PERIODIC TABLE OF THE ELEMENTS

Legend:
- Metals
- Nonmetals
- Metalloids

	IA (1)	IIA (2)	IIIB (3)	IVB (4)	VB (5)	VIB (6)	VIIB (7)	VIIIB (8)	VIIIB (9)	VIIIB (10)	IB (11)	IIB (12)	IIIA (13)	IVA (14)	VA (15)	VIA (16)	VIIA (17)	VIIIA (18)
1	1 H 1.0079																1 H 1.0079	2 He 4.0026
2	3 Li 6.941	4 Be 9.0122											5 B 10.811	6 C 12.011	7 N 14.0067	8 O 15.9994	9 F 18.9984	10 Ne 20.1797
3	11 Na 22.9898	12 Mg 24.3050											13 Al 26.9815	14 Si 28.0855	15 P 30.9738	16 S 32.066	17 Cl 35.4527	18 Ar 39.948
4	19 K 39.0983	20 Ca 40.078	21 Sc 44.9559	22 Ti 47.88	23 V 50.9415	24 Cr 51.9961	25 Mn 54.9380	26 Fe 55.847	27 Co 58.9332	28 Ni 58.69	29 Cu 63.546	30 Zn 65.39	31 Ga 69.723	32 Ge 72.61	33 As 74.9216	34 Se 78.96	35 Br 79.904	36 Kr 83.80
5	37 Rb 85.4678	38 Sr 87.62	39 Y 88.9059	40 Zr 91.224	41 Nb 92.9064	42 Mo 95.94	43 Tc (98)	44 Ru 101.07	45 Rh 102.9055	46 Pd 106.42	47 Ag 107.8682	48 Cd 112.411	49 In 114.82	50 Sn 118.710	51 Sb 121.75	52 Te 127.60	53 I 126.9045	54 Xe 131.29
6	55 Cs 132.9054	56 Ba 137.327	57 La 138.9055 *	72 Hf 178.49	73 Ta 180.9479	74 W 183.85	75 Re 186.207	76 Os 190.2	77 Ir 192.22	78 Pt 195.08	79 Au 196.9665	80 Hg 200.59	81 Tl 204.3833	82 Pb 207.2	83 Bi 208.9804	84 Po (209)	85 At (210)	86 Rn (222)
7	87 Fr (223)	88 Ra (226)	89 Ac (227) **	104 Rf (261)	105 Db (262)	106 Sg (263)	107 Bh (262)	108 Hs (265)	109 Mt (266)									

*Lanthanide Series

58 Ce 140.115	59 Pr 140.9076	60 Nd 144.24	61 Pm (145)	62 Sm 150.36	63 Eu 151.965	64 Gd 157.25	65 Tb 158.9253	66 Dy 162.50	67 Ho 164.9303	68 Er 167.26	69 Tm 168.9342	70 Yb 173.04	71 Lu 174.967

**Actinide Series

90 Th 232.0381	91 Pa 231.0359	92 U 238.0289	93 Np (237)	94 Pu (244)	95 Am (243)	96 Cm (247)	97 Bk (247)	98 Cf (251)	99 Es (252)	100 Fm (257)	101 Md (258)	102 No (259)	103 Lr (260)

Note: Atomic masses are IUPAC values (up to four decimal places). More accurate values for some elements are given on the facing page.

Analytical Chemistry

An Introduction

Seventh Edition

Douglas A. Skoog
Stanford University

Donald M. West
San Jose State University

F. James Holler
University of Kentucky

Stanley R. Crouch
Michigan State University

BROOKS/COLE

THOMSON LEARNING

Australia • Canada • Mexico • Singapore • Spain
United Kingdom • United States

BROOKS/COLE

THOMSON LEARNING ™

Publisher: Emily Barrosse
Acquisitions Editor: John Vondeling
Developmental Editor: Marc Sherman
Production Manager: Charlene Catlett Squibb
Marketing Strategist: Pauline Mula

Production Service: Progressive Publishing Alternatives
Text & Cover Designer: Kathleen Flanagan
Compositor: Progressive Information Technologies
Cover Image: © Cosmo & Action/Photonica

Printed in the United States of America
6 7 05 04 03

For more information about our products, contact us at:
Thomson Learning Academic Resource Center
1-800-423-0563

For permission to use material from this text, contact us by:
Phone: 1-800-730-2214
Fax: 1-800-730-2215
Web: http://www.thomsonrights.com

Library of Congress Catalog-in-Publication Data
Analytical chemistry: an introduction. — 7th ed./Douglas A. Skoog . . . [et al.]
  p.   cm.—(Saunders golden sunburst series)
  Rev. ed. of Analytical chemistry/Douglas A. Skoog, Donald M. West, F. James Holler.
  Includes bibliographical references and index.
  ISBN 0-03-020293-0
  1. Chemistry, Analytic—Quantitative.   1. Skoog, Douglas A.
II. Skoog, Douglas A.   Analytic chemistry.
QD101.2.S55   1999
543—dc21   99-38908
                                           CIP

Asia
Thomson Learning
60 Albert Street, #15-01
Albert Complex
Singapore 189969

Australia
Nelson Thomson Learning
102 Dodds Street
South Melbourne, Victoria 3205
Australia

Canada
Nelson Thomson Learning
1120 Birchmount Road
Toronto, Ontario M1K 5G4
Canada

Europe/Middle East/Africa
Thomson Learning
Berkshire House
168-173 High Holborn
London WC1 V7AA
United Kingdom

Latin America
Thomson Learning
Seneca, 53
Colonia Polanco
11560 Mexico D.F.
Mexico

Spain
Paraninfo Thomson Learning
Calle/Magallanes, 25
28015 Madrid, Spain

Contents Overview

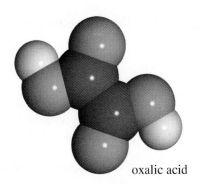

oxalic acid

Contents

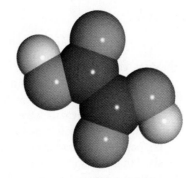

Preface

The seventh edition of *Analytical Chemistry: An Introduction* is more than just an abbreviated version of the text *Fundamentals of Analytical Chemistry* — although it covers many of the same topics in a less detailed and more descriptive manner. The widespread use of computers for instructional purposes has led us to incorporate many spreadsheet applications, examples, and exercises. To accompany the inclusion of spreadsheets, we have added several new topics and revised many older treatments. The primary audience of this text remains students in fields such as biology, medicine, engineering, environmental science, materials science, and other related fields. In many cases the only exposure of these students to analytical chemistry is a one-semester or one-quarter introductory course.

The text should also find use in sophomore level courses for chemistry majors where fewer laboratory experiments and a more descriptive treatment is desired. Finally, earlier editions of the text have also been used in freshman laboratory courses, and this edition should also serve this audience.

OBJECTIVES

A major objective of this text is to provide a strong background in those chemical principles that are particularly important to analytical chemistry. A second goal is to develop an appreciation for the difficult task of judging the accuracy and precision of experimental data and to show how these judgments can be sharpened by the application of statistical methods. A third aim is to introduce a wide range of techniques that are useful in modern analytical chemistry. A fourth goal is to develop the skills needed to solve analytical problems in a quantitative manner, particularly with the aid of the **spreadsheet tools** that are so commonly available. A final goal is to teach those laboratory skills that will give students confidence in their ability to obtain high-quality analytical data.

CHANGES IN THE SEVENTH EDITION

Readers of the sixth edition will find numerous changes in the seventh edition in content as well as in style and format.

Content

Several changes in content have been made to strengthen the book.

*   Many chapters have been strengthened by adding the **spreadsheet exercises, examples, applications, and problems.** Tutorials on using spreadsheets for many statistical calculations have been included. We have incorporated the use of spreadsheets to solve simultaneous equations by basic matrix methods and for iterative solutions to equations. The use of spreadsheets for calculating points on a titration curve and graphically displaying the results is emphasized in the titration chapters.
*   **Section openers** have been written to connect the chapters that fall within a section and to provide an overview of these chapters.
*   The chapter on precipitation titrimetry has been eliminated and some of the material included in Chapter 15 with complexation titrations. The formation of insoluble species is treated as a special case of a complexation reaction.
*   Chapters 16 and 17 on electrochemical cells and cell potentials have been extensively revised in order to clarify the discussion and introduce the free energy of cell processes.
*   Chapter 20 has been revised to include a brief discussion of voltammetry and other modern electroanalytical techniques.
*   Many of the features have been updated and improved to include more environmental, biological, and biomedical applications.
*   We have reduced the number of spectroscopy chapters to three and consolidated the information on spectroscopic instrumentation into one chapter (Chapter 22). The emphasis in the spectroscopy section is on ultraviolet/visible absorption spectrophotometry and molecular fluorescence. Atomic spectrometric methods are included, but in less detail than the previous edition.
*   The chromatography section has been expanded and new information included on supercritical fluid chromatography and capillary electrophoresis.
*   A **Glossary** of important terms used in the text is included.
*   A list of abbreviations and acronyms related to analytical chemistry is provided in Appendix 7.

Style and Format

We have continued to make style and format changes to make the text more readable and student-friendly. The changes include:

*   More active headings and interesting titles are used throughout the book.
*   We have attempted to use short sentences, a more active voice and a more conversational writing style in each chapter.
*   More descriptive figure captions are used whenever appropriate to allow a student to understand the figure and its meaning without alternating between text and caption.
*   **Molecular models** are used liberally in most chapters to stimulate interest in the beauty of molecular structures and to reinforce structural concepts presented in general chemistry and upper-level courses. To represent space-filling molecular models in two colors, we have displayed the most common elements according to the color scheme shown in the margin.
*   We have expanded the use of marginal notes to emphasize important definitions and concepts.

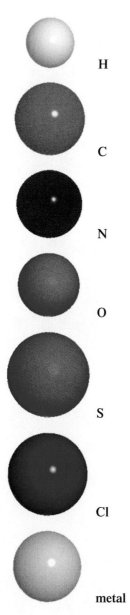

H

C

N

O

S

Cl

metal

COVERAGE AND ORGANIZATION

The material in this text covers both fundamental and practical aspects of chemical analysis. Users of earlier editions will find that we have organized this edition in a somewhat different manner than its predecessors. In particular, we have adopted the use of sections to group related chapters. There are five major sections to the text that follow the brief introduction in Chapter 1.

Section I covers the tools of analytical chemistry and comprises six chapters. Chapter 2 discusses the chemicals and equipment used in analytical laboratories including a tutorial introduction to the use of spreadsheets. Chapters 3 and 4 review basic chemical concepts, including expressions of chemical quantities, stoichiometric relationships, and the principles of chemical equilibrium. Chapters 5, 6, 7 present topics in statistics and data analysis that are important in analytical chemistry and incorporate extensive use of spreadsheet calculations.

Section II covers the principles and application of chemical equilibrium systems in quantitative analysis. Chapter 8 presents a brief overview of the theory and applications of gravimetric methods of analysis. Chapter 9 discusses the effect of electrolytes on equilibrium systems. The systematic approach for attacking equilibrium problems in complex systems is the subject of Chapter 10. In Chapters 11 through 15, we consider the theory and practice of titrimetric methods of analysis, including acid/base titrations, precipitation titrations, and complexometric titrations. Advantage is taken here of the systematic approach to equilibria and the use of spreadsheets in the calculations.

Section III is devoted to electrochemical methods. After an introduction to electrochemistry in Chapter 16, Chapter 17 describes the many uses of electrode potentials. Oxidation/reduction titrations are the subject of Chapter 18, while Chapter 19 presents the use of potentiometric methods to obtain concentrations of molecular and ionic species. The section ends with a presentation of electrolytic methods including electrogravimetry, coulometry, and voltammetry.

Section IV covers spectroscopic methods of analysis. Basic material on the nature of light and its interaction with matter is presented in Chapter 21. Spectroscopic instruments and their components are the topics covered in Chapter 22. The various applications of molecular and atomic spectroscopic methods are covered in some detail in Chapter 23.

Section V includes three chapters dealing with analytical separations. After an introduction to the various ways in which separations are accomplished in Chapter 24, Chapter 25 discusses the techniques of gas chromatography, and high-performance liquid chromatography. The final chapter in this section, Chapter 26, covers supercritical fluid chromatographic methods and capillary electrophoresis, and electrochromatography.

Section VI is the final section of the text and includes just one chapter. Chapter 27 provides detailed procedures for laboratory experiments covering many of the principles and applications discussed in previous chapters.

Because of the organization of the text into sections, there is a good deal of flexibility in the use of the material because many of the sections can stand alone or be taken in a different order. For example, some instructors may want to cover spectroscopic methods prior to electrochemical methods or chromatographic methods prior to spectroscopic methods.

HIGHLIGHTS

We have incorporated in this text many features and methods intended to enhance the learning experience for the student and to provide a versatile teaching tool for the instructor.

Important Equations. Equations that we feel are most important have been highlighted with a color screen for emphasis and ease of review.

Mathematical Level. Generally the principles of chemical analysis developed here are based on college algebra. A few concepts presented are best understood with calculus. However, we tried to give sufficient qualitative descriptions that students can achieve a working knowledge without calculus.

Worked Examples. A large number of worked examples serve as aids in understanding the concepts of analytical chemistry. As in the sixth edition, we follow the practice of including units in chemical calculations and using the factor-label method as a check on their correctness. The examples are also models for the solution of problems found at the end of most of the chapters. Many of these use spreadsheet calculations as described next.

Spreadsheet Calculations. Throughout the book we have introduced spreadsheets for problem solving, graphical analysis and many other applications. Microsoft® Excel has been adapted as the standard for these calculations, but the instructions could be readily adapted to other programs. Several chapters have tutorial discussion of how to enter values, formulas, and built-in functions. Many chapters have spreadsheet examples that are worked in detail and adapted from chemical principles. We have attempted to document each stand-alone spreadsheet with working formulas and entries.

Questions and Problems. An extensive set of questions and problems is included at the end of most chapters. Answers to approximately half of the problems are given at the end of the book; these questions are indicated by an asterisk. Many of the problems are best solved using spreadsheets. These are identified by a spreadsheet icon placed in the margin beside the problem.

Featured Topics. A series of boxed and highlighted Features are found throughout the text. These may contain interesting applications of analytical chemistry to the modern world, derivation of equations, explanations of more difficult theoretical points, and historical notes.

Expanded Figure Captions. Where appropriate, we have attempted to make the figure captions quite descriptive so that reading the caption provides a second level of explanation for many of the concepts. In some cases, the figures can stand by themselves much in the manner of a *Scientific American* figure.

Web Works. In most of the chapters we have included a brief Web Works feature at the end of the chapter. In these we provide addresses (URLs) for sites on the World Wide Web in which the student can find additional information, extended or alternative discussions, or interesting analytical applications. These Web Works are intended to stimulate interest in the student in finding information on the Web or in doing Web searches. The links will be updated regularly on the Brooks/Cole Web Site given later in this Preface.

Appendixes and Endpapers. Included in the appendixes are tables of chemical constants, tables of electrode potentials, a section of the use of logarithms and exponential numbers, a section on normality and equivalents (terms that are not used in the text itself), and a list of acronyms and abbreviations used in analytical chemistry. The endpapers provide a full-color chart of chemical indicators, a table of molar masses of compounds of particular interest in analytical chemistry, a reference card for Microsoft Excel, and a periodic table.

ANCILLARIES

Instructor's Manual. The instructor's manual contains a complete set of worked-out solutions to the problems in the text.
Overhead Transparencies. Approximately 100 transparencies are available for overhead projectors. Many of the transparencies have multiple figures from the text.
Web Site. A site on the World Wide Web is being developed for placing such items as tutorial examples, additional problems, and new Web Works in case URLs in the text become obsolete. See the Brooks/Cole Web Site at http://www.brookscole.com for additional information.

ACKNOWLEDGMENTS

We wish to acknowledge with thanks the comments and suggestions of many others who have reviewed the manuscript in various stages. Professors John Allison, Michigan State University, Kent Clinger, David Lipscomb University, Marcin Majda, University of California at Berkeley, and Thomas Reichel, Miami University all reviewed the 6th edition prior to our revisions.

The reviewers of the current edition did a superb job, and we wish to express our sincere thanks to Julie Brady, Delaware Technical & Community College, Kim Cohn, California State University at Bakersfield, Richard Mitchell, Arkansas State University, Alexander (Raman) Scheeline, University of Illinois at Urbana-Champaign, Jeffery Schneider, State University of New York at Oswego, and John Tyvoll, Salisbury State University. We give a special acknowledgment to Professor David Zellmer of California State University at Fresno for his detailed review and his superb suggestions regarding the presentation and use of spreadsheets.

The authors are grateful to Hypercube, Inc. of Gainesville, FL for providing a copy of HyperChem® 5.1 and HyperChem Raytrace© for building the molecular models in this book. The molecules were rendered using POV-Ray for Windows 3.1, which is available as freeware from the Persistence of Vision Development Team at http://www.povray.org/.

Sincere thanks go to the talented and dedicated staff at Saunders College Publishing: John Vondeling, Publisher; Marc Sherman, Senior Associate Editor;

Kathleen Flanagan, Text and Cover Designer; and Charlene Squibb, Senior Production Manager. Many thanks also go to Donna King and her staff at Progressive Publishing Alternatives for their excellent work.

DOUGLAS A. SKOOG
DONALD M. WEST
F. JAMES HOLLER
STANLEY R. CROUCH

Spreadsheet Activities

Title	Section
• Using Spreadsheets in Analytical Chemistry	2J
• Spreadsheet Exercise: Using Excel to Iterate	4B-3
• Spreadsheet Exercise: A Mean Calculation	5B
• Spreadsheet Exercise: Computing the Standard Deviation With Excel	6B-2
• Spreadsheet Exercise: The Pooled Standard Deviation	6B-4
• Spreadsheet Exercise: Using Excel to Do Least-Squares	7D-2
• Spreadsheet Exercise: Matrix Operations: A Black Box for Solving Simultaneous Linear Equations	8D
• Spreadsheet Exercise: Using Circular References for Iteratively Solving Equations	9B
• Spreadsheet Exercise: Using Microsoft® Excel to Solve the Equations of Complex Equilibrium Systems	10C
• Titrating a Strong Acid with a Strong Base	Example 12-1
• Spreadsheet Exercise: Constructing the Titration Curve for Acetic Acid	Example 12-6
• Calculating α Values for Maleic Acid	13F
• Spreadsheet Exercise: Calculating the pH of a NaH_2PO_4 Solution	Example 13-5
• Spreadsheet Exercise: Constructing the Titration Curve for Maleic Acid	Example 13-6
• Spreadsheet Exercise: Calculating Masses of Acids for Standardizing Bases	Example 14-1
• Spreadsheet Exercise: Calculating the Chromate Concentration Required to Precipitate Ag_2CrO_4 at the Endpoint of Titrations of Chloride, Bromide, and Cyanide	Example 15-3
• Calculating α_4 Values for EDTA	15D-3
• Constructing the Titration Curve of Ca^{2+} with EDTA	Example 15-7
• Constructing the Titration Curve of Fe^{2+} with Ce^{4+}	17C-2
• Spreadsheet Exercise: Constructing the Titration Curve of U^{4+} with Ce^{4+}	Example 17-9
• Spreadsheet Exercise: Constructing Potentiometric Titration Curves and Derivative Plots	Example 19-2
• Calculating Transmittance, %Transmittance, and Absorbance	21C-1
• Analyzing the Data from the Method of Multiple Standard Additions	Example 23-2
• Simultaneous Spectrophotometric Determination of Titanium and Vanadium	23A
• Spreadsheet Exercise: Simulating a Chromatogram	24F
• Spreadsheet Exercise: Quantitative GC Using an Internal Standard Method	Example 25-1
• Spreadsheet Exercise: Calculation of Electrophoretic Mobilities of Ions	Example 26-1

Biological and Environmental Topics

Title	Section
• Deer Kill: Case Study	Feature 1-1
• Lead in Blood	Example 6-1
• Mercury in Fish	Example 6-2
• Alcohol in Blood	Example 7-3
• Immunoassays in the Determination of Drugs	Feature 10-2
• Centers for Disease Control	Web Works Ch. 10
• Buffer Solutions	12C
• The Henderson-Hasselbalch Equation	Feature 12-4
• Acid Rain and the Buffer Capacity of Lakes	Feature 12-5
• Determining the pK Values for Amino Acids	Feature 12-7
• Titration Curves for Amino Acids	Feature 13-4
• Determining Total Serum Protein	Feature 14-1
• Determining Organic and Biochemical Substances	14B-3
• EDTA as a Preservative	Feature 15-3
• Determining Water Hardness	15D-9
• Test Kits for Water Hardness	Feature 15-5
• Biological Redox Systems	Feature 17-1
• Determination of Chromium Species in Water Samples	Feature 18-1
• Antioxidants	Feature 18-2
• The Vitamin E Cycle	Web Works Ch. 18
• Point-of-Care Testing: Blood Gases and Blood Electrolytes with Portable Instrumentation	Feature 19-2
• Clark Oxygen Electrode	20E-2
• Enzyme Sensors	20E-2
• Anodic Stripping Voltammetry	20F
• Firefly Chemiluminescence	21B-2
• UV/Visible Absorption of Biochemical Species	23A-1
• Determination of Phosphate in Urine	Example 23-1
• Determination of Iron in Natural Water	Example 23-2
• Infrared Spectroscopy for Atmospheric Pollutants	Feature 23-1
• Fluorescence Probes in Neurobiology	Feature 23-3
• Chemiluminescence and Bioluminescence	23D-3
• Mercury Determination by Cold Vapor Atomic Absorption Spectroscopy	Feature 23-4
• Infrared Imaging in Biomedical Research	Web Works Ch. 23
• Use of GC/MS for Identifying Drug Metabolites	Feature 25-1
• Capillary Electrophoresis of Pesticides and Biological Molecules	26B-4
• Separation of Nucleotides Using Capillary Zone Electrophoresis	Web Works Ch. 26
• Determination of Water Hardness	Expt. 27E-4
• Determination of Ascorbic Acid in Vitamin C Tablets	Expt. 27I-3
• Determination of Fluoride in Drinking Water and Toothpaste	Expt. 27J-4
• Determination of Iron in Natural Waters	Expt. 27L-2
• Determination of Sodium, Potassium, and Calcium in Mineral Waters	Expt. 27 M-2
• GC Determination of Ethanol in Beverages	Expt. 27O-1

Chapter 1

What Is Analytical Chemistry?

A nalytical chemistry is a measurement science consisting of a set of powerful ideas and methods that are useful in all fields of science and medicine. An exciting illustration of the power and significance of analytical chemistry occurred on July 4, 1997, when the *Pathfinder* spacecraft bounced to a halt on Ares Vallis, Mars, and delivered the *Sojourner* rover from its tetrahedral body to the Martian surface. The world was captivated by the *Pathfinder* mission. As a result, the numerous World Wide Web sites tracking the mission were nearly overwhelmed by millions of Internet surfers who closely monitored the progress of tiny *Sojourner* in its quest for information on the nature of the Red Planet. The key experiment aboard *Sojourner* was the APXS, or alpha proton X-ray spectrometer, which combines the three advanced instrumental techniques of Rutherford backscattering spectroscopy, proton emission spectroscopy, and X-ray fluorescence. The APXS data were collected by *Pathfinder* and transmitted to Earth for further analysis to determine the identity and concentration of most of the elements of the periodic table.[1] The determination of the elemental composition of Martian rocks permitted geologists to rapidly identify them and compare them with terrestrial rocks. The *Pathfinder* mission is a spectacular example illustrating the application of analytical chemistry to practical problems. The experiments aboard the spacecraft and the data from the mission also illustrate how analytical chemistry draws on science and technology in such widely diverse disciplines as nuclear physics and chemistry to identify and determine the relative amounts of the substances in samples of matter.

The *Pathfinder* example demonstrates that both qualitative and quantitative information are required in an analysis. *Qualitative analysis* establishes the

---

[1] For detailed information regarding the instrumentation aboard the *Sojourner,* please refer to the article at http:/mars.sgi.com/mpf/sci_desc.html#APXS, which contains a general description of the instrument package and links to more detailed information, including the article by Economou at http://astro.uchicago.edu/home/web/papers/economou/pathfinder/.

A 360-degree panoramic view of the Mars Pathfinder mission site on July 21, 1997. In the center of photo, the Sojourner rover is exploring a rock nicknamed Yogi using its APXS spectrometer. Photo courtesy of NASA.

Analytes are the components of a sample that are to be determined.

Qualitative analysis reveals the *identity* of the elements and compounds in a sample.

Quantitative analysis indicates the *amount* of each substance in a sample.

chemical identity of the species in the sample. *Quantitative analysis* determines the relative amounts of these species, or *analytes,* in numerical terms. The data from the APXS spectrometer on *Sojourner* contain both types of information. Note that chemical separation of the various elements contained in the rocks was unnecessary in the APXS experiment. More commonly, a separation step is a necessary part of the analytical process. As we shall see, qualitative analysis is often an integral part of the separation step, and the determination of the identity of the analytes is an essential adjunct to quantitative analysis. In this text, we shall explore quantitative methods of analysis, separation methods, and the principles behind their operation.

1A   THE ROLE OF ANALYTICAL CHEMISTRY

Analytical chemistry plays a vital role in the development of science. In 1894, Friedrich Wilhelm Ostwald wrote:

> Analytical chemistry, or the art of recognizing different substances and determining their constituents, takes a prominent position among the applications of science, since the questions which it enables us to answer arise wherever chemical processes are employed for scientific or technical purposes. Its supreme importance has caused it to be assiduously cultivated from a very early period in the history of chemistry, and its records comprise a large part of the quantitative work which is spread over the whole domain of science.

Since Ostwald's time, analytical chemistry has evolved from an art into a science with applications throughout industry, medicine, and all the sciences. To illustrate, consider a few examples. The concentrations of oxygen and of carbon dioxide are determined in millions of blood samples every day and used to diagnose and treat illnesses. Quantities of hydrocarbons, nitrogen oxides, and carbon monoxide present in automobile exhaust gases are measured to determine the effectiveness of smog-control devices. Quantitative measurements of ionized calcium in blood serum help diagnose parathyroid disease in humans. Quantitative

determination of nitrogen in foods establishes their protein content and thus their nutritional value. Analysis of steel during its production permits adjustment in the concentrations of such elements as carbon, nickel, and chromium to achieve a desired strength, hardness, corrosion resistance, and ductility. The mercaptan content of household gas supplies is monitored continually to ensure that the gas has a sufficiently obnoxious odor to warn of dangerous leaks. Farmers tailor fertilization and irrigation schedules to meet changing plant needs during the growing season, gauging these needs from quantitative analyses of the plants and the soil in which they grow.

Quantitative analytical measurements also play a vital role in many research areas in chemistry, biochemistry, biology, geology, physics, and the other sciences. For example, quantitative measurements of potassium, calcium, and sodium ions in the body fluids of animals permit physiologists to study the role these ions play in nerve-signal conduction as well as muscle contraction and relaxation. Chemists unravel the mechanisms of chemical reactions through reaction rate studies. The rate of consumption of reactants or formation of products in a chemical reaction can be calculated from quantitative measurements made at equal time intervals. Materials scientists rely heavily on quantitative analyses of crystalline germanium and silicon in their studies of semiconductor devices. Impurities in these devices are in the concentration range of 1×10^{-6} to 1×10^{-9} percent. Archaeologists identify the source of volcanic glasses (obsidian) by measuring concentrations of minor elements in samples taken from various locations. This knowledge in turn makes it possible to trace prehistoric trade routes for tools and weapons fashioned from obsidian.

Many chemists, biochemists, and medicinal chemists devote a significant fraction of their time in the laboratory gathering quantitative information about systems of interest to them. The central role of analytical chemistry in this enterprise and many others is illustrated in Figure 1-1. All branches of chemistry draw on the ideas and techniques of analytical chemistry. Analytical chemistry plays a similar role with respect to the many other scientific fields listed in the circles and arranged in the periphery of the diagram. Chemistry is often termed the *central science,* and the top center position of chemistry as well as the central position of analytical chemistry in the figure symbolize the importance of

Friedrich Wilhelm Ostwald (1853–1932) was the winner of the 1909 Nobel Prize in Chemistry. He was a pioneer in the field of catalysis and one of the first chemists to recognize the importance of thermodynamics. He was also one of the last great chemists to resist the notion of the existence of atoms. (Oesper Collection of the History of Chemistry, University of Cincinnati)

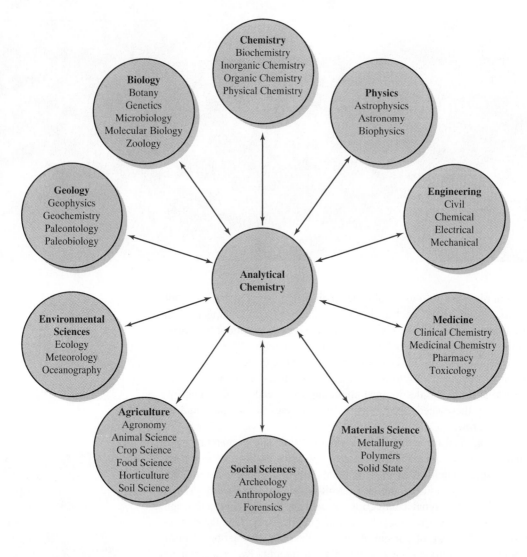

Figure 1-1   The relationship between analytical chemistry, other branches of chemistry, and the other sciences. The central location of analytical chemistry in the diagram signifies its importance and the breadth of its interactions with many other disciplines.

chemistry in the sciences. Analytical chemistry serves as an essential tool in all the fields depicted in the figure. The interdisciplinary nature of chemical analysis makes it a vital tool in medical, industrial, government, and academic laboratories throughout the world.

1B   CLASSIFYING QUANTITATIVE ANALYTICAL METHODS

We compute the results of a typical quantitative analysis from two measurements. One is the mass or the volume of sample to be analyzed. The second is the measurement of some quantity that is proportional to the amount of analyte

"TODAY EVERYONE HAS TO KNOW 'WHAT'S IN THE FOOD?', 'WHAT'S IN THE WATER?', 'WHAT'S IN THE AIR?' THIS IS TRULY THE GOLDEN AGE OF ANALYTICAL CHEMISTRY."

in the sample, such as mass, volume, intensity of light, or electrical charge. This second measurement usually completes the analysis, and we classify analytical methods according to the nature of this final measurement. *Gravimetric methods* determine the mass of the analyte or some compound chemically related to it. In a *volumetric method,* the volume of a solution containing sufficient reagent to react completely with the analyte is measured. *Electroanalytical methods* involve the measurement of such electrical properties as voltage, current, resistance, and quantity of electrical charge. *Spectroscopic methods* are based on measurement of the interaction between electromagnetic radiation and analyte atoms or molecules or on the production of such radiation by analytes. Finally, there is a group of miscellaneous methods that includes the measurement of such quantities as mass-to-charge ratio of molecules by mass spectrometry, rate of radioactive decay, heat of reaction, rate of reaction, sample thermal conductivity, optical activity, and refractive index.

1C STEPPING THROUGH A TYPICAL QUANTITATIVE ANALYSIS

A typical quantitative analysis involves the sequence of steps shown in the flow diagram of Figure 1-2. In some instances, one or more of these steps can be omitted. For example, if the sample is already a liquid, we can avoid the dissolution step. The first 23 chapters of this book focus on the last three steps in Figure 1-2. In the measurement step, we measure one of the physical properties mentioned in Section 1B. In the calculation step, we find the relative amount of the analyte present in the samples. In the final step we evaluate the quality of the results and estimate their reliability.

In the paragraphs that follow, you will find a brief overview of each of the nine steps shown in Figure 1-2. We then present a case study to illustrate these

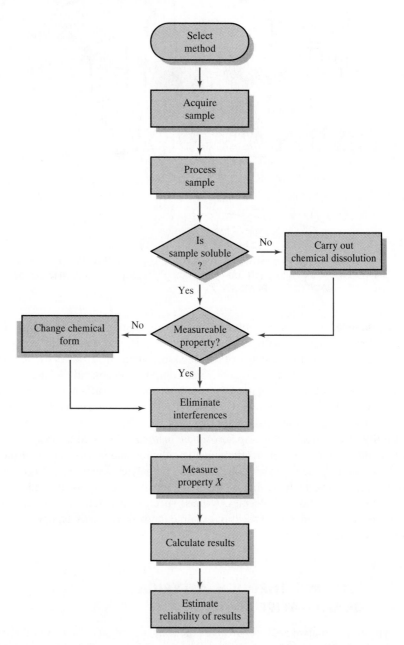

Figure 1-2 Flow diagram showing the steps in a quantitative analysis. There are a number of possible paths through the steps in a quantitative analysis. In the simplest example represented by the central vertical pathway, we select a method, acquire and process the sample, dissolve the sample in a suitable solvent, measure a property of the analyte, calculate the results, and estimate the reliability of the results. Depending on the complexity of the sample and the chosen method, various other pathways may be necessary.

steps in solving an important and practical analytical problem. The details of the case study foreshadow many of the methods and ideas you will explore as you study analytical chemistry.

1C-1   Picking a Method

The essential first step in any quantitative analysis is the selection of a method as depicted in Figure 1-2. The choice is sometimes difficult and requires experience as well as intuition. One of the first questions to be considered in the selection process is the level of accuracy required. Unfortunately, high reliability nearly always requires a large investment of time. The selected method usually represents a compromise between the accuracy needed and the time and money that are available for the analysis.

A second consideration related to economic factors is the number of samples to be analyzed. If there are many samples, we can afford to spend a good deal of time in preliminary operations such as assembling and calibrating instruments and equipment and preparing standard solutions. If we have only a single sample or just a few samples, it may be more appropriate to select a procedure that avoids or minimizes such preliminary steps.

Finally, the complexity of the sample and the number of components in the sample always influence the choice of method to some degree.

1C-2   Acquiring the Sample

As illustrated in Figure 1-2, the next step in a quantitative analysis is to acquire the sample. To produce meaningful information, an analysis must be performed on a sample whose composition faithfully represents that of the bulk of material from which it was taken. Where the bulk is large and heterogeneous, great effort is required to get a representative sample. Consider, for example, a railroad car containing 25 tons of silver ore. Buyer and seller must agree on a price, which will be based primarily on the silver content of the shipment. The ore itself is inherently heterogeneous, consisting of many lumps that vary in size as well as in silver content. The *assay* of this shipment will be performed on a sample that weighs about one gram. For the analysis to have significance, this small sample must have a composition that is representative of the 25 tons (or approximately 22,700,000 g) of ore in the shipment. Isolation of one gram of material that accurately represents the average composition of the nearly 23,000,000 g of bulk sample is a difficult undertaking that requires a careful, systematic manipulation of the entire shipment. *Sampling* involves obtaining a small mass of a material whose composition accurately represents the bulk of the material being sampled.

The collection of specimens from biological sources represents a second type of sampling problem. Sampling of human blood for the determination of blood gases illustrates the difficulty of acquiring a representative sample from a complex biological system. The concentration of oxygen and carbon dioxide in blood depends on a variety of physiological and environmental variables. For example, inappropriate application of a tourniquet or hand flexing by the patient

A material is *heterogeneous* if its constituent parts can be distinguished visually or with the aid of a microscope. Coal, animal tissue, and soil are heterogeneous materials.

An *assay* is the process of determining how much of a given sample is the material indicated by its name. For example, a zinc alloy is assayed for its zinc content, and its assay is a particular numerical value.

We *analyze* samples and we determine *substances*. For example, a blood sample is analyzed to determine the concentrations of various substances such as blood gases and glucose. We therefore speak of the determination of blood gases or glucose, **not** the analysis of blood gases or glucose.

may cause blood oxygen concentration to fluctuate. Because physicians make life-and-death decisions based on results of blood gas analyses, strict procedures have been developed for sampling and transporting specimens to the clinical laboratory. These procedures ensure that the sample is representative of the patient at the time it is collected and that its integrity is preserved until the sample can be analyzed.

Many sampling problems are easier to solve than the two just described. Whether sampling is simple or complex, however, the analyst must be sure that the laboratory sample is representative of the whole before proceeding with an analysis. Sampling is frequently the most difficult step in an analysis and the source of greatest error. The final results of an analysis will never be any more reliable than the reliability of the sampling step.

1C-3   Processing the Sample

The third step in an analysis is to process the sample as shown in Figure 1-2. Under certain circumstances, no sample processing is required prior to the measurement step. For example, once a water sample is withdrawn from a stream, a lake, or an ocean, the pH of the sample can be measured directly. Under most circumstances, we must process the sample in any of a variety of different ways. The first step in processing the sample is often the preparation of a laboratory sample.

Preparing a Laboratory Sample

A solid laboratory sample is ground to decrease particle size, mixed to ensure homogeneity, and stored for various lengths of time before analysis begins. Absorption or desorption of water may occur during each step, depending on the humidity of the environment. Because any loss or gain of water changes the chemical composition of solids, it is a good idea to dry samples just before starting an analysis. Alternatively, the moisture content of the sample can be determined at the time of the analysis in a separate analytical procedure.

Liquid samples present a slightly different but related set of problems during the preparation step. If such samples are allowed to stand in open containers, the solvent may evaporate and change the concentration of the analyte. If the analyte is a gas dissolved in a liquid, as in our blood gas analysis example, the sample container must be kept inside a second sealed container, perhaps during the entire analytical procedure, to prevent contamination by atmospheric gases. Extraordinary measures, including sample manipulation and measurement in an inert atmosphere, may be required to preserve the integrity of the sample.

Defining Replicate Samples

Replicate samples, or *replicates,* are portions of a material of approximately the same size that are carried through an analytical procedure at the same time and in the same way.

We perform most chemical analyses on replicate samples whose masses or volumes have been determined by careful measurements with an analytical balance or with a precise volumetric device. Replication improves the quality of the results and provides a measure of their reliability. Quantitative measurements on replicates are usually averaged, and various statistical tests are performed on the results to establish their reliability.

Excel Shortcut Keystrokes for the PC*

Macintosh equivalents, if different, appear in square brackets

TO ACCOMPLISH THIS TASK	TYPE THESE KEYSTROKES
Alternate between displaying cell values and displaying cell formulas	**Ctrl+`** (Single Left Quotation Mark) **[⌘ +`]**
Calculate all sheets in all open workbooks	**F9**
Calculate the active worksheet	**Shift+F9**
Cancel an entry in a cell or formula bar	**Esc**
Complete a cell entry and move down in the selection	**Enter [Return]**
Complete a cell entry and move to the left in the selection	**Shift+Tab**
Complete a cell entry and move to the right in the selection	**Tab**
Complete a cell entry and move up in the selection	**Shift+Enter**
Copy a formula from the cell above the active cell into the cell or the formula bar	**Ctrl+'** (Apostrophe) **[⌘ +']**
Copy a selection	**Ctrl+C [⌘ +C]**
Copy the value from the cell above the active cell into the cell or the formula bar	**Ctrl+Shift+"** (Quotation Mark) **[⌘ +Shift+"]**
Create names from row and column labels	**Ctrl+Shift+F3**
Cut a selection	**Ctrl+X [⌘ +X]**
Define a name	**Ctrl+F3 [⌘ +F3]**
Delete text to the end of the line	**Ctrl+Delete [Ctrl+Option+Del]**
Delete the character to the left of the insertion point, or delete the selection	**Backspace [Delete]**
Delete the character to the right of the insertion point, or delete the selection	**Delete [Del]**
Display the Formula Palette after you type a valid function name in a formula	**Ctrl+A**
Edit a cell comment	**Shift+F2**
Edit the active cell	**F2 [None]**
Edit the active cell and then clear it, or delete the preceding character in the active cell as you edit the cell contents	**Backspace [Delete]**
Enter a formula as an array formula	**Ctrl+Shift+Enter**
Fill down	**Ctrl+D [⌘ +D]**
Fill the selected cell range with the current entry	**Ctrl+Enter [None]**
Fill to the right	**Ctrl+R [⌘ +R]**
Insert the argument names and parentheses for a function, after you type a valid function name in a formula	**Ctrl+Shift+A**
Insert the AutoSum formula	**Alt+=** (Equal Sign) **[⌘ +Shift+T]**
Move one character up, down, left, or right	**Arrow Keys**
Move to the beginning of the line	**Home**
Paste a name into a formula	**F3 [None]**
Paste a selection	**Ctrl+V [⌘ +V]**
Repeat the last action	**F4 Or Ctrl+Y [⌘ +Y]**
Start a formula	**=** (Equal Sign)
Start a new line in the same cell	**Alt+Enter [⌘ +Option+Enter]**
Undo	**Ctrl+Z [⌘ +Z]**

Microsoft® Excel Toolbars

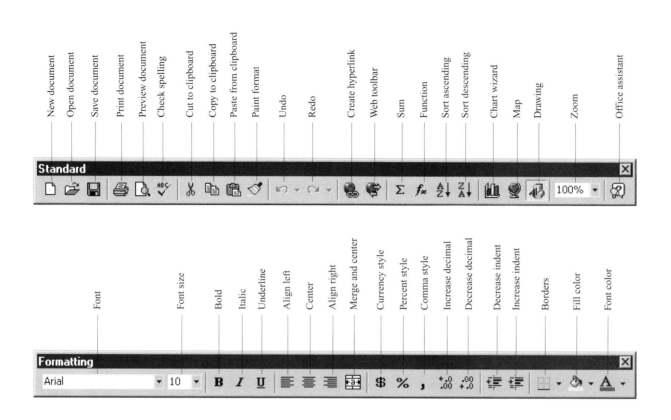

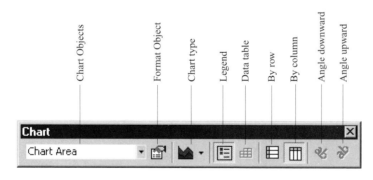

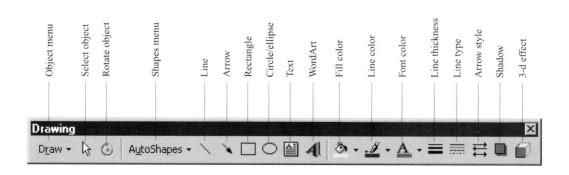

Preparing Solutions: Physical and Chemical Changes

Most analyses are performed on solutions of the sample made with a suitable solvent. Ideally, the solvent should dissolve the entire sample, including the analyte, rapidly and completely. The conditions of dissolution should be sufficiently mild that loss of the analyte cannot occur or is minimized. In our flow diagram of Figure 1-2, we ask whether the sample is soluble in the solvent of choice. Unfortunately, many materials that must be analyzed are insoluble in common solvents. Examples include silicate minerals, high-molecular-weight polymers, and specimens of animal tissue. Under this circumstance, we must follow the flow diagram to the box on the right and carry out some rather harsh chemistry. Conversion of the analyte in such materials into a soluble form is often the most difficult and time-consuming task in the analytical process. The sample may require heating with aqueous solutions of strong acids, strong bases, oxidizing agents, reducing agents, or some combination of such reagents. It may be necessary to ignite the sample in air or oxygen or perform a high-temperature fusion of the sample in the presence of various fluxes. Once the analyte is made soluble, we then ask whether the solution has a property that is proportional to analyte concentration and that we can measure. If it does not, other chemical steps may be necessary to convert the analyte to a form that is suitable for the measurement step as we see in the flow diagram of Figure 1-2. For example, in the determination of manganese in steel, manganese must be oxidized to MnO_4^- before the absorbance of the colored solution is measured (see Chapter 23). At this point in the analysis, it may be possible to proceed directly to the measurement step, but more often than not, we must eliminate interferences in the sample before making measurements as illustrated in the flow diagram.

1C-4   Eliminating Interferences

Once we have gotten the sample into solution and converted the analyte to an appropriate form for the measurement step, the next step is to eliminate substances from the sample that may interfere with the measurement step. This sequence is depicted in the flow diagram of Figure 1-2. Few chemical or physical properties of importance in chemical analysis are unique to a single chemical species. Instead, the reactions used and the properties measured are characteristic of a group of elements or compounds. Species other than the analyte that affect the final measurement are called *interferences,* or *interferents.* A scheme must be devised to isolate the analytes from interferences before the final measurement is made. No hard and fast rules can be given for eliminating interferences; indeed, resolution of this problem can be the most demanding aspect of an analysis. Chapters 24 through 26 describe separation methods.

An *interference* is a species that causes an error in an analysis by enhancing or attenuating (making smaller) the quantity being measured.

1C-5   Calibration and Measurement

All analytical results depend on a final measurement X of a physical or chemical property of the analyte as shown in Figure 1-2. This property must vary in a known and reproducible way with the concentration c_A of the analyte. Ideally, the measurement of the property is directly proportional to the concentration.

Techniques or reactions that work for only one analyte are said to be *specific.* Techniques or reactions that apply for only a few analytes are *selective.*

The *matrix,* or *sample matrix,* is all of the components in the sample containing an analyte.

That is,

$$c_A = kX$$

The process of determining k is thus an important step in most analyses and is termed a *calibration*.

where k is a proportionality constant. With two exceptions, analytical methods require the empirical determination of k with chemical standards for which c_A is known.[2] The process of determining k is thus an important step in most analyses; this step is called a *calibration*.

1C-6   Calculating Results

Computing analyte concentrations from experimental data is usually relatively easy, particularly with modern calculators or computers. This step is depicted in the next-to-last block of the flow diagram of Figure 1-2. These computations are based on the raw experimental data collected in the measurement step, the characteristics of the measurement instruments, and the stoichiometry of the analytical reaction. Samples of these calculations appear throughout this book.

1C-7   Evaluating Results by Estimating Their Reliability

An analytical result without an estimate of reliability is of no value.

As Figure 1-2 implies, analytical results are incomplete without an estimate of their reliability. The experimenter must provide some measure of the uncertainties associated with computed results if the data are to have any value. Chapters 5, 6, and 7 present detailed methods for carrying out this important final step in the analytical process.

1D   AN INTEGRAL ROLE FOR CHEMICAL ANALYSIS: FEEDBACK CONTROL SYSTEMS

Analytical chemistry is usually not an end in itself, but is part of a bigger picture in which we may use analytical results to help control a patient's health, to control the amount of mercury in fish, to control the quality of a product, to determine the status of a synthesis, or to find out whether there is life on Mars. Chemical analysis is the measurement element in all of these examples and in many other cases. Let us consider the role of quantitative analysis in the determination and control of the concentration of glucose in blood. The system flow diagram of Figure 1-3 illustrates the process. Patients suffering from insulin-dependent *diabetes mellitus* develop hyperglycemia, which manifests itself in a blood glucose concentration above the normal concentration of 60 to 95 mg/dL. We begin our example by determining that the desired state is a blood glucose level below 95 mg/dL. Many patients must monitor their blood glucose levels by periodically submitting samples to a clinical laboratory for analysis or by measuring the levels themselves using a handheld electronic glucose monitor.

---

[2]The two exceptions are gravimetric methods, which are discussed in Chapter 7, and coulometric methods, which are considered in Chapter 20. In both these methods, k can be computed from known physical constants.

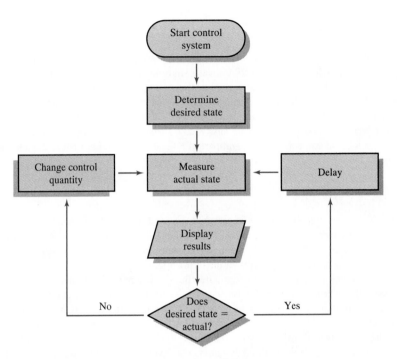

Figure 1-3   Feedback system flow diagram. The desired state is determined, the actual state of the system is measured, and the two states are compared. The difference between the two states is used to change a controllable quantity that results in a change in the state of the system. Quantitative measurements are again performed on the system, and the comparison is repeated. The new difference between the desired state and the actual state is again used to change the state of the system if necessary. The process provides continuous monitoring and feedback to maintain the controllable quantity, and thus the actual state, at the proper level. The text describes the monitoring and control of blood glucose as an example of a feedback control system.

The first step in the monitoring process is to determine the actual state by collecting a blood sample from the patient and measuring the blood glucose level. The results are displayed, and then the actual state is compared to the desired state as shown in Figure 1-3. If the measured blood glucose level is above 95 mg/dL, the patient's insulin level, which is a controllable quantity, is increased by injection or oral administration. After a delay to allow the insulin time to take effect, the glucose level is measured again to determine if the desired state has been achieved. If the level is below the threshold, the insulin level has been maintained, so no insulin is required. After a suitable delay time, the blood glucose level is measured again, and the cycle is repeated. In this way, the insulin level in the patient's blood, and thus the blood glucose level, is maintained at or below the critical threshold, which keeps the metabolism of the patient in control.

The process of continuous measurement and control is often referred to as a *feedback system,* and the cycle of measurement, comparison, and control is called a *feedback loop.* These ideas find wide application in biological and biomedical systems, mechanical systems, and electronics. From the measurement and control of the concentration of manganese in steel to maintaining the proper level of chlorine in a swimming pool, chemical analysis plays a central role in a broad range of systems.

Feature 1-1

Deer Kill: A Case Study
Illustrating the Use of
Analytical Chemistry to
Solve a Problem in
Toxicology

The tools of modern analytical chemistry are widely applied in environmental investigations. In this feature, we describe a case study in which quantitative analysis was used to determine the agent that caused deaths in a population of white-tailed deer in a wildlife area of a national recreational area in Kentucky. We begin with a description of the problem and then show how the steps illustrated in Figure 1-2 were used to solve the analytical problem. This case study also shows how chemical analysis is used in a broad context as an integral part of a feedback control system as depicted in Figure 1-3.

THE PROBLEM

The incident began when a park ranger found a dead white-tailed deer near a pond in the Land Between the Lakes National Recreation Area in western Kentucky. The park ranger enlisted the help of a chemist from the state veterinary diagnostic laboratory to find the cause of death so that further deer kills might be prevented.

The ranger and the chemist investigated the site where the badly decomposed carcass of the deer had been found. Because of the advanced state of decomposition, no fresh organ tissue samples could be gathered. A few days after the original inquiry, the ranger found two more dead deer near the same location. The chemist was summoned to the site of the kill, where he and the ranger loaded the deer onto a truck for transport to the veterinary diagnostic laboratory. The investigators then conducted a careful examination of the surrounding area in an attempt to find clues to establish the cause of death.

The search covered about two acres surrounding the pond. The investigators noticed that grass surrounding nearby power line poles was wilted and discolored. They speculated that a herbicide might have been used on the grass. A common ingredient in herbicides is arsenic in any one of a variety of forms, including arsenic trioxide, sodium arsenite, monosodium methanearsenate, and disodium methanearsenate. The

White-tailed deer, which have proliferated in many parts of the country, shown near Valley Forge, PA. (Photo courtesy of Scott E. Mabry)

last compound is the disodium salt of methanearsenic acid, $CH_3AsO(OH)_2$, which is very soluble in water and thus finds use as the active ingredient in many herbicides. The herbicidal activity of disodium methanearsenate is due to its reactivity with the sulfhydryl (S—H) groups in the amino acid cysteine. When cysteine in plant enzymes reacts with arsenical compounds, the enzyme function is inhibited and the plant eventually dies. Unfortunately, similar chemical effects occur in animals as well. The investigators therefore collected samples of the discolored dead grass for testing along with samples from the organs of the deer. They planned to analyze the samples to confirm the presence of arsenic and, if present, to determine its concentration in the samples.

SELECTING A METHOD

A scheme for the quantitative determination of arsenic in biological samples is found in the published methods of the Association of Official Analytical Chemists (AOAC).[3] This method involves the distillation of arsenic as arsine, which is then determined by colorimetric measurements.

PROCESSING THE SAMPLE: OBTAINING REPRESENTATIVE SAMPLES

Back at the laboratory, the deer were dissected and the kidneys were removed for analysis. The kidneys were chosen because the suspected pathogen (arsenic) is rapidly eliminated from an animal through its urinary tract.

PROCESSING THE SAMPLE: PREPARING A LABORATORY SAMPLE

Each kidney was cut into pieces and homogenized in a high-speed blender. This step served to reduce the size of the pieces of tissue and to homogenize the resulting laboratory sample.

PROCESSING THE SAMPLE: DEFINING REPLICATE SAMPLES

Three 10-g samples of the homogenized tissue from each deer were placed in porcelain crucibles.

DOING CHEMISTRY: DISSOLVING THE SAMPLES

To obtain an aqueous solution of the analyte for analysis, it was necessary to dry ash the sample in air to convert its organic matrix to carbon dioxide and water. This process involved heating each crucible and sample cautiously over an open flame until the sample stopped smoking. The crucible was then placed in a furnace and heated at 555°C for two hours. Dry ashing served to free the analyte from organic material and convert it to arsenic pentoxide. The dry solid in each sample crucible was then dissolved in dilute HCl, which converted the As_2O_5 to soluble H_3AsO_4.

---

[3]*Official Methods of Analysis,* 15th ed., Washington, DC: Association of Official Analytical Chemists, 1990, p. 626.

ELIMINATING INTERFERENCES

Arsenic can be separated from other substances that might interfere in the analysis by converting it to arsine, AsH_3, a toxic, colorless gas that is evolved when a solution of H_3AsO_3 is treated with zinc. The solutions resulting from the deer and grass samples were combined with Sn^{2+}, and a small amount of iodide ion was added to catalyze the reduction of H_3AsO_4 to H_3AsO_3 according to the following reaction:

$$H_3AsO_4 + SnCl_2 + 2HCl \longrightarrow H_3AsO_3 + SnCl_4 + H_2O$$

The H_3AsO_3 was then converted to AsH_3 by the addition of zinc metal as follows:

$$H_3AsO_3 + 3Zn + 6HCl \longrightarrow AsH_3(g) + 3ZnCl_2 + 3H_2O$$

The entire reaction was carried out in flasks equipped with a stopper and delivery tube so that the arsine could be collected in the absorber solution as shown in Figure 1-4. The arrangement ensured that interferences were left in the reaction flask and that only arsine was collected in the absorber in special transparent containers called cuvettes.

Arsine bubbled into the solution in the cuvette, reacted with silver diethyldithiocarbamate to form a colored complex compound according to the following equation:

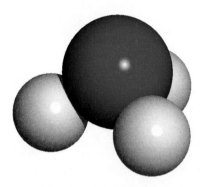

Throughout this text, we will present models of molecules that are important in analytical chemistry. Here we show arsine, AsH_3. Arsine is an extremely toxic colorless gas with a noxious garlic odor. Analytical methods involving the generation of arsine must be carried out with caution and proper ventilation.

$$AsH_3 + 6Ag^+ + 3\left[\begin{array}{c} C_2H_5 \\ \\ C_2H_5 \end{array}\!\!N\!\!-\!\!C\!\!\begin{array}{c} S \\ \\ S \end{array}\right]^- \longrightarrow$$

$$As\left[\begin{array}{c} C_2H_5 \\ \\ C_2H_5 \end{array}\!\!N\!\!-\!\!C\!\!\begin{array}{c} S \\ \\ S \end{array}\right]_3 + 6Ag + 3H^+$$

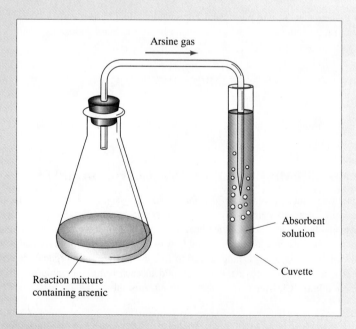

Figure 1-4   An easily constructed apparatus for generating arsine, AsH_3.

Molecular model of diethyldithiocarbamate. This compound is an analytical reagent used in the determination of arsenic as illustrated in this Feature.

MEASURING THE AMOUNT OF THE ANALYTE

The amount of arsenic in each sample was determined by measuring the intensity of the red color formed in the cuvettes with an instrument called a spectrophotometer. As shown in Chapter 23 a spectrophotometer provides a number called absorbance that is directly proportional to the color intensity, which is also proportional to the concentration of the species responsible for the color. To use absorbance for analytical purposes, a calibration curve must be generated by measuring the absorbance of several solutions that contain known concentrations of analyte. The upper part of Figure 1-5 shows that the color becomes more intense as the arsenic content of the standards increases from 0 to 25 parts per million (ppm).

CALCULATING THE CONCENTRATION

The absorbances for the standard solutions containing known concentrations of arsenic are plotted to produce a *calibration curve,* shown in the lower part of Figure 1-5. Each vertical line between the upper and lower parts of Figure 1-5 ties a solution to its corresponding point on the plot. The color intensity of each solution is represented by its absorbance, which is plotted on the vertical axis of the calibration curve. Note that the absorbance increases from 0 to about 0.72 as the concentration of arsenic increases from 0 to 25 parts per million. The concentration of arsenic in each standard solution corresponds to the vertical grid lines of the calibration curve as shown. This curve is then used to determine the concentration of the two unknown solutions shown on the right. We first find the absorbances of the unknowns on the absorbance axis of the plot and then read the corresponding concentrations on the concentration axis. The lines leading from the cuvettes to the calibration curve show that the concentrations of arsenic in the two deer were 16 ppm and 22 ppm, respectively.

Arsenic in kidney tissue of an animal is toxic at levels above about 10 ppm, so it was probable that the deer were killed by ingesting an arsenic compound. The tests also showed that the samples of grass contained about 600 ppm arsenic. This very high level of arsenic suggested that the grass had been sprayed with an arsenical herbicide. The investigators concluded that the deer had probably died as a result of eating the poisoned grass.

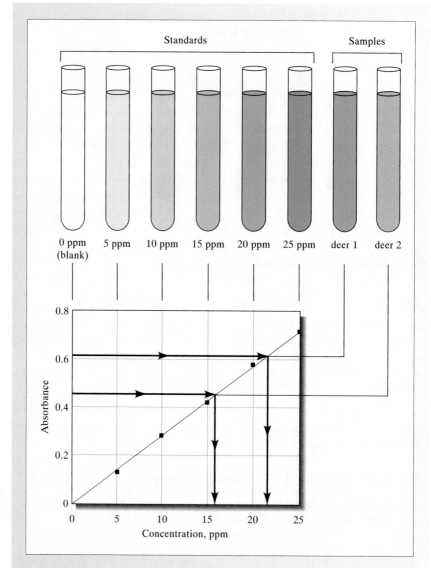

Figure 1-5   Constructing and using a calibration curve to determine the concentration of arsenic. The absorbances of the solutions in the cuvettes are measured using a spectrophotometer. The absorbance values are then plotted against the concentrations of the solutions in the cuvettes, as illustrated in the graph. Finally, the concentrations of the unknown solutions are read from the plot, as shown by the dark arrows.

ESTIMATING THE RELIABILITY OF THE DATA

The data from these experiments were analyzed using the statistical methods described in Chapters 5, 6 and 7. For each of the standard arsenic solutions and the deer samples, the average of the three absorbance measurements was calculated. The average absorbance for the replicates is a more reliable measure of the concentration of arsenic than a single measurement. Least-squares analysis of the standard data (see

Section 7E) was used to find the best straight line among the points and to calculate the concentrations of the unknown samples along with their statistical uncertainties and confidence limits.

In this analysis, the formation of the highly colored product of the reaction served both to confirm the probable presence of arsenic and to provide a reliable estimate of its concentration in the deer and in the grass. Based on their results, the investigators recommended that the use of arsenical herbicides be suspended in the wildlife area to protect the deer and other animals that might eat plants there.

This case study illustrates how chemical analysis is used in the identification and determination of quantities of hazardous chemicals in the environment. Many of the methods and instruments of analytical chemistry are used routinely to provide vital information in environmental and toxicological studies of this type. The system flow diagram of Figure 1-3 may be applied to this case study. The desired state is a concentration of arsenic that is below the toxic level. Chemical analysis is used to determine the actual state, or the concentration of arsenic in the environment, and this value is compared to the desired concentration. The difference is then used to determine appropriate actions (such as decreased use of arsenical pesticides) to ensure that deer are not poisoned by excessive amounts of arsenic in the environment, which in this example is the controlled system.

Section I

The Tools of Analytical Chemistry

Chapter 1 was a survey of the scope and importance of chemical analysis. In the six chapters of Section I, we introduce the methods and operations of analytical chemistry that are essential to begin work in the laboratory and to understand the principles of analysis. Chapter 2 provides a comprehensive treatment of the apparatus, techniques, chemicals, and spreadsheet software necessary for safe, effective, and efficient work in the laboratory. Next, Chapter 3 treats the chemical concepts required for understanding solution preparation and stoichiometry. The basic theory of chemical equilibrium is explored in Chapter 4 as are acid-base chemistry and precipitation equilibria. Chapter 5 introduces some of the fundamental ideas of statistics, Chapter 6 extends these ideas to encompass the descriptive statistics of random errors, and Chapter 7 explores a number of applications of statistics that are useful in the evaluation of analytical data.

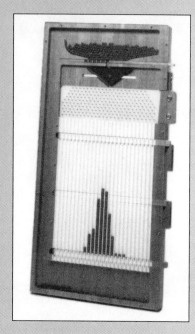

The quincunx is a mechanical device that is used to simulate a normal probability distribution. Small balls fall into the center of the top of the device, which con-tains a regular pattern of pegs to randomly deflect the balls to the right and to the left of the center position. Each time a ball hits a peg, it has a 50-50 chance of falling to the right or to the left. After each ball passes through the array of pegs, it drops into one of the vertical slots of the transpar-ent case. Each slot is called a bin, and the height of the column of balls in each bin is proportional to the probability of a ball falling into a given bin. The center bin is directly below the entry point of the balls above the array of pegs, and so the highest probability of finding a ball is at zero horizontal displacement from the center. The probability falls off to the right and to the left of the center bin as described by the equation for a normal distribution. The quin-cunx is a kind of dynamic me-chanical histogram. We discuss histograms, probability, and nor-mal distributions in Chapters 5 through 7.

It is interesting to note that the precise pattern and location of the pegs in the quincunx determines the type of probability distribution of the balls in the bins, and statis-ticians have worked out the math-ematical formulas that describe the patterns of pegs that yield var-ious different types of distribu-tions. You can learn more about the quincunx and view Java ap-plets that simulate its action at http://www.users.on.net/zhcchz/java/quincunx/quincunx.1.html and http://stad.dsl.nl/~berrie1/index.html. *(Photo courtesy of Lightning Calculator, Troy, MI.)*

Chapter 2

Chemicals and Apparatus: Putting the Tools to Work

I n this chapter, we shall introduce the tools, techniques, and chemicals that are used by analytical chemists. The development of these tools began over two centuries ago and continues today. As the technology of analytical chemistry has improved with the advent of electronic analytical balances, automated titrators, and similar devices, the speed, convenience, accuracy, and precision of analytical methods has generally improved as well. For example, the determination of the mass of a sample that required five to ten minutes 40 years ago is now accomplished in a few seconds. Computations that took ten to twenty minutes using tables of logarithms may now be carried out almost instantaneously with a computer spreadsheet. Our experience with such magnificent technological innovations often elicits impatience with the sometimes tedious techniques of classical analytical chemistry. It is this impatience that drives the quest to develop better methodologies. Indeed, modification of basic methods in the interest of speed or convenience has often been performed without sacrificing accuracy or precision.

We must emphasize, however, that many of the unit operations encountered in the analytical laboratory are timeless. They have gradually evolved over the past two centuries, and they are tried and true. From time to time, the directions given in this chapter may seem somewhat didactic. Although we attempt to explain why unit operations are carried out in the way that we describe, you may be tempted to modify a procedure or skip a step here or there to save time and effort. We must caution you against modifying techniques and procedures unless you have discussed your proposed modification with your instructor and have considered its consequences carefully. Such modifications may cause unanticipated results including unacceptable levels of accuracy and/or precision, or in a worst-case scenario, a serious accident could result. Today, the time required to prepare a carefully standardized solution of sodium hydroxide is about the same as it was 100 years ago.

Mastery of the tools of analytical chemistry will serve you well in chemistry courses and in related scientific fields. In addition, your efforts will be rewarded with the considerable satisfaction of having completed an analysis with high standards of good analytical practice and with levels of accuracy and precision consistent with the limitations of the technique.

2A   SELECTING AND HANDLING REAGENTS AND OTHER CHEMICALS

The purity of reagents has an important bearing on the accuracy that can be attained in any analysis. It is therefore essential that the quality of a reagent be consistent with the use for which it is intended.

2A-1   Classifying Chemicals

Reagent Grade

Reagent-grade chemicals conform to the minimum standards set forth by the Reagent Chemical Committee of the American Chemical Society[1] and are used wherever possible in analytical work. Some suppliers label their products with the maximum limits of impurity allowed by the ACS specifications; others print the actual assay for the various impurities.

Primary-Standard Grade

The qualities required of a *primary standard* — in addition to extraordinary purity — are set forth in Section 11A-3. Primary-standard reagents have been carefully analyzed by the supplier, and the assay is printed on the container label. The National Institute of Standards and Technology is an excellent source for primary standards. This agency also provides *reference standards,* which are complex substances that have been exhaustively analyzed.[2]

The National Institute of Standards and Technology (NIST) is the current name of what was formerly the National Bureau of Standards.

Special-Purpose Reagent Chemicals

Chemicals that have been prepared for a specific application are also available. Included among these are solvents for spectrophotometry and high-performance liquid chromatography. Information pertinent to the intended use is supplied with these reagents. Data provided with a spectrophotometric solvent, for example, might include its absorbance at selected wavelengths and its ultraviolet cutoff wavelength.

---

[1]Committee on Analytical Reagents, *Reagent Chemicals,* 8th ed. Washington, D.C.: American Chemical Society, 1993.
[2]U.S. Department of Commerce, *NIST Standard Reference Materials Catalog, 1998–99, NIST Special Publication 260.* Washington, D.C.: U.S. Government Printing Office, 1998.

2A-2   Rules for Handling Reagents and Solutions

High quality in a chemical analysis requires reagents and solutions of established purity. A freshly opened bottle of a reagent-grade chemical can ordinarily be used with confidence; whether this same confidence is justified when the bottle is half empty depends entirely on the way it has been handled after being opened. The following rules should be observed to prevent the accidental contamination of reagents and solutions.

1. Select the best grade of chemical available for analytical work. Whenever possible, pick the smallest bottle that will supply the desired quantity.
2. Replace the top of every container *immediately* after removal of the reagent; do not rely on someone else to do this.
3. Hold the stoppers of reagent bottles between your fingers; never set a stopper on a desktop.
4. *Unless specifically directed otherwise, never return any excess reagent to a bottle.* The money saved by returning excesses is seldom worth the risk of contaminating the entire bottle.
5. Unless directed otherwise, never insert spatulas, spoons, or knives into a bottle that contains a solid chemical. Instead, shake the capped bottle vigorously or tap it gently against a wooden table to break up any encrustation, then pour out the desired quantity. These measures are occasionally ineffective; in such cases, a clean porcelain spoon should be used.
6. Keep the reagent shelf and the laboratory balance clean and neat. Clean up any spillages immediately, even though someone else is waiting to use the same chemical or reagent.
7. Observe local regulations concerning the disposal of surplus reagents and solutions.

2B   CLEANING AND MARKING LABORATORY WARE

A chemical analysis is ordinarily performed in duplicate or triplicate. Thus, each vessel that holds a sample must be marked so that its contents can be positively identified. Flasks, beakers, and some crucibles have small etched areas on which semipermanent markings can be made with a pencil.

Special marking inks are available for porcelain surfaces. The marking is baked permanently into the glaze by heating at a high temperature. A saturated solution of iron(III) chloride, although not as satisfactory as the commercial preparation, can also be used for marking.

Every beaker, flask, or crucible that will contain the sample must be thoroughly cleaned before being used. The apparatus should be washed with a hot detergent solution and then rinsed, initially with copious amounts of tap water and finally with several small portions of deionized water.[3] Properly cleaned glassware will be coated with a uniform and unbroken film of water. *It is seldom*

Unless you are directed otherwise, do not dry the interior surfaces of glassware or porcelain ware.

---

[3]References to deionized water in this chapter and in Chapter 27 apply equally to distilled water.

Figure 2-1 Arrangement for the evaporation of a liquid.

Bumping is sudden, often violent boiling that tends to spatter solution out of its container.

Wet-ashing is the oxidation of the organic constituents of a sample with oxidizing reagents such as nitric acid, sulfuric acid, hydrogen peroxide, or aqueous bromine, or a combination of these reagents.

An *analytical balance* has a maximum capacity that ranges from 1 g to several kilograms and a precision at maximum capacity of at least 1 part in 10^5.

A *macrobalance* is the most common type of analytical balance; it has a maximum load of 160 to 200 g and a precision of 0.1 mg.

A *semimicroanalytical balance* has a maximum load of 10 to 30 g and a precision of 0.01 mg.

necessary to dry the interior surface of glassware before use; drying is ordinarily a waste of time at best and a potential source of contamination at worst.

An organic solvent, such as benzene or acetone, may be effective in removing grease films. Chemical suppliers also market preparations for the elimination of such films.

2C   EVAPORATING LIQUIDS

It is frequently necessary to decrease the volume of a solution that contains a nonvolatile solute. Figure 2-1 illustrates how this operation is performed. The ribbed cover glass permits vapors to escape and protects the remaining solution from accidental contamination. The use of glass hooks to provide space between the rim of the beaker and a conventional cover glass is less satisfactory than the ribbed cover glass shown.

Evaporation is frequently difficult to control because of the tendency of some solutions to overheat locally. The *bumping* that results can be sufficiently vigorous to cause partial loss of the solution. Careful and gentle heating will minimize the danger of such loss. Where their use is permissible, glass beads will also minimize bumping.

Some unwanted species can be eliminated during evaporation. For example, chloride and nitrate can be removed from a solution by adding sulfuric acid and evaporating until copious white fumes of sulfur trioxide are observed (this operation must be performed in a hood). Urea is effective in removing nitrate ion and nitrogen oxides from acidic solutions. The removal of ammonium chloride is best accomplished by adding concentrated nitric acid and evaporating the solution to a small volume. Ammonium ion is rapidly oxidized upon heating; the solution is then evaporated to dryness.

Organic constituents can frequently be eliminated from a solution by adding sulfuric acid and heating in a hood until sulfur trioxide fumes appear; this process is known as *wet-ashing*. Nitric acid can be added toward the end of heating to hasten oxidation of the last traces of organic matter.

2D   MEASURING MASS

In most analyses, an *analytical balance* must be used to obtain highly accurate masses. Less accurate *laboratory balances* are also employed in the analytical laboratory for mass measurements where the demands for reliability are not critical.

2D-1   Types of Analytical Balances

By definition, an *analytical balance* is a weighing instrument with a maximum capacity that ranges from 1 g to a few kilograms with a precision of at least 1 part in 10^5 at maximum capacity. The precision and accuracy of many modern analytical balances exceed one part in 10^6 at full capacity.

The most commonly encountered analytical balances (*macrobalances*) have a maximum capacity ranging between 160 and 200 g; measurements can be made with a standard deviation of ± 0.1 mg. *Semimicroanalytical balances* have a maximum load of 10 to 30 g with a precision of $\pm$ 0.01 mg. A typical *microana-lytical balance* has a capacity of 1 to 3 g and a precision of $\pm$ 0.001 mg.

The analytical balance has undergone a dramatic evolution over the past several decades. The traditional analytical balance had two pans attached to either end of a lightweight beam that pivoted about a knife edge located in the center of the beam. The object to be weighed was placed on one pan; sufficient standard weights were then added to the other pan to restore the beam to its original position. Weighing with such an *equal-arm* balance was tedious and time consuming.

The first *single-pan analytical balance* appeared on the market in 1946. The speed and convenience of weighing with this balance were vastly superior to what could be realized with the traditional equal-arm balance. Consequently, this balance rapidly replaced the latter in most laboratories. The single-pan balance is still extensively used. The design and operation of a single-pan balance are discussed briefly in Section 2D-3. The single-pan balance is currently being replaced by the *electronic analytical balance,* which has neither a beam nor a knife edge.

2D-2   The Electronic Analytical Balance[4]

Figure 2-2 is a functional diagram of an electronic analytical balance. The pan rides above a hollow metal cylinder that is surrounded by a coil that fits over the inner pole of a cylindrical permanent magnet. An electric current in the coil creates a magnetic field that supports or *levitates* the cylinder, the pan and indicator arm, and whatever load is on the pan. The current is adjusted so that the level of

> A *microanalytical balance* has a maximum load of 1 to 3 g and a precision of 0.001 mg or 1 μg.

> To *levitate* means to cause an object to float in air.

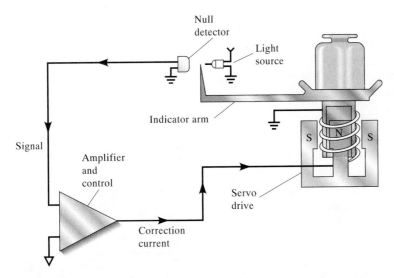

Figure 2-2 Electronic analytical balance. (From R. M. Schoonover, *Anal. Chem.,* **1982,** *54,* 973A. Published 1982 American Chemical Society.)

[4]For a more detailed discussion, see R. M. Schoonover, *Anal. Chem.,* **1982,** *54,* 973A; K. M. Lang, *Amer. Lab.,* **1983,** *15* (3), 72.

the indicator arm is in the null position when the pan is empty. Placing an object on the pan causes the pan and indicator arm to move downward, which increases the amount of light striking the photocell of the null detector. The increased current from the photocell is amplified and fed into the coil, creating a larger magnetic field that returns the pan to its original null position. A device such as this, in which a small electric current causes a mechanical system to maintain a null position, is one type of *servo system*. The current required to keep the pan and object in the null position is directly proportional to the mass of the object and is readily measured, digitized, and displayed. The calibration of an electronic balance involves the use of a standard mass and adjustment of the current so that the mass of the standard is exhibited on the display.

Figure 2-3 shows configurations for two electronic analytical balances. In each, the pan is tethered to a system of constraints known collectively as a *cell*. The cell incorporates several *flexures* that permit limited movement of the pan and prevent torsional forces (resulting from off-center loading) from disturbing the alignment of the balance mechanism. At null, the beam is parallel to the gravitational horizon and each flexure pivot is in a relaxed position.

Figure 2-3a shows an electronic balance with the pan located below the cell. Higher precision is achieved with this arrangement than with the top-loading design shown in Figure 2-3b. Even so, top-loading electronic balances have a precision that equals or exceeds that of the best mechanical balances and additionally provide unencumbered access to the pan.

Electronic balances generally feature an automatic *taring control* that causes the display to read zero with a container (such as a boat or weighing bottle) on the pan. Most balances permit taring to 100% of capacity.

Some electronic balances have dual capacities and dual precisions. These features permit the capacity to be decreased from that of a macrobalance to that of a

> A *tare* is the mass of an empty sample container. Taring is the process of setting a balance to read zero with the tare on the pan.

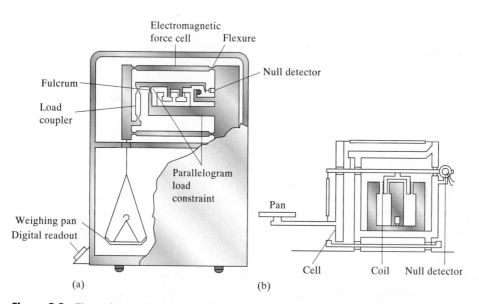

Figure 2-3   Electronic analytical balances. (a) Classical configuration with pan beneath the cell. (From R. M. Schoonover, *Anal. Chem.*, **1982**, *54*, 973A. Published 1982 American Chemical Society.) (b) A top-loading design. Note that the mechanism is enclosed in a windowed case. K. M. Lang, *Amer. Lab.*, **1983**, *15* (3), 72. Copyright 1983 by International Scientific Communications, Inc.

semimicrobalance (30 g) with a concomitant gain in precision of 0.01 mg. Thus, the chemist has effectively two balances in one.

A modern electronic analytical balance provides unprecedented speed and ease of use. For example, one instrument is controlled by touching a single bar at various positions along its length. One position on the bar turns the instrument on or off, another automatically calibrates the balance against a standard weight, and a third zeros the display, either with or without an object on the pan. Reliable weighing data are obtainable with little or no instruction or practice.

A photograph of a modern electronic balance is shown in color plate 23.

2D-3   The Single-Pan Mechanical Analytical Balance

Components

Although they differ considerably in appearance and performance characteristics, all mechanical balances — equal-arm as well as single-pan — have several common components. Figure 2-4 is a diagram of a typical single-pan mechanical balance. Fundamental to this instrument is a lightweight *beam* that is supported on a planar surface by a prism-shaped *knife edge* (*A*). Attached to the left end of the beam is a pan for holding the object to be weighed and a full set of weights held in place by hangers. These weights can be lifted from the beam one at a time by a mechanical arrangement that is controlled by a set of knobs on the exterior of the balance case. The right end of the beam holds a counterweight that just balances the pan and weights on the left end of the beam.

A second knife edge (*B*) is located near the left end of the beam and serves to support a second planar surface, which is located in the inner side of a *stirrup* that couples the pan to the beam. The two knife edges and their planar surfaces

The two *knife edges* in a mechanical balance are prism-shaped agate or sapphire devices that form low-friction bearings with planar surfaces located in the *stirrups* and in the fulcrum of the beam — these surfaces are also of agate or sapphire.

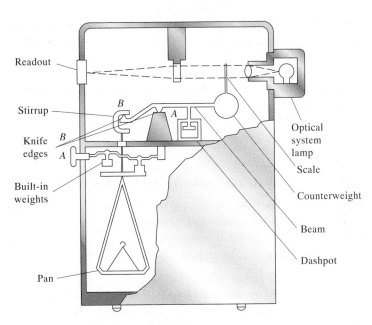

Figure 2-4   Single-pan analytical balance. (From R. M. Schoonover, *Anal. Chem.*, **1982**, *54*, 973A. Published 1982 American Chemical Society.)

are fabricated from extraordinarily hard materials (agate or synthetic sapphire) and form two bearings that permit motion of the beam and pan with a minimum of friction. The performance of a mechanical balance is critically dependent on the perfection of these two bearings.

Single-pan balances are also equipped with a *beam arrest* and a *pan arrest*. The beam arrest is a mechanical device that raises the beam so that the central knife edge no longer touches its bearing surface and simultaneously frees the stirrup from contact with the outer knife edge. The purpose of both arrest mechanisms is to prevent damage to the bearings while objects are being placed on or removed from the pan. When engaged, the pan arrest supports most of the mass of the pan and its contents and thus prevents oscillation. Both arrests are controlled by a lever mounted on the outside of the balance case that should be engaged whenever the balance is not in use.

An *air damper* (also known as a *dashpot*) is mounted near the end of the beam opposite the pan. This device consists of a piston that moves within a concentric cylinder attached to the balance case. Air in the cylinder undergoes expansion and contraction as the beam is set in motion; the beam rapidly comes to rest as a result of this opposition to motion.

Protection from air currents is needed to permit discrimination between small differences in mass (<1 mg). An analytical balance is thus always enclosed in a case equipped with doors to permit the introduction or removal of objects.

> To avoid damage to the knife edges and bearing surfaces, the arrest system for a mechanical balance should be engaged at all times other than during actual weighing.

Weighing with a Single-Pan Balance

The beam of a properly adjusted balance assumes an essentially horizontal position with no object on the pan and all weights in place. When the pan and beam arrests are disengaged, the beam is free to rotate around the knife edge. Placing an object on the pan causes the left end of the beam to move downward. Weights are then removed systematically one by one from the beam until the imbalance is less than 100 mg. The angle of deflection of the beam with respect to its original horizontal position is directly proportional to the milligrams of additional mass that must be removed to restore the beam to its original horizontal position. The optical system shown in the upper part of Figure 2-4 measures this angle of deflection and converts this angle to milligrams. A *reticle,* a small transparent screen mounted on the beam, is scribed with a scale that reads 0 to 100 mg. A beam of light passes through the scale to an enlarging lens, which in turn focuses a small part of the enlarged scale onto a frosted glass plate located on the front of the balance. A vernier makes it possible to read this scale to the nearest 0.1 mg.

2D-4   Precautions in Using an Analytical Balance

An analytical balance is a delicate instrument that you must handle with care. Consult with your instructor for details about weighing with your particular model of balance. Observe the following general rules for working with an analytical balance regardless of make or model:

1. Center the load on the pan as well as possible.
2. Protect the balance from corrosion. Objects to be placed on the pan should be limited to nonreactive metals, nonreactive plastics, and vitreous materials.

3. Observe special precautions (Section 2E-6) for the weighing of liquids.
4. Consult the instructor if the balance appears to need adjustment.
5. Keep the balance and its case scrupulously clean. A camel's-hair brush is useful for the removal of spilled material or dust.
6. Always allow an object that has been heated to return to room temperature before weighing it.
7. Use tongs or finger pads to prevent the uptake of moisture by dried objects.

2D-5   Sources of Error in Weighing

Correction for Buoyancy[5]

A *buoyancy error* will affect data if the density of the object being weighed differs significantly from that of the standard weights. This error has its origin in the difference in the buoyant force exerted by the medium (air) on the object and on the weights. Correction for buoyancy is accomplished with the equation

$$W_1 = W_2 + W_2\left(\frac{d_{air}}{d_{obj}} - \frac{d_{air}}{d_{wts}}\right) \qquad (2\text{-}1)$$

where W_1 is the corrected mass of the object, W_2 is the mass of the standard weights, d_{obj} is the density of the object, d_{wts} is the density of the weights, and d_{air} is the density of the air displaced by them; d_{air} has a value of $0.0012\ \text{g}\cdot\text{cm}^{-3}$.

The consequences of Equation 2-1 are shown in Figure 2-5, in which the relative error due to buoyancy is plotted against the density of objects weighed in air against stainless steel weights. Note that this error is less than 0.1% for objects

> A *buoyancy error* is the weighing error that develops when the object being weighed has a significantly different density than the weights.

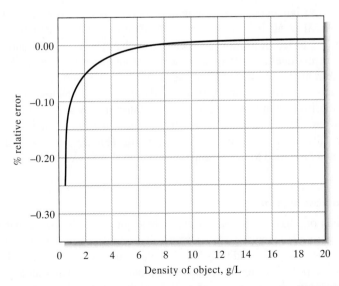

Figure 2-5   Effect of buoyancy on weighing data (density of weights = $8\ \text{g}\cdot\text{cm}^{-3}$). Plot of relative error as a function of the density of the object weighted.

---

[5]For further information, see R. Battino and A. G. Williamson, *J. Chem. Educ.,* **1984,** *64,* 51.

that have a density of 2 g·cm^{-3} or greater. It is thus seldom necessary to apply a correction to the mass of most solids. The same cannot be said for low-density solids, liquids, or gases, however; for these, the effects of buoyancy are significant and a correction must be applied.

The density of weights used in single-pan balances ranges from 7.8 to 8.4 g·cm^{-3}, depending on the manufacturer. Use of 8 g·cm^{-3} is adequate for most purposes. If a greater accuracy is required, the specifications for the balance to be used should be consulted for the necessary density data.

Example 2-1

A bottle weighed 7.6500 g empty and 9.9700 g after introduction of an organic liquid with a density of 0.92 g·cm^{-3}. The balance was equipped with stainless steel weights ($d = 8.0$ g·cm^{-3}). Correct the mass of the sample for the effects of buoyancy.

The apparent mass of the liquid is 9.9700 − 7.6500 = 2.3200 g. The same buoyant force acts on the container during both weighings; thus, we need to consider only the force that acts on the 2.3200 g of liquid. Substitution of 0.0012 g·cm^{-3} for d_{air}, 0.92 g·cm^{-3} for d_{obj}, and 8.0 g·cm^{-3} for d_{wts} in Equation 2-1 gives

$$W_1 = 2.3200 + 2.3200 \left(\frac{0.0012}{0.92} - \frac{0.0012}{8.0} \right) = 2.3227 \text{ g}$$

Temperature Effects

Attempts to weigh an object whose temperature is different from that of its surroundings will result in a significant error. Failure to allow enough time for a heated object to return to room temperature is the most common source of this problem. Errors due to a difference in temperature have two sources. First, convection currents within the balance case exert a buoyant effect on the pan and object. Second, warm air trapped in a closed container weighs less than the same volume at a lower temperature. Both effects cause the apparent mass of the object to be low. This error can amount to as much as 10 or 15 mg for a typical porcelain filtering crucible or a weighing bottle (Figure 2-6). Heated objects must always be cooled to room temperature before being weighed.

Always allow heated objects to return to room temperature before you attempt to weigh them.

Other Sources of Error

A porcelain or glass object will occasionally acquire a static charge that is sufficient to cause a balance to perform erratically; this problem is particularly serious when the relative humidity is low. Spontaneous discharge frequently occurs after a short period. A low-level source of radioactivity (such as a photographer's brush) in the balance case will provide sufficient ions to relieve the charge. Alternatively, the object can be wiped with a faintly damp chamois.

The optical scale of a single-pan balance should be checked regularly for accuracy, particularly under loading conditions that require the full scale range. A standard 100-mg weight is used for this check.

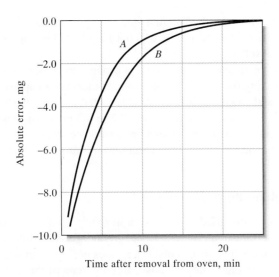

Figure 2-6   Effect of temperature on weighing data. Absolute error in mass as a function of time after the object was removed from a 110°C drying oven. *A*: porcelain filtering crucible. *B*: weighing bottle containing about 7.5 g of KCl.

2D-6   Auxiliary Balances

Balances that are less precise than analytical balances find extensive use in the analytical laboratory. These offer the advantages of speed, ruggedness, large capacity, and convenience; they should be used whenever high sensitivity is not required.

Top-loading auxiliary balances are particularly convenient. A sensitive top-loading balance will accommodate 150 to 200 g with a precision of about 1 mg, an order of magnitude less than a macroanalytical balance. Some balances of this type tolerate loads as great as 25,000 g with a precision of ± 0.05 g. Most are equipped with a taring device that brings the balance reading to zero with an empty container on the pan. Some are fully automatic, require no manual dialing or weight handling, and provide a digital readout of the mass. Modern top-loading balances are electronic (Figure 2-3b).

A triple-beam balance with a sensitivity less than that of a typical top-loading auxiliary balance is also useful. This balance is a single-pan balance with three decades of weights that slide along individual calibrated scales. The precision of a triple-beam balance may be one or two orders of magnitude less than that of a top-loading instrument. This precision is adequate for many weighing operations. The triple-beam balance offers the advantages of simplicity, durability, and low cost.

Use auxiliary laboratory balances for weighings that do not require great accuracy.

2E   THE EQUIPMENT AND MANIPULATIONS ASSOCIATED WITH WEIGHING

The mass of many solids changes with humidity because they tend to absorb weighable amounts of moisture. This effect is especially pronounced when a large surface area is exposed, as with a reagent chemical or a sample that has been ground to a fine powder. The first step in a typical analysis, then, involves drying the sample so that the results will not be affected by the humidity of the surrounding atmosphere.

A sample, a precipitate, or a container is brought to *constant mass* by a cycle that involves heating (ordinarily for one hour or more) at an appropriate temperature, cooling, and weighing. This cycle is repeated as many times as needed to obtain successive masses that agree within 0.2 to 0.3 mg of one another. The establishment of constant mass provides some assurance that the chemical or physical processes that occur during the heating (or ignition) are complete.

Drying or ignition to constant mass is a process in which a solid is cycled through heating, cooling, and weighing steps until its mass becomes constant to within 0.2 to 0.3 mg.

2E-1   Weighing Bottles

Solids are conveniently dried and stored in *weighing bottles,* two common varieties of which are shown in Figure 2-7. The ground-glass portion of the cap-style bottle shown on the left is on the outside and does not come into contact with the contents; this design eliminates the possibility of some of the sample becoming entrained upon and subsequently lost from the ground-glass surface.

Plastic weighing bottles are also available. Ruggedness is the principal advantage of these bottles over their glass counterparts.

2E-2   Desiccators and Desiccants

Oven drying is the most common way of removing moisture from solids. This approach is not appropriate for substances that decompose or for those from which water is not removed at the temperature of the oven.

A *desiccator* is a device for drying substances or objects.

Dried materials are stored in *desiccators* while they cool so as to minimize the uptake of moisture. Figure 2-8 shows the components of a typical desiccator. The base section contains a chemical drying agent, such as anhydrous calcium chloride, calcium sulfate (Drierite®), anhydrous magnesium perchlorate (Anhydrone® or Dehydrite®), or phosphorus pentoxide. The ground-glass surfaces are lightly coated with grease.

When you remove or replace the lid of a desiccator, use a sliding motion to minimize the likelihood of disturbing the sample. An airtight seal is achieved by slight rotation and downward pressure on the positioned lid.

When you place a heated object in a desiccator, the increase in pressure as the enclosed air is warmed may be sufficient to break the seal between lid and base. Conversely, if the seal is not broken, the cooling of heated objects can cause development of a partial vacuum. Both of these conditions can cause the contents of the desiccator to be physically lost or contaminated. Although it defeats the

Figure 2-7   Typical weighing bottles.

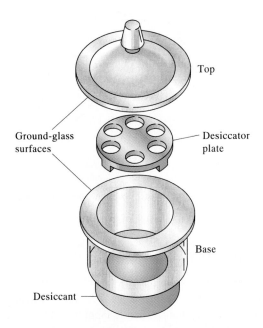

Figure 2-8 Components of a typical desiccator. The base contains a chemical drying agent, which is usually covered with a wire screen and a porcelain plate with holes to accommodate weighing bottles or crucibles.

purpose of the desiccator somewhat, you should allow some cooling to occur before the lid is seated. It is also helpful to break the seal once or twice during cooling to relieve any excessive vacuum that develops. Finally, you should lock the lid in place with your thumbs when moving the desiccator from one place to another.

Very hygroscopic materials should be stored in containers equipped with snug covers, such as weighing bottles; the covers remain in place while in the desiccator. Most other solids can be safely stored uncovered.

2E-3   Manipulating Weighing Bottles

Heating at 105 to 110°C is sufficient to remove the moisture from the surface of most solids. Figure 2-9 depicts the arrangement recommended for drying a sample. The weighing bottle is contained in a labeled beaker with a ribbed cover glass. This arrangement protects the sample from accidental contamination and also allows for the free access of air. Crucibles containing a precipitate that can be freed of moisture by simple drying can be treated similarly. The beaker containing the weighing bottle or crucible to be dried must be carefully marked to permit identification.

You should avoid handling a dried object with your fingers because detectable amounts of water or oil from your skin may be transferred to the object. You can eliminate this problem by using tongs, chamois finger cots, clean cotton gloves, or strips of paper to handle dried objects for weighing. Figure 2-10 shows how to manipulate a weighing bottle with strips of paper.

Figure 2-9   Arrangement for the drying of samples.

Figure 2-10 Method for quantitative transfer of a solid sample. Note the use of a paper strip to avoid contact between glass and skin.

2E-4 Weighing by Difference

Weighing by difference is a simple method for determining a series of sample weights. First the bottle and its contents are weighed. One sample is then transferred from the bottle to a container; gentle tapping of the bottle with its top and slight rotation of the bottle provide control over the amount of sample removed. Following transfer, the bottle and its residual contents are weighed. The mass of the sample is the difference between the two weighings. It is essential that all the solid removed from the weighing bottle be transferred without loss to the container.

2E-5 Weighing Hygroscopic Solids

Hygroscopic substances rapidly absorb moisture from the atmosphere and therefore require special handling. You need a weighing bottle for each sample to be weighed. Place the approximate amount of sample needed in the individual bottles and heat for an appropriate time. When heating is complete, quickly cap the bottles and cool in a desiccator. Before weighing a bottle, open it momentarily to relieve any vacuum. Quickly empty the contents of the bottle into its receiving vessel, cap immediately, and weigh the bottle again (along with any solid that did not get transferred). Repeat for each sample and determine the sample masses by difference.

2E-6 Weighing Liquids

The mass of a liquid is always obtained by difference. Liquids that are noncorrosive and relatively nonvolatile can be transferred to previously weighed containers with snugly fitting covers (such as weighing bottles); the mass of the container is subtracted from the total mass.

A volatile or corrosive liquid should be sealed in a weighed glass ampoule. The ampoule is heated, and the neck is then immersed in the sample; as cooling occurs, the liquid is drawn into the bulb. The ampoule is then inverted and the neck is sealed off with a small flame. The ampoule and its contents, along with any glass removed during sealing, are cooled to room temperature and weighed. The ampoule is then transferred to an appropriate container and broken. A volume correction for the glass of the ampoule may be needed if the receiving vessel is a volumetric flask.

2F   THE EQUIPMENT AND MANIPULATIONS FOR FILTRATION AND IGNITION

2F-1   Apparatus

Simple Crucibles

Simple crucibles serve only as containers. Porcelain, aluminum oxide, silica, and platinum crucibles maintain constant mass — within the limits of experimental error — and are used principally to convert a precipitate into a suitable weighing form. The solid is first collected on a filter paper. The filter and contents are then transferred to a weighed crucible, and the paper is ignited.

Simple crucibles of nickel, iron, silver, and gold are used as containers for the high-temperature fusion of samples that are not soluble in aqueous reagents. Attack by both the atmosphere and the contents may cause these crucibles to suffer mass changes. Moreover, such attack will contaminate the sample with species derived from the crucible. The chemist selects the crucible whose products will offer the least interference in subsequent steps of the analysis.

Filtering Crucibles

Filtering crucibles serve not only as containers but also as filters. A vacuum is used to hasten the filtration; a tight seal between crucible and filtering flask is accomplished with any of several types of rubber adapters (see Figure 2-11; a

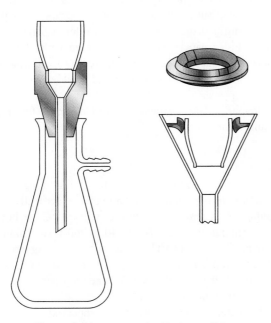

Figure 2-11   Adapters for filtering crucibles.

complete filtration train is shown in Figure 2-16). Collection of a precipitate with a filtering crucible is frequently less time consuming than with paper.

Sintered-glass (also called *fritted-glass*) crucibles are manufactured in fine, medium, and coarse porosities (marked *f, m,* and *c*). The upper temperature limit for a sintered-glass crucible is ordinarily about 200°C. Filtering crucibles made entirely of quartz can tolerate substantially higher temperatures than that without damage. The same is true for crucibles with unglazed porcelain or aluminum oxide frits. The latter are not as costly as quartz.

A *Gooch crucible* has a perforated bottom that supports a fibrous mat. Asbestos was at one time the filtering medium of choice for a Gooch crucible; current regulations concerning this material have virtually eliminated its use. Small circles of glass matting have now replaced asbestos; they are used in pairs to protect against disintegration during the filtration. Glass mats can tolerate temperatures in excess of 500°C and are substantially less hygroscopic than asbestos.

Filter Paper

Paper is an important filtering medium. Ashless paper is manufactured from cellulose fibers that have been treated with hydrochloric and hydrofluoric acids to remove metallic impurities and silica; ammonia is then used to neutralize the acids. The residual ammonium salts in many filter papers may be sufficient to affect the determination of nitrogen by the Kjeldahl method (Section 27C-11).

All papers tend to pick up moisture from the atmosphere, and ashless paper is no exception. It is thus necessary to destroy the paper by ignition if the precipitate collected on it is to be weighed. Typically, 9- or 11-cm circles of ashless paper leave a residue that weighs less than 0.1 mg, an amount that is ordinarily negligible. Ashless paper can be obtained in several porosities.

Gelatinous precipitates, such as hydrous iron(III) oxide, clog the pores of any filtering medium. A coarse-porosity ashless paper is most effective for the filtration of such solids, but even here clogging occurs. This problem can be minimized by mixing a dispersion of ashless filter paper with the precipitate prior to filtration. Filter paper pulp is available in tablet form from chemical suppliers; if necessary, the pulp can be prepared by treating a piece of ashless paper with concentrated hydrochloric acid and washing the disintegrated mass free of acid.

Table 2-1 summarizes the characteristics of common filtering media. None satisfies all requirements.

Heating Equipment

Many precipitates can be weighed directly after being brought to constant mass in a low-temperature drying oven. Such an oven is electrically heated and capable of maintaining a constant temperature to within 1°C (or better). The maximum attainable temperature ranges from 140 to 260°C, depending on make and model; for many precipitates, 110°C is a satisfactory drying temperature. The efficiency of a drying oven is greatly increased by the forced circulation of air. The passage of predried air through an oven designed to operate under a partial vacuum represents an additional improvement.

Microwave laboratory ovens are currently appearing on the market. Where applicable, these greatly shorten drying cycles. For example, slurry samples that

Table 2-1

Comparison of Filtering Media for Gravimetric Analyses

Characteristic	Paper	Gooch Crucible, Glass Mat	Glass Crucible	Porcelain Crucible	Aluminum Oxide Crucible
Speed of filtration	Slow	Rapid	Rapid	Rapid	Rapid
Convenience and ease of preparation	Troublesome, inconvenient	Convenient	Convenient	Convenient	Convenient
Maximum ignition temperature, °C	None	>500	200–500	1100	1450
Chemical reactivity	Carbon has reducing properties	Inert	Inert	Inert	Inert
Porosity	Many available	Several available	Several available	Several available	Several available
Convenience with gelatinous precipitates	Satisfactory	Unsuitable; filter tends to clog	Unsuitable; filter tends to clog	Unsuitable; filter tends to clog	Unsuitable; filter tends to clog
Cost	Low	Low	High	High	High

require 12 to 16 hr for drying in a conventional oven are reported to be dried within 5 to 6 min in a microwave oven.[6]

An ordinary heat lamp can be used to dry a precipitate that has been collected on ashless paper and to char the paper as well. The process is conveniently completed by ignition at an elevated temperature in a muffle furnace.

Burners are convenient sources of intense heat. The maximum attainable temperature depends on the design of the burner and the combustion properties of the fuel. Of the three common laboratory burners, the Meker provides the highest temperatures, followed by the Tirrill and Bunsen types.

A heavy-duty electric furnace (*muffle furnace*) is capable of maintaining controlled temperatures of 1100°C or higher. Long-handled tongs and heat-resistant gloves are needed for protection when transferring objects to or from such a furnace.

2F-2 Filtering and Igniting Precipitates

Preparation of Crucibles

A crucible used to convert a precipitate to a form suitable for weighing must maintain — within the limits of experimental error — a constant mass throughout the drying or ignition. The crucible is first cleaned thoroughly (filtering crucibles are conveniently cleaned by backwashing on a filtration train) and subjected to the same regimen of heating and cooling as that required for the precipitate. This process is repeated until constant mass (page 32) has been achieved, that is, until consecutive weighings differ by 0.3 mg or less.

Backwashing a filtering crucible is done by turning the crucible upside down in the adapter (Figure 2-11) and sucking water through the inverted crucible.

[6]H. M. Kingston and L. B. Jassie, Eds., *Introduction to Microwave Sample Preparation: Theory and Practice,* Washington, D.C.: American Chemical Society, 1988. See also *Anal. Chem.,* **1986,** *58,* 1424A; E. S. Beary, *Anal. Chem.,* **1988,** *60,* 742.

Decantation is the process of pouring a liquid gently so as to not disturb a solid in the bottom of the container.

Filtering and Washing Precipitates

The steps involved in filtering an analytical precipitate are *decantation, washing,* and *transfer.* In decantation, as much supernatant liquid as possible is passed through the filter while the precipitated solid is kept essentially undisturbed in the beaker where it was formed. This procedure speeds the overall filtration rate by delaying the time at which the pores of the filtering medium become clogged with precipitate. A stirring rod is used to direct the flow of liquid (Figure 2-12). When flow ceases, the drop of liquid at the end of the pouring spout is collected with the stirring rod and returned to the beaker. Wash liquid is next added to the beaker and thoroughly mixed with precipitate. The solid is allowed to settle, following which this liquid is also decanted through the filter. Several such washings may be required, depending on the precipitate. Most washing should be carried out *before* the solid is transferred, which results in a more thoroughly washed precipitate and a more rapid filtration.

The transfer process is illustrated in Figure 2-12c and d. The bulk of the precipitate is moved from beaker to filter by suitably directed streams of wash liquid. As in decantation and washing, a stirring rod provides direction for the flow of material to the filtering medium.

The last traces of precipitate that cling to the inside of the beaker are dislodged with a *rubber policeman,* a small section of rubber tubing that has been

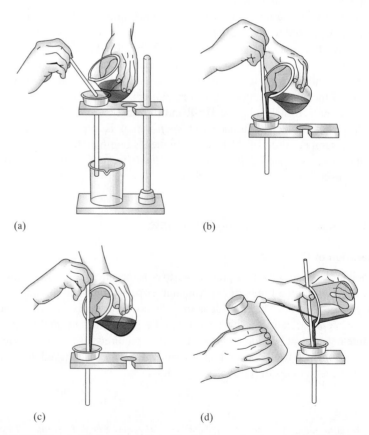

(a)   (b)

(c)   (d)

Figure 2-12   Steps in filtering: (a and b) washing by decantation; (c and d) transfer of the precipitate.

crimped on one end. The open end of the tubing is fitted onto the end of a stirring rod and wetted with wash liquid before use. Any solid collected with it is combined with the main portion on the filter. Small pieces of ashless paper can be used to wipe the last traces of hydrous oxide precipitates from the wall of the beaker; these papers are ignited along with the paper that holds the bulk of the precipitate.

Many precipitates possess the exasperating property of *creeping,* or spreading over a wetted surface against the force of gravity. Filters are never filled to more than three quarters of capacity, since some of the precipitate could be lost as the result of creeping. The addition of a small amount of nonionic detergent, such as Triton X-100, to the supernatant liquid or to the wash liquid can be helpful in minimizing creeping.

A gelatinous precipitate must be completely washed before it is allowed to dry. These precipitates shrink and develop cracks as they dry. Further additions of wash liquid simply pass through these cracks and accomplish little or no washing.

> *Creeping* is a process in which a solid moves up the side of a wetted container or filter paper.

> Do not permit a gelatinous precipitate to dry until it has been washed completely.

2F-3   Directions for Filtering and Igniting Precipitates

Preparing a Filter Paper

Figure 2-13 shows the sequence you should follow to fold a filter paper and seating it in a 60-deg funnel. The paper is folded exactly in half (a), firmly creased, and folded again (b). A triangular piece from one of the corners is torn off parallel to the second fold (c). The paper is then opened so that the untorn quarter

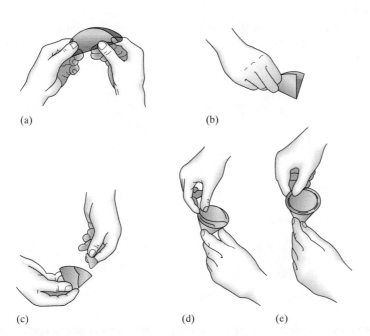

(a)   (b)

(c)   (d)   (e)

Figure 2-13   Folding and seating a filter paper. (a) Fold the paper exactly in half and crease it firmly. (b) Fold the paper a second time. (c) Tear off one of the corners on a line parallel to the second fold. (d) Open the untorn half of the folded paper to form a cone, then (e) seat the cone firmly into the funnel. Moisten the paper slightly and gently pat the paper into place.

forms a cone (d). The cone is fitted into the funnel, and the second fold is creased (e). Seating is completed by dampening the cone with water from a wash bottle and *gently* patting it with a finger. There will be no leakage of air between the funnel and a properly seated cone; in addition, the stem of the funnel will be filled with an unbroken column of liquid.

Transferring Paper and Precipitate to a Crucible

After filtration and washing have been completed, the filter and its contents must be transferred from the funnel to a crucible that has been brought to constant mass. Ashless paper has very low wet strength and must be handled with care during the transfer. The danger of tearing is lessened considerably if the paper is allowed to dry somewhat before it is removed from the funnel.

Figure 2-14 illustrates the transfer process. The triple-thick portion of the filter paper is drawn across the funnel to flatten the cone along its upper edge (a); the corners are next folded inward (b); the top edge is then folded over (c). Finally, the paper and its contents are eased into the crucible (d) so that the bulk of the precipitate is near the bottom.

Ashing Filter Paper

If a heat lamp is to be used, the crucible is placed on a clean, nonreactive surface, such as a wire screen covered with aluminum foil. The lamp is then positioned about 1 cm above the rim of the crucible and turned on. Charring takes place without further attention. The process is considerably accelerated if the paper is moistened with no more than one drop of concentrated ammonium nitrate solution. Elimination of the residual carbon is accomplished with a burner, described next.

Considerably more attention must be paid if you use a burner to ash a filter paper. The burner produces much higher temperatures than a heat lamp. The

You should have a burner for each crucible. You can tend to the ashing of several filter papers at the same time.

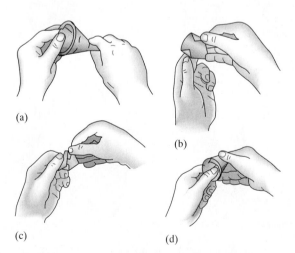

(a)

(b)

(c)

(d)

Figure 2-14   Transferring a filter paper and precipitate to a crucible. (a) Pull the triple-thick portion of the cone to the opposite side of the funnel and flatten the cone along its upper edge. (b) Remove the flattened cone from the funnel and fold the corners inward. (c) Fold the top edge of the cone toward the tip to enclose the precipitate in the paper. (d) Gently ease the folded paper and its contents into the crucible.

possibility thus exists for the mechanical loss of precipitate if moisture is expelled too rapidly in the initial stages of heating or if the paper bursts into flame. Also, partial reduction of some precipitates can occur through reaction with the hot carbon of the charring paper; such reduction is a serious problem if reoxidation following ashing is inconvenient. These difficulties can be minimized by positioning the crucible as illustrated in Figure 2-15. The tilted position allows for the ready access of air; a clean crucible cover should be available to extinguish any flame that might develop.

You should use a small flame at the beginning. The temperature is gradually increased as moisture is evolved and the paper begins to char. The intensity of heating that can be tolerated can be gauged by the amount of smoke given off. Thin wisps are normal. A significant increase in the amount of smoke indicates that the paper is about to flash and that heating should be temporarily discontinued. Any flame that does appear should be immediately extinguished with a crucible cover. (The cover may become discolored, owing to the condensation of carbonaceous products; these products must ultimately be removed from the cover by ignition to confirm the absence of entrained particles of precipitate.) When no further smoking can be detected, heating is increased to eliminate the residual carbon. Strong heating, as necessary, can then be undertaken. This sequence ordinarily precedes the final ignition of a precipitate in a muffle furnace, where a reducing atmosphere is equally undesirable.

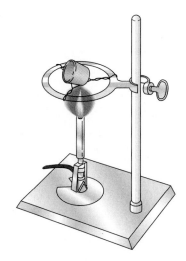

Figure 2-15   Ignition of a precipitate. Proper crucible position for preliminary charring.

The Use of Filtering Crucibles

A vacuum filtration train (Figure 2-16) is used when a filtering crucible is used instead of paper. The trap isolates the filter flask from the source of vacuum.

2F-4   Rules for Manipulating Heated Objects

Careful adherence to the following rules will minimize the possibility of accidental loss of a precipitate.

1. Practice unfamiliar manipulations before putting them to use.
2. *Never* place a heated object on the benchtop; instead, place it on a wire gauze or a heat-resistant ceramic plate.
3. Allow a crucible that has been subjected to the full flame of a burner or to a muffle furnace to cool momentarily (on a wire gauze or ceramic plate) before transferring it to the desiccator.
4. Keep the tongs and forceps used to handle heated objects scrupulously clean. In particular, do not allow the tips to touch the benchtop.

2G   MEASURING VOLUME

The precise measurement of volume is as important to many analytical methods as is the precise measurement of mass.

2G-1   Units of Volume

The unit of volume is the *liter* (L), defined as one cubic decimeter. The *milliliter* (mL) is one one-thousandth of a liter and is used where the liter represents an inconveniently large volume unit.

Figure 2-16   Train for vacuum filtration. The trap isolates the filter flask from the source of vacuum.

2G-2   The Effect of Temperature on Volume Measurements

The volume occupied by a given mass of liquid varies with temperature as does the device that holds the liquid during measurement. Most volumetric measuring devices are made of glass, however, which fortunately has a small coefficient of expansion. Consequently, variations in the volume of a glass container with temperature need not be considered in ordinary analytical work.

The coefficient of expansion for dilute aqueous solutions (approximately 0.025%/°C) is such that a 5°C change has a measurable effect on the reliability of ordinary volumetric measurements.

Example 2-2

A 40.00-mL sample is taken from an aqueous solution at 5°C. What volume does it occupy at 20°C?

$$V_{20°} = V_{5°} + \frac{0.025\%}{100\%}(20 - 5)(40.00) = 40.00 + 0.15 = 40.15 \text{ mL}$$

Volumetric measurements must be referred to some standard temperature; this reference point is ordinarily 20°C. The ambient temperature of most laboratories is sufficiently close to 20°C to eliminate the need for temperature corrections in volume measurements for aqueous solutions. In contrast, the coefficient of expansion for organic liquids may require corrections for temperature differences of 1°C or less.

2G-3   Apparatus for the Precise Measurement of Volume

The reliable measurement of volume is performed with a *pipet*, a *buret*, and a *volumetric flask*. Volumetric equipment as marked by the manufacturer indicates not only the manner of calibration (usually TD for "to deliver" or TC for "to contain") but also the temperature at which the calibration strictly applies. Pipets and burets are ordinarily calibrated to deliver specified volumes, whereas volumetric flasks are calibrated on a to-contain basis.

Pipets

Pipets permit the transfer of accurately known volumes from one container to another. Some common types are shown in Figure 2-17; information concerning their use is given in Table 2-2. A *volumetric*, or *transfer*, pipet (Figure 2-17a) delivers a single, fixed volume between 0.5 and 200 mL. Many such pipets are color coded by volume for convenience in identification and sorting. *Measuring* pipets (Figure 2-17b and c) are calibrated in convenient units to permit delivery of any volume up to a maximum capacity ranging from 0.1 to 25 mL.

Volumetric and measuring pipets are filled to a calibration mark at the outset; the manner in which the transfer is completed depends on the particular type. Because an attraction exists between most liquids and glass, a small amount of liquid tends to remain in the tip after the pipet is emptied. This residual liquid is never blown out of a volumetric pipet or from some measuring pipets; it is blown out of other types of pipets (Table 2-2).

Tolerances, Class A Transfer Pipets

Capacity, mL	Tolerances, mL
0.5	±0.006
1	±0.006
2	±0.006
5	±0.01
10	±0.02
20	±0.03
25	±0.03
50	±0.05
100	±0.08

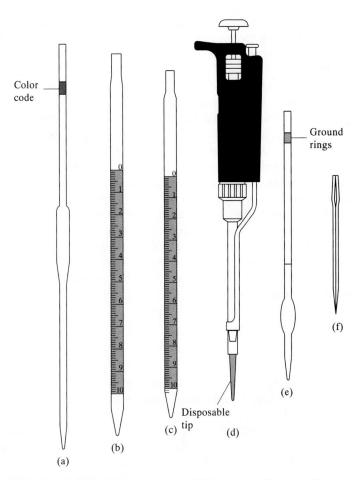

Figure 2-17 Typical pipets: (a) volumetric pipet, (b) Mohr pipet, (c) serological pipet, (d) Eppendorf micropipet, (e) Ostwald–Folin pipet, (f) lambda pipet.

Table 2-2

Characteristics of Pipets

Name	Type of Calibration*	Function	Available Capacity, mL	Type of Drainage
Volumetric	TD	Delivery of fixed volume	1–200	Free
Mohr	TD	Delivery of variable volume	1–25	To lower calibration line
Serological	TD	Delivery of variable volume	0.1–10	Blow out last drop†
Serological	TD	Delivery of variable volume	0.1–10	To lower calibration line
Ostwald–Folin	TD	Delivery of fixed volume	0.5–10	Blow out last drop†
Lambda	TC	Containment of fixed volume	0.001–2	Wash out with suitable solvent
Lambda	TD	Delivery of fixed volume	0.001–2	Blow out last drop†
Eppendorf	TD	Delivery of variable or fixed volume	0.001–1	Tip emptied by air displacement

*TD, to deliver; TC, to contain.
†A frosted ring near the top of recently manufactured pipets indicates that the last drop is to be blown out.

Range and Precision of Typical Eppendorf Micropipets

Volume Range, μL	Standard Deviation, μL
1–20	<0.04 @ 2 μL
	<0.06 @ 20 μL
10–100	<0.10 @ 15 μL
	<0.15 @ 100 μL
20–200	<0.15 @ 25 μL
	<0.30 @ 200 μL
100–1000	<0.6 @ 250 μL
	<1.3 @ 1000 μL
500–5000	<3 @ 1.0 mL
	<8 @ 5.0 mL

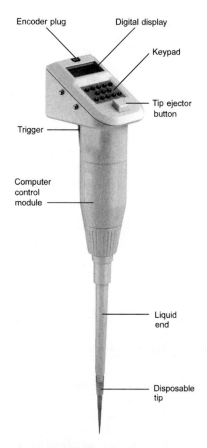

Encoder plug

Digital display

Keypad

Tip ejector button

Trigger

Computer control module

Liquid end

Disposable tip

Figure 2-18 A handheld, battery-operated, computer-controlled, motorized pipet. Courtesy of Rainin Instrument Co. Woburn, MA.

Handheld Eppendorf micropipets (Figure 2-17d) deliver adjustable microliter volumes of liquid. With these pipets, a known and adjustable volume of air is displaced from the disposable plastic tip by depressing the push button on the top of the pipet to a first stop. This button operates a spring-loaded piston that forces air out of the pipet. The volume of displaced air can be varied by a locking digital micrometer adjustment located on the front of the device. The plastic tip is then inserted into the liquid and the pressure on the button is released, causing liquid to be drawn into the tip. The tip is then placed against the walls of the receiving vessel, and the push button is again depressed to the first stop. After 1 s, the push button is depressed further to a second stop, which completely empties the tip. The range of volumes and precision of typical pipets of this type are shown in the margin.

Numerous *automatic* pipets are available for situations that call for the repeated delivery of a particular volume. In addition, motorized, computer-controlled microliter pipets are now available (see Figure 2-18). This device is programmed to function as a pipet, a dispenser of multiple volumes, a buret, and a means for diluting samples. The volume desired is entered on a keyboard and is displayed on a panel. A motor-driven piston dispenses the liquid. Maximum volumes range from 10 to 2500 μL.

Burets

Burets, like measuring pipets, enable the analyst to deliver any volume up to their maximum capacities. You can attain somewhat greater precision with a buret than with a pipet.

A buret consists of a calibrated tube to hold titrant plus a valve arrangement by which the flow of titrant is controlled. This valve is the principal source of difference among burets. The simplest pinchcock valve consists of a close-fitting glass bead inside a short length of rubber tubing that connects the buret and its tip (Figure 2-19a); only when the tubing is deformed does liquid flow past the bead.

Burets are usually equipped with Teflon stopcocks, which are unaffected by most reagents and thus require no lubricant (Figure 2-19b). Burets equipped with glass stopcocks rely on a lubricant between the ground-glass surfaces of stopcock and barrel for a liquid-tight seal. Some solutions, notably bases, cause glass stopcocks to freeze when the solution remains in contact with the ground-glass surfaces for long periods. It is important to thoroughly clean glass stopcocks after each use.

Volumetric Flasks

Volumetric flasks are manufactured with capacities ranging from 5 mL to 5 L and are usually calibrated to contain a specified volume when filled to a line etched on the neck (Figure 2-20). They are used for the preparation of standard solutions and for the dilution of samples to a fixed volume prior to taking aliquots with a pipet. Some are also calibrated on a to-deliver basis; these are readily distinguished by two reference lines on the neck. If delivery of the stated volume is desired, the flask is filled to the upper line.

2G-4 Using Volumetric Equipment

Volume markings are blazed onto clean volumetric equipment by the manufacturer. An equal degree of cleanliness is needed in the laboratory if these markings are to have their stated meanings. Only clean glass surfaces support a uniform film of liquid. Dirt or oil causes breaks in this film; the existence of breaks is a certain indication of an unclean surface.

Cleaning

A brief soaking in a warm detergent solution is usually sufficient to remove the grease and dirt responsible for water breaks. Avoid prolonged soaking because a rough area or ring is likely to develop at a detergent/air interface. This ring cannot be removed and causes a film break that destroys the usefulness of the equipment.

After being cleaned, the apparatus must be thoroughly rinsed with tap water and then with three or four portions of distilled water. It is seldom necessary to dry volumetric ware.

Avoiding Parallax

The top surface of a liquid confined in a narrow tube exhibits a marked curvature, or *meniscus*. It is common practice to use the bottom of the meniscus as the point of reference in calibrating and using volumetric equipment. This minimum can be established more exactly by holding an opaque card or piece of paper behind the graduations (Figure 2-21).

When reading volumes, keep your eye at the level of the liquid surface to avoid an error due to *parallax*, a condition that causes the volume to appear smaller than its actual value if the meniscus is viewed from above (Figure 2-21, position 1) and larger if the meniscus is viewed from below (Figure 2-21, position 2).

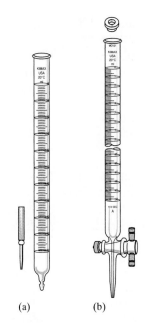

Figure 2-19 Burets: (a) glass-bead valve, (b) Teflon valve.

A *meniscus* is the curved surface of a liquid at its interface with the atmosphere.

Parallax is the apparent displacement of a liquid level or of a pointer as an observer changes position. Parallax occurs when an object is viewed from a position that is not at a right angle to the object.

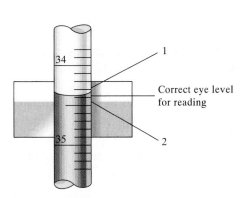

Figure 2-21 Method for reading a buret. The eye should be level with the meniscus. The reading shown is 34.47 mL. If viewed from position 1, the reading appears smaller than 34.47 mL; from position 2, it appears larger.

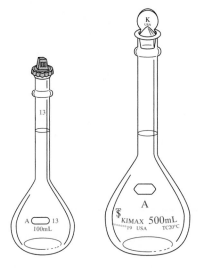

Figure 2-20 Typical volumetric flasks.

Tolerances, Class A Burets

Volume, mL	Tolerances, mL
5	±0.01
10	±0.02
25	±0.03
50	±0.05
100	±0.20

Tolerances, Class A Volumetric Flasks

Capacity, mL	Tolerances, mL
5	±0.02
10	±0.02
25	±0.03
50	±0.05
100	±0.08
250	±0.12
500	±0.20
1000	±0.30
2000	±0.50

2G-5   Directions for the Use of a Pipet

The following directions pertain specifically to volumetric pipets but can also be modified for the use of other types. Liquid is drawn into a pipet through the application of a slight vacuum. *Never use your mouth to draw liquid into a pipet because of the possibility of accidentally ingesting the liquid being pipetted.* Instead, a rubber suction bulb (Figure 2-22a) or a rubber tube connected to a vacuum source should be used.

Cleaning

Use a rubber bulb to draw detergent solution to a level 2 to 3 cm above the calibration mark of the pipet. Drain this solution and then rinse the pipet with several portions of tap water. Inspect for film breaks; repeat this portion of the cleaning cycle if necessary. Finally, fill the pipet with distilled water to perhaps one third of its capacity and carefully rotate it so that the entire interior surface is wetted. Repeat this rinsing step at least twice.

Measuring an Aliquot

Use a rubber bulb to draw a small volume of the liquid to be sampled into the pipet and thoroughly wet the entire interior surface. Repeat with *at least* two additional portions. Then carefully fill the pipet to a level somewhat above the graduation mark (Figure 2-22). Quickly replace the bulb with a *forefinger* to arrest the outflow of liquid (Figure 2-22b). Make certain there are no bubbles in the bulk of the liquid or foam at the surface. Tilt the pipet slightly from the vertical and wipe the exterior free of adhering liquid (Figure 2-22c). Touch the tip of the pipet to the wall of a glass vessel (*not* the container into which the aliquot is to be transferred), and slowly allow the liquid level to drop by partially releasing the forefinger (see note 1 below). Halt further flow as the bottom of the meniscus coincides exactly with the graduation mark. Then place the pipet tip well within the receiving vessel and allow the liquid to drain. When free flow ceases, rest the tip against the inner wall of the receiver for a full 10 s (Figure 2-22d). Finally, withdraw the pipet with a rotating motion to remove any liquid adhering to the tip. *The small volume remaining inside the tip of a volumetric pipet should not be blown or rinsed into the receiving vessel.*

Notes

1. The liquid can best be held at a constant level if the forefinger is *faintly* moist. Too much moisture makes control impossible.
2. Rinse the pipet thoroughly after use.

2G-6   Using a Buret

Before use, a buret must be scrupulously clean; in addition, its valve must be liquid-tight.

Cleaning

Thoroughly clean the tube of the buret with detergent and a long brush. Rinse thoroughly with tap water and then with distilled water. Inspect for water breaks. Repeat the treatment if necessary.

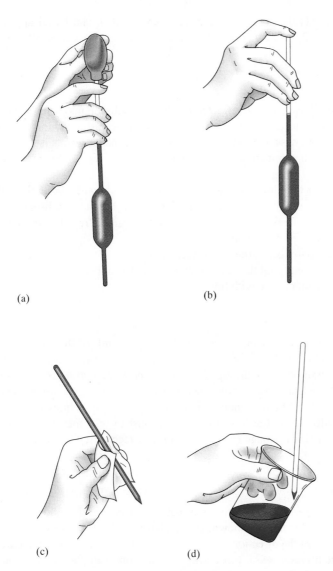

(a)

(b)

(c)

(d)

Figure 2-22 Steps in dispensing an aliquot. (a) Draw a small amount of the liquid into the pipet and wet the interior surface by tilting and rotating the pipet. Repeat this procedure two more times. Manipulate the suction bulb with your left hand if you are right-handed. Then carefully draw the liquid to a point above the calibration mark. (b) Rapidly remove the bulb, and place your index finger over the open end of the pipet to prevent the flow of the liquid out of the tip. (c) Withdraw the tip of the pipet from the liquid and wipe it with a tissue. (d) Hold the pipet vertically with the tip gently pressed against the inside of the beaker as shown, carefully release the pressure of your finger against the end of the pipet, and allow the liquid to flow out of the pipet until the bottom of the meniscus is exactly even with the calibration mark. Stop the flow of liquid, tilt the pipet at a slight angle, and dispense the liquid into a conical flask or other receiving vessel.

Lubrication of a Glass Stopcock

Carefully remove all old grease from a glass stopcock and its barrel with a paper towel and dry both parts completely. Lightly grease the stopcock, taking care to avoid the area adjacent to the hole. Insert the stopcock into the barrel and rotate it vigorously with slight inward pressure. A proper amount of lubricant has been used when (1) the area of contact between stopcock and barrel appears nearly transparent, (2) the seal is liquid-tight, and (3) no grease has worked its way into the tip.

Notes

1. Grease films that are unaffected by cleaning solution may yield to such organic solvents as acetone or benzene. Thorough washing with detergent should follow such treatment. The use of silicone lubricants is not recommended; contamination by such preparations is difficult, if not impossible, to remove.

2. As long as the flow of liquid is not impeded, fouling of a buret tip with stopcock grease is not a serious matter. Removal is best accomplished with organic solvents. A stoppage during a titration can be freed by *gently* warming the tip with a lighted match.

3. Before a buret is returned to service after reassembly, it is advisable to test for leakage. Simply fill the buret with water and establish that the volume reading does not change with time.

Filling

Make certain the stopcock is closed. Add 5 to 10 mL of the solution to be dispensed, and carefully rotate the buret to wet the interior completely. Allow the liquid to drain through the tip. *Repeat this procedure at least two more times.* Then fill the buret well above the zero mark. Free the tip of air bubbles by rapidly rotating the stopcock and permitting small quantities of the titrant to pass. Finally, lower the level of the liquid just to or somewhat below the zero mark. Allow for drainage ($\approx$1 min), and then record the initial volume reading, estimating to the nearest 0.01 mL.

Titration

Figure 2-23 illustrates the preferred method for the manipulation of a stopcock; when the hand is held as shown, any tendency for lateral movement by the stopcock will be in the direction of firmer seating. Be sure the tip of the buret is well within the titration vessel (ordinarily a flask). Introduce the titrant in increments of about 1 mL. Swirl (or stir) constantly to ensure thorough mixing. Decrease the size of the increments as the titration progresses; add titrant dropwise in the immediate vicinity of the end point (note 2). When it is judged that only a few more drops are needed, rinse the walls of the container (note 3). Allow for drainage (at least 30 s) at the completion of the titration. Then record the final volume, again to the nearest 0.01 mL.

Notes

1. When unfamiliar with a particular titration, many chemists prepare an extra sample. No care is lavished on its titration since its functions are to reveal the nature of the end point and to provide a rough estimate of titrant requirements. This deliberate sacrifice of one sample frequently results in an overall saving of time.

2. Increments smaller than one drop can be taken by allowing a small volume of titrant to form on the tip of the buret and then touching the tip to the wall of

Buret readings should be estimated to the nearest 0.01 mL.

Figure 2-23 Recommended method for manipulation of a buret stopcock.

the flask. This partial drop is then combined with the bulk of the liquid as in note 3.

3. Instead of being rinsed toward the end of a titration, the flask can be tilted and rotated so that the bulk of the liquid picks up any drops that adhere to the inner surface.

2G-7   Using Volumetric Flasks

Before use, volumetric flasks should be washed with detergent and thoroughly rinsed. Only rarely do they need to be dried. If required, however, drying is best accomplished by clamping the flask in an inverted position. Insertion of a glass tube connected to a vacuum line hastens the process.

Direct Weighing into a Volumetric Flask

The direct preparation of a standard solution requires the introduction of a known mass of solute to a volumetric flask. Use of a powder funnel minimizes the possibility of loss of solid during the transfer. Rinse the funnel thoroughly; collect the washings in the flask.

The foregoing procedure may be inappropriate if heating is needed to dissolve the solute. Instead, weigh the solid into a beaker or flask, add solvent, heat to dissolve the solute, and allow the solution to cool to room temperature. Transfer this solution quantitatively to the volumetric flask, as described in the next section. *Never heat a volumetric flask.*

Quantitative Transfer of Liquid to a Volumetric Flask

Insert a funnel into the neck of the volumetric flask; use a stirring rod to direct the flow of liquid from the beaker into the funnel. Tip off the last drop of liquid on the spout of the beaker with the stirring rod. Rinse both the stirring rod and the interior of the beaker with distilled water and transfer the washings to the volumetric flask, as before. Repeat the rinsing process *at least* two more times.

Dilution to the Mark

After the solute has been transferred, fill the flask about half full and swirl the contents to hasten solution. Add more solvent and again mix well. Bring the liquid level almost to the mark, and allow time for drainage (≈ 1 min); then use a medicine dropper to make such final additions of solvent as are necessary (see the note below). Firmly stopper the flask, and invert it repeatedly to ensure thorough mixing. Transfer the contents to a storage bottle that either is dry or has been thoroughly rinsed with several small portions of the solution from the flask.

The solute should be completely dissolved *before* you dilute to the mark.

Note

If, as sometimes happens, the liquid level accidentally exceeds the calibration mark, the solution can be saved by correcting for the excess volume. Use a gummed label to mark the location of the meniscus. After the flask has been emptied, carefully refill to the manufacturer's etched mark with water. Use a buret to determine the additional volume needed to fill the flask so that the meniscus is at the gummed-label mark. This volume must be added to the nominal volume of the flask when calculating the concentration of the solution.

2H　CALIBRATING VOLUMETRIC WARE

Volumetric glassware is calibrated by measuring the mass of a liquid (usually distilled water) of known density and temperature that is contained in (or delivered by) the volumetric ware. In carrying out a calibration, a buoyancy correction must be made (Section 2D-4) since the density of water is quite different from that of the weights.

The calculations associated with calibration, although not difficult, are somewhat involved. The raw weighing data are first corrected for buoyancy with Equation 2-1. Next, the volume of the apparatus at the temperature of calibration (T) is obtained by dividing the density of the liquid at that temperature into the corrected mass. Finally, this volume is corrected to the standard temperature of 20°C as in Example 2-2.

Table 2-3 is provided to ease the computational burden of calibration. Corrections for buoyancy with respect to stainless steel or brass masses (the density difference between the two is small enough to be neglected) and for the volume change of water and of glass containers have been incorporated into these data. Multiplication by the appropriate factor from Table 2-3 converts the mass of water at temperature T to (1) the corresponding volume at that temperature or (2) the volume at 20°C.

Example 2-3

A 25-mL pipet delivers 24.976 g of water weighed against stainless steel mass at 25°C. Use the data in Table 2-3 to calculate the volume delivered by this pipet at 25°C and 20°C.

$$\text{At } 25°C: V = 24.976 \text{ g} \times 1.0040 \text{ mL/g} = 25.08 \text{ mL}$$
$$\text{At } 20°C: V = 24.976 \text{ g} \times 1.0037 \text{ mL/g} = 27.07 \text{ mL}$$

2H-1　General Directions for Calibration Work

All volumetric ware should be painstakingly freed of water breaks before being calibrated. Burets and pipets need not be dry; volumetric flasks should be thoroughly drained and dried at room temperature. The water used for calibration should be in thermal equilibrium with its surroundings. This condition is best established by drawing the water well in advance, noting its temperature at frequent intervals, and waiting until no further changes occur.

Although an analytical balance can be used for calibration, weighings to the nearest milligram are perfectly satisfactory for all but the very smallest volumes. Thus, a top-loading balance is more conveniently used. Weighing bottles or small, well-stoppered conical flasks can serve as receivers for the calibration liquid.

Calibrating a Volumetric Pipet

Determine the empty mass of the stoppered receiver to the nearest milligram. Transfer a portion of temperature-equilibrated water to the receiver with the pipet, weigh the receiver and its contents (again, to the nearest milligram), and calculate the mass of water delivered from the difference in these masses.

Calculate the volume delivered with the aid of Table 2-3. Repeat the calibration several times; calculate the mean volume delivered and its standard deviation.

Calibrating a Buret

Fill the buret with temperature-equilibrated water and make sure no air bubbles are trapped in the tip. Allow about 1 min for drainage, then lower the liquid level to bring the bottom of the meniscus to the 0.00-mL mark. Touch the tip to the wall of a beaker to remove any adhering drop. Wait 10 min and recheck the volume; if the stopcock is tight, there should be no perceptible change. During this interval, weigh (to the nearest milligram) a 125-mL conical flask fitted with a rubber stopper.

Once tightness of the stopcock has been established, slowly transfer (at about 10 mL/min) approximately 10 mL of water to the flask. Touch the tip to the wall of the flask. Wait 1 min, record the volume that was apparently delivered, and re-fill the buret. Weigh the flask and its contents to the nearest milligram; the difference between this mass and the initial value gives the mass of water delivered. Use Table 2-3 to convert this mass to the true volume. Subtract the apparent

Table 2-3

Volume Occupied by 1.000 g of Water Weighed in Air Against Stainless Steel Weights*

| | Volume, mL | |
Temperature, T, °C	At T	Corrected to 20°C
10	1.0013	1.0016
11	1.0014	1.0016
12	1.0015	1.0017
13	1.0016	1.0018
14	1.0018	1.0019
15	1.0019	1.0020
16	1.0021	1.0022
17	1.0022	1.0023
18	1.0024	1.0025
19	1.0026	1.0026
20	1.0028	1.0028
21	1.0030	1.0030
22	1.0033	1.0032
23	1.0035	1.0034
24	1.0037	1.0036
25	1.0040	1.0037
26	1.0043	1.0041
27	1.0045	1.0043
28	1.0048	1.0046
29	1.0051	1.0048
30	1.0054	1.0052

*Corrections for buoyancy (stainless steel weights) and change in container volume have been applied.

volume from the true volume. This difference is the correction that should be applied to the apparent volume to give the true volume. Repeat the calibration until agreement within ± 0.02 mL is achieved.

Starting again from the zero mark, repeat the calibration, this time delivering about 20 mL to the receiver. Test the buret at 10-mL intervals over its entire volume. Prepare a plot of the correction to be applied as a function of volume delivered. You can determine the correction associated with any interval from this plot.

Calibrating a Volumetric Flask

Weigh the clean, dry flask to the nearest milligram. Then fill to the mark with equilibrated water and reweigh. Calculate the volume contained with the aid of Table 2-3.

Calibrating a Volumetric Flask Relative to a Pipet

An *aliquot* is a measured fraction of the volume of a liquid sample. A 100-mL volumetric flask calibrated with a 20-mL pipet permits you to deliver a one-fifth aliquot.

The calibration of a volumetric flask relative to a pipet provides an excellent method for partitioning a sample into *aliquots*. These directions pertain to a 50-mL pipet and a 500-mL volumetric flask; other combinations are equally convenient.

Carefully transfer ten 50-mL aliquots from the pipet to a dry 500-mL volumetric flask. Mark the location of the bottom of the meniscus with the top edge of a gummed label. Cover the label with varnish or other transparent medium to ensure that the mark is permanent. Dilution to this mark will enable you to withdraw and deliver precisely a one-tenth aliquot of the dilute solution with the pipet that you used for the calibration. Note that you must recalibrate the flask if you use a different pipet.

2I   THE LABORATORY NOTEBOOK

You will require a laboratory notebook to record measurements, experimental conditions, important observations, and any other significant information you collect during the course of an analysis. The book should be permanently bound with consecutively numbered pages. If your pages are not numbered, hand-number them before you make any entries. Most notebooks have more than ample room for recording a semester's observations. Leave plenty of space for recording your entries.

The first few pages of your notebook should be saved for a table of contents that you can update as entries are made.

2I-1   Maintaining the Laboratory Notebook

Remember that you can discard an experimental measurement *only if you have certain knowledge that you made an experimental error.* Thus, you must carefully record experimental observations in your notebook as soon as they occur.

1. *Record all data and observations directly into the notebook in ink.* Neatness is desirable, but you should not accomplish neatness by transcribing data from a sheet of paper to the notebook or from one notebook to another. The risk of misplacing — or incorrectly transcribing — crucial data and thereby ruining an experiment is too high.
2. Supply each entry or series of entries with a heading or label. A series of weighing data for a set of empty crucibles should carry the heading "empty crucible mass" (or something similar), for example, and the mass of each crucible should be identified by the same number or letter used to label the crucible.

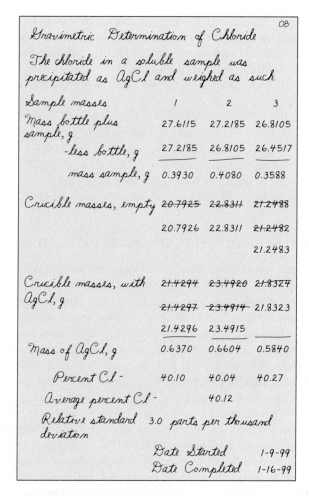

Figure 2-24   Laboratory notebook data page.

3. Date each page of the notebook as it is used.
4. *Never* attempt to erase or obliterate an incorrect entry. Instead, cross it out with a single horizontal line and locate the correct entry as nearby as possible. Do not write over incorrect numbers; with time, it may become impossible to distinguish a correct entry from an incorrect one.
5. Never remove a page from the notebook. Draw diagonal lines across any page that is to be disregarded. Provide a brief rationale for disregarding the page.

An entry in a laboratory notebook should never be erased but should be crossed out instead.

2I-2   Format

Consult your instructor concerning the format to be used in keeping the laboratory notebook.[7] In one common convention, each page is numbered consecutively for the recording of data and observations as they occur. The completed analysis is then summarized on the next available page spread (that is, left and right facing pages). As shown in Figure 2-24, the first of these two facing pages should contain the following entries:

---

[7]See also Howard M. Kanare, *Writing the Laboratory Notebook,* Washington, D.C.: American Chemical Society, 1985.

1. The title of the experiment ("Gravimetric Determination of Chloride").
2. A brief statement of the chemical principles upon which the analysis is based.
3. A complete summary of the weighing, volumetric, and/or instrument response data needed to calculate the results.
4. A report of the best value for the set and a statement of its precision.

The second page should contain the following items:

1. Chemical equations for the principal reactions in the analysis.
2. An algebraic equation showing how the results were calculated.
3. A summary of observations that appear to bear upon the validity of a particular result or the analysis as a whole. *Any such entry must have been originally recorded in the notebook at the time the observation was made.*

2J USING SPREADSHEETS IN ANALYTICAL CHEMISTRY

The spreadsheet is one of the most useful tools that has resulted from the personal computer revolution. Spreadsheets are used for record keeping, mathematical calculations, statistical analysis, curve fitting, data plotting, financial analysis, database management, and a variety of other tasks limited only by the imagination and skill of the user. State-of-the-art spreadsheet programs have many built-in functions to assist in carrying out the computational tasks of analytical chemistry. Throughout this text, we will present examples to illustrate some of these tasks and spreadsheets for performing them. The most popular spreadsheet programs include Microsoft Excel, Lotus 1-2-3, and Quattro Pro. Because of its wide availability and general utility, we chose to illustrate our examples using Microsoft Excel on the PC. Although the syntax and commands for other spreadsheet applications are somewhat different from Excel, the general concepts and principles of operation are similar. The examples that we present may be accomplished using any of the popular spreadsheet applications; the precise instructions must be modified if an application other than Excel is used. In these examples, we assume that Excel is configured with default options as delivered from the manufacturer unless we specifically note otherwise.

It is our feeling that we learn best by doing, not by reading about doing. Although software manufacturers have made great strides in the 1990s in the production of manuals for their products, it is still generally true that when we know enough to read a software manual efficiently, we no longer need the manual. With that in mind, we have tried to provide a series of spreadsheet exercises that evolve in the context of analytical chemistry, not a software manual. We introduce commands and syntax only when they are needed to accomplish a particular task, so if you need more detailed information, please consult the help screens. Help is available at the click of a mouse button from within Excel by clicking on **Help/Microsoft Excel Help.**

Spreadsheet Exercise   Getting Started with Excel

Microsoft
Excel

To illustrate spreadsheet use, we will use Excel to carry out the functions of the laboratory notebook page that is depicted in Figure 2-24. To begin, we must start Excel by double-clicking on its icon, shown in the margin, on the computer desktop. Alternatively, under Windows 98(95), click on **Start/Programs/**

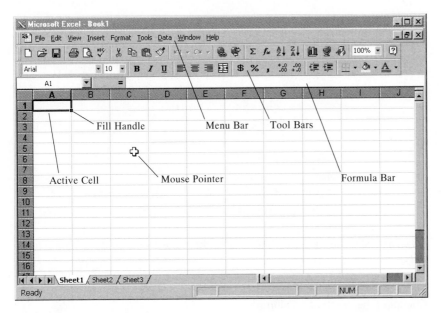

Figure 2-25 The opening window in Microsoft Excel. Note the location of the menu bar, the toolbars, the active cell, and the mouse pointer.

Microsoft Excel on the toolbar, and the opening window of Figure 2-25 appears.[8]

The window contains a *worksheet* consisting of a grid of *cells* arranged in rows and columns. The rows are labeled 1, 2, 3, and so on, and the columns are labeled A, B, C, and so on. Each cell has a unique location specified by its address. For example, the *active cell*, which is surrounded by a dark outline as shown in Figure 2-25, has the address A1. The address of the active cell is always displayed in the box just above the first column of the displayed worksheet in the *formula bar*. You may verify this display of the active cell by clicking on various cells of the worksheet.

Entering Text in the Spreadsheet

Cells may contain text, numbers, or formulas. We will begin by typing some text into the worksheet. Click on cell A1 and type `Gravimetric Determination of Chloride` followed by the Enter key [↵]. Notice that the active cell is now A2, so you may now type `Samples[↵]`. As you type, the data that you enter appear in the formula bar. If you make a mistake, just click the mouse in the formula bar and make necessary corrections. Continue to type text into the cells of column A as shown on the next page.

---

[8]For more information on the use of spreadsheets in chemistry, see D. Diamond and V. C. A. Hanratty, *Spreadsheet Applications in Chemistry Using Microsoft Excel,* New York: John Wiley & Sons, 1997; H. Freiser, *Concepts and Calculations in Analytical Chemistry: A Spreadsheet Approach,* New York: CRC Press, 1992; R. de Levie, *Principles of Quantitative Chemical Analysis,* New York: McGraw-Hill, 1997; R. de Levie, *A Spreadsheet Workbook for Quantitative Chemical Analysis,* New York: McGraw-Hill, 1992.

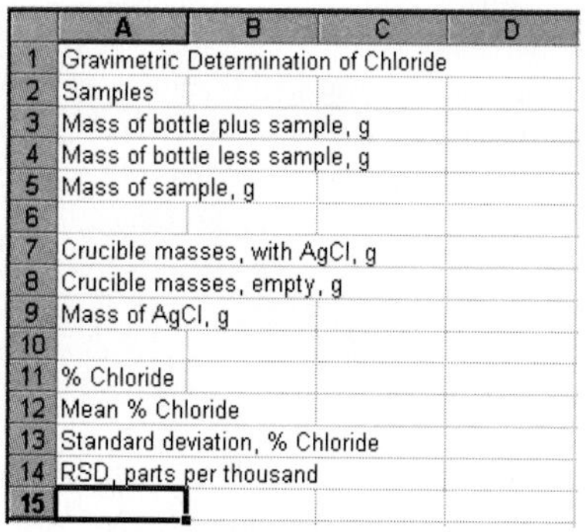

Figure 2-26 The appearance of the worksheet after entering the labels.

```
Mass of bottle plus sample, g[↵]
Mass of bottle less sample, g[↵]
Mass of sample, g[↵]
[↵]
Crucible masses, with AgCl, g[↵]
Crucible masses, empty, g[↵]
Mass of AgCl, g[↵]
[↵]
% Chloride[↵]
Mean % Chloride[↵]
Standard deviation, % Chloride[↵]
RSD, parts per thousand[↵]
```

When you have finished entering the text, the worksheet should appear as shown in Figure 2-26.

Changing the Width of a Column

Notice that the labels that you typed into column A are wider than the column. You can change the width of the column by placing the mouse pointer on the boundary between column A and column B in the column head, as shown in Figure 2-27a, and dragging the boundary to the right so that all the text shows in the column, as in Figure 2-27b.

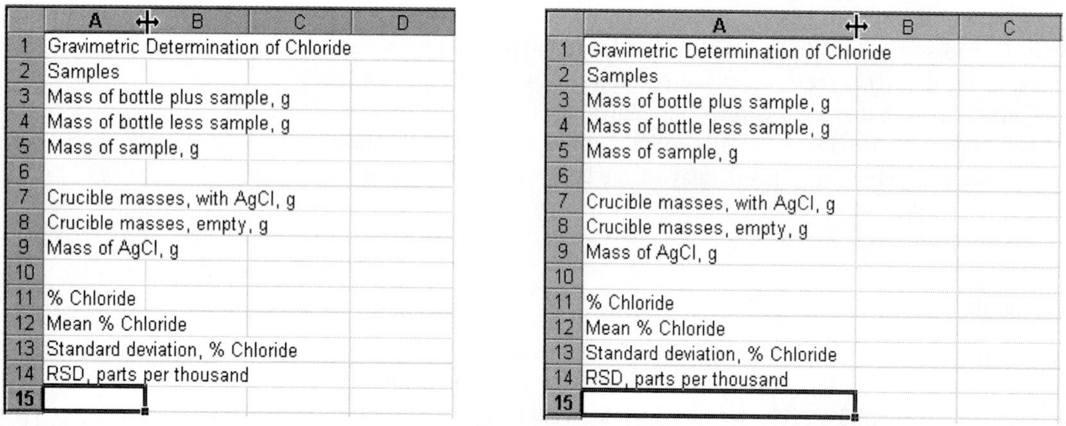

Figure 2-27 Changing the column width. (a) Place the mouse pointer on the boundary between column A and column B, and drag to the right to the position shown in (b).

Entering Numbers into the Spreadsheet

Now let us enter some numerical data into the spreadsheet. Click on cell B2 and type

```
1[↵]
27.6115[↵]
27.2185[↵]
```

	A	B	C	D
1	Gravimetric Determination of Chloride			
2	Samples	1	2	3
3	Mass of bottle plus sample, g	27.6115	27.2185	26.8105
4	Mass of bottle less sample, g	27.2185	26.8105	26.4517
5	Mass of sample, g	0.3930		
6				

Figure 2-28   Sample data entry.

At this point, we wish to calculate the difference between the data in cells B3 and B4, so we type

$$=b3-b4 \; [\dashv]$$

The expression just typed is called a *formula*. In Excel, formulas begin with an equal sign [=] followed by the desired numerical expression. Notice that the difference between the contents of cell B3 and cell B4 is displayed in cell B5. Now continue entering data until your spreadsheet looks like the one illustrated in Figure 2-28.

Filling Cells Using the Fill Handle

The formulas for cells C5 and D5 are identical to the formula in cell B5 except that the cell references for the data are different. In cell C5, we want to compute the difference between the contents of cells C3 and C4, and in cell D5, we want the difference between D3 and D4. We could type the formulas in cells C5 and D5 as we did for cell B5, but Excel provides an easy way to duplicate formulas and automatically changes the cell references to the appropriate values for us. To duplicate a formula in cells adjacent to an existing formula, simply click on the cell containing the formula, which is cell B5 in our example, then click on the fill handle (see Figure 2-25) and drag the corner of the rectangle to the right so that it encompasses the cells where you want the formula to be duplicated. Try it now. Click on cell B5, click on the fill handle, and drag to the right to fill cells C5 and D5. When you let up on the mouse button, the spreadsheet should appear like Figure 2-29. Now click on cell B5, and view the formula in the formula bar. Compare the formula with those in cells C5 and D5.

	A	B	C	D
1	Gravimetric Determination of Chloride			
2	Samples	1	2	3
3	Mass of bottle plus sample, g	27.6115	27.2185	26.8105
4	Mass of bottle less sample, g	27.2185	26.8105	26.4517
5	Mass of sample, g	0.3930	0.4080	0.3588

Figure 2-29   Using the fill handle to copy formulas into adjacent cells of a spreadsheet. In this example, we clicked on cell B2, clicked on the fill handle, and dragged the rectangle to the right to fill cells C5 and D5. The formulas in cells B5, C5, and D5 are identical, but the cell references in the formulas refer to data in columns B, C, and D, respectively.

	A	B	C	D
1	Gravimetric Determination of Chloride			
2	Samples	1	2	3
3	Mass of bottle plus sample, g	27.6115	27.2185	26.8105
4	Mass of bottle less sample, g	27.2185	26.8105	26.4517
5	Mass of sample, g	0.3930	0.4080	0.3588
6				
7	Crucible masses, with AgCl, g	21.4296	23.4915	21.8323
8	Crucible masses, empty, g	20.7926	22.8311	21.2483
9	Mass of AgCl, g			

Figure 2-30 Entering the data into the spreadsheet in preparation for calculating the mass of dry silver chloride in the crucibles.

Now we want to peform the same operations on the data in rows 7, 8, and 9 shown in Figure 2-30. Enter the remaining data into the spreadsheet now.

Now click on cell B9 and type the following formula:

$$=b7-b8 \; [\leftarrow]$$

Again click on cell B9, click on the fill handle, and drag through columns C and D to copy the formula to cells C9 and D9. The mass of silver chloride should now be calculated for all three crucibles.

Making Complex Calculations with Excel

As we shall see in Chapter 7, the equation for finding the % chloride in each of the samples is

$$\% \text{ chloride} = \frac{\dfrac{\text{mass AgCl}}{\text{molar mass AgCl}} \times \text{molar mass Cl}}{\text{mass sample}} \times 100\%$$

$$= \frac{\dfrac{\text{mass AgCl}}{143.321 \text{ g/mol}} \times 35.4527 \text{ g/mol}}{\text{mass sample}} \times 100\%$$

Our task is to translate this equation into an Excel formula and type it into cell B11 as shown below.

$$=B9*35.4527*100/143.321/B5 \; [\leftarrow]$$

Once you have typed the formula, click on cell B11 and drag on the fill handle to copy the formula into cells C11 and D11. The % chloride for samples 2 and 3 should now appear in the spreadsheet as shown in Figure 2-31.

We will complete the spreadsheet in Chapter 6 after we have explored some of the important calculations of statistical analysis. For now, click on **File/Save As...** in the menu bar, enter a file name such as grav_chloride, and save the Excel spreadsheet on a floppy disk for retrieval and editing later. Excel will automatically append the file extension .xls to the file name so that it will appear as grav_chloride.xls on the floppy disk.

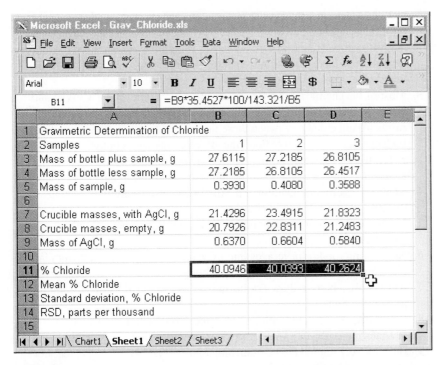

Figure 2-31 Completing the calculation of percent chloride. Type the formula in cell B11, click on the fill handle, and drag to the right through cell D11.

In this section, we have learned some of the basics of spreadsheet operation including typing text into a spreadsheet, changing column widths with the mouse, duplicating cells with the fill handle, and entering formulas into a spreadsheet. In subsequent spreadsheet examples, we will build on the skills that we have learned here and learn many other Excel skills and functions useful in making the calculations of analytical chemistry.

2K SAFETY IN THE LABORATORY

Work in a chemical laboratory necessarily involves a degree of risk; accidents can and do happen. Strict adherence to the following rules will go far toward preventing (or minimizing the effect of) accidents.

1. At the outset, learn the location of the nearest eye fountain, fire blanket, shower, and fire extinguisher. Learn the proper use of each, and do not hesitate to use this equipment should the need arise.
2. *WEAR EYE PROTECTION AT ALL TIMES.* The potential for serious and perhaps permanent eye injury makes it mandatory that all students, instructors, and visitors wear adequate eye protection at all times. Eye protection should be donned before entering the laboratory and should be used continuously until it is time to leave. Serious eye injuries have occurred to people performing such innocuous tasks as computing or writing in a laboratory notebook; such incidents are usually the result of someone else's losing con-

trol of an experiment. Regular prescription glasses are not adequate substitutes for eye protection approved by the Office of Safety and Health Administration (OSHA). Contact lenses should *never* be used in the laboratory because laboratory fumes may react with them and have a harmful effect on the eyes.

3. Most of the chemicals in a laboratory are toxic; some are very toxic, and some — such as concentrated solutions of acids and bases — are highly corrosive. Avoid contact between these liquids and the skin. In the event of such contact, *immediately* flood the affected area with copious quantities of water. If a corrosive solution is spilled on clothing, remove the garment immediately. Time is of the essence; modesty cannot be a matter of concern.

4. *NEVER* perform an unauthorized experiment. Such activity is grounds for disqualification at many institutions.

5. Never work alone in the laboratory; be certain that someone is always within earshot.

6. Never bring food or beverages into the laboratory. Do not drink from laboratory glassware. Do not smoke in the laboratory.

7. Always use a bulb to draw liquids into a pipet; *NEVER* use your mouth to provide suction.

8. Wear adequate foot covering (no sandals). Confine long hair with a net. A laboratory coat or apron will provide some protection and may be required.

9. Be extremely tentative when touching objects that have been heated; hot glass looks just like cold glass.

10. Always fire-polish the ends of freshly cut glass tubing. *NEVER* attempt to force glass tubing through the hole of a stopper. Instead, make sure that both tubing and hole are wet with soapy water. Protect hands with several layers of towel when inserting glass into a stopper.

11. Use fume hoods whenever toxic or noxious gases are likely to be evolved. Be cautious in testing for odors; use your hand to waft vapors above containers toward your nose.

12. Notify the instructor in the event of an injury.

13. Dispose of solutions and chemicals as instructed. It is illegal to flush solutions containing heavy metal ions or organic liquids down the drain in many localities; alternative arrangements are required for the disposal of such liquids.

Chapter 3

Important Chemical Concepts: Expressing Quantities and Concentrations

In this chapter, we explore a number of concepts and skills that are important in analytical chemistry. Although you have probably studied these topics in other chemistry courses, you should find it fruitful to refresh these ideas in your mind before proceeding. We begin by presenting the SI system of units, the important distinction between mass and weight, and the mole. We then investigate the numerous ways of expressing solution concentration and present the basics of chemical stoichiometry.

3A SOME IMPORTANT UNITS OF MEASUREMENT

3A-1 SI Units

Scientists throughout the world are adopting a standardized system of units known as the *International System of Units* (SI). This system is based on the seven fundamental base units shown in Table 3-1. Numerous other useful units, such as volts, hertz, coulombs, and joules, are derived from these base units.[1]

Prefixes are used with these base units and other derived units to express small or large measured quantities in terms of a few simple digits. As shown in Table 3-2, these prefixes multiply the unit by various powers of ten. For example, the wavelength of yellow radiation used for determining sodium by flame photometry is about 5.9×10^{-7} m, which can be expressed more compactly as 590 nm (nanometers); the volume of a liquid injected onto a chromatographic column is often roughly 50×10^{-6} L or 50 μL (microliters); or the amount of memory in some computers is about 64,000,000 bytes or 64 Mbytes (megabytes).

In this course, we often determine the amount of chemical species from mass measurement. For such measurements, metric units of kilograms (kg), grams (g), milligrams (mg), or micrograms (μg) are used. Volumes of liquids are mea-

SI is the acronym for the French "Système International d'Unités."

The *ångstrom unit*, Å, is a non-SI unit of length that is widely used to express the wavelength of very short radiation such as X-rays (1 Å = 0.1 nm = 10^{-10} m). Typical X-radiation lies in the range of 0.1 to 10 Å.

[1]Current information on the definitions of SI base units, conversion factors, and fundamental constants may be found on the National Institute of Standards and Technology (NIST) World Wide Web site at http://physics.nist.gov/cuu/index.html.

Table 3-1

SI Base Units

Physical Quantity	Name of Unit	Abbreviation
Mass	kilogram	kg
Length	meter	m
Time	second	s
Temperature	kelvin	K
Amount of substance	mole	mol
Electric current	ampere	A
Luminous intensity	candela	cd

This French postage stamp commemorates the Meter Convention of 1875. The stamp shows some signatures from the Treaty of the Meter, the seven SI units, and the definition of the meter as 1,650,763.73 wavelengths of the orange light given off by krypton-86. Since the stamp was issued in 1975, the meter has been redefined as the distance that light travels in a vacuum during 1/299,792,458 of a second.

Table 3-2

Prefixes for Units

Prefix	Abbreviation	Multiplier
giga-	G	10^9
mega-	M	10^6
kilo-	k	10^3
deci-	d	10^{-1}
centi-	c	10^{-2}
milli-	m	10^{-3}
micro-	μ	10^{-6}
nano-	n	10^{-9}
pico-	p	10^{-12}
femto-	f	10^{-15}
atto-	a	10^{-18}

sured in units of liters (L), milliliters (mL), and sometimes microliters (μL). The liter is an SI unit that is *defined* as exactly 10^{-3} m³. The milliliter is defined as 10^{-6} m³ or 1 cm³.

3A-2   The Mole

The *mole* (mol) is the SI unit for the amount of a chemical species. It is always associated with a chemical formula and is Avogadro's number (6.022×10^{23}) of particles represented by that formula. The *molar mass* $\mathcal{M}$ of a substance is the mass in grams of one mole of the substance. Molar masses are calculated by summing the atomic masses of all the elements appearing in a chemical formula. For example, the molar mass of formaldehyde CH_2O is

$$\mathcal{M}_{CH_2O} = \frac{1 \text{ mol C}}{\text{mol CH}_2\text{O}} \times \frac{12.0 \text{ g}}{\text{mol C}} + \frac{2 \text{ mol H}}{\text{mol CH}_2\text{O}} \times \frac{1.0 \text{ g}}{\text{mol H}}$$

$$+ \frac{1 \text{ mol O}}{\text{mol CH}_2\text{O}} \times \frac{16.0 \text{ g}}{\text{mol O}}$$

$$= 30.0 \text{ g/mol CH}_2\text{O}$$

It is important to understand the difference between mass and weight. *Mass* is an invariant measure of the amount of matter in an object. *Weight* is the force of attraction between an object and its surroundings, principally Earth. Because gravitational attraction varies with geographic location, the weight of an object depends on where you weigh it. For example, a crucible *weighs* less in Denver than in Atlantic City (both cities are at approximately the same latitude) because the attractive force between the crucible and Earth is smaller at the higher altitude of Denver. Similarly, the crucible *weighs* more in Seattle than in Panama (both cities are at sea level) because Earth is somewhat flattened at the poles, and the force of attraction increases measurably with latitude. The *mass* of the crucible, however, remains constant regardless of where you measure it.

Weight and mass are related by the familiar expression

$$W = mg$$

where W is the weight of an object, m is its mass, and g is the acceleration due to gravity. This equation is one form of *Newton's second law*, which is normally written as $F = ma$. In this instance, the weight of an object W is the force F, and the acceleration due to gravity g corresponds to a in Newton's law.

A chemical analysis is always based on mass so that the results will not depend on locality. A balance is used to compare the weight of an object with the weight of one or more standard masses. Because g affects both unknown and known equally, the mass of the object is identical to the standard masses with which it is compared.

The distinction between mass and weight is often lost in common usage, and the process of comparing masses is ordinarily called *weighing*. In addition, the objects of known mass as well as the results of weighing are frequently called *weights*. Always bear in mind, however, that analytical data are based on mass rather than weight. Therefore, throughout this text we will use mass rather than weight to describe the amounts of substances or objects. On the other hand, for lack of a better word, we will use the verb weigh to describe the act of determining the mass of an object. Also, we will occasionally use the term *weight* to describe the standard masses that are used in weighing.

and that of glucose $C_6H_{12}O_6$ is

$$\mathcal{M}_{C_6H_{12}O_6} = \frac{6\ \text{mol C}}{\text{mol C}_6\text{H}_{12}\text{O}_6} \times \frac{12.0\ \text{g}}{\text{mol C}} + \frac{12\ \text{mol H}}{\text{mol C}_6\text{H}_{12}\text{O}_6} \times \frac{1.0\ \text{g}}{\text{mol H}}$$

$$+ \frac{6\ \text{mol O}}{\text{mol C}_6\text{H}_{12}\text{O}_6} \times \frac{16.0\ \text{g}}{\text{mol O}}$$

$$= 180.0\ \text{g/mol C}_6\text{H}_{12}\text{O}_6$$

Thus, 1 mol of formaldehyde has a mass of 30.0 g and 1 mol of glucose has a mass of 180.0 g.

3A-3   The Millimole

Sometimes it is more convenient to make calculations with millimoles (mmol) rather than moles, where the millimole is 1/1000 of a mole. The mass in grams of a millimole of a substance, the *millimolar mass,* is likewise 1/1000 of the molar mass.

Feature 3-1

Distinguishing between Mass and Weight

Photo of Edwin "Buzz" Aldrin taken by Neil Armstrong in July 1969. Armstrong's reflection may be seen in Aldrin's visor. The suits worn by Armstrong and Aldrin during the *Apollo 11* mission to the Moon appear to be quite massive. But because the mass of the Moon is only 1/81 that of Earth and the acceleration due to gravity is only 1/6 that on Earth, the weight of the suits on the Moon was only 1/6 of their weight on Earth. The mass of the suits, however, was identical in both locations. Photo courtesy of the National Aeronautics and Space Administration.

Mass m is an invariant measure of the amount of matter. Weight W is the force of gravitational attraction between that matter and Earth

$1\ \text{mmol} = 10^{-3}\ \text{mol}$

Feature 3-2

Atomic Mass Units and the Mole

CHALLENGE: Show that the following interesting and useful relationship is correct: 1 mole of atomic mass units = 6.022×10^{23} amu = 1 gram.

A *mole* of a chemical species is 6.022×10^{23} atoms, molecules, ions, electrons, ion pairs, or subatomic particles.

The masses for the elements listed in the table inside the back cover of this text are *relative masses* in terms of *atomic mass units* (amu), or *daltons*. The atomic mass unit is based on a relative scale in which the reference is the ^{12}C carbon isotope, which is *assigned* a mass of exactly 12 amu. Thus, the amu is by definition 1/12 of the mass of one neutral ^{12}C atom. The *molar mass* $\mathcal{M}$ of ^{12}C is then defined as the mass in *grams* of 6.022×10^{23} atoms of the carbon-12 isotope, or exactly 12 g. Likewise, the molar mass of any other element is the mass in grams of 6.022×10^{23} atoms of that element and is numerically equal to the atomic mass of the element in amu units. Thus, the atomic mass of naturally occurring oxygen is 15.9994 amu; its molar mass is 15.9994 g.

Approximately one mole of each of several different elements. Clockwise from the upper left we see 64 g of copper beads, 27 g of crumpled aluminum foil, 207 g of lead shot, 24 g of magnesium chips, 52 g of chromium chunks, and 32 g of sulfur powder. The beakers in the photo have a volume of 50 mL. Photo by C. D. Winters.

3A-4 Calculating the Amount of a Substance in Moles or Millimoles

The number of moles n_x of a species X of molar mass $\mathcal{M}_x$ is given by

$$n_x = \frac{m_x}{\mathcal{M}_x}$$

$$\text{mol X} = \frac{g\text{ X}}{g\text{ X/mol X}} = g\text{ X} \times \frac{\text{mol X}}{g\text{ X}}$$

The number of millimoles is given by

$$\text{mmol} = \frac{g\text{ X}}{g\text{ X/mmol X}} = g\text{ X} \times \frac{\text{mmol X}}{g\text{ X}}$$

In making calculations of this kind, you should include all units as we do throughout this chapter. This practice often reveals errors in setting up equations.

The two examples that follow illustrate how the number of moles or millimoles of a species can be determined from its mass in grams or from the mass of a chemically related species.

Example 3-1

Determine the number of moles and millimoles of benzoic acid ($\mathcal{M}$ = 122.1 g/mol) in 2.00 g of the pure acid.

If we use HBz to represent benzoic acid, we can write that 1 mol of HBz has a mass of 122.1 g. Thus,

$$\text{amount of HBz} = 2.00\ g\ \text{HBz} \times \frac{1\ \text{mol HBz}}{122.1\ g\ \text{HBz}} = 0.0164\ \text{mol HBz}$$

To obtain the number of millimoles, we divide by the millimolar mass (0.1221 g/mol). That is,

$$\text{amount HBz} = 2.00 \; \text{g HBz} \times \frac{1 \; \text{mmol HBz}}{0.1221 \; \text{g HBz}} = 16.4 \; \text{mmol HBz}$$

Example 3-2

Determine the mass in grams of Na^+ (22.99 g/mol) in 25.0 g of Na_2SO_4 (142.0 g/mol).

The chemical formula tells us that 1 mol of Na_2SO_4 contains 2 mol of Na^+. That is,

$$\text{amount Na}^+ = \text{no. mol Na}_2\text{SO}_4 \times \frac{2 \; \text{mol Na}^+}{\text{mol Na}_2\text{SO}_4}$$

To find the number of moles Na_2SO_4 we proceed as in Example 3-1:

$$\text{amount Na}_2\text{SO}_4 = 25.0 \; \text{g Na}_2\text{SO}_4 \times \frac{1 \; \text{mol Na}_2\text{SO}_4}{142.0 \; \text{g Na}_2\text{SO}_4}$$

Combining this equation with the first leads to

$$\text{amount Na}^+ = 25.0 \; \text{g Na}_2\text{SO}_4 \times \frac{1 \; \text{mol Na}_2\text{SO}_4}{142.0 \; \text{Na}_2\text{SO}_4} \times \frac{2 \; \text{mol Na}^+}{\text{mol Na}_2\text{SO}_4}$$

To obtain the mass of sodium in 25.0 g Na_2SO_4, we multiply the number of moles Na^+ by the molar mass of Na^+, or 22.99 g. That is,

$$\text{mass Na}^+ = \text{no. mol Na}^+ \times \frac{22.99 \; \text{g Na}^+}{\text{mol Na}^+}$$

Substituting the previous equation gives the number of grams of Na^+:

$$\text{mass Na}^+ = 25.0 \; \text{g Na}_2\text{SO}_4 \times \frac{1 \; \text{mol Na}_2\text{SO}_4}{142.0 \; \text{g Na}_2\text{SO}_4}$$
$$\times \frac{2 \; \text{mol Na}^+}{\text{mol Na}_2\text{SO}_4} \times \frac{22.99 \; \text{g Na}^+}{\text{mol Na}^+} = 8.10 \; \text{g Na}^+$$

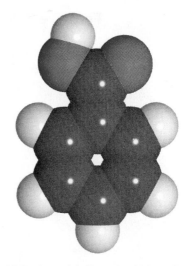

Molecular model of benzoic acid. Benzoic acid occurs widely in nature, particularly in berries. It finds broad use as a preservative in foods, fats, and fruit juices; as a mordant for dying fabric; and as a standard in calorimetry and in acid/base analysis.

3B   SOLUTIONS AND THEIR CONCENTRATIONS

3B-1   Expressing Solution Concentrations

Chemists express the concentration of solutes in solution in several ways. The most important ways are described in this section.

Molar Concentration

The molar concentration of a solution of a chemical species X is the number of moles of the solute species that is contained in one liter of the solution (*not one liter of the solvent*). Molar concentration, or *molarity* M, has the dimensions of mol L^{-1}. Molarity is also equal to the number of millimoles of a solute per

Feature 3-3

Another Approach to
Example 3-2

You may find it advantageous to use the *factor-label method* to solve problems of the type we have been discussing. In this method, we write out the solution to a problem so that units in the denominator of each succeeding factor eliminate the units in the numerator of the preceding one until we arrive at the units of the answer. For example, in Example 3-2, the answer is expressed in units of g Na^+ and the given quantity is expressed in units of g Na_2SO_4. Thus, we can write

$$25.0 \text{ g } Na_2SO_4 \times \frac{\text{mol } Na_2SO_4}{142.0 \text{ g } Na_2SO_4}$$

First eliminate mol Na_2SO_4

$$25.0 \text{ g } Na_2SO_4 \times \frac{\text{mol } Na_2SO_4}{142.0 \text{ g } Na_2SO_4} \times \frac{2 \text{ mol } Na^+}{\text{mol } Na_2SO_4}$$

and then eliminate mol Na^+. That is,

$$25.0 \text{ g } Na_2SO_4 \times \frac{\text{mol } Na_2SO_4}{142.0 \text{ g } Na_2SO_4} \times \frac{2 \text{ mol } Na^+}{\text{mol } Na_2SO_4} \times \frac{22.99 \text{ g } Na^+}{\text{mol } Na^+} = 8.10 \text{ g } Na^+$$

milliliter of solution. Thus, if the number moles of solute is n_x and the volume of the solution is V, then the molar concentration c_x is given by the following equation.

$$c_x = \frac{n_x}{V} \qquad (3\text{-}1)$$

$$\text{molarity} = M = \frac{\text{no. mol solute}}{\text{no. L solution}}$$

$$= \frac{\text{no. mmol solute}}{\text{no. mL solution}}$$

Example 3-3

Calculate the molar concentration of ethanol in an aqueous solution that contains 2.30 g of C_2H_5OH (46.07 g/mol) in 3.50 L of solution.

Because molarity is the number of moles of solute per liter of solution, both the number of moles of solute and the volume of the solution will be needed. The volume V is given as 3.50 L, so all we need to do is to convert the mass of ethanol to the corresponding number of moles.

$$n_{C_2H_5OH} = \text{amount } C_2H_5OH = 2.30 \text{ g } C_2H_5OH \times \frac{1 \text{ mol } C_2H_5OH}{46.07 \text{ g } C_2H_5OH}$$

$$= 0.04992 \text{ mol } C_2H_5OH$$

To obtain the molar concentration, $c_{C_2H_5OH}$, we divide the number of moles of ethanol by the volume. Thus,

$$c_{C_2H_5OH} = \frac{n_{C_2H_5OH}}{V} = \frac{0.04992 \text{ mol } C_2H_5OH}{3.50 \text{ L}}$$

$$= 0.0143 \text{ mol } C_2H_5OH/L = 0.0143 \text{ M}$$

Analytical Molarity.   The *analytical molarity* of a solution gives the *total* number of moles of a solute in 1 L of the solution, or alternatively, the total number of millimoles in 1 mL. That is, the analytical molarity specifies a recipe by which the solution can be prepared. For example, a sulfuric acid solution that has an analytical concentration of 1.0 M can be prepared by dissolving 1.0 mol, or 98 g, of H_2SO_4 in water and diluting to exactly 1.0 L.

Equilibrium Molarity.   The *equilibrium molarity*, or *species molarity*, expresses the molar concentration of a particular species in a solution at equilibrium. To determine the species molarity, it is necessary to know how the solute behaves when it is dissolved in a solvent. For example, the species molarity of H_2SO_4 in a solution with an analytical concentration of 1.0 M is 0.0 M because the sulfuric acid is entirely dissociated into a mixture of H_3O^+, HSO_4^-, and SO_4^{2-} ions; there are essentially no H_2SO_4 molecules as such in this solution. The equilibrium concentrations and thus the species molarities of these three ions are 1.01, 0.99, and 0.01 M, respectively.

Equilibrium molar concentrations are usually symbolized by placing square brackets around the chemical formula for the species, so for our solution of H_2SO_4 with an analytical concentration of 1.0 M, we can write

$$[H_2SO_4] = 0.00 \text{ M} \qquad [H_3O^+] = 1.01 \text{ M}$$
$$[HSO_4^-] = 0.99 \text{ M} \qquad [SO_4^{2-}] = 0.01 \text{ M}$$

> *Analytical molarity* is the total number moles of a solute, regardless of its chemical state, in one liter of solution. The analytical molarity describes how a solution of a given molarity can be prepared.

> *Equilibrium molarity*, or *species molarity*, is the molar concentration of a particular species in a solution at equilibrium.

> Some chemists prefer to distinguish between species and analytical concentrations in a different way. They use *molar concentration* for species concentration and *formal concentration* (F) for analytical concentration. Applying this convention to our example, we can say that the formal concentration of H_2SO_4 is 1.0 F, whereas its equilibrium molar concentration is 0.0 M.

> In this example, the *analytical molarity* of H_2SO_4 is given by
> $c_{H_2SO_4} = [SO_4^{2-}] + [HSO_4^-]$ because these are the only two sulfate-containing species in the solution.

Example 3-4

Calculate the analytical and equilibrium molar concentrations of the solute species in an aqueous solution that contains 285 mg of trichloroacetic acid, Cl_3CCOOH (163.4 g/mol), in 10.0 mL. Trichloroacetic acid is 73% ionized in water.

As in Example 3-3, we calculate the number of moles of Cl_3CCOOH, which we designate as HA, and divide by the volume of the solution, 10.0 mL or 0.0100 L. Thus,

$$\text{amount HA} = 285 \text{ mg HA} \times \frac{1 \text{ g HA}}{1000 \text{ mg HA}} \times \frac{1 \text{ mol HA}}{163.4 \text{ g HA}}$$
$$= 1.744 \times 10^{-3} \text{ mol HA}$$

The molar analytical concentration, c_{HA}, is then

$$c_{HA} = \frac{1.744 \times 10^{-3} \text{ mol HA}}{10.0 \text{ mL}} \times \frac{1000 \text{ mL}}{1 \text{ L}} = 0.174 \frac{\text{mol HA}}{\text{L}} = 0.174 \text{ M}$$

In this solution, 73% of the HA dissociates, giving H^+ and A^-:

$$HA \rightleftharpoons H^+ + A^-$$

The species molarity of HA is then 27% of c_{HA}. Thus,

$$[HA] = c_{HA} \times (100 - 73)/100 = 0.174 \times 0.27$$
$$= 0.047 \text{ M}$$

The species molarity of A^- is equal to 73% of the analytical concentration of HA. That is,

$$[A^-] = \frac{73 \text{ mol } A^-}{100 \text{ mol HA}} \times 0.174 \frac{\text{mol HA}}{L} = 0.127 \text{ M}$$

Because one mole H^+ is formed for each mole A^-, we can also write

$$[H^+] = [A^-] = 0.127 \text{ M}$$

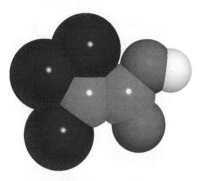

Molecular model of trichloroacetic acid. The rather strong acidity of trichloroacetic acid is usually ascribed to the inductive effect of the three chlorine atoms attached to the end of the molecule opposite the acidic proton. Electron density is withdrawn away from the carboxylate group so that the trichloroacetate anion that is formed when the acid dissociates is stabilized. The acid is used in protein precipitation and in dermatological preparations for the removal of undesirable skin growths.

Example 3-5

Describe the preparation of 2.00 L of 0.108 M $BaCl_2$ from $BaCl_2 \cdot 2H_2O$ (244 g/mol).

To determine the number of grams of solute to be dissolved and diluted to 2.00 L, we note that 1 mol of the dihydrate yields 1 mol of $BaCl_2$. Therefore, to produce this solution we will need

$$2.00 \text{ } L \times \frac{0.108 \text{ mol } BaCl_2 \cdot 2H_2O}{L} = 0.216 \text{ mol } BaCl_2 \cdot 2H_2O$$

The mass of $BaCl_2 \cdot 2H_2O$ is then

$$0.216 \text{ mol } BaCl_2 \cdot 2H_2O \times \frac{244.3 \text{ g } BaCl_2 \cdot 2H_2O}{\text{mol } BaCl_2 \cdot 2H_2O} = 52.8 \text{ g } BaCl_2 \cdot 2H_2O$$

Dissolve 52.8 g of $BaCl_2 \cdot 2H_2O$ in water and dilute to 2.00 L.

Example 3-6

Describe the preparation of 500 mL of 0.0740 M Cl^- solution from solid $BaCl_2 \cdot 2H_2O$ (244 g/mol).

$$\text{mass } BaCl_2 \cdot 2H_2O = \frac{0.0740 \text{ mol } Cl^-}{L} \times 0.500 \text{ } L \times \frac{1 \text{ mol } BaCl_2 \cdot 2H_2O}{2 \text{ mol } Cl^-}$$

$$\times \frac{244.3 \text{ g } BaCl_2 \cdot 2H_2O}{\text{mol } BaCl_2 \cdot 2H_2O}$$

$$= 4.52 \text{ g } BaCl_2 \cdot 2H_2O$$

Dissolve 4.52 g of $BaCl_2 \cdot 2H_2O$ in water and dilute to 0.500 L or 500 mL.

> The number of moles of the species A in a solution of A is given by
> $$n_A = c_A \times V_A$$
> where V_A is the volume of the solution in liters.

Percent Concentration

Chemists frequently express concentrations in terms of percent (parts per hundred). Unfortunately, this practice can be a source of ambiguity because percent composition of a solution can be expressed in several ways. Three common methods are

$$\text{weight percent (w/w)} = \frac{\text{weight solute}}{\text{weight solution}} \times 100\%$$

$$\text{volume percent (v/v)} = \frac{\text{volume solute}}{\text{volume soln}} \times 100\%$$

$$\text{weight/volume percent (w/v)} = \frac{\text{weight solute, g}}{\text{volume soln, mL}} \times 100\%$$

Note that the denominator in each of these expressions refers to the *solution* rather than to the solvent. Note also that the first two expressions do not depend on the units used (provided, of course, that the numerator and denominator are expressed in the same units). In the third expression, units must be defined because the numerator and denominator have different units that do not cancel. Of the three expressions, only weight percent has the virtue of being temperature-independent.

Weight percent is frequently used to express the concentration of commercial aqueous reagents. For example, nitric acid is sold as a 70% solution, which means that the reagent contains 70 g of HNO_3 per 100 g of solution (see Example 3-10).

Volume percent is commonly used to specify the concentration of a solution prepared by diluting a pure liquid solute with another liquid. For example, a 5% aqueous solution of methanol *usually* describes a solution prepared by diluting 5.0 mL of pure methanol with enough water to give 100 mL.

Weight/volume percent is often used to indicate the composition of dilute aqueous solutions of solid reagents. For example, 5% aqueous silver nitrate *often* refers to a solution prepared by dissolving 5 g of silver nitrate in sufficient water to give 100 mL of solution.

To avoid uncertainty, always specify explicitly the type of percent composition being discussed. If this information is missing, the user must decide intuitively which of the several types is involved. The potential error resulting from a wrong choice is considerable. For example, commercial 50% (w/w) sodium hydroxide contains 763 g of the reagent per liter, which corresponds to 76.3% (w/v) sodium hydroxide.

> Weight percent would be more properly called mass percent and be abbreviated m/m. The term *weight percent* is so widely used in the chemical literature, however, that we will use it throughout this text.

> You should always specify the type of percent when reporting concentrations in this way.

Parts Per Million and Parts Per Billion

For very dilute solutions, parts per million (ppm) is a convenient way to express concentration:

$$c_{ppm} = \frac{\text{mass of solute}}{\text{mass of solution}} \times 10^6 \text{ ppm}$$

where c_{ppm} is the concentration in parts per million. Obviously, the units of weight in the numerator and denominator must agree. For even more dilute solutions, 10^9 ppb rather than 10^6 ppm is used in the foregoing equation to give the results in *parts per billion* (ppb). The term *parts per thousand* (ppt) is also encountered, especially in oceanography.

> A handy rule for calculating concentration in parts per million is that for dilute aqueous solutions whose densities are approximately 1.00 g/mL, 1 ppm = 1.00 mg/L. That is,
>
> $$c_{ppm} = \frac{\text{mass solute (mg)}}{\text{volume solution (L)}} \quad (3\text{-}2)$$
>
> For parts per billion, 1 ppb = 1 μg/L.
>
> $$c_{ppb} = \frac{\text{mass solute g}}{\text{mass solution g}} \times 10^9 \text{ ppb}$$

Example 3-7

What is the molarity of K^+ in a solution that contains 63.3 ppm of $K_3Fe(CN)_6$ (329.3 g/mol)?

Because the solution is so dilute, it is reasonable to assume that its density is 1.00 g/mL. Therefore, according to Equation 3-2,

$$63.3 \text{ ppm } K_3Fe(CN)_6 = 63.3 \text{ mg } K_3Fe(CN)_6/L$$

$$c_{K_3Fe(CN)_6} = \frac{63.3 \text{ mg } \cancel{K_3Fe(CN)_6}}{L} \times \frac{1 \text{ g } \cancel{K_3Fe(CN)_6}}{1000 \text{ mg } \cancel{K_3Fe(CN)_6}}$$

$$\times \frac{1 \text{ mol } K_3Fe(CN)_6}{329.3 \text{ g } \cancel{K_3Fe(CN)_6}}$$

$$= 1.922 \times 10^{-4} \frac{\text{mol}}{L} = 1.922 \times 10^{-4} \text{ M}$$

$$[K^+] = \frac{1.922 \times 10^{-4} \text{ mol } \cancel{K_3Fe(CN)_6}}{L} \times \frac{3 \text{ mol } K^+}{1 \text{ mol } \cancel{K_3Fe(CN)_6}}$$

$$= 5.77 \times 10^{-4} \frac{\text{mol } K^+}{L} = 5.77 \times 10^{-4} \text{ M}$$

Solution-Diluent Volume Ratios

The composition of a dilute solution is sometimes specified in terms of the volume of a more concentrated solution and the volume of solvent to be used in diluting it. The volume of the former is separated from that of the latter by a colon. Thus, a 1:4 HCl solution contains four volumes of water for each volume of concentrated hydrochloric acid. This method of notation is frequently ambiguous in that the concentration of the original solution is not always obvious to the reader. Moreover, under some circumstances 1:4 means dilute one volume with three volumes. Because of such uncertainties, you should avoid using solution-diluent ratios.

p-Functions

The most well-known p-function is pH, which is the negative logarithm of [H₃O⁺].

Scientists frequently express the concentration of a species in terms of its *p-function*, or *p-value*. The p-value is the negative base-10 logarithm (log) of the molar concentration of that species. Thus, for the species X,

$$pX = -\log [X]$$

As shown by the following examples, p-values offer the advantage of allowing concentrations that vary over ten or more orders of magnitude to be expressed in terms of small positive numbers.

Example 3-8

Calculate the p-value for each ion in a solution that is 2.00×10^{-3} M in NaCl and 5.4×10^{-4} M in HCl.

$$pH = -\log [H_3O^+] = -\log (5.4 \times 10^{-4}) = 3.27$$

To obtain pNa, we write

$$pNa = -\log (2.00 \times 10^{-3}) = 2.699$$

The total Cl^- concentration is given by the sum of the concentrations of the two solutes:

$$[Cl^-] = 2.00 \times 10^{-3} \text{ M} + 5.4 \times 10^{-4} \text{ M}$$

$$= 2.00 \times 10^{-3} \text{ M} + 0.54 \times 10^{-3} \text{ M} = 2.54 \times 10^{-3} \text{ M}$$

$$pCl = -\log 2.54 \times 10^{-3} = 2.595$$

Molecular model of HCl. Hydrogen chloride is a gas consisting of heteronuclear diatomic molecules. The gas is extremely soluble in water; when a solution is prepared of the gas, only then do the molecules dissociate to form aqueous hydrochloric acid, which consists of H₃O⁺ and Cl⁻ ions.

Note that in Example 3-8, and in the one that follows, the results are rounded according to the rules listed on page 144.

Example 3-9

Calculate the molar concentration of Ag^+ in a solution that has a pAg of 6.372.

$$pAg = -\log [Ag^+] = 6.372$$
$$\log [Ag^+] = -6.372$$
$$[Ag^+] = \text{antilog} (-6.372)$$
$$= 4.246 \times 10^{-7} = 4.25 \times 10^{-7}$$

3B-2 Density and Specific Gravity of Solutions

Density and specific gravity are often encountered in the analytical literature. The *density* of a substance is its mass per unit volume, whereas its *specific gravity* is the ratio of its mass to the mass of an equal volume of water at 4°C. Density has units of kilograms per liter or grams per milliliter in the metric system. Specific gravity is dimensionless and so is not tied to any particular system of units. For this reason, specific gravity is widely used in describing items of commerce. Figure 3-1, which is the label from a bottle of reagent-grade hy-

Density is the mass of a substance per unit volume. In SI units, density is expressed in units of kg/L or alternatively g/mL.

Specific gravity is the ratio of the mass of a substance to the mass of an equal volume of water.

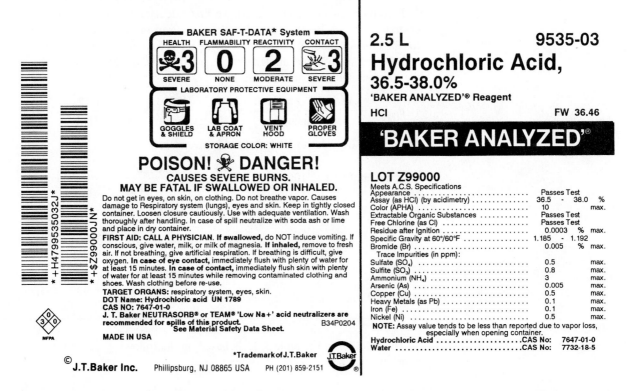

Figure 3-1 Label from a bottle of reagent-grade hydrochloric acid. Note that the specific gravity of the acid over the temperature range of 60 to 80°F is specified on the label.

drochloric acid, illustrates this point. Since the density of water is approximately 1.00 g/mL, and since we use the metric system throughout this text, we use the terms *density* and *specific gravity* interchangeably.

Specific Gravities of Commercial Concentrated Acids and Bases

Reagent	Concentration % (w/w)	Specific Gravity
Acetic acid	99.7	1.05
Ammonia	29.0	0.90
Hydrochloric acid	37.2	1.19
Hydrofluoric acid	49.5	1.15
Nitric acid	70.5	1.42
Perchloric acid	71.0	1.67
Phosphoric acid	86.0	1.71
Sulfuric acid	96.5	1.84

Example 3-10

Calculate the molar concentration of HNO_3 (63.0 g/mol) in a solution that has a specific gravity of 1.42 and is 70% HNO_3 (w/w).

Let us first calculate the grams of acid per liter of concentrated solution.

$$\frac{\text{g HNO}_3}{\text{L reagent}} = \frac{1.42\ \cancel{\text{kg reagent}}}{\text{L reagent}} \times \frac{10^3\ \cancel{\text{g reagent}}}{\cancel{\text{kg reagent}}} \times \frac{70\ \text{g HNO}_3}{100\ \cancel{\text{g reagent}}}$$
$$= \frac{994\ \text{g HNO}_3}{\text{L reagent}}$$

Then

$$c_{\text{HNO}_3} = \frac{994\ \cancel{\text{g HNO}_3}}{\text{L reagent}} \times \frac{1\ \text{mol HNO}_3}{63.0\ \cancel{\text{g HNO}_3}} = \frac{15.8\ \text{mol HNO}_3}{\text{L reagent}} = 16\ \text{M}$$

Example 3-11

Describe the preparation of 100 mL of 6.0 M HCl from a concentrated solution that has a specific gravity of 1.18 and is 37% (w/w) HCl (36.5 g/mol).

Proceeding as in Example 3-10, we first calculate the molarity of the concentrated reagent. We then calculate the number of moles of acid that we need for the diluted solution. Finally, we divide the second figure by the first to obtain the volume of concentrated acid required. Thus, to obtain the molarity of the concentrated reagent, we write

$$c_{\text{HCl}} = \frac{1.18 \times 10^3\ \cancel{\text{g reagent}}}{\text{L reagent}} \times \frac{37\ \cancel{\text{g HCl}}}{100\ \cancel{\text{g reagent}}} \times \frac{1\ \text{mol HCl}}{36.5\ \cancel{\text{g HCl}}} = 12.0\ \text{M}$$

The number of moles of HCl required is given by

$$\text{no. mol HCl} = 100\ \cancel{\text{mL}} \times \frac{1\ \cancel{\text{L}}}{1000\ \cancel{\text{mL}}} \times \frac{6.0\ \text{mol HCl}}{\cancel{\text{L}}} = 0.600\ \text{mol}$$

Finally, to obtain the volume of concentrated reagent, we write

$$\text{vol concd reagent} = 0.600\ \cancel{\text{mol HCl}} \times \frac{1\ \text{L reagent}}{12.0\ \cancel{\text{mol HCl}}}$$
$$= 0.0500\ \text{L or } 50.0\ \text{mL}$$

Thus, dilute 50 mL of the concentrated reagent to 100 mL.

Equation 3-3 can be used with L and mol/L or mL and mmol/mL. Thus,

$$\text{L}_{\text{concd}} \times \frac{\text{mol}_{\text{concd}}}{\text{L}_{\text{concd}}} = \text{L}_{\text{dil}} \times \frac{\text{mol}_{\text{dil}}}{\text{L}_{\text{dil}}}$$

$$\text{mL}_{\text{concd}} \times \frac{\text{mmol}_{\text{concd}}}{\text{mL}_{\text{concd}}} = \text{mL}_{\text{dil}} \times \frac{\text{mmol}_{\text{dil}}}{\text{mL}_{\text{dil}}}$$

The solution to Example 3-11 is based on the following useful relationship, which we will be using countless times:

$$V_{\text{concd}} \times c_{\text{concd}} = V_{\text{dil}} \times c_{\text{dil}} \tag{3-3}$$

where the two terms on the left are the volume and molar concentration of a concentrated solution that is being used to prepare a diluted solution having the volume and concentration given by the corresponding terms on the right. The basis of this equation is that the number of moles of solute in the diluted solution must equal the number of moles in the concentrated reagent. Note that the volumes can be in milliliters or liters as long as the same units are used for both solutions.

3C   CHEMICAL STOICHIOMETRY

Stoichiometry is defined as the mass relationships among reacting chemical species. This section provides a brief review of stoichiometry and its applications to chemical calculations.

> The stoichiometry of a reaction is the relationship among the number of moles of reactants and products as shown by a balanced equation.

3C-1   Empirical Formulas and Molecular Formulas

An *empirical formula* gives the simplest whole-number ratio of atoms in a chemical compound. In contrast, a *molecular formula* specifies the number of atoms in a molecule. Two or more substances may have the same empirical formula but different molecular formulas. For example, CH_2O is both the empirical and the molecular formula for formaldehyde; it is also the empirical formula for such diverse substances as acetic acid ($C_2H_4O_2$), glyceraldehyde ($C_3H_6O_3$), and glucose ($C_6H_{12}O_6$), as well as more than 50 other substances containing 6 or fewer carbon atoms. The empirical formula is obtained from the percent composition of a compound. In addition, the molecular formula requires a knowledge of the molar mass of the species.

A *structural formula* provides additional information. For example, the chemically different ethanol and dimethyl ether share the same molecular formula C_2H_6O. Their structural formulas, C_2H_5OH and CH_3OCH_3, reveal structural differences between these compounds that are not discernible in the molecular formula that they share.

3C-2   Stoichiometric Calculations

A balanced chemical equation is a statement of the combining ratios, or stoichiometry (in units of moles), among the reacting substances and their products. Thus, the equation

$$2NaI(aq) + Pb(NO_3)_2(aq) \rightleftharpoons PbI_2(s) + 2NaNO_3(aq)$$

indicates that 2 mol of aqueous sodium iodide combine with 1 mol of aqueous lead nitrate to produce 1 mol of solid lead iodide and 2 mol of aqueous sodium nitrate.[2]

Example 3-12 demonstrates how the mass in grams of reactants and products in a chemical reaction are related. As shown in Figure 3-2, a calculation of this

> Often the physical state of substances appearing in equations is indicated by the letters (g), (l), (s), or (aq), which refer to gaseous, liquid, solid, and aqueous solution states, respectively

---

[2]Here it is advantageous to depict the reaction in terms of chemical compounds. If we wish to focus on reacting species, the net ionic equation is preferable:

$$2I^-(aq) + Pb^{2+}(aq) \rightleftharpoons PbI_2(s)$$

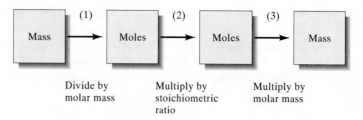

Figure 3-2 Flow diagram for making stoichiometric calculations. (1) When the mass of a reactant or product is given, the mass is first converted to the number of moles using the molar mass. (2) The stoichiometric ratio given by the chemical equation for the reaction is then used to find the number of moles of another reactant that combine with the original substance or the number of moles of product that form. (3) Finally, the mass of the other reactant or the product is computed from its molar mass.

type is a three-step process involving (1) transformation of the known mass of a substance in grams to a corresponding number of moles, (2) multiplication by a factor that accounts for the stoichiometry, and (3) reconversion of the data in moles back to the metric units called for in the answer.

Example 3-12

(a) What mass of $AgNO_3$ (169.9 g/mol) is needed to convert 2.33 g of Na_2CO_3 (106.0 g/mol) to Ag_2CO_3? (b) What mass of Ag_2CO_3 (275.7 g/mol) will be formed?

(a) $Na_2CO_3(aq) + 2AgNO_3(aq) \rightarrow Ag_2CO_3(s) + 2NaNO_3(aq)$

Step 1: $n_{Na_2CO_3} = 2.33 \text{ g Na}_2\text{CO}_3 \times \dfrac{1 \text{ mol Na}_2\text{CO}_3}{106.0 \text{ g Na}_2\text{CO}_3}$

$= 0.02198 \text{ mol Na}_2\text{CO}_3$

Step 2:   The balanced equation reveals that

$$n_{AgNO_3} = 0.02198 \text{ mol Na}_2\text{CO}_3 \times \dfrac{2 \text{ mol AgNO}_3}{1 \text{ mol Na}_2\text{CO}_3}$$

$= 0.04396 \text{ mol AgNO}_3$

Here the stoichiometric factor is (2 mol $AgNO_3$)/(1 mol Na_2CO_3).

Step 3: $m_{AgNO_3} = 0.04396 \text{ mol AgNO}_3 \times \dfrac{169.9 \text{ g AgNO}_3}{\text{mol AgNO}_3}$

$= 7.47 \text{ g AgNO}_3$

(b) $n_{Ag_2CO_3} = n_{Na_2CO_3} = 0.02198 \text{ mol}$

$m_{Ag_2CO_3} = 0.02198 \text{ mol Ag}_2\text{CO}_3 \times \dfrac{275.7 \text{ g Ag}_2\text{CO}_3}{\text{mol Ag}_2\text{CO}_3}$

$= 6.06 \text{ g Ag}_2\text{CO}_3$

Example 3-13

What mass of Ag_2CO_3 (276 g/mol) is formed when 25.0 mL of 0.200 M $AgNO_3$ is mixed with 50.0 mL of 0.0800 M Na_2CO_3?

Mixing these two solutions will result in one (and only one) of three possible outcomes, specifically:

1. An excess of $AgNO_3$ will remain after reaction is complete.
2. An excess of Na_2CO_3 will remain after reaction is complete.
3. An excess of neither reagent will exist (that is, the number of moles of Na_2CO_3 is exactly equal to twice the number of moles of $AgNO_3$).

As a first step, we must establish which of these situations applies by calculating the amounts of reactants (in chemical units) available at the outset.

The initial amounts are

$$n_{AgNO_3} = 25.0 \text{ mL AgNO}_3 \times \frac{1 \text{ L AgNO}_3}{1000 \text{ mL AgNO}_3} \times \frac{0.200 \text{ mol AgNO}_3}{\text{L AgNO}_3}$$

$$= 5.00 \times 10^{-3} \text{ mol AgNO}_3$$

$$n_{Na_2CO_3} = 50.0 \text{ mL Na}_2\text{CO}_3 \times \frac{1 \text{ L Na}_2\text{CO}_3}{1000 \text{ mL Na}_2\text{CO}_3} \times \frac{0.0800 \text{ mol Na}_2\text{CO}_3}{\text{L Na}_2\text{CO}_3}$$

$$= 4.00 \times 10^{-3} \text{ mol Na}_2\text{CO}_3$$

Because each CO_3^{2-} ion reacts with two Ag^+ ions, $2 \times 4.00 \times 10^{-3} = 8.00 \times 10^{-3}$ mol $AgNO_3$ is required to react with the Na_2CO_3. Since we have insufficient $AgNO_3$, situation (2) prevails and the amount of Ag_2CO_3 produced will be limited by the amount of $AgNO_3$ available. Thus,

$$m_{Ag_2CO_3} = 5.00 \times 10^{-3} \text{ mol AgNO}_3 \times \frac{1 \text{ mol Ag}_2\text{CO}_3}{2 \text{ mol AgNO}_3} \times \frac{276 \text{ g Ag}_2\text{CO}_3}{\text{mol Ag}_2\text{CO}_3}$$

$$= 0.690 \text{ g Ag}_2\text{CO}_3$$

Example 3-14

What will be the molar analytical concentration of Na_2CO_3 in the solution produced when 25.0 mL of 0.200 M $AgNO_3$ is mixed with 50.0 mL of 0.0800 M Na_2CO_3?

We have seen in Example 3-13 that formation of 5.00×10^{-3} mol of $AgNO_3$ will require 2.50×10^{-3} mol of Na_2CO_3. The number of moles of unreacted Na_2CO_3 is then given by

$$n_{Na_2CO_3} = 4.00 \times 10^{-3} \text{ mol Na}_2\text{CO}_3$$

$$- 5.00 \times 10^{-3} \text{ mol AgNO}_3 \times \frac{1 \text{ mol Na}_2\text{CO}_3}{2 \text{ mol AgNO}_3}$$

$$= 1.50 \times 10^{-3} \text{ mol Na}_2\text{CO}_3$$

By definition, the molarity is the number of moles Na_2CO_3/L. Thus,

$$c_{Na_2CO_3} = \frac{1.50 \times 10^{-3} \text{ mol Na}_2\text{CO}_3}{(50.0 + 25.0) \text{ mL}} \times \frac{1000 \text{ mL}}{1 \text{ L}} = 0.0200 \text{ M Na}_2\text{CO}_3$$

3D QUESTIONS AND PROBLEMS

3-1. Define
   *(a) millimole.
   (b) molar mass.
   *(c) millimolar mass.
   (d) parts per million.

3-2. What is the difference between species molarity and analytical molarity?

*3-3. Give two examples of units derived from the fundamental base SI units.

3-4. Simplify the following quantities using a unit with an appropriate prefix:
   *(a) 2.5×10^4 Hz.
   (b) 3.72×10^{-8} g.
   *(c) 8.43×10^5 μmol.
   (d) 4.0×10^6 s.
   *(e) 7.44×10^4 nm.
   (f) 97,000 g.

*3-5. How many Na^+ ions are contained in 4.62 g Na_3PO_4?

3-6. How many K^+ ions are contained in 5.96 mol K_3PO_4?

*3-7. Find the number of moles of the indicated species in
   (a) 6.84 g of B_2O_3.
   (b) 296 mg of $Na_2B_4O_7 \cdot 10H_2O$.
   (c) 8.75 g of Mn_3O_4.
   (d) 67.4 mg of CaC_2O_4.

3-8. Find the number of millimoles of the indicated species in
   (a) 64 mg of P_2O_5.
   (b) 12.92 g of CO_2.
   (c) 30.0 g of $NaHCO_3$.
   (d) 764 mg of $MgNH_4PO_4$.

*3-9. Find the number of millimoles of solute in
   (a) 2.00 L of 2.76×10^{-3} M $KMnO_4$.
   (b) 750 mL of 0.0416 M KSCN.
   (c) 250 mL of a solution that contains 4.20 ppm of $CuSO_4$.
   (d) 3.50 L of 0.276 M KCl.

3-10. Find the number of millimoles of solute in
   (a) 175 mL of 0.320 M $HClO_4$.
   (b) 15.0 L of 8.05×10^{-3} M K_2CrO_4.
   (c) 5.00 L of an aqueous solution that contains 6.75 ppm of $AgNO_3$.
   (d) 851 mL of 0.0200 M KOH.

*3-11. What is the mass in milligrams of
   (a) 0.666 mol of HNO_3?
   (b) 300 mmol of MgO?
   (c) 19.0 mol of NH_4NO_3?
   (d) 5.32 mol of $(NH_4)_2Ce(NO_3)_6$ (548.23 g/mol)?

3-12. What is the mass in grams of
   (a) 6.21 mol of KBr?
   (b) 10.2 mmol of PbO?
   (c) 4.92 mol of $MgSO_4$?
   (d) 12.8 mmol of $Fe(NH_4)_2(SO_4)_2 \cdot 6H_2O$?

3-13. What is the mass in milligrams of solute in
   *(a) 26.0 mL of 0.150 M sucrose (342 g/mol)?
   *(b) 2.92 L of 5.23×10^{-3} M H_2O_2?
   (c) 737 mL of a solution that contains 6.38 ppm of $Pb(NO_3)_2$?
   (d) 6.75 mL of 0.0619 M KNO_3?

3-14. What is the mass in grams of solute in
   *(a) 450 mL of 0.164 M H_2O_2?
   *(b) 27.0 mL of 8.75×10^{-4} M benzoic acid (122 g/mol)?
   (c) 3.50 L of a solution that contains 21.7 ppm of $SnCl_2$?
   (d) 21.7 mL of 0.0125 M $KBrO_3$?

3-15. Calculate the p-value for each of the indicated ions in the following:
   *(a) Na^+, Cl^-, and OH^- in a solution that is 0.0235 M in NaCl and 0.0503 M in NaOH.
   (b) Ba^{2+}, Mn^{2+}, and Cl^- in a solution that is 4.62×10^{-3} M in $BaCl_2$ and 1.78 M in $MnCl_2$.
   *(c) H^+, Cl^-, and Zn^{2+} in a solution that is 0.800 M in HCl and 0.101 M in $ZnCl_2$.
   (d) Cu^{2+}, Zn^{2+}, and NO_3^- in a solution that is 3.56×10^{-2} M in $Cu(NO_3)_2$ and 0.104 M in $Zn(NO_3)_2$.
   *(e) K^+, OH^-, and $Fe(CN)_6^{4-}$ in a solution that is 3.79×10^{-7} M in $K_4Fe(CN)_6$ and 4.12×10^{-7} M in KOH.
   (f) H^+, Ba^{2+}, and ClO_4^- in a solution that is 3.62×10^{-4} M in $Ba(ClO_4)_2$ and 6.75×10^{-4} M in $HClO_4$.

3-16. Calculate the molar H_3O^+ ion concentration of a solution that has a pH of
   *(a) 9.21   (b) 4.58
   *(c) 0.45   (d) 14.12
   *(e) 7.32   (f) 6.76
   *(g) -0.21 (h) -0.52

3-17. Calculate the p-functions for each ion in a solution that is:
   *(a) 0.0100 M in NaBr.
   (b) 0.0100 M in $BaBr_2$.
   *(c) 3.5×10^{-3} M in $Ba(OH)_2$.
   (d) 0.040 M in HCl and 0.020 M in NaCl.
   *(e) 5.2×10^{-3} M in $CaCl_2$ and 3.6×10^{-3} M in $BaCl_2$.
   (f) 4.8×10^{-8} M in $Zn(NO_3)_2$ and 5.6×10^{-7} M $Cd(NO_3)_2$.

3-18. Convert the following p-functions to molar concentrations:
   *(a) pH = 8.67   (b) pOH = 0.125
   *(c) pBr = 0.034   (d) pCa = 12.35
   *(e) pLi = -0.321 (f) pNO_3 = 7.77
   *(g) pMn = 0.0025   (h) pCl = 1.020

*3-19. Sea water contains an average of 1.08×10^3 ppm of Na^+ and 270 ppm SO_4^{2-}. Calculate
   (a) the molar concentrations of Na^+ and SO_4^{2-} given that the average density of sea water is 1.02 g/mL.
   (b) the pNa and pSO$_4$ for sea water.

3-20. Average human blood serum contains 18 mg of K^+ and 365 mg of Cl^- per 100 mL. Calculate
   (a) the molar concentration for each of these species; use 1.00 g/mL for the density of serum.
   (b) pK and pCl for human serum.

*3-21. A solution was prepared by dissolving 6.34 g of $KCl \cdot MgCl_2 \cdot 6H_2O$ (277.85 g/mol) in sufficient water to give 2.000 L. Calculate
   (a) the molar analytical concentration of $KCl \cdot MgCl_2$ in this solution.
   (b) the molar concentration of Mg^{2+}.
   (c) the molar concentration of Cl^-.
   (d) the weight/volume percentage of $KCl \cdot MgCl_2 \cdot 6H_2O$.
   (e) the number of millimoles of Cl^- in 25.0 mL of this solution.
   (f) ppm K^+.
   (g) pMg for the solution.
   (h) pCl for the solution.

3-22. A solution was prepared by dissolving 414 mg of $K_3Fe(CN)_6$ (329 g/mol) in sufficient water to give 750 mL of solution. Calculate
   (a) the molar analytical concentration of $K_3Fe(CN)_6$.
   (b) the molar concentration of K^+.
   (c) the molar concentration of $Fe(CN)_6^{3-}$.
   (d) the weight/volume percentage of $K_3Fe(CN)_6$.
   (e) the number of millimoles of K^+ in 50.0 mL of this solution.
   (f) ppm $Fe(CN)_6^{3-}$.
   (g) pK for the solution.
   (h) pFe(CN)$_6$ for the solution.

*3-23. A 6.42% (w/w) $Fe(NO_3)_3$ (241.81 g/mol) solution has a density of 1.059 g/mL. Calculate
   (a) the molar analytical concentration of $Fe(NO_3)_3$ in this solution.
   (b) the molar NO_3^- concentration of the solution.
   (c) the mass in grams of $Fe(NO_3)_3$ in each liter of this solution.

3-24. A 15.0% (w/w) $NiCl_2$ (129.61 g/mol) solution has a density of 1.149 g/mL. Calculate
   (a) the molar concentration of $NiCl_2$ in this solution.
   (b) the molar Cl^- concentration of the solution.
   (c) the grams of $NiCl_2$ contained in each liter of this solution.

*3-25. Describe the preparation of
   (a) 500 mL of 4.75% (w/v) aqueous glycerol (C_2H_5OH, 46.1 g/mol).

(b) 500 g of 4.75% (w/w) aqueous ethanol.
(c) 500 mL of 4.75% (v/v) aqueous ethanol.

3-26. Describe the preparation of
   (a) 2.50 L of 18.0% (w/v) aqueous glycerol ($C_3H_8O_3$, 92.1 g/mol).
   (b) 2.50 kg of 18.0% (w/w) aqueous glycerol.
   (c) 2.50 L of 18.0% (v/v) aqueous glycerol.

*3-27. Describe the preparation of 750 mL of 6.00 M H_3PO_4 from the commercial reagent that is 85% H_3PO_4 (w/w) and has a specific gravity of 1.69.

3-28. Describe the preparation of 900 mL of 3.00 M HNO_3 from the commercial reagent that is 69% HNO_3 (w/w) and has a specific gravity of 1.42.

*3-29. Describe the preparation of
   (a) 500 mL of 0.0650 M $AgNO_3$ from the solid reagent.
   (b) 1.00 L of 0.285 M HCl, starting with a 6.00 M solution of the reagent.
   (c) 400 mL of a solution that is 0.0825 M in K^+, starting with solid $K_4Fe(CN)_6$.
   (d) 600 mL of 3.00% (w/v) aqueous $BaCl_2$ from a 0.400 M $BaCl_2$ solution.
   (e) 2.00 L of 0.120 M $HClO_4$ from the commercial reagent [60% $HClO_4$ (w/w), sp gr 1.60].
   (f) 9.00 L of a solution that is 60.0 ppm in Na^+, starting with solid Na_2SO_4.

3-30. Describe the preparation of
   (a) 5.00 L of 0.0500 M $KMnO_4$ from the solid reagent.
   (b) 4.00 L of 0.250 M $HClO_4$, starting with an 8.00 M solution of the reagent.
   (c) 400 mL of a solution that is 0.0250 M in I^-, starting with MgI_2.
   (d) 200 mL of 1.00% (w/v) aqueous $CuSO_4$, from a 0.365 M $CuSO_4$ solution.
   (e) 1.50 L of 0.215 M NaOH from the concentrated commercial reagent [50% NaOH (w/w), sp gr 1.525].
   (f) 1.50 L of a solution that is 12.0 ppm in K^+, starting with solid $K_4Fe(CN)_6$.

*3-31. What mass of solid $La(IO_3)_3$ (663.6 g/mol) is formed when 50.0 mL of 0.250 M La^{3+} is mixed with 75.0 mL of 0.302 M IO_3^-?

3-32. What mass of solid $PbCl_2$ (278.10 g/mol) is formed when 200 mL of 0.125 M Pb^{2+} is mixed with 400 mL of 0.175 M Cl^-?

*3-33. Exactly 0.1120 g of pure Na_2CO_3 was dissolved in 100.0 mL of 0.0497 M $HClO_4$.
   (a) What mass in grams of CO_2 was evolved?
   (b) What was the molarity of the excess reactant (HCl or Na_2CO_3)?

3-34. Exactly 50.00 mL of a 0.4230 M solution of Na_3PO_4 was mixed with 100.00 mL of 0.5151 M $HgNO_3$.
   (a) What mass of solid Hg_3PO_4 was formed?

(b) What is the molarity of the unreacted species (Na_3PO_4 or $HgNO_3$) after the reaction was complete?

*3-35. Exactly 75.00 mL of a 0.3333 M solution of Na_2SO_3 was treated with 150.0 mL of 0.3912 M $HClO_4$ and boiled to remove the SO_2 formed.

    (a) What was the mass in grams of SO_2 that was evolved?

    (b) What was the concentration of the unreacted reagent (Na_2SO_3 or $HClO_4$) after the reaction was complete?

3-36. What mass of $MgNH_4PO_4$ precipitated when 200.0 mL of a 1.000% (w/v) solution of $MgCl_2$ was treated with 40.0 mL of 0.1753 M Na_3PO_4 and an excess of NH_4^+? What was the molarity of the excess reagent (Na_3PO_4 or $MgCl_2$) after the precipitation was complete?

*3-37. What volume of 0.01000 M $AgNO_3$ would be required to precipitate all the I^- in 200.0 mL of a solution that contained 26.43 ppt KI?

3-38. Exactly 750.0 mL of a solution that contained 480.4 ppm of $Ba(NO_3)_2$ was mixed with 200.0 mL of a solution that was 0.03090 M in $Al_2(SO_4)_3$.

    (a) What mass of solid $BaSO_4$ was formed?

    (b) What was the molarity of the unreacted reagent [$Al_2(SO_4)_3$ or $Ba(NO_3)_2$]?

3-39. Determine for the following (assume densities of 1.00 g/mL in all cases):

    (a) the number of ppm Cr in a 1.54×10^{-5} M $K_2Cr_2O_7$ solution.

    (b) the number of ppm P in a 5.32×10^{-6} M Na_3PO_4 solution.

    (c) the molarity of PO_4^{3-} in a solution that contains 25 mg P/100 mL solution.

    (d) how to prepare 1 L of a solution from $KMnO_4$ that is 25 ppm Mn.

Chapter 4

The Basic Approach to Chemical Equilibrium

Chapter 4 presents a basic approach to aqueous-solution chemistry, including chemical equilibrium and simple equilibrium-constant calculations. This material is treated in most general chemistry courses.

4A THE CHEMICAL COMPOSITION OF AQUEOUS SOLUTIONS

Water is the most plentiful solvent available on Earth, is easily purified, and is not toxic. It therefore finds widespread use as a medium for carrying out chemical analyses.

4A-1 Classifying Solutions of Electrolytes

Most of the solutes we discuss are *electrolytes,* which form ions when dissolved in water (or certain other solvents) and thus produce solutions that conduct electricity. In a solvent, *strong electrolytes* ionize essentially completely, whereas *weak electrolytes* ionize only partially. Thus, a solution of a weak electrolyte will be a poorer conductor than a solution containing an equal concentration of a strong electrolyte. Table 4-1 is a compilation of solutes that act as strong and weak electrolytes in water. Among the strong electrolytes listed are acids, bases, and salts.

A *salt* is produced in the reaction of an acid with a base. Examples include NaCl, Na_2SO_4, and $NaOOCCH_3$ (sodium acetate).

4A-2 Describing Acids and Bases

In 1923, two chemists, J. N. Brønsted in Denmark and J. M. Lowry in England, proposed independently a theory of acid/base behavior that is particularly useful in analytical chemistry. According to the Brønsted–Lowry theory, *an acid is a proton donor* and *a base is a proton acceptor.* For a species to behave as an acid, a proton acceptor (or base) must be present. The reverse is also true.

An *acid* is a substance that donates protons; a *base* is a substance that accepts protons.

An acid donates protons only in the presence of a proton acceptor (a base). Likewise, a base accepts protons only in the presence of a proton donor (an acid).

What are Conjugate Acids and Bases?

An important feature of the Brønsted–Lowry concept is the idea that the species produced when an acid gives up a proton is a potential proton acceptor called the

Table 4-1

Classification of Electrolytes

Strong	Weak
1. Inorganic acids such as HNO_3, $HClO_4$, $H_2SO_4^*$, HCl, HI, HBr, $HClO_3$, $HBrO_3$	1. Many inorganic acids, including H_2CO_3, H_3BO_3, H_3PO_4, H_2S, H_2SO_3
2. Alkali and alkaline-earth hydroxides	2. Most organic acids
3. Most salts	3. Ammonia and most organic bases
	4. Halides, cyanides, and thiocyanates of Hg, Zn, and Cd

*H_2SO_4 is completely dissociated into HSO_4^- and H_3O^+ ions and for this reason is classified as a strong electrolyte. It should be noted, however, that the HSO_4^- ion is a weak electrolyte, being only partially dissociated into SO_4^{2-} and H_3O^+.

A *conjugate base* is the species formed when an acid loses a proton. For example, acetate ion is the conjugate base of acetic acid; similarly, ammonium ion is the conjugate acid of the base ammonia.

A *conjugate acid* is the species formed when a base accepts a proton.

A substance acts as an acid only in the presence of a base and vice versa.

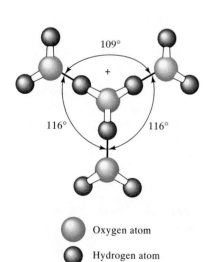

109°

116°   116°

○ Oxygen atom

● Hydrogen atom

The species $H_9O_4^+$ appears to be the predominant form of the hydrated proton. Its structure is shown above.

conjugate base of the parent acid. For example, when the species $acid_1$ gives up a proton, the species $base_1$ is formed as shown by the reaction

$$acid_1 \rightleftharpoons base_1 + proton$$

Here, $acid_1$ and $base_1$ are a conjugate acid/base pair.

Similarly, every base produces a *conjugate acid* as a result of accepting a proton. That is,

$$base_2 + proton \rightleftharpoons acid_2$$

When these two processes are combined, the result is an acid/base, or *neutralization,* reaction:

$$acid_1 + base_2 \rightleftharpoons base_1 + acid_2$$

The extent to which this reaction proceeds depends on the relative tendencies of the two bases to accept a proton (or the two acids to donate a proton). Examples of conjugate acid/base relationships are shown in Equations 4-1 through 4-4.

Many solvents are proton donors or proton acceptors and can thus induce basic or acidic behavior in solutes dissolved in them. For example, in an aqueous solution of ammonia, water can donate a proton and thus acts as an acid with respect to the solute:

$$\underset{base_1}{NH_3} + \underset{acid_2}{H_2O} \rightleftharpoons \underset{\substack{conjugate \\ acid_1}}{NH_4^+} + \underset{\substack{conjugate \\ base_2}}{OH^-} \qquad (4\text{-}1)$$

In this reaction, ammonia ($base_1$) reacts with water, which is labeled $acid_2$, to give the conjugate acid ammonium ion ($acid_1$) and hydroxide ion, which is the conjugate base ($base_2$) of the acid water. In contrast, water acts as a proton acceptor, or base, in an aqueous solution of nitrous acid:

$$\underset{base_1}{H_2O} + \underset{acid_2}{HNO_2} \rightleftharpoons \underset{\substack{conjugate \\ acid_1}}{H_3O^+} + \underset{\substack{conjugate \\ base_2}}{NO_2^-} \qquad (4\text{-}2)$$

The conjugate base of the acid HNO_2 is nitrite ion. The conjugate acid of water is the hydrated proton written as H_3O^+. This species is called the *hydronium* ion and consists of a proton covalently bonded to one water molecule. Higher hydrates such as $H_5O_2^+$ and $H_9O_4^+$ also exist in aqueous solutions of protons. For convenience, however, chemists generally use the notations H_3O^+ or, more simply, H^+ when writing chemical equations in which the proton is involved.

An acid that has donated a proton becomes a conjugate base capable of accepting a proton to reform the original acid; the converse holds equally well. Thus, nitrite ion, the species produced by the loss of a proton from nitrous acid, is a potential acceptor of a proton from a suitable donor. This reaction is the one that causes an aqueous solution of sodium nitrite to be slightly basic:

$$NO_2^- + H_2O \rightleftharpoons HNO_2 + OH^-$$

$$\text{base}_1 \quad \text{acid}_2 \quad \underset{\text{acid}_1}{\text{conjugate}} \quad \underset{\text{base}_2}{\text{conjugate}}$$

Some species have both acidic and basic properties and are called *amphiprotic* solutes, as illustrated in Feature 4-1.

Amphiprotic Solvents

Water is the classic example of an *amphiprotic solvent,* that is, a solvent that can act either as an acid (Equation 4-1) or as a base (Equation 4-2) depending on the solute. Other common amphiprotic solvents are methanol, ethanol, and anhy-

Water can act as either an acid or as a base

Amphiprotic solvents behave as acids in the presence of basic solutes and bases in the presence of acidic solutes.

Feature 4-1

Amphiprotic Species

Solutes that possess both acidic and basic properties are classed as amphiprotic. An example is dihydrogen phosphate ion, $H_2PO_4^-$, which behaves as a base in the presence of a proton donor such as H_3O^+.

$$H_2PO_4^- + H_3O^+ \rightleftharpoons H_3PO_4 + H_2O$$

$$\text{base}_1 \quad \text{acid}_2 \quad \text{acid}_1 \quad \text{base}_2$$

Here, H_3PO_4 is the conjugate acid of the original base. In the presence of a proton acceptor, such as hydroxide ion, however, $H_2PO_4^-$ behaves as an acid and donates a proton to form the conjugate base HPO_4^{2-}.

$$H_2PO_4^- + OH^- \rightleftharpoons HPO_4^{2-} + H_2O$$

$$\text{acid}_1 \quad \text{base}_2 \quad \text{base}_1 \quad \text{acid}_2$$

The simple amino acids are an important class of amphiprotic compounds that contain both a weak acid and a weak base functional group. When dissolved in water, an amino acid such as glycine undergoes a kind of internal acid/base reaction to produce a *zwitterion,* a species that bears both a positive and a negative charge. Thus,

$$\underset{\text{glycine}}{NH_2CH_2COOH} \rightleftharpoons \underset{\text{zwitterion}}{NH_3^+CH_2COO^-}$$

This reaction is analogous to the acid/base reaction one observes between a carboxylic acid and an amine:

$$R'COOH + R''NH_2 \rightleftharpoons R'COO^- + R''NH_3^+$$

$$\text{acid}_1 \quad \text{base}_2 \quad \text{base}_1 \quad \text{acid}_2$$

A *zwitterion* is an ion that bears both a positive and a negative charge.

The *hydronium ion* is the hydrated proton formed when water reacts with an acid. It is usually formulated as H_3O^+, although several higher hydrates exist.

Autoprotolysis (also called autoionization) involves the spontaneous reaction of molecules of a substance to give a pair of ions.

In this text we use the symbol H_3O^+ in the chapters that deal with acid/base equilibria and acid/base equilibrium calculations. In other chapters, we simplify to the more convenient H^+. We should keep in mind, however, that this symbol represents the hydrated proton.

The common strong bases include NaOH, KOH, $Ba(OH)_2$, and the quaternary ammonium hydroxides, R_4NOH, where R is an alkyl group such as CH_3 or C_2H_5.

drous acetic acid. In methanol, for example, the equilibria analogous to those shown in Equations 4-1 and 4-2 are

$$NH_3 + CH_3OH \rightleftharpoons NH_4^+ + CH_3O^- \qquad (4\text{-}3)$$
$$\text{base}_1 \qquad \text{acid}_2 \qquad \text{conjugate} \quad \text{conjugate}$$
$$\text{acid}_1 \qquad \text{base}_2$$

$$CH_3OH + HNO_2 \rightleftharpoons CH_3OH_2^+ + NO_2^- \qquad (4\text{-}4)$$
$$\text{base}_1 \qquad \text{acid}_2 \qquad \text{conjugate} \quad \text{conjugate}$$
$$\text{acid}_1 \qquad \text{base}_2$$

4A-3 Solvents and Autoprotolysis

Amphiprotic solvents undergo self-ionization, or *autoprotolysis,* to form a pair of ionic species. Autoprotolysis is yet another example of acid/base behavior, as illustrated by the following equations.

$$\text{base}_1 \qquad + \text{acid}_2 \qquad \rightleftharpoons \text{acid}_1 \qquad + \text{base}_2$$

$$H_2O \qquad + H_2O \qquad \rightleftharpoons H_3O^+ \qquad + OH^-$$

$$CH_3OH \quad + CH_3OH \rightleftharpoons CH_3OH_2^+ \; + CH_3O^-$$

$$HCOOH + HCOOH \rightleftharpoons HCOOH_2^+ + HCOO^-$$

$$NH_3 \qquad + NH_3 \qquad \rightleftharpoons NH_4^+ \qquad + NH_2^-$$

The extent to which water undergoes autoprotolysis is slight at room temperature. Thus, the hydronium and hydroxide ion concentrations in pure water are only about 10^{-7} M. Despite the small values of these concentrations, this dissociation reaction is very important in understanding the behavior of aqueous solutions.

4A-4 Strong and Weak Acids and Bases

Figure 4-1 shows the dissociation reaction of a few common acids in water. The first two are *strong acids* because they react with water so completely that no undissociated solute molecules remain. The others are *weak acids,* which react incompletely with water to give solutions that contain significant amounts of both the parent acid and its conjugate base. Note that acids can be cationic, anionic, or electrically neutral. The same is true for bases.

The acids in Figure 4-1 become progressively weaker from top to bottom. Perchloric acid and hydrochloric acid are completely dissociated. In contrast, only about 1% of acetic acid ($HC_2H_3O_2$) is dissociated. Ammonium ion is an even weaker acid; only about 0.01% of this ion is dissociated into hydronium ions and ammonia molecules. Another generality illustrated in Figure 4-1 is that the weakest acid forms the strongest conjugate base; that is, ammonia has a much stronger affinity for protons than any base above it. Perchlorate and chloride ions have no affinity for protons in aqueous solution.

The tendency of a solvent to accept or donate protons determines the strength of a solute acid or base dissolved in it. For example, perchloric and

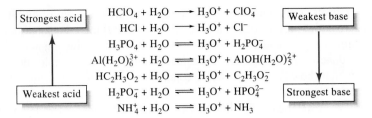

Figure 4-1 Dissociation reactions and relative strengths of some common acids and their conjugate bases. Note that HCl and $HClO_4$ are completely dissociated in water.

The common strong acids include HCl, $HClO_4$, HNO_3, the first proton in H_2SO_4, HBr, HI, and the organic sulfonic acids, RSO_3H.

hydrochloric acids are strong acids in water. If anhydrous acetic acid, a weaker proton acceptor than water, is substituted *as the solvent,* neither of these acids undergoes complete dissociation; instead, equilibria such as the following are established:

$$CH_3COOH + HClO_4 \rightleftharpoons CH_3COOH_2^+ + ClO_4^-$$

$$\text{base}_1 \qquad \text{acid}_2 \qquad \text{acid}_1 \qquad \text{base}_2$$

Perchloric acid is, however, considerably stronger than hydrochloric acid in this solvent; its dissociation is about 5000 times greater. Acetic acid thus acts as a *differentiating* solvent toward the two acids by revealing the inherent differences in their acidities. Water, on the other hand, is a *leveling* solvent for perchloric, hydrochloric, nitric, and sulfuric acids because all four are completely ionized in this solvent and thus exhibit no differences in strength. Differentiating and leveling solvents also exist for bases.

Of all the acids listed in the previous margin note and in Figure 4-1, only perchloric acid is a strong acid in methanol and ethanol. Thus, these two alcohols are also differentiating solvents.

In a *differentiating solvent,* various acids dissociate to different degrees and thus have different strengths. In a *leveling solvent,* several acids are completely dissociated and are thus of the same strength.

4B CHEMICAL EQUILIBRIUM

The reactions we use in analytical chemistry never actually result in complete conversion of reactants to products. Instead, they proceed to a state of *chemical equilibrium* in which the ratio of concentrations of reactants and products is constant. *Equilibrium-constant expressions* are *algebraic* equations that describe the concentration relationships that exist among reactants and products at equilibrium. Among other things, equilibrium-constant expressions permit us to calculate the error that results from any unreacted analyte remaining when equilibrium has been reached.

The discussion that follows deals with use of equilibrium-constant expressions to gain information about analytical systems in which no more than one or two equilibria are present. Chapter 10 extends these methods to systems containing several simultaneous equilibria. Such complex systems are often encountered in analytical chemistry.

4B-1 Describing The Equilibrium State

Consider the chemical equilibrium

$$H_3AsO_4 + 3I^- + 2H^+ \rightleftharpoons H_3AsO_3 + I_3^- + H_2O \qquad (4-5)$$

The rate of this reaction and the extent to which it proceeds to the right can be readily monitored by observing the orange-red color of the triiodide ion I_3^- (the other participants in the reaction are colorless) as it increases with time. If, for example, we add 1 mmol of arsenic acid H_3AsO_4 to 100 mL of a solution containing 3 mmol of potassium iodide, the red color of the triiodide ion appears almost immediately, and within a few seconds the intensity of the color becomes constant, which shows that the triiodide concentration has become constant (see color plate 1b).

A solution of identical color intensity (and hence identical triiodide concentration) can also be produced if we add 1 mmol of arsenous acid H_3AsO_3 to 100 mL of a solution containing 1 mmol of triiodide ion (see color plate 1a). Here, the color intensity is initially greater than in the first solution but rapidly decreases as a result of the reaction

$$H_3AsO_3 + I_3^- + H_2O \rightleftharpoons H_3AsO_4 + 3I^- + 2H^+$$

Ultimately, the color of the two solutions is identical. Many other combinations of the four reactants can be used to yield solutions that are indistinguishable from the two just described.

From the results of the experiments shown in color plates 1a and 1b, we see that the concentration relationship at chemical equilibrium (that is, the *position of equilibrium*) is independent of the route by which the equilibrium state is achieved. This relationship, however, is altered by the application of stress to the system. Such stresses include changes in temperature, pressure (if one of the reactants or products is a gas), or total concentration of a reactant or a product. These effects can be predicted qualitatively from the *principle of Le Châtelier,* which states that the position of chemical equilibrium always shifts in a direction that tends to relieve the effect of an applied stress. Thus, an increase in temperature alters the concentration relationship in the direction that tends to absorb heat, and an increase in pressure favors those participants that occupy a smaller total volume.

In an analysis, the effect of introducing an additional amount of a participating species to the reaction mixture is particularly important. Here, the resulting stress is relieved by a shift in equilibrium in the direction that partially uses up the added substance. Thus, for the equilibrium we have been considering (Equation 4-5), the addition of arsenic acid (H_3AsO_4) or hydrogen ions causes an increase in color as more triiodide ion and arsenous acid are formed; the addition of arsenous acid has the reverse effect. An equilibrium shift brought about by changing the amount of one of the participating species is called a *mass-action effect.*

Theoretical and experimental studies of reacting systems on the molecular level show that reactions among the participating species continue even after equilibrium is achieved. The constant concentration ratio of reactants and products results from an equality in the rates of the forward and reverse processes. In other words, chemical equilibrium is a dynamic state in which the rates of the reactions in the forward and reverse direction are identical.

> The position of a chemical equilibrium is independent of the route by which equilibrium is reached.

> The *Le Châtelier principle* states that the position of an equilibrium always shifts in such a direction as to relieve a stress that is applied to the system.

> The *mass-action effect* is a shift in the position of an equilibrium caused by adding one of the reactants or products to a system.

> Chemical reactions do not cease at equilibrium. Instead, the amounts of reactants and products are constant because the rates of the forward and reverse processes are identical.

> *Thermodynamics* is a branch of chemical science that deals with the flow of heat and energy in chemical reactions. The position of a chemical equilibrium is related to these energy changes.

4B-2   Writing Equilibrium Constants

The influence of concentration (or pressure if the species are gases) on the position of a chemical equilibrium is conveniently described in quantitative terms by means of an *equilibrium-constant expression.* Such expressions are readily derived from thermodynamics. They are important because they permit the chemist to predict the direction and completeness of a chemical reaction. An equilibrium-constant expression, however, yields no information concerning the *rate* at which equilibrium is approached. In fact, we sometimes encounter reactions that have highly favorable equilibrium constants but are of little analytical use because their rates are low. This limitation can often be overcome by the use of a catalyst, which speeds the attainment of equilibrium without changing its position.

Let us consider a generalized equation for a chemical equilibrium

$$wW + xX \rightleftharpoons yY + zZ \tag{4-6}$$

where the capital letters represent the formulas of participating chemical species and the lowercase italic letters are the small whole numbers required to balance the equation. Thus, the equation states that w mol of W react with x mol of X to form y mol of Y and z mol of Z. The equilibrium-constant expression for this reaction is

$$K = \frac{[Y]^y[Z]^z}{[W]^w[X]^x} \tag{4-7}$$

where the bracketed terms have the following meanings:

1. Molar concentration if the species is a dissolved solute.
2. Partial pressure in atmospheres if the species is a gas; in fact, we will often replace the bracketed term (say [Z] in Equation 4-7) with the symbol p_Z, which stands for the partial pressure of the gas Z in atmospheres.

If one (or more) of the species in Equation 4-7 is a pure liquid, a pure solid, or the solvent present in excess, no term for this species appears in the equilibrium-constant expression. For example, if Z in Equation 4-6 is the solvent H_2O, the equilibrium-constant expression simplifies to

$$K = \frac{[Y]^y}{[W]^w[X]^x}$$

The reason for this simplification is discussed in the sections that follow.

The constant K in Equation 4-7 is a temperature-dependent numerical quantity called the *equilibrium constant.* By convention, the concentrations of the products, *as the equation is written,* are always placed in the numerator and the concentrations of the reactants in the denominator.

Equation 4-7 is only an approximate form of a thermodynamic equilibrium-constant expression. The exact form is given by Equation 4-8 in the margin. Generally, we will use the approximate form of this equation because it is less tedious and time consuming than the thermodynamic form. In Section 9B, we show when the use of Equation 4-7 is likely to lead to serious errors in equilibrium calculations and how Equation 4-8 is applied in these cases.

4B-3   Important Equilibrium Constants in Analytical Chemistry

Table 4-2 summarizes the types of chemical equilibria and equilibrium constants that are of importance in analytical chemistry. Simple applications of some of these constants are illustrated in the paragraphs that follow.

Equilibrium-constant expressions provide *no* information as to whether a chemical reaction is fast enough to be useful in an analytical procedure.

Cato Guldberg (1836–1902) and Peter Waage (1833–1900) were Norwegian chemists whose primary interests were in the field of thermodynamics. In 1864, these workers were the first to propose the law of mass action, which is expressed in Equation 4-7. If you would like to learn more about Guldberg and Waage and read a translation of their original paper on the law of mass action, set your browser to http://dbhs.wvusd.k12.ca.us/Chem-History/Concerning-Affinity.html.

$[Z]^Z$ in Equation 4-7 is replaced with p_z in atmospheres if Z is a gas. No term for Z is included in the equation if this species is a pure solid, a pure liquid, or the solvent of a dilute solution.

Remember: Equation 4-7 is only an approximate form of an equilibrium-constant expression. The exact expression takes the form

$$K = \frac{a_Y^y \times a_Z^z}{a_W^w \times a_X^x} \tag{4-8}$$

where a_Y, a_Z, a_W, and a_X are the *activities* of species Y, Z, W, and X (see Section 9B).

Table 4-2

Equilibria and Equilibrium Constants of Importance to Analytical Chemistry

Type of Equilibrium	Name and Symbol of Equilibrium Constant	Typical Example	Equilibrium-Constant Expression
Dissociation of water	Ion-product constant, K_w	$2\,H_2O \rightleftharpoons H_3O^+ + OH^-$	$K_w = [H_3O^+][OH^-]$
Heterogeneous equilibrium between a slightly soluble substance and its ions in a saturated solution	Solubility product, K_{sp}	$BaSO_4(s) \rightleftharpoons Ba^{2+} + SO_4^{2-}$	$K_{sp} = [Ba^{2+}][SO_4^{2-}]$
Dissociation of a weak acid or base	Dissociation constant, K_a or K_b	$CH_3COOH + H_2O \rightleftharpoons$ $H_3O^+ + CH_3COO^-$ $CH_3COO^- + H_2O \rightleftharpoons$ $OH^- + CH_3COOH$	$K_a = \dfrac{[H_3O^+][CH_3COO^-]}{[CH_3COOH]}$ $K_b = \dfrac{[OH^-][CH_3COOH]}{[CH_3COO^-]}$
Formation of a complex ion	Formation constant, β_n	$Ni^{2+} + 4CN^- \rightleftharpoons Ni(CN)_4^{2-}$	$\beta_4 = \dfrac{[Ni(CN)_4^{2-}]}{[Ni^{2+}][CN^-]^4}$
Oxidation/reduction equilibrium	K_{redox}	$MnO_4 + 5Fe^{2+} + 8H^+ \rightleftharpoons$ $Mn^{2+} + 5Fe^{3+} + 4H_2O$	$K_{redox} = \dfrac{[Mn^{2+}][Fe^{3+}]^5}{[MnO_4^-][Fe^{2+}]^5[H^+]^8}$
Distribution equilibrium for a solute between immiscible solvents	K_d	$I_2(aq) \rightleftharpoons I_2(org)$	$K_d = \dfrac{[I_2]_{org}}{[I_2]_{aq}}$

Feature 4-2

Stepwise and Overall Formation Constants for Complex Ions

The formation of $Ni(CN)_4^{2-}$ (Table 4-2) is typical in that it occurs in steps as shown. Note that *stepwise formation* constants are symbolized by K_1, K_2, and so forth.

$$Ni^{2+} + CN^- \rightleftharpoons Ni(CN)^+ \qquad K_1 = \frac{[Ni(CN)^+]}{[Ni^{2+}][CN^-]}$$

$$Ni(CN)^+ + CN^- \rightleftharpoons Ni(CN)_2 \qquad K_2 = \frac{[Ni(CN)_2]}{[Ni(CN)^+][CN^-]}$$

$$Ni(CN)_2 + CN^- \rightleftharpoons Ni(CN)_3^- \qquad K_3 = \frac{[Ni(CN)_3^-]}{[Ni(CN)_2][CN^-]}$$

$$Ni(CN)_3^- + CN^- \rightleftharpoons Ni(CN)_4^{2-} \qquad K_4 = \frac{[Ni(CN)_4^{2-}]}{[Ni(CN)_3^-][CN^-]}$$

Overall constants are designated by the symbol β_n. Thus,

$$Ni^{2+} + 2CN^- \rightleftharpoons Ni(CN)_2 \qquad \beta_2 = K_1K_2 = \frac{[Ni(CN)_2]}{[Ni^{2+}][CN^-]^2}$$

$$Ni^{2+} + 3CN^- \rightleftharpoons Ni(CN)_3^- \qquad \beta_3 = K_1K_2K_3 = \frac{[Ni(CN)_3^-]}{[Ni^{2+}][CN^-]^3}$$

$$Ni^{2+} + 4CN^- \rightleftharpoons Ni(CN)_4^{2-} \qquad \beta_4 = K_1K_2K_3K_4 = \frac{[Ni(CN)_4^{2-}]}{[Ni^{2+}][CN^-]^4}$$

In a dilute aqueous solution, the molar concentration of water is

$$[H_2O] = \frac{1000 \text{ g H}_2\text{O}}{\text{L H}_2\text{O}} \times \frac{1 \text{ mol H}_2\text{O}}{18.0 \text{ g H}_2\text{O}} = 55.6 \text{ M}$$

Let us suppose we have 0.1 mol of HCl in 1 L of water. The presence of this acid will shift the equilibrium shown in Equation 4-9 to the left. Originally, however, there were only 10^{-7} mol/L OH^- to consume the added protons. Thus, even if all the OH^- ions are converted to H_2O, the water concentration will increase only to

$$[H_2O] = 55.6 \frac{\text{mol H}_2\text{O}}{\text{L H}_2\text{O}} + 1 \times 10^{-7} \frac{\text{mol OH}^-}{\text{L H}_2\text{O}} \times \frac{1 \text{ mol H}_2\text{O}}{\text{mol OH}^-} \approx 55.6 \text{ M}$$

The percent change in water concentration is

$$\frac{10^{-7} \text{ M}}{55.6 \text{ M}} \times 100\% = 2 \times 10^{-7}\%$$

which is inconsequential. Thus, $K[H_2O]^2$ in Equation 4-10 is, for all practical purposes, a constant. That is,

$$K(55.6)^2 = K_w = 1.00 \times 10^{-14}$$

The Ion-Product Constant for Water

Aqueous solutions contain small concentrations of hydronium and hydroxide ions as a consequence of the dissociation reaction

$$2H_2O \rightleftharpoons H_3O^+ + OH^- \tag{4-9}$$

An equilibrium constant for this reaction can be formulated as shown in Equation 4-7:

$$K = \frac{[H_3O^+][OH^-]}{[H_2O]^2} \tag{4-10}$$

The concentration of water in dilute aqueous solutions is enormous, however, when compared with the concentration of hydrogen and hydroxide ions. As a consequence, [H_2O] in Equation 4-10 can be taken as constant, and we write

$$K[H_2O]^2 = K_w = [H_3O^+][OH^-] \tag{4-11}$$

where the new constant K_w is given a special name, the *ion-product constant for water*.

At 25°C, the ion-product constant for water is 1.008×10^{-14}. For convenience, we shall use the approximation that at room temperature $K_w \approx 1.00 \times 10^{-14}$. Table 4-3 shows the dependence of this constant on temperature. The ion-product constant for water permits the ready calculation of the hydronium and hydroxide ion concentrations of aqueous solutions as shown in Examples 4-1 and 4-2.

A useful relationship is obtained by taking the negative logarithm of Equation 4-11:

$$-\log K_w = -\log [H_3O^+] - \log [OH^-]$$

By the definition of p-functions,

$$pK_w = pH + pOH$$

At 25°C,

$$pK_w = 14.00$$

Table 4-3

Variation of K_w with Temperature

Temperature, °C	K_w
0	0.114×10^{-14}
25	1.01×10^{-14}
50	5.47×10^{-14}
100	49×10^{-14}

Example 4-1

Calculate the hydronium and hydroxide ion concentrations of pure water at 25°C and 100°C.

Because OH^- and H_3O^+ are formed only from the dissociation of water, their concentrations must be equal:

$$[H_3O^+] = [OH^-]$$

Substitution into Equation 4-11 gives

$$[H_3O^+]^2 = [OH^-]^2 = K_w$$

$$[H_3O^+] = [OH^-] = \sqrt{K_w}$$

At 25°C,

$$[H_3O^+] = [OH^-] = \sqrt{1.00 \times 10^{-14}} = 1.00 \times 10^{-7}\,M$$

At 100°C, from Table 4-3,

$$[H_3O^+] = [OH^-] = \sqrt{49 \times 10^{-14}} = 7.0 \times 10^{-7}\,M$$

Example 4-2

Calculate the hydronium and hydroxide ion concentrations in 0.200 M aqueous NaOH.

Sodium hydroxide is a strong electrolyte, and its contribution to the hydroxide ion concentration in this solution is 0.200 mol/L. As in Example 4-1, hydroxide ions and hydronium ions are formed *in equal amounts* from dissociation of water. Therefore, we write

$$[OH^-] = 0.200 + [H_3O^+]$$

where $[H_3O^+]$ accounts for the hydroxide ions contributed by the solvent. The concentration of OH^- from the water is insignificant, however, when compared with 0.200, so we can write

$$[OH^-] \approx 0.200\,M$$

Equation 4-11 is then used to calculate the hydronium ion concentration:

$$[H_3O^+] = \frac{K_w}{[OH^-]} = \frac{1.00 \times 10^{-14}}{0.200} = 5.00 \times 10^{-14}\,M$$

Note that the approximation

$$[OH^-] = 0.200 + 5.00 \times 10^{-14} \approx 0.200\,M$$

causes no significant error.

What are Solubility-Products?

When we say that a sparingly soluble salt is completely dissociated, *we do not imply* that all the salt dissolves. Instead, the very small amount that *does* go into solution dissociates completely.

Most sparingly soluble salts are essentially completely dissociated in saturated aqueous solution. For example, when an excess of barium iodate is equilibrated with water, the dissociation process is adequately described by the equation

$$Ba(IO_3)_2(s) \rightleftharpoons Ba^{2+}(aq) + 2IO_3^-(aq)$$

Application of Equation 4-7 leads to

$$K = \frac{[Ba^{2+}][IO_3^-]^2}{[Ba(IO_3)_2(s)]}$$

The denominator represents the molar concentration of $Ba(IO_3)_2$ *in the solid*, which is a phase separate from but in contact with the saturated solution. The concentration of a compound in its solid state is, however, constant. In other words, the number of moles of $Ba(IO_3)_2$ divided by the *volume* of the solid $Ba(IO_3)_2$ is constant no matter how much excess solid is present. Therefore, the foregoing equation can be rewritten in the form

$$K[Ba(IO_3)_2(s)] = K_{sp} = [Ba^{2+}][IO_3^-]^2 \qquad (4\text{-}12)$$

where the new constant is called the *solubility-product constant* or the *solubility product*. It is important to appreciate that Equation 4-12 shows that the position of this equilibrium is independent of the *amount* of $Ba(IO_3)_2$ as long as some solid is present; that is, it does not matter whether the amount is a few milligrams or several grams.

A table of solubility product constants for numerous inorganic salts is found in Appendix 1. The examples that follow demonstrate some typical uses of solubility-product expressions. Further applications are considered in Chapter 10.

For Equation 4-12 to apply, it is necessary only that *some solid be present. In the absence of $Ba(IO_3)(s)$, Equation 4-12 is not valid.*

Calculating the Solubility of a Precipitate in Water

The solubility-product expression permits the ready calculation of the solubility of a sparingly soluble substance that ionizes in water as shown in Example 4-3.

Example 4-3

What mass of $Ba(IO_3)_2$ (487 g/mol) can be dissolved in 500.0 mL of water at 25°C?

The solubility-product constant for $Ba(IO_3)_2$ is 1.57×10^{-9} (Appendix 1). The equilibrium between the solid and its ions in solution is described by the equation

$$Ba(IO_3)_2(s) \rightleftharpoons Ba^{2+} + 2IO_3^-$$

and so

$$K_{sp} = [Ba^{2+}][IO_3^-]^2 = 1.57 \times 10^{-9}$$

The equation describing the equilibrium reveals that 1 mol of Ba^{2+} is formed for each mole of $Ba(IO_3)_2$ that dissolves. Therefore,

$$\text{molar solubility of } Ba(IO_3)_2 = [Ba^{2+}]$$

The iodate concentration is twice the barium ion concentration:

$$[IO_3^-] = 2[Ba^{2+}]$$

Substituting this last equation into the equilibrium-constant expression gives

$$[Ba^{2+}](2[Ba^{2+}])^2 = 4[Ba^{2+}]^3 = 1.57 \times 10^{-9}$$

$$[Ba^{2+}] = \left(\frac{1.57 \times 10^{-9}}{4}\right)^{1/3} = 7.32 \times 10^{-4} \text{ M}$$

Note that the molar solubility is equal to $[Ba^{2+}]$ or to $\frac{1}{2}[IO_3^-]$.

Since 1 mol Ba^{2+} is produced for every mole of $Ba(IO_3)_2$,

$$\text{solubility} = 7.32 \times 10^{-4} \text{ M}$$

To compute the number of millimoles of $Ba(IO_3)_2$ dissolved in 500.0 mL of solution, we write

$$\text{no. mmol } Ba(IO_3)_2 = 7.32 \times 10^{-4} \frac{\text{mmol } Ba(IO_3)_2}{\text{mL}} \times 500.0 \text{ mL}$$

The mass of $Ba(IO_3)_2$ in 500.0 mL is given by

$$\text{mass } Ba(IO_3)_2 = (7.32 \times 10^{-4} \times 500.0) \text{ mmol Ba (IO}_3)_2$$

$$\times 0.487 \frac{\text{g } Ba(IO_3)_2}{\text{mmol Ba(IO}_3)_2} = 0.178 \text{ g}$$

The *common-ion effect* is responsible for the reduction in solubility of an ionic precipitate when a soluble compound combining one of the ions of the precipitate is added to the solution in equilibrium with the precipitate.

How Does a Common Ion Affect the Solubility of a Precipitate?   The *common-ion effect* is a mass-action effect predicted from the Le Châtelier principle. It is demonstrated by Examples 4-4 and 4-5.

Example 4-4

Calculate the molar solubility of $Ba(IO_3)_2$ in a solution that is 0.0200 M in $Ba(NO_3)_2$.

The solubility is no longer equal to $[Ba^{2+}]$ because $Ba(NO_3)_2$ is also a source of barium ions. We know, however, that solubility is related to $[IO_3^-]$:

$$\text{molar solubility of } Ba(IO_3)_2 = \tfrac{1}{2}[IO_3^-]$$

There are two sources of barium ions: $Ba(NO_3)_2$ and $Ba(IO_3)_2$. The contribution from the former is 0.0200 M, and that from the latter is equal to the molar solubility, or $\tfrac{1}{2}[IO_3^-]$. Thus,

$$[Ba^{2+}] = 0.0200 + \tfrac{1}{2}[IO_3^-]$$

Substitution of these quantities into the solubility-product expression yields

$$(0.0200 + \tfrac{1}{2}[IO_3^-])[IO_3^-]^2 = 1.57 \times 10^{-9}$$

Since the exact solution for $[IO_3^-]$ requires solving a cubic equation, we seek an approximation that simplifies the algebra. The small numerical value of K_{sp} suggests that the solubility of $Ba(IO_3)_2$ is not large, which is confirmed by the result obtained in Example 4-3. Moreover, barium ions from $Ba(NO_3)_2$ will further repress the limited solubility of $Ba(IO_3)_2$. Thus, it is reasonable to seek a provisional answer to the problem by assuming that 0.0200 is large with respect to $\tfrac{1}{2}[IO_3^-]$. That is, $\tfrac{1}{2}[IO_3^-] \ll 0.0200$, and

$$[Ba^{2+}] = 0.0200 + \tfrac{1}{2}[IO_3^-] \approx 0.0200$$

The original equation then simplifies to

$$0.0200[IO_3^-]^2 = 1.57 \times 10^{-9}$$

$$[IO_3^-] = \sqrt{(1.57 \times 10^{-9})/0.0200} = \sqrt{7.85 \times 10^{-8}} = 2.80 \times 10^{-4} \text{ M}$$

The assumption that $(0.0200 + \frac{1}{2} \times 2.80 \times 10^{-4}) \approx 0.0200$ does not appear to cause serious error because the second term, representing the amount of Ba^{2+} arising from the dissociation of $Ba(IO_3)_2$, is only about 0.7% of 0.0200. Ordinarily, we consider an assumption of this type to be satisfactory if the discrepancy is less than 10%. Finally, then,

$$\text{solubility of } Ba(IO_3)_2 = \tfrac{1}{2}[IO_3^-] = \tfrac{1}{2} \times 2.80 \times 10^{-4} = 1.40 \times 10^{-4} \text{ M}$$

If we compare the solubility of barium iodate in the presence of a common ion (Example 4-4) with its solubility in pure water (Example 4-3), we see that the presence of a small concentration of the common ion has lowered the molar solubility of $Ba(IO_3)_2$ by a factor of about five.

Example 4-5

Calculate the solubility of $Ba(IO_3)_2$ in a solution prepared by mixing 200 mL of 0.0100 M $Ba(NO_3)_2$ with 100 mL of 0.100 M $NaIO_3$.

We must first establish whether either reactant is present in excess at equilibrium. The amounts taken are

$$\text{no. mmol } Ba^{2+} = 200 \text{ mL} \times 0.0100 \text{ mmol/mL} = 2.00$$

$$\text{no. mmol } IO_3^- = 100 \text{ mL} \times 0.100 \text{ mmol/mL} = 10.0$$

If formation of $Ba(IO_3)_2$ is complete,

$$\text{no. mmol excess } NaIO_3 = 10.0 - 2(2.00) = 6.00$$

Thus,

$$[IO_3^-] = \frac{6.00 \text{ mmol}}{200 \text{ mL} + 100 \text{ mL}} = \frac{6.00 \text{ mmol}}{300 \text{ mL}} = 0.0200 \text{ M}$$

The uncertainty in $[IO_3^-]$ is 0.1 part in 6.0 or 1 part in 60. Thus, $0.0200 \times (1/60) = 0.0003$, and we round to 0.0200 M.

As in Example 4-3,

$$\text{molar solubility of } Ba(IO_3)_2 = [Ba^{2+}]$$

Here, however,

$$[IO_3^-] = 0.0200 + 2[Ba^{2+}]$$

where $2[Ba^{2+}]$ represents the iodate contributed by the sparingly soluble $Ba(IO_3)_2$. We can obtain a provisional answer after making the assumption that $[IO_3^-] \approx 0.0200$; thus,

$$\text{solubility of } Ba(IO_3)_2 = [Ba^{2+}] = \frac{K_{sp}}{[IO_3^-]^2} = \frac{1.57 \times 10^{-9}}{(0.0200)^2}$$

$$= 3.93 \times 10^{-6} \text{ M}$$

The approximation appears to be reasonable, and no correction to the provisional answer is needed.

Note that the results from the last two examples demonstrate that an excess of iodate ions is more effective in decreasing the solubility of $Ba(IO_3)_2$ than is the same excess of barium ions.

A 0.02 M excess of Ba^{2+} decreases the solubility of $Ba(IO_3)_2$ by a factor of about five; this same excess of IO_3^- lowers the solubility by a factor of about 200.

Writing Dissociation Constants for Acids and Bases

When a weak acid or a weak base is dissolved in water, partial dissociation occurs. Thus, for nitrous acid, we can write

$$HNO_2 + H_2O \rightleftharpoons H_3O^+ + NO_2^- \qquad K_a = \frac{[H_3O^+][NO_2^-]}{[HNO_2]}$$

where K_a is the *acid dissociation constant* K_b for nitrous acid. In an analogous way, the *base dissociation constant* K_b for ammonia is

$$NH_3 + H_2O \rightleftharpoons NH_4^+ + OH^- \qquad K_b = \frac{[NH_4^+][OH^-]}{[NH_3]}$$

Note that $[H_2O]$ does not appear in the denominator of either equation because the concentration of water is so large relative to the concentration of the weak acid or base that the dissociation does not alter $[H_2O]$ appreciably (see Feature 4-3). Just as in the derivation of the ion-product constant for water, $[H_2O]$ is incorporated in the equilibrium constants K_a and K_b. Dissociation constants for weak acids are found in Appendix 2.

Dissociation Constants for Conjugate Acid/Base Pairs

Consider the base dissociation-constant expression for ammonia and the acid dissociation-constant expression for its conjugate acid, ammonium ion:

$$NH_3 + H_2O \rightleftharpoons NH_4^+ + OH^- \qquad K_b = \frac{[NH_4^+][OH^-]}{[NH_3]}$$

$$NH_4^+ + H_2O \rightleftharpoons NH_3 + H_3O^+ \qquad K_a = \frac{[NH_3][H_3O^+]}{[NH_4^+]}$$

Multiplication of one equilibrium-constant expression by the other gives

$$K_a K_b = \frac{[\cancel{NH_3}][H_3O^+]}{[\cancel{NH_4^+}]} \times \frac{[\cancel{NH_4^+}][OH^-]}{[\cancel{NH_3}]} = [H_3O^+][OH^-]$$

but

$$K_w = [H_3O^+][OH^-]$$

and therefore

$$K_w = K_a K_b \qquad\qquad (4\text{-}13)$$

This relationship is general for all conjugate acid/base pairs. Many compilations of equilibrium constant data list only acid dissociation constants since it is so easy to calculate basic dissociation constants with Equation 4-13 (see Feature 4-4). For example, in Appendix 2, we find no data on the basic dissociation of ammonia (nor for any other bases). Instead, we find the acid dissociation constant for the conjugate acid, ammonium ion. That is,

$$NH_4^+ + H_2O \rightleftharpoons H_3O^+ + NH_3 \qquad K_a = \frac{[H_3O^+][NH_3]}{[NH_4^+]} = 5.70 \times 10^{-10}$$

and we can write

$$NH_3 + H_2O \rightleftharpoons NH_4^+ + OH^-$$

$$K_b = \frac{[NH_4^+][OH^-]}{[NH_3]} = \frac{1.00 \times 10^{-14}}{5.70 \times 10^{-10}} = 1.75 \times 10^{-5}$$

To obtain a dissociation constant for a base, look up in Appendix 2 the dissociation constant for its conjugate acid and then divide this number into 1.00×10^{-14}.

Equation 4-13 confirms the observation in Figure 4-1 that as the acid of a conjugate acid/base pair becomes weaker, its conjugate base becomes stronger and vice versa. Thus, the conjugate base of an acid with a dissociation constant of 10^{-2} will have a basic dissociation constant of 10^{-12}, whereas an acid with a dissociation constant of 10^{-9} has a conjugate base with a dissociation constant of 10^{-5}.

Example 4-6

What is K_b for the equilibrium

$$CN^- + H_2O \rightleftharpoons HCN + OH^-$$

Appendix 2 lists a K_a value of 6.2×10^{-10} for HCN. Thus,

$$K_b = \frac{K_w}{K_{HCN}} = \frac{[HCN][OH^-]}{[CN^-]}$$

$$K_b = \frac{1.00 \times 10^{-14}}{6.2 \times 10^{-10}} = 1.61 \times 10^{-5}$$

Calculating the Hydronium Ion Concentration in Solutions of Weak Acids

When the weak acid HA is dissolved in water, two equilibria are established that yield hydronium ions:

$$HA + H_2O \rightleftharpoons H_3O^+ + A^- \qquad K_a = \frac{[H_3O^+][A^-]}{[HA]}$$

$$2H_2O \rightleftharpoons H_3O^+ + OH^- \qquad K_w = [H_3O^+][OH^-]$$

Ordinarily, the hydronium ions produced from the first reaction suppress the dissociation of water to such an extent that the contribution of hydronium ions from the second equilibrium is negligible. Under these circumstances, one H_3O^+ ion is formed for each A^- ion, and we write

$$[A^-] \approx [H_3O^+] \tag{4-14}$$

Furthermore, the sum of the molar concentrations of the weak acid and its conjugate base must equal the analytical concentration of the acid c_{HA} because the solution contains no other source of A^- ions. Thus,

$$c_{HA} = [A^-] + [HA] \tag{4-15}$$

Substituting $[H_3O^+]$ for $[A^-]$ (Equation 4-14) into Equation 4-15 yields

$$c_{HA} = [H_3O^+] + [HA]$$

which rearranges to

$$[HA] = c_{HA} - [H_3O^+] \tag{4-16}$$

When $[A^-]$ and $[HA]$ are replaced by their equivalent terms from Equations 4-14 and 4-16, the equilibrium-constant expression becomes

$$K_a = \frac{[H_3O^+]^2}{c_{HA} - [H_3O^+]} \tag{4-17}$$

which rearranges to

$$[H_3O^+]^2 + K_a[H_3O^+] - K_ac_{HA} = 0 \qquad (4\text{-}18)$$

The positive solution to this quadratic equation is

$$[H_3O^+] = \frac{-K_a + \sqrt{K_a^2 + 4K_ac_{HA}}}{2} \qquad (4\text{-}19)$$

As an alternative to using Equation 4-19, Equation 4-18 may be solved by successive approximations as shown in Feature 4-5.

Equation 4-16 can frequently be simplified by making the additional assumption that dissociation does not appreciably decrease the molar concentration of HA. Thus, provided that $[H_3O^+] \ll c_{HA}$, $c_{HA} - [H_3O^+] \approx c_{HA}$ and Equation 4-17 reduces to

$$K_a = \frac{[H_3O^+]^2}{c_{HA}} \qquad (4\text{-}20)$$

and

$$[H_3O^+] = \sqrt{K_ac_{HA}} \qquad (4\text{-}21)$$

The magnitude of the error introduced by the assumption that $[H_3O^+] \ll c_{HA}$ increases as the molar concentration of acid becomes smaller and its dissociation constant becomes larger. This statement is supported by the data in Table 4-4. Note that the error introduced by the assumption is about 0.5% when the ratio c_{HA}/K_a is 10^4. The error increases to about 1.6% when the ratio is 10^3, to about 5% when it is 10^2, and to about 17% when it is 10. Figure 4-2 illustrates the effect graphically. It is noteworthy that the hydronium ion concentration

Table 4-4

Error Introduced by Assuming H_3O^+ Concentration Is Small Relative to c_{HA} in Equation 4-16

K_a	c_{HA}	$[H_3O^+]$ Using Assumption	$\dfrac{c_{HA}}{K_a}$	$[H_3O^+]$ Using More Exact Equation	Percent Error
1.00×10^{-2}	1.00×10^{-3}	3.16×10^{-3}	10^{-1}	0.92×10^{-3}	244
	1.00×10^{-2}	1.00×10^{-2}	10^0	0.62×10^{-2}	61
	1.00×10^{-1}	3.16×10^{-2}	10^1	2.70×10^{-2}	17
1.00×10^{-4}	1.00×10^{-4}	1.00×10^{-4}	10^0	0.62×10^{-4}	61
	1.00×10^{-3}	3.16×10^{-4}	10^1	2.70×10^{-4}	17
	1.00×10^{-2}	1.00×10^{-3}	10^2	0.95×10^{-3}	5.3
	1.00×10^{-1}	3.16×10^{-3}	10^3	3.11×10^{-3}	1.6
1.00×10^{-6}	1.00×10^{-5}	3.16×10^{-6}	10^1	2.70×10^{-6}	17
	1.00×10^{-4}	1.00×10^{-5}	10^2	0.95×10^{-5}	5.3
	1.00×10^{-3}	3.16×10^{-5}	10^3	3.11×10^{-5}	1.6
	1.00×10^{-2}	1.00×10^{-4}	10^4	9.95×10^{-5}	0.5
	1.00×10^{-1}	3.16×10^{-4}	10^5	3.16×10^{-4}	0.0

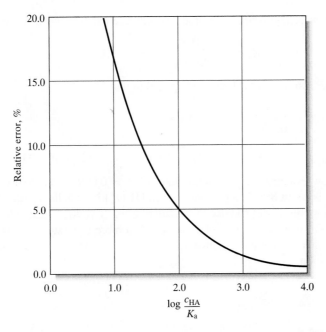

Figure 4-2   Relative error resulting from the assumption that $[H_3O^+] \ll c_{HA}$ in Equation 4-16.

computed with the approximation becomes equal to or greater than the molar concentration of the acid when the ratio is unity or smaller, which is clearly a meaningless result.

In general, it is good practice to make the simplifying assumption and obtain a trial value for $[H_3O^+]$ that can be compared with c_{HA} in Equation 4-16. If the trial value alters $[HA]$ by an amount smaller than the allowable error in the calculation, the solution may be considered satisfactory. Otherwise, the quadratic equation must be solved to obtain a better value for $[H_3O^+]$. Alternatively, the method of successive approximations (Feature 4-5) may be used.

Example 4-7

Calculate the hydronium ion concentration in 0.120 M nitrous acid. The principal equilibrium is

$$HNO_2 + H_2O \rightleftharpoons H_3O^+ + NO_2^-$$

for which (Appendix 2)

$$K_a = 7.1 \times 10^{-4} = \frac{[H_3O^+][NO_2^-]}{[HNO_2]}$$

Substitution into Equations 4-14 and 4-16 gives

$$[NO_2^-] = [H_3O^+]$$

$$[HNO_2] = 0.120 - [H_3O^+]$$

When these relationships are introduced into the expression for K_a, we obtain

$$K_a = \frac{[H_3O^+]^2}{0.120 - [H_3O^+]} = 7.1 \times 10^{-4}$$

If we now assume $[H_3O^+] \ll 0.120$, we find that

$$\frac{[H_3O^+]^2}{0.120} = 7.1 \times 10^{-4}$$

$$[H_3O^+] = \sqrt{0.120 \times 7.1 \times 10^{-4}} = 9.2 \times 10^{-3} \text{ M}$$

We now examine the assumption that $0.120 - 0.0092 \approx 0.120$ and see that the error is about 8%. The relative error in $[H_3O^+]$ is actually smaller than this figure, however, as we can see by calculating $\log (c_{HA}/K_a) = 2.2$, which, from Figure 4-2, suggests an error of about 4%. If a more accurate figure is needed, solution of the quadratic equation yields 8.9×10^{-3} M for the hydronium ion concentration.

Example 4-8

Calculate the hydronium ion concentration in a solution that is 2.0×10^{-4} M in anilinium hydrochloride, $C_6H_5NH_3Cl$.

In aqueous solution, dissociation of the salt to Cl^- and $C_6H_5NH_3^+$ is complete. The weak acid $C_6H_5NH_3^+$ dissociates as follows:

$$C_6H_5NH_3^+ + H_2O \rightleftharpoons C_6H_5NH_2 + H_3O^+ \qquad K_a = \frac{[H_3O^+][C_6H_5NH_2]}{[C_6H_5NH_3^+]}$$

Inspection of Appendix 2 gives K_a for $C_6H_5NH_3^+$ of 2.51×10^{-5}.

Proceeding as in Example 4-7, we have

$$[H_3O^+] = [C_6H_5NH_2]$$

$$[C_6H_5NH_3^+] = 2.0 \times 10^{-4} - [H_3O^+]$$

Let us assume that $[H_3O^+] \ll 2.0 \times 10^{-4}$ and substitute the simplified value for $[C_6H_5NH_3^+]$ into the dissociation-constant expression to obtain (see Equation 4-20)

$$\frac{[H_3O^+]^2}{2.0 \times 10^{-4}} = 2.51 \times 10^{-5}$$

$$[H_3O^+] = \sqrt{5.02 \times 10^{-9}} = 7.09 \times 10^{-5} \text{ M}$$

Comparison of 7.09×10^{-5} with 2.0×10^{-4} suggests that a significant error has been introduced by the assumption that $[H_3O^+] \ll c_{C_6H_5NH_3^+}$ (Figure 4-2 indicates that this error is about 20%). Thus, unless only an approximate value for $[H_3O^+]$ is needed, it is necessary to use the more nearly exact expression (Equation 4-17)

$$\frac{[H_3O^+]^2}{2.0 \times 10^{-4} - [H_3O^+]} = 2.51 \times 10^{-5}$$

which rearranges to

$$[H_3O^+]^2 + 2.51 \times 10^{-5}[H_3O^+] - 5.02 \times 10^{-9} = 0$$

$$[H_3O^+] = \frac{-2.51 \times 10^{-5} + \sqrt{(2.54 \times 10^{-5})^2 + 4 \times 5.02 \times 10^{-9}}}{2}$$

$$= 5.94 \times 10^{-5} \text{ M}$$

The quadratic equation can also be solved by the iterative method shown in Feature 4-5.

Feature 4-5

The Method of Successive Approximations

For convenience, let us write the quadratic equation in Example 4-8 in the form

$$x^2 + 2.51 \times 10^{-5}x - 5.02 \times 10^{-9} = 0$$

where $x = [H_3O^+]$.

As a first step, let us rearrange the equation to the form

$$x = \sqrt{5.02 \times 10^{-9} - 2.51 \times 10^{-5}x}$$

We then assume that x on the right-hand side of the equation is zero and calculate a first value, x_1.

$$x_1 = \sqrt{5.02 \times 10^{-9} - 2.51 \times 10^{-5} \times 0} = 7.09 \times 10^{-5}$$

We then substitute this value into the original equation and derive a second value, x_2. That is,

$$x_2 = \sqrt{5.02 \times 10^{-9} - 2.51 \times 10^{-5} \times 7.09 \times 10^{-5}} = 5.69 \times 10^{-5}$$

Repeating this calculation gives

$$x_3 = \sqrt{5.02 \times 10^{-9} - 2.51 \times 10^{-5} \times 5.69 \times 10^{-5}} = 5.99 \times 10^{-5}$$

Continuing in the same way we obtain

$$x_4 = 5.93 \times 10^{-5}$$

$$x_5 = 5.94 \times 10^{-5}$$

$$x_6 = 5.94 \times 10^{-5}$$

Note that after three iterations x_3 is 5.99×10^{-5}, which is within about 0.8% of the final value of 5.94×10^{-5} M. At this point, we say that the iterations have converged.

Occasionally, the final result will oscillate between a high and a low value. In this case, you may have to solve the complete equation.

The method of successive approximations is particularly useful when cubic or higher power equations need to be solved.

Finding the Hydronium Ion Concentration in Solutions of Weak Bases

The techniques discussed in previous sections are readily adapted to the calculation of the hydroxide or hydronium ion concentration in solutions of weak bases. Aqueous ammonia is basic by virtue of the reaction

$$NH_3 + H_2O \rightleftharpoons NH_4^+ + OH^-$$

The predominant species in such solutions has been clearly demonstrated to be NH_3. Nevertheless, solutions of ammonia are still occasionally called ammonium hydroxide because chemists once thought that NH_4OH rather than NH_3 was the undissociated form of the base. We can write the equilibrium constant for the reaction of NH_3 with H_2O as

$$K_b = \frac{[NH_4^+][OH^-]}{[NH_3]}$$

The examples that follow (4-9 through 4-11) illustrate calculating hydroxide ion concentrations in solutions of weak bases.

Example 4-9

Calculate the hydroxide ion concentration of a 0.0750 M NH_3 solution. The predominant equilibrium is

$$NH_3 + H_2O \rightleftharpoons NH_4^+ + OH^-$$

As shown on page 92,

$$K_b = \frac{[NH_4^+][OH^-]}{[NH_3]} = \frac{1.00 \times 10^{-14}}{5.70 \times 10^{-10}} = 1.75 \times 10^{-5}$$

The chemical equation shows that

$$[NH_4^+] = [OH^-]$$

Both NH_4^+ and NH_3 come from the 0.0750 M solution. Thus,

$$[NH_4^+] + [NH_3] = c_{NH_3} = 0.0750 \text{ M}$$

If we substitute $[OH^-]$ for $[NH_4^+]$ in the second of these equations and rearrange, we find that

$$[NH_3] = 0.0750 - [OH^-]$$

Substituting these quantities into the dissociation-constant expression yields

$$\frac{[OH^-]^2}{7.50 \times 10^{-2} - [OH^-]} = 1.75 \times 10^{-5}$$

which is analogous to Equation 4-17 for weak acids. Provided that $[OH^-] \ll 7.50 \times 10^{-2}$, this equation simplifies to

$$[OH^-]^2 \approx 7.50 \times 10^{-2} \times 1.75 \times 10^{-5}$$

$$[OH^-] = 1.15 \times 10^{-3} \text{ M}$$

When we compare the calculated value for $[OH^-]$ with 7.50×10^{-2}, we see that the error in $[OH^-]$ is less than 2%. If needed, we can obtain a better value for $[OH^-]$ by solving the quadratic equation.

Example 4-10

Calculate the hydroxide ion concentration in a 0.0100 M sodium hypochlorite solution.

The equilibrium between OCl^- and water is

$$OCl^- + H_2O \rightleftharpoons HOCl + OH^-$$

for which

$$K_b = \frac{[HOCl][OH^-]}{[OCl^-]}$$

Appendix 2 reveals that the acid dissociation constant for HOCl is 3.0×10^{-8}. Therefore, we rearrange Equation 4-13 and write

$$K_b = \frac{K_w}{K_a} = \frac{1.00 \times 10^{-14}}{3.0 \times 10^{-8}} = 3.33 \times 10^{-7}$$

Proceeding as in Example 4-9, we have

$$[OH^-] = [HOCl]$$

$$[OCl^-] + [HOCl] = 0.0100$$

$$[OCl^-] = 0.0100 - [OH^-] \approx 0.0100$$

Here we have assumed that $[OH^-] \ll 0.0100$. Substitution into the equilibrium-constant expression gives

$$\frac{[OH^-]^2}{0.0100} = 3.33 \times 10^{-7}$$

$$[OH^-] = 5.8 \times 10^{-5} \, M$$

Note that the error resulting from the approximation is small.

Using Excel to Iterate   *Spreadsheet Exercise*

Excel's properties make it very easy to perform an iterative solution to an equation like the one illustrated in Feature 4-5. Start with a blank worksheet, click on cell A2, and type Equation 4-22 as follows:

```
=sqrt(5.02e-9-2.51e-5*a1)  [↵]
```

The spreadsheet now appears as shown on page 100, and the value in cell A2 is precisely the answer that we obtained in the first iteration of Feature 4-5. Note that since cell A1 is blank, Excel interprets the relative cell reference to mean that the content of A1 is zero. Thus, the second term inside the square-root function is zero, and cell A2 displays the square root of 5.02×10^{-9}.

Now, click on cell A2 and type [Ctrl-c], which copies the contents of the cell into the clipboard. Then, highlight cells A3 to A10 by dragging with the mouse and type [Ctrl-v]. Your spreadsheet now displays the following results.

Click on each of the cells A2 to A10 one at a time and observe how the formula changes in the formula bar. Each formula refers to the result in the cell directly above it, so for example, cell A3 contains the formula =SQRT(0.00000000502-0.0000251*A2). The value of x calculated in A3 results from the previous value of x calculated in A2, the value in A4 is computed from the value in A3, and so on. Note that after four iterations, cell A6 contains the value 5.94×10^{-5}, which is sufficiently precise for virtually any purpose. If you want to see how many iterations are required to reach a six-digit solution to the problem, just copy the formula into more cells in column A and increase the number of displayed digits by highlighting cells, clicking on Format/Cells..., and then clicking on Scientific. Finally, type in the number 6 in the blank labeled Decimal places... and click on OK.

Use this same approach to solve the quadratic equation of Example 4-8.

WEB WORKS

Many professors post lecture slides on the World Wide Web for the benefit of their students. You may wish to review the slides on the following web pages to brush up on chemical equilibrium.

http://www.chem.uncc.edu/faculty/murphy/1252/Chapter16A/index.htm

http://www.chem.uncc.edu/faculty/murphy/1252/Chapter16B/index.htm

http://www.chem.uncc.edu/faculty/murphy/1252/Chapter17A/index.htm

http://www.chem.uncc.edu/faculty/murphy/1252/Chapter17B/index.htm

4C   QUESTIONS AND PROBLEMS

4-1. Briefly describe or define
   *(a) a weak electrolyte.
   (b) a Brønsted–Lowry base.
   *(c) the conjugate acid of a Brønsted–Lowry base.
   (d) neutralization, in terms of the Brønsted–Lowry concepts.
   *(e) an amphiprotic solvent.
   (f) a zwitterion.
   *(g) autoprotolysis.
   (h) a strong acid.
   *(i) the Le Châtelier principle.
   (j) the common-ion effect.

4-2. Briefly describe or define
   *(a) an amphiprotic solute.
   (b) a differentiating solvent.
   *(c) a leveling solvent.
   (d) a mass-action effect.

***4-3.** Briefly explain why there is no term in an equilibrium-constant expression for water or for a pure solid, even though one (or both) appear in the balanced net ionic equation for the equilibrium.

4-4. Identify the acid on the left and its conjugate base on the right in the following equations:
   *(a) $HOCl + H_2O \rightleftharpoons H_3O^+ + OCl^-$
   (b) $HONH_2 + H_2O \rightleftharpoons HONH_3^+ + OH^-$
   *(c) $NH_4^+ + H_2O \rightleftharpoons NH_3 + H_3O^+$
   (d) $2HCO_3^- \rightleftharpoons H_2CO_3 + CO_3^{2-}$
   *(e) $PO_4^{3-} + H_2PO_4^- \rightleftharpoons 2HPO_4^{2-}$

4-5. Identify the base on the left and its conjugate acid on the right in the equations for Problem 4-4.

4-6. Write expressions for the autoprotolysis of
   *(a) H_2O.
   (b) CH_3COOH.
   *(c) CH_3NH_2.
   (d) CH_3OH.

4-7. Write the equilibrium-constant expressions, and obtain numerical values for each constant in

   *(a) the basic dissociation of ethylamine, $C_2H_5NH_2$.
   (b) the acidic dissociation of hydrogen cyanide, HCN.
   *(c) the acidic dissociation of pyridinium hydrochloride, C_5H_5NHCl.
   (d) the basic dissociation of NaCN.
   *(e) the dissociation of H_3AsO_4 to H_3O^+ and AsO_4^{3-}.
   (f) the reaction of CO_3^{2-} with H_2O to give H_2CO_3 and OH^-.

4-8. Generate the solubility-product expression for
   *(a) $AgIO_3$.   *(b) Ag_2SO_3.
   *(c) Ag_3AsO_4.   *(d) $PbClF$.
   (e) CuI.   (f) PbI_2.
   (g) BiI_3.   (h) $MgNH_4PO_4$.

4-9. Express the solubility-product constant for each substance in Problem 4-8 in terms of its molar solubility S.

4-10. Calculate the solubility-product constant for each of the following substances, given that the molar concentrations of their saturated solutions are as indicated:
   *(a) $AgVO_3$ (7.1×10^{-4} mol/L).
   (b) $CuSeO_3$ (1.42×10^{-4} mol/L).
   *(c) $Pb(IO_3)_2$ (4.3×10^{-5} mol/L).
   (d) SrF_2 (8.6×10^{-4} mol/L).
   *(e) $Th(OH)_4$ (3.3×10^{-4} mol/L).
   (f) Ag_3PO_4 (4.7×10^{-6} mol/L).

***4-11.** Calculate the solubility of the solutes in Problem 4-10 for solutions in which the cation concentration is 0.050 M.

***4-12.** Calculate the solubility of the solutes in Problem 4-10 for solutions in which the anion concentration is 0.050 M.

***4-13.** What CrO_4^{2-} concentration is required to
   (a) initiate precipitation of Ag_2CrO_4 from a solution that is 2.12×10^{-3} M in Ag^+?
   (b) lower the concentration of Ag^+ in a solution to 1.00×10^{-6} M?

4-14. What hydroxide concentration is required to
   (a) initiate precipitation of Al^{3+} from a 5.00×10^{-3} M solution of $Al_2(SO_4)_3$?
   (b) lower the Al^{3+} concentration in the foregoing solution to 1.00×10^{-9} M?

*4-15. The solubility-product constant for $Ce(IO_3)_3$ is 3.2×10^{-10}. What is the Ce^{3+} concentration in a solution prepared by mixing 50.0 mL of 0.0500 M Ce^{3+} with 50.00 mL of
   (a) water?
   (b) 0.050 M IO_3^-?
   (c) 0.150 M IO_3^-?
   (d) 0.300 M IO_3^-?

4-16. The solubility-product constant for K_2PdCl_6 is 6.0×10^{-6} ($K_2PdCl_6 \rightleftharpoons 2K^+ + PdCl_6^{2-}$). What is the K^+ concentration of a solution prepared by mixing 50.0 mL of 0.400 M KCl with 50.0 mL of
   (a) 0.100 M $PdCl_6^{2-}$?
   (b) 0.200 M $PdCl_6^{2-}$?
   (c) 0.400 M $PdCl_6^{2-}$?

*4-17. The solubility products for a series of iodides are

$$TlI \quad K_{sp} = 6.5 \times 10^{-8}$$

$$AgI \quad K_{sp} = 8.3 \times 10^{-17}$$

$$PbI_2 \quad K_{sp} = 7.1 \times 10^{-9}$$

$$BiI_3 \quad K_{sp} = 8.1 \times 10^{-19}$$

List these four compounds in order of decreasing molar solubility in
   (a) water.
   (b) 0.10 M NaI.
   (c) a 0.010 M solution of the solute cation.

4-18. The solubility products for a series of hydroxides are

$$BiOOH \quad K_{sp} = 4.0 \times 10^{-10} = [BiO^+][OH^-]$$

$$Be(OH)_2 \quad K_{sp} = 7.0 \times 10^{-22}$$

$$Tm(OH)_3 \quad K_{sp} = 3.0 \times 10^{-24}$$

$$Hf(OH)_4 \quad K_{sp} = 4.0 \times 10^{-26}$$

Which hydroxide has
   (a) the lowest molar solubility in H_2O?
   (b) the lowest molar solubility in a solution that is 0.10 M in NaOH?

4-19. Calculate the hydronium-ion concentration of water at 100°C.

4-20. At 25°C, what are the molar H_3O^+ and OH^- concentrations in
   *(a) 0.0200 M HOCl?
   (b) 0.0800 M butanoic acid?
   *(c) 0.200 M ethylamine?
   (d) 0.100 M trimethylamine?
   *(e) 0.120 M NaOCl?
   (f) 0.0860 M CH_3CH_2COONa?
   *(g) 0.100 M hydroxylamine hydrochloride?
   (h) 0.0500 M ethanolamine hydrochloride?

4-21. For each of the solutions in Problem 4-20, use Excel to find the hydronium ion concentration for 10%, 20%, 30%, 40%, 50%, 60%, 70%, 80%, 90%, 100%, 200%, 300%, 400%, and 500% of the listed molar analytical concentration. Use a separate worksheet for each species, and use a different column to solve for the hydronium ion concentration using the iterative method discussed in the Spreadsheet Exercise. Note any trends in the results as the concentration of the indicated species changes.

4-22. At 25°C, what is the hydronium ion concentration in
   *(a) 0.100 M chloroacetic acid?
   (b) 0.100 M sodium chloroacetate?
   *(c) 0.0100 M methylamine?
   (d) 0.0100 M methylamine hydrochloride?
   *(e) 1.00×10^{-3} anilinium hydrochloride?
   (f) 0.200 M HIO_3?

4-23. For each of the solutions in Problem 4-22, use Excel to find the hydronium ion concentration for 10%, 20%, 30%, 40%, 50%, 60%, 70%, 80%, 90%, 100%, 200%, 300%, 400%, and 500% of the listed molar analytical concentration. Use a separate worksheet for each species, and use a different column to solve for the hydronium ion concentration using the iterative method discussed in the Spreadsheet Exercise. Note any trends in the results as the concentration of the indicated species changes.

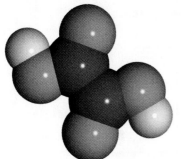

Chapter 5

Errors In
Chemical Analyses:
Assessing the Quality of Results

I t is impossible to perform a chemical analysis in such a way that the results are totally free of errors or uncertainties. We can only hope to minimize these errors and estimate their size with acceptable accuracy.[1] In this and the next two chapters, we explore the nature of experimental errors and their effects on the results of chemical analyses.

The effect of errors in analytical data is illustrated in Figure 5-1, which shows results for the quantitative determination of iron(III). A student analyzed six equal portions of an aqueous solution known to contain exactly 20.00 ppm of iron(III) in exactly the same way. Note that the results range from a low of 19.4 ppm to a high of 20.3 ppm of iron(III). The average, or *mean* value $\bar{x}$, of the data is 19.78 ppm.

> Parts per million (ppm); that is, 20.00 parts of iron(III) per million parts of solution.

Every measurement is influenced by many uncertainties that combine to produce a scatter of results like those shown in Figure 5-1. Measurement uncertainties can never be completely eliminated, so the true value for any quantity is always unknown. The probable magnitude of the error in a measurement can often be evaluated, however. It is then possible to define limits within which the true value of a measured quantity lies at a given level of probability.

> The true value of a measurement is never known exactly.

It is seldom easy to estimate the reliability of experimental data. Nevertheless, we must make such estimates whenever we collect laboratory results *because data of unknown quality are worthless.* On the other hand, results that are not especially accurate may be of considerable value if the limits of uncertainty are known.

Unfortunately, there is no simple and widely applicable method for determining the reliability of data with absolute certainty. As much effort is often required to estimate the quality of experimental results as to collect them.

---

[1]Unfortunately, many people do not understand this truth. For example, when asked by a defense attorney in a celebrated homicide case what the rate of error in a particular blood test was, the assistant district attorney replied that their testing laboratories had no percentage of error because "they have not committed any errors" (*San Francisco Chronicle,* June 29, 1994, p. 4).

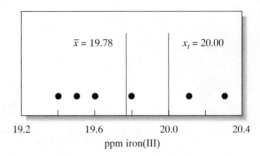

Figure 5-1   Results from six replicate determinations for iron in aqueous samples of a standard solution containing 20.00 ppm of iron(III).

Reliability can be assessed in several ways. Experiments designed to reveal the presence of errors can be performed. Standards of known composition can be analyzed and the results compared with the known composition. A few minutes in the library to consult the literature of analytical chemistry can be profitable. Calibrating equipment enhances the quality of data. Finally, statistical tests can be applied to the data. None of these options is perfect, so in the end we have to make *judgments* as to the probable accuracy of our results. These judgments tend to become harsher and less optimistic with experience.

One of the first questions to answer before beginning an analysis is, What is the maximum error that I can tolerate in the result? The answer to this question determines the time required to do the work. For example, a tenfold increase in accuracy may take hours, days, or even weeks of added labor. *No one can afford to waste time generating data that are more reliable than is needed.*

5A   DEFINING TERMS

Replicates are samples of about the same size that are carried through an analysis in *exactly* the same way.

Chemists usually carry two to five portions (*replicates*) of a sample through an entire analytical procedure. Individual results from a set of measurements are seldom the same (Figure 5-1), so a central or "best" value is used for the set. We justify the extra effort required to analyze several samples in two ways. First, the central value of a set should be more reliable than any of the individual results. Second, variation in the data should provide a measure of the uncertainty associated with the central result. Either the *mean* or the *median* may serve as the central value for a set of replicated measurements.

5A-1   The Mean and Median

Mean, arithmetic mean, and *average* ($\bar{x}$) are synonyms for the quantity obtained by dividing the sum of replicate measurements by the number of measurements in the set:

$$\bar{x} = \frac{\sum_{i=1}^{N} x_i}{N} \tag{5-1}$$

where x_i represents the individual values of x making up a set of N replicate measurements.

The *median* is the middle result when replicate data are arranged in order of size. Equal numbers of results are larger and smaller than the median. For an odd number of data points, the median can be evaluated directly. For an even number, the mean of the middle pair is used (see Example 5-1).

The symbol $\Sigma\, x_i$ means to add all the values x_i for the replicates.

The *mean* of two or more measurements is their average value.

The *median* is the middle value in a set of data that has been arranged in order of size. The median is used advantageously when a set of data contains an *outlier,* a result that differs significantly from others in the set. An outlier can have a significant effect on the mean of the set without affecting the median.

Example 5-1

Calculate the mean and the median for the data shown in Figure 5-1.

mean $= \bar{x}$

$$= \frac{19.4 + 19.5 + 19.6 + 19.8 + 20.1 + 20.3}{6} = 19.78 \approx 19.8 \text{ ppm Fe}$$

Because the set contains an even number of measurements, the median is the average of the central pair:

$$\text{median} = \frac{19.6 + 19.8}{2} = 19.7 \text{ ppm Fe}$$

Ideally, the mean and median are identical. Frequently they are not, however, particularly when the number of measurements in the set is small.

5A-2   What Is Precision?

Precision describes the reproducibility of measurements; in other words, the closeness of results that have been obtained *in exactly the same way.* Generally, the precision of a measurement is readily determined by simply repeating the measurement on replicate samples.

Three terms are widely used to describe the precision of a set of replicate data: *standard deviation, variance,* and *coefficient of variation.* All these terms are a function of the *deviation from the mean d_i,* or just the *deviation,* which is defined as

$$d_i = |x_i - \bar{x}| \tag{5-2}$$

The relationship between deviation from the mean and the three precision terms is given in Section 6B-3.

Precision is the closeness of results to others that have been obtained in exactly the same way.

Note that deviations from the mean are calculated without regard to sign.

5A-3   How About Accuracy?

Figure 5-2 illustrates the difference between accuracy and precision. *Accuracy* indicates the closeness of the measurement to its true or accepted value and is expressed by the *error.* Note the basic difference between accuracy and precision. Accuracy measures agreement between a result and its true value. Precision

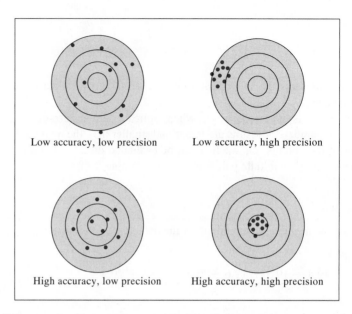

Figure 5-2 Illustration of accuracy and precision using the pattern of darts on a dartboard. Note that we can have very precise results (upper right) that are not accurate and accurate results (lower left) that are imprecise.

describes the agreement among several results that have been obtained in the same way. We may determine precision just by replicating, or repeating, a measurement. On the other hand, we can never determine accuracy exactly because the true value of a measured quantity can never be known exactly. We must use an accepted value instead. Accuracy is expressed in terms of either absolute or relative error.

> *Accuracy* is the closeness of a measurement to the true or accepted value.

Absolute Error

The *absolute error* E in the measurement of a quantity x_i is given by the equation

$$E = x_i - x_t \tag{5-3}$$

where x_t is the true, or accepted, value of the quantity. Returning to the data displayed in Figure 5-1, the absolute error of the result immediately to the left of the true value of 20.00 ppm is -0.2 ppm Fe; the result at 20.10 ppm has an error of $+0.1$ ppm Fe. Note that we retain the sign in stating the error. Thus, the negative sign in the first case shows that the experimental result is smaller than the accepted value.

> The term "absolute" has a different meaning here than it does in mathematics. An absolute value in mathematics means the magnitude of a number *ignoring its sign*. As we shall use it, the absolute error is the difference between an *experimental result and an accepted value including its sign.*

> The *absolute error* of a measurement is the difference between the measured value and the true value. The sign of the absolute error tells you whether the value in question is high or low. If the measurement result is low, the sign is negative; if the measurement result is high, the sign is positive.

Relative Error

Often, the *relative error* E_r is a more useful quantity than the absolute error. The percent relative error is given by the expression

$$E_r = \frac{x_i - x_t}{x_t} \times 100\% \tag{5-4}$$

Relative error is also expressed in parts per thousand (ppt). So, the relative error for the mean of the data in Figure 5-1 is

$$E_r = \frac{19.8 - 20.00}{20.00} \times 100\% = -1\% \text{ or } -10 \text{ ppt}$$

5A-4   Classifying Experimental Errors

The precision of a measurement is readily determined by comparing data from carefully replicated experiments. Unfortunately, an estimate of the accuracy is not as easy to obtain. To determine accuracy, we have to know the true value, and this value is ordinarily exactly what we are seeking in the analysis.

It is tempting to assume that if we know the answer precisely, we also know it accurately. The danger of this assumption is illustrated in Figure 5-3, which summarizes the results for the determination of nitrogen in two pure compounds: benzyl isothiourea hydrochloride and nicotinic acid. The dots show the absolute errors of replicate results obtained by four analysts. Analyst 1 obtained relatively high precision and high accuracy. Analyst 2 had poor precision but good accuracy. The results of analyst 3 are surprisingly common; the precision is excellent, but there is significant error in the numerical average for the data. Both the precision and the accuracy are poor for the results of analyst 4.

Figures 5-1 and 5-3 suggest that chemical analyses are affected by at least two types of errors. One type, called *random* (or *indeterminate*) error, causes data to be scattered more or less symmetrically around a mean value. Refer again to Figure 5-3, and notice that the scatter in the data, and thus the random

Random, or *indeterminate*, *errors* are errors that affect the precision of measurement.

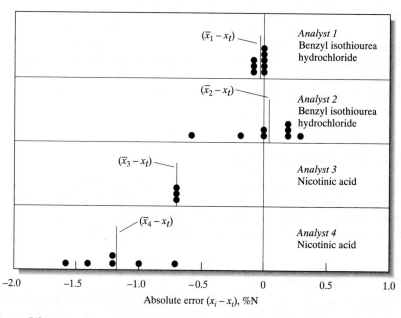

Figure 5-3 Absolute error in the micro-Kjeldahl determination of nitrogen. Each dot represents the error associated with a single determination. Each vertical line labeled $(\bar{x}_i - x_t)$ is the absolute average deviation of the set from the true value. (Data from C. O. Willits and C. L. Ogg, *J. Assoc. Anal. Chem.*, **1949**, *32*, 561. With permission.)

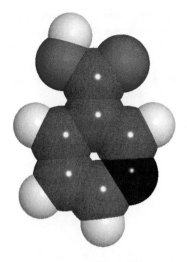

Molecular model of nicotinic acid. Small amounts of this vitamin, which is often called *niacin,* occur in all living cells, and it is essential in the nutrition of mammals. It is used in the prevention and treatment of pellagra.

Systematic, or *determinate, errors* affect the accuracy of results.

An *outlier* is an occasional result in replicate measurements that obviously differs significantly from the rest of the results.

Bias measures the systematic error associated with an analysis. It has a negative sign if it causes the results to be low and a positive sign otherwise.

Of the three types of systematic errors encountered in a chemical analysis, method errors are usually the most difficult to identify and correct.

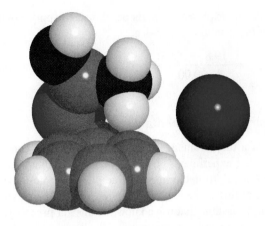

Molecular model of benzyl isothiourea hydrochloride.

error, for analysts 1 and 3 is significantly less than that for analysts 2 and 4. In general, then, the random error in a measurement is reflected by its precision.

A second type of error, called *systematic* (or *determinate*) *error,* causes the mean of a set of data to differ from the accepted value. For example, the mean of the data in Figure 5-1 has a systematic error of about -0.2 ppm Fe. The results of analysts 1 and 2 in Figure 5-3 have little systematic error, but the data of analysts 3 and 4 show systematic errors of about -0.7 and -1.2% nitrogen. In general, a systematic error causes the results in a series of replicate measurements to be all high or all low.

A third type of error is *gross error.* Gross errors differ from indeterminate and determinate errors. They usually occur only occasionally, are often large, and may cause a result to be either high or low. Gross errors lead to *outliers,* results that appear to differ markedly from all other data in a set of replicate measurements. There is no evidence of a gross error in Figures 5-1 and 5-3. Had one of the results shown in Figure 5-1 occurred at, say, 21.2 ppm Fe, it might have been an outlier.

Various statistical tests can be done to determine if a data point is an outlier (see Section 7C).

5B   SYSTEMATIC ERRORS

Systematic errors have a definite value, an assignable cause, and are of about the same magnitude for replicate measurements made in the same way. Systematic errors lead to *bias* in measurement technique. Note that bias affects all the data in a set in approximately the same way and that it bears a sign.

5B-1   How do Systematic Errors Arise?

There are three types of systematic errors. (1) Instrument errors are caused by imperfections in measuring devices and instabilities in their components. (2) Method errors arise from nonideal chemical or physical behavior of analytical systems. (3) Personal errors result from the carelessness, inattention, or personal limitations of the experimenter.

Instrument Errors

All measuring devices are sources of systematic errors. For example, pipets, burets, and volumetric flasks may hold or deliver volumes slightly different from those indicated by their graduations. These differences arise from using glassware at a temperature that differs significantly from the calibration temperature, from distortions in container walls due to heating while drying, from errors in the original calibration, or from contaminants on the inner surfaces of the containers. Calibration eliminates most systematic errors of this type.

Electronic instruments are subject to instrumental systematic errors. These uncertainties have many sources. For example, errors emerge as the voltage of a battery-operated power supply decreases with use. Errors result from increased resistance in circuits because of dirty electrical contacts. Temperature changes cause variation in resistors and standard voltage sources. Currents induced from 110-V power lines affect electronic instruments. Errors from these and other sources are detectable and correctable.

Method Errors

The nonideal chemical or physical behavior of the reagents and reactions upon which an analysis is based often introduce systematic method errors. Such sources of nonideality include the slowness of some reactions, the incompleteness of others, the instability of some species, the nonspecificity of most reagents, and the possible occurrence of side reactions that interfere with the measurement process. For example, a common method error in volumetric methods results from the small excess of reagent required to cause an indicator to undergo the color change that signals completion of the reaction. The accuracy of such an analysis is thus limited by the very phenomenon that makes the titration possible.

Another example of method error is illustrated by the data in Figure 5-3c and 5-3d in which the results by analysts 3 and 4 show a negative bias that can be traced to the chemical nature of the sample, nicotinic acid. The analytical method used involves the decomposition of the organic samples in hot concentrated sulfuric acid that converts the nitrogen in the samples to ammonium sulfate. The amount of ammonia in the ammonium sulfate is then determined in the measurement step. Experiments have shown that compounds containing a pyridine ring such as nicotinic acid (see page 108) are incompletely decomposed by the sulfuric acid unless special precautions are taken. Without these precautions, low results are obtained. It is highly likely the negative errors, $(\bar{x}_3 - x_t)$ and $(\bar{x}_4 - x_t)$ in Figure 5-3 are systematic errors that can be blamed on incomplete decomposition of the samples.

Errors inherent in a method are often difficult to detect and are thus the most serious of the three types of systematic error.

Personal Errors

Many measurements require personal judgments. Examples include estimating the position of a pointer between two scale divisions, the color of a solution at the end point in a titration, or the level of a liquid with respect to a graduation in a pipet or buret (see Figure 6-4). Judgments of this type are often subject to systematic, unidirectional errors. For example, one person may read a pointer consistently high, another may be slightly slow in activating a timer, and a third may

Color blindness is a good example of a disability that amplifies personal errors in a volumetric analysis. A famous color-blind analytical chemist often enlisted his wife to come to the laboratory to help him detect color changes at end points of titrations.

Digital readouts on pH meters, laboratory balances, and other electronic instruments eliminate number bias because no judgment is involved in taking a reading.

Persons who make measurements must guard against personal bias to preserve the integrity of the collected data.

be less sensitive to color changes than others. An analyst who is insensitive to color changes tends to use excess reagent in a volumetric analysis. Physical disabilities are often sources of personal determinate errors.

A universal source of personal error is *prejudice*. Most of us, no matter how honest, have a natural tendency to estimate scale readings in a direction that improves the precision in a set of results. Or, we may have a preconceived notion of the true value for the measurement. We then subconsciously cause the results to fall close to this value. Number bias is another source of personal error that varies considerably from person to person. The most common number bias encountered in estimating the position of a needle on a scale involves a preference for the digits 0 and 5. Also prevalent is a prejudice favoring small digits over large and even numbers over odd.

5B-2   What Effects Do Systematic Errors Have on Analytical Results?

Systematic errors may be either *constant* or *proportional*. The magnitude of a constant error does not depend on the size of the quantity measured. Proportional errors increase or decrease in proportion to the size of the sample taken for analysis.

Constant Errors

Constant errors become more serious as the size of the quantity measured decreases. The effect of solubility losses on the results of a gravimetric analysis, shown in Example 5-2, illustrates this behavior.

Example 5-2

Suppose that 0.50 mg of precipitate is lost as a result of being washed with 200 mL of wash liquid. If the precipitate weighs 500 mg, the relative error due to solubility loss is $-(0.50/500) \times 100\% = -0.1\%$. Loss of the same quantity from 50 mg of precipitate results in a relative error of -1.0%.

The excess of reagent required to bring about a color change during a titration is another example of constant error. This volume, usually small, remains the same regardless of the total volume of reagent required for the titration. Again, the relative error from this source becomes more serious as the total volume decreases. One way of minimizing the effect of constant error is to use as large a sample as possible.

Constant errors are independent of the size of the sample being analyzed. *Proportional errors* decrease or increase in proportion to the size of the sample.

Proportional Errors

A common cause of proportional errors is the presence of interfering contaminants in the sample. For example, a widely used method for the determination of copper is based on the reaction of copper(II) ion with potassium iodide to give iodine. The quantity of iodine is then measured and is proportional to the amount of copper. Iron(III), if present, also liberates iodine from potassium iodide. Unless steps are taken to prevent this interference, high results are observed for the percentage of copper because the iodine produced will be a measure of the copper(II) *and* iron(III) in the sample. The size of this error is fixed by the *fraction* of iron contamination, which is independent of the size of sample taken. If the sample size is doubled, for example, the amount of iodine liberated by both the copper and the iron contaminant is also doubled. Thus, the magnitude of the reported percentage of copper is independent of sample size.

5B-3   Detecting Systematic Instrument and Personal Errors

Systematic instrument errors are usually found and corrected by calibration. Periodic calibration of equipment is always desirable because the response of most instruments changes with time as a result of wear, corrosion, or mistreatment.

Most personal errors can be minimized by care and self-discipline. It is a good habit to check instrument readings, notebook entries, and calculations systematically. Errors that result from a known physical disability can usually be avoided by carefully choosing the method.

After entering an instrument reading into your laboratory notebook, make a second reading, and then check the second reading against the data you entered in your notebook.

5B-4   Detecting Systematic Method Errors

Bias in an analytical method is particularly difficult to detect. We may take one or more of the following steps to recognize and adjust for a systematic error in an analytical method.

Analyzing Standard Samples

The best way to estimate the bias of an analytical method is by analyzing *standard reference materials,* materials that contain one or more analytes at well-known or certified concentration levels. Standard reference materials are obtained in several ways.

Standard materials can sometimes be prepared by synthesis. Here, carefully measured quantities of the pure components of a material are measured out and mixed in such a way as to produce a homogeneous sample whose composition is known from the quantities taken. The overall composition of a synthetic standard material must approximate closely the composition of the samples to be analyzed. Great care must be taken to ensure that the concentration of analyte is known as well as possible. Unfortunately, the synthesis of such standard samples is often impossible or so difficult and time consuming that this approach is not practical.

Standard reference materials can be purchased from a number of governmental and industrial sources. For example, the National Institute of Standards and Technology (NIST) (formerly the National Bureau of Standards) offers over 1300 standard reference materials including rocks and minerals, gas mixtures, glasses, hydrocarbon mixtures, polymers, urban dusts, rainwaters, and river sediments.[2] The concentration of one or more of the components in these materials has been determined in one of three ways: (1) by analysis with a previously vali-

Standard reference materials (*SRMs*) are substances sold by the National Institute of Standards and Technology and certified to contain specified concentrations of one or more analytes.

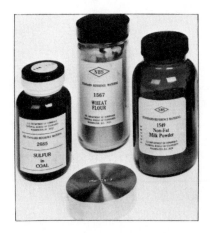

Standard reference materials from NIST. (Photo courtesy of the National Institute of Standards and Technology.)

[2]See U.S. Department of Commerce, *NIST Standard Reference Materials Catalog,* 1998–1999 ed., NIST Special Publication 260 (Washington, D.C.: U.S. Government Printing Office, 1998). For a description of the reference material programs of the NIST, see R. A. Alvarez, S. D. Rasberry, and F. A. Uriano, *Anal. Chem.,* **1982,** *54,* 1226A; F. A. Uriano, *ASTM Standardization News,* **1979,** *7,* 8.

dated reference method; (2) by analysis by two or more independent, reliable measurement methods; or (3) by analysis by a network of cooperating laboratories that are technically competent and thoroughly knowledgeable with the material being tested. Several commercial supply houses also offer analyzed materials for method testing.[3]

Use of a standard to validate an analysis commonly results in values that differ somewhat from theoretical. The question then becomes one of establishing whether such a difference is due to bias or random error. In Section 7B-1, we demonstrate a statistical test that can be applied to aid your judgment in answering this question.

When using SRMs, it is often difficult to separate bias from ordinary random error.

Using an Independent Analytical Method

If standard samples are not available, a second independent and reliable analytical method can be used in parallel with the method being evaluated. The independent method should differ as much as possible from the one under study to minimize the possibility that some common factor in the sample has the same effect on both methods. Here again, a statistical test must be used to determine whether any difference is a result of random errors in the two methods or to bias in the method under study (see Section 7B-2).

Performing Blank Determinations

A *blank* solution contains the solvent and all the reagents in an analysis but usually none of the sample. Occasionally blanks may also contain added constituents to simulate the sample matrix.

Blank determinations are useful for detecting certain types of constant errors. In a blank determination, or *blank,* all steps of the analysis are performed in the absence of a sample. The results from the blank are then applied as a correction to the sample measurements. Blank determinations reveal errors due to interfering contaminants from the reagents and vessels employed in analysis. Blanks also allow the analyst to correct titration data for the volume of reagent needed to cause an indicator to change color at the end point.

Varying the Sample Size

Example 5-2 demonstrated that as the size of a measurement increases, the effect of a constant error decreases. Thus, constant errors can often be detected by varying the sample size.

Spreadsheet Exercise   A Mean Calculation

In this exercise, we learn to calculate the mean of a set of data. First, we define formulas to calculate the mean, and then we use the built-in functions of Excel to accomplish the task.

Entering the Data

Let us begin by starting Excel with a clean spreadsheet. In cell B1, enter the heading Data [↵] . Now enter in column B under the heading the data x_i given in Example 5-1. Click on cell A11 and type

---

[3]For sources of biological and environmental reference materials containing various elements, see C. Veillon, *Anal. Chem.,* **1986,** *58,* 851A.

```
Total[⏎]
N[⏎]
Mean[⏎]
```

Your worksheet should now look like the following:

	A	B	C	D
1		Data		
2		19.4		
3		19.5		
4		19.6		
5		19.8		
6		20.1		
7		20.3		
8				
9				
10				
11	Total			
12	N			
13	Mean			
14				
15				

Finding the Mean

Click on cell B11, and type

$$=\text{SUM(B2:B7)}\ [⏎]$$

This formula calculates the sum of the values in cells B2 through B7 and displays the result in cell B11. Now, in cell B12, type

$$=\text{COUNT(B2:B7)}\ [⏎]$$

The COUNT function counts the number of nonzero cells in the range B2:B7 and displays the result in cell B12. Since we have found the sum of the data and the number N of data, we can find the mean $\bar{x}$ of the data by typing

$$=\text{B11/B12}\ [⏎]$$

in cell B13. Your worksheet should appear as shown on the next page.

Using Excel's Built-In Functions

Excel has built-in functions to compute many of the quantities of interest to us. Now we shall see how to use them to calculate the mean or, in Excel's syntax, the average. Click on cell C13 and type

$$=\text{AVERAGE(B2:B7)}\ [⏎]$$

	A	B	C	D
1		Data		
2		19.4		
3		19.5		
4		19.6		
5		19.8		
6		20.1		
7		20.3		
8				
9				
10				
11	Total	118.7		
12	N	6		
13	Mean	19.78333		
14				
15				

Notice that the mean determined using the built-in AVERAGE function is identical to the value in cell B13 that you determined by typing a formula. Before proceeding or terminating your Excel session, save your file to a floppy disk as `average.xls`.

Finding the Deviations from the Mean

We can now use Excel to determine the deviation from the mean of each of the data in our worksheet using the definition given by Equation 5-2. Click on cell C2 and type

$$=\text{ABS(B2-\$B\$13)}\ [\hookleftarrow]$$

This formula computes the absolute value ABS() of the difference between our first value in cell B2 and the mean value in cell B13. This formula is a bit different than those we have used previously. We have typed a dollar sign, $, before the B and before the 13 in the second cell reference. This type of cell reference is called an *absolute reference*. It means that no matter where we might copy the contents of the cell C2, the reference will always refer to cell B13. The other type of cell reference we consider here is the *relative reference*, which is exemplified by cell B2. We use a relative reference for B2 and an absolute reference for B13 because we want to copy the formula in cell C2 into cells C3 to C7, and we want the mean $B$13 to be subtracted from each of the successive data in column B. Let us now copy the formula by clicking on cell C2, clicking on the fill handle, and dragging the rectangle through cell C7. When you release the mouse button, your worksheet should look like the one on the next page.

Now click on cell C3 and notice that it contains the formula =ABS(B3-$B$13). Compare this formula with the one in cell C2 and cells C4 through C7. The absolute cell reference $B$13 appears in all the cells. As you can see, we have

	A	B	C	D
1		Data	Deviation	
2		19.4	0.383333	
3		19.5	0.283333	
4		19.6	0.183333	
5		19.8	0.016667	
6		20.1	0.316667	
7		20.3	0.516667	
8				
9				
10				
11	Total	118.7		
12	N	6		
13	Mean	19.78333	19.78333	
14				

accomplished our task of calculating the deviation from the mean for all the data. Now we will edit the formula in cell C13 to find the mean deviation of the data.

Editing Formulas

To edit the formula to calculate the mean deviation of the data, click on cell C13 and then click on the formula in the formula bar. Use the arrow keys, [←] and [→], and either the [Backspace] or the [Delete] key to replace both B's in the formula with C's so that it reads =AVERAGE(C2:C7). Finally, type [↵], and the mean deviation will appear in cell C13. Type the label Deviation in cell C1 so that your worksheet appears as shown below. Save the file by clicking on the save icon in the toolbar, by clicking File/Save, or by typing [Ctrl+S].

	A	B	C	D
1		Data	Deviation	
2		19.4	0.383333	
3		19.5	0.283333	
4		19.6	0.183333	
5		19.8	0.016667	
6		20.1	0.316667	
7		20.3	0.516667	
8				
9				
10				
11	Total	118.7		
12	N	6		
13	Mean	19.78333	0.283333	
14				
15				

In this exercise, we have learned to calculate a mean using both the built-in Excel AVERAGE function and a formula of our own design. In Chapter 6, we will use STDEV and other functions to complete our analysis of the data from the gravimetric determination of chloride that we began in Chapter 2. You may now close Excel by clicking on File/Exit or proceed to Chapter 6 to continue with the spreadsheet exercises.

WEB WORKS

Statistics is an important topic not only in chemistry but in all walks of life. Newspapers, magazines, television, and the World Wide Web bombard us with confusing and often misleading, statistics. Browse to http://nilesonline.com/stats/ to see an interesting presentation of statistics for writers. Use the links there to look up the definitions of mean and median. You will find some nice examples using salaries that clarify the distinction between the two measures of central tendency, show the utility of comparing the two, and point out the importance of using the appropriate measure for a particular data set.

5C   QUESTIONS AND PROBLEMS

5-1. Explain the difference between
  *(a) constant and proportional error.
  (b) random and systematic error.
  *(c) mean and median.
  (d) absolute and relative error.

*5-2. Suggest some sources of random error in measuring the width of a 3-m table with a 1-m metal rule.

*5-3. Name three types of systematic errors.

5-4. How are systematic method errors detected?

*5-5. What kind of systematic errors are detected by varying the sample size?

5-6. A method of analysis yields weights for gold that are low by 0.4 mg. Calculate the percent relative error caused by this uncertainty if the weight of gold in the sample is
  *(a) 900 mg.   **(b)** 600 mg.
  *(c) 150 mg.   **(d)** 30 mg.

5-7. The method described in Problem 5-6 is to be used for the analysis of ores that assay about 1.2% gold. What minimum sample weight should be taken if the relative error resulting from a 0.4-mg loss is not to exceed
  *(a) −0.2%?   **(b)** −0.5%?
  *(c) −0.8%?   **(d)** −1.2%?

5-8. The color change of a chemical indicator requires an overtitration of 0.04 mL. Calculate the percent relative error if the total volume of titrant is
  *(a) 50.00 mL.   **(b)** 10.0 mL.
  *(c) 25.0 mL.   **(d)** 40.0 mL.

5-9. A loss of 0.4 mg of Zn occurs in the course of an analysis to determine that element. Calculate the percent relative error due to this loss if the weight of Zn in the sample is
  *(a) 40 mg.
  (b) 175 mg.
  *(c) 400 mg.
  (d) 600 mg.

 5-10. Find the mean and median of each of the following sets of data. Determine the deviation from the mean for each data point within the sets and find the mean deviation for each set.
  *(a) 0.0110   0.0104   0.0105
  (b) 24.53   24.68   24.77   24.81   24.73
  *(c) 188   190   194   187
  (d) 4.52×10^{-3} 4.47×10^{-3}
        4.63×10^{-3} 4.48×10^{-3}
        4.53×10^{-3} 4.58×10^{-3}
  *(e) 39.83   39.61   39.25   39.68
  (f) 850   862   849   869   865

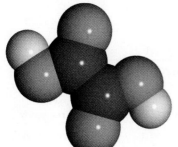

Chapter 6

Random Error: How Certain Can We Be?

All measurements contain random errors. In this chapter, we consider the sources of random errors, the determination of their magnitude, and their effects on computed results of chemical analyses.

6A THE NATURE OF RANDOM ERRORS

Random, or indeterminate, errors occur whenever a measurement is made. This type of error is caused by the many uncontrollable variables that are an inevitable part of every physical or chemical measurement. There are many contributors to random error, but often we cannot positively identify or measure them because they are small enough to avoid individual detection. The accumulated effect of the individual random uncertainties, however, causes the data from a set of replicate measurements to fluctuate randomly around the mean of the set. For example, the scatter of data in Figures 5-1 and 5-3 is a direct result of the accumulation of small random uncertainties. Notice that the random error in the results of analysts 2 and 4 in Figure 5-3 is greater than in the results of analysts 1 and 3.

6A-1 What Are the Sources of Random Errors?

We can get a qualitative idea of the way small undetectable uncertainties produce a detectable random error in the following way. Imagine a situation in which just four small random errors combine to give an overall error. We will assume that each error has an equal probability of occurring and that each can cause the final result to be high or low by a fixed amount $\pm U$.

Table 6-1 shows all the possible ways the four errors can combine to give the indicated deviations from the mean value. Note that only one combination leads to a deviation of $+4U$, four combinations give a deviation of $+2U$, and six give a deviation of $0U$. The negative errors have the same relationship. This ratio of $1:4:6:4:1$ is a measure of the probability for a deviation of each magnitude. Therefore, if we make a sufficiently large number of measurements, we can expect a frequency distribution like that shown in Figure 6-1a. Note that the ordinate in the plot is the relative frequency of occurrence of the five possible combinations.

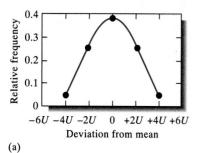

(a)

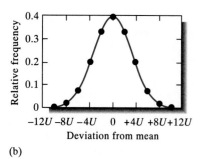

(b)

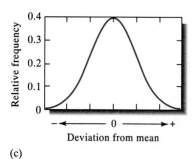

(c)

Figure 6-1 Frequency distribution for measurements containing (a) four random uncertainties, (b) ten random uncertainties, and (c) a very large number of random uncertainties.

In our example, all the uncertainties have the same magnitude. This restriction is not necessary to derive the equation for a Gaussian curve.

The *spread* in a set of replicate measurements is the difference between the highest and lowest result.

Table 6-1

Possible Combinations of Four Equal-Sized Uncertainties

Combinations of Uncertainties	Magnitude of Random Error	Number of Combinations	Relative Frequency
$+U_1 + U_2 + U_3 + U_4$	$+4U$	1	$1/16 = 0.0625$
$-U_1 + U_2 + U_3 + U_4$ $+U_1 - U_2 + U_3 + U_4$ $+U_1 + U_2 - U_3 + U_4$ $+U_1 + U_2 + U_3 - U_4$	$+2U$	4	$4/16 = 0.250$
$-U_1 - U_2 + U_3 + U_4$ $+U_1 + U_2 - U_3 - U_4$ $+U_1 - U_2 + U_3 - U_4$ $-U_1 + U_2 - U_3 + U_4$ $-U_1 + U_2 + U_3 - U_4$ $+U_1 - U_2 - U_3 + U_4$	0	6	$6/16 = 0.375$
$+U_1 - U_2 - U_3 - U_4$ $-U_1 + U_2 - U_3 - U_4$ $-U_1 - U_2 + U_3 - U_4$ $-U_1 - U_2 - U_3 + U_4$	$-2U$	4	$4/16 = 0.250$
$-U_1 - U_2 - U_3 - U_4$	$-4U$	1	$1/16 = 0.0625$

Figure 6-1b shows the theoretical distribution for ten equal-sized uncertainties. Again we see that the most frequent occurrence is zero deviation from the mean. At the other extreme, a maximum deviation of $10U$ occurs only about once in 500 measurements.

When the same procedure is applied to a very large number of individual errors, a bell-shaped curve like that shown in Figure 6-1c results. Such a plot is called a *Gaussian curve* or a *normal error curve*.

6A-2   Describing the Distribution of Experimental Data

We find empirically that the distribution of replicate data from most quantitative analytical experiments approaches that of the Gaussian curve shown in Figure 6-1c. As an example, consider the data in Table 6-2 for the calibration of a 10-mL pipet. In this experiment, a small flask and stopper are weighed. Ten milliliters of water are transferred to the flask with the pipet and the flask is stoppered. The flask, the stopper, and the water are then weighed. The temperature of the water is also measured to establish its density. The mass of the water is then calculated by taking the difference between the two masses; this difference is divided by the density of the water to find the volume delivered by the pipet. The experiment was repeated 50 times.

The data in Table 6-2 are typical of those obtained by an experienced worker weighing to the nearest milligram (which corresponds to 0.001 mL) on a top-loading balance and making every effort to avoid systematic error. Even so, the results vary from a low of 9.969 mL to a high of 9.994 mL. This 0.025 mL *spread* of data results directly from an accumulation of all the random uncertainties in the experiment.

Table 6-2

Replicate Data on the Calibration of a 10-mL Pipet

Trial*	Volume, mL	Trial	Volume, mL	Trial	Volume, mL
1	9.988	18	9.975	35	9.976
2	9.973	19	9.980	36	9.990
3	9.986	20	9.994†	37	9.988
4	9.980	21	9.992	38	9.971
5	9.975	22	9.984	39	9.986
6	9.982	23	9.981	40	9.978
7	9.986	24	9.987	41	9.986
8	9.982	25	9.978	42	9.982
9	9.981	26	9.983	43	9.977
10	9.990	27	9.982	44	9.977
11	9.980	28	9.991	45	9.986
12	9.989	29	9.981	46	9.978
13	9.978	30	9.969‡	47	9.983
14	9.971	31	9.985	48	9.980
15	9.982	32	9.977	49	9.983
16	9.983	33	9.976	50	9.979
17	9.988	34	9.983		

Mean volume = 9.982 mL
Median volume = 9.982 mL
Spread = 0.025 mL
Standard deviation = 0.0056 mL

*Data listed in the order obtained.
†Maximum value.
‡Minimum value.

Table 6-3

Frequency Distribution of Data from Table 6-2

Volume Range, mL	Number in Range	% in Range
9.969 to 9.971	3	6
9.972 to 9.974	1	2
9.975 to 9.977	7	14
9.978 to 9.980	9	18
9.981 to 9.983	13	26
9.984 to 9.986	7	14
9.987 to 9.989	5	10
9.990 to 9.992	4	8
9.993 to 9.995	1	2

The information in Table 6-2 is easier to visualize when the data are rearranged into frequency distribution groups, as in Table 6-3. Here, we tabulate the number of data points falling into a series of adjacent 0.003-mL *cells* and calculate the percentage of measurements falling into each cell. Note that 26% of the data reside in the cell containing the mean and median value of 9.982 mL and that more than half the results are within ± 0.004 mL of this mean.

The frequency distribution data in Table 6-3 are plotted as a bar graph, or *histogram* (labeled *A* in Figure 6-2). We can imagine that as the number of measurements increases, the histogram approaches the shape of the continuous curve shown as plot *B* in Figure 6-2. This curve is a Gaussian curve, or normal error curve, derived for an infinite set of data. These data have the same mean (9.982 mL), the same precision, and the same area under the curve as the histogram.

Variations in replicate results such as those in Table 6-2 come from numerous small and individually undetectable random errors that are attributable to uncontrollable variables in the experiment. Such small errors ordinarily tend to cancel one another and thus have a minimal effect. Occasionally, however, they occur in the same direction to produce a large positive or negative net error.

Sources of random uncertainties in the calibration of a pipet include (1) visual judgments, such as the level of the water with respect to the marking on the pipet and the mercury level in the thermometer; (2) variations in the drainage time and in the angle of the pipet as it drains; (3) temperature fluctuations, which

A *histogram* is a bar graph such as that shown by plot *A* in Figure 6-2.

A *Gaussian*, or *normal error curve*, is a curve that shows the symmetrical distribution of data around the mean of an infinite set of data such as the one in Figure 6-1c.

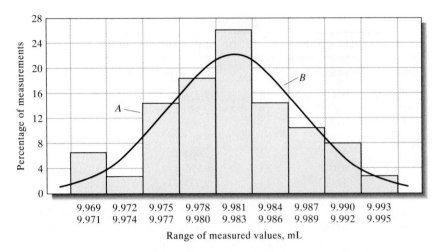

Figure 6-2   A histogram (*A*) showing distribution of the 50 results in Table 6-3 and a Gaussian curve (*B*) for data having the same mean and same standard deviation as the data in the histogram.

affect the volume of the pipet, the viscosity of the liquid, and the performance of the balance; and (4) vibrations and drafts that cause small variations in the balance readings. Undoubtedly, numerous other sources of random uncertainty also operate in this calibration process. Thus, many small and uncontrollable variables affect even the simple process of calibrating a pipet. It is difficult or impossible to determine the influence of any one of the random errors arising from these variables, but the cumulative effect is responsible for the scatter of data points around the mean.

6B   TREATING RANDOM ERRORS WITH STATISTICS

Statistics only reveal information that is already present in a data set. That is, no new information is created by statistics. Statistical analysis does allow us, however, to look at our data in different ways and make objective and intelligent decisions regarding their quality and use.

The random, or indeterminate, errors in the results of an analysis can be evaluated by the methods of statistics. Ordinarily, statistical analysis of analytical data is based on the assumption that random errors follow a Gaussian, or normal, distribution such as that illustrated in curve *B* in Figure 6-2 or in Figure 6-1c. Sometimes analytical data depart seriously from Gaussian behavior, but the normal distribution is the most common. Thus, we base this discussion entirely on normally distributed random errors.

6B-1   Samples and Populations

Do not confuse the statistical sample with the analytical sample. Four analytical samples analyzed in the laboratory represent a single statistical sample. This terminology is an unfortunate duplication of the term sample.

In statistics, a finite number of experimental observations is called a *sample* of data. The sample is treated as a tiny fraction of an infinite number of observations that could in principle be made given infinite time. Statisticians call the theoretical infinite number of data a *population,* or a *universe,* of data. Statistical laws have been derived assuming a population of data; often, they must be modified substantially when applied to a small sample because a few data points may not be representative of the population. In the discussion that follows, we first describe the Gaussian statistics of populations. Then we show how these relationships can be modified and applied to small samples of data.

If you flip a coin ten times, how many heads will you get? Try it and record your results. Repeat the experiment. Are your results the same? Ask friends or members of your class to perform the same experiment and tabulate the results. The table below contains the results obtained by several classes of analytical chemistry students from 1980 to 1998.

Number of Heads	0	1	2	3	4	5	6	7	8	9	10
Frequency	1	1	22	42	102	104	92	48	22	7	1

Add your results to those in the table and plot a histogram similar to the one shown in Figure 6F-1. Find the mean and the standard deviation (see Section 6C-3) for your results and compare them with the values shown in the plot. The smooth curve in the figure is a normal error curve for an infinite number of trials with the same mean and standard deviation as the data. Note that the mean of 5.06 is very close to the value of 5 that you would predict based on the laws of probability. As the number of trials increases, the histogram approaches the shape of the smooth curve and the mean approaches five.

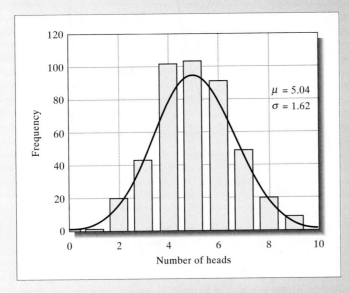

Figure 6F-1 Results of a coin-flipping experiment by 395 students over an 18-year period.

6B-2 Characterizing Gaussian Curves

Figure 6-3a shows two Gaussian curves in which the relative frequency y of occurrence of various deviations from the mean is plotted as a function of the deviation from the mean. As shown in the margin, curves such as these can be described by an equation that contains just two parameters, the *population mean μ* and the *population standard deviation σ*.

The equation for a Gaussian curve has the form

$$y = \frac{e^{-(x-\mu)^2/2\sigma^2}}{\sigma\sqrt{2\pi}}$$

Generally, in making laboratory measurements, the data that we collect are members of a huge population of possible measurements that could be made. It is possible in certain situations to have a population consisting of relatively few data. For example, it is conceivable that in a manufacturing environment measurements might be made on a very small number of unique prototypes. A manufacturer of pH meters might build fifty prototypes and carry out an exhaustive set of measurements on the devices. Statistical analysis of the data would involve the calculation of means and standard deviations of the various measurements. For this population, $N = 50$, and the mean of 50 measurements would represent the population mean μ. Furthermore, the standard deviation of the 50 measurements would be the population standard deviation σ, and Equation 6-1 must be used for the calculation.

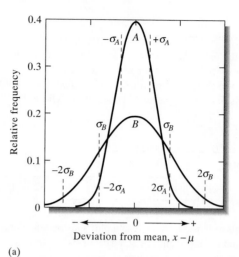

(a)

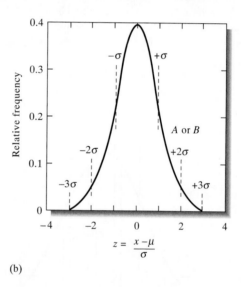

(b)

Figure 6-3   Normal error curves. The standard deviation for curve B is twice that for curve A; that is, $\sigma_B = 2\sigma_A$. (a) The abscissa is the deviation from the mean in the units of measurement. (b) The abscissa is the deviation from the mean in units of σ. Thus, the two curves A and B are identical here.

The Population Mean μ and the Sample Mean $\bar{x}$

Statisticians find it useful to differentiate between a *sample mean* and a *population mean*. The sample mean $\bar{x}$ is the mean of a limited sample drawn from a population of data. It is defined by Equation 5-1, when N is a small number. The population mean μ, in contrast, is the true mean for the population. It is also defined by Equation 5-1 with the added provision that N is so large that it approaches infinity. *In the absence of any systematic error, the population mean is also the true value for the measured quantity.* To emphasize the difference between the two means, the sample mean is symbolized by $\bar{x}$ and the population mean by μ. More often than not, particularly when N is small, $\bar{x}$ differs from μ

In the absence of systematic error, the population mean μ is the true value of a measured quantity.

because a small sample of data does not exactly represent its population. The probable difference between $\bar{x}$ and μ decreases rapidly as the number of measurements making up the sample increases; ordinarily, by the time N reaches 20 to 30, this difference is negligibly small.

The Population Standard Deviation (σ)

The *population standard deviation* σ, which is a measure of the *precision* or scatter of a population of data, is given by the equation

$$\sigma = \sqrt{\frac{\sum_{i=1}^{N} (x_i - \mu)^2}{N}} \qquad (6\text{-}1)$$

where N is the number of data points making up the population.

The two curves in Figure 6-3a are for two populations of data that differ only in their standard deviations. The standard deviation for the data set yielding the broader but lower curve B is twice that for the measurements yielding curve A. The breadth of these curves is a measure of the precision of the two sets of data. Thus, the precision of the data leading to curve A is twice as good as that of the data that are represented by curve B.

Figure 6-3b shows another type of normal error curve in which the abscissa is now a new variable z, which is defined as

$$z = \frac{(x - \mu)}{\sigma} \qquad (6\text{-}2)$$

Note that z is the deviation of a data point from the mean relative to one standard deviation. That is, when $x - \mu = \sigma$, z is equal to one; when $x - \mu = 2\sigma$, z is equal to two; and so forth. Since z is the deviation from the mean relative to the standard deviation, a plot of relative frequency versus this parameter yields a single Gaussian curve that describes all populations of data regardless of standard deviation. Thus, Figure 6-3b is the normal error curve for both sets of data used to plot curves A and B in Figure 6-3a.

Because it appears in the equation for the Gaussian error curve, the square of the standard deviation σ^2 is also of interest. This quantity is called the *variance* (see Section 6B-5).

A normal error curve has several general properties: (a) The mean occurs at the central point of maximum frequency, (2) there is a symmetrical distribution of positive and negative deviations about the maximum, and (3) there is an exponential decrease in frequency as the magnitude of the deviations increases. Thus, small random uncertainties are observed much more often than very large ones.

Areas under a Gaussian Curve

It can be shown that, regardless of its width, 68.3% of the area beneath a Gaussian curve for a population of data lies within one standard deviation ($\pm 1\sigma$) of the mean μ. Thus, 68.3% of the data making up the population will lie within these bounds. Furthermore, approximately 95.4% of all data points are within $\pm 2\sigma$ of the mean and 99.7% within $\pm 3\sigma$. The vertical dashed lines in Figure 6-3 show the areas bounded by $\pm 1\sigma$ and $\pm 2\sigma$.

Sample mean $= \bar{x}$, where

$$\bar{x} = \frac{\sum_{i=1}^{N} x_i}{N} \qquad \text{when } N \text{ is small}$$

Population mean $= \mu$, where

$$\mu = \frac{\sum_{i=1}^{N} x_i}{N} \qquad \text{when } N \to \infty$$

The quantity $(x_i - \mu)$ in Equation 6-1 is the deviation of individual results from the mean of the population; compare with Equation 6-3, which is for a sample of data.

The quantity z represents the deviation of a result from the mean for a population of data measured relative to the standard deviation.

To compute the sample standard deviation statistic we use the σ_{N-1} button on a pocket calculator or.the STDEV() function in Excel. In those rare cases where the entire population is all that there is in the universe, then we would have to compute the population standard deviation using Equation 6-2, or use the σ_N button on a pocket calculator, or the STDEV() function in Excel. Do not confuse the sample standard deviation with the population standard deviation, which is the "true" value of the standard deviation of the very large population that we have sampled.

Because of such area relationships, the standard deviation of a population of data is a useful predictive tool. For example, we can say that the chances are 68.3 in 100 that the random uncertainty of any single measurement is no more than $\pm 1\sigma$. Similarly, the chances are 95.4 in 100 that the error is less than $\pm 2\sigma$, and so forth. The calculation of areas is described in Feature 6-2.

Feature 6-2

How Do You Calculate Those Areas under a Curve?

We often refer to the area under a curve. In the context of statistics, it is important to be able to determine the area under the Gaussian curve between arbitrarily defined limits. The area under the curve between a pair of limits gives the probability of a measured value occurring between the limits. A practical question arises: How do we determine the area under a curve? The margin note on page 121 presents the equation describing the Gaussian curve in terms of the population mean μ and the standard deviation σ. Let us suppose that we want to know the area under the curve between -1σ and $+1\sigma$ of the mean. In other words, we want the area from $\mu - \sigma$ to $\mu + \sigma$.

We can perform this operation using calculus because integration of an equation gives the area under the curve described by the equation. Thus, in this case, we wish to find the definite integral

$$\text{area} = \int_{-\sigma}^{\sigma} \frac{e^{-(x-\mu)^2/2\sigma^2}}{\sigma\sqrt{2\pi}} \, dx$$

To make life easier, let us assume that $\mu = 0$ and $\sigma = 1$. Our equation now becomes

$$\text{area} = \int_{-1}^{1} \frac{e^{-x^2/2}}{\sqrt{2\pi}} \, dx = 0.683$$

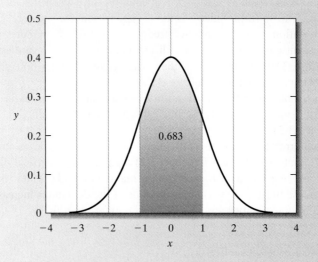

Likewise, if we want to know the area under the Gaussian curve 2σ either side of the mean, we calculate the following integral:

$$\text{area} = \int_{-2}^{2} \frac{e^{-x^2/2}}{\sqrt{2\pi}} \, dx = 0.954$$

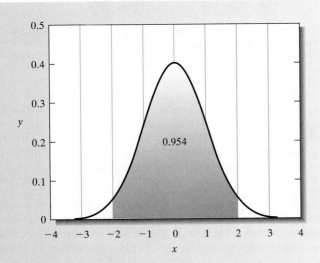

For $\pm 3\sigma$, we have

$$\text{area} = \int_{-3}^{3} \frac{e^{-x^2/2}}{\sqrt{2\pi}}\, dx = 0.997$$

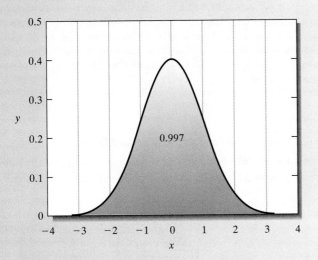

Finally, it is important to know the area under the entire Gaussian curve, so we find the following integral:

$$\text{area} = \int_{-\infty}^{\infty} \frac{e^{-x^2/2}}{\sqrt{2\pi}}\, dx = 1$$

We can see from the integrals that the areas under the Gaussian curve for one, two, and three standard deviations from the mean are, respectively, 68.3%, 95.4%, and 99.7% of the total area under the curve.

6B-3   Finding the Sample Standard Deviation

Equation 6-1 must be modified when it is applied to a small sample of data. Thus, the *sample standard deviation s* is given by the equation

$$s = \sqrt{\frac{\sum\limits_{i=1}^{N} (x_i - \bar{x})^2}{N - 1}} = \sqrt{\frac{\sum\limits_{i=1}^{N} d_i^2}{N - 1}} \tag{6-3}$$

Equation 6-3 applies to small sets of data. It says, "Find the deviations from the mean d_i, square them, sum them, divide the sum by $N - 1$, and take the square root." The quantity $N - 1$ is called the *number of degrees of freedom*. Many scientific calculators have the standard deviation function built in.

Note that Equation 6-3 differs from Equation 6-1 in two ways. First, the sample mean, $\bar{x}$, appears in the numerator of Equation 6-3 in place of the population mean, μ. Second, N in Equation 6-1 is replaced by the *number of degrees of freedom* $(N - 1)$. If this substitution is not used, the calculated s will be less on average than the true standard deviation σ; that is, s will have a negative bias (see Feature 6-3).

The *sample variance* s^2 is also of importance in statistical calculations. It is an estimate of the population variance σ^2, as discussed in Section 6B-5.

Feature 6-3

The Significance of Number of Degrees of Freedom

The number of degrees of freedom indicates the number of *independent* results that go into the computation of a standard deviation. Thus, when μ is unknown, two quantities must be extracted from a set of replicate data: $\bar{x}$ and s. One degree of freedom is used to establish $\bar{x}$ because the sum of the individual deviations, with their signs retained, must add up to zero. Thus, when $N - 1$ deviations have been computed, the final one is known. Consequently, only $N - 1$ deviations provide an *independent* measure of the precision of the set. Failure to use $N - 1$ in calculating the standard deviation for small samples results in values of s, which are on average smaller than the true standard deviation σ.

An Alternative Expression for Sample Standard Deviation

To calculate s with a calculator that does not have a standard deviation key, the following rearrangement is easier to use than directly applying Equation 6-3:

$$s = \sqrt{\frac{\sum\limits_{i=1}^{N} x_i^2 - \dfrac{\left(\sum\limits_{i=1}^{N} x_1\right)^2}{N}}{N - 1}} \tag{6-4}$$

Example 6-1 illustrates the use of Equation 6-4 to calculate s.

Example 6-1

The following results were obtained in the replicate determination of the lead content of a blood sample: 0.752, 0.756, 0.752, 0.751, and 0.760 ppm Pb. Calculate the mean and the standard deviation of this set of data.

To apply Equation 6-4, we calculate $\Sigma\, x_i^2$ and $(\Sigma\, x_i)^2/N$.

Sample	x_i	x_i^2
1	0.752	0.565504
2	0.756	0.571536
3	0.752	0.565504
4	0.751	0.564001
5	0.760	0.577600
	$\Sigma\, x_i = 3.771$	$\Sigma\, x_i^2 = 2.844145$

$$\bar{x} = \frac{\Sigma\, x_i}{N} = \frac{3.771}{5} = 0.7542 \approx 0.754 \text{ ppb Pb}$$

$$\frac{(\Sigma\, x_i)^2}{N} = \frac{(3.771)^2}{5} = \frac{14.220441}{5} = 2.8440882$$

Substituting into Equation 6-4 leads to

$$s = \sqrt{\frac{2.844145 - 2.8440882}{5 - 1}} = \sqrt{\frac{0.0000568}{4}} = 0.00377 \approx 0.004 \text{ ppm Pb}$$

Note in Example 6-1 that the difference between $\Sigma\, x_i^2$ and $(\Sigma\, x_i)^2/N$ is very small. If we had rounded these numbers before subtracting them, a serious error would have appeared in the computed value of s. To avoid this source of error, *never round a standard deviation calculation until the very end.* Furthermore, and for the same reason, never use Equation 6-4 to calculate the standard deviation of numbers containing five or more digits. Use Equation 6-3 instead.[1] Handheld calculators and small computers with a standard deviation function usually employ a version of Equation 6-4. Thus, expect large errors in s when these devices are used to calculate the standard deviation of a data set that has five or more significant figures.

> Any time we subtract two large, approximately equal numbers, the difference will usually have a relatively large uncertainty.

When you make statistical calculations, remember that because of the uncertainty in $\bar{x}$, a sample standard deviation may differ significantly from the population standard deviation.

> When $N \to \infty$, $\bar{x} \to \mu$ and $s \to \sigma$.

What Is the Standard Error of the Mean?

The figures on percentage distribution just quoted refer to the probable error for a *single* measurement. If a series of replicate samples, each containing N measurements, is taken randomly from a population of data, the mean of each set will show less and less scatter as N increases. The standard deviation of each mean is known as the *standard error* of the mean and is given the symbol s_m. It

---

[1] In most cases, the first two or three digits in a set of data are identical to each other. Thus, as an alternative to using Equation 6-3, these identical digits can be dropped and the remaining digits used with Equation 6-4. For example, the standard deviation for the data in Example 6-1 could be based on 0.052, 0.056, 0.052, and so forth (or even 52, 56, 52, etc.).

The *standard error* of a mean is the standard deviation of a set of data divided by the square root of the number of data points in the set.

can be shown that the standard error of the mean is inversely proportional to the square root of the number of data points N used to calculate the mean:

$$s_m = \frac{s}{\sqrt{N}} \tag{6-5}$$

Spreadsheet Exercise

Computing the Standard Deviation

Microsoft
Excel

In this example, we will calculate the standard deviation, the variance, and the relative standard deviation of two sets of data. We begin with the spreadsheet and data from the Spreadsheet Exercise in Chapter 5. The standard deviation s is given by the equation

$$s = \sqrt{\frac{\sum_{i=1}^{N} (x_i - \bar{x})^2}{N - 1}}$$

and the variance is s^2.

Finding the Variance

If you are continuing the Spreadsheet Exercise from Chapter 5, begin with the data on your computer screen. Otherwise, retrieve the file `average.xls` from your floppy disk by clicking on File/Open. Make cell D1 the active cell, and type

```
Deviation^2[↵]
```

Cell D2 should now be the active cell, and your worksheet should appear as follows.

	A	B	C	D	E
1		Data	Deviation	Deviation^2	
2		19.4	0.383333		
3		19.5	0.283333		
4		19.6	0.183333		
5		19.8	0.016667		
6		20.1	0.316667		
7		20.3	0.516667		
8					
9					
10					
11	Total	118.7			
12	N	6			
13	Mean	19.78333	0.283333		

Now type

```
=C2^2[↵]
```

and the square of the deviation in cell C2 appears in D2. Copy this formula into the other cells in column D by once again clicking on cell D2, clicking on the fill handle, and dragging the fill handle through cell D7. You have now calculated the squares of the deviations of each of the data from the mean value in cell B13.

A Shortcut for Performing a Summation

To find the variance, we must find the sum of the squares of the deviations, so now click on cell D11 and then click on the AutoSum icon shown in the margin.

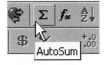

	SUM	▼	✕ ✔ =	=SUM(D2:D10)	
	A	B	C	D	E
1		Data	Deviation	Deviation^2	
2		19.4	0.383333	0.146944	
3		19.5	0.283333	0.080278	
4		19.6	0.183333	0.033611	
5		19.8	0.016667	0.000278	
6		20.1	0.316667	0.100278	
7		20.3	0.516667	0.266944	
8					
9					
10					
11	Total	118.7		=SUM(D2:D10)	
12	N	6			
13	Mean	19.78333	0.283333		

The dashed box shown in the preceding figure now surrounds the column of data in cells D2 through D10, which appear as arguments of the SUM function in cell D11 and in the formula bar. Note that Excel assumes that you want to add all the numerical data above the active cell and automatically completes the formula. When you type [↵], the sum of the squares of the deviations appears in cell D11. Since cells D8 through D10 are blank, they contribute zero to the sum, and so there is no harm in leaving the references to D8 through D10 in the formula. Be aware, however, that references to blank cells could pose difficulty under certain circumstances.

The final step in calculating the variance is to divide the sum of the squares of the deviations by the number of degrees of freedom, which is $N - 1$. We shall type the formula for carrying out this last calculation in cell D12. Before proceeding, type the label Variance in F12. Now click on cell D12, and type

$$=D11/(B12-1) \ [↵]$$

The variance is calculated and appears in the cell. Notice that you must enclose the difference $B12 - 1$ in parentheses so that Excel computes the number of degrees of freedom before the division is carried out. If we had not enclosed the number of degrees of freedom, $B12 - 1$, in parentheses, Excel would have divided D11 by B12 and then subtracted 1, which is incorrect. To illustrate this

Order of Operations

Order	Operator	Description
1	−	Negation
2	%	Percent
3	^	Exponentiation
4	* and /	Multiplication and division
5	+ and −	Addition and subtraction
6	=, <, >, <=, >=, <>	Comparison

point, suppose $D11 = 12$ and $B12 = 3$. If we leave off the parentheses, $D11/B12 - 1 = 3$, but if we put them in, $D11/(B12 - 1) = 6$. The order of mathematical operations in Excel is extremely important. Remember that just as in algebra, Excel performs exponentiation before multiplication and division, and it performs multiplication and division before addition and subtraction. We can change the order of operations by properly placing parentheses. The order that Excel uses in evaluating various mathematical and logical operations is shown in the margin.

Finding the Standard Deviation

Our next step is to calculate the standard deviation by extracting the square root of the variance. Click on D13, and type

$$\texttt{=SQRT(D12) [↵]}$$

Then click on F13, and type

$$\texttt{Standard Deviation [↵]}$$

Your worksheet should then look like the following.

	A	B	C	D	E	F	G
1		Data	Deviation	Deviation^2			
2		19.4	0.383333	0.146944			
3		19.5	0.283333	0.080278			
4		19.6	0.183333	0.033611			
5		19.8	0.016667	0.000278			
6		20.1	0.316667	0.100278			
7		20.3	0.516667	0.266944			
8							
9							
10							
11	Total	118.7		0.628333			
12	N	6		0.125667		Variance	
13	Mean	19.78333	0.283333	0.354495		Standard Deviation	
14							

Notice that we have deliberately left cells E12 and E13 blank. We will now use the built-in variance and standard deviation functions of Excel to check our formulas.

The Built-in Statistical Functions of Excel

Click on cell E12, and then type

$$\texttt{=VAR(}$$

Now click in cell B2, and drag the mouse into cell B7 so that the worksheet appears as follows.

	A	B	C	D	E	F	G
1		Data	Deviation	Deviation^2			
2		19.4	0.383333	0.146944			
3		19.5	0.283333	0.080278			
4		19.6	0.183333	0.033611			
5		19.8	0.016667	0.000278			
6		20.1	0.316667	0.100278			
7		20.3	0.516667	0.266944			
8							
9							
10							
11	Total	118.7		0.628333			
12	N	6		1.26E-01	=VAR(B2:B7)		
13	Mean	19.78333	0.283333	0.35449	0.35449	Standard Deviation	

Notice that the cell references B2:B7 appear in cell E12 and in the formula bar. Now, let up on the mouse button, and type [↵]; the variance appears in cell E12. If you have performed these operations correctly, the values displayed in cells D12 and E12 are identical.

The active cell should now be cell E13. If it is not, click on it, and type

$$\text{STDEV(}$$

and click and drag to highlight cells B2:B7 as you did previously. Let up on the mouse button, type [↵], and the standard deviation appears in cell E13. The computed values in cells D13 and E13 should be equal. Note that the Excel STDEV and VAR functions calculate the *sample standard deviation* and the *sample variance,* not the corresponding population statistics. These built-in functions are quite convenient since your sample will always be sufficiently small that you will want to calculate sample statistics rather than population statistics.

Up to this point, we have paid little attention to the number of decimal places displayed in the cells. To control the number of decimal places in a cell or range of cells, highlight the target cell(s), and click on the Increase Decimal button shown in the margin. Highlight D13:E13 now, and try it. Then click on the Decrease Decimal icon to reverse the process. Excel has no idea how many significant figures to display in a cell; you must control this aspect yourself. Now decrease the number of decimal places until only one significant figure is displayed. Note that Excel conveniently rounds the data for us.

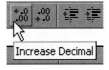

The Coefficient of Variation, or Percent Relative Standard Deviation

Our final goal for this exercise is to calculate the coefficient of variation (CV), or percent relative standard deviation (%RSD) (See Section 6B-5 for an explanation of this term). As shown by Equation 6-7, %RSD is given by

$$\text{CV} = \frac{s}{\bar{x}} \times 100\%$$

Click in E14, and type

=E13*100/B13[↵]

Then click in F13, and type the label CV, %[↵]. Your worksheet should now look something like the one shown below. Note that we have multiplied the ratio of cells E13 to B13 by 100 so that the relative standard deviation is expressed as a percentage. Move the decimal point to indicate only significant figures in the CV.

	A	B	C	D	E	F	G
1		Data	Deviation	Deviation^2			
2		19.4	0.383333	0.146944			
3		19.5	0.283333	0.080278			
4		19.6	0.183333	0.033611			
5		19.8	0.016667	0.000278			
6		20.1	0.316667	0.100278			
7		20.3	0.516667	0.266944			
8							
9							
10							
11	Total	118.7		0.628333			
12	N	6		1.26E-01	0.125667	Variance	
13	Mean	19.78333	0.283333	0.35449	0.35449	Standard Deviation	
14					1.791887	CV, %	
15							

We have now constructed a general-purpose spreadsheet that you may use to make basic statistical calculations. To complete this part of the example, select a convenient location, construct a formula to display the number of degrees of freedom, and then add a label in an adjacent cell to identify this important variable. Save the file for future use in problems and laboratory calculations. Use the spreadsheet now to check the calculations in Example 6-1. To clear the data from your worksheet, just click and drag to highlight cells B2:B7, and then strike [Delete]. Alternatively, you may simply click on B2 and begin typing the data. Terminate each piece of data with [↵]. Be sure to delete the data in cells B7:D7.

As a final exercise, retrieve the spreadsheet that we created in Chapter 5 for the gravimetric determination of chloride, which we called grav_chloride.xls. Enter formulas into cells B12 through B14 to compute the mean, standard deviation, and the RSD in parts per thousand of the percent chloride in the samples. In this example, multiply the relative standard deviation by 1000 in cell B14. Adjust the decimal point in the results to display the proper number of significant figures. The following worksheet shows the results. Save your worksheet so that you can use it as a model for making laboratory calculations.

	A	B	C	D
1	Gravimetric Determination of Chloride			
2	Samples	1	2	3
3	Mass of bottle plus sample, g	27.6115	27.2185	26.8105
4	Mass of bottle less sample, g	27.2185	26.8105	26.4517
5	Mass of sample, g	0.3930	0.4080	0.3588
6				
7	Crucible masses, with AgCl, g	21.4296	23.4915	21.8323
8	Crucible masses, empty, g	20.7926	22.8311	21.2483
9	Mass of AgCl, g	0.6370	0.6604	0.5840
10				
11	%Chloride	40.0947	40.0393	40.2625
12	Mean % Chloride	40.1322		
13	Standard Deviation, % Chloride	0.12		
14	RSD, parts per thousand	2.90		
15				

6B-4   The Reliability of *s* as a Measure of Precision

Most of the statistical tests described in Chapter 7 are based on sample standard deviations, and the probability of correctness of the results of these tests improves as the reliability of *s* becomes greater. Uncertainty in the calculated value of *s* decreases as *N* in Equation 6-4 increases. When *N* is greater than about 20, *s* and σ can be assumed to be identical for all practical purposes. For example, if the 50 measurements in Table 6-2 are divided into ten subgroups of five each, the value of *s* varies widely from one subgroup to another (0.0023 to 0.0079 mL) even though the average of the computed values of *s* is that of the entire set (0.0056 mL). In contrast, the computed values of *s* for two subsets of 25 each are nearly identical (0.0054 and 0.0058 mL).

CHALLENGE: Show that these statements are correct.

When $N > 20$, $s \approx \sigma$.

The rapid improvement in the reliability of *s* with increases in *N* makes it feasible to obtain a good approximation of σ when the method of measurement is not excessively time consuming and when an adequate supply of sample is available. For example, if the pH of numerous solutions is to be measured in the course of an investigation, it is useful to evaluate *s* in a series of preliminary experiments. This measurement is simple, requiring only that a pair of rinsed and dried electrodes be immersed in the test solution and the pH read from a scale or a display. To determine *s*, 20 to 30 portions of a buffer solution of fixed pH can be measured with all steps of the procedure being followed exactly. Normally, it is safe to assume that the random error in this test is the same as that in subsequent measurements. The value of *s* calculated from Equation 6-4 is a valid and accurate measure of the theoretical σ.

Pooling Data to Improve the Reliability of *s*

For analyses that are time consuming, the foregoing procedure is seldom practical. In this situation, data from a series of similar samples accumulated over time can often be pooled to provide an estimate of *s* that is superior to the value for

Feature 6-4

Equation for Calculating
Pooled Standard Deviations

The equation for computing a pooled standard deviation from several sets of data takes the form

$$s_{pooled} = \sqrt{\frac{\sum_{i=1}^{N_1} (x_i - \bar{x}_1)^2 + \sum_{j=1}^{N_2} (x_j - \bar{x}_2)^2 + \sum_{k=1}^{N_3} (x_k - \bar{x}_3)^2 + \cdots}{N_1 + N_2 + N_3 + \cdots - N_t}}$$

where N_1 is the number of results in set 1, N_2 is the number in set 2, and so forth. The term N_t is the number of data sets that are pooled.

any individual subset. Again, we must assume the same sources of random error in all the measurements. This assumption is usually valid if the samples have similar compositions and have been analyzed in exactly the same way.

To obtain a pooled estimate of the standard deviation, s_{pooled}, deviations from the mean for each subset are squared; the squares of all subsets are then summed and divided by an appropriate number of degrees of freedom. The pooled s is obtained by extracting the square root of the quotient. One degree of freedom is lost for each subset. Thus, the number of degrees of freedom for the pooled s is equal to the total number of measurements minus the number of subsets. Feature 6-4 provides a formal equation for obtaining s_{pooled} for t data sets. Example 6-2 illustrates the application of this type of computation.

Example 6-2

The mercury in samples of seven fish taken from Chesapeake Bay was determined by a method based on the absorption of radiation by gaseous elemental mercury. Calculate a pooled estimate of the standard deviation for the method, based on the first three columns of data:

Specimen	Number of Samples Measured	Hg Content, ppm	Mean, ppm Hg	Sum of Squares of Deviations from Mean
1	3	1.80, 1.58, 1.64	1.673	0.0258
2	4	0.96, 0.98, 1.02, 1.10	1.015	0.0115
3	2	3.13, 3.35	3.240	0.0242
4	6	2.06, 1.93, 2.12, 2.16, 1.89, 1.95	2.018	0.0611
5	4	0.57, 0.58, 0.64, 0.49	0.570	0.0114
6	5	2.35, 2.44, 2.70, 2.48, 2.44	2.482	0.0685
7	4	1.11, 1.15, 1.22, 1.04	1.130	0.0170
	Total = 28		Sum of squares =	0.2196

The values in the last two columns for specimen 1 were computed as follows:

x_i	$|(x_i - \bar{x})|$	$(x_i - \bar{x})^2$
1.80	0.127	0.0161
1.58	0.093	0.0086
1.64	0.033	0.0011
5.02	Sum of squares =	0.0258

$$\bar{x} = \frac{5.02}{3} = 1.673$$

The other data in columns 4 and 5 were obtained similarly. Then

s_{pooled}

$$= \sqrt{\frac{0.0258 + 0.0115 + 0.0242 + 0.0611 + 0.0114 + 0.0685 + 0.0170}{28 - 7}}$$

$$= 0.10 \text{ ppm Hg}$$

Note that one degree of freedom is lost for each of the seven samples. Because more than 20 degrees of freedom remain, however, the computed value of s can be considered a good approximation of σ.

Computing the Pooled Standard Deviation

Spreadsheet Exercise

In this spreadsheet exercise, we calculate the pooled standard deviation of the seven data sets of Example 6-2. Let us begin with a clean spreadsheet. Enter the data and column and row headings as shown below.

Microsoft
Excel

	A	B	C	D	E	F	G	H	I	J
1		1	2	3	4	5	6	7		
2	x1	1.80	0.96	3.13	2.06	0.57	2.35	1.11		
3	x2	1.58	0.98	3.35	1.93	0.58	2.44	1.15		
4	x3	1.64	1.02		2.12	0.64	2.70	1.22		
5	x4		1.10		2.16	0.49	2.48	1.04		
6	x5				1.89		2.44			
7	x6				1.95					
8	s^2									
9	N									No. Sets
10	DF									Tot. DF
11	Dev^2									Tot. Dev^2
12								s(pool)		
13										

A Little Mathematical Trickery

As shown in Example 6-2, it is necessary to compute the sum of the squares of the deviations from the mean for each of the seven data sets. Excel does not have a built-in function for the pooled standard deviation, nor does it have a function for computing the sum of the squares of the deviations. We can take advantage of the VARIANCE function, however, if we solve the equation for variance for the sum of the squares of the deviations as follows.

$$s^2 = \frac{\sum\limits_{i=1}^{N} (x_i - \bar{x})^2}{N - 1}$$

$$\sum\limits_{i=1}^{N} (x_i - \bar{x})^2 = s^2(N - 1)$$

So, for each data set, we will compute the variance and multiply it times the number of degrees of freedom $N - 1$, which we can find using the COUNT function. We begin with data set 1.

Click on cell B8, and type

```
=VAR(B2:B7) [↵]
```

Alternatively, you may use the mouse to select cells B2:B7 as illustrated in our previous spreadsheet exercise. Now, in cell B9 type

```
=COUNT(B2:B7) [↵]
=B9-1 [↵]
=B8*B10 [↵]
```

The first two columns of your worksheet should look like those shown below.

	A	B
1		1
2	x1	1.80
3	x2	1.58
4	x3	1.64
5	x4	
6	x5	
7	x6	
8	s^2	0.012933
9	N	3
10	DF	2
11	Dev^2	0.025867
12		

Now, click and drag from cell B8 to cell B11 to highlight those cells. Click on the fill handle, and drag to the right to copy the formulas into columns C through H. When you let up on the mouse button, your worksheet should appear similar to the one at the top of the next page.

	A	B	C	D	E	F	G	H	I	J
1		1	2	3	4	5	6	7		
2	x1	1.80	0.96	3.13	2.06	0.57	2.35	1.11		
3	x2	1.58	0.98	3.35	1.93	0.58	2.44	1.15		
4	x3	1.64	1.02		2.12	0.64	2.70	1.22		
5	x4		1.10		2.16	0.49	2.48	1.04		
6	x5				1.89		2.44			
7	x6				1.95					
8	s^2	0.012933	0.003833	0.0242	0.012217	0.0038	0.01712	0.005667		
9	N	3	4	2	6	4	5	4		No. Sets
10	DF	2	3	1	5	3	4	3		Tot. DF
11	Dev^2	0.025867	0.0115	0.0242	0.061083	0.0114	0.06848	0.017		Tot. Dev^2
12							s(pool)			

Note that the sums of the squares of the deviations (Dev^2) from each of the seven means are displayed in row 11.

Finding the Sum of Squares for All the Data

To find the total sum of squares, click on cell I11, and then click on the AutoSum icon to produce the following on the screen.

	A	B	C	D	E	F	G	H	I	J
8	s^2	0.012933	0.003833	0.0242	0.012217	0.0038	0.01712	0.005667		
9	N	3	4	2	6	4	5	4		No. Sets
10	DF	2	3	1	5	3	4	3		Tot. DF
11	Dev^2	0.025867	0.0115	0.0242	0.061083	0.0114	0.06848	0.017	=SUM(B11:H11)	
12							s(pool)			

Notice that Excel correctly guesses that you want to find the sum of the data in cells B11:H11, so if you type [↵], the total sum of squares appears in cell I11. Now click on I11, and drag the fill handle up to copy the sum formula into cell I10. When you let up on the mouse button, the total number of degrees of freedom appears in cell I10. You have now made all the intermediate calculations necessary to compute the pooled standard deviation. Now click on cell I12, and type

$$\texttt{=SQRT(I11/I10)[↵]}$$

and the pooled standard deviation appears in the cell. For completeness, add a formula to cell I9 to find the number of data sets in the analysis. This is a good time to save your worksheet to a floppy disk under the name `pooled_std_dev.xls`. The completed worksheet follows.

	A	B	C	D	E	F	G	H	I	J
1		1	2	3	4	5	6	7		
2	x1	1.80	0.96	3.13	2.06	0.57	2.35	1.11		
3	x2	1.58	0.98	3.35	1.93	0.58	2.44	1.15		
4	x3	1.64	1.02		2.12	0.64	2.70	1.22		
5	x4		1.10		2.16	0.49	2.48	1.04		
6	x5				1.89		2.44			
7	x6				1.95					
8	s^2	0.012933	0.003833	0.0242	0.012217	0.0038	0.01712	0.005667		
9	N	3	4	2	6	4	5	4	7	No. Sets
10	DF	2	3	1	5	3	4	3	21	Tot. DF
11	Dev^2	0.025867	0.0115	0.0242	0.061083	0.0114	0.06848	0.017	0.21953	Tot. Dev^2
12								s(pool)	0.102244	
13										

In this exercise, we have constructed a worksheet to compute the pooled standard deviation of several sets of data. As extensions of this example, you may use the worksheet to solve some of the pooled standard deviation problems at the end of the chapter. You may also wish to expand the worksheet to accommodate larger sets and larger numbers of sets of data.

6B-5   Alternative Terms for Expressing the Precision of Samples of Data

Chemists ordinarily employ sample standard deviation in reporting the precision of their data. Three other terms are often encountered, however.

Variance (s^2)

The variance is equal to the square of the standard deviation.

The variance is simply the square of the standard deviation:

$$s^2 = \frac{\sum_{i=1}^{N} (x_i - \bar{x})^2}{N - 1} = \frac{\sum_{i=1}^{N} (d_i)^2}{N - 1} \tag{6-6}$$

Note that the standard deviation has the same units as the data, whereas the variance has the units of the data squared. People who do scientific work tend to use standard deviation rather than variance as a measure of precision. It is easier to relate a measurement and its precision if they both have the same units. The advantage of using variance is that variances are additive, as we will see later in this chapter.

Relative Standard Deviation (RSD) and Coefficient of Variation (CV)

Chemists frequently quote standard deviations in relative rather than absolute terms. We calculate the relative standard deviation by dividing the standard deviation by the mean of the set of data. It is often expressed in parts per thousand (ppt) or in percent by multiplying this ratio by 1000 ppt or by 100%. For example,

$$RSD = \frac{s}{\bar{x}} \times 1000 \text{ ppt}$$

The relative standard deviation multiplied by 100% is called the *coefficient of variation* (CV).

$$CV = \frac{s}{\bar{x}} \times 100\% \qquad (6\text{-}7)$$

The *coefficient of variation* is the percent relative standard deviation.

Relative standard deviations often give a clearer picture of data quality than do absolute standard deviations. As an example, suppose that a sample contains about 50 mg of copper and that the standard deviation of a copper determination is 2 mg. The CV for this sample is 4%. For a sample containing only 10 mg, the CV is 20%.

Spread or Range (w).   The *spread,* or *range,* is another term that is sometimes used to describe the precision of a set of replicate results. It is the difference between the largest value in the set and the smallest. Thus, the spread of the data in Figure 5-1 is $(20.3 - 19.4) = 0.9$ ppm Fe.

Example 6-3

For the set of data in Example 6-1, calculate (a) the variance, (b) the relative standard deviation in parts per thousand, (c) the coefficient of variation, and (d) the spread.

In Example 6-1, we found

$$\bar{x} = 0.754 \text{ ppm Pb} \qquad \text{and} \qquad s = 0.0038 \text{ ppm Pb}$$

(a) $s^2 = (0.0038)^2 = 1.4 \times 10^{-5}$

(b) $RSD = \dfrac{0.0038}{0.754} \times 1000 \text{ ppt} = 5.0 \text{ ppt}$

(c) $CV = \dfrac{0.0038}{0.754} \times 100\% = 0.50\%$

(d) $w = 0.760 - 0.751 = 0.009$

6C THE STANDARD DEVIATION OF COMPUTED RESULTS

We often need to estimate the standard deviation of a result that has been computed from two or more experimental data points, each of which has a known sample standard deviation. As shown in Table 6-4, the way such estimates are made depends on the type of arithmetic that is involved.

Table 6-4

Error Propagation in Arithmetic Calculations

Type of Calculation	Example*	Standard Deviation of y	
Addition or subtraction	$y = a + b - c$	$s_y = \sqrt{s_a^2 + s_b^2 + s_c^2}$	(1)
Multiplication or division	$y = a \cdot b/c$	$\dfrac{s_y}{y} = \sqrt{\left(\dfrac{s_a}{a}\right)^2 + \left(\dfrac{s_b}{b}\right)^2 + \left(\dfrac{s_c}{c}\right)^2}$	(2)
Exponentiation	$y = a^x$	$\dfrac{s_y}{y} = x\dfrac{s_a}{a}$	(3)
Logarithm	$y = \log_{10} a$	$s_y = 0.434\dfrac{s_a}{a}$	(4)
Antilogarithm	$y = \text{antilog}_{10}\, a$	$\dfrac{s_y}{y} = 2.303 s_a$	(5)

*a, b, and c are experimental variables whose standard deviations are s_a, s_b, and s_c, respectively.

6C-1   The Standard Deviation of Sums and Differences

Consider the summation

$$
\begin{array}{ll}
+0.50 & (\pm 0.02) \\
+4.10 & (\pm 0.03) \\
-1.97 & (\pm 0.05) \\
\hline
2.63 &
\end{array}
$$

where the numbers in parentheses are absolute standard deviations. If the three individual standard deviations happen by chance to have the same sign, the standard deviation of the sum could be as large as $+0.02 + 0.03 + 0.05 = +0.10$ or $-0.02 - 0.03 - 0.05 = -0.10$. On the other hand, it is possible that the three standard deviations could combine to give an accumulated value of zero: $-0.02 - 0.03 + 0.05 = 0$ or $+0.02 + 0.03 - 0.05 = 0$. More likely, however, the standard deviation of the sum will lie between these two extremes. It can be shown by propagation of error mathematics that the variance of a sum or difference is equal to the sum of the individual variances. Thus, the most probable value for a standard deviation of a sum or difference can be found by taking the square root of the sum of the squares of the individual absolute standard deviations. So, for the computation

$$y = a(\pm s_a) + b(\pm s_b) - c(\pm s_c)$$

The variance of y, s_y^2, is given by

> The variance of a sum or difference is equal to the *sum* of the variances of the numbers making up that sum or difference.

$$s_y^2 = s_a^2 + s_b^2 + s_c^2$$

Hence, the standard deviation of the result s_y is given by

$$s_y = \sqrt{s_a^2 + s_b^2 + s_c^2} \tag{6-8}$$

where s_a, s_b, and s_c are the standard deviations of the three terms making up the result. Substituting the standard deviations from the example gives

> For a sum or a difference, the *absolute standard deviation of the answer* is the square root of the sum of the squares of the *absolute standard deviations* of the numbers used to calculate the sum or difference.

$$s_y = \sqrt{(\pm 0.02)^2 + (\pm 0.03)^2 + (\pm 0.05)^2} = \pm 0.06$$

and the sum should be reported as 2.63 (± 0.06).

6C-2   The Standard Deviation of Products and Quotients

Consider the following computation, where the numbers in parentheses are again absolute standard deviations:

$$\frac{4.10(\pm 0.02) \times 0.0050(\pm 0.0001)}{1.97(\pm 0.04)} = 0.010406(\pm ?)$$

In this situation, the standard deviations of two of the numbers in the calculation are larger than the result itself. Evidently, we need a different approach for multiplication and division. As shown in Table 6-4, the *relative standard deviation* of a product or quotient is determined by the *relative standard deviations* of the numbers forming the computed result. For example, in the case of

$$y = \frac{a \times b}{c} \tag{6-9}$$

we obtain the relative standard deviation s_y/y of the result y by summing the squares of the relative standard deviations of a, b, and c and extracting the square root of the sum:

$$\frac{s_y}{y} = \sqrt{\left(\frac{s_a}{a}\right)^2 + \left(\frac{s_b}{b}\right)^2 + \left(\frac{s_c}{c}\right)^2} \tag{6-10}$$

> For multiplication or division, the *relative standard deviation of the answer* is the square root of the sum of the squares of the *relative standard deviations* of the numbers that are multiplied or divided.

Applying this equation to the numerical example gives

$$\frac{s_y}{y} = \sqrt{\left(\frac{\pm 0.02}{4.10}\right)^2 + \left(\frac{\pm 0.0001}{0.005}\right)^2 + \left(\frac{\pm 0.04}{1.97}\right)^2}$$

$$= \sqrt{(0.0049)^2 + (0.0200)^2 + (0.0203)^2} = \pm 0.0289$$

To complete the calculation, we must find the absolute standard deviation of the result,

> To find the absolute standard deviation in a product or a quotient, first find the relative standard deviation in the result and then multiply it by the result.

$$s_y = y \times (\pm 0.0289) = 0.0104 \times (\pm 0.0289) = \pm 0.000301$$

and we can write the answer and its uncertainty as $0.0104(\pm 0.0003)$.

Example 6-4 demonstrates the calculation of the standard deviation of the result for a more complex calculation.

Example 6-4

Calculate the standard deviation of the result of

$$\frac{[14.3(\pm 0.2) - 11.6(\pm 0.2)] \times 0.050(\pm 0.001)}{[820(\pm 10) + 1030(\pm 5)] \times 42.3(\pm 0.4)} = 1.725(\pm ?) \times 10^{-6}$$

First, we must calculate the standard deviation of the sum and the difference. For the difference in the numerator,

$$s_a = \sqrt{(\pm 0.2)^2 + (\pm 0.2)^2} = \pm 0.283$$

and for the sum in the denominator,

$$s_b = \sqrt{(\pm 10)^2 + (\pm 5)^2} = 11.2$$

Figure 6-4 Buret section showing the liquid level and meniscus.

We may then rewrite the equation as

$$\frac{2.7(\pm 0.283) \times 0.050(\pm 0.001)}{1850(\pm 11.2) \times 42.3(\pm 0.4)} = 1.725 \times 10^{-6}$$

The equation now contains only products and quotients, and Equation 6-10 applies. Thus,

$$\frac{s_y}{y} = \sqrt{\left(\pm \frac{0.283}{2.7}\right)^2 + \left(\pm \frac{0.001}{0.050}\right)^2 + \left(\pm \frac{11.2}{1850}\right)^2 + \left(\pm \frac{0.4}{42.3}\right)^2} = 0.107$$

To obtain the absolute standard deviation, we write

$$s_y = y \times 0.107 = 1.725 \times 10^{-6} \times (\pm 0.107) = \pm 0.185 \times 10^{-6}$$

and round the answer to $1.7(\pm 0.2) \times 10^{-6}$.

6D   REPORTING COMPUTED DATA

A numerical result is worthless unless the user knows something about its accuracy and/or precision. Therefore, it is always essential for you to indicate your best estimate of the reliability of your data. One of the best ways of indicating reliability is to give a confidence interval at the 90% or 95% confidence level as we describe in Section 7A-2. Another method is to report the absolute standard deviation or the coefficient of variation of the data. In this case, it is a good idea to indicate the number of data points that were used to obtain the standard deviation so that the user has some idea of the probable reliability of s. A less satisfactory but more common indicator of the quality of data is the *significant figure convention.*

6D-1   The Significant Figure Convention

> The *significant figures* in a number are all the certain digits plus the first uncertain digit.

A simple way of indicating the probable uncertainty associated with an experimental measurement is to round the result so that it contains only *significant figures.* By definition, the significant figures in a number are all the certain digits *and the first uncertain digit.* For example, when you read the 50-mL buret section shown in Figure 6-4, you can easily tell that the liquid level is greater than 30.2 mL and less than 30.3 mL. You can also estimate the position of the liquid between the graduations to about ± 0.02 mL. So, using the significant figure convention, you should report the volume delivered as 30.24 mL, which is four significant figures. Note that the first three digits are certain and that the last digit (4) is uncertain.

A zero may or may not be significant depending on its location in a number. A zero that is surrounded by other digits is always significant (such as in 30.24 mL) because it is read directly and with certainty from a scale or instrument readout. On the other hand, zeros that only locate the decimal point for us are not. If we write 30.24 mL as 0.03024 L, the number of significant figures is the same. The only function of the zero before the 3 is to locate the decimal point, so it is not significant. Terminal, or final, zeros may or may not be significant.

> Express data in scientific notation to avoid confusion in determining whether terminal zeros are significant.

For example, if the volume of a beaker is expressed as 2.0 L, the presence of the zero tells us that the volume is known to a few tenths of a liter, so both the 2 and the zero are significant figures. If this same volume is reported as 2000 mL, the situation becomes confusing. The last two zeros are not significant because the uncertainty is still a few tenths of a liter or a few hundred milliliters. To follow the significant figure convention in such a case, use scientific notation and report the volume as 2.0×10^2 mL.

Rules for determining the number of significant figures:

1. Disregard all initial zeros.
2. Disregard all final zeros *unless they follow a decimal point.*
3. All remaining digits including zeros between nonzero digits are significant.

6D-2   Significant Figures in Numerical Computations

Care is required to determine the appropriate number of significant figures in the result of an arithmetic combination of two or more numbers.[2]

Sums and Differences

For addition and subtraction, the number of significant figures can be found by visual inspection. For example, in the expression

$$3.4 + 0.020 + 7.31 = 10.73 = 10.7$$

the second and third decimal places in the answer cannot be significant because 3.4 is uncertain in the first decimal place. Note that the result contains three significant digits even though two of the numbers involved have only two significant figures.

You have probably heard it said that a chain is only as strong as its weakest link. For addition and subtraction, the weak link is the number of decimal places in the number with the *smallest* number of decimal places.

Products and Quotients

A rule of thumb sometimes suggested for multiplication and division is that the answer should be rounded so that it contains the same number of significant digits as the original number with the smallest number of significant digits. Unfortunately, this procedure often leads to incorrect rounding. For example, consider the two calculations

$$\frac{24 \times 4.52}{100.0} = 1.08 \quad \text{and} \quad \frac{24 \times 4.02}{100.0} = 0.965$$

By this rule, the first answer would be rounded to 1.1 and the second to 0.96. If we assume a unit uncertainty in the last digit of each number in the first quotient, however, the relative uncertainties associated with each of these numbers are $\frac{1}{24}, \frac{1}{452}$, and $\frac{1}{1000}$. Because the first relative uncertainty is much larger than the other two, the relative uncertainty in the result is also $\frac{1}{24}$; the absolute uncertainty is then

$$1.08 \times \frac{1}{24} = 0.045 = 0.04$$

By the same argument, the absolute uncertainty of the second answer is given by

$$0.965 \times \frac{1}{24} = 0.040 = 0.04$$

When adding and subtracting numbers in scientific notation, express the numbers to the same power of ten. For example,

$$
\begin{aligned}
2.432 \times 10^6 &= 2.432 \times 10^6 \\
+6.512 \times 10^4 &= +0.06512 \times 10^6 \\
-1.227 \times 10^5 &= -0.1227 \times 10^6 \\
2.37442 \times 10^6 &= 2.374 \times 10^6
\end{aligned}
$$

[2]For an extensive discussion of propagation of significant figures, see L. M. Schwartz, *J. Chem. Educ.*, **1985**, *62*, 693.

The weak link for multiplication and division is the number of *significant figures* in the number with the smallest number of significant figures. *Use this rule of thumb with caution.*

Therefore, the first result should be rounded to three significant figures or 1.08, but the second should be rounded to only two; that is 0.96.

Logarithms and Antilogarithms

Be especially careful in rounding the results of calculations involving logarithms. The following rules apply to most situations:[3]

1. In a logarithm of a number, keep as many digits to the right of the decimal point as there are significant figures in the original number.
2. In an antilogarithm of a number, keep as many digits as there are digits to the right of the decimal point in the original number.

The number of significant figures in the *mantissa,* or the digits to the right of the decimal point of a logarithm, is the same as the number of significant figures in the original number. Thus,

$$\log(9.57 \times 10^4) = 4.981$$

Example 6-5

Round the following answers so that only significant digits are retained: (a) log $4.000 \times 10^{-5} = -4.3979400$ and (b) antilog $12.5 = 3.162277 \times 10^{12}$.

(a) Following rule 1, we retain 4 digits to the right of the decimal point

$$\log 4.000 \times 10^{-5} = -4.3979$$

(b) Following rule 2, we may retain only 1 digit

$$\text{antilog } 12.5 = 3 \times 10^{12}$$

6D-3   Rounding Data

Note that 0.635 rounds to 0.64 and 0.625 rounds to 0.62.

In rounding a number ending in 5, always round so that the result ends with an even number.

Always round the computed results of a chemical analysis in an appropriate way. For example, consider the replicate results: 61.60, 61.46, 61.55, and 61.61. The mean of these data is 61.555, and the standard deviation is 0.069. When we round the mean, do we take 61.55 or 61.56? A good guide to follow when rounding a 5 is always to round to the nearest even number. In this way, we eliminate any tendency to round in a set direction. In other words, there is an equal likelihood that the nearest even number will be the higher or the lower in any given situation. Accordingly, we might choose to report the result as 61.56 ± 0.07. If we had reason to doubt the reliability of the estimated standard deviation, we might report the result as 61.6 ± 0.1.

We should note that it is seldom justifiable to keep more than one significant figure in the standard deviation because the standard deviation contains error as well. For certain specialized purposes, it may be useful to keep two significant figures, and there is certainly nothing wrong with including a second digit in the standard deviation. It is important to recognize that the uncertainty lies in the first digit, however.[4]

---

[3]D. E. Jones, *J. Chem. Educ.,* **1971,** *49,* 753.

[4]For more details on this topic, direct your Web browser to http://www.chem.uky.edu/courses/ che226/download/CI_for_sigma.html.

6D-4   Rounding the Results from Chemical Computations

Throughout this text and others, you are asked to perform calculations with data whose precision is indicated only by the significant figure convention. In these circumstances, common sense assumptions must be made as to the uncertainty in each number. The uncertainty of the result is then estimated using the techniques presented in Section 6C. Finally, the result is rounded so that it contains only significant digits. *It is especially important to postpone rounding until the calculation is completed.* At least one extra digit beyond the significant digits should be carried through all the computations to avoid a *rounding error.* This extra digit is sometimes called a "guard" digit. Modern calculators generally retain several extra digits that are not significant, and the user must be careful to round final results properly so that only significant figures are included. Example 6-6 illustrates this procedure.

Example 6-6

A 3.4842-g sample of a solid mixture containing benzoic acid, C_6H_5COOH (122.123 g/mol), was dissolved and titrated with base to a phenolphthalein end point. The acid consumed 41.36 mL of 0.2328 M NaOH. Calculate the percent benzoic acid (HBz) in the sample.

As shown in Section 11C-3, the computation takes the following form:

$$\% \text{ HBz} = \frac{41.36 \text{ mL} \times 0.2328 \dfrac{\text{mmol NaOH}}{\text{mL NaOH}} \times \dfrac{1 \text{ mmol HBz}}{\text{mmol NaOH}} \times \dfrac{122.123 \text{ g HBz}}{1000 \text{ mmol HBz}}}{3.4842 \text{ g sample}} \times 100\%$$

$$= 33.749\%$$

Since all operations are either multiplication or division, the relative uncertainty of the answer is determined by the relative uncertainties of the experimental data. Let us estimate what these uncertainties are.

1. The position of the liquid level in a buret can be estimated to ± 0.02 mL (Figure 6-4). Initial and final readings must be made, however, so that the standard deviation of the volume will be

$$\sqrt{(0.02)^2 + (0.02)^2} = \pm 0.028 \text{ mL} \qquad \text{(Equation 6-8)}$$

The relative uncertainty is then

$$\frac{\pm 0.028}{41.36} \times 1000 \text{ ppt} = \pm 0.68 \text{ ppt}$$

2. Generally, the absolute uncertainty of a mass obtained with an analytical balance will be on the order of ± 0.0001 g. Thus, the relative uncertainty of the denominator is

$$\frac{0.0001}{3.4842} \times 1000 \text{ ppt} = 0.029 \text{ ppt}$$

3. Usually, we can assume that the absolute uncertainty molarity of a reagent solution is 0.0001, and so

$$\frac{0.0001}{0.2328} \times 1000 \text{ ppt} = 0.43 \text{ ppt}$$

4. The relative uncertainty in the molar mass of HBz is several orders of magnitude smaller than any of the three experimental data and is of no consequence. Note, however, that we should retain enough digits in the calculation so that the molar mass is given to at least one more digit (the guard digit) than any of the experimental data. Thus, in the calculation, we use 122.123 for the molar mass (here we are carrying two extra digits).
5. No uncertainty is associated with 100% and the 1000-mmol HBz since these are exact numbers.

Substituting the three relative uncertainties into Equation 6-10, we obtain

$$\frac{s_y}{y} = \sqrt{\left(\frac{0.028}{41.36}\right)^2 + \left(\frac{0.0001}{3.4842}\right)^2 + \left(\frac{0.0001}{0.2328}\right)^2}$$
$$= \sqrt{(0.68)^2 + (0.029)^2 + (0.43)^2}$$
$$= 8.02 \times 10^{-4}$$

$$s_y = 8.02 \times 10^{-4} \times y = 8.02 \times 10^{-4} \times 33.749 = 0.027$$

Thus, the uncertainty in the calculated result is $\pm 0.03\%$ HBz, and we should report the result as 33.75% HBz or, better, 33.75 (± 0.03)% HBz.

There is no relationship between the number of digits displayed on a calculator or a computer screen and the true number of significant figures.

It is important to emphasize that rounding decisions are an important part of *every calculation* and that such decisions *cannot* be based on the number of digits displayed on the readout of a calculator or on a computer screen.

WEB WORKS

The National Institute of Standards and Technology maintains web pages of statistical data for testing software. Direct your web browser to http://www.itl.nist.gov/div898/strd/index.html and browse the site to see what kinds of data are available for testing. We use two of the NIST data sets in Problems 6-13 and 6-14.

6E   QUESTIONS AND PROBLEMS

6-1. Define
   *(a) spread or range.
   (b) coefficient of variation.
   *(c) significant figures.
   (d) Gaussian distribution.

6-2. Differentiate between
   *(a) sample standard deviation and sample variance.
   (b) population mean and sample mean.
   *(c) accuracy and precision.
   (d) random and systematic error.

6-3. Distinguish between
   *(a) the meaning of the word "sample" as it is used in a chemical and in a statistical sense.
   (b) the sample standard deviation and the population standard deviation.

6-4. What is the standard error of a mean? Why is the standard deviation of the mean lower than the standard deviation of the data points in a set?

6-5. Consider the following sets of replicate measurements:

*A	B	*C	D	*E	F
3.5	70.24	0.812	2.7	70.65	0.514
3.1	70.22	0.792	3.0	70.63	0.503
3.1	70.10	0.794	2.6	70.64	0.486
3.3		0.900	2.8	70.21	0.497
2.5			3.2		0.472

For each set, calculate the

(a)   mean.

(b)   median.

(c)   spread or range.

(d)   standard deviation.

(e)   coefficient of variation.

6-6. The accepted values for the sets of data in Problem 6-5 are: *set A, 3.0; set B, 70.05; *set C, 0.830; set D, 3.4; *set E, 70.05; set F, 0.525. For the mean of each set, calculate the

(a)   absolute error.

(b)   relative error in parts per thousand.

6-7. Estimate the absolute deviation and the coefficient of variation for the results of the following calculations. Round each result so that it contains only significant digits. The numbers in parentheses are absolute standard deviations.

*(a) $y = 5.75(\pm0.03) + 0.833(\pm0.001)$
        $- 8.021(\pm0.001) = -1.4381$

(b) $y = 18.97(\pm0.04) + 0.0025(\pm0.0001)$
        $+ 2.29(\pm0.08) = 21.2625$

*(c) $y = 66.2(\pm0.3) \times 1.13(\pm0.02) \times 10^{-17}$
        $= 7.4806 \times 10^{-16}$

(d) $y = 251(\pm1) \times \dfrac{860(\pm2)}{1.673(\pm0.006)}$
        $= 129,025.70$

*(e) $y = \dfrac{157(\pm6) - 59(\pm3)}{1220(\pm1) + 77(\pm8)}$
        $= 7.5539 \times 10^{-2}$

(f) $y = \dfrac{1.97(\pm0.01)}{243(\pm3)} = 8.106996 \times 10^{-3}$

6-8. Estimate the absolute standard deviation and the coefficient of variation for the results of the following calculations. Round each result to include only significant figures. The numbers in parentheses are absolute standard deviations.

*(a) $y = -1.02(\pm0.02) \times 10^{-8} - 3.54(\pm0.2)$
        $\times 10^{-9} = -1.374 \times 10^{-8}$

(b) $y = 90.31(\pm0.08) - 89.32(\pm0.06)$
        $+ 0.200(\pm0.004) = 0.790$

*(c) $y = 0.0020(\pm0.0005) \times 20.20(\pm0.02)$
        $\times 300(\pm1) = 12.12$

(d) $y = \dfrac{1.63(\pm0.03) \times 10^{-14}}{1.03(\pm0.04) \times 10^{-16}} = 158.25243$

*(e) $y = \dfrac{100(\pm1)}{2(\pm1)} = 50$

(f) $y = \dfrac{2.45(\pm0.02) \times 10^{-2} - 5.06(\pm0.06) \times 10^{-3}}{23.2(\pm0.7) + 9.11(\pm0.08)}$
        $= 6.0167 \times 10^{-4}$

*6-9.   Analysis of several plant-food preparations for potassium ion yielded the following data:

Sample	Mean Percent K^+	Number of Observations	Deviation of Individual Results from Mean
1	5.12	5	0.13, 0.09, 0.08, 0.06, 0.08
2	7.09	3	0.09, 0.08, 0.12
3	3.98	4	0.02, 0.17, 0.05, 0.12
4	4.73	4	0.12, 0.06, 0.05, 0.11
5	5.96	5	0.08, 0.06, 0.14, 0.10, 0.08

(a)   Evaluate the standard deviation s for each sample.

(b)   Obtain a pooled estimate for s.

6-10. Six bottles of wine were analyzed for residual sugar, with the following results:

Bottle	Percent (w/v) Residual Sugar	Number of Observations	Deviation of Individual Results from Mean
1	0.94	3	0.050, 0.10, 0.08
2	1.08	4	0.060, 0.050, 0.090, 0.060
3	1.20	5	0.05, 0.12, 0.07, 0.00, 0.08
4	0.67	4	0.05, 0.10, 0.06, 0.09
5	0.83	3	0.07, 0.09, 0.10
6	0.76	4	0.06, 0.12, 0.04, 0.03

(a)   Evaluate the standard deviation s for each set of data.

(b)   Pool the data to establish an absolute standard deviation for the method.

*6-11.   Nine samples of illicit heroin preparations were analyzed in duplicate by a gas-chromatographic technique. Pool the following data to establish an absolute standard deviation for the procedure:

Sample	Heroin, %	Sample	Heroin, %
1	2.24, 2.27	6	1.07, 1.02
2	8.4, 8.7	7	14.4, 14.8
3	7.6, 7.5	8	21.9, 21.1
4	11.9, 12.6	9	8.8, 8.4
5	4.3, 4.2		

6-12. Calculate a pooled estimate of s from the following spectrophotometric determination of NTA (nitrilotriacetic acid) in water from the Ohio River:

Sample	NTA, ppb
1	12, 17, 15, 8
2	32, 31, 32
3	25, 29, 23, 29, 26

6-13. Direct your web browser to the NIST page at http://www.itl.nist.gov/div898/strd/univ/michels.html. After reading the page, click on the link labeled Data file (*ASCII Format*). The page that you see contains 100 values for the speed of light as measured by E. N. Dorsey, *Transactions of the American Philosophical Society,* **1944,** *34,* pp. 1–110, Table 22. Once you have the data on the screen, use your mouse to highlight only the 100 values of the speed of light. Copy the text to the clipboard. Then start Excel with a clean spreadsheet, and click on Copy/Paste to insert the data in column A. Now, find the mean and standard deviation, and compare your values to those presented when you click on Certified Values on the NIST web page. These values appear at http://www.itl.nist.gov/div898/strd/univ/certvalues/michelso.html. Be sure to increase the displayed number of digits in your spreadsheet so that you can compare all the results.

6-14. Direct your web browser to the NIST page at http://www.itl.nist.gov/div898/strd/anova/Ag_Atomic_Wtt.dat. This page contains data for the determination of the atomic mass of silver as presented by L. J. Powell, T. J. Murphy, and J. W. Gramlich, "The Absolute Isotopic Abundance and Atomic Weight of a Reference Sample of Silver," *NBS Journal of Research,* **1982,** *87,* pp. 9–19. The page that you see contains 48 values for the atomic mass of silver: 24 determined by one instrument and 24 determined by another. Once you have the data on the screen, click on File/Save As..., and Ag_Atomic_Wtt.dat will appear in the File name blank. Click on Save. Then start Excel, click on File/Open, and be sure that the Files of type: blank contains All Files (*.*). Find Ag_Atomic_Wtt.dat, highlight the file name, and click on Open. After the Test Import Wizard appears, click on Delimited and then Next. In the next window, be sure that only Space is checked, and scroll down to the bottom of the file to be sure that Excel draws vertical lines to separate the two columns of atomic mass data; then click on Finish. The data should then appear in the spreadsheet. The data in the first 60 rows will look a bit disorganized, but beginning in row 61, the atomic mass data should appear in two columns of the spreadsheet. Now, find the mean and standard deviation of the two sets of data, and then find the pooled standard deviation of the two sets of data. Compare your value with the value for the certified residual standard deviation presented when you click on Certified Values on the NIST web page. These values appear at http://www.itl.nist.gov/div898/strd/anova/Ag_Atomic_Wt_cv.html. Be sure to increase the displayed number of digits in your spreadsheet so that you can compare all the results. You may also want to compare your sum of squares of the deviations from the two means with the NIST value for the certified sum of squares (within instrument).

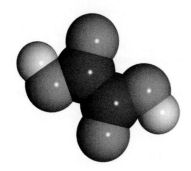

Chapter 7

Statistical Analysis: Evaluating the Data

Experimentalists use statistical calculations to sharpen their judgments concerning the quality of experimental measurements. In this chapter, we consider several of the most common applications of statistical tests to the treatment of analytical results. These applications include:

1. Defining a numerical interval around the mean of a set of replicate analytical results within which the population mean can be expected to lie with a certain probability. This interval is called the *confidence interval*.

2. Determining the number of replicate measurements required to ensure at a given probability that an experimental mean falls within a certain confidence interval.

3. Estimating the probability that (a) an experimental mean and a true value or (b) two experimental means are different, that is, whether the difference is real or simply the result of random error. This test is particularly important for discovering systematic errors in a method and for determining whether two samples come from the same source.

4. Deciding whether what appears to be an outlier in a set of replicate measurements is with a certain probability the result of a gross error and can thus be rejected or whether it is a legitimate result that must be retained in calculating the mean of the set.

5. Using the least-squares method for constructing calibration curves.

7A   CONFIDENCE LIMITS

The true value of the mean μ for a population of data can never be determined exactly because such a determination requires that an infinite number of measurements be made. Statistical theory, however, allows us to set limits around an experimentally determined mean $\bar{x}$ within which the population mean μ lies with a given degree of probability. These limits are called *confidence limits,* and the interval they define is known as the *confidence interval.*

The size of the confidence interval, which is computed from the sample standard deviation, depends on how accurately we know s, that is, how closely we

Confidence limits define a numerical interval around $\bar{x}$ that contains μ with a certain probability.

A *confidence interval* is the numerical magnitude of the confidence limit.

149

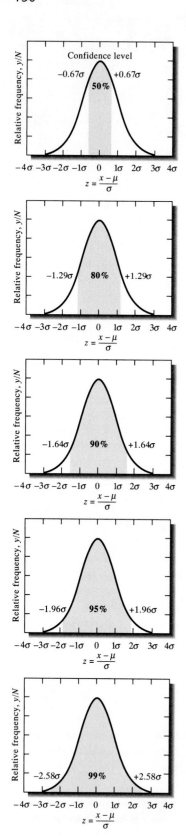

think our sample standard deviation is to the population standard deviation σ. If we have reason to believe that s is a good estimate of σ, then the confidence interval can be significantly narrower than if s is based on only two or three replicates.

7A-1   Finding the Confidence Interval When s Is a Good Estimate of σ

Figure 7-1 shows a series of five normal error curves. In each, the relative frequency is plotted as a function of the quantity z (Equation 6-2, page 123), which is the deviation from the mean normalized to *the population standard deviation*. The shaded areas in each plot lie between the values of $-z$ and $+z$ that are indicated to the left and right of the curves. The numbers within the shaded areas are the percentage of the total area under the curve that is included within these values of z. For example, as shown in the top curve, 50% of the area under any Gaussian curve is located between -0.67σ and $+0.67\sigma$. Proceeding on down, we see that 80% of the total area lies between -1.29σ and $+1.29\sigma$ and that 90% lies between -1.64σ and $+1.64\sigma$. Relationships such as these allow us to define a range of values around a measurement within which the true mean is likely to lie with a certain probability *provided we have a reasonable estimate of* σ. For example, we may assume that 90 times out of 100, the true mean, μ, will be within $\pm 1.64\sigma$ of any measurement that we make (see Figure 7-1c). Here, the *confidence level* is 90% and the *confidence interval* is $\pm z\sigma = \pm 1.64\sigma$.

We find a general expression for the confidence limits (CL) of a single measurement by rearranging Equation 6-2 (remember that z can take positive or negative values). Thus,

$$CL = x \pm z\sigma \tag{7-1}$$

For the mean of N measurements, the standard error of the mean, $\sigma/\sqrt{N}$ (Equation 6-5), is used in place of σ. That is,

$$CL \text{ for } \mu = \bar{x} \pm \frac{z\sigma}{\sqrt{N}} \tag{7-2}$$

Values for z at various confidence levels are found in Table 7-1. Sample calculations of confidence limits and confidence intervals are given in Examples 7-1 and 7-2.

Example 7-1

Calculate the 80% and 95% confidence limits for (a) the first entry (1.80 ppm Hg) in Example 6-2 (page 134) and (b) the mean value (1.67 ppm Hg) for specimen 1 in the example. Assume that in each part $s = 0.1$ is a good estimate of σ ($s \rightarrow \sigma$).

(a) From Table 7-1, we see that $z = 1.29$ and 1.96 for the two confidence levels. Substituting into Equation 7-2 gives

Figure 7-1   Areas under a Gaussian curve for various values of $\pm z$.

$$80\% \text{ CL} = 1.80 \pm \frac{1.29 \times 0.10}{\sqrt{1}} = 1.80 \pm 0.13$$

$$95\% \text{ CL} = 1.80 \pm \frac{1.96 \times 0.10}{\sqrt{1}} = 1.80 \pm 0.20$$

From these calculations, we conclude that it is 80% probable that μ, the population mean (and, *in the absence of determinate error,* the true value) lies in the interval between 1.67 and 1.93 ppm Hg. Furthermore, there is a 95% chance that it lies in the interval between 1.60 and 2.00 ppm Hg.

(b) For the three measurements,

$$80\% \text{ CL} = 1.67 \pm \frac{1.29 \times 0.10}{\sqrt{3}} = 1.67 \pm 0.07$$

$$95\% \text{ CL} = 1.67 \pm \frac{1.96 \times 0.10}{\sqrt{3}} = 1.67 \pm 0.11$$

Thus, there is an 80% probability that the population mean is located in the interval from 1.60 to 1.74 ppm Hg and a 95% probability that it lies between 1.56 and 1.78 ppm.

The *confidence level* is the probability and is often expressed as a percentage.

Table 7-1

Confidence Levels for Various Values of z

Confidence Levels, %	z
50	0.67
68	1.00
80	1.29
90	1.64
95	1.96
95.4	2.00
99	2.58
99.7	3.00
99.9	3.29

Example 7-2

How many replicate measurements of specimen 1 in Example 7-1 are needed to decrease the 95% confidence interval to ± 0.07 ppm Hg?

The confidence interval (CI) is given by the second term on the right-hand side of Equation 7-2 (CI $= \pm z\sigma/\sqrt{N}$). Thus,

$$\text{CI} = 0.07 = \pm \frac{zs}{\sqrt{N}} = \pm \frac{1.96 \times 0.10}{\sqrt{N}}$$

$$\sqrt{N} = \pm \frac{1.96 \times 0.10}{0.07} = \pm 2.80$$

$$N = (\pm 2.8)^2 = 7.8$$

We conclude that eight measurements would provide a slightly better than 95% chance of the population mean lying within ± 0.07 ppm of the experimental mean.

Number of Measurements Averaged, N	Relative Size of Confidence Interval
1	1.00
2	0.71
3	0.58
4	0.50
5	0.45
6	0.41
10	0.32

Equation 7-2 tells us that the confidence interval for an analysis can be halved by carrying out four measurements. Sixteen measurements will narrow the interval by a factor of 4, and so on. We rapidly reach a point of diminishing returns in acquiring additional data. Ordinarily, we take advantage of the relatively large gain attained by averaging two to four measurements, but seldom can we afford the time required for additional increases in confidence.

It is essential to keep in mind at all times that confidence intervals based on Equation 7-2 apply only *in the absence of bias and only if we can assume that s is a good estimate of σ ($s \rightarrow \sigma$).*

7A-2   Finding the Confidence Interval When σ Is Unknown

Often, we are faced with limitations in time or the amount of available sample that prevent us from accurately estimating σ. In such cases, a single set of replicate measurements must provide not only a mean but also an estimate of precision. As indicated earlier, s calculated from a small set of data may be quite uncertain. Thus, confidence limits are necessarily broader when a good estimate of σ is not available.

To account for the variability of s, we use the important statistical parameter t, which is defined in the same way as z (Equation 6-2) except that s is substituted for σ:

$$t = \frac{x - \mu}{s} \tag{7-3}$$

Like z in Equation 7-2, t depends on the desired confidence level, but t also depends on the number of degrees of freedom in the calculation of s. Table 7-2 provides values for t for a few degrees of freedom. More extensive tables are found in various mathematical and statistical handbooks. Note that t approaches z as the number of degrees of freedom approaches infinity.

The confidence limits for the mean $\bar{x}$ of N replicate measurements can be calculated from t by an equation similar to Equation 7-2:

$$\text{CL for } \mu = \bar{x} \pm \frac{ts}{\sqrt{N}} \tag{7-4}$$

The use of the t statistic for confidence limits is illustrated in Example 7-3.

> The t statistic is often called *Student's t*. Student was the name used by W. S. Gossett when he wrote the classic paper on t that appeared in *Biometrika,* **1908,** *6,* 1. Gossett was employed by the Guinness Brewery to provide a statistical analysis of the results of alcohol determinations on their products. As a result of this work, he discovered the now-famous statistical treatment of small sets of data. To avoid the disclosure of any trade secrets of his employer, Gossett published the paper under the name Student.

> Remember that the number of degrees of freedom is not equal to N but $N - 1$ instead.

Table 7-2

Values of t for Various Levels of Probability

Degrees of Freedom	Factor for Confidence Interval				
	80%	90%	95%	99%	99.9%
1	3.08	6.31	12.7	63.7	637
2	1.89	2.92	4.30	9.92	31.6
3	1.64	2.35	3.18	5.84	12.9
4	1.53	2.13	2.78	4.60	8.60
5	1.48	2.02	2.57	4.03	6.86
6	1.44	1.94	2.45	3.71	5.96
7	1.42	1.90	2.36	3.50	5.40
8	1.40	1.86	2.31	3.36	5.04
9	1.38	1.83	2.26	3.25	4.78
10	1.37	1.81	2.23	3.17	4.59
11	1.36	1.80	2.20	3.11	4.44
12	1.36	1.78	2.18	3.06	4.32
13	1.35	1.77	2.16	3.01	4.22
14	1.34	1.76	2.14	2.98	4.14
∞	1.29	1.64	1.96	2.58	3.29

Example 7-3

You have obtained the following data for the alcohol content of a sample of blood: % C_2H_5OH: 0.084, 0.089, and 0.079. Calculate the 95% confidence limits for the mean assuming (a) that you know nothing about the precision of the method and (b) that on the basis of previous experience, you know that $s = 0.005\%$ C_2H_5OH and that s is a good estimate of σ.

(a) $\Sigma x_i = 0.084 + 0.089 + 0.079 = 0.252$

$\Sigma x_i^2 = 0.007056 + 0.007921 + 0.006241 = 0.021218$

$$s = \sqrt{\frac{0.021218 - (0.252)^2/3}{3 - 1}} = 0.0050\% \; C_2H_5OH$$

Here, $\bar{x} = 0.252/3 = 0.084$. Table 7-2 indicates that $t = 4.30$ for two degrees of freedom and 95% confidence. Thus,

$$95\% \; CL = \bar{x} \pm \frac{ts}{\sqrt{N}} = 0.084 \pm \frac{4.30 \times 0.0050}{\sqrt{3}} = 0.084 \pm 0.012\% \; C_2H_5OH$$

(b) Because $s = 0.005\%$ is a good estimate of σ,

$$95\% \; CL = \bar{x} \pm \frac{z\sigma}{\sqrt{N}} = 0.084 \pm \frac{1.96 \times 0.0050}{\sqrt{3}} = 0.084 \pm 0.006\% \; C_2H_5OH$$

Note that a sure knowledge of σ decreases the confidence interval by a significant amount.

7B   STATISTICAL AIDS TO HYPOTHESIS TESTING

Much of scientific and engineering endeavor is based on hypothesis testing. Thus, to explain an observation, a hypothetical model is advanced and tested experimentally to determine its validity. If the results from these experiments do not support the model, we reject it and seek a new hypothesis. If agreement is found, the hypothetical model serves as the basis for further experiments. When the hypothesis is supported by sufficient experimental data, it becomes recognized as a useful theory until such time as data are obtained that refute it.

Experimental results seldom agree exactly with those predicted from a theoretical model. Consequently, scientists and engineers frequently must judge whether a numerical difference is a manifestation of the random errors inevitable in all measurements. Certain statistical tests are useful in sharpening these judgments.

Tests of this kind make use of a *null hypothesis*, which assumes that the numerical quantities being compared are, in fact, the same. The probability of the observed differences appearing as a result of random error is then computed from a probability distribution. Usually, if the observed difference is greater than or equal to the difference that would occur 5 times in 100 (the 5% probability level), the null hypothesis is considered questionable and the difference is

In statistics, a *null hypothesis* postulates that two observed quantities are the same.

judged to be significant. Other probability levels, such as 1 in 100 or 10 in 100, may also be adopted, depending on the certainty desired in the judgment. These probability levels are often called *significance levels* and are given the symbol α in statistics. The confidence level as a percentage is related to α and is given by $(1 - \alpha) \, 100\%$.

The kinds of testing that chemists use most often include the comparison of (1) the mean of an experimental data set $\bar{x}$ with what is believed to be the true value μ; (2) the means $\bar{x}_1$ and $\bar{x}_2$ or the standard deviations σ_1 and σ_2 from two sets of data; and (3) the mean $\bar{x}$ to a predicted or theoretical value. The sections that follow consider some of the methods for making these comparisons.

7B-1   Comparing an Experimental Mean with the True Value

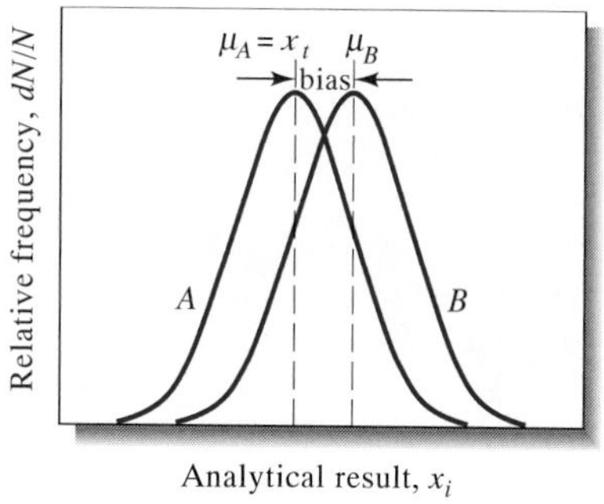

Figure 7-2   Illustration of bias: bias $= \mu_B - x_t = \mu_B - \mu_A$

A common way of testing for bias in an analytical method is to use the method to analyze a sample whose composition is accurately known. Bias in an analytical method is illustrated by the two curves shown in Figure 7-2, which show the frequency distribution of replicate results in the analysis of identical samples by two analytical methods having random errors of exactly the same size. Method A has no bias, so the population mean μ_A is the true value x_t. Method B has a systematic error, or bias, that is given by

$$\text{bias} = \mu_B - x_t = \mu_B - \mu_A \tag{7-5}$$

Note that bias effects all the data in the set in the same way and that it can be either positive or negative.

In testing for bias by analyzing a sample whose analyte concentration is known exactly, it is likely that the experimental mean $\bar{x}$ will differ from the accepted value x_t as shown in the figure; the judgment must then be made whether this difference is the consequence of random error or, alternatively, a systematic error.

In treating this type of problem statistically, the difference $\bar{x} - x_t$ is compared with the difference that could be caused by random error. If the observed difference is less than that computed for a chosen probability level, the null hypothesis that $\bar{x}$ and x_t are the same cannot be rejected; that is, no significant systematic error has been demonstrated. It is important to realize, however, that this statement does not say that there is no systematic error; it says only that whatever systematic error is present is so small that it cannot be distinguished from random error. If $\bar{x} - x_t$ is significantly larger than either the expected or the critical value, we may assume that the difference is real and that the systematic error is significant.

The critical value for rejecting the null hypothesis is calculated by rewriting Equation 7-4 in the form

$$\bar{x} - x_t = \pm \frac{ts}{\sqrt{N}} \tag{7-6}$$

where N is the number of replicate measurements used in the test. If a good estimate of σ is available, Equation 7-6 can be modified by replacing t with z and s with σ. Example 7-4 illustrates the use of an hypothesis test to determine whether there is bias in a method.

Example 7-4

A new procedure for the rapid determination of sulfur in kerosenes was tested on a sample known from its method of preparation to contain 0.123% S (x_t). The results were % S = 0.112, 0.118, 0.115, and 0.119. Do the data indicate that there is bias in the method?

$$\Sigma x_i = 0.112 + 0.118 + 0.115 + 0.119 = 0.464$$

$$\bar{x} = \frac{0.464}{4} = 0.116\% \text{ S}$$

$$\bar{x} - x_t = 0.116 - 0.123 = -0.007\% \text{ S}$$

$$\Sigma x_i^2 = 0.012544 + 0.013924 + 0.013225 + 0.014161 = 0.053854$$

$$s = \sqrt{\frac{0.053854 - (0.464)^2/4}{4-1}} = \sqrt{\frac{0.000030}{3}} = 0.0032$$

From Table 7-2, we find that at the 95% confidence level, t has a value of 3.18 for three degrees of freedom. Thus, we can calculate a test value of t from our data and compare it to the values given in the tables at the desired confidence level. The values of t from the tables are often called critical values and symbolized t_{crit}. The test value is calculated from

$$t = \frac{\bar{x} - x_t}{s/\sqrt{N}}$$

If $|t| \geq t_{crit}$, we reject the null hypothesis at the confidence level chosen. The absolute value of t is used because we are interested in testing only that there is a difference between our mean and the true value and do not care about the sign of the difference. This type of test is often called a two-tailed test. In our case

$$t = \frac{-0.007}{0.0032/\sqrt{4}} = -4.375$$

Since 4.375 > 3.18, the critical value of t at the 95% confidence level, we conclude that a difference this large is significant and reject the null hypothesis.

At the 99% confidence level, $t_{crit} = 5.84$ (Table 7-2). Since 4.375 < 5.84, we would accept the null hypothesis at the 99% confidence level and conclude that there is no difference between the results. Note that the probability (significance) level (0.05 or 0.01) is the probability of making an error by rejecting the null hypothesis.

The probability of a difference this large occurring because of only random errors can be obtained from the Excel function TDIST(x, deg_freedom, tails), where x is the test value of t(4.375), deg_freedom is 3 for our case, and tails = 2. The result is TDIST(4.375,3,2) = 0.022. Hence it is only 2.2% probable to get a value this large because of random errors. The critical value of t for a given confidence level can be obtained in Excel from TINV(probability,deg_freedom). In our case TINV(0,05,3) = 3.1825.

If it was confirmed by further experiments that the method always gave low results, we would say that the method had a *negative bias*.

Even if a mean value is shown to be equal to the true value at a given confidence level, we cannot conclude that there is *no* systematic error in the data.

7B-2   Comparing Two Experimental Means

The results of chemical analyses are frequently used to determine whether two materials are identical. Here, the chemist must judge whether a difference in the means of two sets of identical analyses is real and constitutes evidence that the samples are different or whether the discrepancy is simply a consequence of random errors in the two sets. To illustrate, let us assume that N_1 replicate analyses of material 1 yielded a mean value of $\bar{x}_1$ and that N_2 analyses of material 2 obtained by the same method gave a mean of $\bar{x}_2$. If the data were collected in an identical way, it is usually safe to assume that the standard deviations of the two sets of measurements are the same. We can then modify Equation 7-6 to take into account that one set of results is being compared with a second rather than with the true mean of the data, x_t.

In this case, as with the previous one, we invoke the null hypothesis that the samples are identical and that the observed difference in the results, $(\bar{x}_1 - \bar{x}_2)$, is the result of random errors. To test this hypothesis statistically, we modify Equation 7-6 in the following way. First, we substitute $\bar{x}_2$ for x_t, thus making the left side of the equation the numerical difference between the two means $\bar{x}_1 - \bar{x}_2$. Since we know from Equation 6-5 that the standard deviation of the mean $\bar{x}_1$ is

$$s_{m1} = \frac{s_1}{\sqrt{N_1}}$$

and likewise for $\bar{x}_2$,

$$s_{m2} = \frac{s_2}{\sqrt{N_2}}$$

Thus, the variance s_d^2 of the difference $(d = x_1 - x_2)$ between the means is given by

$$s_d^2 = s_{m1}^2 + s_{m2}^2$$

By substituting the values of s_d, s_{m1}, and s_{m2} into this equation, we have

$$\left(\frac{s_d}{\sqrt{N}}\right)^2 = \left(\frac{s_{m1}}{\sqrt{N_1}}\right)^2 + \left(\frac{s_{m2}}{\sqrt{N_2}}\right)^2$$

If we then assume that the pooled standard deviation s_{pooled} is a good estimate of both s_{m1} and s_{m2}, then

$$\left(\frac{s_d}{\sqrt{N}}\right)^2 = \left(\frac{s_{\text{pooled}}}{\sqrt{N_1}}\right)^2 + \left(\frac{s_{\text{pooled}}}{\sqrt{N_2}}\right)^2 = s_{\text{pooled}}^2 \left(\frac{N_1 + N_2}{N_1 N_2}\right)$$

and

$$\frac{s_d}{\sqrt{N}} = s_{\text{pooled}} \sqrt{\frac{N_1 + N_2}{N_1 N_2}}$$

Substituting this equation into Equation 7-6 (and also $\bar{x}_2$ for x_t), we find that

$$\bar{x}_1 - \bar{x}_2 = \pm t s_{\text{pooled}} \sqrt{\frac{N_1 + N_2}{N_1 N_2}} \tag{7-7}$$

or the test value of t is given by

$$t = \frac{\bar{x}_1 - \bar{x}_2}{s_{\text{pooled}} \sqrt{\dfrac{N_1 + N_2}{N_1 N_2}}} \tag{7-8}$$

We then compare our test value of t with the critical value obtained from the table for the particular confidence level desired. The number of degrees of freedom for finding the critical value of t in Table 7-2 is $N_1 + N_2 - 2$. If the absolute value of the test statistic is smaller than the critical value, the null hypothesis is accepted and no significant difference between the means has been demonstrated. A test value of t greater than the critical value of t indicates that there is a significant difference between the means.

If a good estimate of σ is available, Equation 7-7 can be modified by inserting z for t and σ for s.

Example 7-5

Two barrels of wine were analyzed for their alcohol content to determine whether they were from different sources. On the basis of six analyses, the average content of the first barrel was established to be 12.61% ethanol. Four analyses of the second barrel gave a mean of 12.53% alcohol. The ten analyses yielded a pooled value of $s = 0.070\%$. Do the data indicate a difference between the wines?

Here we employ Equation 7-8 to calculate the test statistic t.

$$t = \frac{\bar{x}_1 - \bar{x}_2}{s_{pooled}\sqrt{\dfrac{N_1 + N_2}{N_1 N_2}}} = \frac{12.61 - 12.53}{0.07\sqrt{\dfrac{6 + 4}{6 \times 4}}} = 1.771$$

The critical value of t at the 95% confidence level for $10 - 2 = 8$ degrees of freedom is 2.31. Since $1.771 < 2.31$, we accept the null hypothesis at the 95% confidence level and conclude that there is no difference in the alcohol content of the wines. The probability of getting a t value of 1.771 may be calculated using the Excel function TDIST() and is TDIST(1.771,8,2) = 0.11. Hence there is a better than 10% chance that a value this large could occur just because of random error.

In Example 7-5, no significant difference between the alcohol content of the two wines was indicated at the 95% confidence level. Note that this statement is equivalent to saying that $\bar{x}_1$ is equal to $\bar{x}_2$ with a certain probability, but the tests do not prove that the wines come from the same source. Indeed, it is conceivable that one wine is a red and the other is a white. To establish with a reasonable probability that the two wines are from the same source would require extensive testing of other characteristics, such as taste, color, odor, and refractive index as well as tartaric acid, sugar, and trace element content. If no significant differences are revealed by all these tests and by others, it might be possible to judge the two wines as having a common origin. In contrast, the finding of *one* significant difference in any test would clearly show that the two wines are different. Thus, the establishment of a significant difference by a single test is much more revealing than the establishment of an absence of difference.

7C   DETECTING GROSS ERRORS

A data point that differs excessively from the mean in a data set is termed an *outlier*. When a set of data contains an outlier, the decision must be made whether to retain or reject it. The choice of criterion for the rejection of a sus-

Outliers are the result of gross errors.

pected result has its perils. If we set a stringent standard that makes the rejection of a questionable measurement difficult, we run the risk of retaining results that are spurious and have an inordinate effect on the mean of the data. If we set lenient limits on precision and thereby make the rejection of a result easy, we are likely to discard measurements that rightfully belong in the set, thus introducing a bias to the data. It is an unfortunate fact that no universal rule can be invoked to settle the question of retention or rejection.[1]

7C-1   Using the Q Test

The Q test is a simple and widely used statistical test.[2] In this test, the absolute value of the difference between the questionable result x_q and its nearest neighbor x_n is divided by the spread w of the entire set to give the quantity Q_{exp}:

$$Q_{exp} = \frac{|x_q - x_n|}{w} = \frac{|x_q - x_n|}{|x_{high} - x_{low}|} \tag{7-9}$$

This ratio is then compared with rejection values Q_{crit} found in Table 7-3. If Q_{exp} is greater than Q_{crit}, the questionable result can be rejected with the indicated degree of confidence (See Figure 7-3.).

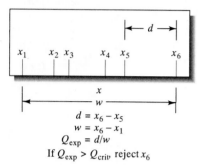

$$d = x_6 - x_5$$
$$w = x_6 - x_1$$
$$Q_{exp} = d/w$$
If $Q_{exp} > Q_{crit}$, reject x_6

Figure 7-3   The Q test for outliers.

Table 7-3

Critical Values for the Rejection Quotient Q

Number of Observations	Q_{crit} (Reject if $Q_{exp} > Q_{crit}$)		
	90% Confidence	**95% Confidence**	**99% Confidence**
3	0.941	0.970	0.994
4	0.765	0.829	0.926
5	0.642	0.710	0.821
6	0.560	0.625	0.740
7	0.507	0.568	0.680
8	0.468	0.526	0.634
9	0.437	0.493	0.598
10	0.412	0.466	0.568

Source: Reproduced from D. B. Rorabacher, *Anal. Chem.,* **1991,** *63,* 139. By courtesy of the American Chemical Society.

Example 7-6

The analysis of a calcite sample yielded CaO percentages of 55.95, 56.00, 56.04, 56.08, and 56.23. The last value appears anomalous; should it be retained or rejected?

The difference between 56.23 and 56.08 is 0.15%. The spread (56.23 − 55.95) is 0.28%. Thus,

$$Q_{exp} = \frac{0.15}{0.28} = 0.54$$

---

[1]J. Mandel, in *Treatise on Analytical Chemistry,* 2nd ed., I. M. Kolthoff and P. J. Elving, Eds., Part I, Vol. 1 (New York: Wiley, 1978), pp. 282–289.

[2]R. B. Dean and W. J. Dixon, *Anal. Chem.,* **1951,** *23,* 636.

For five measurements, Q_{crit} at the 90% confidence level is 0.64. Because $0.54 < 0.64$, we must retain the outlier at the 90% confidence level.

7C-2   A Word of Caution about Rejecting Outliers

Several other statistical tests have been developed to provide criteria for rejection or retention of outliers. Such tests, like the Q test, assume that the distribution of the population data is normal, or Gaussian. Unfortunately, this condition cannot be proved or disproved for samples that have many fewer than 50 results. Consequently, statistical rules, which are perfectly reliable for normal distributions of data, should be *used with extreme caution* when applied to samples containing only a few data. J. Mandel, in discussing treatment of small sets of data, writes, "Those who believe that they can discard observations with statistical sanction by using statistical rules for the rejection of outliers are simply deluding themselves." [3] Thus, statistical tests for rejection should be used only as aids to common sense when small samples are involved.

The blind application of statistical tests to retain or reject a suspect measurement in a small set of data is not likely to be much more fruitful than an arbitrary decision. The application of good judgment based on broad experience with an analytical method is usually a sounder approach. In the end, the only valid reason for rejecting a result from a small set of data is the sure knowledge that a mistake was made in the measurement process. Without this knowledge, *a cautious approach to rejection of an outlier is wise.*

Use extreme caution when rejecting data for any reason.

7C-3   How Do We Deal with Outliers?

Recommendations for the treatment of a small set of results that contains a suspect value are:

1. Reexamine carefully all data relating to the outlying result to see if a gross error could have affected its value. This recommendation demands a *properly kept laboratory notebook containing careful notations of all observations* (see Section 2I).
2. If possible, estimate the precision that can be reasonably expected from the procedure to be sure that the outlying result actually is questionable.
3. Repeat the analysis if sufficient sample and time are available. Agreement between the newly acquired data and those of the original set that appear to be valid will lend weight to the notion that the outlying result should be rejected. Furthermore, if retention is still indicated, the questionable result will have a smaller effect on the mean of the larger set of data.
4. If more data cannot be obtained, apply the Q test to the existing set to see if the doubtful result should be retained or rejected on statistical grounds.
5. If the Q test indicates retention, consider reporting the median of the set rather than the mean. The median has the great virtue of allowing inclusion of

---

[3] J. Mandel, in *Treatise on Analytical Chemistry,* 2nd ed., I. M. Kolthoff and P. J. Elving, Eds., Part I, Vol. 1 (New York: Wiley, 1978), p. 282.

all data in a set without undue influence from an outlying value. In addition, the median of a normally distributed set containing three measurements provides a better estimate of the correct value than the mean of the set after the outlying value has been discarded.

7D   ANALYZING TWO-DIMENSIONAL DATA: THE LEAST-SQUARES METHOD

Many analytical methods are based on a calibration curve in which a measured quantity y is plotted as a function of the known concentration x of a series of standards. The typical calibration curve shown in Figure 7-4 was constructed for the determination of isooctane in a hydrocarbon sample. The ordinate (the dependent variable) is the area under the chromatographic peak for isooctane, and the abscissa (the independent variable) is the mole percent of isooctane. As is typical (and desirable), the plot approximates a straight line. Note, however, that because of the indeterminate errors in the measurement process, not all the data

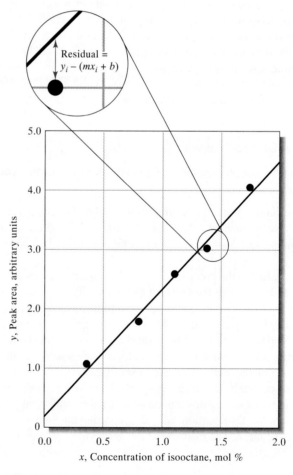

Figure 7-4   Calibration curve for the determination of isooctane in a hydrocarbon mixture.

fall exactly on the line. Thus, the investigator must try to draw the "best" straight line among the points. A statistical technique called *regression analysis* provides the means for objectively obtaining such a line and also for specifying the uncertainties associated with its subsequent use. We consider here only the basic procedure for two-dimensional data that conform to a linear regression model: the *method of least squares.*

7D-1   Assumptions of the Least-Squares Method

When the method of least squares is used to generate a calibration curve, two assumptions are required. The first is that there is actually a linear relationship between the measured variable (*y*) and the analyte concentration (*x*). The mathematical relationship that describes this assumption is called the regression model, which may be represented as

$$y = mx + b$$

where *b* is the *y* intercept (the value of *y* when *x* is zero) and *m* is the slope of the line (see Figure 7-5). We also assume that any deviation of individual points from the straight line results from error in the *measurement*. That is, we must assume that there is no error in the *x* values of the points. For the example of a calibration curve, we assume that exact concentrations of the standards are known. Both of these assumptions are appropriate for many analytical methods, but bear in mind that whenever there is significant uncertainty associated with *x* data, basic linear least-squares analysis may not give the best straight line among a set of points. In such a case, a more complex regression analysis may be necessary to treat the data in question.

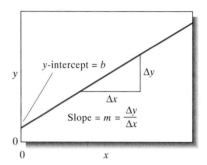

Figure 7-5   The slope-intercept form of a straight line.

Linear least-squares analysis gives you the equation for the best straight line among a set of *x*, *y* data pairs when there is a linear relationship between the two variables and when the *x* data contain negligible uncertainty.

7D-2   Computing the Regression Coefficients and Finding the Least-Squares Line

As illustrated in Figure 7-4, the vertical deviation of each point from the straight line is called a *residual*. The line generated by the least-squares method is the one that minimizes the sum of the squares of the residuals for all the points. In addition to providing the best fit between the experimental points and the straight line, the method gives the standard deviations for *m* and *b*.[4]

For convenience, we define three quantities S_{xx}, S_{yy}, and S_{xy} as follows:

$$S_{xx} = \Sigma\,(x_i - \bar{x})^2 = \Sigma\,x_i^2 - \frac{(\Sigma\,x_i)^2}{N} \qquad (7\text{-}10)$$

$$S_{yy} = \Sigma\,(y_i - \bar{y})^2 = \Sigma\,y_i^2 - \frac{(\Sigma\,y_i)^2}{N} \qquad (7\text{-}11)$$

$$S_{xy} = \Sigma\,(x_i - \bar{x})(y_i - \bar{y}) = \Sigma\,x_i y_i - \frac{\Sigma\,x_i\,\Sigma\,y_i}{N} \qquad (7\text{-}12)$$

The equations for S_{xx} and S_{yy} are similar in form to the numerator in the equation for standard deviation.

[4]R. L. Anderson, *Practical Statistics for Analytical Chemists* (New York: Van Nostrand-Reinhold, 1987), pp. 89–121.

where x_i and y_i are individual pairs of data for x and y, N is the number of pairs of data used in preparing the calibration curve, and $\bar{x}$ and $\bar{y}$ are the average values for the variables; that is,

$$\bar{x} = \frac{\sum x_i}{N} \quad \text{and} \quad \bar{y} = \frac{\sum y_i}{N}$$

Note that S_{xx} and S_{yy} are the sum of the squares of the deviations from the mean for the individual values of x and y. The equivalent expressions shown to the far right in Equations 7-9 to 7-12 are more convenient when a calculator without a built-in regression function is being used.

Six useful quantities can be derived from S_{xx}, S_{yy}, and S_{xy}, as follows.

1. The slope of the line m:

$$m = \frac{S_{xy}}{S_{xx}} \tag{7-13}$$

2. The intercept b:

$$b = \bar{y} - m\bar{x} \tag{7-14}$$

3. The standard deviation about regression s_r:

$$s_r = \sqrt{\frac{S_{yy} - m^2 S_{xx}}{N - 2}} \tag{7-15}$$

4. The standard deviation of the slope s_m:

$$s_m = \sqrt{\frac{s_r^2}{S_{xx}}} \tag{7-16}$$

5. The standard deviation of the intercept s_b:

$$s_b = s_r \sqrt{\frac{\sum x_i^2}{N \sum x_i^2 - (\sum x_i)^2}} = s_r \sqrt{\frac{1}{N - (\sum x_i)^2 / \sum x_i^2}} \tag{7-17}$$

6. The standard deviation for results obtained from the calibration curve s_c:

$$s_c = \frac{s_r}{m} \sqrt{\frac{1}{M} + \frac{1}{N} + \frac{(\bar{y}_c - \bar{y})^2}{m^2 S_{xx}}} \tag{7-18}$$

Equation 7-18 gives us a way to calculate the standard deviation from the mean $\bar{y}_c$ of a set of M replicate analyses of unknowns when a calibration curve that contains N points is used; recall that $\bar{y}$ is the mean value of y for the N calibration data.

The standard deviation about regression s_r (Equation 7-15) is the standard deviation for y when the deviations are measured not from the mean of y (as is usally the case) but from the straight line that results from the least-squares analysis:

$$s_r = \sqrt{\frac{\sum\limits_{i=1}^{N} [y_i - (b + mx_i)]^2}{N - 2}} \tag{7-19}$$

In this equation, the number of degrees of freedom is $N - 2$ since one degree is lost in calculating m and one in determining b. The standard deviation about regression is often called the *standard error of the estimate* or the *standard error in y*. Examples 7-7 and 7-8 illustrate how these quantities are calculated and used. A spreadsheet application is given in Example 7-9.

The standard deviation about regression is analogous to the standard deviation for one-dimensional data.

Example 7-7

Carry out a least-squares analysis of the experimental data provided in the first two columns in Table 7-4 and plotted in Figure 7-5.

Columns 3, 4, and 5 of the table contain computed values for x_i^2, y_i^2, and x_iy_i, with their sums appearing as the last entry in each column. Note that the number of digits carried in the computed values should be the *maximum allowed by the calculator or computer;* that is, *rounding should not be performed until the calculation is complete.*

We now substitute into Equations 7-10, 7-11, and 7-12 and obtain

$$S_{xx} = \Sigma x_i^2 - \frac{(\Sigma x_i)^2}{N} = 6.90201 - \frac{(5.365)^2}{5} = 1.14537$$

$$S_{yy} = \Sigma y_i^2 - \frac{(\Sigma y_i)^2}{N} = 36.3775 - \frac{(12.51)^2}{5} = 5.07748$$

$$S_{xy} = \Sigma x_iy_i - \frac{\Sigma x_i \Sigma y_i}{N} = 15.81992 - \frac{5.365 \times 12.51}{5} = 2.39669$$

Substitution of these quantities into Equations 7-13 and 7-14 yields

$$m = \frac{2.39669}{1.14537} = 2.0925 \approx 2.09$$

$$b = \frac{12.51}{5} - 2.0925 \times \frac{5.365}{5} = 0.2567 \approx 0.26$$

Thus, the equation for the least-squares line is

$$y = 2.09x + 0.26$$

Substitution into Equation 7-15 yields the standard deviation about regression,

Table 7-4

Calibration Data for a Chromatographic Method for the Determination of Isooctane in a Hydrocarbon Mixture

Mole Percent Isooctane, x_i	Peak Area y_i	x_i^2	y_i^2	x_iy_i
0.352	1.09	0.12390	1.1881	0.38368
0.803	1.78	0.64481	3.1684	1.42934
1.08	2.60	1.16640	6.7600	2.80800
1.38	3.03	1.90440	9.1809	4.18140
1.75	4.01	3.06250	16.0801	7.01750
5.365	12.51	6.90201	36.3775	15.81992

$$s_r = \sqrt{\frac{S_{yy} - m^2 S_{xx}}{N - 2}} = \sqrt{\frac{5.07748 - (2.0925)^2 \times 1.14537}{5 - 2}} = 0.144 \approx 0.14$$

and substitution into Equation 7-16 gives the standard deviation of the slope,

$$s_m = \sqrt{\frac{s_r^2}{S_{xx}}} = \sqrt{\frac{(0.144)^2}{1.14537}} = 0.13$$

Finally, we find the standard deviation of the intercept from Equation 7-17:

$$s_b = 0.144 \sqrt{\frac{1}{5 - (5.365)^2/6.90201}} = 0.16$$

Example 7-8

The calibration curve determined in Example 7-7 was used for the chromatographic determination of isooctane in a hydrocarbon mixture. A peak area of 2.65 was obtained. Calculate the mole percent of isooctane and the standard deviation for the result if the area was (a) the result of a single measurement and (b) the mean of four measurements.

In either case, substituting into Equation 7-14 and rearranging gives

$$x = \frac{y - b}{m} = \frac{y - 0.26}{2.09} = \frac{2.65 - 0.26}{2.09} = 1.14 \text{ mol } \%$$

(a) Substituting into Equation 7-18, we obtain

$$s_c = \frac{0.14}{2.09} \sqrt{\frac{1}{1} + \frac{1}{5} + \frac{(2.65 - 12.51/5)^2}{(2.09)^2 \times 1.145}} = 0.074 \text{ mol } \%$$

(b) For the mean of four measurements,

$$s_c = \frac{0.14}{2.09} \sqrt{\frac{1}{4} + \frac{1}{5} + \frac{(2.65 - 12.51/5)^2}{(2.09)^2 \times 1.145}} = 0.046 \text{ mol } \%$$

Spreadsheet Exercise   Using Excel to Do Least-Squares

Microsoft
Excel

Linear least-squares analysis is quite easy with Excel. This type of analysis may be accomplished in several ways: by using the equations presented in this chapter, by employing the basic built-in functions of Excel, or by using the regression data analysis tool. Because the built-in functions are the easiest of these options, we explore them in detail and see how they may be used to evaluate analytical data. In addition, we use Excel to construct our first graphical representation of data.

The Slope and Intercept

As usual, we begin with a blank worksheet. Enter the data from Table 7-4 into the worksheet so that it appears as follows.

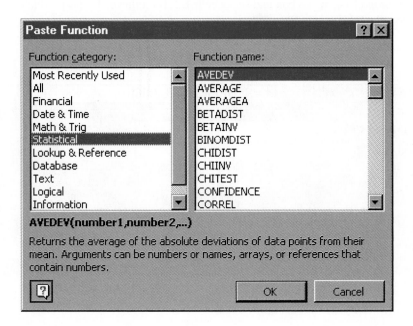

Now, click on cell B8, then click on the Paste Function icon shown in the margin so that the Paste Function window appears, and then click on Statistical. The window appears as follows.

Note that numerous statistical functions appear in the window labeled Function name:. Use the mouse to scroll down the list of functions until you come to the SLOPE function, and the click on it. The function appears in bold under the left-hand window, and a description of the function appears below it. Read the description of the slope function, and then click OK. The following window appears just below the formula bar.

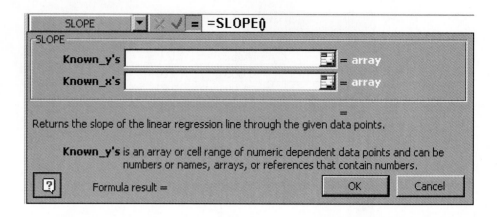

Look at the information that is provided in the window and in the formula bar. The SLOPE() function appears in the formula bar with no arguments, so we must select the data that Excel will use to determine the slope of the line. Now click on the selection button at the right end of the Known_y's box, use the mouse to select cells C2:C6, and type [↵]. Similarly, click on the selection button for the Known_x's box and select cells B2:B8 followed by [↵], which produces the following in the window.

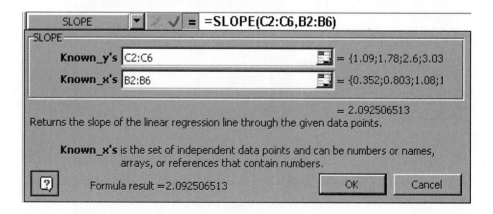

The window shows not only the cell references for the *x* and *y* data, but it also shows the first few of the data points to the right and displays the result of the slope calculation. Now click on OK, and the slope of the line appears in cell B8.

Click on cell B9 followed by the Paste Function icon, and repeat the process that we just carried out, except that now you should select the INTERCEPT function. When the intercept function window appears, select the Known_y's and the Known_x's as before, and click OK. When you have finished, the worksheet will have the following appearance.

	A	B	C
1		x	y
2		0.352	1.09
3		0.803	1.78
4		1.08	2.6
5		1.38	3.03
6		1.75	4.01
7			
8	slope	2.092507	
9	intercept	0.256741	
10			

At this point you may wish to compare these results with those obtained for the slope and intercept in Example 7-7. We should note at this point that Excel provides many digits that are not significant. We shall see how many figures are significant after we find the standard deviations of the slope and intercept.

Using LINEST

Excel features other individual functions related to least-squares analysis, but let us now see how the LINEST function can accomplish many important functions in a single procedure. Begin by using the mouse to select an array of cells two cells wide and five cells high, such as E2:F6. Then click on the Paste Function icon, select STATISTICAL and LINEST in the left and right windows, respectively, and click on OK. Select the Known_y's and Known_x's as before, then click on the box labeled Const and type `true`. Also type `true` in the box labeled Stats. When you click on each of the latter two boxes, notice that a description of the meaning of these logical variables appears below the box. To activate the LINEST function, you must now type the rather unusual keystroke combination `Ctrl+Shift+[↵]`. This keystroke combination must be used whenever you perform a function on an array of cells. The worksheet should now appear as follows.

E2	▼	= {=LINEST(C2:C6,B2:B6,TRUE,TRUE)}			
	A	B	C	D	E	F
1		x	y			
2		0.352	1.09		2.092507	0.256741
3		0.803	1.78		0.134749	0.158318
4		1.08	2.6		0.987712	0.144211
5		1.38	3.03		241.1465	3
6		1.75	4.01		5.015089	0.062391
7						
8	slope	2.092507				
9	intercept	0.256741				

As you can see, cells E2 and F2 contain the slope and intercept of the least-squares line. Cells E3 and F3 are the respective standard deviations of the slope and intercept. Cell E4 contains the square of the correlation coefficient, or the *coefficient of determination*. This value indicates the fraction of the total variation in *y* that is explained by the linear model. The higher the value of the coefficient of determination, the more successful is the model in explaining the variation in *y*. The correlation coefficient is often used as a criterion for goodness of fit of the least-squares line to the points, but in fact, this statistic is not a sensitive measure of goodness of fit. For any reasonable analytical data, the correlation coefficient almost always has a value near unity, which is the maximum value that it can have. A value near zero indicates no correlation between the two variables, and a value near -1 indicates a strong negative correlation; that is, as *x* becomes large, *y* becomes small. An important statistic for indicating the error in estimated values of *y* is the standard error of the estimate, or standard deviation about regression, which is located in cell F4. The smaller the standard error of the estimate, the closer the points are to the line. The value in cell E5 is the *F* statistic, which is useful but is not considered here. Cell F5 contains a familiar statistical variable, which you can look up using the Excel Help/Microsoft Excel Help menu. Before you look it up, try to deduce what variable it is. Finally, cells E6 and F6 contain the sum of the squares of the regression and the sum of the squares of the residuals, respectively. The significance of these quantities is beyond the scope of our discussion.

The number of significant figures that are kept in a least-squares analysis depends on the use for which the data are intended. If the results are to be used to carry out further spreadsheet computations, wait until final results are computed before rounding to an appropriate number of significant figures. Excel provides 15 digits of numerical precision, so, in general, spreadsheet computations will not contribute to the uncertainty in the results. Final answers must be rounded to be consistent with the uncertainty in the original data, which is reflected in the standard deviations of the slope and intercept and the standard error of the estimate. The standard deviations of the slope and intercept in our example suggest that, at most, we should express both the slope and the intercept to only two decimal places. Thus, the least-squares results for the slope and intercept may be expressed as 2.09 ± 0.13 and 0.26 ± 0.16, respectively, or as 2.1 ± 0.1 and 0.3 ± 0.2.

Plotting a Graph of the Data and the Least-Squares Fit

It is customary and useful to plot a graph of the data and the least-squares line similar to Figure 7-4. The built-in Chart Wizard of Excel makes creating graphs relatively easy. Let us begin the process by typing `y(calc)` in cell D1. Now type the following formula in cell D2:

```
=$E$2*B2+$F$2
```

This formula finds the *y* value corresponding to the experimental value of *x* using the formula $y = mx + b$, where *m* and *b* are the least-squares slope and intercept in cells E2 and F2, respectively. Use the fill handle to copy the formula in cell D2 into cells D3:D6. At this point, we have the experimental values for y in column C and the calculated values for y in column D. The next step is to select cells B2:D6, and then click on the Chart Wizard icon shown in the margin. The following window then appears on the screen.

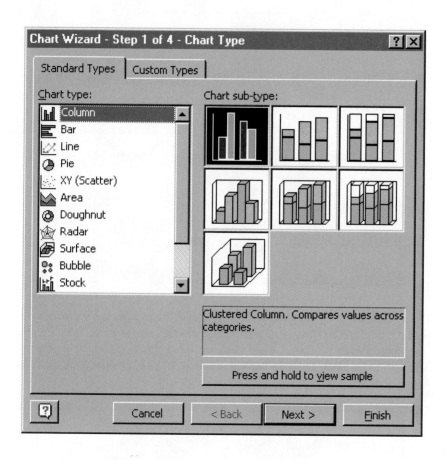

Now select XY (Scatter), and click on Next > and then the Series tab to produce the following window:

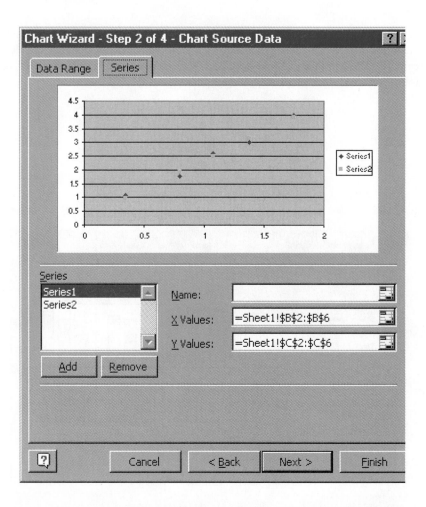

Enter Data in the Name: blank, click on Series2, and enter Fit in the Name: blank; then click on Next >. Click on the Gridlines tab, and check Major grid-lines under Value (X) axis. Then click on the Titles tab, and enter x in the Value (X) axis blank and y in the Value (Y) axis blank. Finally, click on Finish to produce the following graph of the data and the fit.

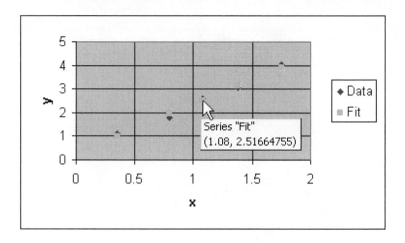

Note that if you click on one of the plotted points, the x and y data are highlighted in your spreadsheet with a colored box. You can include or exclude
points in your graph by simply dragging the fill handle of the colored highlight
box.

Our next step is to plot the fit as a line rather than as individual points. Move
your mouse cursor over one of the fitted points, and notice that a small window
appears as shown in the plot above. The window tells you which series of data
you are pointing at and the x and y coordinates of the point. Move the mouse
around to various points and verify this action. Then place the mouse cursor over
one of the Series "Fit" points as shown above, and right-click on the point. Click
on Format Data Series... in the window that appears, and another window appears as follows.

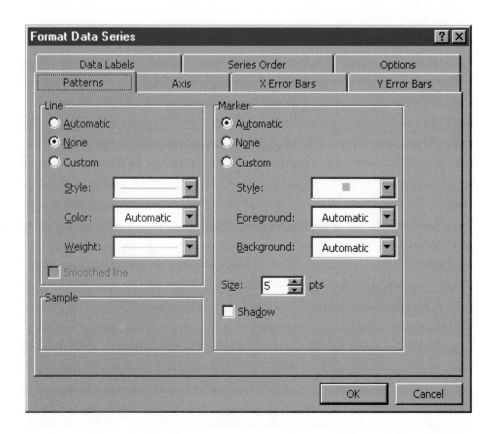

Click on Automatic under Line, None under Marker, and OK to remove the
points on the graph and plot just the least-squares fitted line as follows.

Excel also has a built-in feature for plotting a linear least-squares line as well
as other types of trend lines. Before trying this feature, right-click on the line,
select Format Data Series..., and click on None under Line. Now, right-click
on a data point, click on Add Trendline..., select Linear and the desired data
series, and click OK. A trendline will then appear on your plot. You may also
wish to find a way to plot a trend line that extends across the entire range of
x values.

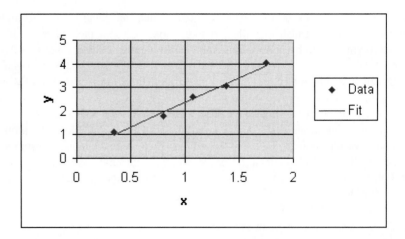

In this exercise, we have given precise instructions to produce a particular type of graph. You no doubt have noticed that there are many options on the Excel menus that we have not investigated. We encourage you to explore the options available in Excel for plotting and presenting data in various ways.

As an extension to this exercise, modify your spreadsheet to include a column for the residuals, that is, $y_i - (b + mx_i)$. Then create a residual plot or a plot of the residuals as a function of x. Residual plots will assist you in detecting any systematic deviations of the experimental points from the least-squares fitted curve. Be sure to save your spreadsheet in a file for reference and for use in analyzing laboratory data.

To test the LINEST function further, obtain the NIST standard data set described in Web Works and perform a least-squares analysis of the data using Excel.

WEB WORKS

Use your web browser to view the NIST web site described in the Web Works of Chapter 6. The universal resource locator (URL) for the page containing certification data for linear regression is http://www-09.nist.gov/div898/strd/lls/data/norris11.shtml, which presents a standard data set compiled by J. Norris of NIST. These data were collected in the calibration of ozone monitors. Click on Data file (ASCII Format), and download the file as described in the Web Works of Chapter 6. Read the data file into Excel using the Text Import Wizard. Remember to select Fixed Width to ensure that the columns of x and y data appear in separate columns of the spreadsheet. Use the LINEST function of Excel to carry out a least-squares analysis of the data, and compare the results with the certified values that you find at http://www-09.nist.gov/div898/strd/lls/data/LINKS/v-norris11.shtml.

7E   QUESTIONS AND PROBLEMS

7-1.   Consider the following sets of replicate measurements:

*A	B	*C	D	*E	F
2.4	69.94	0.0902	2.3	69.65	0.624
2.1	69.92	0.0884	2.6	69.63	0.613
2.1	69.80	0.0886	2.2	69.64	0.596
2.3		0.1000	2.4	69.21	0.607
1.5			2.9		0.582

Calculate the mean and the standard deviation for each of these six data sets. Calculate the 95% confidence limit for each set of data. What do these confidence limits mean?

*7-2.   Calculate the 95% confidence limit for each set of data in Problem 7-1 if $s \rightarrow \sigma$ and has a value of *set A, 0.20; set B, 0.050; *set C, 0.0070; set D, 0.50; *set E, 0.15; set F, 0.015.

*7-3.   The last result in each set of data in Problem 7-1 may be an outlier. Apply the Q test (95% confidence level) to determine whether or not there is a statistical basis for rejection.

*7-4.   An atomic absorption method for the determination of the amount of iron present in used jet engine oil was found, from pooling 30 triplicate analyses, to have a standard deviation $s \rightarrow \sigma = 2.4$ μg Fe/mL. Calculate the 80% and 95% confidence limits for the result, 18.5 μg Fe/mL, if based on
(a)   a single analysis.
(b)   the mean of two analyses.
*(c)   the mean of four analyses.

7-5.   An atomic absorption method for determination of copper in fuels yielded a pooled standard deviation of $s \rightarrow \sigma = 0.32$ μg Cu/mL. The analysis of an oil from a reciprocating aircraft engine showed a copper content of 8.53 μg Cu/mL. Calculate the 90% and 99% confidence limit for the result if it was based on
(a)   a single analysis.
(b)   the mean of 4 analyses.
(c)   the mean of 16 analyses.

*7-6.   How many replicate measurements are needed to decrease the 95% and 99% confidence intervals for the analysis described in Problem 7-4 to ± 1.5 μg Fe/mL?

7-7.   How many replicate measurements are necessary to decrease the 95% and 99% confidence intervals for the analysis described in Problem 7-5 to ± 0.2 μg Cu/mL?

*7-8.   A volumetric calcium determination on triplicate samples of the blood serum of a patient believed to be suffering from a hyperparathyroid condition produced the following data: meq Ca/L = 3.15, 3.25, 3.26. What is the 95% confidence limit for the mean of the data, assuming

(a)   no prior information about the precision of the analysis?
(b) $s \rightarrow \sigma = 0.056$ meq Ca/L?

7-9.   A chemist obtained the following data for percent lindane in the triplicate analysis of an insecticide preparation: 7.47, 6.98, 7.27. Calculate the 90% confidence limit for the mean of the three data, assuming that
(a)   the only information about the precision of the method is the precision for the three data points.
(b)   on the basis of long experience with the method, it is believed that $s \rightarrow \sigma = 0.28\%$ lindane.

7-10.   A standard method for the determination of glucose in serum is reported to have a standard deviation of 0.40 mg/dL. If $s \rightarrow \sigma = 0.40$, how many replicate determinations should be made for the mean for the analysis of a sample to be within
*(a) ± 0.3 mg/dL of the true mean 99% of the time?
(b) ± 0.3 mg/dL of the true mean 95% of the time?
(c) ± 0.2 mg/dL of the true mean 90% of the time?

*7-11.   A titrimetric method for the determination of calcium in limestone was tested by analysis of an NIST limestone containing 30.15% CaO. The mean result of four analyses was 30.26% CaO, with a standard deviation of 0.085%. By pooling data from several analyses, it was established that $s \rightarrow \sigma = 0.094\%$ CaO.
(a)   Do the data indicate the presence of a systematic error at the 95% confidence level?
(b)   Do the data indicate the presence of a systematic error at the 95% confidence level if no pooled value for σ was available?

7-12.   To test the quality of the work of a commercial laboratory, duplicate analyses of a purified benzoic acid (68.8% C, 4.953% H) sample was requested. It is assumed that the relative standard deviation of the method is $s_r \rightarrow \sigma_r = 4$ ppt for carbon and 6 ppt for hydrogen. The means of the reported results are 68.5% C and 4.882% H. At the 95% confidence level, is there any indication of systematic error in either analysis?

*7-13.   A prosecuting attorney in a criminal case presented as principal evidence small fragments of glass found imbedded in the coat of the accused. The attorney claimed that the fragments were identical in composition to a rare Belgian stained-glass window broken during the crime. The average of triplicate analyses for five elements in the glass are shown below. On the basis of these data, does the defendant have grounds for claiming reasonable doubt as to guilt? Use the 99% confidence level as a criterion for doubt.

	Concentration, ppm		Standard Deviation
Element	From Clothes	From Window	$s \to \sigma$
As	129	119	9.5
Co	0.53	0.60	0.025
La	3.92	3.52	0.20
Sb	2.75	2.71	0.25
Th	0.61	0.73	0.043

7-14. The homogeneity of a standard chloride sample was tested by analyzing portions of the material from the top and the bottom of the container, with the following results:

% Chloride	
Top	Bottom
26.32	26.28
26.33	26.25
26.38	26.38
26.39	

(a) Is nonhomogeneity indicated at the 95% confidence level?

(b) Is nonhomogeneity indicated at the 95% level if it is known that
$s \to \sigma = 0.03\%$ Cl?

*7-15. Lord Rayleigh prepared nitrogen samples by several different methods. The density of each sample was measured as the mass of gas required to fill a particular flask at a certain temperature and pressure. Masses of nitrogen samples prepared by decomposition of various nitrogen compounds were 2.29890, 2.29940, 2.29849, and 2.30054 g. Masses of "nitrogen" prepared by removing oxygen from air in various ways were 2.31001, 2.31163, and 2.31028 g. Is the density of nitrogen prepared from nitrogen compounds significantly different from that prepared from air? What are the chances of the conclusion being in error? (Study of this difference led to the discovery of the inert gases by Sir William Ramsey, Lord Rayleigh.)

*7-16. Apply the Q test to the following data sets to determine whether the outlying result should be retained or rejected at the 95% confidence level.
(a) 41.27, 41.61, 41.84, 41.70
(b) 7.295, 7.284, 7.388, 7.292

7-17. Apply the Q test to the following data sets to determine whether the outlying result should be retained or rejected at the 95% confidence level.
(a) 85.10, 84.62, 84.70
(b) 85.10, 84.62, 84.65, 84.70

 7-18. The sulfate ion concentration in natural water can be determined by measuring the turbidity that results when an excess of $BaCl_2$ is added to a measured quantity of the sample. A turbidimeter, the instrument used for this analysis, was calibrated with a series of standard Na_2SO_4 solutions. The following data were obtained in the calibration:

mg SO_4^{2-}/L, C_x	Turbidimeter Reading, R
0.00	0.06
5.00	1.48
10.00	2.28
15.0	3.98
20.0	4.61

Assume that a linear relationship exists between the instrument reading and concentration.
(a) Plot the data and draw a straight line through the points by eye.
(b) Compute the least-squares coefficients of the equation for the best straight line among the points.
(c) Compare the straight line from the relationship determined in part (b) with that in part (a).
(d) Calculate the concentration of sulfate in a sample yielding a turbidimeter reading of 3.67. Calculate the absolute standard deviation of the result and the coefficient of variation.
(e) Repeat the calculations in part (d) assuming that the 3.67 was a mean of six turbidimeter readings.

 *7-19. The following data were obtained in calibrating a calcium ion electrode for the determination of pCa. A linear relationship between the potential E and pCa is known to exist.

pCa	E, mV
5.00	-53.8
4.00	-27.7
3.00	$+2.7$
2.00	$+31.9$
1.00	$+65.1$

(a) Plot the data and draw a line through the points by eye.
(b) Find the least-squares expression for the best straight line through the points. Plot this line.

(c) Calculate the pCa of a serum solution in which the electrode potential was 20.3 mV. Calculate the absolute and relative standard deviations for pCa if the result was from a single voltage measurement.

(d) Calculate the absolute and relative standard deviations for pCa if the millivolt reading in part (c) was the mean of two replicate measurements. Repeat the calculation based on the mean of eight measurements.

 7-20. The following are relative peak areas for chromatograms of standard solutions of methyl vinyl ketone (MVK).

Concentration MVK, mmol/L	Relative Peak Area
0.500	3.76
1.50	9.16
2.50	15.03
3.50	20.42
4.50	25.33
5.50	31.97

(a) Determine the coefficients of the best-fit line using the least-squares method.

(b) Plot the least-squares line as well as the experimental points.

(c) A sample containing MVK yielded relative peak area of 6.3. Calculate the concentration of MVK in the solution.

(d) Assume that the result in part (c) represents a single measurement as well as the mean of four measurements. Calculate the respective absolute and relative standard deviations.

(e) Repeat the calculations in parts (c) and (d) for a sample that gave a peak area of 27.5.

Section II

Principles and Applications of Chemical Equilibria

In Section II, we present the theory and methodology associated with many of the so-called *classical methods of analysis.* Prominent among these techniques are *gravimetry* and *titrimetry,* which are discussed in considerable detail in this section. These methods, whose roots lie in the work of the early chemists Black, Gay-Lussac, Descroxzilles, and others, have evolved over a period of over two hundred years as relevant science and technology has improved. The full potential of classical gravimetry and titrimetry was not realized, however, until thermodynamics was developed in the late nineteenth and early twentieth centuries. Prior to this development, classical analysis was largely empirical in nature. The principles of equilibrium thermodynamics placed quantitative analysis on firm theoretical ground and provided impetus for developing the vast array of accurate and precise techniques that is now at our disposal. To fully comprehend how these methods are developed and put into practice, it is essential that you develop an understanding of solution equilibrium. For this reason, we have woven discussions, examples, problems, and exercises related to the principles of chemical equilibrium into the fabric of this section.

Chapter 8 presents the theory and practice of gravimetric methods of analysis including equilibrium problems related to the solubility of precipitates. Chapter 9 introduces the concepts of activity, ionic strength, and the Debye-Hückel equation and explores activity effects in chemical equilibria. In Chapter 10, we investigate the mass balance and charge balance relationships and develop a systematic approach to the solution of complex equilibrium systems. Chapter 11 presents the fundamentals of titrimetric analysis and applies the principles of stoichiometry to the calcula-

tions of titrimetry. We investigate acids, bases, and acid/base titrations in Chapter 12, along with the principles of equilibrium necessary for understanding titrations and acid/base indicators. In Chapter 13, the theory of polyfunctional acids and bases and the titration of these compounds are examined. We present the practical details of numerous acid/base titrations in Chapter 14. Finally, Section II concludes with the discussion of complexometric titrations and a brief treatment of precipitation titrations in Chapter 15. *Several spreadsheet examples are given in this section to illustrate applications to equilibrium calculations and titrations.*

The photo to the left shows the beautiful natural formation called *Frozen Niagra* in Mammoth Cave National Park in Kentucky. As water seeps over the limestone surface of the cave, calcium carbonate dissolves in the water according to the chemical equilibrium: $CaCO_3(s) + CO_2(g) + H_2O(l) \rightleftharpoons Ca^{2+}(aq) + 2HCO_3^-(aq)$. The flowing water becomes saturated with calcium carbonate so that as carbon dioxide is swept away, the equilibrium reverses, and limestone is again deposited in formations whose shape is governed by the path of the flowing water. Stalactites and stalagmites are examples of similar formations found where water saturated with calcium carbonate drips from the ceiling to the floor of caves over aeons of time. *Photo by Roshan Photo, Bowling Green, KY.*

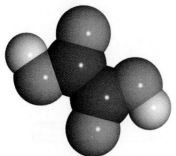

Chapter 8

Gravimetric Methods of Analysis

Gravimetric methods of analysis are based on the measurement of mass.[1] There are two major types of gravimetric methods: precipitation methods and volatilization methods. In *precipitation methods,* the analyte is converted to a sparingly soluble precipitate. This precipitate is then filtered, washed free of impurities, and converted to a product of known composition by suitable heat treatment, and the product is weighed. For example, the Association of Official Analytical Chemists recommends a precipitation method for determining calcium in natural waters. In this method, an excess of oxalic acid, $H_2C_2O_4$, is added to a carefully measured volume of the sample. Ammonia is then added to neutralize the solution and cause the calcium in the sample to precipitate as calcium oxalate. The reactions are

$$2NH_3 + H_2C_2O_4 \rightarrow 2NH_4^+ + C_2O_4^{2-}$$

$$Ca^{2+}(aq) + C_2O_4^{2-}(aq) \rightarrow CaC_2O_4(s)$$

The precipitate is filtered using a weighed filtering crucible and is dried and ignited. This process converts the precipitate entirely to calcium oxide. The reaction is

$$CaC_2O_4(s) \xrightarrow{\Delta} CaO(s) + CO(g) + CO_2(g)$$

After cooling, the crucible and precipitate are weighed and the mass of calcium oxide is determined by subtraction of the known mass of the crucible. The calcium content of the sample is then computed as shown later in Example 8-1.

In *volatilization methods,* the analyte or its decomposition products are volatilized at a suitable temperature. The volatile product is then collected and weighed, or, alternatively, the mass of the product is determined indirectly from the loss in mass of the sample. An example of a gravimetric volatilization procedure is the determination of the sodium hydrogen carbonate content of antacid

> *Gravimetric methods* are quantitative methods based on determining the mass of a pure compound to which the analyte is chemically related.

Theodore W. Richards (1868–1928) and his graduate students at Harvard University developed or refined many of the techniques of gravimetric analysis involving silver and chlorine. These techniques were used to determine the atomic weights of 25 of the elements by preparing pure samples of the chlorides of the elements, decomposing known weights of these compounds, and determining their chloride content by gravimetric methods. For this work, in 1914 Richards became the first American to receive the Nobel Prize in chemistry.

---

[1] For an extensive treatment of gravimetric methods, see C. L. Rulfs, in *Treatise on Analytical Chemistry,* I. M. Kolthoff and P. J. Elving, Eds. (New York: Wiley, 1975), Part I, Vol. 11, Chapter 13.

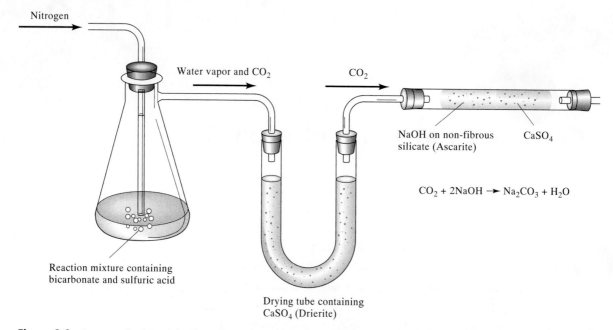

Figure 8-1   Apparatus for determining the sodium hydrogen carbonate content of antacid tablets by a gravimetric volatilization procedure.

Gravimetric methods of analysis are based on mass measurements with an analytical balance, an instrument that yields highly accurate and precise data. In fact, if you perform a gravimetric chloride determination in the laboratory, you may make some of the most accurate and precise measurements of your life.

tablets. Here, a weighed sample of the finely ground tablets is treated with dilute sulfuric acid to convert the sodium hydrogen carbonate to carbon dioxide:

$$NaHCO_3(aq) + H_2SO_4(aq) \rightarrow CO_2(g) + H_2O(l) + NaHSO_4(aq)$$

As shown in Figure 8-1, this reaction is carried out in a flask connected to a weighed absorption tube that contains an absorbent (see Section 8D-5 for a description of the absorbent) that retains the carbon dioxide selectively as it is removed from the solution by heating. The difference in mass of the tube before and after absorption is used to calculate the amount of sodium hydrogen carbonate.

8A   PROPERTIES OF PRECIPITATES AND PRECIPITATING REAGENTS

Ideally, a gravimetric precipitating agent should react *specifically,* or if not, then *selectively* with the analyte. Specific reagents, which are rare, react only with a single chemical species. Selective reagents, which are more common, react with only a limited number of species. In addition to specificity or selectivity, the ideal precipitating reagent would react with the analyte to give a product that is

1. Readily filtered and washed free of contaminants
2. Of sufficiently low solubility so that no significant loss of the solid occurs during filtration and washing
3. Unreactive with constituents of the atmosphere
4. Of known composition after it is dried or, if necessary, ignited

Few if any reagents produce precipitates that possess all these desirable properties perfectly.

The variables that influence solubility (the second property in the foregoing list) are discussed in Section 10B. In the section that follows, we shall be concerned with methods for obtaining pure and easily filtered solids of known composition.[2]

8A-1   Particle Size and Filterability of Precipitates

Precipitates made up of large particles are generally desirable in gravimetric work because large particles are easy to filter and wash free of impurities. In addition, such precipitates are usually purer than are precipitates made up of fine particles.

What Factors Determine Particle Size?

The particle size of solids formed by precipitation varies enormously. At one extreme are *colloidal suspensions,* whose tiny particles are invisible to the naked eye (10^{-7} to 10^{-4} cm in diameter). Colloidal particles show no tendency to settle from solution, nor are they easily filtered. At the other extreme are particles with dimensions on the order of tenths of a millimeter or greater. The temporary dispersion of such particles in the liquid phase is called a *crystalline suspension.* The particles of a crystalline suspension tend to settle spontaneously and are readily filtered.

Precipitate formation has been studied for nearly two centuries, but the mechanism of the process is still not fully understood. It is certain, however, that the particle size of a precipitate is influenced by such experimental variables as precipitate solubility, temperature, reactant concentrations, and the rate at which reactants are mixed. The net effect of these variables can be accounted for, at least qualitatively, by assuming that the particle size is related to a single property of the system called its *relative supersaturation,* where

$$\text{relative supersaturation} = \frac{Q - S}{S} \tag{8-1}$$

In this equation, Q is the concentration of the solute at any instant and S is its equilibrium solubility.

Precipitation reactions are often so slow that even with dropwise addition of a precipitating agent, some supersaturation is likely. Experimental evidence indicates that the particle size of a precipitate varies inversely with the average relative supersaturation during the time the reagent is being introduced. Thus, when $(Q - S)/S$ is large, the precipitate tends to be colloidal; when $(Q - S)/S$ is small, a crystalline solid is more likely.

An example of a selective reagent is $AgNO_3$. The only common ions that it precipitates from acidic solution are Cl^-, Br^-, I^-, and SCN^-. Dimethylglyoxime, which is discussed in Section 8D-3, is a specific reagent that precipitates only Ni^{2+} from alkaline solutions.

A *colloid* is a solid made up of particles having diameters that are less than 10^{-4} cm.

In diffuse light, colloidal suspensions may be perfectly clear and appear to contain no solid. The presence of the second phase can be detected, however, by shining the beam of a flashlight into the solution. Because particles of colloidal dimensions scatter visible radiation, the path of the beam through the solution can be seen by the eye. This phenomenon is called the *Tyndall effect.*

The particles of colloidal suspension are not easily filtered. To trap these particles, the pore size of the filtering medium must be so small that filtrations take too long. With suitable treatment, however, the individual colloidal particles can be made to stick together, thus giving a filterable mass.

Equation 8-1 is known as the Von Weimarn equation in recognition of the scientist who proposed it in 1925.

A *supersaturated solution* is an unstable solution that contains more solute than a saturated solution. With time, supersaturation is relieved by precipitation of the excess solute.

To increase the particle size of a precipitate, minimize the relative supersaturation during precipitate formation.

---

[2]For a more detailed treatment of precipitates, see H. A. Laitinen and W. E. Harris, *Chemical Analysis,* 2nd ed. (New York: McGraw-Hill, 1975), Chapters 8 and 9; A. E. Nielsen, in *Treatise on Analytical Chemistry,* 2nd ed., I. M. Kolthoff and P. J. Elving, Eds. (New York: Wiley, 1983), Part I, Vol. 3, Chapter 27.

How Do Precipitates Form?

Nucleation is a process in which a minimum number of atoms, ions, or molecules join together to produce a stable solid.

The effect of relative supersaturation on particle size can be explained if we assume that precipitates form in two ways, namely by *nucleation* and by *particle growth*. The particle size of a freshly formed precipitate is determined by which way is faster.

In nucleation, a few ions, atoms, or molecules (perhaps as few as four or five) come together to form a stable solid. Often, these nuclei form on the surface of suspended solid contaminants, such as dust particles. Further precipitation then involves a competition between additional nucleation and growth on existing nuclei (particle growth). If nucleation predominates, a precipitate containing a large number of small particles results; if growth predominates, a smaller number of larger particles is produced.

Precipitates form by nucleation and by particle growth. If nucleation predominates, a large number of very fine particles results; if particle growth predominates, a smaller number of larger particles is obtained.

The rate of nucleation is believed to increase enormously with increasing relative supersaturation. In contrast, the rate of particle growth is only moderately enhanced by high relative supersaturations. Thus, when a precipitate is formed at high relative supersaturation, nucleation is the major precipitation mechanism and a large number of small particles is formed. At low relative supersaturations, on the other hand, the rate of particle growth tends to predominate and deposition of solid on existing particles occurs to the exclusion of further nucleation; a crystalline suspension results.

Controlling Particle Size

Experimental variables that minimize supersaturation and thus lead to crystalline precipitates include elevated temperatures to increase the solubility of the precipitate (S in Equation 8-1), dilute solutions (to minimize Q), and slow addition of the precipitating agent with good stirring. The last two measures also minimize the concentration of the solute (Q) at any given instant.

Larger particles can also be obtained by pH control, provided the solubility of the precipitate depends on pH. For example, large, easily filtered crystals of calcium oxalate are obtained by forming the bulk of the precipitate in a mildly acidic environment in which the salt is moderately soluble. The precipitation is then completed by slowly adding aqueous ammonia until the acidity is sufficiently low for removal of substantially all the calcium oxalate. The additional precipitate produced during this step forms on the solid particles formed in the first step.

Precipitates having very low solubilities, such as many sulfides and hydroxides, generally form as colloids.

Unfortunately, many precipitates cannot be formed as crystals under practical laboratory conditions. A colloidal solid is generally encountered when a precipitate has such a low solubility that S in Equation 8-1 always remains negligible relative to Q. The relative supersaturation thus remains enormous throughout precipitate formation, and a colloidal suspension results. For example, under ordinary conditions for an analysis, the hydroxides of iron(III), aluminum, and chromium(III) and the sulfides of most heavy-metal ions form only as colloids because of their very low solubilities.[3]

---

[3]Silver chloride illustrates that the relative supersaturation concept is imperfect. This compound ordinarily forms as a colloid, yet its molar solubility is not significantly different from that of other compounds, such as $BaSO_4$, which generally form as crystals.

8A-2   Colloidal Precipitates

Individual colloidal particles are so small that they are not retained by ordinary filters. Moreover, Brownian motion prevents their settling out of solution under the influence of gravity. Fortunately, however, we can coagulate, or agglomerate, the individual particles of most colloids to give a filterable, amorphous mass that will settle out of solution.

Coagulation of Colloids

Coagulation can be hastened by heating, stirring, and adding an electrolyte to the medium. To understand the effectiveness of these measures, we need to look into why colloidal suspensions are stable and do not coagulate spontaneously.

Colloidal suspensions are stable because all the particles present are either positively or negatively charged. This charge results from cations or anions that are bound to the surface of the particles. The process by which ions are retained *on the surface of a solid* is known as *adsorption.* We can readily demonstrate that colloidal particles are charged by observing their migration when placed in an electrical field.

The adsorption of ions on an ionic solid originates in the normal bonding forces that are responsible for crystal growth. For example, a silver ion at the surface of a silver chloride particle has a partially unsatisfied bonding capacity for anions because of its surface location. Negative ions are attracted to this site by the same forces that hold chloride ions in the silver chloride lattice. Chloride ions at the surface of the solid exert an analogous attraction for cations dissolved in the solvent.

The kind of ions retained on the surface of a colloidal particle and their number depend, in a complex way, on several variables. For a suspension produced in the course of a gravimetric analysis, however, the species adsorbed, and hence the charge on the particles, can be readily predicted because lattice ions are generally more strongly held than others. For example, when silver nitrate is first added to a solution containing chloride ion, the colloidal particles of the precipitate are negatively charged due to adsorption of some of the excess chloride ions. This charge, however, becomes positive when enough silver nitrate has been added to provide an excess of silver ions. The surface charge is at a minimum when the supernatant liquid contains an excess of neither ion.

The extent of adsorption, and thus the charge on a given particle, increases rapidly as the concentration of a common ion becomes greater. Eventually, however, the surface of the particles becomes covered with the adsorbed ions and the charge becomes constant and independent of the concentration of the common ion.

Figure 8-2 shows a colloidal silver chloride particle in a solution that contains an excess of silver nitrate. Attached directly to the solid surface is the *primary adsorption layer,* which consists mainly of adsorbed silver ions. Surrounding the charged particle is a layer of solution, called the *counter-ion layer,* which contains sufficient excess of negative ions (principally nitrate) to just balance the charge on the surface of the particle. The primarily adsorbed silver ions and the negative counter-ion layer constitute an *electrical double layer* that imparts stability to the colloidal suspension. As colloidal particles approach one another, this double layer exerts an electrostatic repulsive force that prevents particles from colliding and adhering.

Adsorption is a process in which a substance (gas, liquid, or solid) is held *on the surface* of a solid. In contrast, *absorption* involves retention of a substance *within* the pores of a solid.

The charge on a colloidal particle formed in a gravimetric analysis is determined by the charge of the lattice ion that is in excess when the precipitation is complete.

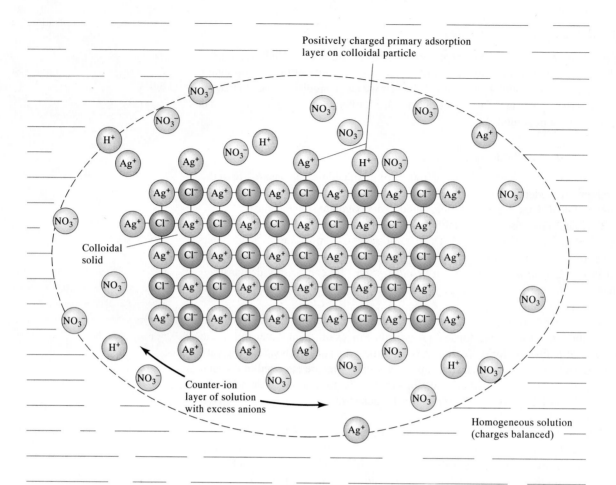

Figure 8-2   A colloidal silver chloride particle suspended in a solution of silver nitrate.

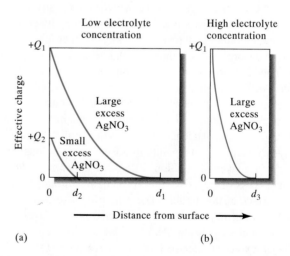

Figure 8-3   Effect of $AgNO_3$ and electrolyte concentration on the thickness of the double layer surrounding a colloidal AgCl particle in a solution containing excess of $AgNO_3$.

Figure 8-3a shows the effective charge on two silver chloride particles. The upper curve is for a particle in a solution that contains a reasonably large excess of silver nitrate, whereas the lower curve represents a particle in a solution that has a much lower silver nitrate content. The effective charge can be thought of as a measure of the repulsive force the particle exerts on like particles in the solution. Note that the effective charge falls off rapidly as the distance from the surface increases and approaches zero at points d_1 or d_2. These decreases in effective charge (in both cases positive) are caused by the negative charge of the excess counter ions in the double layer surrounding each particle. At points d_1 and d_2, the number of counter ions in the layer is approximately equal to the number of primarily adsorbed ions on the surfaces of the particles, so the effective charge of the particles approaches zero at this point.

The upper schematic in Figure 8-4 depicts two silver chloride particles and their counter-ion layers as they approach one another in the concentrated silver nitrate solution just considered. Note that the effective charge on the particles prevents them from approaching one another more closely than about $2d_1$, a distance that is too great for coagulation to occur. As shown in the lower part of Figure 8-4 in the more dilute silver nitrate solution, the two particles can approach within $2d_2$ of one another. Ultimately, as the concentration of silver nitrate is further decreased, the distance between particles becomes small enough for the forces of agglomeration to take effect and for a coagulated precipitate to appear.

Coagulation of a colloidal suspension can often be brought about by a short period of heating, particularly if accompanied by stirring. Heating decreases the number of adsorbed ions and thus the thickness d_i of the double layer. The particles may also gain enough kinetic energy at the higher temperature to overcome the barrier to close approach posed by the double layer.

An even more effective way to coagulate a colloid is to increase the electrolyte concentration of the solution. If we add a suitable ionic compound to a colloidal suspension, the concentration of counter ions increases in the vicinity of each particle. As a result, the volume of solution that contains sufficient counter ions to balance the charge of the primary adsorption layer decreases. The net effect of adding an electrolyte is thus a shrinkage of the counter-ion layer as shown in Figure 8-3b. The particles can then approach one another more closely and agglomerate.

Peptization of Colloids

Peptization refers to the process by which a coagulated colloid reverts to its original dispersed state. When a coagulated colloid is washed, some of the electrolyte responsible for its coagulation is leached from the internal liquid in contact with the solid particles. Removal of this electrolyte has the effect of increasing the volume of the counter-ion layer. The repulsive forces responsible for the original colloidal state are then reestablished, and particles detach themselves from the coagulated mass. The washings become cloudy as the freshly dispersed particles pass through the filter.

The chemist is thus faced with a dilemma in working with coagulated colloids. On one hand, washing is needed to minimize contamination; on the other hand, there is the risk of losses resulting from peptization if pure water is used. The problem is commonly resolved by washing the precipitate with a solution containing an electrolyte that volatilizes when the precipitate is dried during the

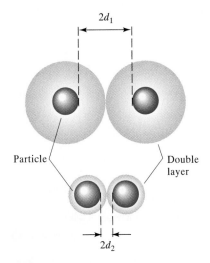

Figure 8-4   The electric double layer of a colloid consists of a layer of charge adsorbed on the surface of the particle (the primary adsorption layer) and a layer of opposite charge (the counter-ion layer) in the solution surrounding the particle. Increasing the electrolyte concentration has the effect of decreasing the volume of the counter-ion layer, thereby increasing the chances for coagulation.

Colloidal suspensions can often be coagulated by heating, stirring, and adding an electrolyte.

Peptization is a process by which a coagulated colloid returns to its dispersed state.

subsequent drying or ignition step. For example, silver chloride is ordinarily washed with a dilute solution of nitric acid. Although the precipitate undoubtedly becomes contaminated with the acid, no harm results, since the nitric acid is volatilized during the ensuing drying step.

Treatment of Colloidal Precipitates

Colloids are best precipitated from hot, stirred solutions containing sufficient electrolyte to ensure coagulation. The filterability of a coagulated colloid frequently improves if it is allowed to stand in contact with the hot solution from which it was formed for an hour or more. During this process, which is known as *digestion,* weakly bound water appears to be lost from the precipitate; the result is a denser mass that is easier to filter.

Digestion is a process in which a precipitate is heated for an hour or more in the solution from which it was formed (the mother liquor).

Mother liquor is the solution from which a precipitate forms.

8A-3   Crystalline Precipitates

Crystalline precipitates are generally more easily filtered and purified than coagulated colloids. In addition, the size of individual crystalline particles, and thus their filterability, can be controlled to a degree.

Improving Particle Size and Filterability

The particle size of crystalline solids can often be improved significantly by minimizing Q, maximizing S, or both in Equation 8-1. Minimization of Q is generally accomplished by using dilute solutions and adding the precipitating reagent slowly and with good mixing. Often, S is increased by precipitating from hot solution or by adjusting the pH of the precipitation medium.

Digestion improves the purity and filterability of both colloidal and crystalline precipitates.

Digestion of crystalline precipitates (without stirring) for some time after formation frequently yields a purer, more filterable product. The improvement in filterability undoubtedly results from the dissolution and recrystallization that occur continuously and at an enhanced rate at elevated temperatures. Recrystallization apparently results in bridging between adjacent particles, a process that yields larger and more easily filtered crystalline aggregates. This view is supported by the observation that little improvement in filtering characteristics occurs if the mixture is stirred during digestion.

8A-4   Coprecipitation

Coprecipitation is the phenomenon in which *otherwise soluble* compounds are removed from solution during precipitate formation. It is important to understand that contamination of a precipitate by a second substance whose solubility product has been exceeded *does not constitute coprecipitation.*

Coprecipitation is a process in which *normally soluble* compounds are carried out of solution by a precipitate.

There are four types of coprecipitation: *surface adsorption, mixed-crystal formation, occlusion,* and *mechanical entrapment.*[4] Surface adsorption and mixed-crystal formation are equilibrium processes, whereas occlusion and mechanical entrapment arise from the kinetics of crystal growth.

---

[4]Several systems of classification of coprecipitation phenomena have been suggested. We follow the simple system proposed by A. E. Nielsen, in *Treatise on Analytical Chemistry,* 2nd ed., I. M. Kolthoff and P. J. Elving, Eds. (New York: Wiley, 1983), Part I, Vol. 3, p. 333.

Surface Adsorption

Adsorption is a common source of coprecipitation that is likely to cause significant contamination of precipitates with large specific surface areas, that is, coagulated colloids (see Feature 8-1 for the definition of specific area). Although adsorption does occur in crystalline solids, its effects on purity are usually undetectable because of the relatively small specific surface area of these solids.

Adsorption is often the major source of contamination in coagulated colloids but is of no significance in crystalline precipitates.

Feature 8-1

Specific Surface Area of Colloids

Specific surface area is defined as the surface area per unit mass of solid and is ordinarily expressed in terms of square centimeters per gram. For a given mass of solid, the specific surface area increases dramatically as particle size decreases and becomes enormous for colloids. For example, the solid cube shown in Figure 8F-1, which has dimensions of 1 cm on a side, has a surface area of 6 cm². If this cube weighs 2 g, its specific surface area is 6 cm²/2g = 3 cm²/g. This cube could be divided into 1000 cubes, each being 0.1 cm on a side. The surface area of each face of these cubes is now 0.1 cm × 0.1 cm = 0.01 cm², and the total area for the six faces of the cube is 0.06 cm². Because there are 1000 of these cubes, the total surface area for the 2 g of solid is now 60 cm²; the specific surface area is then 30 cm²/g. Continuing in this way we find that the specific surface area becomes 300 cm²/g when we have 10^6 cubes that are 0.01 cm on a side. The particle size of a typical crystalline suspension lies in the region of 0.1 and 0.01 cm, so a typical crystalline precipitate has a specific surface area between 30 and 300 cm²/g. Contrast these figures with that for 2 g of a colloid made up of 10^{18} particles with dimensions of 10^{-6} cm. Here, the specific area is 3×10^6 cm²/g, which converts to over 3000 ft²/g. Thus, 1 g of a colloidal suspension has a surface area that is equivalent to the floor area of a good-sized home.

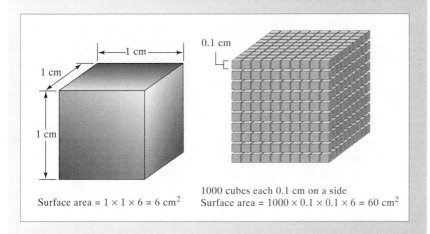

0.1 cm

1 cm

1 cm

1 cm

1 cm

Surface area = $1 \times 1 \times 6 = 6$ cm²

1000 cubes each 0.1 cm on a side
Surface area = $1000 \times 0.1 \times 0.1 \times 6 = 60$ cm²

Figure 8F-1   Increase in surface area per unit mass with decrease in particle size.

Coagulation of a colloid does not significantly decrease the amount of adsorption because the coagulated solid still contains large internal surface areas that remain exposed to the solvent (Figure 8-5). The coprecipitated contaminant on the coagulated colloid consists of the lattice ion originally adsorbed on the surface before coagulation and the counter ion of opposite charge held in the film of solution immediately adjacent to the particle. *The net effect of surface adsorption is therefore the carrying down of an otherwise soluble compound as*

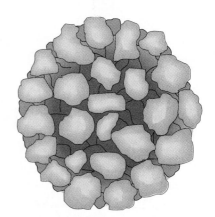

Figure 8-5   A coagulated colloid. This figure suggests that a coagulated colloid continues to expose a large surface area to the solution from which it was formed.

In adsorption, a normally soluble compound is carried out of solution on the surface of a coagulated colloid. This compound consists of the primarily adsorbed ion and an ion of opposite charge from the counter-ion layer.

a surface contaminant. For example, the coagulated silver chloride formed in the gravimetric determination of chloride ion is contaminated with primarily adsorbed silver ions and nitrate or other anions in the counter-ion layer. As a consequence, silver nitrate, a normally soluble compound, is coprecipitated with the silver chloride.

Minimizing Adsorbed Impurities on Colloids.   The purity of many coagulated colloids is improved by digestion. During this process, water is expelled from the solid to give a denser mass that has a smaller specific surface area for adsorption.

Washing a coagulated colloid with a solution containing a volatile electrolyte may also be helpful because any nonvolatile electrolyte added earlier to cause coagulation is displaced by the volatile species. Washing generally does not remove much of the primarily adsorbed ions because the attraction between these ions and the surface of the solid is too strong. Exchange occurs, however, between existing *counter ions* and ions in the wash liquid. For example, in the determination of silver by precipitation with chloride ion, the primarily adsorbed species is chloride. Washing with an acidic solution converts the counter-ion layer largely to hydrogen ions so that both chloride and hydrogen ions are retained by the solid. Volatile HCl is then given off when the precipitate is dried.

Regardless of the method of treatment, a coagulated colloid is always contaminated to some degree, even after extensive washing. The error introduced into the analysis from this source can be as low as 1 to 2 ppt, as in the coprecipitation of silver nitrate on silver chloride. In contrast, coprecipitation of heavy-metal hydroxides on the hydroxides of trivalent iron or aluminum can result in errors as large as several percent, which is generally intolerable.

Reprecipitation.   A drastic but effective way to minimize the effects of adsorption is *reprecipitation,* or *double precipitation.* Here, the filtered solid is redissolved and reprecipitated. The first precipitate ordinarily carries down only a fraction of the contaminant present in the original solvent. Thus, the solution containing the redissolved precipitate has a significantly lower contaminant concentration than the original, and even less adsorption occurs during the second precipitation. Reprecipitation adds substantially to the time required for an analysis but is often necessary for such precipitates as the hydroxides of iron(III) and aluminum, which have extraordinary tendencies to adsorb the hydroxides of heavy-metal cations, such as zinc, cadmium, and manganese.

Mixed-Crystal Formation

In mixed-crystal formation, one of the ions in the crystal lattice of a solid is replaced by an ion of another element. For this exchange to occur, it is necessary that the two ions have the same charge and that their sizes differ by no more than about 5%. Furthermore, the two salts must belong to the same crystal class. For example, barium sulfate formed by adding barium chloride to a solution containing sulfate, lead, and acetate ions is found to be severely contaminated by lead sulfate even though acetate ions normally prevent precipitation of lead sulfate by complexing the lead. Here, lead ions replace some of the barium ions in the barium sulfate crystals. Other examples of coprecipitation by mixed-crystal formation include $MgKPO_4$ in $MgNH_4PO_4$, $SrSO_4$ in $BaSO_4$, and MnS in CdS.

The extent of mixed-crystal contamination is governed by the law of mass action and increases as the ratio of contaminant to analyte concentration increases. Mixed-crystal formation is a particularly troublesome type of coprecipitation because little can be done about it when certain combinations of ions are present in a sample matrix. This problem is encountered with both colloidal suspensions and crystalline precipitates. When mixed-crystal formation occurs, separation of the interfering ion may have to be carried out before the final precipitation step. Alternatively, a different precipitating reagent that does not give mixed crystals with the ions in question may be used.

Mixed-crystal formation is a type of coprecipitation in which a contaminant ion replaces an ion in the lattice of a crystal.

Occlusion and Mechanical Entrapment

When a crystal is growing rapidly during precipitate formation, foreign ions in the counter-ion layer may become trapped, or *occluded,* within the growing crystal. Because supersaturation, and thus growth rate, decreases as a precipitation progresses, the amount of occluded material is greatest in that part of a crystal that forms first.

Mechanical entrapment occurs when crystals lie close together during growth. Here, several crystals grow together and in so doing trap a portion of the solution in a tiny pocket.

Both occlusion and mechanical entrapment are at a minimum when the rate of precipitate formation is low, that is, under conditions of low supersaturation. In addition, digestion is often remarkably helpful in reducing these types of coprecipitation. Undoubtedly, the rapid solution and reprecipitation that goes on at the elevated temperature of digestion opens up the pockets and allows the impurities to escape into the solution.

Occlusion is a type of coprecipitation in which a compound is trapped within a pocket formed during rapid crystal growth.

Mixed-crystal formation may occur in both colloidal and crystalline precipitates, whereas occlusion and mechanical entrapment are confined to crystalline precipitates.

Coprecipitation Errors

Coprecipitated impurities may cause either negative or positive errors in an analysis. If the contaminant is not a compound of the ion being determined, a positive error will always result. Thus, a positive error is observed whenever colloidal silver chloride adsorbs silver nitrate during a chloride analysis. In contrast, when the contaminant does contain the ion being determined, either positive or negative errors may be observed. For example, in the determination of barium by precipitation as barium sulfate, occlusion of other barium salts occurs. If the occluded contaminant is barium nitrate, a positive error is observed because this compound has a larger molar mass than the barium sulfate that would have formed had no coprecipitation occurred. If barium chloride is the contaminant, the error is negative because its molar mass is less than that of the sulfate salt.

Coprecipitation can cause either negative or positive errors.

8A-5 Precipitation from Homogeneous Solution

Precipitation from homogeneous solution is a technique in which a precipitating agent is generated in a solution of the analyte by a slow chemical reaction.[5] Local reagent excesses do not occur because the precipitating agent appears gradually and homogeneously throughout the solution and reacts immediately

Homogeneous precipitation is a process in which a precipitate is formed by slow generation of a precipitating reagent homogeneously throughout a solution.

---

[5]For a general reference on this technique, see L. Gordon, M. L. Salutsky, and H. H. Willard, *Precipitation from Homogeneous Solution* (New York: Wiley, 1959).

Figure 8-6 Aluminum hydroxide formed by the direct addition of ammonia (left) and the homogeneous production of hydroxide ion (right).

Solids formed by homogeneous precipitation are generally purer and more easily filtered than precipitates generated by direct addition of a reagent to the analyte solution.

with the analyte. As a result, the relative supersaturation is kept low during the entire precipitation. In general, homogeneously formed precipitates, both colloidal and crystalline, are better suited for analysis than a solid formed by direct addition of a precipitating reagent.

Urea is often used for the homogeneous generation of hydroxide ion. The reaction can be expressed by the equation

$$(H_2N)_2CO + 3H_2O \rightarrow CO_2 + 2NH_4^+ + 2OH^-$$

This hydrolysis proceeds slowly at temperatures just below 100°C, and 1 to 2 hr is needed to complete a typical precipitation. Urea is particularly valuable for the precipitation of hydroxides or basic salts. For example, hydroxides of iron(III) and aluminum, formed by direct addition of base, are bulky and gelatinous masses that are heavily contaminated and difficult to filter. In contrast, when these same products are produced by homogeneous generation of hydroxide ion, they are dense and readily filtered and have considerably higher purity. Figure 8-6 shows hydroxide precipitates of aluminum formed by direct addition of base and by homogeneous precipitates with urea. Similar pictures for hydroxides of iron(III) are shown in color plate 7. Homogeneous precipitation of crystalline precipitates also results in marked increases in crystal size as well as improvements in purity.

Representative methods based on precipitation by homogeneously generated reagents are given in Table 8-1.

Table 8-1

Methods for Homogeneous Generation of Precipitating Agents

Precipitating Agent	Reagent	Generation Reaction	Elements Precipitated
OH^-	Urea	$(NH_2)_2CO + 3H_2O \rightarrow CO_2 + 2NH_4^+ + 2OH^-$	Al, Ga, Th, Bi, Fe, Sn
PO_4^{3-}	Trimethyl phosphate	$(CH_3O)_3PO + 3H_2O \rightarrow 3CH_3OH + H_3PO_4$	Zr, Hf
$C_2O_4^{2-}$	Ethyl oxalate	$(C_2H_5)_2C_2O_4 + 2H_2O \rightarrow 2C_2H_5OH + H_2C_2O_4$	Mg, Zn, Ca
SO_4^{2-}	Dimethyl sulfate	$(CH_3O)_2SO_2 + 4H_2O \rightarrow 2CH_3OH + SO_4^{2-} + 2H_3O^+$	Ba, Ca, Sr, Pb
CO_3^{2-}	Trichloroacetic acid	$Cl_3CCOOH + 2OH^- \rightarrow CHCl_3 + CO_3^{2-} + H_2O$	La, Ba, Ra
H_2S	Thioacetamide*	$CH_3CSNH_2 + H_2O \rightarrow CH_3CONH_2 + H_2S$	Sb, Mo, Cu, Cd
DMG†	Biacetyl + hydroxylamine	$CH_3COCOCH_3 + 2H_2NOH \rightarrow DMG + 2H_2O$	Ni
HOQ‡	8-Acetoxyquinoline§	$CH_3COOQ + H_2O \rightarrow CH_3COOH + HOQ$	Al, U, Mg, Zn

$\overset{\displaystyle S}{\overset{\displaystyle \|}{^*CH_3-C-NH_2}}$

†DMG = Dimethylglyoxime = $CH_3-\overset{\displaystyle \overset{OH}{\overset{|}{N}}}{\underset{}{C}}=\overset{\displaystyle \overset{OH}{\overset{|}{N}}}{\underset{}{C}}-CH_3$

‡HOQ = 8-Hydroxyquinoline =

§$CH_3-\overset{O}{\overset{\|}{C}}-O$

8B   DRYING AND IGNITION OF PRECIPITATES

After filtration, a gravimetric precipitate is heated until its mass becomes constant. Heating removes the solvent and any volatile species carried down with the precipitate. Some precipitates are also ignited to decompose the solid and form a compound of known composition. This new compound is often called the *weighing form.*

The temperature required to produce a suitable weighing form varies from precipitate to precipitate. Figure 8-7 shows mass loss as a function of temperature for several common analytical precipitates. These data were obtained with an automatic thermobalance,[6] an instrument that records the mass of a substance continuously as its temperature is increased at a constant rate in a furnace (see Figure 8-8). Heating three of the precipitates — silver chloride, barium sulfate, and aluminum oxide — simply causes removal of water and perhaps volatile electrolytes. Note the vastly different temperatures required to produce an anhydrous precipitate of constant mass. Moisture is completely removed from silver chloride above 110°C, but dehydration of aluminum oxide is not complete until a temperature greater than 1000°C is achieved. It is of interest to note that aluminum oxide formed homogeneously with urea can be completely dehydrated at about 650°C.

The thermal curve for calcium oxalate is considerably more complex than the others shown in Figure 8-7. Below about 135°C, unbound water is eliminated to give the monohydrate $CaC_2O_4 \cdot H_2O$. This compound is then converted to the anhydrous oxalate CaC_2O_4 at 225°C. The abrupt change in mass at about 450°C signals the decomposition of calcium oxalate to calcium carbonate and carbon

> The temperature required to dehydrate a precipitate completely may be as low as 100°C or as high as 1000°C.

> Recording thermal decomposition curves is often called *thermogravimetry* or *thermal gravimetric analysis,* and the mass vs. temperature curves are called *thermograms.*

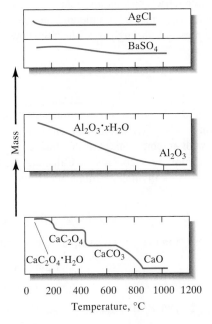

Figure 8-7   Effect of temperature on precipitate mass.

---

[6]For descriptions of thermobalances, see W. W. Wendlandt, *Thermal Methods of Analysis,* 2nd ed. (New York: Wiley, 1974).

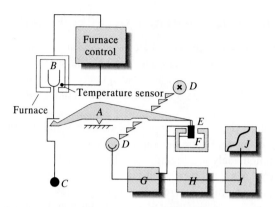

Figure 8-8 Schematic of a thermobalance: *A*: beam; *B*: sample cup and holder; *C*: counterweight; *D*: lamp and photodiodes; *E*: coil; *F*: magnet; *G*: control amplifier; *H*: tare calculator; *I*: amplifier; and *J*: recorder. (Courtesy of Mettler Instrument Corp., Hightstown, NJ.)

monoxide. The final step in the curve depicts the conversion of the carbonate to calcium oxide and carbon dioxide. It is evident that the compound finally weighed in a gravimetric calcium determination based on precipitation as oxalate is highly dependent on the ignition temperature.

8C   CALCULATING RESULTS FROM GRAVIMETRIC DATA

The results of a gravimetric analysis are generally computed from two experimental measurements: the mass of sample and the mass of a product of known composition. Examples 8-1 through 8-3 illustrate how such computations are carried out.

Example 8-1

The calcium in a 200.0-mL sample of a natural water was determined by precipitating the cation as CaC_2O_4. The precipitate was filtered, washed, and ignited in a crucible with an empty mass of 26.6002 g. The mass of the crucible plus CaO (56.077 g/mol) was 26.7134 g. Calculate the concentration of Ca (40.078 g/mol) in the water in units of grams per 100 mL.

The mass of CaO is

$$26.7134 \text{ g} - 26.6002 \text{ g} = 0.1132 \text{ g}$$

The number of moles Ca in the sample is equal to the number of moles CaO or

$$\text{amount of Ca} = 0.1132 \text{ g CaO} \times \frac{1 \text{ mol CaO}}{56.077 \text{ g CaO}} \times \frac{1 \text{ mol Ca}}{\text{mol CaO}}$$

$$= 2.0186 \times 10^{-3} \text{ mol Ca}$$

$$\text{mass Ca/100 mL} = \frac{2.0186 \times 10^{-3} \text{ mol Ca} \times 40.078 \text{ g Ca/mol Ca}}{200 \text{ mL sample}} \times 100 \text{ mL}$$

$$= 0.04045 \text{ g/100 mL}$$

Example 8-2

An iron ore was analyzed by dissolving a 1.1324-g sample in concentrated HCl. The resulting solution was diluted with water, and the iron(III) was precipitated as the hydrous oxide $Fe_2O_3 \cdot xH_2O$ by the addition of NH_3. After filtration and washing, the residue was ignited at a high temperature to give 0.5394 g of pure Fe_2O_3 (159.69 g/mol). Calculate (a) the % Fe (55.847 g/mol) and (b) the % Fe_3O_4 (231.54 g/mol) in the sample.

For both parts of this problem, we need to calculate the number of moles of Fe_2O_3. Thus,

$$\text{amount } Fe_2O_3 = 0.5394 \text{ g } Fe_2O_3 \times \frac{1 \text{ mol } Fe_2O_3}{159.69 \text{ g } Fe_2O_3}$$

$$= 3.3778 \times 10^{-3} \text{ mol } Fe_2O_3$$

(a) The number of moles of Fe is twice the number of moles of Fe_2O_3, and

$$\text{mass Fe} = 3.3778 \times 10^{-3} \text{ mol } Fe_2O_3 \times \frac{2 \text{ mol Fe}}{\text{mol } Fe_2O_3} \times 55.847 \frac{\text{g Fe}}{\text{mol Fe}}$$

$$= 0.37728 \text{ g Fe}$$

$$\% \text{ Fe} = \frac{0.37728 \text{ g Fe}}{1.1324 \text{ g sample}} \times 100\% = 33.317 \approx 33.32\%$$

(b) As shown by the following balanced equation, 3 mol of Fe_2O_3 are chemically equivalent to 2 mol of Fe_3O_4. That is,

$$3Fe_2O_3 \rightarrow 2Fe_3O_4 + \frac{1}{2}O_2$$

$$\text{mass } Fe_3O_4 = 3.3778 \times 10^{-3} \text{ mol } Fe_2O_3 \times \frac{2 \text{ mol } Fe_3O_4}{3 \text{ mol } Fe_2O_3} \times \frac{231.54 \text{ g } Fe_3O_4}{\text{mol } Fe_3O_4}$$

$$= 0.52140 \text{ g } Fe_3O_4$$

$$\% \text{ } Fe_3O_4 = \frac{0.5140 \text{ g } Fe_3O_4}{1.1324 \text{ g sample}} \times 100\% = 46.044 \approx 46.04\%$$

Example 8-3

A 0.2356-g sample containing *only* NaCl (58.44 g/mol) and $BaCl_2$ (208.23 g/mol) yielded 0.4637 g of dried AgCl (143.32 g/mol). Calculate the percent of each halogen compound in the sample.

If we let x be the mass of NaCl in grams and y be the mass of $BaCl_2$ in grams, we can write as a first equation

$$x + y = 0.2356 \text{ g sample}$$

To obtain the mass of AgCl from the NaCl, we write an expression for the number of moles of AgCl formed from the NaCl. That is,

$$\text{amount AgCl from NaCl} = x \text{ g NaCl} \times \frac{1 \text{ mol NaCl}}{58.44 \text{ g NaCl}} \times \frac{1 \text{ mol AgCl}}{\text{mol NaCl}}$$

$$= 0.017111x \text{ mol AgCl}$$

The mass of AgCl from this source is

$$\text{mass AgCl from NaCl} = 0.017111x \text{ mol AgCl} \times 143.32 \frac{\text{g AgCl}}{\text{mol AgCl}}$$

$$= 2.4524x \text{ g AgCl}$$

Proceeding in the same way, we can write that the number of moles of AgCl from the $BaCl_2$ is given by

$$\text{amount AgCl from BaCl}_2 = y \text{ g BaCl}_2 \times \frac{1 \text{ mol BaCl}_2}{208.23 \text{ g BaCl}_2} \times \frac{2 \text{ mol AgCl}}{\text{mol BaCl}_2}$$

$$= 9.605 \times 10^{-3} y \text{ mol AgCl}$$

$$\text{amount AgCl from BaCl}_2 = 9.605 \times 10^{-3} y \text{ mol AgCl} \times 143.32 \frac{\text{g AgCl}}{\text{mol AgCl}}$$

$$= 1.3766y \text{ g AgCl}$$

Because 0.4637 g of AgCl comes from the two compounds we can write

$$2.4524x + 1.3766y = 0.4637$$

The first equation can be rewritten as

$$y = 0.2356 - x$$

Substituting into the previous equation gives

$$2.4524x + 1.3766 (0.2356 - x) = 0.4637$$

which rearranges to

$$1.0758x = 0.13942$$

$$x = \text{mass NaCl} = 0.12960 \text{ g NaCl}$$

$$\% \text{ NaCl} = \frac{0.12956 \text{ g NaCl}}{0.2356 \text{ g sample}} \times 100\% = 55.01\%$$

$$\% \text{ BaCl}_2 = 100.00\% - 55.01\% = 44.99\%$$

8D APPLICATIONS OF GRAVIMETRIC METHODS

Gravimetric methods have been developed for most inorganic anions and cations as well as for such neutral species as water, sulfur dioxide, carbon dioxide, and iodine. A variety of organic substances can also be readily determined gravimetrically. Examples include lactose in milk products, salicylates in drug preparations, phenolphthalein in laxatives, nicotine in pesticides, cholesterol in cereals, and benzaldehyde in almond extracts. Indeed, gravimetric methods are among the most widely applicable of all analytical procedures.

8D-1 Inorganic Precipitating Agents

Table 8-2 lists common inorganic precipitating agents. These reagents typically form slightly soluble salts or hydrous oxides with the analyte. As you can see from the many entries for each reagent, most inorganic reagents are not very selective.

Gravimetric methods do not require a calibration or standardization step because the results are calculated directly from the experimental data and molar masses. Thus, when only one or two samples are to be analyzed, a gravimetric procedure may be the method of choice because it requires less time and effort than a procedure that requires preparation of standards and calibration. All analytical procedures except gravimetry and coulometry require standardization or calibration.

Table 8-2

Some Inorganic Precipitating Agents

Precipitating Agent	Element Precipitated*
$NH_3(aq)$	**Be** (BeO), **Al** (Al_2O_3), **Sc** (Sc_2O_3), Cr (Cr_2O_3),† **Fe** (Fe_2O_3), Ga (Ga_2O_3), Zr (ZrO_2), **In** (In_2O_3), Sn (SnO_2), U (U_3O_8)
H_2S	Cu (CuO),† **Zn** (ZnO, or $ZnSO_4$), **Ge** (GeO_2), As (As_2O_3, or As_2O_5), Mo (MoO_3), Sn (SnO_2),† Sb (Sb_2O_3, or $\underline{Sb_2O_5}$), Bi (Bi_2S_3)
$(NH_4)_2S$	Hg (HgS), Co (Co_3O_4)
$(NH_4)_2HPO_4$	**Mg** ($\underline{Mg_2P_2O_7}$), Al ($AlPO_4$), Mn ($Mn_2P_2O_7$), Zn ($Zn_2P_2O_7$), Zr ($Zr_2P_2O_7$), Cd ($Cd_2P_2O_7$), Bi ($BiPO_4$)
H_2SO_4	Li, Mn, **Sr, Cd, Pb, Ba** (all as sulfates)
H_2PtCl_6	K (K_2PtCl_6, or Pt), Rb ($\underline{Rb_2PtCl_6}$), Cs ($\underline{Cs_2PtCl_6}$)
$H_2C_2O_4$	Ca (CaO), Sr (SrO), **Th** ($\underline{ThO_2}$)
$(NH_4)_2MoO_4$	Cd ($CdMoO_4$),† Pb ($PbMoO_4$)
HCl	**Ag** (AgCl), Hg (Hg_2Cl_2), Na (as NaCl from butyl alcohol), Si (SiO_2)
$AgNO_3$	**Cl** (AgCl), Br ($\underline{AgBr}$), I($\underline{AgI}$)
$(NH_4)_2CO_3$	**Bi** (Bi_2O_3)
NH_4SCN	Cu [$Cu_2(SCN)_2$]
$NaHCO_3$	Ru, Os, Ir (precipitated as hydrous oxides; reduced with H_2 to metallic state)
HNO_3	Sn (SnO_2)
H_5IO_6	Hg [$Hg_5(IO_6)_2$]
NaCl, $Pb(NO_3)_2$	F (PbClF)
$BaCl_2$	SO_4^{2-} ($BaSO_4$)
$MgCl_2$, NH_4Cl	PO_4^{3-} ($Mg_2P_2O_7$)

*Boldface type indicates that gravimetric analysis is the preferred method for the element or ion. The weighed form is indicated in parentheses.

†A dagger indicates that the gravimetric method is seldom used. An underscore indicates the most reliable gravimetric method.

Source: From W. F. Hillebrand, G. E. F. Lundell, H. A. Bright, and J. I. Hoffman, *Applied Inorganic Analysis* (New York: Wiley, 1953). By permission of John Wiley & Sons, Inc.

Table 8-3

Some Reducing Agents Employed in Gravimetric Methods

Reducing Agent	Analyte
SO_2	Se, Au
$SO_2 + H_2NOH$	Te
H_2NOH	Se
$H_2C_2O_4$	Au
H_2	Re, Ir
HCOOH	Pt
$NaNO_2$	Au
$SnCl_2$	Hg
Electrolytic reduction	Co, Ni, Cu, Zn, Ag, In, Sn, Sb, Cd, Re, Bi

8D-2   Reducing Agents

Table 8-3 lists several reagents that convert an analyte to its elemental form for weighing.

8D-3   Organic Precipitating Agents

Numerous organic reagents have been developed for the gravimetric determination of inorganic species. Some of these reagents are significantly more selective in their reactions than are most of the inorganic reagents listed in Table 8-2.

We encounter two types of organic reagents. One forms slightly soluble non-ionic products called *coordination compounds;* the other forms products in which the bonding between the inorganic species and the reagent is largely ionic.

Chelates are cyclical metal-organic compounds in which the metal is a part of one or more five- or six-membered rings. The chelate pictured below is heme, which is a part of hemoglobin, the oxygen carrying molecule in human blood.

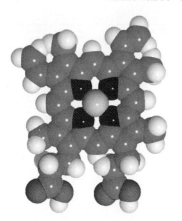

Notice the four six-membered rings that are formed with Fe^{2+}.

Organic reagents that yield sparingly soluble coordination compounds typically contain at least two functional groups. Each of these groups is capable of bonding with a cation by donating a pair of electrons. The functional groups are located in the molecule such that a five- or six-membered ring results from the reaction. Reagents that form compounds of this type are called *chelating agents,* and their products are called *chelates.*

Metal chelates are relatively nonpolar and, as a consequence, have solubilities that are low in water but high in organic liquids. Usually such compounds possess low densities, and they are often intensely colored. Because they are not wetted by water, coordination compounds are readily freed of moisture at low temperatures. Two widely used chelating reagents are described in the paragraphs that follow.

8-Hydroxyquinoline

Approximately two dozen cations form sparingly soluble chelates with 8-hydroxyquinoline. The structure of magnesium 8-hydroxyquinolate is typical of these chelates:

The solubilities of metal 8-hydroxyquinolates vary widely from cation to cation and are pH-dependent because 8-hydroxyquinoline is always deprotonated during chelation reaction. Therefore, we can achieve a considerable degree of selectivity in the use of 8-hydroxyquinoline by controlling pH.

Dimethylglyoxime

Dimethylglyoxime is an organic precipitating agent of unparalleled specificity. Only nickel(II) is precipitated from a weakly alkaline solution. The reaction is

8-Hydroxyquinoline

This precipitate is so bulky that only small amounts of nickel can be handled conveniently. It also has an exasperating tendency to creep up the sides of the container as it is filtered and washed. The solid is readily dried at 110°C and has the composition indicated by its formula.

Sodium Tetraphenylborate

Sodium tetraphenylborate, $(C_6H_5)_4B^-Na^+$, is an important example of an organic precipitating reagent that forms salt-like precipitates. In cold mineral-acid solutions, it is a near-specific precipitating agent for potassium and ammonium ions. The precipitates have stoichiometric composition and contain 1 mole of potassium or ammonium ion for each mole of tetraphenylborate ion; these ionic compounds are readily filtered and can be brought to constant mass at 105 to 120°C. Only mercury(II), rubidium, and cesium interfere and must be removed by prior treatment.

Nickel dimethylglyoxime is spectacular in appearance. As shown in color plate 8, it has a beautiful and vivid red color.

8D-4   Organic Functional Group Analysis

Several reagents react selectively with certain organic functional groups and thus can be used for the determination of most compounds containing these groups. A list of gravimetric functional-group reagents is given in Table 8-4. Many of the reactions shown can also be used for volumetric and spectrophotometric determinations.

8D-5   Volatilization Methods

The two most common gravimetric methods based on volatilization are those for water and carbon dioxide.

Table 8-4

Gravimetric Methods for Organic Functional Groups

Functional Group	Basis for Method	Reaction and Product Weighed*
Carbonyl	Mass of precipitate with 2,4-dinitrophenylhydrazine	$RCHO + H_2NNHC_6H_3(NO_2)_2 \rightarrow$ $\underline{R\text{—}CH = NNHC_6H_3(NO_2)_2}(s) + H_2O$ (RCOR′ reacts similarly)
Aromatic carbonyl	Mass of CO_2 formed at 230°C in quinoline; CO_2 distilled, absorbed, and weighed	$ArCHO \xrightarrow[CuCO_3]{230°C} Ar + \underline{CO_2}(g)$
Methoxyl and ethoxyl	Mass of AgI formed after distillation and decomposition of CH_3I or C_2H_5I	$\left.\begin{array}{l} ROCH_3 \;\;+ HI \rightarrow ROH \;\;\;\;+ CH_3I \\ RCOOCH_3 + HI \rightarrow RCOOH + CH_3I \\ ROC_2H_5 \;\;+ HI \rightarrow ROH \;\;\;\;+ C_2H_5I \end{array}\right\} \begin{array}{l} CH_3I + Ag^+ + H_2O \rightarrow \\ \underline{AgI}(s) + CH_3OH \end{array}$
Aromatic nitro	Mass loss of Sn	$RNO_2 + \frac{3}{2}\underline{Sn}(s) + 6H^+ \rightarrow RNH_2 + \frac{3}{2}Sn^{4+} + 2H_2O$
Azo	Mass loss of Cu	$RN = NR' + 2\underline{Cu}(s) + 4H^+ \rightarrow RNH_2 + R'NH_2 + 2Cu^{2+}$
Phosphate	Mass of Ba salt	$\underset{\underset{ROP(OH)_2}{}}{\overset{O}{\overset{\|}{}}} + Ba^{2+} \rightarrow \underset{}{\overset{O}{\overset{\|}{\underline{ROPO_2Ba}}}}(s) + 2H^+$
Sulfamic acid	Mass of $BaSO_4$ after oxidation with HNO_2	$RNHSO_3H + HNO_2 + Ba^{2+} \rightarrow ROH + \underline{BaSO_4}(s) + N_2 + 2H^+$
Sulfinic acid	Mass of Fe_2O_3 after ignition of Fe(III) sulfinate	$3ROSOH + Fe^{3+} \rightarrow (ROSO)_3Fe(s) + 3H^+$ $(ROSO)_3Fe \underset{O_2}{\rightarrow} CO_2 + H_2O + SO_2 + \underline{Fe_2O_3}(s)$

*The substance weighed is underlined.

Water is quantitatively eliminated from many inorganic samples by ignition. In the direct determination, it is collected on any of several solid desiccants and its mass is determined from the mass gain of the desiccant.

The indirect method in which the amount of water is determined by the loss of mass of the sample during heating is less satisfactory because it must be assumed that water is the only component volatilized. This assumption is frequently unjustified, however, because heating of many substances results in their decomposition and a consequent change in mass, irrespective of the presence of water. Nevertheless, the indirect method has found wide use for the determination of water in items of commerce. For example, a semiautomated instrument for the determination of moisture in cereal grains can be purchased. It consists of a platform balance on which a 10-g sample is heated with an infrared lamp. The percent residue is read directly.

Carbonates are ordinarily decomposed by acids to give carbon dioxide, which is readily evolved from solution by heat. As in the direct analysis for water, the mass of carbon dioxide is established from the increase in the mass of a solid absorbent. Ascarite II,[7] which consists of sodium hydroxide on a nonfibrous silicate, retains carbon dioxide by the reaction

$$2NaOH + CO_2 \rightarrow Na_2CO_3 + H_2O$$

The absorption tube must also contain a desiccant to prevent loss of the evolved water.

Sulfides and sulfites can also be determined by volatilization. Hydrogen sulfide or sulfur dioxide evolved from the sample after treatment with acid is collected in a suitable absorbent.

Finally, the classical method for the determination of carbon and hydrogen in organic compounds is a gravimetric procedure in which the combustion products (H_2O and CO_2) are collected selectively on weighed absorbents. The increase in mass serves as the analytical parameter.

Automatic instruments for the routine determination of water in various products of agriculture and commerce are marketed by several instrument manufacturers.

Microsoft
Excel

Spreadsheet Exercise

Matrix Operations: A Black Box for Solving Simultaneous Linear Equations

Example 8-3 involved the solution of two simultaneous linear equations to find the relative amounts of two compounds in a mixture. For two equations in two unknowns, we may solve for the unknowns in straightforward fashion by substitution. If the number of equations and unknowns exceeds three, however, solution of the equations by substitution becomes time consuming and fraught with error. Excel affords an extremely easy way to solve systems of equations of virtually any size using matrix algebra. The details of matrix algebra are beyond the scope of this presentation, but we can demonstrate the Excel matrix functions that are necessary for solving simultaneous equations. Let us begin by considering the following pair of equations:

[7]®Thomas Scientific, Swedesboro, NJ.

$$3x - 7y = 5$$

$$2x + y = 9$$

Solve these equations by substitution, and you will find that $x = 4$ and $y = 1$. From the two equations, we form two arrays of numbers: the *coefficient matrix* and the *constant matrix*. The coefficient matrix in matrix notation is

$$\mathbf{A} = \begin{bmatrix} 3 & -7 \\ 2 & 1 \end{bmatrix}$$

and the constant matrix is

$$\mathbf{C} = \begin{bmatrix} 5 \\ 9 \end{bmatrix}$$

The system of two equations may be written as

$$\mathbf{AX} = \mathbf{C}$$

That is, the product of the coefficient matrix and the solution matrix is the constant matrix, where the solution matrix for the system of equations is written

$$\mathbf{X} = \mathbf{A}^{-1}\mathbf{C} = \begin{bmatrix} 4 \\ 1 \end{bmatrix}$$

This equation says that if we find the inverse of the coefficient matrix and multiply it by the constant matrix, we obtain the solution matrix. Thus, we require two matrix operations to find the solution matrix: the matrix inverse function *minverse( )* and the matrix multiplication function *mmult( )*. Now let's try out these Excel functions using the previous equations. Beginning with a clean spreadsheet, type the arrays of coefficients and constants as shown below.

	A	B	C	D	E
1	Coefficient Matrix			Constant Matrix	
2	3	-7		5	
3	2	1		9	

In cell A5, type

```
Inverse of Coefficient Matrix[↵]
```

In cell D5, type

```
Solution Matrix[↵]
```

Then highlight cells A6:B7, and type

```
=minverse(a2:b3) [Ctrl+Shift+↵]
```

Recall that you must simultaneously depress the Control key, the Shift key, and the Enter key to accomplish array operations. Your spreadsheet should now appear as follows:

	A	B	C	D	E
1	Coefficient Matrix			Constant Matrix	
2	3	-7		5	
3	2	1		9	
4					
5	Inverse of Coefficient Matrix			Solution Matrix	
6	0.058824	0.411765			
7	-0.11765	0.176471			

Now to determine the solution matrix, highlight cells D6:D7, and type

```
=mmult(a6:b7,d2:d3)[Ctrl+Shift+↵]
```

and your spreadsheet will appear as follows:

	A	B	C	D	E
1	Coefficient Matrix			Constant Matrix	
2	3	-7		5	
3	2	1		9	
4					
5	Inverse of Coefficient Matrix			Solution Matrix	
6	0.058824	0.411765		4	
7	-0.11765	0.176471		1	

Excel has confirmed that the solutions to the equations are $x = 4$ and $y = 1$. Now, to further test the *minverse( )* and *mmult( )* functions, confirm the results of Example 8-3. You can simply enter the x and y coefficients in cells A2:B3 and the constants in cells D2:D3, and Excel will recalculate the solution matrix for you. You do not need to reenter the matrix functions as long as you have the same number of equations and unknowns. Further examples of this type may be found in the problems at the end of the chapter. You may wish to use this spreadsheet to solve some of the problems involving two equations and two unknowns.

Systems with Large Numbers of Equations and Unknowns

Consider the following system of four equations in four unknowns:

$$2w + 5x - y + 4z = 13$$

$$-3w - 2y + 6z = 12$$

$$2x + 3y + 10z = 0$$

$$5w + 3x + 2y = -5$$

Use the matrix functions of Excel to solve these four equations for w, x, y, and z as shown in the following spreadsheet. Note that if a coefficient does not appear in an equation, it must be entered as a zero in the coefficient matrix.

	A	B	C	D	E	F	G
1	Coefficient Matrix					Constant Matrix	
2	2	5	-1	4		13	
3	-3	0	-2	6		12	
4	0	2	3	10		0	
5	5	3	2	0		-5	
6							
7	Inverse Matrix					Solution Matrix	
8	-0.3	0.5	-0.18	0.62		-1	
9	0.433333	-0.5	0.126667	-0.47333		2	
10	0.1	-0.5	0.26	-0.34		-3	
11	-0.11667	0.25	-0.00333	0.196667		0.5	

The solution matrix gives $w = -1$, $x = 2$, $y = -3$, and $z = 0.5$. For systems of three or more equations and unknowns, the Excel matrix functions are most useful.

WEB WORKS

Direct your web browser to http://www.14.nist.gov/nist839/839.01/pubs.htm, where you will find two articles by C. M. Beck of NIST entitled "Classical Analysis: Now and Then" (http://www.14.nist.gov/nist839/839.01/beck1.htm) and "Toward a Revival of Classical Analysis" (http://www.14.nist.gov/nist839/839.01/beck2.htm). In these articles, which were originally published in the scientific literature,[8] Beck makes a strong case for the revival of classical analysis. What is Beck's definition of classical analysis? Why does Beck maintain that classical analysis should be cultivated in this age of automated, computerized instrumentation? What solution does he propose for the problem of dwindling numbers of qualified classical analysts? List three reasons why, in Beck's view, a supply of classical analysts must be maintained.

8E   QUESTIONS AND PROBLEMS

*8-1. Explain the difference between
   *(a) a colloidal and a crystalline precipitate.
   (b) a gravimetric precipitation and volatilization method.
   *(c) precipitation and coprecipitation.
   (d) peptization and coagulation.
   *(e) occlusion and mixed-crystal formation.
   (f) nucleation and particle growth.

8-2. Define
   *(a) digestion.
   (b) adsorption.
   *(c) reprecipitation.
   (d) precipitation from homogeneous solution.
   *(e) counter-ion layer.
   (f) mother liquor.
   *(g) supersaturation.

---

[8]C. M. Beck, *Anal. Chem.,* **1991,** *63* (20), 993A–1003A, and C. M. Beck, *Metrologia,* **1997,** *34* (1), 19–30.

*8-3. What are the structural characteristics of a chelating agent?

8-4. How can the relative supersaturation during precipitate formation be varied?

*8-5. An aqueous solution contains $NaNO_3$ and KSCN. The thiocyanate ion is precipitated as AgSCN by addition of $AgNO_3$. After an excess of the precipitating reagent has been added,
   (a) what is the charge on the surface of the coagulated colloidal particles?
   (b) what is the source of the charge?
   (c) what ions make up the counter-ion layer?

8-6. Suggest a method by which Cu^{2+} can be precipitated homogeneously as CuS.

*8-7. What is peptization and how is it avoided?

8-8. Suggest a precipitation method for separation of K^+ from Na^+ and Li^+.

8-9. Write an equation showing how the mass of the substance on the left can be converted to the mass of the substance on the right.

	Sought	Weighed
*(a)	SO_3	$BaSO_4$
(b)	Zn	$Zn_2P_2O_7$
*(c)	In	In_2O_3
(d)	K	K_2PtCl_6
*(e)	CuO	$Cu_2(SCN)_2$
(f)	Mn_2O_3	Mn_3O_4
(g)	Pb_3O_4	PbO_2
(h)	$U_2P_2O_{11}$	P_2O_5
*(i)	$Na_2B_4O_7 \cdot 10H_2O$	B_2O_3
(j)	Na_2O	$NaZn(UO_2)_3$ $(C_2H_3O_2)_9 \cdot 6H_2O$

*8-10. Treatment of a 0.4000-g sample of impure potassium chloride with an excess of $AgNO_3$ resulted in the formation of 0.7332 g of AgCl. Calculate the percentage of KCl in the sample.

8-11. The aluminum in a 1.200-g sample of impure ammonium aluminum sulfate was precipitated with aqueous ammonia as the hydrous $Al_2O_3 \cdot xH_2O$. The precipitate was filtered and ignited at 1000°C to give anhydrous Al_2O_3, which weighed 0.1798 g. Express the result of this analysis in terms of
   (a) % $NH_4Al(SO_4)_2$. **(b)** % Al_2O_3. **(c)** % Al.

*8-12. What mass of $Cu(IO_3)_2$ can be formed from 0.400 g of $CuSO_4 \cdot 5H_2O$?

8-13. What mass of KIO_3 is needed to convert the copper in 0.4000 g of $CuSO_4 \cdot 5H_2O$ to $Cu(IO_3)_2$?

*8-14. What mass of AgI can be produced from a 0.240-g sample that assays 30.6% MgI_2?

8-15. Precipitates used in the gravimetric determination of uranium include $Na_2U_2O_7$ (634.0 g/mol), $(UO_2)_2P_2O_7$ (714.0 g/mol), and $V_2O_5 \cdot 2UO_3$ (753.9 g/mol). Which of these weighing forms provides the greatest mass of precipitate from a given quantity of uranium?

*8-16. A 0.7406-g sample of impure magnesite, $MgCO_3$, was decomposed with HCl; the liberated CO_2 was collected on calcium oxide and found to weigh 0.1881 g. Calculate the percentage of magnesium in the sample.

8-17. The hydrogen sulfide in a 50.0-g sample of crude petroleum was removed by distillation and collected in a solution of $CdCl_2$. The precipitated CdS was then filtered, washed, and ignited to $CdSO_4$. Calculate the percentage of H_2S in the sample if 0.108 g of $CdSO_4$ was recovered.

*8-18. A 0.1799-g sample of an organic compound was burned in a stream of oxygen, and the CO_2 produced was collected in a solution of barium hydroxide. Calculate the percentage of carbon in the sample if 0.5613 g of $BaCO_3$ was formed.

8-19. A 5.000-g sample of a pesticide was decomposed with metallic sodium in alcohol, and the liberated chloride ion was precipitated as AgCl. Express the results of this analysis in terms of percent DDT ($C_{14}H_9Cl_5$, 354.50 g/mol) based on the recovery of 0.1606 g of AgCl.

*8-20. The mercury in a 0.7152-g sample was precipitated with an excess of paraperiodic acid, H_5IO_6:

$$5Hg^{2+} + 2H_5IO_6 \rightarrow Hg_5(IO_6)_2 + 10H^+$$

The precipitate was filtered, washed free of precipitating agent, dried, and weighed, 0.3408 g being recovered. Calculate the percentage of Hg_2Cl_2 in the sample.

8-21. The iodide in a sample that also contained chloride was converted to iodate by treatment with an excess of bromine:

$$3H_2O + 3Br_2 + I^- \rightarrow 6Br^- + IO_3^- + 6H^+$$

The unused bromine was removed by boiling; an excess of barium ion was then added to precipitate the iodate:

$$Ba^{2+} + 2IO_3^- \rightarrow Ba(IO_3)_2$$

In the analysis of a 2.72-g sample, 0.0720 g of barium iodate was recovered. Express the results of this analysis as percent potassium iodide.

*8-22. Ammoniacal nitrogen can be determined by treatment of the sample with chloroplatinic acid; the product is slightly soluble ammonium chloroplatinate:

$$H_2PtCl_6 + 2NH_4^+ \rightarrow (NH_4)_2PtCl_6 + 2H^+$$

The precipitate decomposes on ignition, yielding metallic platinum and gaseous products:

$$(NH_4)_2PtCl_6 \rightarrow$$
$$Pt(s) + 2Cl_2(g) + 2NH_3(g) + 2HCl(g)$$

Calculate the percentage of ammonia in a sample if 0.2213 g gave rise to 0.5881 g of platinum.

8-23. A 0.6447-g portion of manganese dioxide was added to an acidic solution in which 1.1402 g of a chloride-containing sample was dissolved. Evolution of chlorine took place as a consequence of the following reaction:

$$MnO_2(s) + 2Cl^- + 4H^+ \rightarrow$$
$$Mn^{2+} + Cl_2(g) + 2H_2O$$

After the reaction was complete, the excess MnO_2 was collected by filtration, washed, and weighed, 0.3521 g being recovered. Express the results of this analysis in terms of percent aluminum chloride.

*8-24. A series of sulfate samples is to be analyzed by precipitation as $BaSO_4$. If it is known that the sulfate content in these samples ranges between 20 and 55%, what minimum sample mass should be taken to ensure that a precipitate mass no smaller than 0.300 g is produced? What is the maximum precipitate weight to be expected if this quantity of sample is taken?

8-25. The addition of dimethylglyoxime, $H_2C_4H_6O_2N_2$, to a solution containing nickel(II) ion gives rise to a precipitate:

$$Ni^{2+} + 2H_2C_4H_6O_2N_2 \rightarrow$$
$$2H^+ + Ni(HC_4H_6O_2N_2)_2$$

Nickel dimethylglyoxime is a bulky precipitate that is inconvenient to manipulate in amounts greater than 175 mg. The amount of nickel in a type of permanent-magnet alloy ranges between 24 and 35%. Calculate the sample size that should not be exceeded when analyzing these alloys for nickel.

*8-26. The success of a particular catalyst is highly dependent on its zirconium content. The starting material for this preparation is received in batches that assay between 68 and 84% $ZrCl_4$. Routine analysis based on precipitation of AgCl is feasible, it having been established that there are no sources of chloride ion other than the $ZrCl_4$ in the sample.

(a) What sample mass should be taken to ensure a AgCl precipitate that weighs at least 0.400 g?

(b) If this sample mass is used, what is the maximum weight of AgCl that can be expected in this analysis?

(c) To simplify calculations, what sample mass should be taken to have the percentage of $ZrCl_4$ exceed the mass of AgCl produced by a factor of 100?

8-27. A 0.8720-g sample of a mixture consisting solely of sodium bromide and potassium bromide yields 1.505 g of silver bromide. What are the percentages of the two salts in the sample?

*8-28. A 0.6407-g sample containing chloride and iodide ions gave a silver halide precipitate weighing 0.4430 g. This precipitate was then strongly heated in a stream of Cl_2 gas to convert the AgI to AgCl; on completion of this treatment, the precipitate weighed 0.3181 g. Calculate the percentage of chloride and iodide in the sample.

8-29. The phosphorus in a 0.2374-g sample was precipitated as the slightly soluble $(NH_4)_3PO_4 \cdot 12MoO_3$. This precipitate was filtered, washed, and then redissolved in acid. Treatment of the resulting solution with an excess of Pb^{2+} resulted in the formation of 0.2752 g of $PbMoO_4$. Express the results of this analysis in terms of percent P_2O_5.

*8-30. How many grams of CO_2 are evolved from a 1.204-g sample that is 36.0% $MgCO_3$ and 44.0% K_2CO_3 by mass?

8-31. A 6.881-g sample containing magnesium chloride and sodium chloride was dissolved in sufficient water to give 500 mL of solution. Analysis for the chloride content of a 50.0-mL aliquot resulted in the formation of 0.5923 g of AgCl. The magnesium in a second 50.0-mL aliquot was precipitated as $MgNH_4PO_4$; on ignition, 0.1796 g of $Mg_2P_2O_7$ was found. Calculate the percentage of $MgCl_2 \cdot 6H_2O$ and of NaCl in the sample.

*8-32. A 50.0-mL portion of a solution containing 0.200 g of $BaCl_2 \cdot 2H_2O$ is mixed with 50.0 mL of a solution containing 0.300 g of $NaIO_3$. Assume that the solubility of $Ba(IO_3)_2$ in water is negligibly small and calculate

(a) the mass of the precipitated $Ba(IO_3)_2$.

(b) the mass of the unreacted compound that remains in solution.

8-33. When a 100.0-mL portion of a solution containing 0.500 g of $AgNO_3$ is mixed with 100.0 mL of a solution containing 0.300 g of K_2CrO_4, a bright red precipitate of Ag_2CrO_4 forms.

(a) Assuming that the solubility of Ag_2CrO_4 is negligible, calculate the mass of the precipitate.

(b) Calculate the mass of the unreacted component that remains in solution.

Chapter 9

Electrolyte Effects:
Activity or Concentration?

Ve noted that Equation 4-7 (page 85) is an approximate form of the thermodynamic equilibrium-constant expression. In this chapter, we describe when the use of this approximate form is likely to lead to error. We also show how the approximate equation is modified to give a better description of a system at chemical equilibrium.

9A EFFECTS OF ELECTROLYTES ON CHEMICAL EQUILIBRIA

Experimentally, we find that the position of most solution equilibria depends on the electrolyte concentration of the medium, even when the added electrolyte contains no ion in common with those involved in the equilibrium. For example, consider again the oxidation of iodide ion by arsenic acid that we described in Section 4B-1:

$$H_3AsO_4 + 3I^- + 2H^+ \rightleftharpoons H_3AsO_3 + I_3^- + H_2O$$

If an electrolyte, such as barium nitrate, potassium sulfate, or sodium perchlorate, is added to this solution, the color of the triiodide ion becomes less intense. This decrease in color intensity indicates that the concentration of I_3^- has decreased and that the equilibrium has been shifted to the left by the added electrolyte.

Figure 9-1 further illustrates the effect of electrolytes. Curve A is a plot of the product of the molar hydronium and hydroxide ion *concentration* ($\times 10^{14}$) as a function of the concentration of sodium chloride. This *concentration*-based ion product is designated K'_w. At low sodium chloride concentrations, K'_w becomes independent of the electrolyte concentration and is equal to 1.00×10^{-14}, which is the *thermodynamic* ion-product constant for water, K_w. A relationship that approaches a constant value as some variable (here, the electrolyte concentration) approaches zero is called a *limiting law;* the constant numerical value observed at this limit is referred to as a *limiting value.*

The vertical axis for curve B in Figure 9-1 is the product of the molar concentrations of barium and sulfate ions ($\times 10^{10}$) in saturated solutions of barium sul-

Concentration-based equilibrium constants are often indicated by adding a prime mark. Examples are K'_w, K'_{sp}, and K'_a.

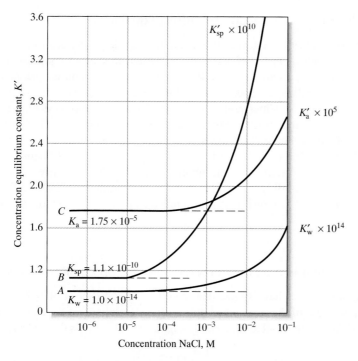

Figure 9-1   Effect of electrolyte concentration on concentration-based equilibrium constants.

fate. This concentration-based solubility product is designated as K'_{sp}. At low electrolyte concentrations, K'_{sp} has a limiting value of 1.1×10^{-10}, which is the accepted thermodynamic value of K_{sp} for barium sulfate.

Curve C is a plot of K'_a ($\times 10^5$), the concentration quotient for the equilibrium involving the dissociation of acetic acid, as a function of electrolyte concentration. Here again, the ordinate function approaches a limiting value K_a, which is the thermodynamic acid dissociation constant for acetic acid.

The dashed lines in Figure 9-1 represent ideal behavior of the solutes. Note that departures from ideality can be significant. For example, the product of the molar concentrations of hydrogen and hydroxide ion increases from 1.0×10^{-14} in pure water to about 1.7×10^{-14} in a solution that is 0.1 M in sodium chloride. The effect is even more pronounced with barium sulfate; here, in 0.1 M sodium chloride K'_{sp} is more than double its limiting value.

The electrolyte effect shown in Figure 9-1 is not peculiar to sodium chloride. Indeed, we would see identical curves when potassium nitrate or sodium perchlorate is substituted for sodium chloride. In each case, the effect has as its origin the electrostatic attraction between the ions of the electrolyte and the ions of reacting species of opposite charge. Since the electrostatic forces associated with all singly charged ions are approximately the same, the three salts exhibit essentially identical effects on equilibria.

We must now consider how we can take the electrolyte effect into account when we wish to make more accurate equilibrium calculations.

As the electrolyte concentration becomes very small, concentration-based equilibrium constants approach their thermodynamic values: K_w, K_{sp}, K_a.

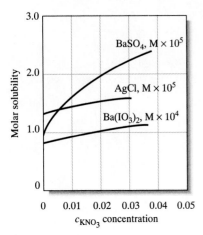

Figure 9-2 Effect of electrolyte concentration on the solubility of some salts.

9A-1   How Do Ionic Charges Affect Equilibria?

Extensive studies have revealed that the magnitude of the electrolyte effect is highly dependent on the charges of the participants in an equilibrium. When only neutral species are involved, the position of equilibrium is essentially independent of electrolyte concentration. With ionic participants, the magnitude of the electrolyte effect increases with charge. This generality is demonstrated by the three solubility curves in Figure 9-2. Note, for example, that in a 0.02 M solution of potassium nitrate, the solubility of barium sulfate, with its pair of doubly charged ions, is larger than it is in pure water by a factor of 2. This same change in electrolyte concentration increases the solubility of barium iodate by a factor of only 1.25 and that of silver chloride by 1.2. The enhanced effect due to doubly charged ions is also reflected in the greater slope of curve B in Figure 9-1.

9A-2   What Is the Effect of Ionic Strength on Equilibria?

Systematic studies have shown that the effect of added electrolyte on equilibria is *independent* of the chemical nature of the electrolyte but depends on a property of the solution called the *ionic strength*. This quantity is defined as

$$\text{ionic strength} = \mu = \frac{1}{2}([A]\,Z_A^2 + [B]\,Z_B^2 + [C]\,Z_C^2 + \cdots) \qquad (9\text{-}1)$$

where [A], [B], [C], . . . represent the molar species concentrations of ions A, B, C, . . . and $Z_A, Z_B, Z_C,$ . . . are their ionic charges. Examples 9-1 and 9-2 illustrate the calculation of ionic strength.

Example 9-1

Calculate the ionic strength of (a) a 0.1 M solution of KNO_3 and (b) a 0.1 M solution of Na_2SO_4.

(a) For the KNO_3 solution, $[K^+]$ and $[NO_3^-]$ are 0.1 M and

$$\mu = \frac{1}{2}(0.1 \times 1^2 + 0.1 \times 1^2) = 0.1$$

(b) For the Na_2SO_4 solution, $[Na^+] = 0.2$ and $[SO_4^{2-}] = 0.1$. Therefore,

$$\mu = \frac{1}{2}(0.2 \times 1^2 + 0.1 \times 2^2) = 0.3$$

Example 9-2

What is the ionic strength of a solution that is 0.05 M in KNO_3 and 0.1 M in Na_2SO_4?

$$\mu = \frac{1}{2}(0.05 \times 1^2 + 0.05 \times 1^2 + 0.2 \times 1^2 + 0.1 \times 2^2) = 0.35\ \text{M}$$

It is apparent from these examples that the ionic strength of a solution of a strong electrolyte consisting solely of singly charged ions is identical with its total molar salt concentration. The ionic strength is, however, greater than the molar concentration if the solution contains ions with multiple charges.

For solutions with ionic strengths of 0.1 M or less, the electrolyte effect is *independent* of the *kind* of ions and dependent *only on the ionic strength*. Thus, the solubility of barium sulfate is the same in aqueous sodium iodide, potassium nitrate, or aluminum chloride provided that the concentrations of these species are such that the ionic strengths are identical. Note that this independence with respect to electrolyte species disappears at high ionic strengths.

9A-3   The Salt Effect

The electrolyte effect, which we have just described, results from the electrostatic attractive and repulsive forces that exist between the ions of an electrolyte and the ions involved in an equilibrium. These forces cause each ion from the dissociated reactant to be surrounded by a sheath of solution that contains a slight excess of electrolyte ions of opposite charge. For example, when a barium sulfate precipitate is equilibrated with a sodium chloride solution, each dissolved barium ion is surrounded by an ionic atmosphere that, because of electrostatic attraction and repulsion, carries a small net negative charge on the average due to repulsion of sodium ions and an attraction of chloride ions. Similarly, each sulfate ion is surrounded by an ionic atmosphere that tends to be slightly positive. These charged layers make the barium ions somewhat less positive and the sulfate ions somewhat less negative than they would be in the absence of electrolyte. The consequence of this effect is a decrease in overall attraction between barium and sulfate ions and an increase in solubility, which becomes greater as the number of electrolyte ions in the solution becomes larger. That is, the *effective concentration* of barium ions and of sulfate ions becomes less as the ionic strength of the medium becomes greater.

9B   ACTIVITY COEFFICIENTS

Chemists use the term *activity, a,* to account for the effects of electrolytes on chemical equilibria. The activity, or effective concentration, of species X depends on the ionic strength of the medium and is defined as

$$a_X = \gamma_X[X] \tag{9-2}$$

where a_X is the activity of X, $[X]$ is its molar concentration, and γ_X is a dimensionless quantity called the *activity coefficient*. The activity coefficient and thus the activity of X vary with ionic strength such that substitution of a_X for $[X]$ in any equilibrium-constant expression frees the numerical value of the constant from dependence on the ionic strength. To illustrate, if X_mY_n is a precipitate, the thermodynamic solubility product expression is defined by the equation

$$K_{sp} = a_X^m \cdot a_Y^n \tag{9-3}$$

Effect of Charge on Ionic Strength

Type Electrolyte	Example	Ionic Strength*
1:1	NaCl	c
1:2	Ba(NO$_3$)$_2$, Na$_2$SO$_4$	$3c$
1:3	Al(NO$_3$)$_3$, Na$_3$PO$_4$	$6c$
2:2	MgSO$_4$	$4c$

*c = molarity of the salt.

Applying Equation 9-2 gives

$$K_{sp} = \gamma_X^m \gamma_Y^n \cdot [X]^m[Y]^n = \gamma_X^m \gamma_Y^n \cdot K'_{sp} \qquad (9\text{-}4)$$

Here K'_{sp} is the *concentration solubility product constant* and K_{sp} is the thermodynamic equilibrium constant.[1] The activity coefficients γ_X and γ_Y vary with ionic strength in such a way as to keep K_{sp} numerically constant and independent of ionic strength (in contrast to the concentration constant K'_{sp}).

9B-1   Properties of Activity Coefficients

Activity coefficients have the following properties:

1. The activity coefficient of a species is a measure of the effectiveness with which that species influences an equilibrium in which it is a participant. In very dilute solutions, where the ionic strength is minimal, this effectiveness becomes constant and the activity coefficient becomes unity. Under such circumstances, the activity and the molar concentration are identical (as are thermodynamic and concentration equilibrium constants). As the ionic strength increases, however, an ion loses some of its effectiveness and its activity coefficient decreases. We may summarize this behavior in terms of Equations 9-2 and 9-3. At moderate ionic strengths, $\gamma_X < 1$; as the solution approaches infinite dilution, however, $\gamma_X \rightarrow 1$ and thus $a_X \rightarrow [X]$ and $K'_{sp} \rightarrow K_{sp}$. At high ionic strengths ($\mu > 0.1$ M), activity coefficients often increase and may even become greater than unity. Because interpretation of the behavior of solutions in this region is difficult, we shall confine our discussion to regions of low or moderate ionic strength (that is, where $\mu \leq 0.1$ M). The variation of typical activity coefficients as a function of ionic strength is shown in Figure 9-3.

> The activity of a species is a measure of its effective concentration as determined by the lowering of the freezing point of water, by electrical conductivity, and by the mass-action effect.

> As $\mu \rightarrow 0$, $\gamma_X \rightarrow 1$, $a_X \rightarrow [X]$, and $K'_{sp} \rightarrow K_{sp}$.

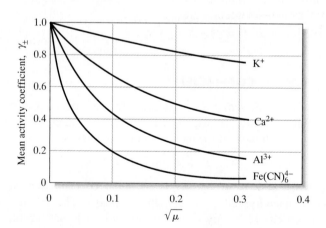

Figure 9-3 Effect of ionic strength on activity coefficients.

---

[1]In the chapters that follow, we use the prime notation only when it is necessary to distinguish between thermodynamic and concentration equilibrium constant.

2. In solutions that are not too concentrated, the activity coefficient for a given species is independent of the nature of the electrolyte and dependent only on the ionic strength.

3. For a given ionic strength, the activity coefficient of an ion departs farther from unity as the charge carried by the species increases. This effect is shown in Figure 9-2. The activity coefficient of an uncharged molecule is approximately unity, regardless of ionic strength.

4. At any given ionic strength, the activity coefficients of ions of the same charge are approximately equal. The small variations that do exist can be correlated with the effective diameter of the hydrated ions.

5. The activity coefficient of a given ion describes its effective behavior in all equilibria in which it participates. For example, at a given ionic strength, a single activity coefficient for cyanide ion describes the influence of that species on any of the following equilibria:

$$HCN + H_2O \rightleftharpoons H_3O^+ + CN^-$$

$$Ag^+ + CN^- \rightleftharpoons AgCN(s)$$

$$Ni^{2+} + 4CN^- \rightleftharpoons Ni(CN)_4^{2-}$$

9B-2 The Debye–Hückel Equation

In 1923, P. Debye and E. Hückel used the ionic atmosphere model, described in Section 9A-3, to derive a theoretical expression that permits the calculation of activity coefficients of ions from their charge and their average size.[2] This equation, which has become known as the *Debye–Hückel equation*, takes the form

$$-\log \gamma_X = \frac{0.51 Z_X^2 \sqrt{\mu}}{1 + 3.3 \alpha_X \sqrt{\mu}} \qquad (9\text{-}5)$$

where

γ_X = activity coefficient of the species X

Z_X = charge on the species X

μ = ionic strength of the solution

α_X = effective diameter of the hydrated ion X in nanometers (10^{-9} m)

The constants 0.51 and 3.3 are applicable to aqueous solutions at 25°C; other values must be used at other temperatures.

Unfortunately, considerable uncertainty exists regarding the magnitude of α_X in Equation 9-5. Its value appears to be approximately 0.3 nm for most singly charged ions; for these species, then, the denominator of the Debye–Hückel equation simplifies to approximately $1 + \sqrt{\mu}$. For ions with higher charge, α_X may be as large as 1.0 nm. This increase in size with increase in charge makes good chemical sense. The larger the charge on an ion, the larger the number of polar water molecules that will be held in the solvation shell about the ion. The

Peter Debye (1884–1966) was born and educated in Europe but became Professor of Chemistry at Cornell University in 1940. He was noted for his work in several different areas of chemistry including electrolyte solutions, X-ray diffraction, and the properties of polar molecules. He received the 1936 Nobel Prize in Chemistry. (Cornell University, courtesy of AIP Emilio Segrè Visual Archives)

[2]P. Debye and E. Hückel, *Physik. Z.,* **1923**, *24*, 185.

Table 9-1

Activity Coefficients for Ions at 25°C

Ion	α_X, nm	Activity Coefficient at Indicated Ionic Strength				
		0.001	0.005	0.01	0.05	0.1
H_3O^+	0.9	0.967	0.934	0.913	0.85	0.83
Li^+, $C_6H_5COO^-$	0.6	0.966	0.930	0.907	0.83	0.80
Na^+, IO_3^-, HSO_3^-, HCO_3^-, $H_2PO_4^-$, $H_2AsO_4^-$, OAc^-	0.4–0.45	0.965	0.927	0.902	0.82	0.77
OH^-, F^-, SCN^-, HS^-, ClO_3^-, ClO_4^-, BrO_3^-, IO_4^-, MnO_4^-	0.35	0.965	0.926	0.900	0.81	0.76
K^+, Cl^-, Br^-, I^-, CN^-, NO_2^-, NO_3^-, $HCOO^-$	0.3	0.965	0.925	0.899	0.81	0.75
Rb^+, Cs^+, Tl^+, Ag^+, NH_4^+	0.25	0.965	0.925	0.897	0.80	0.75
Mg^{2+}, Be^{2+}	0.8	0.872	0.756	0.690	0.52	0.44
Ca^{2+}, Cu^{2+}, Zn^{2+}, Sn^{2+}, Mn^{2+}, Fe^{2+}, Ni^{2+}, Co^{2+}, Phthalate^{2-}	0.6	0.870	0.748	0.676	0.48	0.40
Sr^{2+}, Ba^{2+}, Cd^{2+}, Hg^{2+}, S^{2-}	0.5	0.869	0.743	0.668	0.46	0.38
Pb^{2+}, CO_3^{2-}, SO_3^{2-}, $C_2O_4^{2-}$	0.45	0.868	0.741	0.665	0.45	0.36
Hg_2^{2+}, SO_4^{2-}, $S_2O_3^{2-}$, CrO_4^{2-}, HPO_4^{2-}	0.40	0.867	0.738	0.661	0.44	0.35
Al^{3+}, Fe^{3+}, Cr^{3+}, La^{3+}, Ce^{3+}	0.9	0.737	0.540	0.443	0.24	0.18
PO_4^{3-}, $Fe(CN)_6^{3-}$	0.4	0.726	0.505	0.394	0.16	0.095
Th^{4+}, Zr^{4+}, Ce^{4+}, Sn^{4+}	1.1	0.587	0.348	0.252	0.10	0.063
$Fe(CN)_6^{4-}$	0.5	0.569	0.305	0.200	0.047	0.020

Source: J. Kielland, *J. Am. Chem. Soc.*, **1937**, *59*, 1675. Courtesy of the American Chemical Society.

When μ is less than 0.01, $1 + \sqrt{\mu} \approx 1$, and Equation 9-5 becomes

$$-\log \gamma_X = 0.51 Z_X^2 \sqrt{\mu}$$

This equation is referred to as the Debye–Hückel limiting law (DHLL). Thus, in solutions of very low ionic strength, the DHLL can be used to calculate approximate activity coefficients.

second term of the denominator is small with respect to the first when the ionic strength is less than 0.01, so that at these ionic strengths, uncertainties in α_A are of little significance in calculating activity coefficients.

J. Kielland[3] has calculated values of α_X for numerous ions from a variety of experimental data. His best values for effective diameters are given in Table 9-1. Also presented are activity coefficients calculated from Equation 9-5 using these values for the size parameter.

Experimental determination of single-ion activity coefficients such as those shown in Table 9-1 is unfortunately impossible because all experimental methods give only a mean activity coefficient for the positively and negatively charged ions in a solution. In other words, it is impossible to measure the properties of individual ions, which necessarily exist, in the presence of counter ions of opposite charge and solvent molecules. Mean activity coefficients calculated from the data in Table 9-1 agree satisfactorily with the experimental values, however.

Example 9-3 demonstrates how activity coefficients may be calculated.

Example 9-3

Use Equation 9-5 to calculate the activity coefficient for Hg^{2+} in a solution that has an ionic strength of 0.085. Use 0.5 nm for the effective diameter of the ion. Compare the calculated value with $\gamma_{Hg^{2+}}$ obtained by interpolation of data from Table 9-1.

---

[3]J. Kielland, *J. Amer. Chem. Soc.*, **1937**, *59*, 1675.

$$-\log \gamma_{Hg^{2+}} = \frac{(0.51)(2)^2\sqrt{0.085}}{1 + (3.3)(0.5)\sqrt{0.085}} = 0.4016$$

$$\gamma_{Hg^{2+}} = antilog(-0.4016)$$

$$\gamma_{Hg^{2+}} = 0.397 \approx 0.40$$

Table 9-1 indicates that $\gamma_{Hg^{2+}} = 0.38$ when $\mu = 0.100$ and 0.46 when $\mu = 0.050$. Thus, for $\mu = 0.085$,

$$\gamma_{Hg^{2+}} = 0.38 + \frac{0.015}{0.050}(0.46 - 0.38) = 0.404 \approx 0.40$$

The Debye–Hückel relationship and the data in Table 9-1 give satisfactory activity coefficients for ionic strengths up to about 0.1. Beyond this value, the equation fails, and experimentally determined mean activity coefficients must be used.

Values for activity coefficients at ionic strengths not listed in Table 9-1 can be approximated by interpolation as shown in Example 9-3.

Feature 9-1

Mean Activity Coefficients

The mean activity coefficient of the electrolyte A_mB_n is defined as

$$\gamma_\pm = \text{mean activity coefficient} = (\gamma_A^m \gamma_B^n)^{1/(m+n)}$$

The mean activity coefficient can be measured in any of several ways, but it is impossible experimentally to resolve this term into the individual activity coefficients for γ_A and γ_B. For example, if

$$K_{sp} = \gamma_A^m \gamma_B^n \cdot [A]^m[B]^n = \gamma_\pm^{m+n}[A]^m[B]^n$$

we can obtain K_{sp} by measuring the solubility of A_mB_n in a solution in which the electrolyte concentration approaches zero (that is, where both γ_A and $\gamma_B \rightarrow 1$). A second solubility measurement at some ionic strength μ_1 gives values for [A] and [B]. These data then permit the calculation of $\gamma_A^m \gamma_B^n = \gamma_\pm^{m+n}$ for ionic strength μ_1.

It is important to understand that this procedure does not provide enough experimental data to permit the calculation of the *individual* quantities γ_A and γ_B and that there appears to be no additional experimental information that would permit evaluation of these quantities. This situation is general, and the *experimental* determination of an individual activity coefficient is impossible.

9B-3   Equilibrium Calculations with Activity Coefficients

Equilibrium calculations with activities yield values that are better in agreement with experimental results than those obtained with molar concentrations. Unless otherwise specified, equilibrium constants found in tables are generally based on activities and are thus thermodynamic equilibrium constants. Examples 9-4 and 9-5 illustrate how activity coefficients from Table 9-1 are applied to such data.

Example 9-4

Find the relative error introduced by neglecting activities in calculating the solubility of $Ba(IO_3)_2$ in a 0.033 M solution of $Mg(IO_3)_2$. The thermodynamic solubility product for $Ba(IO_3)_2$ is 1.57×10^{-9} (Appendix 1).

At the outset, we write the solubility-product expression in terms of activities:

$$a_{Ba^{2+}} \cdot a_{IO_3^-}^2 = K_{sp} = 1.57 \times 10^{-9}$$

where $a_{Ba^{2+}}$ and $a_{IO_3^-}$ are the activities of barium and iodate ions. Replacing activities in this equation by activity coefficients and concentrations from Equation 9-2 yields

$$[Ba^{2+}]\gamma_{Ba^{2+}} \cdot [IO_3^-]^2\gamma_{IO_3^-}^2 = K_{sp}$$

where $\gamma_{Ba^{2+}}$ and $\gamma_{IO_3^-}$ are the activity coefficients for the two ions. Rearranging this expression gives

$$K'_{sp} = \frac{K_{sp}}{\gamma_{Ba^{2+}}\gamma_{IO_3^-}^2} = [Ba^{2+}][IO_3^-]^2 \tag{9-6}$$

where K'_{sp} is the *concentration-based solubility product*.

The ionic strength of the solution is obtained by substituting into Equation 9-1:

$$\mu = \frac{1}{2}([Mg^{2+}] \times 2^2 + [IO_3^-] \times 1^2)$$

$$= \frac{1}{2}(0.033 \times 4 + 0.066 \times 1) = 0.099 \approx 0.1$$

In calculating μ, we have assumed that the Ba^{2+} and IO_3^- ions from the precipitate do not significantly affect the ionic strength of the solution. This simplification seems justified, considering the low solubility of barium iodate and the relatively high concentration of $Mg(IO_3)_2$. In situations where it is not possible to make such an assumption, the concentrations of the two ions can be approximated by a solubility calculation in which activities and concentrations are assumed to be identical (as in Examples 4-4 and 4-5). These concentrations can then be introduced to give a better value for μ.

Turning now to Table 9-1, we find that at an ionic strength of 0.1,

$$\gamma_{Ba^{2+}} = 0.38 \qquad \gamma_{IO_3^-} = 0.77$$

If the calculated ionic strength did not match that of one of the columns in the table, $\gamma_{Ba^{2+}}$ and $\gamma_{IO_3^-}$ could be calculated from Equation 9-5.

Substituting into the thermodynamic solubility-product expression gives

$$K'_{sp} = \frac{1.57 \times 10^{-9}}{(0.38)(0.77)^2} = 6.8 \times 10^{-9}$$

$$[Ba^{2+}][IO_3^-]^2 = 6.8 \times 10^{-9}$$

Proceeding now as in earlier solubility calculations yields

$$\text{solubility} = [Ba^{2+}]$$

$$[IO_3^-] = 2 \times 0.033 + 2[Ba^{2+}] \approx 0.066$$

$$[Ba^{2+}](0.066)^2 = 6.8 \times 10^{-9}$$

$$[Ba^{2+}] = \text{solubility} = 1.56 \times 10^{-6}\,M$$

If we neglect activities, the solubility is

$$[Ba^{2+}](0.066)^2 = 1.57 \times 10^{-9}$$

$$[Ba^{2+}] = \text{solubility} = 3.60 \times 10^{-7}\,M$$

$$\text{relative error} = \frac{3.60 \times 10^{-7} - 1.56 \times 10^{-6}}{1.56 \times 10^{-6}} \times 100\% = -77\%$$

Example 9-5

Use activities to calculate the hydronium ion concentration in a 0.120 M solution of HNO_2 that is also 0.050 M in NaCl.

The ionic strength of this solution, neglecting dissociation of HNO_2, is

$$\mu = \frac{1}{2}(0.0500 \times 1^2 + 0.0500 \times 1^2) = 0.0500$$

In Table 9-1, at ionic strength 0.050, we find

$$\gamma_{H_3O^+} = 0.85 \qquad \gamma_{NO_2^-} = 0.81$$

Also, from rule 3 (page 209), we can write

$$\gamma_{HNO_2} = 1.0$$

These three values for activity coefficients permit the calculation of a concentration-based dissociation constant from the thermodynamic constant of 7.1×10^{-4} (Appendix 2):

$$K_a' = \frac{[H_3O^+][NO_2^-]}{[HNO_2]} = \frac{K_a \cdot \gamma_{HNO_2}}{\gamma_{H_3O^+}\gamma_{NO_2^-}} = \frac{7.1 \times 10^{-4} \times 1.0}{0.85 \times 0.81} = 10.3 \times 10^{-4}$$

Proceeding as in Example 8-7, we write

$$[H_3O^+] = \sqrt{0.120 \times 10.3 \times 10^{-4}} = 11.1 \times 10^{-3}$$

Note that assuming activity coefficients are unity gives $[H_3O^+] = 9.2 \times 10^{-3}$.

9B-4   Omitting Activity Coefficients in Equilibrium Calculations

We shall ordinarily neglect activity coefficients and simply use molar concentrations in applications of the equilibrium law. This recourse simplifies the calculations and greatly decreases the amount of data needed. For most purposes, the error introduced by the assumption of unity for the activity coefficient is not large enough to lead to false conclusions. It should be apparent from the preceding examples, however, that disregard of activity coefficients may introduce a significant numerical error in calculations of this kind. Note, for example, that neglect of activities in Example 9-4 resulted in an error of about -77% relative. The reader should be alert to the conditions under which the substitution of concentration for activity is likely to lead to the largest error. Significant discrepancies occur when the ionic strength is large (0.01 or larger) or when the ions involved have multiple charges (Table 9-1). With dilute solutions (ionic strength <0.01) of nonelectrolytes or of singly charged ions, the use of concentrations in a mass-law calculation often provides reasonably accurate results.

It is also important to note that the decrease in solubility resulting from the presence of an ion common to the precipitate is in part counteracted by the larger electrolyte concentration associated with the presence of the salt containing the common ion.

Spreadsheet Exercise   Using Circular References for Iteratively Solving Equations

Microsoft
Excel

In this spreadsheet exercise, we use circular references to solve equilibrium problems by iteration and study the effects of diverse ions on solubility. We will begin by finding the solubility of TlCl in 0.1 M NaCl without activity corrections. We have chosen TlCl because its solubility is sufficiently high ($K_{sp} = 1.7 \times 10^{-4}$) so that the ions of the salt have an appreciable and measurable effect on the activity coefficients. Let us first look at the solubility expressions for TlCl without activity corrections:

$$K_{sp} = [Tl^+][Cl^-] = [Tl^+](c_{NaCl} + [Tl^+]) \qquad \text{and}$$

$$[Tl^+] = \frac{K_{sp}}{c_{NaCl} + [Tl^+]} = \frac{K_{sp}}{0.1 + [Tl^+]}$$

We could solve this equation iteratively for $[Tl^+]$ as we did in the Spreadsheet Exercise in Chapter 4, but Excel provides an even better way to accomplish the task in a single spreadsheet cell.

Begin with a clean spreadsheet, click on Tools/Options..., make sure there is a check mark in Iteration, and enter 500 in Maximum iterations: and 0.000001 in Maximum change:. Then enter the following in cells A1:A2:

```
Ksp[↵]
1.7e-4[↵]
```

Then in cells A4:A5 enter

```
c(NaCl)[↵]
0.1[↵]
```

Enter the following formula for $[Tl^+]$ in cells B4:B5:

```
s=[Tl+][↵]
=$a$2/(b5+a5)[↵]
```

Notice that this formula, which is in cell B5, refers to cell B5. This reference is called a *circular reference*. In ordinary calculations, you would not want to use the result of a computation to compute a value, but in this case, we want Excel to iterate and find a value for cell B5 that satisfies our equation. If you have not set the Excel options as suggested above to iterate, an error message will appear as shown below.

Whenever a circular reference is used to perform an iterative calculation, the result should be checked by an independent calculation. This calculation ensures that the result has been computed to the desired level of precision.

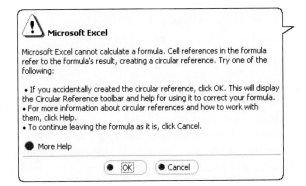

If you receive such a message, click on Cancel to leave the intended circular reference in place, and check to make sure that you have set the options correctly. If you have done this, the following should appear in the spreadsheet.

	A	B
1	Ksp	
2	1.70E-04	
3		
4	c(NaCl)	s=[Tl+]
5	0.1	0.001672

Excel has iteratively solved the quadratic equation to find the solubility of TlCl in 0.1 M NaCl to be 1.7×10^{-3} M.

Making Activity Corrections

Now we will calculate the solubility using activity corrections based on the Debye–Hückel equation. The thermodynamic solubility product constant is given by

$$K_{sp} = a_{Tl^+} a_{Cl^-} = \gamma_{Tl^+} \gamma_{Cl^-} [Tl^+][Cl^-]$$

$$[Tl^+] = \frac{K_{sp}}{\gamma_{Tl^+} \gamma_{Cl^-} [Cl^-]} = \frac{K_{sp}}{\gamma_{Tl^+} \gamma_{Cl^-} (c_{NaCl} + [Tl^+])}$$

We determine the activity coefficients from the Debye–Hückel equation as follows:

$$\log \gamma_{Tl^+} = \frac{-0.51 Z_{Tl^+}^2 \sqrt{\mu}}{1 + 3.3 \alpha_{Tl^+} \sqrt{\mu}} \quad \text{or} \quad \gamma_{Tl^+} = 10^{-0.51\sqrt{\mu}/[1+3.3(0.25)\sqrt{\mu}]}$$

and

$$\log \gamma_{Cl^-} = \frac{-0.51 Z_{Cl^-}^2 \sqrt{\mu}}{1 + 3.3 \alpha_{Cl^-} \sqrt{\mu}} \qquad \text{or} \qquad \gamma_{Cl^-} = 10^{-0.51\sqrt{\mu}/[1+3.3(0.3)\sqrt{\mu}]}$$

where

$$\mu = 0.5([Tl^+]Z_{Tl^+}^2 + [Na^+]Z_{Na^+}^2 + [Cl^-]Z_{Cl^-}^2)$$

$$= 0.5\{[Tl^+]Z_{Tl^+}^2 + c_{NaCl}Z_{Na^+}^2 + (c_{NaCl} + [Tl^+])Z_{Cl^-}^2\}$$

$$= [Tl^+] + c_{NaCl} \qquad \text{when} \qquad Z_{Tl^+}^2 = Z_{Na^+}^2 = Z_{Cl^-}^2 = 1$$

Since the charges of all the ions are unity and the α_X values are 0.25 and 0.3 for thallium and chloride, respectively, we have substituted these values into the final expressions above. We have now written expressions for the solubility of TlCl in terms of the activity coefficients and c_{NaCl}, the activity coefficients in terms of ionic strength, and the ionic strength in terms of the solubility and c_{NaCl}. In other words, because the ionic strength depends on the solubility and the solubility depends on ionic strength, we have to use Excel to iteratively solve the set of equations.

Enter the following labels in cells D4:G4. Type the right arrow key [→] to move laterally:

```
s=[Tl+] [→] mu [→] gamma(Tl) [→] gamma(Cl)
```

Now click on cell E5, and type the expressions for the ionic strength:

```
=d5+a5 [→]
=10^(-0.51*sqrt(e5)/(1+3.3*0.25*sqrt(e5))) [→]
=10^(-0.51*sqrt(e5)/(1+3.3*0.3*sqrt(e5))) [↵]
```

Your spreadsheet should now appear as follows.

	A	B	C	D	E	F	G
1	Ksp						
2	1.70E-04						
3							
4	c(NaCl)	s = [Tl+]		s = [Tl+]	mu	gamma(Tl)	gamma(Cl)
5	0.1	0.001672			0.1	0.744892	0.753661

At this point in our analysis, Excel has calculated the activity coefficients based on just the concentration of NaCl. As soon as we type the formula for solubility in cell D5, the TlCl contribution to the ionic strength will appear and the activity coefficients will be recalculated, which will produce a new value for the solubility. This iterative process will continue until Excel finds consistent values in cells D5:G5. Let us now enter the following formula in cell D5:

```
=$a$2/(f5*g5*(a5+d5))
```

Note that Excel immediately performs the iterative solution, and your spreadsheet should appear as follows.

	A	B	C	D	E	F	G
1	Ksp						
2	1.70E-04						
3							
4	c(NaCl)	s = [Tl+]		s = [Tl+]	mu	gamma(Tl)	gamma(Cl)
5	0.1	0.001672		0.00296	0.10296	0.742347	0.751287

There are several interesting results from this calculation. First, notice that the ionic strength is only about 3% larger than the concentration of NaCl. In addition, the activity coefficients have not changed much from the values obtained when we had not yet included the concentrations of Tl^+ and Cl^-. What is surprising is that the solubility has increased by nearly a factor of 2 relative to our original estimate in cell B5.

Studying Solubility as a Function
of Ionic Strength

Now change c_{NaCl} to 1.0, and note the effect. Change the value successively to 0.1, 0.01, 0.001, 0.0001, and so on, and note the various effects. Notice that it takes Excel longer to achieve an iterative solution as the ionic strength becomes smaller. Let us now create a table of solubility as a function of ionic strength that will enable us to make plots of various combinations of the variables in this system.

We begin by typing the following data in cells A5:A8.

$$0.0001 [\dashv]$$
$$0.0002 [\dashv]$$
$$0.0004 [\dashv]$$
$$0.0008 [\dashv]$$

Then in cells A9:A12 type

$$=a5*10 [\dashv]$$
$$=a6*10 [\dashv]$$
$$=a7*10 [\dashv]$$
$$=a8*10 [\dashv]$$

Select cells A9:A12, and drag the fill handle to cell A25. This method is an elegant way of copying a series of cells that contains a regular series of numbers or formulas. Excel detects the pattern and duplicates the cells according to the pattern. Finally, select cells B5:G5, and drag the fill handle to row 21. When Excel finishes the iteration process for all 17 rows, your spreadsheet should look like the following.

	A	B	C	D	E	F	G
1	Ksp						
2	1.70E-04						
3							
4	c(NaCl)	s = [Tl+]		s = [Tl+]	mu	gamma(Tl)	gamma(Cl)
5	0.0001	0.012989		0.014784	0.014884	0.877951	0.880006
6	0.0002	0.012939		0.014737	0.014937	0.877765	0.879826
7	0.0004	0.01284		0.014644	0.015044	0.877393	0.879467
8	0.0008	0.012645		0.014461	0.015261	0.876647	0.878746
9	0.001	0.012548		0.01437	0.01537	0.876273	0.878384
10	0.002	0.012077		0.013926	0.015926	0.874394	0.876569
11	0.004	0.011191		0.01309	0.01709	0.870602	0.872909
12	0.008	0.009638		0.011608	0.019608	0.862941	0.865526
13	0.01	0.008964		0.010956	0.020956	0.859108	0.861838
14	0.02	0.006432		0.00843	0.02843	0.840444	0.843934
15	0.04	0.003875		0.005668	0.045668	0.807881	0.812913
16	0.08	0.002071		0.003487	0.083487	0.760336	0.768099
17	0.1	0.001672		0.002961	0.102961	0.742347	0.751287
18	0.2	0.000846		0.001784	0.201784	0.680535	0.694106
19	0.4	0.000425		0.001091	0.401091	0.613553	0.633109
20	0.8	0.000212		0.000678	0.800678	0.546336	0.572815
21	1	0.00017		0.000583	1.000583	0.525416	0.554219

Observe the various trends in the data. Note what happens to the solubility of TlCl as the ionic strength approaches zero. What happens to the activity coefficients as the ionic strength approaches zero? How do solubilities calculated without activity corrections compare to activity-based solubilities? Although you can see these trends in a table of numbers, plotting the data in various ways is useful and informative.

Visualizing the Solubility Data

The Excel Chart Wizard was used to produce the following plots of the data in the spreadsheet. Create these same plots yourself. To do so, you will need to tabulate six more columns of results in the spreadsheet: $\log \gamma_{Tl^+}$, $\log \gamma_{Cl^-}$, $\sqrt{\mu}$, $ps = -\log s$ (activity based), $ps = -\log s$ (concentration based), and $pc_{NaCl} = -\log c_{NaCl}$. Once you have created these results, use the Chart Wizard to create the plots. If you see other combinations of variables that seem interesting, plot those as well.

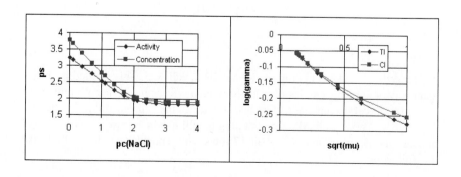

This spreadsheet exercise may be extended in several ways. First, look up in Appendix 1 the solubility product for another uni-univalent salt such as AgCl, and type its K_{sp} in cell A2. All the other cells in the spreadsheet will recalculate automatically, and the plots will be updated. If you have made changes in the scaling on either of the axes, you must change the scaling to automatic by right-clicking on the axis, selecting Format Axis..., selecting the Scale tab, and checking Auto in all the boxes.

If you wish to extend the exercise to salts that are not uni-univalent, such as $Ca_3(PO_4)_2$, you must rewrite the solubility equations, the equation for ionic strength, and the Debye–Hückel expressions and then edit the formulas in the spreadsheet accordingly. It is important to note that the calculations you just completed in this exercise would have required many hours of work if you had performed them manually. The advent of spreadsheets and other similar computer-based computational aids has greatly extended our ability to carry out sophisticated and intensive numerical analysis of chemical systems and to present the results in graphical form.

WEB WORKS

Point your web browser to http://maple.lemoyne.edu/~giunta/papers.html. This site presents a comprehensive set of links to *Selected Classic Papers from the History of Chemistry*. Locate the link to the famous 1923 paper[2] by Debye and Hückel on the theory of electrolytic solutions, and click on it. Read the paper, and compare the notation in their paper to the notation in this chapter. What symbol do the authors use for the activity coefficient? What important phenomena do the authors relate to their theory? Note that the mathematical details are missing from the translation of the paper.

9C   QUESTIONS AND PROBLEMS

*9-1. Make a distinction between
   (a) activity and activity coefficient.
   (b) thermodynamic and concentration equilibrium constants.

9-2. List general properties of activity coefficients.

*9-3. Neglecting any effects caused by volume changes, would you expect the ionic strength to (1) increase, (2) decrease, or (3) remain essentially unchanged by the addition of NaOH to a dilute solution of
   (a) magnesium chloride [$Mg(OH)_2(s)$ forms]?
   (b) hydrochloric acid?
   (c) acetic acid?

9-4. Neglecting any effects caused by volume changes, would you expect the ionic strength to (1) increase, (2) decrease, or (3) remain essentially unchanged by the addition of iron(III) chloride to
   (a) HCl?

   (b) NaOH?
   (c) $AgNO_3$?

*9-5. Why is the slope of the curve for Ca^{2+} in Figure 9-3 greater than that for K^+?

9-6. What is the numerical value of the activity coefficient of aqueous ammonia (NH_3) at an ionic strength of 0.1?

9-7. Calculate the ionic strength of a solution that is
   *(a) 0.040 M in $FeSO_4$.
   (b) 0.20 M in $(NH_4)_2CrO_4$.
   *(c) 0.10 M in $FeCl_3$ and 0.20 M in $FeCl_2$.
   (d) 0.060 M in $La(NO_3)_3$ and 0.030 M in $Fe(NO_3)_2$.

9-8. Use Equation 9-5 to calculate the activity coefficient of
   *(a) Fe^{3+} at $\mu = 0.075$.
   (b) Pb^{2+} at $\mu = 0.012$.
   *(c) Ce^{4+} at $\mu = 0.080$.
   (d) Sn^{4+} at $\mu = 0.060$.

9-9. Calculate activity coefficients for the species in Problem 9-8 by linear interpolation of the data in Table 9-1.

9-10. For a solution in which μ is 5.0×10^{-2}, calculate K'_{sp} for

*(a) AgSCN.   (b) PbI$_2$.

*(c) La(IO$_3$)$_3$.   (d) MgNH$_4$PO$_4$.

*9-11. Use activities to calculate the molar solubility of Zn(OH)$_2$ in

(a) 0.0100 M KCl.

(b) 0.0167 M K$_2$SO$_4$.

(c) the solution that results when you mix 20.0 mL of 0.250 M KOH with 80.0 mL of 0.0250 M ZnCl$_2$.

(d) the solution that results when you mix 20.0 mL of 0.100 M KOH with 80.0 mL of 0.0250 M ZnCl$_2$.

*9-12. Calculate the solubilities of the following compounds in a 0.0333 M solution of Mg(ClO$_4$)$_2$ using (1) activities and (2) molar concentrations:

(a) AgSCN.

(b) PbI$_2$.

(c) BaSO$_4$.

(d) Cd$_2$Fe(CN)$_6$

$[Cd_2Fe(CN)_6(s) \rightleftharpoons 2Cd^{2+} + Fe(CN)_6^{4-}$

$K_{sp} = 3.2 \times 10^{-17}]$.

*9-13. Calculate the solubilities of the following compounds in a 0.0167 M solution of Ba(NO$_3$)$_2$ using (1) activities and (2) molar concentrations:

(a) AgIO$_3$.

(b) Mg(OH)$_2$.

(c) BaSO$_4$.

(d) La(IO$_3$)$_2$.

9-14. (a) Repeat the computations of Problem 9-13 using a spreadsheet. Vary the concentration of Ba(NO$_3$)$_2$ from 0.0001 M to 1 M in a manner similar to that used in the spreadsheet exercise.

(b) Plot ps versus pc, where pc is the negative logarithm of the concentration of Ba(NO$_3$)$_2$.

9-15. Design and construct a spreadsheet to calculate activity coefficients in a format similar to Table 9-1. Enter values of α_X in cells A3, A4, A5, and so forth, and enter ionic charges in cells B3, B4, B4, and so forth. Enter in cells C2:G2 the same set of values for ionic strength listed in Table 9-1. Enter the formula for the activity coefficients in cells C3:G3. Be sure to use absolute cell references for ionic strength in your formulas for the activity coefficients. Finally, copy the formulas for the activity coefficients into the rows below row C by highlighting C3:G3 and dragging the fill handle downward. Compare the activity coefficients that you calculate to those in Table 9-1. Do you find any discrepancies? If so, explain how they arise.

Chapter 10

How Equilibrium Calculations Can Be Applied to Complex Systems

A queous solutions encountered in the laboratory often contain several species that interact with one another and water to yield two or more equilibria that function simultaneously. For example, when water is saturated with the sparingly soluble barium sulfate, three equilibria develop:

$$BaSO_4(s) \rightleftharpoons Ba^{2+} + SO_4^{2-} \tag{10-1}$$

$$SO_4^{2-} + H_3O^+ \rightleftharpoons HSO_4^- + H_2O \tag{10-2}$$

$$2H_2O \rightleftharpoons H_3O^+ + OH^- \tag{10-3}$$

If hydronium ions are added to this system, the second equilibrium is shifted to the right by the common-ion effect. The resulting decrease in sulfate concentration causes the first equilibrium to shift to the right as well, which increases the solubility of the barium sulfate.

The solubility of barium sulfate is also increased when acetate ions are added to an aqueous suspension of barium sulfate because acetate ions tend to form a soluble complex with barium ions as shown by the reaction

$$Ba^{2+} + OAc^- \rightleftharpoons BaOAc^+ \tag{10-4}$$

Again, the common-ion effect causes both this equilibrium and the solubility equilibrium to shift to the right; an increase in solubility results.

If we wish to calculate the solubility of barium sulfate in a system containing hydronium and acetate ions, we must take into account not only the solubility equilibrium, but also the other three equilibria. We find, however, that employing four equilibrium-constant expressions to calculate a solubility is much more difficult and complex than the rather simple procedure illustrated in Examples 4-3, 4-4, and 4-5. To solve this type of problem, we will find it helpful to use a systematic approach that is described in Section 10A. We then use this approach to illustrate the effect of pH and complex formation on the solubility of typical analytical precipitates. In subsequent chapters, we will use this same systematic method for solution of problems involving multiple equilibria of several types.

The introduction of a new equilibrium system into a solution does not change the equilibrium constants for any existing equilibria.

221

10A   SOLVING MULTIPLE-EQUILIBRIUM PROBLEMS BY A SYSTEMATIC METHOD

Solution of a multiple-equilibrium problem requires us to develop as many independent equations as there are participants in the system being studied. For example, if we wish to compute the solubility of barium sulfate in a solution of acid, we need to be able to calculate the concentration of all the species present in the solution. There are five species: Ba^{2+}, SO_4^{2-}, HSO_4^-, H_3O^+, and OH^-. To calculate the solubility of barium sulfate in this solution rigorously, it would then be necessary to develop five independent algebraic equations containing these five unknowns and solve them simultaneously.

Three types of algebraic equations are used in solving multiple-equilibrium problems: (1) equilibrium-constant expressions, (2) *mass-balance* equations, and (3) a single *charge-balance* equation. We have already shown in Section 4B how equilibrium-constant expressions are written; we must now turn our attention to the development of the other two types of equations.

10A-1   Mass-Balance Equations

The term *mass-balance equation,* although widely used, is somewhat misleading because such equations are really based on balancing *concentrations* rather than *masses.* On the other hand, since all species are in the same volume of solvent, equating masses to concentrations does not create a problem.

Mass-balance or material-balance equations relate the *equilibrium* concentrations of various species in a solution to one another and to the *analytical* concentrations of the various solutes. They are derived from information about how the solution was prepared and from a knowledge of the kinds of equilibria established in the solution.

Example 10-1

Write mass-balance expressions for a 0.0100 M solution of HCl that is in equilibrium with an excess of solid $BaSO_4$.

From our general knowledge of the behavior of aqueous solutions, we can write equations for three equilibria that must be present in this solution.

$$BaSO_4(s) \rightleftharpoons Ba^{2+} + SO_4^{2-}$$

$$SO_4^{2-} + H_3O^+ \rightleftharpoons HSO_4^- + H_2O$$

$$2H_2O \rightleftharpoons H_3O^+ + OH^-$$

For a slightly soluble salt with a 1:1 stoichiometry, the equilibrium concentration of the cation is equal to equilibrium concentration of the anion. This equality is the mass-balance expression.

Because the only source for the two sulfate species is the dissolved $BaSO_4$, the barium ion concentration must equal the total concentration of sulfate-containing species, and a mass-balance equation can be written that expresses this equality. Thus,

$$[Ba^{2+}] = [SO_4^{2-}] + [HSO_4^-]$$

The hydronium ion concentration in this solution has two sources: one from the HCl and the other from the dissociation of the solvent. A second mass-balance expression is thus

$$[H_3O^+] + [HSO_4^-] = c_{HCl} + [OH^-] = 0.0100 + [OH^-]$$

Since the only source of hydroxide is the dissociation of water, $[OH^-]$ is equal to the hydronium ion concentration from the dissociation of water.

Example 10-2

Write mass-balance expressions for the system formed when a 0.010 M NH_3 solution is saturated with AgBr.

Here, equations for the pertinent equilibria in the solution are

$$AgBr(s) \rightleftharpoons Ag^+ + Br^-$$

$$Ag^+ + 2NH_3 \rightleftharpoons Ag(NH_3)_2^+$$

$$NH_3 + H_2O \rightleftharpoons NH_4^+ + OH^-$$

$$2H_2O \rightleftharpoons H_3O^+ + OH^-$$

Because the only source of Br^-, Ag^+, and $Ag(NH_3)_2^+$ is AgBr and because silver and bromide ions are present in a $1:1$ ratio in that compound, it follows that one mass-balance equation is

$$[Ag^+] + [Ag(NH_3)_2^+] = [Br^-]$$

where the bracketed terms are molar species concentrations. Also, we know that the only source of ammonia-containing species is the 0.010 M NH_3. Therefore,

$$c_{NH_3} = [NH_3] + [NH_4^+] + 2[Ag(NH_3)_2^+] = 0.010$$

From the last two equilibria, we see that one hydroxide ion is formed for each NH_4^+ and each hydronium ion. Therefore,

$$[OH^-] = [NH_4^+] + [H_3O^+]$$

For slightly soluble salts with stoichiometries other than $1:1$, the mass-balance expression is obtained by multiplying the concentration of one of the ions by the stoichiometric ratio. For example, in a solution saturated with Ag_2CrO_4 the silver ion concentration is twice that of the CrO_4^{2-}. That is,

$$[Ag^+] = 2[CrO_4^{2-}]$$

10A-2 Charge-Balance Equation

We know that electrolyte solutions are electrically neutral even though they may contain many millions of charged ions. Solutions are neutral because the *molar concentration of positive charge* in an electrolyte solution always equals the *molar concentration of negative charge*. That is, for any solution containing electrolytes, we may write

$$\text{no. mol/L positive charge} = \text{no. mol/L negative charge}$$

This equation represents the charge-balance condition and is called the *charge-balance equation*. To be useful for equilibrium calculations, the equality must be expressed in terms of the molar concentrations of the species that carry a charge in the solution.

How much charge is contributed to a solution by 1 mol of Na^+? Or, how about 1 mol of Mg^{2+} or 1 mol of PO_4^{3-}? The concentration of charge contributed to a solution by an ion is equal to the molar concentration of that ion multiplied by its charge. Thus, the molar concentration of positive charge in a solution due to the presence of sodium ions is the molar sodium ion concentration. That is,

$$\frac{\text{mol positive charge}}{L} = \frac{1 \text{ mol positive charge}}{\text{mol } Na^+} \times \left(\frac{\text{mol } Na^+}{L} \right)$$

$$= 1 \times [Na^+]$$

The concentration of positive charge due to magnesium ions is

$$\frac{\text{mol positive charge}}{L} = \frac{2 \text{ mol positive charge}}{\cancel{\text{mol } Mg^{2+}}} \times \left(\frac{\cancel{\text{mol } Mg^{2+}}}{L} \right)$$

$$= 2 \times [Mg^{2+}]$$

since each mole of magnesium ion contributes 2 mol of positive charge to the solution. Similarly, we may write for phosphate ion that

$$\frac{\text{mol negative charge}}{L} = \frac{3 \text{ mol negative charge}}{\cancel{\text{mol } PO_4^{3-}}} \times \left(\frac{\cancel{\text{mol } PO_4^{3-}}}{L} \right)$$

$$= 3 \times [PO_4^{3-}]$$

Now, consider how we would write a charge-balance equation for a 0.100 M solution of sodium chloride. Positive charges in this solution are supplied by Na^+ and H_3O^+ (from the dissociation of water). Negative charges come from Cl^- and OH^-. The molarities of positive and negative charges are

$$\text{mol/L positive charge} = [Na^+] + [H_3O^+] = 0.100 + 1 \times 10^{-7}$$

$$\text{mol/L negative charge} = [Cl^-] + [OH^-] = 0.100 + 1 \times 10^{-7}$$

We write the charge-balance equation by equating the concentrations of positive and negative charges. That is,

$$[Na^+] + [H_3O^+] = [Cl^-] + [OH^-] = 0.100 + 1 \times 10^{-7}$$

Let us now consider a solution of magnesium chloride that has an analytical concentration of 0.100 M. Here the molarities of positive and negative charge are given by

$$\text{mol/L positive charge} = 2[Mg^{2+}] + [H_3O^+] = 2 \times 0.100 + 1 \times 10^{-7}$$

$$\text{mol/L negative charge} = [Cl^-] + [OH^-] = 2 \times 0.100 + 1 \times 10^{-7}$$

In the first equation, the molar concentration of magnesium ion is multiplied by two (2×0.100) because 1 mol of that ion contributes 2 mol of positive charge to the solution. In the second equation, the molar chloride ion concentration is twice that of the magnesium chloride concentration of 2×0.100. To obtain the charge-balance equation, we equate the concentration of positive charge with the concentration of negative charge to obtain

$$2[Mg^{2+}] + [H_3O^+] = [Cl^-] + [OH^-] = 0.200 + 1 \times 10^{-7}$$

For a neutral solution, $[H_3O^+]$ and $[OH^-]$ are very small and equal so that we can ordinarily simplify the charge-balance equation to

$$2[Mg^{2+}] = [Cl^-] = 0.200$$

Always remember that a charge-balance equation is based on the equality in molar charge concentrations and that to obtain the charge concentration of an ion you must multiply the molar concentration of the ion by its charge.

Example 10-3

Write a charge-balance equation for the system in Example 10-2.

$$[Ag^+] + [Ag(NH_3)_2^+] + [H_3O^+] + [NH_4^+] = [OH^-] + [Br^-]$$

Example 10-4

Neglecting the dissociation of water, write a charge-balance equation for a solution that contains NaCl, $Ba(ClO_4)_2$, and $Al_2(SO_4)_3$.

$$[Na^+] + 2[Ba^{2+}] + 3[Al^{3+}] = [ClO_4^-] + [NO_3^-] + 2[SO_4^{2-}]$$

10A-3   Steps for Solving Problems Involving Several Equilibria

Step 1. Write a set of balanced chemical equations for all pertinent equilibria.

Step 2. State in terms of equilibrium concentrations what quantity is being sought.

Step 3. Write equilibrium-constant expressions for all equilibria developed in step 1, and find numerical values for the constants in tables of equilibrium constants.

Step 4. Write mass-balance expressions for the system.

Step 5. If possible, write a charge-balance expression for the system.

Step 6. Count the number of unknown concentrations in the equations developed in steps 3, 4, and 5, and compare this number with the number of independent equations. If the number of equations is equal to the number of unknowns, proceed to step 7. If the number is not, seek additional equations. If enough equations cannot be developed, try to eliminate unknowns by suitable approximations regarding the concentration of one or more of the unknowns. If such approximations cannot be found, the problem cannot be solved.

Step 7. Make suitable approximations to simplify the algebra.

Step 8. Solve the algebraic equations for the equilibrium concentrations needed to give a provisional answer as defined in step 2.

Step 9. Check the validity of the approximations made in step 7 using the provisional concentrations computed in step 8.

These steps are shown in Figure 10-1.

Step 6 is critical because it shows whether or not an exact solution to the problem is possible. If the number of unknowns is identical to the number of equations, the problem has been reduced to one of *algebra* alone. That is, answers can be obtained with sufficient perseverance. On the other hand, if there are not enough equations even after approximations are made, the problem should be abandoned.

10A-4   Making Approximations to Solve Equilibrium Equations

When step 6 of the systematic approach is complete, we have a *mathematical* problem of solving several nonlinear simultaneous equations. This job is often formidable, tedious, and time consuming unless a suitable computer program is available or unless approximations can be found that decrease the number of unknowns and equations. In this section, we consider in general terms how equations describing equilibrium relationships can be simplified by suitable approximations.

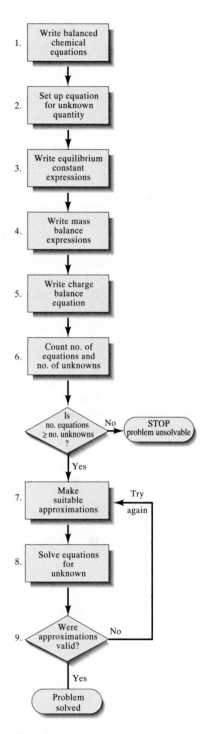

Figure 10-1 A systematic method for solving multiple equation problems.

Bear in mind that *only* the mass-balance and charge-balance equations can be simplified because only in these equations do the concentration terms appear as sums or differences rather than as products or quotients. It is always possible to assume that one (or more) of the terms in a sum or difference is so much smaller than the others that it can be ignored without significantly affecting the equality. The assumption that a concentration term in an equilibrium-constant expression is zero makes the expression meaningless.

The assumption that a given term in a mass- or charge-balance equation is sufficiently small that it can be neglected is generally based on a knowledge of the chemistry of the system. For example, in a solution containing a reasonable concentration of an acid, the hydroxide concentration will often be negligible with respect to many other species in the solution and the term for the hydroxide concentration can usually be neglected in a mass- or charge-balance expression without introducing a significant error in a calculation.

Many students find step 7 to be the most troublesome because they fear that making invalid approximations will lead to serious errors in their computed results. *Such fears are groundless.* Experienced scientists are often as puzzled as beginners when making an approximation that simplifies an equilibrium calculation. Nonetheless, they make such approximations without fear because they know that the effects of an invalid assumption will become obvious by the time a computation is completed. Generally, questionable assumptions should be tried at the outset and provisional answers computed. If the assumption leads to an intolerable error (which is easily recognized), a recalculation without the faulty approximation is then performed. Usually, it is more efficient to try a questionable assumption at the outset than to make a more time-consuming and tedious calculation without the assumption.

10B   CALCULATING SOLUBILITIES BY THE SYSTEMATIC METHOD

The use of the systematic method is illustrated in this section with examples involving the solubility of precipitates under various conditions. In later chapters, we will apply this method to other types of equilibria.

10B-1   Solubility Calculations

Example 10-5 involves calculating the solubilities of metal hydroxides. This example illustrates making an approximation and checking its validity.

Example 10-5

Calculate the molar solubility of $Mg(OH)_2$ in water.

Step 1.   **Pertinent Equilibria**

Two equilibria that need to be considered are

$$Mg(OH)_2(s) \rightleftharpoons Mg^{2+} + 2OH^-$$

$$2H_2O \rightleftharpoons H_3O^+ + OH^-$$

Step 2. **Definition of the Unknown**

Since 1 mol of Mg^{2+} is formed for each mole of $Mg(OH)_2$ dissolved,

$$\text{solubility } Mg(OH)_2 = [Mg^{2+}]$$

Step 3. **Equilibrium-Constant Expressions**

$$K_{sp} = [Mg^{2+}][OH^-]^2 = 7.1 \times 10^{-12} \qquad (10\text{-}5)$$

$$K_w = [H_3O^+][OH^-] = 1.00 \times 10^{-14} \qquad (10\text{-}6)$$

Step 4. **Mass-Balance Expression**

As shown by the two equilibrium equations, there are two sources of hydroxide ions: $Mg(OH)_2$ and H_2O. The hydroxide ion resulting from dissociation of $Mg(OH)_2$ is twice the magnesium ion concentration and that from the dissociation of water is equal to the hydronium ion concentration. Thus,

$$[OH^-] = 2[Mg^{2+}] + [H_3O^+] \qquad (10\text{-}7)$$

Step 5. **Charge-Balance Expression**

$$[OH^-] = 2[Mg^{2+}] + [H_3O^+]$$

Note that this equation is identical to Equation 10-7. Often a mass-balance and a charge-balance equation are the same.

Step 6. **Number of Independent Equations and Unknowns**

We have developed three independent algebraic equations (Equations 10-5, 10-6, and 10-7) and have three unknowns ($[Mg^{2+}]$, $[OH^-]$, and $[H_3O^+]$). Therefore, the problem can be solved rigorously.

Step 7. **Approximations**

We can make approximations only in Equation 10-7. Since the solubility-product constant for $Mg(OH)_2$ is relatively large, the solution will be somewhat basic. Therefore, it is reasonable to assume that $[H_3O^+] \ll [OH^-]$. Equation 10-7 then simplifies to

$$2[Mg^{2+}] \approx [OH^-] \qquad (10\text{-}8)$$

Step 8. **Solution to Equations**

Substitution of Equation 10-8 into Equation 10-5 gives

$$[Mg^{2+}](2[Mg^{2+}])^2 = 7.1 \times 10^{-12}$$

$$[Mg^{2+}]^3 = \frac{7.1 \times 10^{-12}}{4} = 1.78 \times 10^{-12}$$

$$[Mg^{2+}] = \text{solubility} = 1.21 \times 10^{-4} \text{ or } 1.2 \times 10^{-4} \text{ M}$$

Step 9. **Check of Assumptions**

Substitution into Equation 10-8 yields

$$[OH^-] = 2 \times 1.21 \times 10^{-4} = 2.42 \times 10^{-4}$$

We arrived at Equation 10-7 with the following reasoning. If $[OH^-]_{H_2O}$ and $[OH^-]_{Mg(OH)_2}$ are the concentrations of OH^- produced from H_2O and $Mg(OH)_2$, respectively, then

$$[OH^-]_{H_2O} = [H_3O^+]$$
$$[OH^-]_{Mg(OH)_2} = 2[Mg^{2+}]$$

$$[OH^-]_{total} = [OH^-]_{H_2O} + [OH^-]_{Mg(OH)_2}$$
$$= [H_3O^+] + 2[Mg^{2+}]$$

$$[H_3O^+] = \frac{1.00 \times 10^{-14}}{2.42 \times 10^{-4}} = 4.1 \times 10^{-11}$$

Thus, our assumption that $4.1 \times 10^{-11} \ll 1.6 \times 10^{-4}$ is certainly valid.

10B-2   How pH Influences Solubility

All precipitates that contain an anion that is the conjugate base of a weak acid are more soluble at low than at high pH.

The solubility of a precipitate containing an anion with basic properties, a cation with acidic properties, or both will depend on pH. Example 10-6 illustrates how the effect of pH on solubility can be treated in quantitative terms.

Solubility Calculations at Constant pH

A *buffer* keeps the pH of a solution nearly constant (see Section 12C-2).

Analytical precipitations are frequently performed in buffered solutions in which the pH is fixed at some predetermined and known value. The calculation of solubility under this circumstance is illustrated by Example 10-6.

Example 10-6

Calculate the molar solubility of calcium oxalate in a solution that has been buffered so that its pH is constant and equal to 4.00.

Step 1.   **Pertinent Equilibria**

$$CaC_2O_4(s) \rightleftharpoons Ca^{2+} + C_2O_4^{2-} \tag{10-9}$$

Oxalate ions react with water to form $HC_2O_4^-$ and $H_2C_2O_4$. Thus, there are three other equilibria present in this solution:

$$H_2C_2O_4 + H_2O \rightleftharpoons H_3O^+ + HC_2O_4^- \tag{10-10}$$

$$HC_2O_4^- + H_2O \rightleftharpoons H_3O^+ + C_2O_4^{2-} \tag{10-11}$$

$$2H_2O \rightleftharpoons H_3O^+ + OH^-$$

Step 2.   **Definition of the Unknown**

Calcium oxalate is a strong electrolyte so that its molar analytical concentration is equal to the equilibrium calcium ion concentration. That is,

$$\text{solubility} = [Ca^{2+}] \tag{10-12}$$

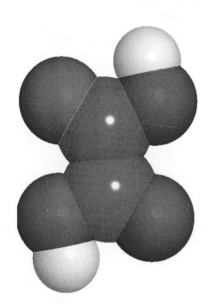

The molecular structure of oxalic acid. Oxalic acid occurs naturally in many plants as the potassium or calcium salt, and molds produce oxalic acid as the calcium salt. The sodium salt is used as a primary standard in redox titrimetry (see Chapter 18). The acid is widely used in the dye industry as a cleaning agent in a variety of applications, including the cleaning and restoration of wood surfaces, in the ceramics industry, in metallurgy, in the paper industry, and in photography. It is quite poisonous if ingested and may cause severe gastroenteritis or kidney damage. It may be prepared by passing carbon monoxide into concentrated sodium hydroxide.

Step 3.   **Equilibrium-Constant Expressions**

$$[Ca^{2+}][C_2O_4^{2-}] = K_{sp} = 1.7 \times 10^{-9} \tag{10-13}$$

$$\frac{[H_3O^+][HC_2O_4^-]}{[H_2C_2O_4]} = K_1 = 5.60 \times 10^{-2} \tag{10-14}$$

$$\frac{[H_3O^+][C_2O_4^{2-}]}{[HC_2O_4^-]} = K_2 = 5.42 \times 10^{-5} \tag{10-15}$$

$$[H_3O^+][OH^-] = K_w = 1.0 \times 10^{-14}$$

Step 4.   **Mass-Balance Expressions**

Because CaC_2O_4 is the only source of Ca^{2+} and the three oxalate species,

$$[Ca^{2+}] = [C_2O_4^{2-}] + [HC_2O_4^-] + [H_2C_2O_4] \qquad (10\text{-}16)$$

Moreover, the problem states that the pH is 4.00. Thus,

$$[H_3O^+] = 1.00 \times 10^{-4} \quad \text{and} \quad [OH^-] = \frac{K_{sp}}{[H_3O^+]} = 1.00 \times 10^{-10}$$

Step 5.   **Charge-Balance Expressions**

A buffer is required to maintain the pH at 4.00. The buffer most likely consists of some weak acid HA and its conjugate base A^- (Section 12C-1). The identity of the three species and their concentrations have not been specified, however, so we do not have enough information to write a charge-balance equation.

Step 6.   **Number of Independent Equations and Unknowns**

We have four unknowns ($[Ca^{2+}]$, $[C_2O_4^{2-}]$, $[HC_2O_4^-]$, and $[H_2C_2O_4]$) as well as four independent algebraic relationships (Equations 10-13, 10-14, 10-15, and 10-16). Therefore, an exact solution can be obtained, and the problem becomes one of algebra.

Step 7.   **Approximations**

An exact solution is so readily obtained in this case that we will not bother with approximations.

Step 8.   **Solution to Equations**

A convenient way to solve the problem is to substitute Equations 10-14 and 10-15 into Equation 10-16 in such a way as to develop a relationship between $[Ca^{2+}]$, $[C_2O_4^{2-}]$, and $[H_3O^+]$. Thus, we rearrange Equation 10-15 to give

$$[HC_2O_4^-] = \frac{[H_3O^+][C_2O_4^{2-}]}{K_2}$$

Substituting numerical values for $[H_3O^+]$ and K_2 gives

$$[HC_2O_4^-] = \frac{1.00 \times 10^{-4}[C_2O_4^{2-}]}{5.42 \times 10^{-5}} = 1.85\,[C_2O_4^{2-}]$$

Substituting this relationship into Equation 10-14 and rearranging gives

$$[H_2C_2O_4] = \frac{[H_3O^+][C_2O_4^{2-}] \times 1.85}{K_1}$$

Substituting numerical values for $[H_3O^+]$ and K_1 yields

$$[H_2C_2O_4] = \frac{1.85 \times 10^{-4}[C_2O_4^{2-}]}{5.60 \times 10^{-2}} = 3.30 \times 10^{-3}\,[C_2O_4^{2-}]$$

Substituting these expressions for $[HC_2O_4^-]$ and $[H_2C_2O_4]$ into Equation 10-16 gives

$$[Ca^{2+}] = [C_2O_4^{2-}] + 1.85\,[C_2O_4^{2-}] + 3.30 \times 10^{-3}\,[C_2O_4^{2-}]$$
$$= 2.85\,[C_2O_4^{2-}]$$

or

$$[C_2O_4^{2-}] = \frac{[Ca^{2+}]}{2.85}$$

Substituting into Equation 10-13 gives

$$\frac{[Ca^{2+}][Ca^{2+}]}{2.85} = 1.7 \times 10^{-9}$$

$$[Ca^{2+}] = \text{solubility} = \sqrt{2.85 \times 1.7 \times 10^{-9}} = 7.0 \times 10^{-5}\ M$$

Solubility Calculations When the pH Is Variable

Computing the solubility of a precipitate such as calcium oxalate in a solution in which the pH is not fixed and known is considerably more complicated than the example just considered. Thus, to determine the solubility of CaC_2O_4 in pure water, we must take into account the change in OH^- and H_3O^+ that accompanies the solution process. Here four equilibria need to be considered:

$$CaC_2O_4(s) \rightleftharpoons Ca^{2+} + C_2O_4^{2-}$$

$$C_2O_4^{2-} + H_2O \rightleftharpoons HC_2O_4^- + OH^-$$

$$HC_2O_4^- + H_2O \rightleftharpoons H_2C_2O_4 + OH^-$$

$$2H_2O \rightleftharpoons H_3O^+ + OH^-$$

In contrast to Example 10-6, the hydroxide ion concentration now becomes an unknown and an additional algebraic equation must therefore be developed if the solubility of calcium carbonate is to be calculated.

It is not difficult to write six algebraic equations needed to calculate the solubility of CaC_2O_4 (see Feature 10-1). Solving the six algebraic equations, however, is tedious and time consuming.

10B-3   The Solubility of Precipitates in the Presence of Complexing Agents

The solubility of a precipitate may increase dramatically in the presence of reagents that form complexes with the anion or the cation of the precipitate. For example, fluoride ions prevent the quantitative precipitation of aluminum hydroxide even though the solubility product of this precipitate is very small (2×10^{-32}). The cause of this increase in solubility is shown by

$$Al(OH)_3(s) \rightleftharpoons Al^{3+} + 3OH^-$$
$$+$$
$$6F^-$$
$$\updownarrow$$
$$AlF_6^{3-}$$

The solubility of a precipitate always increases in the presence of a complexing agent that reacts with the cation of the precipitate.

The fluoride complex is sufficiently stable to permit fluoride ions to compete successfully with hydroxide ions for aluminum ions.

Feature 10-1

Algebraic Expressions
Needed to Calculate the
Solubility of CaC_2O_4 in
Water

Here, as in Example 10-6, the solubility is equal to the cation concentration. Thus,

$$\text{solubility} = [Ca^{2+}] = [C_2O_4^{2-}] + [HC_2O_4^-] + [H_2C_2O_4]$$

Equilibrium-constant expressions for the four equilibria shown above are

$$K_{sp} = [Ca^{2+}][C_2O_4^{2-}] = 1.7 \times 10^{-9} \tag{10-17}$$

$$K_2 = \frac{[H_3O^+][C_2O_4^{2-}]}{[HC_2O_4^-]} = 5.42 \times 10^{-5} \tag{10-18}$$

$$K_1 = \frac{[H_3O^+][HC_2O_4^-]}{[H_2C_2O_4]} = 5.60 \times 10^{-2} \tag{10-19}$$

$$K_w = [H_3O^+][OH^-] = 1.00 \times 10^{-14} \tag{10-20}$$

The mass-balance equation is

$$[Ca^{2+}] = [C_2O_4^{2-}] + [HC_2O_4^-] + [H_2C_2O_4] \tag{10-21}$$

The charge-balance equation is

$$2[Ca^{2+}] + [H_3O^+] = 2[C_2O_4^{2-}] + [HC_2O_4^-] + [OH^-] \tag{10-22}$$

We now have six unknowns $[Ca^{2+}]$, $[C_2O_4^{2-}]$, $[HC_2O_4^-]$, $[H_2C_2O_4]$, $[H_3O^+]$, and $[OH^-]$ and six equations (10-17 through 10-22). Thus, in principle the problem can be solved exactly.

Complex Formation with a Common Ion

Many precipitates react with the precipitating reagent to form soluble complexes. In a gravimetric analysis, this tendency may have the undesirable effect of diminishing the recovery of analytes if too large an excess of reagent is used. For example, silver is often determined by precipitation of silver ion by addition of an excess of a potassium chloride solution. The amount of excess reagent must be carefully controlled because of the formation of soluble chloro complexes as shown by the reactions

$$AgCl(s) + Cl^- \rightleftharpoons AgCl_2^-$$

$$AgCl_2^- + Cl^- \rightleftharpoons AgCl_3^{2-}$$

The effect of these complexes is illustrated in Figure 10-2 in which the experimentally determined solubility of silver chloride is plotted against the logarithm of the potassium chloride concentration. For low anion concentrations, the experimental solubilities do not differ greatly from those calculated with the solubility-product constant for silver chloride; beyond a chloride ion concentration of about 10^{-3} M, however, the calculated solubilities approach zero whereas the measured values rise steeply. Note that the solubility of silver chloride is about the same in 0.3 M KCl as in pure water and is about eight times that figure in a 1 M solution. We can describe these effects quantitatively if the compositions of the complexes and their formation constants are known.

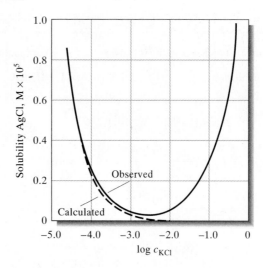

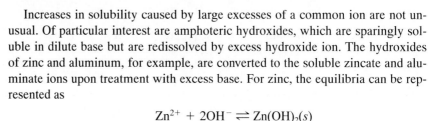

Figure 10-2   Solubility of silver chloride in potassium chloride solutions. The dashed curve is calculated from K_{sp}; the solid curve is plotted from experimental data of A. Pinkus and A. M. Timmermans, *Bull. Soc. Belges*, **1937**, *46*, 46–73.

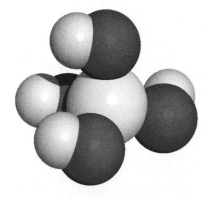

The tetrahedral molecular structure of tetrahydroxozincate(II) ion. The complex ion forms in basic solutions of Zn^{2+}.

In gravimetric procedures, a small excess of precipitating agent minimizes solubility losses, but a large excess often causes increased losses due to complex formation.

Increases in solubility caused by large excesses of a common ion are not unusual. Of particular interest are amphoteric hydroxides, which are sparingly soluble in dilute base but are redissolved by excess hydroxide ion. The hydroxides of zinc and aluminum, for example, are converted to the soluble zincate and aluminate ions upon treatment with excess base. For zinc, the equilibria can be represented as

$$Zn^{2+} + 2OH^- \rightleftharpoons Zn(OH)_2(s)$$

$$Zn(OH)_2(s) + 2OH^- \rightleftharpoons Zn(OH)_4^{2-}$$

As with silver chloride, the solubilities of amphoteric hydroxides pass through minima and then increase rapidly with increasing concentrations of base. The hydroxide ion concentration at which the solubility is a minimum can be calculated, provided equilibrium constants for the reactions are available.

Quantitative Treatment of the Effect of Complex Formation on Solubility

Solubility calculations for a precipitate in the presence of a complexing reagent are similar in principle to those discussed in the previous section. Formation constants for the complexes involved must be available.[1]

10C   SEPARATING IONS BY pH CONTROL: SULFIDE SEPARATIONS

Several precipitating agents permit separation of ions based on solubility differences. Such separations require close control of the active reagent concentration

---

[1]For an example of such a calculation, see D. A. Skoog, D. M. West, and F. J. Holler, *Fundamentals of Analytical Chemistry*, 7th ed., pp. 177–179. (Philadelphia: Saunders College Publishing, 1996.)

at a suitable and predetermined level. Most often, such control is achieved by controlling the pH of the solution with suitable buffers. This method is applicable to anionic reagents in which the anion is the conjugate base of a weak acid. A well-known example of this technique involves the separation of heavy-metal cations by selective precipitation as sulfides.

Sulfide ion forms precipitates with heavy-metal cations that have solubility products that vary widely. The concentration of S^{2-} in a solution can be varied over a range of about 0.1 M to 10^{-22} M by controlling the pH of a saturated solution of hydrogen sulfide. These two properties make possible a number of useful cation separations. To illustrate the use of hydrogen sulfide to separate cations based on pH control, let us consider the precipitation of the divalent cation M^{2+} from a solution that is kept saturated with hydrogen sulfide by bubbling the gas continuously through the solution. The important equilibria in this solution are

$$MS(s) \rightleftharpoons M^{2+} + S^{2-} \qquad [M^{2+}][S^{2-}] = K_{sp}$$

$$H_2S + H_2O \rightleftharpoons H_3O^+ + HS^- \qquad \frac{[H_3O^+][HS^-]}{[H_2S]} = K_1 = 9.6 \times 10^{-8}$$

$$HS^- + H_2O \rightleftharpoons H_3O^+ + S^{2-} \qquad \frac{[H_3O^+][S^{2-}]}{[HS^-]} = K_2 = 1.3 \times 10^{-14}$$

We may also write

$$\text{solubility} = [M^{2+}]$$

The concentration of hydrogen sulfide in a saturated solution of the gas is approximately 0.1 M. Thus, we may write as a mass-balance expression

$$0.1 = [S^{2-}] + [HS^-] + [H_2S]$$

Because we are using pH control, we know the hydronium ion concentration and thus have four unknowns: the concentration of the metal ion and the three sulfide species.

We can simplify the calculation greatly by assuming that $([S^{2-}] + [HS^-]) \ll [H_2S]$ so that

$$[H_2S] \approx 0.10 \text{ mol/L}$$

The two dissociation-constant expressions for hydrogen sulfide may be multiplied together to give an expression for the overall dissociation of hydrogen sulfide to sulfide ion:

$$H_2S + 2H_2O \rightleftharpoons 2H_3O^+ + S^{2-} \qquad \frac{[H_3O^+]^2[S^{2-}]}{[H_2S]} = K_1K_2 = 1.2 \times 10^{-21}$$

The constant for this overall reaction is simply the product of K_1 and K_2.

Substituting the numerical value for $[H_2S]$ into this equation gives

$$\frac{[H_3O^+]^2[S^{2-}]}{0.10} = 1.2 \times 10^{-21}$$

Upon rearranging this equation, we obtain

$$[S^{2-}] = \frac{1.2 \times 10^{-22}}{[H_3O^+]^2} \qquad (10\text{-}23)$$

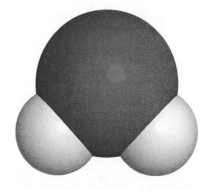

Hydrogen sulfide is a colorless, flammable gas with important chemical and toxicological properties. It is the product of a number of natural processes including the decay of sulfur-containing organic material. Its noxious odor of rotten eggs permits its detection at extremely low concentration (0.02 ppm), but because the olfactory sense is dulled by its action, higher concentrations may be tolerated and the lethal concentration of 100 ppm may be exceeded. Aqueous solutions of the gas were used traditionally as a source of sulfide ion for the precipitation of metals, but because of the toxicity of H_2S, this role has been taken over by other sulfur-containing compounds such as thioacetamide.

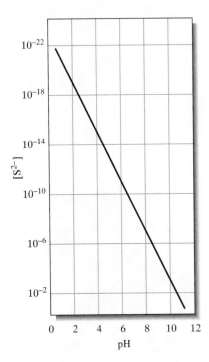

Figure 10-3 Sulfide ion concentration as a function of pH in a saturated H_2S solution.

Thus, we see that the sulfide ion concentration of a saturated hydrogen sulfide solution varies inversely as the square of the hydronium ion concentration. Figure 10-3, which was obtained with this equation, reveals that the sulfide ion concentration of an aqueous solution can be varied by over 20 orders of magnitude by varying the pH from 1 to 11.

Substituting Equation 10-23 into the solubility-product expression gives

$$\frac{[M^{2+}] \times 1.2 \times 10^{-22}}{[H_3O^+]^2} = K_{sp}$$

$$[M^{2+}] = \text{solubility} = \frac{[H_3O^+]^2 K_{sp}}{1.2 \times 10^{-22}}$$

Thus, the solubility of a divalent metal sulfide increases as the square of the hydronium ion concentration.

Example 10-7

Cadmium sulfide is less soluble than thallium(I) sulfide. Find the conditions under which Cd^{2+} and Tl^+ can, in theory, be separated quantitatively with H_2S from a solution that is 0.1 M in each cation.

The constants for the two solubility equilibria are

$$CdS(s) \rightleftharpoons Cd^{2+} + S^{2-} \qquad [Cd^{2+}][S^{2-}] = 1 \times 10^{-27}$$

$$Tl_2S(s) \rightleftharpoons 2Tl^+ + S^{2-} \qquad [Tl^+]^2[S^{2-}] = 6 \times 10^{-22}$$

Since CdS precipitates at a lower $[S^{2-}]$ than does Tl_2S, we first compute the sulfide ion concentration necessary for quantitative removal of Cd^{2+} from solution. To make such a calculation, we must first specify what constitutes a quantitative removal. The decision is arbitrary and depends on the purpose of the separation. In this example, we shall consider a separation to be quantitative when all but 1 part in 1000 of the Cd^{2+} has been removed; that is, the concentration of the cation has been lowered to 1.00×10^{-4} M. Substituting this value into the solubility-product expression gives

$$10^{-4} [S^{2-}] = 1 \times 10^{-27}$$

$$[S^{2-}] = 2 \times 10^{-23}$$

Thus, if we maintain the sulfide concentration at this level or greater, we may assume that quantitative removal of the cadmium will take place. Next, we compute the $[S^{2-}]$ needed to initiate precipitation of Tl_2S from a 0.1 M solution. Precipitation will begin when the solubility product is just exceeded. Since the solution is 0.1 M in Tl^+,

$$(0.1)^2[S^{2-}] = 6 \times 10^{-22}$$

$$[S^{2-}] = 6 \times 10^{-20}$$

These two calculations show that quantitative precipitation of Cd^{2+} takes place if $[S^{2-}]$ is made greater than 1×10^{-23}. No precipitation of Tl^+ occurs, however, until $[S^{2-}]$ becomes greater than 6×10^{-20} M.

Substituting these two values for $[S^{2-}]$ into Equation 10-23 permits calculation of the $[H_3O^+]$ range required for the separation.

$$[H_3O^+]^2 = \frac{1.2 \times 10^{-22}}{1 \times 10^{-23}} = 12$$

$$[H_3O^+] = 3.5$$

and

$$[H_3O^+]^2 = \frac{1.2 \times 10^{-22}}{6 \times 10^{-20}} = 2.0 \times 10^{-3}$$

$$[H_3O^+] = 0.045$$

By maintaining $[H_3O^+]$ between approximately 0.045 and 3.5 M, we can in theory separate CdS quantitatively from Tl_2S.

Feature 10-2

Immunoassay: Equilibria in the Specific Determination of Drugs

The determination of drugs in the human body is a matter of great importance in drug therapy and in the detection and prevention of drug abuse. The diversity of drugs and their low concentration levels in body fluids make it difficult to identify them and measure their concentrations. Fortunately, it is possible to harness one of nature's own mechanisms, the immune response, to determine a variety of therapeutic and illicit drugs quantitatively.

When a foreign substance, or antigen (Ag), shown schematically in (a) of Figure 10F-2A is introduced into the body of a mammal, the immune system synthesizes protein-based molecules (b) in Figure 10F-2A called antibodies (Ab), which specifically bind to the antigen molecules via electrostatic interactions, hydrogen bonding, and other noncovalent short-range forces. These massive molecules (molar mass $\approx$ 150,000) form a complex (see Figure 10F-2D) with antigens as shown in the reaction below and in (c) Figure 10F-2A.

$$Ag + Ab \rightleftharpoons AgAb \qquad K = \frac{[AgAb]}{[Ag][Ab]}$$

The immune system does not recognize relatively small molecules, so we must use a trick to prepare antibodies with binding sites that are specific for a particular drug. As shown in (d) of Figure 10F-2A, we attach the drug covalently to an antigenic carrier molecule such as bovine serum albumin (BSA), a protein that is obtained from the blood of cattle.

$$D + Ag \rightarrow D\text{-}Ag$$

When the resulting drug-antigen conjugate (D-Ag) is injected into the bloodstream of a rabbit, the immune system of the rabbit synthesizes antibodies with binding sites that are specific for the drug, as illustrated in (e) of Figure 10F-2A. Approximately three weeks following injection of the antigen, blood is drawn from the rabbit, the serum is separated from the blood, and the antibodies of interest are separated from the serum and other antibodies, usually by chromatographic methods (see Chapter 24). Once the drug-specific antibody has been synthesized by the immune system of the rabbit, the drug can bind directly to the antibody without the aid of the carrier

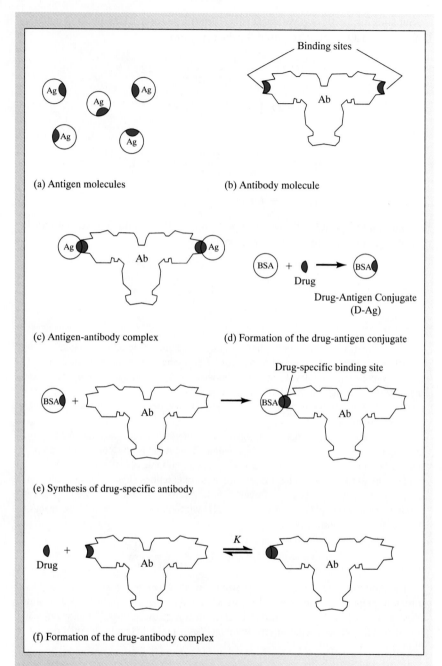

(a) Antigen molecules

(b) Antibody molecule

(c) Antigen-antibody complex

(d) Formation of the drug-antigen conjugate

(e) Synthesis of drug-specific antibody

(f) Formation of the drug-antibody complex

Figure 10F-2A

molecule, as shown in (f) Figure 10F-2A. This direct drug-antibody forms the basis for the specific determination of the drug.

The measurement step of the immunoassay is accomplished by mixing the sample containing the drug with a measured amount of drug-specific antibody. At this point, the quantity of Ab-D must be determined by adding a standard sample of the drug that has been chemically altered to contain a detectable *label*. Typical labels are enzymes,

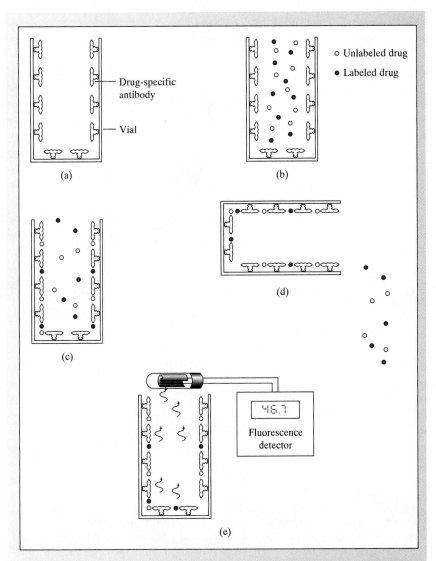

Figure 10F-2B   Procedure for determining drugs by immunoassay with fluorescence-labeling. (a) vial lined with drug-specific antibodies; (b) vial is filled with solution containing both labeled and unlabeled drug; (c) labeled and unlabeled drug binds to antibodies; (d) solution is discarded leaving bound drug behind; (e) fluorescence of the bound, labeled drug is measured. The concentration of drug is determined by using the dose-response curve of Figure 10F-2C.

fluorescent or chemiluminescent molecules, or radioactive atoms. For our example, we will assume that a fluorescent molecule has been attached to the drug to produce the labeled drug D*.[2] If the amount of the antibody is somewhat less than the sum of the amounts of D and D*, then D and D* compete for the antibody as shown in the following equilibria.

---

[2]For a discussion of molecular fluorescence, see Chapter 23.

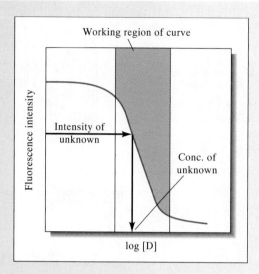

Figure 10F-2C   Dose-response curve for determining drugs by fluorescence-based immunoassay.

$$D^* + Ab \rightleftharpoons Ab\text{-}D^* \qquad K^* = \frac{[Ab\text{-}D^*]}{[D^*][Ab]}$$

$$D + Ab \rightleftharpoons Ab\text{-}D \qquad K = \frac{[Ab\text{-}D]}{[D][Ab]}$$

It is important to select a label that does not substantially alter the affinity of the drug for the antibody so that the labeled and unlabeled drugs bind with the antibody equally well. If this is true, then $K = K^*$. Typical values for equilibrium constants of this type, called *binding constants*, range from 10^7 to 10^{12}. The larger the concentration of the unknown, unlabeled drug, the smaller the concentration of Ab-D*, and vice versa. This inverse relationship between D and Ab-D* forms the basis for the quantitative determination of the drug. We can find the amount of D if we measure *either* Ab-D* or D*. To differentiate between bound drug and unbound labeled drug, it is necessary to separate them prior to measurement. The amount of Ab-D* can then be found by using a fluorescence detector to measure the intensity of the fluorescence resulting from the Ab-D*. A determination of this type using a fluorescent drug and radiation detection is called a *fluorescence immunoassay*. Determinations of this type are very sensitive and selective.

One convenient way to separate D* and Ag-D* is to prepare polystyrene vials that are coated on the inside with antibody molecules, as illustrated in (a) of Figure 10F-2B. A sample of blood serum, urine, or other body fluid containing an unknown concentration of D along with a volume of solution containing labeled drug D* are added to the vial as depicted in (b) of Figure 10F-2B. After equilibrium is achieved in the vial ((c) of Figure 10F-2B), the solution containing residual D and D* is decanted

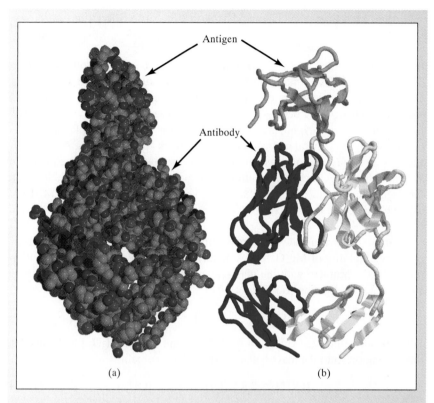

Figure 10F-2D   The molecular structure of an antigen-antibody complex. Pictures above are two representations of the complex formed between a digestion fragment of intact mouse antibody A6 and genetically engineered human interferon-gamma receptor alpha chain. (a) A space filling model of the molecular structure of the complex. (b) A ribbon diagram showing the protein chains in the complex. (From the Brookhaven Protein Data Base, Structure lJRH, S. Sogabe, F. Stuart, C. Henke, A. Bridges, G. Williams, A Birch, F. K. Winkler, and J. A. Robinson, 1997; http://www.pdb.bnl.gov.

and the vial is rinsed, leaving an amount of D* bound to the antibody that is inversely proportional to the concentration of D in the sample ((d) in Figure 10F-2B). Finally, the fluorescence intensity of the bound D* is determined using a fluorometer as shown in (e) of Figure 10F-2B. This procedure is repeated for several standard solutions of D to produce a nonlinear working curve called a *dose-response curve* similar to the curve of Figure 10F-2C. The fluorescence intensity for an unknown solution of D is located on the calibration curve, and the concentration is read from the concentration axis. Immunoassay is a powerful tool in the clinical laboratory and is one of the most widely used of all analytical techniques. Reagent kits for many different immunoassays are available commercially as are automated instruments for carrying out fluorescent immunoassays and immunoassays of other types. In addition to drugs, vitamins, proteins, growth hormones, pregnancy hormones, cancer and other disease indicators, and pesticide residues in natural waters and food are determined by immunoassay. The structure of an antigen-antibody complex is shown in Figure 10F-2D.

Spreadsheet Exercise

Using Microsoft® Excel to Solve the Equations
of Complex Equilibrium Systems

Microsoft
Excel

In previous chapters, we have shown how Excel can be used to solve systems of linear equations. Excel may also be used to solve polynomial and other nonlinear equations, but systems of equations describing chemical equilibria are a special challenge. Other general-purpose programs such as Mathcad®, Mathematica®, and Maple® are certainly useful for solving multiple-equilibrium problems, but no general approach appears to work for all systems. A number of special computer programs have been written to solve equilibrium systems, but each system requires some customization, and solutions are far from automatic.[3] In this exercise, we show how a feature of Excel called Goal Seek can be used to advantage in solving some multiple equilibrium problems.

Example 10-5 illustrates the application of the systematic method to determine the solubility of $Mg(OH)_2$ in water. As in Example 10-5, we begin our approach to the problem with the mass-balance equation, which is identical to the charge-balance equation (Equation 10-7).

$$[OH^-] = 2[Mg^{2+}] + [H_3O^+]$$

We then solve the K_{sp} expression for $[Mg^{2+}]$ and K_w for $[H_3O^+]$ and substitute these equations into the mass-balance expression to obtain

$$[OH^-] = 2 K_{sp}/[OH^-]^2 + K_w/[OH^-] \tag{10-24}$$

This cubic equation in $[OH^-]$ may be solved in a variety of different ways such as the method illustrated in Example 10-5, but it can be solved numerically using the Goal Seek function without making chemical approximations.

What is Goal Seek, and How Does it Work?

Goal seek is a tool for finding a value of a cell referenced by a formula that will give a specific result for the formula. To accomplish this task, Excel uses an iterative process not unlike the method of successive approximations. A common example of the use of Goal Seek might be to find out what final exam score you need in order to earn an A in one of your courses. You would enter a formula into a spreadsheet for calculating your grade from the individual scores on exams, homework, etc. that are entered into various cells of the spreadsheet. You would then use Goal Seek to find the value for the final exam that would yield a final score for the course that would earn you an A. You may want to try this example after we have used Goal Seek to solve our chemical problem. The grade problem can be solved algebraically, but for complex problems, there may be no straightforward algebraic solution, or the algebraic solution may require more time and effort than it is worth. Hence, Goal Seek can be quite a valuable tool in your computational arsenal. Let us now use Goal Seek to solve Equation 10-24.

---

[3]D. P. Cobranchi and E. M. Eyring, *J. Chem. Educ.*, **1991**, *68*, 40.

To begin, open Excel, and click on Tools/Options... Then click on the Calculation tab, and make sure that you click on Automatic and Iteration. Enter 500 in the Maximum iterations: box and 0.0001 in the Maximum change: box. Finally, click on OK. Goal Seek will attain the most accurate results when it is allowed to iterate a large number of times and when the maximum change in consecutive solutions is set to a very small number. The value of 0.0001 for the maximum change is the absolute amount by which the results of consecutive iterations change before Excel concludes that it has found the solution of the equation. The values selected for the maximum number of iterations and the maximum change depend on the type of problem at hand and the magnitude of the goal that is sought. In this case our Goal will be 1.000, so a maximum change of 0.0001 seems appropriate. Some experimentation may be required to achieve an acceptable solution for a given equation. Large numbers of iterations and small values for the maximum change generally result in longer computation times.

In this exercise, we present for the first time a method for documenting spreadsheets that we shall use throughout the remainder of this book. The spreadsheet below is designed to solve Equation 10-24 using the Goal Seek function.

	A	B	C	D	E
1	The Solubility of Mg(OH)2				
2	Ksp	Kw			
3	7.10E-12	1.00E-14			
4					
5	Left Side	Right Side		Right Side/Left Side	
6	1.00E+00	1.42E-11		7.04E+10	
7					
8	OH-	H3O+	Mg2+		
9	1.00E+00	1.00E-14	7.10E-12		
10					
11	Spreadsheet Documentation				
12	Cell B6=2*A3/(A6^2)+B3/A6				
13	Cell D6=A6/B6				
14	Cell A9=A6				
15	Cell B9=B3/A6				
16	Cell C9=A3/(A6^2)				

We will discuss the easiest way to document the spreadsheet later, but for now, just start Excel, and type the headings and the values for K_{sp} and K_w in cells A3 and B3. Now type the formulas (beginning with =) that you see in cells A12 to A16 in the indicated cells. To give the Goal Seek function a starting value, type 1 in cell A6, which is also duplicated in cell A9 for display purposes. Compare the formula that you typed in cell B6 to the right-hand-side of Equation 10-24, and note that they are identical in form. Now, we are ready to solve the equation. Our goal is to find a value of the hydroxide ion concentration for which the right side of the equation is equal to the left side. When the two sides of the equation are equal, their ratio located in cell D6 will be one. Goal Seek uses a systematic method to vary the contents of cell A6 until this goal is met. Now make cell D6 the active cell by clicking on it, and click on Tools/Goal Seek... to bring up the following menu.

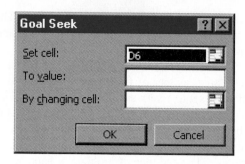

Note that D6 appears in the Set cell: blank. Click on the To value: blank, and enter 1, and enter A6 in the By changing cell: blank. Before clicking on OK, consider what Goal Seek is about to do. It is going to try to set cell D6 to a value of one by systematically changing cell A6. Now click on OK to start Goal Seek, and observe the results. If everything works properly, your spreadsheet should appear as shown below.

	A	B	C	D	E	F	G	H
1	The Solubility of Mg(OH)2							
2	Ksp	Kw						
3	7.10E-12	1.00E-14						
4								
5	Left Side	Right Side		Right Side/Left Side				
6	2.42E-04	2.42E-04		1.000066				
7								
8	OH-	H3O+	Mg2+					
9	2.42E-04	4.13E-11	1.21E-04					
10								
11	Spreadsheet Documentation							
12	Cell B6=2*A3/(A6^2)+B3/A6							
13	Cell D6=A6/B6							
14	Cell A9=A6							
15	Cell B9=B3/A6							
16	Cell C9=A3/(A6^2)							

Goal Seek Status ? X
Goal Seeking with Cell D6
found a solution.

Target value: 1
Current value: 1.000066144

[OK] [Cancel] [Step] [Pause]

The Goal Seek Status window shows us that it found a value in cell A6 that caused cell D6 to become 1 within the tolerance that we set for calculation options at the beginning of this exercise. The left-hand-side of the equation (cell A6) is $[OH^-]$, which is displayed in cell A9. From this value, we find that $[H_3O^+] = 4.13 \times 10^{-11}$ and $[Mg^{2+}] = 1.21 \times 10^{-4}$ in cells B9 and C9, respectively. Now, you may click on OK. As an additional check on the validity of the solution, you may wish to enter a formula to calculate the solubility product constant for magnesium hydroxide, and compare the result with the value in cell A3.

Documenting and Formatting Spreadsheets

In the spreadsheet that we developed above, you saw how we display formulas in one part of the spreadsheet as text that appear as numerical results of the formulas in another part of the spreadsheet. Documenting spreadsheets in this way makes it easy to see all of the important formulas while working on a specific cell. In addition, the formulas are displayed when you print a hard copy of the spreadsheet. To create a documentation line, click on the cell whose formula you would like to display; for example, click on cell C9 now. Then click in the formula bar at the top of the spreadhseet, highlight the characters in the formula with your mouse, and type `[Ctrl+c]`. You have just copied the characters, or text, of the formula to the clipboard. Alternatively, you could have clicked on Edit/Copy on the menu rather than using the shortcut key. Now click on the icon containing the red X in the formula bar to deselect cell C9, click on cell A16, and type `Cell C9` followed by `[Ctrl+v]` to paste the formula text into the cell. Clicking on Edit/Paste works just as well. Since we didn't type an equals sign at the beginning of the text string, Excel interprets the contents of A16 as text rather than a formula, and it displays the formula so that we can read it. We will use this method of documentation from now on.

It is often useful to be able to format the contents of spreadsheet cells to make them easy to read or to help distinguish text from numerical results. For example, in the spreadsheet on page 242, we have made the headings bold. This is accomplished quite easily by clicking on the cell that you wish to format and then clicking on the bold capital **B** in the format toolbar. Underlining and italics may be selected in a similar way.

For displaying chemical and algebraic formulas, it is useful to be able to format subscripts and superscripts. Let us format H_3O^+ in cell B8 so that the subscript 3 and the superscript + appear properly. Click on cell B8, and then click on the formula bar. Using your mouse, highlight only the 3 in H_3O^+, and then click on Format/Cells..., and the following window appears.

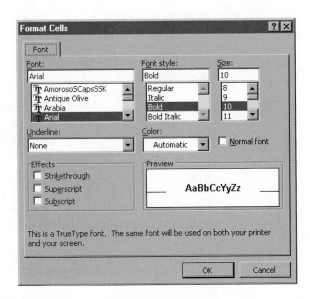

Now click on the subscript button, click on OK, and the 3 should now appear as a subscript. Notice that you could have selected a different font, such as, for example, the Symbol font, if you wished to format a character as a mathematical symbol or as a Greek character. You may also select combinations of italic and bold fonts. Now format all of the subscripts and superscripts in the spreadsheet using the same technique. Type some text in a blank cell, and format it using the

Symbol font if it is available to you, and note the result. These techniques make it possible to format the contents of spreadsheets to make them more readable and to ensure that the results are interpreted as they are intended.

In this exercise, we have solved a relatively straightforward solubility problem to illustrate the power and ease of use of Goal Seek. Goal Seek may be used to solve problems of considerably greater complexity, however, and it is a worthy addition to your Excel repertoire. In addition, we have explored some of the options for documenting and formatting spreadsheets, which should serve you well in completing spreadsheet assignments in your analytical chemistry course and other courses in your curriculum.

WEB WORKS

The Centers for Disease Control and Prevention (CDC) maintains a web site for providing information related to AIDS and HIV. Set your browser to this site at http://www.cdcnac.org/ and locate pages on testing for HIV. You will find that various types of immunoassay are useful for HIV testing. What are these immunoassay types? Then use one of the web search engines such as HotBot, Lycos, Alta Vista, or MetaCrawler to search the Web for these various types of immunoassay. What physical or chemical properties are used in these immunoassays? What are the principles behind these methods?

10D   QUESTIONS AND PROBLEMS

10-1. Demonstrate how the sulfide ion concentration is related to the hydronium ion concentration of a solution that is kept saturated with hydrogen sulfide.

10-2. Why are simplifying assumptions restricted to relationships that are sums or differences?

*10-3. Why do molar concentrations of some species appear as multiples in charge-balance equations?

10-4. Write the mass-balance expressions for a solution that is
*(a) 0.10 M in H_3PO_4.
(b) 0.10 M in Na_2HPO_4.
*(c) 0.100 M in HNO_2 and 0.0500 M in $NaNO_2$.
(d) 0.25 M in NaF and saturated with CaF_2.
*(e) 0.100 M in NaOH and saturated with $Zn(OH)_2$ (which undergoes the reaction $Zn(OH)_2 + 2OH^- \rightleftharpoons Zn(OH)_4^{2-}$).
(f) saturated with $MgCO_3$.
*(g) saturated with CaF_2.

10-5. Write the charge-balance equations for the solutions in Problem 10-4.

10-6. Calculate the molar solubility of Ag_2CO_3 in a solution that has a fixed H_3O^+ concentration of
*(a) 1.0×10^{-6} M.   ***(c)** 1.0×10^{-9} M.
(b) 1.0×10^{-7} M.   **(d)** 1.0×10^{-11} M.

10-7. Calculate the molar solubility of $BaSO_4$ in a solution in which $[H_3O^+]$ is
*(a) 2.0 M.   ***(c)** 0.50 M.
(b) 1.0 M.   **(d)** 0.10 M.

*10-8. Calculate the molar solubility of CuS in a solution in which the $[H_3O^+]$ concentration is held constant at **(a)** 1.0×10^{-1} M and **(b)** 1.0×10^{-4} M.

10-9. Calculate the concentration of CdS in a solution in which the $[H_3O^+]$ is held constant at **(a)** 1.0×10^{-1} M and **(b)** 1.0×10^{-4} M.

10-10. Calculate the molar solubility of MnS in a solution with a constant $[H_3O^+]$ of ***(a)** 1.00×10^{-5} and **(b)** 1.00×10^{-7}. (Use $K_{sp} = 3 \times 10^{-14}$)

*10-11. Calculate the molar solubility of $PbCO_3$ in a solution buffered to a pH of 7.00.

10-12. Calculate the molar solubility of Ag_2SO_3 ($K_{sp} = 1.5 \times 10^{-14}$) in a solution buffered to a pH of 7.00.

*10-13. Dilute NaOH is introduced into a solution that is 0.050 M in Cu^{2+} and 0.040 M in Mn^{2+}.
(a) Which hydroxide precipitates first?
(b) What OH^- concentration is needed to initiate precipitation of the first hydroxide?
(c) What is the concentration of the cation forming the less soluble hydroxide when the more soluble hydroxide begins to form?

10-14. A solution is 0.040 M in Na_2SO_4 and 0.050 M in $NaIO_3$. To this is added a solution containing Ba^{2+}. Assuming that no HSO_4^- is present in the original solution:

(a) Which barium salt will precipitate first?

(b) What is the Ba^{2+} concentration as the first precipitate forms?

(c) What is the concentration of the anion that forms the less soluble barium salt when the more soluble precipitate begins to form?

*10-15. Silver ion is being considered as a reagent for separating I^- from SCN^- in a solution that is 0.060 M in KI and 0.070 M in NaSCN.

(a) What Ag^+ concentration is needed to lower the I^- concentration to 1.0×10^{-6} M?

(b) What is the Ag^+ concentration of the solution when AgSCN begins to precipitate?

(c) What is the ratio of SCN^- to I^- when AgSCN begins to precipitate?

(d) What is the ratio of SCN^- to I^- when the Ag^+ concentration is 1.0×10^{-3} M?

10-16. Using 1.0×10^{-6} M as the criterion for quantitative removal, determine whether it is feasible to use

(a) SO_4^{2-} in separate Ba^{2+} and Sr^{2+} in a solution that is initially 0.10 M in Sr^{2+} and 0.25 M in Ba^{2+}.

(b) SO_4^{2-} to separate Ba^{2+} and Ag^+ in a solution that is initially 0.040 M in each cation. For Ag_2SO_4, $K_{sp} = 1.6 \times 10^{-5}$.

(c) OH^- to separate Ce^{3+} and Hf^{4+} in a solution that is initially 0.020 M in Ce^{3+} and 0.010 M in Hf^{4+}. For $Ce(OH)_3$, $K_{sp} = 7.0 \times 10^{-22}$; for $Hf(OH)_4$, $K_{sp} = 4.0 \times 10^{-26}$.

(d) IO_3^- to separate In^{3+} and Tl^+ in a solution that is initially 0.11 M in In^{3+} and 0.060 M in Tl^+. For $In(IO_3)_3$, $K_{sp} = 3.3 \times 10^{-11}$; for $TlIO_3$, $K_{sp} = 3.1 \times 10^{-6}$.

*10-17. What weight of AgBr dissolves in 200 mL of 0.100 M NaCN?

$$Ag^+ + 2CN^- \rightleftharpoons Ag(CN)_2^-$$

$$\beta_2 = 1.3 \times 10^{21}$$

10-18. The equilibrium constant for formation of $CuCl_2^-$ is given by

$$Cu^+ + 2Cl^- \rightleftharpoons CuCl_2^-$$

$$\beta_2 = \frac{[CuCl_2^-]}{[Cu^+][Cl^-]^2} = 7.9 \times 10^4$$

What is the solubility of CuCl in solutions having an analytical NaCl concentration of

(a) 1.0 M?

(b) 1.0×10^{-1} M?

(c) 1.0×10^{-2} M?

(d) 1.0×10^{-3} M?

(e) 1.0×10^{-4} M?

*10-19. In contrast to many salts, calcium sulfate is only partially dissociated in aqueous solution:

$$CaSO_4(aq) \rightleftharpoons Ca^{2+} + SO_4^{2-}$$

$$K_d = 5.2 \times 10^{-3}$$

The solubility-product constant for $CaSO_4$ is 2.6×10^{-5}. Calculate the solubility of $CaSO_4$ in (a) water and (b) 0.0100 M Na_2SO_4. In addition, calculate the percent of undissociated $CaSO_4$ in each solution.

10-20. Calculate the molar solubility of CdS as a function of pH from pH 10 to pH 1. Find values for every 0.5 pH unit and plot solubility vs. pH.

10-21. Calculate the molar solubility of TlS as a function of pH over the range of pH 10 to pH 1. Find values every 0.5 pH unit and use the charting function to plot solubility vs. pH.

Chapter 11

Titrations: Taking Advantage of Stoichiometric Reactions

Titrimetry includes a group of analytical methods based on determining the quantity of a reagent of known concentration that is required to react completely with the analyte. The reagent may be a standard solution of a chemical or an electric current of known magnitude.

Volumetric titrimetry is a type of titrimetry in which the standard reagent is measured volumetrically.

Coulometric titrimetry is a type of titrimetry in which the quantity of charge in coulombs required to complete a reaction with the analyte is measured.

T itrimetric methods include a large and powerful group of quantitative procedures based on measuring the amount of a reagent of known concentration that is consumed by an analyte. *Volumetric titrimetry* involves measuring the volume of a solution of known concentration that is needed to react essentially completely with the analyte. *Gravimetric titrimetry* differs only in that the mass of the reagent is measured instead of its volume. In *coulometric titrimetry,* the "reagent" is a constant direct electrical current of known magnitude that consumes the analyte; here, the time required to complete the electrochemical reaction is measured.

Titrimetric methods are widely used for routine determinations because they are rapid, convenient, accurate, and readily automated. This chapter provides introductory material on volumetric titrimetry. The next three chapters then consider the theory and applications of titrimetric methods based on neutralization reactions. Neutralization titrations are used for the determination of naturally occurring and synthetic acids and bases and, in addition, molecular species that can be converted to acids or bases by chemical reaction. Chapter 15 treats titrimetric methods based on complex formation reactions that are used for the determination of a multitude of different cations. Titrimetric methods based on formation of precipitates of limited solubility are considered briefly in Section 15B-2. Chapter 18 deals with volumetric methods in which the analyte is consumed by an oxidizing or a reducing reagent of known concentration. These methods are applicable to the determination of a wide variety of inorganic, organic, and biochemical species. Spectrophotometric titrations are considered in Section 23A-2. Coulometric titrimetry, which serves as a basis for a number of commercial automated instruments, is described in Section 20D-5, and amperometric titrations are discussed briefly in Section 20E-2.

11A   SOME GENERAL ASPECTS OF VOLUMETRIC TITRIMETRY[1]

11A-1   Defining Some Terms

A *standard solution* (or a *standard titrant*) is a reagent of known concentration that is used to carry out a titrimetric analysis. A *titration* is performed by adding a standard solution from a buret or other liquid-dispensing device to a solution of the analyte until the reaction between the two is judged complete. The volume of reagent needed to complete the titration is determined from the difference between the initial and final volume readings. The titration process is depicted in Figure 11-1.

The *equivalence point* in a titration is reached when the amount of added titrant is chemically equivalent to the amount of analyte in the sample. For example, the equivalence point in the titration of sodium chloride with silver nitrate occurs after exactly 1 mol of silver ion has been added for each mole of chloride ion in the sample. The equivalence point in the titration of sulfuric acid with sodium hydroxide is reached after introduction of 2 mol of base for each mole of acid.

It is sometimes necessary to add an excess of the standard titrant and then determine the excess amount by *back-titration* with a second standard titrant. Here, the equivalence point corresponds to the point where the amount of initial titrant is chemically equivalent to the amount of analyte plus the amount of back-titrant.

11A-2   Equivalence Points and End Points

The equivalence point of a titration cannot be determined experimentally. Instead, we can only estimate its position by observing some physical change associated with the condition of equivalence. This change is called the *end point* for the titration. Every effort is made to ensure that any volume or mass difference between the equivalence point and the end point is small. Such differences do exist, however, as a result of inadequacies in the physical changes and in our ability to observe them. The difference in volume or mass between the equivalence point and the end point is the *titration error*.

Indicators are often added to the analyte solution to give an observable physical change (the end point) at or near the equivalence point. We shall see that large changes in the relative concentration of analyte or titrant occur in the equivalence-point region. These concentration changes cause the indicator to change in appearance. Typical indicator changes include the appearance or disappearance of a color, a change in color, or the appearance or disappearance of turbidity.

We often use instruments to detect end points. These instruments respond to certain properties of the solution that change in a characteristic way during the titration. Instruments for detecting end points include colorimeters, turbidimeters, temperature monitors, voltmeters, current meters, and conductivity meters.

> A *standard solution* is a reagent of known concentration that is used in a titrimetric analysis.

> *Titration* is a process in which a standard reagent is added to a solution of an analyte until the reaction between the analyte and reagent is judged to be complete.

> *Back-titration* is a process in which the excess of a standard solution used to consume an analyte is determined by titration with a second standard solution. Back-titrations are often required when the rate of reaction between the analyte and reagent is slow or when the standard solution lacks stability.

> The *equivalence point* is the point in a titration when the amount of added standard reagent is equivalent to the amount of analyte.

> The *end point* is the point in a titration when a physical change occurs that is associated with the condition of chemical equivalence.

> In volumetric methods, the *titration error* E_t is given by
> $$E_t = V_{ep} - V_{eq}$$
> where V_{eq} is the theoretical volume of reagent required to reach the equivalence point and V_{ep} is the actual volume used to arrive at the end point.

---

[1]For a detailed discussion of volumetric methods, see J. I. Watters, in *Treatise on Analytical Chemistry*, I. M. Kolthoff and P. J. Elving, Eds., Part I, Vol. 11 (New York: Wiley, 1975), Chapter 114.

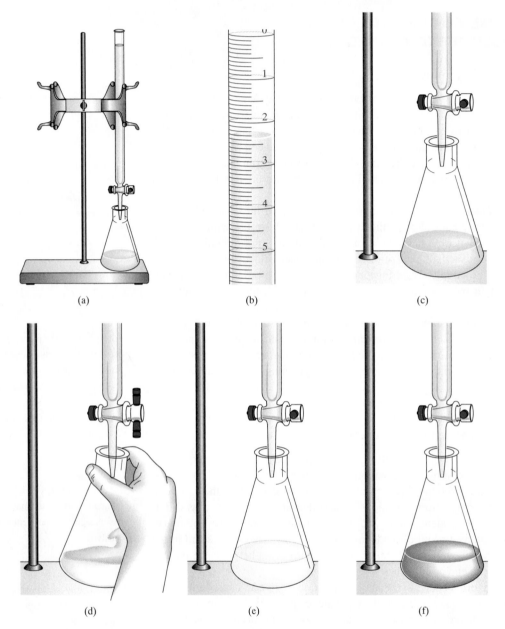

(a)   (b)   (c)

(d)   (e)   (f)

Figure 11-1   The titration process. This sequence of drawings shows the titration of an acid with sodium hydroxide titrant using phenolphthalein as indicator. (a) A typical setup for carrying out a titration. The apparatus consists of a buret, a buret stand and clamp with a white porcelain base to provide an appropriate background for viewing indicator color changes, and a wide-mouth Erlenmeyer flask containing a precisely known volume of the solution to be titrated. This solution is normally delivered into the flask using a pipet as shown in Figure 2-22. (b) Detail of the buret graduations. Normally, the buret is filled with titrant solution to within one or two milliliters of the zero position at the top of the buret. The initial volume of the buret is read to the nearest ±0.01 mL. The reference point on the meniscus and the proper position of the eye for reading are depicted in Figure 2-21. (c) Before titration begins. The solution to be titrated, an acid in this example, is placed into the flask. (d) During titration. The titrant is added to the flask with swirling until the color of the indicator persists. (e) Titration end point. The indicator turns pink in basic solution. The end point is achieved when the barely perceptible pink color of phenolphthalein is observed. A final reading of the buret is made at this time, and the volume of base delivered is calculated from the difference in the two buret readings. (f) Excess of base. When an excess of base is added to the flask, the solution turns a deep pink and the end point has been exceeded.

11A-3   Primary Standards

A *primary standard* is a highly purified compound that serves as a reference material in all volumetric and mass titrimetric methods. The accuracy of a method is critically dependent on the properties of this compound. Important requirements for a primary standard are

1. High purity (established methods for confirming purity should be available)
2. Atmospheric stability
3. Absence of hydrate water so that the composition of the solid does not change with variations in relative humidity
4. Ready availability at modest cost
5. Reasonable solubility in the titration medium
6. Reasonably large molar mass so that the relative error associated with weighing the standard is minimized

Compounds that meet or even approach these criteria are very few, and only a limited number of primary standard substances are available to the chemist. As a consequence, less pure compounds must sometimes be used in lieu of a primary standard. The purity of such a *secondary standard* must be established by careful analysis.

> A *primary standard* is an ultrapure compound that serves as the reference material for a titrimetric method of analysis.

> A *secondary standard* is a compound whose purity has been established by chemical analysis and that serves as the reference material for a titrimetric method of analysis.

11B   STANDARD SOLUTIONS

Standard solutions play a central role in all titrimetric methods of analysis. Therefore, we need to consider the desirable properties for such solutions, how they are prepared, and how their concentrations are expressed.

11B-1   Desirable Properties of Standard Solutions

The ideal standard solution for a titrimetric method will

1. Be sufficiently stable so that it is only necessary to determine its concentration once
2. React rapidly with the analyte so that the time required between additions of reagent is minimized
3. React completely with the analyte so that satisfactory end points are realized
4. Undergo a selective reaction with the analyte that can be described by a balanced equation

Few reagents meet all these ideals perfectly.

11B-2   Establishing the Concentration of Standard Solutions

The accuracy of a titrimetric method can be no better than the accuracy of the concentration of the standard solution used in the titration. Two basic methods are used to establish the concentration of such solutions. The first is the *direct method* in which a carefully weighed quantity of a primary standard is dissolved in a suitable solvent and diluted to a known volume in a volumetric flask. The

In a *standardization,* the concentration of a volumetric solution is determined by titrating it against a carefully measured quantity of a primary or secondary standard or an accurately known volume of another standard solution.

second is by *standardization* in which the titrant to be standardized is used to titrate (1) a weighed quantity of a primary standard, (2) a weighed quantity of a secondary standard, or (3) a measured volume of another standard solution. A titrant that is standardized against a secondary standard or against another standard solution is sometimes referred to as a *secondary standard solution.* The concentration of a secondary standard solution is subject to a larger uncertainty than that for a primary standard solution. Thus, if there is a choice, solutions are best prepared by the direct method. On the other hand, many reagents lack the properties required for a primary standard and therefore require standardization.

11B-3   Expressing the Concentration of Standard Solutions

The concentrations of standard solutions are generally expressed in units of either *molarity c* or *normality* c_N. The first gives the number of moles of reagent contained in one liter of solution, and the second gives the number of *equivalents* of reagent in the same volume.

Throughout this book we base volumetric calculations exclusively on molarity and molar masses. We have included in Appendix 6, however, a discussion of how volumetric calculations are carried out based on normality and equivalent weights because you may encounter these terms and their uses in the industrial and health science literature.

11C   VOLUMETRIC CALCULATIONS

Chemists express the concentration of solutes in several ways. The most important of these are described in this section.

11C-1   Some Useful Algebraic Relationships

Most volumetric calculations are based on two pairs of fundamental equations that are derived from definitions of millimole, mole, and molar concentration. For the chemical species A, we may write

$$n_A = \frac{m_A}{\mathcal{M}_A}$$

$$\text{amount A (mol)} = \frac{\text{mass A (g)}}{\text{molar mass A (g/mol)}} \tag{11-1}$$

$$\text{amount A (mmol)} = \frac{\text{mass A (g)}}{\text{millimolar mass A (g/mmol)}} \tag{11-2}$$

We may derive a second pair from the definition of molar concentration. That is,

$$c_A = \frac{n_A}{V} \quad \text{or} \quad n_A = V \times c_A$$

$$\text{amount A (mol)} = \text{volume (L)} \times \text{concentration A} \left(\frac{\text{mol}}{\text{L}} \right) \tag{11-3}$$

$$\text{amount A (mmol)} = \text{volume (mL)} \times \text{concentration A} \left(\frac{\text{mmol}}{\text{mL}} \right) \tag{11-4}$$

You should use Equation 11-3 when volumes are measured in liters and Equation 11-4 when the units are milliliters.

11C-2   Calculating the Molarity of Standard Solutions

Examples 11-1 through 11-3 illustrate how volumetric reagents are prepared.

Example 11-1

Describe the preparation of 5.000 L of 0.1000 M Na_2CO_3 (105.99 g/mol) from the primary standard solid.

Since the volume is in liters, we base our calculations on the mole rather than the millimole. Thus, to obtain the amount of Na_2CO_3 needed, we write

$$\text{amount Na}_2\text{CO}_3 = n_{Na_2CO_3}(\text{mol}) = V_{soln}(\text{L}) \times c_{Na_2CO_3}(\text{mol/L})$$

$$= 5.000 \; \cancel{L} \times \frac{0.1000 \text{ mol Na}_2\text{CO}_3}{\cancel{L}} = 0.5000 \text{ mol Na}_2\text{CO}_3$$

To obtain the mass of Na_2CO_3, we rearrange Equation 11-2 to give

$$\text{mass Na}_2\text{CO}_3 = m_{Na_2CO_3} = 0.5000 \; \cancel{\text{mol Na}_2\text{CO}_3} \times \frac{105.99 \text{ g Na}_2\text{CO}_3}{\cancel{\text{mol Na}_2\text{CO}_3}}$$

$$= 53.00 \text{ g Na}_2\text{CO}_3$$

Therefore, the solution is prepared by dissolving 53.00 g of Na_2CO_3 in water and diluting to 5.000 L.

Example 11-2

A standard 0.0100 M solution of Na^+ is required for calibrating a flame photometric method for determining the element. Describe how 500.0 mL of this solution can be prepared from primary standard Na_2CO_3.

We wish to compute the mass of reagent required to give a species molarity of 0.0100 mol/L. Here we will use millimoles since the volume is in milliliters. Because Na_2CO_3 dissociates to give two Na^+ ions, we can write that the number of millimoles of Na_2CO_3 needed is

$$\text{amount Na}_2\text{CO}_3 = 500.0 \; \cancel{\text{mL}} \times \frac{0.0100 \; \cancel{\text{mmol Na}^+}}{\cancel{\text{mL}}} \times \frac{1 \text{ mmol Na}_2\text{CO}_3}{2 \; \cancel{\text{mmol Na}^+}}$$

$$= 2.50 \text{ mmol}$$

From the definition of millimole, we write

$$\text{mass Na}_2\text{CO}_3 = 2.50 \; \cancel{\text{mmol Na}_2\text{CO}_3} \times 0.10599 \; \frac{\text{g Na}_2\text{CO}_3}{\cancel{\text{mmol Na}_2\text{CO}_3}}$$

$$= 0.265 \text{ g}$$

The solution is therefore prepared by dissolving 0.265 g of Na_2CO_3 in water and diluting to 500.0 mL.

It is useful to know that any combination of grams, moles, and liters can be replaced with any analogous combination expressed in milligrams, millimoles, and milliliters. For example, a 0.1 M solution contains 0.1 mol of a species per liter or 0.1 mmol per milliliter. Similarly, the number of moles of a compound is equal to the mass in grams of that compound divided by its molar mass in grams or the mass in milligrams divided by its millimolar mass in milligrams.

Example 11-3

How would you prepare 50.0-mL portions of standard solutions that are 0.00500 M, 0.00200 M, and 0.00100 M in Na^+ from the solution in Example 11-2?

The number of millimoles of Na^+ taken from the concentrated solution must equal the number in the diluted solutions. Thus,

$$\text{amount } Na^+ \text{ from concd soln} = \text{amount } Na^+ \text{ in dil soln}$$

Recall that the number of millimoles is equal to number of millimoles per milliliter times the number of milliliters. Therefore,

$$V_{concd} \times c_{concd} = V_{dil} \times c_{dil}$$

where V_{concd} and V_{dil} are the volumes in milliliters of the concentrated and diluted solutions respectively and c_{concd} and c_{dil} are their molar Na^+ concentrations. For the first solution, this equation rearranges to

$$V_{concd} = \frac{V_{dil} \times c_{dil}}{c_{concd}} = \frac{50.0 \text{ mL} \times 0.00500 \text{ mmol } Na^+/mL}{0.0100 \text{ mmol } Na^+/mL} = 25.0 \text{ mL}$$

Thus, to produce 50.0 mL of 0.00500 M Na^+, 25.0 mL of the concentrated solution should be diluted to 50.0 mL.

Repeat the calculation for the other two molarities to confirm that diluting 10.0 and 5.00 mL of the concentrated solution to 50.0 mL produces the desired solutions.

> A useful algebraic relationship is
> $$V_{concd} \times c_{concd} = V_{dil} \times c_{dil}$$

11C-3   Treating Titration Data

In this section, we describe two types of volumetric calculations. The first involves computing the molarity of solutions that have been standardized against either a primary standard or another standard solution. The second involves calculating the amount of analyte in a sample from titration data. Both types are based on three algebraic relationships. Two of these are Equations 11-2 and 11-4, which are based on millimoles and milliliters. The third relationship is the stoichiometric ratio of the number of millimoles of the analyte to the number of millimoles of titrant.

Calculating Molarities from Standardization Data

Examples 11-4 and 11-5 illustrate how standardization data are treated.

Example 11-4

A 50.00 mL volume of HCl solution required 29.71 mL of 0.01963 M $Ba(OH)_2$ to reach an end point with bromocresol green indicator. Calculate the molarity of the HCl.

In the titration, 1 mmol of $Ba(OH)_2$ reacts with 2 mmol of HCl.

$$Ba(OH)_2 + 2HCl \rightarrow BaCl_2 + 2H_2O$$

Thus the stoichiometric ratio is

$$\text{stoichiometric ratio} = \frac{2 \text{ mmol HCl}}{1 \text{ mmol Ba(OH)}_2}$$

The number of millimoles of the standard is obtained by substituting into Equation 11-3:

$$\text{amount Ba(OH)}_2 = 29.71 \text{ mL Ba(OH)}_2 \times 0.01963 \frac{\text{mmol Ba(OH)}_2}{\text{mL Ba(OH)}_2}$$

To obtain the number of millimoles of HCl, we multiply this result by the stoichiometric ratio determined initially:

$$\text{amount HCl} = (29.71 \times 0.01963) \text{ mmol Ba(OH)}_2 \times \frac{2 \text{ mmol HCl}}{1 \text{ mmol Ba(OH)}_2}$$

To obtain the number of millimoles of HCl per mL, we divide by the volume of the acid. Thus,

$$c_{\text{HCl}} = \frac{(29.71 \times 0.01963 \times 2) \text{ mmol HCl}}{50.0 \text{ mL solution}}$$

$$= 0.023328 \frac{\text{mmol HCl}}{\text{mL solution}} = 0.0233 \text{ M}$$

In determining the number of significant figures to retain in volumetric calculations, the stoichiometric ratio is assumed to be known exactly without uncertainty.

Example 11-5

Titration of 0.2121 g of pure $Na_2C_2O_4$ (134.00 g/mol) required 43.31 mL of $KMnO_4$. What is the molarity of the $KMnO_4$ solution? The chemical reaction is

$$2MnO_4^- + 5C_2O_4^{2-} + 16H^+ \rightarrow 2Mn^{2+} + 10CO_2 + 8H_2O$$

From this equation, we see that the stoichiometric ratio is

$$\text{stoichiometric ratio} = \frac{2 \text{ mmol KMnO}_4}{5 \text{ mmol Na}_2\text{C}_2\text{O}_4}$$

The amount of primary standard $Na_2C_2O_4$ is given by Equation 11-1:

$$\text{amount Na}_2\text{C}_2\text{O}_4 = 0.2121 \text{ g Na}_2\text{C}_2\text{O}_4 \times \frac{1 \text{ mmol Na}_2\text{C}_2\text{O}_4}{0.13400 \text{ g Na}_2\text{C}_2\text{O}_4}$$

To obtain the number of millimoles of $KMnO_4$, we multiply this result by the stoichiometric factor:

$$\text{amount KMnO}_4 = \frac{0.2121}{0.1340} \text{ mmol Na}_2\text{C}_2\text{O}_4 \times \frac{2 \text{ mmol KMnO}_4}{5 \text{ mmol Na}_2\text{C}_2\text{O}_4}$$

The molarity is then obtained by dividing by the volume of $KMnO_4$ consumed. Thus,

$$c_{\text{KMnO}_4} = \frac{\left(\dfrac{0.2121}{0.1340} \times \dfrac{2}{5} \right) \text{ mmol KMnO}_4}{43.31 \text{ mL KMnO}_4} = 0.01462 \text{ M}$$

Note that units are carried through all calculations as a check on the correctness of the relationships used in Examples 11-4 and 11-5.

Calculating the Quantity of Analyte from Titration Data

As shown by Examples 11-6 through 11-9 and Feature 11-1 that follow, the same systematic approach just described is also used to compute analyte concentrations from titration data.

Example 11-6

A 0.8040-g sample of an iron ore is dissolved in acid. The iron is then reduced to Fe^{2+} and titrated with 47.22 mL of 0.02242 M $KMnO_4$ solution. Calculate the results of this analysis in terms of (a) % Fe (55.847 g/mol) and (b) % Fe_3O_4 (231.54 g/mol). The reaction of the analyte with the reagent is described by the equation

$$MnO_4^- + 5Fe^{2+} + 8H^+ \rightarrow Mn^{2+} + 5Fe^{3+} + 4H_2O$$

(a) stoichiometric ratio $= \dfrac{5 \text{ mmol } Fe^{2+}}{1 \text{ mmol } KMnO_4}$

amount $KMnO_4 = 47.22 \text{ mL } \cancel{KMnO_4} \times \dfrac{0.02242 \text{ mmol } KMnO_4}{\cancel{mL \ KMnO_4}}$

amount $Fe^{2+} = (47.22 \times 0.02242) \cancel{\text{mmol } KMnO_4} \times \dfrac{5 \text{ mmol } Fe^{2+}}{1 \cancel{\text{mmol } KMnO_4}}$

The mass of Fe^{2+} is then given by

mass $Fe^{2+} = (47.22 \times 0.02242 \times 5) \cancel{\text{mmol } Fe^{2+}} \times 0.055847 \dfrac{\text{g } Fe^{2+}}{\cancel{\text{mmol } Fe^{2+}}}$

The percent Fe^{2+} is

$$\% \ Fe^{2+} = \frac{(47.22 \times 0.02242 \times 5 \times 0.055847) \text{ g } Fe^{2+}}{0.8040 \text{ g sample}} \times 100\% = 36.77\%$$

(b) To determine the correct stoichiometric ratio, we note that

$$5 \ Fe^{2+} \equiv 1 \ MnO_4^-$$

Therefore,

$$5Fe_3O_4 \equiv 15Fe^{2+} \equiv 3MnO_4^-$$

and

$$\text{stoichiometric ratio} = \frac{5 \text{ mmol } Fe_3O_4}{3 \text{ mmol } KMnO_4}$$

As in part (a),

amount $KMnO_4 = 47.22 \text{ mL } \cancel{KMnO_4} \times 0.02242 \dfrac{\text{mmol } KMnO_4}{\cancel{mL \ KMnO_4}}$

$$\text{amount Fe}_3\text{O}_4 = (47.22 \times 0.02242) \text{ mmol KMnO}_4^- \times \frac{5 \text{ mmol Fe}_3\text{O}_4}{3 \text{ mmol KMnO}_4}$$

$$\text{mass Fe}_3\text{O}_4 = \left(47.22 \times 0.02242 \times \frac{5}{3}\right) \text{ mmol Fe}_3\text{O}_4$$

$$\times \, 0.23154 \, \frac{\text{g Fe}_3\text{O}_4}{\text{mmol Fe}_3\text{O}_4}$$

$$\% \, \text{Fe}_3\text{O}_4 = \frac{(47.22 \times 0.02242 \times \frac{5}{3}) \times 0.23154 \text{ g Fe}_3\text{O}_4}{0.8040 \text{ g sample}} \times 100\%$$

$$= 50.81\%$$

Feature 11-1

The Factor-Label Approach to Example 11-6(a)

Some people find it easier to write out the solution to a problem in such a way that the units in the denominator of each succeeding term eliminate the units in the numerator of the preceding one until the units of the answer are obtained. For example, the solution to part (a) of Example 11-6 can be written

$$47.22 \text{ mL KMnO}_4 \times \frac{0.02242 \text{ mmol KMnO}_4}{\text{mL KMnO}_4} \times \frac{5 \text{ mmol Fe}}{1 \text{ mmol KMnO}_4}$$

$$\times \frac{0.05585 \text{ g Fe}}{\text{mmol Fe}} \times \frac{1}{0.8040 \text{ g sample}} \times 100\% = 36.77\% \text{ Fe}$$

Example 11-7

The organic matter in a 3.776-g sample of a mercuric ointment is decomposed with HNO_3. After dilution, the Hg^{2+} is titrated with 21.30 mL of a 0.1144 M solution of NH_4SCN. Calculate the percent Hg (200.59 g/mol) in the ointment.

This titration involves the formation of a stable neutral complex, $Hg(SCN)_2$:

$$Hg^{2+} + 2SCN^- \rightarrow Hg(SCN)_2(aq)$$

At the equivalence point,

$$\text{stoichiometric ratio} = \frac{1 \text{ mmol Hg}^{2+}}{2 \text{ mmol NH}_4\text{SCN}}$$

$$\text{amount NH}_4\text{SCN} = 21.30 \text{ mL NH}_4\text{SCN} \times 0.1144 \, \frac{\text{mmol NH}_4\text{SCN}}{\text{mL NH}_4\text{SCN}}$$

$$\text{amount Hg}^{2+} = (21.30 \times 0.1144) \text{ mmol NH}_4\text{SCN} \times \frac{1 \text{ mmol Hg}^{2+}}{2 \text{ mmol NH}_4\text{SCN}}$$

$$\text{mass Hg}^{2+} = \left(21.30 \times 0.1144 \times \frac{1}{2}\right) \text{mmol Hg}^{2+} \times \frac{0.20059 \text{ g Hg}^{2+}}{\text{mmol Hg}^{2+}}$$

$$\% \text{ Hg} = \frac{(21.30 \times 0.1144 \times \frac{1}{2}) \times 0.20059 \text{ g Hg}^{2+}}{3.776 \text{ g sample}} \times 100\%$$

$$= 6.472\%$$

Example 11-8

A 0.4755-g sample containing $(NH_4)_2C_2O_4$ and inert materials was dissolved in H_2O and made strongly alkaline with KOH, which converted NH_4^+ to NH_3. The liberated NH_3 was distilled into 50.00 mL of 0.05035 M H_2SO_4. The excess H_2SO_4 was back-titrated with 11.13 mL of 0.1214 M NaOH. Calculate (a) the % N (14.007 g/mol) and (b) the % $(NH_4)_2C_2O_4$ (124.10 g/mol) in the sample.

(a) The H_2SO_4 reacts with both NH_3 and NaOH, and two stoichiometric ratios can be determined. These are

$$\frac{2 \text{ mmol NH}_3}{1 \text{ mmol H}_2SO_4} \quad \text{and} \quad \frac{1 \text{ mmol H}_2SO_4}{2 \text{ mmol NaOH}}$$

$$\text{total amount H}_2SO_4 = 50.00 \text{ mL H}_2SO_4 \times 0.05035 \frac{\text{mmol H}_2SO_4}{\text{mL H}_2SO_4}$$

$$= 2.5175 \text{ mmol H}_2SO_4$$

The amount of H_2SO_4 consumed by the NaOH in the back-titration is

$$\text{amount H}_2SO_4 = (11.13 \times 0.1214) \text{ mmol NaOH} \times \frac{1 \text{ mmol H}_2SO_4}{2 \text{ mmol NaOH}}$$

$$= 0.6756 \text{ mmol H}_2SO_4$$

Note that we keep all figures until the final calculation is made and then round off the answer to the appropriate number of significant figures.

The amount of H_2SO_4 that reacted with NH_3 is then

$$\text{amount H}_2SO_4 = (2.5175 - 0.6756) \text{ mmol H}_2SO_4 = 1.8419 \text{ mmol H}_2SO_4$$

The amount of NH_3, which is equal to the number of millimoles of N, is

$$\text{amount N} = \text{no. mmol NH}_3 = 1.8419 \text{ mmol H}_2SO_4 \times \frac{2 \text{ mmol N}}{1 \text{ mmol H}_2SO_4}$$

$$= 3.6838 \text{ mmol N}$$

$$\% \text{ N} = \frac{3.6838 \text{ mmol N} \times 0.014007 \text{ g N/mmol N}}{0.4755 \text{ g sample}} \times 100\% = 10.85\%$$

(b) Since each millimole of $(NH_4)_2C_2O_4$ produces 2 mmol of NH_3, which reacts with 1 mmol of H_2SO_4,

$$\text{stoichiometric ratio} = \frac{1 \text{ mmol } (NH_4)_2C_2O_4}{1 \text{ mmol } H_2SO_4}$$

$$\text{amount } (NH_4)_2C_2O_4 = 1.8419 \cancel{\text{ mmol } H_2SO_4} \times \frac{1 \text{ mmol } (NH_4)_2C_2O_4}{1 \cancel{\text{ mmol } H_2SO_4}}$$

$$\text{mass } (NH_4)_2C_2O_4 = 1.8419 \cancel{\text{ mmol } (NH_4)_2C_2O_4} \times \frac{0.12410 \text{ g } (NH_4)_2C_2O_4}{\cancel{\text{mmol } (NH_4)_2C_2O_4}}$$

$$= 0.22858 \text{ g}$$

$$\% (NH_4)_2C_2O_4 = \frac{0.22858 \text{ g } (NH_4)_2C_2O_4}{0.4755 \text{ g sample}} \times 100\% = 48.07\%$$

Example 11-9

The CO in a 20.3-L sample of gas was converted to CO_2 by passing the gas over iodine pentoxide heated to 150°C:

$$I_2O_5(s) + 5CO(g) \rightarrow 5CO_2(g) + I_2(g)$$

The iodine was distilled at this temperature and was collected in an absorber containing 8.25 mL of 0.01101 M $Na_2S_2O_3$:

$$I_2(aq) + 2S_2O_3^{2-}(aq) \rightarrow 2I^-(aq) + S_4O_6^{2-}(aq)$$

The excess $Na_2S_2O_3$ was back-titrated with 2.16 mL of 0.00947 M I_2 solution. Calculate the mass in milligrams of CO (28.01 g/mol) per liter of sample.

Based on the two reactions, the stoichiometric ratios are

$$\frac{5 \text{ mmol CO}}{1 \text{ mmol } I_2} \quad \text{and} \quad \frac{2 \text{ mmol } Na_2S_2O_3}{1 \text{ mmol } I_2}$$

We divide the first ratio by the second to get a third useful ratio:

$$\frac{5 \text{ mmol CO}}{2 \text{ mmol } Na_2S_2O_3}$$

This relationship reveals that 5 mmol of CO are responsible for the consumption of 2 mmol $Na_2S_2O_3$. The total amount of $Na_2S_2O_3$ is

$$\text{amount } Na_2S_2O_3 = 8.25 \cancel{\text{ mL } Na_2S_2O_3} \times 0.01101 \frac{\text{mmol } Na_2S_2O_3}{\cancel{\text{mL } Na_2S_2O_3}}$$

$$= 0.09083 \text{ mmol } Na_2S_2O_3$$

The amount of $Na_2S_2O_3$ consumed in the back-titration is

$$\text{amount } Na_2S_2O_3 = 2.16 \cancel{\text{ mL } I_2} \times 0.00947 \frac{\cancel{\text{mmol } I_2}}{\cancel{\text{mL } I_2}} \times \frac{2 \text{ mmol } Na_2S_2O_3}{\cancel{\text{mmol } I_2}}$$

$$= 0.04091 \text{ mmol } Na_2S_2O_3$$

The number of millimoles of CO can then be obtained by using the third stoichiometric ratio:

$$\text{amount CO} = (0.09083 - 0.04091) \, \text{mmol Na}_2\text{S}_2\text{O}_3 \times \frac{5 \text{ mmol CO}}{2 \text{ mmol Na}_2\text{S}_2\text{O}_3}$$

$$= 0.1248 \text{ mmol CO}$$

$$\text{mass CO} = 0.1248 \, \text{mmol CO} \times \frac{28.01 \text{ mg CO}}{\text{mmol CO}} = 3.4956 \text{ mg}$$

$$\frac{\text{mass CO}}{\text{vol sample}} = \frac{3.4956 \text{ mg CO}}{20.3 \text{ L sample}} = 0.172 \, \frac{\text{mg CO}}{\text{L}}$$

Note again that we keep figures until the final calculation. Only three significant figures are justified in the final result.

11D   TITRATION CURVES

As noted in Section 11A-2, an end point is an observable physical change that occurs near the equivalence point of a titration. The two most widely used end points involve (1) changes in color due to the reagent, the analyte, or an indicator and (2) a change in potential of an electrode that responds to the concentration of the reagent or the analyte.

To help us understand the detection of end points and the sources of titration errors, we will construct *titration curves* for the system under consideration. Titration curves consist of a plot of reagent volume on the horizontal axis and some function of the analyte or reagent concentration on the vertical axis.

11D-1   Types of Titration Curves

Two general types of titration curves (and thus two general types of end points) are encountered in titrimetric methods. In the first type, called a *sigmoidal curve*, important observations are confined to a small region (typically ± 0.1 to ± 0.5 mL) surrounding the equivalence point. A sigmoidal curve, in which the p-function of analyte (or sometimes the reagent) is plotted as a function of reagent volume, is shown in Figure 11-2a. In the second type of curve, called a *linear segment curve*, measurements are made on both sides of but well away from the equivalence point. Measurements near equivalence are avoided. In this type of curve, the vertical axis is an instrument reading that is directly proportional to the concentration of the analyte or the reagent. A typical linear segment curve is found in Figure 11-2b. The sigmoidal type offers the advantages of speed and convenience. The linear segment type is advantageous for reactions that are complete only in the presence of a considerable excess of the reagent or analyte.

In this chapter and the several chapters that follow, we deal exclusively with sigmoidal titration curves. Linear-segment curves are considered in Section 23A-2.

> *Titration curves* are plots of a concentration-related variable as a function of reagent volume.

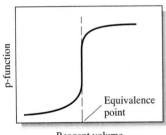

(a) Sigmoidal curve

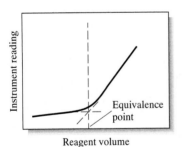

(b) Linear-segment curve

Figure 11-2 Two types of titration curves.

> The vertical axis in a sigmoidal titration curve is either the p-function of the analyte or reagent or else the potential of an analyte- or reagent-sensitive electrode. Remember that $\text{pAnalyte} = -\log c_{\text{Analyte}}$.

> The vertical axis for a linear-segment titration curve is an instrumental signal that is proportional to the concentration of the analyte or reagent.

Table 11-1

Concentration Changes during a Titration of 50.00 mL of 0.1000 M HCl

Volume of 0.1000 M NaOH, mL	$[H_3O^+]$ mol/L	Volume of NaOH to Cause a Tenfold Decrease in $[H_3O^+]$, mL	pH	pOH
0.00	1.000×10^{-1}		1.00	13.00
40.91	1.000×10^{-2}	40.91	2.00	12.00
49.01	1.000×10^{-3}	8.11	3.00	11.00
49.90	1.000×10^{-4}	0.89	4.00	10.00
49.99	1.000×10^{-5}	0.09	5.00	9.00
49.999	1.000×10^{-6}	0.009	6.00	8.00
50.00	1.000×10^{-7}	0.001	7.00	7.00
50.001	1.000×10^{-8}	0.001	8.00	6.00
50.01	1.000×10^{-9}	0.009	9.00	5.00
50.10	1.000×10^{-10}	0.09	10.00	4.00
51.01	1.000×10^{-11}	0.91	11.00	3.00
61.11	1.000×10^{-12}	10.10	12.00	2.00

11D-2   Concentration Changes During Titrations

The equivalence point in a titration is characterized by major changes in the *relative* concentrations of reagent and analyte. Table 11-1 illustrates this phenomenon. The data in the second column of the table show the changes in the hydronium ion concentration as a 50.00-mL aliquot of a 0.1000 M solution of hydrochloric acid is titrated with a 0.1000 M solution of sodium hydroxide. The neutralization reaction is described by the equation

$$H_3O^+ + OH^- \rightleftharpoons 2H_2O \tag{11-5}$$

To emphasize the changes in *relative* concentration that occur in the equivalence region, the volume increments computed are those required to cause tenfold decreases in the concentration of H_3O^+ (or tenfold increases in hydroxide ion concentration). Thus, we see in the third column that an addition of 40.91 mL of base is needed to decrease the concentration of the hydronium ion by one order of magnitude from 0.100 M to 0.0100 M. An addition of only 8.11 mL is required to lower the concentration by another factor of 10 to 0.00100 M; 0.89 mL causes yet another tenfold decrease. Corresponding increases in hydroxide concentration occur at the same time. End-point detection, then, is based on this large change in the *relative* concentration of the analyte (or the reagent) that occurs at the equivalence point for every type of titration. Feature 11-2 describes how the volumes in the first column of Table 11-1 are obtained.

The large relative concentration changes that occur in the region of chemical equivalence are shown by plotting the negative logarithm of the analyte or the reagent concentration (the p-function) against reagent volume, as has been done in Figure 11-3. The data for this plot are found in the fourth column of Table 11-1. Titration curves for reactions involving complex formation, precipitation, and oxidation/reduction all exhibit the same sharp increase or decrease in p-function as those shown in Figure 11-3 in the equivalence-point region. Titration curves define the properties required of an indicator and allow us to estimate the error associated with titration methods.

At the beginning of the titration described in Table 11-1, about 41 mL of reagent brings about a tenfold decrease in the concentration of the titrant; only 0.001 mL is required to cause this same change at the equivalence point.

Feature 11-2

How Did We Calculate
the Volumes of NaOH Shown
in the First Column of
Table 11-1?

Up until the equivalence point, $[H_3O^+]$ will equal the concentration of unreacted HCl (c_{HCl}). The concentration of HCl is equal to the original number of millimole HCl (50.00×0.1000) minus the number of millimoles NaOH added ($V_{NaOH} \times 0.1000$) divided by the total volume of the solution. That is,

$$c_{HCl} = [H_3O^+] = \frac{50.00 \times 0.1000 - V_{NaOH} \times 0.1000}{50.00 + V_{NaOH}}$$

where V_{NaOH} is the volume of 0.1000 M NaOH added. This equation reduces to

$$50.00[H_3O^+] + V_{NaOH}[H_3O^+] = 5.000 - 0.1000\,V_{NaOH}$$

Collecting the terms containing V_{NaOH} gives

$$V_{NaOH}(0.1000 + [H_3O^+]) = 5.000 - 50.00[H_3O^+]$$

or

$$V_{NaOH} = \frac{5.000 - 50.00[H_3O^+]}{0.1000 + [H_3O^+]}$$

Thus, to obtain $[H_3O^+] = 1.000 \times 10^{-2}$, we find

$$V_{NaOH} = \frac{5.000 - 50.00 \times 1.000 \times 10^{-2}}{0.1000 + 1.000 \times 10^{-2}} = 40.91 \text{ mL}$$

Challenge: Use the same reasoning to show that beyond the equivalence point,

$$V_{NaOH} = \frac{50.00[OH^-] + 5.000}{0.1000 - [OH^-]}$$

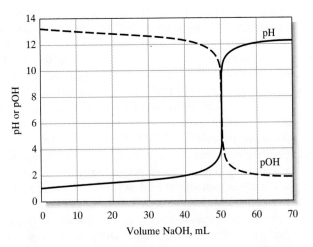

Figure 11-3 Titration curves of pH and pOH versus volume of base for the titration of 0.1000 M HCl with 0.1000 M NaOH.

WEB WORKS

To get an overview of the titration process and the care and use of a buret, direct your web browser to the web pages http://genchem.chem.wisc.edu/labdocs/modules/buret/buret.htm and http://genchem.chem.wisc.edu/labdocs/modules/titrate/titrate.htm. You will find links to a number of QuickTime® movies depicting various titration operations. These movie files are quite large (2–3 Mbytes), so you will want to use a fast internet connection to download them. Be sure that the QuickTime® plugin is installed for your browser.

11E   QUESTIONS AND PROBLEMS

11-1. Write two equations that, along with the stoichiometric ratio, form the basis for the calculations of volumetric titrimetry.

11-2. Define
  *(a) millimole.
  (b) titration.
  *(c) stoichiometric ratio.
  (d) titration error.

11-3. Distinguish between
  *(a) the equivalence point and the end point of a titration.
  (b) a direct titration and a back titration.
  *(c) a primary standard and a secondary standard.

*11-4. Briefly explain why milligrams of solute per liter and parts per million can be used interchangeably to describe the concentration of a dilute aqueous solution.

11-5. Calculations of volumetric analysis ordinarily consist of transforming the quantity of titrant used (in chemical units) to a chemically equivalent quantity of analyte (also in chemical units) through use of a stoichiometric ratio. Use chemical formulas (no calculations required) to express this ratio to calculate the percentage of
  *(a) hydrazine in rocket fuel by titration with standard iodine. Reaction:

$$H_2NNH_2 + 2I_2 \rightarrow N_2(g) + 4I^- + 4H^+$$

  (b) hydrogen peroxide in a cosmetic preparation by titration with standard permanganate. Reaction:

$$5H_2O_2 + 2MnO_4^- + 6H^+ \rightarrow \\ 2Mn^{2+} + 5O_2(g) + 8H_2O$$

  *(c) boron in a sample of borax, $Na_2B_4O_7 \cdot 10H_2O$ by titration with standard acid. Reaction:

$$B_4O_7^{2-} + 2H^+ + 5H_2O \rightarrow 4H_3BO_3$$

  (d) sulfur in an agricultural spray that was con-

verted to thiocyanate with an unmeasured excess of cyanide. Reaction:

$$S(s) + CN^- \rightarrow SCN^-$$

After removal of the excess cyanide, the thiocyanate was titrated with a standard potassium iodate solution in strong HCl. Reaction:

$$2SCN^- + 3IO_3^- + 2H^+ + 6Cl^- \rightarrow \\ 2SO_4^{2-} + 2CN^- + 3ICl_2^- + H_2O$$

*11-6. How many millimoles of solute are contained in
  (a) 2.00 L of 2.76×10^{-3} M $KMnO_4$?
  (b) 750.0 mL of 0.0416 M KSCN?
  (c) 250.0 mL of a solution that contains 4.20 ppm of $CuSO_4$?
  (d) 3.50 L of 0.276 M KCl?

11-7. How many millimoles of solute are contained in
  (a) 4.25 mL of 0.0917 M KH_2PO_4?
  (b) 0.1020 L of 0.0643 M $HgCl_2$?
  (c) 2.81 L of a 49.0 ppm solution of $Mg(NO_3)_2$?
  (d) 79.8 mL of 0.1379 M NH_4VO_3 (116.98 g/mol)?

11-8. What mass of solute in milligrams is contained in
  *(a) 26.0 mL of 0.150 M sucrose (342 g/mol)?
  *(b) 2.92 L of 5.23×10^{-3} M H_2O_2?
  (c) 737 mL of a solution that contains 6.38 ppm of $Pb(NO_3)_2$ (331.20 g/mol)?
  (d) 6.75 mL of 0.0619 M KNO_3?

11-9. What mass of solute in grams is contained in
  *(a) 450.0 mL of 0.164 M H_2O_2?
  *(b) 27.0 mL of 8.75×10^{-4} M benzoic acid (122 g/mol)?
  (c) 3.50 L of a solution that contains 21.7 ppm of $SnCl_2$?
  (d) 21.7 mL of 0.0125 M $KBrO_3$?

*11-10. Calculate the molar concentration of a solution that is 50.0% in NaOH (w/w) and has a specific gravity of 1.52.

11-11. Calculate the molar concentration of a 20.0% solution (w/w) of KCl that has a specific gravity of 1.13.

11-12. Calculate the molar analytical concentration of solute in an aqueous solution that is
  *(a) 11.00% (w/w) NH_3 and has a density of 0.9538.
  (b) 18.00% (w/w) KBr and has a density of 1.149.
  *(c) 28.00% (w/w) ethylene glycol (62.07 g/mol) and has a density of 1.0350.
  (d) 15.00% (w/w) sucrose (342.5 g/mol) and has a density of 1.0592.

*11-13. How would you prepare
  (a) 500.0 mL of 16.0% (w/v) aqueous ethanol (46.1 g/mol)?
  (b) 500.0 mL of 16.0% (v/v) aqueous ethanol?
  (c) 500.0 g of 16.0% (w/w) aqueous ethanol?

11-14. Describe the preparation of
  (a) 250.0 mL of 20.0% (w/v) aqueous acetone (58.05 g/mol).
  (b) 250.0 mL of 20.0% (v/v) aqueous acetone.
  (c) 250.0 mL of 20.0% (w/w) aqueous acetone.

*11-15. Describe how you would prepare 2.00 L of 0.150 M perchloric acid from a concentrated solution that has a specific gravity of 1.66 and is 70% $HClO_4$ (w/w).

11-16. Describe how you would prepare 800 mL of 0.400 M aqueous ammonia from a concentrated solution that has a specific gravity of 0.90 and is 27% NH_3 (w/w).

*11-17. Describe the preparation of
  (a) 500.0 mL of 0.0750 M $AgNO_3$ from the solid reagent.
  (b) 1.00 L of 0.315 M HCl, starting with a 6.00 M solution of the reagent.
  (c) 600.0 mL of a solution that is 0.0825 M in K^+, starting with solid $K_4Fe(CN)_6$.
  (d) 400.0 mL of 3.00% (w/v) aqueous $BaCl_2$ from a 0.400 M $BaCl_2$ solution.
  (e) 2.00 L of 0.120 M $HClO_4$ from the commercial reagent [60% $HClO_4$ (w/w), sp gr 1.60].
  (f) 9.00 L of a solution that is 60.0 ppm in Na^+, starting with solid Na_2SO_4.

11-18. Describe the preparation of
  (a) 5.00 L of 0.150 M $KMnO_4$ from the solid reagent.
  (b) 4.00 L of 0.175 M $HClO_4$, starting with an 8.00 M solution of the reagent.
  (c) 400.0 mL of a solution that is 0.0500 M in I^-, starting with MgI_2.
  (d) 200.0 mL of 1.00% (w/v) aqueous $CuSO_4$ from a 0.218 M $CuSO_4$ solution.
  (e) 1.50 L of 0.215 M NaOH from the concentrated commercial reagent [50% NaOH (w/w), sp gr 1.525].
  (f) 1.50 L of a solution that is 12.0 ppm in K^+, starting with solid $K_4Fe(CN)_6$.

*11-19. A solution of $HClO_4$ was standardized by dissolving 0.3745 g of primary-standard-grade HgO in a solution of KBr:

$$HgO(s) + 4Br^- + H_2O \rightarrow HgBr_4^{2-} + 2OH^-$$

The liberated OH^- required 37.79 mL of the acid to be neutralized. Calculate the molarity of the $HClO_4$.

11-20. A 0.3367-g sample of primary-standard-grade Na_2CO_3 required 28.66 mL of a H_2SO_4 solution to reach the end point in the reaction

$$CO_3^{2-} + 2H^+ \rightarrow H_2O + CO_2(g)$$

What is the molarity of the H_2SO_4?

*11-21. A 0.3396-g sample that assayed 96.4% Na_2SO_4 required 37.70 mL of a barium chloride solution. Reaction:

$$Ba^{2+} + SO_4^{2-} \rightarrow BaSO_4(s)$$

Calculate the analytical molarity of $BaCl_2$ in the solution.

*11-22. A 0.4793-g sample of primary-standard-grade Na_2CO_3 was treated with 40.00 mL of dilute perchloric acid. The solution was boiled to remove CO_2, following which the excess $HClO_4$ was back-titrated with 8.70 mL of dilute NaOH. In a separate experiment, it was established that 27.43 mL of the $HClO_4$ neutralized the NaOH in a 25.00-mL portion. Calculate the molarities of the $HClO_4$ and NaOH.

11-23. Titration of 50.00 mL of 0.05251 M $Na_2C_2O_4$ required 38.71 mL of a potassium permanganate solution:

$$2MnO_4^- + 5C_2O_4^{2-} + 16H^+ \rightarrow$$
$$2Mn^{2+} + 10CO_2(g) + 8H_2O$$

Calculate the molarity of the $KMnO_4$ solution.

*11-24. Titration of the I_2 produced from 0.1238 g of primary-standard-grade KIO_3 required 41.27 mL of sodium thiosulfate:

$$IO_3^- + 5I^- + 6H^+ \rightarrow 3I_2 + 3H_2O$$
$$I_2 + 2S_2O_3^{2-} \rightarrow 2I^- + S_4O_6^{2-}$$

Calculate the concentration of the $Na_2S_2O_3$.

*11-25. A 4.476-g sample of a petroleum product was burned in a tube furnace, and the SO_2 produced was collected in 3% H_2O_2. Reaction:

$$SO_2(g) + H_2O_2 \rightarrow H_2SO_4$$

A 25.00-mL portion of 0.00923 M NaOH was introduced into the solution of H_2SO_4, following which the excess base was back-titrated with 13.33 mL of 0.01007 M HCl. Calculate the parts per million of sulfur in the sample.

11-26. A 100.0-mL sample of spring water was treated to convert any iron present to Fe^{2+}. Addition of 25.00 mL of 0.002107 M $K_2Cr_2O_7$ resulted in the reaction

$$6Fe^{2+} + Cr_2O_7^{2-} + 14H^+ \rightarrow$$
$$6Fe^{3+} + 2Cr^{3+} + 7H_2O$$

The excess $K_2Cr_2O_7$ was back-titrated with 7.47 mL of a 0.00979 M Fe^{2+} solution. Calculate the parts per million of iron in the sample.

*11-27. The arsenic in a 1.223-g sample of a pesticide was converted to H_3AsO_4 by suitable treatment. The acid was then neutralized, and 40.00 mL of 0.07891 M $AgNO_3$ was added to precipitate the arsenic quantitatively as Ag_3AsO_4. The excess Ag^+ in the filtrate and washings from the precipitate was titrated with 11.27 mL of 0.1000 M KSCN; the reaction was

$$Ag^+ + SCN^- \rightarrow AgSCN(s)$$

Calculate the percent As_2O_3 in the sample.

*11-28. The thiourea in a 1.455-g sample of organic material was extracted into a dilute H_2SO_4 solution and titrated with 37.31 mL of 0.009372 M Hg^{2+} via the reaction

$$4(NH_2)_2CS + Hg^{2+} \rightarrow [(NH_2)_2CS]_4Hg^{2+}$$

Calculate the percent $(NH_2)_2CS$ (76.12 g/mol) in the sample.

11-29. The ethyl acetate concentration in an alcoholic solution was determined by diluting a 10.00-mL sample to 100.0 mL. A 20.00-mL portion of the diluted solution was refluxed with 40.00 mL of 0.04672 M KOH:

$$CH_3COOC_2H_5 + OH^- \rightarrow CH_3COO^- + C_2H_5OH$$

After cooling, the excess OH^- was back-titrated with 3.41 mL of 0.05042 M H_2SO_4. Calculate the number of grams of ethyl acetate (88.11 g/mol) per 100 mL of the original sample.

*11-30. A solution of $Ba(OH)_2$ was standardized against 0.1016 g of primary-standard-grade benzoic acid C_6H_5COOH (122.12 g/mol). An end point was observed after addition of 44.42 mL of base.
 (a) Calculate the molarity of the base.
 (b) Calculate the standard deviation of the molarity if the standard deviation for weighing was ± 0.2 mg and that for the volume measurement was ± 0.03 mL.
 (c) Assuming an error of -0.3 mg in the weighing, calculate the absolute and relative systematic error in the molarity.

11-31. A 0.1475 M solution of $Ba(OH)_2$ was used to titrate the acetic acid (60.05 g/mol) in a dilute aqueous solution. The following results were obtained.

Sample	Sample Volume, mL	$Ba(OH)_2$ Volume, mL
1	50.00	43.17
2	49.50	42.68
3	25.00	21.47
4	50.00	43.33

 (a) Calculate the mean w/v percentage of acetic acid in the sample.
 (b) Calculate the standard deviation for the results.
 (c) Calculate the 90% confidence interval for the mean.
 (d) At the 90% confidence level, could any of the results be discarded?
 (e) Assume that the buret used to measure the acetic acid had a systematic error of -0.05 mL at all volumes delivered. Calculate the systematic error in the mean result.

*11-32. (a) A 0.3147-g sample of primary-standard grade $Na_2C_2O_4$ was dissolved in dilute H_2SO_4 and titrated with 31.67 mL of dilute $KMnO_4$:

$$2MnO_4^- + 5C_2O_4^{2-} + 16H^+ \rightarrow$$
$$2Mn^{2+} + 10CO_2(g) + 8H_2O$$

Calculate the molarity of the $KMnO_4$ solution.
 (b) The iron in a 0.6656-g ore sample was reduced quantitatively to the +2 state and then titrated with 26.75 mL of the $KMnO_4$ solution. Calculate the percent Fe_2O_3 in the sample.

11-33. (a) A 0.1752-g sample of primary-standard-grade $AgNO_3$ was dissolved in 502.3 mL of distilled water. Calculate the molarity of Ag^+ in this solution.
 (b) The standard solution described in part (a) was used to titrate a 25.171-g sample of a KSCN solution. An end point was obtained after adding 23.76 mL of the $AgNO_3$ solution. Calculate the molarity of the KSCN solution.
 (c) The solutions described in parts (a) and (b) were used to determine the $BaCl_2 \cdot 2H_2O$ in a 0.7120-g sample. A 20.10 mL volume of the $AgNO_3$ was added to a solution of the sample, and the excess $AgNO_3$ was back-titrated with 7.54 mL of the KSCN solution. Calculate the percent $BaCl_2 \cdot 2H_2O$ in the sample.

*11-34. A solution was prepared by dissolving 10.12 g of $KCl \cdot MgCl_2 \cdot 6H_2O$ (277.85 g/mol) in sufficient water to give 2.000 L. Calculate
 (a) the molar analytical concentration of $KCl \cdot MgCl_2$ in this solution.
 (b) the molar concentration of Mg^{2+}.
 (c) the molar concentration of Cl^-.

(d) the weight/volume percentage of
      $KCl \cdot MgCl_2 \cdot 6H_2O$.

(e) the number of millimoles of Cl^- in 25.0 mL of
      this solution.

(f) ppm K^+.

11-35. A solution was prepared by dissolving 367 mg of
$K_3Fe(CN)_6$ (329.2 g/mol) in sufficient water to give
750 mL. Calculate

(a) the molar analytical concentration of
      $K_3Fe(CN)_6$.

(b) the molar concentration of K^+.

(c) the molar concentration of $Fe(CN)_6^{3-}$.

(d) the weight/volume percentage of $K_3Fe(CN)_6$.

(e) the number of millimoles of K^+ in 50.0 mL of
      this solution.

(f) ppm $Fe(CN)_6^{3-}$.

Chapter 12

Principles of Neutralization Titrations: Determining Acids, Bases, and the pH of Buffer Solutions

The equilibria that exist in solutions of acids and bases are of paramount importance throughout chemistry and throughout science in general. For example, you will find that the material in this chapter as well as that in Chapters 13 and 14 is directly relevant to the acid/base reactions that are important in biochemistry and the other biological sciences.

Standard solutions of strong acids and strong bases are used extensively for determining analytes that are themselves acids or bases or analytes that can be converted to such species by chemical treatment. This chapter deals with the principles of titrations with such reagents. In addition, buffer solutions that are vital in maintaining a nearly constant pH are explored and several examples of pH calculations are presented.

12A WHAT SOLUTIONS AND INDICATORS ARE USED?

Like any titration, neutralization titrations depend on a chemical reaction between the analyte and a standard reagent. The point of chemical equivalence is indicated by a chemical indicator or an instrumental measurement. The discussion here focuses on the types of standard solutions and the chemical indicators that are used for neutralization titrations.

12A-1 Standard Solutions

The standard solutions employed in neutralization titrations are strong acids or strong bases because these substances react more completely with an analyte than their weaker counterparts do and thus yield sharper end points. Standard solutions of acids are prepared by diluting concentrated hydrochloric, perchloric, or sulfuric acid. Nitric acid is seldom used because its oxidizing properties offer the potential for undesirable side reactions. Hot concentrated perchloric

The standard reagents used in acid/base titrations are always strong acids or strong bases, most commonly HCl, HClO$_4$, H$_2$SO$_4$, NaOH, and KOH. Weak acids and weak bases are never used as standard reagents because they react incompletely with analytes.

and sulfuric acids are potent oxidizing agents and are therefore hazardous. Fortunately, cold dilute solutions of these reagents are relatively benign and can be used in the analytical laboratory without any special precautions other than eye protection.

Standard solutions of bases are ordinarily prepared from solid sodium or potassium and occasionally barium hydroxides. The concentrations of these bases must be established by standardization. Again, eye protection should always be used when handling dilute solutions of these reagents.

12A-2   Acid/Base Indicators

Many substances, both naturally occurring and synthetic, display colors that depend on the pH of the solutions in which they are dissolved. Some of these substances, which have been used for centuries to indicate the acidity or alkalinity of water, are still applied as acid/base indicators.

An acid/base indicator is a weak organic acid or a weak organic base whose undissociated form differs in color from its conjugate form. For example, the behavior of an acid-type indicator, HIn, is described by the equilibrium

$$HIn + H_2O \rightleftharpoons In^- + H_3O^+$$

acid color   base color

For a list of common acid/base indicators and their colors, look inside the front cover of this book.

Here, internal structural changes accompany dissociation and cause the color change (for example, see Figure 12-1). The equilibrium for a base-type indicator, In, is

$$In + H_2O \rightleftharpoons InH^+ + OH^-$$

base color   acid color

In the paragraphs that follow, we focus on the behavior of acid-type indicators. The principles, however, can be easily extended to base-type indicators as well.

The equilibrium-constant expression for the dissociation of an acid-type indicator takes the form

$$K_a = \frac{[H_3O^+][In^-]}{[HIn]} \tag{12-1}$$

Rearranging leads to

$$[H_3O^+] = K_a \frac{[HIn]}{[In^-]} \tag{12-2}$$

We see then that the hydronium ion concentration determines the ratio of the acid to the conjugate base form of the indicator and thus determines the color developed by the solution.

The human eye is not very sensitive to color differences in a solution containing a mixture of In$^-$ and HIn, particularly when the ratio [HIn]/[In$^-$] is greater than about 10 or smaller than about 0.1. Consequently, the color imparted to a solution by a typical indicator appears to the average observer to change rapidly only within the limited concentration ratio of approximately 10 to 0.1. At greater or smaller ratios, the color becomes essentially constant to the eye and is independent of the ratio. Hence, we can write that the average indicator HIn exhibits its pure acid color when

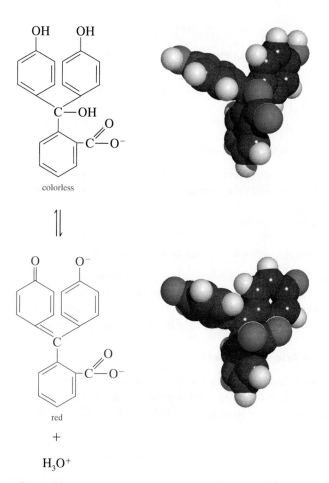

colorless

red

+

H_3O^+

Figure 12-1   Color change and molecular model for phenolphthalein.

$$\frac{[\text{HIn}]}{[\text{In}^-]} \geq \frac{10}{1}$$

and its base color when

$$\frac{[\text{HIn}]}{[\text{In}^-]} \leq \frac{1}{10}$$

The color appears to be intermediate for ratios between these two values. These ratios vary considerably from indicator to indicator, of course. Furthermore, people differ significantly in their ability to distinguish between colors; indeed, a color-blind person may be unable to distinguish any color change at all.

If the two concentration ratios are substituted into Equation 12-2, the range of hydronium ion concentrations needed to change the indicator color can be evaluated. Thus, for the full acid color,

$$[\text{H}_3\text{O}^+] = 10K_a$$

and similarly for the full base color,

pH

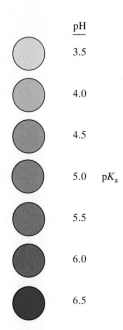

pH	
3.5	
4.0	
4.5	
5.0	pK_a
5.5	
6.0	
6.5	

Figure 12-2 Indicator color as a function of pH (pK_a = 5.0).

The approximate pH transition range of most acid type indicators is roughly $pK_a \pm 1$.

$$[H_3O^+] = 0.1\,K_a$$

To obtain the indicator pH range, we take the negative logarithms of the two expressions:

$$\text{pH(acid color)} = -\log(10K_a) = pK_a + 1$$

$$\text{pH(basic color)} = -\log(0.1K_a) = pK_a - 1$$

$$\text{indicator pH range} = pK_a \pm 1 \qquad (12\text{-}3)$$

Thus, an indicator with an acid dissociation constant of 1×10^{-5} ($pK_a = 5$) typically shows a complete color change when the pH of the solution in which it is dissolved changes from 4 to 6 (see Figure 12-2). A similar relationship is easily derived for a basic-type indicator.

What Variables Influence the Behavior of Indicators?

The pH interval over which a given indicator exhibits a color change is influenced by temperature, by the ionic strength of the medium, and by the presence of organic solvents and colloidal particles. Some of these effects, particularly the last two, can cause the transition range to shift by one or more pH units.[1]

The Common Acid/Base Indicators

The list of acid/base indicators is large and includes a number of organic compounds. Indicators are available for almost any desired pH range. A few common indicators and their properties are listed in Table 12-1. Note that the transi-

Table 12-1

Some Important Acid/Base Indicators

Common Name	Transition Range, pH	pK_a*	Color Change†	Indicator Type‡
Thymol blue	1.2–2.8	1.65§	R–Y	1
	8.0–9.6	8.96§	Y–B	
Methyl yellow	2.9–4.0		R–Y	2
Methyl orange	3.1–4.4	3.46§	R–O	2
Bromocresol green	3.8–5.4	4.66§	Y–B	1
Methyl red	4.2–6.3	5.00§	R–Y	2
Bromocresol purple	5.2–6.8	6.12§	Y–P	1
Bromothymol blue	6.2–7.6	7.10§	Y–B	1
Phenol red	6.8–8.4	7.81§	Y–R	1
Cresol purple	7.6–9.2		Y–P	1
Phenolphthalein	8.3–10.0		C–R	1
Thymolphthalein	9.3–10.5		C–B	1
Alizarin yellow GG	10–12		C–Y	2

*At ionic strength of 0.1.
†B = blue; C = colorless; O = orange; P = purple; R = red; Y = yellow.
‡(1) Acid type: $HIn + H_2O \rightleftharpoons H_3O^+ + In^-$; (2) base type: $In + H_2O \rightleftharpoons InH^+ + OH^-$.
§For the reaction $InH^+ + H_2O \rightleftharpoons H_3O^+ + In$.

[1]For a discussion of these effects, see H. A. Laitinen and W. E. Harris, *Chemical Analysis,* 2nd ed., pp. 48–51. New York: McGraw-Hill, 1975.

tion ranges vary from 1.1 to 2.2, with the average being about 1.6 units. These indicators and several more are shown along with their transition ranges in the colored chart inside the front cover of this book.

12B CALCULATING pH IN TITRATIONS OF STRONG ACIDS AND STRONG BASES

The hydronium ions in an aqueous solution of a strong acid have two sources: (1) the reaction of the acid with water and (2) the dissociation of water itself. In all but the most dilute solutions, however, the contribution from the strong acid far exceeds that from the solvent. Thus, for a solution of HCl with a concentration greater than about 10^{-6} M, we can write

$$[H_3O^+] = c_{HCl} + [OH^-] \approx c_{HCl}$$

where $[OH^-]$ represents the contribution of hydronium ions from the dissociation of water. An analogous relationship applies for a solution of a strong base, such as sodium hydroxide. That is,

$$[OH^-] = c_{NaOH} + [H_3O^+] \approx c_{NaOH}$$

> In solutions of a strong acid that are more concentrated than about 1×10^{-6} M, it is safe to assume that the equilibrium concentration of H_3O^+ is equal to the analytical concentration of the acid. The same is true for the hydroxide concentration in solutions of strong bases.

12B-1 Titrating a Strong Acid with a Strong Base

We will be interested here, and in the next several chapters, in calculating *hypothetical* titration curves of pH versus volume of titrant. We distinguish between hypothetical curves constructed by computing the values of pH and *experimental* titration curves observed in the laboratory. Three types of calculations must be done to construct the hypothetical titration curve for titrating a solution of a strong acid with a strong base. Each calculation corresponds to a distinct stage in the titration: (1) preequivalence, (2) equivalence, and (3) postequivalence. In the preequivalence stage, we compute the concentration of the acid from its starting concentration and the amount of base that has been added. At the equivalence point, the hydronium and hydroxide ions are present in equal concentrations, and the hydronium ion concentration is derived directly from the ion-product constant for water. In the postequivalence stage, the analytical concentration of the excess base is computed, and the hydroxide ion concentration is assumed to be equal to or a multiple of the analytical concentration.

A convenient way of converting hydroxide concentration to pH can be developed by taking the negative logarithm of both sides of the ion-product constant expression for water. Thus,

> Before the equivalence point, we calculate the pH from the molar concentration of unreacted acid.

> At the equivalence point, the solution is neutral and pH = 7.00.

> Beyond the equivalence point, we first calculate pOH and then pH. Remember that pH = pK_w − pOH = 14.00 − pOH.

$$K_w = [H_3O^+][OH^-]$$

$$-\log K_w = -\log [H_3O^+][OH^-] = -\log [H_3O^+] - \log [OH^-]$$

$$pK_w = pH + pOH$$

$$-\log 10^{-14} = 14.00 = pH + pOH$$

Example 12-1

Do the calculations needed to generate the hypothetical titration curve for the titration of 50.00 mL of 0.0500 M HCl with 0.1000 M NaOH.

Initial Point

Before any base is added, the solution is 0.0500 M in H_3O^+, and

$$pH = -\log[H_3O^+] = -\log 0.0500 = 1.30$$

After Addition of 10.00 mL of Reagent

The hydronium ion concentration is decreased as a result of both reaction with the base and dilution. Thus, the analytical concentration of HCl is

$$c_{HCl} = \frac{\text{no. mmol HCl remaining after addition of NaOH}}{\text{total volume soln}}$$

$$= \frac{\text{original no. mmol HCl} - \text{no. mmol NaOH added}}{\text{total volume soln}}$$

$$= \frac{(50.00 \text{ mL} \times 0.0500 \text{ M}) - (10.00 \text{ mL} \times 0.1000 \text{ M})}{50.00 \text{ mL} + 10.00 \text{ mL}}$$

$$= \frac{(2.500 \text{ mmol} - 1.000 \text{ mmol})}{60.00 \text{ mL}} = 2.500 \times 10^{-2} \text{ M}$$

$$[H_3O^+] = 2.500 \times 10^{-2}$$

$$pH = -\log[H_3O^+] = -\log(2.500 \times 10^{-2}) = 1.602$$

Additional points defining the curve in the region before the equivalence point are obtained in the same way. The results of such calculations are shown in the second column of Table 12-2.

Table 12-2

Changes in pH During the Titration of a Strong Acid With a Strong Base

	pH	
Volume of NaOH, mL	50.00 mL of 0.0500 M HCl with 0.100 M NaOH	50.00 mL of 0.000500 M HCl with 0.00100 M NaOH
0.00	1.30	3.30
10.00	1.60	3.60
20.00	2.15	4.15
24.00	2.87	4.87
24.90	3.87	5.87
25.00	7.00	7.00
25.10	10.12	8.12
26.00	11.12	9.12
30.00	11.80	9.80

Equivalence Point

At the equivalence point, neither HCl nor NaOH is in excess, so the concentrations of hydronium and hydroxide ions must be equal. Substituting this equality into the ion-product constant for water yields

$$[H_3O^+] = \sqrt{K_w} = \sqrt{1.00 \times 10^{-14}} = 1.00 \times 10^{-7}$$

$$pH = -\log(1.00 \times 10^{-7}) = 7.00$$

After Addition of 25.10 mL of Reagent

The solution now contains an excess of NaOH, and we can write

$$c_{NaOH} = \frac{\text{no. mmol NaOH added } - \text{ original no. mmol HCl}}{\text{total volume soln}}$$

$$c_{NaOH} = \frac{25.10 \times 0.100 - 50.00 \times 0.0500}{75.10} = 1.33 \times 10^{-4} \text{ M}$$

and the equilibrium concentration of hydroxide ion is

$$[OH^-] = c_{NaOH} = 1.33 \times 10^{-4} \text{ M}$$

$$pOH = -\log(1.33 \times 10^{-4}) = 3.88$$

and

$$pH = 14.00 - 3.88 = 10.12$$

Additional data defining the curve beyond the equivalence point are computed in the same way. The results of such computations are shown in Table 12-2. The calculations of Table 12-2 are done in spreadsheet format in Figure 12-3.

How Do Concentrations Influence Titration Curves?

The effects of reagent and analyte concentration on the neutralization titration curves for strong acids are shown by the two sets of data in Table 12-2 and the plots in Figure 12-3. Note that with 0.1 M NaOH as the titrant, the change in pH in the equivalence-point region is large. With 0.001 M NaOH, the change is markedly less but still pronounced.

Choosing an Indicator

Figure 12-4 shows that the selection of an indicator is not critical when the reagent concentration is approximately 0.1 M. Here, the volume differences in titrations with the three indicators shown are of the same magnitude as the uncertainties associated with reading the buret and are thus negligible. Note, however, that bromocresol green is clearly unsuited for a titration involving the 0.001 M reagent because the color change occurs over a 5-mL range well before the equivalence point. The use of phenolphthalein is subject to similar objections. Of the three indicators, then, only bromothymol blue provides a satisfac-

	A	B	C	D	E	F	G	H	I
1	50.00 mL 0.0500 M HCl with 0.100 M NaOH					50.00 mL 0.000500 M HCl with 0.00100 M NaOH			
2	K_w	1.00E-14	Vol. HCL, mL	50.00		Vol HCl, mL	50.00	c_{NaOH}	0.00100
3	Initial c_{HCl}	0.05	c_{NaOH}	0.100		Initial c_{HCl}	0.000500		
4	Vol. NaOH, mL	[H$_2$O$^+$]	[OH$^-$]	pOH	pH	[H$_2$O$^+$]	[OH$^-$]	pOH	pH
5	0.00	0.05			1.30	0.0005			3.30
6	5.00	0.036363636			1.44	0.0003636			3.44
7	10.00	0.025			1.60	0.00025			3.60
8	15.00	0.015384615			1.81	0.0001538			3.81
9	20.00	0.007142857			2.15	7.143E-05			4.15
10	24.00	0.001351351			2.87	1.351E-05			4.87
11	24.20	0.001078167			2.97	1.078E-05			4.97
12	24.40	0.000806452			3.09	8.065E-06			5.09
13	24.60	0.000536193			3.27	5.362E-06			5.27
14	24.80	0.00026738			3.57	2.674E-06			5.57
15	24.90	0.000133511			3.87	1.335E-06			5.87
16	24.99	1.33351E-05			4.88	1.334E-07			6.88
17	25.00	1.00E-07			7.00	1.00E-07			7.00
18	25.01		1.33316E-05	4.88	9.12		1.33316E-07	6.88	7.12
19	25.10		0.000133156	3.88	10.12		1.33156E-06	5.88	8.12
20	25.20		0.000265957	3.58	10.42		2.65957E-06	5.58	8.42
21	25.40		0.000530504	3.28	10.72		5.30504E-06	5.28	8.72
22	25.60		0.000793651	3.10	10.90		7.93651E-06	5.10	8.90
23	25.80		0.001055409	2.98	11.02		1.05541E-05	4.98	9.02
24	26.00		0.001315789	2.88	11.12		1.31579E-05	4.88	9.12
25	30.00		0.00625	2.20	11.80		0.0000625	4.20	9.80
26	35.00		0.011764706	1.93	12.07		0.000117647	3.93	10.07
27	40.00		0.016666667	1.78	12.22		0.000166667	3.78	10.22
28	45.00		0.021052632	1.68	12.32		0.000210526	3.68	10.32
29	50.00		0.025	1.60	12.40		0.00025	3.60	10.40
30	Spreadsheet Documentation								
31	Cell B5=($B$3*$D$2-A5*$D$3)/($D$2+A5)					Cell F5=($G$2*$G$3-A5*$I$2)/($D$2+A5)			
32	Cell B17=SQRT(B2)					Cell F17=SQRT(B2)			
33	Cell E5=-LOG10(B5)					Cell I5=-LOG10(F5)			
34	Cell C18=(A18*$D$3-$B$3*$D$2)/($D$2+A18)					Cell G18=(A18*$I$2-$G$2*$G$3)/($D$2+A18)			
35	Cell D18=-LOG10(C18)					Cell H18=-LOG10(G18)			
36	Cell E18=14-D18					Cell I18=14-H18			

Figure 12-3 *See figure caption on facing page.*

Figure 12-3   Spreadsheet for titration curve data of Example 12-1. The dissociation constant for water, K_w is entered into cell B2. The initial volume of HCl is entered into D2 for the first case and G2 for the second case. Initial concentrations of HCl are entered into B3 and G3. The concentration of titrant is entered into D3 and I2. Appropriate labels are placed in the cells next to these parameter entries. Labels as shown are entered into the cells in row 4. The volume of NaOH is entered in cells A5 through A29. The H_3O^+ concentration is calculated directly in the preequivalance point region in cells B5 through B16 and F5 through F16. The formulas for calculating the H_3O^+ concentration are entered into B5 and F5 as shown in the documentation (cells C31 and F31). The B5 formula is copied into cells B6:B16, and the F5 formula is copied into cells F6:F16. You should verify these entries. The pH values for these concentrations are calculated in cells E5 through E16 and I5 through I16. The formula `-log10(B5)` is entered into cell E5 and copied into cells E6:E16. Likewise `-log10(F5)` is entered in cell I5 and copied into cells I6:I16. In cells B17 and F17, the H_3O^+ concentration is calculated at the equivalence point as $\sqrt{K_w}$ as shown in the documentation (cells A32 and F32), and the pH in cells E17 and I17 from `=-LOG10(B15)` and `=-LOG10(F15)`. In the postequivalance point region, cells 18 through 29, the OH^- concentration is calculated and the pH obtained from this. The formulas for the first postequivalence points, cell C18 and G18, are shown in the documentation section (Cells A34 and F34). These are then copied into the remaining C and G cells. The pOH is calculated in cells D18 and H18 as given in Cells A35 and F35. These formulas are copied into the remaining D and H cells. Finally in cells E18 and I18, the pH is calculated from `=14-D18` and `14-H18`, respectively, and the formulas are copied into the remaining E and I cells. The chart that results from plotting pH vs. Volume NaOH is shown below the spreadsheet. This chart is constructed by plotting cells E5:E29 and I5:I29 against cells A5:A29.

---

tory end point with a minimal systematic error in the titration of the most dilute solution.

12B-2   Titrating a Strong Base with a Strong Acid

Titration curves for strong bases are constructed in an analogous way to those for strong acids. Short of the equivalence point, the solution is highly basic; the hydroxide ion concentration is numerically related to the analytical molarity of the base. The solution is neutral at the equivalence point and becomes acidic in the region beyond the equivalence point; here, the hydronium ion concentration is equal to the analytical concentration of the excess strong acid.

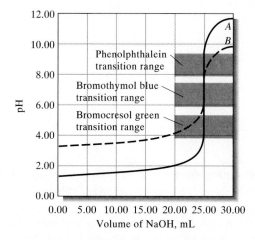

Figure 12-4   Titration curves for HCl with NaOH. Curve *A*: 50.00 mL of 0.0500 M HCl with 0.1000 M NaOH. Curve *B*: 50.00 mL of 0.000500 M HCl with 0.001000 M NaOH.

Feature 12-1

Constructing Titration Curves
from the Charge-Balance
Equation

In Example 12-1, we generated an acid/base titration curve from the reaction stoichiometry. We can show that all points on the curve can also be derived from the charge-balance equation.

For the system treated in Example 12-1, the charge-balance equation is given by

$$[H_3O^+] + [Na^+] = [OH^-] + [Cl^-]$$

where the sodium and chloride ion concentrations are given by

$$[Na^+] = \frac{V_{NaOH}c_{NaOH}}{V_{NaOH} + V_{HCl}}$$

$$[Cl^-] = \frac{V_{HCl}c_{HCl}}{V_{NaOH} + V_{HCl}}$$

For volumes of NaOH short of the equivalence point, $[OH^-] \ll [Cl^-]$, and we can rewrite the first equation in the form

$$[H_3O^+] \approx [Cl^-] - [Na^+]$$

and

$$[H_3O^+] = \frac{V_{HCl}c_{HCl}}{V_{HCl} + V_{NaOH}} - \frac{V_{NaOH}c_{NaOH}}{V_{HCl} + V_{NaOH}} = \frac{V_{HCl}c_{HCl} - V_{NaOH}c_{NaOH}}{V_{HCl} + V_{NaOH}}$$

At the equivalence point, $[Na^+] = [Cl^-]$ and

$$[H_3O^+] = [OH^-]$$

$$[H_3O^+] = \sqrt{K_w}$$

Beyond the equivalence point, $[H_3O^+] \ll [Na^+]$, and the original equation rearranges to

$$[OH^-] \approx [Na^+] - [Cl^-]$$

$$= \frac{V_{NaOH}c_{NaOH}}{V_{NaOH} + V_{HCl}} - \frac{V_{HCl}c_{HCl}}{V_{NaOH} + V_{HCl}} = \frac{V_{NaOH}c_{NaOH} - V_{HCl}c_{HCl}}{V_{NaOH} + V_{HCl}}$$

Example 12-2

Calculate the pH during the titration of 50.00 mL of 0.0500 M NaOH with 0.1000 M HCl after the addition of the following volumes of reagent: (a) 24.50 mL, (b) 25.00 mL, (c) 25.50 mL.

(a) At 24.50 mL added, $[H_3O^+]$ is very small and cannot be computed from stoichiometric considerations but can be obtained readily from $[OH^-]$.

$$[OH^-] = c_{NaOH} = \frac{\text{original no. mmol NaOH} - \text{no. mmol HCl added}}{\text{total volume of solution}}$$

$$= \frac{(50.00 \times 0.0500) - (24.50 \times 0.100)}{50.00 + 24.50} = 6.71 \times 10^{-4} \text{ M}$$

$$[H_3O^+] = \frac{K_w}{6.71 \times 10^{-4}} = \frac{1.00 \times 10^{-14}}{6.71 \times 10^{-4}}$$

$$= 1.49 \times 10^{-11} \text{ M}$$

$$pH = -\log(1.49 \times 10^{-11}) = 10.83$$

(b) This is the equivalence point where $[H_3O^+] = [OH^-]$.

$$[H_3O^+] = \sqrt{K_w} = \sqrt{1.00 \times 10^{-14}} = 1.00 \times 10^{-7} \text{ M}$$

$$pH = -\log(1.00 \times 10^{-7}) = 7.00$$

(c) At 25.50 mL added,

$$[H_3O^+] = c_{HCl} = \frac{(25.50 \times 0.100) - (50.00 \times 0.0500)}{75.50}$$

$$= 6.62 \times 10^{-4} \text{ M}$$

$$pH = -\log(6.62 \times 10^{-4}) = 3.18$$

Curves for the titration of 0.0500 M and 0.00500 M NaOH with 0.1000 M and 0.0100 M HCl are shown in Figure 12-5. Indicator selection is based on the same considerations described for the titration of a strong acid with a strong base.

Feature 12-2

How Many Significant Figures Should We Retain in Titration Curve Calculations?

Concentrations calculated in the equivalence-point region of titration curves are generally of low precision because they are based on small differences between large numbers. For example, in the calculation of c_{NaOH} after introduction of 25.10 mL of NaOH in Example 12-1, the numerator $(2.510 - 2.500 = 0.010)$ is known to only two significant digits. To minimize rounding error, however, three digits were retained in c_{NaOH} (1.33×10^{-4}), and rounding was postponed until pOH and pH were computed.

In rounding the calculated values for p-functions, you should remember (Section 5E-2) that it is the *mantissa of a logarithm* (that is, the number to the right of the decimal point) *that should be rounded to include only significant figures* because the characteristic (the number to the left of the decimal point) serves merely to locate the decimal point. Fortunately, the large changes in p-functions characteristic of most equivalence points are not obscured by the limited precision of the calculated data. Generally, in deriving data for titration curves, we will round p-functions to two places to the right of the decimal point whether or not such rounding is called for.

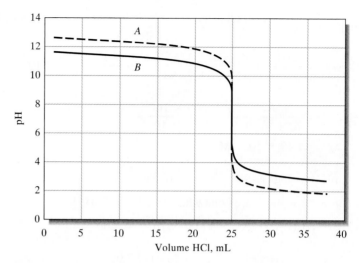

Figure 12-5   Titration curves for NaOH with HCl. Curve *A*: 50.00 mL of 0.0500 M NaOH with 0.1000 M HCl. Curve *B*: 50.00 mL of 0.00500 M NaOH with 0.0100 M HCl.

A *buffer* is a mixture of a weak acid and its conjugate base or a weak base and its conjugate acid that resists changes in pH of a solution.

Buffers are used in all types of chemistry whenever it is desirable to maintain the pH of a solution at a constant and predetermined level.

Buffered aspirin contains buffers to help prevent stomach irritation from the acidity of the carboxylic acid group in aspirin.

Molecular model and structure of aspirin. The analgesic action is thought to arise because aspirin interferes with the synthesis of prostaglandins, which are hormones involved in the transmission of pain signals.

12C   BUFFER SOLUTIONS

Whenever a weak acid is titrated with a strong base or a weak base with a strong acid, a *buffer solution* consisting of a conjugate acid/base pair is formed. Thus, before we can show how titration curves for weak acids and weak bases are derived we must investigate in some detail the properties and behavior of buffer solutions. By definition, a buffer solution is a solution of a conjugate acid/base pair that resists changes in pH. Chemists employ buffers whenever they need to maintain the pH of a solution at a constant and predetermined level. You will find many references to the use of buffers throughout this text.

12C-1   Calculating the pH of Buffer Solutions

Buffers Formed from a Weak Acid and Its Conjugate Base

A solution containing a weak acid, HA and its conjugate base A^-, may be acidic, neutral, or basic, depending on the position of two competitive equilibria:

$$HA + H_2O \rightleftharpoons H_3O^+ + A^- \qquad K_a = \frac{[H_3O^+][A^-]}{[HA]} \qquad (12\text{-}4)$$

$$A^- + H_2O \rightleftharpoons OH^- + HA \qquad K_b = \frac{K_w}{K_a} = \frac{[OH^-][HA]}{[A^-]} \qquad (12\text{-}5)$$

If the first equilibrium lies farther to the right than the second, the solution is acidic. If the second equilibrium is more favorable, the solution is basic. These two equilibrium-constant expressions show that the relative concentrations of the hydronium and hydroxide ions depend not only on the magnitudes of K_a and K_b but also on the ratio between the concentrations of the acid and its conjugate base.

Equations 12-6 and 12-7 can also be derived from mass- and charge-balance expressions. Mass-balance considerations require that

$$c_{HA} + c_{NaA} = [HA] + [A^-]$$

Electrical neutrality considerations require that

$$[Na^+] + [H_3O^+] = [A^-] + [OH^-]$$

But

$$[Na^+] = c_{NaA}$$

Therefore, the charge-balance equation is

$$c_{NaA} + [H_3O^+] = [A^-] + [OH^-]$$

which rearranges to Equation 12-7:

$$[A^-] = c_{NaA} + [H_3O^+] - [OH^-]$$

When we subtract the first equation from the fourth and rearrange the resulting equation, we obtain

$$[HA] = c_{HA} - [H_3O^+] + [OH^-]$$

which is identical to Equation 12-6.

To compute the pH of a solution containing both an acid, HA, and its salt, NaA, we need to express the equilibrium concentrations of HA and NaA in terms of their analytical concentrations, c_{HA} and c_{NaA}. An examination of the two equilibria shown in Equations 12-4 and 12-5 reveals that the first reaction decreases the concentration of HA by an amount equal to $[H_3O^+]$, whereas the second increases the HA concentration by an amount equal to $[OH^-]$. Thus, the species concentration of HA is related to its analytical concentration by the mass-balance equation

$$[HA] = c_{HA} - [H_3O^+] + [OH^-] \tag{12-6}$$

Similarly, the first equilibrium will increase the concentration of A^- by an amount equal to $[H_3O^+]$, and the second will decrease this concentration by the amount $[OH^-]$. Thus the equilibrium concentration of A^- is given by a second mass-balance equation

$$[A^-] = c_{NaA} + [H_3O^+] - [OH^-] \tag{12-7}$$

Because of the inverse relationship between $[H_3O^+]$ and $[OH^-]$, it is *always* possible to eliminate one or the other from Equations 12-6 and 12-7. Moreover, the *difference* in concentration between these two species is often so small relative to the molar concentrations of acid and conjugate base that Equations 12-6 and 12-7 simplify to

$$[HA] \approx c_{HA} \tag{12-8}$$

$$[A^-] \approx c_{NaA} \tag{12-9}$$

The product formed during neutralization of the weak acid HA is the ionic salt NaA; the anion A^- is the conjugate base of HA.

Feature 12-4

The Henderson–Hasselbalch
Equation

The Henderson–Hasselbalch equation is an alternative form of Equation 12-10 that is frequently encountered in biological literature and biochemical texts. It is obtained by expressing each term in the equation in the form of its negative logarithm and inverting the concentration ratio to keep all signs positive:

$$-\log [H_3O^+] = -\log K_a - \log \frac{c_{HA}}{c_{NaA}}$$

Therefore,

$$pH = pK_a + \log \frac{c_{NaA}}{c_{HA}} \qquad (12\text{-}11)$$

If the assumptions leading to Equation 12-10 are not valid, the values for [HA] and [A⁻] are given by Equations 12-6 and 12-7, respectively. If negative logarithms are then taken, we may derive an extended Henderson–Hasselbalch equation.

Substitution of Equations 12-8 and 12-9 into the dissociation-constant expression and rearrangement yields

$$[H_3O^+] = K_a \frac{c_{HA}}{c_{NaA}} \qquad (12\text{-}10)$$

The assumption leading to Equations 12-8 and 12-9 sometimes breaks down with acids or bases that have dissociation constants greater than about 10^{-3} or when the molar concentration of either the acid or its conjugate base (or both) is very small (often <1 mM). In these circumstances, either [OH⁻] or [H₃O⁺] must be retained in Equations 12-6 and 12-7, depending on whether the solution is acidic or basic. In any case, Equations 12-8 and 12-9 should always be used initially. The provisional values for [H₃O⁺] and [OH⁻] can then be employed to test the assumptions.

Within the limits imposed by the assumptions made in its derivation, Equation 12-10 says that the hydronium ion concentration of a solution containing a weak acid and its conjugate base is dependent only on the ratio of the molar concentrations of these two solutes. Furthermore, this ratio is *independent of dilution* because the concentration of each component changes proportionately when the volume changes.

Example 12-3

What is the pH of a solution that is 0.400 M in formic acid and 1.00 M in sodium formate?

The pH of this solution will be affected by the K_a of formic acid and the K_b of formate ion.

$$HCOOH + H_2O \rightleftharpoons H_3O^+ + HCOO^- \qquad K_a = 1.80 \times 10^{-4}$$

$$HCOO^- + H_2O \rightleftharpoons HCOOH + OH^- \qquad K_b = \frac{K_w}{K_a} = 5.56 \times 10^{-11}$$

Since the K_a for formic acid is orders of magnitude larger than the K_b for formate, the solution will be acidic and K_a will determine the H_3O^+ concentration. We can thus write

$$K_a = \frac{[H_3O^+][HCOO^-]}{[HCOOH]} = 1.80 \times 10^{-4}$$

$$[HCOO^-] \approx c_{HCOO^-} = 1.00 \text{ M}$$

$$[HCOOH] \approx c_{HCOOH} = 0.400 \text{ M}$$

Substitution into Equation 12-10 gives, with rearrangement,

$$[H_3O^+] = 1.80 \times 10^{-4} \times \frac{0.400}{1.00} = 7.20 \times 10^{-5}$$

Note that the assumption that $[H_3O^+] \ll c_{HCOOH}$ and $[H_3O^+] \ll c_{HCOO^-}$ is valid. Thus,

$$pH = -\log (7.20 \times 10^{-5}) = 4.14$$

Buffers Formed from a Weak Base and Its Conjugate Acid

As shown in Example 12-4, Equations 12-6 and 12-7 also apply to buffers consisting of a weak base and its conjugate acid. In most cases it is possible to simplify these equations so that Equation 12-10 can be used.

Example 12-4

Calculate the pH of a solution that is 0.200 M in NH_3 and 0.300 M in NH_4Cl. In Appendix 2, we find that the acid dissociation constant K_a for NH_4^+ is 5.70×10^{-10}.

The equilibria we must consider are

$$NH_4^+ + H_2O \rightleftharpoons NH_3 + H_3O^+ \quad K_a = 5.70 \times 10^{-10}$$

$$NH_3 + H_2O \rightleftharpoons NH_4^+ + OH^- \quad K_b = \frac{K_w}{K_a} = \frac{1.00 \times 10^{-14}}{5.70 \times 10^{-10}} = 1.75 \times 10^{-5}$$

Using the arguments that led to Equations 12-6 and 12-7, we obtain

$$[NH_4^+] = c_{NH_4Cl} + [OH^-] - [H_3O^+] \approx c_{NH_4Cl} + [OH^-]$$

$$[NH_3] = c_{NH_3} + [H_3O^+] - [OH^-] \approx c_{NH_3} - [OH^-]$$

Here, on the basis of the equilibrium constants, we have assumed that the solution will be basic and that $[OH^-]$ is much larger than $[H_3O^+]$. Thus, we have neglected the concentration of H_3O^+ in the approximations above.

Let us also assume that $[OH^-]$ is much smaller than c_{NH_4Cl} and c_{NH_3} so that

$$[NH_4^+] \approx c_{NH_4Cl} = 0.300 \text{ M}$$

$$[NH_3] \approx c_{NH_3} = 0.200 \text{ M}$$

Substituting into the acid-dissociation constant for NH_4^+, we obtain a relationship similar to Equation 12-10. That is,

$$[H_3O^+] = \frac{K_a \times [NH_4^+]}{[NH_3]} = \frac{5.70 \times 10^{-10} \times c_{NH_4Cl}}{c_{NH_3}}$$

$$= \frac{5.70 \times 10^{-10} \times 0.300}{0.200} = 8.55 \times 10^{-10}$$

To check the validity of our approximations, we calculate $[OH^-]$. Thus,

$$[OH^-] = \frac{1.00 \times 10^{-14}}{8.55 \times 10^{-10}} = 1.17 \times 10^{-5}$$

which is certainly much smaller than c_{NH_4Cl} or c_{NH_3}. Thus we may write

$$pH = -\log(8.55 \times 10^{-10}) = 9.07$$

12C-2   What Are the Unique Properties of Buffer Solutions?

In this section we illustrate the resistance of buffers to changes of pH brought about by dilution or addition of strong acids or bases.

The Effect of Dilution

The pH of a buffer solution remains essentially independent of dilution until the concentrations of the species it contains are decreased to the point where the approximations used to develop Equations 12-8 and 12-9 become invalid. Figure 12-6 contrasts the behavior of buffered and unbuffered solutions with dilution. For each, the initial solute concentration is 1.00 M. The resistance of the buffered solution to changes in pH during dilution is clear.

The Effect of Added Acids and Bases

Example 12-5 illustrates a second property of buffer solutions: their resistance to pH change after addition of small amounts of strong acids or bases. It is of interest to contrast the behavior of an unbuffered solution with a pH of 9.07 to that of the buffer in Example 12-5. Adding the same quantity of base to the unbuffered solution would increase the pH to 12.00, a pH change of 2.93 units. Adding the acid would decrease the pH by slightly more than 7 units.

What Is the Buffer Capacity?

Figure 12-6 and Example 12-5 demonstrate that a solution containing a conjugate acid/base pair possesses remarkable resistance to changes in pH. The ability of a buffer to prevent a significant change in pH is directly related to the total concentration of the buffering species as well as to their concentration ratio. For example, the pH of a 400-mL portion of a buffer formed by diluting the solution described in Example 12-4 by 10 would change by about 0.4 to 0.5 unit when treated with 100 mL of 0.0500 M sodium hydroxide or 0.0500 M hydrochloric acid. We show in Example 12-5 that the change is only about 0.04 to 0.05 unit for the more concentrated buffer.

Buffers do not maintain pH at an absolutely constant value, but changes in pH are relatively small when small amounts of acid or base are added.

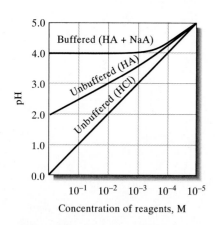

Figure 12-6 The effect of dilution of the pH of buffered and unbuffered solutions. The dissociation constant for HA is 1.00×10^{-4}. Initial solute concentrations are 1.00 M.

Example 12-5

Calculate the pH change that takes place when a 100-mL portion of (a) 0.0500 M NaOH and (b) 0.0500 M HCl is added to 400 mL of the buffer solution described in Example 12-4.

(a) Addition of NaOH converts part of the NH_4^+ in the buffer to NH_3:

$$NH_4^+ + OH^- \rightleftharpoons NH_3 + H_2O$$

The analytical concentrations of NH_3 and NH_4Cl then become

$$c_{NH_3} = \frac{400 \times 0.200 + 100 \times 0.0500}{500} = \frac{85.0}{500} = 0.170 \text{ M}$$

$$c_{NH_4Cl} = \frac{400 \times 0.300 - 100 \times 0.0500}{500} = \frac{115}{500} = 0.230 \text{ M}$$

When substituted into the acid-dissociation constant expression for NH_4^+, these values yield

$$[H_3O^+] = 5.70 \times 10^{-10} \times \frac{0.230}{0.170} = 7.71 \times 10^{-10}$$

$$pH = -\log(7.71 \times 10^{-10}) = 9.11$$

and the change in pH is

$$\Delta pH = 9.11 - 9.07 = 0.04$$

(b) Addition of HCl converts part of the NH_3 to NH_4^+; thus,

$$NH_3 + H_3O^+ \rightleftharpoons NH_4^+ + H_2O$$

$$c_{NH_3} = \frac{400 \times 0.200 - 100 \times 0.0500}{500} = \frac{75.0}{500} = 0.150 \text{ M}$$

$$c_{NH_4Cl} = \frac{400 \times 0.300 + 100 \times 0.0500}{500} = \frac{125}{500} = 0.250 \text{ M}$$

$$[H_3O^+] = 5.70 \times 10^{-10} \times \frac{0.250}{0.150} = 9.50 \times 10^{-10}$$

$$pH = -\log(9.50 \times 10^{-10}) = 9.02$$

$$\Delta pH = 9.02 - 9.07 = -0.05$$

The *buffer capacity* of a buffer is the number of moles of strong acid or strong base that causes one liter of the buffer to change pH by one unit.

The *buffer capacity* of a solution is defined as the number of moles of a strong acid or a strong base that causes 1.00 L of the buffer to undergo a 1.00-unit change in pH. The capacity of a buffer depends not only on the total concentration of the two buffer components but also on their concentration ratio. Buffer capacity falls off moderately rapidly as the concentration ratio of acid to conjugate base departs from unity (see Figure 12-7). For this reason, the pK_a of the acid chosen for a given application should lie within ±1 unit of the desired pH for the buffer to have a reasonable capacity.

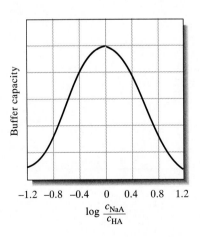

Figure 12-7 Buffer capacity as a function of the ratio c_{NaA}/c_{HA}.

Preparing Buffers

In principle, a buffer solution of any desired pH can be prepared by combining calculated quantities of a suitable conjugate acid/base pair. In practice, however, the pH values of buffers prepared from theoretically generated recipes differ from the predicted values because of uncertainties in the numerical values of many dissociation constants and from the simplifications used in calculations. It is important to realize that often the ionic strength of a buffer is so high that good values for the activity coefficients of the ions in the solution cannot be obtained from the Debye–Hückel relationship. Because of these uncertainties, we prepare buffers by making up a solution of approximately the desired pH and then adjust by adding acid or conjugate base until the required pH is indicated by a pH meter. Alternatively, empirically derived recipes for preparing buffer solutions of known pH are available in chemical handbooks and reference works.[2]

Buffers are of tremendous importance in biological and biochemical studies where a low but constant concentration of hydronium ions (10^{-6} to 10^{-10} M) must be maintained throughout experiments. Several biological supply houses offer a variety of such buffers.

Feature 12-5

Acid Rain and the Buffer Capacity of Lakes

Acid rain has been the subject of considerable controversy since the 1970s. Acid rain forms when the gaseous oxides of nitrogen and sulfur dissolve in water droplets in the air. These gases form at high temperatures in power plants, automobiles, and other combustion sources. The combustion products pass into the atmosphere where they react with water to form nitric acid and sulfuric acid. These reactions are shown by the equations

$$4NO_2(g) + 2H_2O(l) + O_2(g) \rightarrow 4HNO_3(aq)$$

$$SO_3(g) + H_2O(l) \rightarrow H_2SO_4(aq)$$

Eventually, the droplets coalesce with other droplets to form acid rain. The profound effects of acid rain have been highly publicized. Stone buildings and monuments literally dissolve away as acid rain flows over their surfaces. Forests are slowly being killed off in some locations. To illustrate the effects on aquatic life, consider the changes in pH that have occurred in the lakes of the Adirondack Mountains area of New York, illustrated in the bar graphs of Figure 12F-5A.

The graphs show the distribution of pH in these lakes, which were studied first in the 1930s and then again in 1975.[3] The shift in pH of the lakes over a 40-year period is dramatic. The average pH of the lakes changed from 6.4 to about 5.1, which represents a 20-fold change in the hydronium ion concentration. Such changes in pH have a profound effect on aquatic life, as shown by a study of the fish population in lakes in the same area.[4] In the graph of Figure 12F-5B, the number of lakes is plotted as a

---

[2]See, for example, J. A. Dean, *Analytical Chemistry Handbook,* pp. 14.26–14.34. New York: McGraw-Hill, 1995.
[3]R. F. Wright and E. T. Gjessing, *Ambio,* **1976,** *5,* 219.
[4]C. L. Schofield, *Ambio,* **1976,** *5,* 228.

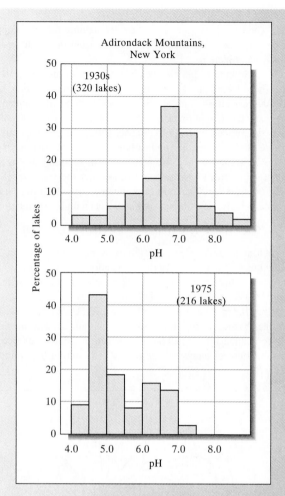

Figure 12F-5A   Changes in pH of lakes between 1930 and 1975.

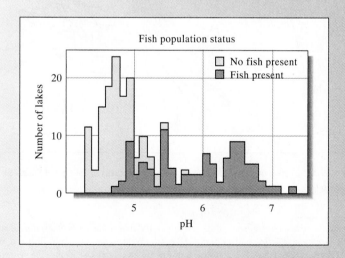

Figure 12F-5B   Effect of pH of lakes on their fish population.

function of pH. The darker bars represent lakes containing fish; the lighter bars refer to lakes with no fish. There is a distinct correlation between pH changes in the lakes and diminished fish population.

Many factors contribute to pH changes in groundwater and lakes in a given geographical area. These include the prevailing wind patterns and weather, types of soils, water sources, nature of the terrain, characteristics of plant life, human activity, and the geological characteristics of the area. The susceptibility of natural water to acidification is largely determined by its buffer capacity, and the principal buffer of natural water is a mixture of bicarbonate ion and carbonic acid. Recall that the buffer capacity of a solution is proportional to the concentration of the buffering agent. So, the higher the concentration of dissolved bicarbonate, the greater the capacity of the water to neutralize acid from acid rain. The most important source of bicarbonate ion in natural water is limestone, or calcium carbonate, which reacts with hydronium ion:

$$CaCO_3(s) + H_3O^+(aq) \rightleftharpoons HCO_3^-(aq) + Ca^{2+}(aq)$$

Limestone-rich areas have lakes with relatively high concentrations of dissolved bicarbonate and thus low susceptibility to acidification. Granite, sandstone, shale, and other rocks containing little or no calcium carbonate are associated with lakes having high susceptibility to acidification.

The map of the United States shown in Figure 12F-5C vividly illustrates the correlation between the absence of limestone-bearing rocks and the acidification of groundwaters.[5] Areas containing little limestone are shaded in color; areas rich in limestone are white. Contour lines of equal pH for groundwater during the period 1978–1979

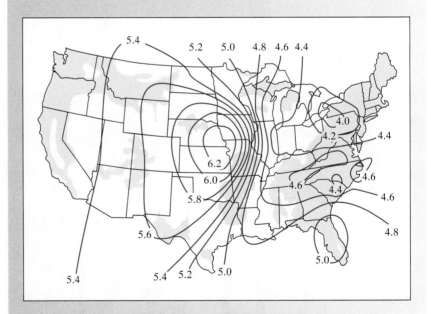

Figure 12F-5C   Effect of presence of limestone on pH of lakes in the United States. Shaded areas contain little limestone

[5]J. Root et al., cited in *The Effects of Air Pollution and Acid Rain on Fish, Wildlife, and Their Habitats—Introduction*, M. A. Peterson, Ed., p. 63. (Washington, DC. U.S. Fish and Wildlife Service, Biological Services Program, Eastern Energy and Land Use Team, U.S. Government Publication FWS/OBS-80/40.3).

are superimposed on the map. The Adirondack Mountains area, located in northeastern New York, contains little limestone and exhibits pH values in the range of 4.2 to 4.4. The low buffer capacity of the lakes in this region combined with the low pH of precipitation appear to have caused the decline of fish populations. Similar correlations among acid rain, buffer capacity of lakes, and wildlife decline occur throughout the industrialized world.

Although natural sources such as volcanos produce sulfur trioxide and lightning discharges in the atmosphere generate nitrogen dioxide, large quantities of these compounds come from the burning of high-sulfur coal and from automobile emissions. To minimize emissions of these pollutants, some states have enacted legislation imposing strict standards on automobiles sold and operated within their borders. Some states have required the installation of scrubbers to remove oxides of sulfur from the emissions of coal-fired power plants. To minimize the effects of acid rain on lakes, powdered limestone is dumped on their surfaces to increase the buffer capacity of the water. Solutions to these problems require the expenditure of much time, energy, and money. We must sometimes make difficult economic decisions to preserve the quality of our environment and to reverse trends that have operated for many decades.

The 1990 Clean Air Act Amendments provided a dramatic new way to regulate sulfur dioxide. Congress issued specific emission limits to power plant operators as shown in Figure 12F-5D, but no specific methods were proposed for meeting the standards. In addition, Congress established an emissions trading system by which power plants could buy, sell, and trade rights to pollute. Although detailed scientific and economic analysis of the effects of these congressional measures is still under way, it is clear from the results so far that the Clean Air Act Amendments have had a profound positive effect on the causes and effects of acid rain.[6]

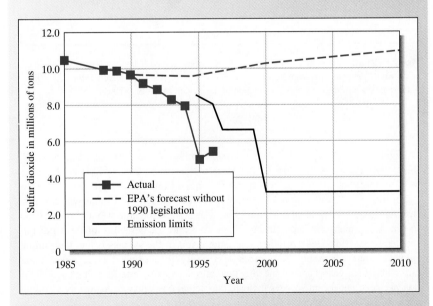

Figure 12F-5D   Sulfur dioxide emissions from selected plants in the United States have dropped below the levels required by law. From R. A. Kerr, *Science,* **1998,** *282,* 1024, with permission of the American Association of the Advancement of Science.

---

[6]R. A. Kerr, *Science,* **1998,** *282,* 1024.

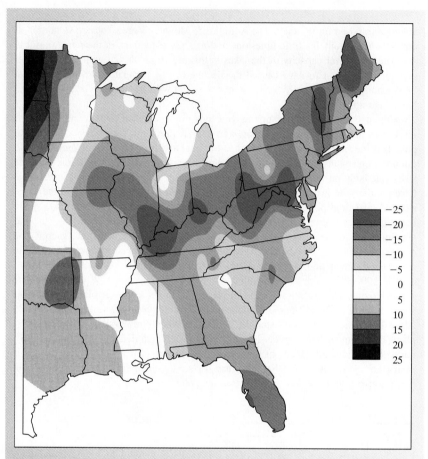

Figure 12F-5E   Precipitation over much of the eastern United States has become less acidic, as shown by the percent change from 1983 to 1994. From R. A. Kerr, *Science*, **1998**, *282*, 1024, with permission of the American Association of the Advancement of Science.

Figure 12F-5D shows that sulfur dioxide emissions have decreased dramatically since 1990 and are well below levels forecasted by the EPA and within the limits set by Congress. The effects of these measures on acid rain are depicted in the map of Figure 12F-5E. The map shows the percent change in acidity in various regions of the eastern United States from 1983 to 1994. The significant improvement in acid rain shown on the map has been attributed tentatively to the flexibility of the regulatory statutes imposed in 1990. Another surprising result of the statutes is that their implementation has apparently been much less costly than originally projected. Initial estimates of the yearly cost of meeting the emission standards were as high as $10 billion per year, but recent surveys indicate that actual costs may be as low as $1 billion per year.[6]

For a comprehensive discussion of the problem of acid rain, refer to the book by Park.[7]

[7]C. C. Park, *Acid Rain*. New York: Methuen, 1987.

12D CALCULATING pH IN WEAK ACID TITRATIONS

Four distinctly different types of calculations are needed to derive a titration curve for a weak acid or a weak base:

1. At the beginning, the solution contains only a weak acid or a weak base, and the pH is calculated from the concentration of that solute and its dissociation constant.
2. After various increments of titrant have been added (in quantities up to, but not including, an equivalent amount), the solution consists of a series of buffers. The pH of each buffer can be calculated from the analytical concentrations of the conjugate base or acid and the residual concentrations of the weak acid or base.
3. At the equivalence point, the solution contains only the conjugate of the weak acid or base being titrated (that is, a salt), and the pH is calculated from the concentration of this product.
4. Beyond the equivalence point, the excess of strong acid or base titrant represses the acidic or basic character of the reaction product to such an extent that the pH is governed largely by the concentration of the excess titrant.

| Example 12-6 | *Spreadsheet Exercise* |

Generate a curve for the titration of 50.00 mL of 0.1000 M acetic acid with 0.1000 M sodium hydroxide. Find the pH after adding 0.00. 5.00, 10.00, 20.00, 25.00, 30.00, 40.00, 45.00, 49.00, 49.95, 49.90, 49.99, 50.00, 50.01, 50.05. 50.10, 51.00, 55.00, 60.00, 70.00, 80.00, and 90.00 mL of titrant. Plot the titration curve in the spreadsheet.

Initial Entry Values

The spreadsheet is shown in Figure 12-8. Let us label cells A2 and A3, K_a and Vol HOAc, mL, respectively. The values of 1.75E-5 and 50.00 are then entered in cells B2 and B3. Cells C2 and C3 are labeled Initial c_{HOAc} and Initial c_{NaOH}, and the values 0.1000 for both are entered into D2 and D3. Cell E2 is labeled K_w and 1.00×10^{-14} is entered into cell F2. The titrant volumes are entered into cells A5:A26 with A4 being labeled **Vol. NaOH, mL.** The remaining cells in row 4 are given the labels shown in the spreadsheet. We are now ready to enter the appropriate formulas in cells B5:D26 and F5:I26. Cells B5:B26 will contain the analytical concentration of HOAc during the titration, while cells C5:C26 will contain the analytical concentration of OAc⁻. The excess NaOH added is calculated in cells D5:D26. Note that cells E5:E26 contain a qualitative description of the system after each volume of titrant.

Initial pH

To check our spreadsheet entries, we will calculate a few values manually as well as with the spreadsheet. The concentration of HOAc in cells B5 is given by the number of mmoles of HOAc initially present minus the number of mmoles of NaOH added divided by the total volume. Hence we enter into cell B5, the formula shown in the documentation section in cell A28. Likewise the analytical concentration of OAc⁻ produced is given by the number of mmoles of NaOH added divided by the total volume. The formula shown in cell A29 is thus en-

	A	B	C	D	E	F	G	H	I
1	50.00 mL 0.1000 M HOAc with 0.100 M NaOH								
2	K_a	1.75E-05	Initial c_{HOAc}	0.1000	K_w	1.00E-14			
3	Vol. HOAC, mL	50.00	Initial c_{NaOH}	0.1000					
4	Vol. NaOH, mL	c_{HOAC}	c_{OAC^-}	C_{NaOH}	System	$[H_3O^+]$	$[OH^-]$	pOH	pH
5	0.00	0.1000	0.0000	0.0000	weak acid	0.001322876			2.88
6	5.00	0.0818	0.0091	0.0000	buffer	1.58E-04			3.80
7	10.00	0.0667	0.0167	0.0000	buffer	7.00E-05			4.15
8	20.00	0.0429	0.0286	0.0000	buffer	2.63E-05			4.58
9	25.00	0.0333	0.0333	0.0000	buffer	1.75E-05			4.76
10	30.00	0.0250	0.0375	0.0000	buffer	1.17E-05			4.93
11	40.00	0.0111	0.0444	0.0000	buffer	4.38E-06			5.36
12	45.00	0.0053	0.0474	0.0000	buffer	1.94E-06			5.71
13	49.00	0.0010	0.0495	0.0000	buffer	3.57E-07			6.45
14	49.90	0.0001	0.0499	0.0000	buffer	3.51E-08			7.46
15	49.95	0.0001	0.0500	0.0000	buffer	1.75E-08			7.76
16	49.99	0.0000	0.0500	0.0000	buffer	2.10E-09			8.68
17	50.00	0.0000	0.0500	0.0000	weak base		5.35E-06	5.27	8.73
18	50.01	0.0000	0.0500	9.999E-06	strong base		1.00E-05	5.00	9.00
19	50.05	0.0000	0.0500	4.998E-05	strong base		5.00E-05	4.30	9.70
20	50.10	0.0000	0.0500	9.990E-05	strong base		9.99E-05	4.00	10.00
21	51.00	0.0000	0.0495	9.901E-04	strong base		9.90E-04	3.00	11.00
22	55.00	0.0000	0.0476	4.762E-03	strong base		4.76E-03	2.32	11.68
23	60.00	0.0000	0.0455	9.091E-03	strong base		9.09E-03	2.04	11.96
24	70.00	0.0000	0.0417	1.667E-02	strong base		1.67E-02	1.78	12.22
25	80.00	0.0000	0.0385	2.308E-02	strong base		2.31E-02	1.64	12.36
26	90.00	0.0000	0.0357	2.857E-02	strong base		2.86E-02	1.54	12.46
27	**Spreadsheet Documentation**								
28	Cell B5=($D$2*$B$3-$D$3*A5)/($B$3+A5)				Cell G17=SQRT(($B$3*$D$3)/(A17+$B$3)*($F$2/$B$2))				
29	Cell C5=$D$3*A5/($B$3+A5)				Cell H17=-LOG10(G17)				
30	Cell D5=0				Cell I17=14-H17				
31	Cell F5=SQRT($B$2*$D$2)				Cell C18=($D$2*$B$3)/($B$3+A18)				
32	Cell I5=-LOG10(F5)				Cell D18=($D$3*A18-$D$3*$B$3)/($B$3+A18)				
33	Cell F6=$B$2*($B$3*$D$2-A6*$D$3)/(A6*$D$3)				Cell G18=D18				
34	Cell B18=0								

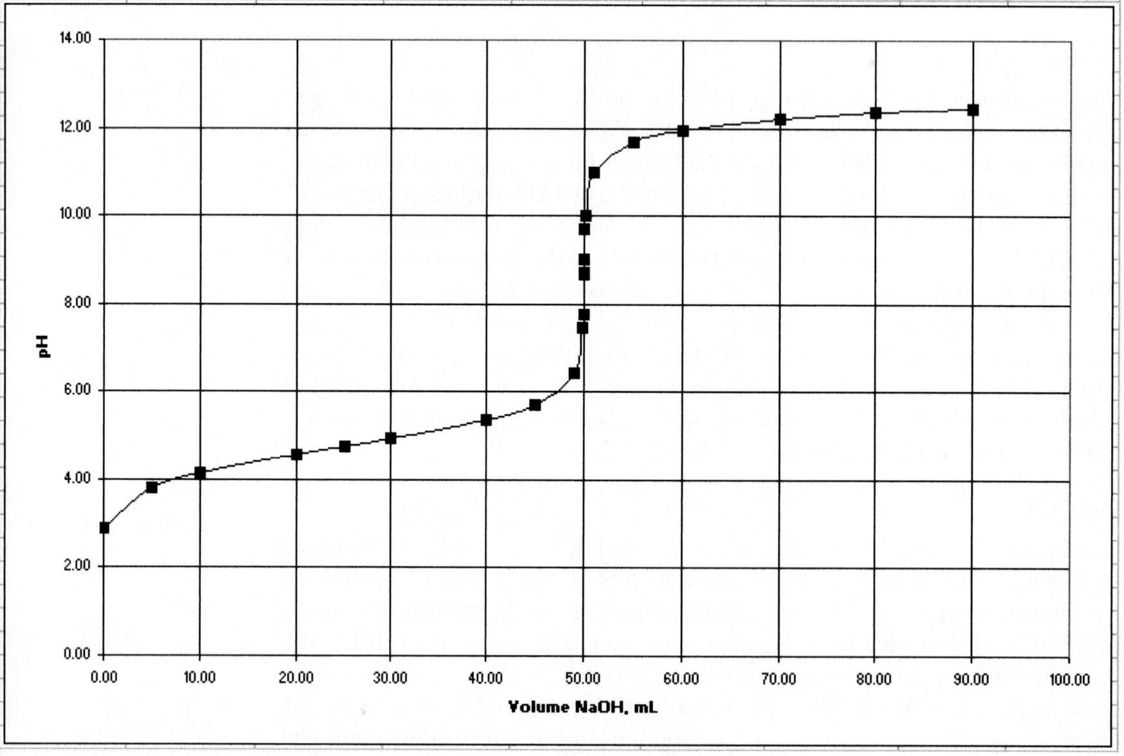

Figure 12-8 *See figure caption on facing page.*

tered into cell C5. Until after the equivalence point, there is no excess NaOH and so 0.00 is entered into cell D5. The system is initially just a weak acid and so we enter that into cell E5. Initially, we must calculate the pH of a 0.1000 M solution of HOAc using Equation 4-21 (page 94). Manually, this is

$$[H_3O^+] = \sqrt{K_a c_{HOAc}} = \sqrt{1.75 \times 10^{-5} \times 0.100} = 1.32 \times 10^{-3} \text{ M}$$

$$pH = -\log(1.32 \times 10^{-3}) = 2.88$$

We find $[H_3O^+]$ in the spreadsheet by entering =SQRT($B$2*$D$2) in cell F5. Since we do not need to calculate $[OH^-]$ and pOH to obtain the pH of this solution, cells G5 and H5 are left blank. To find the solution pH, we enter =-LOG10(F5) in cell I5. The results should agree with those calculated manually above.

pH after Addition of 5.00 mL of Reagent

A buffer solution consisting of NaOAc and HOAc has now been produced. The analytical concentrations of the two constituents are

$$c_{HOAc} = \frac{50.00 \text{ mL} \times 0.100 \text{ M} - 5.00 \text{ mL} \times 0.100 \text{ M}}{60.00 \text{ mL}} = \frac{4.500}{60.00} \text{ M}$$

$$c_{NaOAc} = \frac{5.00 \text{ mL} \times 0.100 \text{ M}}{60.00 \text{ mL}} = \frac{0.500}{60.00} \text{ M}$$

Since we calculated the HOAc and OAc^- concentrations in cells B5 and C5 from this same relationship, the concentrations can be calculated in the spreadsheet by copying the formulas in cells B5 and C5 into B6 and C6 respectively. Note that these same formulas apply up to and including the equivalence point. Thus, we can copy the content of cell B5 into B6:B17 and the contents of C5 into C6:C17.

Now for the 5.00 mL volume, we substitute the concentrations of HOAc and OAc^- into the dissociation-constant expression for acetic acid and obtain

$$\frac{[H_3O^+](0.500/60.00)}{4.500/60.00} = K_a = 1.75 \times 10^{-5}$$

$$[H_3O^+] = 1.58 \times 10^{-4}$$

$$pH = 3.80$$

Note that the total volume of solution is present in both numerator and denominator and thus cancels in the expression for $[H_3O^+]$. We take advantage of this to simplify the spreadsheet entry in cell F6 as shown in the documentation section in cell A33. We then copy this formula into cells F7:F16 which completes the preequivalence point region. For the pH, we copy cell I5 into I6:I16.

◀ **Figure 12-8**   Spreadsheet to calculate pH during titration of 50.00 mL 0.1000 M acetic acid with 0.1000 M NaOH.

Equivalence Point pH

At the equivalence point, all the acetic acid has been converted to sodium acetate. The solution is therefore similar to one formed by dissolving that salt in water, and the pH calculation is identical to that shown in Example 4-10 (page 99) for a weak base. In the present example, the NaOAc concentration is 0.0500 M. Thus,

$$OAc^- + H_2O \rightleftharpoons HOAc + OH^-$$

$$[OH^-] = [HOAc]$$

$$[OAc^-] = 0.0500 - [OH^-] \approx 0.0500$$

Substituting in the base dissociation-constant expression for OAc^- gives

$$\frac{[OH^-]^2}{0.0500} = \frac{K_w}{K_a} = \frac{1.00 \times 10^{-14}}{1.75 \times 10^{-5}} = 5.71 \times 10^{-10}$$

$$[OH^-] = \sqrt{0.0500 \times 5.71 \times 10^{-10}} = 5.34 \times 10^{-6}$$

$$pH = 14.00 - [-\log (5.34 \times 10^{-6})] = 8.73$$

Note that the pH at the equivalence point of this titration is greater than 7. The solution is alkaline.

In the spreadsheet, we characterize the system as a weak base in cell E17 and enter into cell G17 the formula shown in the documentation section in cell E28. In H17, we calculate pOH by entering -LOG10(G17). The pH is calculated in I17 by entering 14-H17.

pH after Addition of 50.01 mL of Base

After the addition of 50.01 mL of NaOH, both the excess base and the acetate ion are sources of hydroxide ion. The contribution of the latter is small, however, because the excess of strong base represses the reaction. This fact becomes evident when we consider that the hydroxide ion concentration is only 5.35×10^{-6} at the equivalence point; once an excess of strong base is added, the contribution from the reaction of the acetate will be even smaller. Thus,

$$[OH^-] \approx c_{NaOH} = \frac{50.01 \text{ mL} \times 0.1000 \text{ M} - 50.00 \text{ mL} \times 0.1000 \text{ M}}{100.01 \text{ mL}}$$

$$= 1.00 \times 10^{-5} \text{ M}$$

$$pH = 14.00 - (-\log 1.00 \times 10^{-5}) = 9.00$$

Since the HOAc concentration is now essentially 0.00 M, we enter 0 into cell B18. The OAc^- concentration is the number of mmoles produced divided by the total volume and so the formula documented in cell E31 is entered into cell C18. The excess NaOH in cell D18 is calculated as shown above from the number of mmoles of NaOH added minus the number of mmoles of HOAc initially present over the total volume. Hence, the formula documented in cell E32 is entered in D18. Since the OH^- concentration is essentially equal to this excess NaOH, we set cell G18 equal to cell D18. Note that these same formulas hold throughout the postequivalence point region. We can then copy the formulas in cells B18, C18, D18, and G18 into the remaining B, C, D and G cells. Likewise the pOH and pH calculations in cells H17 and I17 are copied into the remaining H and I

Table 12-3

Changes in pH During the Titration of a
Weak Acid With a Strong Base

Volume of NaOH, mL	pH	
	50.00 mL of 0.1000 M HOAc with 0.1000 M NaOH	50.00 mL of 0.001000 M HOAc with 0.001000 M NaOH
0.00	2.88	3.91
10.00	4.15	4.30
25.00	4.76	4.80
40.00	5.36	5.38
49.00	6.45	6.46
49.90	7.46	7.47
50.00	8.73	7.73
50.10	10.00	8.09
51.00	11.00	9.00
60.00	11.96	9.96
70.00	12.22	10.25

Challenge: Calculate the pH values given in the third column of Table 12-3.

cells. Now we draw the titration curve by selecting an x:y plot with A5:A26 as the x axis and I5:I26 as the y axis. The resulting plot is also shown in Figure 12-8. Note that the titration curve for a weak acid with a strong base is identical with that for a strong acid with a strong base in the region slightly beyond the equivalence point.

Table 12-3 compares some selected values from the spreadsheet with a more dilute titration. The effect of concentration is discussed in Section 12D-1.

Titration curves for strong and weak acids are identical just slightly beyond the equivalence point. The same is true for strong and weak bases.

Note from Figure 12-8 that the analytical concentrations of acid and conjugate base are identical when an acid has been half neutralized (in Example 12-6, after the addition of exactly 25.00 mL of base). Thus, these terms cancel in the equilibrium-constant expression, and the hydronium ion concentration is numerically equal to the dissociation constant. Likewise, in the titration of a weak base, the hydroxide ion concentration is numerically equal to the dissociation constant of the base at the midpoint in the titration curve. In addition, the buffer capacities of each of the solutions is at a maximum at this point.

At the half-neutralization point in the titration of a weak acid, $[H_3O^+] = K_a$ or $pH = pK_a$.

At the half-neutralization point in the titration of a weak base, $[OH^-] = K_b$ or $pOH = pK_b$.

Feature 12-6

Determining Dissociation Constants for Weak Acids and Weak Bases

The dissociation constants of weak acids or weak bases are often determined by monitoring the pH of the solution while the acid or base is being titrated. A pH meter with a glass pH electrode (see Section 19D-3) is used for the measurements. For an acid, the measured pH when the acid is exactly half neutralized is numerically equal to pK_a. For a weak base, the pH at half neutralization must be converted to pOH, which is then equal to pK_b.

12D-1   The Effect of Concentration

The second and third columns of Table 12-3 contain pH data for the titration of 0.1000 M and of 0.001000 M acetic acid with sodium hydroxide solutions of the same two concentrations. In calculating the values for the more dilute acid, none of the approximations shown in Example 12-6 was valid, and solution of a quadratic equation was necessary until after the equivalence point. In the post-equivalence point region, the excess OH$^-$ predominates and the simple calculation suffices.

Figure 12-9 is a plot of the data in Table 12-3. Note that the initial pH values are higher and the equivalence-point pH is lower for the more dilute solution. At intermediate titrant volumes, however, the pH values differ only slightly because of the buffering action of the acetic acid/sodium acetate system that exists in this region. Figure 12-9 is graphical confirmation that the pH of buffers is largely independent of dilution. Note that the change in pH in the equivalence-point region becomes smaller with lower analyte and reagent concentrations. This effect is analogous to the effect for the titration of a strong acid with a strong base (see Figure 12-4).

12D-2   The Effect of Reaction Completeness

Titration curves for 0.1000 M solutions of acids with different dissociation constants are shown in Figure 12-10. Note that the pH change in the equivalence-point region becomes smaller as the acid becomes weaker, that is, as the reaction between the acid and the base becomes less complete.

12D-3   Choosing an Indicator; the Feasibility of Titration

Figures 12-9 and 12-10 show clearly that the choice of indicator for the titration of a weak acid is more limited than that for a strong acid. For example, from Figure 12-9, we see that bromocresol green is totally unsuited for titration of 0.1000 M acetic acid. Bromothymol blue is also unsatisfactory because its full color change occurs over a range from about 47 mL to 50 mL of 0.1000 M base. On the other

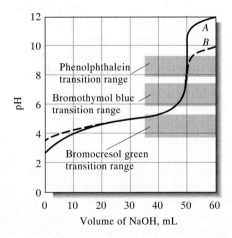

Figure 12-9   Curve for the titration of acetic acid with sodium hydroxide. Curve *A*: 0.1000 M acid with 0.1000 M base. Curve *B*: 0.001000 M acid with 0.001000 M base.

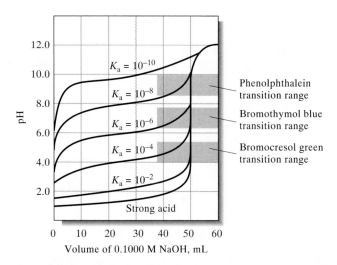

Figure 12-10   The effect of acid strength (dissociation constant, K_a) on titration curves. Each curve represents the titration of 50.00 mL of 0.1000 M acid with 0.1000 M base.

hand, an indicator exhibiting a color change in the basic region, such as phenolphthalein, should provide a sharp end point with a minimal titration error.

The end-point pH change associated with the titration of 0.001000 M acetic acid (curve *B*, Figure 12-9) is so small that a significant titration error is likely to be introduced regardless of indicator. But the use of an indicator with a transition range between that of phenolphthalein and that of bromothymol blue in conjunction with a suitable color comparison standard makes it possible to establish the end point in this titration with a reproducibility of a few percent relative.

Figure 12-10 illustrates that similar problems exist as the strength of the acid being titrated decreases. Precision on the order of ± 2 ppt can be achieved in the titration of a 0.1000 M acid solution with a dissociation constant of 10^{-8} provided a suitable color comparison standard is available. With more concentrated solutions, somewhat weaker acids can be titrated with reasonable precision.

12E   CALCULATING pH IN TITRATIONS OF WEAK BASES

The calculations needed to draw the titration curve for a weak base are analogous to those of a weak acid.

Example 12-7

A 50.00-mL aliquot of 0.0500 M NaCN is titrated with 0.1000 M HCl. The reaction is

$$CN^- + H_3O^+ \rightleftharpoons HCN + H_2O$$

Calculate the pH after the addition of (a) 0.00, (b) 10.00, (c) 25.00, and (d) 26.00 mL of acid.

(a) 0.00 mL of reagent: The pH of a solution of NaCN can be derived by the method shown in Example 4-10, (page 99):

$$CN^- + H_2O \rightleftharpoons HCN + OH^-$$

$$K_b = \frac{[OH^-][HCN]}{[CN^-]} = \frac{K_w}{K_a} = \frac{1.00 \times 10^{-14}}{6.2 \times 10^{-10}} = 1.61 \times 10^{-5}$$

$$[OH^-] = [HCN]$$

$$[CN^-] = c_{NaCN} - [OH^-] \approx c_{NaCN} = 0.050 \text{ M}$$

Substitution into the dissociation-constant expression gives, after rearrangement,

$$[OH^-] = \sqrt{K_b c_{NaCN}} = \sqrt{1.61 \times 10^{-5} \times 0.0500} = 8.97 \times 10^{-4}$$

$$pH = 14.00 - [-\log (8.97 \times 10^{-4})] = 10.95$$

(b) 10.00 mL of reagent: Addition of acid produces a buffer with a composition given by

$$c_{NaCN} = \frac{50.00 \times 0.0500 - 10.00 \times 0.1000}{60.00} = \frac{1.500}{60.00} \text{ M}$$

$$c_{HCN} = \frac{10.00 \times 0.1000}{60.00} = \frac{1.000}{60.00} \text{ M}$$

These values are then substituted into the expression for the acid dissociation constant of HCN to give $[H_3O^+]$ directly (see margin note):

Challenge: Show that the pH of the buffer can be calculated with K_a for HCN, as was done here, or equally well with K_b. We used K_a because it gives $[H_3O^+]$ directly; K_b gives $[OH^-]$.

$$[H_3O^+] = \frac{6.2 \times 10^{-10} \times (1.000/60.00)}{1.500/60.00} = 4.13 \times 10^{-10}$$

$$pH = -\log 4.13 \times 10^{-10} = 9.38$$

(c) 25.00 mL of reagent: This volume corresponds to the equivalence point, where the principal solute species is the weak acid HCN. Thus,

$$c_{HCN} = \frac{25.00 \times 0.1000}{75.00} = 0.03333 \text{ M}$$

Applying Equation 4-21 gives

$$[H_3O^+] = \sqrt{K_a c_{HCN}} = \sqrt{6.2 \times 10^{-10} \times 0.03333} = 4.45 \times 10^{-6} \text{ M}$$

Since the principal solute species at the equivalence point is HCN, the pH is acidic.

$$pH = -\log 4.45 \times 10^{-6} = 5.34$$

(d) 26.00 mL of reagent: The excess of strong acid now present represses the dissociation of the HCN to the point where its contribution to the pH is negligible. Thus,

$$[H_3O^+] = c_{HCl} = \frac{26.00 \times 0.1000 - 50.00 \times 0.0500}{76.00} = 1.32 \times 10^{-3} \text{ M}$$

$$pH = -\log 1.32 \times 10^{-3} = 2.88$$

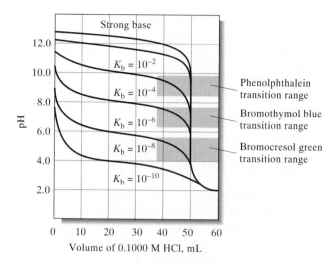

Figure 12-11 The effect of base strength (K_b) on titration curves. Each curve represents the titration of 50.00 mL of 0.1000 M base with 0.1000 M HCl.

Figure 12-11 shows hypothetical titration curves for a series of weak bases of different strengths. Clearly, indicators with *acidic* transition ranges must be employed for weak bases.

When titrating a weak base, use an indicator with an acidic transition range. When titrating a weak acid, use an indicator with a basic transition range.

Feature 12-7

Amino acids contain both an acidic and a basic group. For example, the structure of alanine is represented by

Determining the pK Values for Amino Acids

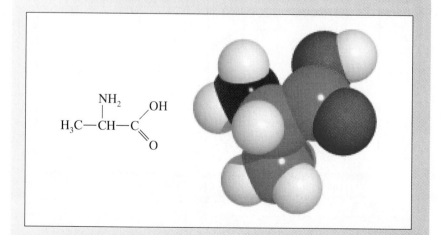

Structure and molecular model of alanine. Alanine is an amino acid. It can exist in two mirror image forms the left-handed (L) form and the right-handed (D) form. All naturally occurring amino acids are left handed.

The amine group behaves as a base, whereas the carboxyl group acts as an acid. In aqueous solution, the amino acid exists as an internally ionized molecule or "zwitterion" in which the amine group acquires a proton and becomes positively charged while the carboxyl group, having lost a proton, becomes negatively charged.

The pK values for amino acids can conveniently be determined by the general procedure described in Feature 12-6. Since the zwitterion has both acidic and basic character, two pKs can be determined. The pK for deprotonation of the protonated amine group can be determined by adding base, whereas the pK for protonating the carboxyl group can be determined by adding acid. In practice, a solution is prepared containing a known concentration of the amino acid. Hence, the scientist knows the amount of base or acid to add to reach halfway to the equivalence point. A curve of pH versus volume of acid or base added is shown in Figure 12F-7. Here the titration starts in the middle of the plot (0.00 mL added) and is only taken to a point that is half the volume required for equivalence. Note that in this example for alanine, a volume of 20.00 mL of HCl is needed to completely protonate the carboxyl group. By adding acid to the zwitterion, the curve on the left is obtained. At a volume of 10.00 mL of HCl added, the pH is equal to the pK_a for the carboxyl group, 2.35.

By adding NaOH to the zwitterion, the pK for deprotonating the $-NH_3^+$ group can be determined. Now 20.00 mL of base is required for complete deprotonation. At a volume of 10.00 mL of NaOH added, the pH is equal to the pK_a for the amine group or 9.89. The pK_a values for other amino acids and more complicated biomolecules such as peptides and proteins can often be obtained in a similar manner. Some amino acids have more than one carboxyl or amine group. Aspartic acid is an example.

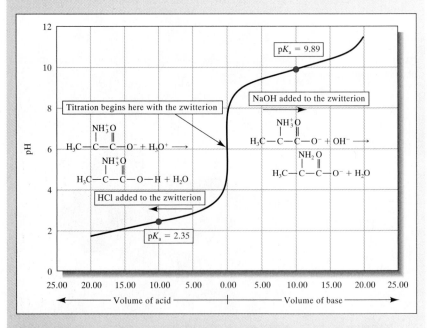

Figure 12F-7   Curves for the titration of 20.00 mL of 0.1000 M alanine with 0.1000 M NaOH and 0.1000 M HCl. Note that the zwitterion exists before any acid or base has been added. Adding acid protonates the carboxylate group with a pK_a of 2.35. Adding base deprotonates the protonated amine group with a pK_a of 9.89.

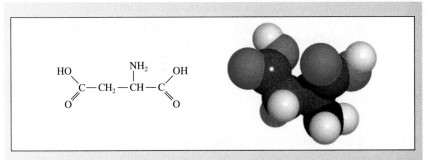

Structure and molecular model of aspartic acid.

In general, amino acids cannot be quantitatively determined by direct titration because end points for completely protonating or deprotonating the zwitterion are usually indistinct. Amino acids are normally determined by high-performance liquid chromatography (see Chapter 25) or spectroscopic methods (see Section 23A).

Aspartic acid is an amino acid with two carboxyl groups. It can be combined with phenylalanine to make the artificial sweetener aspartame, which is sweeter and less fattening than ordinary sugar (sucrose).

12F HOW DO BUFFER SOLUTIONS CHANGE AS A FUNCTION OF pH?

The changes in composition that occur while a solution of a weak acid or a weak base is being titrated are sometimes of interest. They can be visualized by plotting the *relative* equilibrium concentration of the weak acid as well as the relative equilibrium concentration of the conjugate base as a function of the pH of the solution. These relative concentrations are called *alpha values*. For example, if we let c_T be the sum of the analytical concentrations of acetic acid and sodium acetate at any point in the titration curve derived in Example 12-6, we may write

$$c_T = c_{HOAc} + c_{NaOAc} \qquad (12\text{-}12)$$

We then define α_0 as

$$\alpha_0 = \frac{[HOAc]}{c_T} \qquad (12\text{-}13)$$

and α_1 as

$$\alpha_1 = \frac{[OAc^-]}{c_T} \qquad (12\text{-}14)$$

Alpha values are unitless ratios whose sum must equal unity. That is,

$$\alpha_0 + \alpha_1 = 1$$

Alpha values depend only on $[H_3O^+]$ and K_a and are independent of c_T. To obtain expressions for α_0, we rearrange the dissociation-constant expression to

$$[OAc^-] = \frac{K_a[HOAc]}{[H_3O^+]}$$

Mass-balance requires that

$$c_T = [HOAc] + [OAc^-] \qquad (12\text{-}15)$$

Substituting the previous equation into Equation 12-15 gives

$$c_T = [\text{HOAc}] + \frac{K_a[\text{HOAc}]}{[\text{H}_3\text{O}^+]} = [\text{HOAc}]\left(\frac{[\text{H}_3\text{O}^+] + K_a}{[\text{H}_3\text{O}^+]}\right)$$

Upon rearrangement we obtain

$$\frac{[\text{HOAc}]}{c_T} = \frac{[\text{H}_3\text{O}^+]}{[\text{H}_3\text{O}^+] + K_a}$$

But by definition, $[\text{HOAc}]/c_T = \alpha_0$ (Equation 12-13), or

$$\alpha_0 = \frac{[\text{HOAc}]}{c_T} = \frac{[\text{H}_3\text{O}^+]}{[\text{H}_3\text{O}^+] + K_a} \qquad (12\text{-}16)$$

To obtain an expression for α_1, we rearrange the dissociation-constant expression to

$$[\text{HOAc}] = \frac{[\text{H}_3\text{O}^+][\text{OAc}^-]}{K_a}$$

and substitute into Equation 12-16:

$$c_T = \frac{[\text{H}_3\text{O}^+][\text{OAc}^-]}{K_a} + [\text{OAc}^-] = [\text{OAc}^-]\left(\frac{[\text{H}_3\text{O}^+] + K_a}{K_a}\right)$$

Rearranging this equation gives α_1 as defined by Equation 12-14:

$$\alpha_1 = \frac{[\text{OAc}^-]}{c_T} = \frac{K_a}{[\text{H}_3\text{O}^+] + K_a} \qquad (12\text{-}17)$$

Note that the denominator is the same in Equations 12-16 and 12-17.

The solid straight lines labeled α_0 and α_1 in Figure 12-12 were calculated with Equations 12-16 and 12-17 using values for $[\text{H}_3\text{O}^+]$ shown in column 2 of Table 12-3. The actual titration curve is shown as the curved line in Figure 12-12. Note that at the beginning of the titration α_0 is nearly one (0.987), meaning that 98.7% of the acetate containing species is present as HOAc and only

Alternatively,

$$\alpha_1 = 1 - \alpha_0$$

$$= 1 - \frac{[\text{H}_3\text{O}^+]}{[\text{H}_3\text{O}^+] + K_a}$$

$$= \frac{[\text{H}_3\text{O}^+] + K_a - [\text{H}_3\text{O}^+]}{[\text{H}_3\text{O}^+] + K_a}$$

$$= \frac{K_a}{[\text{H}_3\text{O}^+] + K_a}$$

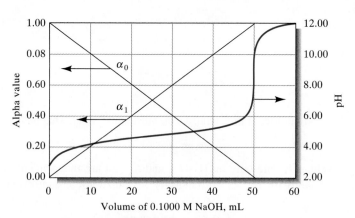

Figure 12-12 Plots of relative amounts of acetic acid and acetate ion during a titration. The straight lines show the change in relative amounts of HOAc (α_0) and OAc$^-$ (α_1) during the titration of 50.00 mL of 0.1000 M acetic acid. The curved line is the titration curve for the system.

1.3% is present as OAc^-. At the equivalence point, α_0 decreases to 1.1×10^{-4} and α_1 approaches one. Thus, only about 0.011% of the acetate containing species is HOAc. Note that when the acid is half neutralized (25.00 mL), α_0 and α_1 are each 0.5.

WEB WORKS

For a web-based presentation of neutralization titrations including titration curves and choosing indicators, set your browser to http://www.cem.msu.edu/~cem262/Neutralization.html.

12G QUESTIONS AND PROBLEMS

In this chapter, round all calculated values for pH and pOH to two figures to the right of the decimal point unless otherwise instructed.

*12-1. Consider curves for the titration of 0.10 M NaOH and 0.010 M NH_3 with 0.10 M HCl.
    (a) Briefly account for the differences between curves for the two titrations.
    (b) In what respect will the two curves be indistinguishable?

12-2. What factors affect end-point sharpness in an acid/base titration?

*12-3. Why does the typical acid/base indicator exhibit its color change over a range of about 2 pH units?

12-4. What variables can cause the pH range of an indicator to shift?

*12-5. Why are the standard reagents used in neutralization titrations generally strong acids and bases rather than weak acids and weak bases?

12-6. What is a buffer solution and what are its properties?

*12-7. Define buffer capacity.

12-8. Which has the greater buffer capacity: (a) a mixture containing 0.100 mol of NH_3 and 0.200 mol of NH_4Cl or (b) a mixture containing 0.0500 mol of NH_3 and 0.100 mol of NH_4Cl?

*12-9. Consider solutions prepared by
    (a) dissolving 8.00 mmol of NaOAc in 200 mL of 0.100 M HOAc.
    (b) adding 100 mL of 0.0500 M NaOH to 100 mL of 0.175 M HOAc.
    (c) adding 40.0 mL of 0.1200 M HCl to 160.0 mL of 0.0420 M NaOAc.
    In what respects do these solutions resemble one another? How do they differ?

12-10. Consult Appendix 2 and pick out a suitable acid/base pair to prepare a buffer with a pH of
    *(a) 3.5   **(b)** 7.6
    *(c) 9.3   **(d)** 5.1

12-11. Which solute would provide the sharper end point in a titration with 0.10 M HCl:
    *(a) 0.10 M NaOCl or 0.10 M hydroxylamine?
    (b) 0.10 M NH_3 or 0.10 M sodium phenolate?
    *(c) 0.10 M methylamine or 0.10 M hydroxylamine?
    (d) 0.10 M hydrazine or 0.10 M NaCN?

12-12. Which solute would provide the sharper end point in a titration with 0.10 M NaOH:
    *(a) 0.10 M nitrous acid or 0.10 M iodic acid?
    (b) 0.10 M anilinium hydrochloride ($C_6H_5NH_3Cl$) or 0.10 M benzoic acid?
    *(c) 0.10 M hypochlorous acid or 0.10 M pyruvic acid?
    (d) 0.10 M salicylic acid or 0.10 M acetic acid?

12-13. Before glass electrodes and pH meters became widely used, pH was often determined by measuring the concentration of the acid and base forms of the indicator colorimetrically. If bromothymol blue is introduced into a solution and the concentration ratio of acid to base form is found to be 1.43, what is the pH of the solution?

*12-14. The procedure described in Problem 12-13 was used to determine pH with methyl orange as the indicator. The concentration ratio of the acid to base form of the indicator was 1.64. Calculate the pH of the solution.

12-15. Values for K_w at 0, 50, and 100°C are 1.14×10^{-15}, 5.47×10^{-14}, and 4.9×10^{-13}, respectively. Calculate the pH for a neutral solution at each of these temperatures.

12-16. Using the data in Problem 12-15, calculate pK_w at
  *(a) 0°C   **(b)** 50°C   **(c)** 100°C

12-17. Using the data in Problem 12-15, calculate the pH of a 1.00×10^{-2} M NaOH solution at
  *(a) 0°C   **(b)** 50°C   **(c)** 100°C

*12-18. What is the pH of an aqueous solution that is 14.0% HCl by weight and has a density of 1.054 g/mL?

12-19. Calculate the pH of a solution that contains 9.00% (w/w) NaOH and has a density of 1.098 g/mL.

*12-20. What is the pH of a solution that is 2.00×10^{-8} M in NaOH? (Hint: In such a dilute solution, you must take into account the contribution of H_2O to the hydroxide ion concentration.)

12-21. What is the pH of a 2.00×10^{-8} M HCl solution?

12-22. What is the pH of the solution that results when 0.102 g of $Mg(OH)_2$ is mixed with
  (a) 75.0 mL of 0.0600 M HCl?
  (b) 15.0 mL of 0.0600 M HCl?
  (c) 30.0 mL of 0.0600 M HCl?
  (d) 30.0 mL of 0.0600 M $MgCl_2$?

*12-23. Calculate the pH of the solution that results upon mixing 20.0 mL of 0.2000 M HCl with 25.0 mL of
  (a) distilled water.
  (b) 0.132 M $AgNO_3$.
  (c) 0.132 M NaOH.
  (d) 0.132 M NH_3.
  (e) 0.232 M NaOH.

*12-24. Calculate the hydronium ion concentration and pH of a solution that is 0.0500 M in HCl
  (a) neglecting activities.
  (b) using activities.

12-25. Calculate the hydroxide ion concentration and the pH of a 0.0167 M $Ba(OH)_2$ solution
  (a) neglecting activities.
  (b) using activities.

*12-26. Calculate the pH of a HOCl solution that is **(a)** 1.00×10^{-1} M, **(b)** 1.00×10^{-2} M, **(c)** 1.00×10^{-4} M.

12-27. Calculate the pH of a NaOCl solution that is **(a)** 1.00×10^{-1} M, **(b)** 1.00×10^{-2} M, **(c)** 1.00×10^{-4} M.

*12-28. Calculate the pH of an ammonia solution that is **(a)** 1.00×10^{-1}, **(b)** 1.00×10^{-2} M, **(c)** 1.00×10^{-4} M.

12-29. Calculate the pH of an NH_4Cl solution that is **(a)** 1.00×10^{-1} M, **(b)** 1.00×10^{-2} M, **(c)** 1.00×10^{-4} M.

*12-30. Calculate the pH of a solution in which the concentration of the piperidine is **(a)** 1.00×10^{-1} M, **(b)** 1.00×10^{-2} M, **(c)** 1.00×10^{-4} M.

12-31. Calculate the pH of an iodic acid solution that is **(a)** 1.00×10^{-1} M, **(b)** 1.00×10^{-2} M, **(c)** 1.00×10^{-4} M.

*12-32. Calculate the pH of a solution prepared by
  (a) dissolving 43.0 g of lactic acid in water and diluting to 500.0 mL.
  (b) diluting 25.0 mL of the solution in part (a) to 250.0 mL.
  (c) diluting 10.0 mL of the solution in part (b) to 1.00 L.

12-33. Calculate the pH of a solution prepared by
  (a) dissolving 1.05 g of picric acid, $(NO_2)_3C_6H_2OH$ (229.11 g/mol), in 100.0 mL of water.
  (b) diluting 10.0 mL of the solution in part (a) to 100.0 mL.
  (c) diluting 10.0 mL of the solution in part (b) to 1.00 L.

*12-34. Calculate the pH of the solution that results when 20.0 mL of 0.200 M formic acid is
  (a) diluted to 45.0 mL with distilled water.
  (b) mixed with 25.0 mL of 0.160 M NaOH solution.
  (c) mixed with 25.0 mL of 0.200 M NaOH solution.
  (d) mixed with 25.0 mL of 0.200 sodium formate solution.

12-35. Calculate the pH of the solution that results when 40.0 mL of 0.100 M NH_3 is
  (a) diluted to 20.0 mL with distilled water.
  (b) mixed with 20.0 mL of 0.200 M HCl solution.
  (c) mixed with 20.0 mL of 0.250 M HCl solution.
  (d) mixed with 20.0 mL of 0.200 M NH_4Cl solution.
  (e) mixed with 20.0 mL of 0.100 M HCl solution.

12-36. A solution is 0.0500 M in NH_4Cl and 0.0300 M in NH_3. Calculate its OH^- concentration and its pH
  (a) neglecting activities.
  (b) taking activities into account.

*12-37. What is the pH of a solution that is
  (a) prepared by dissolving 9.20 g of lactic acid (90.08 g/mol) and 11.15 g of sodium lactate (112.06 g/mol) in water and diluting to 1.00 L?
  (b) 0.0550 M in acetic acid and 0.0110 M in sodium acetate?

(c) prepared by dissolving 3.00 g of salicylic acid, $C_6H_4(OH)COOH$ (138.12 g/mol) in 50.0 mL of 0.1130 M NaOH and diluting to 500.0 mL? Hint: The simplified Henderson–Hasselbalch equation will not work in this case.

(d) 0.0100 M in picric acid and 0.100 M in sodium picrate?

12-38. What is the pH of a solution that is

(a) prepared by dissolving 3.30 g of $(NH_4)_2SO_4$ in water, adding 125.0 mL of 0.1011 M NaOH, and diluting to 500.0 mL?

(b) 0.120 M in piperidine and 0.080 M in its chloride salt?

(c) 0.050 M in ethylamine and 0.167 M in its chloride salt?

(d) prepared by dissolving 2.32 g of aniline (93.13 g/mol) in 100 mL of 0.0200 M HCl and diluting to 250.0 mL?

12-39. Calculate the change in pH that occurs in each of the solutions listed below as a result of a tenfold dilution with water. Round calculated values for pH to three figures to the right of the decimal point.

*(a) H_2O.

(b) 0.0500 M HCl.

*(c) 0.0500 M NaOH.

(d) 0.0500 M CH_3COOH.

*(e) 0.0500 M CH_3COONa.

(f) 0.0500 M CH_3COOH + 0.0500 M CH_3COONa.

*(g) 0.500 M CH_3COOH + 0.500 M CH_3COONa.

12-40. Calculate the change in pH that occurs when 1.00 mmol of a strong acid is added to 100 mL of the solutions listed in Problem 12-39.

12-41. Calculate the change in pH that occurs when 1.00 mmol of a strong base is added to 100 mL of the solutions listed in Problem 12-39. Calculate values to three decimal places.

12-42. Calculate the change in pH to three decimal places that occurs when 0.50 mmol of a strong acid is added to 100 mL of

(a) 0.0200 M lactic acid + 0.0800 M sodium lactate.

*(b) 0.0800 M lactic acid + 0.0200 M sodium lactate.

(c) 0.0500 M lactic acid + 0.0500 M sodium lactate.

*12-43. What weight of sodium formate must be added to 400.0 mL of 1.00 M formic acid to produce a buffer solution that has a pH of 3.50?

12-44. What weight of sodium glycolate should be added to 300.0 mL of 1.00 M glycolic acid to produce a buffer solution with a pH of 4.00?

*12-45. What volume of 0.200 M HCl must be added to 250.0 mL of 0.300 M sodium mandelate to produce a buffer solution with a pH of 3.37?

12-46. What volume of 2.00 M NaOH must be added to 300.0 mL of 1.00 M glycolic acid to produce a buffer solution having a pH of 4.00?

*12-47. A 50.00-mL aliquot of 0.1000 M NaOH is titrated with 0.1000 M HCl. Calculate the pH of the solution after the addition of 0.00, 10.00, 25.00, 40.00, 45.00, 49.00, 50.00, 51.00, 55.00, and 60.00 mL of acid and prepare a titration curve from the data.

*12-48. In a titration of 50.00 mL of 0.05000 M formic acid with 0.1000 M KOH, the titration error must be smaller than 0.05 mL. What indicator can be chosen to realize this goal?

12-49. In a titration of 50.00 mL of 0.1000 M ethylamine with 0.1000 M $HClO_4$, the titration error must be no more than 0.05 mL. What indicator can be chosen to realize this goal?

12-50. Calculate the pH after addition of 0.00, 5.00, 15.00, 25.00, 40.00, 45.00, 49.00, 50.00, 51.00, 55.00, and 60.00 mL of 0.1000 M NaOH in the titration of 50.00 mL of

*(a) 0.1000 M HNO_2.

(b) 0.1000 M lactic acid.

*(c) 0.1000 M pyridinium chloride.

12-51. Calculate the pH after addition of 0.00, 5.00, 15.00, 25.00, 40.00, 45.00, 49.00, 50.00, 51.00, 55.00, and 60.00 mL of 0.1000 M HCl in the titration of 50.00 mL of

*(a) 0.1000 M ammonia.

(b) 0.1000 M hydrazine.

(c) 0.1000 M sodium cyanide.

12-52. Calculate the pH after addition of 0.00, 5.00, 15.00, 25.00, 40.00, 49.00, 50.00, 51.00, 55.00, and 60.00 mL of reagent in the titration of 50.0 mL of

*(a) 0.1000 M anilinium chloride with 0.1000 M NaOH.

(b) 0.01000 M chloroacetic acid with 0.01000 M NaOH.

*(c) 0.1000 M hypochlorous acid with 0.1000 M NaOH.

(d) 0.1000 M hydroxylamine with 0.1000 M HCl.

(e) Construct titration curves from the data.

12-53. Calculate α_0 and α_1 for

*(a) acetic acid species in a solution with a pH of 5.320.

(b) picric acid species in a solution with a pH of 1.250.

*(c) hypochlorous acid species in a solution with a pH of 7.000.

(d) hydroxylamine acid species in a solution with a pH of 5.120.

*(e) piperidine species in a solution with a pH of 10.080.

If you perform these calculations using a spreadsheet, compute data for α_0 and α_1 over the pH range 0 to 14 with points every 0.5 pH unit and construct a plot of α_0 and α_1 versus pH.

*12-54. Calculate the equilibrium concentration of undissociated HCOOH in a formic acid solution with an analytical formic acid concentration of 0.0850 and a pH of 3.200.

12-55. Calculate the equilibrium concentration of methylammonia in a solution that has a molar analytical CH_3NH_2 concentration of 0.120 and a pH of 11.471.

12-56. Supply the missing data in the table below.

Acid	Molar Analytical Concentration, c_T ($c_T = c_{HA} + c_{A^-}$)	pH	[HA]	[A⁻]	α_0	α_1
Lactic	0.120				0.640	
Iodic	0.200					0.765
Butanoic		5.00	0.0644			
Hypochlorous	0.280	7.00				
Nitrous				0.105	0.413	0.587
Hydrogen cyanide			0.145	0.221		
Sulfamic	0.250	1.20				

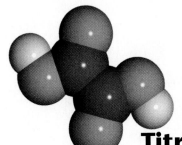

Chapter 13

Titrating Polyfunctional Acids and Bases

I n this chapter, we describe methods for computing the necessary data to construct titration curves for acids and bases that have two or more acidic or basic functional groups. Solutions of amphiprotic salts and the composition of polyprotic acid solutions as a function of pH are also discussed.

Throughout the remainder of this chapter, we will find it useful to use K_{a1}, K_{a2}, and so forth to represent the first and second dissociation constants of acids and K_{b1}, K_{b2}, and so forth for the stepwise constants for bases.

Generally, $K_{a1} > K_{a2}$, often by a factor of 10^4 to 10^5 because of electrostatic forces. That is, the first dissociation involves separating a single positively charged hydronium ion from a singly charged anion. In the second step, a hydronium ion is separated from a doubly charged anion, a process that requires considerably more energy.

A second reason that $K_{a1} > K_{a2}$ is a statistical one. For a dibasic acid H_2A, a proton can be removed from two locations in the first step and only one in the second. Thus, the first dissociation is twice as probable as the second.

13A   POLYFUNCTIONAL ACIDS

Phosphoric acid is a typical polyfunctional acid. In aqueous solution, it undergoes the following three dissociation reactions:

$$H_3PO_4 + H_2O \rightleftharpoons H_2PO_4^- + H_3O^+ \quad K_{a1} = \frac{[H_3O^+][H_2PO_4^-]}{[H_3PO_4]} = 7.11 \times 10^{-3}$$

$$H_2PO_4^- + H_2O \rightleftharpoons HPO_4^{2-} + H_3O^+ \quad K_{a2} = \frac{[H_3O^+][HPO_4^{2-}]}{[H_2PO_4^-]} = 6.32 \times 10^{-8}$$

$$HPO_4^{2-} + H_2O \rightleftharpoons PO_4^{3-} + H_3O^+ \quad K_{a3} = \frac{[H_3O^+][PO_4^{3-}]}{[HPO_4^{2-}]} = 4.5 \times 10^{-13}$$

With this acid, as with other polyprotic acids, $K_{a1} > K_{a2} > K_{a3}$.

13B   DESCRIBING POLYFUNCTIONAL BASES

Polyfunctional bases, such as sodium carbonate, are also common. Carbonate ion, the conjugate base of the hydrogen carbonate ion, is involved in the stepwise equilibria:

$$CO_3^{2-} + H_2O \rightleftharpoons HCO_3^- + OH^- \quad K_{b1} = \frac{[HCO_3^-][OH^-]}{[CO_3^{2-}]} = \frac{K_w}{K_{a2}}$$

$$= \frac{1.00 \times 10^{-14}}{4.69 \times 10^{-11}} = 2.13 \times 10^{-4}$$

Feature 13-1

Combining Equilibrium-
Constant Expressions

When two adjacent stepwise equilibria are added, the equilibrium constant for the resulting overall reaction is the product of the two constants. Thus, for the first two dissociation equilibria for H_3PO_4, we may write

$$H_3PO_4 + H_2O \rightleftharpoons H_2PO_4^- + H_3O^+$$
$$H_2PO_4^- + H_2O \rightleftharpoons HPO_4^{2-} + H_3O^+$$
$$\overline{H_3PO_4 + 2H_2O \rightleftharpoons HPO_4^{2-} + 2H_3O^+}$$

and

$$K_{a1}K_{a2} = \frac{[H_3O^+]^2[HPO_4^{2-}]}{[H_3PO_4]}$$

$$= 7.11 \times 10^{-3} \times 6.32 \times 10^{-8} = 4.49 \times 10^{-10}$$

Similarly, for the reaction

$$H_3PO_4 + 3H_2O \rightleftharpoons PO_4^{3-} + 3H_3O^+$$

we may write

$$K_{a1}K_{a2}K_{a3} = \frac{[H_3O^+]^3[PO_4^{3-}]}{[H_3PO_4]}$$

$$= 7.11 \times 10^{-3} \times 6.32 \times 10^{-8} \times 4.5 \times 10^{-13} = 2.0 \times 10^{-22}$$

$$HCO_3^- + H_2O \rightleftharpoons H_2CO_3 + OH^- \qquad K_{b2} = \frac{[H_2CO_3][OH^-]}{[HCO_3^-]} = \frac{K_w}{K_{a1}}$$

$$= \frac{1.00 \times 10^{-14}}{1.5 \times 10^{-4}} = 6.7 \times 10^{-11}$$

where K_{a1} and K_{a2} are the first and second dissociation constants for carbonic acid and K_{b1} and K_{b2} are the first and second dissociation constants of the base CO_3^{2-}. In some cases, we must also consider a third equilibrium, namely, $H_2CO_3 \rightleftharpoons CO_2(aq) + H_2O$. Because of the limited solubility of CO_2, this last reaction can limit the concentration of H_2CO_3 in solution. For our purposes, however, we will neglect this reaction.

The overall basic dissociation reaction of sodium carbonate is described by the equations

$$CO_3^{2-} + 2H_2O \rightleftharpoons H_2CO_3 + 2OH^- \qquad K_{b1}K_{b2} = \frac{[H_2CO_3][OH^-]^2}{[CO_3^{2-}]}$$

$$= 2.13 \times 10^{-4} \times 6.7 \times 10^{-11}$$

$$= 1.4 \times 10^{-14}$$

The pH of polyfunctional systems, such as phosphoric acid or sodium carbonate, can be computed rigorously through use of the systematic approach to multiequilibrium problems described in Chapter 10. Solving the several simultaneous equations that are involved can be difficult and time consuming, particularly if a computer is not handy. Fortunately, simplifying assumptions can be invoked when the successive equilibrium constants for the acid (or base) differ by a factor of about 10^3 (or more). With one exception, which is described in Section 13C, these assumptions make it possible to generate pH data for titration curves by the techniques discussed in earlier chapters.

Challenge: Write a sufficient number of equations to permit you to calculate the concentration of each species in a solution that contains known molar analytical concentrations of Na_2CO_3 and $NaHCO_3$.

13C   FINDING THE pH OF SOLUTIONS OF AMPHIPROTIC SALTS

Thus far, we have not considered how to calculate the pH of solutions of salts that have both acidic and basic properties, that is, salts that are amphiprotic. Such salts are formed during neutralization titration of polyfunctional acids and bases. For example, when 1 mol of NaOH is added to a solution containing 1 mol of the acid H_2A, 1 mol of NaHA is formed. The pH of this solution is determined by two equilibria established between HA^- and water. The solution could be acidic because of

$$HA^- + H_2O \rightleftharpoons A^{2-} + H_3O^+$$

or basic because of

$$HA^- + H_2O \rightleftharpoons H_2A + OH^-$$

Whether a solution of NaHA is acidic or basic depends on the relative magnitude of the equilibrium constants for these processes:

$$K_{a2} = \frac{[H_3O^+][A^{2-}]}{[HA^-]} \tag{13-1}$$

$$K_{b2} = \frac{K_w}{K_{a1}} = \frac{[H_2A][OH^-]}{[HA^-]} \tag{13-2}$$

where K_{a1} and K_{a2} are the acid dissociation constants for H_2A and K_{b2} is the basic dissociation constant for HA^-. If K_{b2} is greater than K_{a2}, the solution is basic; otherwise, it is acidic.

A solution of the amphiprotic salt NaHA contains five species H_3O^+, OH^-, H_2A, HA^-, and A^{2-}. Thus, five independent equations are needed to compute the hydronium ion concentration rigorously. As we can see in Feature 13-2, solution of these equations yields an approximate value of $[H_3O^+]$ that is given by the equation

> An amphiprotic salt is a species that can act as an acid and as a base when dissolved in a suitable solvent.

$$[H_3O^+] = \sqrt{\frac{K_{a2}c_{NaHA} + K_w}{1 + c_{NaHA}/K_{a1}}} \tag{13-3}$$

where c_{NaHA} is the molar concentration of the salt NaHA. This equation should be used when NaHA is the *only* species that contributes significantly to the pH of a solution. It is not valid for very dilute solutions of NaHA or for systems in which K_{a2} or K_w/K_{a1} is relatively large.

Frequently, the ratio c_{NaHA}/K_{a1} is much larger than unity in the denominator of Equation 13-3 and $K_{a2}c_{NaHA}$ is considerably greater than K_w in the numerator. When these assumptions are valid, the equation simplifies to

$$[H_3O^+] \cong \sqrt{K_{a1}K_{a2}} \tag{13-4}$$

Make sure that you always check the assumptions that are inherent in Equation 13-4. The ratio c_{NaHA}/K_{a1} must be much greater than unity and the product $K_{a2}c_{NaHA}$ must be greater than K_w.

Note that Equation 13-4 does not contain c_{NaHA}, which implies that the pH of solutions of this type remains constant over a considerable range of solute concentrations.

Feature 13-2

How Was Equation 13-3 Generated?

We first write a mass-balance expression. That is,

$$c_{NaHA} = [H_2A] + [HA^-] + [A^{2-}] \tag{13-5}$$

The charge-balance equation takes the form

$$[Na^+] + [H_3O^+] = [HA^-] + 2[A^{2-}] + [OH^-]$$

Since the sodium ion concentration is equal to the molar analytical concentration of the salt, the last equation can be rewritten as

$$c_{NaHA} + [H_3O^+] = [HA^-] + 2[A^{2-}] + [OH^-] \tag{13-6}$$

We now have four algebraic equations (Equations 13-5 and 13-6 and the two dissociation constant expressions for H_2A) and need one additional to solve for the five unknowns. The ion-product constant for water serves this purpose:

$$K_w = [H_3O^+][OH^-]$$

The rigorous computation of the hydronium ion concentration from these five equations is difficult. A reasonable approximation, applicable to solutions of most acid salts, however, can be obtained as follows.

We first subtract the mass-balance equation from the charge-balance equation.

$$
\begin{aligned}
c_{NaHA} + [H_3O^+] &= [HA^-] + 2[A^{2-}] + [OH^-] \quad \text{charge balance} \\
c_{NaHA} &= [H_2A] + [HA^-] + [A^{2-}] \quad \text{mass balance} \\
\hline
[H_3O^+] &= [A^{2-}] + [OH^-] - [H_2A] \quad\quad\quad\quad\quad (13\text{-}7)
\end{aligned}
$$

We then rearrange the acid-dissociation constant expressions for H_2A to obtain

$$[H_2A] = \frac{[H_3O^+][HA^-]}{K_{a1}}$$

and for HA^- to give

$$[A^{2-}] = \frac{K_{a2}[HA^-]}{[H_3O^+]}$$

Substituting these expressions and the expression for K_w into Equation 13-7 yields

$$[H_3O^+] = \frac{K_{a2}[HA^-]}{[H_3O^+]} + \frac{K_w}{[H_3O^+]} - \frac{[H_3O^+][HA^-]}{K_{a1}}$$

Multiplication through by $[H_3O^+]$ gives

$$[H_3O^+]^2 = K_{a2}[HA^-] + K_w - \frac{[H_3O^+]^2[HA^-]}{K_{a1}}$$

We collect terms to obtain

$$[H_3O^+]^2\left(\frac{[HA^-]}{K_{a1}} + 1\right) = K_{a2}[HA^-] + K_w$$

Finally, this equation rearranges to

$$[H_3O^+] = \sqrt{\frac{K_{a2}[HA^-] + K_w}{1 + [HA^-]/K_{a1}}} \qquad (13\text{-}8)$$

Under most circumstances, we can make the approximation that

$$[HA^-] \approx c_{NaHA} \qquad (13\text{-}9)$$

Introduction of this relationship into Equation 13-9 gives

$$[H_3O^+] = \sqrt{\frac{K_{a2}c_{NaHA} + K_w}{1 + c_{NaHA}/K_{a1}}} \qquad (13\text{-}10)$$

It is important to understand that the approximation used to obtain Equation 13-10 requires that $[HA^-]$ be much larger than any of the other equilibrium concentrations in Equations 13-5 and 13-6. This assumption is not valid for very dilute solutions of NaHA or when K_{a2} or K_w/K_{a1} is relatively large.

Example 13-1

Calculate the hydronium ion concentration of a 0.100 M $NaHCO_3$ solution.

We first examine the assumptions leading to Equation 13-4. The dissociation constants for H_2CO_3 are $K_{a1} = 1.5 \times 10^{-4}$ and $K_{a2} = 4.69 \times 10^{-11}$. Clearly, c_{NaHA}/K_{a1} in the denominator is much larger than unity; in addition, $K_{a2}c_{NaHA}$ has a value of 4.69×10^{-12}, which is substantially greater than K_w. Thus, Equation 13-4 applies and

$$[H_3O^+] = \sqrt{1.5 \times 10^{-4} \times 4.69 \times 10^{-11}} = 8.4 \times 10^{-8} \text{ M}$$

Example 13-2

Calculate the hydronium ion concentration of a 1.00×10^{-3} M Na_2HPO_4 solution.

The pertinent dissociation constants are K_{a2} and K_{a3}, which both contain $[HPO_4^{2-}]$. Their values are $K_{a2} = 6.32 \times 10^{-8}$ and $K_{a3} = 4.5 \times 10^{-13}$. Considering again the assumptions that led to Equation 13-4, we find that the ratio $(1.00 \times 10^{-3})/(6.32 \times 10^{-8})$ is much larger than 1, so the denominator can be simplified. The product $K_{a3}c_{Na_2HPO_4}$ is by no means much larger than K_w, however. We therefore use a partially simplified version of Equation 13-3:

$$[H_3O^+] = \sqrt{\frac{4.5 \times 10^{-13} \times 1.00 \times 10^{-3} + 1.00 \times 10^{-14}}{(1.00 \times 10^{-3})/(6.32 \times 10^{-8})}} = 8.1 \times 10^{-10} \text{ M}$$

Use of Equation 13-4 yields a value of 1.7×10^{-10} M.

Example 13-3

Find the hydronium ion concentration of a 0.0100 M NaH_2PO_4 solution.

The two dissociation constants of importance (those containing $[H_2PO_4^-]$) are $K_{a1} = 7.11 \times 10^{-3}$ and $K_{a2} = 6.32 \times 10^{-8}$. We see that the denominator of Equation 13-3 cannot be simplified, but the numerator reduces to $K_{a2}c_{NaH_2PO_4}$. Thus, Equation 13-3 becomes

$$[H_3O^+] = \sqrt{\frac{6.32 \times 10^{-8} \times 1.00 \times 10^{-2}}{1.00 + (1.00 \times 10^{-2})/(7.11 \times 10^{-3})}} = 1.62 \times 10^{-5} \text{ M}$$

This problem is solved using a spreadsheet and the full Equation 13-3 in Spreadsheet Example 13-5.

13D   CONSTRUCTING TITRATION CURVES FOR POLYFUNCTIONAL ACIDS

Compounds with two or more acid functional groups yield multiple end points in a titration, provided the functional groups differ sufficiently in strengths as acids. The computational techniques described in Chapter 12 permit construction of reasonably accurate theoretical titration curves for polyprotic acids if the ratio K_{a1}/K_{a2} is somewhat greater than 10^3. If this ratio is smaller, the error, particularly in the region of the first equivalence point, becomes large, and a more rigorous treatment of the equilibrium relationships is required.

Figure 13-1 shows the titration curve for a diprotic acid H_2A with dissociation constants of $K_{a1} = 1.00 \times 10^{-3}$ and $K_{a2} = 1.00 \times 10^{-7}$. Because the ratio K_{a1}/K_{a2} is significantly greater than 10^3, we can construct this curve (except for the equivalence points and the region just surrounding them) using the techniques developed in Chapter 12 for simple monoprotic weak acids. Thus, to obtain the initial pH (point A), we treat the system as if it contained a single monoprotic acid with a dissociation constant of $K_{a1} = 1.00 \times 10^{-3}$. In region B we have the equivalent of a simple buffer solution consisting of the weak acid H_2A and its conjugate base NaHA. That is, we assume that the concentration of A^{2-} is negligible with respect to the other two A-containing species and employ Equation 12-10 (page 278) to obtain $[H_3O^+]$. Just prior to the first equivalence point (region C), we have a solution consisting of primarily HA^- with a small amount of H_2A remaining and some A^{2-} being formed by dissociation. At the first equivalence point, we have a solution of an acid salt and use Equation 13-3 or one of its simplifications to compute the hydronium ion concentration. In region D, we have a second buffer consisting of a weak acid HA^- and its conjugate base Na_2A, and we calculate the pH employing the second dissociation constant, $K_{a2} = 1.00 \times 10^{-7}$. Just prior to the second equivalence point (region E), the solution consists primarily of A^{2-} with a small amount of HA^- remaining. At the second equivalence point, the solution consists of A^{2-} and the hydroxide concentration of the solution is determined by the reaction of A^{2-} with water to form HA^- and OH^-. Just after this equivalence point, the excess OH^- must also be considered. Finally, in region F, we compute the hydroxide concentration from the molarity of the excess NaOH and obtain the pH from this quantity.

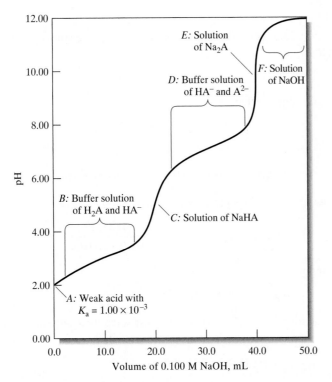

Figure 13-1   Titration of 20.00 mL of 0.1000 M H$_2$A with 0.100 M NaOH. For H$_2$A, K_{a1} = 1.00 × 10^{-3} and K_{a2} = 1.00 × 10^{-7}. The method of pH calculation is shown for several points and regions on the titration curve.

Example 13-4 demonstrates these calculations for maleic acid, a typical diprotic acid. As noted earlier, the regions just prior to and just after the equivalence points involve fairly complex calculations, which are beyond the scope of this book.

Example 13-4

Construct a curve for the titration of 25.00 mL of 0.1000 M maleic acid, HOOC—CH=CH—COOH, with 0.1000 M NaOH. Consider the initial pH, the first buffer region, the first equivalence point, the second buffer region, the second equivalence point, and the region beyond the second equivalence point. The calculations are done manually here. Later in Spreadsheet Example 13-6, we show the spreadsheet approach to this same problem.

Symbolizing the acid as H$_2$M, we can write the two dissociation equilibria as

$$H_2M + H_2O \rightleftharpoons H_3O^+ + HM^- \qquad K_{a1} = 1.3 \times 10^{-2}$$

$$HM^- + H_2O \rightleftharpoons H_3O^+ + M^{2-} \qquad K_{a2} = 5.9 \times 10^{-7}$$

Because the ratio K_{a1}/K_{a2} is large (2 × 10^4), we proceed as just described for constructing Figure 13-1.

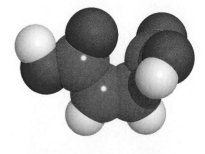

Molecular models of maleic acid, or (Z)-butenedioic acid, (top) and fumaric acid, or (E)-butenedioic acid (bottom). These geometric isomers exhibit striking differences in both their physical and their chemical properties. Because the *cis* isomer (maleic acid) has both carboxyl groups on the same side of the molecule, the compound eliminates water to form cyclic maleic anhydride, which is a very reactive precursor widely used in plastics, dyes, pharmaceuticals, and agrichemicals. Fumaric acid, which is essential to animal and vegetable respiration, is used industrially as an antioxidant, to synthesize resins, and to fix colors in dyeing. It is interesting to compare the pK_a values for the two acids; for fumaric acid, pK_{a1} = 3.05 and pK_{a2} = 4.49; for maleic acid, pK_{a1} = 1.89 and pK_{a2} = 6.23. Challenge: Explain the differences in the pK_a values based on differences in the molecular structures.

Initial pH

Initially, only the first dissociation makes an appreciable contribution to $[H_3O^+]$; thus,

$$[H_3O^+] \approx [HM^-]$$

Mass balance requires that

$$c_{H_2M} \approx [H_2M] + [HM^-] = 0.1000$$

or

$$[H_2M] = 0.1000 - [HM^-] = 0.1000 - [H_3O^+]$$

Substituting these relationships into the expression for K_{a1} gives

$$K_{a1} = 1.3 \times 10^{-2} = \frac{[H_3O^+]^2}{0.1000 - [H_3O^+]}$$

Rearranging yields

$$[H_3O^+]^2 + 1.3 \times 10^{-2}[H_3O^+] - 1.3 \times 10^{-3} = 0$$

Because K_{a1} for maleic acid is large, we must solve the quadratic equation exactly or by successive approximations. The result is

$$[H_3O^+] = 3.01 \times 10^{-2}$$

$$pH = -\log(3.01 \times 10^{-2}) = 1.52$$

First Buffer Region

The addition of 5.00 mL of base results in the formation of a buffer consisting of the weak acid H_2M and its conjugate base HM^-. To the extent that dissociation of HM^- to give M^{2-} is negligible, the solution can be treated as a simple buffer system. Thus, applying Equations 12-8 and 12-9 (page 277) gives

$$c_{NaHM} \approx [HM^-] = \frac{5.00 \times 0.1000}{30.00} = 1.67 \times 10^{-2} \text{ M}$$

$$c_{H_2M} \approx [H_2M] = \frac{25.00 \times 0.1000 - 5.00 \times 0.1000}{30.00} = 6.67 \times 10^{-2} \text{ M}$$

Substitution of these values into the equilibrium-constant expression for K_{a1} yields a tentative value of 5.2×10^{-2} M for $[H_3O^+]$. It is clear, however, that the approximation $[H_3O^+] \ll c_{H_2M}$ or c_{HM^-} is not valid; therefore, Equations 12-6 and 12-7 must be used, and

$$[HM^-] = 1.67 \times 10^{-2} + [H_3O^+] - [OH^-]$$

$$[H_2M] = 6.67 \times 10^{-2} - [H_3O^+] - [OH^-]$$

Because the solution is quite acidic, the approximation that $[OH^-]$ is very small is surely justified. Substitution of these expressions into the dissociation-constant relationship gives

$$K_{a1} = \frac{[H_3O^+](1.67 \times 10^{-2} + [H_3O^+])}{6.67 \times 10^{-2} - [H_3O^+]} = 1.3 \times 10^{-2}$$

$$[H_3O^+]^2 + 2.97 \times 10^{-2}[H_3O^+] - 8.67 \times 10^{-4} = 0$$

$$[H_3O^+] = 1.81 \times 10^{-2} \, M$$

$$pH = -\log(1.81 \times 10^{-2}) = 1.74$$

Additional points in the first buffer region are computed in a similar way. One complication occurs just prior to the first equivalence point where the concentration of H_2M becomes so small that it is comparable to the concentration of M^{2-}, and the second equilibrium must also be considered. We will not consider this complication here.

First Equivalence Point

At the first equivalence point,

$$[HM^-] \approx c_{NaHM} = \frac{25.00 \times 0.1000}{50.00} = 5.00 \times 10^{-2} \, M$$

Simplification of the numerator in Equation 13-3 is clearly justified. On the other hand, the second term in the denominator is not $\ll 1$. Hence,

$$[H_3O^+] = \sqrt{\frac{K_{a2}c_{NaHM}}{1 + c_{NaHM}/K_{a1}}}$$

$$= \sqrt{\frac{5.9 \times 10^{-7} \times 5.00 \times 10^{-2}}{1 + (5.00 \times 10^{-2})/(1.3 \times 10^{-2})}} = 7.8 \times 10^{-5}$$

$$pH = -\log(7.8 \times 10^{-5}) = 4.11$$

In the region just after the first equivalence point, calculation of the pH is again complicated because we have not yet formed enough M^{2-} to make a good buffer solution.

Second Buffer Region

Further additions of base to the solution create a new buffer system consisting of HM^- and M^{2-}. When enough base has been added so that the reaction of HM^- with water to give OH^- can be neglected (a few tenths of a milliliter beyond the first equivalence point), the pH of the mixture is readily obtained from K_{a2}. With the introduction of 25.50 mL of NaOH, for example,

$$[M^{2-}] \approx c_{Na_2M} \approx \frac{(25.50 - 25.00)(0.1000)}{50.50} = \frac{0.050}{50.50} = 9.90 \times 10^{-4}$$

and the molar concentration of NaHM is

$$[HM^-] \approx c_{NaHM} \approx \frac{(25.00 \times 0.1000) - (25.50 - 25.00)(0.1000)}{50.50}$$

$$= \frac{2.45}{50.50} = 0.04851$$

Substituting these values into the expression for K_{a2} gives

$$[H_3O^+] = \frac{K_{a2} \times c_{NaHM}}{c_{Na_2M}} = \frac{5.9 \times 10^{-7} \times 0.04851}{9.90 \times 10^{-4}} = 2.89 \times 10^{-5} \text{ M}$$

The assumption that $[H_3O^+]$ is small relative to c_{HM^-} and $c_{M^{2-}}$ is valid, and pH = 4.54.

The other values in the second buffer region are calculated in a similar manner. Just prior to the second equivalence point (49.90 mL), the ratio $[M^{2-}]/[HM^-]$ becomes large, and the simple buffer equation no longer applies.

Second Equivalence Point

After the addition of 50.00 mL of 0.1000 M sodium hydroxide, the solution is 0.0333 M in Na_2M (2.5 mmol/75.00 mL). Reaction of the base M^{2-} with water is the predominant equilibrium in the system and the only one that we need to take into account. Thus,

$$M^{2-} + H_2O \rightleftharpoons OH^- + HM^-$$

$$K_{b1} = \frac{K_w}{K_{a2}} = \frac{[OH^-][HM^-]}{[M^{2-}]} = 1.69 \times 10^{-8}$$

$$[OH^-] \approx [HM^-]$$

$$[M^{2-}] = 0.0333 - [OH^-] \approx 0.0333$$

$$\frac{[OH^-]^2}{0.0333} = 1.69 \times 10^{-8}$$

$$[OH^-] = 2.38 \times 10^{-5} \quad \text{and} \quad pOH = -\log(2.38 \times 10^{-5}) = 4.62$$

$$pH = 14.00 - 4.62 = 9.38$$

pH Beyond the Second Equivalence Point

In the region just beyond the second equivalence point (50.01 mL, for example), we still need to take into account the reaction of M^{2-} with water in addition to the excess OH^- that has been added.

Further additions of sodium hydroxide repress the basic dissociation of M^{2-}. The pH is calculated from the concentration of NaOH added in excess of that required for the complete neutralization of H_2M. Thus, when 51.00 mL of NaOH have been added, we have 1.00-mL excess of 0.1000 M NaOH and

$$[OH^-] = \frac{1.00 \times 0.100}{76.00} = 1.32 \times 10^{-3}$$

$$pOH = 2.88$$

$$pH = 14.00 - 2.88 = 11.12$$

Figure 13-2 is the titration curve for 0.1000 M maleic acid generated as shown in Example 13-4. Two end points are apparent, either of which could in principle be used as a measure of the concentration of the acid. The second end point is clearly more satisfactory, however, because the pH change is more pronounced.

Figure 13-3 shows titration curves for three other polyprotic acids. These curves illustrate that a well-defined end point corresponding to the first equivalence point is observed only when the degree of dissociation of the two acids is sufficiently different. The ratio K_{a1}/K_{a2} for oxalic acid (curve B) is approximately 1000. The curve for this titration shows an inflection corresponding to the first equivalence point. The magnitude of the pH change is too small to permit precise location of equivalence with an indicator, however. But the second end point provides a means for the accurate determination of oxalic acid.

Curve A in Figure 13-3 is the theoretical titration curve for triprotic phosphoric acid. Here, the ratio K_{a1}/K_{a2} is approximately 10^5, as is K_{a2}/K_{a3}. This ratio results in two well-defined end points, either of which is satisfactory for analytical purposes. An acid-range indicator will provide a color change when 1 mol of base has been introduced for each mole of acid; a base-range indicator will require 2 mol of base per mole of acid. The third hydrogen of phosphoric acid is so slightly dissociated ($K_{a3} = 4.5 \times 10^{-13}$) that no practical end point is associated with its neutralization. The buffering effect of the third dissociation is noticeable, however, and causes the pH for curve A to be lower than the pH for the other two curves in the region beyond the second equivalence point.

Curve C is the titration curve for sulfuric acid, a substance that has one fully dissociated proton and one that is dissociated to a relatively large extent ($K_{a2} = 1.02 \times 10^{-2}$). Because of the similarity in strengths of the two acids, only a single end point, corresponding to the titration of both protons, is observed.

In general, the titration of acids or bases that have two reactive groups yields individual end points that are of practical value only when the ratio between the two dissociation constants is at least 10^4. If the ratio is much smaller than this, the pH change at the first equivalence point will prove less satisfactory for an analysis.

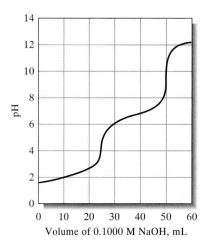

Figure 13-2 Titration curve for 25.00 mL of 0.1000 M maleic acid, H_2M, with 0.1000 M NaOH.

In titrating a polyprotic acid or base, two usable end points are obtained if the ratio of dissociation constants is greater than 10^4 and if the weaker acid or base has a dissociation constant greater than 10^{-8}.

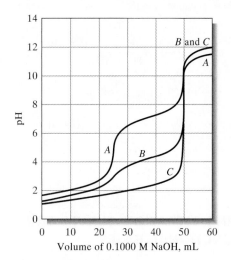

Figure 13-3 Curves for the titration of polyprotic acids. A 0.1000 M NaOH solution is used to titrate 25.00 mL of 0.1000 M H_3PO_4 (curve A), 0.1000 M oxalic acid (curve B), and 0.1000 M H_2SO_4 (curve C).

Feature 13-3

The Dissociation of Sulfuric Acid

Sulfuric acid is unusual in that one of its protons behaves as a strong acid in water and the other as a weak acid ($K_{a2} = 1.02 \times 10^{-2}$). Let us consider how the hydronium ion concentration of sulfuric acid solutions is computed using a 0.0400 M solution as an example.

We will first assume that the dissociation of HSO_4^- is negligible because of the large excess of H_3O^+ resulting from the complete dissociation of H_2SO_4. Therefore,

$$[H_3O^+] \cong [HSO_4^-] \approx 0.0400 \text{ M}$$

An estimate of $[SO_4^{2-}]$ based on this approximation and the expression for K_{a2}, however, reveals that

$$\frac{0.0400[SO_4^-]}{0.0400} = 1.02 \times 10^{-2}$$

This result shows that $[SO_4^{2-}]$ is *not* small relative to $[HSO_4^-]$, and a more rigorous solution is required.

From stoichiometric considerations, it is necessary that

$$[H_3O^+] = 0.0400 + [SO_4^{2-}]$$

The first term on the right is the concentration of H_3O^+ from dissociation of the H_2SO_4 to HSO_4^-. The second term is the contribution of the dissociation of HSO_4^-. Rearrangement yields

$$[SO_4^{2-}] = [H_3O^+] - 0.0400$$

Mass-balance considerations require that

$$c_{H_2SO_4} = 0.0400 = [HSO_4^-] + [SO_4^{2-}]$$

Combining the last two equations and rearranging yields

$$[HSO_4^-] = 0.0800 - [H_3O^+]$$

Introduction of these equations for $[SO_4^{2-}]$ and HSO_4^- into the expression for K_{a2} yields

$$\frac{[H_3O^+]([H_3O^+] - 0.0400)}{0.0800 - [H_3O^+]} = 1.02 \times 10^{-2}$$

$$[H_3O^+]^2 - (0.0298[H_3O^+]) - 8.16 \times 10^{-4} = 0$$

$$[H_3O^+] = 0.0471$$

Challenge: Construct a titration curve for 50.0 mL of 0.0500 M H_2SO_4 with 0.1000 M NaOH.

13E   DRAWING TITRATION CURVES FOR POLYFUNCTIONAL BASES

The construction of a titration curve for a polyfunctional base involves no new principles. To illustrate, consider the titration of a sodium carbonate solution with standard hydrochloric acid. The important equilibrium constants are

$$CO_3^{2-} + H_2O \rightleftharpoons OH^- + HCO_3^-$$

$$K_{b1} = \frac{K_w}{K_{a2}} = \frac{1.00 \times 10^{-14}}{4.69 \times 10^{-11}} = 2.13 \times 10^{-4}$$

$$HCO_3^- + H_2O \rightleftharpoons OH^- + H_2CO_3$$

$$K_{b2} = \frac{K_w}{K_{a1}} = \frac{1.00 \times 10^{-14}}{1.5 \times 10^{-4}} = 6.7 \times 10^{-11}$$

The reaction of carbonate ion with water governs the initial pH of the solution, which can be computed by the method shown for the second equivalence point in Example 13-4. With the first additions of acid, a carbonate/hydrogen carbonate buffer is established. In this region, the pH can be obtained from *either* the hydroxide ion concentration calculated from K_{b1} *or* the hydronium ion concentration calculated from K_{a2}. Because we are usually interested in calculating $[H_3O^+]$ and pH, the expression for K_{a2} is the easier one to use.

Sodium hydrogen carbonate is the principal solute species at the first equivalence point, and Equation 13-4 is used to compute the hydronium ion concentration (see Example 13-1). With the addition of more acid, a new buffer consisting of sodium hydrogen carbonate and carbonic acid is formed. The pH of this buffer is readily obtained from either K_{b2} or K_{a1}.

At the second equivalence point, the solution consists of carbonic acid and sodium chloride. The carbonic acid can be treated as a simple weak acid having a dissociation constant K_{a1}. Finally, after excess hydrochloric acid has been introduced, the dissociation of the weak acid is repressed to a point where the hydronium ion concentration is essentially that of the molar concentration of the strong acid.

Figure 13-4 illustrates that two end points are observed in the titration of sodium carbonate, the second being appreciably sharper than the first. It is apparent that the individual components in mixtures of sodium carbonate and sodium hydrogen carbonate can be determined by neutralization methods.

Challenge: Show that either K_{b2} or K_{a1} can be used to calculate the pH of a buffer that is 0.100 M in Na_2CO_3 and 0.100 M in $NaHCO_3$.

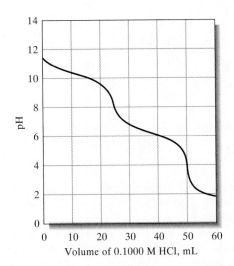

Figure 13-4   Curve for the titration of 25.00 mL of 0.1000 M Na_2CO_3 with 0.1000 M HCl.

Feature 13-4

Titration Curves for Amino Acids

Amino acids are often titrated to determine pK values as discussed in Feature 12-7. The internal proton transfer from the carboxyl group to the amine group of an amino acid is very favorable in aqueous solution so that amino acids exist as zwitterions as shown below for glycine.

$$NH_2CH_2COOH \rightleftharpoons NH_3^+CH_2COO^-   \text{zwitterion formation}$$

The conjugate acid of the zwitterion can be treated as a polyprotic acid.

$$\underset{\textbf{conjugate acid}}{NH_3^+CH_2COOH} + H_2O \rightleftharpoons H_3O^+ + \underset{\textbf{zwitterion}}{NH_3^+CH_2COO^-}   K_{a1} = 5 \times 10^{-3}$$

$$NH_3^+CH_2COO^- + H_2O \rightleftharpoons H_3O^+ + \underset{\textbf{conjugate base}}{NH_2CH_2COO^-}   K_{a2} = 2 \times 10^{-10}$$

The zwitterion can act as an acid because of the second proton transfer or as a base because of

$$NH_3^+CH_2COO^- + H_2O \rightleftharpoons NH_3^+CH_2COOH + OH^-   K_b = \frac{K_w}{K_{a1}} = 2 \times 10^{-12}$$

The zwitterion of an amino acid, containing as it does a positive and a negative charge, has no tendency to migrate in an electric field, whereas the singly charged anionic and cationic species are attracted to electrodes of opposite charge. No *net* migration of the amino acid occurs in an electric field when the pH of the solvent is such that the concentration of the anionic and cationic forms are identical. The pH at which no net migration occurs is called the *isoelectric point;* this point is an important physical constant for characterizing amino acids. The isoelectric point is readily related to the ionization constants for the species. Thus, for glycine,

$$K_a = \frac{[NH_2CH_2COO^-][H_3O^+]}{[NH_3^+CH_2COO^-]}   \text{and}   K_b = \frac{[NH_3^+CH_2COOH][OH^-]}{[NH_3^+CH_2COO^-]}$$

At the isoelectric point,

$$[NH_2CH_2COO^-] = [NH_3^+CH_2COOH]$$

Division of K_a by K_b gives

$$\frac{[H_3O^+][NH_2CH_2COO^-]}{[OH^-][NH_3^+CH_2COOH]} = \frac{[H_3O^+]}{[OH^-]} = \frac{K_a}{K_b}$$

If we substitute $K_w/[H_3O^+]$ for $[OH^-]$ and rearrange, we get

$$[H_3O^+] = \sqrt{\frac{K_aK_w}{K_b}}$$

The isoelectric point for glycine occurs at a pH of 6.0. That is,

$$[H_3O^+] = \sqrt{\left(\frac{2 \times 10^{-10}}{2 \times 10^{-12}}\right) \times 1 \times 10^{-14}} = 1 \times 10^{-6}$$

A titration curve for the zwitterion of alanine was presented in Figure 12F-4 to illustrate the determination of pK values. For simple amino acids, K_a and K_b are generally so small that their quantitative determination by neutralization titrations is impossible. Amino acids that contain more than one carboxyl or amine group can sometimes be determined. If the K_a values are different enough (10^4 or more), stepwise end points can be obtained just like other polyfunctional acids or bases as long as the K_a values

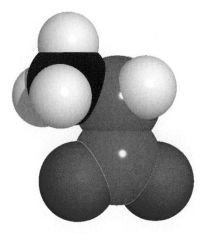

The molecular structure of glycine. Glycine is one of the so-called nonessential amino acids; it is nonessential in the sense that it is synthesized in the bodies of mammals and so is not generally essential in the diet. Because of its compact structure, glycine acts as a versatile building block in protein synthesis and in the biosynthesis of hemoglobin. A significant fraction of the collagen, or the fibrous protein constituent of bone, cartilage, tendon, and other connective tissue, in the human body is made up of glycine. Glycine is also an *inhibitory neurotransmitter* and, as a result has been suggested as a possible therapeutic agent in diseases of the central nervous system such as multiple sclerosis and epilepsy. The calming effects of glycine are currently being investigated to assess its utility in the treatment of schizophrenia.

are large enough. In some cases, addition of formaldehyde is used to remove the amine group and leave the carboxylic acid available for titration with a standard base. For example, with glycine,

$$NH_3^+CH_2COO^- + CH_2O \rightarrow CH_2{=}NCH_2COOH + H_2O$$

The titration curve for the product is that of a typical carboxylic acid.

13F   THE COMPOSITION OF POLYPROTIC ACID SOLUTIONS AS A FUNCTION OF pH

In Section 12F, we showed how alpha values are useful in visualizing the changes in the concentration of various species that occur in a titration of a simple weak acid. Alpha values can also be calculated for polyfunctional acids and bases. For example, if we let c_T be the sum of the molar concentrations of the maleate-containing species in the solution throughout the titration described in Example 13-4, the alpha value for the free acid α_0 is defined as

$$\alpha_0 = \frac{[H_2M]}{c_T}$$

where

$$c_T = [H_2M] + [HM^-] + [M^{2-}] \tag{13-11}$$

The alpha values for HM^- and M^{2-} are given by similar equations:

$$\alpha_1 = \frac{[HM^-]}{c_T}$$

$$\alpha_2 = \frac{[M^{2-}]}{c_T}$$

As noted earlier, the sum of the alpha values for a system must equal unity:

$$\alpha_0 + \alpha_1 + \alpha_2 = 1$$

The alpha values for the maleic acid system are readily expressed in terms of $[H_3O^+]$, K_{a1}, and K_{a2}. We follow the method used to obtain Equations 12-16 and 12-17 in Section 12F with the results

$$\alpha_0 = \frac{[H_3O^+]^2}{[H_3O^+]^2 + K_{a1}[H_3O^+] + K_{a1}K_{a2}} \tag{13-12}$$

$$\alpha_1 = \frac{K_{a1}[H_3O^+]}{[H_3O^+]^2 + K_{a1}[H_3O^+] + K_{a1}K_{a2}} \tag{13-13}$$

$$\alpha_2 = \frac{K_{a1}K_{a2}}{[H_3O^+]^2 + K_{a1}[H_3O^+] + K_{a1}K_{a2}} \tag{13-14}$$

Challenge: Derive Equations 13-12, 13-13, and 13-14.

Note that the denominator is the same for each expression. Note also that the fractional amount of each species is fixed at any pH and is *independent* of the total concentration, c_T.

A spreadsheet for calculating the alpha values of maleic acid as a function of pH is given in Figure 13-5. The three curves plotted show the alpha value for each maleate-containing species. Note the regions where each species predominates. The solid curves in Figure 13-6 depict the same alpha values but now plotted as a function of volume of sodium hydroxide as the acid is titrated. The titration curve is also shown by the dashed line in Figure 13-6. Consideration of these curves gives a clear picture of all concentration changes that occur during the titration. For example, Figure 13-6 reveals that before the addition of any base, α_0 for H_2M is roughly 0.7 and α_1 for HM^- is approximately 0.3. For all practical purposes, α_2 is zero. Thus, approximately 70% of the maleic acid exists as H_2M and 30% as HM^-. With addition of base, the pH rises, as does the fraction of HM^-. At the first equivalence point (pH = 4.11), essentially all the maleate is present as HM^- ($\alpha_1 \rightarrow 1$). Beyond the first equivalence point, α_0 is negligible, α_1 decreases and α_2 increases. At the second equivalence point (pH = 9.38) and beyond, essentially all the maleate exists as M^{2-}.

Feature 13-5

A General Expression for Alpha Values

For the weak acid H_nA, the denominator in all alpha-value expressions takes the form:

$$[H_3O^+]^n + K_{a1}[H_3O^+]^{(n-1)} + K_{a1}K_{a2}[H_3O^+]^{(n-2)} + \cdots + K_{a1}K_{a2} \cdots K_{an}$$

The numerator for α_0 is the first term in the denominator; for α_1, it is the second term; and so forth. Thus, if we let D be the denominator, $\alpha_0 = [H_3O^+]^n/D$ and $\alpha_1 = K_{a1}[H_3O^+]^{(n-1)}/D$.

Alpha values for polyfunctional bases are generated in an analogous way, with the equations written in terms of base dissociation constants and $[OH^-]$.

Spreadsheet Exercise   Example 13-5

Set up a spreadsheet to find the hydronium ion concentration of a 0.0100 M NaH_2PO_4 solution. Use the full Equation 13-3, the full equation neglecting K_w in the numerator of Equation 13-3, and the simplified Equation 13-4. Compare the results to those obtained in Example 13-3.

Figure 13-5 Spreadsheet to calculate alpha values for maleic acid. The dissociation constants are entered into cells B1 and D1, and the pH values for which alpha values are desired are entered into cells A4 through A24. The H_3O^+ concentration is calculated in cell B4 as `=10^-A4`. This formula is copied into cells B5 through B24. The value for D (see Feature 13-5) is calculated in cell C4 by the formula shown in the documentation section, cell A27. The calculations to obtain α_0, α_1, and α_2 are shown in the documentation, cells A28 through A30. On the right, the chart gives the composition as a function of pH.

	A	B	C	D	E	F
1	Spreadsheet to calculate alpha values for maleic acid					
2	K_{a1}=	1.30E-03		K_{a2}=	5.90E-07	
3	pH	[H_3O^+]	Denominator, D	α_0	α_1	α_2
4	0	1.00E+00	1.00E+00	9.99E-01	1.30E-03	7.66E-10
5	0.5	3.16E-01	1.00E-01	9.96E-01	4.09E-03	7.64E-09
6	1	1.00E-01	1.01E-02	9.87E-01	1.28E-02	7.57E-08
7	1.5	3.16E-02	1.04E-03	9.61E-01	3.95E-02	7.37E-07
8	2	1.00E-02	1.13E-04	8.85E-01	1.15E-01	6.79E-06
9	2.5	3.16E-03	1.41E-05	7.09E-01	2.91E-01	5.44E-05
10	3	1.00E-03	2.30E-06	4.35E-01	5.65E-01	3.33E-04
11	3.5	3.16E-04	5.12E-07	1.95E-01	8.03E-01	1.50E-03
12	4	1.00E-04	1.41E-07	7.10E-02	9.24E-01	5.45E-03
13	4.5	3.16E-05	4.29E-08	2.33E-02	9.59E-01	1.79E-02
14	5	1.00E-05	1.39E-08	7.21E-03	9.37E-01	5.53E-02
15	5.5	3.16E-06	4.89E-09	2.05E-03	8.41E-01	1.57E-01
16	6	1.00E-06	2.07E-09	4.84E-04	6.29E-01	3.71E-01
17	6.5	3.16E-07	1.18E-09	8.49E-05	3.49E-01	6.51E-01
18	7	1.00E-07	8.97E-10	1.11E-05	1.45E-01	8.55E-01
19	7.5	3.16E-08	8.08E-10	1.24E-06	5.09E-02	9.49E-01
20	8	1.00E-08	7.80E-10	1.28E-07	1.67E-02	9.83E-01
21	8.5	3.16E-09	7.71E-10	1.30E-08	5.33E-03	9.95E-01
22	9	1.00E-09	7.68E-10	1.30E-09	1.69E-03	9.98E-01
23	9.5	3.16E-10	7.67E-10	1.30E-10	5.36E-04	9.99E-01
24	10	1.00E-10	7.67E-10	1.30E-11	1.69E-04	1.00E+00
25	Spreadsheet Documentation					
26	Cell B4=10^-A4					
27	Cell C4=(B4^2+$B$2*B4+$B$2*$D$2)					
28	Cell D4=(B4^2)/C4					
29	Cell E4=($B$2*B4)/C4					
30	Cell F4=($B$2*$D$2)/C4					

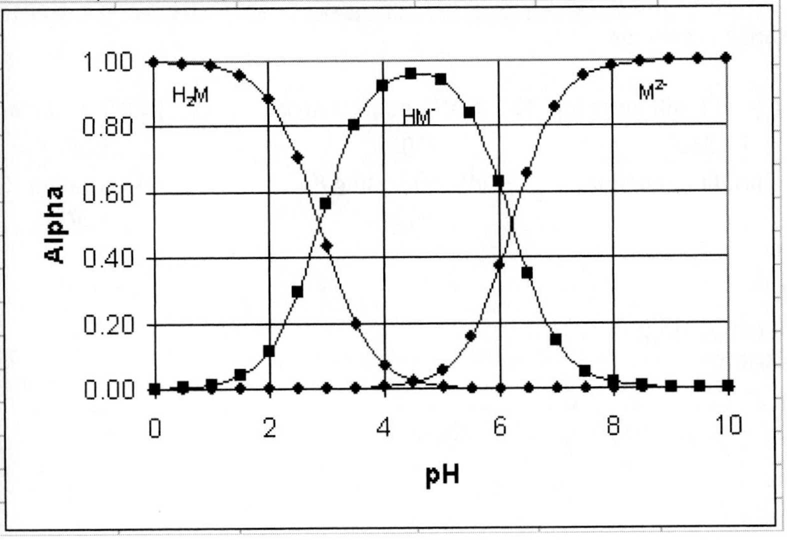

Figure 13-5 *See figure caption on facing page.*

The value for K_w is entered into cell B2 of the spreadsheet shown in Figure 13-7. The two dissociation constants of importance (those containing $[H_2PO_4^-]$) are $K_{a1} = 7.11 \times 10^{-3}$ and $K_{a2} = 6.32 \times 10^{-8}$. These are entered in cells B3 and B4. The value for $c_{NaH_2PO_4}$ is entered into cell B5. It can be seen from the results in cells D4, F4, and I4 that neglecting K_w is a good approximation, whereas the simplified Equation 13-4 leads to a result that is approximately 31% too high.

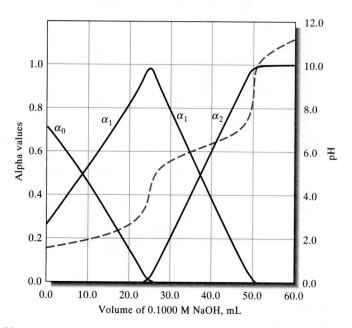

Figure 13-6   Titration of 25.00 mL of 0.1000 M maleic acid with 0.1000 M NaOH. The solid curves are plots of alpha values as a function of volume. The broken curve is the titration curve of pH as a function of volume.

	A	B	C	D	E	F	G	H	I	J
1	**Spreadsheet to calculate pH of a NaH₂PO₄ solution**									
2	K_w	1.00E-14								
3	K_{a1}	7.11E-03		**[H₃O⁺], full equation 13-3**		**[H₃O⁺], neglecting K_w**			**[H₃O⁺], equation 13-4**	
4	K_{a2}	6.32E-08		1.62E-05		1.62E-05			2.12E-05	
5	$c_{NaH2PO4}$	0.0100		**pH, full equation**		**pH, neglecting K_w**			**pH, equation 13-4**	
6				4.79		4.79			4.67	
7										
8										
9	**Spreadsheet Documentation**									
10	Cell D4=SQRT(($B$4*$B$5+$B$2)/(1+($B$5/$B$3)))									
11	Cell F4=SQRT(($B$4*$B$5)/(1+($B$5/$B$3)))									
12	Cell I4=SQRT(($B$4*$B$5)/($B$5/$B$3))									
13	Cell D6=-LOG($D$4)									
14	Cell F6=-LOG($F$4)									
15	Cell I6=-LOG($I$4)									

Figure 13-7   Spreadsheet to calculate the pH of 0.0100 M NaH₂PO₄ solution.

| Example 13-6 | Spreadsheet Exercise |

Use a spreadsheet to generate a curve for the titration of 25.00 mL of 0.1000 M maleic acid, $HOOC-CH=CH-COOH$, with 0.1000 M NaOH as was done in Example 13-4. Compare the results generated by the spreadsheet approach to those obtained manually in the earlier Example.

The spreadsheet is shown in Figure 13-8. As we work through the example, we will explain the spreadsheet entries. Let us begin the spreadsheet by placing in cells A2–A4 the labels K_{a1}, K_{a2}, and K_w. In cells C2–C4 we type the labels K_{b1}, Initial c_{H_2M} and Initial c_{NaOH}. The label mL H_2M is placed in cell E2. The values for these constants and starting conditions are entered into the appropriate cells. Note that K_{b1} can be calculated in cell D2 from K_w/K_{a2}. That is, we enter into cell D2 the formula =$B$4/$B$3. Because the ratio K_{a1}/K_{a2} is large (2×10^4), we proceed as described for constructing Figure 13-1.

Let us label the 5th row as shown in Figure 13-8: **Vol. NaOH, mL** (A5), c_{H_2M} (B5), c_{HM^-} (C5), $c_{M^{2-}}$ (D5), $[H_3O^+]$ (E5), **pH** (F5), $[OH^-]$ (G5), and **pOH** (H5). The volumes for which the pH is to be calculated are entered into cells A6–A29 as shown.

Initial pH

In the spreadsheet we can calculate the analytical concentrations of H_2M (cell B6) and HM^- (cell C6) as shown in the documentation section, cells A31 and A32. These values hold until the first equivalence point so we can copy cell B6 into B7 through B17 and cell C6 into C7 through C17.

As shown in Example 13-4, a quadratic equation must be solved to calculate the initial pH. The equation derived previously is

$$[H_3O^+]^2 + 1.3 \times 10^{-2}[H_3O^+] - 1.3 \times 10^{-3} = 0$$

Because K_{a1} for maleic acid is large, we must solve the quadratic equation exactly or by iteration. To solve the quadratic in the spreadsheet we enter the formula given in the documentation for cell E6(A33). The result is identical to that calculated on page 310, $[H_3O^+] = 3.01 \times 10^{-2}$. In cell F6, we enter =-LOG10(E6), which returns pH = 1.52. Since pH is calculated in the same way for all values prior to the second equivalence point, we copy F6 into F7 through F24.

First Buffer Region

The addition of 5.00 mL of base results in the formation of a buffer consisting of the weak acid H_2M and its conjugate base HM^-. We have previously entered the formulas for c_{NaHM} and c_{H_2M} into the B7:B17 cells and the C7:C17 cells. We note that for the 5.00 mL point, the values returned in cells B7 and C7 are identical to those calculated manually on page 310. As before, calculation of the pH requires solving a quadratic. Hence, we enter into cell E7 the formula shown in the documentation section, cell A34. The returns from this calculation and the one in F7 are $[H_3O^+] = 1.81 \times 10^{-2}$ M and pH = 1.74 as calculated manually on page 311.

Additional points in the first buffer region are computed in a similar way. Hence we copy the results of E7 into cells E8 through E16. The simplified buffer formula used gives large errors very near the equivalence point. You can see that in cells B16 and C16, the ratio c_{HM^-}/c_{H_2M} becomes large indicating that the

	A	B	C	D	E	F	G	H	I
1	**Spreadsheet to calculate titration curve for maleic acid.**								
2	K_{a1}	1.30E-02	K_{b1}	1.69E-08	mL H$_2$M	25.00			
3	K_{a2}	5.90E-07	Initial c_{H2M}	0.1000					
4	K_w	1.00E-14	Initial c_{NaOH}	0.1000					
5	**Vol. NaOH, mL**	c_{H2M}	c_{HM-}	c_{M2-}	**[H$_3$O$^+$]**	**pH**	**[OH$^-$]**	**pOH**	
6	0.00	0.10000	0.00000		3.01E-02	1.52			
7	5.00	0.06667	0.01667		1.81E-02	1.74			
8	8.19	0.05067	0.02467		1.30E-02	1.89			
9	10.00	0.04286	0.02857		1.07E-02	1.97			
10	12.50	0.03333	0.03333		7.98E-03	2.10			
11	15.00	0.02500	0.03750		5.78E-03	2.24			
12	17.00	0.01905	0.04048		4.29E-03	2.37			
13	18.00	0.01628	0.04186		3.62E-03	2.44			
14	19.00	0.01364	0.04318		3.00E-03	2.52			
15	24.00	2.041E-03	0.04898		4.25E-04	3.37			
16	24.50	1.010E-03	0.04949		2.09E-04	3.68			
17	25.00	0.0000	0.05		7.80E-05	4.11			
18	25.50		0.048515	9.9010E-04	2.89E-05	4.54			
19	26.00		0.047059	1.9608E-03	1.42E-05	4.85			
20	30.00		0.036364	9.0909E-03	2.36E-06	5.63			
21	40.00		0.015385	2.3077E-02	3.93E-07	6.41			
22	45.00		0.007143	2.8571E-02	1.48E-07	6.83			
23	49.00		0.001351	3.2432E-02	2.46E-08	7.61			
24	49.50		0.000671	3.2886E-02	1.20E-08	7.92			
25	50.00		0.000000	3.3333E-02	4.21E-10	9.38	2.38E-05	4.62	
26	50.50			3.3113E-02	1.51E-11	10.82	6.62E-04	3.18	
27	51.00			3.2895E-02	7.60E-12	11.12	1.32E-03	2.88	
28	55.00			3.1250E-02	1.60E-12	11.80	6.25E-03	2.20	
29	60.00			2.9412E-02	8.50E-13	12.07	1.18E-02	1.93	
30	**Spreadsheet Documentation**								
31	Cell B6=($D$3*$F$2-$D$4*A6)/($F$2+A6)				Cell E17=SQRT(($B$3*C17+$B$4)/(1+C17/$B$2))				
32	Cell C6=$D$4*A6/($F$2+A6)				Cell E18=$B$3*C18/D18				
33	Cell E6=(-($B$2)+SQRT(($B$2)^2+(4*B6*$B$2)))/2				Cell D26=(($A$25-$A$17)*$D$4)/(A26+$F$2)				
34	Cell E7=(-(C7+$B$2)+SQRT((C7+$B$2)^2+(4*B7*$B$2)))/2				Cell G25=SQRT($D$2*D25)				
35	Cell F6=-LOG10(E6)				Cell G26=(A26-2*$F$2)*$D$4/(A26+$F$2)				
36	Cell C17=$D$4*A17/($F$2+A17)				Cell H25=-LOG(G25)				
37	Cell C18=(($D$4*$A$17)-((A18-$F$2)*$D$4))/($F$2+A18)				Cell F25=14-H25				
38	Cell D18=(($D$4*$A$17)-((A18-$F$2)*$D$4))/($F$2+A18)				Cell E25=10^-F25				

Figure 13-8   Spreadsheet for calculating the maleic acid titration curve.

solution is no longer a good buffer. If we were to go closer to the equivalence point (24.90 mL, for example), a more complex expression would need to be developed.

First Equivalence Point

At the first equivalence point, $[HM^-] \approx c_{NaHM} = 0.0500$ M and $c_{H_2M} = 0.0000$ M. The concentration of H_3O^+ must be calculated from Equation 13-3 with K_w negligible. In spreadsheet cell E17, we thus enter the formula shown in the documentation section, cell D33.

Second Buffer Region

Addition of a small volume of base beyond that required to reach the first equivalence point creates a new buffer system consisting of HM^- and M^{2-}. When enough base has been added so that the reaction of HM^- with water to give OH^- can be neglected (a few tenths of a milliliter beyond the first equivalence point), the pH of the mixture is readily obtained from K_{a2} as shown on page 312. With the introduction of 25.50 mL of NaOH, for example, the $[M^{2-}]$ and $[HM^-]$ are readily calculated from the stoichiometry. The formulas for these are shown in cell C18 and D18. These same values hold until the second equivalence point and so the formula in C18 is copied into cells C19:C25. Likewise the formula in cell D18 is copied into D19:D25. Substituting these values into the expression for K_{a2} gives the equation for $[H_3O^+]$ and we thus enter into cell E18 the formula shown in the documentation section, cell D34.

The other values in the second buffer region are calculated in a similar manner. Thus, the formula in cell E18 is is copied into cells E19 through E25. Again if we were to do the calculation of pH very close to the second equivalence point (49.90 mL for example), the simple buffer equation used would break down because the approximations used to derive the equation are no longer valid.

Second Equivalence Point

After the addition of 50.00 mL of 0.1000 M sodium hydroxide, the solution is 0.0333 M in Na_2M (2.5 mmol/75.00 mL) and there is no longer any HM^- left. Reaction of the base M^{2-} with water is the predominant equilibrium in the system and the only one that we need to take into account. Thus, we calculate $[OH^-]$ from K_{b1} and the analytical concentration of Na_2M. The formula shown in the documentation section, cell D36 is entered into cell G25. This returns $[OH^-] = 2.38 \times 10^{-5}$, pOH = 4.62. The pH in cell F25 is calculated from pH = 14 − pOH. This gives pH = 9.38. From the pH, we can calculate $[H_3O^+] = 10^{-pH}$. The pH and $[H_3O^+]$ formulas hold for the rest of the spreadsheet and so we copy cell F25 into cells F26:F29 and cell E25 into cells E26:E29.

pH Beyond the Second Equivalence Point

As long as we are a few tenths of a milliliter beyond the second equivalence point, the pH is calculated from the concentration of NaOH added in excess of that required for the complete neutralization of H_2M. Thus when 50.50 mL of NaOH have been added, we have 0.50-mL excess 0.1000 M NaOH, and we enter into G26 the formula shown in the documentation section, cell H31. The result is 1.32×10^{-3} and pOH = 2.88, pH = 11.12. The formula in G26 is

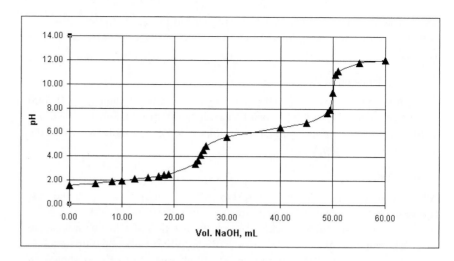

Figure 13-9   Titration curve for 25.00 mL of 0.1000 M maleic acid H_2M with 0.1000 M NaOH generated from the spreadsheet of Figure 13-8.

copied into the remaining G cells. Note that we can calculate the concentration of M^{2-} by the formula shown in the documentation section, cell D35. The titration curve that results from the spreadsheet calculation is plotted in Figure 13-9. Note the similarity to the curve calculated manually in Figure 13-2.

WEB WORKS

Using your web browser, connect to the Virtual Titrator at http://hamers.chem.wisc.edu/chapman/Titrator/. Click on the indicated frame to invoke the Virtual Titrator Java applet and display two windows, the Menu Panel and the Virtual Titrator main window. To begin, click on Acids on the main window menu bar and select the diprotic acid *o*-phthalic acid. Examine the titration curve that results. Then click on Graphs/Alpha Plot vs. pH and observe the result. Click on Graphs/Alpha Plot vs. mL base. Repeat the process for several monoprotic and polyprotic acids and note the results.

13G   QUESTIONS AND PROBLEMS

*13-1.   As its name implies, NaHA is an "acid salt" because it has a proton available to donate to a base. Briefly explain why a pH calculation for a solution of NaHA differs from that for a weak acid of the type HA.

13-2.   Briefly explain why the use of Equation 13-3 is limited to calculation of the hydronium ion concentration of solutions in which NaHA is the only solute that determines the pH.

*13-3.   Why is it impossible to titrate all three protons of phosphoric acid in aqueous solution?

13-4.   Indicate whether an aqueous solution of the fol-

lowing compounds is acidic, neutral, or basic. Explain your answer.

*(a) NH_4OAc **(b)** $NaNO_2$
*(c) $NaNO_3$ **(d)** $NaHC_2O_4$
*(e) $Na_2C_2O_4$ **(f)** Na_2HPO_4
*(g) NaH_2PO_4 **(h)** Na_3PO_4

13-5.   Suggest an indicator that could be used to provide an end point for the titration of the first proton in H_3AsO_4.

*13-6.   Suggest an indicator that would give an end point for the titration of the first two protons in H_3AsO_4.

13-7. Suggest a method for the determination of the amounts of H_3PO_4 and NaH_2PO_4 in an aqueous solution.

13-8. Suggest a suitable indicator for a titration based upon the following reactions; use 0.05 M if an equivalence point concentration is needed.

*(a) $H_2CO_3 + NaOH \rightarrow NaHCO_3 + H_2O$

(b) $H_2P + 2NaOH \rightarrow Na_2P + 2H_2O$
(H_2P = o-phthalic acid)

*(c) $H_2T + 2NaOH \rightarrow Na_2T + 2H_2O$
(H_2T = tartaric acid)

(d) $NH_2C_2H_4NH_2 + HCl \rightarrow NH_2C_2H_4NH_3Cl$

*(e) $NH_2C_2H_4NH_2 + 2HCl \rightarrow$
$ClNH_3C_2H_4NH_3Cl$

(f) $H_2SO_3 + NaOH \rightarrow NaHSO_3 + H_2O$

*(g) $H_2SO_3 + 2NaOH \rightarrow Na_2SO_3 + 2H_2O$

13-9. Calculate the pH of a solution that is 0.0400 M in

*(a) H_3PO_4.   (b) $H_2C_2O_4$.

*(c) H_3PO_3.   (d) H_2SO_3.

*(e) H_2S.   (f) $H_2NO_2H_4NH_2$.

13-10. Calculate the pH of a solution that is 0.0400 M in

*(a) NaH_2PO_4.   (b) $NaHC_2O_4$.

*(c) NaH_2PO_3.   (d) $NaHSO_3$.

*(e) NaHS   (f) $H_2NC_2H_4NH_3^+Cl^-$.

13-11. Calculate the pH of a solution that is 0.0400 M in

*(a) Na_3PO_4.   (b) $Na_2C_2O_4$.

*(c) Na_2HPO_3.   (d) Na_2SO_3.

*(e) Na_2S.   (f) $C_2H_4(NH_3^+Cl^-)_2$.

*13-12. Calculate the pH of a solution that is made up to contain the following analytical concentrations:

(a) 0.0500 M in H_3AsO_4 and 0.0200 M in NaH_2AsO_4.

(b) 0.0300 M in NaH_2AsO_4 and 0.0500 M in Na_2HAsO_4.

(c) 0.0600 M in Na_2CO_3 and 0.0300 M in $NaHCO_3$.

(d) 0.0400 M in H_3PO_4 and 0.0200 M in Na_2HPO_4.

(e) 0.0500 M in $NaHSO_4$ and 0.0400 M in Na_2SO_4.

13-13. Calculate the pH of a solution that is made up to contain the following analytical concentrations:

(a) 0.240 M in H_3PO_4 and 0.480 M in NaH_2PO_4.

(b) 0.0670 M in Na_2SO_3 and 0.0315 M in $NaHSO_3$.

(c) 0.640 M in $HOC_2H_4NH_2$ and 0.750 M in $HOC_2H_4NH_3Cl$.

(d) 0.0240 in $H_2C_2O_4$ (oxalic acid) and 0.0360 M in $Na_2C_2O_4$.

(e) 0.0100 M in $Na_2C_2O_4$ and 0.0400 M in $NaHC_2O_4$.

*13-14. Calculate the pH of a solution that is

(a) 0.0100 M in HCl and 0.0200 M in picric acid.

(b) 0.0100 M in HCl and 0.0200 M in benzoic acid.

(c) 0.0100 M in NaOH and 0.100 M in Na_2CO_3.

(d) 0.0100 M in NaOH and 0.100 M in NH_3.

13-15. Calculate the pH of a solution that is

(a) 0.0100 M in $HClO_4$ and 0.0300 M in monochloroacetic acid.

(b) 0.0100 M in HCl and 0.0150 M in H_2SO_4.

(c) 0.0100 M in NaOH and 0.0300 M in Na_2S.

(d) 0.0100 M in NaOH and 0.0300 M in sodium acetate.

*13-16. Identify the principal conjugate acid/base pair and calculate the ratio between them in a solution that is buffered to pH 6.00 and contains

(a) H_2SO_3.   (b) citric acid.

(c) malonic acid.   (d) tartaric acid.

13-17. Identify the principal conjugate acid/base pair and calculate the ratio between them in a solution that is buffered to pH 9.00 and contains

(a) H_2S.   (b) ethylenediamine dihydrochloride.

(c) H_3AsO_4.   (d) H_2CO_3.

*13-18. How many grams of $Na_2HPO_4 \cdot 2H_2O$ must be added to 400 mL of 0.200 M H_3PO_4 to give a buffer of pH 7.30?

13-19. How many grams of dipotassium phthalate must be added to 750 mL of 0.0500 M phthalic acid to give a buffer of pH 5.75?

*13-20. What is the pH of the buffer formed by mixing 50.0 mL of 0.200 M NaH_2PO_4 with

(a) 50.0 mL of 0.120 M HCl?

(b) 50.0 mL of 0.120 M NaOH?

13-21. What is the pH of the buffer formed by adding 100 mL of 0.150 M potassium hydrogen phthalate to

(a) 100 mL of 0.0800 M NaOH?

(b) 100 mL of 0.0800 M HCl?

*13-22. How would you prepare 1.00 L of a buffer with a pH of 9.60 from 0.300 M Na_2CO_3 and 0.200 M HCl?

13-23. How would you prepare 1.00 L of a buffer with a pH of 7.00 from 0.200 M H_3PO_4 and 0.160 M NaOH?

*13-24. How would you prepare 1.00 L of a buffer with a pH of 6.00 from 0.500 M Na_3AsO_4 and 0.400 M HCl?

13-25. Identify by letter the curve you would expect in the titration of a solution containing

(a) disodium maleate, Na_2M, with standard acid.

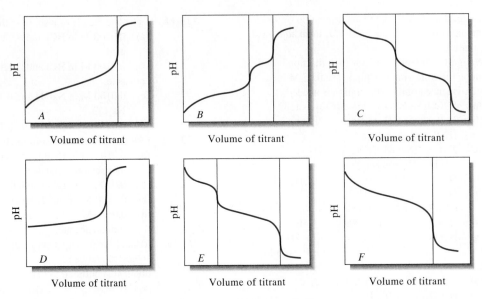

Titration curves for Problem 13-25.

(b) pyruvic acid, HP, with standard base.

(c) sodium carbonate, Na_2CO_3, with standard acid.

13-26. Describe the composition of a solution that would be expected to yield a curve resembling (see Problem 13-25)

(a) curve B.   **(b)** curve A.   **(c)** curve E.

13-27. Briefly explain why curve B cannot describe the titration of a mixture consisting of H_3PO_4 and NaH_2PO_4.

13-28. Derive a curve for the titration of 50.00 mL of a 0.1000 M solution of compound A with a 0.2000 M solution of compound B in the following list. For each titration, calculate the pH after the addition of 0.00, 12.50, 20.00, 24.00, 25.00, 26.00, 37.50, 45.00, 49.00, 50.00, 51.00, and 60.00 mL of compound B:

	A	B
*(a)	Na_2CO_3	HCl
(b)	ethylenediamine	HCl
*(c)	H_2SO_4	NaOH
(d)	H_2SO_3	NaOH

*13-29. Generate a curve for the titration of 50.00 mL of a solution in which the analytical concentration of NaOH is 0.1000 M and that for hydrazine is 0.0800 M. Calculate the pH after addition of 0.00, 10.00, 20.00, 24.00, 25.00, 26.00, 35.00, 44.00, 45.00, 46.00, and 50.00 mL of 0.2000 M $HClO_4$.

13-30. Generate a curve for the titration of 50.00 mL of a solution in which the analytical concentration of $HClO_4$ is 0.1000 M and that for formic acid is 0.0800 M. Calculate the pH after addition of 0.00, 10.00, 20.00, 24.00, 25.00, 26.00, 35.00, 44.00, 45.00, 46.00, and 50.00 mL of 0.2000 M KOH.

13-31. Formulate equilibrium constants for the following equilibria, giving numerical values for the constants:

*(a) $H_2AsO_4^- + H_2AsO_4^- \rightleftharpoons$
$$H_3AsO_4 + HAsO_4^{2-}$$

(b) $HAsO_4^{2-} + HAsO_4^{2-} \rightleftharpoons$
$$AsO_4^{3-} + H_2AsO_4^-$$

*13-32. Derive a numerical value for the equilibrium constant for the reaction

$$NH_4^+ + OAc^- \rightleftharpoons NH_3 + HOAc$$

13-33. For pH values of 2.00, 6.00, and 10.00, calculate the alpha value for each species in an aqueous solution of

*(a) phthalic acid.   **(b)** phosphoric acid.
*(c) citric acid.   **(d)** arsenic acid.
*(e) phosphorous acid.   **(f)** oxalic acid.

If you perform these calculations using a spreadsheet, compute alpha values for each species over the pH range 0 to 14 with points every 0.5 pH unit, and construct a plot of alpha values versus pH.

13-34. Derive equations that define α_0, α_1, α_2, and α_3 for the acid H_3AsO_4.

Chapter 14

Applying Neutralization Titrations

Neutralization titrations are widely used to determine the concentration of analytes that are themselves acids or bases or that are convertible to such species by suitable treatment.[1] Water is the usual solvent for neutralization titrations because it is readily available, inexpensive, and non-toxic. Its low coefficient of expansion with temperature is an added virtue. Some analytes, however, are not titratable in aqueous media because their solubilities are too low or because their strengths as acids or bases are not sufficient to provide satisfactory end points. Nonaqueous solvents such as methyl and ethyl alcohol, glacial acetic acid, and methyl isobutyl ketone often make it possible to titrate such analytes in a solvent other than water.[2] We shall restrict our discussions in this chapter to aqueous systems.

14A REAGENTS FOR NEUTRALIZATION TITRATIONS

In Chapter 12, we noted that strong acids and strong bases cause the most pronounced change in pH at the equivalence point. For this reason, standard solutions for neutralization titrations are always prepared from these reagents.

14A-1 Preparing Standard Acid Solutions

Hydrochloric acid is widely used for titration of bases. Dilute solutions of HCl are stable indefinitely and do not cause troublesome precipitation reactions with most cations. It is reported that 0.1 M solutions of HCl can be boiled for as long as one hour without loss of acid, provided that the water lost by evaporation is periodically replaced; 0.5 M solutions can be boiled for at least 10 min without significant loss.

---

[1]For a review of applications of neutralization titrations, see D. Rosenthal and P. Zuman, in *Treatise on Analytical Chemistry,* 2nd ed., I. M. Kolthoff and P. J. Elving, Eds. (New York: Wiley, 1979), Part I, Vol. 2, Chapter 18.
[2]For a review of nonaqueous acid/base titrimetry, see *Treatise on Analytical Chemistry,* 2nd ed., I. M. Kolthoff and P. J. Elving, Eds. (New York: Wiley, 1979), Part I, Vol. 2, Chapters 19A–19E.

Solutions of HCl, $HClO_4$, and H_2SO_4 are stable indefinitely. Restandardization is never required.

Solutions of perchloric acid and sulfuric acid are also stable and are useful for titrations where chloride ion interferes by forming precipitates. Standard solutions of nitric acid are seldom encountered because of their oxidizing properties. Standard acid solutions are ordinarily prepared by diluting an approximate volume of the concentrated reagent and subsequently standardizing the diluted solution against a primary-standard base.

14A-2   Standardizing Acids

Sodium Carbonate

Sodium carbonate occurs naturally in large deposits as *washing soda*, $Na_2CO_3 \cdot 10H_2O$ and as *trona*, $Na_2CO_3 \cdot NaHCO_3 \cdot 2H_2O$. These minerals are widely used in many industries, such as the glass industry. Primary standard sodium carbonate is manufactured by extensive purification of these minerals.

Acids are frequently standardized against weighed quantities of sodium carbonate. The primary-standard-grade reagent is available commercially or can be prepared by heating purified sodium hydrogen carbonate between 270 and 300°C for 1 hr.

$$2NaHCO_3(s) \rightarrow Na_2CO_3(s) + H_2O(g) + CO_2(g)$$

A high mass per proton consumed is desirable in a primary standard because a larger mass of reagent must be used, thus decreasing the relative weighing error. TRIS reacts in a 1:1 molar ratio with hydronium ions.

As shown in Figure 13-4, two end points are observed in the titration of sodium carbonate. The first, corresponding to the conversion of carbonate to hydrogen carbonate, occurs at about pH 8.3; the second, involving the formation of carbonic acid and carbon dioxide, is observed at about pH 3.8. The second end point is always used for standardization because the change in pH is greater than that of the first. An even sharper end point can be achieved by boiling the solution briefly to eliminate the reaction products, carbonic acid and carbon dioxide. The sample is titrated to the first appearance of the acid color of the indicator (such as bromocresol green or methyl orange). At this point, the solution contains a large amount of dissolved carbon dioxide and small amounts of carbonic acid and unreacted hydrogen carbonate. Boiling effectively destroys this buffer by eliminating the carbonic acid:

$$H_2CO_3(aq) \rightarrow CO_2(g) + H_2O(l)$$

The solution then becomes alkaline again due to the residual hydrogen carbonate ion. The titration is completed after the solution has cooled. Now, however, a substantially larger decrease in pH occurs during the final additions of acid, thus giving a more abrupt color change (see Figure 14-1).

As an alternative, the acid can be introduced in an amount slightly in excess of that needed to convert the sodium carbonate to carbonic acid. The solution is boiled as before to remove carbon dioxide and cooled; the excess acid is then back-titrated with a dilute solution of base. Any indicator suitable for a strong acid/strong base titration is satisfactory. Of course, the volume ratio of acid to base must be established by an independent titration.

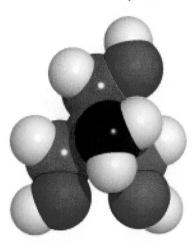

$$\begin{array}{c} CH_2OH \\ | \\ H_2N-C-CH_2OH \\ | \\ CH_2OH \end{array}$$

Molecular model and structure of TRIS.

Borax, $Na_2B_4O_7 \cdot 10H_2O$, a mineral mined in the desert, is widely used in cleaning preparations. A highly purified form of borax is used as a primary standard for acids.

Other Primary Standards for Acids

Tris-(hydroxymethyl)aminomethane, $(HOCH_2)_3CNH_2$, known also as TRIS or THAM, is available in primary-standard purity from commercial sources. It possesses the advantage of a substantially greater mass per mole of protons consumed (121.1) than sodium carbonate (53.0) (see Spreadsheet Example 14-1).

Sodium tetraborate decahydrate and mercury(II) oxide have also been recommended as primary standards. The reaction of an acid with the tetraborate is

$$B_4O_7^{2-} + 2H_3O^+ + 3H_2O \rightarrow 4H_3BO_3$$

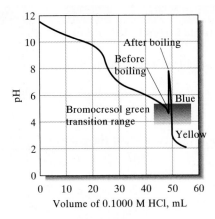

Figure 14-1 Titration of 25.00 mL of 0.1000 M Na_2CO_3 with 0.1000 M HCl. After about 49 mL of HCl have been added, the solution is boiled, causing the increase in pH shown. The change in pH on further addition of HCl is much larger.

Example 14-1

Spreadsheet Exercise

Use a spreadsheet to compare the masses of (a) TRIS (121 g/mol), (b) Na_2CO_3 (106 g/mol), and (c) $Na_2B_4O_7 \cdot 10H_2O$ (381 g/mol) that should be taken to standardize an approximately 0.020 molar solution of HCl for the following volumes of HCl: 20.00, 30.00, 40.00, and 50.00 mL. If the standard deviation associated with weighing out the primary standard bases is 0.1 mg, use the spreadsheet to calculate the percent relative standard deviation that this uncertainty would introduce into each of the calculated molarities.

The spreadsheet is shown in Figure 14-2. We enter the molarity of HCl in cell B2 and the molecular weights of the three primary standards in cells B3, B4, and B5. Appropriate labels are entered into columns A and C. The volumes of HCl for which the calculations are desired are entered in cells A8 through A11. We will do a sample calculation for the 20.00-mL volume here and show the spreadsheet entry. In each case, the number of mmoles of HCl is calculated from

$$\text{mmol HCl} = \text{mL HCl} \times 0.020 \, \frac{\text{mmol HCl}}{\text{mL HCl}}$$

(a) For TRIS,

$$\text{g TRIS} = \text{mmol HCl} \times \frac{1 \text{ mmol TRIS}}{\text{mmol HCl}} \times \frac{121 \text{ g TRIS/mol TRIS}}{1000 \text{ mmol TRIS/mol TRIS}}$$

For the 20.00-mL volume of HCl, the appropriate entry in cell B8 is =$B$2*A8*1*$B$3/1000 as shown in the documentation section of Figure 14-2. The result returned is 0.048 g. The formula in cell B8 is then copied into cells B9 through B11 to complete the column. The relative uncertainty in the molarity due to weighing would be equal to the relative uncertainty in the weighing process. For 0.048 g of TRIS, the percent relative standard deviation (%RSD) is (0.0001g/0.048 g) $\times$ 100%, so the entry in cell C8 is as shown in Figure 14-2. This formula in cell C8 is then copied into C9:C11.

	A	B	C	D	E	F	G
1	**Spreadsheet to compare masses required for various bases in the standardization of 0.020 M HCl**						
2	M HCl	0.020					
3	MW TRIS	121	g/mol	**Note**: All weighings have standard deviations of 0.1 mg (0.0001g)			
4	MW Na$_2$CO$_3$	106	g/mol				
5	MW Na$_2$B$_4$O$_7$•H$_2$O	381	g/mol				
6							
7	**mL HCl**	**g TRIS**	**%RSD TRIS**	**g Na$_2$CO$_3$**	**%RSD Na$_2$CO$_3$**	**g Na$_2$B$_4$O$_7$•H$_2$O**	**%RSD Na$_2$B$_4$O$_7$•H$_2$O**
8	20.00	0.048	0.21	0.021	0.47	0.08	0.13
9	30.00	0.073	0.14	0.032	0.31	0.11	0.09
10	40.00	0.097	0.10	0.042	0.24	0.15	0.07
11	50.00	0.121	0.08	0.053	0.19	0.19	0.05
12							
13							
14	**Documentation**						
15	Cell B8=$B$2*A8*1*$B$3/1000						
16	Cell C8 =(0.0001/B8)*100						
17	Cell D8 =$B$2*A8*1/2*$B$4/1000						
18	Cell E8 = (0.0001/D8)*100						
19	Cell F8 = $B$2*A8*1/2*$B$5/1000						
20	Cell G8 = (0.0001/F8)*100						

Figure 14-2 Spreadsheet to compare masses and relative errors associated with using different primary standard bases to standardize HCl solutions.

(b) For Na$_2$CO$_3$,

$$\text{mass Na}_2\text{CO}_3 = \text{mmol HCl} \times \frac{1 \text{ mmol Na}_2\text{CO}_3}{2 \text{ mmol HCl}}$$

$$\times \frac{106 \text{ g Na}_2\text{CO}_3/\text{mol Na}_2\text{CO}_3}{1000 \text{ mmol Na}_2\text{CO}_3/\text{mol Na}_2\text{CO}_3}$$

This result is entered into cell D8 as shown in Figure 14-2 and copied into D9:D11. The percent relative standard deviation in cell E8 is calculated as (0.0001/D8) × 100. The formula in E8 is copied into E9:E11.

(c) Similarly, for Na$_2$B$_4$O$_7 \cdot$ 10H$_2$O,

$$\text{mass borax} = \text{mmol HCl} \times \frac{1 \text{ mmol borax}}{2 \text{ mmol HCl}}$$

$$\times \frac{381 \text{ g borax/mol borax}}{1000 \text{ mmol borax/mol borax}}$$

Note in Figure 14-2 that the relative standard deviation for TRIS is 0.10% or less if the volume of HCl taken is more than 40.00 mL. For Na$_2$CO$_3$, more than 50.00 mL of HCl would be required for this level of uncertainty. For borax, any volume above about 26.00 mL would suffice.

14A-3 Preparing Standard Base Solutions

Sodium hydroxide is the most common base for preparing standard solutions, although potassium hydroxide and barium hydroxide are also encountered. None of these is obtainable in primary-standard purity, so standardization is required after preparation.

The Effect of Carbon Dioxide upon Standard Base Solutions

In solution as well as in the solid state, the hydroxides of sodium, potassium, and barium react rapidly with atmospheric carbon dioxide to produce the corresponding carbonate:

$$CO_2 + 2OH^- \rightarrow CO_3^{2-} + H_2O$$

Although production of each carbonate ion uses up two hydroxide ions, the uptake of carbon dioxide by a solution of base does not necessarily alter its combining capacity for hydronium ions. If the end point of a titration occurs in acidic solution and thus requires an acid-range indicator (such as bromocresol green), each carbonate ion produced from sodium or potassium hydroxide will have reacted with two hydronium ions of the acid (Figure 14-1):

$$CO_3^{2-} + 2H_3O^+ \rightarrow H_2CO_3 + 2H_2O$$

Because the amount of hydronium ion consumed by this reaction is identical to the amount of hydroxide lost during formation of the carbonate ion, no error is incurred.

Unfortunately, most titrations that make use of a standard base have basic end points and require an indicator with a basic transition range (phenolphthalein, for example). In these basic solutions, each carbonate ion has reacted with only one hydronium ion when the color change of the indicator is observed:

$$CO_3^{2-} + H_3O^+ \rightarrow HCO_3^- + H_2O$$

The effective concentration of the base is thus diminished by absorption of carbon dioxide, and a systematic error (called a *carbonate error*) results (see Example 14-2).

Absorption of carbon dioxide by a standardized solution of sodium or potassium hydroxide leads to a negative systematic error in analyses in which an indicator with a basic range is used; no systematic error is incurred if an indicator with an acidic range is used.

Example 14-2

A carbonate-free NaOH solution was found to be 0.05118 M immediately after preparation. If exactly 1.000 L of this solution was exposed to air for some time and absorbed 0.1962 g CO_2, calculate the relative carbonate error that would arise in the determination of acetic acid with the contaminated solution if phenolphthalein were used as an indicator.

$$2NaOH + CO_2 \rightarrow Na_2CO_3 + H_2O$$

$$c_{Na_2CO_3} = 0.1962 \text{ g } \cancel{CO_2} \times \frac{1 \text{ mol } \cancel{CO_2}}{44.01 \text{ g } \cancel{CO_2}} \times \frac{1 \text{ mol } Na_2CO_3}{\text{mol } \cancel{CO_2}}$$

$$\times \frac{1}{1.000 \text{ L soln}} = 4.458 \times 10^{-3} \text{ M}$$

The effective concentration c_{NaOH} of NaOH for acetic acid is

$$c_{NaOH} = \frac{0.05118 \text{ mol NaOH}}{L} - \frac{4.458 \times 10^{-3} \text{ mol } \cancel{Na_2CO_3}}{L}$$

$$\times \frac{1 \text{ mol } \cancel{HCl}}{\text{mol } \cancel{Na_2CO_3}} \times \frac{1 \text{ mol NaOH}}{\text{mol } \cancel{HCl}} = 0.04672 \text{ M}$$

$$\text{percent rel error} = \frac{0.04672 - 0.05118}{0.05118} \times 100\% = -8.7\%$$

Generally, carbonate ion in standard solutions of bases is undesirable because it decreases the sharpness of end points.

WARNING: Concentrated solutions of NaOH (and KOH) are extremely corrosive to the skin. In making up standard solutions of NaOH, a *face shield, rubber gloves, and protective clothing must be worn at all times.*

Water that is in equilibrium with atmospheric constituents contains only about 1.5×10^{-5} mol CO_2/L, an amount that has a negligible effect on the strength of most standard bases. As an alternative to boiling to remove CO_2 from supersaturated solutions of CO_2, the excess gas can be removed by bubbling air through the water for several hours. This process, called *sparging*, produces a solution that contains the equilibrium concentration of CO_2.

Sparging is the process of removing a gas from a solution by bubbling an inert gas through the solution.

Solutions of bases should be stored in polyethylene bottles rather than glass because of the reaction between bases and glass. Such solutions should never be stored in glass-stoppered bottles; after standing for a period, removal of the stopper often becomes impossible.

The solid reagents used to prepare standard solutions of base are always contaminated by significant amounts of carbonate ion. The presence of this contaminant does not cause a carbonate error provided the same indicator is used for both standardization and analysis. It does, however, lead to a decreased sharpness of end points. Consequently, steps are usually taken to remove carbonate ion before a solution of a base is standardized.

The best method for preparing carbonate-free sodium hydroxide solutions takes advantage of the very low solubility of sodium carbonate in concentrated solutions of the base. An approximately 50% aqueous solution of sodium hydroxide is prepared (or purchased from commercial sources). The solid sodium carbonate is allowed to settle to give a clear liquid that is decanted and diluted to give the desired concentration (alternatively, the solid is removed by vacuum filtration).

Water that is used to prepare carbonate-free solutions of base must also be free of carbon dioxide. Distilled water, which is sometimes supersaturated with carbon dioxide, should be boiled briefly to eliminate the gas. The water is then allowed to cool to room temperature before the introduction of base because hot alkali solutions rapidly absorb carbon dioxide. Deionized water ordinarily does not contain significant amounts of carbon dioxide.

A tightly capped polyethylene bottle usually provides adequate short-term protection against the uptake of atmospheric carbon dioxide. Before capping, the bottle is squeezed to minimize the interior air space. Care should also be taken to keep the bottle closed except during the brief periods when the contents are being transferred to a buret. Sodium hydroxide solutions will ultimately cause a polyethylene bottle to become brittle.

The concentration of a sodium hydroxide solution will decrease slowly (0.1 to 0.3% per week) if the base is stored in glass bottles. The loss in strength is caused by the reaction of the base with the glass to form sodium silicates. For this reason, standard solutions of base should not be stored for extended periods (longer than 1 or 2 weeks) in glass containers. In addition, bases should never be kept in glass-stoppered containers because the reaction between the base and the stopper may cause the latter to "freeze" after a brief period. Finally, to avoid the same type of freezing, burets with glass stopcocks should be promptly drained and thoroughly rinsed with water after use with standard base solutions. This problem is avoided with burets equipped with Teflon stopcocks.

14A-4 Standardizing Solutions of Bases

Several excellent primary standards are available for the standardization of bases. Most are weak organic acids that require the use of an indicator with a basic transition range.

Potassium Hydrogen Phthalate, $KHC_8H_4O_4$

Potassium hydrogen phthalate is an ideal primary standard. It is a nonhygroscopic crystalline solid with a high molar mass (204.2 g/mol). For most purposes, the commercial analytical-grade salt can be used without further purification. For the most exacting work, potassium hydrogen phthalate of certified purity is available from the National Institute of Standards and Technology.

Other Primary Standards for Bases

Benzoic acid is obtainable in primary-standard purity and can be used for the standardization of bases. Because its solubility in water is limited, this reagent is ordinarily dissolved in ethanol prior to dilution with water and titration. A solvent blank should always be titrated during this standardization process because commercial alcohol is sometimes slightly acidic.

Potassium hydrogen iodate, $KH(IO_3)_2$, is an excellent primary standard with a high molecular mass per mole of protons. It is also a strong acid that can be titrated using virtually any indicator with a transition range between pH 4 and 10.

14B TYPICAL APPLICATIONS OF NEUTRALIZATION TITRATIONS

Neutralization titrations are used to determine the innumerable inorganic, organic, and biological species that possess inherent acidic or basic properties. Equally important, however, are the many applications that involve conversion of an analyte to an acid or base by suitable chemical treatment, followed by titration with a standard strong base or acid.

Two major types of end points find widespread use in neutralization titrations. The first is a visual end point based on indicators such as those described in Section 12A-2. The second is a *potentiometric* end point in which the potential of a glass/calomel electrode system is determined with a voltage-measuring device. The measured potential is directly proportional to pH. Potentiometric end points are described in Section 19G-1.

14B-1 Elemental Analysis

Several important elements that occur in organic and biological systems are conveniently determined by methods that involve an acid/base titration as the final step. Generally, the elements susceptible to this type of analysis are nonmetallic and include carbon, nitrogen, chlorine, bromine, and fluorine as well as a few other less common species. Pretreatment converts the element to an inorganic acid or base that is then titrated. A few examples follow.

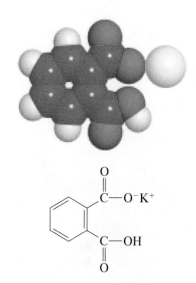

Molecular model and structure of potassium hydrogen phthalate.

Standard solutions of strong bases cannot be prepared directly by mass and must always be standardized against a primary standard with acidic properties.

$KH(IO_3)_2$, in contrast to all other primary standards for bases, has the advantage of being a strong acid, thus making the choice of indicator less critical.

Neutralization titrations are still among the most widely used of all analytical methods.

Kjeldahl is pronounced *Kyell' dahl.* Hundreds of thousands of Kjeldahl nitrogen determinations are performed each year, primarily to provide a measure of the protein content of meats, grains, and animal feeds.

The Kjeldahl method was developed by a Danish chemist who first described it in 1883. J. Kjeldahl, *Z. Anal. Chem.* **1883,** *22,* 366.

Nitrogen

Nitrogen occurs in a wide variety of substances of interest in research, industry, and agriculture. Examples include amino acids, proteins, synthetic drugs, fertilizers, explosives, soils, potable water supplies, and dyes. Thus, analytical methods for the determination of nitrogen, particularly in organic substrates, are of singular importance.

The most common method for determining organic nitrogen is the *Kjeldahl method,* which is based on a neutralization titration. The procedure is straightforward, requires no special equipment, and is readily adapted to the routine analysis of large numbers of samples. It (or one of its modifications) is the standard means for determining the protein content of grains, meats, and other biological materials. Since most proteins contain approximately the same percentage of nitrogen, multiplication of this percentage by a suitable factor (6.25 for meats, 6.38 for dairy products, and 5.70 for cereals) gives the percentage of protein in a sample.

In the Kjeldahl method, the sample is decomposed in hot, concentrated sulfuric acid to convert the bound nitrogen to ammonium ion. The resulting solution is then cooled, diluted, and made basic. The liberated ammonia is distilled, collected in an acidic solution, and determined by a neutralization titration.

The critical step in the Kjeldahl method is the decomposition with sulfuric acid, which oxidizes the carbon and hydrogen in the sample to carbon dioxide and water. The fate of the nitrogen, however, depends on its state of combination in the original sample. Amine and amide nitrogens are quantitatively converted to ammonium ion. In contrast, nitro, azo, and azoxy groups are likely to yield the element or its various oxides, which are all lost from the hot acidic medium. This loss can be avoided by pretreating the sample with a reducing agent to form products that behave as amide or amine nitrogen. In one such prereduction scheme, salicylic acid and sodium thiosulfate are added to the concentrated sulfuric acid solution containing the sample. After a brief period, the digestion is performed in the usual way.

Certain aromatic heterocyclic compounds, such as pyridine (C_6H_5N) and its derivatives, are particularly resistant to complete decomposition by sulfuric acid.

Feature 14-1

Determining Total Serum Protein

The determination of total serum protein is an important clinical measurement used in diagnosing liver malfunctions. Although the Kjeldahl method is capable of high precision and accuracy, it is too slow and cumbersome to be used routinely in determining total serum protein. The Kjeldahl procedure, however, has historically been the reference method against which other methods are compared. Methods commonly used include the *biuret method* and the *Lowry method.* In the biuret method, a reagent containing cupric ions is used, and a violet-colored complex is formed between the Cu^{2+} ions and peptide bonds. The increase in the absorption of visible radiation is used to measure serum protein. This method is readily automated. In the Lowry procedure, the serum sample is pretreated with an alkaline copper solution followed by a phenolic reagent. A color develops because of reduction of phosphotungstic acid and phosphomolybdic acid to a blue heteropoly acid. Both the biuret and Lowry methods use spectrophotometry (see Chapter 22) for quantitative measurements.

Several other methods are used to determine the nitrogen content of organic materials. In the *Dumas method,* the sample is mixed with powdered copper(II) oxide and ignited in a combustion tube to give carbon dioxide, water, nitrogen, and small amounts of nitrogen oxides. A stream of carbon dioxide carries these products through a packing of hot copper, which reduces any oxides of nitrogen to elemental nitrogen. The mixture then is passed into a gas buret filled with concentrated potassium hydroxide. The only component not absorbed by the base is nitrogen, and its volume is measured directly.

The newest method for determining organic nitrogen involves combusting the sample at 1100°C for a few minutes to convert the nitrogen to nitric oxide, NO. Ozone is then introduced into the gaseous mixture, which oxidizes the nitric oxide to nitrogen dioxide. This reaction gives off visible radiation (*chemiluminescence*), the intensity of which is measured and is proportional to the nitrogen content of the sample. An instrument based on this phenomenon is available from commercial sources.

Such compounds yield low results as a consequence (Figure 5-3) unless special precautions are taken.

The decomposition step is frequently the most time-consuming aspect of a Kjeldahl determination. Some samples may require heating periods in excess of one hour. Numerous modifications of the original procedure have been proposed with the aim of shortening the digestion time. In the most widely used modification, a neutral salt, such as potassium sulfate, is added to increase the boiling point of the sulfuric acid solution and thus the temperature at which the decomposition occurs. In another modification, a solution of hydrogen peroxide is added to the mixture as an additional oxidant after the digestion has decomposed most of the organic matrix.

Many substances catalyze the decomposition of organic compounds by sulfuric acid. Mercury, copper, and selenium, either combined or in the elemental state, are effective. Mercury(II), if present, must be precipitated with hydrogen sulfide prior to distillation to prevent retention of ammonia as a mercury(II) ammine complex.

$$-NO_2 \qquad -N{=}N- \qquad \begin{matrix} -N_+{=}N- \\ | \\ O^- \end{matrix}$$

nitro group   azo group   azoxy group

Example 14-3

A 0.7121-g sample of a wheat flour was analyzed by the Kjeldahl method. The ammonia formed by addition of concentrated base after digestion with H_2SO_4 was distilled into 25.00 mL of 0.04977 M HCl. The excess HCl was then back-titrated with 3.97 mL of 0.04012 M NaOH. Calculate the percent protein in the flour.

$$\text{no. mmol HCl} = 25.00 \text{ mL HCl} \times 0.04977 \frac{\text{mmol HCl}}{\text{mL HCl}} = 1.2443$$

$$\text{no. mmol NaOH} = 3.97 \text{ mL NaOH} \times 0.04012 \frac{\text{mmol NaOH}}{\text{mL NaOH}} = 0.1593$$

$$\text{no. mmol N} = 1.2443 - 0.1593 = 0.10850$$

$$\text{percent N} = \frac{1.0850 \; \cancel{\text{mmol N}} \times 0.014007 \; \text{g N/} \cancel{\text{mmol N}}}{0.7121 \; \text{g sample}} \times 100\%$$

$$= 2.1341\%$$

$$\text{percent protein} = 2.1341\% \; \text{N} \times \frac{5.70\% \; \text{protein}}{\% \; \text{N}} = 12.16\%$$

Sulfur

Sulfur dioxide in the atmosphere is often determined by drawing a sample through a hydrogen peroxide solution and then titrating the sulfuric acid that is produced.

Sulfur in organic and biological materials is conveniently determined by burning the sample in a stream of oxygen. The sulfur dioxide (as well as the sulfur trioxide) formed during the oxidation is collected by distillation into a dilute solution of hydrogen peroxide:

$$SO_2(g) + H_2O_2 \rightarrow H_2SO_4$$

The sulfuric acid is then titrated with standard base.

Other Elements

Table 14-1 lists other elements that can be determined by neutralization methods.

14B-2   Determining Inorganic Substances

Numerous inorganic species can be determined by titration with strong acids or bases. A few examples follow.

Ammonium Salts

Ammonium salts are conveniently determined by conversion to ammonia with strong base followed by distillation. The ammonia is collected and titrated as in the Kjeldahl method.

Table 14-1

Elemental Analyses Based on Neutralization Titrations

Element	Converted to	Adsorption or Precipitation Products	Titration
N	NH_3	$NH_3(g) + H_3O^+ \rightarrow NH_4^+ + H_2O$	Excess HCl with NaOH
S	SO_2	$SO_2(g) + H_2O_2 \rightarrow H_2SO_4$	NaOH
C	CO_2	$CO_2(g) + Ba(OH)_2 \rightarrow Ba(CO)_3(s) + H_2O$	Excess $Ba(OH)_2$ with HCl
Cl(Br)	HCl	$HCl(g) + H_2O \rightarrow Cl^- + H_3O^+$	NaOH
F	SiF_4	$SiF_4(g) + H_2O \rightarrow H_2SiF_6$	NaOH
P	H_3PO_4	$12H_2MoO_4 + 3NH_4^+ + H_3PO_4 \rightarrow$ $(NH_4)_3PO_4 \cdot 12MoO_3(s) + 12H_2O + 3H^+$ $(NH_4)_3PO_4 \cdot 12MoO_3(s) + 26OH^- \rightarrow$ $HPO_4^{2-} + 12MoO_4^{2-} + 14H_2O + 3NH_3(g)$	Excess NaOH with HCl

Nitrates and Nitrites

The method just described for ammonium salts can be extended to the determination of inorganic nitrate or nitrite. These ions are first reduced to ammonium ion by Devarda's alloy (50% Cu, 45% Al, 5% Zn). Granules of the alloy are introduced into a strongly alkaline solution of the sample in a Kjeldahl flask. The ammonia is distilled after reaction is complete. Arnd's alloy (60% Cu, 40% Mg) has also been used as the reducing agent.

Carbonate and Carbonate Mixtures

The qualitative and quantitative determination of the constituents in a solution containing sodium carbonate, sodium hydrogen carbonate, and sodium hydroxide, either alone or admixed, provides interesting examples of how neutralization titrations can be employed to analyze mixtures. No more than two of these three constituents can exist in appreciable amount in any solution because reaction eliminates the third. Thus, mixing sodium hydroxide with sodium hydrogen carbonate results in the formation of sodium carbonate until one or the other (or both) of the original reactants is exhausted. If the sodium hydroxide is used up, the solution will contain sodium carbonate and sodium hydrogen carbonate; if sodium hydrogen carbonate is depleted, sodium carbonate and sodium hydroxide will remain; if equimolar amounts of sodium hydrogen carbonate and sodium hydroxide are mixed, the principal solute species will be sodium carbonate.

The analysis of such mixtures requires two titrations, one with an alkaline-range indicator, such as phenolphthalein, and the other with an acid-range indicator, such as bromocresol green. The composition of the solution can then be deduced from the relative volumes of acid needed to titrate equal volumes of the sample (Table 14-2 and Figure 14-3). Once the composition of the solution has been established, the volume data can be used to determine the concentration of each component in the sample.

Table 14-2

Volume Relationships in the Analysis of Mixtures Containing Hydroxide, Carbonate, and Hydrogen Carbonate Ions

Constituents in Sample	Relationship between V_{phth} and V_{beg} in the Titration of an Equal Volume of Sample*
NaOH	$V_{phth} = V_{beg}$
Na_2CO_2	$V_{phth} = \frac{1}{2}V_{beg}$
$NaHCO_3$	$V_{phth} = 0; V_{beg} > 0$
NaOH, Na_2CO_2	$V_{phth} > \frac{1}{2}V_{beg}$
Na_2CO_2, $NaHCO_3$	$V_{phth} < \frac{1}{2}V_{beg}$

*V_{phth} = volume of acid needed for a phenolphthalein end point; V_{beg} = volume of acid needed for a bromocresol green end point.

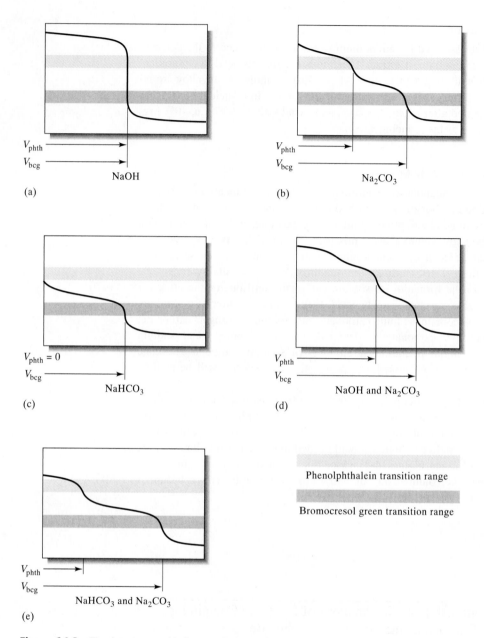

Figure 14-3   Titration curves and indicator transition ranges for the analysis of mixtures containing hydroxide, carbonate, and hydrogen carbonate ions.

Example 14-4

A solution contains $NaHCO_3$, Na_2CO_3, and NaOH, either alone or in permissible combination. Titration of a 50.0-mL portion to a phenolphthalein end point requires 22.1 mL of 0.100 M HCl. A second 50.0-mL aliquot requires 48.4 mL of the HCl when titrated to a bromocresol green end point. Deduce the composition, and calculate the molar solute concentrations of the original solution.

If the solution contained only NaOH, the volume of acid required would be the same regardless of indicator (see Figure 14-3a). Similarly, we can rule out the presence of Na_2CO_3 alone because titration of this compound to a bromocresol green end point would consume just twice the volume of acid required to reach the phenolphthalein end point (see Figure 14-3b). In fact, however, the second titration requires 48.4 mL. Because less than half of this amount is involved in the first titration, the solution must contain some $NaHCO_3$ in addition to Na_2CO_3 (see Figure 14-3e). We can now calculate the concentration of the two constituents.

When the phenolphthalein end point is reached, the CO_3^{2-} originally present is converted to HCO_3^-. Thus,

$$\text{no. mmol } Na_2CO_3 = 22.1 \; \cancel{mL} \times 0.100 \; \text{mmol/}\cancel{mL} = 2.21$$

The titration from the phenolphthalein to the bromocresol green end point (48.4 − 22.1 = 26.3 mL) involves both the hydrogen carbonate originally present and that formed by titration of the carbonate. Thus,

$$\text{no. mmol } NaHCO_3 + \text{no. mmol } Na_2CO_3 = 26.3 \times 0.100 = 2.63$$

Hence,

$$\text{no. mmol } NaHCO_3 = 2.63 - 2.21 = 0.42$$

The molar concentrations are readily calculated from these data:

$$c_{Na_2CO_3} = \frac{2.21 \; \text{mmol}}{50.0 \; \text{mL}} = 0.0442 \; M$$

$$c_{NaHCO_3} = \frac{0.42 \; \text{mmol}}{50.0 \; \text{mL}} = 0.0084 \; M$$

Compatible mixtures containing two of the following can also be analyzed in a similar way: HCl, H_3PO_4, NaH_2PO_4, Na_2HPO_4, Na_3PO_4, and NaOH.

The method described in Example 14-4 is not entirely satisfactory because the pH change corresponding to the hydrogen carbonate equivalence point is not sufficient to give a sharp color change with a chemical indicator (Figure 13-6). Relative errors of 1% or more must be expected as a consequence.

The accuracy of methods for analyzing solutions containing mixtures of carbonate and hydrogen carbonate ions or carbonate and hydroxide ions can be greatly improved by taking advantage of the limited solubility of barium carbonate in neutral and basic solutions. For example, in the *Winkler method* for the analysis of carbonate/hydroxide mixtures, both components are titrated with a standard acid to the end point with an acid-range indicator, such as bromocresol green (the end point is established after the solution is boiled to remove carbon dioxide). An unmeasured excess of neutral barium chloride is then added to a second aliquot of the sample solution to precipitate the carbonate ion; then the hydroxide ion is titrated to a phenolphthalein end point. The presence of the sparingly soluble barium carbonate does not interfere as long as the concentration of barium ion is greater than 0.1 M.

Carbonate and hydrogen carbonate ions can be accurately determined in mixtures by first titrating both ions with standard acid to an end point with an acid-range indicator (with boiling to eliminate carbon dioxide). The hydrogen

How could you analyze a mixture of HCl and H_3PO_4? A mixture of Na_3PO_4 and Na_2HPO_4? See Figure 13-5, curve A.

carbonate in a second aliquot is converted to carbonate by the addition of a known excess of standard base. After a large excess of barium chloride has been introduced, the excess base is titrated with standard acid to a phenolphthalein end point. The presence of solid barium carbonate does not hamper end-point detection in either of these methods.

14B-3   Determining Organic and Biochemical Substances

Neutralization titrations provide convenient methods for the direct or indirect determination of numerous organic species containing acidic or basic functional groups, such as carboxylic and sulfonic acids and various types of amines. A widely used method for the analysis of esters involves heating the sample in the presence of an excess of standard base. The resulting hydrolysis, which is called *saponification*, consumes one mole of the base per mole of ester. After saponification is judged to be complete, the excess base is titrated with a standard acid solution.

Saponification is the process by which an ester is hydrolyzed in alkaline solution to give an alcohol and a conjugate base. For example,

$$CH_3\overset{\overset{O}{\|}}{C}OCH_3 + OH^- \rightarrow$$

$$CH_3\overset{\overset{O}{\|}}{C}-O^- + CH_3OH$$

WEB WORKS

Nitrogen in organic compounds and ammonia is often determined by the Kjeldahl procedure. Nitrogen in these forms comes from human and animal waste and is a good indicator of pollution, particularly from sites that have agricultural uses. To learn about the results of nitrogen determinations in the Santa Rosa, California area, direct your browser to *http://www.sonoma.edu/RCHS/biobeyond/diskel.html*. The sampling sites for the determinations are near Sonoma State University and are given on a site map located at *http://www.sonoma.edu/RCHS/biobeyond/map.jpg*. Discover which of the sampling sites had the highest nitrogen amounts.

14C   QUESTIONS AND PROBLEMS

*14-1. The boiling points of HCl and CO_2 are nearly the same (-85 and $-78°C$). Explain why CO_2 can be removed from an aqueous solution by boiling briefly while essentially no HCl is lost even after boiling for 1 hr or more.

14-2. Why is HNO_3 seldom used to prepare standard acid solutions?

*14-3. Explain how Na_2CO_3 of primary-standard grade can be prepared from primary standard $NaHCO_3$.

14-4. Why is it common practice to boil the solution near the equivalence point in the standardization of acid with Na_2CO_3?

*14-5. Give two reasons why $KH(IO_3)_2$ would be preferred over benzoic acid as a primary standard for a 0.010 M NaOH solution.

14-6. Briefly describe the circumstance where the molarity of a sodium hydroxide solution will apparently be unaffected by the absorption of carbon dioxide.

14-7. What types of organic nitrogen-containing compounds tend to yield low results with the Kjeldahl method unless special precautions are taken?

*14-8. How would you prepare 2.00 L of
(a) 0.15 M KOH from the solid?
(b) 0.015 M $Ba(OH)_2 \cdot 8H_2O$ from the solid?
(c) 0.200 M HCl from a reagent that has a density of 1.0579 g/mL and is 11.50% HCl (w/w)?

14-9. How would you prepare 500.0 mL of
(a) 0.250 M H_2SO_4 from a reagent that has a density of 1.1539 g/mL and is 21.8% H_2SO_4 (w/w)?
(b) 0.30 M NaOH from the solid?
(c) 0.08000 M Na_2CO_3 from the pure solid?

*14-10. Standardization of a sodium hydroxide solution against potassium hydrogen phthalate (KHP) yielded these results:

Mass KHP, g	Volume NaOH, mL
0.7987	38.29
0.8365	39.96
0.8104	38.51
0.8039	38.29

Calculate
(a) the average molarity of the base.
(b) the standard deviation and the coefficient of variation for the data.
(c) the spread of the data and determine if the first value is an outlier.

14-11. The molarity of a perchloric acid solution was established by titration against primary standard sodium carbonate (product: CO_2); the following data were obtained.

Mass Na_2CO_3, g	Volume $HClO_4$, mL
0.2068	36.31
0.1997	35.11
0.2245	39.00
0.2137	37.54

(a) Calculate the average molarity of the acid.
(b) Calculate the standard deviation for the data as well as the coefficient of variation for the data.
(c) Does statistical justification exist for disregarding the outlying result?

*14-12. If 1.000 L of 0.1500 M NaOH was unprotected from the air after standardization and absorbed 11.2 mmol of CO_2, what is its new molarity when it is standardized against a standard solution of HCl using
(a) phenolphthalein?
(b) bromocresol green?

14-13. A NaOH solution was 0.1019 M immediately after standardization. Exactly 500.0 mL of the reagent was left exposed to air for several days and absorbed 0.652 g of CO_2. Calculate the relative carbonate error in the determination of acetic acid with this solution if the titrations were performed with phenolphthalein.

*14-14. Calculate the molar concentration of a dilute HCl solution if
(a) a 50.00-mL aliquot yielded 0.6010 g of AgCl.
(b) titration of 25.00 mL of 0.04010 M $Ba(OH)_2$ required 19.92 mL of the acid.
(c) titration of 0.2694 g of primary standard Na_2CO_3 required 38.77 mL of the acid (products: CO_2 and H_2O).

14-15. Calculate the molarity of a dilute $Ba(OH)_2$ solution if
(a) 50.00 mL yielded 0.1684 g of $BaSO_4$.
(b) titration of 0.4815 g of primary standard potassium hydrogen phthalate (KHP) required 29.41 mL of the base.
(c) addition of 50.00 mL of the base to 0.3614 g of benzoic acid required a 4.13-mL back-titration with 0.05317 M HCl.

14-16. Suggest a range of sample masses for the indicated primary standard if it is desired to use between 35 and 45 mL of titrant:
*(a) 0.150 M $HClO_4$ titrated against Na_2CO_3 (CO_2 product).
(b) 0.075 M HCl titrated against $Na_2C_2O_4$.

$$Na_2C_2O_4 \rightarrow Na_2CO_3 + CO$$
$$CO_3^{2-} + 2H^+ \rightarrow H_2O + CO_2$$

*(c) 0.20 M NaOH titrated against benzoic acid.
(d) 0.030 M $Ba(OH)_2$ titrated against $KH(IO_3)_2$.
*(e) 0.040 M $HClO_4$ titrated against TRIS.
(f) 0.080 M H_2SO_4 titrated against $Na_2B_4O_7 \cdot 10H_2O$. Reaction:

$$B_4O_7^{2-} + 2H_3O^+ + 3H_2O \rightarrow 4H_3BO_3$$

*14-17. Calculate the relative standard deviation in the computed molarity of 0.0200 M HCl if this acid were standardized against the masses derived in Example 14-1 for (a) TRIS, (b) Na_2CO_3, and (c) $Na_2B_4O_7 \cdot 10H_2O$. Assume that the absolute standard deviation in the mass measurement is 0.0001 g and that this measurement limits the precision of the computed molarity.

14-18. (a) Compare the masses of potassium hydrogen phthalate (204.22 g/mol), potassium hydrogen iodate (389.91 g/mol), and benzoic acid (122.12 g/mol) needed for a 30.00-mL standardization of 0.0400 M NaOH.
(b) What would be the relative standard deviation in the molarity of the base if the standard deviation in the measurement of mass in part (a) is 0.002 g and this uncertainty limits the precision of the calculation?

*14-19. A 50.00-mL sample of a white dinner wine required 21.48 mL of 0.03776 M NaOH to achieve a phenolphthalein end point. Express the acidity of the wine in terms of grams of tartaric acid ($H_2C_4H_4O_6$; 150.09 g/mol) per 100 mL. (Assume that two hydrogens of the acid are titrated.)

14-20. A 25.0-mL aliquot of vinegar was diluted to 250.0 mL in a volumetric flask. Titration of 50.0-mL aliquots of the diluted solution required an average of 34.88 mL of 0.09600 M NaOH. Express the acidity of the vinegar in terms of the percentage (w/v) of acetic acid.

*14-21. Titration of a 0.7439-g sample of impure $Na_2B_4O_7$ required 31.64 mL of 0.1081 M HCl [see Problem 14-16(f) for the reaction]. Express the results of this analysis in terms of percent
   (a) $Na_2B_4O_7$.
   (b) $Na_2B_4O_7 \cdot 10H_2O$.
   (c) B_2O_3.
   (d) B.

14-22. A 0.6334-g sample of impure mercury(II) oxide was dissolved in an unmeasured excess of potassium iodide. Reaction:

$$HgO(s) + 4I^- + H_2O \rightarrow HgI_4^{2-} + 2OH^-$$

Calculate the percentage of HgO in the sample if titration of the liberated hydroxide required 42.59 mL of 0.1178 M HCl.

*14-23. The formaldehyde content of a pesticide preparation was determined by weighing 0.3124 g of the liquid sample into a flask containing 50.0 mL of 0.0996 M NaOH and 50 mL of 3% H_2O_2. Upon heating, the following reaction took place:

$$OH^- + HCHO + H_2O_2 \rightarrow HCOO^- + 2H_2O$$

After cooling, the excess base was titrated with 23.3 mL of 0.05250 M H_2SO_4. Calculate the percentage of HCHO (30.026 g/mol) in the sample.

14-24. The benzoic acid extracted from 106.3 g of catsup required a 14.76-mL titration with 0.0514 M NaOH. Express the results of this analysis in terms of percent sodium benzoate (144.10 g/mol).

*14-25. The active ingredient in Antabuse, a drug used for the treatment of chronic alcoholism, is tetraethylthiuram disulfide,

$$(C_2H_5)_2N\overset{\overset{S}{\|}}{C}SS\overset{\overset{S}{\|}}{C}N(C_2H_5)_2$$

(296.54 g/mol). The sulfur in a 0.4329-g sample of an Antabuse preparation was oxidized to SO_2, which was absorbed in H_2O_2 to give H_2SO_4. The acid was titrated with 22.13 mL of 0.03736 M base. Calculate the percentage of active ingredient in the preparation.

14-26. A 25.00-mL sample of a household cleaning solution was diluted to 250.0 mL in a volumetric flask. A 50.00-mL aliquot of this solution required 40.38 mL of 0.2506 M HCl to reach a bromocresol green end point. Calculate the weight/volume percentage of NH_3 in the sample. (Assume that all the alkalinity results from the ammonia.)

*14-27. A 0.1401-g sample of a purified carbonate was dissolved in 50.00 mL of 0.1140 M HCl and boiled to eliminate CO_2. Back-titration of the excess HCl required 24.21 mL of 0.09802 M NaOH. Identify the carbonate.

14-28. A dilute solution of an unknown weak acid required a 28.62-mL titration with 0.1084 M NaOH to reach a phenolphthalein end point. The titrated solution was evaporated to dryness. Calculate the equivalent weight (the equivalent weight of an acid or base is the mass of the acid or base in grams that reacts with or contains one mole of protons) of the acid if the sodium salt was found to weigh 0.2110 g.

*14-29. A 3.00-L sample of urban air was bubbled through a solution containing 50.0 mL of 0.0116 M $Ba(OH)_2$, which caused the CO_2 in the sample to precipitate as $BaCO_3$. The excess base was back-titrated to a phenolphthalein end point with 23.6 mL of 0.0108 M HCl. Calculate the parts per million of CO_2 in the air (that is, mL $CO_2/10^6$ mL air); use 1.98 g/L for the density of CO_2.

14-30. Air was bubbled at a rate of 30.0 L/min through a trap containing 75 mL of 1% H_2O_2 ($H_2O_2 + SO_2 \rightarrow H_2SO_4$). After 10.0 min, the H_2SO_4 was titrated with 11.1 mL of 0.00204 M NaOH. Calculate the parts per million of SO_2 (that is, mL $SO_2/10^6$ mL air) if the density of SO_2 is 0.00285 g/mL.

*14-31. The digestion of a 0.1417-g sample of a phosphorus containing compound in a mixture of HNO_3 and H_2SO_4 resulted in the formation of CO_2, H_2O, and H_3PO_4. Addition of ammonium molybdate yielded a solid having the composition $(NH_4)_3PO_4 \cdot 12MoO_3$ (1876.3 g/mol). This precipitate was filtered, washed, and dissolved in 50.00 mL of 0.2000 M NaOH:

$$(NH_4)_3PO_4 \cdot 12MoO_3(s) + 26OH^- \rightarrow$$
$$HPO_4^{2-} + 12MoO_4^{2-} + 14H_2O + 3NH_3(g)$$

After the solution was boiled to remove the NH_3, the excess NaOH was titrated with 14.17

mL of 0.1741 M HCl to a phenolphthalein end point. Calculate the percentage of phosphorus in the sample.

*14-32. A 0.8160-g sample containing dimethylphthalate, $C_6H_4(COOCH_3)_2$ (194.19 g/mol), and unreactive species was refluxed with 50.00 mL of 0.1031 M NaOH to hydrolyze the ester groups (this process is called saponification).

$$C_6H_4(COOCH_3)_2 + 2OH^- \rightarrow$$
$$C_6H_4(COO)_2^{2-} + 2CH_3OH$$

After the reaction was complete, the excess NaOH was back-titrated with 24.27 mL of 0.1644 M HCl. Calculate the percentage of dimethylphthalate in the sample.

*14-33. Neohetramine, $C_{16}H_{21}ON_4$ (285.37 g/mol), is a common antihistamine. A 0.1247-g sample containing this compound was analyzed by the Kjeldahl method. The ammonia produced was collected in H_3BO_3; the resulting $H_2BO_3^-$ was titrated with 26.13 mL of 0.01477 M HCl. Calculate the percentage of neohetramine in the sample.

14-34. The Merck Index indicates that 10 mg of guanidine, CH_5N_3, may be administered for each kilogram of body weight in the treatment of myasthenia gravis. The nitrogen in a four-tablet sample that weighed a total of 7.50 g was converted to ammonia by a Kjeldahl digestion, followed by distillation into 100.0 mL of 0.1750 M HCl. The analysis was completed by titrating the excess acid with 11.37 mL of 0.1080 M NaOH. How many of these tablets represent a proper dose for patients that weigh (a) 100 lbs, (b) 150 lbs, and (c) 200 lbs?

*14-35. A 1.047-g sample of canned tuna was analyzed by the Kjeldahl method; 24.61 mL of 0.1180 M HCl were required to titrate the liberated ammonia. Calculate the percentage of nitrogen in the sample.

14-36. Calculate the grams of protein in a 6.50-oz can of tuna in Problem 14-35 (1 oz = 28.3 g).

*14-37. A 0.5843-g sample of a plant food preparation was analyzed for its nitrogen content by the Kjeldahl method, the liberated NH_3 being collected in 50.00 mL of 0.1062 M HCl. The excess acid required an 11.89 mL back-titration with 0.0925 M NaOH. Express the results of this analysis in terms of percent

  *(a) N.   **(b)** urea, H_2NCONH_2.
  *(c) $(NH_4)_2SO_4$. **(d)** $(NH_4)_3PO_4$.

14-38. A 0.9092-g sample of a wheat flour was analyzed by the Kjeldahl procedure. The ammonia formed was distilled into 50.00 mL of 0.05063 M HCl; a 7.46-mL back-titration with 0.04917 M NaOH was required. Calculate the percentage of protein in the flour.

*14-39. A 1.219-g sample containing $(NH_4)_2SO_4$, NH_4NO_3, and nonreactive substances was diluted to 250 mL in a volumetric flask. A 50.00-mL aliquot was made basic with strong alkali, and the liberated NH_3 was distilled into 30.00 mL of 0.08421 M HCl. The excess HCl required 10.17 mL of 0.08802 M NaOH. A 25.00-mL aliquot of the sample was made alkaline after the addition of Devarda's alloy, and the NO_3^- was reduced to NH_3. The NH_3 from both NH_4^+ and NO_3^- was then distilled into 30.00 mL of the standard acid and back-titrated with 14.16 mL of the base. Calculate the percentage of $(NH_4)_2SO_4$ and NH_4NO_3 in the sample.

*14-40. A 1.217-g sample of commercial KOH contaminated by K_2CO_3 was dissolved in water, and the resulting solution was diluted to 500.0 mL. A 50.00-mL aliquot of this solution was treated with 40.00 mL of 0.05304 M HCl and boiled to remove CO_2. The excess acid consumed 4.74 mL of 0.04983 M NaOH (phenolphthalein indicator). An excess of neutral $BaCl_2$ was added to another 50.00-mL aliquot to precipitate the carbonate as $BaCO_3$. The solution was then titrated with 28.56 mL of the acid (also to a phenolphthalein end point). Calculate the percentage of KOH, K_2CO_3, and H_2O in the sample, assuming that these are the only compounds present.

14-41. A 0.5000-g sample containing $NaHCO_3$, Na_2CO_3, and H_2O was dissolved and diluted to 250.0 mL. A 25.00-mL aliquot was then boiled with 50.00 mL of 0.01255 M HCl. After cooling, the excess acid in the solution required 2.34 mL of 0.01063 M NaOH when titrated to a phenolphthalein end point. A second 25.00-mL aliquot was then treated with an excess of $BaCl_2$ and 25.00 mL of the base; precipitation of all the carbonate resulted, and 7.63 mL of the HCl was required to titrate the excess base. Calculate the composition of the mixture.

*14-42. Calculate the volume of 0.06122 M HCl needed to titrate

  (a) 10.00, 15.00, 25.00, and 40.00 mL of 0.05555 M Na_3PO_4 to a thymolphthalein end point.

  (b) 10.00, 15.00, 20.00, and 25.00 mL of 0.05555 M Na_3PO_4 to a bromocresol green end point.

(c)   20.00, 25.00, 30.00, and 40.00 mL of a solution that is 0.02102 M in Na_3PO_4 and 0.01655 M in Na_2HPO_4 to a bromocresol green end point.

(d)   15.00, 20.00, 35.00, and 40.00 mL of a solution that is 0.02102 M in Na_3PO_4 and 0.01655 M in NaOH to a thymolphthalein end point.

14-43.   Calculate the volume of 0.07731 M NaOH needed to titrate

(a)   25.00 mL of a solution that is 0.03000 M in HCl and 0.01000 M in H_3PO_4 to a bromocresol green end point.

(b)   the solutions in part (a) to a thymolphthalein end point.

(c)   10.00, 20.00, 30.00, and 40.00 mL of 0.06407 M NaH_2PO_4 to a thymolphthalein end point.

(d)   20.00, 25.00, and 30.00 mL of a solution that is 0.02000 M in H_3PO_4 and 0.03000 M in NaH_2PO_4 to a thymolphthalein end point.

*14-44.   A series of solutions containing NaOH, Na_2CO_3, and $NaHCO_3$, alone or in compatible combination, was titrated with 0.1202 M HCl. Tabulated below are the volumes of acid needed to titrate 25.00-mL portions of each solution to a (1) phenolphthalein and (2) bromocresol green end point. Use this information to deduce the composition of the solutions. In addition, calculate the number of milligrams of each solute per milliliter of solution.

	(1)	(2)
(a)	22.42	22.44
(b)	15.67	42.13
(c)	29.64	36.42
(d)	16.12	32.23
(e)	0.00	33.33

14-45.   A series of solutions containing NaOH, Na_3AsO_4, and Na_2HAsO_4, alone or in compatible combination, was titrated with 0.08601 M HCl. Tabulated below are the volumes of acid needed to titrate 25.00-mL portions of each solution to a (1) phenolphthalein and (2) bromocresol green end point. Use this information to deduce the composition of the solutions. In addition, calculate the number of milligrams of each solute per milliliter of solution.

	(1)	(2)
(a)	0.00	18.15
(b)	21.00	28.15
(c)	19.80	39.61
(d)	18.04	18.03
(e)	16.00	37.37

Chapter 15

Complexation and Precipitation Titrations: Taking Advantage of Complexing and Precipitating Agents

Complex-formation reagents are widely used in analytical chemistry. One of the first uses of these reagents was for titrating cations, the major topic of this chapter. In addition, many complexes are colored or absorb ultraviolet radiation; the formation of these complexes is often the basis for spectrophotometric determinations (see Chapter 23). Some complexes are sparingly soluble and can be used in gravimetric analysis (see Chapter 8) or for precipitation titrations. Complexes are also widely used for extracting cations from one solvent to another and for dissolving insoluble precipitates. The most useful complex forming reagents are organic compounds that contain several electron-donor groups that form multiple covalent bonds with metal ions. Inorganic complexing agents are also used to control solubility, form colored species, or form precipitates.

15A FORMING COMPLEXES

Most metal ions react with electron-pair donors to form coordination compounds or complexes. The donor species, or *ligand,* must have at least one pair of unshared electrons available for bond formation. Water, ammonia, and halide ions are common inorganic ligands. In fact, most metal ions in aqueous solution actually exist as aquo complexes. Copper(II), for example, is readily complexed in aqueous solution by water molecules to form species such as $Cu(H_2O)_4^{2+}$. We often simplify such complexes in chemical equations by writing the metal ion as if it were uncomplexed Cu^{2+}. We should remember, however, that such ions are actually aquo complexes in aqueous solution.

The number of covalent bonds that a cation tends to form with electron donors is its coordination number. Typical values for coordination numbers are two, four, and six. The species formed as a result of coordination can be electri-

> A ligand is an ion or a molecule that forms a covalent bond with a cation or a neutral metal atom by donating a pair of electrons that are then shared by the two.

cally positive, neutral, or negative. For example, copper(II), which has a coordination number of four, forms a cationic ammine complex, $Cu(NH_3)_4^{2+}$; a neutral complex with glycine, $Cu(NH_2CH_2COO)_2$; and an anion complex with chloride ion, $CuCl_4^{2-}$.

Titrimetric methods based on complex formation (sometimes called *complexometric methods*) have been used for more than a century. The truly remarkable growth in their analytical application, based on a particular class of coordination compounds called *chelates*, began in the 1940s. A chelate is produced when a metal ion coordinates with two or more donor groups of a single ligand to form a five- or six-membered heterocyclic ring. The copper complex of glycine mentioned earlier is an example. Here, copper bonds to both the oxygen of the carboxyl group and the nitrogen of the amine group:

Chelate, pronounced kee'late, is derived from the Greek word for claw.

Dentate (Latin) means having toothlike projections.

A ligand that has a single donor group, such as ammonia, is called *unidentate* (single-toothed), whereas one such as glycine, which has two groups available for covalent bonding, is called *bidentate*. *Tridentate, tetradentate, pentadentate,* and *hexadentate* chelating agents are also known.

18-crown-6   dibenzo-18-crown-6   cryptand 2,2,2

The selectivity of a ligand for one metal ion over another relates to the stability of the complexes formed. The higher the formation constant of a metal-ligand complex, the better the selectivity of the ligand for the metal relative to similar complexes formed with other metals.

Another important type of complex, a *macrocycle*, is formed between a metal ion and a cyclic organic compound. Macrocycles contain nine or more atoms in the cycle and include at least three heteroatoms, usually oxygen, nitrogen, or sulfur. Some macrocyclic compounds form three-dimensional cavities that can just accommodate appropriately sized metal ions. Selectivity occurs to a large extent because of the size and shape of the cycle or cavity relative to that of the metal, although the nature of the heteroatoms and their electron densities, the compatibility of the donor atoms with the metal, and several other factors also play important roles.

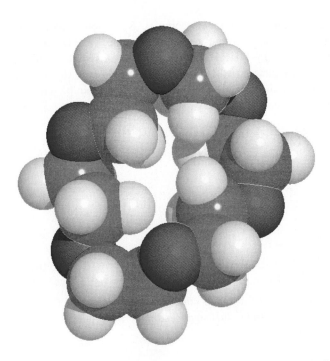

Molecular model of 18-crown-6. This crown ether can form strong complexes with alkali metal ions. The formation constants of the Na^+, K^+ and Rb^+ complexes with 18-crown-6 are in the 10^5-10^6 range in organic solvents.

15A-1   Producing Soluble Complexes

Complexation reactions involve a metal ion M reacting with a ligand L to form a complex ML, as shown in Equation 15-1:

$$M + L \rightleftharpoons ML \tag{15-1}$$

where the charges on the ions have been omitted so as to be general. Complexation reactions occur in a stepwise fashion, and the reaction above is often followed by additional reactions:

$$ML + L \rightleftharpoons ML_2 \tag{15-2}$$

$$ML_2 + L \rightleftharpoons ML_3 \tag{15-3}$$

$$\vdots \qquad \vdots$$

$$ML_{n-1} + L \rightleftharpoons ML_n \tag{15-4}$$

Unidentate ligands invariably add in a series of steps as shown above. With multidentate ligands, the maximum coordination number of the cation may be satisfied with only one or a few added ligands. For example, Cu(II), with a maximum coordination number of 4, can form complexes with ammonia that have the formulas $Cu(NH_3)^{2+}$, $Cu(NH_3)_2^{2+}$, $Cu(NH_3)_3^{2+}$, and $Cu(NH_3)_4^{2+}$. With the bidentate ligand glycine (gly), the only complexes that form are $Cu(gly)^{2+}$ and $Cu(gly)_2^{2+}$.

The equilibrium constants for complex formation reactions are generally written as formation constants, as discussed in Chapter 3. Thus, each of Equations 15-1 to 15-4 is associated with a stepwise formation constant K_1 through K_4. For example, $K_1 = [ML]/[M][L]$, $K_2 = [ML_2]/[ML][L]$, and so on. Reactions that are the sum of two or more steps can also be written with overall formation constants designated by the symbol β_n. Thus,

$$M + 2L \rightleftharpoons ML_2 \qquad \beta_2 = \frac{[ML_2]}{[M][L]^2} = K_1 K_2 \tag{15-5}$$

$$M + 3L \rightleftharpoons ML_3 \qquad \beta_3 = \frac{[ML_3]}{[M][L]^3} = K_1 K_2 K_3 \tag{15-6}$$

$$\vdots \qquad\qquad \vdots$$

$$M + nL \rightleftharpoons ML_n \qquad \beta_n = \frac{[ML_n]}{[M][L]^n} = K_1 K_2 \cdots K_n \tag{15-7}$$

As shown, the overall formation constants are products of the stepwise formation constants for the individual steps leading to the product.

15A-2   Forming Insoluble Species

In the cases discussed in Section 15A-1, the complexes formed are soluble in solution. The addition of ligands to a metal ion, however, may result in insoluble species, such as the familiar nickel-dimethylglyoxime precipitate. In many cases, the intermediate uncharged complexes in the stepwise formation scheme may be sparingly soluble, whereas the addition of more ligand molecules may result in soluble species. For example, adding Cl^- to Ag^+ first results in the insoluble AgCl precipitate. Addition of a large excess of Cl^- produces soluble species $AgCl_2^-$, $AgCl_3^{2-}$, and $AgCl_4^{3-}$.

In contrast to complexation equilibria, which are most often treated as formation reactions, solubility equilibria are usually treated as dissociation reactions, as discussed in Chapter 4. In general, for a sparingly soluble salt M_xA_y in a saturated solution, we can write

$$M_xA_y(s) \rightleftharpoons xM^{y+}(aq) + yA^{x-}(aq) \qquad K_{sp} = [M^{y+}]^x[A^{x-}]^y \tag{15-8}$$

where K_{sp} is the solubility product. Hence, for BiI_3, the solubility product is written $K_{sp} = [Bi^{3+}][I^-]^3$.

The formation of soluble complexes can be used to control the concentration of free metal ions in solution and thus control their reactivity. For example, we can prevent a metal ion from precipitating or taking part in another reaction by forming a stable complex, which decreases the free metal ion concentration. Example 15-1 shows how formation of complexes can be used in separating Zn from Ni. Another use of complexing agents is to dissolve an already formed precipitate. For example, AgBr is sparingly soluble with $K_{sp} = 5.0 \times 10^{-13}$. AgBr can be dissolved if sufficient ammonia is added, however, because the Ag^+ ions are complexed to form the soluble species $Ag(NH_3)^+$ and $Ag(NH_3)_2^+$.

Example 15-1

It is desired to separate Ni and Zn by precipitation as the sulfides NiS and ZnS. The solubility products are $(K_{sp})_{NiS} = 4.0 \times 10^{-20}$ and $(K_{sp})_{ZnS} = 3.0 \times 10^{-25}$. If excess KCN is added to a solution initially containing 0.01 M each of Ni^{2+} and Zn^{2+} and the volume remains approximately constant, both ions form the $M(CN)_4^{2-}$ complex almost exclusively. If KCN is added until the $[CN^-]$ is 1.0 M and the $[S^{2-}]$ is maintained at 0.5 M, can we precipitate ZnS while leaving most of the Ni in solution? Adding CN^- causes the formation of the Zn^{2+} and Ni^{2+} complexes according to

$$Ni^{2+} + 4CN^- \rightleftharpoons Ni(CN)_4^{2-} \qquad \beta_4 = \frac{[Ni(CN)_4^{2-}]}{[Ni^{2+}][CN^-]^4} = 1.6 \times 10^{30}$$

$$Zn^{2+} + 4CN^- \rightleftharpoons Zn(CN)_4^{2-} \qquad \beta_4 = \frac{[Zn(CN)_4^{2-}]}{[Zn^{2+}][CN^-]^4} = 4.2 \times 10^{19}$$

From mass balance, $(c_T)_{Ni} = 0.01$ M $= [Ni(CN)_4^{2-}] + [Ni^{2+}]$ and $(c_T)_{Zn} = 0.01$ M $= [Zn(CN)_4^{2-}] + [Zn^{2+}]$. Since the formation constants of the complexes are so large, we can estimate that most of the metals will be in the form of the complex ion. Thus, $[Ni(CN)_4^{2-}] \approx 0.01$ M and $[Zn(CN)_4^{2-}] \approx 0.01$ M. We can then estimate the free $[Ni^{2+}]$ and $[Zn^{2+}]$ concentrations from the β_4 values:

$$[Ni^{2+}] = \frac{[Ni(CN)_4^{2-}]}{\beta_4[CN^-]^4} = \frac{0.01}{1.6 \times 10^{30}(1.0)^4} = 6.25 \times 10^{-33}$$

and
$$[Zn^{2+}] = \frac{[Zn(CN)_4^{2-}]}{\beta_4[CN^-]^4} = \frac{0.01}{4.2 \times 10^{19}(1.0)^4} = 2.38 \times 10^{-22}$$

Now the product of the ion concentrations for NiS ($[Ni^{2+}][S^{2-}] = 6.25 \times 10^{-33} \times 0.5 = 3.1 \times 10^{-33}$) is seen to be much smaller than K_{sp} (4.0×10^{-20}) so that the Ni^{2+} remains in solution. The product of the ion concentrations for ZnS ($[Zn^{2+}][S^{2-}] = 2.38 \times 10^{-22} \times 0.5 = 1.2 \times 10^{-22}$), however, exceeds K_{sp} (3.0×10^{-25}) so that ZnS readily precipitates, which then allows a good separation of Zn from Ni.

15A-3   Ligands That Can Protonate

Example 15-1 showed that side reactions of the metal can influence solubility. Side reactions of the *ligands* can also complicate complexation or solubility equilibria. Such side reactions make it possible to exert even more control over the complexes that form or the solubility of a sparingly soluble species. One of the most common side reactions is of a ligand that can protonate; that is, the ligand is a weak acid.

Complexation with Protonating Ligands

We will first consider the case of the formation of soluble complexes between the metal M and the ligand L, where the ligand L is a weak acid. For completeness, we will assume that L is the conjugate base of a polyprotic acid and forms $HL, H_2L, \ldots, H_nL$, where again the charges have been omitted for generality. Adding acid to a solution containing M and L reduces the concentration of free L available to complex with M and thus decreases the effectiveness of L as a complexing agent (Le Châtelier's principle). For example, ferric ions (Fe^{3+}) form complexes with oxalate ($C_2O_4^{2-}$, abbreviated Ox^{2-}) with formulas $(FeOx)^+$, $(FeOx_2)^-$, and $(FeOx_3)^{3-}$. Oxalate can protonate to form HOx^- and H_2Ox. In basic solution, where most of the oxalate is present as Ox^{2-} before complexation with Fe^{3+}, the ferric/oxalate complexes are very stable. Adding acid, however, protonates the oxalate, which in turn causes dissociation of the ferric complexes.

For a diprotic acid, like oxalic acid, the fraction of the total oxalate containing species in any given form, Ox^{2-}, HOx^-, and H_2Ox, is given by an alpha value (recall Section 13F). Since

$$c_T = [H_2Ox] + [HOx^-] + [Ox^{2-}] \tag{15-9}$$

we can write the alpha values α_0, α_1, and α_2 as

$$\alpha_0 = \frac{[H_2Ox]}{c_T} = \frac{[H^+]^2}{[H^+]^2 + K_{a1}[H^+] + K_{a1}K_{a2}} \tag{15-10}$$

$$\alpha_1 = \frac{[HOx^-]}{c_T} = \frac{K_{a1}[H^+]}{[H^+]^2 + K_{a1}[H^+] + K_{a1}K_{a2}} \tag{15-11}$$

$$\alpha_2 = \frac{[Ox^{2-}]}{c_T} = \frac{K_{a1}K_{a2}}{[H^+]^2 + K_{a1}[H^+] + K_{a1}K_{a2}} \tag{15-12}$$

Since we are interested in the free oxalate concentration, we will be most concerned with the α_2 value. From Equation 15-12 we can write

$$[Ox^{2-}] = c_T\alpha_2 \tag{15-13}$$

Note that as the solution gets more acidic, the first two terms in the denominator of Equation 15-12 dominate and α_2 and the free oxalate concentration decrease. You may recall that the alpha value is a fraction and varies between 0 and 1. On the other hand, when the solution is very basic, α_2 becomes nearly unity and $[Ox^{2-}] \approx c_T$, indicating that nearly all the oxalate is in the Ox^{2-} form in basic solution.

Accounting for pH Effects with Conditional Formation Constants

To take into account the effect of pH on the free ligand concentration in a complexation reaction, it is useful to introduce a *conditional* or *effective* formation constant. Such constants are pH-dependent equilibrium constants that apply at a single pH only. For the reaction of Fe^{3+} with oxalate, for example, we can write the formation constant K_1 for the first complex as

$$K_1 = \frac{[(FeOx)^+]}{[Fe^{3+}][Ox^{2-}]} = \frac{[(FeOx)^+]}{[Fe^{3+}]\alpha_2 c_T} \tag{15-14}$$

At a particular pH value, α_2 is constant, and we can combine K_1 and α_2 to yield a new conditional constant K_1':

$$K_1' = \alpha_2 K_1 = \frac{[(FeOx)^+]}{[Fe^{3+}]c_T} \qquad (15\text{-}15)$$

The use of conditional constants greatly simplifies calculations because c_T is often known or readily computed, whereas the free ligand concentration is not as easily determined. The overall formation constants, beta values, for the higher complexes, $(FeOx_2)^-$ and $(FeOx_3)^{3-}$, can also be written as conditional constants.

15B   TITRATIONS WITH INORGANIC COMPLEXING AGENTS

Complexation reactions have many uses in analytical chemistry, but their classical application is in complexometric titrations. Here, a metal ion reacts with a suitable ligand to form a complex, and the equivalence point is determined by an indicator or a suitable instrumental method. The formation of soluble inorganic complexes is not widely used for titrations as discussed later, but the formation of precipitates, particularly with silver nitrate as the titrant, is the basis for many important determinations.

15B-1   Complexation Titrations

The progress of a complexometric titration is generally illustrated by a titration curve, which is usually a plot of pM $= -\log[M]$ as a function of the volume of titrant added. Most often in complexometric titrations the ligand is the titrant and the metal ion the analyte, although occasionally the reverse is true. Many precipitation titrations, as discussed in Section 15B-2, use the metal ion as the titrant. Most simple inorganic ligands are unidentate, which can lead to low complex stability and indistinct titration end points. As titrants, multidentate ligands, particularly those having four or six donor groups, have two advantages over their unidentate counterparts. First, they generally react more completely with cations and thus provide sharper end points. Second, they ordinarily react with metal ions in a single-step process, whereas complex formation with unidentate ligands usually involves two or more intermediate species (recall Equations 15-1 to 15-4).

The advantage of a single-step reaction is illustrated by the titration curves shown in Figure 15-1. Each of the titrations involves a reaction that has an overall equilibrium constant of 10^{20}. Curve A is derived for a reaction in which a metal ion M with a coordination number of 4 reacts with a tetradentate ligand D to form the complex MD (we have again omitted the charges on the two reactants for convenience). Curve B is for the reaction of M with a hypothetical bidentate ligand B to give MB_2 in two steps. The formation constant for the first step is 10^{12} and for the second 10^8. Curve C involves a unidentate ligand, A, that forms MA_4 in four steps with successive formation constants of 10^8, 10^6, 10^4, and 10^2. These curves demonstrate that a much sharper end point is obtained with a reaction that takes place in a single step. For this reason, multidentate ligands are ordinarily preferred for complexometric titrations.

The most widely used complexometric titration employing a unidentate ligand is the titration of cyanide with silver nitrate, a method introduced by Justus

Tetradentate or hexadentate ligands are more satisfactory as titrants than ligands with a lesser number of donor groups because their reactions with cations are more complete and because they tend to form 1 : 1 complexes.

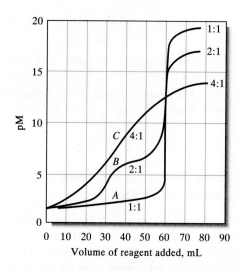

Figure 15-1 Titration curves for complexometric titrations. Titration of 60.0 mL of a solution that is 0.020 M in M with (curve *A*) a 0.020-M solution of the tetradentate ligand D to give MD as the product; (curve *B*) a 0.040-M solution of the bidentate ligand B to give MB_2; and (curve *C*) a 0.080-M solution of the unidentate ligand A to give MA_4. The overall formation constant for each product is 10^{20}.

von Liebig in the 1850s. This method involves the formation of the soluble $Ag(CN)_2^-$ as discussed in Feature 15-1. Other common inorganic complexing agents and their applications are listed in Table 15-1.

15B-2 Precipitation Titrations

Precipitation titrimetry, which is based on reactions that yield ionic compounds of limited solubility, is one of the oldest analytical techniques, dating to the mid-1800s. The slow rate of formation of most precipitates, however, limits the number of precipitating agents that can be used in titrations to a handful. We will limit our discussion here to the most widely used and important precipitating

Table 15-1

Typical Inorganic Complex-Forming Titrations

Titrant	Analyte	Remarks
$Hg(NO_3)_2$	Br^-, Cl^-, SCN^-, CN^-, thiourea	Products are neutral Hg(II) complexes; various indicators used
$AgNO_3$	CN^-	Product is $Ag(CN)_2^-$; indicator is I^-; titrate to first turbidity of AgI
$NiSO_4$	CN^-	Product is $Ni(CN)_4^{2-}$; indicator is I^-; titrate to first turbidity of AgI
KCN	Cu^{2+}, Hg^{2+}, Ni^{2+}	Products are $Cu(CN)_4^{2-}$, $Hg(CN)_2$, and $Ni(CN)_4^{2-}$; various indicators used

reagent, silver nitrate, which is used for the determination of the halogens, the halogen-like anions, mercaptans, fatty acids, and several divalent inorganic anions. Titrations with silver nitrate are sometimes called *argentometric titrations*.

Feature 15-1

Determination of Hydrogen Cyanide in Acrylonitrile Plant Streams

Acrylonitrile, $CH_2=CH-C\equiv N$, is an extremely important chemical in the production of polyacrylonitrile. This thermoplastic is drawn out into fine threads and woven into synthetic fabrics such as Orlon, Acrilan, and Creslan. Hydrogen cyanide is an impurity in the plant streams that carry aqueous acrylonitrile. The cyanide is commonly determined by titration with $AgNO_3$. The titration reaction is

$$Ag^+ + 2CN^- \rightarrow Ag(CN)_2^-$$

To determine the end point of the titration, the aqueous sample is mixed with a basic solution of potassium iodide before the titration. Before the equivalence point, cyanide is in excess and all the Ag^+ is complexed. As soon as all the cyanide has been reacted, the first excess of Ag^+ causes a permanent turbidity to appear in the solution because of the formation of the AgI precipitate according to

$$Ag^+ + I^- \rightarrow AgI(s)$$

Example 15-2

Calculate the pAg of the solution during the titration of 50.00 mL of 0.0500 M NaCl with 0.1000 M $AgNO_3$ after the addition of the following volumes of reagent: (a) 0.00 mL, (b) 24.50 mL, (c) 25.00 mL, (d) 25.50 mL.

(a) Because no $AgNO_3$ has been added, $[Ag^+] = 0$ and pAg is indeterminate.

(b) At 24.5 mL, $[Ag^+]$ is very small and cannot be computed from stoichiometric considerations, but $[Cl^-]$ can be obtained readily.

$$[Cl^-] \approx c_{NaCl} = \frac{\text{original no. mmol } Cl^- - \text{no. mol } AgNO_3}{\text{total volume of solution}}$$

$$= \frac{(50.00 \times 0.0500 - 24.50 \times 0.100)}{50.00 + 24.50} = 6.71 \times 10^{-4}$$

$$[Ag^+] = \frac{K_{sp}}{6.71 \times 10^{-4}} = \frac{1.82 \times 10^{-10}}{6.71 \times 10^{-4}} = 2.71 \times 10^{-7}$$

$$pAg = -\log(2.71 \times 10^{-7}) = 6.57$$

(c) This volume corresponds to the equivalence point where $[Ag^+] = [Cl^-]$ and

$$[Ag^+] = \sqrt{K_{sp}} = \sqrt{1.82 \times 10^{-10}} = 1.35 \times 10^{-5}$$

$$pAg = -\log(1.35 \times 10^{-5}) = 4.87$$

$$(d) \ [Ag^+] = c_{AgNO_3} = \frac{(25.50 \times 0.1000 - 50.00 \times 0.0500)}{75.50} = 6.62 \times 10^{-4}$$

$$pAg = -\log(6.62 \times 10^{-4}) = 3.18$$

This titration curve can also be derived from the charge-balance equation that was described in Feature 12-1.

The Shapes of Titration Curves

Titration curves for precipitation reactions are derived in a completely analogous way to the methods described in Section 12B for titrations involving strong acids and strong bases except that the solubility product of the precipitate is substituted for the ion-product constant for water. Example 15-2 illustrates how p-functions are derived for the preequivalence-point region, the postequivalence-point region, and the equivalence point for a typical precipitation titration.

Most indicators for argentometric titrations respond to changes in the concentration of silver ions. As a consequence, titration curves for precipitation reactions usually consist of a plot of pAg versus volume of $AgNO_3$.

Figure 15-2 shows titration curves for chloride ion that have been derived employing the techniques shown in Example 15-2. Note the effect of analyte and reagent concentrations on the magnitude of the change in pAg in the equivalence-point region. The effect here is analogous to that illustrated for acid/base titrations in Figure 12-4. As shown by the shaded region in Figure 15-2, an indicator with a pAg range of 4 to 6 should provide a sharp end point in the titration of the 0.05 M chloride solution. For the more dilute solution, in contrast, the end point would be drawn out over a large enough volume of reagent to make accurate establishment of the end point impossible. That is, the color change would begin at about 24 mL and be complete at about 26 mL.

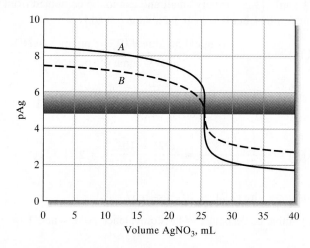

Figure 15-2   Effect of titrant concentration on precipitation titration curves. Curve A shows 50.00 mL of 0.0500 M NaCl with 0.1000 M $AgNO_3$, and curve B shows 50.00 mL of 0.00500 M NaCl with 0.01000 M $AgNO_3$. Note the increased sharpness of the break for the more concentrated solution, A.

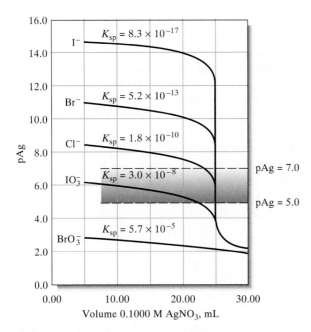

Figure 15-3   Effect of reaction completeness on precipitation titration curves. For each curve, 50.00 mL of a 0.0500 M solution of the anion was titrated with 0.1000 M $AgNO_3$. Note that smaller values of K_{sp} give much sharper breaks at the end point.

Figure 15-3 illustrates the effect of solubility product on the sharpness of the end point in titrations with 0.1 M silver nitrate. Clearly the change in pAg at the equivalence point becomes greater as the solubility products become smaller, that is, as the reaction between the analyte and silver nitrate becomes more complete. By careful choice of indicator — one that changes color in the region of pAg from 4 to 6 — titration of chloride ions should be possible with a minimal titration error. Note that ions forming precipitates with solubility products much larger than about 10^{-10} do not yield satisfactory end points.

Signaling the End Point for Argentometric Titrations

Three types of end points are encountered in titrations with silver nitrate: (1) chemical, (2) potentiometric, and (3) amperometric. Some chemical indicators are described with specific methods later in this section. Potentiometric end points are obtained by measuring the potential between a silver electrode and a reference electrode whose potential is constant and independent of the added reagent. Titration curves similar to those shown in Figures 15-2 and 15-3 are obtained. Potentiometric end points are discussed in Section 17E. To obtain an amperometric end point, the current generated between a pair of silver microelectrodes in the solution of the analyte is measured and plotted as a function of reagent volume. Amperometric methods are considered in Section 20E.

The end point produced by a chemical indicator usually consists of a color change or, occasionally, the appearance or disappearance of turbidity in the solution being titrated. The requirements for an indicator for a precipitation titration are analogous to those for an indicator for a neutralization titration: (1) the color change should occur over a limited range in the p-function of the reagent or the

A useful relationship can be derived by taking the negative logarithm of both sides of a solubility-product expression. Thus, for silver chloride,

$$-\log K_{sp} = -\log([Ag^+][Cl^-])$$
$$= -\log[Ag^+] - \log[Cl^-]$$

$$pK_{sp} = pAg + pCl$$

Compare this expression with the one in marginal note in Chapter 4 on page 87.

analyte, and (2) the color change should take place within the steep portion of the titration curve for the analyte. For example, turning to Figure 15-3, we see that the indicator shown should provide a satisfactory end point for the titration of iodide and bromide ions, but not chloride. An indicator with a pAg range of 6.0 to 4.0, however, should be satisfactory for chloride ions (Figure 15-2). Three indicators that have found extensive use for argentometric titrations are discussed in the sections that follow.

Formation of a Colored Precipitate: The Mohr Method. Sodium chromate can serve as an indicator for the argentometric determination of chloride, bromide, and cyanide ions by reacting with silver ion to form a brick-red silver chromate (Ag_2CrO_4) precipitate in the equivalence-point region. The reactions involved in the determination of chloride and bromide (X^-) are

titration reaction: $Ag^+ + X^- \rightleftharpoons AgX(s)$
                                            white

indicator reaction: $2Ag^+ + CrO_4^{2-} \rightleftharpoons Ag_2CrO_4(s)$
                                                 red

The solubility of silver chromate is several times greater than that of silver chloride or silver bromide. For the titration of chloride and bromide, it should be possible to select a chromate ion concentration such that the Ag_2CrO_4 appears at a silver ion concentration corresponding to the equivalence point of the titration, as given in Spreadsheet Example 15-3. In practice, however, the high chromate ion concentration needed imparts such an intense yellow color to the solution that it obscures the red color of the silver chromate. For this reason, lower concentrations of chromate ion are generally used, and, as a consequence, excess silver nitrate is required before precipitation begins. An additional excess of the reagent must also be added to produce enough silver chromate to be seen. These two factors create a positive systematic error in the Mohr method that becomes significant in magnitude at reagent concentrations lower than about 0.1 M. A good way to correct for this error is to standardize the silver nitrate solution against primary-standard-grade sodium chloride using the same conditions that are to be used in the analysis. This technique not only compensates for the overconsumption of reagent but also for the acuity of the analyst in detecting the appearance of the color.

The Mohr titration must be carried out at a pH of 7 to 10 because chromate ion is the conjugate base of the weak chromic acid. Consequently, in acidic solutions, where the pH is less than 7, the chromate ion concentration is too low to produce the precipitate at the equivalence point. Normally, a suitable pH is achieved by saturating the analyte solution with sodium hydrogen carbonate.

Adsorption Indicators: The Fajans Method. An adsorption indicator is an organic compound that tends to be adsorbed onto the surface of the solid in a precipitation titration. Ideally, the adsorption (or desorption) occurs near the equivalence point and results not only in a color change but also in a transfer of color from the solution to the solid (or the reverse).

Fluorescein is a typical adsorption indicator useful for the titration of chloride ion with silver nitrate. In aqueous solution, fluorescein partially dissociates into hydronium ions and negatively charged fluoresceinate ions that are

The Mohr method was first described in 1865 by K. F. Mohr, a German pharmaceutical chemist who did much pioneering work in the development of titrimetry. Because Cr(VI) is a carcinogen, the Mohr method is no longer in common use.

Adsorption indicators were first described by K. Fajans, a Polish chemist, in 1926. His name is pronounced *Fay yahns.*

Example 15-3   **Spreadsheet Exercise**

	A	B	C	D	E	F	G
1	**Spreadsheet to calculate chromate concentration needed to precipitate**						
2	Ag_2CrO_4 at the endpoint of titrations of chloride, bromide, and cyanide						
3	K_{sp} for Ag_2CrO_4	1.2E-12					
4							
5		$(K_{sp})_{AgX}$	$[Ag^+]_{e.p.} = \sqrt{K_{sp}}$		$[CrO_4^{2-}] = (K_{sp})_{Ag_2CrO_4} / [Ag^+]^2$		
6	Ion						
7	Chloride	1.82E-10	1.35E-05		0.0066	M	
8	Bromide	5.60E-13	7.48E-07		2.1	M	
9	Cyanide	2.20E-16	1.48E-08		5.5E+03	M (Impossibly high)	
10			**Documentation**				
11			Cell C7=SQRT(B7)		Cell E7=$B$3/C7^2		
12			Cell C8=SQRT(B8)		Cell E8=$B$3/C8^2		
13			Cell C9=SQRT(B9)		Cell E9=$B$3/C9^2		

yellow-green. The fluoresceinate ion forms an intensely red silver salt. Whenever this dye is used as an indicator, however, *its concentration is never large enough to precipitate as silver fluoresceinate.*

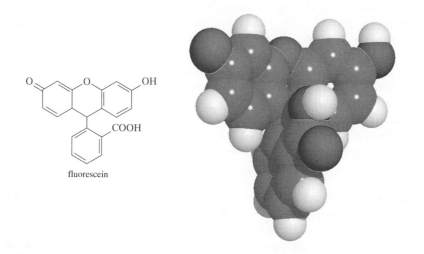

fluorescein

Fluorescein structural formula and molecular model. This strongly fluorescent dye has many applications. It is widely used to study retinal circulation and various diseases involving the retina. The technique is known as fluorescein angiography. Fluorescein can be bound to DNA and other proteins and its fluorescence used as probe of these molecules and their interactions. Fluorescein is also used for water tracing to provide information on the contamination of underground wells. It has also been used as a laser dye.

In the early stages of the titration of chloride ion with silver nitrate, the colloidal silver chloride particles are negatively charged because of adsorption of excess chloride ions (see Section 8A). The dye anions are repelled from this surface by electrostatic repulsion and impart a yellow-green color to the solution. Beyond the equivalence point, however, the silver chloride particles strongly adsorb silver ions and thereby acquire a positive charge. Fluoresceinate anions are now attracted *into the counter-ion layer* that surrounds each colloidal silver chloride particle. The net result is the appearance of the red color of silver fluoresceinate *in the surface layer of the solution surrounding the solid.* It is important to emphasize that the color change is an *adsorption* (and not a precipitation) process because the solubility product of the silver fluoresceinate is never exceeded. The adsorption is reversible.

Titrations involving adsorption indicators are rapid, accurate, and reliable. Their application, however, is limited to the relatively few precipitation reactions in which a colloidal precipitate is formed rapidly.

Forming a Colored Complex: The Volhard Method.   In the Volhard method, silver ions are titrated with a standard solution of thiocyanate ion:

$$Ag^+ + SCN^- \rightleftharpoons AgSCN(s)$$

Iron(III) serves as the indicator. The solution turns red with the first slight excess of thiocyanate ion:

$$Fe^{3+} + SCN^- \rightleftharpoons \underset{red}{Fe(SCN)^{2+}} \quad K_f = \frac{[Fe(SCN)^{2+}]}{[Fe^{3+}][SCN^-]} = 1.05 \times 10^3$$

The titration must be carried out in acidic solution to prevent precipitation of iron(III) as the hydrated oxide.

The indicator concentration is not critical in the Volhard titration. In fact, calculations demonstrate that a titration error of one part in a thousand or less is possible if the iron(III) concentration is held between 0.002 and 1.6 M. In practice, it is found that an indicator concentration greater than 0.2 M imparts sufficient color to the solution to make detection of the thiocyanate complex difficult because of the yellow color of Fe^{3+}. Therefore, lower concentrations (usually about 0.01 M) of iron (III) are preferred.

The most important application of the Volhard method is for the indirect determination of halide ions. A measured excess of standard silver nitrate solution is added to the sample, and the excess silver ion is determined by back-titration with a standard thiocyanate solution. The strong acid environment required for the Volhard procedure represents a distinct advantage over other titrimetric methods of halide analysis because such ions as carbonate, oxalate, and arsenate (which form slightly soluble silver salts in neutral media but not in acidic media) do not interfere.

Silver chloride is more soluble than silver thiocyanate. As a consequence, in chloride determinations by the Volhard method, the reaction

$$AgCl(s) + SCN^- \rightleftharpoons AgSCN(s) + Cl^-$$

occurs to a significant extent near the end of the back-titration of the excess silver ion. This reaction causes the end point to fade and results in an overconsumption of thiocyanate ion, which in turn leads to low values for the chloride analysis. This error can be circumvented by filtering the silver chloride before undertaking the back-titration. Filtration is not required in the determination of other halides because they all form silver salts that are less soluble than silver thiocyanate.

Applications of Precipitation Titrations.   Table 15-2 lists some typical applications of precipitation titrations in which silver nitrate is the standard solution. In most of these methods, the analyte is precipitated with a measured excess of silver nitrate and the excess determined by a Volhard titration with standard potassium thiocyanate.

Both silver nitrate and potassium thiocyanate are obtainable in primary-standard quality. The latter compound is somewhat hygroscopic, however, and thiocyanate solutions are ordinarily standardized against silver nitrate. Both silver nitrate and potassium thiocyanate solutions are stable indefinitely.

The Volhard method was first described by Jacob Volhard, a German chemist, in 1874.

Volhard method for chloride:

$$\underset{excess}{Ag^+} + Cl^- \rightleftharpoons \underset{white}{AgCl(s)}$$

$$SCN^- + Ag^+ \rightleftharpoons \underset{white}{AgSCN(s)}$$

$$Fe^{3+} + SCN^- \rightleftharpoons \underset{red}{Fe(SCN)^{2+}}$$

The Volhard procedure requires that the analyte solution be distinctly acidic.

Table 15-2

Typical Argentometric Precipitation Methods

Substance Being Determined	End Point	Remarks
AsO_4^{3-}, Br^-, I^-, CNO^-, SCN^-	Volhard	Removal of silver salt not required
CO_3^{2-}, CrO_4^{2-}, CN^-, Cl^-, $C_2O_4^{2-}$, PO_4^{3-}, S^{2-}, NCN^{2-}	Volhard	Removal of silver salt required before back-titration of excess Ag^+
BH_4^-	Modified Volhard	Titration of excess Ag^+ following $$BH_4^- + 8Ag^+ + 8OH^- \rightarrow$$ $$8Ag(s) + H_2BO_3^- + 5H_2O$$
Epoxide	Volhard	Titration of excess Cl^- following hydrohalogenation
K^+	Modified Volhard	Precipitation of K^+ with known excess of $B(C_6H_5)_4^-$, addition of excess Ag^+ giving $AgB(C_6H_5)_4(s)$, and back-titration of the excess
Br^-, Cl^-	$2Ag^+ + CrO_4^{2-} \rightarrow Ag_2CrO_4(s)$ red	In neutral solution
Br^-, Cl^-, I^-, SeO_3^{2-}	Adsorption indicator	
$V(OH)_4^+$, fatty acids, mercaptans	Electroanalytical	Direct titration with Ag^+
Zn^{2+}	Modified Volhard	Precipitation as $ZnHg(SCN)_4$, filtration, dissolution in acid, addition of excess Ag^+, back-titration of excess Ag^+
F^-	Modified Volhard	Precipitation as $PbClF$, filtration, dissolution in acid, addition of excess Ag^+, back-titration of excess Ag^+

15C   ORGANIC COMPLEXING AGENTS

Many different organic complexing agents have become important in analytical chemistry because of their inherent sensitivity and potential selectivity in reacting with metal ions. Such reagents are particularly useful in precipitating metals and in extracting metals from one solvent to another. The most useful organic reagents form chelate complexes with metal ions.

15C-1   Reagents for Precipitating Metals

One important type of reaction involving an organic complexing agent is that in which an insoluble, uncharged complex is formed. Usually, it is necessary to consider stepwise formation of soluble species in addition to the formation of the insoluble species. Thus, a metal ion M^{n+} reacts with a complexing agent X^- to form $MX^{(n-1)+}$, $MX_2^{(n-2)+}$, MX_{n-1}^+, and $MX_n(soln)$. If we consider the overall formation of the soluble, uncharged species, we can write

$$M^{n+} + nX^- \rightleftharpoons MX_n(soln) \qquad \beta_n = \frac{[MX_n]}{[M^{n+}][X^-]^n} = K_1K_2 \cdots K_n \quad (15\text{-}16)$$

Precipitation occurs when the solubility of the species MX_n has been exceeded. We can write the solubility equilibrium as

$$MX_n(solid) \rightleftharpoons MX_n(soln) \qquad K_{eq} = [MX_n] \qquad (15\text{-}17)$$

Table 15-3

Organic Reagents for Precipitating Metals

Reagent	Metal(s) Precipitated	Species Formed
8-Hydroxyquinoline	Mg^{2+}, Zn^{2+}, Al^{3+}, several others	Sparingly soluble chelates
Dimethylglyoxime	Ni^{2+} (very selectively)	Bulky, insoluble chelate
Cupferron	Zr^{4+}, Sn^{4+}, highly charged metal ions	Chelates
1-Nitroso-2-naphthol	Co^{2+}	Chelate, unsuitable for weighing
$NaB(C_6H_5)_4$	K^+	Insoluble salt

and the solubility product expression as

$$K_{sp} = [M^{n+}][X^-]^n = \frac{K_{eq}}{\beta_n} \qquad (15\text{-}18)$$

where the constant K_{eq} is sometimes referred to as the *intrinsic solubility* of the species MX_n.

Several of the most common organic reagents for precipitating metals were described in Chapter 8 and are listed in Table 15-3 along with the metals with which they commonly react.

15C-2   Forming Soluble Complexes for Extractions and Other Uses

Many organic reagents are useful in converting metal ions into forms that can be readily extracted from water into an immiscible organic phase. Extractions are widely used to separate metals of interest from potential interfering ions and for achieving a concentrating effect by extracting into a phase of smaller volume. Extractions are applicable to much smaller amounts of metals than precipitations, and they avoid problems associated with coprecipitation. Separations by extraction are considered in detail in Chapter 24.

Several of the most widely used organic complexing agents for extractions are listed in Table 15-4. Note that some of the same reagents as used for precipi-

Table 15-4

Organic Reagents for Extracting Metals

Reagent	Metal Ions Extracted	Solvents
8-Hydroxyquinoline	Zn^{2+}, Cu^{2+}, Ni^{2+}, Al^{3+}, many others	Water $\rightarrow$ chloroform ($CHCl_3$)
Diphenylthiocarbazone (dithizone)	Cd^{2+}, Co^{2+}, Cu^{2+}, Pb^{2+}, many others	Water $\rightarrow$ $CHCl_3$ or CCl_4
Acetylacetone	Fe^{3+}, Cu^{2+}, Zn^{2+}, U(VI), many others	Water $\rightarrow$ $CHCl_3$, (CCl_4, or C_6H_6)
Ammonium pyrrolidine dithiocarbamate	Transition metals	Water $\rightarrow$ methyl isobutyl ketone
Tenoyltrifluoroacetone	Ca^{2+}, Sr^{2+}, La^{3+}, Pr^{3+} other rare earths	Water $\rightarrow$ benzene (C_6H_6)
Dibenzo-18-crown-6	Alkali metals, some alkaline earths	Water $\rightarrow$ benzene

Note: Carbon tetrachloride and benzene are known carcinogens

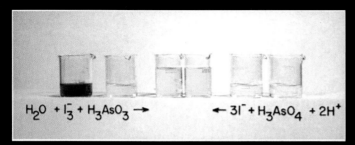

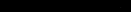

a   b

Color Plate 1
Chemical Equilibrium 1: Reaction between iodine and arsenic(III) at pH 1. (a) One mmol I_3^- added to one mmol H_3AsO_3. (b) Three mmol I^- added to one mmol H_3AsO_4 (Section 4B-1, page 83).

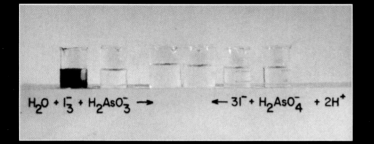

Color Plate 2
Chemical Equilibrium 2: The same reaction as in color plate 1 carried out at pH 7 (Section 4B-1, page 83).

a   b

Color Plate 3
Chemical Equilibrium 3: Reaction between iodine and ferrocyanide. One mmol I_3^- added to two mmol $Fe(CN)_6^{4-}$. (b) Three mmol I^- added to two mmol $Fe(CN)_6^{3-}$ (Section 4B-1).

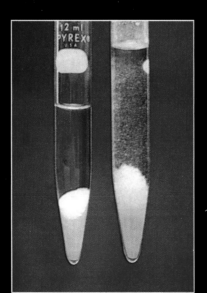

Color Plate 4
The common ion effect. The test tube on the left contains a saturated solution of silver acetate, AgOAc. The following equilibrium is established in the test tube:

$$AgOAc(s) \rightleftharpoons Ag^+(aq) + OAc^-(aq)$$

When $AgNO_3$ is added to the test tube, the equilibrium shifts to the left to form more AgOAc as shown in the test tube on the right (Section 4B-3, page 90).

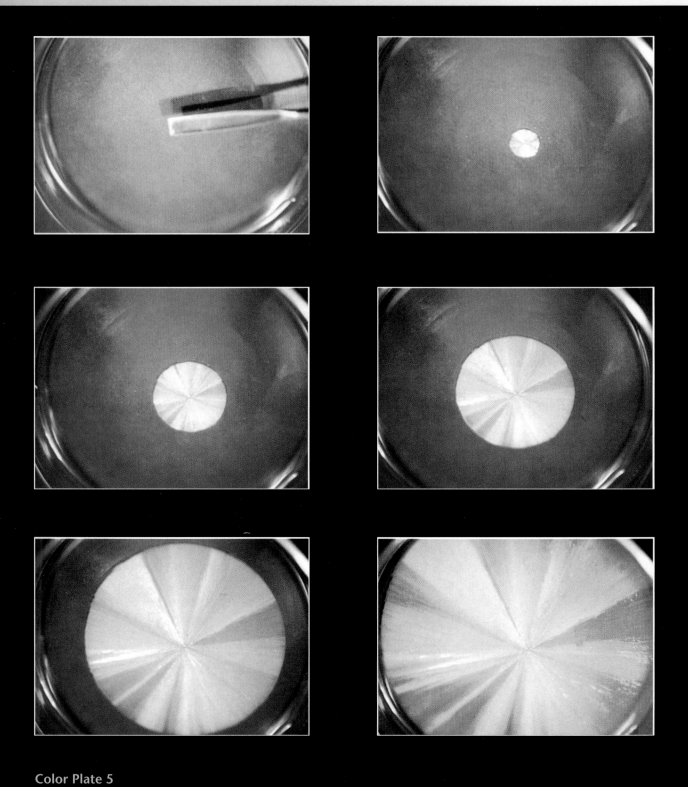

Color Plate 5
Crystallization of sodium acetate from a supersaturated solution (Section 8A-1).

Color Plate 6
The Tyndall effect (see margin note, page 181).

Color Plate 7
Precipitation of 0.1041 g of Fe(III) with NH_3 (left), and homogeneous precipitation of the same amount of iron with urea (right) (see Section 8A-5, page 190).

Color Plate 8
When dimethylglyoxime is added to a slightly basic solution of $Ni^{2+}(aq)$ shown on the left, a bright red precipitate of $Ni(C_4H_7N_2O_2)_2$ is formed as seen in the beaker on the right (Section 8D-3, page 196).

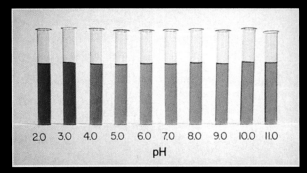

Methyl orange (3.1–4.4)

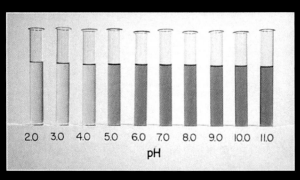

Bromocresol green (3.8–5.4)

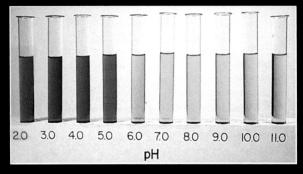

Methyl red (4.2–6.3)

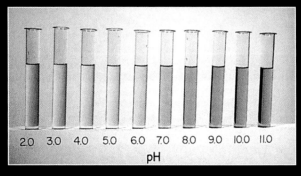

Chlorophenol red (5.2–8.8)

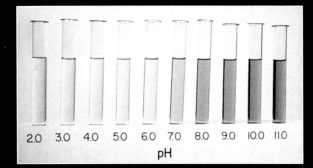

Bromothymol blue (6.0–7.6)

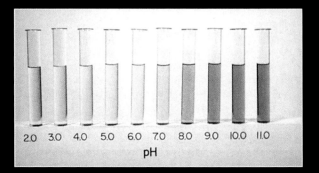

Cresol red (7.2–8.8)

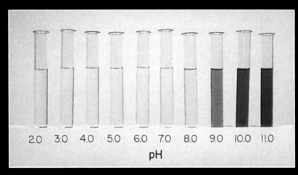

Phenolphthalein (8.3–10.0)

Color Plate 9
Acid-base indicators and
their transition pH
ranges (Section 12A-2).

a   b   c   d

Color Plate 10
Argentometric determination of chloride: Fajans method (Section 15B-2, page 356). (a) aqueous 2′,7′-dichlorofluorescein; (b) same, plus 1 mL 0.10 M Ag^+. Note the absence of a precipitate; (c) same, plus AgCl and an excess of Cl^-; (d) same, plus AgCl and the first slight excess of Ag^+.

a   b   c   d

Color Plate 11
In (c) and (d) the AgCl has coagulated. Note that in (d) the dye has been carried down on the precipitate and that the supernatant solution is clear while in (c) the dye remains in solution. (Section 15B-2, page 356).

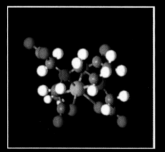

Color Plate 12
Model of an EDTA/cation chelate (Section 15D-2).

Color Plate 13
Reduction of silver(I) by direct reaction with copper: the "silver tree" (Section 16A-2, page 392).

Color Plate 14
Reduction of silver(I) by copper in an electrochemical cell (Section 16A-2, page 392).

Color Plate 15
Reaction between iron(III) and iodide (see margin note, Section 16C-6, page 410).

$$2Fe^{3+} + 3I^- \rightleftharpoons 2Fe^{2+} + I_3^-$$

a   b   c   d

Color Plate 16
Starch/iodine end point (Section 18B-2, page 448): (a) iodine solution; (b) same, within a few drops of the equivalence point; (c) same as (b), with starch added; (d) same as (c), at the equivalence point.

Color Plate 17
The time dependence of the reaction between permanganate and oxalate (Section 18C-1, page 454).

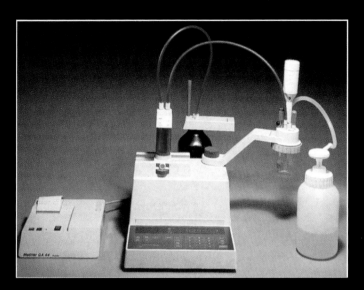

Color Plate 18
Computer-controlled automatic potentiometric titrator. The sample is pipetted into the sample cup. The titrator adds titrant, measures the pH of the titration mixture as titrant is added, and locates the equivalence point. The results of the titration are printed to provide a permanent record of the analysis.
Photo courtesy of Mettler-Toledo, Inc., Hightstown, NJ.

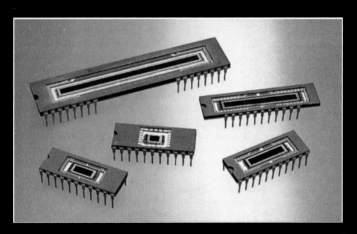

Color Plate 19
Photodiode arrays of various sizes. The arrays contain 256, 512, 1024, 2048, and 4096 diodes (Section 22A-4, page 582).
Photo courtesy of EG&G Reticon, Sunnyvale, CA.

Color Plate 20
Laminar flow burner for atomic absorption spectroscopy (Section 23F-1).
Photo courtesy of Varian Instruments, Sunnyvale, CA.

a

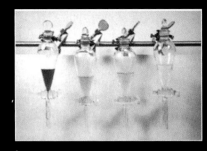

b

(a) Extraction of iodine with chloroform (Section 24D). Left to right: iodine solution; after single 40-mL extraction; after two 20-ml extractions; after four 10-ml extractions.

(b) Extraction of iodine with chloroform, aqueous layers only.

c

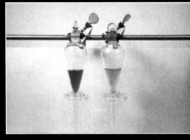

d

(c) Extraction of iodine with chloroform (Section 24D). Original solution (left); after a single 40-mL extraction (right).

(d) Extraction of iodine with chloroform. Original solution (left); aqueous phase only (right).

e

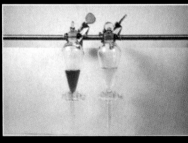

f

(e) Extraction of iodine with chloroform (Section 24D). Original solution (left); after two 20-mL extractions (right).

(f) Extraction of iodine with chloroform. Original solution (left); aqueous phase only (right).

g

h

(g) Extraction of iodine with chloroform (Section 24D). Original solution (left); after four 10-mL extractions (right).

(h) Extraction of iodine with chloroform. Original solution (left); aqueous phase only (right).

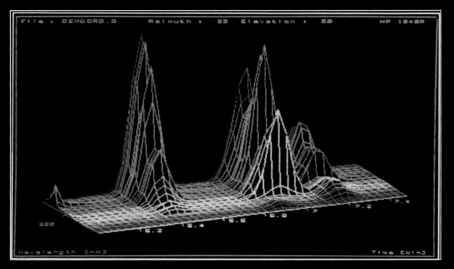

Color Plate 22
High performance liquid chromatogram (Section 25B) obtained with a multichannel diode array spectrometer (Section 22B-3, page 588) as detector. The chromatogram is displayed on the terminal of a computer, which controls the instrument. Retention time is displayed horizontally, and various detected wavelengths are presented in different colors.
Photo courtesy of Hewlett-Packard Company, Palo Alto, CA.

Color Plate 23
Electronic analytical balance with 0.1 mg precision (Section 2D-2).
Photo courtesy of Mettler-Toledo, Inc., Hightstown, NY.

Color Plate 24

Stamps from various countries celebrating scientists and events of importance to analytical chemistry. The numbers following each citation refer to the Scott Standard Postage Stamp Catalogue. Stamps Commemorating (1) the Centennial of the Metric Convention of 1875, France, #1475; (2) T. W. Richards, Sweden, #1104; (3) Svante Arrhenius, Sweden, #549; (4) Cato Guldberg and Peter Waage, Norway, #453; (5) Walther Nernst, Sweden, #1344; (6) Allessandro Volta, Italy, #527; (7) André Marie Ampère, Monaco, #1001; (8) Gustav Robert Kirchhoff, Germany, #9N345; (9) spectroscopic analysis, Vatican, #655; (10) Jöns Jacob Berzelius, Sweden, #1293. (11) Stamp honoring Archer J. P. Martin and Richard L. M. Synge for their work in chromatography, Great Britain, #808. (12) Stamp depicting the visible spectrum and atomic lines, Canada, #613. (13) Stamp announcing an international spectroscopy colloquium in Madrid, Spain, #1570.

These stamps are from the private collection of Professor C. M. Lang.

tations can also be used for extractions. In such cases, the solubility of the metal chelate in the organic phase keeps the complex from precipitating in the aqueous phase. In many cases, the pH of the aqueous phase is used to achieve some control over the extraction process, since most of the reactions are pH dependent as shown in Equation 15-19.

$$nHX(org) + M^{n+}(aq) \rightleftharpoons MX_n(org) + nH^+(aq) \qquad (15\text{-}19)$$

15D   TITRATIONS WITH AMINOCARBOXYLIC ACIDS

Tertiary amines that also contain carboxylic acid groups form remarkably stable chelates with many metal ions.[1] Gerold Schwarzenbach first recognized their potential as analytical reagents in 1945. Since this original work, investigators throughout the world have described applications of these compounds to the volumetric determination of most of the metals in the periodic table.

15D-1   Versatile Ethylenediaminetetraacetic Acid (EDTA)

Ethylenediaminetetraacetic acid [also called (ethylenedinitrilo)tetraacetic acid], which is commonly shortened to EDTA, is the most widely used complexometric titrant. Fully protonated EDTA has the structure

> EDTA, a hexadentate ligand, is among the most important and widely used reagents in titrimetry.

$$
\begin{array}{c}
\text{HOOC}-\text{CH}_2 \\
\hspace{2.5em}\diagdown \\
\text{N}-\text{CH}_2-\text{CH}_2-\text{N} \\
\hspace{2.5em}\diagup \\
\text{HOOC}-\text{CH}_2 \\
\end{array}
\quad
\begin{array}{c}
\text{CH}_2-\text{COOH} \\
\diagup \\
\\
\diagdown \\
\text{CH}_2-\text{COOH} \\
\end{array}
$$

The EDTA molecule has six potential sites for bonding a metal ion: the four carboxyl groups and the two amino groups, each of the latter with an unshared pair of electrons. Thus, EDTA is a hexadentate ligand.

EDTA Is a Tetrabasic Acid

The dissociation constants for the acidic groups in EDTA are $K_1 = 1.02 \times 10^{-2}$, $K_2 = 2.14 \times 10^{-3}$, $K_3 = 6.92 \times 10^{-7}$, and $K_4 = 5.50 \times 10^{-11}$. It is of interest that the first two constants are of the same order of magnitude, which suggests that the two protons involved dissociate from opposite ends of the rather long molecule. As a consequence of their physical separation, the negative charge created by the first dissociation does not greatly affect the removal of the second proton. The same cannot be said for the dissociation of the other two protons, however, which are much closer to the negatively charged carboxylate ions created by the initial dissociations.

The various EDTA species are often abbreviated H_4Y, H_3Y^-, H_2Y^{2-}, HY^{3-}, and Y^{4-}. Figure 15-4 illustrates how the relative amounts of these five species vary as a function of pH. It is apparent that H_2Y^{2-} predominates in moderately acidic medium (pH 3 to 6).

---

[1]See, for example, R. Pribil, *Applied Complexometry* (New York: Pergamon, 1982); A. Ringbom and E. Wanninen, in *Treatise on Analytical Chemistry,* 2nd ed., I. M. Kolthoff and P. J. Elving, Eds. (New York: Wiley, 1979), Part I, Vol. 2, Chapter 11.

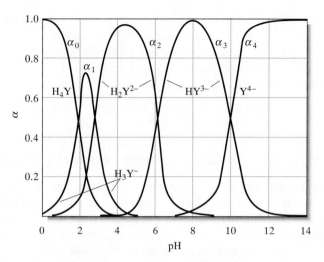

Figure 15-4 Composition of EDTA solutions as a function of pH. Note that the fully protonated form, H_4Y, is only a major component in very acidic solutions (pH $<$ 3). Throughout the pH range of 3 to 10, the species H_2Y^{2-} and HY^{3-} are predominant. The fully unprotonated form Y^{4-} is a significant component only in very basic solutions (pH $>$ 10).

Reagents for EDTA Titrations

Standard EDTA solutions are ordinarily prepared by dissolving weighed quantities of $Na_2H_2Y \cdot 2H_2O$ and diluting to the mark in a volumetric flask.

The free acid H_4Y and the dihydrate of the sodium salt, $Na_2H_2Y \cdot 2H_2O$, are commercially available in reagent quality. The former can serve as a primary standard after it has been dried for several hours at 130 to 145°C. It is then dissolved in the minimum amount of base required for complete solution.

Under normal atmospheric conditions, the dihydrate, $Na_2H_2Y \cdot 2H_2O$, contains 0.3% moisture in excess of the stoichiometric amount. For all but the most exacting work, this excess is sufficiently reproducible to permit use of a corrected weight of the salt in the direct preparation of a standard solution. If necessary, the pure dihydrate can be prepared by drying at 80°C for several days in an atmosphere of 50% relative humidity.

Nitrilotriacetic acid (NTA) is the second most common aminopolycarboxylic acid used for titrations. It is a tetradentate chelating agent and has the structure

A number of compounds that are chemically related to EDTA have also been investigated but do not appear to offer significant advantages. We thus limit our discussion here to the properties and applications of EDTA.

$$COOH-CH_2 \diagdown \diagup CH_2-COOH$$
$$N$$
$$CH_2-COOH$$

Feature 15-2

Species Present in a Solution of EDTA

When it is dissolved in water, EDTA behaves like an amino acid, such as glycine (see Features 12-7 and 13-4). Here, however, a double zwitterion forms, which has the structure shown in Figure 15-5a. Note that the net charge on this species is zero and that it contains four dissociable protons, two associated with two of the carboxyl groups and the other two with the two amine groups. For simplicity, we generally formulate the double zwitterion as H_4Y, where Y^{4-} is the fully deprotonated form of Figure 15-5e. The first and second steps in the dissociation process involve successive loss of protons from the two carboxylic acid groups; the third and fourth steps involve dissociation of the protonated amine groups. The structures of H_3Y^-, H_2Y^{2-}, and HY^{3-} are shown in Figure 15-5b, c, and d.

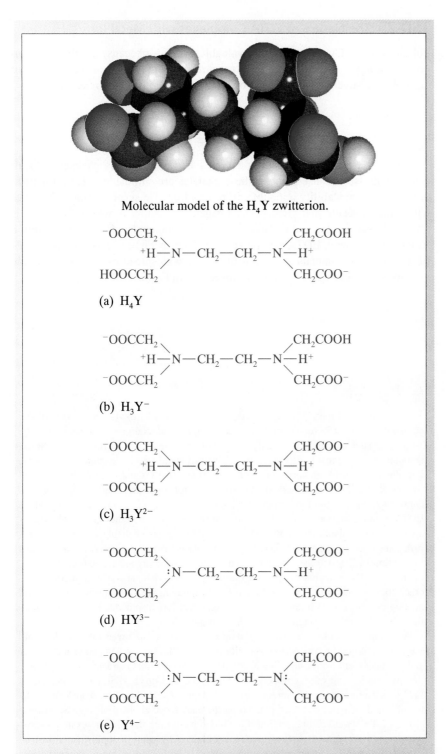

Molecular model of the H_4Y zwitterion.

(a) H_4Y

(b) H_3Y^-

(c) H_3Y^{2-}

(d) HY^{3-}

(e) Y^{4-}

Figure 15-5   Structure of H_4Y and its dissociation products. Note that the fully protonated species H_4Y exist as the double zwitterion with the amine nitrogens and two of the carboxylic acid groups protonated. The first two protons dissociate from the carboxyl groups, whereas the last two come from the amine groups

15D-2   The Nature of EDTA Complexes with Metal Ions

Solutions of EDTA are particularly valuable as titrants because the reagent *combines with metal ions in a 1:1 ratio regardless of the charge on the cation.* For example, the silver and aluminum complexes are formed by the reactions

$$Ag^+ + Y^{4-} \rightleftharpoons AgY^{3-}$$

$$Al^{3+} + Y^{4-} \rightleftharpoons AlY^-$$

In general, we can write the reaction of the EDTA anion with a metal ion M^{n+} as

$$M^{n+} + Y^{4-} \rightleftharpoons MY^{(n-4)+}$$

EDTA is a remarkable reagent not only because it forms chelates with all cations but also because most of these chelates are sufficiently stable for titrations. This great stability undoubtedly results from the several complexing sites within the molecule that give rise to a cagelike structure in which the cation is effectively surrounded and isolated from solvent molecules. One of the common structures for metal/EDTA complexes is shown in Figure 15-6. The ability of EDTA to complex metals is responsible for its widespread use as a preservative in foods and in biological samples as discussed in Feature 15-3.

Feature 15-3

EDTA as a Preservative

Trace quantities of metal ions can effectively catalyze the air oxidation of many of the compounds present in foods and biological samples (for example, proteins in blood). To prevent such oxidation reactions, it is important to inactivate or remove even trace amounts of metal ions. In processed foods, we can readily get trace quantities of metal ions due to contact with various metallic containers (kettles and vats) during the processing stages. EDTA is an excellent preservative for foods and is a common ingredient of such commercial food products as mayonnaise, salad dressings, and oils. When EDTA is added to foods, it so tightly binds most metal ions that they are unable to catalyze the air oxidation reaction. EDTA and other similar chelating agents are often called *sequestering agents* because of their ability to remove or inactivate metal ions. In addition to EDTA, other common sequestering agents are salts of citric and phosphoric acid. These agents can protect the unsaturated side chains of triglycerides and other components against air oxidation. Such oxidation reactions are responsible for making fats and oils turn rancid. Sequestering agents are also added to prevent oxidation of easily oxidized compounds, such as ascorbic acid.

With biological samples, it is important to add EDTA as a preservative if the sample is to be stored for any period. As with foods, EDTA tightly complexes metal ions and prevents them from catalyzing air oxidation reactions that can lead to decomposition of proteins and other compounds. During the trial of O. J. Simpson, the use of EDTA as a preservative became an important item. The prosecution team contended that if blood evidence had been planted on the back fence at Nicole Brown Simpson's home, EDTA should be present, but if the blood were from the perpetrator no preservative should be seen. Analytical evidence, obtained by using a sophisticated instrumental system (liquid chromatography combined with tandem mass spectrometry), did show traces of EDTA, but the amounts were very small and subject to differing interpretations.

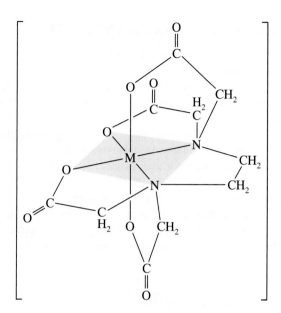

Figure 15-6 Structure of a metal/EDTA complex. Note that EDTA behaves here as a hexadentate ligand in that six donor atoms are involved in bonding the divalent metal cation.

Table 15-5 lists formation constants K_{MY} for common EDTA complexes. Note that the constant refers to the equilibrium involving the fully unprotonated species Y^{4-} with the metal ion:

$$M^{n+} + Y^{4-} \rightleftharpoons MY^{(n-4)+} \qquad K_{MY} = \frac{[MY^{(n-4)+}]}{[M^{n+}][Y^{4-}]} \qquad (15\text{-}20)$$

Table 15-5

Formation Constants for EDTA Complexes

Cation	K_{MY}*	$\log K_{MY}$	Cation	K_{MY}	$\log K_{MY}$
Ag^+	2.1×10^7	7.32	Cu^{2+}	6.3×10^{18}	18.80
Mg^{2+}	4.9×10^8	8.69	Zn^{2+}	3.2×10^{16}	16.50
Ca^{2+}	5.0×10^{10}	10.70	Cd^{2+}	2.9×10^{16}	16.46
Sr^{2+}	4.3×10^8	8.63	Hg^{2+}	6.3×10^{21}	21.80
Ba^{2+}	5.8×10^7	7.76	Pb^{2+}	1.1×10^{18}	18.04
Mn^{2+}	6.2×10^{13}	13.79	Al^{3+}	1.3×10^{16}	16.13
Fe^{2+}	2.1×10^{14}	14.33	Fe^{3+}	1.3×10^{25}	25.1
Co^{2+}	2.0×10^{16}	16.31	V^{3+}	7.9×10^{25}	25.9
Ni^{2+}	4.2×10^{18}	18.62	Th^{4+}	1.6×10^{23}	23.2

Source: G. Schwarzenbach, Complexometric Titrations, (London: Chapman and Hall, 1957), p. 8. With permission.
* Constants are valid at 20°C and ionic strength of 0.1.

15D-3   Equilibrium Calculations Involving EDTA

A titration curve for the reaction of a cation M^{n+} with EDTA consists of a plot of pM versus reagent volume. Values for pM are readily computed in the early stage of a titration by assuming that the equilibrium concentration of M^{n+} is equal to its analytical concentration, which in turn is readily derived from stoichiometric data.

Calculation of M^{n+} at and beyond the equivalence point requires the use of Equation 15-20. The computation in this region is troublesome and time-consuming if the pH is unknown and variable because both $[MY^{(n-4)+}]$ and $[M^{n+}]$ are pH-dependent. Fortunately, EDTA titrations are always performed in solutions that are buffered to a known pH to avoid interferences by other cations or to ensure satisfactory indicator behavior. Calculating $[M^{n+}]$ in a buffered solution containing EDTA is a relatively straightforward procedure provided the pH is known. In this computation, we use the alpha value for EDTA. Recall from Section 13F that α_4 for EDTA can be defined as

$$\alpha_4 = \frac{[Y^{4-}]}{c_T} \tag{15-21}$$

where c_T is the total molar concentration of *uncomplexed* EDTA:

$$c_T = [Y^{4-}] + [HY^{3-}] + [H_2Y^{2-}] + [H_3Y^-] + [H_4Y]$$

Conditional Formation Constants

To obtain the conditional formation constant for the equilibrium shown in Equation 15-20, we substitute $\alpha_4 c_T$ from Equation 15-21 for $[Y^{4-}]$ in the formation constant expression (Equation 15-20):

$$M^{n+} + Y^{4-} \rightleftharpoons MY^{(n-4)+} \qquad K_{MY} = \frac{[MY^{(n-4)+}]}{[M^{n+}]\alpha_4 c_T} \tag{15-22}$$

Combining the two constants α_4 and K_{MY} yields the conditional formation constant K'_{MY}:

$$K'_{MY} = \alpha_4 K_{MY} = \frac{[MY^{(n-4)+}]}{[M^{n+}]c_T} \tag{15-23}$$

where K'_{MY} is a constant *only at the pH for which α_4 is applicable.*

Conditional constants are readily computed and provide a simple means by which the equilibrium concentration of the metal ion and the complex can be calculated at the equivalence point and where there is an excess of reactant. Note that replacement of $[Y^{4-}]$ with c_T in the equilibrium-constant expression greatly simplifies calculations because c_T is easily determined from the reaction stoichiometry whereas $[Y^{4-}]$ is not.

> Conditional formation constants are pH-dependent.

Computing α_4 Values for EDTA Solutions

An expression for calculating α_4 at a given hydrogen ion concentration is derived by the method demonstrated in Section 13F (see Feature 13-5). Thus, α_4 for EDTA is

$$\alpha_4 = \frac{K_1 K_2 K_3 K_4}{[H^+]^4 + K_1[H^+]^3 + K_1 K_2[H^+]^2 + K_1 K_2 K_3[H^+] + K_1 K_2 K_3 K_4} \tag{15-24}$$

$$\alpha_4 = \frac{K_1 K_2 K_3 K_4}{D} \qquad (15\text{-}25)$$

where K_1, K_2, K_3, and K_4 are the four dissociation constants for H_4Y and D is the denominator of Equation 15-24.

Figure 15-7 is an Excel spreadsheet to calculate α_4 at selected pH values according to Equations 15-24 and 15-25. Note from the results that only about 4×10^{-12} percent of EDTA exists as Y^{4-} at pH 2.00. Example 15-4 illustrates how Y^{4-} is calculated for a solution of known pH.

The alpha values for the other EDTA species are calculated in a similar manner and are found to be

$$\alpha_0 = \frac{[H^+]^4}{D} \qquad \alpha_1 = \frac{K_1[H^+]^3}{D}$$

$$\alpha_2 = \frac{K_1 K_2[H^+]^2}{D} \qquad \alpha_3 = \frac{K_1 K_2 K_3[H^+]}{D}$$

Only α_4 is needed in deriving titration curves.

Example 15-4

Calculate the molar Y^{4-} concentration in a 0.0200 M EDTA solution buffered to a pH of 10.00.

At pH 10.00, α_4 is 0.35 (Figure 15-7). Thus,

$$[Y^{4-}] = \alpha_4 c_T = 0.35 \times 0.0200 = 7.00 \times 10^{-3} \text{ M}$$

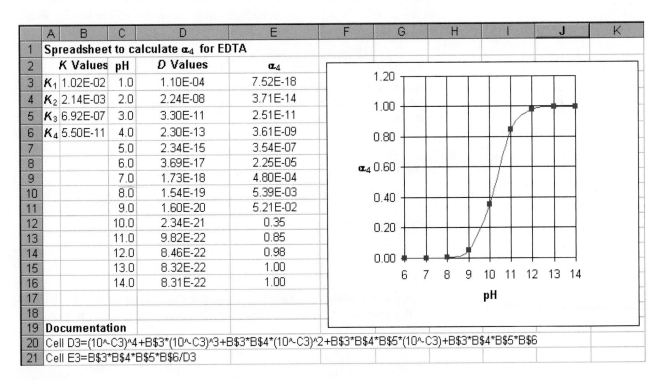

	A	B	C	D	E
1	**Spreadsheet to calculate α_4 for EDTA**				
2		**K Values**	**pH**	**D Values**	**α_4**
3	K_1	1.02E-02	1.0	1.10E-04	7.52E-18
4	K_2	2.14E-03	2.0	2.24E-08	3.71E-14
5	K_3	6.92E-07	3.0	3.30E-11	2.51E-11
6	K_4	5.50E-11	4.0	2.30E-13	3.61E-09
7			5.0	2.34E-15	3.54E-07
8			6.0	3.69E-17	2.25E-05
9			7.0	1.73E-18	4.80E-04
10			8.0	1.54E-19	5.39E-03
11			9.0	1.60E-20	5.21E-02
12			10.0	2.34E-21	0.35
13			11.0	9.82E-22	0.85
14			12.0	8.46E-22	0.98
15			13.0	8.32E-22	1.00
16			14.0	8.31E-22	1.00
17					
18					
19	**Documentation**				
20	Cell D3=(10^C3)^4+B$3*(10^C3)^3+B$3*B$4*(10^C3)^2+B$3*B$4*B$5*(10^C3)+B$3*B$4*B$5*B$6				
21	Cell E3=B$3*B$4*B$5*B$6/D3				

Figure 15-7   Spreadsheet to calculate α_4 for EDTA at selected pH values. Note that the acid dissociation constants for EDTA are entered in column B (labels in column A). Next, the pH values for which the calculations are to be done are entered in column C. The formula for calculating the denominator D in Equation 15-24 is placed into cell D3 and copied into cells D4 through D16. The final column E contains the equation for calculating the α_4 values as given in Equation 15-25. The graph shows an XY (Scatter) plot of α_4 versus pH.

Calculating the Cation Concentration in EDTA Solutions

In an EDTA titration, we are interested in finding the cation concentration as a function of the amount of titrant (EDTA) added. Prior to the equivalence point, the cation is in excess and its concentration can be found from the reaction stoichiometry. At the equivalence point and in the postequivalence-point region, however, the conditional formation constant of the complex must be used to calculate the cation concentration. Example 15-5 demonstrates how the cation concentration can be calculated in a solution of an EDTA complex. Example 15-6 illustrates this calculation when excess EDTA is present.

Example 15-5

Calculate the equilibrium concentration of Ni^{2+} in a solution with an analytical NiY^{2-} concentration of 0.0150 M at pH (a) 3.0 and (b) 8.0.

From Table 15-5,

$$Ni^{2+} + Y^{4-} \rightleftharpoons NiY^{2-} \qquad K_{NiY} = \frac{[NiY^{2-}]}{[Ni^{2+}][Y^{4-}]} = 4.2 \times 10^{18}$$

The equilibrium concentration of NiY^{2-} is equal to the analytical concentration of the complex minus the concentration lost by dissociation. The latter is identical to the equilibrium Ni^{2+} concentration. Thus,

$$[NiY^{2-}] = 0.0150 - [Ni^{2+}]$$

If we assume that $[Ni^{2+}] \ll 0.0150$, an assumption that is almost certainly valid in light of the large formation constant of the complex, this equation simplifies to

$$[NiY^{2-}] \approx 0.0150$$

Since the complex is the only source of both Ni^{2+} and the EDTA species,

$$[Ni^{2+}] = [Y^{4-}] + [HY^{3-}] + [H_2Y^{2-}] + [H_3Y^-] + [H_4Y] = c_T$$

Substitution of this equality into Equation 15-23 gives

$$K'_{NiY} = \frac{[NiY^{2-}]}{[Ni^{2+}]c_T} = \frac{[NiY^{2-}]}{[Ni^{2+}]^2} = \alpha_4 K_{NiY}$$

(a) The spreadsheet in Figure 15-7 indicates that α_4 is 2.5×10^{-11} at pH 3.0. If we substitute this value and the concentration of NiY^{2-} into the equation for K'_{MY}, we get

$$\frac{0.0150}{[Ni^{2+}]^2} = 2.5 \times 10^{-11} \times 4.2 \times 10^{18} = 1.05 \times 10^8$$

$$[Ni^{2+}] = \sqrt{1.43 \times 10^{-10}} = 1.2 \times 10^{-5} \text{ M}$$

(b) At pH 8.0, the conditional constant is much larger. Thus,

$$K'_{NiY} = 5.4 \times 10^{-3} \times 4.2 \times 10^{18} = 2.27 \times 10^{16}$$

and, after we substitute this into the equation for K'_{NiY}, we get

$$[Ni^{2+}] = \sqrt{0.0150/(2.27 \times 10^{16})} = 8.1 \times 10^{-10} \text{ M}$$

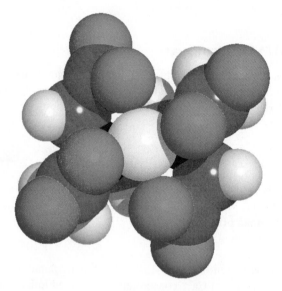

Molecular model of NiY^{2-}. This complex is typical of the strong complexes that EDTA forms with metal ions. The formation constant of the Ni^{2+} complex is 4.2×10^{18}.

Example 15-6

Calculate the concentration of Ni^{2+} in a solution that was prepared by mixing 50.0 mL of 0.0300 M Ni^{2+} with 50.0 mL of 0.0500 M EDTA. The mixture was buffered to a pH of 3.0.

Here, the solution has an excess of EDTA, and the analytical concentration of the complex is determined by the amount of Ni^{2+} originally present. Thus,

$$c_{NiY^{2+}} = 50.0 \text{ mL} \times \frac{0.0300 \text{ M}}{100.0 \text{ mL}} = 0.0150 \text{ M}$$

$$c_{EDTA} = \frac{(50.0 \times 0.0500) \text{ mmol} - (50.0 \times 0.0300) \text{ mmol}}{100.0 \text{ mL}} = 0.0100 \text{ M}$$

Again, let us assume that $[Ni^{2+}] \ll [NiY^{2-}]$ so that

$$[NiY^{2-}] = 0.0150 - [Ni^{2+}] \cong 0.0150 \text{ M}$$

At this point, the total concentration of uncomplexed EDTA is given by its molarity:

$$c_T = 0.0100 \text{ M}$$

The value for K'_{NiY} was found in Example 15-5 to be 1.05×10^8 at pH 3.00.

If we substitute this value into Equation 15-23, we get

$$K'_{NiY} = \frac{0.0150}{[Ni^{2+}]\,0.0100} = \alpha_4 K_{NiY}$$

$$[Ni^{2+}] = \frac{0.0150}{0.0100 \times 1.05 \times 10^8} = 1.4 \times 10^{-8}\ M$$

15D-4   EDTA Titration Curves

The principles illustrated in Examples 15-5 and 15-6 can be used in the derivation of the titration curve for a metal ion with EDTA in a solution of fixed pH. Spreadsheet Example 15-7 demonstrates how the titration curve is constructed.

Spreadsheet Exercise   Example 15-7

Use a spreadsheet to construct the titration curve of pCa vs. volume of EDTA for 50.0 mL of 0.00500 M Ca^{2+} being titrated with 0.0100 M EDTA in a solution buffered to a constant pH of 10.0.

Initial Entries

The spreadsheet is shown in Figure 15-8. The initial volume of Ca^{2+} is entered into cell B3 and the initial Ca^{2+} concentration is entered in E2. The EDTA con-

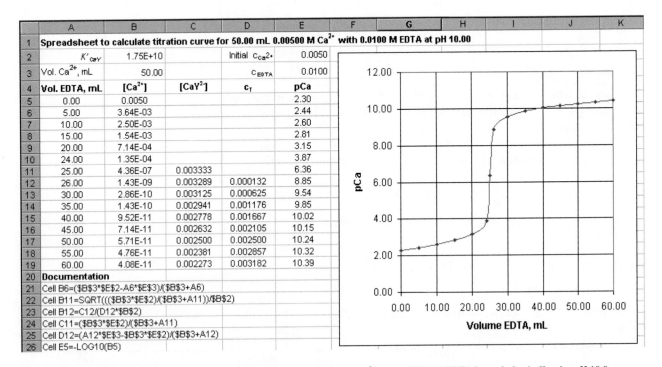

Figure 15-8   Spreadsheet for the titration of 50.00 mL of 0.00500 M Ca^{2+} with 0.0100 M EDTA in a solution buffered at pH 10.0.

centration is entered into cell E3. The volumes for which pCa values are to be calculated are entered into cells A5 through A19.

Calculating the Conditional Constant

The conditional formation constant for the calcium-EDTA complex at pH 10 is obtained from the formation constant of the complex (Table 15-4) and the α_4 value for EDTA at pH 10 (Figure 15-7). Thus, if we substitute into Equation 15-23, we get

$$K'_{\text{CaY}} = \frac{[\text{CaY}^{2-}]}{[\text{Ca}^{2+}]c_{\text{T}}} = \alpha_4 K_{\text{CaY}}$$

$$= 0.35 \times 5.0 \times 10^{10} = 1.75 \times 10^{10}$$

This value is entered into cell B2.

Preequivalence-Point Values for pCa

The initial $[\text{Ca}^{2+}]$ at 0.00 mL titrant is just the value in cell E2. Hence =E2 is entered into cell B5. The initial pCa is calculated from the initial $[\text{Ca}^{2+}]$ by taking the negative logarithm as shown in the documentation for cell E5 (cell A26). This formula is copied into cells E6 through E19. For the other entries prior to the equivalence point, the equilibrium concentration of Ca^{2+} is equal to the untitrated excess of the cation plus any dissociation of the complex, the latter being equal numerically to c_{T}. Usually, c_{T} is small relative to the analytical concentration of the uncomplexed calcium ion. Thus, for example, after 5.00 mL of EDTA has been added,

$$[\text{Ca}^{2+}] = \frac{50.0 \text{ mL} \times 0.00500 \text{ M} - 5.00 \text{ mL} \times 0.0100 \text{ M}}{(50 + 5.00) \text{ mL}} + c_{\text{T}}$$

$$\cong \frac{50.0 \text{ mL} \times 0.00500 \text{ M} - 5.00 \text{ mL} \times 0.0100 \text{ M}}{55.00 \text{ mL}}$$

We thus enter into cell B6 the formula shown in the documentation section of the spreadsheet (cell A21). The reader should verify that the spreadsheet formula is equivalent to the expression for $[\text{Ca}^{2+}]$ given above. The volume of titrant (A6) is the only value that changes in this preequivalence-point region. Hence, other preequivalence-point values of pCa are calculated by copying the formula in cell B6 into cells B7 through B10.

The Equivalence-Point pCa

At the equivalence point (25.00 mL of EDTA), we follow the method shown in Example 15-5 and first compute the analytical concentration of CaY^{2-}:

$$c_{\text{CaY}^{2-}} = \frac{(50.0 \times 0.00500) \text{ mmol}}{(50.0 + 25.0) \text{ mL}}$$

The only source of Ca^{2+} ions is the dissociation of the complex. It also follows that the Ca^{2+} concentration must be equal to the sum of the concentrations of the uncomplexed EDTA, c_{T}. Thus,

$$[\text{Ca}^{2+}] = c_{\text{T}} \quad \text{and} \quad [\text{CaY}^{2-}] = c_{\text{CaY}^{2-}} - [\text{Ca}^{2+}] \approx c_{\text{CaY}^{2-}}$$

The formula for $[\text{CaY}^{2-}]$ is thus entered into cell C11 as shown in the documentation in cell A24. The reader should again be able to verify this formula. To obtain $[Ca^{2+}]$, we substitute into the expression for K'_{CaY},

$$K'_{CaY} = \frac{[CaY^{2-}]}{[Ca^{2+}]c_T} \approx \frac{c_{CaY^{2-}}}{[Ca^{2+}]^2}$$

$$[Ca^{2+}] = \sqrt{\frac{c_{CaY^{2-}}}{K'_{CaY}}}$$

We thus enter into cell B11 the formula corresponding to this expression as shown in cell A22.

Postequivalence-Point pCa

Beyond the equivalence point, analytical concentrations of CaY^{2-} and EDTA are obtained directly from the stoichiometric data. Since there is now excess EDTA, a calculation similar to that in Example 15-6 is then performed. Thus, after the addition of 26.0 mL of EDTA, we can write

$$c_{CaY^{2-}} = \frac{(50.0 \times 0.00500) \text{ mmol}}{(50.0 + 26.0) \text{ mL}}$$

$$c_{EDTA} = \frac{(26.0 \times 0.0100) \text{ mL} - (50.0 \times 0.00500) \text{ mL}}{76.0 \text{ mL}}$$

As an approximation,

$$[CaY^{2-}] = c_{CaY^{2-}} - [Ca^{2+}] \approx c_{CaY^{2-}} \approx \frac{(50.0 \times 0.00500) \text{ mmol}}{(50.0 + 26.0) \text{ mL}}$$

Since this expression is the same as that previously entered into cell C11, we copy that equation into cell C12. We also note that $[CaY^{2-}]$ will be given by this same expression (with the volume varied) throughout the remainder of the titration. Hence the formula in cell C12 is copied into cells C13 through C19. Also, we approximate

$$c_T = c_{EDTA} + [Ca^{2+}] \approx c_{EDTA}$$

$$= \frac{(26.0 \times 0.0100) \text{ mL} - (50.0 \times 0.00500) \text{ mL}}{76.0 \text{ mL}}$$

We enter this formula into cell D12 as shown in the documentation (cell A25) and copy it into cells D13 through D16.

To calculate $[Ca^{2+}]$, we then substitute into the conditional formation-constant expression, and obtain

$$K'_{CaY} = \frac{[CaY^{2-}]}{[Ca^{2+}] \times c_T} \approx \frac{c_{CaY^{2-}}}{[Ca^{2+}] \times c_{EDTA}}$$

$$[Ca^{2+}] = \frac{c_{CaY^{2-}}}{c_{EDTA} \times K'_{CaY}}$$

Hence the $[Ca^{2+}]$ in cell B12 is computed from the values in cells C12 and D12 as shown in cell A23. We copy this formula into cells B13 through B19, and plot the titration curve shown in Figure 15-8.

Curve *A* in Figure 15-9 is a plot of data for the titration in Spreadsheet Exercise Example 15-7. Curve *B* is for the titration of magnesium ion under identical conditions. The formation constant for the EDTA complex of magnesium is smaller than that of the calcium complex, which results in a smaller change in the p-function in the equivalence-point region.

Figure 15-10 provides titration curves for calcium ion in solutions buffered to various pH levels. Recall that α_4, and hence K'_{CaY}, becomes smaller as the pH decreases. The less favorable equilibrium constant leads to a smaller change in pCa in the equivalence-point region. It is apparent from Figure 15-10 that an adequate end point in the titration of calcium requires a pH of about 8 or greater. As shown in Figure 15-11, however, cations with larger formation constants provide good end points even in acidic media. Figure 15-12 shows the minimum permissible pH for a satisfactory end point in the titration of various metal ions in the absence of competing complexing agents. Note that a moderately acidic environment is satisfactory for many divalent heavy-metal cations and that a strongly acidic medium can be tolerated in the titration of such ions as iron(III) and indium(III).

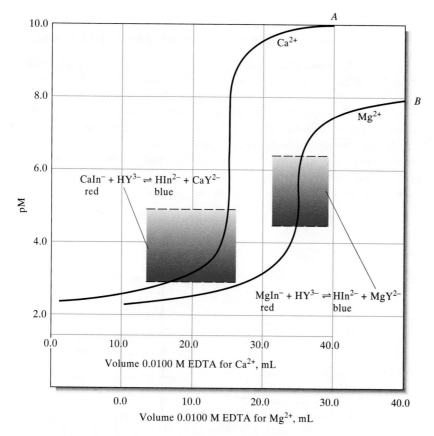

Figure 15-9   EDTA titration curves for 50.0 mL of 0.00500 M Ca^{2+} ($K'_{CaY} = 1.75 \times 10^{10}$) and Mg^{2+} ($K'_{MgY} = 1.72 \times 10^{8}$) at pH 10.0. Note that because of the larger formation constant, the reaction of calcium ion with EDTA is more complete, and a larger change occurs in the equivalence-point region. The shaded areas show the transition range for the indicator Eriochrome Black T.

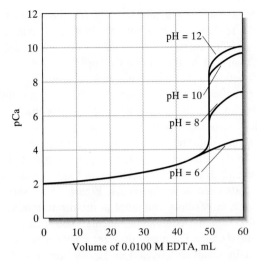

Figure 15-10 Influence of pH on the titration of 0.0100 M Ca^{2+} with 0.0100 M EDTA. Note that the end point becomes less sharp as the pH decreases because the complex-formation reaction is less complete under these circumstances.

15D-5   How Do Other Complexing Agents Affect EDTA Titrations?

Many cations form precipitates with hydroxide when the pH is raised to the level required for their successful titration with EDTA. When this problem is encountered, an auxiliary complexing agent is needed to keep the cation in solution. For example, zinc(II) is ordinarily titrated in a medium that has fairly high concentrations of ammonia and ammonium chloride. These species buffer the solution to a pH that ensures complete reaction between cation and titrant; in addition, ammonia forms ammine complexes with zinc(II) and prevents formation of the sparingly soluble zinc hydroxide, particularly in the early stages of the titration. A somewhat more realistic description of the reaction is then

$$Zn(NH_3)_4^{2+} + HY^{3-} \rightarrow ZnY^{2-} + 3NH_3 + NH_4^+$$

Auxiliary complexing agents must often be used in EDTA titrations to prevent precipitation of the analyte as a hydroxide or oxide. Such reagents cause the end points to be less sharp.

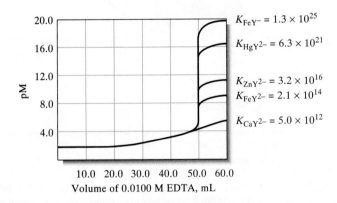

Figure 15-11 Titration curves for 50.0 mL of 0.0100 M solutions of various cations at pH 6.0.

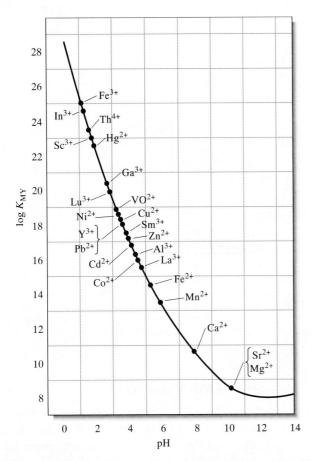

Figure 15-12 Minimum pH needed for satisfactory titration of various cations with EDTA. (From C. N. Reilley and R. W. Schmid, *Anal. Chem.,* **1958,** *30,* 947. With permission of the American Chemical Society.)

The solution also contains such other zinc-ammonia species as $Zn(NH_3)_3^{2+}$, $Zn(NH_3)_2^{2+}$, and $Zn(NH_3)^{2+}$. Calculation of pZn in a solution that contains ammonia must take these species into account.[2] Qualitatively, complexation of a cation by an auxiliary complexing reagent causes preequivalence point pM values to be larger than in a comparable solution with no such reagent.

Figure 15-13 shows two theoretical curves for the titration of zinc(II) with EDTA at pH 9.00. The equilibrium concentration of ammonia was 0.100 M for one titration and 0.0100 M for the other. Note that the presence of ammonia decreases the change in pZn near the equivalence point. For this reason, the concentration of an auxiliary complexing reagent should always be kept to the minimum required to prevent precipitation of the analyte. Note that the auxiliary complexing agent does not affect pZn beyond the equivalence point. On the other hand, keep in mind that α_4, and thus pH, plays an important role in defining this part of the titration curve (Figure 15-10).

---

[2]See, for example, D. A. Skoog, D. M. West, and F. J. Holler, *Fundamentals of Analytical Chemistry,* 7th ed. (Philadelphia: Saunders, 1996), pp. 291–294.

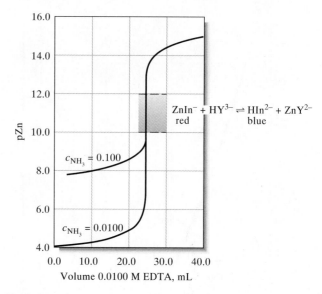

Figure 15-13   Influence of ammonia concentration on the end point for the titration of 50.0 mL of 0.00500 M Zn^{2+}. Solutions are buffered to pH 9.00. The shaded region shows the transition range for Eriochrome Black T. Note that ammonia decreases the change in pZn in the equivalence-point region.

15D-6   Indicators for EDTA Titrations

Reilley and Barnard[3] have listed nearly 200 organic compounds that have been investigated as indicators for metal ions in EDTA titrations. In general, these indicators are organic dyes that form colored chelates with metal ions in a pM range that is characteristic of the particular cation and dye. The complexes are often intensely colored and are discernible to the eye at concentrations in the range of 10^{-6} to 10^{-7} M.

Eriochrome Black T is a typical metal-ion indicator used in the titration of several common cations. The structural formula of Eriochrome Black T is shown in Figure 15-14. Its behavior as a weak acid is described by the equations

$$H_2O + H_2In^- \rightleftharpoons HIn^{2-} + H_3O^+ \qquad K_1 = 5 \times 10^{-7}$$
$$\text{red} \text{blue}$$

$$H_2O + HIn^{2-} \rightleftharpoons In^{3-} + H_3O^+ \qquad K_2 = 2.8 \times 10^{-12}$$
$$\text{blue} \phantom{^{2-} \rightleftharpoons} \text{orange}$$

Note that the acids and their conjugate bases have different colors. Thus, Eriochrome Black T behaves as an acid/base indicator as well as a metal-ion indicator.

The metal complexes of Eriochrome Black T are generally red, as is H_2In^-. Thus, for metal-ion detection, it is necessary to adjust the pH to 7 or above so that the blue form of the species, HIn^{2-}, predominates in the absence of a metal ion. Until the equivalence point in a titration, the indicator complexes the excess

---

[3]C. N. Reilley and A. J. Barnard, Jr., in *Handbook of Analytical Chemistry*, L. Meites, Ed. (New York: McGraw-Hill, 1963), p. 3-77.

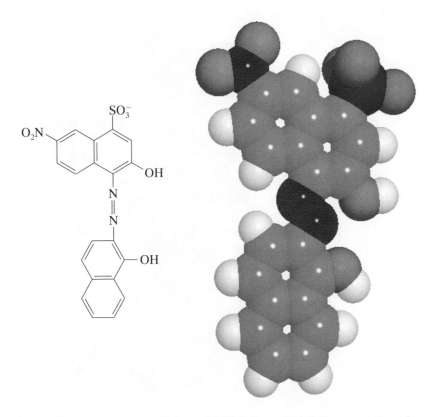

Figure 15-14   Structure and molecular model of Eriochrome Black T. The compound contains a sulfonic acid group that completely dissociates in water and two phenolic groups that only partially dissociate.

metal ion so that the solution is red. With the first slight excess of EDTA, the solution turns blue as a consequence of the reaction

$$\text{MIn}^- + \text{HY}^{3-} \rightleftharpoons \text{HIn}^{2-} + \text{MY}^{2-}$$
$$\quad\ \textbf{red} \qquad\qquad\qquad\ \textbf{blue}$$

Eriochrome Black T forms red complexes with more than two dozen metal ions, but the formation constants of only a few are appropriate for end-point detection. The applicability of a given indicator to an EDTA titration can be determined from the change in pM in the equivalence-point region, provided the formation constant for the metal-indicator complex is known.[4] Transition ranges for magnesium and calcium are indicated on the titration curves in Figure 15-9. The indicator is clearly ideal for the titration of magnesium and zinc (Figure 15-13); for calcium, however, it is unsatisfactory. Eriochrome Black T can be used in the determination of total calcium plus magnesium (total hardness) because both metals have similar formation constants with EDTA and are titrated together.

A limitation of Eriochrome Black T is that its solutions decompose slowly with standing. It is claimed that solutions of Calmagite (see Figure 15-15), an indicator that for all practical purposes is identical in behavior to Eriochrome Black T, do not suffer this disadvantage. Many other metal indicators have been

[4]C. N. Reilley and R. W. Schmid, *Anal. Chem.,* **1959,** *31,* 887.

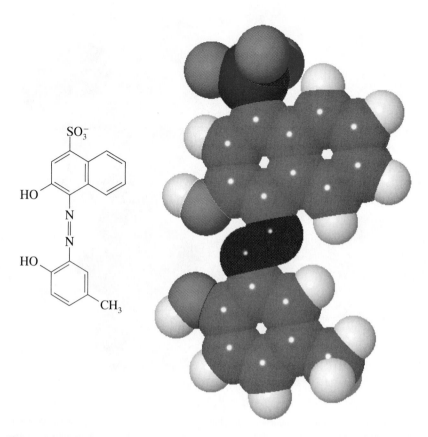

Figure 15-15   Structural formula and molecular model of Calmagite. Note the similarity to Eriochrome Black T (Figure 15-14).

developed for EDTA titrations.[5] In contrast to Eriochrome Black T, some of these indicators can be used in strongly acidic media.

15D-7   Titration Methods Employing EDTA

Several different types of titration methods can be used with EDTA as described below.

Direct Titration

Direct titration procedures with a metal-ion indicator that responds to the analyte are the easiest and most convenient methods to use. Methods that use an added metal ion are also widely employed.

Many of the metals in the periodic table can be determined by titration with standard EDTA solutions. Some methods are based on indicators that respond to the analyte itself, whereas others are based on an added metal ion.

Methods Based on Indicators for the Analyte.   Reilley and Barnard[6] list 40 elements that can be determined by direct titration with EDTA using metal-ion

[5]See, for example, J. A. Dean, *Analytical Chemistry Handbook* (New York: McGraw-Hill, 1995), pp. 3.94–3.96.
[6]C. N. Reilley and A. J. Barnard, Jr., in *Handbook of Analytical Chemistry,* L. Meites, Ed. (New York: McGraw-Hill, 1963), pp. 3-166 to 3-200.

indicators. Indicators that respond to the metal directly cannot be used in all cases either because no indicator with an appropriate transition range is available or because the reaction between the metal ion and EDTA is so slow as to make titration impractical.

Methods Based on Indicators for an Added Metal Ion.   In cases where a good, direct indicator for the analyte is unavailable, a small amount of a metal ion for which a good indicator is available can be added. The metal ion must form a complex that is less stable than the analyte complex. For example, indicators for calcium ion are generally less satisfactory than those we have described for magnesium ion. Consequently, a small amount of magnesium chloride is often added to an EDTA solution that is to be used for the determination of calcium. In this case, Eriochrome Black T can be used in the titration. In the initial stages, magnesium ions are displaced from the EDTA complex by calcium ions and are free to combine with the Eriochrome Black T, thus imparting a red color to the solution. When all the calcium ions have been complexed, however, the liberated magnesium ions again combine with the EDTA until the end point is observed. This procedure requires standardization of the EDTA solution against primary-standard calcium carbonate.

Potentiometric Methods.   Potential measurements can be used for end-point detection in the EDTA titration of those metal ions for which specific ion electrodes are available. Electrodes of this type are described in Section 19D. In addition, a mercury electrode can be made sensitive to EDTA ions and used in titrations with this reagent.

Spectrophotometric Methods.   Measurement of UV/visible absorption can also be used to determine the end points of titrations (see Section 23A-2). In these cases, an instrument responds to the color change in the titration rather than relying on a visual determination of the end point.

Back-Titration Methods

Back-titrations are useful for the determination of cations that form stable EDTA complexes and for which a satisfactory indicator is not available; the determination of thallium is an extreme example. The method is also useful for cations such as Cr(III) and Co(III) that react only slowly with EDTA. A measured excess of standard EDTA solution is added to the analyte solution. After the reaction is judged complete, the excess EDTA is back-titrated with a standard magnesium or zinc ion solution to an Eriochrome Black T or Calmagite end point.[7] For this procedure to be successful, it is necessary that the magnesium or zinc ions form an EDTA complex that is less stable than the corresponding analyte complex.

Back-titration is also useful for analyzing samples that contain anions that would otherwise form sparingly soluble precipitates with the analyte under the analytical conditions. Here, the excess EDTA prevents precipitate formation.

Back-titration procedures are used when no suitable indicator is available, when the reaction between analyte and EDTA is slow, or when the analyte forms precipitates at the pH required for its titration.

[7]For a discussion of the back-titration procedure, see C. Macca and M. Fiorana, *J. Chem. Educ.*, **1986,** *63,* 121.

Displacement Methods

In displacement titrations, an unmeasured excess of a solution containing the magnesium or zinc complex of EDTA is introduced into the analyte solution. If the analyte forms a more stable complex than that of magnesium or zinc, the following displacement reaction occurs:

$$MgY^{2-} + M^{2+} \rightarrow MY^{2-} + Mg^{2+}$$

where M^{2+} represents the analyte cation. The liberated Mg^{2+} or, in some cases Zn^{2+}, is then titrated with a standard EDTA solution. Displacement titrations are used when no indicator for an analyte is available.

15D-8   The Scope of EDTA Titrations

Complexometric titrations with EDTA have been applied to the determination of virtually every metal cation with the exception of the alkali metal ions. Because EDTA complexes most cations, the reagent might appear at first glance to be totally lacking in selectivity. In fact, however, considerable control over interferences can be realized by pH regulation. For example, trivalent cations can usually be titrated without interference from divalent species by maintaining the solution at a pH of about 1 (Figure 15-12). At this pH, the less stable divalent chelates do not form to any significant extent, but the trivalent ions are quantitatively complexed.

Similarly, ions such as cadmium and zinc, which form more stable EDTA chelates than does magnesium, can be determined in the presence of the latter ion by buffering the mixture to pH 7 before titration. Eriochrome Black T serves as an indicator for the cadmium or zinc end points without interference from magnesium because the indicator chelate with magnesium is not formed at this pH.

Finally, interference from a particular cation can sometimes be eliminated by adding a suitable *masking agent,* an auxiliary ligand that preferentially forms highly stable complexes with the potential interfering ion.[8] Thus, cyanide ion is often employed as a masking agent to permit the titration of magnesium and calcium ions in the presence of ions such as cadmium, cobalt, copper, nickel, zinc, and palladium. The latter all form sufficiently stable cyanide complexes to prevent reaction with EDTA. Feature 15-4 illustrates how masking and demasking reagents are used to improve the selectivity of EDTA reactions.

> A *masking agent* is a complexing agent that reacts selectively with a component in a solution to prevent that component from interfering in a determination.

15D-9   Determining Water Hardness

Historically, water "hardness" was defined in terms of the capacity of cations in the water to replace the sodium or potassium ions in soaps and form sparingly soluble products, which cause "scum" in the sink or bathtub. Most multiply charged cations share this undesirable property. In natural waters, however, the concentrations of calcium and magnesium ions generally far exceed that of any

> Hard water contains calcium, magnesium, and heavy metal ions that form precipitates with soap (but not detergents).

---

[8]For further information, see D. D. Perrin, *Masking and Demasking of Chemical Reactions* (New York: Wiley-Interscience, 1970); J. A. Dean, *Analytical Chemistry Handbook,* (New York: McGraw-Hill, 1995), pp. 3.92 to 3.111.

other metal ion. Consequently, hardness is now expressed in terms of the concentration of calcium carbonate that is equivalent to the total concentration of all the multivalent cations in the sample.

The determination of hardness is a useful analytical test that provides a measure of the quality of water for household and industrial uses. The test is important to industry because hard water, when heated, precipitates calcium carbonate, which then clogs boilers and pipes.

Water hardness is ordinarily determined by an EDTA titration after the sample has been buffered to pH 10. Magnesium, which forms the least stable EDTA complex of all the common multivalent cations in typical water samples, does not form a stable EDTA complex until enough EDTA has been added to complex all the other cations in the sample. Therefore, a magnesium ion indicator, such as Calmagite or Eriochrome Black T, can serve as indicator in water-hardness titrations. Often, a small concentration of the magnesium-EDTA chelate is incorporated in the buffer or in the titrant to ensure the presence of sufficient magnesium ions for satisfactory indicator action.

Feature 15-4

How Can Masking and Demasking Agents Be Used to Enhance the Selectivity of EDTA Titrations?

Lead, magnesium, and zinc can be determined on a single sample by two titrations with standard EDTA and one titration with standard Mg^{2+}. The sample is first treated with an excess of NaCN, which masks Zn^{2+} and prevents it from reacting with EDTA.

$$Zn^{2+} + 4CN^- \rightleftharpoons Zn(CN)_4^{2-}$$

The Pb^{2+} and Mg^{2+} are then titrated with standard EDTA. After the equivalence point has been reached, a solution of the complexing agent BAL (2-3-dimercapto-1-propanol, $CH_2SHCHSHCH_2OH$), which we will write as $R(SH)_2$, is added to the solution. This bidentate ligand reacts selectively to form a complex with Pb^{2+} that is much more stable than PbY^{2-}.

$$PbY^{2-} + 2R(SH)_2 \rightarrow Pb(RS)_2 + 2H^+ + Y^{4-}$$

The liberated Y^{4-} is then titrated with a standard solution of Mg^{2+}. Finally, the zinc is demasked by adding formaldehyde.

$$Zn(CN)_4^{2-} + 4HCHO + 4H_2O \rightarrow Zn^{2+} + 4HOCH_2CN + 4OH^-$$

The liberated Zn^{2+} is then titrated with the standard EDTA solution.

Suppose the initial titration of Mg^{2+} and Pb^{2+} required 42.22 mL of 0.02064 M EDTA. Titration of the Y^{4-} liberated by the BAL consumed 19.35 mL of 0.007657 M Mg^{2+}. Finally, after addition of formaldehyde, the liberated Zn^{2+} was titrated with 28.63 mL of the EDTA. Calculate the percent of the three elements if a 0.4085-g sample was used.

The initial titration reveals the number of millimoles of Pb^{2+} and Mg^{2+} present. That is,

$$mmol\ (Pb^{2+} + Mg^{2+}) = 42.22 \times 0.02064 = 0.87142$$

The second titration gives the number of millimoles of Pb^{2+}. Thus,

$$\text{mmol Pb}^{2+} = 19.35 \times 0.007657 = 0.14816$$

$$\text{mmol Mg}^{2+} = 0.87142 - 0.14816 = 0.72326$$

Finally, from the third titration we obtain

$$\text{mmol Zn}^{2+} = 28.63 \times 0.02064 = 0.59092$$

To obtain the percentages, we write

$$\frac{0.014816 \text{ mmol Pb} \times 0.2072 \text{ g Pb/mmol Pb}}{0.4085 \text{ g sample}} \times 100\% = 7.515\% \text{ Pb}$$

$$\frac{0.72326 \text{ mmol Mg} \times 0.024305 \text{ g Mg/mmol Mg}}{0.4085 \text{ g sample}} \times 100\% = 4.303\% \text{ Mg}$$

$$\frac{0.59095 \text{ mmol Zn} \times 0.06539 \text{ g Zn/mmol Zn}}{0.4085 \text{ g sample}} \times 100\% = 9.459\% \text{ Zn}$$

Feature 15-5

Test Kits for Water Hardness

Test kits for determining the hardness of household water are available at stores selling water softeners and plumbing supplies. The kits usually consist of a vessel calibrated to contain a known volume of water, a measuring scoop to deliver an appropriate amount of a solid buffer mixture, an indicator solution, and a bottle of standard EDTA equipped with a medicine dropper. The drops of standard reagent needed to cause a color change are counted. The concentration of the EDTA solution is ordinarily such that 1 drop corresponds to 1 grain (about 0.065 g) of calcium carbonate per gallon of water.

WEB WORKS

There are several good educational sites available that give additional help on complexation equilibria and complexometric titrations. For additional information on chemistry as applied to aquatic systems, see *http://nexus.chemistry.duq.edu/snes/esm/Course_Material/ESM552/Notes/Chapter3/chapter3.html*. Complexation equilibria are described near the end of the file. There are also good discussions of acid/base equilibria, buffers and other systems of interest in aquatic chemistry.

Several Web sites describe experiments that can be done in the laboratory based on complexation methods. For an abstract of two of these see *http://jchemed.chem.wisc.edu/Journal/Issues/1997/Dec/abs1422.html* and *http://jchemed.chem.wisc.edu/Journal/Issues/1997/Dec/abs1463.html*. The first describes the determination of water hardness, while the second experiment involves the determination of zinc by a complexometric titration. The papers appear in the December, 1997 issue of the Journal of Chemical Education.

15E QUESTIONS AND PROBLEMS

15-1. Define

*(a) chelate.

(b) tetradentate chelating agent.

*(c) ligand.

(d) coordination number.

*(e) conditional formation constant.

(f) NTA.

*(g) water hardness.

(h) EDTA displacement titration.

*15-2. Describe three general methods for performing EDTA titrations. What are the advantages of each?

15-3. Why are multidentate ligands preferable to unidentate ligands for complexometric titrations?

15-4. Write chemical equations and equilibrium-constant expressions for the stepwise formation of

*(a) $Ag(S_2O_3)_2^{3-}$.   (b) $Ni(SCN)_3^-$.

*15-5. Explain how stepwise and overall formation constants are related.

*15-6. In what respect is the Fajans method superior to the Volhard method for the titration of chloride ion?

15-7. Briefly explain why the sparingly soluble product must be removed by filtration before you back-titrate the excess silver ion in the Volhard determination of

(a) chloride ion.

(b) cyanide ion.

(c) carbonate ion.

15-8. Why does the charge on the surface of precipitate particles change sign at the equivalence point in a titration?

*15-9. Outline a method for the determination of K^+ based on argentometry. Write balanced equations for the reactions.

15-10. Propose a complexometric method for the determination of the individual components in a solution containing In^{3+}, Zn^{2+}, and Mg^{2+}.

15-11. Why is a small amount of MgY^{2-} often added to a water specimen that is to be titrated for hardness?

*15-12. An EDTA solution was prepared by dissolving 3.853 g of purified and dried $Na_2H_2Y \cdot H_2O$ in sufficient water to give 1.000 L. Calculate the molar concentration, given that the solute contained 0.3% excess moisture (page 362).

15-13. A solution was prepared by dissolving about 3.0 g of $Na_2H_2Y \cdot 2H_2O$ in approximately 1 L of water and standardizing against 50.00-mL aliquots of 0.004517 M Mg^{2+}. An average titra-

tion volume of 32.22 mL was required. Calculate the molar concentration of the EDTA.

15-14. Calculate the volume of 0.0500 M EDTA needed to titrate

*(a) 26.37 mL of 0.0741 M $Mg(NO_3)_2$.

(b) the Ca in 0.2145 g of $CaCO_3$.

*(c) the Ca in a 0.4397-g mineral specimen that is 81.4% brushite, $CaHPO_4 \cdot 2H_2O$ (172.09 g/mol).

(d) the Mg in a 0.2080-g sample of the mineral hydromagnesite, $3MgCO_3Mg(OH)_2 \cdot 3H_2O$ (365.3 g/mol).

*(e) the Ca and Mg in a 0.1557-g sample that is 92.5% dolomite, $CaCO_3 \cdot MgCO_3$ (184.4 g/mol).

15-15. A solution contains 1.694 mg of $CoSO_4$ (155.0 g/mol) per milliliter. Calculate

(a) the volume of 0.08640 M EDTA needed to titrate a 25.00-mL aliquot of this solution.

(b) the volume of 0.009450 M Zn^{2+} needed to titrate the excess reagent after addition of 50.00 mL of 0.008640 M EDTA to a 25.00-mL aliquot of this solution.

(c) the volume of 0.008640 M EDTA needed to titrate the Zn^{2+} displaced by Co^{2+} following addition of an unmeasured excess of ZnY^{2-} to a 25.00-mL aliquot of the $CoSO_4$ solution. The reaction is

$$Co^{2+} + ZnY^{2-} \rightarrow CoY^{2-} + Zn^{2+}$$

*15-16. The Zn in a 0.7556-g sample of foot powder was titrated with 21.27 mL of 0.01645 M EDTA. Calculate the percentage of Zn in this sample.

*15-17. The Cr plating on a surface that measured 3.00×4.00 cm was dissolved in HCl. The pH was suitably adjusted, following which 15.00 mL of 0.01768 M EDTA were introduced. The excess reagent required a 4.30-mL back-titration with 0.008120 M Cu^{2+}. Calculate the average weight of Cr on each square centimeter of surface.

15-18. A silver nitrate solution contains 14.77 g of primary-standard $AgNO_3$ in 1.00 L. What volume of this solution will be needed to react with

*(a) 0.2631 g of NaCl.

(b) 0.1799 g of Na_2CrO_4.

*(c) 64.13 mg of Na_3AsO_4.

(d) 381.1 mg of $BaCl_2 \cdot 2H_2O$.

*(e) 25.00 mL of 0.05361 M Na_3PO_4.

(f) 50.00 mL of 0.01808 M H_2S.

15-19. What is the molar analytical concentration of a silver nitrate solution if a 25.00-mL aliquot reacts with the amount of solutes listed in Problem 15-18?

15-20. What minimum volume of 0.09621 M $AgNO_3$ will be needed to assure an excess of silver ion in the titration of

  *(a) an impure NaCl sample that weighs 0.2513 g?

  (b) a 0.3462-g sample that is 74.52% (w/w) $ZnCl_2$?

  *(c) 25.00 mL of 0.01907 M $AlCl_3$?

15-21. A Fajans titration of a 0.7908-g sample required 45.32 mL of 0.1046 M $AgNO_3$. Express the results of this analysis in terms of the percentage of

  (a) Cl^-.

  (b) $BaCl_2 \cdot 2H_2O$.

  (c) $ZnCl_2 \cdot 2NH_4Cl$ (243.28 g/mol).

*15-22. The Tl in a 9.76-g sample of rodenticide was oxidized to the trivalent state and treated with an unmeasured excess of Mg/EDTA solution. The reaction is

$$Tl^{3+} + MgY^{2-} \rightarrow TlY^- + Mg^{2+}$$

Titration of the liberated Mg^{2+} required 13.34 mL of 0.03560 M EDTA. Calculate the percentage of Tl_2SO_4 (504.8 g/mol) in the sample.

15-23. An EDTA solution was prepared by dissolving approximately 4 g of the disodium salt in approximately 1 L of water. An average of 42.35 mL of this solution was required to titrate 50.00-mL aliquots of a standard that contained 0.7682 g of $MgCO_3$ per liter. Titration of a 25.00-mL sample of mineral water at pH 10 required 18.81 mL of the EDTA solution. A 50.00-mL aliquot of the mineral water was rendered strongly alkaline to precipitate the magnesium as $Mg(OH)_2$. Titration with a calcium specific indicator required 31.54 mL of the EDTA solution. Calculate

  (a) the molarity of the EDTA solution.

  (b) the ppm of $CaCO_3$ in the mineral water.

  (c) the ppm of $MgCO_3$ in the mineral water.

*15-24. A 50.00-mL aliquot of a solution containing iron(II) and iron(III) required 13.73 mL of 0.01200 M EDTA when titrated at pH 2.0 and 29.62 mL when titrated at pH 6.0. Express the concentration of the solution in terms of the parts per million of each solute.

15-25. A 24-hr urine specimen was diluted to 2.000 L. After the solution was buffered to pH 10, a 10.00-mL aliquot was titrated with 26.81 mL of 0.003474 M EDTA. The calcium in a second 10.00-mL aliquot was isolated as $CaC_2O_4(s)$, redissolved in acid, and titrated with 11.63 mL of the EDTA solution. Assuming that 15 to 300 mg of magnesium and 50 to 400 mg of calcium per day are normal, did this specimen fall within these ranges?

*15-26. A 1.509-g sample of a Pb/Cd alloy was dissolved in acid and diluted to exactly 250.0 mL in a volumetric flask. A 50.00-mL aliquot of the diluted solution was brought to a pH of 10.0 with a NH_3/NH_4^+ buffer; the subsequent titration involved both cations and required 28.89 mL of 0.06950 M EDTA. A second 50.00-mL aliquot was brought to a pH of 10.0 with an HCN/NaCN buffer, which also served to mask the Cd^{2+}; 11.56 mL of the EDTA solution were needed to titrate the Pb^{2+}. Calculate the percentages of Pb and Cd in the sample.

15-27. A 0.6004-g sample of Ni/Cu condenser tubing was dissolved in acid and diluted to 100.0 mL in a volumetric flask. Titration of both cations in a 25.00-mL aliquot of this solution required 45.81 mL of 0.05285 M EDTA. Mercaptoacetic acid and NH_3 were then introduced; production of the Cu complex with the former resulted in the release of an equivalent amount of EDTA, which required a 22.85-mL titration with 0.07238 M Mg^{2+}. Calculate the percentages of Cu and Ni in the alloy.

*15-28. Calamine, which is used for relief of skin irritations, is a mixture of zinc and iron oxides. A 1.022-g sample of dried calamine was dissolved in acid and diluted to 250.0 mL. Potassium fluoride was added to a 10.00-mL aliquot of the diluted solution to mask the iron; after suitable adjustment of the pH, Zn^{2+} consumed 38.71 mL of 0.01294 M EDTA. A second 50.00-mL aliquot was suitably buffered and titrated with 2.40 mL of 0.002727 M ZnY^{2-} solution:

$$Fe^{3+} + ZnY^{2-} \rightarrow FeY^- + Zn^{2+}$$

Calculate the percentages of ZnO and Fe_2O_3 in the sample.

15-29. The potassium ion in a 250.0-mL sample of mineral water was precipitated with sodium tetraphenylborate:

$$K^+ + B(C_6H_4)_4^- \rightarrow KB(C_6H_4)(s)$$

The precipitate was filtered, washed, and redissolved in an organic solvent. An excess of the mercury(II)/EDTA chelate was added:

$$4HgY^{2-} + B(C_6H_4)_4^- + 4H_2O \rightarrow$$
$$H_3BO_3 + 4C_6H_5Hg^+ + 4HY^{3-} + OH^-$$

The liberated EDTA was titrated with 29.64 mL of 0.05581 M Mg^{2+}. Calculate the potassium ion concentration in parts per million.

*15-30. Chromel is an alloy composed of nickel, iron, and chromium. A 0.6472-g sample was dissolved and diluted to 250.0 mL. When a 50.00-mL aliquot of 0.05182 M EDTA was mixed with an equal volume of the diluted sample, all three ions were chelated and a 5.11-mL back-titration with 0.06241 M copper(II) was required. The chromium in a second 50.0-mL aliquot was masked through the addition of hexamethylenetetramine; titration of the Fe and Ni required 36.28 mL of 0.05182 M EDTA. Iron and chromium were masked with pyrophosphate in a third 50.0-mL aliquot, and the nickel was titrated with 25.91 mL of the EDTA solution. Calculate the percentages of nickel, chromium, and iron in the alloy.

15-31. A 0.3284-g sample of brass (containing lead, zinc, copper, and tin) was dissolved in nitric acid. The sparingly soluble $SnO_2 \cdot 4H_2O$ was removed by filtration, and the combined filtrate and washings were then diluted to 500.0 mL. A 10.00-mL aliquot was suitably buffered; titration of the lead, zinc, and copper in this aliquot required 37.56 mL of 0.002500 M EDTA. The copper in a 25.00-mL aliquot was masked with thiosulfate; the lead and zinc were then titrated with 27.67 mL of the EDTA solution. Cyanide ion was used to mask the copper and zinc in a 100-mL aliquot; 10.80 mL of the EDTA solution was needed to titrate the lead ion. Determine the composition of the brass sample and evaluate the percentage of tin by difference.

*15-32. Calculate conditional constants for the formation of the EDTA complex of Fe^{2+} at a pH of
 (a) 6.0
 (b) 8.0
 (c) 10.0

15-33. Calculate conditional constants for the formation of the EDTA complex of Ba^{2+} at a pH of
 (a) 7.0
 (b) 9.0
 (c) 11.0

*15-34. Construct a titration curve for 50.00 mL of 0.01000 M Sr^{2+} with 0.02000 M EDTA in a so-

lution buffered to pH 11.0. Calculate pSr values after the addition of 0.00, 10.00, 24.00, 24.90, 25.00, 25.10, 26.00, and 30.00 mL of titrant.

15-35. Construct a titration curve for 50.00 mL of 0.0150 M Fe^{2+} with 0.0300 M EDTA in a solution buffered to pH 7.0. Calculate pFe values after the addition of 0.00, 10.00, 24.00, 24.90, 25.00, 25.10, 26.00, and 30.00 mL of titrant.

15-36. For each of the following precipitation titrations, calculate the cation and anion concentrations at equivalence as well as at reagent volumes corresponding to ±20.00 mL, ±10.00 mL, and ±1.00 mL of equivalence. Construct a titration curve from the data, plotting the p-function of the cation versus reagent volume.
 *(a) 25.00 mL of 0.05000 M $AgNO_3$ with 0.02500 M NH_4SCN.
 (b) 20.00 mL of 0.06000 M $AgNO_3$ with 0.03000 M KI.
 *(c) 30.00 mL of 0.07500 M $AgNO_3$ with 0.07500 M NaCl.
 (d) 35.00 mL of 0.4000 M Na_2SO_4 with 0.2000 M $Pb(NO_3)_2$.
 *(e) 40.00 mL of 0.02500 M $BaCl_2$ with 0.05000 M Na_2SO_4.
 (f) 50.00 mL of 0.2000 M NaI with 0.4000 M $TlNO_3$ (K_{sp} for TlI = 6.5×10^{-8}).

15-37. Calculate the silver ion concentration after the addition of 5.00*, 15.00, 25.00, 30.00, 35.00, 39.00, 40.00*, 41.00, 45.00*, and 50.00 mL of 0.05000 M $AgNO_3$ to 50.0 mL of 0.0400 M KBr. Construct a titration curve from these data plotting pAg as a function of titrant volume.

15-38. Titration of Ca^{2+} and Mg^{2+} in a 50.00-mL sample of hard water required 22.35 mL of 0.01115 M EDTA. A second 50.00-mL aliquot was made strongly basic with NaOH to precipitate the Mg^{2+} as $Mg(OH)_2(s)$. The supernatant liquid was titrated with 15.19 mL of the EDTA solution. Calculate
 (a) the total hardness of the water sample, expressed as ppm $CaCO_3$.
 *(b) the concentration in ppm of $CaCO_3$ in the sample.
 (c) the concentration in ppm of $MgCO_3$ in the sample.

Section III

Electrochemical Methods

Outline

Analytical methods based on electrochemistry are widely used in many scientific disciplines. The measurement of pH with the glass/calomel electrode pair is the one analytical measurement that is nearly ubiquitous in scientific laboratories. Hundreds of thousands of pH and other electrochemical measurements are made daily in clinical, environmental, research, and industrial laboratories. The five chapters in this section treat the most popular and most common electrochemical techniques.

Chapter 16 introduces the principles of electrochemistry and lays the foundation for the remaining chapters in this section. Chapter 17 describes how electrode potentials can be used to calculate equilibrium constants of oxidation/reduction reactions and to construct curves for oxidation/reduction titrations. Chapter 18 presents the applications of oxidation/reduction titrations and some practical information about reagents and indicators. The focus of Chapter 19 is on the use of potentiometry for measuring concentrations of ionic and molecular species. The glass electrode and several other ion-selective electrodes are described in this chapter. Chapter 20 completes this section with a discussion of several electrolytic analytical methods, including electrogravimetry, coulometry and voltammetry.

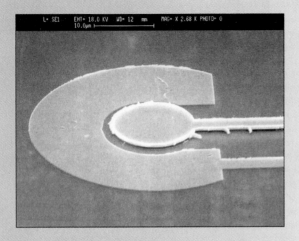

Electron photomicrograph of an immobilized enzyme-based biosensor fabricated by microlithographic techniques.[1] The same technology used to construct computer microchips was used to produce the glucose oxidase-coated biosensor shown in the photo. A silicon nitride layer is first laid down on the surface of a silicon wafer. Then, after suitable application of a polymer coating to mask all but the outer loop, a layer of chromium and a layer of silver are vapor deposited only on the surface of the loop. The polymer is then removed, and the entire area except for the central disk is coated with polymer. A layer of chromium and a layer of platinum are then vapor deposited on the surface of the disk. After removal of the polymer, glucose oxidase is deposited through a microstencil onto the central disk by a laser desorption ionization technique. The resulting biosensor can be used to measure the concentration of glucose in a variety of analyte solutions using voltammetric techniques as described in Section 20E-2. The distance scale indicated in the legend to the photo shows the sensor to be only about 30 μm in diameter. Because of its small size, the sensor can be used to analyze extremely small samples or to make *in vivo* glucose determinations. (Photo courtesy of P. Morales. Reproduced by permission of the American Institute of Physics.)

[1] E. Di Fabrizio, M. Gentili, P. Morales, R. Pilloton, J. Mela, S. Santucci, and A. Sese, *Appl. Phys. Lett.*, **1996**, *69*(21), 3280.

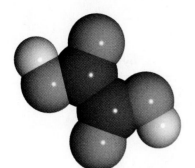

Chapter 16

Elements of Electrochemistry

We now turn our attention to several analytical methods that are based on oxidation/reduction reactions. These methods, which are described in Chapters 16 through 20, include oxidation/reduction titrimetry, potentiometry, coulometry, electrogravimetry, and voltammetry. Fundamentals of electrochemistry necessary for understanding the theory of these procedures are presented in the current chapter.

16A CHARACTERIZING OXIDATION/ REDUCTION REACTIONS

Oxidation is the loss of electrons by a species while *reduction* is the gain of electrons. An *oxidation/reduction reaction* is one in which electrons are transferred from one reactant to another. An example is the oxidation of iron(II) ions by cerium(IV) ions. The reaction is described by the equation

$$Ce^{4+} + Fe^{2+} \rightleftharpoons Ce^{3+} + Fe^{3+} \tag{16-1}$$

In this reaction, an electron is transferred from Fe^{2+} to Ce^{4+} to form Ce^{3+} and Fe^{3+} ions. A substance, such as Ce^{4+}, that accepts electrons during the course of a reaction is called an *oxidizing agent,* or an *oxidant.* A *reducing agent,* or *reductant,* is a species, such as Fe^{2+}, that readily donates electrons to another species. With regard to Equation 16-1, we say that Fe^{2+} is oxidized by Ce^{4+}; similarly, Ce^{4+} is reduced by Fe^{2+}.

We can split any oxidation/reduction equation into two half-reactions that show clearly which species gains electrons and which loses them. For example, Equation 16-1 is the sum of the two half-reactions

$$Ce^{4+} + e^- \rightleftharpoons Ce^{3+} \quad \text{(reduction of } Ce^{4+})$$

$$Fe^{2+} \rightleftharpoons Fe^{3+} + e^- \quad \text{(oxidation of } Fe^{2+})$$

The rules for balancing half-reactions are the same as those for other reaction types; that is, the number of atoms of each element as well as the net charge on each side of the equation must be the same. Thus, for the oxidation of Fe^{2+} by MnO_4^-, the half-reactions are

> *Oxidation/reduction reactions* are sometimes called *redox* reactions.

> A *reducing agent* is an electron donor. An *oxidizing agent* is an electron acceptor.

> It is important to understand that although we can readily write an equation for a half-reaction in which electrons are consumed or generated, we cannot observe an isolated half-reaction experimentally because there must always be present a second half-reaction that serves as a source of electrons or a recipient of electrons; that is, an individual half-reaction is a theoretical concept.

Copyright 1993, Johnny Hart and Creator's Syndicate.

$$MnO_4^- + 8H^+ + 5e^- \rightleftharpoons Mn^{2+} + 4H_2O$$

$$5Fe^{2+} \rightleftharpoons 5Fe^{3+} + 5e^-$$

In the first half-reaction, the net charge on the left side is $(-1 - 5 + 8) = +2$, which is the same as that on the right. Note also that we have multiplied the second half-reaction by 5 so that the number of electrons lost by Fe^{2+} equals the number gained by MnO_4^-. A balanced net-ionic equation (see Feature 16-1) for the overall reaction is then obtained by adding the two half-reactions:

$$MnO_4^- + 5Fe^{2+} + 8H^+ \rightleftharpoons Mn^{2+} + 5Fe^{3+} + 4H_2O$$

16A-1   Comparing Redox Reactions to Acid/Base Reactions

Recall that in the Brønsted–Lowry concept an acid/base reaction is described by the equation

$$acid_1 + base_2 \rightleftharpoons base_1 + acid_2$$

Oxidation/reduction reactions can be viewed in a way that is analogous to the Brønsted–Lowry concept of acid/base reactions (Section 4A-2). Both involve the transfer of one or more charged particles from a donor to an acceptor, the particles being electrons in oxidation/reduction and protons in neutralization. When an acid donates a proton, it becomes a conjugate base that is capable of accepting a proton. By analogy, when a reducing agent donates an electron, it becomes an oxidizing agent that can accept an electron. This product could be called a conjugate oxidant, but this terminology is seldom, if ever, used. We may then write a generalized equation for a redox reaction as

$$A_{red} + B_{ox} \rightleftharpoons A_{ox} + B_{red} \tag{16-2}$$

Here, B_{ox}, the oxidized form of species B, accepts electrons from A_{red} to form the new reductant, B_{red}. At the same time, reductant A_{red}, having given up electrons, becomes an oxidizing agent, A_{ox}. If we know from chemical evidence that the equilibrium in Equation 16-2 lies to the right, we can state that B_{ox} is a better electron acceptor (stronger oxidant) than A_{ox}. Furthermore, A_{red} is a more effective electron donor (better reductant) than B_{red}. Several specific cases are described in Example 16-1.

A knowledge of how to balance oxidation/reduction reactions is essential to understanding all the concepts covered in this chapter. Although you probably remember this technique from your general chemistry course, a quick review here will remind you of how the process works. For practice, let us complete and balance the following equation after adding H^+, OH^-, or H_2O as needed.

$$MnO_4^- + NO_2^- \rightleftharpoons Mn^{2+} + NO_3^-$$

First, we write and balance the two half-reactions involved. For MnO_4^-, we first write

$$MnO_4^- \rightleftharpoons Mn^{2+}$$

To account for the four oxygen atoms on the left-hand side of the equation, we add $4H_2O$ on the right-hand side of the equation, which means that we must provide $8H^+$ on the left:

$$MnO_4^- + 8H^+ \rightleftharpoons Mn^{2+} + 4H_2O$$

To balance the charge, we need to add five electrons to the left side of the equation. Thus,

$$MnO_4^- + 8H^+ + 5e^- \rightleftharpoons Mn^{2+} + 4H_2O$$

For the other half-reaction,

$$NO_2^- \rightleftharpoons NO_3^-$$

we add one H_2O to the left side of the equation to supply the needed oxygen and $2H^+$ on the right.

$$NO_2^- + H_2O \rightleftharpoons NO_3^- + 2H^+$$

Then we add two electrons to the right-hand side to balance the charge.

$$NO_2^- + H_2O \rightleftharpoons NO_3^- + 2H^+ + 2e^-$$

Before combining the two equations, we must multiply the first by 2 and the second by 5 so that the electrons will cancel. We then add the two equations to obtain

$$2MnO_4^- + 16H^+ + 10e^- + 5NO_2^- + 5H_2O \rightleftharpoons$$
$$2Mn^{2+} + 8H_2O + 5NO_3^- + 10H^+ + 10e^-$$

which then rearranges to the balanced equation

$$2MnO_4^- + 6H^+ + 5NO_2^- \rightleftharpoons 2Mn^{2+} + 5NO_3^- + 3H_2O$$

Example 16-1

The following reactions are spontaneous and thus proceed to the right, as written:

$$2H^+ + Cd(s) \rightleftharpoons H_2 + Cd^{2+}$$

$$2Ag^+ + H_2(g) \rightleftharpoons 2Ag(s) + 2H^+$$

$$Cd^{2+} + Zn(s) \rightleftharpoons Cd(s) + Zn^{2+}$$

What can we deduce regarding the strengths of H^+, Ag^+, Cd^{2+}, and Zn^{2+} as electron acceptors (or oxidizing agents)?

Figure 16-1   Photograph of a "silver tree."

For an interesting illustration of this reaction, immerse a piece of copper in a solution of silver nitrate. The result is the deposition of silver on the copper in the form of a "silver tree." See Figure 16-1 and color plate 13?

Salt bridges are widely used to prevent mixing of the contents of the two electrolyte solutions making up electrochemical cells. Ordinarily, the two ends of the bridge are equipped with sintered glass disks or other devices to prevent liquid from siphoning from one part of the cell to the other.

When the $CuSO_4$ and $AgNO_3$ solutions are 0.0200 M, the cell develops a potential of 0.412 V as shown in Figure 16-2a.

The second reaction establishes that Ag^+ is a more effective electron acceptor than H^+; the first reaction demonstrates that H^+ is more effective than Cd^{2+}. Finally, the third equation shows that Cd^{2+} is more effective than Zn^{2+}. Thus, the order of oxidizing strength is $Ag^+ > H^+ > Cd^{2+} > Zn^{2+}$.

16A-2   Oxidation/Reduction Reactions in Electrochemical Cells

Many oxidation/reduction reactions can be carried out in either of two ways; the two methods are physically quite different. In one, the reaction is performed by direct contact between the oxidant and the reductant in a suitable container. In the second, the reactants do not come in direct contact with one another. An example of the first process involves immersing a strip of copper in a solution containing silver nitrate. Here, silver ions migrate to the metal and are reduced:

$$Ag^+ + e^- \rightleftharpoons Ag(s)$$

At the same time, an equivalent quantity of copper is oxidized:

$$Cu(s) \rightleftharpoons Cu^{2+} + 2e^-$$

Multiplication of the silver half-reaction by two and addition yields a net-ionic equation for the overall process.

$$2Ag^+ + Cu(s) \rightleftharpoons 2Ag(s) + Cu^{2+} \tag{16-3}$$

A unique aspect of oxidation/reduction reactions is that the transfer of electrons — and thus an identical net reaction — can often be brought about in an *electrochemical cell* in which the oxidizing agent and the reducing agent are physically separated from one another. Figure 16-2a shows such an arrangement. Note that a *salt bridge* isolates the reactants but maintains electrical contact between the two halves of the cell. When a voltmeter of high internal resistance is connected as shown or when the electrodes are not connected externally, the cell is said to be at *open circuit* and delivers the full cell potential. When the circuit is open, no net reaction occurs in the cell although, as we shall show, the cell has the *potential* for doing work. The voltmeter measures the potential difference, or *voltage,* between the two electrodes at any instant. This voltage, also called the *cell potential,* is a measure of the tendency of the cell reaction to proceed toward equilibrium.

In Figure 16-2b, the cell is connected so that electrons can flow in the external circuit. The potential energy of the cell can now be converted into electrical energy to light a lamp or do some other type of electrical work. In the cell in Figure 16-2b, metallic copper is oxidized at the left-hand electrode, silver ions are reduced at the right-hand electrode, and electrons flow through the external circuit to the silver electrode. As the reaction goes on, the cell potential, initially 0.412 V when the circuit is open, decreases continuously and approaches zero as the state of equilibrium for the overall reaction is approached. When the cell is at equilibrium, the cell voltage is zero. A cell with zero voltage cannot perform work, as anyone who has experienced a dead battery in a flashlight or a laptop computer can attest.

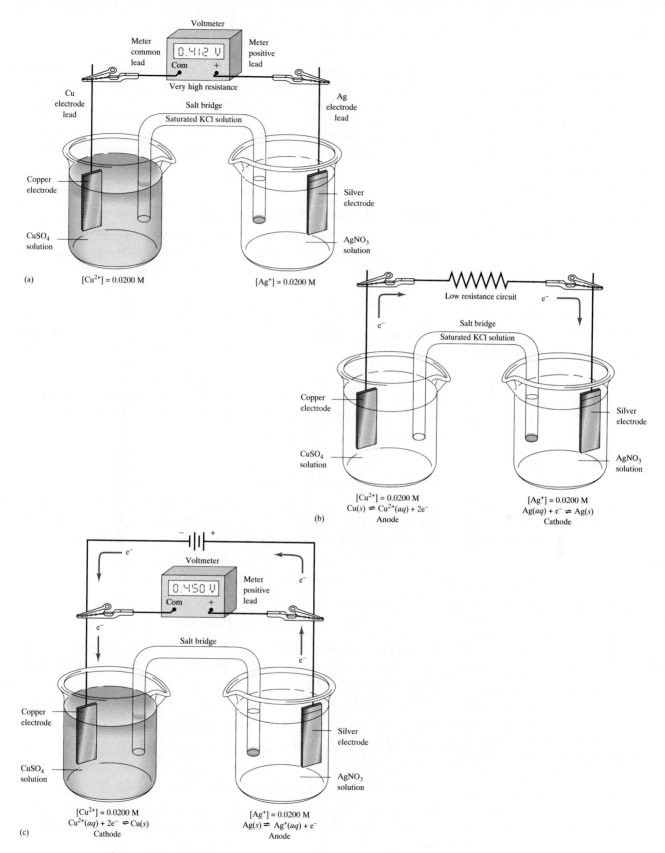

Voltmeter

0.4112 V

Meter common lead   Com   +   Meter positive lead

Very high resistance

Cu electrode lead

Ag electrode lead

Salt bridge
Saturated KCl solution

Copper electrode

Silver electrode

CuSO$_4$ solution

AgNO$_3$ solution

(a)   [Cu^{2+}] = 0.0200 M   [Ag$^+$] = 0.0200 M

e$^-$ Low resistance circuit   e$^-$

Salt bridge
Saturated KCl solution

Copper electrode

Silver electrode

CuSO$_4$ solution

AgNO$_3$ solution

[Cu^{2+}] = 0.0200 M   [Ag$^+$] = 0.0200 M
Cu(s) $\rightleftharpoons$ Cu^{2+}(aq) + 2e$^-$ Ag(aq) + e$^-$ $\rightleftharpoons$ Ag(s)
Anode   Cathode

(b)

$-$ |||| $+$

e$^-$ e$^-$

Voltmeter

0.450 V

Com   +   Meter positive lead

e$^-$ Salt bridge   e$^-$

Copper electrode

Silver electrode

CuSO$_4$ solution

AgNO$_3$ solution

[Cu^{2+}] = 0.0200 M   [Ag$^+$] = 0.0200 M
Cu^{2+}(aq) + 2e$^-$ $\rightleftharpoons$ Cu(s)   Ag(s) $\rightleftharpoons$ Ag$^+$(aq) + e$^-$
Cathode   Anode

(c)

Figure 16-2   (a) A galvanic cell at open circuit, (b) a galvanic cell doing work, and (c) an electrolytic cell.

393

The equilibrium-constant expression for the reaction shown in Equation 16-3 is

$$K_{eq} = \frac{[Cu^{2+}]}{[Ag^+]^2} = 4.1 \times 10^{15} \quad (16\text{-}4)$$

This expression applies regardless of whether the reaction occurs directly between reactants or within an electrochemical cell. The measured value of K_{eq} is completely independent of the pathway by which equilibrium is achieved.

At equilibrium, the two half-reactions in a cell continue, but their rates become equal.

The electrodes in some cells share a common electrolyte; these cells are known as *cells without liquid junction.* For an example of such a cell, see Figure 17-2 and Example 17-4.

A *cathode* is an electrode where reduction occurs. An *anode* is an electrode where oxidation occurs.

The reaction $2H^+ + 2e^- \rightleftharpoons H_2(g)$ occurs at the cathode when an aqueous solution contains no easily reduced species. Silver ions are reduced at a silver electrode. Reduction of iron(III) and nitrate ions will occur at an inert platinum electrode.

The Fe^{2+} half-reaction may seem somewhat unusual because a cation rather than an anion migrates to the electrode and gives up an electron. As we shall see, oxidation of a cation at an anode or reduction of an anion at a cathode is not uncommon.

The reaction $2H_2O \rightleftharpoons O_2(g) + 4H^+ + 4e^-$ occurs at the anode when an aqueous solution contains no more easily oxidized species.

When zero voltage is reached in the cell of Figure 16-2b, the concentrations of Cu(II) and Ag(I) ions will have values that satisfy the equilibrium-constant expression shown in Equation 16-4 (see margin). At this point, no further net flow of electrons will occur. *Note that the overall reaction and its position of equilibrium are totally independent of the way the reaction is carried out,* whether it is by direct reaction in a solution or by indirect action in an electrochemical cell.

16B   ELECTROCHEMICAL CELLS

Oxidation/reduction equilibria are conveniently studied by measuring the potentials of electrochemical cells in which the two half-reactions making up the equilibrium are participants. For this reason, we need to consider some of the characteristics of cells.

An electrochemical cell consists of two conductors called *electrodes,* each of which is immersed in an electrolyte solution. In most of the cells that will be of interest to us, the solutions surrounding the two electrodes are different and must be separated to avoid direct reaction between the reactants. The most common way of avoiding mixing is to insert a *salt bridge,* such as that shown in Figure 16-2, between the solutions. Conduction of electricity from one electrolyte solution to the other then occurs by migration of potassium ions in the bridge in one direction and chloride ions in the other. Direct contact between copper metal and silver ions, however, is prevented.

16B-1   Cathodes and Anodes

The *cathode* in an electrochemical cell is the electrode at which a reduction reaction occurs. The *anode* is the electrode at which an oxidation takes place. Examples of typical cathodic reactions include

$$Ag^+ + e^- \rightleftharpoons Ag(s)$$

$$Fe^{3+} + e^- \rightleftharpoons Fe^{2+}$$

$$NO_3^- + 10H^+ + 8e^- \rightleftharpoons NH_4^+ + 3H_2O$$

We can force a desired reaction to occur by applying a suitable potential to an inert electrode such as platinum. Note that both anions and cations can be involved in the cathodic reaction.

Typical anodic reactions include

$$Cu(s) \rightleftharpoons Cu^{2+} + 2e^-$$

$$2Cl^- \rightleftharpoons Cl_2(g) + 2e^-$$

$$Fe^{2+} \rightleftharpoons Fe^{3+} + e^-$$

The first reaction requires a copper anode, but the other two can be carried out at the surface of an inert platinum electrode that serves to provide or accept electrons.

16B-2   Two Types of Electrochemical Cells

Electrochemical cells are either galvanic or electrolytic. They can also be classified as reversible or irreversible.

Galvanic, or *voltaic,* cells store electrical energy. *Batteries* are usually made from several such cells connected in series to produce higher voltages than a single cell can produce. The reactions at the two electrodes in such cells tend to proceed spontaneously and produce a flow of electrons from the anode to the cathode via an external conductor. The cell shown in Figure 16-2a is a galvanic cell that develops a potential of about 0.412 V when no current is being drawn from it. The silver electrode is positive with respect to the copper electrode in this cell. The copper electrode, which is negative with respect to the silver electrode, is a potential source of electrons to the external circuit when the cell is discharged. The cell in Figure 16-2b is the same galvanic cell, but it is now under discharge so that electrons move through the external circuit from the copper electrode to the silver electrode. While being discharged, the silver electrode is the *cathode* since the reduction of Ag^+ occurs here; the copper electrode is the *anode* since the oxidation of $Cu(s)$ occurs at this electrode. Galvanic cells operate spontaneously, and the net reaction during discharge is called the *spontaneous cell reaction.* For the cell of Figure 16-2b, the spontaneous cell reaction is that given by Equation 16-3, namely $2Ag^+ + Cu(s) \rightleftharpoons 2Ag(s) + Cu^{2+}$. Another example of a galvanic cell is described in Feature 16-2.

An *electrolytic cell,* in contrast, requires an external source of electrical energy for operation. The cell just considered can be operated electrolytically by connecting the positive terminal of an external voltage source having a potential of somewhat greater than 0.412 V to the silver electrode and the negative terminal to the copper electrode as shown in Figure 16-2c. Now since the negative terminal of the external voltage source is electron rich, electrons flow from this terminal to the copper electrode where reduction of Cu^{2+} to $Cu(s)$ occurs. The current is sustained by $Ag(s)$ being oxidized to Ag^+ at the right-hand electrode, producing electrons that flow to the positive terminal of the voltage source. Note that in the electrolytic cell the direction of the current is reversed over the galvanic cell of Figure 16-2b and the reactions at the electrodes are reversed as well. The silver electrode is forced to become the *anode,* whereas the copper electrode is forced to become the *cathode.* The net reaction that occurs when a higher voltage than the galvanic cell voltage is applied is the opposite of the spontaneous cell reaction. That is,

$$2Ag(s) + Cu^{2+} \rightleftharpoons 2Ag^+ + Cu(s)$$

The cell in Figure 16-2 is an example of a reversible cell in which the direction of the electrochemical reaction is reversed when the direction of electron flow is changed. In an irreversible cell, changing the direction of current causes entirely different half-reactions to occur at one or both electrodes. The lead-acid storage battery in an automobile is another example of a reversible cell. When the battery is being charged by the generator or an external charger, it operates as an electrolytic cell. When it is used to operate the headlights, the radio, or the ignition, it behaves as a galvanic cell.

Alessandro Volta (1745–1827), Italian physicist, was the inventor of the first battery, the so-called voltaic pile. It consisted of alternating disks of copper and zinc separated by disks of cardboard soaked with salt solution. In honor of Volta's many contributions to electrical science, the unit of potential difference, the volt, is named for him. In fact, in modern usage we often call the quantity the voltage instead of the potential difference.

Galvanic cells store and supply electrical energy; *electrolytic cells* consume electricity from an external source.

For both galvanic and electrolytic cells, remember that (1) reduction always takes place at the cathode, and (2) oxidation always takes place at the anode.

In a *reversible cell,* reversing the current reverses the cell reaction. In an *irreversible cell,* reversing the current causes a different half-reaction to occur at one or both of the electrodes.

Feature 16-2

The Daniell Gravity Cell

The Daniell gravity cell was one of the earliest galvanic cells that found widespread practical application. It was used in the mid-1800s as a battery to power telegraphic communication systems. As shown in Figure 16-3, the cathode was a piece of copper immersed in a saturated solution of copper sulfate. A much less dense solution of dilute zinc sulfate was layered on top of the copper sulfate, and a massive zinc electrode was located in this solution. The electrode reactions were

$$Zn(s) \rightleftharpoons Zn^{2+} + 2e^-$$

$$Cu^{2+} + 2e^- \rightleftharpoons Cu(s)$$

This cell develops an initial voltage of 1.18 V, which gradually decreases with use.

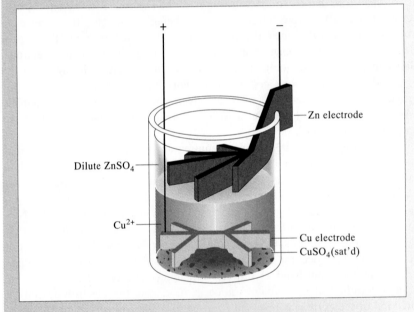

Figure 16-3   A Daniell gravity cell.

16B-3   Representing Cells Schematically

Chemists frequently use a shorthand notation to describe electrochemical cells. The cell in Figure 16-2a, for example, is described by

$$Cu \,|\, Cu^{2+}(0.0200 \text{ M}) \,\|\, Ag^+(0.0200 \text{ M}) \,|\, Ag \qquad (16\text{-}5)$$

By convention, a single vertical line indicates a phase boundary, or interface, at which a potential develops. For example, the first vertical line in this schematic indicates that a potential develops at the phase boundary between the copper electrode and the copper sulfate solution. The double vertical line represents two phase boundaries, one at each end of the salt bridge. A *liquid-junction potential* develops at each of these interfaces. This potential results from differences in rates with which the ions in the cell compartments and the salt bridge migrate across the interfaces. A liquid-junction potential can amount to as much as several hundredths of a volt, but it can be made negligibly small if the electrolyte in the salt bridge has an anion and a cation that migrate at nearly the same rate. A saturated solution of potassium chloride, KCl, is the electrolyte that is most widely used; it can reduce the junction potential to a few millivolts or less. For our purposes, we will neglect the contribution of liquid-junction potentials to the total potential of the cell. There are also several examples of cells that are without liquid junction and therefore do not require a salt bridge.

An alternative way of writing the cell shown in Figure 16-2a is

$$Cu \mid CuSO_4(0.0200\ M) \parallel AgNO_3(0.0200\ M) \mid Ag$$

Here, the compounds used to prepare the cell are indicated rather than the active participants in the cell half-reactions.

16B-4   Describing Currents in Electrochemical Cells

Figure 16-4 shows the movement of various charge carriers in a galvanic cell during discharge. The electrodes are connected with a wire so that the spontaneous cell reaction occurs. Charge is transported through such an electrochemical cell by three mechanisms:

1. Electrons carry the charge within the electrodes as well as the external conductor.
2. Anions and cations are the charge carriers within the cell. At the left-hand electrode, copper is oxidized to copper ions and gives up electrons to the electrode. As shown in Figure 16-4, the copper ions formed move away from the copper electrode into the bulk of solution, whereas anions, such as sulfate and hydrogen sulfate ions, migrate toward the copper anode. Within the salt bridge, chloride ions migrate toward and into the copper compartment, and potassium ions move in the opposite direction. In the right-hand compartment, silver ions move toward the silver electrode where they are reduced to silver metal, and the nitrate ions move away from the electrode into the bulk of solution.
3. The ionic conduction of the solution is coupled to the electronic conduction in the electrodes by the reduction reaction at the cathode and the oxidation reaction at the anode.

> In a cell, electricity is carried by movement of anions toward the anode and cations toward the cathode.

> The phase boundary between an electrode and its solution is called an *interface*.

16C   ELECTRODE POTENTIALS

The potential difference that develops between the electrodes of the cell in Figure 16-5a is a measure of the tendency for the reaction

$$2Ag^+ + Cu(s) \rightleftharpoons 2Ag(s) + Cu^{2+}$$

to proceed from a nonequilibrium state to the condition of equilibrium. The cell potential E_{cell} is related to the free energy of the reaction ΔG by

$$\Delta G = -nFE_{cell} \qquad (16\text{-}6)$$

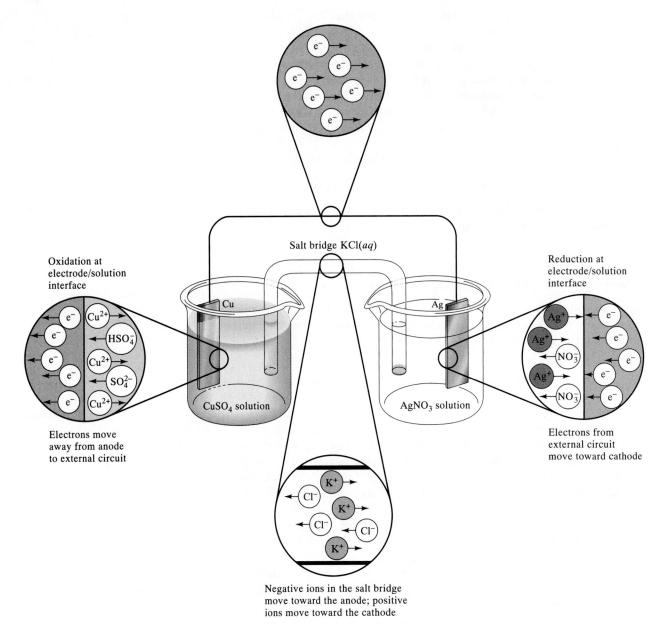

Figure 16-4   Movement of charge in a galvanic cell.

The standard state of a substance is a reference state that allows us to obtain relative values of such thermodynamic quantities as free energy, activity, enthalpy, and entropy. All substances are assigned unit activity in their standard state.

where n is the number of electrons transferred in the reaction, and F is Faraday's constant. If the reactants and products are in their *standard states*, the resulting cell potential is called the *standard cell potential*. This quantity is related to the standard free energy change for the reaction and thus to the equilibrium constant by

$$\Delta G^0 = -nFE_{cell}^0 = -RT \ln K_{eq} \tag{16-7}$$

where R is the gas constant, and T is the absolute temperature.

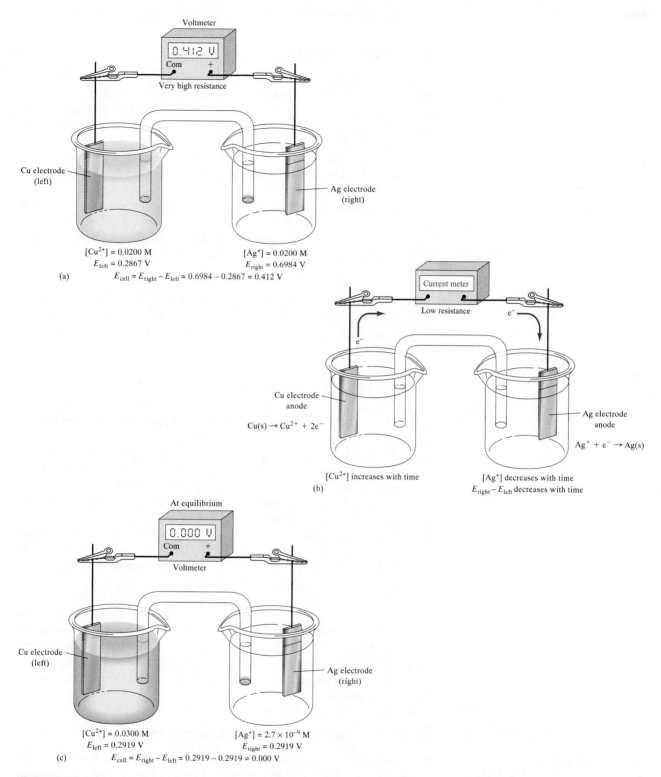

Figure 16-5 Change in cell potential after passage of current until equilibrium is reached. In (a), the high-resistance voltmeter prevents any significant electron flow, and the full open circuit cell potential is measured. For the concentrations shown, this value is + 0.412 V. In (b), the voltmeter is replaced with a low-resistance current meter and the cell discharges with time until eventually equilibrium is reached. In (c), after equilibrium is reached, the cell potential is again measured with a voltmeter and found to be 0.000 V. The concentrations in the cell are now those at equilibrium as shown.

16C-1   Cell Potential Sign Convention

By the International Union of Pure and Applied Chemistry (IUPAC) sign convention, when we consider an electrochemical cell and its resulting potential we consider the cell reaction to occur in a certain direction, just as when we consider a normal chemical reaction we speak of the reaction occurring from reactants on the left side of the arrow to products on the right side. The convention for cells is called the *plus right* rule, and it implies that we always measure the cell potential by connecting the positive lead (red lead) of the voltmeter to the right-hand electrode in the schematic or cell drawing (Ag electrode in Figure 16-5) and the common (or ground) lead (black lead) of the voltmeter to the left-hand electrode (Cu electrode in Figure 16-5). If we always follow this convention, the value of E_{cell} is a measure of the tendency of the cell reaction to occur spontaneously in the direction written below from left to right.

The leads of voltmeters are color coded. The positive lead is red and the common, or ground, lead is black.

$$Cu \,|\, Cu^{2+}(0.0200\ M) \,\|\, Ag^+(0.0200\ M) \,|\, Ag$$

That is, the direction considered is copper metal being oxidized to Cu^{2+} in the left-hand compartment and Ag^+ being reduced to silver metal in the right-hand compartment. Or, the reaction being considered is $Cu(s) + 2Ag^+ \rightleftharpoons Cu^{2+} + 2Ag(s)$.

What Does the IUPAC Convention Imply?

Several implications of the sign convention may not be apparent. First, if the measured value of E_{cell} is positive, the right-hand electrode is positive with respect to the left-hand electrode and the free energy change for the reaction in the direction being considered is negative by Equation 16-6. Hence, the reaction being considered would occur spontaneously if the cell were short-circuited or were caused to perform work (light a lamp, power a radio, start a car, etc.). On the other hand, if E_{cell} is negative, the right-hand electrode is negative with respect to the left-hand electrode, the free energy change is positive, and the reaction as considered (oxidation on the left, reduction on the right) is *not* the spontaneous cell reaction. For our cell of Figure 16-5, $E_{cell} = +0.412$ V and the oxidation of Cu and reduction of Ag^+ occurs spontaneously when allowed to do so.

The IUPAC convention is in accordance with the signs that the electrodes actually assume in a galvanic cell. That is, in the Cu/Ag cell as shown in Figure 16-5, the Cu electrode becomes electron rich (negative) due to the tendency of Cu to be oxidized to Cu^{2+}, whereas the Ag electrode becomes electron deficient (positive) because of the tendency for Ag^+ to be reduced to Ag. In the spontaneous galvanic cell under discharge, the silver electrode is the cathode, whereas the Cu electrode is the anode.

Note that for the same cell written in the opposite direction,

$$Ag \,|\, AgNO_3(0.0200\ M) \,\|\, CuSO_4(0.0200\ M) \,|\, Cu$$

the measured cell potential would be $E_{cell} = -0.412$ V, and the reaction considered is $2Ag(s) + Cu^{2+} \rightleftharpoons 2Ag^+ + Cu(s)$. This reaction is *not* the spontaneous cell reaction since E_{cell} is negative and thus ΔG is positive. It does not matter to the cell which electrode is written in the schematic on the right and which on the left. The spontaneous cell reaction is always $Cu(s) + 2Ag^+ \rightleftharpoons Cu^{2+} + 2Ag(s)$. By convention, we just measure the cell potential in a consistent manner and consider the cell reaction in a certain direction. Finally, we must emphasize that no matter how we may write the cell schematic or arrange the cell in the laboratory, if we connect a wire or a low-resistance circuit to the cell, the spontaneous cell reaction will occur. The only way to achieve the reverse reaction is to connect an external voltage source and cause the electrolytic reaction $2Ag(s) + Cu^{2+} \rightleftharpoons 2Ag^+ + Cu(s)$ to occur.

What Are Half-Cell Potentials?

The potential of a cell such as that shown in Figure 16-5a is the difference between two half-cell or single-electrode potentials, one associated with the half-reaction at the right-hand electrode (E_{right}), the other associated with the half-reaction at the left-hand electrode (E_{left}). Hence, according to the IUPAC sign convention, as long as the liquid-junction potential is negligible, or there is no liquid junction, we may write the cell potential E_{cell} as

$$E_{cell} = E_{right} - E_{left} \qquad (16-8)$$

Although we cannot determine absolute potentials of electrodes such as these (see Feature 16-3), we can readily determine relative electrode potentials. For example, if we replace the copper electrode in the cell in Figure 16-2 with a cadmium electrode immersed in a cadmium sulfate solution, the voltmeter reads about 0.7 V more positive than the original cell. Since the right-hand compartment remains unaltered, we conclude that the half-cell potential for cadmium is about 0.7 V less than that for copper (that is, cadmium is a stronger reductant than copper). Substituting other electrodes while keeping one of the electrodes unchanged allows us to construct a table of relative electrode potentials as discussed in Section 16C-3.

Discharging a Galvanic Cell

The galvanic cell of Figure 16-5a is in a nonequilibrium state because the very high resistance of the voltmeter prevents the cell from discharging significantly.

Feature 16-3

Why We Cannot Measure Absolute Electrode Potentials

Although it is not difficult to measure *relative* half-cell potentials, it is impossible to determine absolute half-cell potentials because all voltage-measuring devices measure only *differences* in potential. To measure the potential of an electrode, one contact of a voltmeter is connected to the electrode in question. The other contact from the meter must then be brought into electrical contact with the solution in the electrode compartment via another conductor. This second contact, however, inevitably involves a solid/solution interface that acts as a second half-cell when the potential is measured. Thus, an absolute half-cell potential is not obtained; rather, the difference between the half-cell potential of interest and a half-cell made up of the second contact and the solution is found.

Our inability to measure absolute half-cell potentials presents no real obstacle because relative half-cell potentials are just as useful provided they are all measured against the same reference half-cell. Relative potentials can be combined to give cell potentials. We can also use them to calculate equilibrium constants and generate titration curves.

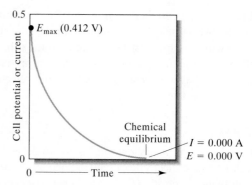

Figure 16-6   Cell potential in the galvanic cell of Figure 16-5b as a function of time. The current, which is directly related to the cell potential, also decreases with the same time behavior.

Hence, when we measure the cell potential, no reaction occurs and what we measure is the tendency of the reaction to occur should we allow it to proceed. For the Cu/Ag cell of Figure 16-5a with the concentrations shown, the cell potential measured under open circuit conditions is $+0.412$ V, as previously noted. If we now allow the cell to discharge by replacing the voltmeter with a low-resistance current meter as shown in Figure 16-5b, the spontaneous cell reaction occurs. The current, initially high, decreases exponentially with time as shown in Figure 16-6. As shown in Figure 16-5c, when equilibrium is reached there is no net current in the cell and the cell potential is 0.000 V. The copper ion concentration at equilibrium is then 0.0300 M, whereas the silver ion concentration falls to 2.7×10^{-9} M.

16C-2   The Standard Hydrogen Reference Electrode

The standard hydrogen electrode is sometimes called the *normal hydrogen electrode.*

SHE is the acronym for standard hydrogen electrode.

Platinum black is a layer of finely divided platinum that is formed on the surface of a smooth platinum electrode by electrolytic deposition of the metal from a solution of chloroplatinic acid H_2PtCl_6. The platinum black provides a large specific surface area of platinum at which the H^+/H_2 reaction can occur. Platinum black catalyzes the reaction shown in Equation 16-9. Remember that catalysts do not change the position of equilibrium but simply shorten the time it takes to reach equilibrium.

For relative electrode potential data to be widely applicable and useful, we must have a generally agreed-upon reference half-cell against which all others are compared. Such an electrode must be easy to construct, reversible, and highly reproducible in its behavior. The *standard hydrogen electrode* (SHE), although limited in practical utility, has been used throughout the world for many years as a universal reference electrode. It is a typical *gas electrode.*

Figure 16-7 shows how a hydrogen electrode is constructed. The metal conductor is a piece of platinum that has been coated, or *platinized,* with finely divided platinum (platinum black) to increase its specific surface area. This electrode is immersed in an aqueous acid solution of known, constant hydrogen ion activity. The solution is kept saturated with hydrogen by bubbling the gas at constant pressure over the surface of the electrode. The platinum does not take part in the electrochemical reaction and serves only as the site where electrons are transferred. The half-reaction responsible for the potential that develops at this electrode is

$$2H^+(aq) + 2e^- \rightleftharpoons H_2(g) \tag{16-9}$$

The hydrogen electrode shown in Figure 16-7 can be represented schematically as

$$Pt, H_2(p = 1.00 \text{ atm}) \,|\, ([H^+] = x \text{ M}) \,\|$$

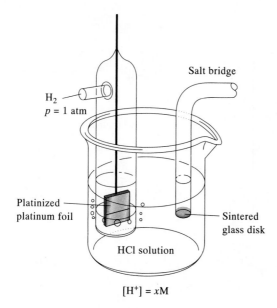

Figure 16-7   The hydrogen gas electrode.

Here, the hydrogen is specified as having a partial pressure of one atmosphere and the concentration of hydrogen ions in the solution is x M. The hydrogen electrode is reversible.

The potential of a hydrogen electrode depends on temperature and the activities of hydrogen ion and molecular hydrogen in the solution. The molecular hydrogen activity, in turn, is proportional to the pressure of the gas that is used to keep the solution saturated in hydrogen. For the standard hydrogen electrode (SHE), the activity of hydrogen ions is specified as unity and the partial pressure of the gas is specified as 1 atm. *By convention, the potential of the standard hydrogen electrode is assigned a value of 0.000 V at all temperatures.* As a consequence of this definition, any potential developed in a galvanic cell consisting of a standard hydrogen electrode and some other electrode is attributed entirely to the other electrode.

Several other reference electrodes that are more convenient for routine measurements have been developed. Some of these are described in Section 19B.

16C-3   Defining Electrode Potential and Standard Electrode Potential

An *electrode potential* is defined as the potential of a cell consisting of the electrode in question acting as the right-hand electrode and the standard hydrogen electrode acting as the left-hand electrode. Thus, if we desire to obtain the potential of a silver electrode in contact with a solution of Ag^+, we would construct a cell as shown in Figure 16-8. Here, the half-cell on the right consists of a strip of pure silver in contact with a solution containing silver ions; the electrode on the left is the SHE. The cell potential is defined as in Equation 16-8.

Because the left-hand electrode is the standard hydrogen electrode with a potential assigned a value of 0.000 V, we can write

The reaction shown as Equation 16-9 involves two equilibria:

$$2H^+ + 2e^- \rightleftharpoons H_2(aq)$$

$$H_2(aq) \rightleftharpoons H_2(g)$$

The continuous stream of gas at constant pressure provides the solution with a constant molecular hydrogen concentration.

At $p_{H_2} = 1.00$ atm and $a_{H^+} = 1.00$ M, the potential of the hydrogen electrode is assigned a value of exactly 0.000 V at all temperatures.

The SHE is not convenient to use in the laboratory because it requires hydrogen gas and because the Pt electrode surface is easily contaminated.

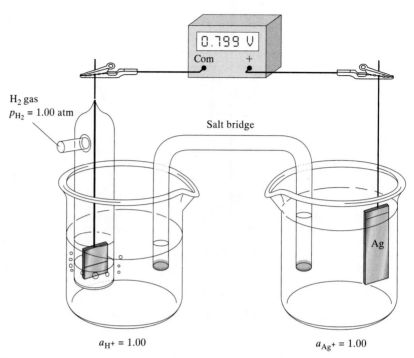

Figure 16-8 Measurement of the electrode potential for a Ag electrode. If the silver ion activity in the right-hand compartment is 1.00, the cell potential is the standard electrode potential of the Ag^+/Ag half-reaction.

$$E_{cell} = E_{right} - E_{left} = E_{Ag} - E_{SHE} = E_{Ag} - 0.000 = E_{Ag}$$

An electrode potential is the potential of a cell that has a standard hydrogen electrode as the left electrode (reference).

where E_{Ag} is the potential of the silver electrode. Despite its name, an electrode potential is, in fact, the potential of an electrochemical cell involving a carefully defined reference electrode. Often the potential of an electrode, such as the silver electrode in Figure 16-8, is referred to as E_{Ag} versus SHE to emphasize that it is the potential of a complete cell measured against the standard hydrogen electrode as a reference.

For gases, the standard state has the properties of an ideal gas, but at one atmosphere pressure. It is thus said to be a *hypothetical* state. For pure liquids and solvents, the standard states are *real* states and are the pure substances at a specified temperature and pressure. For solutes in dilute solution, the standard state is a hypothetical state that has the properties of an infinitely dilute solute, but at unit concentration (molarity, molality, or mole fraction). The standard state of a solid is a real state and is the pure solid in its most stable crystalline form.

The *standard electrode potential, E^0*, of a half-reaction is defined as its electrode potential when the activities of the reactants and products are all unity. For the cell in Figure 16-8, the E^0 value for the half-reaction

$$Ag^+ + e^- \rightleftharpoons Ag(s)$$

can be obtained by measuring E_{cell} with the activity of Ag^+ equal to 1.00. In this case, the cell shown in Figure 16-8 can be represented schematically as

$$Pt, H_2(p = 1.00\ atm)\,|\,H^+(a_{H^+} = 1.00)\,\|\,Ag^+(a_{Ag^+} = 1.00)\,|\,Ag$$

or alternatively as

$$SHE\,\|\,Ag^+(a_{Ag^+} = 1.00)\,|\,Ag$$

This galvanic cell develops a potential of $+0.799$ V with the silver electrode on the right; that is, the spontaneous cell reaction is oxidation in the left-hand compartment and reduction in the right-hand compartment:

$$2Ag^+ + H_2(g) \rightleftharpoons 2Ag(s) + 2H^+$$

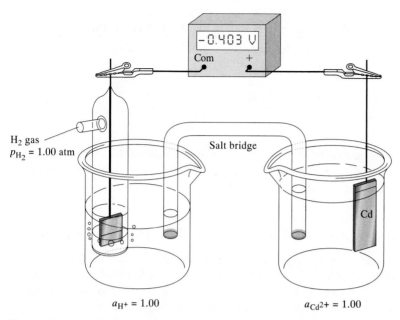

Figure 16-9   Measurement of the standard electrode potential for $Cd^{2+} + 2e^- \rightleftharpoons Cd(s)$.

Because the silver electrode is on the right, the measured potential is, by definition, the standard electrode potential for the silver half-reaction, or the Ag^+/Ag *couple*. Note that the silver electrode is positive with respect to the standard hydrogen electrode. Therefore, the standard electrode potential is given a positive sign, and we write

$$Ag^+ + e^- \rightleftharpoons Ag(s) \qquad E^0_{Ag^+/Ag} = +0.799 \text{ V}$$

Figure 16-9 illustrates a cell to measure the standard electrode potential for the half-reaction

$$Cd^{2+} + 2e^- \rightleftharpoons Cd(s)$$

In contrast to the silver electrode, the cadmium electrode is negative with respect to the standard hydrogen electrode. Consequently, the standard electrode potential of the Cd^2/Cd^+ couple is *by convention* given a negative sign, and $E^0_{Cd^{2+}/Cd} = -0.403$ V. Because the cell potential is negative, the spontaneous cell reaction is not the reaction as written, that is, oxidation on the left and reduction on the right. Rather, the spontaneous reaction is in the opposite direction.

A half-cell is sometimes called a *couple*.

$$Cd(s) + 2H^+ \rightleftharpoons Cd^{2+} + H_2(g)$$

A zinc electrode immersed in a solution having a zinc ion activity of unity develops a potential of -0.763 V when it is the right-hand electrode paired with a standard hydrogen electrode on the left. Thus we can write

$$E^0_{Zn^{2+}/Zn} = -0.763 \text{ V}$$

The standard electrode potentials for the four half-cells just described can be arranged in the following order:

Half-Reaction	Standard Electrode Potential, V
$Ag^+ + e^- \rightleftharpoons Ag(s)$	$+0.799$
$2H^+ + 2e^- \rightleftharpoons H_2(g)$	0.000
$Cd^{2+} + 2e^- \rightleftharpoons Cd(s)$	-0.403
$Zn^{2+} + 2e^- \rightleftharpoons Zn(s)$	-0.763

The magnitude of these electrode potentials indicate the relative strength of the four ionic species as electron acceptors (oxidizing agents); that is, in decreasing strength, $Ag^+ > H^+ > Cd^{2+} > Zn^{2+}$.

16C-4   Additional Implications of the IUPAC Sign Convention

The sign convention just described was adopted at the International Union of Pure and Applied Chemistry (IUPAC) meeting in Stockholm in 1953 and is now generally accepted. Prior to this agreement, chemists had not always used the same convention, which caused controversy and confusion in the development and routine use of electrochemistry.

Any sign convention must be based on expressing half-cell processes in a single way, that is, either as oxidations or as reductions. According to the IUPAC convention, the term *electrode potential* (or, more exactly, *relative electrode potential*) *is reserved exclusively to describe half-reactions written as reductions*. There is no objection to the use of the term *oxidation potential* to indicate a process written in the opposite sense, but it is not proper to refer to such a potential as an electrode potential.

The sign of an electrode potential is determined by the sign of the half-cell in question when it is coupled to a standard hydrogen electrode. When the half-cell of interest exhibits a positive potential relative to the SHE (Figure 16-8), it will behave spontaneously as the cathode when the cell is discharging. When the half-cell of interest is negative relative to the SHE (Figure 16-9), it will behave spontaneously as the anode when the cell is discharging.

> An electrode potential is, by definition, a reduction potential. An oxidation potential is the potential for the half-reaction written in the opposite way. The sign of an oxidation potential is therefore opposite that for a reduction potential, but the magnitude is the same.

> The IUPAC sign convention is based on the actual sign of the half-cell of interest when it is part of a cell containing the standard hydrogen electrode as the other half-cell.

16C-5   The Nernst Equation: How Does Concentration Affect Electrode Potentials?

An electrode potential is a measure of the extent to which the concentrations of the species in a half-cell differ from their equilibrium values. Thus, for example, there is a greater tendency for the process

$$Ag^+ + e^- \rightleftharpoons Ag(s)$$

to occur in a concentrated solution of silver(I) than in a dilute solution of that ion. It follows that the magnitude of the electrode potential for this process must also become larger (more positive) as the silver ion concentration of a solution is increased. We now examine the quantitative relationship between concentration and electrode potential.

Consider the reversible half-reaction

$$a\mathrm{A} + b\mathrm{B} + \cdots + ne^- \rightleftharpoons c\mathrm{C} + d\mathrm{D} + \cdots \qquad (16\text{-}10)$$

where the capital letters represent formulas for the participating species (atoms, molecules, or ions), e^- represents the electron, and the lowercase italic letters indicate the number of moles of each species appearing in the half-reaction as it has been written. The electrode potential for this process is given by the equation

$$E = E^0 - \frac{RT}{nF} \ln \frac{[C]^c[D]^d \cdots}{[A]^a[B]^b \cdots} \qquad (16\text{-}11)$$

where

E^0 = the *standard electrode potential*, which is characteristic for each half-reaction

R = the gas constant, $8.314 \text{ J K}^{-1} \text{ mol}^{-1}$

T = temperature, K

n = number of moles of electrons that appear in the half-reaction for the electrode process as written

F = the faraday = 96,485 C (coulombs)

ln = natural logarithm = 2.303 log

If we substitute numerical values for the constants, convert to base 10 logarithms, and specify 25°C for the temperature, we get

$$E = E^0 - \frac{0.0592}{n} \log \frac{[C]^c[D]^d \cdots}{[A]^a[B]^b \cdots} \qquad (16\text{-}12)$$

The letters in brackets strictly represent activities, but we shall ordinarily follow the practice of substituting molar concentrations for activities in most calculations. Thus, if some participating species A is a solute, [A] is the concentration of A in moles per liter. If A is a gas, [A] in Equation 16-12 is replaced by p_A, the partial pressure of A in atmospheres. If A is a pure liquid, a pure solid, or the solvent, its activity is unity and no term for A is included in the equation. The rationale for these assumptions is the same as that described in Section 4B-2, which deals with equilibrium-constant expressions. Equation 16-12 is known as the Nernst equation in honor of the German chemist Walther Nernst who was responsible for its development. Example 16-2 contains typical applications of Equation 16-12.

Strictly the bracketed terms in Equations 16-11 and 16-12 are activities. In practice, we often substitute concentrations for activities. Hence, for a solute A, [A] = molar concentration, for a gas B, [B] = p_B = partial pressure in atmospheres. If one or more of the species appearing in Equation 16-11 is a pure liquid, pure solid, or the solvent present in excess, then no bracketed term for this species appears in the quotient because the activities of these are unity.

Walther Nernst (1864–1941) received the Nobel Prize in chemistry in 1920 for his numerous contributions to the field of chemical thermodynamics.

Example 16-2

Typical half-cell reactions and their corresponding Nernst expressions follow.

1. $Zn^{2+} + 2e^- \rightleftharpoons Zn(s)$ $\qquad E = E^0 - \dfrac{0.0592}{2} \log \dfrac{1}{[Zn^{2+}]}$

No term for elemental zinc is included in the logarithmic term because it is a pure solid. Thus, the electrode potential varies linearly with the logarithm of the reciprocal of the zinc ion concentration.

2. $Fe^{3+} + e^- \rightleftharpoons Fe^{2+}(s)$ $\qquad E = E^0 - \dfrac{0.0592}{1} \log \dfrac{[Fe^{2+}]}{[Fe^{3+}]}$

The potential for this couple can be measured with an inert metallic electrode immersed in a solution containing both iron species. The potential depends on the logarithm of the ratio between the molar concentrations of these ions.

3. $2H^+ + 2e^- \rightleftharpoons H_2(g)$ $E = E^0 - \dfrac{0.0592}{2} \log \dfrac{p_{H_2}}{[H^+]^2}$

In this example, p_{H_2} is the partial pressure of hydrogen (in atmospheres) at the surface of the electrode. Ordinarily, its value will be the same as the atmospheric pressure.

4. $MnO_4^- + 5e^- + 8H^+ \rightleftharpoons Mn^{2+} + 4H_2O$

$$E = E^0 - \frac{0.0592}{5} \log \frac{[Mn^{2+}]}{[MnO_4^-][H^+]^8}$$

Here, the potential depends not only on the concentrations of the manganese species but also on the pH of the solution.

5. $AgCl(s) + e^- \rightleftharpoons Ag(s) + Cl^-$ $E = E^0 - \dfrac{0.0592}{1} \log[Cl^-]$

The Nernst expression in item 5 of Example 16-2 requires an excess of solid AgCl so that the solution is saturated with the compound at all times.

This half-reaction describes the behavior of a silver electrode immersed in a chloride solution that is *saturated* with AgCl. To ensure this condition, an excess of the solid must always be present. Note that this electrode reaction is the sum of two reactions, namely,

$$AgCl(s) \rightleftharpoons Ag^+ + Cl^-$$

$$Ag^+ + e^- \rightleftharpoons Ag(s)$$

Note also that the electrode potential is independent of the amount of AgCl present as long as there is some present to keep the solution saturated.

16C-6   The Standard Electrode Potential, E^0

The *standard electrode potential* for a half-reaction, E^0, is defined as the electrode potential when all reactants and products of a half-reaction are at unit activity

Examination of Equations 16-11 and 16-12 reveals that the constant E^0 is the electrode potential whenever the concentration quotient (strictly activity quotient) has a value of one. This constant is by definition the standard electrode potential for the half-reaction. Note that the quotient is always unity when the activities of the reactants and products of a half-reaction are unity.

The standard electrode potential is an important physical constant that provides quantitative information regarding the driving force for a half-cell reaction.[1] The important characteristics of these constants are:

1. The standard electrode potential is a relative quantity in the sense that it is the potential of an electrochemical cell in which the reference electrode (left-hand electrode) is the standard hydrogen electrode whose potential has been arbitrarily set at 0 V.
2. The standard electrode potential for a half-reaction refers exclusively to a reduction reaction; that is, it is a relative reduction potential.

---

[1]For further reading on standard electrode potentials, see R. G. Bates, in *Treatise on Analytical Chemistry*, 2nd ed., I. M. Kolthoff and P. J. Elving, Eds. (New York: Wiley, 1978), Part I, Vol. 1, Chapter 13.

3. The standard electrode potential measures the relative force tending to drive the half-reaction from a state in which the reactants and products are at unit activity to a state in which the reactants and products are at their equilibrium activities relative to the standard hydrogen electrode.

4. The standard electrode potential is independent of the number of moles of reactant and product shown in the balanced half-reaction. Thus, the standard electrode potential for the half-reaction

$$Fe^{3+} + e^- \rightleftharpoons Fe^{2+} \qquad E^0 = +0.771 \text{ V}$$

does not change if we choose to write the reaction as

$$5Fe^{3+} + 5e^- \rightleftharpoons 5Fe^{2+} \qquad E^0 = +0.771 \text{ V}$$

Note, however, that the Nernst equation must be consistent with the half-reaction as written. For the first case, it will be

$$E = 0.771 - \frac{0.0592}{1} \log \frac{[Fe^{2+}]}{[Fe^{3+}]}$$

and for the second,

$$E = 0.771 - \frac{0.0592}{5} \log \frac{[Fe^{2+}]^5}{[Fe^{3+}]^5} = 0.771 - \frac{0.0592}{5} \log \left(\frac{[Fe^{2+}]}{[Fe^{3+}]} \right)^5$$

$$= 0.771 - \frac{\cancel{5} \times 0.0592}{\cancel{5}} \log \frac{[Fe^{2+}]}{[Fe^{3+}]}$$

Note that the two log terms have identical values. That is,

$$\frac{0.0592}{1} \log \frac{[Fe^{2+}]}{[Fe^{3+}]} = \frac{0.059}{\cancel{5}} \log \frac{[Fe^{2+}]^{\cancel{5}}}{[Fe^{3+}]^{\cancel{5}}}$$

5. A positive electrode potential indicates that the half-reaction in question is spontaneous with respect to the standard hydrogen electrode half-reaction. That is, the oxidant in the half-reaction is a stronger oxidant than hydrogen ion. A negative sign indicates just the opposite.

6. The standard electrode potential for a half-reaction is temperature dependent.

Standard electrode potential data are available for an enormous number of half-reactions. Many have been determined directly from electrochemical measurements, and others have been computed from equilibrium studies of oxidation/reduction systems and from thermochemical data associated with such reactions. Table 16-1 contains standard electrode data from several half-reactions that we consider in the pages that follow. A more extensive listing is found in Appendix 4.[2]

Table 16-1 and Appendix 4 illustrate the two common ways for tabulating standard potential data. In Table 16-1, potentials are listed in decreasing numerical order. As a consequence, the species in the upper left are the most effective electron acceptors, as evidenced by their large positive values. They are therefore the strongest oxidizing agents. As we proceed down the left side of such a table, each succeeding species is less effective as an electron acceptor than the one above it. The half-cell reactions at the bottom of the table have little or no

---

[2]Comprehensive sources for standard electrode potentials include A. J. Bard, R. Parsons, and J. Jordan, Eds., *Standard Electrode Potentials in Aqueous Solutions* (New York: Marcel Dekker, 1985); G. Milazzo and S. Caroli, *Tables of Standard Electrode Potentials* (New York: Wiley-Interscience, 1977); M. S. Antelman and F. J. Harris, *Chemical Electrode Potentials* (New York: Plenum Press, 1982). Some compilations are arranged alphabetically by element; others are tabulated according to the numerical value of E^0.

Table 16-1

Standard Electrode Potentials

Reaction	E^0 at 25°C, V
$Cl_2(g) + 2e^- \rightleftharpoons 2Cl^-$	$+1.359$
$O_2(g) + 4H^+ + 4e^- \rightleftharpoons 2H_2O$	$+1.229$
$Br_2(aq) + 2e^- \rightleftharpoons 2Br^-$	$+1.087$
$Br_2(l) + 2e^- \rightleftharpoons 2Br^-$	$+1.065$
$Ag^+ + e^- \rightleftharpoons Ag(s)$	$+0.799$
$Fe^{3+} + e^- \rightleftharpoons Fe^{2+}$	$+0.771$
$I_3^- + 2e^- \rightleftharpoons 3I^-$	$+0.536$
$Cu^{2+} + 2e^- \rightleftharpoons Cu(s)$	$+0.337$
$UO_2^{2+} + 4H^+ + 2e^- \rightleftharpoons U^{4+} + 2H_2O$	$+0.334$
$Hg_2Cl_2(s) + 2e^- \rightleftharpoons 2Hg(l) + 2Cl^-$	$+0.268$
$AgCl(s) + e^- \rightleftharpoons Ag(s) + Cl^-$	$+0.222$
$Ag(S_2O_3)_2^{3-} + e^- \rightleftharpoons Ag(s) + 2S_2O_3^{2-}$	$+0.017$
$2H^+ + 2e^- \rightleftharpoons H_2(g)$	**0.000**
$AgI(s) + e^- \rightleftharpoons Ag(s) + I^-$	-0.151
$PbSO_4 + 2e^- \rightleftharpoons Pb(s) + SO_4^{2-}$	-0.350
$Cd^{2+} + 2e^- \rightleftharpoons Cd(s)$	-0.403
$Zn^{2+} + 2e^- \rightleftharpoons Zn(s)$	-0.763

See Appendix 4 for a more extensive list.

tendency to take place as they are written. On the other hand, they do tend to occur in the opposite sense. The most effective reducing agents, then, are those species that appear in the lower right of the table.

Compilations of electrode-potential data such as that shown in Table 16-1 provide the chemist with qualitative insights into the extent and direction of electron-transfer reactions. For example, the standard potential for Ag^+/Ag $(+0.799$ V) is more positive than that for Cu^{2+}/Cu $(+0.337$ V). We therefore conclude that a piece of copper immersed in a silver(I) solution will cause the reduction of that ion and the oxidation of the copper. On the other hand, we would expect no reaction if we place a piece of silver in a copper(II) solution.

In contrast to the data in Table 16-1, standard potentials in Appendix 4 are arranged alphabetically by element. This arrangement makes it easy to locate data for a given electrode reaction.

Based on the E^0 values in Table 16-1 for Fe^{3+} and I_3^-, which species would you expect to predominate in a solution produced by mixing iron(III) and iodide ions (see color plate 15)?

Systems Involving Precipitates or Complex Ions

In Table 16-1, we find several entries involving Ag(I), including

$$Ag^+ + e^- \rightleftharpoons Ag(s) \qquad\qquad E^0_{Ag^+/Ag} = +0.799 \text{ V}$$

$$AgCl(s) + e^- \rightleftharpoons Ag(s) + Cl^- \qquad\qquad E^0_{AgCl/Ag} = +0.222 \text{ V}$$

$$Ag(S_2O_3)_2^{3-} + e^- \rightleftharpoons Ag(s) + 2S_2O_3^{2-} \qquad E^0_{Ag(S_2O_3)_2^{3-}/Ag} = +0.017 \text{ V}$$

Each gives the potential of a silver electrode in a different environment. Let's see how the three potentials are related.

Feature 16-4

Sign Conventions in the
Older Literature

Reference works, particularly those published before 1953, often contain tabulations of electrode potentials that are not in accord with the IUPAC recommendations. For example, in a classic source of standard-potential data complied by Latimer,[3] one finds

$$Zn(s) \rightleftharpoons Zn^{2+} + 2e^-   \quad E = +0.76 \text{ V}$$

$$Cu(s) \rightleftharpoons Cu^{2+} + 2e^-   \quad E = +0.34 \text{ V}$$

In converting these oxidation potentials to electrode potentials as defined by the IUPAC convention, one must mentally (1) express the half-reactions as reductions and (2) change the signs of the potentials.

The sign convention used in a tabulation of electrode potentials may not be explicitly stated. This information can be readily deduced, however, by noting the direction and sign of the potential for a half-reaction with which one is familiar. If the sign agrees with the IUPAC convention, the table can be used as is; if not, the signs of all of the data must be reversed. For example, the reaction

$$O_2(g) + 4H^+ + 4e^- \rightleftharpoons 2H_2O   \quad E = +1.229 \text{ V}$$

occurs spontaneously with respect to the standard hydrogen electrode and thus carries a positive sign. If the potential for this half-reaction is negative in a tabulation, it and all the other potentials should be multiplied by -1.

[3]W. M. Latimer, *The Oxidation States of the Elements and Their Potentials in Aqueous Solutions*, 2nd ed. (Englewood Cliffs, N.J.: Prentice-Hall, 1952).

The Nernst expression for the first half-reaction is

$$E = E^0_{Ag^+/Ag} - \frac{0.0592}{1} \log \frac{1}{[Ag^+]}$$

Since for AgCl, $K_{sp} = [Ag^+][Cl^-]$, we can replace $[Ag^+]$ with $K_{sp}/[Cl^-]$, and obtain

$$E = E^0_{Ag^+/Ag} - \frac{0.0592}{1} \log \frac{[Cl^-]}{K_{sp}}$$
$$= E^0_{Ag^+/Ag} + 0.0592 \log K_{sp} - 0.0592 \log [Cl^-]$$

By definition, the standard potential for the second half-reaction is the potential where $[Cl^-] = 1.00$. That is, when $[Cl^-] = 1.00, E = E^0_{AgCl/Ag}$. Substituting these values gives

$$E^0_{AgCl/Ag} = E^0_{Ag^+/Ag} + 0.0592 \log 1.82 \times 10^{-10} - 0.0592 \log (1.00)$$
$$= 0.799 + (-0.577) - 0.000 = 0.222 \text{ V}$$

Figure 16-10 illustrates measurement of the standard electrode potential for the Ag/AgCl electrode.

If we proceed in the same way, we can obtain an expression for the standard electrode potential for the reduction of the thiosulfate complex of silver ion shown by the third equilibrium at the start of this section. Here the standard potential is given by

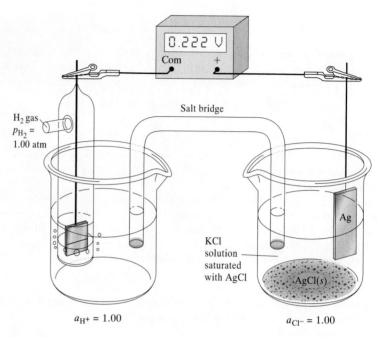

Figure 16-10 Measurement of the standard electrode potential for a Ag/AgCl electrode.

CHALLENGE: Derive Equation 16-13.

$$E^0_{Ag(S_2O_3)_2^{3-}/Ag} = E^0_{Ag^+/Ag} - 0.0592 \log \beta_2 \qquad (16\text{-}13)$$

where β_2 is the formation constant for the complex. That is,

$$\beta_2 = \frac{[Ag(S_2O_3)_2^{3-}]}{[Ag^+][S_2O_3^{2-}]^2}$$

Example 16-3

Calculate the electrode potential of a silver electrode immersed in a 0.0500 M solution of NaCl using (a) $E^0_{Ag^+/Ag} = 0.799$ V and (b) $E^0_{AgCl/Ag} = 0.222$ V.

(a) $Ag^+ + e^- \rightleftharpoons Ag(s)$ $E^0_{Ag^+/Ag} = +0.799$ V

The Ag^+ concentration of this solution is given by

$$[Ag^+] = \frac{K_{sp}}{[Cl^-]} = \frac{1.82 \times 10^{-10}}{0.0500} = 3.64 \times 10^{-9} \text{ M}$$

Substituting into the Nernst expression gives

$$E = 0.799 - 0.0592 \log \frac{1}{3.64 \times 10^{-9}} = 0.299 \text{ V}$$

(b) Here we may write

$$E = 0.222 - 0.0592 \log [Cl^-] = 0.222 - 0.0592 \log 0.0500$$
$$= 0.299 \text{ V}$$

Feature 16-5

Why Are There Two
Electrode Potentials for
Br_2 in Table 16-1?

In Table 16-1, we find the following data for Br_2:

$$Br_2(aq) + 2e^- \rightleftharpoons 2Br^- \qquad E^0 = +1.087 \text{ V}$$

$$Br_2(l) + 2e^- \rightleftharpoons 2Br^- \qquad E^0 = +1.065 \text{ V}$$

The second standard potential applies only to a solution that is saturated with Br_2 and not to undersaturated solutions. You should use 1.065 V to calculate the electrode potential of a 0.0100 M solution of KBr that is saturated with Br_2 and in contact with an excess of the liquid. In such a case,

$$E = 1.065 - \frac{0.0592}{2} \log [Br^-]^2 = 1.065 - \frac{0.0592}{2} \log (0.0100)^2$$

$$= 1.065 - \frac{0.0592}{2}(-4.00) = 1.183 \text{ V}$$

In this calculation, no term for Br_2 appears in the logarithmic term because it is a pure liquid present in excess (unit activity).

The standard electrode potential shown in the first entry for $Br_2(aq)$ is a hypothetical standard potential because the solubility of Br_2 at 25°C is only about 0.18 M. Thus, the recorded value of 1.087 V is based on a system that — in terms of our definition of E^0 — cannot be realized experimentally. Nevertheless, the hypothetical potential does permit us to calculate electrode potentials for solutions that are undersaturated in Br_2. For example, if we wish to calculate the electrode potential for a solution that was 0.0100 M in KBr and 0.00100 M in Br_2, we would write

$$E = 1.087 - \frac{0.0592}{2} \log \frac{[Br^-]^2}{[Br_2(aq)]} = 1.087 - \frac{0.0592}{2} \log \frac{(0.0100)^2}{0.00100}$$

$$= 1.087 - \frac{0.0592}{2} \log 0.100 = 1.117 \text{ V}$$

16C-7   Limitations to the Use of Standard Electrode Potentials

We shall be using standard electrode potentials throughout the rest of this text to calculate cell potentials and equilibrium constants for redox reactions as well as to derive data for redox titration curves. You should be aware that such calculations sometimes lead to results that are significantly different from those you would obtain experimentally in the laboratory. The main sources of these differences are (1) the necessity of using concentrations in place of activities in the Nernst equation and (2) failure to take into account other equilibria such as dissociation, association, complex formation, and solvolysis.

Use of Concentrations Instead of Activities

Most analytical oxidation/reduction reactions are carried out in solutions that have such high ionic strengths that activity coefficients cannot be obtained via the Debye–Hückel equation (Equation 9-5, Section 9B-2). Significant errors may result, however, if concentrations are used instead of activities in the Nernst equation. For example, the standard potential for the half-reaction

$$Fe^{3+} + e^- \rightleftharpoons Fe^{2+} \qquad E^0 = +0.771 \text{ V}$$

is $+0.771$ V. When the potential of a platinum electrode, immersed in a solution that is 10^{-4} M in iron(III) ion, iron(II) ion, and perchloric acid, is measured against a standard hydrogen electrode, a reading of close to $+0.77$ V is obtained as predicted by theory. If, however, perchloric acid is added to this mixture until the acid concentration is 0.1 M, the potential is found to decrease to about $+0.75$ V. This difference is because the activity coefficient of iron(III) is considerably smaller than that of iron(II) (0.4 versus 0.18) at the high ionic strength of the 0.1 M perchloric acid medium (see Table 9-1, page 210). As a consequence, the ratio of activities of the two species ($[Fe^{2+}]/[Fe^{3+}]$) in the Nernst equation is greater than unity, a condition that leads to a decrease in the electrode potential. In 1 M $HClO_4$, the electrode potential is still smaller (≈ 0.73 V).

Effect of Other Equilibria

The application of standard electrode potential data to many systems of interest in analytical chemistry is further complicated by association, dissociation, complex formation, and solvolysis equilibria involving the species that appear in the Nernst equation. These phenomena can be taken into account only if their existence is known and appropriate equilibrium constants are available. More often than not, neither of these requirements is met and significant discrepancies arise as a consequence. For example, the presence of 1 M hydrochloric acid in the iron(II)/iron(III) mixture we have just discussed leads to a measured potential of $+0.70$ V; in 1 M sulfuric acid, a potential of $+0.68$ V is observed; and in a 2 M phosphoric acid, the potential is $+0.46$ V. In each of these cases, the iron(II)/iron(III) activity ratio is larger because the complexes of iron(III) with chloride, sulfate, and phosphate ions are more stable than those of iron(II); thus, the ratio of the species concentrations, $[Fe^{2+}]/[Fe^{3+}]$, in the Nernst equation is greater than unity and the measured potential is less than the standard potential. If formation constants for these complexes were available, it would be possible to make appropriate corrections. Unfortunately, such data are often not available or, if they are, are not very reliable.

What Are Formal Potentials?

Formal potentials are empirically derived potentials that compensate for the types of activity and competing equilibria effects we have just described. The formal potential $E^{0'}$ of a system is the potential of the half-cell with respect to the standard hydrogen electrode measured under conditions such that the ratio of analytical concentrations of reactants and products as they appear in the Nernst equation is exactly unity and the concentrations of other species in the system are all carefully specified. For example, the formal potential for the half-reaction

$$Ag^+ + e^- \rightleftharpoons Ag(s) \qquad E^{0'} = 0.792 \text{ V}$$

A *formal potential* is the electrode potential when the ratio of *analytical concentrations* of reactants and products of a half-reaction are exactly 1.00 and the molar concentrations of any other solutes are specified.

in 1.00 M $HClO_4$ could be obtained by measuring the potential of the cell shown in Figure 16-11. Here, the right-hand electrode is a silver electrode immersed in a solution that is 1.00 M in $AgNO_3$ and 1.00 M in $HClO_4$; the reference electrode on the left is a standard hydrogen electrode. This cell develops a potential of $+0.792$ V, which is the formal potential of the Ag^+/Ag couple in 1.00 M $HClO_4$. Note that the standard potential for this couple is $+0.799$ V.

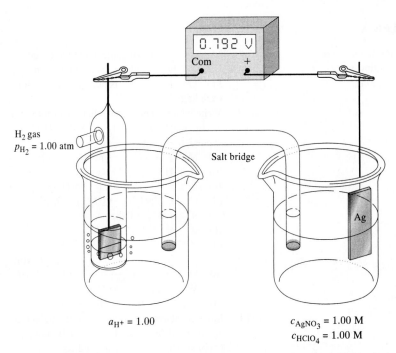

Figure 16-11   Measurement of the formal potential of the Ag^+/Ag couple in 1 M $HClO_4$.

Formal potentials for many half-reactions are listed in Appendix 4. Note that large differences exist between the formal and standard potentials for some half-reactions. For example, the formal potential for

$$Fe(CN)_6^{3-} + e^- \rightleftharpoons Fe(CN)_6^{4-} \qquad E^0 = +0.36 \text{ V}$$

is 0.72 V in 1 M perchloric or sulfuric acids, which is 0.36 V greater than the standard electrode potential for the half-reaction. The reason for this difference is that in the presence of high concentrations of hydrogen ion, ferrocyanide ions, $Fe(CN)_6^{4-}$, and ferricyanide ions, $Fe(CN)_6^{3-}$, combine with one or more protons to form hydroferrocyanic and hydroferricyanic acid species. Because the former are weaker, the ratio of the species concentrations, $[Fe(CN)_6^{4-}]/[Fe(CN)_6^{3-}]$, in the Nernst equation is less than unity, thus making the observed potentials greater.

Substitution of formal potentials for standard electrode potentials in the Nernst equation yields better agreement between calculated and experimental results, provided, of course, that the electrolyte concentration of the solution approximates that for which the formal potential is applicable. Not surprisingly, attempts to apply formal potentials to systems that differ substantially in type and in concentration of electrolyte can result in errors that are larger than those associated with the use of standard electrode potentials. In this text, we use whichever is more appropriate.

WEB WORKS

Set your browser to the following site and explore the use of fuel cells as batteries of the future: http://www.sciam.com/explorations/122396explorations.html. Describe a proton-exchange membrane from the information and links given.

16D   QUESTIONS AND PROBLEMS

Note: Numerical data are molar analytical concentrations where the full formula of a species is provided. Molar equilibrium concentrations are supplied for species displayed as ions.

16-1. Briefly describe or define
 *(a) oxidation.
 (b) oxidizing agent.
 *(c) salt bridge.
 (d) liquid junction.
 *(e) Nernst equation.

16-2. Briefly describe or define
 *(a) electrode potential.
 (b) formal potential.
 *(c) standard electrode potential.
 (d) liquid-junction potential.
 *(e) oxidation potential.

16-3. Make a clear distinction between
 *(a) reduction and reducing agent.
 (b) a galvanic cell and an electrolytic cell.
 *(c) the anode and the cathode of an electrochemical cell.
 (d) a reversible electrochemical cell and an irreversible electrochemical cell.
 *(e) the standard electrode potential and formal potential.

*16-4. The following entries are found in a table of standard electrode potentials:

$$I_2(s) + 2e^- \rightleftharpoons 2I^- \qquad E^0 = 0.5355 \text{ V}$$

$$I_2(aq) + 2e^- \rightleftharpoons 2I^- \qquad E^0 = 0.615 \text{ V}$$

What is the significance of the difference between these two standard potentials?

*16-5. Why is it necessary to bubble hydrogen through the electrolyte in a hydrogen electrode?

16-6. The standard electrode potential for the reduction of Ni^{2+} to Ni is -0.25 V. Would the potential of a nickel electrode immersed in a 1.00 M NaOH solution saturated with $Ni(OH)_2$ be more negative than $E^0_{Ni^{2+}/Ni}$ or less? Explain.

16-7. Write balanced net-ionic equations for the following reactions. Supply H^+ and/or H_2O, as needed, to obtain balance.
 *(a) $Fe^{3+} + Sn^{2+} \rightarrow Fe^{2+} + Sn^{4+}$
 (b) $Cr(s) + Ag^+ \rightarrow Cr^{3+} + Ag(s)$
 *(c) $NO_3^- + Cu(s) \rightarrow NO_2(g) + Cu^{2+}$
 (d) $MnO_4^{2-} + H_2SO_3 \rightarrow Mn^{2+} + SO_4^{2-}$
 *(e) $Ti^{3+} + Fe(CN)_6^{3-} \rightarrow TiO^{2+} + Fe(CN)_6^{4-}$
 (f) $H_2O_2 + Ce^{4+} \rightarrow O_2(g) + Ce^{3+}$
 *(g) $Ag(s) + I^- + Sn^{4+} \rightarrow AgI(s) + Sn^{2+}$
 (h) $UO_2^{2+} + Zn(s) \rightarrow U^{4+} + Zn^{2+}$
 *(i) $HNO_2 + MnO_4^- \rightarrow NO_3^- + Mn^{2+}$
 (j) $HN_2NNH_2 + IO_3^- + Cl^- \rightarrow N_2(g) + ICl_2^-$

*16-8. Identify the oxidizing agent and the reducing agent on the left side of each equation in Problem 16-7; write a balanced equation for each half-reaction.

16-9. Write balanced net-ionic equations for the following reactions. Supply H^+ and/or H_2O, as needed, to obtain balance.
 *(a) $MnO_4^- + VO^{2+} \rightarrow Mn^{2+} + V(OH)_4^+$
 (b) $I_2 + H_2S(g) \rightarrow I^- + S(s)$
 *(c) $Cr_2O_7^{2-} + U^{4+} \rightarrow Cr^{3+} + UO_2^{2+}$
 (d) $Cl^- + MnO_2(s) \rightarrow Cl_2(g) + Mn^{2+}$
 *(e) $IO_3^- + I^- \rightarrow I_2(aq)$
 (f) $IO_3^- + I^- + Cl^- \rightarrow ICl_2^-$
 *(g) $HPO_3^{2-} + MnO_4^- + OH^- \rightarrow$
 $\qquad\qquad\qquad\qquad PO_4^{3-} + MnO_4^{2-}$
 (h) $SCN^- + BrO_3^- \rightarrow Br^- + SO_4^{2-} + HCN$
 *(i) $V^{2+} + V(OH)_4^+ \rightarrow VO^{2+}$
 (j) $MnO_4^- + Mn^{2+} + OH^- \rightarrow MnO_2(s)$

16-10. Identify the oxidizing agent and the reducing agent on the left side of each equation in Problem 16-9; write a balanced equation for each half-reaction.

*16-11. Consider the following oxidation/reduction reactions:

$$AgBr(s) + V^{2+} \rightarrow Ag(s) + V^{3+} + Br^-$$

$$Tl^{3+} + 2Fe(CN)_6^{4-} \rightarrow Tl^+ + 2Fe(CN)_6^{3-}$$

$$2V^{3+} + Zn(s) \rightarrow 2V^{2+} + Zn^{2+}$$

$$Fe(CN)_6^{3-} + Ag(s) + Br^- \rightarrow$$
$$Fe(CN)_6^{4-} + AgBr(s)$$

$$S_2O_8^{2-} + Tl^+ \rightarrow 2SO_4^{2-} + Tl^{3+}$$

 (a) Write each net process in terms of two balanced half-reactions.
 (b) Express each half-reaction as a reduction.
 (c) Arrange the half-reactions in part (b) in order of decreasing effectiveness as electron acceptors.

16-12. Consider the following oxidation/reduction reactions:

$$2H^+ + Sn(s) \rightarrow H_2(g) + Sn^{2+}$$

$$Ag^+ + Fe^{2+} \rightarrow Ag(s) + Fe^{3+}$$

$$Sn^{4+} + H_2(g) \rightarrow Sn^{2+} + 2H^+$$

$$2Fe^{3+} + Sn^{2+} \rightarrow 2Fe^{2+} + Sn^{4+}$$

$$Sn^{2+} + Co(s) \rightarrow Sn(s) + Co^{2+}$$

(a) Write each net process in terms of two balanced half-reactions.

(b) Express each half-reaction as a reduction.

(c) Arrange the half-reactions in part (b) in order of decreasing effectiveness as electron acceptors.

*16-13. Calculate the potential of a copper electrode immersed in

    (a) 0.0440 M $Cu(NO_3)_2$.

    (b) 0.0750 M in NaCl and saturated with CuCl.

    (c) 0.0400 M in NaOH and saturated with $Cu(OH)_2$.

    (d) 0.0250 M in $Cu(NH_3)_4^{2+}$ and 0.128 M in NH_3. [β_4 for $Cu(NH_3)_4^{2+}$ is 5.62×10^{11}].

    (e) a solution in which the molar analytical concentration of $Cu(NO_3)_2$ is 4.00×10^{-3} M, that for H_2Y^{2-} is 2.90×10^{-2} M (Y = EDTA), and the pH is fixed at 4.00.

16-14. Calculate the potential of a zinc electrode immersed in

    (a) 0.0600 M $Zn(NO_3)_2$.

    (b) 0.01000 M in NaOH and saturated with $Zn(OH)_2$.

    (c) 0.0100 M in $Zn(NH_3)_4^{2+}$ and 0.250 M in NH_3, [β_4 for $Zn(NH_3)_4^{2+}$ is 7.76×10^8].

    (d) a solution in which the molar analytical concentration of $Zn(NO_3)_2$ is 5.00×10^{-3}, that for H_2Y^{2-} is 0.0445 M, and the pH is fixed at 9.00.

16-15. Use activities to calculate the electrode potential of a hydrogen electrode in which the electrolyte is 0.0100 M HCl and the activity of H_2 is 1.00 atm.

*16-16. Calculate the potential of a platinum electrode immersed in a solution that is

    (a) 0.0263 M in K_2PtCl_4 and 0.1492 M in KCl.

    (b) 0.0750 M in $Sn(SO_4)_2$ and 2.5×10^{-3} M in $SnSO_4$.

    (c) buffered to a pH of 6.00 and saturated with $H_2(g)$ at 1.00 atm.

    (d) 0.0353 M in $VOSO_4$, 0.0586 M in $V_2(SO_4)_3$, and 0.100 M in $HClO_4$.

    (e) prepared by mixing 25.00 mL of 0.0918 M $SnCl_2$ with an equal volume of 0.1568 M $FeCl_3$.

    (f) prepared by mixing 25.00 mL of 0.0832 M $V(OH)_4^+$ with 50.00 mL of 0.01087 M $V_2(SO_4)_3$ and has a pH of 1.00.

16-17. Calculate the potential of a platinum electrode immersed in a solution that is

    (a) 0.0813 M in $K_4Fe(CN)_6$ and 0.00566 M in $K_3Fe(CN)_6$.

    (b) 0.0400 M in $FeSO_4$ and 0.00845 M in $Fe_2(SO_4)_3$.

    (c) buffered to a pH of 5.55 and saturated with H_2 at 1.00 atm.

    (d) 0.1996 M in $V(OH)_4^+$, 0.0789 M in VO^{2+}, and 0.0800 M in $HClO_4$.

    (e) prepared by mixing 50.00 mL of 0.0607 M $Ce(SO_4)_2$ with an equal volume of 0.100 M $FeCl_2$. Assume solutions were 1.00 M in H_2SO_4 and use formal potentials.

    (f) prepared by mixing 25.00 mL of 0.0832 M $V_2(SO_4)_3$ with 50.00 mL of 0.00628 M $V(OH)_4^+$ and has a pH of 1.00.

*16-18. If the following half-cells are the right-hand electrode in a galvanic cell with a standard hydrogen electrode on the left, calculate the cell potential. If the cell were shorted, indicate whether the electrodes shown would act as the anode or cathode.

    (a) $Ni \,|\, Ni^{2+}(0.0943\ M)$.

    (b) $Ag \,|\, AgI(sat'd), KI(0.0922\ M)$.

    (c) $Pt, O_2(780\ torr)\ HCl(1.50 \times 10^{-4}\ M)$

    (d) $Pt \,|\, Sn^{2+}(0.0944\ M), Sn^{4+}(0.350\ M)$

    (e) $Ag \,|\, Ag(S_2O_3)_2^{3-}(0.00753\ M),$ $Na_2S_2O_3(0.1439\ M)$

16-19. The following half-cells are on the left and coupled with the standard hydrogen electrode on the right to form a galvanic cell. Calculate the cell potential. Indicate which electrode would be the cathode if each cell were short-circuited.

    (a) $Cu \,|\, Cu^{2+}(0.0897\ M)$

    (b) $Cu \,|\, CuI(sat'd), KI(0.1214\ M)$

    (c) $Pt, H_2(0.984\ atm) \,|\, HCl(1.00 \times 10^{-4}\ M)$

    (d) $Pt \,|\, Fe^{3+}(0.0906\ M), Fe^{2+}(0.1628\ M)$

    (e) $Ag \,|\, Ag(CN)_2^-(0.0827\ M),$ $KCN(0.0699\ M)$

*16-20. The solubility-product constant for Ag_2SO_3 is 1.5×10^{-14}. Calculate E^0 for the process

$$Ag_2SO_3(s) + 2e^- \rightleftharpoons 2Ag + SO_3^{2-}$$

16-21. The solubility-product constant for $Ni_2P_2O_7$ is 1.7×10^{-13}. Calculate E^0 for the process

$$Ni_2P_2O_7(s) + 4e^- \rightleftharpoons 2Ni(s) + P_2O_7^{4-}$$

*16-22. The solubility-product constant for Tl_2S is 6×10^{-22}. Calculate E^0 for the reaction

$$Tl_2S(s) + 2e^- \rightleftharpoons 2Tl(s) + S^{2-}$$

16-23. The solubility product for $Pb_3(AsO_4)_2$ is 4.1×10^{-36}. Calculate E^0 for the reaction

$$Pb_3(AsO_4)_2(s) + 6e^- \rightleftharpoons 3Pb(s) + 2AsO_4^{2-}$$

*16-24. Compute E^0 for the process

$$ZnY^{2-} + 2e^- \rightleftharpoons Zn(s) + Y^{4-}$$

where Y^{4-} is the completely deprotonated anion of EDTA. The formation constant for ZnY^{2-} is 3.2×10^{16}.

*16-25. Given the formation constants

$$Fe^{3+} + Y^{4-} \rightleftharpoons FeY^- \qquad K_f = 1.3 \times 10^{25}$$

$$Fe^{2+} + Y^{4-} \rightleftharpoons FeY^{2-} \qquad K_f = 2.1 \times 10^{14}$$

calculate E^0 for the process

$$FeY^- + e^- \rightleftharpoons FeY^{2-}$$

16-26. Calculate E^0 for the process

$$Cu(NH_3)_4^{2+} + e^- \rightleftharpoons Cu(NH_3)_2^+ + 2NH_3$$

given that

$$Cu^+ + 2NH_3 \rightleftharpoons Cu(NH_3)_2^+ \qquad \beta_2 = 7.2 \times 10^{10}$$

$$Cu^{2+} + 4NH_3 \rightleftharpoons Cu(NH_3)_4^{2+} \qquad \beta_4 = 5.62 \times 10^{11}$$

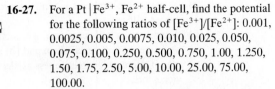

16-27. For a $Pt \,|\, Fe^{3+}, Fe^{2+}$ half-cell, find the potential for the following ratios of $[Fe^{3+}]/[Fe^{2+}]$: 0.001, 0.0025, 0.005, 0.0075, 0.010, 0.025, 0.050, 0.075, 0.100, 0.250, 0.500, 0.750, 1.00, 1.250, 1.50, 1.75, 2.50, 5.00, 10.00, 25.00, 75.00, 100.00.

16-28. For a $Pt \,|\, Ce^{4+}, Ce^{3+}$ half-cell, find the potential for the same ratios of $[Ce^{4+}]/[Ce^{3+}]$ as given in Problem 16-27 for $[Fe^{3+}]/[Fe^{2+}]$.

16-29. Plot the half-cell potential as a function of concentration ratio for the half-cells of Problems 16-27 and 16-28. How would the plot look if potential were plotted against log (concentration ratio)?

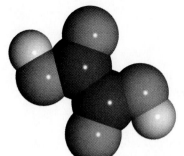

Chapter 17

Using Electrode Potentials

I n this chapter, we show how standard electrode potentials can be used for (1) calculating thermodynamic cell potentials, (2) calculating equilibrium constants for redox reactions, and (3) deriving redox titration curves.

17A CALCULATING POTENTIALS OF ELECTROCHEMICAL CELLS

We can use standard electrode potentials and the Nernst equation to calculate the potential obtainable from a galvanic cell or the potential required to operate an electrolytic cell. The calculated potentials (sometimes called thermodynamic potentials) are theoretical in the sense that they refer to cells in which there is no current. As we show in Chapter 20, additional factors must be taken into account if a current is involved.

The thermodynamic potential of an electrochemical cell is the difference between the electrode potential of the right-hand electrode and the electrode potential of the left-hand electrode. That is,

$$E_{cell} = E_{right} - E_{left} \qquad (17\text{-}1)$$

where E_{right} and E_{left} are the electrode potentials of the right-hand and left-hand electrodes, respectively.

It is important to note that E_{right} and E_{left} in Equation 17-1 are both *electrode potentials* as defined at the beginning of Section 16C-3.

Example 17-1

Calculate the thermodynamic potential of the following cell and the free energy change associated with the cell reaction.

$$Cu \,|\, Cu^{2+}(0.0200\ M) \,\|\, Ag^+(0.0200\ M) \,|\, Ag$$

Note that this cell is the galvanic cell shown in Figure 16-2a.

The two half-reactions and standard potentials are

$$Ag^+ + e^- \rightleftharpoons Ag(s) \qquad E^0 = 0.799\ V \qquad (17\text{-}2)$$

$$Cu^{2+} + 2e^- \rightleftharpoons Cu(s) \qquad E^0 = 0.337\ V \qquad (17\text{-}3)$$

The electrode potentials are

$$E_{Ag^+/Ag} = 0.799 - 0.0592 \log \frac{1}{0.0200} = 0.6984 \text{ V}$$

$$E_{Cu^{2+}/Cu} = 0.337 - \frac{0.0592}{2} \log \frac{1}{0.0200} = 0.2867 \text{ V}$$

We see from the cell diagram that the silver electrode is the right-hand electrode and the copper electrode is the left-hand electrode. Therefore, application of Equation 17-1 gives

$$E_{cell} = E_{right} - E_{left} = E_{Ag^+/Ag} - E_{Cu^{2+}/Cu} = 0.6984 - 0.2867 = +0.412 \text{ V}$$

The free energy change ΔG for the reaction $Cu(s) + 2Ag^+ \rightleftharpoons Cu^{2+} + Ag(s)$ is found from

$$\Delta G = -nFE_{cell} = -2 \times 96,485 \text{ C} \times 0.412 \text{ V} = -79,500 \text{ J} (18.99 \text{ kcal})$$

Example 17-2

Calculate the potential for the cell

$$Ag \,|\, Ag^+ (0.0200 \text{ M}) \,\|\, Cu^{2+} (0.0200 \text{ M}) \,|\, Cu$$

The electrode potentials for the two half-reactions are identical to the electrode potentials calculated in Example 17-1. That is,

$$E_{Ag^+/Ag} = 0.6984 \text{ V} \qquad \text{and} \qquad E_{Cu^{2+}/Cu} = 0.2867 \text{ V}$$

In contrast to Example 17-1, however, the silver electrode is on the left and the copper electrode is on the right. Substituting these electrode potentials into Equation 17-1 gives

$$E_{cell} = E_{right} - E_{left} = E_{Cu^{2+}/Cu} - E_{Ag^+/Ag} = 0.2867 - 0.6984 = -0.412 \text{ V}$$

Examples 17-1 and 17-2 illustrate an important fact. The magnitude of the potential difference between the two electrodes is 0.412 V regardless of which electrode is considered the left, or reference, electrode. If the Ag electrode is the left electrode as in Example 17-2, the cell potential has a negative sign, whereas if the Cu electrode is the reference as in Example 17-2, the cell potential has a positive sign. No matter how the cell is arranged, however, the spontaneous cell reaction is oxidation of Cu and reduction of Ag^+, and the free energy change is $-79,500$ J. Examples 17-3 and 17-4 illustrate other types of electrode reactions.

Example 17-3

Calculate the potential of the following cell and indicate the reaction that would occur spontaneously if the cell were short-circuited (see Figure 17-1).

$$Pt \,|\, U^{4+} (0.200 \text{ M}), UO_2^{2+} (0.0150 \text{ M}), H^+ (0.0300 \text{ M}) \,\|$$

$$Fe^{2+} (0.0100 \text{ M}), Fe^{3+} (0.0250 \text{ M}) \,|\, Pt$$

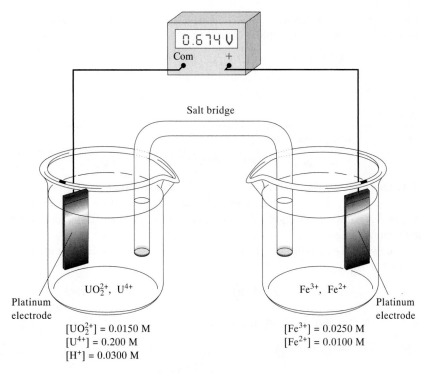

Figure 17-1 Cell for Example 17-3.

The two half-reactions are

$$Fe^{3+} + e^- \rightleftharpoons Fe^{2+} \qquad\qquad E^0 = +0.771 \text{ V}$$

$$UO_2^{2+} + 4H^+ + 2e^- \rightleftharpoons U^{4+} + 2H_2O \qquad E^0 = +0.334 \text{ V}$$

The electrode potential for the right-hand electrode is

$$E_{\text{right}} = 0.771 - 0.0592 \log \frac{[Fe^{2+}]}{[Fe^{3+}]}$$

$$= 0.771 - 0.0592 \log \frac{0.0100}{0.0250} = 0.771 - (-0.0236) = 0.7946 \text{ V}$$

The electrode potential for the left-hand electrode is

$$E_{\text{left}} = 0.334 - \frac{0.0592}{2} \log \frac{[U^{4+}]}{[UO_2^{2+}][H^+]^4}$$

$$= 0.334 - \frac{0.0592}{2} \log \frac{0.200}{(0.0150)(0.0300)^4}$$

$$= 0.334 - 0.2136 = 0.1204 \text{ V}$$

and

$$E_{\text{cell}} = E_{\text{right}} - E_{\text{left}} = 0.7946 - 0.1204 = 0.674 \text{ V}$$

The positive sign means that the spontaneous reaction is the oxidation of U^{4+} on the left and the reduction of Fe^{3+} on the right, or

$$U^{4+} + 2Fe^{3+} + 2H_2O \rightleftharpoons UO_2^{2+} + 2Fe^{2+} + 4H^+$$

Example 17-4

Calculate the cell potential for

$$Ag \mid AgCl(sat'd), HCl(0.0200\ M) \mid H_2(0.800\ atm), Pt$$

Note that this cell does not require two compartments (nor a salt bridge) because molecular H_2 has little tendency to react directly with the low concentration of Ag^+ in the electrolyte solution. This example is of a cell *without liquid junction* (see Figure 17-2).

The two half-reactions and their corresponding standard electrode potentials are (Table 16-1)

$$2H^+ + 2e^- \rightleftharpoons H_2(g) \qquad\qquad E^0_{H^+/H_2} = 0.000\ V$$

$$AgCl(s) + e^- \rightleftharpoons Ag(s) + Cl^- \qquad E^0_{AgCl/Ag} = 0.222\ V$$

The two electrode potentials are

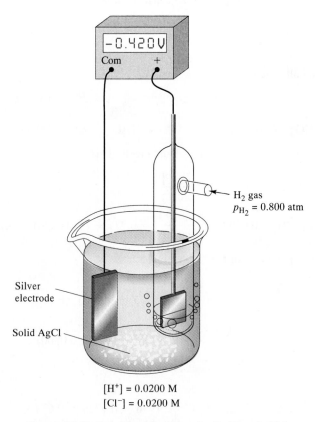

$[H^+] = 0.0200\ M$
$[Cl^-] = 0.0200\ M$

Figure 17-2   Cell without liquid junction for Example 17-4.

$$E_{right} = 0.000 - \frac{0.0592}{2} \log \frac{p_{H_2}}{[H^+]^2} = \frac{0.0592}{2} \log \frac{0.800}{(0.0200)^3}$$

$$= -0.0977 \text{ V}$$

$$E_{left} = 0.222 - 0.0592 \log [Cl^-] = 0.222 - 0.0592 \log 0.0200$$

$$= 0.3226 \text{ V}$$

The cell potential is thus

$$E_{cell} = E_{right} - E_{left} = -0.0977 - 0.3226 = -0.420 \text{ V}$$

The negative sign indicates that the cell reaction as considered,

$$2H^+ + 2Ag(s) \rightleftharpoons H_2(g) + 2AgCl(s)$$

is nonspontaneous. To get this reaction to occur, we would have to apply an external voltage and construct an electrolytic cell.

17B   CALCULATING REDOX EQUILIBRIUM CONSTANTS

Let us again consider the equilibrium that is established when a piece of copper is immersed in a solution containing a dilute solution of silver nitrate:

$$Cu(s) + 2Ag^+ \rightleftharpoons Cu^{2+} + 2Ag(s) \tag{17-4}$$

The equilibrium constant for this reaction is

$$K_{eq} = \frac{[Cu^{2+}]}{[Ag^+]^2} \tag{17-5}$$

As we showed in Example 17-1, this reaction can be carried out in the galvanic cell

$$Cu \,|\, Cu^{2+}(xM) \,\|\, Ag^+(yM) \,|\, Ag$$

A sketch of a cell similar to this is shown in Figure 16-2a on page 393. Its cell potential at any instant is given by Equation 17-1:

$$E_{cell} = E_{right} - E_{left} = E_{Ag} - E_{Cu}$$

As the reaction proceeds, the concentration of Cu(II) ions increases and the concentration of Ag(I) ions decreases. These changes make the potential of the copper electrode more positive and that of the silver electrode less positive. As shown in Figure 16-6 on page 402, the net effect of these changes is a continuous decrease in the potential of the cell as it discharges. Ultimately, the concentrations of Cu(II) and Ag(I) attain their equilibrium values as determined by Equation 17-5, and the current ceases. Under these conditions, *the potential of the cell becomes zero. Thus, at chemical equilibrium,* we may write

$$E_{cell} = 0 = E_{right} - E_{left} = E_{Ag} - E_{Cu}$$

or

$$\boxed{E_{right} = E_{left} = E_{Ag} = E_{Cu}} \tag{17-6}$$

For simplicity in many texts and in the electrochemical literature, the potential of the right-hand electrode in Example 17-1 is symbolized E_{Ag} and that of the left-hand electrode in Example 17-1 E_{Cu}. A completely unambiguous description that specifies the redox couple that determines the potential of these electrodes is to symbolize these potentials as $E_{Ag^+/Ag}$ and $E_{Cu^{2+}/Cu}$. Throughout this text the more unambiguous description is used except for some simple metal ion/metal couples, such as Ag^+/Ag and Cu^{2+}/Cu, where the redox couple is clear from the context or cell schematic.

Feature 17-1

Biological Redox Systems

There are many redox systems of importance in biology and biochemistry. The cytochromes are excellent examples of such systems. Cytochromes are iron-heme proteins in which a porphyrin ring is coordinated through nitrogens to an iron atom. These undergo one-electron redox reactions. Cytochromes function physiologically to facilitate electron transport. In the respiratory chain, the cytochromes are intimately involved in the formation of water from H_2. Reduced pyridine nucleotides deliver hydrogen to flavoproteins. The reduced flavoproteins are reoxidized by the Fe^{3+} of cytochrome b or c. The result is the formation of H^+ and the transport of electrons. The chain is completed when cytochrome oxidase transfers electrons to oxygen. The resulting superoxide ion (O^{2-}) is unstable and immediately picks up two H^+ ions to produce H_2O. The scheme is illustrated in Figure 17F-1.

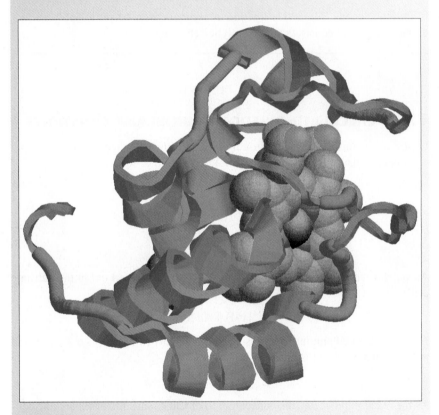

Model of cytochrome c

Most biological redox systems are pH-dependent. It has become standard practice to compile the electrode potentials of these systems at pH 7.0 to make comparisons of oxidizing or reducing powers. The compiled values are typically formal potentials at pH 7.0 and are sometimes symbolized $E_7^{0'}$.

Other redox systems of importance in biochemistry include the NADH/NAD system, the flavins, the pyruvate/lactate system, the oxalacetate/malate system, and the quinone/hydroquinone system.

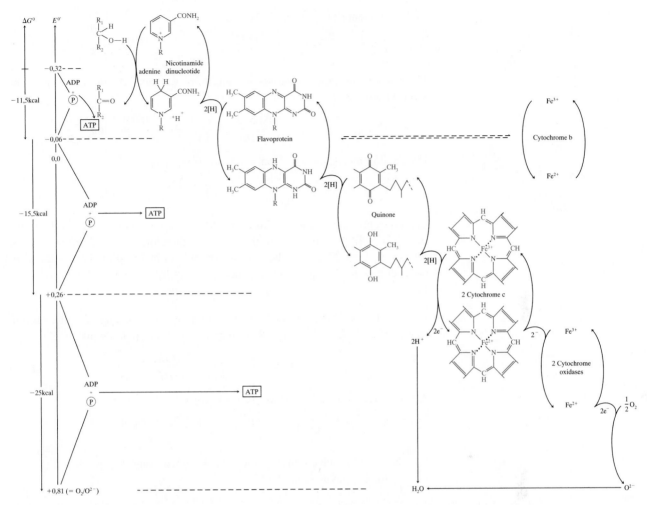

Figure 17F-1 Redox systems in the respiratory chain (P = phosphate ion. From P. Karlson, *Introduction to Modern Biochemistry* (New York: Academic Press, 1963) with permission).

We can generalize Equation 17-6 by stating that *at equilibrium, the electrode potentials for all half-reactions in an oxidation/reduction system are equal.* This generalization applies regardless of the number of half-reactions present in the system because interactions among all must take place until the electrode potentials are identical. For example, if we have four oxidation/reduction systems in a solution, interaction among all four takes place until the potentials of all four redox couples are equal.

Returning to the reaction shown in Equation 17-4, let us substitute Nernst expressions for the two electrode potentials in Equation 17-6, which gives

$$E^0_{Ag^+/Ag} - \frac{0.0592}{2}\log\frac{1}{[Ag^+]^2} = E^0_{Cu^{2+}/Cu} - \frac{0.0592}{2}\log\frac{1}{[Cu^{2+}]} \quad (17\text{-}7)$$

In Equation 17-7, the Nernst equation is applied to the silver half-reaction as it appears in the balanced reaction (Equation 17-4):

$$2Ag^+ + 2e^- \rightleftharpoons 2Ag(s) \qquad E^0 = 0.799 \text{ V}$$

Remember: *When redox systems are at equilibrium, the electrode potentials of all redox couples that are present are identical.* This generality applies whether the reactions take place directly in solution or indirectly in a galvanic cell.

Rearrangement of Equation 17-7 gives

$$E^0_{Ag^+/Ag} - E^0_{Cu^{2+}/Cu} = \frac{0.0592}{2} \log \frac{1}{[Ag^+]^2} - \frac{0.0592}{2} \log \frac{1}{[Cu^{2+}]}$$

If we invert the ratio in the second log term, we must change the sign of the term. This gives

$$E^0_{Ag^+/Ag} - E^0_{Cu^{2+}/Cu} = \frac{0.0592}{2} \log \frac{1}{[Ag^+]^2} + \frac{0.0592}{2} \log \frac{[Cu^{2+}]}{1}$$

Finally, combining the log terms and rearranging gives

$$\frac{2(E^0_{Ag^+/Ag} - E^0_{Cu^{2+}/Cu})}{0.0592} = \log \frac{[Cu^{2+}]}{[Ag^+]^2} = \log K_{eq} \qquad (17\text{-}8)$$

The concentration terms in Equation 17-8 are *equilibrium concentrations;* the ratio $[Cu^{2+}]/[Ag^+]^2$ in the logarithmic term is therefore the *equilibrium constant for the reaction.* Note that the term in parenthesis in Equation 17-8 is the standard cell potential E^0_{cell}, which in general is given by

$$E^0_{cell} = E^0_{right} - E^0_{left}$$

We can also obtain Equation 17-8 from the free energy change for the reaction as was given in Equation 16-7. Rearrangement of this equation gives

$$\ln K_{eq} = -\frac{\Delta G^0}{RT} = \frac{nFE^0_{cell}}{RT} \qquad (17\text{-}9)$$

At 25°C after conversion to base 10 logarithms, we can write

$$\log K_{eq} = \frac{nE^0_{cell}}{0.0592} = \frac{n(E^0_{right} - E^0_{left})}{0.0592}$$

For the reaction given in Equation 17-4, substituting $E^0_{Ag^+/Ag}$ for E^0_{right} and $E^0_{Cu^{2+}/Cu}$ for E^0_{left} gives Equation 17-8.

Example 17-5

Calculate the equilibrium constant for the reaction shown in Equation 17-4.
  Substituting numerical values into Equation 17-8 yields

In making calculations of the sort shown in Example 17-5, you should follow the rounding rule for antilogs that is given on page 144.

$$\log K_{eq} = \log \frac{[Cu^{2+}]}{[Ag^+]^2} = \frac{2(0.799 - 0.337)}{0.0592} = 15.61$$
$$K_{eq} = \text{antilog } 15.61 = 4.1 \times 10^{15}$$

Example 17-6

Calculate the equilibrium constant for the reaction

$$2Fe^{3+} + 3I^- \rightleftharpoons 2Fe^{2+} + I_3^-$$

In Appendix 4, we find

$$2Fe^{3+} + 2e^- \rightleftharpoons 2Fe^{2+} \qquad E^0 = 0.771 \text{ V}$$

$$I_3^- + 2e^- \rightleftharpoons 3I^- \qquad E^0 = 0.536 \text{ V}$$

We have multiplied the first half-reaction by 2 so that the number of moles of Fe^{3+} and Fe^{2+} are the same as in the balanced overall equation. We write the Nernst equation for Fe^{3+} based on the half-reaction for a two-electron transfer. That is,

$$E_{Fe^{3+}/Fe^{2+}} = E^0_{Fe^{3+}/Fe^{2+}} - \frac{0.0592}{2} \log \frac{[Fe^{2+}]^2}{[Fe^{3+}]^2}$$

and

$$E_{I_3^-/I^-} = E^0_{I_3^-/I^-} - \frac{0.0592}{2} \log \frac{[I^-]^3}{[I_3^-]}$$

At equilibrium, the electrode potentials are equal and

$$E_{Fe^{3+}/Fe^{2+}} = E_{I_3^-/I^-}$$

$$E^0_{Fe^{3+}/Fe^{2+}} - \frac{0.0592}{2} \log \frac{[Fe^{2+}]^2}{[Fe^{3+}]^2} = E^0_{I_3^-/I^-} - \frac{0.0592}{2} \log \frac{[I^-]^3}{[I_3^-]}$$

This equation rearranges to

$$\frac{2(E^0_{Fe^{3+}/Fe^{2+}} - E^0_{I_3^-/I^-})}{0.0592} = \log \frac{[Fe^{2+}]^2}{[Fe^{3+}]^2} - \log \frac{[I^-]^3}{[I_3^-]} = \log \frac{[Fe^{2+}]^2}{[Fe^{3+}]^2} + \log \frac{[I_3^-]}{[I^-]^3}$$

$$= \log \frac{[Fe^{2+}]^2[I_3^-]}{[Fe^{3+}]^2[I^-]^3}$$

Notice that we have changed the sign of the second logarithmic term by inverting the fraction. Further arrangement gives

$$\log \frac{[Fe^{2+}]^2[I_3^-]}{[Fe^{3+}]^2[I^-]^3} = \frac{2(E^0_{Fe^{3+}/Fe^{2+}} - E^0_{I_3^-/I^-})}{0.0592}$$

Recall, however, that here the concentration terms are *equilibrium concentrations* and that

$$\log K_{eq} = \frac{2(E^0_{Fe^{3+}/Fe^{2+}} - E^0_{I_3^-/I^-})}{0.0592} = \frac{2(0.771 - 0.536)}{0.0592} = 7.94$$

$$K_{eq} = \text{antilog } 7.94 = 8.7 \times 10^7$$

We round the answer to two figures because $\log K_{eq}$ contains only two significant figures (the two to the right of the decimal point).

Example 17-7

Calculate the equilibrium constant for the reaction

$$2MnO_4^- + 3Mn^{2+} + 2H_2O \rightleftharpoons 5MnO_2(s) + 4H^+$$

In Appendix 4, we find

$$2MnO_4^- + 8H^+ + 6e^- \rightleftharpoons 2MnO_2(s) + 4H_2O \qquad E^0 = +1.695 \text{ V}$$

$$3MnO_2(s) + 12H^+ + 6e^- \rightleftharpoons 3Mn^{2+} + 6H_2O \qquad E^0 = +1.23 \text{ V}$$

Again we have multiplied both equations by integers so that the number of electrons are equal. When this system is at equilibrium,

$$E_{MnO_4^-/MnO_2} = E_{MnO_2/Mn^{2+}}$$

$$1.695 - \frac{0.0592}{6} \log \frac{1}{[MnO_4^-]^2[H^+]^8} = 1.23 - \frac{0.0592}{6} \log \frac{[Mn^{2+}]^3}{[H^+]^{12}}$$

If we invert the log term on the right and rearrange, we obtain

$$\frac{6(1.695 - 1.23)}{0.0592} = \log \frac{1}{[MnO_4^-]^2[H^+]^8} + \log \frac{[H^+]^{12}}{[Mn^{2+}]^3}$$

Adding the two log terms gives

$$\frac{6(1.695 - 1.23)}{0.0592} = \log \frac{[H^+]^{12}}{[MnO_4^-]^2[Mn^{2+}]^3[H^+]^8}$$

$$47.1 = \log \frac{[H^+]^4}{[MnO_4^-]^2[Mn^{2+}]^3} = \log K_{eq}$$

$$K_{eq} = \text{antilog } 47.1 = 1 \times 10^{47}$$

Note that the final result has only one significant figure.

17C   CONSTRUCTING REDOX TITRATION CURVES

Because most redox indicators respond to changes in electrode potential, the vertical axis in oxidation/reduction titration curves is generally an electrode potential instead of the logarithmic p-functions that were used for complex formation and neutralization titration curves. We have seen in Chapter 16 that there is a logarithmic relationship between electrode potential and concentration of the analyte or titrant; as a result, redox titration curves are similar in appearance to those for other types of titrations in which a p-function is plotted as the ordinate.

17C-1   Electrode Potentials during Redox Titrations

Let us now consider the redox titration of iron(II) with a standard solution of cerium(IV). This reaction is widely used for the determination of iron in various kinds of samples. The titration reaction is

$$Fe^{2+} + Ce^{4+} \rightleftharpoons Fe^{3+} + Ce^{3+}$$

This reaction is rapid and reversible so that the system *is at equilibrium at all times throughout the titration.* Consequently, the electrode potentials for the two half-reactions are always identical (Equation 17-6); that is,

$$E_{Ce^{4+}/Ce^{3+}} = E_{Fe^{3+}/Fe^{2+}} = E_{system}$$

where we have termed E_{system} *the potential of the system.* If a redox indicator has been added to this solution, the ratio of the concentrations of its oxidized and reduced forms must adjust so that the electrode potential for the indicator, E_{In}, is also equal to the system potential; thus, employing Equation 17-6, we may write

$$E_{In} = E_{Ce^{4+}/Ce^{3+}} = E_{Fe^{3+}/Fe^{2+}} = E_{system}$$

The electrode potential of a system is readily derived from standard potential data. Thus, for the reaction under consideration, the titration mixture is treated as if it were part of the hypothetical cell

$$SHE \parallel Ce^{4+}, Ce^{3+}, Fe^{3+}, Fe^{2+} \mid Pt$$

where SHE symbolizes the standard hydrogen electrode. Note that when we monitor the system potential with a cell such as that shown, the *titration reaction* is at equilibrium, but the *overall cell reaction* (including the SHE) is not. The potential of the platinum electrode with respect to the standard hydrogen electrode is determined by the tendencies of iron(III) and cerium(IV) to accept electrons, that is, by the tendencies of the following half-reactions to occur:

$$Fe^{3+} + e^- \rightleftharpoons Fe^{2+}$$

$$Ce^{4+} + e^- \rightleftharpoons Ce^{3+}$$

At equilibrium, the concentration ratios of the oxidized and reduced forms of the two species are such that their attraction for electrons (and thus their electrode potentials) are identical. Note that these concentration ratios vary continuously throughout the titration, as must E_{system}. End points are determined from the characteristic variation in E_{system} that occurs during the titration.

Because $E_{Ce^{4+}/Ce^{3+}} = E_{Fe^{3+}/Fe^{2+}} = E_{system}$, data for a titration curve can be obtained by applying the Nernst equation for either the cerium(IV) half-reaction or the iron(III) half-reaction. It turns out, however, that one or the other will be more convenient, depending on the stage of the titration. For example, the iron(III) potential is easier to compute in the region short of the equivalence point because here the species concentrations of iron(II) and iron(III) are appreciable and are approximately equal to their analytical concentrations. In contrast, the concentration of cerium(IV), which is negligible prior to the equivalence point because of the large excess of iron(II), can be obtained at this stage only by calculations based on the equilibrium constant for the reaction. Beyond the equivalence point, the analytical concentrations of cerium(IV) and cerium(III) are readily computed directly from the volumetric data, whereas that for iron(II) is not. In this region, then, the cerium(IV) electrode potential is the easier to use. Equivalence-point potentials are derived by the method shown in the next section.

Equivalence-Point Potentials

At the equivalence point, the concentrations of cerium(IV) and iron(II) are minute and cannot be obtained from the stoichiometry of the reaction. Fortunately, equivalence-point potentials are readily obtained by taking advantage of the two reactant species and the two product species having known concentration ratios at chemical equivalence.

At the equivalence point in the titration of iron(II) with cerium(IV), the potential of the system E_{eq} is controlled by both half reactions:

Remember: When redox systems are at equilibrium, *the electrode potentials of all half-reactions are identical.* This generality applies whether the reactions take place directly in solution or indirectly in a galvanic cell.

Most end points in oxidation/reduction titrations are based on the rapid changes in E_{system} that occur at or near chemical equivalence.

Before the equivalence point, E_{system} calculations are easiest to make using the Nernst equation for the analyte. Beyond the equivalence point, the Nernst equation for the titrant is more convenient.

$$E_{eq} = E^0_{Ce^{4+}/Ce^{3+}} - \frac{0.0592}{1} \log \frac{[Ce^{3+}]}{[Ce^{4+}]}$$

and

$$E_{eq} = E^0_{Fe^{3+}/Fe^{2+}} - \frac{0.0592}{1} \log \frac{[Fe^{2+}]}{[Fe^{3+}]}$$

Adding these two expressions gives

The concentration quotient in Equation 17-10 is not the usual ratio of product concentrations and reactant concentrations that appears in equilibrium-constant expressions.

$$2E_{eq} = E^0_{Fe^{3+}/Fe^{2+}} + E^0_{Ce^{4+}/Ce^{3+}} - \frac{0.0592}{1} \log \frac{[Ce^{3+}][Fe^{2+}]}{[Ce^{4+}][Fe^{3+}]} \quad (17\text{-}10)$$

The definition of equivalence point requires that

$$[Fe^{3+}] = [Ce^{3+}]$$

$$[Fe^{2+}] = [Ce^{4+}]$$

Substitution of these equalities into Equation 17-10 results in the concentration quotient becoming unity and the logarithmic term becoming zero:

$$2E_{eq} = E^0_{Fe^{3+}/Fe^{2+}} + E^0_{Ce^{4+}/Ce^{3+}} - \frac{0.0592}{1} \log \frac{[Ce^{3+}][Ce^{4+}]}{[Ce^{4+}][Ce^{3+}]}$$

$$= E^0_{Fe^{3+}/Fe^{2+}} + E^0_{Ce^{4+}/Ce^{3+}}$$

$$E_{eq} = \frac{E^0_{Fe^{3+}/Fe^{2+}} + E^0_{Ce^{4+}/Ce^{3+}}}{2} \quad (17\text{-}11)$$

Example 17-8 illustrates how the equivalence-point potential is derived for a more complex reaction.

Example 17-8

Obtain an expression for the equivalence-point potential in the titration of 0.0500 M U^{4+} with 0.1000 M Ce^{4+}. Assume that both solutions are 1.0 M in H_2SO_4.

$$U^{4+} + 2Ce^{4+} + 2H_2O \rightleftharpoons UO_2^{2+} + 2Ce^{3+} + 4H^+$$

In Appendix 5, we find that

$$UO_2^{2+} + 4H^+ + 2e^- \rightarrow U^{4+} + 2H_2O \qquad E^0 = 0.334 \text{ V}$$

$$Ce^{4+} + e^- \rightleftharpoons Ce^{3+} \qquad E^{0'} = 1.44 \text{ V}$$

Recall that we use the prime notation to indicate formal potentials. Thus, the formal potential for the Ce^{4+}/Ce^{3+} in 1.0 M H_2SO_4 is symbolized by $E^{0'}$.

Here we use the formal potential for Ce^{4+} in 1.0 M H_2SO_4.

Proceeding as in the cerium(IV)/iron(II) equivalence-point calculation, we write

$$E_{eq} = E^0_{UO_2^{2+}/U^{4+}} - \frac{0.0592}{2} \log \frac{[U^{4+}]}{[UO_2^{2+}][H^+]^4}$$

$$E_{eq} = E^{0'}_{Ce^{4+}/Ce^{3+}} - \frac{0.0592}{1} \log \frac{[Ce^{3+}]}{[Ce^{4+}]}$$

To combine the log terms, we must multiply the first equation by 2 to give

$$2E_{eq} = 2E^0_{UO_2^{2+}/U^{4+}} - 0.0592 \log \frac{[U^{4+}]}{[UO_2^{2+}][H^+]^4}$$

Adding this value to the previous equation leads to

$$3E_{eq} = 2E^0_{UO_2^{2+}/U^{4+}} + E^{0'}_{Ce^{4+}/Ce^{3+}} - 0.0592 \log \frac{[U^{4+}][Ce^{3+}]}{[UO_2^{2+}][Ce^{4+}][H^+]^4}$$

But at equivalence

$$[U^{4+}] = \frac{[Ce^{4+}]}{2}$$

and

$$[UO_2^{2+}] = \frac{[Ce^{3+}]}{2}$$

Substituting these equations and rearranging, gives,

$$E_{eq} = \frac{2E^0_{UO_2^{2+}/U^{4+}} + E^{0'}_{Ce^{4+}/Ce^{3+}}}{3} - \frac{0.0592}{3} \log \frac{2[Ce^{4+}][Ce^{3+}]}{2[Ce^{3+}][Ce^{4+}][H^+]^4}$$

$$= \frac{2E^0_{UO_2^{2+}/U^{4+}} + E^{0'}_{Ce^{4+}/Ce^{3+}}}{3} - \frac{0.0592}{3} \log \frac{1}{[H^+]^4}$$

We see that the equivalence-point potential for this titration is pH-dependent.

17C-2   The Titration Curve

Let us first consider the titration of 50.00 mL of 0.0500 M Fe^{2+} with 0.1000 M Ce^{4+} in a medium that is 1.0 M in H_2SO_4 at all times. Formal potential data for both half-cell processes are available in Appendix 4 and are used for these calculations. That is,

$$Ce^{4+} + e^- \rightleftharpoons Ce^{3+} \qquad E^{0'} = 1.44 \text{ V}(1 \text{ M } H_2SO_4)$$

$$Fe^{3+} + e^- \rightleftharpoons Fe^{2+} \qquad E^{0'} = 0.68 \text{ V}(1 \text{ M } H_2SO_4)$$

Initial Potential

The solution contains no cerium species at the outset. In all likelihood, there is a small but unknown amount of Fe^{3+} present due to air oxidation of Fe^{2+}. In any event, we lack sufficient information to calculate an initial potential.

Why is it impossible to calculate the potential of the system before titrant is added?

Potential after the Addition of 5.00 mL of Cerium(IV)

When oxidant is added, Ce^{3+} and Fe^{3+} are formed and the solution contains appreciable and readily calculated concentrations of three of the participants; that of the fourth, Ce^{4+}, is vanishingly small. Therefore, it is more convenient to use the concentrations of the two iron species to calculate the electrode potential of the system.

The equilibrium concentration of Fe(III) is equal to its analytical concentration less the equilibrium concentration of the unreacted Ce(IV):

Remember: The equation for this reaction is

$$Fe^{3+} + Ce^{4+} \rightleftharpoons Fe^{3+} + Ce^{3+}$$

$$[Fe^{3+}] = \frac{5.00 \times 0.1000}{50.00 + 5.00} - [Ce^{4+}] = \frac{0.500}{55.00} - [Ce^{4+}]$$

Similarly, the Fe^{2+} concentration is given by its molarity plus the equilibrium concentration of unreacted Ce(IV):

$$[Fe^{2+}] = \frac{50.00 \times 0.0500 - 5.00 \times 0.1000}{55.00} + [Ce^{4+}] = \frac{2.00}{55.00} + [Ce^{4+}]$$

The quantity $[Ce^{4+}]$ is equal to the concentration of $[Fe^{2+}]$ that *does not react* with Ce^{4+}. Thus, $[Fe^{3+}]$ must be decreased by this amount, and $[Fe^{2+}]$ must be increased by this amount. Ordinarily, $[Ce^{4+}]$ is so small that we can neglect it in both cases.

Generally, redox reactions used in titrimetry are sufficiently complete that the equilibrium concentration of one of the species (in this case $[Ce^{4+}]$) is minuscule with respect to the other species present in the solution. Thus, the foregoing two equations can be simplified to

$$[Fe^{3+}] = \frac{0.500}{55.00} \quad \text{and} \quad [Fe^{2+}] = \frac{2.00}{55.00}$$

Substitution for $[Fe^{2+}]$ and $[Fe^{3+}]$ in the Nernst equation gives

$$E_{system} = +0.68 - \frac{0.0592}{1} \log \frac{2.00/55.00}{0.20/55.00} = 0.64 \text{ V}$$

Note that the volumes in the numerator and denominator cancel, which indicates that the potential is independent of dilution. This independence persists until the solution becomes so dilute that the two assumptions made in the calculation become invalid.

It is worth emphasizing again that the use of the Nernst equation for the Ce(IV)/Ce(III) system would yield the same value for E_{system}, but to do so would require computing $[Ce^{4+}]$ by means of the equilibrium constant for the reaction.

Additional potentials needed to define the titration curve short of the equivalence point can be obtained similarly. Such data are given in Table 17-1. You may want to confirm one or two of these values.

Table 17-1

Electrode Potential versus SHE in Titrations with 0.100 M Ce^{4+}

Reagent Volume, mL	Potential, V vs. SHE		
	50.00 mL of 0.0500 M Fe^{2+}		50.00 mL of 0.02500 M U^{4+}
5.00	0.64		0.316
15.00	0.69		0.339
20.00	0.72		0.352
24.00	0.76		0.375
24.90	0.82		0.405
25.00	**1.06**	← Equivalence → Point	**0.703**
25.10	1.30		1.30
26.00	1.36		1.36
30.00	1.40		1.40

Note: H_2SO_4 concentration is such that $[H^+] = 1.0$ M throughout.

Equivalence-Point Potential

Substitution of the two formal potentials into Equation 17-11 yields

$$E_{eq} = \frac{E^{0\prime}_{Ce^{4+}/Ce^{3+}} + E^{0\prime}_{Fe^{3+}/Fe^{2+}}}{2} = \frac{1.44 + 0.68}{2} = 1.06 \text{ V}$$

Potential after the Addition of 25.10 mL of Cerium(IV)

The molar concentrations of Ce(III), Ce(IV), and Fe(III) are readily computed at this point, but that for Fe(II) is not. Therefore, E_{system} computations based on the cerium half-reaction are more convenient. The concentrations of the two cerium ion species are

$$[Ce^{3+}] = \frac{25.00 \times 0.1000}{75.10} - [Fe^{2+}] \approx \frac{2.500}{75.10}$$

$$[Ce^{4+}] = \frac{25.10 \times 0.1000 - 50.00 \times 0.0500}{75.10} + [Fe^{2+}] \approx \frac{0.010}{75.10}$$

Here, the iron(II) concentration is negligible with respect to the analytical concentrations of the two cerium species. Substitution into the Nernst equation for the cerium couple gives

$$E = +1.44 - \frac{0.0592}{1} \log \frac{[Ce^{3+}]}{[Ce^{4+}]}$$

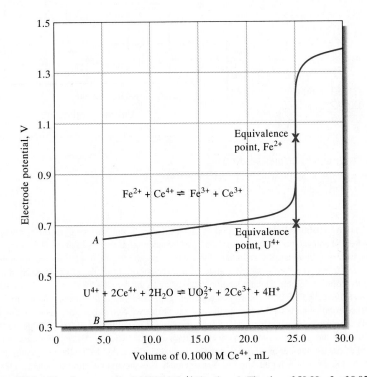

Figure 17-3   Titration curves for 0.1000 M Ce^{4+} titration. A: Titration of 50.00 mL of 0.05000 M Fe^{2+}. B: Titration of 50.00 mL of 0.02500 M U^{4+}.

$$E = +1.44 - \frac{0.0592}{1} \log\frac{2.500/\cancel{75.10}}{0.010/\cancel{75.10}} = 1.30 \text{ V}$$

The additional postequivalence potentials in Table 17-1 were derived in a similar fashion.

The titration curve of iron(II) with cerium(IV) appears as A in Figure 17-3. This plot closely resembles the curves encountered in neutralization, precipitation, and complex-formation titrations with the equivalence point being signaled by a rapid change in the ordinate function. A titration involving 0.00500 M iron(II) and 0.01000 M cerium(IV) yields a curve that, for all practical purposes, is identical to the one we have derived, since the electrode potential of the system is independent of dilution. A spreadsheet to calculate E_{system} as a function of the volume of Ce(IV) added is shown in Figure 17-4. Example 17-9 illustrates the construction of a spreadsheet for the titration of U^{4+} with Ce^{4+}.

> In contrast to other titration curves we have encountered, oxidation/reduction curves are *independent* of reactant concentration except for very dilute solutions.

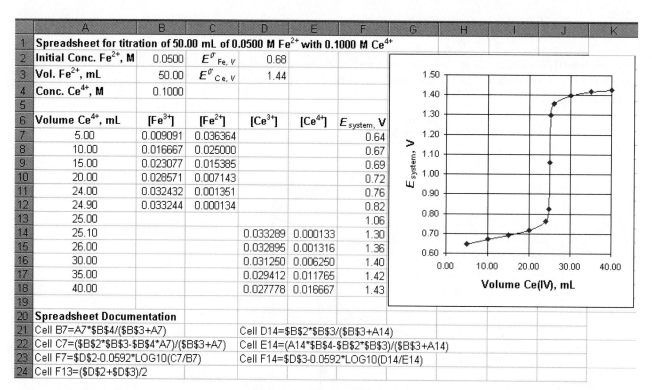

Figure 17-4 spreadsheet:

	A	B	C	D	E	F
1	Spreadsheet for titration of 50.00 mL of 0.0500 M Fe^{2+} with 0.1000 M Ce^{4+}					
2	Initial Conc. Fe^{2+}, M	0.0500	$E^{0'}_{Fe, V}$	0.68		
3	Vol. Fe^{2+}, mL	50.00	$E^{0'}_{Ce, V}$	1.44		
4	Conc. Ce^{4+}, M	0.1000				
5						
6	Volume Ce^{4+}, mL	$[Fe^{3+}]$	$[Fe^{2+}]$	$[Ce^{3+}]$	$[Ce^{4+}]$	E_{system}, V
7	5.00	0.009091	0.036364			0.64
8	10.00	0.016667	0.025000			0.67
9	15.00	0.023077	0.015385			0.69
10	20.00	0.028571	0.007143			0.72
11	24.00	0.032432	0.001351			0.76
12	24.90	0.033244	0.000134			0.82
13	25.00					1.06
14	25.10			0.033289	0.000133	1.30
15	26.00			0.032895	0.001316	1.36
16	30.00			0.031250	0.006250	1.40
17	35.00			0.029412	0.011765	1.42
18	40.00			0.027778	0.016667	1.43
19						
20	Spreadsheet Documentation					
21	Cell B7=A7*$B$4/($B$3+A7)			Cell D14=$B$2*$B$3/($B$3+A14)		
22	Cell C7=($B$2*$B$3-$B$4*A7)/($B$3+A7)			Cell E14=(A14*$B$4-$B$2*$B$3)/($B$3+A14)		
23	Cell F7=$D$2-0.0592*LOG10(C7/B7)			Cell F14=$D$3-0.0592*LOG10(D14/E14)		
24	Cell F13=($D$2+$D$3)/2					

Figure 17-4 Spreadsheet and plot for titration of 50.00 mL of 0.0500 M Fe^{2+} with 0.1000 M Ce^{4+}. Prior to the equivalence point the system potential is calculated from the Fe^{3+} and Fe^{2+} concentrations. After the equivalence point the Ce^{4+} and Ce^{3+} concentrations are used in the Nernst equation. The Fe^{3+} concentration in cell B7 is calculated from the number of millimoles of Ce^{4+} added divided by the total volume of solution. The formula used for the first volume is shown in documentation cell A21. In cell C7 the $[Fe^{2+}]$ is calculated as the initial number of millimoles of Fe^{2+} present minus that number of millimoles of Fe^{3+} formed divided by the total solution volume. Documentation cell A22 gives the formula for the 5.00 mL volume. The system potential prior to the equivalence point is calculated in cells F7:F12 by the Nernst equation, expressed for the first volume by the formula shown in documentation cell A23. In cell F13, the equivalence point potential is found from the average of the two formal potentials as shown in documentation cell A24. After the equivalence point, the Ce(III) concentration (cell D14) is found from the number of millimoles of Fe^{2+} initially present divided by the total solution volume as shown for the 25.10 mL volume by the formula in documentation cell D21. The Ce(IV) concentration (E14) is found from the total number of millimoles of Ce(IV) added minus the number of millimoles of Fe^{2+} initially present divided by the total solution volume as shown in documentation cell D22. The system potential in cell F14 is found from the Nernst equation as shown in documentation cell D23. The chart is then the resulting titration curve plot.

Example 17-9   ***Spreadsheet***

Construct a spreadsheet for the calculations and plot a titration curve for the reaction of 50.00 mL of 0.02500 M U^{4+} with 0.1000 M Ce^{4+}. The solution is 1.0 M in H_2SO_4 throughout the titration. For the sake of simplicity, assume that $[H^+]$ for this solution is also about 1.0 M. The spreadsheet is shown in Figure 17-5. In this example, the reasoning is shown first, and then the spreadsheet solution is presented.

The analytical reaction is

$$U^{4+} + 2Ce^{4+} + 2H_2O \rightleftharpoons UO_2^{2+} + 2Ce^{3+} + 4H^+$$

and in Appendix 4 we find

$$UO_2^{2+} + 4H^+ + 2e^- \rightleftharpoons U^{4+} + 2H_2O \qquad E^0 = 0.334 \text{ V}$$

$$Ce^{4+} + e^- \rightleftharpoons Ce^{3+} \qquad\qquad E^{0'} = 1.44 \text{ V}$$

Potential after Adding 5.00 mL of Ce^{4+}

$$\text{original amount } U^{4+} = 50.00 \text{ mL } U^{4+} \times 0.02500 \frac{\text{mmol } U^{4+}}{\text{mL } U^{4+}}$$

$$= 1.250 \text{ mmol } U^{4+}$$

$$\text{amount } Ce^{4+} \text{ added} = 5.00 \text{ mL } Ce^{4+} \times 0.1000 \frac{\text{mmol } Ce^{4+}}{\text{mL } Ce^{4+}}$$

$$= 0.5000 \text{ mmol } Ce^{4+}$$

$$\text{total volume of solution} = (50.00 + 5.00) \text{ mL} = 55.00 \text{ mL}$$

In cell B7, we will place the concentration of UO_2^{2+} formed as a result of adding Ce^{4+} to U^{4+}. We reason that

$$\text{concentration } UO_2^{2+} \text{ formed} = \frac{0.5000 \text{ mmol } Ce^{4+} \times \dfrac{1 \text{ mmol } UO_2^{2+}}{2 \text{ mmol } Ce^{4+}}}{55.00 \text{ mL}}$$

The reader should verify that the formula in documentation cell A21 is identical to that shown above for calculating the concentration of UO_2^{2+} formed. In cell C7, we reason that

concentration U^{4+} remaining

$$= \frac{1.250 \text{ mmol } U^{4+} - 0.2500 \text{ mmol } UO_2^{2+} \times \dfrac{1 \text{ mmol } U^{4+}}{\text{mmol } UO_2^{2+}}}{55.00 \text{ mL}}$$

and place the formula shown in documentation cell A22. The expressions in B7 and C7 are copied into B8:B12 and C8:C12 respectively.

The number of millimoles UO_2^{2+} formed per millimole Ce^{4+} reacted (1/2) occurs because the Ce^{4+}/Ce^{3+} reaction involves only one electron, whereas the UO_2^{2+}/U^{4+} reaction involves two electrons.

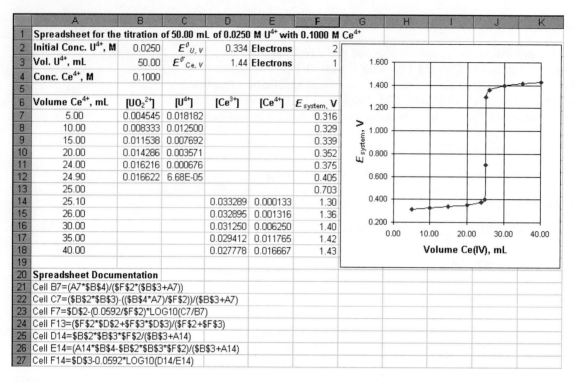

Figure 17-5   Spreadsheet for the titration of 50.00 mL of 0.0250 M U^{4+} with 0.1000 M Ce^{4+}. The solution is assumed to be 1.0 M in H^+ throughout the titration. Example 17-9 gives the details of the calculations.

Applying the Nernst equation for UO_2^{2+}, we obtain

$$E = 0.334 - \frac{0.0592}{2} \log \frac{[U^{4+}]}{[UO_2^{2+}][H^+]^4}$$

$$= 0.334 - \frac{0.0592}{2} \log \frac{[U^{4+}]}{[UO_2^{2+}](1.00)^4}$$

Substituting concentrations of the two uranium species gives, for cell F7,

$$E = 0.334 - \frac{0.0592}{2} \log \frac{1.000 \text{ mmol } U^{4+}/55.00 \text{ mL}}{0.2500 \text{ mmol } UO_2^{2+}/55.00 \text{ mL}}$$

Hence, the formula shown in documentation cell A23 is entered into cell F7, and the contents are copied into cells F8:F12. These values are also tabulated in the third column in Table 17-1.

Equivalence-Point Potential

Following the procedure shown in Example 17-8, we obtain

$$E_{eq} = \frac{(2E^0_{UO_2^{2+}/U^{4+}} + E^{0'}_{Ce^{4+}/Ce^{3+}})}{3} - \frac{0.0592}{3} \log \frac{1}{[H^+]^4}$$

Substituting gives, for cell F13,

$$E_{eq} = \frac{2 \times 0.334 + 1.44}{3} - \frac{0.0592}{3} \log \frac{1}{(1.00)^4} = \frac{2 \times 0.334 + 1.44}{3}$$

Potential after Adding 25.10 mL of Ce^{4+}

$$\text{total volume of solution} = 75.10 \text{ mL}$$

$$\text{original amount } U^{4+} = 50.00 \text{ mL } U^{4+} \times 0.02500 \frac{\text{mmol } U^{4+}}{\text{mL } U^{4+}}$$

$$= 1.250 \text{ mmol } U^{4+}$$

$$\text{amount } Ce^{4+} \text{ added} = 25.10 \text{ mL } Ce^{4+} \times 0.1000 \frac{\text{mmol } Ce^{4+}}{\text{mL } Ce^{4+}}$$

$$= 2.510 \text{ mmol } Ce^{4+}$$

concentration of Ce^{3+} formed

$$= \frac{1.250 \text{ mmol } U^{4+} \times 2 \text{ mmol } Ce^{3+}/\text{mmol } U^{4+}}{75.10 \text{ mL}}$$

In cell D14, we thus write the expression shown in documentation cell A25.

concentration of Ce^{4+} remaining

$$= \frac{2.510 \text{ mmol } Ce^{4+} - 2.500 \text{ mmol } Ce^{3+} \times 1 \text{ mmol } Ce^{4+}/\text{mmol } Ce^{3+}}{75.10 \text{ mL}}$$

In cell E14, we place the formula given in documentation cell A26.
  Substituting into the expression for the formal potential gives

$$E = 1.44 - 0.0592 \log \frac{2.500/75.10}{0.010/75.10}$$

In cell F14, we thus write the formula shown in documentation cell A27 and copy this result into F15 through F18. This completes the spreadsheet. The plot is shown alongside the spreadsheet in Figure 17-5.

Table 17-1 contains other postequivalence-point data obtained in this same way. The data in the third column of Table 17-1 are plotted as curve *B* in Figure 17-3 to compare the two titrations. The two curves are seen to be identical for volumes greater than 25.10 mL because the concentrations of the two cerium species are identical in this region. It is also interesting that the curve for iron(II) is symmetric about the equivalence point, whereas the curve for uranium(IV) is not. In general, symmetric curves are always obtained when the analyte and titrant react in a 1:1 molar ratio.

Redox titration curves are symmetric when the reactants combine in a 1:1 ratio. Otherwise, they are asymmetric.

17C-3   Effect of Variables on Redox Titration Curves

In earlier chapters, we have considered the effects of reactant concentrations and completeness of the reaction on titration curves. Here, we describe the effects of these variables on oxidation/reduction titration curves.

Reactant Concentration

As we have just seen, E_{system} for an oxidation/reduction titration is ordinarily independent of dilution. Consequently, titration curves for oxidation/reduction reactions are usually independent of analyte and reagent concentrations. This independence is in distinct contrast to that observed in the other types of titration curves we have encountered.

Completeness of the Reaction

The change in the equivalence-point region of an oxidation/reduction titration becomes larger as the reaction becomes more complete. This effect is demonstrated by the two curves in Figure 17-3. The equilibrium constant for the reaction of cerium(IV) with iron(II) is 7×10^{12}, whereas for U(IV) it is 2×10^{37}. The effect of reaction completeness is further demonstrated in Figure 17-6, which shows curves for the titration of a hypothetical reductant with a standard electrode potential of 0.20 V with several hypothetical oxidants with standard potentials ranging from 0.40 to 1.20 V; the corresponding equilibrium constants lie between about 2×10^3 and 8×10^{16}. Clearly, the greatest change in potential of the system is associated with the reaction that is most complete. In this respect, then, oxidation/reduction titration curves are similar to those involving other types of reactions.

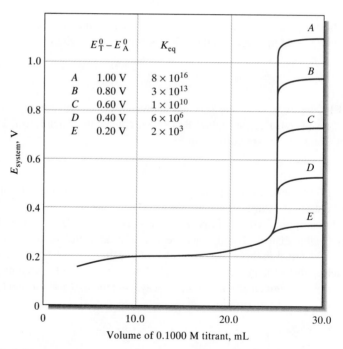

	$E_{\text{T}}^0 - E_{\text{A}}^0$	K_{eq}
A	1.00 V	8×10^{16}
B	0.80 V	3×10^{13}
C	0.60 V	1×10^{10}
D	0.40 V	6×10^6
E	0.20 V	2×10^3

Figure 17-6 Effect of titrant electrode potential on reaction completeness. The standard electrode potential for the analyte (E_{A}^0) is 0.200 V; starting with curve A, standard electrode potentials for the titrant (E_{T}^0) are 1.20, 1.00, 0.80, 0.60, and 0.40, respectively. Both analyte and titrant undergo a one-electron change.

17D   OXIDATION/REDUCTION INDICATORS

Two types of chemical indicators are used for obtaining end points for oxidation/reduction titrations: general redox indicators and specific indicators.

17D-1   General Redox Indicators

General oxidation/reduction indicators are substances that change color upon being oxidized or reduced. In contrast to specific indicators, the color changes of true redox indicators are largely independent of the chemical nature of the analyte and titrant and depend instead on the changes in the electrode potential of the system that occur as the titration progresses.

> Color changes for general redox indicators depend only on the potential of the system.

The half-reaction responsible for color change in a typical general oxidation/reduction indicator can be written as

$$\text{In}_{\text{ox}} + n\text{e}^- \rightleftharpoons \text{In}_{\text{red}}$$

If the indicator reaction is reversible, we can write

$$E = E^0_{\text{In}_{\text{ox}}/\text{In}_{\text{red}}} - \frac{0.0592}{n} \log \frac{[\text{In}_{\text{red}}]}{[\text{In}_{\text{ox}}]} \qquad (17\text{-}12)$$

Typically, a change from the color of the oxidized form of the indicator to the color of the reduced form requires a change of about 100 in the ratio of reactant concentrations; that is, a color change is seen when

$$\frac{[\text{In}_{\text{red}}]}{[\text{In}_{\text{ox}}]} \leq \frac{1}{10}$$

changes to

$$\frac{[\text{In}_{\text{red}}]}{[\text{In}_{\text{ox}}]} \geq 10$$

The potential change required to produce the full color change of a typical general indicator can be found by substituting these two values into Equation 17-12, which gives

$$E = E^0_{\text{In}} \pm \frac{0.0592}{n}$$

This equation shows that a typical general indicator exhibits a detectable color change when a titrant causes the system potential to shift from $E^0_{\text{In}} + 0.0592/n$ to $E^0_{\text{In}} - 0.0592/n$, or about $(0.118/n)$ V. For many indicators, $n = 2$, and a change of 0.059 V is thus sufficient.

> Protons are involved in the reduction of many indicators, and so the range of potentials over which a color change occurs (the *transition potential*) is often pH-dependent.

Table 17-2 lists transition potentials for several redox indicators. Note that indicators functioning in any desired potential range up to about + 1.25 V are available. Structures for and reactions of a few of the indicators listed in the table are considered in the paragraphs that follow.

Iron(II) Complexes of Orthophenanthrolines

A class of organic compounds known as 1,10-phenanthrolines, or orthophenanthrolines, form stable complexes with iron(II) and certain other ions. The parent compound has a pair of nitrogen atoms located in such positions that each can

The compound 1,10-phenanthroline is an excellent complexing agent for Fe(II).

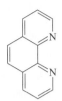

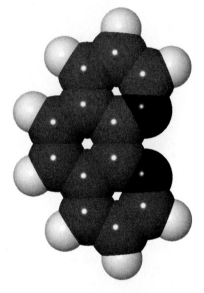

Model of 1,10-phenanthroline.

Table 17-2

Selected Oxidation/Reduction Indicators

Indicator	Color		Transition Potential, V	Conditions
	Oxidized	Reduced		
5-Nitro-1,10-phenanthroline iron(II) complex	Pale blue	Red-violet	+ 1.25	1 M H_2SO_4
2,3'-Diphenylamine dicarboxylic acid	Blue-violet	Colorless	+ 1.12	7–10 M H_2SO_4
1,10-Phenanthroline iron(II) complex	Pale blue	Red	+ 1.11	1 M H_2SO_4
5-Methyl-1,10-phenanthroline iron(II) complex	Pale blue	Red	+ 1.02	1 M H_2SO_4
Erioglaucin A	Blue-red	Yellow-green	+ 0.98	0.5 M H_2SO_4
Diphenylamine sulfonic acid	Red-violet	Colorless	+ 0.85	Dilute acid
Diphenylamine	Violet	Colorless	+ 0.76	Dilute acid
p-Ethoxychrysoidine	Yellow	Red	+ 0.76	Dilute acid
Methylene blue	Blue	Colorless	+ 0.53	1 M acid
Indigo tetrasulfonate	Blue	Colorless	+ 0.36	1 M acid
Phenosafranine	Red	Colorless	+ 0.28	1 M acid

Source: Data in part from I. M. Kolthoff and V. A. Stenger, *Volumetric Analysis,* 2nd ed., Vol. 1 (New York: Interscience, 1942), p. 140.

form a covalent bond with the iron(II) ion. Three orthophenanthroline molecules combine with each iron ion to yield a complex sometimes called ferroin, which is conveniently formulated as $(phen)_3Fe^{2+}$.

The complexed iron in the ferroin undergoes a reversible oxidation/reduction reaction that can be written as

$$(phen)_3Fe^{3+} + e^- \rightleftharpoons (phen)_3Fe^{2+}$$

pale blue   **red**

In practice, the color of the oxidized form is so slight as to go undetected, and the color change associated with this reduction is thus from nearly colorless to red. Because of the difference in color intensity, the end point is usually taken when only about 10% of the indicator is in the iron(II) form. The transition potential is thus approximately +1.11 V in 1 M sulfuric acid.

Of all the oxidation/reduction indicators, ferroin approaches most closely the ideal substance. It reacts rapidly and reversibly, its color change is pronounced, and its solutions are stable and readily prepared. In contrast to many indicators, the oxidized form of ferroin is remarkably inert toward strong oxidizing agents. At temperatures above 60°C, ferroin decomposes.

A number of substituted phenanthrolines have been investigated for their indicator properties, and some have proved to be as useful as the parent compound. Among these, the 5-nitro and 5-methyl derivatives are noteworthy, with transition potentials of +1.25 V and +1.02 V, respectively.

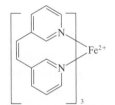

ferroin $(phen)_3Fe^{2+}$

5-nitro-1, 10-phenanthroline   5-methyl-1, 10-phenanthroline

Starch / Iodine Solutions

Starch, which forms a blue complex with triiodide ion, is a widely used specific indicator in oxidation/reduction reactions involving iodine as an oxidant or iodide ion as a reductant. A starch solution containing a little triiodide or iodide ion can also function as a true redox indicator, however. In the presence of excess oxidizing agent, the concentration ratio of iodine to iodide is high, giving a blue color to the solution. With excess reducing agent, on the other hand, iodide ion predominates and the blue color is absent. Thus, the indicator system changes from colorless to blue in the titration of many reducing agents with various oxidizing agents. This color change is quite independent of the chemical composition of the reactants, depending only on the potential of the system at the equivalence point.

The Choice of Redox Indicator

It is apparent from Figure 17-6 that all the indicators in Table 17-2 except for the first and the last could be employed with titrant A. In contrast, with titrant D, only indigo tetrasulfonate could be employed. The change in potential with titrant E is too small to be satisfactorily detected by an indicator.

17D-2   Specific Indicators

Perhaps the best-known specific indicator is starch, which forms a dark blue complex with triiodide ion as discusssed above. This complex signals the end point in titrations in which iodine is either produced or consumed.

Another specific indicator is potassium thiocyanate, which may be employed, for example, in the titration of iron(III) with solutions of titanium(III) sulfate. The end point involves the disappearance of the red color of the iron(III)/thiocyanate complex as a result of the marked decrease in the iron(III) concentration at the equivalence point.

17E   POTENTIOMETRIC END POINTS

End points for many oxidiation/reduction titrations are readily observed by making the solution of the analyte part of the cell:

$$\text{reference electrode} \parallel \text{analyte solution} \mid \text{Pt}$$

By measuring the potential of this cell during a titration, data for curves analogous to those shown in Figures 17-3 and 17-6 can be generated. End points are readily estimated from such curves. Potentiometric end points are considered in detail in Chapter 19.

WEB WORKS

Set your browser to http://www.iupac.org/divisions/commissions.html and then click on the Electroanalytical Chemistry Commission of the International Union of Pure and Applied Chemistry (IUPAC). Describe one of the current projects of this commission or a paper published by a member.

17F   QUESTIONS AND PROBLEMS

*17-1. Briefly define the electrode potential of a system that contains two or more redox couples.

17-2. For an oxidation/reduction reaction, briefly distinguish between

 *(a) equilibrium and equivalence.

 (b) a true oxidation/reduction indicator and a specific indicator.

17-3. What is unique about the condition of equilibrium in an oxidation/reduction reaction?

*17-4. How is an oxidation/reduction titration curve generated through the use of standard electrode potentials for the analyte species and the volumetric titrant?

17-5. How does calculation of the electrode potential of the system at the equivalence point differ from that for any other point of an oxidation/reduction titration?

*17-6. Under what circumstance is the curve for an oxidation/reduction titration asymmetric about the equivalence point?

17-7. Calculate the theoretical potential of the following cells. Indicate whether the reaction will proceed spontaneously in the direction considered (oxidation on the left, reduction on the right) or whether an external voltage source is needed to force this reaction to occur.

 (a) $Pb \mid Pb^{2+}(0.1393\ M) \parallel Cd^{2+}(0.0511) \mid Cd$

 (b) $Zn \mid Zn^{2+}(0.0364\ M) \parallel$ $Tl^{3+}(9.06 \times 10^{-3}\ M), Tl^{+}(0.0620\ M) \mid Pt$

 (c) $Pt, H_2(765\ torr) \mid HCl(1.00 \times 10^{-4}\ M) \parallel$ $Ni^{2+}(0.0214\ M) \mid Ni$

 (d) $Pb \mid PbI_2\ (sat'd), (0.0120\ M) \parallel$ $Hg^{2+}(4.59 \times 10^{-3}\ M) \mid Hg$

 (e) $Pt, H_2(1.00\ atm) \mid NH_3(0.438\ M), NH_4^+$ $(0.379\ M) \parallel SHE$

 (f) $Pt \mid TiO^{2+}(0.0790\ M), Ti^{3+}(0.00918\ M),$ $H^+(1.47 \times 10^{-2}\ M) \parallel VO^{2+}(0.1340\ M),$ $V^{3+}(0.0784\ M), H^+(0.0538\ M) \mid Pt$

*17-8. Calculate the theoretical cell potential of the following cells. If the cell is short-circuited, indicate the direction of the spontaneous cell reaction.

 (a) $Zn \mid Zn^{2+}(0.0955\ M) \parallel Co^{2+}(6.78 \times 10^{-3}\ M) \mid Co$

 (b) $Pt \mid Fe^{3+}(0.1310\ M), Fe^{2+}(0.0681\ M) \parallel$ $Hg^{2+}(0.0671\ M) \mid Hg$

 (c) $Ag \mid Ag^{+}(0.1544\ M) \mid H^{+}(0.0794\ M) \mid$ $O_2(1.12\ atm), Pt$

 (d) $Cu \mid Cu^{2+}(0.0601\ M) \parallel (0.1350\ M),$ $AgI\ (sat'd) \mid Ag$

 (e) $SHE \parallel HCOOH(0.1302\ M),$ $HCOO^-(0.0764\ M) \mid H_2(1.00\ atm), Pt$

 (f) $Pt \mid UO_2^{2+}(7.93 \times 10^{-3}\ M),$ $U^{4+}(6.37 \times 10^{-2}\ M),$ $H^+(1.16 \times 10^{-3}\ M) \parallel Fe^{3+}(0.003876\ M),$ $Fe^{2+}(0.1134\ M) \mid Pt$

17-9. Calculate the potential of the following cells:

 *(a) a galvanic cell consisting of a lead electrode immersed in 0.0848 M Pb^{2+} and a zinc electrode in contact with 0.1364 M Zn^{2+}.

 (b) a galvanic cell with two platinum electrodes, one immersed in a solution that is 0.0301 M in Fe^{3+} and 0.0760 M in Fe^{2+}, the other in a solution that is 0.00309 M in $Fe(CN)_6^{4-}$ and 0.1564 M in $Fe(CN)_6^{3-}$.

 *(c) a galvanic cell consisting of a standard hydrogen electrode and a platinum electrode immersed in a solution that is 1.46 × 10^{-3} M in TiO^{2+}, 0.02723 M in Ti^{3+}, and buffered to a pH of 3.00.

17-10. Use the shorthand notation (page 396) to describe the cells in Problem 17-9. Each cell is supplied with a salt bridge to provide electrical contact between the solutions in the two cell compartments.

17-11. Generate equilibrium constant expressions for the following titration reactions. Calculate numerical values for K_{eq}.

 *(a) $Fe^{3+} + V^{2+} \rightleftharpoons Fe^{2+} + V^{3+}$

 (b) $Fe(CN)_6^{3-} + Cr^{2+} \rightleftharpoons Fe(CN)_6^{4-} + Cr^{3+}$

 *(c) $2V(OH)_4^+ + U^{4+} \rightleftharpoons$ $2VO^{2+} + UO_2^{2+} + 4H_2O$

 (d) $Tl^{3+} + 2Fe^{2+} \rightleftharpoons Tl^+ + 2Fe^{3+}$

 *(e) $2Ce^{4+} + H_3AsO_3 + H_2O \rightleftharpoons$ $2Ce^{3+} + H_3AsO_4 + 2H^+ \ (1\ M\ HClO_4)$

 (f) $2V(OH)_4^+ + H_2SO_3 \rightleftharpoons$ $SO_4^{2-} + 2VO^{2+} + 5H_2O$

 *(g) $VO^{2+} + V^{2+} + 2H^+ \rightleftharpoons 2V^{3+} + H_2O$

 (h) $TiO^{2+} + Ti^{2+} + 2H^+ \rightleftharpoons 2Ti^{3+} + H_2O$

*17-12. Calculate the electrode potential of the system at the equivalence point for each of the reactions in Problem 17-11. Use 0.100 M where a value for $[H^+]$ is needed and is not otherwise specified.

17-13. If you start with 0.1000 M solutions and the first-named species is the titrant, what will be the concentration of each reactant and product at the equivalence point of the titrations in Problem 17-11? Assume that there is no change in $[H^+]$ during the titration.

*17-14. Select an indicator from Table 17-2 that might be suitable for each of the titrations in Problem 17-11. Write "none" if no indicator listed in Table 17-2 is suitable.

17-15. Use a spreadsheet and construct curves for the following titrations. Calculate potentials after the addition of 10.00, 25.00, 49.00, 49.90, 50.00, 50.10, 51.00, and 60.00 mL of the reagent. Where necessary, assume that $[H^+] = 1.00$ throughout.

*(a)   50.00 mL of 0.1000 M V^{2+} with 0.05000 M Sn^{4+}.

(b)   50.00 mL of 0.1000 M $Fe(CN)_6^{3-}$ with 0.1000 M Cr^{2+}.

*(c)   50.00 mL of 0.1000 M $Fe(CN)_6^{4-}$ with 0.05000 M Tl^{3+}.

(d)   50.00 mL of 0.1000 M Fe^{3+} with 0.05000 M Sn^{2+}.

*(e)   50.00 mL of 0.05000 M U^{4+} with 0.02000 M MnO_4^-.

Chapter 18

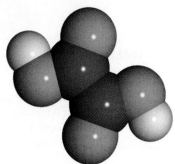

Applying Oxidation/ Reduction Titrations

This chapter is concerned with the preparation of standard solutions of oxidants and reductants and with their applications in analytical chemistry. In addition, auxiliary reagents that convert an analyte to a single oxidation state are described.[1]

18A AUXILIARY OXIDIZING AND REDUCING REAGENTS

The analyte in an oxidation/reduction titration must be in a single oxidation state at the outset. Often, however, the steps that precede the titration (dissolution of the sample and separation of interferences) convert the analyte to a mixture of oxidation states. For example, the solution formed when an iron-containing sample is dissolved usually contains a mixture of iron(II) and iron(III) ions. If we choose to use a standard oxidant for determining iron, we must first treat the sample solution with an auxiliary reducing agent to convert all the iron to iron(II). On the other hand, if we plan to titrate with a standard reductant, pretreatment with an auxiliary oxidizing reagent is needed.[2]

To be useful as a preoxidant or a prereductant, a reagent must react quantitatively with the analyte. In addition, any reagent excess must be readily removable because the excess reagent usually interferes by reacting with the standard solution.

18A-1 Auxiliary Reducing Reagents

A number of metals are good reducing agents and have been used for the prereduction of analytes. Included among these are zinc, aluminum, cadmium, lead, nickel, copper, and silver (in the presence of chloride ion). Sticks or coils of the metal can be immersed directly in the analyte solution. After reduction is judged

---

[1]For further reading on redox titrimetry, see J. A. Dean, *Analytical Chemistry Handbook* (New York: McGraw-Hill, 1995), Section 3, pp. 3.65–3.75.

[2]For a brief summary of auxiliary reagents, see J. A. Goldman and V. A. Stenger, in *Treatise on Analytical Chemistry,* I. M. Kolthoff and P. J. Elving, Eds., Part I, Vol. 11 (New York: Wiley, 1975), pp. 7204–7206.

complete, the solid is removed manually and rinsed with water. Filtration of the analyte solution is needed to remove granular or powdered forms of the metal. An alternative to filtration is the use of a *reductor,* such as that shown in Figure 18-1.[3] Here, the finely divided metal is held in a vertical glass tube through which the solution is drawn under a mild vacuum. The metal in a reductor is ordinarily sufficient for hundreds of reductions.

A typical *Jones reductor* has a diameter of about 2 cm and holds a 40- to 50-cm column of amalgamated zinc. Amalgamation is accomplished by allowing zinc granules to stand briefly in a solution of mercury(II) chloride, where the following reaction occurs:

$$2Zn(s) + Hg^{2+} \rightarrow Zn^{2+} + Zn(Hg)(s)$$

Zinc amalgam is nearly as effective for reductions as the pure metal and has the important virtue of inhibiting the reduction of hydrogen ions by zinc. This side reaction needlessly uses up the reducing agent and also contaminates the sample solution with a large amount of zinc(II) ions. Solutions that are quite acidic can be passed through a Jones reductor without significant hydrogen formation.

Table 18-1 lists the principal applications of the Jones reductor. Also listed in this table are reductions that can be accomplished with a *Walden reductor,* in which granular metallic silver held in a narrow glass column is the reductant. Silver is not a good reducing agent unless chloride or some other ion that forms a silver salt of low solubility is present. For this reason, prereductions with a Walden reductor are generally carried out from hydrochloric acid solutions of the analyte. The coating of silver chloride produced on the metal is removed periodically by dipping a zinc rod into the solution that covers the packing. Table 18-1 suggests that the Walden reductor is somewhat more selective in its action than is the Jones reductor.

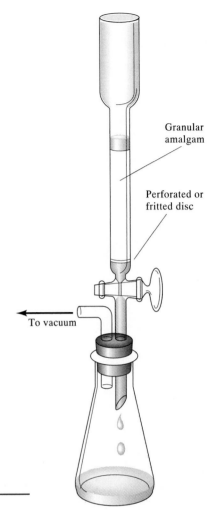

Granular amalgam

Perforated or fritted disc

To vacuum

Figure 18-1 A Jones reductor.

Table 18-1

Uses of the Walden Reductor and the Jones Reductor

Walden $Ag(s) + Cl^- \rightarrow AgCl(s) + e^-$	Jones $Zn(Hg)(s) \rightarrow Zn^{2+} + Hg + 2e^-$
$Fe^{3+} + e^- \rightarrow Fe^{2+}$	$Fe^{3+} + e^- \rightleftharpoons Fe^{2+}$
$Cu^{2+} + e^- \rightarrow Cu^+$	$Cu^{2+} + 2e^- \rightleftharpoons Cu(s)$
$H_2MoO_4 + 2H^+ + e^- \rightarrow MoO_2^+ + 2H_2O$	$H_2MoO_4 + 6H^+ + 3e^- \rightleftharpoons Mo^{3+} + 3H_2O$
$UO_2^{2+} + 4H^+ + 2e^- \rightarrow U^{4+} + 2H_2O$	$UO_2^{2+} + 4H^+ + 2e^- \rightleftharpoons U^{4+} + 2H_2O$
	$UO_2^{2+} + 4H^+ + 3e^- \rightleftharpoons U^{3+} + 2H_2O*$
$V(OH)_4^+ + 2H^+ + e^- \rightarrow VO^{2+} + 3H_2O$	$V(OH)_4^+ + 4H^+ + 3e^- \rightleftharpoons V^{2+} + 4H_2O$
TiO^{2+} not reduced	$TiO^{2+} + 2H^+ + e^- \rightleftharpoons Ti^{2+} + H_2O$
Cr^{3+} not reduced	$Cr^{3+} + e^- \rightleftharpoons Cr^{2+}$

Source: From I. M. Kolthoff and R. Belcher, *Volumetric Analysis,* Vol. 3 (New York: Interscience, 1957), p. 12. With permission.

*A mixture of oxidation states is obtained. The Jones reductor may still be used for the determination of uranium, however, because any U^{2+} formed can be converted to U^{3+} by shaking the solution with air for a few minutes.

[3]For a discussion of reductors, see F. Hecht, in *Treatise on Analytical Chemistry,* I. M. Kolthoff and P. J. Elving, Eds., Part I, Vol. 11 (New York: Wiley, 1975), pp. 6703–6707.

18A-2   Auxiliary Oxidizing Reagents

Sodium Bismuthate

Sodium bismuthate is a powerful oxidizing agent; it is capable, for example, of converting manganese(II) quantitatively to permanganate ion. This bismuth salt is a sparingly soluble solid with a formula that is usually written as $NaBiO_3$, although its exact composition is somewhat uncertain. Oxidations are performed by suspending the bismuthate in the analyte solution and boiling for a brief period. The unused reagent is then removed by filtration. The half-reaction for the reduction of sodium bismuthate can be written as

$$NaBiO_3(s) + 4H^+ + 2e^- \rightleftharpoons BiO^+ + Na^+ + 2H_2O$$

Ammonium Peroxydisulfate

Ammonium peroxydisulfate, $(NH_4)_2S_2O_8$, is also a powerful oxidizing agent. In acidic solution, it converts chromium(III) to dichromate, cerium(III) to cerium(IV), and manganese(II) to permanganate. The half-reaction is

$$S_2O_8^{2-} + 2e^- \rightleftharpoons 2SO_4^{2-}$$

The oxidations are catalyzed by traces of silver ion. The excess reagent is readily decomposed by a brief period of boiling:

$$2S_2O_8^{2-} + 2H_2O \rightarrow 4SO_4^{2-} + O_2(g) + 4H^+$$

Sodium Peroxide and Hydrogen Peroxide

Peroxide is a convenient oxidizing agent either as the solid sodium salt or as a dilute solution of the acid. The half-reaction for hydrogen peroxide in acidic solution is

$$H_2O_2 + 2H^+ + 2e^- \rightleftharpoons 2H_2O \qquad E^0 = 1.78 \text{ V}$$

After oxidation is complete, the solution is freed of excess reagent by boiling:

$$2H_2O_2 \rightarrow 2H_2O + O_2(g)$$

18B   APPLYING STANDARD REDUCING AGENTS

Standard solutions of most reductants tend to react with atmospheric oxygen. For this reason, reductants are seldom used for the direct titration of oxidizing analytes; indirect methods are used instead. The two most common reductants, iron(II) and thiosulfate ions, are discussed in the paragraphs that follow.

18B-1   Iron(II) Solutions

Solutions of iron(II) are readily prepared from iron(II) ammonium sulfate, $Fe(NH_4)_2(SO_4)_2 \cdot 6H_2O$ (Mohr's salt), or from the closely related iron(II) ethylenediamine sulfate, $FeC_2H_4(NH_3)_2(SO_4)_2 \cdot 4H_2O$ (Oesper's salt). Air oxidation of iron(II) takes place rapidly in neutral solutions but is inhibited in the presence of acids, with the most stable preparations being about 0.5 M in H_2SO_4. Such solutions are stable for no more than one day, if that long. Numerous oxidizing agents are conveniently determined by treatment of the analyte solution with a measured excess of standard iron(II) followed by immediate titration of the excess with a standard solution of potassium dichromate or cerium(IV) (Sections 18C-1 and 18C-2). Just before or just after the analyte is titrated, the volumetric ratio between the standard oxidant and the iron(II) solution is established by titrating two or three aliquots of the latter with the former. This procedure has been applied to the determination of organic peroxides; hydroxylamine; chromium(VI); cerium(IV); molybdenum(VI); nitrate, chlorate, and perchlorate ions; and numerous other oxidants (see, for example, Problems 18-37 and 18-39).

18B-2 Sodium Thiosulfate

Thiosulfate ion ($S_2O_3^{2-}$) is a moderately strong reducing agent that has been widely used to determine oxidizing agents by an indirect procedure that involves iodine as an intermediate. With iodine, thiosulfate ion is oxidized quantitatively to tetrathionate ion ($S_4O_6^{2-}$) according to the half-reaction

$$2S_2O_3^{2-} \rightleftharpoons S_4O_6^{2-} + 2e^-$$

In its reaction with iodine, each thiosulfate ion loses one electron.

The quantitative reaction with iodine is unique. Other oxidants can oxidize the tetrathionate ion to sulfate ion.

The scheme used to determine oxidizing agents involves adding an unmeasured excess of potassium iodide to a slightly acidic solution of the analyte. Reduction of the analyte produces a stoichiometrically equivalent amount of iodine. The liberated iodine is then titrated with a standard solution of sodium thiosulfate, $Na_2S_2O_3$, one of the few reducing agents that is stable toward air oxidation. An example of this procedure is the determination of sodium hypochlorite in bleaches. The reactions are

Sodium thiosulfate is one of the few reducing agents that is not oxidized by air.

$$OCl^- + 2I^- + 2H^+ \rightarrow Cl^- + I_2 + H_2O \quad \text{(unmeasured excess KI)}$$

$$I_2 + 2S_2O_3^{2-} \rightarrow 2I^- + S_4O_6^{2-} \quad (18\text{-}1)$$

Very dilute solutions of iodine have a pale yellow color that is actually due to the triiodide ion I_3^-, which is described on page 461.

Molecular model of thiosulfate ion. Sodium thiosulfate, formerly called sodium hyposulfite, or *hypo,* is used to "fix" photographic images, to extract silver from ore, as an antidote in cyanide poisoning, as a mordant in the dye industry, as a bleaching agent in a variety of applications, as the solute in the supersaturated solution of hot packs, and of course, as an analytical reducing agent. The action of thiosulfate as a photographic fixer is based on its capacity to form complexes with silver and thus dissolve unexposed silver bromide from the surface of photographic film and paper. Thiosulfate is often used as a dechlorinating agent to make aquarium water safe for fish and other aquatic life.

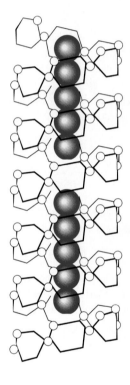

$n > 1000$

(a)

(b)

Figure 18-2   Thousands of glucose molecules polymerize to form huge molecules of β-amylose as shown schematically in (a). Molecules of β-amylose tend to assume a helical structure. The iodine species I_5^- as shown in (b) is incorporated into the amylose helix. For further details, see R. C. Teitelbaum, S. L. Ruby and T. J. Marks, *J. Amer. Chem. Soc.*, **1980**, *102*, 3322.

Starch undergoes decomposition in solutions with high I_2 concentrations. In titrations of excess I_2 with $Na_2S_2O_3$, addition of the indicator must be deferred until most of the I_2 has been reduced.

The quantitative conversion of thiosulfate ion to tetrathionate ion shown in Equation 18-1 requires a pH smaller than 7. If strongly acidic solutions must be titrated, air oxidation of the excess iodide must be prevented by blanketing the solution with an inert gas, such as carbon dioxide or nitrogen.

Detecting End Points in Iodine / Thiosulfate Titrations

A solution that is about 5×10^{-6} M in I_2 has a discernible color, which corresponds to less than one drop of a 0.05 M iodine solution in 100 mL. Thus, provided the analyte solution is colorless, the disappearance of the iodine color can serve as the indicator in titrations with sodium thiosulfate.

More commonly, titrations involving iodine are performed with a suspension of starch as an indicator. The deep blue color that develops in the presence of iodine is believed to arise from the absorption of iodine into the helical chain of β-amylose (see Figure 18-2), a macromolecular component of most starches. The closely related α-amylose forms a red adduct with iodine. This reaction is not readily reversible and is thus undesirable. In commercially available *soluble starch,* the alpha fraction has been removed to leave principally β-amylose; indicator solutions are readily prepared from this product.

Aqueous starch suspensions decompose within a few days, primarily because of bacterial action. The decomposition products tend to interfere with the indicator properties of the preparation and may also be oxidized by iodine. The rate of decomposition can be inhibited by preparing and storing the indicator under sterile conditions and by adding mercury(II) iodide or chloroform as a bacteriostat. Perhaps the simplest alternative is to prepare a fresh suspension of the indicator, which takes only a few minutes, on the day it is to be used.

Starch decomposes irreversibly in solutions containing large concentrations of iodine. Therefore, in titrating solutions of iodine with thiosulfate ion, as in the indirect determination of oxidants, addition of the indicator is delayed until the color of the solution changes from red-brown to yellow; at this point, the titration is nearly complete. The indicator can be introduced at the outset when thiosulfate solutions are being titrated directly with iodine.

How Stable Are Sodium Thiosulfate Solutions?

Although sodium thiosulfate solutions are resistant to air oxidation, they do tend to decompose to give sulfur and hydrogen sulfite ion:

$$S_2O_3^{2-} + H^+ \rightleftharpoons HSO_3^- + S(s)$$

Variables that influence the rate of this reaction include pH, the presence of microorganisms, the concentration of the solution, the presence of copper(II) ions, and exposure to sunlight. These variables may cause the concentration of a thiosulfate solution to change by several percent over a period of a few weeks. On the other hand, proper attention to detail will yield solutions that need only occasional restandardization. The rate of the decomposition reaction increases markedly as the solution becomes acidic.

The most important single cause for the instability of neutral or slightly basic thiosulfate solutions is bacteria that metabolize thiosulfate ion to sulfite and sulfate ions as well as to elemental sulfur. To minimize this problem, standard solutions of the reagent are prepared under reasonably sterile conditions. Bacterial activity appears to be at a minimum at a pH between 9 and 10, which accounts, at least in part, for the reagent's greater stability in slightly basic solutions. The presence of a bactericide, such as chloroform, sodium benzoate, or mercury(II) iodide, also slows decomposition.

<aside>When sodium thiosulfate is added to a strongly acidic medium, a cloudiness develops almost immediately as a consequence of the precipitation of elemental sulfur. Even in neutral solution, this reaction proceeds at such a rate that standard sodium thiosulfate must be restandardized periodically</aside>

Standardizing Thiosulfate Solutions

Potassium iodate is an excellent primary standard for thiosulfate solutions. In this application, weighed amounts of primary-standard-grade reagent are dissolved in water containing an excess of potassium iodide. When this mixture is acidified with a strong acid, the reaction

$$IO_3^- + 5I^- + 6H^+ \rightleftharpoons 3I_2 + 2H_2O$$

occurs instantaneously. The liberated iodine is then titrated with the thiosulfate solution. The stoichiometry of the reactions is

$$1 \text{ mol } IO_3^- = 3 \text{ mol } I_2 = 6 \text{ mol } S_2O_3^{2-}$$

The use of KIO_3 to standardize a thiosulfate solution is illustrated in Example 18-1.

Example 18-1

A solution of sodium thiosulfate was standardized by dissolving 0.1210 g of KIO_3 (214.00 g/mol) in water, adding a large excess of KI, and acidifying with HCl. The liberated iodine required 41.64 mL of the thiosulfate solution to decolorize the blue starch/iodine complex. Calculate the molarity of the $Na_2S_2O_3$.

$$\text{amount } Na_2S_2O_3 = 0.1210 \text{ g } \cancel{KIO_3} \times \frac{1 \text{ mmol } \cancel{KIO_3}}{0.21400 \text{ g } \cancel{KIO_3}} \times \frac{6 \text{ mmol } Na_2S_2O_3}{\text{mmol } \cancel{KIO_3}}$$

$$= 3.3925 \text{ mmol } Na_2S_2O_3$$

$$c_{Na_2S_2O_3} = \frac{3.3925 \text{ mmol}}{41.64 \text{ mL}} = 0.08147 \text{ M}$$

Other primary standards for sodium thiosulfate are potassium dichromate, potassium bromate, potassium hydrogen iodate, potassium ferricyanide, and metallic copper. All these compounds liberate stoichiometric amounts of iodine when treated with excess potassium iodide.

Applications of Sodium Thiosulfate Solutions

Numerous substances can be determined by the indirect method involving titration with sodium thiosulfate. Typical applications are summarized in Table 18-2.

Table 18-2

Some Applications of Sodium Thiosulfate as a Reductant

Analyte	Half-Reaction	Special Conditions
IO_4^-	$IO_4^- + 8H^+ + 7e^- \rightleftharpoons \frac{1}{2}I_2 + 4H_2O$	Acidic solution
	$IO_4^- + 2H^+ + 2e^- \rightleftharpoons IO_3^- + H_2O$	Neutral solution
IO_3^-	$IO_3^- + 6H^+ + 5e^- \rightleftharpoons \frac{1}{2}I_2 + 3H_2O$	Strong acid
BrO_3^-, ClO_3^-	$XO_3^- + 6H^+ + 6e^- \rightleftharpoons X^- + 3H_2O$	Strong acid
Br_2, Cl_2	$X_2 + 2I^- \rightleftharpoons I_2 + 2X^-$	
NO_2^-	$HNO_2 + H^+ + e^- \rightleftharpoons NO(g) + H_2O$	
Cu^{2+}	$Cu^{2+} + I^- + e^- \rightleftharpoons CuI(s)$	
O_2	$O_2 + 4Mn(OH)_2(s) + 2H_2O \rightleftharpoons 4Mn(OH)_3(s)$	Basic solution
	$Mn(OH)_3(s) + 3H^+ + e^- \rightleftharpoons Mn^{2+} + 3H_2O$	Acidic solution
O_3	$O_3(g) + 2H^+ + 2e^- \rightleftharpoons O_2(g) + H_2O$	
Organic peroxide	$ROOH + 2H^+ + 2e^- \rightleftharpoons ROH + H_2O$	

18C   APPLYING STANDARD OXIDIZING AGENTS

Table 18-3 summarizes the properties of five of the most widely used volumetric oxidizing reagents. Note that the standard potentials for these reagents vary from 0.5 to 1.5 V. The choice among them depends on the strength of the analyte as a reducing agent, the rate of reaction between oxidant and analyte, the stability of the standard oxidant solutions, the cost, and the availability of a satisfactory indicator.

Table 18-3

Some Common Oxidants Used as Standard Solutions

Reagent and Formula	Reduction Product	Standard Potential, V	Standardized with	Indicator*	Stability†
Potassium permanganate, $KMnO_4$	Mn^{2+}	1.51‡	$Na_2C_2O_4$, Fe, As_2O_3	MnO_4^-	(b)
Potassium bromate, $KBrO_3$	Br^-	1.44‡	$KBrO_3$	(1)	(a)
Cerium(IV), Ce^{4+}	Ce^{3+}	1.44‡	$Na_2C_2O_4$, Fe, As_2O_3	(2)	(a)
Potassium dichromate, $K_2Cr_2O_7$	Cr^{3+}	1.33‡	$K_2Cr_2O_7$, Fe	(3)	(a)
Iodine, I_2	I^-	0.536‡	$BaS_2O_3 \cdot H_2O$, $Na_2S_2O_3$	starch	(c)

*(1) α-Napthoflavone; (2) 1,10-phenanthroline iron(II) complex (ferroin); (3) diphenylamine sulfonic acid.

†(a) indefinitely stable; (b) moderately stable, requires periodic standardization; (c) somewhat unstable, requires frequent standardization.

‡$E^{0'}$ in 1 M H_2SO_4.

18C-1   The Strong Oxidants: Potassium Permanganate and Cerium(IV)

Solutions of permanganate ion and cerium(IV) ion are strong oxidizing reagents whose applications closely parallel one another. Half-reactions for the two are

$$MnO_4^- + 8H^+ + 5e^- \rightleftharpoons Mn^{2+} + 4H_2O \qquad E^0 = 1.51 \text{ V}$$

$$Ce^{4+} + e^- \rightleftharpoons Ce^{3+} \qquad E^{0'} = 1.44 \text{ V}(1 \text{ M } H_2SO_4)$$

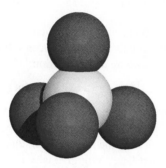

Molecular model of permanganate ion, MnO_4^-. In addition to its use as an analytical reagent, usually in the form of its potassium salt, permanganate is very useful as an oxidizing agent in synthetic organic chemistry. It is used as a bleaching agent with fats, oils, cotton, silk, and other fibers. It has also been used as an antiseptic and anti-infective, as a component in outdoor survival kits, for destroying organic matter in fish ponds, in manufacturing printed wiring boards, for neutralizing the effects of the pesticide rotenone, and for scrubbing flu gases in the determination of mercury. Solid potassium permanganate reacts violently with organic matter, and this effect is often used as a demonstration in general chemistry courses. To further explore these and other uses of permanganate, go to *http://www.google.com/*. Use *permanganate uses* as your search term.

The formal potential shown for the reduction of cerium(IV) is for solutions that are 1 M in sulfuric acid. In 1 M perchloric acid and 1 M nitric acid, the potentials are 1.70 and 1.61 V, respectively. Solutions of cerium(IV) in the latter two acids are not very stable and thus find limited application.

The half-reaction shown for permanganate ion occurs only in solutions that are 0.1 M or greater in strong acid. In less acidic media, the product may be Mn(III), Mn(IV), or Mn(VI), depending on conditions.

How Do the Two Reagents Compare?

For all practical purposes, the oxidizing strengths of permanganate and cerium(IV) solutions are comparable. Solutions of cerium(IV) in sulfuric acid, however, are stable indefinitely, whereas permanganate solutions decompose slowly and thus require occasional restandardization. Furthermore, cerium(IV) solutions in sulfuric acid do not oxidize chloride ion and can be used to titrate hydrochloric acid solutions of analytes; in contrast, permanganate ion cannot be used with hydrochloric acid solutions unless special precautions are taken to prevent the slow oxidation of chloride ion that leads to overconsumption of the standard reagent. A further advantage of cerium(IV) is that a primary-standard-grade salt of the reagent is available, which makes it possible to prepare standard solutions directly.

Despite these advantages of cerium solutions over permanganate solutions, the latter are more widely used. One reason is the color of permanganate solutions, which is intense enough to serve as an indicator in titrations. A second reason for the popularity of permanganate solutions is their modest cost. The cost of one liter of 0.02 M $KMnO_4$ solution is about \$0.10, whereas one liter of a comparable strength Ce(IV) solution costs more than \$2.00 (\$4.00 if reagent of primary-standard grade is used). Another disadvantage of cerium(IV) solutions is their tendency to form precipitates of basic salts in solutions that are less than 0.1 M in strong acid.

Detecting the End Points

A useful property of a potassium permanganate solution is its intense purple color, which is sufficient to serve as an indicator for most titrations. As little as 0.01 to 0.02 mL of a 0.02 M solution imparts a perceptible color to 100 mL of water. If the permanganate solution is very dilute, diphenylamine sulfonic acid or the 1,10-phenanthroline complex of iron(II) (Table 17-2, page 440) will provide a sharper end point.

The permanganate end point is not permanent because excess permanganate ions react slowly with the relatively large concentration of manganese(II) ions present at the end point, according to the reaction

$$2MnO_4^- + 3Mn^{2+} + 2H_2O \rightleftharpoons 5MnO_2(s) + 4H^+$$

The equilibrium constant for this reaction is about 10^{47}, which indicates that the equilibrium concentration of permanganate ion is vanishingly small even in highly acidic media. Fortunately, the rate at which this equilibrium is approached is so slow that the end point fades only gradually over a period of perhaps 30 s.

Solutions of cerium(IV) are yellow-orange, but the color is not intense enough to act as an indicator in titrations. Several oxidation/reduction indicators are available for titrations with standard solutions of cerium(IV). The most widely used of these is the iron(II) complex of 1,10-phenanthroline or one of its substituted derivatives (Table 17-2).

The Preparation and Stability of Standard Solutions

Aqueous solutions of permanganate are not entirely stable because the ion tends to oxidize water:

$$4MnO_4^- + 2H_2O \rightarrow 4MnO_2(s) + 3O_2(g) + 4OH^-$$

Although the equilibrium constant for this reaction indicates that the products are favored, permanganate solutions, when properly prepared, are reasonably stable because the decomposition reaction is slow. It is catalyzed by light, heat, acids, bases, manganese(II), and manganese dioxide.

Moderately stable solutions of permanganate ion can be prepared if the effects of these catalysts, particularly manganese dioxide, are minimized. Manganese dioxide is a contaminant in even the best grade of solid potassium permanganate. Furthermore, this compound forms in freshly prepared solutions of the reagent as a consequence of the reaction of permanganate ion with organic matter and dust present in the water used to prepare the solution. Removal of manganese dioxide by filtration before standardization markedly improves the stability of standard permanganate solutions. Before filtration, the reagent solution is allowed to stand for about 24 hr or is heated for a brief period to hasten oxidation of the organic species generally present in small amounts in distilled and deionized water. Paper cannot be used for filtering because permanganate ion reacts with it to form additional manganese dioxide.

Chromium is an important metal to monitor in environmental samples. Not only is the total amount of chromium of interest, but the oxidation state in which the chromium is found is quite important. In water, chromium can exist as Cr(III) or as Cr(VI) species. Chromium (III) is an essential nutrient and nontoxic. Chromium (VI), however, is a known carcinogen. Hence, the determination of the amount of chromium in each of these oxidation states is often of more interest than the total amount of chromium. There are several good methods available for determining Cr(VI) selectively. One of the most popular involves the oxidation of the reagent 1,5-diphenylcarbohydrazide (diphenylcarbazide) by Cr(VI) in acid solution. The reaction produces a red-purple chelate of Cr(III) and diphenylcarbazide that can be monitored colorimetrically (see Section 23A). The direct reaction of Cr(III) itself and the reagent is so slow that essentially only the Cr(VI) is measured. To determine Cr(III), the sample is oxidized with excess permanganate in alkaline solution to convert all the Cr(III) to Cr(VI). The excess oxidant is destroyed with sodium azide. A new colorimetric measurement is made that now determines total chromium (the original Cr(VI) plus that formed by oxidation of Cr(III)). The amount of Cr(III) present is then obtained by subtracting the amount of Cr(VI) obtained in the original measurement from the amount of total chromium obtained after permanganate oxidation. Note that here permanganate is being used as an auxiliary oxidizing agent.

Standardized permanganate solutions should be stored in the dark. Filtration and restandardization are required if any solid is detected in the solution or on the walls of the storage bottle. In any event, restandardization every 1 or 2 weeks is a good precautionary measure.

Hot, aqueous permanganate will oxidize the water in which it is dissolved. This decomposition cannot be compensated for with a blank. On the other hand, it is possible to titrate hot, acidic solutions of reductants with permanganate without error provided the reagent is added slowly enough that large excesses do not accumulate.

Permanganate solutions are moderately stable provided they are freed of manganese dioxide and stored in a dark container.

Example 18-2

Describe how you would prepare 2.0 L of an approximately 0.010 M solution of KMnO$_4$ (158.03 g/mol).

$$\text{mass KMnO}_4 \text{ needed} = 2.0 \text{ L} \times 0.010 \, \frac{\text{mol KMnO}_4}{\text{L}} \times 158.03 \, \frac{\text{g KMnO}_4}{\text{mol KMnO}_4}$$

$$= 3.16 \text{ g}$$

Dissolve about 3.2 g of KMnO$_4$ in a little water. After complete dissolution, add water to bring the volume to about 2.0 L. Heat the solution to boiling for a brief period and let stand until it is cool. Filter through a glass filtering crucible and store in a clean dark bottle.

The most widely used compounds for the preparation of solutions of cerium(IV) are listed in Table 18-4. Primary-standard-grade cerium ammonium nitrate is available commercially and can be used to prepare standard solutions of the cation directly by weight. More commonly, less expensive reagent-grade cerium(IV)

Table 18-4

Analytically Useful Cerium(IV) Compounds

Name	Formula	Molar Mass
Cerium(IV) ammonium nitrate	$Ce(NO_3)_4 \cdot 2NH_4NO_3$	548.2
Cerium(IV) ammonium sulfate	$Ce(SO_4)_2 \cdot 2(NH_4)_2SO_4 \cdot 2H_2O$	632.6
Cerium(IV) hydroxide	$Ce(OH)_4$	208.1
Ce(IV) hydrogen sulfate	$Ce(HSO_4)_4$	528.4

ammonium nitrate or ceric hydroxide is used to prepare solutions that are subsequently standardized. In either case, the reagent must be dissolved in a solution that is at least 0.1 M in sulfuric acid to prevent the precipitation of basic salts.

Sulfuric acid solutions of cerium(IV) are remarkably stable. They can be stored for months or heated at 100°C for prolonged periods without change in concentration.

Standardizing Permanganate and Ce(IV) Solutions

Sodium oxalate is a primary standard widely used for standardizing solutions of potassium permanganate and cerium(IV). In acidic solutions, the oxalate ion is converted to the undissociated acid. Thus, its reaction with permanganate can be described by

$$2MnO_4^- + 5H_2C_2O_4 + 6H^+ \rightarrow 2Mn^{2+} + 10CO_2(g) + 8H_2O$$

The reaction between permanganate ion and oxalic acid is complex and proceeds slowly even at elevated temperature unless manganese(II) is present as a catalyst. Thus, when the first few milliliters of standard permanganate are added to a hot solution of oxalic acid, several seconds are required before the color of the permanganate ion disappears. As the concentration of manganese(II) builds up, however, the reaction proceeds more and more rapidly as a result of autocatalysis.

It has been found that when solutions of sodium oxalate are titrated at 60 to 90°C, the consumption of permanganate is from 0.1 to 0.4% less than theoretical, probably due to the air oxidation of a fraction of the oxalic acid. This small error can be avoided by adding 90 to 95% of the required permanganate to a cool solution of the oxalate. After the added permanganate is completely consumed (as indicated by the disappearance of color), the solution is heated to about 60°C and titrated to a pink color that persists for about 30 s. The disadvantage of this procedure is that it requires prior knowledge of the approximate concentration of the permanganate solution so that a proper initial volume can be added. For most purposes, direct titration of the hot oxalic acid solution is adequate (usually 0.2 to 0.3% high). If greater accuracy is required, a direct titration of the hot solution of one portion of the primary standard can be followed by titration of two or three portions in which the solution is not heated until the end.

Sodium oxalate is also widely used to standardize Ce(IV) solutions. The reaction between Ce^{4+} and $H_2C_2O_4$ is

$$2Ce^{4+} + H_2C_2O_4 \rightarrow 2Ce^{3+} + 2CO_2(g) + 2H^+$$

Autocatalysis is a type of catalysis in which the product of a reaction catalyses the reaction. This phenomenon causes the rate of the reaction to increase as the reaction proceeds.

Solutions of $KMnO_4$ and Ce^{4+} can also be standardized with electrolytic iron wire or with potassium iodide.

Cerium(IV) standardizations against sodium oxalate are usually performed at 50°C in a hydrochloric acid solution containing iodine monochloride as a catalyst.

Example 18-3

If you wish the standardization of the permanganate solution in Example 18-2 to require 30 to 45 mL of the reagent, what range of masses of primary-standard $Na_2C_2O_4$ (134.00 g/mol) should you weigh out?

For a 30-mL titration:

$$\text{amount KMnO}_4 = 30 \text{ mL KMnO}_4 \times 0.010 \frac{\text{mmol KMnO}_4}{\text{mL KMnO}_4}$$

$$= 0.30 \text{ mmol KMnO}_4$$

$$\text{mass Na}_2\text{C}_2\text{O}_4 = 0.30 \text{ mmol KMnO}_4 \times \frac{5 \text{ mmol Na}_2\text{C}_2\text{O}_4}{2 \text{ mmol KMnO}_4}$$

$$\times 0.134 \frac{\text{g Na}_2\text{C}_2\text{O}_4}{\text{mmol Na}_2\text{C}_2\text{O}_4}$$

$$= 0.101 \text{ g Na}_2\text{C}_2\text{O}_4$$

Proceeding in the same way, we find for a 45-mL titration,

$$\text{mass Na}_2\text{C}_2\text{O}_4 = 45 \times 0.010 \times \frac{5}{2} \times 0.134 = 0.151 \text{ g Na}_2\text{C}_2\text{O}_4$$

Thus, you should weigh between 0.10- and 0.15-g samples of the primary standard.

Example 18-4

A 0.1278-g sample of primary-standard $Na_2C_2O_4$ required exactly 33.31 mL of the permanganate solution in Example 18-2 to reach the end point. What was the molarity of the $KMnO_4$ reagent?

$$\text{amount Na}_2\text{C}_2\text{O}_4 = 0.1278 \text{ g Na}_2\text{C}_2\text{O}_4 \times \frac{1 \text{ mmol Na}_2\text{C}_2\text{O}_4}{0.13400 \text{ g Na}_2\text{C}_2\text{O}_4}$$

$$= 0.95373 \text{ mmol Na}_2\text{C}_2\text{O}_4$$

$$c_{\text{KMnO}_4} = 0.95373 \text{ mmol Na}_2\text{C}_2\text{O}_4 \times \frac{2 \text{ mmol KMnO}_4}{5 \text{ mmol Na}_2\text{C}_2\text{O}_4} \times \frac{1}{33.31 \text{ mL KMnO}_4}$$

$$= 0.01145 \text{ M}$$

Using Potassium Permanganate and Cerium(IV) Solutions

Table 18-5 lists some of the many applications of permanganate and cerium(IV) solutions to the volumetric determination of inorganic species. Both reagents have also been applied to the determination of organic compounds with oxidizable functional groups.

Table 18-5

Some Applications of Potassium Permanganate and Cerium(IV) Solutions

Substance Sought	Half-Reaction	Conditions
Sn	$Sn^{2+} \rightleftharpoons Sn^{4+} + 2e^-$	Prereduction with Zn
H_2O_2	$H_2O_2 \rightleftharpoons O_2(g) + 2H^+ + 2e^-$	
Fe	$Fe^{2+} \rightleftharpoons Fe^{3+} + e^-$	Prereduction with $SnCl_2$ or with Jones or Walden reductor
$Fe(CN)_6^{4-}$	$Fe(CN)_6^{4-} \rightleftharpoons Fe(CN)_6^{3-} + e^-$	
V	$VO^{2+} + 3H_2O \rightleftharpoons V(OH)_4^+ + 2H^+ + e^-$	Prereduction with Bi amalgam or SO_2
Mo	$Mo^{3+} + 4H_2O \rightleftharpoons MoO_4^{2-} + 8H^+ + 3e^-$	Prereduction with Jones reductor
W	$W^{3+} + 4H_2O \rightleftharpoons WO_4^{2-} + 8H^+ + 3e^-$	Prereduction with Zn or Cd
U	$U^{4+} + 2H_2O \rightleftharpoons UO_2^{2+} + 4H^+ + 2e^-$	Prereduction with Jones reductor
Ti	$Ti^{3+} + H_2O \rightleftharpoons TiO^{2+} + 2H^+ + e^-$	Prereduction with Jones reductor
$H_2C_2O_4$	$H_2C_2O_4 \rightleftharpoons 2CO_2 + 2H^+ + 2e^-$	
Mg, Ca, Zn, Co, Pb, Ag	$H_2C_2O_4 \rightleftharpoons 2CO_2 + 2H^+ + 2e^-$	Sparingly soluble metal oxalates filtered, washed, and dissolved in acid; liberated oxalic acid titrated
HNO_2	$HNO_2 + H_2O \rightleftharpoons NO_3^- + 3H^+ + 2e^-$	15-min reaction time; excess $KMnO_4$ back-titrated
K	$K_2NaCo(NO_2)_6 + 6H_2O \rightleftharpoons Co^{2+} + 6NO_3^-$ $+ 12H^+ + 2K^+ + Na^+ + 11e^-$	Precipitated as $K_2NaCo(NO_2)_6$; filtered and dissolved in $KMnO_4$; excess $KMnO_4$ back-titrated
Na	$U^{4+} + 2H_2O \rightleftharpoons UO_2^{2+} + 4H^+ + 2e^-$	Precipitated as $NaZn(UO_2)_3(OAc)_9$; filtered, washed, dissolved; U determined as above

Example 18-5

Aqueous solutions containing approximately 3% (w/w) H_2O_2 are sold in drug-stores as a disinfectant. Propose a method for determining the peroxide content of such a preparation using the standard solution described in Examples 18-3 and 18-4. Assume that you wish to use between 30 and 45 mL of the reagent for a titration. The reaction is

$$5H_2O_2 + 2MnO_4^- + 6H^+ \rightarrow 5O_2 + 2Mn^{2+} + 8H_2O$$

The amount of $KMnO_4$ in 35 to 45 mL of the reagent is between

$$\text{amount } KMnO_4 = 35 \text{ mL } KMnO_4 \times 0.01145 \frac{\text{mmol } KMnO_4}{\text{mL } KMnO_4}$$

$$= 0.401 \text{ mmol } KMnO_4$$

and

$$\text{amount } KMnO_4 = 45 \times 0.01145 = 0.515 \text{ mmol } KMnO_4$$

The amount of H_2O_2 consumed by 0.401 mmol of $KMnO_4$ is

$$\text{amount } H_2O_2 = 0.401 \text{ mmol } KMnO_4 \times \frac{5 \text{ mmol } H_2O_2}{2 \text{ mmol } KMnO_4} = 1.00 \text{ mmol } H_2O_2$$

and

$$amount\ H_2O_2 = 0.515 \times \frac{5}{2} = 1.29\ mmol\ H_2O_2$$

We therefore need to take samples that contain from 1.00 to 1.29 mmol H_2O_2.

$$mass\ sample = 1.00\ mmol\ H_2O_2 \times 0.03401\ \frac{g\ H_2O_2}{mmol\ H_2O_2} \times \frac{100\ g\ sample}{3\ g\ H_2O_2}$$

$$= 1.1\ g\ sample$$

to

$$mass\ sample = 1.29 \times 0.03401 \times \frac{100}{3} = 1.5\ g\ sample$$

Thus, our sample should weigh between 1.1 and 1.5 g. These samples should be diluted to perhaps 75 to 100 mL with water and made slightly acidic with dilute H_2SO_4 before titration.

18C-2 Potassium Dichromate

In its analytical applications, dichromate ion is reduced to green chromium(III) ion:

$$Cr_2O_7^{2-} + 14H^+ + 6e^- \rightleftharpoons 2Cr^{3+} + 7H_2O \qquad E^0 = 1.33\ V$$

Dichromate titrations are generally carried out in solutions that are about 1 M in hydrochloric or sulfuric acid. In these media, the formal potential for the half-reaction is 1.0 to 1.1 V.

Molecular model of dichromate ion. For many years, dichromate in the form of its ammonium, potassium, or sodium salts was used in nearly all areas of chemistry as a powerful oxidizing agent. In addition to its use as a primary standard in analytical chemistry, it has been used as an oxidizing agent in synthetic organic chemistry, as a pigment in the paint, dye, and photographic industries, as a bleaching agent, and as a corrosion inhibitor. Chromic acid solution made from sodium dichromate and sulfuric acid was once the reagent of choice for thorough cleaning of glassware. Dichromate has been used as the analytical reagent in alcohol Breathalyzer®, but in recent years these devices have largely been replaced by analyzers based on the absorption of infrared radiation. Early color photography utilized the colors produced by chromium compounds in the so-called *gum bichromate* process, but this process has been replaced by silver bromide-based processes. The use of chromium compounds in general and dichromate in particular has decreased over the last decade because of the discovery that chromium compounds are carcinogenic. In spite of this, many millions of pounds of chromium compounds are manufactured and consumed by industry each year. Before using dichromate in laboratory work, read the MSDS for potassium dichromate (*http://msds.pdc.cornell.edu/ msds/siri/q315/q129.html* or *http://msds.pdc.cornell.edu/msds/siri/q283/q234.html*), and explore its chemical, toxicological, and carcinogenic properties. Observe all precautions in handling this useful but potentially hazardous chemical either in the solid form or in solution.

Feature 18-2

Antioxidants[4]

Oxidation can have deleterious effects on the cells and tissues of the human body. There is considerable evidence that reactive oxygen and nitrogen species, such as superoxide ion O_2^-, hydroxyl radical $OH\cdot$, peroxyl radicals $RO_2\cdot$, alkoxyl radicals $RO\cdot$, nitric oxide $NO\cdot$, and nitrogen dioxide $NO_2\cdot$, damage cells and other body components. A group of compounds known as *antioxidants* can help counteract the influence of reactive oxygen and nitrogen species. Antioxidants are reducing agents that are so easily oxidized that they can protect other compounds in the body from oxidation. Typical antioxidants include vitamins A, C, and E; minerals such as selenium; and herbs such as ginkgo, rosemary, and milk thistle.

There are several proposed mechanisms for antioxidant action. Their presence may result in the decreased formation of the reactive oxygen and nitrogen species in the first place. Antioxidants may also scavenge the reactive species or their precursors. Vitamin E is an example of this latter behavior in its inhibition of lipid oxidation by its reaction with radical intermediates generated from polyunsaturated fatty acids. Some antioxidants can bind the metal ions needed to catalyze the formation of the reactive oxidants. Other antioxidants can repair oxidative damage to biomolecules or can influence enzymes that catalyze repair mechanisms.

Vitamin E, or α-tocopherol, is thought to deter atherosclerosis, accelerate wound healing, and protect lung tissue from inhaled pollutants. It may also reduce the risk for heart disease and prevent premature skin aging. Researchers suspect that vitamin E has several other beneficial effects ranging from alleviating rheumatoid arthritis to preventing cataracts. Most of us get enough vitamin E through our diet and do not require supplements. Dark-green leafy vegetables, nuts, vegetable oils, seafood, eggs, and avocados are food sources rich in vitamin E.

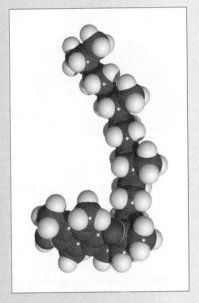

Molecular model of vitamin E.

[4]See B. Halliwell, *Nutr. Rev.,* **1997,** *55,* S44.

Selenium has antioxidant effects that complement those of vitamin E. Selenium is a required constituent of several enzymes that remove reactive oxidants. The metal may support the immune function and neutralize some heavy metal poisons. It may also aid in deterring heart disease and some cancers. Good sources of selenium in the diet are whole grains, asparagus, garlic, eggs, mushrooms, lean meats, and seafood. Usually, diet alone provides sufficient selenium for good health. Supplements should be taken only if prescribed by a doctor because high doses can be toxic.

Potassium dichromate solutions are indefinitely stable, can be boiled without decomposition, and do not react with hydrochloric acid. Moreover, primary-standard reagent is available commercially and at a modest cost. The disadvantages of potassium dichromate compared with cerium(IV) and permanganate ion are its lower electrode potential and the slowness of its reaction with certain reducing agents.

Standard solutions of $K_2Cr_2O_7$ have the great advantage that they are indefinitely stable and do not oxidize HCl. Furthermore, primary-standard grade is inexpensive and readily available commercially.

Preparing Dichromate Solutions

For most purposes, reagent-grade potassium dichromate is sufficiently pure to permit the direct preparation of standard solutions; the solid is simply dried at 150°C to 200°C before being weighed.

The orange color of a dichromate solution is not intense enough for use in end-point detection. Diphenylamine sulfonic acid (Table 17-2), however, is an excellent indicator for titrations with this reagent. The oxidized form of the indicator is violet, and its reduced form is essentially colorless; thus, the color change observed in a direct titration is from the green of chromium(III) to violet.

Applying Potassium Dichromate Solutions

The principal use of dichromate is for the volumetric titration of iron(II) based on the reaction

$$Cr_2O_7^{2-} + 6Fe^{2+} + 14H^+ \rightarrow 2Cr^{3+} + 6Fe^{3+} + 7H_2O$$

Often, this titration is performed in the presence of moderate concentrations of hydrochloric acid.

The reaction of dichromate with iron(II) has been widely used for the indirect determination of a variety of oxidizing agents. In these applications, a measured excess of an iron(II) solution is added to an acidic solution of the analyte. The excess iron(II) is then back-titrated with standard potassium dichromate (Section 18B-1). Standardization of the iron(II) solution by titration with the dichromate is performed concurrently with the determination because solutions of iron(II) tend to be air oxidized. This method has been applied to the determination of nitrate, chlorate, permanganate, and dichromate ions as well as organic peroxides and several other oxidizing agents.

Example 18-6

A 5.00-mL sample of brandy was diluted to 1.000 L in a volumetric flask. The ethanol (C_2H_5OH) in a 25.00-mL aliquot of the diluted solution was distilled into 50.00 mL of 0.02000 M $K_2Cr_2O_7$ and oxidized to acetic acid with heating. Reaction:

$$3C_2H_5OH + 2Cr_2O_7^{2-} + 16H^+ \rightarrow 4Cr^{3+} + 3CH_3COOH + 11H_2O$$

After cooling, 20.00 mL of 0.1253 M Fe^{2+} was pipetted into the flask. The excess Fe^{2+} was then titrated with 7.46 mL of the standard $K_2Cr_2O_7$ to a diphenylamine sulfonic acid end point. Calculate the percent (w/v) C_2H_5OH (46.07 g/mol) in the brandy.

total amount $K_2Cr_2O_7$

$$= (50.00 + 7.46) \text{ mL } K_2Cr_2O_7 \times 0.02000 \frac{\text{mmol } K_2Cr_2O_7}{\text{mL } K_2Cr_2O_7}$$

$$= 1.1492 \text{ mmol } K_2Cr_2O_7$$

amount $K_2Cr_2O_7$ consumed by Fe^{2+}

$$= 20.00 \text{ mL } Fe^{2+} \times 0.1253 \frac{\text{mmol } Fe^{2+}}{\text{mL } Fe^{2+}} \times \frac{1 \text{ mmol } K_2Cr_2O_7}{6 \text{ mmol } Fe^{2+}}$$

$$= 0.41767 \text{ mmol } K_2Cr_2O_7$$

amount $K_2Cr_2O_7$ consumed by $C_2H_5OH = (1.1492 - 0.41767) \text{ mmol } K_2Cr_2O_7$
$$= 0.73153 \text{ mmol } K_2Cr_2O_7$$

mass C_2H_5OH

$$= 0.73153 \text{ mmol } K_2Cr_2O_7 \times \frac{3 \text{ mmol } C_2H_5OH}{2 \text{ mmol } K_2Cr_2O_7} \times 0.04607 \frac{\text{g } C_2H_5OH}{\text{mmol } C_2H_5OH}$$

$$= 0.050552 \text{ g } C_2H_5OH$$

$$\text{percent } C_2H_5OH = \frac{0.050552 \text{ g } C_2H_5OH}{5.00 \text{ mL sample} \times 25.00 \text{ mL}/1000 \text{ mL}} \times 100\%$$

$$= 40.4\% \text{ } C_2H_5OH$$

18C-3   Iodine

Iodine is a weak oxidizing agent used primarily for the determination of strong reductants. The most accurate description of the half-reaction for iodine in these applications is

$$I_3^- + 2e^- \rightleftharpoons 3I^- \qquad E^0 = 0.536 \text{ V}$$

where I_3^- is the triiodide ion.

Standard iodine solutions have relatively limited application compared with the other oxidants we have described because of their significantly smaller electrode potential. Occasionally, however, this low potential is advantageous because it imparts a degree of selectivity that makes it possible to determine strong reducing agents in the presence of weak ones. An important advantage of iodine is the availability of a sensitive and reversible indicator for the titrations. On the other hand, iodine solutions lack stability and must be restandardized regularly.

What Are the Properties of Iodine Solutions?

Iodine is not very soluble in water (0.001 M). To obtain solutions having analytically useful concentrations of the element, iodine is dissolved in moderately concentrated solutions of potassium iodide. In this medium, iodine is reasonably soluble as a consequence of the reaction

$$I_2(s) + I^- \rightleftharpoons I_3^- \qquad K = 7.1 \times 10^2$$

Iodine dissolves only very slowly in solutions of potassium iodide, particularly if the iodide concentration is low. To ensure complete solution, the iodine is always dissolved in a small volume of concentrated potassium iodide, with care being taken to avoid dilution of the concentrated solution until the last trace of solid iodine has disappeared; otherwise, the molarity of the diluted solution gradually increases with time. This problem can be avoided by filtering the solution through a sintered glass crucible before standardization.

Iodine solutions lack stability for several reasons, one being the volatility of the solute. Losses of iodine from an open vessel occur in a relatively short time even in the presence of an excess of iodide ion. In addition, iodine slowly attacks most organic materials. Consequently, cork or rubber stoppers are never used to close containers of the reagent, and precautions must be taken to protect standard solutions from contact with organic dusts and fumes.

Air oxidation of iodide ion also causes changes in the molarity of an iodine solution:

$$4I^- + O_2(g) + 4H^+ \rightarrow 2I_2 + 2H_2O$$

In contrast to the other effects, this reaction causes the molarity of the iodine to increase. Air oxidation is promoted by acids, heat, and light.

> Solutions prepared by dissolving iodine in a concentrated solution of potassium iodide are properly called *triiodide solutions*. In practice, however, they are often called *iodine solutions* because this terminology accounts for the stoichiometric behavior of these solutions $(I_2 + 2e^- \rightarrow 2I^-)$.

Standardizing and Applying Iodine Solutions

Iodine solutions can be standardized against anhydrous sodium thiosulfate or barium thiosulfate monohydrate, both of which are available commercially. The reaction between iodine and sodium thiosulfate was discussed in detail in Section 18B-2. Often, solutions of iodine are standardized against solutions of sodium thiosulfate that have in turn been standardized against potassium iodate or potassium dichromate (Section 18B-2). Table 18-6 summarizes methods that use iodine as an oxidizing agent.

18C-4 Potassium Bromate as a Source of Bromine

Primary-standard potassium bromate is available from commercial sources and can be used directly to prepare standard solutions that are stable indefinitely. Direct titrations with potassium bromate are relatively few. Instead, the reagent

Table 18-6

Some Applications of Iodine Solutions

Substance Determined	Half-Reaction
As	$H_3AsO_3 + H_2O \rightleftharpoons H_3AsO_4 + 2H^+ + 2e^-$
Sb	$H_3SbO_3 + H_2O \rightleftharpoons H_3SbO_4 + 2H^+ + 2e^-$
Sn	$Sn^{2+} \rightleftharpoons Sn^{4+} + 2e^-$
H_2S	$H_2S \rightleftharpoons S(s) + 2H^+ + 2e^-$
SO_2	$SO_3^{2-} + H_2O \rightleftharpoons SO_4^{2-} + 2H^+ + 2e^-$
$S_2O_3^{2-}$	$2S_2O_3^{2-} \rightleftharpoons S_4O_6^{2-} + 2e^-$
N_2H_4	$N_2H_4 \rightleftharpoons N_2(g) + 4H^+ + 4e^-$
Ascorbic acid	$C_6H_8O_6 \rightleftharpoons C_6H_6O_6 + 2H^+ + 2e^-$

is a convenient and widely used stable source of bromine.[5] In this application, an unmeasured excess of potassium bromide is added to an acidic solution of the analyte. Upon introduction of a measured volume of standard potassium bromate, a stoichiometric quantity of bromine is produced.

$1 \text{ mol } KBrO_3 \equiv 3 \text{ mol } Br_2$

$$BrO_3^- + 5Br^- + 6H^+ \rightarrow 3Br_2 + 3H_2O$$

standard   excess
solution

This indirect generation circumvents the problems associated with the use of standard bromine solutions, which lack stability.

The primary use of standard potassium bromate is for the determination of organic compounds that react with bromine. Bromine is incorporated into an organic molecule either by substitution or by addition. Few of these reactions are rapid enough to make direct titration feasible. Instead, a measured excess of standard bromate is added to the solution that contains the sample plus an excess of potassium bromide. After acidification, the mixture is allowed to stand in a glass-stoppered vessel until the bromine/analyte reaction is judged complete. To determine the excess bromine, an excess of potassium iodide is introduced so that the following reaction occurs:

$$2I^- + Br_2 \rightarrow I_2 + 2Br^-$$

The liberated iodine is then titrated with standard sodium thiosulfate (Equation 18-1).

Substitution Reactions

Halogen substitution involves the replacement of hydrogen in an aromatic ring by a halogen. Substitution methods have been successfully applied to the determination of aromatic compounds that contain strong ortho-para-directing groups, particularly amines and phenols.

---

[5]For a discussion of bromate solutions and their applications, see M. R. F. Ashworth, *Titrimetric Organic Analysis,* Part I (New York: Interscience, 1964), pp. 118–130.

Example 18-7

A 0.2981-g sample of an antibiotic powder was dissolved in HCl and the solution diluted to 100.0 mL. A 20.00-mL aliquot was transferred to a flask followed by 25.00 mL of 0.01767 M KBrO$_3$. An excess of KBr was added to form Br$_2$, and the flask was stoppered. After 10 min, during which time the Br$_2$ brominated the sulfanilamide, an excess of KI was added. The liberated iodine was titrated with 12.92 mL of 0.1215 M sodium thiosulfate. The reactions are

$$BrO_3^- + 5Br^- + 6H^+ \longrightarrow 3Br_2 + 3H_2O$$

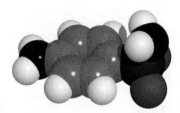

sulfanilamide

Molecular model of sulfanilamide. In the 1930s, sulfanilamide was found to be an effective antibacterial agent. In an effort to provide a solution of the drug that could be conveniently administered to patients, drug companies distributed sulfanilamide elixir containing a high concentration of ethylene glycol, which is toxic to the kidneys. Consequently, over 100 people died from the effects of the solvent. This event led to the rapid passage of the *1938 Federal Food, Drug, and Cosmetic Act,* which required toxicity testing prior to marketing and listing of active ingredients on product labels. For more information on the history of drug laws, see *http://www.fda.gov/fdac/special/newdrug/ benlaw.html.*

$$Br_2 + 2I^- \longrightarrow 2Br^- + I_2 \quad \text{(excess KI)}$$

$$I_2 + 2S_2O_3^{2-} \longrightarrow 2S_4O_6^{2-} + 2I^-$$

Calculate the percent sulfanilamide (NH$_2$C$_6$H$_4$SO$_2$NH$_2$, 172.21 g/mol) in the powder.

total amount Br$_2$

$$= 25.00 \text{ mL KBrO}_3 \times 0.01767 \frac{\text{mmol KBrO}_3}{\text{mL KBrO}_3} \times \frac{3 \text{ mmol Br}_2}{\text{mmol KBrO}_3}$$

$$= 1.32525 \text{ mmol Br}_2$$

We next calculate how much Br$_2$ was in excess over that required to brominate the analyte:

amount excess Br_2 = amount I_2

$$= 12.92 \text{ mL Na}_2S_2O_3 \times 0.1215 \frac{\text{mmol Na}_2S_2O_3}{\text{mL Na}_2S_2O_3} \times \frac{1 \text{ mmol } I_2}{2 \text{ mmol Na}_2S_2O_3}$$

$$= 0.78489 \text{ mmol Br}_2$$

The amount of Br_2 consumed by the sample is given by

$$\text{amount Br}_2 = 1.32525 - 0.78489 = 0.54036 \text{ mmol Br}_2$$

$$\text{mass analyte} = 0.54036 \text{ mmol Br}_2 \times \frac{1 \text{ mmol analyte}}{2 \text{ mmol Br}_2}$$

$$\times 0.17221 \frac{\text{g analyte}}{\text{mmol analyte}}$$

$$= 0.046528 \text{ g analyte}$$

$$\text{percent analyte} = \frac{0.046528 \text{ g analyte}}{0.2891 \text{ g sample} \times 20.00 \text{ mL}/100 \text{ mL}} \times 100\%$$

$$= 80.47\% \text{ sulfanilamide}$$

An important example of the use of a bromine substitution reaction is the determination of 8-hydroxyquinoline:

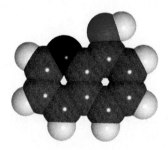

In contrast to most bromine substitutions, this reaction takes place rapidly enough in hydrochloric acid solution to make direct titration feasible. The titration of 8-hydroxyquinoline with bromine is particularly significant because the

Molecular model of 8-hydroxyquinoline.

former is an excellent precipitating reagent for cations (Section 8D-3). For example, aluminum can be determined according to the sequence

$$Al^{3+} + 3HOC_9H_6N \xrightarrow{\text{pH 4–9}} Al(OC_9H_6N)_3(s) + 3H^+$$

$$Al(OC_9H_6N)_3(s) \xrightarrow{\text{hot 4 M HCl}} 3HOC_9H_6N + Al^{3+}$$

$$3HOC_9H_6N + 6 Br_2 \longrightarrow 3HOC_9H_4NBr_2 + 6HBr$$

The stoichiometric relationships in this case are

$$1 \text{ mol } Al^{3+} = 3 \text{ mol } HOC_9H_6N = 6 \text{ mol } Br_2 = 2 \text{ mol } KBrO_3$$

Addition Reactions

Addition reactions involve the opening of an olefinic double bond. For example, 1 mol of ethylene reacts with 1 mol of bromine in the reaction

The literature contains numerous references to the use of bromine for the estimation of olefinic unsaturation in fats, oils, and petroleum products. A method for the determination of ascorbic acid in vitamin C tablets is given in Section 27I-3.

18C-5   Determining Water with the Karl Fischer Reagent

In industry and commerce, one of the most widely used analytical methods is the Karl Fischer titration procedure for the determination of water in various types of solids and organic liquids. This important titrimetric method is based on an oxidation/reduction reaction that is relatively specific for water.[6]

Describing the Reaction Stoichiometry

Karl Fischer reagent usually consists of iodine, sulfur dioxide, and an organic base, such as pyridine or imidazole, dissolved in methanol or some other low-molecular-mass alcohol. For a mixture containing pyridine (C_5H_5N) and methanol, the analytical reactions can be formulated as

$$C_5H_5N \cdot I_2 + C_5H_5N \cdot SO_2 + C_5H_5N + H_2O \longrightarrow$$
$$2C_5H_5N \cdot HI + C_5H_5N \cdot SO_3 \qquad (18\text{-}2)$$

$$C_5H_5N \cdot SO_3 + CH_3OH \longrightarrow C_5H_5N(H)SO_4CH_3 \qquad (18\text{-}3)$$

where I_2, SO_2, and SO_3 are shown as complexed by the pyridine.

---

[6]For a review of the composition and uses of the Karl Fischer reagent, see S. K. MacLeod, *Anal. Chem.*, **1991**, *63*, 557A; J. D. Mitchell, Jr., and D. M. Smith, *Aquametry*, 2nd ed., Vol. 3 (New York: Wiley, 1977).

Note that the reaction involves the consumption of 1 mol of iodine, 1 mol of sulfur dioxide, and 3 mol of base for each mole of water. In practice, the reagent contains excesses of both sulfur dioxide and base; the combining capacity for water is determined by its iodine content.

Detecting the End Point

An end point in a Karl Fischer titration can be observed visually based on the brown color of the excess reagent. More commonly, however, end points are obtained by electroanalytical measurements. Several instrument manufacturers offer automatic or semiautomatic instruments for performing Karl Fischer titrations. These are all based on electrometric end-point detection.

Reagent Properties

Karl Fischer reagent decomposes on standing. Because decomposition is particularly rapid immediately after preparation, it is common practice to prepare the reagent a day or two before it is to be used. Ordinarily, its strength must be established at least daily against a standard solution of water in methanol. A proprietary commercial Karl Fischer reagent reported to require only occasional restandardization is now available.

It is obvious that great care must be exercised to keep atmospheric moisture from contaminating the Karl Fischer reagent and the sample. All glassware must be carefully dried before use, and the standard solution must be stored out of contact with air. It is also necessary to minimize contact between the atmosphere and the solution during the titration.

Applications

Karl Fischer reagent has been applied to the determination of water in numerous types of samples. There are several variations of the basic technique, depending on the solubility of the material, the state in which the water is retained, and the physical state of the sample. If the sample can be dissolved completely in methanol, a direct and rapid titration is usually feasible. This method has been applied to the determination of water in many organic acids, alcohols, esters, ethers, anhydrides, and halides. The hydrated salts of most organic acids as well as the hydrates of a number of inorganic salts that are soluble in methanol can also be determined by direct titration.

Direct titration of samples that are only partially dissolved in the reagent usually leads to incomplete recovery of the water. Satisfactory results with this type of sample are often obtained, however, by the addition of excess reagent and back-titration with a standard solution of water in methanol after a suitable reaction time. An effective alternative is to extract the water from the sample by refluxing with anhydrous methanol or other organic solvents. The resulting solution is then titrated directly with the Karl Fischer solution.

WEB WORKS

Set your browser to http://packer.berkeley.edu and learn about the research interests of the Packer lab at the University of California at Berkeley. Investigate the nature of leukocytes and their role in wound healing. Discover the vitamin E cycle.

18D   QUESTIONS AND PROBLEMS

*18-1. Write balanced net-ionic equations to describe
   (a) the oxidation of Mn^{2+} to MnO_4^- by ammonium peroxydisulfate.
   (b) the oxidation of Ce^{3+} to Ce^{4+} by sodium bismuthate.
   (c) the oxidation of U^{4+} to UO_2^{2+} by H_2O_2.
   (d) the reaction of $V(OH)_4^+$ in a Walden reductor.
   (e) the titration of H_2O_2 with $KMnO_4$.
   (f) the reaction between KI and ClO_3^- in acidic solution.

18-2. Write balanced net-ionic equations to describe
   (a) the reduction of Fe^{3+} to Fe^{2+} by SO_2.
   (b) the reaction of H_2MoO_4 in a Jones reductor.
   (c) the oxidation of HNO_2 by a solution of MnO_4^-.
   (d) the reaction of aniline ($C_6H_4NH_2$) with a mixture of $KBrO_3$ and KBr in acidic solution.
   (e) the air oxidation of $HAsO_3^{2-}$ to $HAsO_4^{2-}$.
   (f) the reaction of KI with HNO_2 in acidic solution.

*18-3. Why is a Walden reductor always used with solutions that contain appreciable concentrations of HCl?

18-4. Why is zinc amalgam preferable to pure zinc in a Jones reductor?

*18-5. Write a balanced net-ionic equation for the reduction of UO_2^{2+} in a Walden reductor.

18-6. Write a balanced net-ionic equation for the reduction of TiO^{2+} in a Jones reductor.

*18-7. Why are standard solutions of reductants less often used for titrations than standard solutions of oxidants?

*18-8. Why are standard $KMnO_4$ solutions seldom used for the titration of solutions containing HCl?

18-9. Why are Ce^{4+} solutions never used for the titration of reductants in basic solutions?

*18-10. Write a balanced net-ionic equation showing why $KMnO_4$ end points fade.

18-11. Why are $KMnO_4$ solutions filtered before they are standardized?

18-12. Why are solutions of $KMnO_4$ and $Na_2S_2O_3$ generally stored in dark reagent bottles?

*18-13. When a solution of $KMnO_4$ was left standing in a buret for 3 hr, a brownish ring formed at the surface of the liquid. Write a balanced net-ionic equation to account for this observation.

18-14. What is the primary use of standard $K_2Cr_2O_7$ solutions?

*18-15. Why are iodine solutions prepared by dissolving I_2 in concentrated KI?

18-16. A standard solution of I_2 increased in molarity with standing. Write a balanced net-ionic equation that accounts for the increase.

*18-17. When a solution of $Na_2S_2O_3$ is introduced into a solution of HCl, a cloudiness develops almost immediately. Write a balanced net-ionic equation explaining this phenomenon.

18-18. Suggest a way in which a solution of KIO_3 could be used as a source of known quantities of I_2.

*18-19. Write balanced equations showing how $KBrO_3$ could be used as a primary standard for solutions of $Na_2S_2O_3$.

18-20. Write balanced equations showing how $K_2Cr_2O_7$ could be used as a primary standard for solutions of $Na_2S_2O_3$.

*18-21. Write a balanced net-ionic equation describing the titration of hydrazine (N_2H_4) with standard iodine.

18-22. In the titration of I_2 solutions with $Na_2S_2O_3$, starch indicator is never added until just before chemical equivalence. Why?

18-23. A solution prepared by dissolving a 0.2464-g sample of electrolytic iron wire in acid was passed through a Jones reductor. The iron(II) in the resulting solution required a 39.31-mL titration. Calculate the molar oxidant concentration if the titrant used was
   *(a) Ce^{4+} (product: Ce^{3+}).
   (b) $Cr_2O_7^{2-}$ (product: Cr^{3+}).
   *(c) MnO_4^- (product: Mn^{2+}).
   (d) $V(OH)_4^+$ (product: VO^{2+}).
   *(e) IO_3^- (product: ICl_2^-).

*18-24. How would you prepare 250.0 mL of 0.03500 M $K_2Cr_2O_7$?

18-25. How would you prepare 2.000 L of 0.02500 M $KBrO_3$?

*18-26. How would you prepare 1.5 L of approximately 0.1 M $KMnO_4$?

18-27. How would you prepare 2.0 L of approximately 0.05 M I_3^-?

*18-28. Titration of 0.1467 g of primary-standard $Na_2C_2O_4$ required 28.85 mL of a potassium permanganate solution. Calculate the molar concentration of $KMnO_4$ in this solution.

18-29. A 0.1809-g sample of pure iron wire was dissolved in acid, reduced to the +2 state, and titrated with 31.33 mL of cerium(IV). Calculate the molar concentration of the Ce^{4+} solution.

*18-30. The iodine produced when an excess of KI was added to a solution containing 0.1518 g of

$K_2Cr_2O_7$ required a 46.13-mL titration with $Na_2S_2O_3$. Calculate the molar concentration of the thiosulfate solution.

18-31. A 0.1017-g sample of $KBrO_3$ was dissolved in dilute HCl and treated with an unmeasured excess of KI. The liberated iodine required 39.75 mL of a sodium thiosulfate solution. Calculate the molar concentration of the $Na_2S_2O_3$.

*18-32. The Sb(III) in a 1.080-g ore sample required a 41.67-mL titration with 0.03134 M I_2 [reaction product: Sb(V)]. Express the results of this analysis in terms of
  (a) percentage of Sb.
  (b) percentage of stibnite (Sb_2S_3).

18-33. Calculate the percentage of MnO_2 in a mineral specimen if the I_2 liberated by a 0.1344-g sample in the net reaction

$$MnO_2(s) + 4H^+ + 2I^- \rightarrow Mn^{2+} + I_2 + 2H_2O$$

required 32.30 mL of 0.07220 M $Na_2S_2O_3$.

*18-34. Under suitable conditions, thiourea is oxidized to sulfate by solutions of bromate

$$3CS(NH_2)_2 + 4BrO_3^- + 3H_2O \rightarrow$$
$$3CO(NH_2)_2 + 3SO_4^{2-} + 4Br^- + 6H^+$$

A 0.0715-g sample of a material was found to consume 14.1 mL of 0.00833 M $KBrO_3$. What was the percent purity of the thiourea sample?

*18-35. A 0.7120-g specimen of iron ore was brought into solution and passed through a Jones reductor. Titration of the Fe(II) produced required 39.21 mL of 0.02086 M $KMnO_4$. Express the results of this analysis in terms of percent
  (a) Fe.
  (b) Fe_2O_3.

18-36. The Sn in a 0.4352-g mineral specimen was reduced to the +2 state with Pb and titrated with 29.77 mL of 0.01735 M $K_2Cr_2O_7$. Calculate the results of this analysis in terms of percent
  (a) Sn.
  (b) SnO_2.

*18-37. Treatment of hydroxylamine (H_2NOH) with an excess of Fe(III) results in the formation of N_2O and an equivalent amount of Fe(II):

$$2H_2NOH + 4Fe^{3+} \rightarrow$$
$$N_2O(g) + 4Fe^{2+} + 4H^+ + H_2O$$

Calculate the molar concentration of an H_2NOH solution if the Fe(II) produced by treatment of a 50.00-mL aliquot required 23.61 mL of 0.02170 M $K_2Cr_2O_7$.

18-38. The organic matter in a 0.9280-g sample of burn ointment was eliminated by ashing, following

which the solid residue of ZnO was dissolved in acid. Treatment with $(NH_4)_2C_2O_4$ resulted in the formation of the sparingly soluble ZnC_2O_4. The solid was filtered, washed, and then redissolved in dilute acid. The liberated $H_2C_2O_4$ required 37.81 mL of 0.01508 M $KMnO_4$. Calculate the percentage of ZnO in the medication.

*18-39. The $KClO_3$ in a 0.1342-g sample of an explosive was determined by reaction with 50.00 mL of 0.09601 M Fe^{2+}:

$$ClO_3^- + 6Fe^{2+} + 6H^+ \rightarrow$$
$$Cl^- + 3H_2O + 6Fe^{3+}$$

When the reaction was complete, the excess Fe^{2+} was back-titrated with 12.99 mL of 0.08362 M Ce^{4+}. Calculate the percentage of $KClO_3$ in the sample.

18-40. The tetraethyl lead [$Pb(C_2H_5)_4$] in a 25.00-mL sample of aviation gasoline was shaken with 15.00 mL of 0.02095 M I_2. The reaction is

$$Pb(C_2H_5)_4 + I_2 \rightarrow Pb(C_2H_5)_3I + C_2H_5I$$

After the reaction was complete, the unused I_2 was titrated with 6.09 mL of 0.03465 M $Na_2S_2O_3$. Calculate the weight (in milligrams) of $Pb(C_2H_5)_4$ (323.4 g/mol) in each liter of the gasoline.

*18-41. An 8.13-g sample of an ant-control preparation was decomposed by wet-ashing with H_2SO_4 and HNO_3. The As in the residue was reduced to the trivalent state with hydrazine. After removal of the excess reducing agent, the As(III) required a 23.77-mL titration with 0.02425 M I_2 in a faintly alkaline medium. Express the results of this analysis in terms of percentage of As_2O_3 in the original sample.

18-42. A sample of alkali metal chlorides was analyzed for sodium by dissolving a 0.800-g sample in water and diluting to exactly 500 mL. A 25.0-mL aliquot of this was treated in such a way as to precipitate the sodium as $NaZn(UO_2)_3(OAc)_9 \cdot 6H_2O$. The precipitate was filtered, dissolved in acid, and passed through a lead reductor that converted the uranium to U^{4+}. Oxidation of U^{4+} to UO_2^{2+} required 19.9 mL of 0.100 M $K_2Cr_2O_7$. Calculate the percent NaCl in the sample.

*18-43. The ethyl mercaptan concentration in a mixture was determined by shaking a 1.657-g sample with 50.0 mL of 0.01194 M I_2 in a tightly stoppered flask:

$$2C_2H_5SH + I_2 \rightarrow C_2H_5SSC_2H_5 + 2I^- + 2H^+$$

The excess I_2 was back-titrated with 16.77 mL of 0.01325 M $Na_2S_2O_3$. Calculate the percentage of C_2H_5SH (62.13 g/mol).

18-44. A 4.971-g sample containing the mineral tellurite was dissolved and then treated with 50.00 mL of 0.03114 M $K_2Cr_2O_7$:

$$3TeO_2 + Cr_2O_7^{2-} + 8H^+ \rightarrow$$
$$3H_2TeO_4 + 2Cr^{3+} + H_2O$$

Upon completion of the reaction, the excess required a 10.05-mL back-titration with 0.1135 M Fe^{2+}. Calculate the percentage of TeO_2 in the sample.

*18-45. A sensitive method for I^- in the presence of Cl^- and Br^- entails oxidation of the I^- to IO_3^- with Br_2. The excess Br_2 is then removed by boiling or by reduction with formate ion. The IO_3^- produced is determined by addition of excess I^- and titration of the resulting I_2. A 1.204-g sample of mixed halides was dissolved and analyzed by the foregoing procedure; 20.66 mL of 0.05551 M thiosulfate was required. Calculate the percentage of KI in the sample.

*18-46. A 1.065-g sample of stainless steel was dissolved in HCl (this treatment converts the Cr present to Cr^{3+}) and diluted to 500.0 mL in a volumetric flask. One 50.00-mL aliquot was passed through a Walden reductor and then titrated with 13.72 mL of 0.01920 M $KMnO_4$. A 100.0-mL aliquot was passed through a Jones reductor into 50 mL of 0.10 M Fe^{3+}. Titration of the resulting solution required 36.43 mL of the $KMnO_4$ solution. Calculate the percentages of Fe and Cr in the alloy.

18-47. A 2.559-g sample containing both Fe and V was dissolved under conditions that converted the elements to Fe(III) and V(V). The solution was diluted to 500.0 mL, and a 50.00-mL aliquot was passed through a Walden reductor and titrated with 17.74 mL of 0.1000 M Ce^{4+}. A second 50.00-mL aliquot was passed through a Jones reductor and required 44.67 mL of the same Ce^{4+} solution to reach an end point. Calculate the percentage of Fe_2O_3 and V_2O_5 in the sample.

*18-48. A 25.0-mL aliquot of a solution containing Tl(I) ion was treated with K_2CrO_4. The Tl_2CrO_4 was filtered, washed free of excess precipitating agent, and dissolved in dilute H_2SO_4. The $Cr_2O_7^{2-}$ produced was titrated with 40.60 mL of 0.1004 M Fe^{2+} solution. What was the mass of Tl in the sample? The reactions are

$$2Tl^+ + CrO_4^{2-} \rightarrow Tl_2CrO_4(s)$$

$$2Tl_2CrO_4(s) + 2H^+ \rightarrow 4Tl^+ + Cr_2O_7^{2-} + H_2O$$

$$Cr_2O_7^{2-} + 6Fe^{2+} + 14H^+ \rightarrow$$
$$6Fe^{3+} + 2Cr^{3+} + 7H_2O$$

*18-49. A gas mixture was passed at the rate of 2.50 L/min through a solution of sodium hydroxide for a total of 64.00 min. The SO_2 in the mixture was retained as sulfite ion:

$$SO_2(g) + 2OH^- \rightarrow SO_3^{2-} + H_2O$$

After acidification with HCl, the sulfite was titrated with 4.98 mL of 0.003125 M KIO_3:

$$IO_3^- + 2H_2SO_3 + 2Cl^- \rightarrow$$
$$ICl_2^- + SO_4^{2-} + 2H^+$$

Use 1.20 g/L for the density of the mixture and calculate the concentration of SO_2 in ppm.

18-50. A 24.7-L sample of air drawn from the vicinity of a coking oven was passed over iodine pentoxide at 150°C, where CO was converted to CO_2 and a chemically equivalent quantity of I_2 was produced:

$$I_2O_5(s) + 5CO(g) \rightarrow 5CO_2(g) + I_2(g)$$

The I_2 distilled at this temperature and was collected in a solution of KI. The I_3^- produced was titrated with 7.76 mL of 0.00221 M $Na_2S_2O_3$. Does the air in this space comply with federal regulations that mandate a maximum CO level no greater than 50 ppm?

*18-51. A 30.00-L air sample was passed through an absorption tower containing a solution of Cd^{2+}, where H_2S was retained as CdS. The mixture was acidified and treated with 10.00 mL of 0.01070 M I_2. After the reaction

$$S^{2-} + I_2 \rightarrow S(s) + 2I^-$$

was complete, the excess iodine was titrated with 12.85 mL of 0.01344 M thiosulfate. Calculate the concentration of H_2S in ppm; use 1.20 g/L for the density of the gas stream.

18-52. A square of photographic film 2.0 cm on an edge was suspended in a 5% solution of $Na_2S_2O_3$ to dissolve the silver halides. After removal and washing of the film, the solution was treated with an excess of Br_2 to oxidize the iodide present to IO_3^- and destroy the excess thiosulfate ion. The solution was boiled to remove the bromine, and an excess of iodide was added. The liberated iodine was titrated with 13.7 mL of 0.0352 M thiosulfate solution.

(a) Write balanced equations for the reactions involved in the method.

(b) Calculate the milligrams of AgI per square centimeter of film.

*18-53. The Winkler method for dissolved oxygen in water is based on the rapid oxidation of solid $Mn(OH)_2$ to $Mn(OH)_3$ in alkaline medium. When acidified, the Mn(III) readily releases iodine from iodide. A 250-mL water sample, in a stoppered vessel, was treated with 1.00 mL of a concentrated solution of NaI and NaOH and 1.00 mL of a manganese(II) solution. Oxidation of the $Mn(OH)_2$ was complete in about 1 min. The precipitates were then dissolved by addition of 2.00 mL of concentrated H_2SO_4, whereupon an amount of iodine equivalent to the $Mn(OH)_3$ (and hence to the dissolved O_2) was liberated. A 25.0-mL aliquot (of the 254 mL) was titrated with 12.7 mL of 0.00962 M thiosulfate. Calculate the milligrams O_2 per milliliter sample (assume that the concentrated reagents are O_2 free and take the dilutions of the sample into account).

18-54. Use a spreadsheet to do the calculations and plot the titration curves for the following titrations. Calculate potentials after the addition of titrant corresponding to 10, 20, 30, 40, 50, 60, 70, 80, 90, 95, 99, 99.9, 100, 101, 105, 110, and 120% of the equivalence point volume.

(a) 25.00 mL of 0.025 M $SnCl_2$ with 0.050 M $FeCl_3$.

(b) 25.00 mL of 0.08467 M $Na_2S_2O_3$ with 0.10235 M I_3^-.

(c) 0.1250 g of primary-standard grade $Na_2C_2O_4$, dissolved in 25.0 mL of solution, with 0.01035 M $KMnO_4$.

(d) 20.00 mL of 0.1034 M Fe^{2+} with 0.01500 M $K_2Cr_2O_7$.

Note: For (c) and (d), assume $[H^+] = 1.00$ M throughout. For (c), $P_{CO_2} = 1.00$ atm.

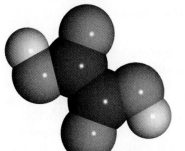

Chapter 19

Potentiometry: Measuring Concentrations of Ions and Molecules

I n Chapter 16, we showed that the potential of an electrode, relative to the standard hydrogen electrode, is determined by the activity of one or more of the species in the solution in which the electrode is immersed. This chapter deals with how electrode potential data are used to determine the concentration of analytes.[1] Analytical methods that are based on potential measurements are called *potentiometric methods,* or *potentiometry.*

19A GENERAL PRINCIPLES

In Feature 16-3, we showed that absolute values for individual half-cell potentials cannot be determined in the laboratory. That is, only cell potentials can be obtained experimentally. Figure 19-1 shows a typical cell for potentiometric analysis. This cell can be depicted as

$$\underbrace{\text{reference electrode}}_{E_{ref}} \,|\, \underbrace{\text{salt bridge}}_{E_j} \,|\, \underbrace{\text{analyte solution} \,|\, \text{indicator electrode}}_{E_{ind}}$$

The *reference electrode* in this figure is a half-cell with an accurately known electrode potential, E_{ref}, that is independent of the concentration of the analyte or any other ions in the solution under study. Although it can be a standard hydrogen electrode, it seldom is because a standard hydrogen electrode is somewhat troublesome to maintain and use. By convention, the reference electrode is always treated as the left-hand electrode in potentiometric measurements. The indicator electrode, which is immersed in a solution of the analyte, develops a potential, E_{ind}, that depends on the activity of the analyte. Most indicator electrodes used in potentiometry are highly selective in their responses. The third component of a potentiometric cell is a salt bridge that prevents the components

A *reference electrode* is a half-cell having a known electrode potential that remains constant and is independent of the composition of the analyte solution.

Reference electrodes are *always* treated as the left-hand electrode in this text.

An *indicator electrode* has a potential that varies in a known way with variations in the concentration of an analyte.

---

[1]For further reading on potentiometric methods, see E. P. Serjeant, *Potentiometry and Potentiometric Titrations* (New York: Wiley, 1984). Reprinted by Krieger Publishing, Malabar, FL, 1991.

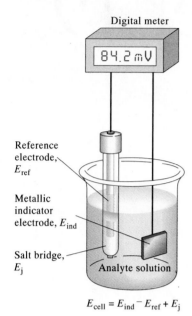

Digital meter

84.2 mV

Reference
electrode,
E_{ref}

Metallic
indicator
electrode, E_{ind}

Salt bridge,
E_j

Analyte solution

$$E_{cell} = E_{ind} - E_{ref} + E_j$$

Figure 19-1   A cell for potentiometric
determinations.

of the analyte solution from mixing with those of the reference electrode. As noted in Chapter 16, a potential develops across the liquid junctions at each end of the salt bridge. These two potentials tend to cancel one another if the mobilities of the cation and the anion in the bridge solution are approximately the same. Potassium chloride is a nearly ideal electrolyte for the salt bridge because the mobilities of the K^+ ion and the Cl^- ion are nearly equal. The net potential across the salt bridge E_j is thereby reduced to a few millivolts or less. For most electroanalytical methods, this net junction potential is small enough to be neglected. In the potentiometric methods discussed in this chapter, however, the junction potential and its uncertainty can be factors that limit the measurement accuracy and precision.

The potential of the cell we have just considered is given by the equation

$$E_{cell} = E_{ind} - E_{ref} + E_j \qquad (19\text{-}1)$$

The first term in this equation, E_{ind}, contains the information we seek about the concentration of the analyte. A potentiometric determination of an analyte, then, involves measuring a cell potential, correcting this potential for the reference and junction potentials, and computing the analyte concentration from the indicator electrode potential.

The sections that follow deal with the sources of the three potentials shown on the right side of Equation 19-1. We begin by considering the reference electrode.

19B   WHAT REFERENCE ELECTRODES ARE USED?

The ideal reference electrode has a potential that is accurately known, constant, and completely insensitive to the composition of the analyte solution. In addition, this electrode should be rugged, be easy to assemble, and be able to maintain a constant potential while passing small currents.

19B-1   Calomel Reference Electrodes

A hydrogen electrode is seldom used as a reference electrode for day-to-day potentiometric measurements because it is somewhat inconvenient. It is also a fire hazard.

A calomel electrode can be represented schematically as

$$Hg\,|\,Hg_2Cl_2(\text{sat'd}), KCl(x\ M)\,\|$$

Calomel, Hg_2Cl_2, has a limited solubility in water.

Molecular model of Hg_2Cl_2 (calomel).

The "saturated" in a saturated calomel electrode refers to the KCl concentration and not the calomel concentration. All calomel electrodes are saturated with Hg_2Cl_2 (calomel).

where x represents the molar concentration of potassium chloride in the solution. Three concentrations of potassium chloride are commonly 0.1 M, 1 M, and saturated (about 4.6 M). The saturated calomel electrode (SCE) is the most widely used of the calomel electrodes because it is easily prepared. Its main disadvantage is its somewhat larger temperature coefficient than the other two. This disadvantage is important only in those rare circumstances where substantial temperature changes occur during a measurement. The *electrode potential* of the saturated calomel electrode is 0.2444 V at 25°C. The electrode reaction in calomel half-cells is

$$Hg_2Cl_2(s) + 2e^- \rightleftharpoons 2Hg(l) + 2Cl^-(aq)$$

Table 19-1 lists the compositions and electrode potentials for the three most common calomel electrodes. Note that the electrodes differ only in their potas-

Table 19-1

Electrode Potentials for Reference Electrodes as a Function of
Composition and Temperature

Temperature, °C	Potential vs. SHE, V				
	0.1 M Calomel*	3.5 M Calomel†	Sat'd Calomel*	3.5 M Ag/AgCl†	Sat'd Ag/AgCl†
12	0.3362		0.2528		
15	0.3362	0.254	0.2511	0.212	0.209
20	0.3359	0.252	0.2479	0.208	0.204
25	0.3356	0.250	0.2444	0.205	0.199
30	0.3351	0.248	0.2411	0.201	0.194
35	0.3344	0.246	0.2376	0.197	0.189

*From R. G. Bates, in *Treatise on Analytical Chemistry,* 2nd ed., I. M. Kolthoff and P. J. Elving, Eds., Part I, Vol. 1 (New York: Wiley, 1978), p. 793.
†From D. T. Sawyer and J. L. Roberts, Jr., *Experimental Electrochemistry for Chemists* (New York: Wiley, 1974), p. 42.

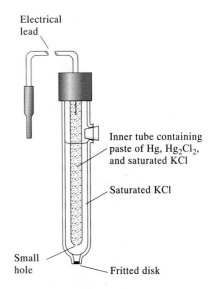

Figure 19-2 Diagram of a typical commercial saturated calomel electrode.

sium chloride concentrations; all are saturated with calomel (Hg_2Cl_2). Figure 19-2 illustrates a typical commercial saturated calomel electrode. It consists of a 5- to 15-cm-long tube that is 0.5 to 1.0 cm in diameter. A mercury/mercury(I) chloride paste in saturated potassium chloride is contained in an inner tube and is connected to the saturated potassium chloride solution in the outer tube through a small opening. An inert metal electrode is immersed in the paste. Contact with the analyte solution is made through a fritted disk, a porous fiber, or a piece of porous Vycor® ("thirsty glass") sealed in the end of the outer tubing.

Figure 19-3 shows a saturated calomel electrode that you can easily construct from materials available in most laboratories. A salt bridge (Section 16B-2) provides electrical contact with the analyte solution.

19B-2   Silver/Silver Chloride Reference Electrodes

A system analogous to the saturated calomel electrode consists of a silver electrode immersed in a solution that is saturated in both potassium chloride and silver chloride:

$$Ag \,|\, AgCl(sat'd), KCl(sat'd) \,\|$$

The half-reaction is

$$AgCl + e^- \rightleftharpoons Ag(s) + Cl^-$$

The potential of this electrode is 0.199 V at 25°C.

Silver/silver chloride electrodes of various sizes and shapes are on the market (Table 19-1). A simple and easily constructed electrode of this type is shown in Figure 19-4.

A salt bridge is readily constructed by filling a U-tube with a conducting gel prepared by heating about 5 g of agar in 100 mL of an aqueous solution containing about 35 g of potassium chloride. When the liquid cools, it sets into a gel that prevents the two solutions at either end of the tube from mixing and is also a good conductor.

Agar, which is available as translucent flakes, is a heteropolysaccharide extracted from certain East Indian seaweed. Solutions of agar in hot water set to a gel upon cooling.

At 25°C, the potential of the saturated calomel electrode versus the standard hydrogen electrode is 0.244 V; for the saturated silver/silver chloride electrode it is 0.199 V.

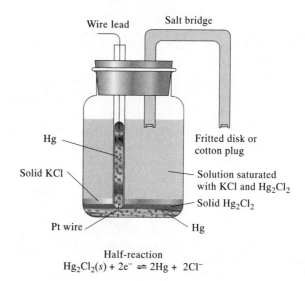

Half-reaction
$Hg_2Cl_2(s) + 2e^- \rightleftharpoons 2Hg + 2Cl^-$

Figure 19-3   A saturated calomel electrode made from materials readily available in any laboratory.

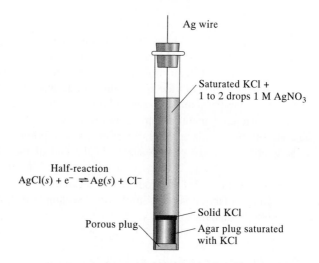

Half-reaction
$AgCl(s) + e^- \rightleftharpoons Ag(s) + Cl^-$

Figure 19-4   Diagram of a silver/silver chloride electrode.

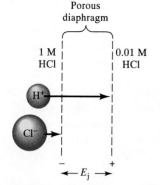

Figure 19-5   Schematic representation of a liquid junction, showing the source of the junction potential, E_j. The length of the arrows corresponds to the relative mobilities of the ions.

19C   DEALING WITH LIQUID-JUNCTION POTENTIALS

A liquid-junction potential develops across the boundary between two electrolyte solutions that have different compositions. Figure 19-5 shows a very simple liquid junction consisting of a 1 M hydrochloric acid solution in contact with a solution that is 0.01 M in that acid. An inert porous barrier, such as a fritted glass plate, prevents the two solutions from mixing. Both hydrogen ions and chloride ions tend to diffuse across this boundary from the more concentrated to the more dilute solution. The driving force for each ion is proportional to the activity difference between the two solutions. In this example, hydrogen ions are substantially more mobile than chloride ions. Thus, hydrogen ions diffuse more rapidly than chloride ions, and, as shown in Figure 19-5, a separation of charge results. The more dilute side of the boundary becomes positively charged because of the more rapid diffusion of hydrogen ions. The concentrated side therefore acquires a negative charge from the excess of slower moving chloride ions. The charge developed tends to counteract the differences in diffusion rates of the two ions so that a condition of steady state is attained rapidly. The potential difference resulting from this charge separation may be several hundredths of a volt.

The magnitude of the liquid-junction potential can be minimized by placing a salt bridge between the two solutions. The salt bridge is most effective if the mobilities of the negative and positive ions in the bridge are nearly equal and if their concentrations are large. A saturated solution of potassium chloride is good from both standpoints. The net-junction potential with such a bridge is typically a few millivolts.

> The net-junction potential across a typical salt bridge is a few millivolts.

19D INDICATOR ELECTRODES

An ideal indicator electrode responds rapidly and reproducibly to changes in the concentration of an analyte ion (or group of analyte ions). Although no indicator electrode is absolutely specific in its response, a few are now available that are remarkably selective. Indicator electrodes are of three types: metallic, membrane, and ion-sensitive field effect transistors.

> The results of potentiometric determinations are the activities of analytes, in contrast to most analytical methods that give the concentrations of analytes. Recall that the activity of a species a_X is related to the molar concentration of X by Equation 9-2:
>
> $$a_X = \gamma_X [X]$$
>
> where γ_X is the activity coefficient of X, a parameter that varies with the ionic strength of the solution. Because potentiometric data are dependent on activities, in most cases we will not make the usual approximation that $a_X \approx [X]$ in this chapter.

19D-1 Describing Metallic Indicator Electrodes

It is convenient to classify metallic indicator electrodes as *electrodes of the first kind, electrodes of the second kind,* and *inert redox electrodes.*

Electrodes of the First Kind

An electrode of the first kind is a pure metal that is in direct equilibrium with its cation in the solution. A single reaction is involved. For example, the equilibrium between a metal X and its cation X^{n+} is

$$X^{n+}(aq) + ne^- \rightleftharpoons X(s)$$

for which

$$E_{ind} = E^0_{X^{n+}/X} - \frac{0.0592}{n} \log \frac{1}{a_{X^{n+}}} = E^0_{X^{n+}/X} + \frac{0.0592}{n} \log a_{X^{n+}} \quad (19\text{-}2)$$

where E_{ind} is the electrode potential of the metal and $a_{X^{n+}}$ is the activity of the ion (or approximately its molar concentration, $[X^{n+}]$).

We often express the electrode potential of the indicator electrode in terms of the p-function of the cation ($pX = -\log [a_{X^{n+}}]$). Thus, substituting this definition of pX into Equation 19-2 gives

$$E_{ind} = E^0_{X^{n+}/X} + \frac{0.0592}{n} \log a_{X^{n+}} = E^0_{X^{n+}/X} - \frac{0.0592}{n} pX \quad (19\text{-}3)$$

This function is plotted in Figure 19-6.

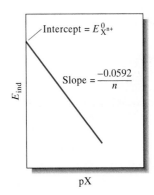

Figure 19-6 A plot of Equation 19-3 for an electrode of the first kind.

Electrode systems of the first kind are not widely used for potentiometric determinations for several reasons. For one, metallic indicator electrodes are not very selective and respond not only to their own cations but also other more easily reduced cations. For example, a copper electrode cannot be used for the determination of copper(II) ions in the presence of silver(I) ions because the electrode potential is also a function of the Ag^+ concentration. In addition, many metal electrodes, such as zinc and cadmium, can only be used in neutral or basic solutions because they dissolve in the presence of acids. Third, other metals are so easily oxidized that their use is restricted to solutions that have been deaerated. Finally, certain harder metals, such as iron, chromium, cobalt, and nickel, do not provide reproducible potentials. Moreover, for these electrodes, plots of E_{ind} versus pX yield slopes that differ significantly and irregularly from the theoretical $(-0.0592/n)$. For these reasons, the only electrode systems of the first kind that have been used in potentiometry are Ag/Ag^+ and Hg/Hg_2^{2+} in neutral solutions and Cu/Cu^{2+}, Zn/Zn^{2+}, Cd/Cd^{2+}, Bi/Bi^{3+}, Tl/Tl^+, and Pb/Pb^{2+} in deaerated solutions.

Electrodes of the Second Kind

Metals not only serve as indicator electrodes for their own cations but also respond to the activities of anions that form sparingly soluble precipitates or stable complexes with such cations. The potential of a silver electrode, for example, correlates reproducibly with the activity of chloride ion in a solution saturated with silver chloride. Here, the electrode reaction can be written as

$$AgCl(s) + e^- \rightleftharpoons Ag(s) + Cl^-(aq) \qquad E^0 = 0.222 \text{ V}$$

The Nernst expression for this process is

$$E_{ind} = E^0_{AgCl/Ag} - 0.0592 \log a_{Cl^-} = E^0_{AgCl/Ag} + 0.0592 \text{ pCl} \qquad (19\text{-}4)$$

Equation 19-4 shows that the potential of a silver electrode is proportional to pCl, the negative logarithm of the chloride ion activity. Thus, in a solution saturated with silver chloride, a silver electrode can serve as an indicator electrode of the second kind for chloride ion. Note that the sign of the log term for an electrode of this type is opposite that for an electrode of the first kind (see Equation 19-3). A plot of the potential of the silver electrode versus pCl is shown in Figure 19-7.

Mercury serves as an indicator electrode of the second kind for the EDTA anion Y^{4-}. For example, when a small amount of HgY^{2-} is added to a solution containing Y^{4-}, the half-reaction at a mercury electrode is

$$HgY^{2-} + 2e^- \rightleftharpoons Hg(l) + Y^{4-} \qquad E^0 = 0.21 \text{ V}$$

for which

$$E_{ind} = 0.21 - \frac{0.0592}{2} \log \frac{a_{Y^{4-}}}{a_{HgY^{2-}}}$$

The formation constant for HgY^{2-} is very large (6.3×10^{21}), so the concentration of the complex remains essentially constant over a large range of Y^{4-} concentrations. The Nernst equation for the process can therefore be written as

$$E = K - \frac{0.0592}{2} \log a_{Y^{4-}} = K + \frac{0.0592}{2} \text{ pY} \qquad (19\text{-}5)$$

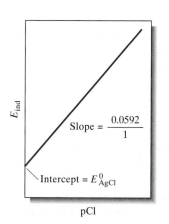

Figure 19-7   A plot of Equation 19-4 for an electrode of the second kind for Cl^-.

where

$$K = 0.21 - \frac{0.0592}{2} \log \frac{1}{a_{HgY^{2-}}}$$

The mercury electrode is thus a valuable electrode of the second kind for EDTA titrations.

Inert Metallic Electrodes for Redox Systems

As noted in Chapter 16, several inert conductors respond to redox systems. Such materials as platinum, gold, palladium, and carbon can be used to monitor redox systems. For example, the potential of a platinum electrode immersed in a solution containing cerium(III) and cerium(IV) is

$$E_{ind} = E^0_{Ce^{4+}/Ce^{3+}} - 0.0592 \log \frac{a_{Ce^{3+}}}{a_{Ce^{4+}}}$$

A platinum electrode is a convenient indicator electrode for titrations involving standard cerium(IV) solutions.

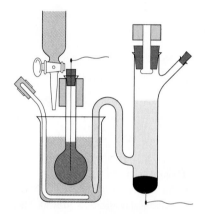

Figure 19-8 The first practical glass electrode (From Haber and Klemensiewicz, *Z. Phys. Chem.,* **1909,** *65,* 385.)

19D-2 Characterizing Membrane Electrodes[2]

For many years, the most convenient method for determining pH has involved measurement of the potential that develops across a thin glass membrane that separates two solutions with different hydrogen ion concentrations. A diagram of the first practical glass electrode is shown in Figure 19-8. The phenomenon upon which the measurement is based was first reported in 1906 and by now has been extensively studied by many investigators. As a result, the sensitivity and selectivity of glass membranes toward hydrogen ions are reasonably well understood. Furthermore, this understanding has led to the development of other types of membranes that respond selectively to more than two dozen other ions.

Membrane electrodes are sometimes called *p-ion electrodes* because the data obtained from them are usually presented as p-functions, such as pH, pCa, or pNO_3. In this section, we consider several types of p-ion membranes.

Membrane electrodes are fundamentally different from metal electrodes both in design and in principle. We shall use the glass electrode for pH measurements to illustrate these differences.

19D-3 The Glass Electrode for Measuring pH

Figure 19-9 shows a typical *cell* for measuring pH. The cell consists of a glass indicator electrode and a saturated calomel reference electrode immersed in the solution of unknown pH. The indicator electrode consists of a thin, pH-sensitive glass membrane sealed onto one end of a heavy-walled glass or plastic tube. A

---

[2]Some suggested sources for additional information on this topic are A. Evans, *Potentiometry and Ion-Selective Electrodes* (New York: Wiley, 1987); J. Koryta, *Ions, Electrodes, and Membranes,* 2nd ed. (New York: Wiley, 1991); R. S. Hutchins and L. G. Bachas, in *Handbook of Instrumental Techniques for Analytical Chemistry,* F. A. Settle, Ed. (Upper Saddle River, N.J.: Prentice-Hall, 1997).

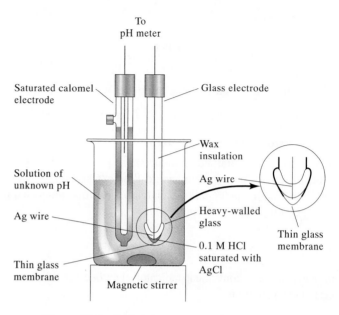

Figure 19-9   Typical electrode system for measuring pH.

The membrane of a typical glass electrode (with a thickness of 0.03 to 0.1 mm) has an electrical resistance of 50 to 500 MΩ.

small volume of dilute hydrochloric acid saturated with silver chloride is contained in the tube (the inner solution in some electrodes is a buffer containing chloride ion). A silver wire in this solution forms a silver/silver chloride reference electrode, which is connected to one of the terminals of a potential-measuring device. The calomel electrode is connected to the other terminal.

Figure 19-9 and the schematic representation of this cell in Figure 19-10 show that a glass electrode system contains two reference electrodes: (1) the external calomel electrode and (2) the internal silver/silver chloride electrode. Although the internal reference electrode is a part of the glass electrode, it is not the pH-sensing element. Instead, *it is the thin glass membrane at the tip of the electrode that responds to pH.*

The Composition and Structure of Glass Membranes

A good deal of research has been devoted to the effects of glass composition on the sensitivity of membranes to protons and other cations, and a number of formulations are now used for the manufacture of electrodes. Corning 015 glass, which has been widely used for membranes, consists of approximately 22%

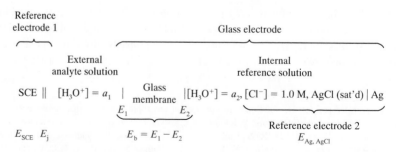

Figure 19-10   Diagram of a glass/calomel cell for the measurement of pH.

Na_2O, 6% CaO, and 72% SiO_2. This membrane shows excellent specificity toward hydrogen ions up to a pH of about 9. At higher pH values, however, the glass becomes somewhat responsive to sodium as well as to other singly charged cations. Other glass formulations are now in use in which sodium and calcium ions are replaced to various degrees by barium and lithium ions. These membranes have superior selectivity and lifetime.

As shown in Figure 19-11, a silicate glass used for membranes consists of an infinite three-dimensional network of groups in which each silicon is bonded to four oxygens and each oxygen is shared by two silicons. Within the interstices of this structure are sufficient cations to balance the negative charge of the silicate groups. Singly charged cations, such as sodium and lithium, are mobile in the lattice and are responsible for electrical conduction within the membrane.

The two surfaces of a glass membrane must be hydrated before it will function as a pH electrode. Nonhygroscopic glasses show no pH function. Even hygroscopic glasses lose their pH sensitivity after dehydration. The effect is reversible, however, and the response of a glass electrode is restored by soaking it in water.

The hydration of a pH-sensitive glass membrane involves an ion-exchange reaction between singly-charged cations in the interstices of the glass lattice and protons from the solution. The process involves univalent cations exclusively because di- and trivalent cations are too strongly held within the silicate structure to exchange with ions in the solution. Typically, then, the ion-exchange reaction can be written as

$$H^+ + Na^+Gl^- \rightleftharpoons Na^+ + H^+Gl^- \qquad (19\text{-}6)$$
$$\text{soln} \qquad \text{glass} \qquad \text{soln} \qquad \text{glass}$$

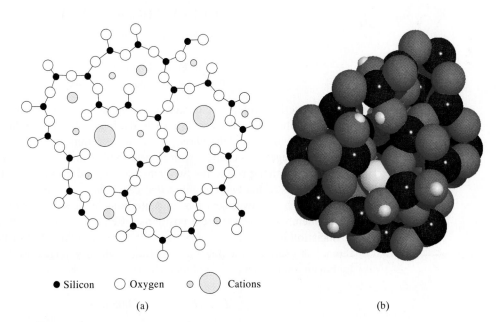

● Silicon   ○ Oxygen   ○ ◯ Cations

(a)   (b)

Figure 19-11   (a) Cross-sectional view of a silicate glass structure. In addition to the three Si—O bonds shown, each silicon is bonded to an additional oxygen atom, either above or below the plane of the paper. (Adapted with permission from G. A. Perley, *Anal. Chem.,* **1949,** *21,* 395. Copyright 1949 American Chemical Society.) (b) model showing three-dimensional structure of amorphous silica with Na^+ ion incorporated.

Oxygen atoms attached to only one silicon atom are the negatively charged Gl^- sites shown in this equation. The equilibrium constant for this process is so large that the surfaces of a hydrated glass membrane ordinarily consist entirely of silicic acid (H^+Gl^-). An exception to this situation exists in highly alkaline media, where the hydrogen-ion concentration is extremely small and the sodium-ion concentration is large; here, a significant fraction of the sites are occupied by sodium ions.

Glasses and other substances that absorb water are said to be *hygroscopic*.

Membrane Potentials

The lower part of Figure 19-10 shows four potentials that develop in a cell when pH is being determined with a glass electrode. Two of these, $E_{Ag,AgCl}$ and E_{SCE}, are reference electrode potentials that are constant. A third is the net potential across the salt bridge that separates the calomel electrode from the analyte solution. This junction and its associated *junction potential, E_j*, are found in all cells used for the potentiometric measurement of ion concentration. The fourth, and most important, potential shown in Figure 19-10 is the boundary potential, E_b, *which varies with the pH of the analyte solution*. The two reference electrodes simply provide electrical contacts with the solutions so that changes in the boundary potential can be measured.

The Boundary Potential

In Figure 19-10, the boundary potential is shown as being made up of two potentials, E_1 and E_2, that develop at the two *surfaces* of the glass membrane. The source of these two potentials is the charge that develops as a consequence of the reactions

$$\underset{\text{glass}_1}{H^+Gl^-(s)} \rightleftharpoons \underset{\text{soln}_1}{H^+(aq)} + \underset{\text{glass}_1}{Gl^-(s)} \qquad (19\text{-}7)$$

$$\underset{\text{glass}_2}{H^+Gl^-(s)} \rightleftharpoons \underset{\text{soln}_2}{H^+(aq)} + \underset{\text{glass}_2}{Gl^-(s)} \qquad (19\text{-}8)$$

where subscript 1 refers to the interface between the exterior of the glass and the analyte solution and subscript 2 refers to the interface between the internal solution and the interior of the glass. These two reactions cause the two glass surfaces to be negatively charged with respect to the solutions in which they are immersed, thus giving rise to the two potentials E_1 and E_2 shown in Figure 19-10. The positions of the two equilibria that cause the two potentials to develop are determined by the hydrogen-ion concentrations in the solutions on the two sides of the membrane. Where these positions differ, the surface at which the greater dissociation has occurred is negative with respect to the other surface. This difference in charge is the boundary potential, which is related to the activities of hydrogen ion in each of the solutions by the Nernst-like equation

$$E_b = E_1 - E_2 = 0.0592 \log \frac{a_1}{a_2} \qquad (19\text{-}9)$$

where a_1 is the hydrogen-ion activity of the analyte solution and a_2 is that of the internal solution. For a glass pH electrode, the hydrogen ion activity of the internal solution is held constant so that Equation 19-9 simplifies to

$$E_b = L' + 0.0592 \log a_1 = L' - 0.0592 \text{ pH} \qquad (19\text{-}10)$$

where

$$L' = -0.0592 \log a_2$$

The boundary potential is then a measure of the hydrogen ion activity of the external solution.

The Asymmetry Potential

When identical solutions and reference electrodes are placed on the two sides of a glass membrane, the boundary potential should in principle be zero. In fact, however, a small asymmetry potential that changes gradually with time is frequently encountered.

The sources of the asymmetry potential are obscure but undoubtedly include such causes as differences in strain on the two surfaces of the membrane imparted during manufacture, mechanical abrasion on the outer surface during use, and chemical etching of the outer surface. To eliminate the bias caused by the asymmetry potential, all membrane electrodes must be calibrated against one or more standard analyte solutions. Such calibrations should be carried out at least daily and more often when the electrode receives heavy use.

The Glass Electrode Potential

The potential of a glass indicator electrode E_{ind} has three components: (1) the boundary potential, given by Equation 19-9; (2) the potential of the internal Ag/AgCl reference electrode; and (3) the small asymmetry potential, E_{asy}, which changes slowly with time. In equation form, we may write

$$E_{ind} = E_b + E_{Ag/AgCl} + E_{asy}$$

Substitution of Equation 19-10 for E_b gives

$$E_{ind} = L' + 0.0592 \log a_1 + E_{Ag/AgCl} + E_{asy}$$

or

$$E_{ind} = L + 0.0592 \log a_1 = L - 0.0592 \text{ pH} \qquad (19\text{-}11)$$

where L is a combination of the three constant terms. Compare Equations 19-11 and 19-3. Although these two equations are similar in form, remember that the mechanism for the development of the potential of the electrodes that they *describe are totally different.*

The Alkaline Error

In basic solutions, glass electrodes respond to the concentration of both hydrogen ion and alkali metal ions. The magnitude of the resulting *alkaline error* for four different glass membranes is shown in Figure 19-12 (curves C to F). These curves refer to solutions in which the sodium ion concentration was held constant at 1 M while the pH was varied. Note that the error is negative (that is, the measured pH values are lower than the true values), which suggests that the electrode is responding to sodium ions as well as to protons. This observation is confirmed by data obtained for solutions containing different sodium ion concentrations. Thus, at pH 12, the electrode with a Corning 015 membrane (curve

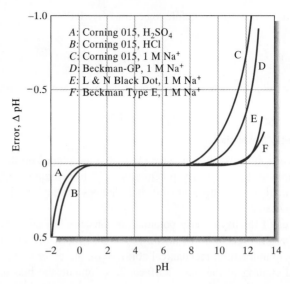

Figure 19-12   Acid and alkaline error for selected glass electrodes at 25°C. [From R. G. Bates, *Determination of pH,* 2nd ed. (New York: Wiley, 1973), p. 365. With permission.]

C in Figure 19-12) registered a pH of 11.3 when immersed in a solution having a sodium ion concentration of 1 M but 11.7 in a solution that was 0.1 M in this ion. All singly charged cations induce an alkaline error whose magnitude depends on both the cation in question and the composition of the glass membrane.

The alkaline error can be satisfactorily explained by assuming an exchange equilibrium between the hydrogen ions on the glass surface and the cations in solution. This process is simply the reverse of that shown in Equation 19-6:

$$H^+Gl^- + B^+ \rightleftharpoons B^+Gl^- + H^+$$

$$\textbf{glass} \qquad \textbf{soln} \qquad \textbf{glass} \qquad \textbf{soln}$$

where B^+ represents some singly charged cation, such as sodium ion.

The equilibrium constant for this reaction is

In Equation 19-12, b_1 represents the activity of some singly charged cation such as Na^+ or K^+.

$$K_{ex} = \frac{a_1 b_1'}{a_1' b_1} \tag{19-12}$$

where a_1 and b_1 represent the activities of H^+ and B^+ in solution and a_1' and b_1' are the activities of these ions on the glass surface. Equation 19-12 can be rearranged to give ratio of the activities of B^+ to H^+ on the glass surface:

$$\frac{b_1'}{a_1'} = \frac{b_1}{a_1} K_{ex}$$

For the glasses used for pH electrodes, K_{ex} is so small that the activity ratio b_1'/a_1' is ordinarily minuscule. The situation differs in strongly alkaline media, however. For example, b_1'/a_1' for an electrode immersed in a pH 11 solution that is 1 M in sodium ions (Figure 19-12) is $10^{11} \times K_{ex}$. Here, the activity of the sodium ions relative to that of the hydrogen ions becomes so large that the electrode responds to both species.

Describing Selectivity

The effect of an alkali metal ion on the potential across a membrane can be accounted for by inserting an additional term in Equation 19-10 to give

$$E_b = L' + 0.0592 \log (a_1 + k_{H,B}b_1) \tag{19-13}$$

where $k_{H,B}$ is the *selectivity coefficient* for the electrode. Equation 19-13 applies not only to glass indicator electrodes for hydrogen ion but also to all other types of membrane electrodes. Selectivity coefficients range from zero (no interference) to values greater than unity. Thus, if an electrode for ion A responds 20 times more strongly to ion B than to ion A, $k_{H,B}$ has a value of 20. If the response of the electrode to ion C is 0.001 of its response to A (a much more desirable situation), $k_{H,B}$ is 0.001.[3]

The product $k_{H,B}b_1$ for a glass pH electrode is ordinarily small relative to a_1 provided the pH is less than 9; under these conditions, Equation 19-13 simplifies to Equation 19-10. At high pH values and high concentrations of a singly charged ion, however, the second term in Equation 19-13 assumes a more important role in determining E_b, and an alkaline error is encountered. For electrodes specifically designed for work in highly alkaline media (curve *E* in Figure 19-12), the magnitude of $k_{H,B}b_1$ is appreciably smaller than for ordinary glass electrodes.

> The *selectivity coefficient* is a measure of the response of an ion-selective electrode to other ions.

The Acid Error

As shown in Figure 19-12, the typical glass electrode exhibits an error, opposite in sign to the alkaline error, in solutions of pH less than about 0.5; pH readings tend to be too high in this region. The magnitude of the error depends on a variety of factors and is generally not very reproducible. All the causes of the acid error are not well understood, but one source is a saturation effect that occurs when all the surface sites on the glass are occupied with H^+ ions. Under these conditions, the electrode no longer responds to further increases in the H^+ concentration and the pH readings are too high.

19D-4 Glass Electrodes for Other Cations

The alkaline error in early glass electrodes led to investigations concerning the effect of glass composition on the magnitude of this error. One consequence has been the development of glasses for which the alkaline error is negligible below about pH 12 (see curves *E* and *F* in Figure 19-12). Other studies have discovered glass compositions that permit the determination of cations other than hydrogen. Incorporation of Al_2O_3 or B_2O_3 in the glass has the desired effect. Glass electrodes that permit the direct potentiometric measurement of such singly charged species as Na^+, K^+, NH_4^+, Rb^+, Cs^+, Li^+, and Ag^+ have been developed. Some of these glasses are reasonably selective toward particular singly charged cations. Glass electrodes for Na^+, Li^+, NH_4^+, and total concentration of univalent cations are available from commercial sources.

---

[3]For tables of selectivity coefficients for a variety of membranes and ionic species, see Y. Umezawa, *CRC Handbook of Ion Selective Electrodes: Selectivity Coefficients* (Boca Raton, FL: CRC Press, 1990).

19D-5   Liquid-Membrane Electrodes

The potential of liquid-membrane electrodes develops across the interface between the solution containing the analyte and a liquid-ion exchanger that selectively bonds with the analyte ion. These electrodes have been developed for the direct potentiometric measurement of numerous polyvalent cations as well as certain anions.

Figure 19-13 is a schematic of a liquid-membrane electrode for calcium. It consists of a conducting membrane that selectively binds calcium ions, an internal solution containing a fixed concentration of calcium chloride, and a silver electrode that is coated with silver chloride to form an internal reference electrode. Notice the similarities between the liquid-membrane electrode and the glass electrode as shown in Figure 19-14. The active membrane ingredient is an ion exchanger consisting of a calcium dialkyl phosphate that is nearly insoluble in water. In the electrodes shown in Figures 19-13 and 19-14, the ion exchanger is dissolved in an immiscible organic liquid that is forced by gravity into the pores of a hydrophobic porous disk. This disk then serves as the membrane that separates the internal solution from the analyte solution. In a more recent design, the ion exchanger is immobilized in a tough polyvinyl chloride gel cemented to the end of a tube that holds the internal solution and reference electrode. In either design, a dissociation equilibrium develops at each membrane interface that is analogous to Equations 19-7 and 19-8:

> *Hydrophobic* means water hating. The hydrophobic disk is porous toward organic liquids but repels water.

$$[(RO)_2POO]_2Ca \rightleftharpoons 2(RO)_2POO^- + Ca^{2+}$$
$$\textbf{organic} \qquad\qquad \textbf{organic} \qquad\quad \textbf{aqueous}$$

where R is a high-molecular-weight aliphatic group. As with the glass electrode, a potential develops across the membrane when the extent of dissociation of the ion exchanger at one surface differs from that at the other surface. This potential is a result of differences in the calcium ion activity of the internal and external solutions. The relationship between the membrane potential and the calcium ion activities is given by an equation that is similar to Equation 19-9:

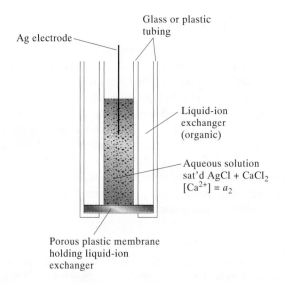

Figure 19-13   Diagram of a liquid-membrane electrode for Ca^{2+}.

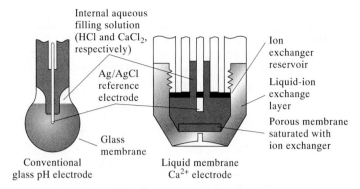

Figure 19-14   Comparison of a liquid-membrane calcium ion electrode with a glass pH electrode. (Courtesy of Orion Research, Boston, MA.)

$$E_b = E_1 - E_2 = \frac{0.0592}{2} \log \frac{a_1}{a_2} \qquad (19\text{-}14)$$

where a_1 and a_2 are the activities of calcium ion in the external and internal solutions, respectively. Since the calcium ion activity of the internal solution is constant,

$$E_b = N + \frac{0.0592}{2} \log a_1 = N - \frac{0.0592}{2} \text{pCa} \qquad (19\text{-}15)$$

where N is a constant (compare Equations 19-15 and 19-10). Because calcium is divalent, note that a 2 appears in the denominator of the coefficient of the log term.

Figure 19-14 compares the structural features of a glass-membrane electrode and a commercially available liquid-membrane electrode for calcium ion. The sensitivity of the liquid-membrane electrode for calcium ion is reported to be 50 times greater than for magnesium ion and 1000 times greater than for sodium or potassium ions. Calcium ion activities as low as 5×10^{-7} M can be measured. Performance of the electrode is independent of pH in the range of 5.5 to 11. At lower pH levels, hydrogen ions undoubtedly replace some of the calcium ions on the exchanger; the electrode then becomes sensitive to pH as well as to pCa.

The calcium ion liquid-membrane electrode is a valuable tool for physiological investigations because this ion plays important roles in such processes as nerve conduction, bone formation, muscle contraction, cardiac expansion and contraction, renal tubular function, and perhaps hypertension. Most of these processes are more influenced by the activity than the concentration of the calcium ion; activity, of course, is the parameter measured by the membrane electrode. Thus, the calcium ion electrode (and the potassium ion electrode and others) is an important tool in studying physiological processes.

A liquid-membrane electrode specific for potassium ion is also of great value for physiologists because the transport of neural signals appears to involve movement of this ion across nerve membranes. Investigation of this process requires an electrode that can detect small concentrations of potassium ion in media that contain much larger concentrations of sodium ion. Several liquid-membrane electrodes show promise in meeting this requirement. One is based on the antibiotic valinomycin, a cyclic ether that has a strong affinity for potassium ion.

Ion-selective microelectrodes can be used to make measurements of ion activities within a living organism.

Table 19-2

Characteristics of Liquid-Membrane Electrodes

Analyte Ion	Concentration Range, M	Major Interferences
Ca^{2+}	10^0 to 5×10^{-7}	Pb^{2+}, Fe^{2+}, Ni^{2+}, Hg^{2+}, Sr^{2+}
Cl^-	10^0 to 5×10^{-6}	I^-, OH^-, SO_4^{2-}
NO_3^-	10^0 to 7×10^{-6}	ClO_4^-, I^-, ClO_3^-, CN^-, Br^-
ClO_4^-	10^0 to 7×10^{-6}	I^-, ClO_3^-, CN^-, Br^-
K^+	10^0 to 1×10^{-6}	Cs^+, NH_4^+, Tl^+
Water hardness (Ca^{2+} + Mg^{2+})	10^0 to 6×10^{-6}	Cu^{2+}, Zn^{2+}, Ni^{2+}, Fe^{2+}, Sr^{2+}, Ba^{2+}

Source: *Orion Guide to Ion Analysis* (Boston, MA: Orion Research, 1992). With permission.

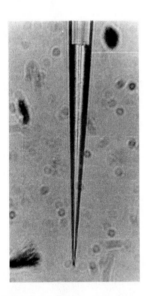

Figure 19-15 Photograph of a potassium liquid-ion exchanger microelectrode with 125 μm of ion exchanger inside the tip. The magnification of the original photo was 400×. (From J. L. Walker, *Anal. Chem.,* **1971,** *43* (3), 91A. Reproduced by permission of the American Chemical Society.)

Of equal importance is the observation that a liquid membrane consisting of valinomycin in diphenyl ether is about 10^4 times as responsive to potassium ion as it is to sodium ion.[4] Figure 19-15 is a photomicrograph of a tiny electrode used for determining the potassium content of a single cell.

Table 19-2 lists some liquid-membrane electrodes available from commercial sources. The anion-sensitive electrodes shown make use of a solution containing an anion-exchange resin in an organic solvent. Liquid-membrane electrodes in which the exchange liquid is held in a polyvinyl chloride gel have been developed for Ca^{2+}, K^+, NO_3^-, and BF_4^-. These have the appearance of crystalline electrodes, which are considered in the following section. A homemade liquid-membrane ion-selective electrode is described in Feature 19-1.

19D-6   Crystalline-Membrane Electrodes

Considerable work has been devoted to the development of solid membranes that are selective toward anions in the same way that some glasses respond to cations. We have seen that anionic sites on a glass surface account for the selectivity of a membrane toward certain cations. By analogy, a membrane with cationic sites might be expected to respond selectively toward anions.

Membranes prepared from cast pellets of silver halides have been used successfully in electrodes for the selective determination of chloride, bromide, and iodide ions. In addition, an electrode based on a polycrystalline Ag_2S membrane is offered by one manufacturer for the determination of sulfide ion. In both types of membranes, silver ions are sufficiently mobile to conduct electricity through the solid medium. Mixtures of PbS, CdS, and CuS with Ag_2S provide membranes that are selective for Pb^{2+}, Cd^{2+}, and Cu^{2+}, respectively. Silver ion must be present in these membranes to conduct electricity because divalent ions are immobile in crystals. The potential that develops across crystalline solid state electrodes is described by a relationship similar to Equation 19-10.

A crystalline electrode for fluoride ion is available from commercial sources. The membrane consists of a slice of a single crystal of lanthanum fluoride that has been doped with europium(II) fluoride to improve its conductivity. The

---

[4]M. S. Frant and J. W. Ross, Jr., *Science,* **1970,** *167,* 987.

You can make a liquid-membrane ion-selective electrode with glassware and chemicals available in most laboratories.[5] All you need are a pH meter, a pair of reference electrodes, a fritted-glass filter crucible or tube, trimethylchlorosilane, and a liquid ion exchanger.

First, cut the filter crucible (or alternatively, a fritted tube) as shown in Figure 19F-1. Carefully clean and dry the crucible, and then draw a small amount of trimethylchlorosilane into the frit. This coating makes the glass in the frit hydrophobic. Rinse the frit with water, dry it, and apply a commercial liquid-ion exchanger to it. After a minute, remove the excess exchanger. Add a few milliliters of a 10^{-2} M solution of the ion of interest to the crucible, insert a reference electrode into the solution, and voilá, you have a very nice ion-selective electrode. The exact details of washing, drying, and preparing the electrode are provided in the original article.

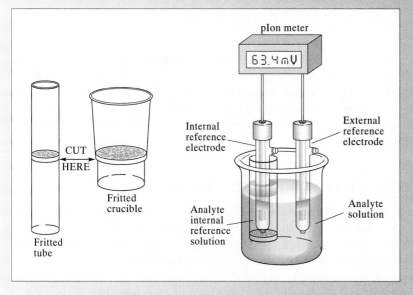

Figure 19F-1   A homemade liquid-membrane electrode.

Connect the ion-selective electrode and the second reference electrode to the pH meter as shown in Figure 19F-1. Prepare a series of standard solutions of the ion of interest, measure the cell potential for each concentration, plot a working curve of E_{cell} versus log c, and perform a least-squares analysis on the data (Section 7D). Compare the slope of the line with the theoretical slope of $(0.0592 \text{ V})/n$. Measure the potential for an unknown solution of the ion and calculate the concentration from the least-squares parameters.

---

[5]See T. K. Christopoulus and E. P. Diamandis, *J. Chem. Educ.,* **1988,** *65,* 648.

Table 19-3

Characteristics of Solid State Crystalline Electrodes

Analyte Ion	Concentration Range, M	Major Interferences
Br^-	10^0 to 5×10^{-6}	CN^-, I^-, S^{2-}
Cd^{2+}	10^{-1} to 1×10^{-7}	$Fe^{2+}, Pb^{2+}, Hg^{2+}, Ag^+, Cu^{2+}$
Cl^-	10^0 to 5×10^{-5}	$CN^-, I^-, Br^-, S^{2-}, OH^-, NH_3$
Cu^{2+}	10^{-1} to 1×10^{-8}	Hg^{2+}, Ag^+, Cd^{2+}
CN^-	10^{-2} to 1×10^{-6}	S^{2-}, I^-
F^-	Sat'd to 1×10^{-6}	OH^-
I^-	10^0 to 5×10^{-8}	CN^-
Pb^{2+}	10^{-1} to 1×10^{-6}	Hg^{2+}, Ag^+, Cu^{2+}
Ag^+/S^{2-}	Ag^+: 10^0 to 1×10^{-7} S^{2-}: 10^0 to 1×10^{-7}	Hg^{2+}
SCN^-	10^0 to 5×10^{-6}	I^-, Br^-, CN^-, S^{2-}

Source: *Orion Guide to Ion Analysis* (Boston, MA: Orion Research, 1992). With permission.

membrane, supported between a reference solution and the solution to be measured, shows a theoretical response to changes in fluoride ion activity from 10^0 to 10^{-6} M. The electrode is selective for fluoride ion over other common anions by several orders of magnitude; only hydroxide ion appears to offer serious interference. Some solid state electrodes available from commercial sources are listed in Table 19-3.

19D-7   Ion-Sensitive Field Effect Transistors (ISFETs)

The field effect transistor, or the metal oxide field effect transistor (MOSFET), is a tiny solid state semiconductor device widely used in computers and other electronic circuits as a switch to control current. One problem in employing this type of device in electronic circuits has been its pronounced sensitivity to ionic surface impurities, and a great deal of money and effort has been expended by the electronic industry in minimizing or eliminating this sensitivity to produce stable transistors.

Beginning in 1970, a number of scientists have attempted to exploit the sensitivities of MOSFETs to surface ionic impurities for the selective potentiometric determination of various ions. These studies have led to the development of a number of different ion-selective field effect transistors called ISFETs. The theory of their selective ion sensitivity is well understood, but it is beyond the scope of this book.[6]

ISFETs offer a number of significant advantages over membrane electrodes, including ruggedness, small size, inertness toward harsh environments, rapid

The term *ISFETs* stands for ion-sensitive field effect transistors.

---

[6]For an explanation of the theory, see D. A. Skoog, D. M. West, and F. J. Holler, *Fundamentals of Analytical Chemistry*, 7th ed. (Philadelphia, Saunders College Publishing, 1996), pp. 406–408.

response, and low electrical impedance. In contrast to membrane electrodes, ISFETs do not require hydration before use and can be stored indefinitely in the dry states. Despite these many advantages, ISFETs did not appear on the market until the early 1990s, over 20 years after their invention. The reason for this delay is that manufacturers were unable to develop the technology of encapsulating the devices to give a product that did not exhibit drift and instability. Several companies now produce ISFETs for the determination of pH, but they are certainly not as routinely used as the glass pH electrode at this time.

19D-8   Gas-Sensing Probes

Figure 19-16 illustrates the essential features of a potentiometric gas-sensing probe, which consists of a tube containing a reference electrode, a specific ion electrode, and an electrolyte solution. A thin, replaceable, gas-permeable membrane attached to one end of the tube serves as a barrier between the internal and analyte solutions. As can be seen from Figure 19-16, this device is a complete electrochemical cell and is more properly referred to as a probe rather than an electrode, a term that is frequently encountered in advertisements by instrument manufacturers. Gas-sensing probes have found widespread use for determining dissolved gases in water and other solvents.

A *gas-sensing probe* is a galvanic *cell* whose potential is related to the concentration of a gas in a solution. Often, in instrument brochures these devices are called gas-sensing electrodes, which is a misnomer.

Membrane Composition

A *microporous membrane* is fabricated from a hydrophobic polymer. As the name implies, the membrane is highly porous (the average pore size is less than 1 μm) and allows the free passage of gases; at the same time, the water-repellent polymer prevents water and solute ions from entering the pores. The thickness of the membrane is about 0.1 mm.

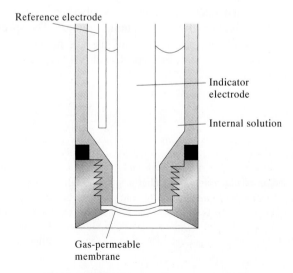

Reference electrode

Indicator electrode

Internal solution

Gas-permeable membrane

Figure 19-16   Diagram of a gas-sensing probe.

The Mechanism of Response

Using carbon dioxide as an example, we can represent the transfer of gas to the internal solution in Figure 19-16 by the following set of equations:

$$CO_2(aq) \rightleftharpoons CO_2(g)$$

analyte   membrane
solution   pores

$$CO_2(g) \rightleftharpoons CO_2(aq)$$

membrane   internal
pores   solution

$$CO_2(aq) + 2H_2O \rightleftharpoons HCO_3^- + H_3O^+$$

internal   internal
solution   solution

The last equilibrium causes the pH of the internal surface film to change. This change is then detected by the internal glass/calomel electrode system. A description of the overall process is obtained by adding the equations for the three equilibria to give

$$CO_2(aq) + 2H_2O \rightleftharpoons HCO_3^- + H_3O^+$$

analyte   internal
solution   solution

It can be shown that the potential of the internal cell is given by

$$E_{cell} = L + 0.0592 \log [CO_2(aq)]_{ext} \qquad (19\text{-}16)$$

where L is a constant. Thus, the potential between the glass electrode and the reference electrode in the internal solution is determined by the CO_2 concentration in the external solution. Note that no electrode comes in direct contact with the analyte solution. Therefore, these devices are gas-sensing cells, or probes, rather than gas-sensing electrodes. Nevertheless, they continue to be called electrodes in some literature and many advertising brochures.

The only species that interfere are other dissolved gases that permeate the membrane and then affect the pH of the internal solution. The specificity of gas probes depends only on the permeability of the gas membrane. Gas-sensing probes for CO_2, NO_2, H_2S, SO_2, HF, HCN, and NH_3 are now available from commercial sources.

Although sold as gas-sensing electrodes, these devices are complete electrochemical cells and should be called *gas-sensing probes.*

Feature 19-2

Point-of-Care Testing: Blood Gases and Blood Electrolytes with Portable Instrumentation

Modern medicine relies heavily on analytical measurements for diagnosis and treatment in emergency rooms, operating rooms, and intensive care units. Prompt reporting of blood gas and blood electrolyte concentrations is especially important to physicians in these areas. There is seldom sufficient time to transport blood samples to the clinical laboratory, perform required analyses, and transmit the results back to the bedside. In this feature, we describe an automated blood gas and electrolyte monitor, designed specifically to be attached to a patient.[7] The monitor setup is similar to a conventional

[7]VIA Medical Corporation, 11425 Sorrento Valley Rd., San Diego, CA 92121.

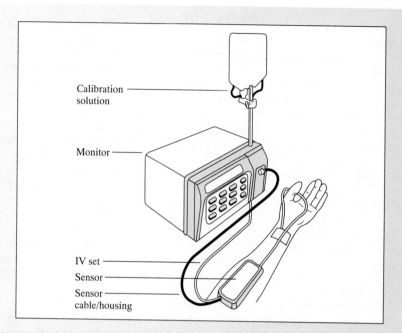

Figure 19F-2A   Layout of patient-attached blood gas and electrolyte monitor (Courtesy of VIA Medical Corp., San Diego, CA.)

intravenous (IV) infusion pump except that an in-line sensor array is included and located near the entry point of the arterial line, as illustrated in Figure 19F-2A. The microcomputer-based monitor determines oxygen, carbon dioxide, sodium, potassium, and calcium concentrations as well as pH and hematocrit in whole blood. The analytes, measurement ranges, and measurement resolutions of the instrument are as follows:

The hematocrit is the ratio of the volume of red blood cells to the total volume of a blood sample expressed as a percent.

Analyte	Range	Resolution
pO_2	20–699 mm Hg	1 mm Hg
pCO_2	10–99 mm Hg	1 mm Hg
Na^+	80–190 mM	1 mM
K^+	0.2–9.9 mM	0.1 mM
	10–20 mM	1 mM
Ca^{2+}	0.2–4.25 mM	0.01 mM
pH	6.80–7.70	0.01
Hematocrit	12–70%	1%

Most of the measured quantities (pCO_2, Na^+, K^+, Ca^{2+}, and pH) are determined by potentiometry using membrane-based ion-selective electrode technology. The hematocrit is measured by electrolytic conductivity detection and pO_2 is determined with a Clark voltammetric sensor (see Section 20-E2).

The central component of the monitor is the disposable electrochemical sensor array depicted in Figure 19F-2B. The individual sensors are located along a narrow flow channel. Materials for the array were selected to be sterile, nontoxic, blood compatible, and robust enough to be used for hundreds of measurements over a 3 to 4 day period of time.

Each new sensor array is calibrated prior to attachment by a two-point calibration procedure that requires about 15 min. The calibration procedure establishes the slopes and intercepts of the calibration equations for each sensor. After the sensor array is attached to the patient, an infusible, isotonic calibration solution containing known concentrations of the analytes is pumped continuously at a slow rate ($\approx$ 5 mL/hr) over the sensors. Every 10 min or so, a one-point calibration is performed on the isotonic solution to compensate for drift in the response of the sensor array.

Blood measurements can be initiated automatically at a preset time interval, or on demand, by pressing a key. The measurement cycle consists of the following steps: (1) the monitor's pump reverses direction and withdraws a 1.5-mL blood sample across the sensors; (2) the levels of blood gases, electrolytes, and hematocrit are determined at 37°C; (3) the results are displayed while the blood sample is pumped back to the patient; and (4) the sensor array is thoroughly cleaned by the infusible calibration solution to maintain the electrodes in a friendly environment. The entire measurement cycle takes less than 90 s and may be repeated as frequently as every 10 min. Accuracy and precision of the instrument on whole blood samples are comparable to dedicated instruments in the clinical laboratory. The monitor virtually eliminates sampling and sample handling errors. It minimizes risks of infection and involves no loss of blood whatsoever.

This feature shows how modern ion-selective electrode technology coupled with microcomputer control of the measurement process and data reporting can be used to provide essential measurements of analyte concentrations in whole blood on a nearly real-time basis at the patient's bedside.

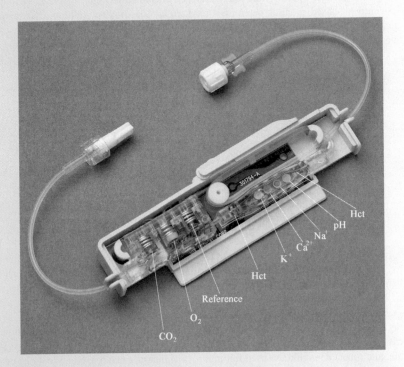

Figure 19F-2B   Miniature sensor array. (Courtesy of VIA Medical Corp., San Diego, CA.)

19E   INSTRUMENTS FOR MEASURING CELL POTENTIAL

Most cells containing a membrane electrode have very high electrical resistance (as much as 10^8 ohms or more). To measure potentials of such high-resistance circuits accurately, it is necessary that the voltmeter have an electrical resistance that is several orders of magnitude greater than the resistance of the cell being measured. If the meter resistance is too low, current is drawn from the cell, which has the effect of lowering its output potential, thus creating a negative *loading error.* When the meter and the cell have the same resistance, a relative error of − 50% results. When this ratio is 10, the error is about − 9%. When it is 1000, the error is less than 0.1% relative.

Numerous high-resistance, direct-reading digital voltmeters with internal resistances of more than 10^{11} ohms are now on the market. These meters are commonly called *pH meters,* but they could more properly be referred to as *pIon meters* or *ion meters* since they are frequently used for the measurement of concentrations of other ions as well.

Modern ion meters are digital; some are capable of a precision on the order of 0.001 to 0.005 pH unit. Seldom is it possible to measure pH with a comparable degree of *accuracy.* Inaccuracies of ± 0.02 to ± 0.03 pH unit are typical.

19F   DIRECT POTENTIOMETRY

Direct potentiometric measurements provide a rapid and convenient method for determining the activity of a variety of cations and anions. The technique requires only a comparison of the potential developed in a cell containing the indicator electrode in the analyte solution with its potential when immersed in one or more standard solutions of known analyte concentration. If the response of the electrode is specific for the analyte, as it often is, no preliminary separation steps are required. Direct potentiometric measurements are also readily adapted to applications requiring continuous and automatic recording of analytical data.

19F-1   Equations Governing Direct Potentiometry

The sign convention for potentiometry is consistent with the convention described in Chapter 16 for standard electrode potentials.[8] In this convention, the indicator electrode is always treated as the right-hand electrode and the reference electrode as the left-hand electrode. For direct potentiometric measurements, the potential of a cell can then be expressed in terms of the potentials developed by the indicator electrode, the reference electrode, and a junction potential, as described in Section 19A:

$$E_{cell} = E_{ind} - E_{ref} + E_j \qquad (19\text{-}17)$$

---

[8]According to Bates, the convention being described here has been endorsed by standardizing groups in the United States and Great Britain as well as IUPAC. See R. G. Bates, in *Treatise on Analytical Chemistry,* 2nd ed., I. M. Kolthoff and P. J. Elving, Eds., Part I, Vol. 1 (New York: Wiley, 1978), pp. 831–832.

In Section 19D, we described the response of various types of indicator electrodes to analyte activities. For the cation X^{n+} at 25°C, the electrode response takes the general *Nernstian* form

$$E_{ind} = L - \frac{0.0592}{n} pX = L + \frac{0.0592}{n} \log a_X \qquad (19\text{-}18)$$

where L is a constant and a_X is the activity of the cation. For metallic indicator electrodes, L is ordinarily the standard electrode potential; for membrane electrodes, L is the summation of several constants, including the time-dependent asymmetry potential of uncertain magnitude.

Substitution of Equation 19-18 into Equation 19-17 and rearrangement yields

$$pX = -\log a_X = -\frac{E_{cell} - (E_j - E_{ref} + L)}{0.0592/n} \qquad (19\text{-}19)$$

The constant terms in parentheses can be combined to give a new constant K.

$$pX = -\log a_X = -\frac{(E_{cell} - K)}{0.0592/n} = -\frac{n(E_{cell} - K)}{0.0592} \qquad (19\text{-}20)$$

For an anion A^{n-}, the sign of Equation 19-20 is reversed:

$$pA = \frac{(E_{cell} - K)}{0.0592/n} = \frac{n(E_{cell} - K)}{0.0592} \qquad (19\text{-}21)$$

All direct potentiometric methods are based on Equation 19-20 or 19-21. The difference in sign in the two equations has a subtle but important consequence in the way that ion-selective electrodes are connected to pH meters and pIon meters. When the two equations are solved for E_{cell}, we find that for cations,

$$E_{cell} = K - \frac{0.0592}{n} pX \qquad (19\text{-}22)$$

and for anions,

$$E_{cell} = K + \frac{0.0592}{n} pA \qquad (19\text{-}23)$$

Equation 19-22 shows that for a cation-selective electrode, an increase in pX results in a *decrease* in E_{cell}. Thus, when a high-resistance voltmeter is connected to the cell in the usual way, with the indicator electrode attached to the positive terminal, the meter reading decreases as pX increases. To eliminate this problem, instrument manufacturers generally reverse the leads so that cation-sensitive electrodes are connected to the negative terminal of the voltage-measuring device. Meter readings then increase with increases of pX. Anion-selective electrodes, on the other hand, are connected to the positive terminal of the meter so that increases in pA also yield larger readings.

19F-2   Applying the Electrode-Calibration Method

As we have seen from our discussions in Section 16D, the constant K in Equations 19-20 and 19-21 is made up of several constants at least one of which, the junction potential, cannot be computed from theory nor measured directly. Thus, before these equations can be used for the determination of pX or pA, K must be evaluated experimentally with a standard solution of the analyte.

In the electrode-calibration method, K in Equations 19-20 and 19-21 is determined by measuring E_{cell} for one or more standard solutions of known pX or pA. The assumption is then made that K is unchanged when the standard is replaced by the analyte solution. The calibration is ordinarily performed at the time pX or pA for the unknown is determined. With membrane electrodes, recalibration may be required if measurements extend over several hours because of slow changes in the asymmetry potential.

The electrode-calibration method offers the advantages of simplicity, speed, and applicability to the continuous monitoring of pX or pA. It suffers, however, from a somewhat limited accuracy because of uncertainties in junction potentials.

What Is the Inherent Error in the Electrode-Calibration Procedure?

A serious disadvantage of the electrode-calibration method is the inherent error that results from the assumption that K in Equations 19-20 and 19-21 remains constant after calibration. This assumption can seldom, if ever, be exactly true because the electrolyte composition of the unknown almost inevitably differs from that of the solution employed for calibration. The junction potential term contained in K varies slightly as a consequence, even when a salt bridge is used. This error is frequently on the order of 1 mV or more. Unfortunately, because of the nature of the potential/activity relationship, such an uncertainty has an amplified effect on the inherent accuracy of the analysis. The relative error in analyte concentration can be estimated from the following equation:[9]

$$\text{percent relative error} = \frac{\Delta a_1}{a_1} \times 100\% = 3.89 \times 10^3 n \, \Delta K\% \approx 4000 n K\%$$

The quantity $\Delta a_1/a_1$ is the relative error in a_1 associated with an absolute uncertainty ΔK in K. If, for example, ΔK is ± 0.001 V, a relative error in activity of about $\pm 4n\%$ can be expected. *This error is characteristic of all measurements involving cells that contain a salt bridge, and this error cannot be eliminated by even the most careful measurements of cell potentials or the most sensitive and precise measuring devices.*

Electrode Response: Activity or Concentration?

Electrode response is related to analyte activity rather than analyte concentration. We are usually interested in concentration, however, and the determination of this quantity from a potentiometric measurement requires activity coefficient data. Activity coefficients are seldom available because the ionic strength of the solution is either unknown or else is so large that the Debye–Hückel equation is not applicable.

The difference between activity and concentration is illustrated in Figure 19-17, in which the response of a calcium ion electrode is plotted against a logarithmic function of calcium chloride concentration. The nonlinearity is due to the increase in ionic strength — and the consequent decrease in the activity of calcium ion — with increasing electrolyte concentration. The upper curve is

---

[9]For a derivation of this equation, see D. A. Skoog, D. M. West, and F. J. Holler, *Fundamentals of Analytical Chemistry*, 7th ed. (Philadelphia: Saunders College Publishing, 1996), p. 417.

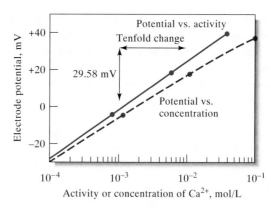

Figure 19-17 Response of a liquid-membrane electrode to variations in the concentration and activity of calcium ion. (Courtesy of Orion Research, Boston, MA.)

obtained when these concentrations are converted to activities. This straight line has the theoretical slope of 0.0296 (0.0592/2).

Activity coefficients for singly charged species are less affected by changes in ionic strength than are the coefficients for ions with multiple charges. Thus, the effect shown in Figure 19-17 is less pronounced for electrodes that respond to H^+, Na^+, and other univalent ions.

In potentiometric pH measurements, the pH of the standard buffer used for calibration is generally based on the activity of hydrogen ions. Thus, the results are also on an activity scale. If the unknown sample has a high ionic strength, the hydrogen ion *concentration* will differ appreciably from the activity measured.

> Many chemical reactions of physiological importance depend on the activity of metal ions rather than their concentration.

An obvious way to convert potentiometric measurements from activity to concentration is to make use of an empirical calibration curve, such as the lower plot in Figure 19-17. For this approach to be successful, it is necessary to make the ionic composition of the standards essentially the same as that of the analyte solution. Matching the ionic strength of standards to that of samples is often difficult, particularly for samples that are chemically complex.

> A *total ionic strength adjustment buffer (TISAB)* is used to control the ionic strength and the pH of samples and standards, in ion-selective electrode measurements.

Where electrolyte concentrations are not too great, it is often useful to swamp both samples and standards with a measured excess of an inert electrolyte. The added effect of the electrolyte from the sample matrix becomes negligible under these circumstances, and the empirical calibration curve yields results in terms of concentration. This approach has been used, for example, in the potentiometric determination of fluoride ion in drinking water. Both samples and standards are diluted with a solution that contains sodium chloride, an acetate buffer, and a citrate buffer; the diluent is sufficiently concentrated that the samples and standards have essentially identical ionic strengths. This method provides a rapid means for measuring fluoride concentrations in the part-per-million range with a relative accuracy of about 5%.

19F-3   Using the Standard-Addition Method

The standard-addition method involves determining the potential of the electrode system before and after one or more measured volumes of a standard have been added to a known volume of the analyte solution. Often, an excess of an electrolyte is incorporated into the analyte solution at the outset to prevent any major shift in ionic strength that might accompany the addition of standard. It is also necessary to assume that the junction potential remains constant during the two measurements. Although Example 19-1 shows the addition of a single volume of standard solution, a better practice is to use multiple additions. This provides a check of the calibration curve linearity. The multiple standard addition method is discussed in Section 23A-2.

Example 19-1

A cell consisting of a saturated calomel electrode and a lead ion electrode developed a potential of -0.4706 V when immersed in 50.00 mL of a sample. A 5.00-mL addition of standard 0.02000 M lead solution caused the potential to shift to -0.4490 V. Calculate the molar concentration of lead in the sample.

We shall assume that the activity of Pb^{2+} is approximately equal to $[Pb^{2+}]$ and apply Equation 19-20. Thus,

$$pPb = -\log [Pb^{2+}] = -\frac{E'_{cell} - K}{0.0592/2}$$

where E'_{cell} is the initial measured potential (-0.4706 V).

After the standard solution is added, the potential becomes E''_{cell} (-0.4490 V), and

$$-\log \frac{50.00 \times [Pb^{2+}] + 5.00 \times 0.02000}{50.00 + 5.00} = -\frac{E''_{cell} - K}{0.0592/2}$$

$$-\log(0.9091 [Pb^{2+}] + 1.818 \times 10^{-3}) = -\frac{E''_{cell} - K}{0.0592/2}$$

Subtracting this equation from the first leads to

$$-\log\frac{[Pb^{2+}]}{0.9091[Pb^{2+}] + 1.818 \times 10^{-3}} = \frac{2(E''_{cell} - E'_{cell})}{0.0592}$$

$$= \frac{2(-0.4490 - 0.4706)}{0.0592}$$

$$= 0.7297$$

$$\frac{[Pb^{2+}]}{0.9091[Pb^{2+}] + 1.818 \times 10^{-3}} = \text{antilog}(-0.7297) = 0.1863$$

$$[Pb^{2+}] = 4.08 \times 10^{-4} \text{ M}$$

19F-4   Making pH Measurements with the Glass Electrode[10]

The glass electrode is unquestionably the most important indicator electrode for hydrogen ion. It is convenient to use and is subject to few of the interferences that affect other pH-sensing electrodes.

The glass/calomel electrode system is a remarkably versatile tool for the measurement of pH under many conditions. It can be used without interference in solutions containing strong oxidants, strong reductants, proteins, and gases; the pH of viscous or even semisolid fluids can be determined. Electrodes for

---

[10]For a detailed discussion of potentiometric pH measurements, see R. G. Bates, *Determination of pH*, 2nd ed. (New York: Wiley, 1973).

special applications are available. Included among these are small electrodes for pH measurements in one drop (or less) of solution, in a tooth cavity, or in sweat on skin; microelectrodes that permit the measurement of pH inside a living cell; rugged electrodes for insertion in a flowing liquid stream to provide a continuous monitoring of pH; and small electrodes that can be swallowed to measure the acidity of the stomach contents (the calomel electrode is kept in the mouth).

Errors Affecting pH Measurements

The ubiquity of the pH meter and the general applicability of the glass electrode tend to lull the chemist into the attitude that any measurement obtained with such equipment is surely correct. There are, however, distinct limitations to the electrode, some of which were discussed in earlier sections:

1. *The alkaline error.* The ordinary glass electrode becomes somewhat sensitive to alkali metal ions and gives low readings at pH values greater than 9.
2. *The acid error.* Values registered by the glass electrode tend to be somewhat high when the pH is less than about 0.5.
3. *Dehydration.* Dehydration may cause erratic electrode performance.
4. *Errors in low-ionic-strength solutions.* It has been found that significant errors (as much as 1 or 2 pH units) may occur when the pH of samples of low ionic strength, such as water from a lake or stream, is measured with a glass/calomel electrode system.[11] The prime source of such errors has been shown to be non-reproducible junction potentials, that apparently result from partial clogging of the fritted plug or porous fiber used to restrict the flow of liquid from the salt bridge into the analyte solution. To overcome this problem, free diffusion junctions of various types have been designed, and one is produced commercially.
5. *Variation in junction potential.* A fundamental source of uncertainty for which a correction cannot be applied is the junction-potential variation resulting from differences in the composition of the standard and the unknown solution.
6. *Error in the pH of the standard buffer.* Any inaccuracies in the preparation of the buffer used for calibration or any changes in its composition during storage cause an error in subsequent pH measurements. The action of bacteria on organic buffer components is a common cause for deterioration.

> Particular care must be taken in measuring the pH of approximately neutral unbuffered solutions, such as samples from lakes and streams.

Giving an Operational Definition for pH

The utility of pH as a measure of the acidity and alkalinity of aqueous media, the wide availability of commercial glass electrodes, and the relatively recent proliferation of inexpensive pH meters have made the potentiometric measurement of pH perhaps the most common analytical technique in all science. It is thus extremely important that pH be defined in a manner that is easily duplicated at various times and in various laboratories throughout the world. To meet this requirement, it is necessary to define pH in operational terms, that is, by the way the measurement is made. Only then will the pH measured by one worker be the same as that by another.

> Perhaps the most common instrumental measurement is that of pH.

The operational definition of pH endorsed by the National Institute of Standards and Technology (NIST), similar organizations in other countries, and

---

[11]See W. Davison and C. Woof, *Anal Chem.,* **1985,** *57,* 2567; T. R. Harbinson and W. Davison, *Anal. Chem.,* **1987,** *59,* 2450; A. Kopelove, S. Franklin, and G. M. Miller, *Amer. Lab.,* **1989,** *21* (6), 40.

the IUPAC is based on the direct calibration of the meter with carefully prescribed standard buffers followed by potentiometric determination of the pH of unknown solutions. Consider, for example, the glass/calomel electrode pair of Figure 19-10. When these electrodes are immersed in a standard buffer, Equation 19-20 applies and we can write

$$pH_S = \frac{E_S - K}{0.0592}$$

where E_S is the cell potential when the electrodes are immersed in the buffer. Similarly, if the cell potential is E_U when the electrodes are immersed in a solution of unknown pH, we have

$$pH_U = \frac{E_U - K}{0.0592}$$

By subtracting the first equation from the second and solving for pH_U, we find

$$pH_U = pH_S - \frac{(E_U - E_S)}{0.0592} \tag{19-24}$$

Equation 19-24 has been adopted throughout the world as the *operational definition of pH*.

Workers at NIST and elsewhere have used cells without liquid junctions to study primary-standard buffers extensively. Some of the properties of these buffers are discussed in detail elsewhere.[12] Note that the NIST buffers are described by their molal concentrations (mol solute/kg solvent) for accuracy and precision of preparation. For general use, the buffers can be prepared from relatively inexpensive laboratory reagents; for careful work, however, certified buffers can be purchased from the NIST.

The strength of the operational definition of pH is that it provides a coherent scale for the determination of acidity or alkalinity. Measured pH values, however, cannot be expected to yield a detailed picture of solution composition that is entirely consistent with solution theory. This uncertainty stems from our fundamental inability to measure single ion activities. That is, the operational definition of pH does not yield the exact pH as defined by the equation

$$pH = -\log \gamma_{H^+}[H^+]$$

> By definition, pH is what you measure with a glass electrode and a pH meter. It is approximately equal to the theoretical definition of pH $= -\log a_{H^+}$.

> An operational definition of a quantity defines the quantity in terms of how it is measured.

19G FOLLOWING TITRATIONS BY POTENTIOMETRY

A *potentiometric titration* involves measurement of the potential of a suitable indicator electrode as a function of titrant volume. The information provided by a potentiometric titration is not the same as that obtained from a direct potentiometric measurement. For example, the direct measurement of 0.100 M solutions of hydrochloric and acetic acids would yield two substantially different hydrogen ion concentrations because the latter is only partially dissociated. In contrast, the potentiometric titration of equal volumes of the two acids would require the same amount of standard base because both solutes have the same number of titratable protons.

> Automatic *titrators* for potentiometric titrations are available from several manufacturers.

---

[12]R. G. Bates, *Determination of pH,* 2nd ed. (New York: Wiley, 1973), Chapter 4.

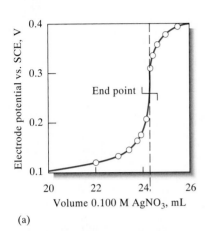

(a)

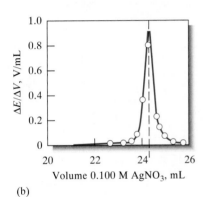

(b)

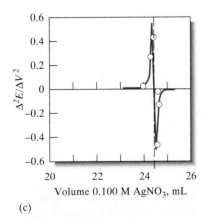

(c)

Figure 19-19   Titration of 2.433 mmol of chloride ion with 0.1000 M silver nitrate. (a) Titration curve. (b) First-derivative curve. (c) Second-derivative curve.

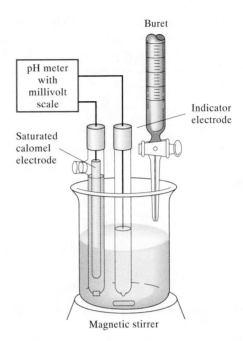

Figure 19-18   Apparatus for a potentiometric titration.

Potentiometric titrations provide data that are more reliable than data from titrations that use chemical indicators. Potentiometric titrations are particularly useful with colored or turbid solutions and for detecting the presence of unsuspected species. Such titrations are also readily automated. Manual titrations of this kind, on the other hand, suffer from the disadvantage of being more time consuming than those involving indicators.

Potentiometric titrations offer additional advantages over direct potentiometry. Because the measurement is based on the titrant volume that causes a rapid *change* in potential near the equivalence point, potentiometric titrations are not dependent on measuring absolute values of E_{cell}. Thus, the titration is relatively free from junction potential uncertainties because the junction potential remains approximately constant during the titration. Titration results instead depend most highly on having a titrant of accurately known concentration. The potentiometric instrument merely signals the end point and thus behaves in an identical fashion to a chemical indicator. Problems with electrodes fouling or not displaying Nernstian response are not nearly as serious when the electrode system is used to monitor a titration. Likewise, the reference electrode potential does not need to be known accurately in a potentiometric titration. Another advantage of a titration is that the results are obtained in terms of the analyte concentration even though the electrode responds to activity. Hence, ionic strength effects are unimportant in the titration procedure.

Figure 19-18 illustrates a typical apparatus for performing a manual potentiometric titration. Its use involves measuring and recording the cell potential (in units of millivolts or pH, as appropriate) after each addition of reagent. The titrant is added in large increments at the outset and in smaller and smaller increments as the end point is approached (as indicated by larger changes in response per unit volume as shown in Figure 19-19a).

19G-1   Detecting the End Point

Several methods can be used to determine the end point of a potentiometric titration. The most straightforward method involves a direct plot of potential as a function of reagent volume, as in Figure 19-19a; the inflection point in the steeply rising portion of the curve is estimated visually and taken as the end point.

A second approach to end-point detection is to calculate the change in potential per unit volume of titrant (that is, $\Delta E/\Delta V$). A plot of these data as a function of the average volume V produces a curve with a maximum that corresponds to the point of inflection (Figure 19-19b). Alternatively, this ratio can be evaluated during the titration and recorded in lieu of the potential. From the plot, it can be seen that the maximum is about 24.30 mL. If the titration curve is symmetrical, the point of maximum slope coincides with the equivalence point. For the asymmetrical titration curves that are observed when the titrant and analyte half-reactions involve different numbers of electrons, a small titration error occurs if the point of maximum slope is used.

Figure 19-19c shows that the second derivative for the data changes sign at the point of inflection. This change is used as the analytical signal in some automatic titrators. The point at which the second derivative crosses zero is the inflection point that is taken as the end point of the titration; this point can be located quite precisely, as discussed in Spreadsheet Example 19-2.

Example 19-2   *Spreadsheet Exercise*

Use the data in the table below to construct a spreadsheet showing the titration curve data, the first-derivative data, and the second-derivative data. Plot each of these results versus titrant volume and determine the end point of the titration.

Volume AgNO$_3$, mL	E vs. SCE, V
5.00	0.062
15.00	0.085
20.00	0.107
22.00	0.123
23.00	0.138
23.50	0.146
23.80	0.161
24.00	0.174
24.10	0.183
24.20	0.194
24.30	0.233
24.40	0.316
24.50	0.340
24.60	0.351
24.70	0.358
25.00	0.373
25.50	0.385
26.00	0.396
28.00	0.426

We will first enter the data into the spreadsheet and then construct the ordinary titration curve and the derivative plots. Enter the data from the table above into columns A and B of the spreadsheet shown in Figure 19-20. At this point the ordinary titration curve could be plotted as shown in Figure 19-21.

Now we will compute the first and second derivatives. We will first compute in column C, the average or midpoint volume to be associated with the calculated first derivatives. Enter in cell C5 the formula shown in documentation cell A24 which computes the average of the volumes in cells A4 and A5. Copy this formula into the remaining cells in column C. In column D we will compute the difference in potential, ΔE, that corresponds to this average volume. Enter into cell D5 the formula shown in documentation cell A25 and copy it into the remaining D cells. In column E we calculate the change in volume, ΔV, corresponding to this change in potential. Enter into cell E5 the formula shown in documentation cell A26 and copy the formula into the remaining cells in column E. The first derivative is then calculated in column F by dividing the ΔE values in column D by the ΔV values in column E as shown in documentation cell A27. This same formula can be copied into the rest of the cells in column F.

The first derivative plot can be constructed as described in Figure 19-22. Note that the spike at about 24.3 mL makes it easier to locate the end point than

	A	B	C	D	E	F	G	H	I	J
1	**Derivative End Points**									
2	**Data**		**First Derivative**				**Second Derivative**			
3	Volume AgNO$_3$, mL	E vs. SCE, Volts	Midpoint Vol.	ΔE	ΔVol	$\Delta E/\Delta V$	Midpoint Vol.	$\Delta(\Delta E/\Delta V)$	ΔVol	$\Delta^2 E/\Delta V^2$
4	5.00	0.062								
5	15.00	0.085	10.00	0.023	10.000	0.002				
6	20.00	0.107	17.50	0.022	5.000	0.004	13.75	0.002	7.500	0.000
7	22.00	0.123	21.00	0.016	2.000	0.008	19.25	0.004	3.500	0.001
8	23.00	0.138	22.50	0.015	1.000	0.015	21.75	0.007	1.500	0.005
9	23.50	0.146	23.25	0.008	0.500	0.016	22.88	0.001	0.750	0.001
10	23.80	0.161	23.65	0.015	0.300	0.050	23.45	0.034	0.400	0.085
11	24.00	0.174	23.90	0.013	0.200	0.065	23.78	0.015	0.250	0.060
12	24.10	0.183	24.05	0.009	0.100	0.090	23.98	0.025	0.150	0.167
13	24.20	0.194	24.15	0.011	0.100	0.110	24.10	0.020	0.100	0.200
14	24.30	0.233	24.25	0.039	0.100	0.390	24.20	0.280	0.100	2.800
15	24.40	0.316	24.35	0.083	0.100	0.830	24.30	0.440	0.100	4.400
16	24.50	0.340	24.45	0.024	0.100	0.240	24.40	-0.590	0.100	-5.900
17	24.60	0.351	24.55	0.011	0.100	0.110	24.50	-0.130	0.100	-1.300
18	24.70	0.358	24.65	0.007	0.100	0.070	24.60	-0.040	0.100	-0.400
19	25.00	0.373	24.85	0.015	0.300	0.050	24.75	-0.020	0.200	-0.100
20	25.50	0.385	25.25	0.012	0.500	0.024	25.05	-0.026	0.400	-0.065
21	26.00	0.396	25.75	0.011	0.500	0.022	25.50	-0.002	0.500	-0.004
22	28.00	0.426	27.00	0.030	2.000	0.015	26.38	-0.007	1.250	-0.006
23	**Documentation**									
24	Cell C5=(A5+A4)/2		Cell G6=(C6+C5)/2							
25	Cell D5=B5-B4		Cell H6=F6-F5							
26	Cell E5=A5-A4		Cell I6=C6-C5							
27	Cell F5=D5/E5		Cell J6=H6/I6							

Figure 19-20   Spreadsheet for constructing titration curves and derivative plots.

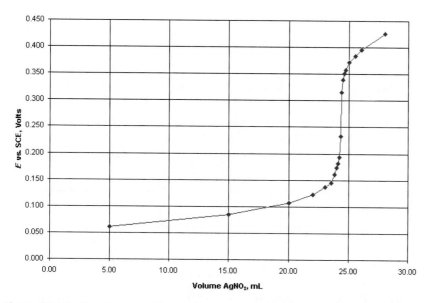

Figure 19-21   Plot of the normal titration curve. An XY (Scatter) plot is made of the electrode potential (column B in Figure 19-20) vs. titrant volume (column A in Figure 19-20).

with the normal titration curve. We could even expand this plot to do a more precise location. However, expansion of this plot shows that more data points would be needed in the region of 24.30 to 24.40 mL in order to locate the end point more precisely. Any small differences or uncertainties in the electrode potential will cause ΔE to change. This is one of the drawbacks of the derivative

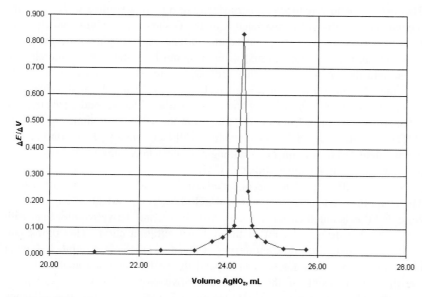

Figure 19-22   Plot of the first derivative vs. volume of titrant. This plot was made as an XY (Scatter) plot with the first derivative data in column F of Figure 19-20 plotted against the midpoint volume in column C.

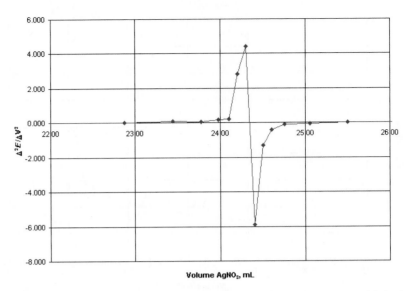

Figure 19-23   Second derivative plot. Here an XY (Scatter) plot is made of the second derivative data in column J of Figure 19-20 vs. the midpoint volume in column G.

method for locating the end point. Any noise in the data can cause an error because the rate of change of the noise can be higher than that of the signal. Hence, derivatives are said to amplify noise. Computer averaging and smoothing of the data can be used to reduce this problem and is often employed with automatic titrators.

Now we will compute the second derivative and enter the results in column J. The first step is to calculate another average or midpoint volume with which the second derivative is to be associated. Column G shows this volume which is obtained from the average midpoint volume in column C as shown by the documentation in cell C24 for the first value. Column H contains the numerator for the second derivative. These values are the change in the first derivative and computed by the formula in documentation cell C25 for the first value. The change in volumes to be used in the denominator of the second derivative are calculated in column I from the differences in the midpoint volumes of column C. The formula is shown in documentation cell C26. Finally, the second derivative is calculated in column J by dividing the values in column H by those in column I as shown in documentation cell C27.

Now we will prepare the second derivative plot as shown in Figure 19-23. The endpoint is taken as the point at which the second derivative crosses zero. Note that this is much easier to locate with the second derivative plot than with either the first derivative plot or the ordinary titration curve. This plot could again be expanded to allow more precise location of the end point, but again noise often limits how far we can expand the scale. Some automatic titrators use the zero crossing of the second derivative plot to locate the end point precisely. They often use noise averaging techniques to reduce the influence of noise.

WEB WORKS

Use a Web search engine to find sites dealing with potentiometric titrators. This search should turn up such companies as Metrohm, Mettler-Toledo, and Orion. Set your browser to one or two of these and explore the types of titrators that are available commercially.

19H   QUESTIONS AND PROBLEMS

19-1. Briefly describe or define
  *(a) indicator electrode.
  (b) reference electrode.
  *(c) electrode of the first kind.
  (d) electrode of the second kind.

19-2. Briefly describe or define
  *(a) liquid-junction potential.
  (b) boundary potential.
  (c) asymmetry potential.

19-3. Describe how a mercury electrode could function as
  (a) an electrode of the first kind for Hg(II).
  (b) an electrode of the second kind for EDTA.

19-4. What is meant by Nernstian behavior in an indicator electrode?

19-5. Describe the source of pH dependence in a glass membrane electrode.

19-6. Why is it necessary for the glass in the membrane of a pH-sensitive electrode to be appreciably hygroscopic?

19-7. List several sources of uncertainty in pH measurements with a glass/calomel electrode system.

19-8. What experimental factor places a limit on the number of significant figures in the response of a membrane electrode?

19-9. Describe the alkaline error in the measurement of pH. Under what circumstances is this error appreciable? How are pH data affected by alkaline error?

19-10. How does a gas-sensing probe differ from other membrane electrodes?

19-11. What is the source of
  (a) the asymmetry potential in a membrane electrode?
  *(b) the boundary potential in a membrane electrode?
  (c) a junction potential in a glass/calomel electrode system?
  *(d) the potential of a crystalline membrane electrode used to determine the concentration of F^-?

*19-12. How does information supplied by a direct potentiometric measurement of pH differ from that obtained from a potentiometric acid/base titration?

19-13. Give several advantages of a potentiometric titration over a direct potentiometric measurement.

19-14. What is the operational definition of pH? Why is it used?

*19-15. (a) Calculate E^0 for the process

$$AgIO_3(s) + e^- \rightleftharpoons Ag(s) + IO_3^-$$

  (b) Use the shorthand notation to describe a cell consisting of a saturated calomel reference electrode and a silver indicator electrode that could be used to measure pIO_3.
  (c) Develop an equation that relates the potential of the cell in part (b) to pIO_3.
  (d) Calculate pIO_3 if the cell in part (b) has a potential of 0.294 V.

19-16. (a) Calculate E^0 for the process

$$PbI_2(s) + e^- \rightleftharpoons Pb(s) + 2I^-$$

  (b) Use the shorthand notation to describe a cell consisting of a saturated calomel reference electrode and a lead indicator electrode that could be used to measure pI.
  (c) Develop an equation that relates the potential of the cell in part (b) to pI.
  (d) Calculate pI if the cell in part (b) has a potential of -0.348 V.

19-17. Use the shorthand notation to describe a cell consisting of a saturated calomel reference electrode and a silver indicator electrode for the measurement of
  *(a) pSCN.
  (b) pI.
  *(c) pSO_3.
  (d) pPO_4.

19-18. Generate an equation that relates pAnion to E_{cell} for each of the cells in Problem 19-17. (For Ag_2SO_3, $K_{sp} = 1.5 \times 10^{-14}$; for Ag_3PO_4, $K_{sp} = 1.3 \times 10^{-20}$.)

19-19. Calculate
*(a) pSCN if the cell in Problem 19-17, part (a), has a potential of 0.122 V.
(b) pI if the cell in Problem 19-17, part (b), has a potential of −211 mV.
*(c) pSO_3 if the cell in Problem 19-17, part (c), has a potential of 300 mV.
(d) pPO_4 if the cell in Problem 19-17, part (d), has a potential of 0.244 V.

*19-20. The cell

$$SCE \parallel Ag_2CrO_4(sat'd), CrO_4^{2-}(x\ M) \mid Ag$$

is employed for the determination of $pCrO_4$. Calculate $pCrO_4$ when the cell potential is 0.402 V.

*19-21. The cell

$$SCE \parallel H^+(a = x) \mid glass\ electrode$$

has a potential of 0.2094 V when the solution in the right-hand compartment is a buffer of pH 4.006. The following potentials are obtained when the buffer is replaced with unknowns. Calculate the pH and the hydrogen ion activity of each unknown.
(a) − 0.3011 V.
(b) + 0.1163 V.
(c) Assuming an uncertainty of 0.002 V in the junction potential, what is the range of hydrogen ion activities within which the true value might be expected to lie?

*19-22. A 0.3798-g sample of a purified organic acid was dissolved in water and titrated potentiometrically. A plot of the data revealed a single end point after 27.22 mL of 0.1025 M NaOH had been introduced. Calculate the molecular mass of the acid.

19-23. Calculate the potential of a silver indicator electrode versus the standard calomel electrode after the addition of 5.00, 15.00, 25.00, 30.00, 35.00, 39.00, 39.50, 39.60, 39.70, 39.80, 39.90, 39.95, 39.99, 40.00, 40.01, 40.05, 40.10, 40.20, 40.30, 40.40, 40.50, 41.00, 45.00, 50.00, 55.00 mL, and 70.00 of 0.1000 M $AgNO_3$ to 50.00 mL of 0.0800 M KSeCN. Construct a titration curve and a first- and second-derivative plot from these data. (K_{sp} for AgSeCN = 4.20×10^{-16}.)

19-24. A 40.00-mL aliquot of 0.05000 M HNO_2 is diluted to 75.00 mL and titrated with 0.0800 M

Ce^{4+}. The pH of the solution is maintained at 1.00 throughout the titration; the formal potential of the cerium system is 1.44 V.
*(a) Calculate the potential of the indicator electrode with respect to a saturated calomel reference electrode after the addition of 5.00, 10.00, 15.00, 25.00, 40.00, 49.00, 49.50, 49.60, 49.70, 49.80, 49.90, 49.95, 49.99, 50.00, 50.01, 50.05, 50.10, 50.20, 50.30, 50.40, 50.50, 51.00, 60.00, 75.00, and 90.00 mL of cerium (IV).
(b) Draw a titration curve for these data.
(c) Generate a first- and second-derivative curve for these data. Does the volume at which the second-derivative curve crosses zero correspond to the theoretical equivalence point? Why or why not?

19-25. The titration of Fe(II) with permanganate yields a particularly asymmetrical titration curve because of the different number of electrons involved in the two half-reactions. Consider the titration of 25.00 mL of 0.1 M Fe(II) with 0.1 M MnO_4^-. The H^+ concentration is maintained at 1.0 M throughout the titration. Use a spreadsheet to generate a theoretical titration curve and a first- and second-derivative plot. Do the inflection points obtained from the maximum of the first-derivative plot or the zero crossing of the second-derivative plot correspond to the equivalence point? Explain why or why not.

*19-26. The Na^+ concentration of a solution was determined by measurements with a sodium ion-selective electrode. The electrode system developed a potential of − 0.2331 V when immersed in 10.0 mL of a solution of unknown concentration. After addition of 1.00 mL of 2.00×10^{-2} M NaCl, the potential changed to − 0.1846 V. Calculate the Na^+ concentration of the original solution.

19-27. The F^- concentration of a solution was determined by measurements with a liquid-membrane electrode. The electrode system developed a potential of 0.4965 V when immersed in 25.00 mL of the sample, and 0.4117 V after the addition of 2.00 mL of 5.45×10^{-2} M NaF. Calculate pF for the sample.

19-28. A lithium ion-selective electrode gave the potentials shown below for four standard solutions of LiCl and 3 samples of unknown concentration.

Solution (a_{Li^+})	Potential vs. SCE, mV
0.100 M	+ 1.0
0.050 M	− 30.0
0.010 M	− 60.0
0.001 M	− 138.0
Unknown 1	− 48.5
Unknown 2	− 75.3

Solution Containing F⁻	Potential vs. SCE, mV
5.00×10^{-4} M	0.02
1.00×10^{-4} M	41.4
5.00×10^{-5} M	61.5
1.00×10^{-5} M	100.2
Unknown 1	38.9
Unknown 2	55.3

(a) Draw a calibration curve of electrode potential vs. $\log a_{Li^+}$ and determine if the electrode follows the Nernst equation.

*(b) Use a linear least squares procedure to determine the concentrations of the two unknowns.

19-29. A fluoride electrode was used to determine the amount of fluoride in drinking water samples. The results given in the table below were obtained for four standards and two unknowns. Constant ionic strength and pH conditions were used.

(a) Plot a calibration curve of potential vs. $\log[F^-]$. Determine whether the electrode system shows Nernstian response.

(b) Determine the concentration of F^- in the two unknown samples by a linear least-squares procedure.

Chapter 20

A Brief Look at Some Other Electroanalytical Methods

Electrogravimetry and coulometry often have relative errors of a few parts per thousand.

André Marie Ampère (1775–1836), French mathematician and physicist, was the first to apply mathematics to the study of electrical current. Consistent with Benjamin Franklin's definitions of positive and negative charge, Ampère defined a positive current to be the direction of flow of positive charge. Although we now know that negative electrons carry current in metals, Ampère's definition has survived to the present. The unit of current, the ampere, is named in his honor.

The potentiometric methods described in Chapter 19 involved potential measurements under conditions in which no current was present in the cell, and there was thus no net cell reaction. In contrast, the methods discussed in this chapter are electrolytic methods in which there is a net current and a net cell reaction.[1] Electrogravimetry and coulometry are two related methods based on an electrolysis that is carried out for a sufficient length of time to ensure complete oxidation or reduction of the analyte to a product of known composition. In electrogravimetry, the product is weighed as a deposit on one of the electrodes. In coulometric procedures, the quantity of electrical charge needed to complete the electrolysis is measured. On the other hand, voltammetric techniques measure the current in an electrochemical cell as a function of applied voltage and use a working microelectrode under conditions in which minimal consumption of the analyte occurs.

Electrogravimetry and coulometry are moderately sensitive and among the most accurate and precise techniques available to the chemist. Like gravimetry, these methods require no preliminary calibration against standards because the functional relationship between the quantity measured and the analyte concentration can be derived from theory and atomic-mass data.

Voltammetry is not only of analytical utility, but it is also widely used by inorganic, physical, and biochemists for fundamental studies of oxidation and reduction processes in various media, adsorption processes on surfaces, and electron transfer mechanisms at chemically modified electrode surfaces.

---

[1] For further information concerning the methods in this chapter, see J. A. Dean, *Analytical Chemistry Handbook* (New York: McGraw-Hill, 1995), Section 14, pp. 14.57–14.118.

20A   HOW DOES CURRENT AFFECT THE POTENTIAL OF ELECTROCHEMICAL CELLS?

When a current develops in an electrochemical cell, the measured potential across the two electrodes is no longer simply the difference between the two electrode potentials (the thermodynamic cell potential). Two additional phenomena, *IR drop* and *polarization,* require application of potentials greater than the thermodynamic potential to operate an electrolytic cell and result in the development of potentials smaller than theoretical in a galvanic cell. Before proceeding, we will examine these two phenomena in detail. To do so, we will employ as an example an electrolytic cell that has found application in determining such metal ions as copper(II), cadmium(II), and zinc(II) in hydrochloric acid solutions of the metals by electrogravimetry and coulometry. A typical cell for the determination of cadmium ion takes the following form:

$$Ag\,|\,AgCl(s),\ Cl^-(0.200\ M),\ Cd^{2+}(0.00500\ M)\,|\,Cd$$

Here, the right-hand electrode is a metal electrode that has been coated with a layer of cadmium. Because this is the electrode at which the reduction of Cd^{2+} ions occurs, it is called the *working electrode.* The left-hand electrode is a silver/silver chloride electrode whose electrode potential remains more or less constant during the analysis. The left-hand electrode is thus the *reference electrode.* Note that this is an example of a cell without liquid junction. As shown in Example 20-1, this cell, as written, has a thermodynamic potential of -0.734 V. Here, the negative sign for the cell potential indicates that the spontaneous reaction is *not* the reduction of Cd^{2+} on the right and the oxidation of Ag on the left. To reduce Cd^{2+} to Cd, we must construct an electrolytic cell and *apply* a voltage somewhat more negative than -0.734 V in order to make the cadmium electrode electron-rich enough to reduce Cd^{2+} to Cd. Such a cell is shown in Figure 20-1a. With this cell, we force the Cd electrode to become the cathode so that we cause the net reaction shown in Equation 20-1 to occur.

$$Cd^{2+} + 2Ag(s) + 2Cl^- \rightleftharpoons Cd(s) + AgCl(s) \qquad (20\text{-}1)$$

Note that the cell being considered is reversible so that in the absence of the external voltage source shown in the figure, the spontaneous cell reaction is the reverse of Equation 20-1. If the spontaneous reaction is allowed to occur by discharging the galvanic cell, the Cd electrode is the anode.

20A-1   Ohmic Potential; *IR* Drop

Electrochemical cells, like metallic conductors, resist the flow of charge. Ohm's law describes the effect of this resistance on the magnitude of the current in the cell. The product of the resistance R of a cell in ohms (Ω) and the current I in amperes (A) is called the *ohmic potential* or the *IR* drop of the cell. In Figure 20-1b, we have used a resistor R to represent the cell resistance in Figure 20-1a. To generate a current of I amperes in this cell, we must apply a potential that is *IR* volts more negative than the thermodynamic cell potential, $E_{cell} = E_{right} - E_{left}$. That is,

$$E_{applied} = E_{cell} - IR \qquad (20\text{-}2)$$

Ohm's law: $E = IR$ or $I = E/R$.

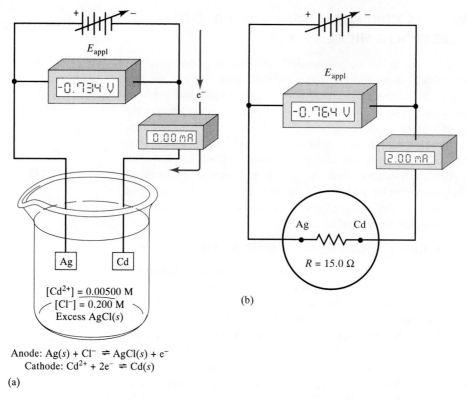

Anode: $Ag(s) + Cl^- \rightleftharpoons AgCl(s) + e^-$
Cathode: $Cd^{2+} + 2e^- \rightleftharpoons Cd(s)$

(a)

Figure 20-1   An electrolytic cell for determining Cd^{2+}. (a) Current $= 0.00$ mA. (b) Schematic of cell in (a) with the internal resistance of the cell represented by a 15.0-Ω resistor and $E_{applied}$ increased to give a current of 2.00 mA.

Usually, we try to minimize the IR drop in the cell by having a very small cell resistance (high ionic strength) or by using a special *three-electrode* cell (see Section 20C-2) in which the current passes between the working electrode and an *auxiliary* (or *counter*) electrode. With this arrangement, only a very small current passes between the working electrode and the reference electrode, which minimizes the IR drop.

Example 20-1

The following cell has been used for the determination of cadmium in the presence of chloride ions by both electrogravimetry and coulometry.

$$Ag \,|\, AgCl(s), Cl^-(0.200 \text{ M}), Cd^{2+}(0.00500 \text{ M}) \,|\, Cd$$

Calculate the potential that (a) must be applied to prevent a current from developing in the cell when the two electrodes are connected and (b) must be applied to cause an electrolytic current of 2.00 mA to develop. Assume that the internal resistance of the cell is 15.0 Ω.

(a) In Appendix 4, we find that

$$Cd^{2+} + 2e^- \rightleftharpoons Cd(s) \qquad E^0 = -0.403 \text{ V}$$

$$AgCl(s) + e^- \rightleftharpoons Ag(s) + Cl^- \qquad E^0 = 0.222 \text{ V}$$

The potential of the cadmium electrode is

$$E_{right} = -0.403 - \frac{0.0592}{2} \log \frac{1}{0.00500} = -0.471 \text{ V}$$

and that of the silver electrode is

$$E_{left} = 0.222 - 0.0592 \log (0.200) = 0.263 \text{ V}$$

Since the current is to be 0.00 mA, we find, from Equations 20-2 and 17-1 that

$$E_{applied} = E_{cell} = E_{right} - E_{left}$$
$$= -0.471 - 0.263 = -0.734 \text{ V}$$

Hence, to prevent the passage of current in this cell, we would need to apply a voltage of -0.734 V as shown in Figure 20-1a. Note that to obtain a current of 0.00 mA, the applied voltage must exactly match the galvanic cell potential. This match is the basis for a very precise null comparison measurement of the galvanic cell potential. We use a variable, standard voltage source as the applied voltage and adjust its output until a current of 0.00 mA is obtained as indicated by a very sensitive current meter. At this *null point*, the standard voltage is read on a voltmeter to obtain the value of E_{cell}. Since there is no current at the null point, this type of voltage measurement prevents the loading error mentioned in Section 19E.

(b) To calculate the applied potential needed to develop a current of 2.00 mA, or 2.00×10^{-3} A, we substitute into Equation 20-2 to obtain

$$E_{applied} = E_{cell} - IR$$
$$= -0.734 - 2.00 \times 10^{-3} \text{ A} \times 15 \text{ }\Omega$$
$$= -0.734 - 0.030 = -0.764 \text{ V}$$

Thus, to obtain a 2.00-mA current as shown in Figure 20-1b, an applied potential of -0.764 V would be required.

20A-2 Polarization Effects

According to Equation 20-2, a plot of current in an electrolytic cell as a function of applied potential should be a straight line with a slope equal to the negative reciprocal of the resistance. As shown in Figure 20-2, the plot is indeed linear with small currents. (In this experiment, the measurements were made in a brief enough time so that neither electrode potential is changed significantly as a consequence of the electrolytic reaction.) As the applied voltage increases, the current ultimately begins to deviate from linearity.

Cells that exhibit nonlinear behavior at higher currents are said to be *polarized,* and the degree of polarization is given by an *overvoltage,* or overpotential, symbolized by Π in Figure 20-2. Note that polarization requires the application of a potential greater than the theoretical value to give a current of the expected magnitude. Thus, the overpotential required to achieve a current of 7.00 mA in the electrolytic cell in Figure 20-2 is about -0.23 V. For an electrolytic cell affected by overvoltage, Equation 20-2 then becomes

Overvoltage is the potential difference between the theoretical cell potential from Equation 20-2 and the actual cell potential at a given level of current.

$$E_{applied} = E_{cell} - IR - \Pi \tag{20-3}$$

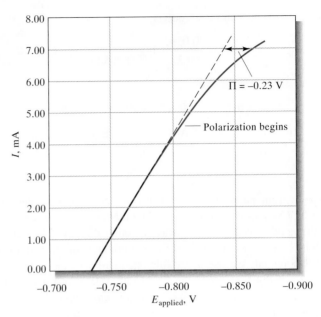

Figure 20-2   Experimental current–voltage curve for operation of the cell shown in Figure 20-1. Dashed line is the theoretical curve assuming no polarization. Overvoltage Π is the potential difference between the theoretical curve and the experimental.

Factors that influence polarization are (1) the electrode size, shape, and composition; (2) the composition of the electrolyte solution; (3) the temperature and stirring rate; (4) the current level; and (5) the physical state of species involved in the cell reaction.

Polarization is an electrode phenomenon that may affect either or both electrodes in a cell. The degree of polarization of an electrode varies widely. In some instances, it approaches zero, but in others, it can be so large that the current in the cell becomes independent of potential. Under this circumstance, polarization is said to be complete. Polarization phenomena are conveniently divided into two categories: *concentration polarization* and *kinetic polarization.*

Concentration Polarization

Mass transfer is the movement of material, such as ions, from one location to another.

Concentration polarization occurs because of the finite rate of mass transfer from the solution to the electrode surface. Electron transfer between a reactive species in a solution and an electrode can take place only from a thin film of solution located immediately adjacent to the surface of the electrode; this film is only a fraction of a nanometer in thickness and contains a limited number of reactive ions or molecules. For there to be a steady current in a cell, this film must be continuously replenished with reactant from the bulk of the solution. That is, as reactant ions or molecules are consumed by the electrochemical reaction, more must be transported into the surface film at a rate that is sufficient to maintain the current. For example, to have a current of 2.0 mA in the cell described in Figure 20-1b, it is necessary to transport cadmium ions to the cathode surface at a rate of about 1×10^{-8} mol/s or 6×10^{15} cadmium ions/s. (Similarly, silver ions must be removed from the surface film of the anode at a rate of 2×10^{-8} mol/s.)

Reactants can be transported to an electrode by (1) diffusion, (2) migration, and (3) convection.

Concentration polarization occurs when reactant species do not arrive at the surface of the electrode or product species do not leave the surface of the electrode fast enough to maintain the desired current. When this happens, the current is limited to values less than that predicted by Equation 20-2.

Reactants are transported to the surface of an electrode by three mechanisms: (1) *diffusion,* (2) *migration,* and (3) *convection.* Products are removed from electrode surfaces in the same ways. We will focus on mass-transport processes to the cathode, but our discussion applies equally to anodes.

Diffusion. When there is a concentration difference between two regions of a solution, ions or molecules move from the more concentrated region to the more dilute. This process, called diffusion, ultimately leads to a disappearance of the concentration difference. The rate of diffusion is directly proportional to the concentration difference. For example, when cadmium ions are deposited at a cathode by a current as illustrated in Figure 20-3, the concentration of Cd^{2+} at the electrode surface $[Cd^{2+}]_0$ becomes lower than the bulk concentration. The difference between the concentration at the surface and the concentration in the bulk solution $[Cd^{2+}]$ creates a concentration gradient that causes cadmium ions to diffuse from the bulk of the solution to the surface film. The rate of diffusion is given by

$$\text{rate of diffusion to cathode surface} = k'([Cd^{2+}] - [Cd^{2+}]_0) \quad (20\text{-}4)$$

where $[Cd^{2+}]$ is the reactant concentration in the bulk of the solution, $[Cd^{2+}]_0$ is its equilibrium concentration at the surface of the cathode, and k' is a proportionality constant. The value of $[Cd^{2+}]_0$ at any instant is fixed by the potential of the electrode and can be calculated from the Nernst equation. In the present example, we find the surface cadmium ion concentration from the relationship

$$E_{\text{cathode}} = E^0_{Cd^{2+}/Cd} - \frac{0.0592}{2} \log \frac{1}{[Cd^{2+}]_0}$$

where E_{cathode} is the potential applied to the cathode. As the applied potential becomes more and more negative, $[Cd^{2+}]_0$ becomes smaller and smaller. The result is that the rate of diffusion and the current become correspondingly larger until the surface concentration falls to zero, and the maximum or *limiting current* is reached.

Migration. The process by which ions move under the influence of an electric field is called migration. This process, shown schematically in Figure 20-4, is the primary cause of mass transfer in the bulk of the solution in a cell. The rate at which ions migrate to or away from an electrode surface generally increases as the electrode potential increases. This charge movement constitutes a current, which also increases with potential. Migration of analyte species is undesirable in most types of electrochemistry, and migration can be minimized by having a high concentration of an inert electrolyte, called a *supporting electrolyte,* present in the cell. The current in the cell is then primarily due to charges carried by ions from the supporting electrolyte. The supporting electrolyte also serves to reduce the resistance of the cell, which decreases the *IR* drop.

Convection. Reactants can also be transferred to or from an electrode by mechanical means. *Forced convection,* such as stirring or agitation, will tend to decrease the thickness of the diffusion layer at the surface of an electrode and thus decrease concentration polarization. *Natural convection* resulting from temperature or density differences also contributes to the transport of molecules to and from an electrode.

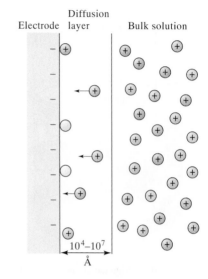

Diffusion layer

Electrode   Bulk solution

10^4–10^7
Å

–   Electrons

◯   Freshly deposited Cd atoms

⊕   Cd ions

Figure 20-3 Concentration changes at the surface of a cadmium cathode. As Cd^{2+} ions are reduced to Cd atoms at the electrode surface, the concentration of Cd^{2+} at the surface becomes smaller than the bulk concentration. Ions then diffuse from the bulk of the solution to the surface as a result of the concentration gradient. The higher the current, the larger the concentration gradient until the surface concentration falls to zero, its lowest possible value. At this point, the maximum possible current, called the *limiting current,* is obtained.

Diffusion is the movement of a species under the influence of a concentration gradient. It is the process that causes ions or molecules to move from a more concentrated part of a solution to a more dilute.

Migration involves the movement of ions through a solution as a result of electrostatic attraction between the ions and the electrodes.

Convection is the transport of ions or molecules through a solution as a result of stirring, vibration, or temperature gradients.

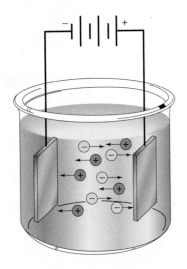

Figure 20-4 Migration is the movement of ions through a solution as a result of electrostatic attraction between the ions and the electrodes.

The experimental variables that influence the degree of concentration polarization are (1) reactant concentration, (2) total electrolyte concentration, (3) mechanical agitation, and (4) electrode size.

The current in a kinetically polarized cell is governed by the rate of electron transfer rather than the rate of mass transfer.

Current density is defined as amperes per square centimeter (A/cm^2) of electrode surface.

Kinetic polarization is most commonly encountered when the reactant or product in an electrochemical cell is a gas.

The Importance of Concentration Polarization.   As noted earlier, concentration polarization sets in when the effects of diffusion, migration, and convection are insufficient to transport a reactant to or from an electrode surface at a rate that produces a current of the magnitude given by Equation 20-2. Concentration polarization requires applied potentials that are larger than calculated from Equation 20-2 to maintain a given current in an electrolytic cell (Figure 20-2). Similarly, the phenomenon causes a galvanic cell potential when the cell is discharging, to be smaller than the value predicted on the basis of the theoretical potential and the *IR* drop.

Concentration polarization is important in several electroanalytical methods. In some applications, its effects are undesirable, and steps are taken to eliminate it; in others, it is essential to the analytical method, and every effort is made to promote its occurrence.

Kinetic Polarization

In kinetic polarization, the magnitude of the current is limited by the rate of one or both of the electrode reactions, that is, by the rate of electron transfer between the reactants and the electrodes. To offset kinetic polarization, an additional potential, or overvoltage, is required to overcome the activation energy of the half-reaction.

Kinetic polarization is most pronounced for electrode processes that yield gaseous products and is often negligible for reactions that involve the deposition or solution of a metal. Kinetic effects usually decrease with increasing temperature and decreasing current density. These effects also depend on the composition of the electrode and are most pronounced with softer metals, such as lead, zinc, and particularly mercury. The magnitude of overvoltage effects cannot be predicted from present theory and can only be estimated from empirical information in the literature.[2] In common with *IR* drop, overvoltage effects cause the application of voltages more negative than calculated to operate an electrolytic cell at a desired current. Kinetic polarization also causes the potential of a galvanic cell to be smaller than the value calculated from the Nernst equation and the *IR* drop (Equation 20-2).

The overvoltages associated with the formation of hydrogen and oxygen are often 1 V or more and are of considerable importance because these molecules are frequently produced by electrochemical reactions. Of particular interest is the high overvoltage of hydrogen on such metals as copper, zinc, lead, and mercury. These metals and several others can therefore be deposited without interference from hydrogen evolution. In theory, it is not possible to deposit zinc from a neutral aqueous solution because hydrogen forms at a potential that is considerably less than that required for zinc deposition. In fact, zinc can be deposited on a copper electrode with no significant hydrogen formation because the rate at which the gas forms on both zinc and copper is negligible, as shown by the high hydrogen overvoltage associated with these metals.

---

[2]Overvoltage data for various gaseous species on different electrode surfaces have been compiled in J. A. Dean, *Analytical Chemistry Handbook* (New York: McGraw-Hill, 1995), Section 14, pp. 14.96–14.97.

If it were not for the high overvoltage of hydrogen on lead and lead oxide electrodes, the lead-acid storage batteries found in automobiles and trucks would not operate because of hydrogen formation at the cathode both during charging and use. Certain trace metals in the system lower this overvoltage and eventually lead to gassing, or hydrogen formation, which limits the lifetime of the battery. The basic difference between a battery with a 48-month warranty and a 72-month warranty is the concentration of these trace metals in the system.

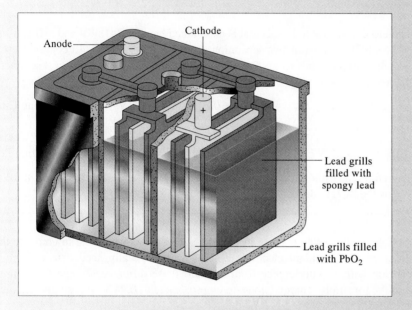

Figure 20F-1 The lead-acid storage battery.

20B   ARE ELECTROLYTIC METHODS SELECTIVE?

In principle, electrolytic methods offer a reasonably selective means for separating and determining a number of ions. The feasibility of and theoretical conditions for accomplishing a given separation can be readily derived from the standard electrode potentials of the species of interest as illustrated in Example 20-2.

Example 20-2

Is a quantitative separation of Cu^{2+} and Pb^{2+} by electrolytic deposition feasible in principle? If so, what range of cathode potentials (versus SCE) can be used? Assume that the sample solution is initially 0.1000 M in each ion and that quantitative removal of an ion is realized when only 1 part in 10,000 remains undeposited.

In Appendix 4, we find that

$$Cu^{2+} + 2e^- \rightleftharpoons Cu(s) \qquad E^0 = 0.337 \text{ V}$$

$$Pb^{2+} + 2e^- \rightleftharpoons Pb(s) \qquad E^0 = -0.126 \text{ V}$$

It is apparent that copper will begin to deposit before lead. Let us first calculate the potential required to decrease the Cu^{2+} concentration to 10^{-4} of its original concentration (that is, to 1.00×10^{-5} M). Substituting into the Nernst equation, we obtain

$$E = 0.337 - \frac{0.0592}{2} \log \frac{1}{1.00 \times 10^{-5}} = 0.189 \text{ V}$$

Similarly, we can derive the potential at which lead begins to deposit:

$$E = -0.126 - \frac{0.0592}{2} \log \frac{1}{0.1000} = -0.156 \text{ V}$$

Therefore, if the cathode potential is maintained between 0.189 V and -0.156 V (versus SHE), a quantitative separation should in theory occur. To convert these potentials to potentials relative to a saturated calomel electrode, we must subtract the reference electrode potential and write

$$E_{cell} = E_{cathode} - E_{SCE} = 0.189 - 0.244 = -0.055 \text{ V} \qquad \text{for depositing Cu}$$

and

$$E_{cell} = E_{cathode} - E_{SCE} = -0.156 - 0.244 = -0.400 \text{ V} \qquad \text{for depositing Pb}$$

Therefore, the cathode potential should be kept between -0.055 and -0.400 V versus the SCE to deposit Cu without depositing Pb.

Calculations such as those in Example 20-2 make it possible to compute the differences in standard electrode potentials theoretically needed to determine one ion without interference from another; these differences range from about 0.04 V for triply charged ions to approximately 0.24 V for singly charged species.

These theoretical separation limits can be approached only by maintaining the potential of the working electrode (usually the cathode, at which a metal deposits) at the level required. The potential of this electrode can be controlled only by variation of the potential applied to the cell, however. From Equation 20-3, it is evident that variations in $E_{applied}$ affect not only the cathode potential but also the anode potential, the *IR* drop, and the overpotential. As a consequence, the only practical way of achieving separation of species whose electrode potentials differ by a few tenths of a volt is to monitor the cathode potential continuously against a reference electrode whose potential is known; the applied cell potential can then be adjusted to maintain the cathode potential at the desired level. An analysis performed in this way is called a *controlled-potential electrolysis* or a *potentiostatic electrolysis*. Controlled-potential methods are discussed in Sections 20C-2 and 20D-4.

20C   ELECTROGRAVIMETRIC METHODS: WEIGHING THE DEPOSIT

Electrolytic precipitation has been used since the late nineteenth century for the gravimetric determination of metals. In most applications, the metal is deposited on a weighed platinum cathode, and the increase in mass is determined.

Important exceptions to this procedure include the anodic deposition of lead as lead dioxide on platinum and of chloride as silver chloride on silver.

There are two general types of electrogravimetric methods. In one, no control of the potential of the working electrode is exercised and the applied cell potential is held at a more or less constant level that provides a large enough current to complete the electrolysis in a reasonable length of time. The second type of electrogravimetric method is the controlled-potential or potentiostatic method.

> A *working electrode* is the electrode at which the analytical reaction occurs.

> A *potentiostatic method* is an electrolytic procedure in which the potential of the working electrode is maintained at a constant level versus a reference electrode, such as a SCE.

20C-1   Electrogravimetry without Potential Control

Electrolytic procedures in which no effort is made to control the potential of the working electrode make use of simple and inexpensive equipment and require little operator attention. In these procedures, the potential applied across the cell is maintained at a more or less constant level throughout the electrolysis.

Instrumentation

As shown in Figure 20-5, the apparatus for an analytical electrodeposition without cathode potential control consists of a suitable cell and a 6- to 12-V dc power supply. The voltage applied to the cell is controlled by the variable resistor, R. An ammeter and a voltmeter indicate the approximate current and applied voltage. In performing an electrolysis with this apparatus, the applied voltage is adjusted with potentiometer R to give a current of several tenths of an ampere. The voltage is then maintained at about the initial level until the deposition is judged to be complete.

Electrolysis Cells

Figure 20-5 shows a typical cell for the deposition of a metal on a solid electrode. Ordinarily, the working electrode is a platinum gauze cylinder 2 or 3 cm in diameter and perhaps 6 cm in length. Often, as shown, the anode takes the form of a solid platinum stirring paddle that is located inside and connected to the cathode through the external circuit.

Describing the Physical Properties of Electrolytic Precipitates

Ideally, an electrolytically deposited metal should be strongly adherent, dense, and smooth so that it can be washed, dried, and weighed without mechanical loss or reaction with the atmosphere. Good metallic deposits are fine grained and have a metallic luster; spongy, powdery, or flaky precipitates are likely to be less pure and less adherent than fine-grained deposits.

The principal factors that influence the physical characteristics of deposits are current density, temperature, and the presence of complexing agents. Ordinarily, the best deposits are formed at current densities that are less than $0.1 \ A/cm^2$. Stirring generally improves the quality of a deposit. The effects of temperature are unpredictable and must be determined empirically.

Many metals form smoother and more adherent films when deposited from solutions in which their ions exist primarily as complexes. Cyanide and ammonia complexes often provide the best deposits. The reasons for this effect are not obvious.

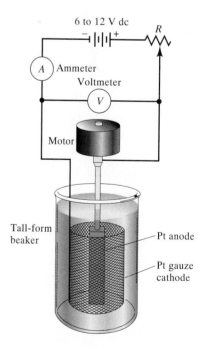

Figure 20-5 Apparatus for electrodeposition of metals without cathode-potential control.

Applying Electrogravimetric Methods

In practice, electrolysis at a constant cell potential is limited to the separation of easily reduced cations from those that are more difficult to reduce than hydrogen ion or nitrate ion. The reason for this limitation is illustrated in Figure 20-6, which shows the changes of current, IR drop, and cathode potential during an electrolysis in the cell in Figure 20-5. The analyte here is copper(II) ions in a solution containing an excess of acid. Initially, R is adjusted so that the potential applied to the cell is about -2.5 V, which, as shown in Figure 20-6a, leads to a current of about 1.5 A. The electrolytic deposition of copper is then completed at this applied potential.

As shown in Figure 20-6b, the IR drop decreases continually as the reaction proceeds, primarily because concentration polarization at the cathode limits the rate at which copper ions are brought to the electrode surface, which limits the current. From Equation 20-3, it is apparent that the decrease in the IR drop must be offset by an increase in the cathode potential since the applied cell potential is constant.

Ultimately, the decrease in current and the increase in cathode potential is slowed at point B by the reduction of hydrogen ions. Because the solution contains a large excess of acid, the current is now no longer limited by concentration polarization, and codeposition of copper and hydrogen occurs simultane-

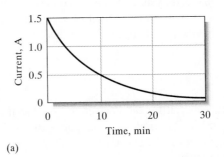

(a)

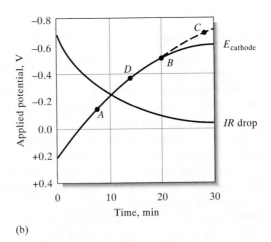

(b)

Figure 20-6   (a) Current; (b) IR drop and cathode potential change during electrolytic deposition of copper at a constant applied cell potential.

ously until the remainder of the copper ions is deposited. Under these conditions, the cathode is said to be *depolarized* by hydrogen ions.

Consider now the fate of some metal ion, such as lead(II), which begins to deposit at point *A* on the cathode potential curve. Clearly, it would codeposit well before copper deposition was complete, and an interference would result. In contrast, a metal ion, such as cobalt(II), which reacts at a cathode potential corresponding to point *C* on the curve, would not interfere because depolarization by hydrogen gas formation prevents the cathode from reaching this potential.

Codeposition of hydrogen during electrolysis often leads to formation of nonadherent deposits, which are unsatisfactory for analytical purposes. This problem can be resolved by introducing another species that is reduced at a less negative potential than hydrogen ion and that does not adversely affect the physical properties of the deposit. One such cathode *depolarizer* is nitrate ion.

Electrolytic methods performed without electrode-potential control, although limited by their lack of selectivity, do have several applications of practical importance. Table 20-1 lists the common analytes that are determined by this procedure.

> A *cathode depolarizer* is a chemical that is easily reduced (or oxidized) and stabilizes the potential of a working electrode by minimizing concentration polarization.

20C-2 Potentiostatic Gravimetry

In the discussion that follows, we shall assume that the working electrode is a cathode at which an analyte is deposited as a metal. The discussion may be extended to an anodic working electrode and nonmetallic deposits.

Instrumentation

To separate species with electrode potentials that differ by only a few tenths of a volt, it is necessary to use a more elaborate approach than the one just described. Such techniques are required because concentration polarization at the cathode, if unchecked, causes the potential of that electrode to become so negative that codeposition of the other species present begins before the analyte is completely deposited (Figure 20-6). A large negative drift in the cathode potential can be avoided by employing a three-electrode system, such as that shown in Figure 20-7.

The controlled-potential apparatus shown in Figure 20-7 is made up of two independent electrical circuits that share a common electrode, the working

Table 20-1

Typical Applications of Electrogravimetry without Potential Control

Analyte	Weighed As	Cathode	Anode	Conditions
Ag^+	Ag	Pt	Pt	Alkaline CN^- solution
Br^-	AgBr (on anode)	Pt	Ag	
Cd^{2+}	Cd	Cu on Pt	Pt	Alkaline CN^- solution
Cu^{2+}	Cu	Pt	Pt	H_2SO_4/HNO_3 solution
Mn^{2+}	MnO_2 (on anode)	Pt	Pt dish	HCOOH/HCOONa solution
Ni^{2+}	Ni	Cu on Pt	Pt	Ammoniacal solution
Pb^{2+}	PbO_2 (on anode)	Pt	Pt	Strong HNO_3 solution
Zn^{2+}	Zn	Cu on Pt	Pt	Acidic citrate solution

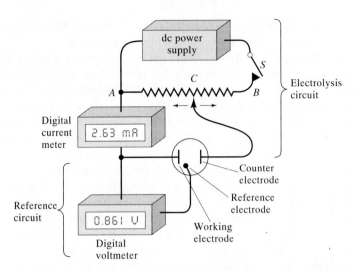

Figure 20-7 Apparatus for controlled-potential, or potentiostatic, electrolysis. Contact *C* is adjusted as necessary to maintain the working electrode (cathode in this example) at a constant potential. The current in the reference electrode is essentially zero at all times. Modern potentiostats are fully automatic and frequently computer controlled.

A *counter electrode* has no effect on the reaction at the working electrode. It simply serves to feed electrons to the working electrode.

A *potentiostat* is an instrument that maintains the working electrode potential at a constant value.

electrode at which the analyte is deposited. The electrolysis circuit consists of a dc source, a potentiometer (*ACB*) that permits continuous variation in the potential applied across the working electrode and a counter electrode, and a current meter. The control circuit is made up of a reference electrode (often a SCE), a high-resistance digital voltmeter, and the working electrode. The electrical resistance of the control circuit is so large that the electrolysis circuit supplies essentially all the current for the electrolysis. The purpose of the control circuit is to monitor continuously the potential between the working electrode and the reference electrode and to maintain it at a constant value.

The current and the cell potential changes that occur in a typical constant-cathode-potential electrolysis are depicted in Figure 20-8. Note that the applied cell potential has to be decreased continuously throughout the electrolysis. Manual adjustment of this potential is tedious (particularly at the outset) and, above all, time consuming. To avoid such waste of operator time, controlled-potential electrolyses are generally performed with automated instruments called *potentiostats*, which maintain a constant cathode potential electronically.

Applying Controlled-Potential Methods

The controlled-cathode-potential method is a potent tool for separating and determining metallic species having standard potentials that differ by only a few tenths of a volt. An example illustrating the power of the method involves determining copper, bismuth, lead, cadmium, zinc, and tin in mixtures by successive deposition of the metals on a weighed platinum cathode. The first three elements are deposited from a nearly neutral solution containing tartrate ion to complex the tin(IV) and prevent its deposition. Copper is first reduced quantitatively by

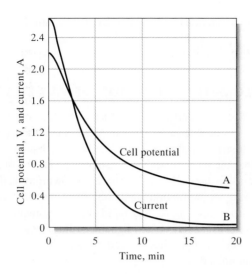

Figure 20-8   Changes in cell potential (A) and current (B) during a controlled-cathode-potential deposition of copper. The cathode is maintained at -0.36 V (versus SCE) throughout the experiment. (Data from J. J. Lingane, *Anal. Chim. Acta,* **1948,** *2,* 590.)

maintaining the cathode potential at -0.2 V with respect to a saturated calomel electrode. After being weighed, the copper-plated cathode is returned to the solution and bismuth is removed at a potential of -0.4 V. Lead is then deposited quantitatively by increasing the cathode potential to -0.6 V. When lead deposition is complete, the solution is made strongly ammoniacal and cadmium and zinc are deposited successively at -1.2 and -1.5 V. Finally, the solution is acidified to decompose the tin/tartrate complex by the formation of undissociated tartaric acid. Tin is then deposited at a cathode potential of -0.65 V. A fresh cathode must be used here because the zinc redissolves under these conditions. A procedure such as this is particularly attractive for use with a potentiostat because little operator time is required for the complete analysis. Table 20-2

Table 20-2

Some Applications of Controlled-Cathode-Potential Electrolysis

Element Determined	Other Elements That Can Be Present
Ag	Cu and heavy metals
Cu	Bi, Sb, Pb, Sn, Ni, Cd, Zn
Bi	Cu, Pb, Zn, Sb, Cd, Sn
Sb	Pb, Sn
Sn	Cd, Zn, Mn, Fe
Pb	Cd, Sn, Ni, Zn, Mn, Al, Fe
Cd	Zn
Ni	Zn, Al, Fe

lists some other separations performed by the controlled-cathode-potential method.

20D   COULOMETRIC METHODS OF ANALYSIS

Coulometric methods are performed by measuring the quantity of electrical charge (electrons) required to convert a sample of an analyte quantitatively to a different oxidation state. Coulometric and gravimetric methods share the common advantage that the proportionality constant between the quantity measured and the analyte mass is derived from accurately known physical constants, thus eliminating the need for calibration standards. In contrast to gravimetric methods, coulometric procedures are usually rapid and do not require that the product of the electrochemical reaction be a weighable solid. Coulometric methods are as accurate as conventional gravimetric and volumetric procedures; in addition, they are readily automated.[3]

20D-1   Determining the Quantity of Electrical Charge

1 coulomb = 1 ampere · second = 1 A · s

Units for the quantity of charge include the coulomb (C) and the faraday (F). *The coulomb is the quantity of electrical charge transported by a constant current of one ampere in one second.* Thus, the number of coulombs (Q) resulting from a constant current of I amperes operated for t seconds is

$$Q = It \tag{20-5}$$

For a variable current i, the number of coulombs is given by the integral

$$Q = \int_0^t i\,dt \tag{20-6}$$

A *faraday* of charge is equivalent to one mole of electrons or 6.022×10^{23} electrons.

The faraday is the quantity of charge that corresponds to one mole or 6.022×10^{23} electrons. The faraday also equals 96,485 C. As shown in Example 20-3, we can use these definitions to calculate the mass of a chemical species that is formed at an electrode by a current of known magnitude.

Example 20-3

A constant current of 0.800 A is used to deposit copper at the cathode and oxygen at the anode of an electrolytic cell. Calculate the number of grams of each product formed in 15.2 min, assuming no other redox reaction.

The two half-reactions are

$$Cu^{2+} + 2e^- \rightarrow Cu(s)$$

---

[3]For additional information about coulometric methods, see J. A. Dean, *Analytical Chemistry Handbook* (New York: McGraw-Hill, 1995), Section 14, pp. 14.118–14.133; D. J. Curran, in *Laboratory Techniques in Electroanalytical Chemistry,* P. T. Kissinger and W. R. Heineman, Eds., (New York: Marcel Dekker, 1984), pp. 539–568; J. A. Plambeck, *Electroanalytical Chemistry* (New York: Wiley, 1982), Chapter 12.

$$2H_2O \rightarrow 4e^- + O_2(g) + 4H^+$$

Thus 1 mol of copper is equivalent to 2 mol of electrons, and 1 mol of oxygen corresponds to 4 mol of electrons.

Substituting into Equation 20-5 yields

$$Q = 0.800 \text{ A} \times 15.2 \text{ min} \times 60 \text{ s/min} = 729.6 \text{ A} \cdot \text{s} = 729.6 \text{ C}$$

$$\text{no. F} = \frac{729.6 \text{ C}}{96,485 \text{ C/F}} = 7.56 \times 10^{-3} \text{ F} = 7.56 \times 10^{-3} \text{ mol of electrons}$$

The masses of Cu and O_2 are given by

$$\text{mass Cu} = 7.56 \times 10^{-3} \text{ mol e}^- \times \frac{1 \text{ mol Cu}}{2 \text{ mol e}^-} \times \frac{63.55 \text{ g Cu}}{1 \text{ mol Cu}} = 0.240 \text{ g Cu}$$

$$\text{mass O}_2 = 7.56 \times 10^{-3} \text{ mol e}^- \times \frac{1 \text{ mol O}_2}{4 \text{ mol e}^-} \times \frac{32.00 \text{ g O}_2}{1 \text{ mol O}_2} = 0.0605 \text{ g O}_2$$

20D-2   Characterizing Coulometric Methods

Two methods have been developed that are based on measuring the quantity of charge: *potentiostatic coulometry* and *amperostatic coulometry,* or *coulometric titrimetry.* Potentiostatic methods are performed in much the same way as controlled-potential gravimetric methods, with the potential of the working electrode being maintained at a constant level relative to a reference electrode throughout the electrolysis. Here, however, the electrolysis current is recorded as a function of time to give a curve similar to curve B in Figure 20-8. The analysis is then completed by integrating the current-time curve to obtain the number of coulombs and thus the number of faradays of charge consumed or produced by the analyte.

Coulometric titrations are similar to other titrimetric methods in that analyses are based on measuring the combining capacity of the analyte with a standard reagent. In a coulometric procedure, the "reagent" is composed of electrons, and the "standard solution" is a constant current of known magnitude. Electrons are added to the analyte (via the direct current) or to some species that immediately reacts with the analyte until the point of chemical equivalence is indicated. At that point, the electrolysis is discontinued. The amount of analyte is determined from the magnitude of the current and the time required to complete the titration. The magnitude of the current in amperes is analogous to the molarity of a standard solution, and the time measurement is analogous to the volume measurement in conventional titrimetry.

20D-3   Current-Efficiency Requirements

A fundamental requirement for all coulometric methods is 100% current efficiency; that is, each faraday of electricity must bring about one equivalent of chemical change in the analyte. Note that 100% current efficiency can be

Michael Faraday (1791–1867) was one of the foremost chemists and physicists of his time. Among his most important discoveries were Faraday's laws of electrolysis. Faraday, a simple man who lacked mathematical sophistication, was a superb experimentalist and an inspiring teacher and lecturer. The quantity of charge equal to a mole of electrons is named in his honor.

Amperostatic coulometry is also called coulometric titrimetry.

Electrons are the reagent in a coulometric titration.

One equivalent of chemical change is the change that corresponds to 1 mol of electrons. Thus, for the two half-reactions in Example 20-3, one equivalent of chemical change involves production of 1/2 mol of Cu or 1/4 mol of O_2.

achieved without direct participation of the analyte in electron transfer at an electrode. For example, chloride ions are readily determined by potentiostatic coulometry or by coulometric titrations by generating silver ions at a silver anode. These ions then react with the analyte to form a precipitate or deposit of silver chloride. The quantity of electricity required to complete the silver chloride formation serves as the analytical parameter. In this instance, 100% current efficiency is realized because the number of moles of electrons is exactly equal to the number of moles of chloride ion in the sample, even though these ions do not react directly at the electrode surface.

20D-4   Controlled-Potential Coulometry

In controlled-potential coulometry, the potential of the working electrode is maintained at a constant level such that only the analyte is responsible for the conduction of charge across the electrode-solution interface. The number of coulombs required to convert the analyte to its reaction product is then determined by recording and integrating the current-versus-time curve for the electrolysis.

Instrumentation

The instrumentation for potentiostatic coulometry consists of an electrolysis cell, a potentiostat, and a device for determining the number of coulombs consumed by the analyte.

Cells.   Figure 20-9 illustrates two types of cells that are used for potentiostatic coulometry. The first consists of a platinum-gauze working electrode, a platinum-wire counter electrode, and a saturated calomel reference electrode. The counter electrode is separated from the analyte solution by a salt bridge that usually contains the same electrolyte as the solution containing the analyte. This bridge is needed to prevent the reaction products formed at the counter electrode from diffusing into the analyte solution and interfering. For example, hydrogen gas is a common product at a cathodic counter electrode. Unless this species is physically isolated from the analyte solution by the bridge, it will react directly with many of the analytes that are determined by oxidation at the working anode.

The second type of cell, shown in Figure 20-9b, contains a mercury pool, which functions as the cathode. A mercury cathode is particularly useful for separating easily reduced elements as a preliminary step in an analysis. In addition, it has found considerable use for the coulometric determination of several cations that when reduced form metals that are soluble in mercury. In these applications, little or no hydrogen evolution occurs even at high applied potentials because of the large overvoltage effects. A coulometric cell such as that shown in Figure 20-9b is also useful for the coulometric determination of certain types of organic compounds.

Potentiostats and Coulometers.   For controlled-potential coulometry, a potentiostat similar in design to that shown in Figure 20-7 is required. Generally, however, the potentiostat is automated and is equipped with a computer, or an elec-

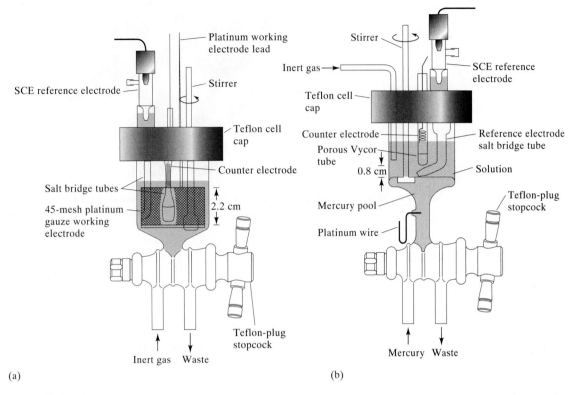

Figure 20-9   Electrolysis cells for potentiostatic coulometry. Working electrode: (a) platinum gauze, (b) mercury pool. (Reprinted with permission from J. E. Harrar and C. L. Pomernacki, *Anal. Chem.*, **1973**, *45*, 57. Copyright 1973 American Chemical Society.)

tronic current integrator that gives the number of coulombs necessary to complete the reaction, as shown in Figure 20-10.

Applications of Controlled-Potential Coulometry

Controlled-potential coulometric methods have been applied to the determination of many elements in inorganic compounds.[4] Mercury appears to be favored as the cathode, and methods for depositing dozens of metals at this electrode have been described. The method has found widespread use in the nuclear energy field for the relatively interference-free determination of uranium and plutonium.

Controlled-potential coulometry also offers possibilities for the electrolytic determination (and synthesis) of organic compounds. For example, trichloroacetic acid and picric acid are quantitatively reduced at a mercury cathode whose potential is suitably controlled:

$$Cl_3CCOO^- + H^+ + 2e^- \rightarrow Cl_2HCCOO^- + Cl^-$$

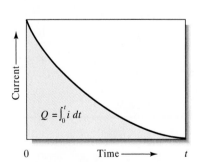

$$Q = \int_0^t i \, dt$$

Figure 20-10   For a current that varies with time, the quantity of charge Q in a time t is the shaded area under the curve, obtained by integration of the current–time curve.

---

[4]For a summary of applications, see J. A. Dean, *Analytical Chemistry Handbook* (New York: McGraw-Hill, 1995), Section 14, pp. 14.119–14.123.

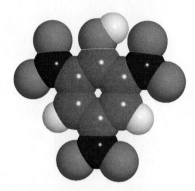

Molecular model of picric acid. Picric acid (2,4,6-trinitrophenol) is a close relative of trinitrotoluene (TNT). It is an explosive compound and has military applications. Picric acid has also been used as a yellow dye and staining agent and as an antiseptic.

$$O_2N\text{—}\underset{\underset{NO_2}{|}}{\overset{\overset{OH}{|}}{\bigcirc}}\text{—}NO_2 + 18H^+ + 18e^- \rightarrow H_2N\text{—}\underset{\underset{NH_2}{|}}{\overset{\overset{OH}{|}}{\bigcirc}}\text{—}NH_2 + 6H_2O$$

Coulometric measurements permit the determination of these compounds with a relative error of a few tenths of a percent.

20D-5   Coulometric Titrations[5]

Coulometric titrations are carried out with a constant-current source called an *amperostat,* which senses decreases in current in a cell and responds by increasing the potential applied to the cell until the current is restored to its original level. Because of the effects of concentration polarization, 100% current efficiency with respect to the analyte can be maintained only by having a large excess of an auxiliary reagent that is oxidized or reduced at the electrode to give a product that in turn reacts with the analyte. As an example, consider the coulometric titration of iron(II) at a platinum anode. At the beginning of the titration, the primary anodic reaction is

$$Fe^{2+} \rightarrow Fe^{3+} + e^-$$

As the concentration of iron(II) decreases, however, the requirement of a constant current results in an increase in the applied cell potential. Because of con-

> Constant current generators are called *amperostats* or *galvanostats.*

---

[5]For further details on this technique, see D. J. Curran, in *Laboratory Techniques in Electroanalytical Chemistry,* P. T. Kissinger and W. R. Heineman, Eds. (New York: Marcel Dekker, 1984), Chapter 20.

centration polarization, this increase in potential causes the anode potential to increase to the point where the decomposition of water becomes a competing process:

$$2H_2O \rightarrow O_2(g) + 4H^+ + 4e^-$$

The quantity of electricity required to complete the oxidation of iron(II) then exceeds that demanded by theory, and the current efficiency is less than 100%. The lowered current efficiency is prevented, however, by introducing at the outset an unmeasured excess of cerium(III), which is oxidized at a lower potential than is water:

$$Ce^{3+} \rightarrow Ce^{4+} + e^-$$

With stirring, the cerium(IV) produced is rapidly transported from the surface of the electrode to the bulk of the solution, where it oxidizes an equivalent amount of iron(II):

$$Ce^{4+} + Fe^{2+} \rightarrow Ce^{3+} + Fe^{3+}$$

The net effect is an electrochemical oxidation of iron(II) with 100% current efficiency, even though only a fraction of that species is directly oxidized at the electrode surface.

> Auxiliary reagents are essential in coulometric titrations.

Detecting the End Point in a Coulometric Titration

Coulometric titrations, like their volumetric counterparts, require a means for determining when the reaction between analyte and reagent is complete. Generally, the end point detection techniques described in Chapters 11–15 and Chapters 18 and 19 on volumetric methods are applicable to coulometric titrations as well. Thus, for the titration of iron(II) just described, an oxidation/reduction indicator, such as 1,10-phenanthroline, can be used; as an alternative, the end point can be determined potentiometrically. Some coulometric titrations use a photometric end point (see Section 23A-2).

Instrumentation

As shown in Figure 20-11, the equipment required for a coulometric titration includes a source of constant current that provides one to several hundred milliamperes, a titration vessel, a switch, a timer, and a device for monitoring current. Movement of the switch to position 1 simultaneously starts the timer and initiates a current in the titration cell. When the switch is moved to position 2, the electrolysis and the timing are discontinued. With the switch in this position, however, electricity continues to be drawn from the source and passes through a dummy resistor R_D that has about the same electrical resistance as the cell. This arrangement ensures continuous operation of the source, which aids in maintaining the current at a constant level.

The constant-current source for a coulometric titration is often an amperostat, an electronic device capable of maintaining a current of 210 mA or more that is constant to a few hundredths percent. Amperostats are available from several instrument manufacturers. The electrolysis time can be measured very accurately with a digital timer or a computer-based timing system.

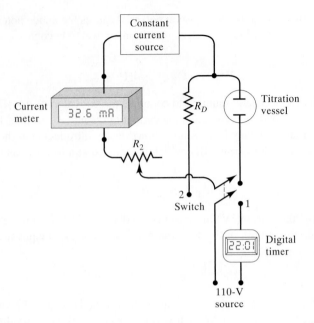

Figure 20-11   Conceptual diagram of a coulometric titration apparatus. Commercial coulometric titrators are totally electronic and usually computer controlled.

Cells for Coulometric Titrations.   Figure 20-12 shows a typical coulometric titration cell consisting of a generator electrode at which the reagent is produced and a counter electrode to complete the circuit. The generator electrode — usually a platinum rectangle, a coil of wire, or a gauze cylinder — has a relatively large surface area to minimize polarization effects. Ordinarily, the counter electrode is isolated from the reaction medium by a sintered disk or some other porous medium to prevent interference by the reaction products from this electrode.

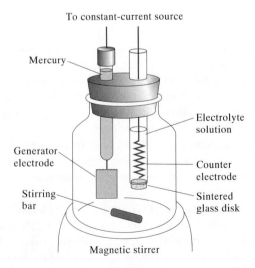

Figure 20-12   A typical coulometric titration cell.

Comparing Coulometric and Conventional Titrations

The various components of the titrator in Figure 20-11 have their counterparts in the reagents and apparatus required for a volumetric titration. The constant-current source of known magnitude serves the same function as the standard solution in a volumetric method. The electronic timer and switch correspond to the buret and stopcock, respectively. Electricity is passed through the cell for relatively long periods of time at the outset of a coulometric titration, but the time intervals are made smaller and smaller as chemical equivalence is approached. Note that these steps are analogous to the operation of a buret in a conventional titration.

A coulometric titration offers several significant advantages over a conventional volumetric procedure. Principal among these is the elimination of the problems associated with the preparation, standardization, and storage of standard solutions. This advantage is particularly significant with labile reagents such as chlorine, bromine, and titanium(III) ion, which are sufficiently unstable in aqueous solution to seriously limit their value as volumetric reagents. In contrast, their use in a coulometric determination is straightforward because they are consumed as soon as they are generated.

Coulometric methods also excel when small amounts of analyte have to be titrated because tiny quantities of reagent are generated with ease and accuracy through the proper choice of current. In contrast, the use of very dilute solutions and the accurate measurement of small volumes are inconvenient at best.

A further advantage of the coulometric procedure is that a single constant-current source provides reagents for precipitation, complex formation, neutralization, or oxidation/reduction titrations. Finally, coulometric titrations are more readily automated since currents are easier to control than liquid flow.

The current-time measurements required for a coulometric titration are inherently as accurate as or more accurate than the comparable volume/molarity measurements of a conventional volumetric method, particularly where small quantities of reagent are involved. When the accuracy of a titration is limited by the sensitivity of the end point, the two titration methods have comparable accuracies.

> Coulometric methods are as accurate and precise as comparable volumetric methods.

Applying the Coulometric Titration Procedure

Coulometric titrations have been developed for all types of volumetric reactions.[6] Selected applications are described in this section.

Neutralization Titrations.   Hydroxide ion can be generated at the surface of a platinum cathode immersed in a solution containing the analyte acid:

$$2H_2O + 2e^- \rightarrow 2OH^- + H_2(g)$$

The platinum anode must be isolated by a diaphragm to eliminate potential interference from the hydrogen ions produced by anodic oxidation of water. As a convenient alternative, a silver wire can be substituted for the platinum anode,

---

[6]For a summary of applications, see J. A. Dean, *Analytical Chemistry Handbook* (New York: McGraw-Hill, 1995), Section 14, pp. 14.127–14.133.

provided chloride or bromide ions are added to the analyte solution. The anode reaction then becomes

$$Ag(s) + Br^- \rightarrow AgBr(s) + e^-$$

Silver bromide does not interfere with the neutralization reaction.

Coulometric titrations of acids are much less susceptible to the carbonate error encountered in volumetric methods (Section 14A-3). The only measure required to avoid this error is to remove the carbon dioxide from the solvent by boiling or by bubbling an inert gas, such as nitrogen, through the solution for a brief period.

Hydrogen ions generated at the surface of a platinum anode can be used for the coulometric titration of strong as well as weak bases:

$$2H_2O \rightarrow O_2 + 4H^+ + 4e^-$$

Here, the cathode must be isolated from the analyte solution to prevent interference from hydroxide ion.

Precipitation and Complex-Formation Reactions. Coulometric titrations with EDTA use the reduction of the ammine mercury(II) EDTA chelate at a mercury cathode:

$$HgNH_3Y^{2-} + NH_4^+ + 2e^- \rightarrow Hg(l) + 2NH_3 + HY^{3-} \qquad (20\text{-}7)$$

Because the mercury chelate is more stable than the corresponding complexes of such cations as calcium, zinc, lead, or copper, complexation of these ions occurs only after the ligand has been freed by the electrode process.

As shown in Table 20-3, several precipitating reagents can be generated coulometrically. The most widely used of these is silver ion, which is generated at a silver anode.

Oxidation/Reduction Titrations. Table 20-4 reveals that a variety of redox reagents can be generated coulometrically. For example, the coulometric generation of bromine forms the basis for a large number of coulometric methods. Of interest as well are reagents not ordinarily encountered in conventional volumetric analysis owing to the instability of their solutions; silver(II), manganese(III), and the chloride complex of copper(I) are examples.

Table 20-3

Summary of Coulometric Titrations Involving Neutralization, Precipitation, and Complex-Formation Reactions

Species Determined	Generator Electrode Reaction	Secondary Analytical Reaction
Acids	$2H_2O + 2e^- \rightleftharpoons 2OH^- + H_2$	$OH^- + H^+ \rightleftharpoons H_2O$
Bases	$H_2O \rightleftharpoons 2H^+ + \frac{1}{2}O_2 + 2e^-$	$H^+ + OH^- \rightleftharpoons H_2O$
Cl^-, Br^-, I^-	$Ag \rightleftharpoons Ag^+ + e^-$	$Ag^+ + X^- \rightleftharpoons AgX(s)$
Mercaptans	$Ag \rightleftharpoons Ag^+ + e^-$	$Ag^+ + RSH \rightleftharpoons AgSR(s) + H^+$
Cl^-, Br^-, I^-	$2Hg \rightleftharpoons Hg_2^{2+} + 2e^-$	$Hg_2^{2+} + 2X^- \rightleftharpoons Hg_2X_2(s)$
Zn^{2+}	$Fe(CN)_6^{3-} + e^- \rightleftharpoons Fe(CN)_6^{4-}$	$3Zn^{2+} + 2K^+ + 2Fe(CN)_6^{4-} \rightleftharpoons$ $K_2Zn_3[Fe(CN)_6]_2(s)$
$Ca^{2+}, Cu^{2+}, Zn^{2+}, Pb^{2+}$	See Equation 20-7	$HY^{3-} + Ca^{2+} \rightleftharpoons CaY^{2-} + H^+$, etc.

Table 20-4

Summary of Coulometric Titration Involving Oxidation/Reduction Reactions

Reagent	Generator Electrode Reaction	Substance Determined
Br_2	$2Br^- \rightleftharpoons Br_2 + 2e^-$	As(III), Sb(III), U(IV), Ti(I), I^-, SCN^-, NH_3, N_2H_4, NH_2OH, phenol, aniline, mustard gas, mercaptans, 8-hydroxyquinoline, olefins
Cl_2	$2Cl^- \rightleftharpoons Cl_2 + 2e^-$	As(III), I^-, styrene, fatty acids
I_2	$2I^- \rightleftharpoons I_2 + 2e^-$	As(III), Sb(III), $S_2O_3^{2-}$, H_2S, ascorbic acid
Ce^{4+}	$Ce^{3+} \rightleftharpoons Ce^{4+} + e^-$	Fe(II), Ti(III), U(IV), As(III), I^-, $Fe(CN)_6^{4-}$
Mn^{3+}	$Mn^{2+} \rightleftharpoons Mn^{3+} + e^-$	$H_2C_2O_4$, Fe(II), As(III)
Ag^{2+}	$Ag^+ \rightleftharpoons Ag^{2+} + e^-$	Ce(III), V(IV), $H_2C_2O_4$, As(III)
Fe^{2+}	$Fe^{3+} + e^- \rightleftharpoons Fe^{2+}$	Cr(VI), Mn(VII), V(V), Ce(IV)
Ti^{3+}	$TiO^{2+} + 2H^+ + e^- \rightleftharpoons Ti^{3+} + H_2O$	Fe(III), V(V), Ce(IV), U(VI)
$CuCl_3^{2-}$	$Cu^{2+} + 3Cl^- + e^- \rightleftharpoons CuCl_3^{2-}$	V(V), Cr(VI), IO_3^-
U^{4+}	$UO_2^{2+} + 4H^+ + 2e^- \rightleftharpoons U^{4+} + 2H_2O$	Cr(VI), Ce(IV)

Automating Coulometric Titrations

A number of instrument manufacturers offer automatic coulometric titrators, most of which employ a potentiometric end point. Some of these instruments are multipurpose and can be used for the determination of a variety of species. Others are designed for a single type of analysis. Examples of the latter are chloride titrators, in which silver ion is generated coulometrically; sulfur dioxide monitors, where anodically generated bromine oxidizes the analyte to sulfate ions; carbon dioxide monitors, in which the gas, absorbed in monoethanolamine, is titrated with coulometrically generated base; and water titrators, in which Karl Fischer reagent (see Section 18C-5) is generated electrolytically.

20E   VOLTAMMETRY

Voltammetry consists of a group of electroanalytical methods in which information about the analyte is derived from measurement of current as a function of applied potential under conditions that encourage polarization of the indicator or working electrode. Generally, to enhance polarization, the working electrodes in voltammetry are *microelectrodes* that have surface areas from a few square millimeters at the most to, in some applications, only a few square micrometers.

Voltammetry is based on the measurement of current in an electrochemical cell under conditions of complete concentration polarization in which the rate of oxidation or reduction of the analyte is limited by the rate of mass transfer of the analyte to the electrode surface. Voltammetry differs from electrogravimetry and coulometry in that with the latter two methods, measures are taken to minimize or compensate for the effects of concentration polarization. Furthermore, a minimal consumption of analyte takes place in voltammetry; in electrogravimetry and coulometry, essentially all the analyte is converted to another state.

Historically, the field of voltammetry developed from *polarography,* a type of voltammetry that was discovered by Czechoslovakian chemist Jaroslav

Voltammetric methods are based on measurement of current as a function of the potential applied to a microelectrode.

Polarography is voltammetry at the dropping mercury electrode.

Heyrovsky in the early 1920s.[7] Polarography, which is still an important branch of voltammetry, differs from other types of voltammetry in that a *dropping mercury electrode* (DME) is used as the working microelectrode.

In voltammetry,[8] the voltage of the working microelectrode is varied while the current response is measured. Several different voltage-time functions can be applied to the electrode. The simplest of these is a linear scan in which the potential of the working electrode is changed linearly with time. Typically, the potential of the working electrode is varied over a 1- or 2-V range. The current in the cell is then measured as a function of the applied potential. Pulsed waveforms and triangular waveforms can also be applied. We will consider here only linear-scan voltammetry.

20E-1   Linear-Scan Voltammetry

In the earliest and simplest voltammetric methods, the potential of the working electrode is increased or decreased at a typical rate of 2 to 5 mV/s. The current, usually in microamperes, is then recorded to give a *voltammogram,* which is a plot of current as a function of applied potential. Linear-scan voltammetry is of two types: *hydrodynamic voltammetry* and *polarography.*

Voltammetric Instrumentation

Typically, a three-electrode potentiostat, an all-electronic version of the one shown in Figure 20-7, is employed in linear-scan voltammetry. The potential sweep is provided by an electronic *linear-sweep generator* or, in modern instruments, a computer. The cell is made up of the three electrodes immersed in a solution containing the analyte and an excess of a *supporting electrolyte.* The potential of the microelectrode or *working electrode* is varied linearly with time. The second electrode is a reference electrode whose potential remains invariant throughout the experiment. The third electrode is the *auxiliary* or *counter electrode,* which is often a coil of platinum wire or a pool of mercury that simply serves to conduct electricity from the source through the solution to the microelectrode. The electrical resistance in the circuit containing the reference electrode is so large ($> 10^{11}$ Ω) that essentially no current is present in it. Thus, the entire current from the source is carried from the counter electrode to the microelectrode.

What Microelectrodes Are Used?

The microelectrodes used in voltammetry take a variety of shapes and forms. Often, they are small, flat disks of a conductor that are press-fitted into a rod of an inert material, such as Teflon or Kel-F, that has a wire contact imbedded in it (see Figure 20-13a). The conductor may be an inert metal, such as platinum or gold; pyrolytic graphite or glassy carbon; a semiconductor, such as tin or indium

---

[7]J. Heyrovsky, *Chem. Listy,* **1922,** *16,* 256.

[8]For further information on voltammetric methods, see J. A. Dean, *Analytical Chemistry Handbook* (New York: McGraw-Hill, 1995), Section 14, pp. 14.57–14.93; M. R. Smyth and F. G. Vos, Eds., *Analytical Voltammetry* (New York: Elsevier, 1992); A. J. Bard and L. R. Faulkner, *Electrochemical Methods* (New York: Wiley, 1980).

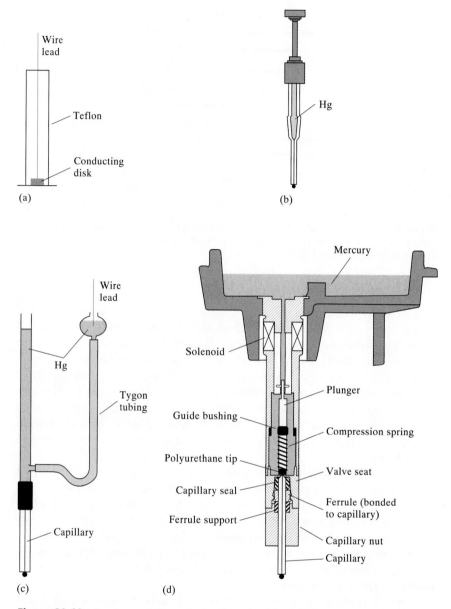

Figure 20-13   Some common types of microelectrodes: (a) a disk electrode, (b) a hanging mercury drop electrode, (c) a dropping mercury electrode, (d) a static mercury dropping electrode.

oxide; or a metal coated with a film of mercury. As shown in Figure 20-14, the range of potentials that can be used with these electrodes in aqueous solutions varies; it depends not only on the electrode material but also on the composition of the solution in which it is immersed.

Mercury microelectrodes have been widely employed in voltammetry for several reasons. One is the relatively large negative potential range that can be achieved with mercury because of the large overvoltage of hydrogen on mercury (see Section 20A-2). This overvoltage allows the determination of metals such as Cd, Zn, and Pb that are reduced at negative potentials to amalgams at the surface

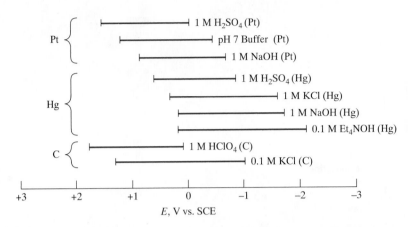

Figure 20-14 Potential ranges for three types of electrodes in various supporting electrolytes. [Adapted from A. J. Bard and L. R. Faulkner, *Electrochemical Methods* (New York: Wiley, 1980), back cover. With permission.]

Large negative potentials can be used with mercury electrodes because of the large overvoltage associated with the reduction of hydrogen ions to hydrogen gas.

of a mercury electrode. Furthermore, a fresh metallic surface is readily formed by simply producing a new mercury drop. Mercury microelectrodes take several forms. The simplest of these is a mercury film electrode formed by electrodeposition of the metal onto a disk electrode, such as that shown in Figure 20-13a. Figure 20-13b illustrates a *hanging mercury drop electrode* (HMDE). This electrode, which is available from several commercial sources, consists of a very fine capillary tube connected to a reservoir containing mercury. The metal is forced out of the capillary by a piston arrangement driven by a micrometer screw, which permits formation of drops having surface areas reproducible to 5% or better.

Figure 20-13c shows a typical *dropping mercury electrode* (DME), which was used in nearly all early polarographic experiments. It consists of roughly 10 cm of a fine capillary tubing (inside diameter ≈ 0.05 mm) through which mercury is forced by a mercury head of perhaps 50 cm. The diameter of the capillary is such that a new drop forms, grows, and falls every 2 to 6 s. The diameter of the drop is 0.5 to 1 mm and is highly reproducible. In some applications, the drop time is controlled by a mechanical *drop knocker* that dislodges the drop at a fixed time after it begins to form.

Figure 20-13d shows a commercially available mercury electrode, which can be operated either as a dropping mercury electrode or as a hanging mercury drop electrode. This system has the advantage that the drop forms quickly allowing current measurements to be delayed until the surface area is stable and constant. This procedure largely eliminates the large current fluctuations encountered with the classical dropping mercury electrode.

Describing the Voltammogram

Figure 20-15 illustrates the appearance of a typical linear-scan voltammogram for an electrolysis involving the reduction of an analyte species A to give a product P at a mercury film microelectrode. Here, the microelectrode is assumed to be connected to the negative terminal of the linear-sweep generator so that the applied potentials are given a negative sign as shown. By convention, cathodic

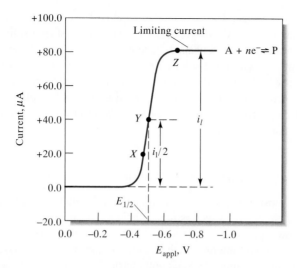

Figure 20-15 Linear-scan voltammogram for the reduction of a hypothetical species A to give a product P. The limiting current i_l is proportional to the analyte concentration and is used for quantitative analysis. The half-wave potential $E_{1/2}$ is related to the standard potential for the half-reaction and is often used for qualitative identification of species. The half-wave potential is the applied potential at which the current i is $i_l/2$.

(reduction) currents are treated as being positive, whereas anodic currents are given a negative sign.

Linear-scan voltammograms under slow sweep conditions (a few mV/s) with the solution moving past the electrode generally have the shape of a sigmoidal ($\int$-shaped) curve called a *voltammetric wave*. The constant current beyond the steep rise (point Z on Figure 20-15) is called the *limiting current* i_l, because it is limited by the rate at which the reactant can be brought to the electrode surface by mass transport processes. Limiting currents are generally directly proportional to reactant concentration. Thus, we may write

$$i_l = kc_A$$

where c_A is the analyte concentration and k is a constant. Quantitative linear-scan voltammetry is based on this relationship.

The potential at which the current is equal to one-half the limiting current is called the *half-wave potential* and is given the symbol $E_{1/2}$. The half-wave potential is closely related to the standard potential for the half-reaction but is usually not identical to that constant. Half-wave potentials are sometimes useful for identification of the components of a solution.

To obtain reproducible limiting currents rapidly, it is necessary that either (1) the solution or the microelectrode be in continuous and reproducible motion or (2) a dropping mercury electrode be used. Linear-scan voltammetry in which the solution is stirred or the electrode is rotated is called *hydrodynamic voltammetry*. Voltammetry with the dropping mercury electrode is called *polarography*.

In the type of linear-scan voltammetry discussed above, the potential is changed slowly enough and mass transfer is rapid enough that a steady state is reached at the electrode surface. Hence, the mass transport rate of analyte A to the electrode just balances its reduction rate at the electrode. Likewise, the mass transport of product P away from the electrode is just equal to its production rate

A *voltammetric wave* is an $\int$-shaped wave obtained in current-voltage plots in voltammetry.

The *half-wave potential* occurs when the current is equal to one-half the limiting value.

The *limiting current* in voltammetry is the current plateau observed at the top of the voltammetric wave. It occurs because the surface concentration of the analyte falls to zero. At this point, the mass transfer rate is its maximum value. The limiting-current plateau is an example of complete concentration polarization.

at the electrode surface. There is another type of linear-sweep voltammetry in which fast scan rates (1 V/s or greater) are used with unstirred solutions. In this type of voltammetry, a peak-shaped current-time signal is obtained because of depletion of the analyte in the solution near the electrode. *Cyclic voltammetry* is an example in which forward and reverse linear scans are applied. With cyclic voltammetry, products formed on the forward scan can be detected on the reverse scan if they have not moved away from the electrode or have not been altered by a chemical reaction.

20E-2   Applying Voltammetric Methods

Currently, the most important uses of voltammetry include (1) detection and determination of chemical species as they exit from chromatographic columns or a continuous-flow apparatus; (2) routine determination of oxygen and certain species of biochemical interest, such as glucose, lactate, and sucrose; (3) detection of end points in coulometric and volumetric titrations; and (4) determination of metals (polarography).

Voltammetric Detectors

Voltammetric detectors are widely used for detection and determination of oxidizable or reducible compounds or ions in flowing streams. Compounds that have been separated by liquid chromatography (see Chapter 25) or are present in flow injection analyzers are typical examples. The manner in which most voltammetric detectors are used is to apply a constant potential between the working electrode and a reference electrode; the potential corresponds to the limiting current region for the analyte of interest. Because the potential is maintained at a constant value and the current is measured, the technique is sometimes called *amperometry*. In these applications, a thin-layer cell is used with a volume that is typically 0.1 to 1.0 μL. Detection limits as low as 10^{-9} to 10^{-10} M of analyte can be obtained.

Voltammetric Sensors

A number of voltammetric systems are produced commercially for the determination of specific species that are of interest in industry and research. These devices are sometimes called electrodes but are, in fact, complete voltammetric cells that are better referred to as sensors. Two of these devices are described here.

The determination of dissolved oxygen in a variety of aqueous environments, such as seawater, blood, sewage, effluents from chemical plants, and soils, is of tremendous importance. One of the most common and convenient methods for making such measurements is with the *Clark oxygen sensor*, which was patented by L. C. Clark, Jr., in 1956.[9] A schematic of the Clark oxygen sensor is shown in Figure 20-16. The cell consists of a cathodic platinum disk working electrode imbedded in a centrally located cylindrical insulator. Surrounding the lower end of this insulator is a ring-shaped silver anode. The tubular insulator and elec-

The Clark oxygen sensor is widely used in clinical laboratories for the determination of dissolved O_2 in blood and other body fluids

---

[9]For a detailed discussion of the Clark oxygen sensor, see M. L. Hitchman, *Measurement of Dissolved Oxygen* (New York: Wiley, 1978), Chapters 3–5.

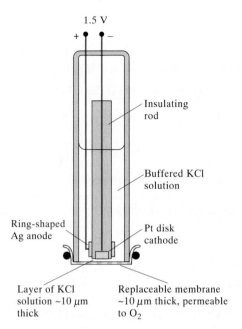

Figure 20-16   The Clark voltammetric oxygen sensor. Cathode reaction: $O_2 + 4H^+ + 4e^- \rightleftharpoons$ H_2O. Anodic reaction: $Ag + Cl^- \rightleftharpoons AgCl(s) + e^-$.

trodes are mounted inside a second cylinder that contains a buffered solution of potassium chloride. A thin ($\approx 20\ \mu m$), replaceable, oxygen-permeable membrane of Teflon or polyethylene is held in place at the bottom end of the tube by an O-ring. The thickness of the electrolyte solution between the cathode and the membrane is approximately 20 μm.

When the oxygen sensor is immersed in a flowing or stirred solution of the analyte, oxygen diffuses through the membrane into the thin layer of electrolyte immediately adjacent to the disk cathode, where it diffuses to the electrode and is immediately reduced to water. Two diffusion processes are involved, one through the membrane and the other through the solution between the membrane and the electrode surface. For a steady state condition to be reached in a reasonable time (10 to 20 s), the thickness of the membrane and the electrolyte film must be 20 μm or less. Under these conditions, the rate of equilibration of oxygen transfer across the membrane determines the steady state current that is reached. This rate is directly proportional to the dissolved oxygen concentration in solution.

A number of enzyme-based voltammetric sensors are available commercially. An example is a glucose sensor that is widely used in clinical laboratories. This device is similar in construction to the oxygen electrode of Figure 20-16, but the membrane in this case is more complex and consists of three layers. The outer layer is a polycarbonate film that is permeable to glucose but impermeable to proteins and other constituents of blood. The middle layer is an immobilized enzyme, here glucose oxidase. The inner layer is a cellulose acetate membrane, which is permeable to small molecules such as hydrogen peroxide. When this device is immersed in a glucose-containing solution, glucose diffuses through the outer membrane into the immobilized enzyme, where the following catalytic reaction occurs:

Enzyme-based sensors can be based on detecting hydrogen peroxide, oxygen, or H^+, depending on the analyte and enzyme. Voltammetric sensors are used for H_2O_2 and O_2, whereas a potentiometric pH electrode is used for H^+.

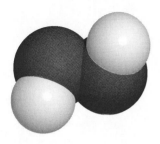

Molecular model of hydrogen peroxide. Hydrogen peroxide is a strong oxidizing agent that plays an important role in air pollution. Peroxide radicals occur in smog and can attack unburned fuel molecules. Hydrogen peroxide is used as a bleaching agent for hair and to oxidize and destroy other pigments. It is produced in enzyme reactions involving the oxidation of sugar molecules. The antibiotic action that was once responsible for honey being used to treat wounds results from the presence of trace amounts of H_2O_2 produced by the oxidation of glucose.

$$glucose + O_2 \xrightarrow{\text{glucose oxidase}} H_2O_2 + \text{gluconic acid}$$

The hydrogen peroxide then diffuses through the inner layer of membrane and to the electrode surface, where it is oxidized to give oxygen. That is,

$$H_2O_2 + 2OH^- \rightarrow O_2 + 2H_2O + 2e^-$$

The resulting current is directly proportional to the glucose concentration of the analyte solution.

Several other sensors are available that are based on the voltammetric measurement of hydrogen peroxide produced by enzymatic oxidations of other species of clinical interest. These analytes include sucrose, lactose, ethanol, and L-lactate. A different enzyme is, of course, required for each species. In some cases, enzyme electrodes can be based on measuring oxygen or on measuring pH.

Amperometric Titrations

Hydrodynamic voltammetry can be used to estimate the equivalence point of titrations, provided at least one of the participants or products of the reaction involved is electroactive. Here, the current at some fixed potential in the limiting current region is measured as a function of reagent volume (or time, if the reagent is generated coulometrically).

Amperometric titration curves are typically one of the shapes shown in Figure 20-17. Figure 20-17a represents a titration in which the analyte is reduced at the electrode, but the reagent is not electroactive. Figure 20-17b is typical of the case in which the reagent reacts at the electrode, but the analyte does not. Figure 20-17c corresponds to a titration in which both analyte and titrant are reduced at the electrode.

Two types of amperometric electrode systems are encountered. One employs a single polarizable microelectrode coupled to a reference electrode; the other uses a pair of identical solid state microelectrodes immersed in a stirred solution. For the first, a rotating platinum electrode is often used. Titrations involving dual polarized electrodes use a pair of metallic microelectrodes, usually platinum.[10]

Amperometric end points are widely used in various types of titrations. The most common end-point detection method for the Karl Fischer titration discussed in Section 18-C5 is the amperometric method with dual polarized electrodes. Several instrument manufacturers offer fully automated instruments for performing these titrations.

Determining Metals by Polarography

Polarographic methods are still widely employed for determining metals and in some cases organic species.[11] The classical linear-scan methods are giving way to techniques based on more complex pulsed waveforms. Quantitative determi-

---

[10]Further information about the amperometric titration method can be found in J. A. Dean, *Analytical Chemistry Handbook* (New York: McGraw-Hill, 1995), Section 14, pp. 14.87–14.91; D. A. Skoog, D. M. West, and F. J. Holler, *Fundamentals of Analytical Chemistry*, 7th ed. (Philadelphia: Saunders College Publishing, 1996), pp. 479–480.

[11]References dealing with polarography include R. C. Kapoor and B. S. Aggarwal, *Principles of Polarography* (New York: Wiley, 1991); A. M. Bond, *Modern Polarographic Methods in Analytical Chemistry* (New York: Marcel Dekker, 1980).

nations are most often based on calibration curves of limiting current versus analyte concentration.

Most metallic cations can be determined at the dropping mercury electrode. It is even feasible to determine mixtures of cations as long as their half-wave potentials differ by 100 mV or more. Some of the metals commonly determined by polarographic methods include Cd^{2+}, Zn^{2+}, Pb^{2+}, Ni^{2+}, Tl^+, Co^{2+}, and Cu^{2+}. Even the alkali and alkaline-earth metals are reducible at a mercury surface, provided that the supporting electrolyte does not react at the high potentials required.

20F   SOME ADDITIONAL ELECTROANALYTICAL METHODS

There are several other important electroanalytical methods. *Stripping methods* encompass a variety of electrochemical procedures having a common characteristic initial step.[12] In all these procedures, the analyte is first deposited on a microelectrode, usually from a stirred solution. After an accurately measured deposition period, the electrolysis is discontinued, the stirring is stopped, and the deposited analyte is determined by one of the voltammetric procedures described in the preceding section. During this second step, the analyte is redissolved or stripped from the microelectrode; hence, the name attached to these methods.

In *anodic stripping methods,* the microelectrode behaves as a cathode during the electrodeposition step and as an anode during the stripping step. In *cathodic stripping methods,* the working electrode is an anode during deposition and a cathode during stripping. The deposition step amounts to an electrochemical preconcentration of the analyte; that is, the concentration of the analyte in the surface of the microelectrode is greater than it is in the bulk of solution. Because of this, stripping methods can be extremely sensitive; they can improve the detection limits for selected species by factors of up to 10^6.

Anodic stripping is widely applied in the environmental area because of its extreme sensitivity. Elements commonly determined by anodic stripping include Ag, Au, Bi, Cd, Cu, Ga, Ge, Hg, In, Pb, Sb, Sn, Tl, and Zn. Cathodic stripping methods are used to determine anions that form insoluble mercury or silver salts. Some species that can be determined by cathodic stripping include arsenate, chloride, bromide, iodide, chromate, sulfate, sulfide, mercaptans, and thiocyanate.

A classical bulk electrochemical method that has again become important because of its use in detecting the effluents from ion chromatography columns (see Chapter 25) is *electrolytic conductivity.* Here, ions migrate to electrodes under the influence of an electric field. The amount of current is directly proportional to the applied potential and the resistance (reciprocal of the conductance) of the solution by Ohm's law. Net electrolysis is usually prevented by having the field alternate in sign (ac field) or by pulsing the electrode potential. Conductance measurements are also employed to detect the end points in titrations involving electrolytes.

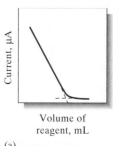

(a)

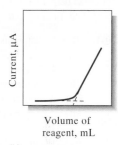

(b)

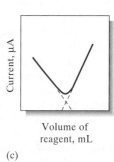

(c)

Figure 20-17 Typical amperometric titration curves: (a) analyte is reduced, reagent is not; (b) reagent is reduced, analyte is not; (c) both reagent and analyte are reduced.

A major advantage of stripping analysis is the capability of preconcentrating the analyte prior to the measurement step.

---

[12]For a discussion of stripping methods, see J. Wang, *Stripping Analysis* (Deerfield Beach, FL: VCH Publishers, 1985).

WEB WORKS

Set your browser to http://www.bioanalytical.com and investigate the electrochemical instruments produced by one instrument company. In addition, determine if this company makes amperometric detectors for liquid chromatography and voltammetric sensors for other applications.

20G   QUESTIONS AND PROBLEMS

Note: Numerical data are molar analytical concentrations where the full formula of a species is provided. Molar equilibrium concentrations are supplied for species displayed as ions.

20-1. Briefly distinguish between
   *(a)   concentration polarization and kinetic polarization.
   (b)   an amperostat and a potentiostat.
   *(c)   a coulomb and a faraday.
   (d)   a working electrode and a counter electrode.
   *(e)   the electrolysis circuit and the control circuit for controlled-potential methods.

20-2. Briefly define
   *(a)   current density.
   (b)   ohmic potential.
   *(c)   coulometric titration.
   (d)   controlled-potential electrolysis.
   *(e)   current efficiency.
   (f)   an electrochemical equivalent.

*20-3. Describe three mechanisms responsible for the transport of dissolved species to and from an electrode surface.

20-4. How does the current in an electrochemical cell affect its potential?

*20-5. How do concentration polarization and kinetic polarization resemble one another? How do they differ?

20-6. What experimental variables affect concentration polarization in an electrochemical cell?

*20-7. Describe conditions that favor kinetic polarization in an electrochemical cell.

20-8. How do electrogravimetric and coulometric methods differ from potentiometric methods?

*20-9. Identify three factors that influence the physical characteristics of an electrolytic deposit.

20-10. What is the purpose of a cathode depolarizer?

*20-11. What is the function of
   (a)   an amperostat?
   (b)   a potentiostat?

*20-12. Differentiate between amperostatic coulometry and potentiostatic coulometry.

*20-13. Why is it ordinarily necessary to isolate the working electrode from the counter electrode in a controlled-potential coulometric analysis?

20-14. Why is an auxiliary reagent always required in a coulometric titration?

20-15. Distinguish between
   *(a)   voltammetry and polarography.
   (b)   linear-scan and pulse voltammetry.
   (c)   a hanging mercury drop electrode and a dropping mercury electrode.
   (d)   the standard electrode potential and the half-wave potential for a reversible couple at a microelectrode.

20-16. Define
   *(a)   voltammogram.
   (b)   hydrodynamic voltammetry.
   (c)   limiting current.
   (d)   mercury film electrode.
   *(e)   half-wave potential.

20-17. List the advantages and limitations of the dropping mercury electrode compared with platinum or carbon microelectrodes.

20-18. Calculate the number of ions involved at the surface of an electrode during each second that an electrochemical cell is operated at 0.020 A and the ions involved are
   (a)   univalent.
   *(b)   divalent.
   (c)   trivalent.

20-19. Calculate the theoretical potential needed to initiate the deposition of
   *(a)   copper from a solution that is 0.150 M in Cu^{2+} and buffered to a pH of 3.00. Oxygen is evolved at the anode at 1.00 atm.
   (b)   tin from a solution that is 0.120 M in Sn^{2+} and buffered to a pH of 4.00. Oxygen is evolved at the anode at 770 torr.
   *(c)   silver bromide on a silver anode from a solution that is 0.0864 M in Br^- and buffered to a pH of 3.40. Hydrogen is evolved at the cathode at 765 torr.
   (d) Tl_2O_3 from a solution that is 4.00×10^{-3} M in Tl^+ and buffered to a pH of 8.00. The solution is also 0.010 M in Cu^{2+}, which

acts as a cathode depolarizer. The reduction process is

$$Tl_2O_3 + 3H_2O + 4e^- \rightleftharpoons$$
$$2Tl^+ + 6OH^- \qquad E^0 = 0.020 \text{ V}$$

*20-20.   Calculate the initial potential needed for a current of 0.078 A in the cell

$$Co \mid Co^{2+}(6.40 \times 10^{-3} \text{ M}) \parallel$$
$$Zn^{2+}(3.75 \times 10^{-3} \text{ M}) \mid Zn$$

if this cell has a resistance of 5.00 Ω.

*20-21.   The cell

$$Sn \mid Sn^{2+}(8.22 \times 10^{-4} \text{ M}) \parallel$$
$$Cd^{2+}(7.50 \times 10^{-2} \text{ M}) \mid Cd$$

has a resistance of 3.95 Ω. Calculate the initial potential that will be needed for a current of 0.072 A in this cell.

20-22.   Copper is to be deposited from a solution that is 0.200 M in Cu(II) and buffered to a pH of 4.00. Oxygen is evolved from the anode at a partial pressure of 740 torr. The cell has a resistance of 3.60 Ω; the temperature is 25°C. Calculate

  (a)   the theoretical potential needed to initiate the deposition of copper from this solution.

  (b)   the *IR* drop associated with a current of 0.10 A in this cell.

  (c)   the initial applied potential, given that the overvoltage of oxygen is 0.50 V under these conditions.

  (d)   the potential of the cell when $[Cu^{2+}]$ is 8.00×10^{-6}, assuming that the *IR* drop and O_2 overvoltage remain unchanged.

20-23.   Nickel is to be deposited on a platinum cathode (area = 120 cm²) from a solution that is 0.200 M in Ni^{2+} and buffered to a pH of 2.00. Oxygen is evolved at a partial pressure of 1.00 atm at a platinum anode with an area of 80 cm². The cell has a resistance of 3.15 Ω; the temperature is 25°C. Calculate

  (a)   the theoretical potential needed to initiate the deposition of nickel from this solution.

  (b)   the *IR* drop for a current of 1.10 A in this cell.

  (c)   the current density at the anode and the cathode.

  (d)   the initial applied potential, given that the overvoltage of oxygen on platinum is approximately 0.52 V under these conditions.

  (e)   the applied potential when the nickel concentration has decreased to 2.00×10^{-4} M

(assume that all variables other than $[Ni^{2+}]$ remain constant).

*20-24.   Silver is to be deposited from a solution that is 0.150 M in $Ag(CN)_2^-$, 0.320 M in KCN, and buffered to a pH of 10.00. Oxygen is evolved at the anode at a partial pressure of 1.00 atm. The cell has a resistance of 2.90 Ω; the temperature is 25°C. Calculate

  (a)   the theoretical potential needed to initiate the deposition of silver from this solution.

  (b)   the *IR* drop associated with a current of 0.12 A in this cell.

  (c)   the initial applied potential, given that the O_2 overvoltage is 0.80 V under these conditions.

  (d)   the applied potential when $[Ag(CN)_2^-]$ is 1.00×10^{-5} M, assuming no changes in *IR* drop and O_2 overvoltage.

*20-25.   A solution is 0.150 M in Co^{2+} and 0.0750 M in Cd^{2+}. Calculate

  (a)   the Co^{2+} concentration in the solution as the first cadmium starts to deposit.

  (b)   the cathode potential needed to lower the Co^{2+} concentration to 1.00×10^{-5} M.

20-26.   A solution is 0.0500 M in BiO^+ and 0.0400 M in Co^{2+} and has a pH of 2.50.

  (a)   What is the concentration of the more readily reduced cation at the onset of deposition of the less reducible one?

  (b)   What is the potential of the cathode when the concentration of the more easily reduced species is 1.00×10^{-6} M?

20-27.   Electrogravimetric analysis involving control of the cathode potential is proposed as a means for separating Bi^{3+} and Sn^{2+} in a solution that is 0.200 M in each ion and buffered to pH 1.50.

  (a)   Calculate the theoretical cathode potential at the start of deposition of the more readily reduced ion.

  (b)   Calculate the residual concentration of the more readily reduced species at the outset of the deposition of the less readily reduced species.

  (c)   Propose a range (versus SCE), if such exists, within which the cathode potential should be maintained; consider a residual concentration less than 10^{-6} M as constituting quantitative removal.

*20-28.   Halide ions can be deposited on a silver anode via the reaction

$$Ag(s) + X^- \rightarrow AgX(s) + e^-$$

  (a)   If 1.00×10^{-5} M is used as the criterion for quantitative removal, is it theoretically feasible to separate Br^- from I^- through control of the anode potential in a solution that is initially 0.250 M in each ion?

(b) Is a separation of Cl^- and I^- theoretically feasible in a solution that is initially 0.250 M in each ion?

(c) If a separation is feasible in either part (a) or part (b), what range of anode potential (versus SCE) should be used?

20-29. A solution is 0.100 M in each of two reducible cations, A and B. Removal of the more reducible species A is deemed complete when [A] has been decreased to 1.00×10^{-5} M. What minimum difference in standard electrode potentials will permit the isolation of A without interference from B when

	A is	B is
*(a)	univalent	univalent
(b)	divalent	univalent
*(c)	trivalent	univalent
(d)	univalent	divalent
*(e)	divalent	divalent
(f)	trivalent	divalent
*(g)	univalent	trivalent
(h)	divalent	trivalent
*(i)	trivalent	trivalent

*20-30. Calculate the time needed for a constant current of 0.961 A to deposit 0.500 g of Co(II) as

(a) elemental cobalt on the surface of a cathode.

(b) Co_3O_4 on an anode.

20-31. Calculate the time needed for a constant current of 1.20 A to deposit 0.500 g of

(a) Tl(III) as the element on a cathode.

(b) Tl(I) as Tl_2O_3 on an anode.

(c) Tl(I) as the element on a cathode.

*20-32. A 0.1516-g sample of a purified organic acid was neutralized by the hydroxide ion produced in 5 min and 24 s by a constant current of 0.401 A. Calculate the equivalent weight of the acid.

20-33. The concentration of 10.0 mL of a plating solution was determined by titration with electro-generated hydrogen ion to a methyl orange end point. A color change occurred after 3 min and 22 s with a current of 43.4 mA. Calculate the number of grams of NaCN per liter of solution.

*20-34. An excess of $HgNH_3Y^{2-}$ was introduced to 25.00 mL of well water. Express the hardness of the water in terms of ppm $CaCO_3$ if the EDTA needed for the titration was generated at a mercury cathode (Equation 20-7) in 2.02 min by a constant current of 31.6 mA.

20-35. Electrolytically generated I_2 was used to determine the amount of H_2S in 100.0 mL of brackish water. Following addition of excess KI, a titration required a constant current of 36.32 mA for 10.12 min. The reaction was

$$H_2S + I_2 \rightarrow S(s) + 2H^+ + 2I^-$$

Express the results of the analysis in terms of ppm H_2S.

*20-36. The nitrobenzene in 210 mg of an organic mixture was reduced to phenylhydroxylamine at a constant potential of -0.96 V (versus SCE) applied to a mercury cathode:

$$C_6H_5NO_2 + 4H^+ + 4e^- \rightarrow$$
$$C_6H_5NHOH + H_2O$$

The sample was dissolved in 100 mL of methanol; after electrolysis for 30 min, the reaction was judged complete. An electronic coulometer in series with the cell indicated that the reduction required 26.74 C. Calculate the percentage of $C_6H_5NO_2$ in the sample.

20-37. The phenol content of water downstream from a coking furnace was determined by coulometric analysis. A 100-mL sample was rendered slightly acidic and an excess of KBr was introduced. To produce Br_2 for the reaction

$$C_6H_5OH + 3Br_2 \rightarrow Br_3C_6H_2OH(s) + 3HBr$$

a steady current of 0.0313 A for 7 min and 33 s was required. Express the results of this analysis in terms of parts of C_6H_5OH per million parts of water. (Assume that the density of water is 1.00 g/mL.)

*20-38. At a potential of -1.0 V (versus SCE), CCl_4 in methanol is reduced to $CHCl_3$ at a mercury cathode:

$$2CCl_4 + 2H^+ + 2e^- + 2Hg(l) \rightarrow$$
$$2CHCl_3 + Hg_2Cl_2(s)$$

At -1.80 V, the $CHCl_3$ further reacts to give CH_4:

$$2CHCl_3 + 6H^+ + 6e^- + 6Hg(l) \rightarrow$$
$$2CH_4 + 3Hg_2Cl_2(s)$$

A 0.750-g sample containing CCl_4, $CHCl_3$, and inert organic species was dissolved in methanol and electrolyzed at -1.0 V until the current approached zero. A coulometer indicated that 11.63 C were required to complete the reaction. The potential of the cathode was adjusted to -1.8 V. Completion of the titration at this potential required an additional 68.6 C. Calculate the percentage of CCl_4 and $CHCl_3$ in the mixture.

20-39. A 0.1309-g sample containing only $CHCl_3$ and CH_2Cl_2 was dissolved in methanol and was electrolyzed in a cell containing a mercury cathode; the potential of the cathode was held constant at -1.80 V (versus SCE). Both compounds were reduced to CH_4 (see Problem 20-38 for the reaction). Calculate the percentage of $CHCl_3$ and CH_2Cl_2 if 306.7 C was required to complete the reduction.

*20-40. The voltammogram for 20.00 mL of solution that was 3.65×10^{-3} M in Cd^{2+} gave a wave for that ion with a limiting current of 31.3 μA. Calculate the percentage change in concentration of the solution if the current in the limiting current region were allowed to continue for
(a) 5 min.
(b) 10 min.
(c) 30 min.

20-41. Sulfate ion can be determined by an amperometric titration procedure using Pb^{2+} as the titrant. If the potential of a Hg electrode is adjusted to -1.00 V versus SCE, the current can be used to monitor the Pb^{2+} concentration during the titration. In a calibration experiment, the limiting current, after correction for background and residual currents, was found to be related to the Pb^{2+} concentration by $i_l = 10c_{Pb^{2+}}$, where i_l is the limiting current (μA) and $c_{Pb^{2+}}$ is the Pb^{2+} concentration (mM). The titration reaction is

$$SO_4^{2-} + Pb^{2+} \rightleftharpoons PbSO_4(s)$$

$$K_{sp} = 1.6 \times 10^{-8}$$

If 25 mL of 0.025 M Na_2SO_4 is titrated with 0.04 M $Pb(NO_3)_2$, develop the titration curve in spreadsheet format and plot the limiting current versus volume of titrant.

*20-42. Why are stripping methods more sensitive than other voltammetric procedures?

20-43. What is the purpose of the electrodeposition step in stripping analysis?

Section IV

Spectrochemical Analysis

Outline

Spectrochemical methods are among the most popular of the instrumental analytical techniques. Ultraviolet and visible absorption and molecular fluorescence spectroscopy are applied for quantitative determinations in many areas of the physical and life sciences. Likewise infrared absorption methods are frequently used to identify molecules and to supply structural information. Atomic spectrometric methods can be used for qualitative identification of elements, but the most frequent use is for quantitative elemental analysis.

Chapter 21 introduces the principles of spectrochemical methods and discusses the absorption law and its limitations. Chapter 22 describes the instruments and instrumental components that are used for spectroscopic measurements.

Radiation sources and detectors and complete spectrometers are presented in this chapter. Chapter 23 completes this section by presenting the most important applications of spectrochemical methods, including those involving UV-visible and infrared absorption, molecular fluorescence, atomic absorption, and atomic emission methods.

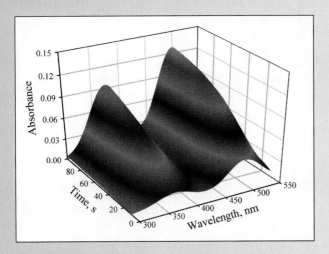

This 3d plot shows an absorption spectrum for a two component mixture undergoing a time-dependent chemical reaction. Two products are shown forming at different rates. Product 1 has an absorption maximum near 375 nm, while product 2 has an absorption maximum near 500 nm. Product 2 forms at a faster rate than product 1. This type of information is obtained from a diode array spectrophotometer and is useful for determining multiple analytes in mixtures without prior separation.

Chapter 21

Spectroscopic Methods of Analysis: Making Measurements with Light

Measurements based on light and other forms of electromagnetic radiation are widely used throughout analytical chemistry. The interactions of radiation and matter are the subject of the science called *spectroscopy*. Spectroscopic analytical methods are based on measuring the amount of radiation produced or absorbed by molecular or atomic species of interest.[1] We can classify spectroscopic methods according to the region of the electromagnetic spectrum involved in the measurement. The regions that have been used include γ-ray, X-ray, ultraviolet (UV), visible, infrared (IR), microwave, and radio frequency (RF). Indeed, current usage extends the meaning of spectroscopy yet further to include techniques that do not even involve electromagnetic radiation. Examples of this last usage include acoustic, mass, and electron spectroscopy.

Spectroscopy has played a vital role in the development of modern atomic theory. In addition, *spectrochemical methods* have provided perhaps the most widely used tools for the elucidation of molecular structure as well as the quantitative and qualitative determination of both inorganic and organic compounds.

In this chapter, the basic principles necessary for understanding measurements made with electromagnetic radiation are discussed, particularly those dealing with the absorption of UV, visible, and IR radiation. The nature of electromagnetic radiation and its interactions with matter are stressed. The following two chapters are devoted to spectroscopic instruments (Chapter 22) and to specific applications of spectroscopy (Chapter 23).

Other analytically useful types of electromagnetic radiation include γ-ray, X-ray, microwave, and RF radiation. Optical spectroscopic methods involve UV, visible, or IR radiation.

---

[1]For further study, see E. J. Meehan, in *Treatise on Analytical Chemistry,* 2nd ed., P. J. Elving, E. J. Meehan, and I. M. Kolthoff, Eds. (New York: Wiley, 1981), Part I, Vol. 7, Chapters 1–3; J. D. Ingle, Jr., and S. R. Crouch, *Spectrochemical Methods of Analysis* (Englewood Cliffs, N.J.: Prentice-Hall, 1988), J. E. Crooks, *The Spectrum in Chemistry* (New York: Academic Press, 1978).

21A   LIGHT: A PARTICLE, A WAVE, OR BOTH?

Electromagnetic radiation is a form of energy that is transmitted through space at enormous velocities. We will call electromagnetic radiation in the UV/visible and sometimes in the IR region *light,* although strictly speaking the term refers only to visible radiation. Electromagnetic radiation can be described as a wave with properties of wavelength, frequency, velocity, and amplitude. In contrast to sound waves, light requires no supporting medium for its transmission; thus, it readily passes through a vacuum. Light also travels much faster than sound.

The wave model fails to account for phenomena associated with the absorption and emission of radiant energy. For these processes, electromagnetic radiation can be treated as discrete packets of energy or particles called *photons* or *quanta.* These dual views of radiation as particles and waves are not mutually exclusive but complementary. In fact, the energy of a photon is directly proportional to its frequency, as we shall see. Similarly, this duality applies to streams of electrons, protons, and other elementary particles, that can produce interference and diffraction effects typically associated with wave behavior.

21A-1   Wave Properties

In dealing with such phenomena as reflection, refraction, interference, and diffraction, electromagnetic radiation is conveniently modeled as waves consisting of perpendicularly oscillating electric and magnetic fields as shown in Figure 21-1a. The electric field for a single-frequency wave oscillates sinusoidally in space and time as shown in Figure 21-1b. Here, the electric field is represented as a vector whose length is proportional to the field strength. The x-axis in this plot is either time as the radiation passes a fixed point in space or distance at a fixed time. Note that the direction in which the field oscillates is perpendicular to the direction in which the radiation is being propagated.

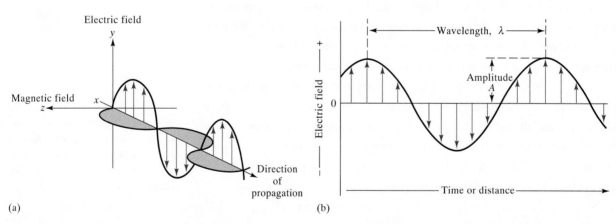

Figure 21-1   Wave nature of a beam of single-frequency electromagnetic radiation. In (a), a plane-polarized wave is shown propagating along the x-axis. The electric field oscillates in a plane perpendicular to the magnetic field. If the radiation were unpolarized, a component of the electric field would be seen in all planes. In (b), only the electric field oscillations are shown. The amplitude of the wave is the length of the electric field vector at the wave maximum, whereas the wavelength is the distance between successive maxima.

Wave Characteristics

In Figure 21-1b, the *amplitude* of the sine wave is shown and the wavelength is defined. The time required for successive maxima (or minima) to pass through a fixed point in space is called the *period p* of the wave. The *frequency ν* is the number of oscillations of the electric field vector per unit time and is equal to $1/p$.

The frequency of a light wave, or any wave of electromagnetic radiation, is determined by the source that emits it and remains constant regardless of the medium traversed. In contrast, the *velocity v* of the wave front through a medium depends on both the medium and the frequency. The *wavelength λ* is the linear distance between successive maxima or minima of a wave as shown in Figure 21-1b. Multiplication of the frequency in waves per unit time by the wavelength in distance per wave gives the velocity of the wave in distance per unit time (centimeters per second, cm s^{-1} or meters per second, m s^{-1}) as shown in Equation 21-1. Note that both the velocity and the wavelength depend on the medium.

$$v = \nu\lambda \tag{21-1}$$

> The unit of frequency is the *hertz* (Hz), which corresponds to one cycle per second. That is, 1 Hz = 1 s^{-1}. The frequency of a beam of electromagnetic radiation does not change as it passes through different media.

Wavelength Units for Various Spectral Regions

Region	Unit	Definition
X-ray	Angstrom unit, Å	10^{-10} m
Ultraviolet/visible	Nanometer, nm	10^{-9} m
Infrared	Micrometer, μm	10^{-6} m

> The refractive index η of a medium measures the extent of interaction between electromagnetic radiation and the medium through which it passes and is defined by $\eta = c/v$. For example, the refractive index of water at room temperature is 1.33, which means that radiation passes through water at a rate of $c/1.33$ or 2.26×10^{10} cm s^{-1}. In other words, light travels 1.33 times slower in water than it does in vacuum. The velocity and wavelength of radiation become proportionally smaller as the radiation passes from a vacuum, or from air, to a denser medium while the frequency remains constant.

> Note in Equation 21-1 that v (distance/time) $= \nu$ (waves/time) $\times \lambda$ (distance/wave).

The Speed of Light

In a vacuum, light travels at its maximum velocity. This velocity, which is given the special symbol c, is 2.99792×10^8 m s^{-1}. The velocity of light in air is only about 0.03% less than its velocity in vacuum. Thus, for a vacuum, or for air, Equation 21-2 conveniently gives the velocity of light.

$$c = \nu\lambda = 3.00 \times 10^8 \text{ m s}^{-1} = 3.00 \times 10^{10} \text{ cm s}^{-1} \tag{21-2}$$

> To three significant figures, Equation 21-2 is equally applicable in air or vacuum.

In a medium containing matter, light travels with a velocity less than c because of interaction between the electromagnetic field and electrons in the atoms or molecules of the medium. Since the frequency of the radiation is constant, the wavelength must decrease as the light passes from a vacuum to a medium containing matter (see Equation 21-1). This effect is illustrated in Figure 21-2 for a beam of visible radiation. Note that the effect can be quite large.

The wavenumber $\bar{\nu}$ is another way to describe electromagnetic radiation. It is defined as the number of waves per centimeter and is equal to $1/\lambda$. By definition, $\bar{\nu}$ has the units of reciprocal centimeters (cm^{-1}).

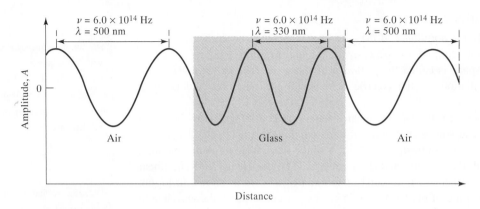

Figure 21-2   Change in wavelength as radiation passes from air into a dense glass and back to air. Note that the wavelength shortens by nearly 200 nm, or more than 30%, as it passes into glass; a reverse change occurs as the radiation again enters air.

The wavenumber $\bar{\nu}$ in cm^{-1} (Kayser) can be used to describe radiation throughout the electromagnetic spectrum. It is most often used in the infrared region. The most useful part of the infrared spectrum for the detection and determination of organic species is from 2.5 to 15 μm, which corresponds to a wavenumber range of 4000 to 667 cm^{-1}. As shown later, the wavenumber of a beam of electromagnetic radiation is directly proportional to its energy and thus its frequency.

Equation 21-3 gives the energy of radiation in SI units of *joules,* where one joule (J) is the work done by a force of one newton (N) acting over a distance of one meter. Note that both frequency and wavenumber are directly proportional to the energy of a photon.

Radiant Power and Intensity

The *radiant power P* in watts (W) is the energy of a beam that reaches a given area per unit time. The *intensity* is the radiant power-per-unit solid angle.[2] Both quantities are proportional to the square of the amplitude of the electric field (see Figure 21-1b). Although it is not strictly correct, radiant power and intensity are frequently used interchangeably.

21A-2   The Particle Nature of Light: Photons

In many radiation/matter interactions, it is useful to consider light as consisting of photons or quanta. We can relate the energy of a photon to its wavelength, frequency, and wavenumber by

$$E = h\nu = \frac{hc}{\lambda} = hc\bar{\nu} \tag{21-3}$$

where h is Planck's constant (6.63×10^{-34} J s). Note that the wavenumber $\bar{\nu}$ and frequency ν, in contrast to the wavelength λ, are directly proportional to the photon energy E. The radiant power of a beam of radiation is directly proportional to the number of photons per second.

21B   HOW RADIATION AND MATTER INTERACT

The types of interactions of most interest in spectroscopy involve transitions between different energy levels of chemical species. Other types of interactions, such as reflection, refraction, elastic scattering, interference, and diffraction, are often related to the bulk properties of materials rather than to energy levels of specific molecules or atoms. Although these bulk interactions are also of interest

---

[2]Solid angle is the three-dimensional spread at the vertex of a cone measured as the area intercepted by the cone on a unit sphere whose center is at the vertex. The angle is measured in stereradians (sr).

in spectroscopy, we will restrict the discussion here to those interactions that involve energy level transitions. The specific types of interactions observed depend strongly on the energy of the radiation used and the mode of detection.

21B-1   The Electromagnetic Spectrum

The electromagnetic spectrum covers an enormous range of energies (frequencies) and thus wavelengths. Useful frequencies vary from $>10^{19}$ Hz (γ-ray) to 10^3 Hz (radio waves). An X-ray photon ($\nu \approx 3 \times 10^{18}$ Hz, $\lambda \approx 10^{-10}$ m), for example, is approximately 10,000 times as energetic as a photon emitted by an ordinary light bulb ($\nu \approx 3 \times 10^{14}$ Hz, $\lambda \approx 10^{-6}$ m) and 10^{15} times as energetic as a radio-frequency photon ($\nu \approx 3 \times 10^3$ Hz, $\lambda \approx 10^5$ m).

The major divisions of the spectrum are shown in color in the inside front cover of this book. Note that the visible region, to which our eyes respond, is only a minute region of the entire spectrum. Such different types of radiation as gamma (γ) rays or radio waves differ from visible light only in the energy (frequency) of their photons.

Figure 21-3 shows the regions of electromagnetic spectrum that are used for spectroscopic analyses. Also shown are the types of atomic and molecular transitions that result from interactions of the radiation with a sample. Note that the low-energy radiation used in nuclear magnetic resonance (NMR) and electron spin resonance (ESR) spectroscopy causes subtle changes, such as changes in spin; the high-energy radiation used in γ-ray spectroscopy can cause much more dramatic effects, such as nuclear configuration changes.

Regions of the UV/Visible and IR Spectrum

Region	Wavelength Range
UV	$180 - 380$ nm
Visible	$380 - 780$ nm
Near-IR	$0.78 - 2.5 \ \mu$m
Mid-IR	$2.5 - 50 \ \mu$m

You can recall the order of the colors in the spectrum by the mnemonic ROY G BIV, which is short for Red, Orange, Yellow, Green, Blue, Indigo, and Violet.

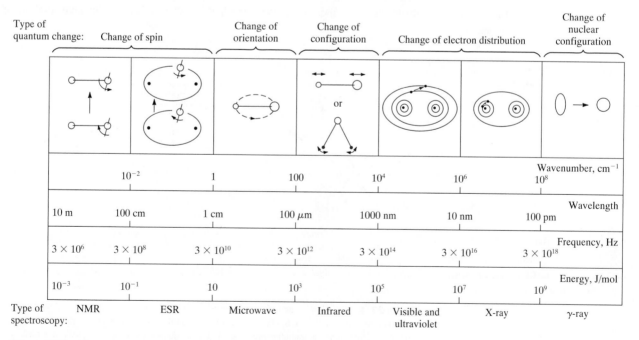

Figure 21-3   The regions of the electromagnetic spectrum. Interaction of an analyte with electromagnetic radiation can result in the types of changes shown. Note that changes in electron distributions occur in the UV/visible region. The wavenumber, wavelength, frequency, and energy are characteristics that describe electromagnetic radiation. [From C. N. Banwell, *Fundamentals of Molecular Spectroscopy*, 3rd ed. (New York; McGraw-Hill, 1983), p. 7.]

21B-2   What Do Spectroscopists Measure?

Spectroscopists use the interactions of radiation with matter to obtain information about a sample. Several of the elements were discovered by spectroscopy (see Feature 21-1). The sample is usually stimulated in some way, by applying energy in the form of heat, electrical energy, light, particles, or a chemical reaction. Prior to the application of this stimulus, the analyte is predominately in its lowest-energy or *ground state*. The stimulus then causes some analyte species to undergo a transition to a higher-energy or *excited state*. We obtain information about the analyte by measuring the electromagnetic radiation emitted as it returns to the ground state or by measuring the amount of electromagnetic radiation absorbed as a result of excitation.

Figure 21-4 illustrates the processes involved in emission and chemiluminescence spectroscopy. Here, the analyte is stimulated by the application of heat, electrical energy, or a chemical reaction. The term *emission spectroscopy* usually refers to methods in which the stimulus is heat or electrical energy, whereas *chemiluminescence spectroscopy* refers to excitation of the analyte by a chemical reaction. In both cases, measurement of the radiant power emitted as the analyte returns to the ground state can give information about its identity and concentration. The results of such a measurement are often expressed graphically by a *spectrum*, which is a plot of the emitted radiation as a function of frequency or wavelength.

When the sample is stimulated by application of an external electromagnetic radiation source, several processes are possible. For example, the radiation can be scattered or reflected. What is important to us is that some of the incident radiation can be absorbed and promote some of the analyte species to an excited state as shown in Figure 21-5. In *absorption spectroscopy,* the amount of light absorbed as a function of wavelength is measured, which can give qualitative

A familiar example of chemiluminescence is found in the light emitted by a firefly. In the firefly reaction, an enzyme luciferase catalyzes the oxidative phosphorylation reaction of luciferin with adenosine triphosphate (ATP) to produce oxyluciferin, carbon dioxide, adenosine monophosphate (AMP), and light. Chemiluminescence involving a biological or enzyme reaction is often called *bioluminescence.* The popular light stick is another familiar example of chemiluminescence.

Figures 21-4, 21-5 and 21-6 are from J. D. Ingle, Jr. and S. R. Crouch, *Spectrochemical Analysis,* (Englewood Cliffs, N.J.: Prentice Hall, 1988) with permission.

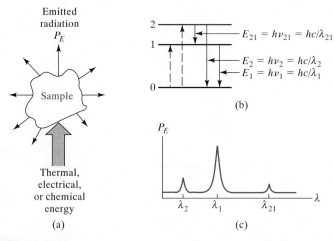

Figure 21-4 Emission or chemiluminescence processes. In (a), the sample is excited by the application of thermal, electrical, or chemical energy. These processes do not involve radiant energy and are therefore called nonradiative processes. In the energy-level diagram (b), the dashed lines with upward-pointing arrows symbolize these nonradiative excitation processes, whereas the solid lines with downward-pointing arrows indicate that the analyte loses its energy by emission of a photon. In (c), the resulting spectrum is shown as a measurement of the radiant power emitted P_E as a function of wavelength, λ.

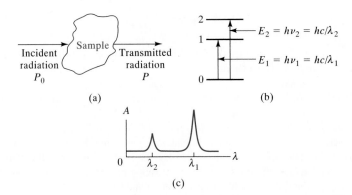

(a)   (b)   (c)

Figure 21-5   Absorption methods. Radiation of incident radiant power P_0 can be absorbed by the analyte, resulting in a transmitted beam of lower radiant power P. For absorption to occur, the energy of the incident beam must correspond to one of the energy differences shown in (b). The resulting absorption spectrum is shown in (c).

and quantitative information about the sample. In *photoluminescence spectroscopy* (Figure 21-6), the emission of photons is measured following absorption. The most important forms of photoluminescence for analytical purposes are *fluorescence* and *phosphorescence spectroscopy.*

We will focus on absorption spectroscopy in the UV/visible region of the spectrum because it is so widely used in chemistry, biology, forensic science, engineering, agriculture, clinical chemistry, and many other fields. Chapter 23 briefly treats IR absorption, photoluminescence, and atomic spectroscopic methods. Note that the processes shown in Figures 21-4 to 21-6 can occur in any re-

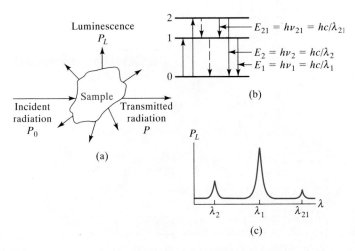

(a)   (b)   (c)

Figure 21-6   Photoluminescence methods (fluorescence and phosphorescence). Fluorescence and phosphoresecence result from absorption of electromagnetic radiation and then dissipation of the energy by emission of radiation (a). In (b), the absorption can cause excitation of the analyte to state 1 or state 2. Once excited, the excess energy can be lost by emission of a photon (luminescence shown as solid line) or by nonradiative processes (dashed lines). The emission occurs over all angles, and the wavelengths emitted (c) correspond to energy differences between levels. The major distinction between fluorescence and phosphorescence is the time scale of emission, with fluorescence being prompt emission and phosphorescence being delayed emission.

Feature 21-1

Spectroscopy and the
Discovery of Elements

The modern era of spectroscopy began with the observation of the prism spectrum of the sun by Sir Isaac Newton in 1672. In Newton's experiment, rays from the sun passed through a small opening into a dark room where they struck a prism and dispersed into the colors of the spectrum. The first description of spectral features beyond the simple observation of colors was by William Hyde Wollaston in 1802, who noted dark lines while observing the solar spectrum with a spectroscope. These lines were later described in detail by Joseph Fraunhofer, beginning in 1817, who gave the prominent lines letters starting with "A" at the red end of the spectrum. It remained for Gustav Robert Kirchhoff and Robert Wilhelm Bunsen in 1859 and 1860, however, to explain the origin of the Fraunhofer lines. Bunsen had invented his famous burner (see Figure 21F-1) a few years earlier, which made possible spectral observations of emission and absorption phenomena in a nearly transparent flame. Kirchhoff concluded that the Fraunhofer "D" lines were due to sodium in the sun's atmosphere and that the "A" and "B" lines were due to potassium. Still today, we call the emission lines of sodium the sodium "D" lines. These lines are responsible for the familiar yellow color seen in flames containing sodium or in sodium vapor lamps. The absence of lithium in the sun's spectrum led Kirchhoff to conclude that there was little lithium present in the sun. During these studies, Kirchhoff also developed his famous laws relating the absorption and emission of light from bodies and at interfaces. Together with Bunsen, Kirchhoff observed that different elements could impart different colors to flames and produce spectra exhibiting differently colored bands or lines. Kirchhoff and Bunsen are thus credited with discovering the use of spectroscopy for chemical analysis. The method was soon put to many practical uses, including the discovery of new elements. In 1860, the elements cesium and rubidium were discovered, followed in 1861 by thallium and in 1864 by indium. The age of spectroscopic analysis had clearly begun.

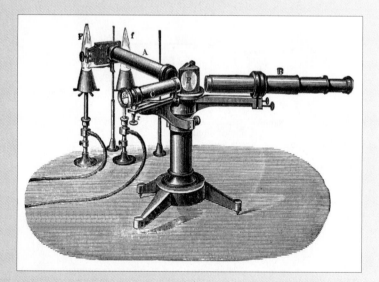

Figure 21F-1   Bunsen burner of the type used in early spectroscopic studies with a prism spectroscope of type used by Kirchhoff. [From H. Kayser, *Handbuch der Spectroscopie* Vol. 1, p. 492. (Stuttgart, Germany: S. Hirzel Verlag GmbH and Co., 1900.]

gion of the electromagnetic spectrum; the different energy levels can be nuclear levels, electronic levels, vibrational levels, or spin levels.

21C   ABSORBING LIGHT

Every molecular species is capable of absorbing its own characteristic frequencies of electromagnetic radiation as described in Figure 21-5. This process transfers energy to the molecule and results in a decrease in the intensity of the incident electromagnetic radiation. Absorption of the radiation thus *attenuates* the beam in accordance with the absorption law as described below.

21C-1   The Absorption Law: Describing the Absorption Process

The absorption law, also known as the Beer–Lambert law or just Beer's law, tells us quantitatively how the amount of attenuation depends on the concentration of the absorbing molecules and the pathlength over which absorption occurs. As light traverses a medium containing an absorbing analyte, decreases in intensity occur as the analyte becomes excited. For an analyte solution of a given concentration, the longer the length of the medium through which the light passes (pathlength of light), the more absorbers are in the path and the greater the attenuation. Also, for a given pathlength of light, the higher the concentration of absorbers, the stronger the attenuation.

Figure 21-7 depicts the attenuation of a parallel beam of monochromatic radiation as it passes through an absorbing solution of thickness b cm and concentration c moles per liter. Because of interactions between the photons and absorbing particles (recall Figure 21-5), the radiant power of the beam decreases from P_0 to P. The *transmittance T* of the solution is the fraction of incident radiation transmitted by the solution as shown in Equation 21-4. Transmittance is often expressed as a percentage and called the *percent transmittance*.

$$T = \frac{P}{P_0} \qquad (21\text{-}4)$$

Percent transmittance =

$$\%T = \frac{P}{P_0} \times 100\%$$

The *absorbance A* of a solution is related to the transmittance in a logarithmic manner as shown in Equation 21-5. Note that increases in the absorbance of a solution are accompanied by decreases in transmittance. The relationship be-

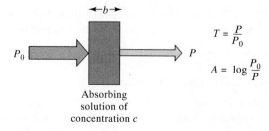

$$T = \frac{P}{P_0}$$

$$A = \log \frac{P_0}{P}$$

Figure 21-7 Attenuation of a beam of radiation by an absorbing solution. The larger arrow on the incident beam signifies a higher radiant power than is transmitted by the solution. The pathlength of the absorbing solution is b, and the concentration is c.

	A	B	C	D	E
1	**Calculation of absorbance from transmittance**				
2	T	%T	A	A	
3	0.001	0.1	3.000	3.000	
4	0.010	1.0	2.000	2.000	
5	0.050	5.0	1.301	1.301	
6	0.075	7.5	1.125	1.125	
7	0.100	10.0	1.000	1.000	
8	0.200	20.0	0.699	0.699	
9	0.300	30.0	0.523	0.523	
10	0.400	40.0	0.398	0.398	
11	0.500	50.0	0.301	0.301	
12	0.600	60.0	0.222	0.222	
13	0.700	70.0	0.155	0.155	
14	0.800	80.0	0.097	0.097	
15	0.900	90.0	0.046	0.046	
16	1.000	100.0	0.000	0.000	
17					
18	**Spreadsheet Documentation**				
19	Cell B3=100*A3				
20	Cell C3=-LOG10(A3)				
21	Cell D3=2-LOG10(B3)				

Figure 21-8 Conversion spreadsheet relating transmittance T, percent transmittance $\%T$ and absorbance A. The transmittance data to be converted are entered in cells A3 through A16. The percent transmittance is calculated in cells B3 by the formula shown in the documentation section, cell A19. This formula is copied into cells B4 through B16. The absorbance is calculated from $-\log T$ in cells C3 through C16 and from $2 - \log \%T$ in cells D3 through D16. The formulas for the first cell in the C and D columns are shown in cells A20 and A21.

tween transmittance and absorbance is illustrated by the conversion spreadsheet shown in Figure 21-8. The scales on earlier instruments were linear in transmittance; modern instruments have linear absorbance scales or a computer that calculates absorbance from measured quantities.

$$A = -\log T = \log \frac{P_0}{P} \tag{21-5}$$

Since $\%T = T \times 100\%$, A can also be calculated from $A = -\log T = 2 - \log \%T$, where $2 = \log 100$. The equivalence of the two calculations of A is shown in Figure 21-8.

Measuring Transmittance and Absorbances

Ordinarily, transmittance and absorbance, as defined by Equations 21-4 and 21-5 and Figure 21-7, cannot be measured as shown because the solution to be studied must be held in some sort of container (cell or cuvette). Reflection and scattering losses can occur at the cell walls as shown in Figure 21-9. These losses

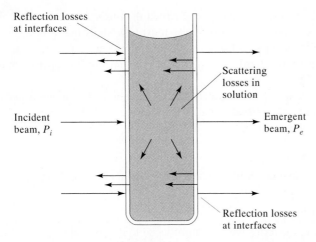

Figure 21-9   Reflection and scattering losses with a solution contained in a typical glass cell. Losses by reflection can occur at all the boundaries that separate the different materials. In this example, the light passes through the following boundaries, called interfaces: air-glass, glass-solution, solution-glass, and glass-air.

can be substantial. For example, about 8.5% of a beam of yellow light is lost by reflection when it passes through a glass cell. Light can also be scattered in all directions from the surface of large molecules or particles, such as dust, in the solvent, and this can also cause further attenuation of the beam as it passes through the solution.

To compensate for these effects, the power of the beam transmitted through a cell containing the analyte solution is compared with one that traverses an identical cell containing only the solvent or a reagent blank. An experimental absorbance that closely approximates the true absorbance for the solution is thus obtained; that is,

$$A = \log \frac{P_0}{P} \approx \log \frac{P_{\text{solvent}}}{P_{\text{solution}}} \qquad (21\text{-}6)$$

The terms P_0 and P will henceforth refer to the power of a beam that has passed through cells containing the blank (solvent) and the analyte, respectively.

Beer's Law[3]

According to Beer's law, absorbance is linearly related to the concentration of the absorbing species c and the pathlength b of the absorbing medium as expressed by Equation 21-7.

$$A = \log \frac{P_0}{P} = abc \qquad (21\text{-}7)$$

---

[3]For a derivation of Beer's law, see D. J. Swinehart, *J. Chem. Ed.,* **1972,** *32,* 333; J. D. Ingle, Jr., and S. R. Crouch, *Spectrochemical Analysis* (Englewood Cliffs, N.J.: Prentice-Hall, 1988), pp. 34–35.

Here a is a proportionality constant called the *absorptivity*. Because absorbance is a unitless quantity, the absorptivity must have units that cancel the units of b and c. If, for example, c has the units of grams per liter (g L^{-1}) and b has the units of centimeters (cm), absorptivity has the units of liters per gram centimeter (L g^{-1} cm^{-1}).

When we express the concentration in Equation 21-7 in moles per liter and b in centimeters, the proportionality constant, called the *molar absorptivity*, is given the special symbol ε. Thus,

$$A = \varepsilon bc \qquad (21\text{-}8)$$

where ε has the units of liters per mole centimeter (L mol^{-1} cm^{-1}).

> The molar absorptivity of a species at an absorption maximum is characteristic of that species. Peak molar absorptivities for many organic compounds range from 10 or less to 10,000 or more. Some transition metal complexes have molar absorptivities of 10,000 to 50,000. High molar absorptivities are desirable for quantitative analysis because they lead to high analytical sensitivity.

Coming to Terms

In addition to the terms we have introduced to describe absorption of radiant energy, you may encounter other terms in the literature or with older instruments. The terms, symbols, and definitions given in Table 21-1 are recommended by the American Society for Testing Materials (ASTM) as well as by the American Chemical Society. The third column contains the older names and symbols. Because a standard nomenclature is highly desirable to avoid ambiguities, we urge you to learn and use the recommended terms and symbols and to avoid the older terms.

Using Beer's Law

Beer's law, as expressed in Equations 21-7 and 21-8, can be used in several different ways. We can calculate molar absorptivities of species if their con-

Table 21-1

Important Terms and Symbols Employed in Absorption Measurements

Term and Symbol*	Definition	Alternative Name and Symbol
Incident radiant power, P_0	Radiant power in watts incident on sample	Incident intensity, I_0
Transmitted radiant power, P	Radiant power transmitted by sample	Transmitted intensity, I
Absorbance, A	$\log(P_0/P)$	Optical density, D; extinction, E
Transmittance, T	P/P_0	Transmission, T
Pathlength of sample, b	Length over which attenuation occurs	l, d
Absorptivity, $a\dagger$	$A/(bc)$	Extinction coefficient, k
Molar absorptivity, $\varepsilon\ddagger$	$A/(bc)$	Molar extinction coefficient

*Terminology recommended by the American Chemical Society (*Anal. Chem.*, **1990**, *62*, 91).
$\dagger c$ may be expressed in g L^{-1} or in other specified concentration units; b may be expressed in cm or other units of length.
$\ddagger c$ is expressed in mol L^{-1}; b is expressed in cm.

centrations are known, as shown in Example 21-1. We can use the measured value of absorbance to obtain concentration if absorptivity and pathlength are known. Absorptivities, however, are functions of such variables as solvent, solution composition, and temperature. Because of variations in absorptivities with conditions, it is never wise to depend on literature values for quantitative work. Hence, a standard solution of the analyte in the same solvent and at a similar temperature is used to obtain the absorptivity at the time of the analysis. Most often, we use a series of standard solutions of the analyte to construct a calibration curve, or working curve, of A versus c (see Figure 23-4) or to obtain a linear regression equation (see Section 7D). A spreadsheet is given in Example 21-2 for determining the concentration of a solution of permanganate. It may also be necessary to duplicate closely the overall composition of the analyte solution to compensate for matrix effects. Alternatively, the method of standard additions (see Section 23A-2) is used for the same purpose.

Example 21-1

A 7.50×10^{-5} M solution of potassium permanganate has a transmittance of 36.4% when measured in a 1.05-cm cell at a wavelength of 525 nm. Calculate (a) the absorbance of this solution and (b) the molar absorptivity of $KMnO_4$.

   (a) $A = -\log T = -\log 0.364 = -(-0.4389) = 0.439$

   (b) From Equation 21-8,

$$\varepsilon = \frac{A}{bc} = \frac{0.4389}{1.05 \text{ cm} \times 7.50 \times 10^{-5} \text{ mol L}^{-1}} = 5.57 \times 10^{3} \text{ L mol}^{-1} \text{ cm}^{-1}$$

Example 21-2

A series of permanganate solutions was used to prepare a calibration curve and determine the concentration of a solution. The absorbances given in cells B3 through B7 were obtained for the permanganate concentrations given in A3 through A7. The absorbance of the solution of unknown permanganate concentration is given in cell B8. A linear least-squares analysis was done as described in Section 7D. Cells B10 and B11 show the slope and intercept. These were used to calculate the concentration of permanganate in the unknown. Note that the prediction function in Excel predicts the value of y (Absorbance) given a known value of x (concentration). We desire, however, the value of x for a measured value of y. Hence, we calculate the unknown concentration from the absorbance of the unknown minus the intercept divided by the slope. The statistical quantities needed to obtain the standard deviation in the concentration of the unknown, s_c, are calculated in the error analysis section, cells B14 through B18. The equation given in Chapter 7 (Equation 7-18) is used to calculate s_c in cell B19. The complete spreadsheed documentation is given in cells A21 through A28. From the values in cells B12 and B19, the concentration of the unknown should be reported as 97.3 ± 0.9 (1 std dev) μM.

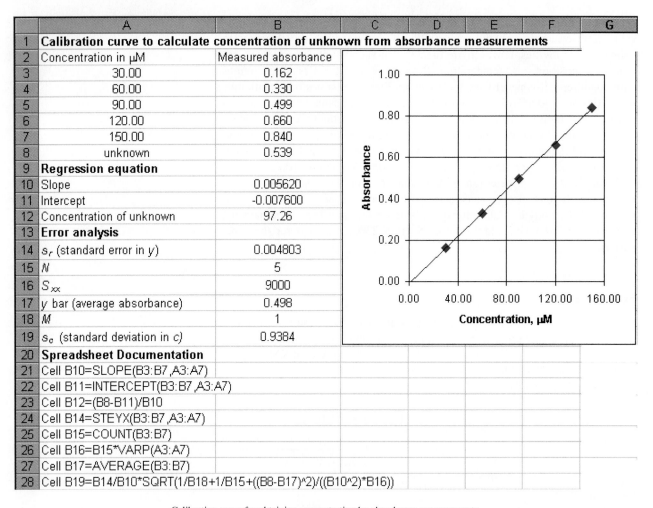

	A	B	C	D	E	F	G
1	**Calibration curve to calculate concentration of unknown from absorbance measurements**						
2	Concentration in μM	Measured absorbance					
3	30.00	0.162					
4	60.00	0.330					
5	90.00	0.499					
6	120.00	0.660					
7	150.00	0.840					
8	unknown	0.539					
9	**Regression equation**						
10	Slope	0.005620					
11	Intercept	-0.007600					
12	Concentration of unknown	97.26					
13	**Error analysis**						
14	s_r (standard error in y)	0.004803					
15	N	5					
16	S_{xx}	9000					
17	y bar (average absorbance)	0.498					
18	M	1					
19	s_c (standard deviation in c)	0.9384					
20	**Spreadsheet Documentation**						
21	Cell B10=SLOPE(B3:B7,A3:A7)						
22	Cell B11=INTERCEPT(B3:B7,A3:A7)						
23	Cell B12=(B8-B11)/B10						
24	Cell B14=STEYX(B3:B7,A3:A7)						
25	Cell B15=COUNT(B3:B7)						
26	Cell B16=B15*VARP(A3:A7)						
27	Cell B17=AVERAGE(B3:B7)						
28	Cell B19=B14/B10*SQRT(1/B18+1/B15+((B8-B17)^2)/((B10^2)*B16))						

Calibration curve for obtaining concentration by absorbance measurements.

Beer's law also applies to solutions containing more than one kind of absorbing substance. Provided that there are no interactions among the various species, the total absorbance for a multicomponent system is the sum of the individual absorbances. In other words,

$$A_{total} = A_1 + A_2 + \cdots + A_n = \varepsilon_1 bc_1 + \varepsilon_2 bc_2 + \cdots + \varepsilon_n bc_n \quad (21\text{-}9)$$

where the subscripts refer to absorbing components 1, 2, . . . , n.

21C-2   Limits to Beer's Law

We find few exceptions to the linear relationship between absorbance and pathlength at a fixed concentration. On the other hand, we frequently observe deviations from the direct proportionality between absorbance and concentration where b is a constant. Some of these deviations, called *real deviations*, are fundamental and represent real limitations to the law. Others occur as a consequence of the manner in which the absorbance measurements are made or as a result of chemical changes associated with concentration changes. These deviations are called *instrumental deviations* and *chemical deviations,* respectively.

Real Limitations to Beer's Law

Beer's law describes the absorption behavior of dilute solutions only and in this sense is a *limiting law.* At concentrations exceeding about 0.01 M, the average distances between ions or molecules of the absorbing species are diminished to the point where each particle affects the charge distribution, and thus the extent of absorption, of its neighbors. Because the extent of interaction depends on concentration, the occurrence of this phenomenon causes deviations from the linear relationship between absorbance and concentration. A similar effect sometimes occurs in dilute solutions of absorbers that contain high concentrations of other species, particularly electrolytes. When ions are in close proximity, the molar absorptivity of the analyte can be altered because of electrostatic interactions, which can lead to departures from Beer's law.

> Limiting laws in science are those that hold under limiting conditions such as dilute solutions. In addition to Beer's law, other limiting laws in chemistry include the Debye–Hückel law (see Chapter 9) and the law of independent migration that describes the conductance of electricity by ions.

Chemical Deviations

As shown in Example 21-3, deviations from Beer's law appear when the absorbing species undergoes association, dissociation, or reaction with the solvent to give products that absorb differently from the analyte. The extent of such departures can be predicted from the molar absorptivities of the absorbing species and the equilibrium constants for the equilibria involved. Unfortunately, we are usually unaware that such processes are affecting the analyte, so compensation is often impossible. Typical equilibria that give rise to this effect include monomer-dimer equilibria, metal complexation equilibria where more than one complex is present, acid/base equilibria, and solvent-analyte association equilibria.

Example 21-3

Consider a solution of an indicator with an acid dissociation constant of 1.42×10^{-5} and a molar analytical concentration c_{total}. Mass balance requires that

$$c_{total} = [HIn] + [In^-]$$

For all practical purposes the indicator is entirely in the acid form in 0.1 M HCl; that is, $c_{total} = [HIn]$. Likewise, in 0.1 M NaOH, the indicator is completely in its conjugate base form, and $c_{total} = [In^-]$. The molar absorptivities of the indicator species at 430 nm and 570 nm are

	ε_{430}	ε_{570}
HIn (in 0.1 M HCl)	6.30×10^2	7.12×10^3
In$^-$ (in 0.1 M NaOH)	2.06×10^4	9.60×10^2

Although both species individually obey Beer's law, plots of absorbance versus concentration of unbuffered solutions of the indicator are nonlinear, as shown in Figure 21-10, because the indicator equilibrium shifts as its concentration

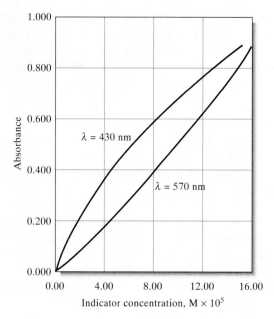

Figure 21-10   Chemical deviations from Beer's law for unbuffered solutions of the indicator HIn. The absorbance values were calculated at various indicator concentrations as shown in Example 21-3. Note that there are positive deviations at 430 nm and negative deviations at 570 nm. At 430 nm, the absorbance is primarily due to the ionized In^- form of the indicator and is in fact proportional to the fraction ionized. The fraction ionized varies nonlinearly with total concentration. At lower total concentrations ($[HIn] + [In^-]$), the fraction ionized is larger than at high total concentrations. Hence, a positive error occurs. At 570 nm, the absorbance is due principally to the undissociated acid HIn. The fraction in this form begins as a low amount and increases nonlinearly, with the total concentration giving rise to the negative deviation shown.

changes. However, if we know the equilibrium concentrations of HIn and In^-, we may calculate the observed absorbances. Thus, for a solution with $c_{total} = 2.00 \times 10^{-5}$ M and using 1.00 cm-cells,

$$[H_3O^+] = [In^-]$$

$$[HIn] = 2.00 \times 10^{-5} - [In^-]$$

$$K_{HIn} = \frac{[H_3O^+][In^-]}{[HIn]} = \frac{[In^-]^2}{2.00 \times 10^{-5} - [In^-]} = 1.42 \times 10^{-5}$$

$$[In^-] = 1.12 \times 10^{-5}$$

$$[HIn] = 2.00 \times 10^{-5} - 1.12 \times 10^{-5} = 0.88 \times 10^{-5}$$

Finally, if 1.00-cm cells are used,

$$A_{430} = (6.30 \times 10^2)(1.00)(0.88 \times 10^{-5})$$
$$+ (2.06 \times 10^4)(1.00)(1.12 \times 10^{-5}) = 0.236$$

and

$$A_{570} = (7.12 \times 10^3)(1.00)(0.88 \times 10^{-5})$$
$$+ (9.60 \times 10^2)(1.00)(1.12 \times 10^{-5}) = 0.073$$

CHALLENGE: Perform calculations to confirm that $A_{430} = 0.596$ and $A_{570} = 0.401$ for a solution in which the analytical concentration of HIn is 8.00×10^{-5} M.

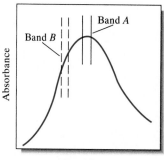

Instrumental Deviations

The need for monochromatic radiation and the absence of stray radiation are practical factors that limit the applicability of Beer's law. Beer's law strictly applies only when measurements are made with monochromatic source radiation. In practice, polychromatic sources that have a continuous distribution of wavelengths are used in conjunction with a grating or a filter to isolate a nearly symmetric band of wavelengths around the wavelength to be employed (see Section 22A-3). If the band selected corresponds to a region in which the absorptivity of the analyte is essentially constant, departures from Beer's law will be minimal. Many molecular bands in the UV/visible region fit this description. For these, Beer's law is obeyed as demonstrated by band A in Figure 21-11. On the other hand, some absorption bands in the UV/visible region and many in the IR region are very narrow, and departures from Beer's law are common, as illustrated for band B in Figure 21-11. Hence, to avoid deviations, it is advisable to select a wavelength band near the wavelength of maximum absorption where the analyte absorptivity changes little with wavelength. Atomic absorption lines are so narrow that they require special sources to obtain adherence to Beer's law as discussed in Section 23F.

Stray radiation, commonly called *stray light,* is defined as radiation from the instrument that is outside the nominal wavelength band chosen for the determination. This stray radiation is often the result of scattering and reflection off the surfaces of gratings, lenses or mirrors, filters, and windows. When measurements are made in the presence of stray light, the observed absorbance is given by

$$A' = \log \frac{P_0 + P_s}{P + P_s}$$

where P_s is the radiant power of the stray light. Figure 21-12 shows a plot of the apparent absorbance A' versus concentration for various levels of P_s, relative to P_0. Stray light always causes the apparent absorbance to be lower than the true absorbance. The deviations due to stray light are most significant at high absorbance values. Because stray radiation levels can be as high as 0.5% in modern instruments, absorbance levels above 2.0 are rarely measured unless special precautions are taken or special instruments with extremely low stray light levels are used. Some inexpensive filter instruments can exhibit deviations from Beer's law at absorbances as low as 1.0 because of high stray light levels or the presence of polychromatic light.

Another almost trivial, but important, deviation is caused by mismatched cells. If the cells holding the analyte and blank solutions are not of equal pathlength and equivalent in optical characteristics, an intercept will occur in the calibration curve, and $A = \varepsilon bc + k$ will be the actual equation instead of Equation 21-8. This error can be avoided either by using matched cells or by using a linear regression procedure (recall Example 21-2) to calculate both

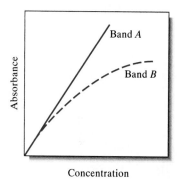

Figure 21-11   The effect of polychromatic radiation on Beer's law. In the absorption spectrum at the top, the absorptivity of the analyte is seen to be nearly constant over band A from the source. Note in the Beer's law plot at the bottom that employing band A gives a linear relationship. In the spectrum, band B coincides with a region of the spectrum over which the absorptivity of the analyte changes. Note the marked Beer's law deviation that results in the lower plot.

Polychromatic light, literally multicolored light, is light of many wavelengths, such as that from a tungsten light bulb.

Monochromatic light is light of a single wavelength, such as that from a laser, or a single small band of wavelengths.

Monochromatic light can be produced by filtering, diffracting, or refracting polychromatic light, as discussed in Section 22A-3.

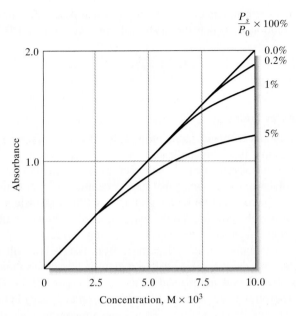

Figure 21-12 Deviation from Beer's law caused by various levels of stray light. Note that absorbance begins to level off with concentration at high stray light levels. Stray light always limits the maximum absorbance that can be obtained because when the absorbance is high, the radiant power transmitted through the sample can become comparable to or lower than the stray light level.

the slope and intercept of the calibration curve. In most cases, the linear regression strategy is best because an intercept can also occur if the blank solution does not totally compensate for interferences. Another way to avoid the mismatched cell problem with single-beam instruments is to use only one cell and keep it in the same position for both blank and analyte measurements. After obtaining the blank reading, the cell is emptied by aspiration, washed, and rinsed and filled with analyte solution.

21C-3 Absorption Spectra

An *absorption spectrum* is a plot of absorbance versus wavelength as illustrated in Figure 21-13a. Absorbance could also be plotted against wavenumber or frequency. Most modern scanning instruments can produce such an absorption spectrum directly. Older instruments often display transmittance and produce plots of %T versus wavelength as shown in Figure 21-13b. Occasionally, plots with log A as the ordinate are used. The logarithmic axis leads to a loss of spectral detail, but it is convenient for comparing solutions of widely different concentrations. A plot of molar absorptivity ε as a function of wavelength is independent of concentration. This type of spectral plot is characteristic for a given molecule and is sometimes used to aid in identifying or confirming the identity of a particular species. The color of a solution is related to the absorption spectrum (see Feature 21-2).

A bit of Latin: One plot of absorbance versus wavelength is called a *spectrum;* two or more plots are called *spectra.*

Why Is a Red Solution Red?

A solution such as $Fe(SCN)^{2+}$ is red not because the complex adds red radiation to the solvent. Instead, it absorbs green from the incoming white radiation and transmits the red component unaltered (see Figure 21F-2). Thus, in a colorimetric determination of iron based on its thiocyanate complex, the maximum change in absorbance with concentration occurs with green radiation; the absorbance change with red radiation is negligible. In general, then, the radiation used for a colorimetric analysis should be the complementary color of the analyte solution. The table below shows this relationship for various parts of the visible spectrum.

The Visible Spectrum

Wavelength Region Absorbed, nm	Color of Light Absorbed	Complementary Transmitted
400–435	Violet	Yellow-green
435–480	Blue	Yellow
480–490	Blue-green	Orange
490–500	Green-blue	Red
500–560	Green	Purple
560–580	Yellow-green	Violet
580–595	Yellow	Blue
595–650	Orange	Blue-green
650–750	Red	Green-blue

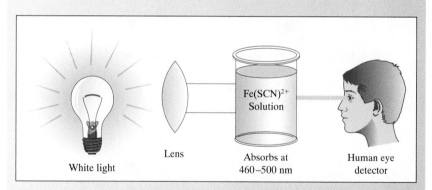

Figure 21F-2 Color of a solution. White light from a lamp or the sun strikes the solution of $Fe(SCN)^{2+}$. The fairly broad absorption spectrum shows a maximum absorbance in the 460 to 500 nm range. The complementary red color is transmitted.

21C-4 Molecular Absorption Transitions

Three types of energy changes occur when *molecules* are excited by ultraviolet, visible, and infrared radiation. For UV/visible radiation, when we excite a molecule, an electron residing in a low-energy molecular or atomic orbital is promoted to a higher-energy orbital. The change in energy levels is called a *transition*. We can only get such a transition when the energy $h\nu$ of the photon is exactly the same as the energy difference between the two orbital energies (recall Figure 21-5). The transition of an electron between different energy levels

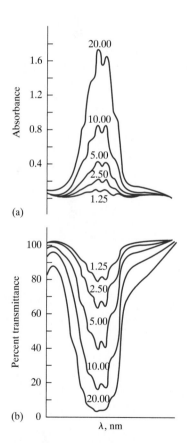

Figure 21-13 Methods for plotting absorption spectra. The numbers for the curves indicate ppm of $KMnO_4$; $b = 2.00$ cm. [From M. G. Mellon, *Analytical Absorption Spectroscopy* (New York: Wiley, 1950), pp. 104–106. With permission.]

An electronic absorption transition involves promotion of an electron to an empty or partially filled molecular orbital.

is called an *electronic transition,* and the absorption process is called *electronic absorption.*

Molecules exhibit two additional types of radiation-induced transitions, namely *vibrational transitions* and *rotational transitions.* The vibrational energy of a molecule is associated with the bonds that hold the molecule together. The vibrational energy of a molecule is quantized and can only assume certain discrete levels. Transitions between vibrational levels are vibrational transitions.

To get an idea of the nature of vibrational states, picture a chemical bond as a flexible spring with atoms attached at both ends. In Figure 21-14a, two types of stretching vibrations are shown. With each vibration, atoms first approach and then move away from one another. The potential energy of the system at any instant depends on the extent to which the spring-like bond is stretched or compressed. For an ordinary spring, the potential energy of the system varies continuously and reaches a maximum when the spring is fully stretched or fully compressed. In contrast, the energy of a spring system with atomic dimensions can assume only discrete levels. Figure 21-14b shows several additional types of molecular vibrations. The energy associated with each of these vibrational states differs from others and from the energies associated with stretching vibrations.

In addition to quantized vibrational states, a molecule has a host of quantized rotational states that are associated with the rotational motion of a molecule around its center of mass. The overall energy E associated with a molecule in a given state can be written as

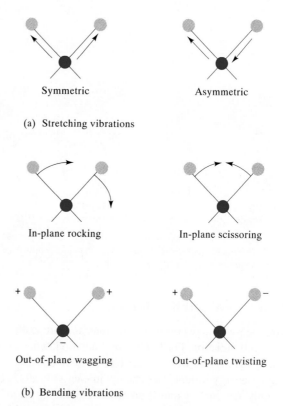

Symmetric   Asymmetric

(a) Stretching vibrations

In-plane rocking   In-plane scissoring

Out-of-plane wagging   Out-of-plane twisting

(b) Bending vibrations

Figure 21-14 Types of molecular vibrations. The plus sign indicates motion from the page toward the reader; the minus sign indicates motion in the opposite direction.

$$E = E_{\text{electronic}} + E_{\text{vibrational}} + E_{\text{rotational}} \qquad (21\text{-}10)$$

where $E_{\text{electronic}}$ is the electronic energy of the molecule, $E_{\text{vibrational}}$ is its vibrational energy, and $E_{\text{rotational}}$ is its rotational energy.

Energy-Level Diagrams for Molecules

Figure 21-15 is a partial energy-level diagram depicting typical processes that occur when a polyatomic molecule absorbs infrared, visible, and ultraviolet radiation. The energies E_1 and E_2 of the first two of the several electronically excited states of a molecule are shown relative to the energy of its ground state E_0. In addition, the relative energies of a few of the many vibrational states associated with each electronic state are indicated by the lighter horizontal lines labeled 1, 2, 3, and 4. The lowest vibrational levels are labeled 0. Note that the differences in energy among the vibrational states are typically an order of magnitude smaller than among energy levels of the electronic states.

Although they are not shown in Figure 21-15, a set of rotational energy states is superimposed on each of the vibrational states shown in the diagram. The energy differences among these states are even smaller than those among vibrational states. Thus, pure rotational transitions can be stimulated by low-energy radiation in the microwave region of the spectrum.

Infrared Absorption.   Infrared radiation is generally not of sufficient energy to cause electronic transitions but can induce transitions in the vibrational and rotational states associated with the *ground electronic state* of the molecule. Four of these transitions are depicted in the lower left of Figure 21-15. For absorption to occur, the analyte must be irradiated with frequencies corresponding exactly to the energies indicated by the lengths of the four arrows (wavelengths $\lambda_1 - \lambda_4$ as shown).

Ultraviolet and Visible Absorption.   The center arrows in Figure 21-15 suggest that the molecules under consideration absorb visible radiation of five wavelengths (λ_1' to λ_5'), thereby promoting electrons to the five vibrational levels of the excited electronic level E_1. Ultraviolet photons that are more energetic are required to produce the absorption to level E_2 shown by the five arrows on the right (wavelengths λ_1'' to λ_5'').

As suggested by the energy levels of Figure 21-15, molecular absorption in the UV/visible region consists of absorption *bands* made up of closely spaced lines. A real molecule will have many more vibrational levels than shown here and will necessarily have a more complex spectrum. Such spectra can be observed in the gas phase with a high-resolution instrument. In solution, interaction between the absorbing species and solvent molecules results in a spreading of quantum states and causes the spectrum to become smooth and continuous.

Figure 21-16 shows visible spectra for 1,2,4,5-tetrazine that were obtained under three different conditions. Note that, in the gas phase, the individual tetrazine molecules are sufficiently separated from one another to vibrate and rotate freely, and many individual absorption peaks resulting from transitions among the various vibrational and rotational states are clearly evident. In the condensed state and in solution, however, freedom to rotate is largely lost, and rotational structure is rarely observed. Furthermore, frequent collisions and

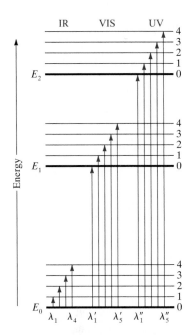

Figure 21-15   Energy-level diagram showing some of the energy changes that occur during absorption of infrared (IR), visible (VIS), and ultraviolet (UV) radiation. Note that with some molecules a transition from E_0 to E_1 may require UV radiation instead of visible radiation shown. With other molecules, the transition from E_0 to E_2 may occur with visible radiation instead of UV radiation.

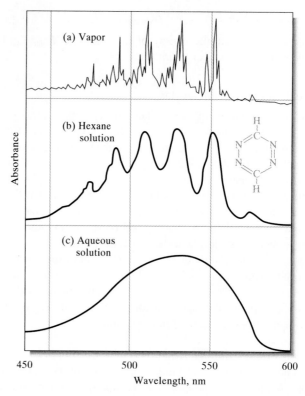

Figure 21-16 Typical ultraviolet absorption spectra. The compound is 1,2,4,5-tetrazine. In (a), the spectrum is shown in the gas phase where many lines due to electronic, vibrational, and rotational transitions are seen. In a nonpolar solvent (b), the electronic transitions can be observed, but the vibrational and rotational structure have been lost. In a polar solvent (c), the strong intermolecular forces have caused the electronic peaks to blend together to give only a single, smooth absorption peak. (From S. F. Mason, *J. Chem. Soc.,* **1959,** 1265. With permission.)

interactions with solvent molecules cause the vibrational levels to be modified energetically in an irregular way.

WEB WORKS

To learn more about Beer's law, set your browser to *http://www.scimedia. com/chem-ed/spec/beerslaw.htm.* Find how the molar absorptivity of a compound (ε) relates to the absorption cross section (σ).

21D   QUESTIONS AND PROBLEMS

*21-1. Why is a solution of $Cu(NH_3)_4^{2+}$ blue?

21-2. What is the relationship between
  *(a) absorbance and transmittance?
  (b) absorptivity a and molar absorptivity ε?

*21-3. Identify factors that cause the Beer's law relationship to depart from linearity.

21-4. Describe the difference between "real" deviations from Beer's law and those due to instrumental or chemical factors.

21-5. How does an electronic transition resemble a vibrational transition? How do they differ?

21-6. Calculate the frequency in hertz of
  *(a) an X-ray beam with a wavelength of 2.65 Å.
  (b) an emission line for copper at 211.0 nm.
  *(c) the line at 694.3 nm produced by a ruby laser.
  (d) the output of a CO_2 laser at 10.6 μm.
  *(e) an infrared absorption peak at 19.6 μm.
  (f) a microwave beam at 1.86 cm.

21-7. Calculate the wavelength in centimeters of
  *(a) an airport tower transmitting at 118.6 MHz.
  (b) a VOR (radio navigation aid) transmitting at 114.10 kHz.
  *(c) an NMR signal at 105 MHz.
  (d) an infrared absorption peak with a wavenumber of 1210 cm^{-1}.

21-8. A typical simple infrared spectrophotometer covers a wavelength range from 3 to 15 μm. Express its range
  (a) in wavenumbers. (b) in hertz.

*21-9. A sophisticated ultraviolet/visible/near-IR instrument has a wavelength range of 185 to 3000 nm. What are its wavenumber and frequency ranges?

21-13. What are the units for absorptivity when the path length is given in centimeters and the concentration is expressed in
  *(a) parts per million?
  (b) micrograms per liter?
  *(c) weight-volume percent?
  (d) grams per liter?

21-14. Express the following absorbances in terms of percent transmittance:
  *(a) 0.0510 (b) 0.918
  *(c) 0.379 (d) 0.261
  *(e) 0.485 (f) 0.702

21-15. Convert the accompanying transmittance data to absorbances:
  *(a) 25.5% (b) 0.567
  *(c) 32.8% (d) 3.58%
  *(e) 0.085 (f) 53.8%

21-16. Calculate the percent transmittance of solutions that have twice the absorbance of the solutions in Problem 21-14.

21-17. Calculate the absorbances of solutions with half the transmittance of those in Problem 21-15.

21-18. Evaluate the missing quantities in the accompanying table. Where needed, use 200 for the molar mass of the analyte.

	A	%T	ε L mol^{-1} cm^{-1}	a cm^{-1} ppm^{-1}	b cm	c M	c ppm
*(a)	0.172		4.23×10^3		1.00		
(b)		44.9		0.0258		1.35×10^{-4}	
*(c)	0.520		7.95×10^3		1.00		
(d)		39.6		0.0912			1.76
*(e)			3.73×10^3		0.100	1.71×10^{-3}	
(f)		83.6			1.00	8.07×10^{-6}	
*(g)	0.798				1.50		33.6
(h)		11.1	1.35×10^4			7.07×10^{-5}	
*(i)		5.23	9.78×10^3				5.24
(j)	0.179				1.00	7.19×10^{-5}	

21-10. Calculate the frequency in hertz and the energy in joules of an X-ray photon with a wavelength of 2.70 Å (1 eV = 1.062 J).

*21-11. Calculate the wavelength in cm and the energy in joules associated with a signal at 220 MHz.

21-12. Calculate the wavelength of
  *(a) the sodium line at 589 nm in an aqueous solution with a refractive index of 1.35.
  (b) the output of a ruby laser at 694.3 nm when it is passing through a piece of quartz that has a refractive index of 1.55.

*21-19. A solution containing 4.48 ppm $KMnO_4$ has a transmittance of 0.309 in a 1.00-cm cell at 520 nm. Calculate the molar absorptivity of $KMnO_4$.

21-20. Beryllium(II) forms a complex with acetylacetone (166.2 g/mol). Calculate the molar absorptivity of the complex, given that a 1.34 ppm solution has a transmittance of 55.7% when measured in a 1.00-cm cell at 295 nm, the wavelength of maximum absorption.

*21-21. At 580 nm, the wavelength of its maximum absorption, the complex $Fe(SCN)^{2+}$ has a molar absorptivity of 7.00×10^3 L mol^{-1} cm^{-1}. Calculate

(a) the absorbance of a 2.50×10^{-5} M solution of the complex at 580 nm in a 1.00-cm cell.

(b) the absorbance of a solution in which the concentration of the complex is twice that in part (a).

(c) the transmittance of the solutions described in parts (a) and (b).

(d) the absorbance of a solution that has half the transmittance of that described in part (a).

*21-22. A 2.50-mL aliquot of a solution that contains 3.8 ppm iron(III) is treated with an appropriate excess of KSCN and diluted to 50.0 mL. What is the absorbance of the resulting solution at 580 nm in a 2.50-cm cell? See Problem 21-21 for absorptivity data.

21-23. A solution containing the complex formed between Bi(III) and thiourea has a molar absorptivity of 9.32×10^3 L mol^{-1} cm^{-1} at 470 nm.

(a) What is the absorbance of a 6.24×10^{-5} M solution of the complex at 470 nm in a 1.00-cm cell?

(b) What is the percent transmittance of the solution described in part (a)?

(c) What is the molar concentration of the complex in a solution that has the absorbance described in part (a) when measured at 470 nm in a 5.00-cm cell?

*21-24. The complex formed between Cu(I) and 1,10-phenanthroline has a molar absorptivity of 7000 L mol^{-1} cm^{-1} at 435 nm, the wavelength of maximum absorption. Calculate

(a) the absorbance of an 8.50×10^{-5} M solution of the complex when measured in a 1.00-cm cell at 435 nm.

(b) the percent transmittance of the solution in part (a).

(c) the concentration of a solution that in a 5.00-cm cell has the same absorbance as the solution in part (a).

(d) the pathlength through a 3.40×10^{-5} M solution of the complex that is needed for an absorbance that is the same as the solution in part (a).

21-25. A solution with a "true" absorbance $[A = \log(P_0/P)]$ of 2.10 was placed in a spectrophotometer with a stray light level (P_s/P_0) of 0.75%. What absorbance A' would be measured? What percentage error would result?

*21-26. A compound X is to be determined by UV/Visible spectrophotometry. A calibration curve is constructed from standard solutions of X with the following results: 0.50 ppm, $A =$ 0.24; 1.5 ppm, $A = 0.36$; 2.5 ppm, $A = 0.44$; 3.5 ppm, $A = 0.59$; 4.5 ppm, $A = 0.70$. Find the slope and intercept of the calibration curve, the standard error in y, the concentration of the solution of unknown X concentration and the standard deviation in the concentration of X. Construct a plot of the calibration curve and determine the unknown concentration by hand from the plot.

21-27. One common way to determine phosphorus in urine is to treat the sample, after removing the protein, with molybdenum (VI) and then reducing the resulting 12-molybdophosphate complex with ascorbic acid to give an intense blue colored species called molybdenum blue. The absorbance of molybdenum blue can be measured at 650 nm. A 24-hour urine sample was collected and the patient produced 1122 mL in 24 hours. A 1.00 mL aliquot of the sample was treated with Mo(VI) and ascorbic acid and diluted to a volume of 50.00 mL. A calibration curve was prepared by treating 1.00 mL aliquots of phosphate standard solutions in the same manner as the urine sample. The absorbances of the standards and the urine sample were obtained at 650 nm and the following results obtained:

Solution	Absorbance at 650 nm
1.00 ppm P	0.230
2.00 ppm P	0.436
3.00 ppm P	0.638
4.00 ppm P	0.848
Urine sample	0.518

(a) Find the slope and intercept of the calibration curve. Construct a plot of the calibration curve. Determine the number of ppm P in the urine sample and the standard deviation from the least-squares equation of the line. Compare the unknown concentration to that obtained manually from a calibration curve.

(b) How many grams of phosphorus were eliminated per day by the patient?

(c) What is the phosphate concentration in urine in mM?

21-28. Nitrite is commonly determined by a colorimetric procedure using a reaction called the Griess reaction. In this reaction, the sample containing nitrite is reacted with sulfanilimide and N-(1-Napthyl)ethylenediamine to form a colored species that absorbs at 550 nm. Using an automated flow analysis instrument, the following results were obtained for standard solutions of nitrite and for a sample containing an unknown amount.

Solution	Absorbance at 550 nm
2.00 μM	0.065
6.00 μM	0.205
10.00 μM	0.338
14.00 μM	0.474
18.00 μM	0.598
Unknown	0.402

*(a)   Obtain the slope and intercept of the calibration curve, and the standard error in y.

(b)   Construct a plot of the calibration curve.

*(c)   Determine the concentration of nitrite in the sample and its standard deviation.

Chapter 22

Instruments for Measuring Absorption: Is It a Photometer, a Spectrophotometer, or a Spectrometer?

The basic components of analytical instruments for absorption, as well as for emission and fluorescence spectroscopy, are remarkably alike in function and in general performance requirements regardless of whether the instruments are designed for UV, visible, or IR radiation. Because of the similarities, such instruments are frequently referred to as *optical instruments* even though the human eye is only sensitive to the visible region. In this chapter, we first examine the characteristics of the components common to all optical instruments. We then consider the characteristics of typical instruments designed for UV, visible, and IR absorption spectroscopy.

22A INSTRUMENT COMPONENTS

Most spectroscopic instruments in the UV/visible and IR regions are made up of five components, including (1) a stable source of radiant energy; (2) a wavelength selector that isolates a limited region of the spectrum for measurement; (3) one or more sample containers; (4) a radiation detector, which converts radiant energy to a measurable electrical signal; and (5) a signal processing and readout unit usually consisting of electronic hardware and, invariably today, a computer. Figure 22-1 illustrates the three ways these components are configured for carrying out optical spectroscopic measurements. As can be seen in the figure, components (3), (4), and (5) have similar configurations for each type of measurement.

The first two instrumental configurations, for absorption and fluorescence, require an external source of radiation. In absorption measurements (Figure 22-1a), the attenuation of the source radiation at the selected wavelength is measured. In fluorescence measurements (Figure 22-1b), the source excites the analyte and causes the emission of characteristic radiation, which is usually measured at a 90-degree angle with respect to the incident source beam. In emission

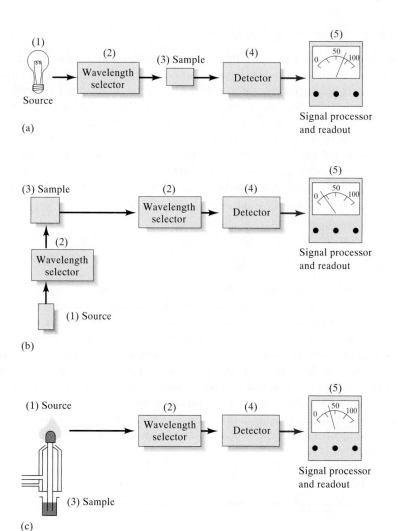

Figure 22-1 Components of various types of instruments for optical spectroscopy. In (a), the arrangement for absorption measurements is shown. Note that source radiation of the selected wavelength is sent through the sample and that the transmitted radiation is measured by the detector/signal processing/readout unit. With some instruments, the position of the sample and wavelength selector is reversed. In (b), the configuration for fluorescence measurements is shown. Here, two wavelength selectors are needed to select the excitation and the emission wavelengths. The selected source radiation is incident on the sample and the radiation emitted is measured, usually at right angles to avoid scattering. In (c), the configuration for emission spectroscopy is shown. Here, a source of thermal energy, such as a flame, produces an analyte vapor that emits radiation that is isolated by the wavelength selector and converted to an electrical signal by the detector.

We often call the UV/visible and IR regions of the spectrum the optical region. Even though the optic nerve is responsive only to visible radiation, the other regions are included because the lenses, mirrors, prisms, and gratings used are similar and function in a comparable manner. Spectroscopy in the UV/visible and IR regions is therefore often called *optical spectroscopy*.

spectroscopy (Figure 22-1c), the sample itself is the emitter and no external radiation source is needed. In emission methods, the sample is usually fed into a plasma or a flame that provides enough thermal energy to cause the analyte to emit characteristic radiation. Emission and fluorescence methods are described in more detail in Sections 23D and 23F.

22A-1 Selecting Optical Materials

The cells, windows, lenses, mirrors, and wavelength selecting elements in an optical spectroscopic instrument must transmit radiation in the wavelength region being employed. Figure 22-2 shows the usable wavelength range for several optical materials that find use in the UV, visible, and IR regions of the spectrum. Ordinary silicate glass is completely adequate for the visible region and has the considerable advantage of low cost. In the UV region, at wavelengths shorter than about 380 nm, glass begins to absorb and fused silica or quartz must be substituted. Also, glass, quartz, and fused silica all absorb in the IR region at wavelengths longer than about 2.5 μm. Hence, optical elements for IR spectrometry are typically made from halide salts or in some cases polymeric materials.

22A-2 The Source of It All

To be suitable for spectroscopic studies, a source must generate a beam of radiation that is sufficiently powerful for easy detection and measurement. In addition, its output power should be stable for reasonable periods. For good stability, a well-regulated power supply must provide electrical power for the source. Spectroscopic sources are of two types: *continuum sources,* which emit radiation that changes in intensity only slowly as a function of wavelength, and *line sources,* which emit a limited number of spectral lines, each of which spans a very limited wavelength range. The distinction between these sources is illustrated in Figure 22-3. Sources can also be classified as *continuous sources,* which emit radiation continuously with time, or *pulsed sources,* which emit radiation in bursts.

> A continuum source provides a broad distribution of wavelengths within a particular spectral range. This distribution is known as a spectral continuum.

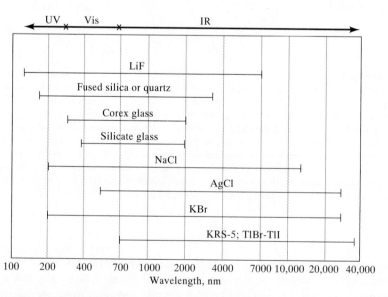

Figure 22-2 Transmittance ranges for various optical materials. Simple glasses are fine in the visible region, whereas fused silica or quartz is necessary in the UV region (<380 nm). Halide salts (KBr, NaCl, AgCl) are often used in the IR, but they have the disadvantages of being expensive and being somewhat water soluble.

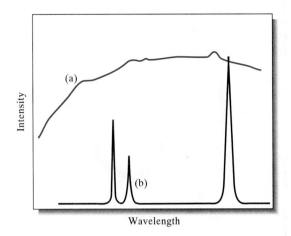

Figure 22-3 Spectral source types. The spectrum of a continuum source (a) is much broader than that of a line source (b).

Continuum Sources in the UV/Visible Region

The most widely used continuum sources are listed in Table 22-1. An ordinary tungsten filament lamp provides a distribution of wavelengths from 320 to 2500 nm (Figure 22-4). Generally, these lamps are operated at a temperature of around 2900 K, which produces useful radiation from about 350 to 2200 nm.

Tungsten/halogen lamps, also called quartz/halogen lamps, contain a small amount of iodine within the quartz envelope that houses the filament. Quartz allows the filament to be operated at a temperature of about 3500 K, which leads to higher intensities and extends the range of the lamp well into the UV region. The lifetime of a tungsten/halogen lamp is more than double that of an ordinary tungsten lamp because the life of the latter is limited by sublimation of tungsten from the filament. In the presence of iodine, the sublimed tungsten reacts to give gaseous WI_2 molecules, which then diffuse back to the hot filament where they decompose and redeposit as W atoms. These lamps are finding ever-increasing use in modern spectroscopic instruments because of their extended wavelength range, great intensity, and long life.

Table 22-1

Continuum Sources for Optical Spectroscopy

Source	Wavelength Region, nm	Type of Spectroscopy
Xenon arc lamp	250–600	Molecular fluorescence
H_2 and D_2 lamps	160–380	UV molecular absorption
Tungsten/halogen lamp	240–2500	UV/visible/near-IR molecular absorption
Tungsten lamp	350–2200	Visible/near-IR molecular absorption
Nernst glower	400–20,000	IR molecular absorption
Nichrome wire	750–20,000	IR molecular absorption
Globar	1200–40,000	IR molecular absorption

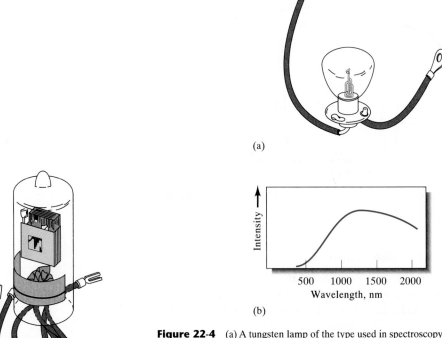

Figure 22-4 (a) A tungsten lamp of the type used in spectroscopy and (b) its spectrum. Intensity of the tungsten source is usually quite low at wavelengths shorter than about 350 nm. Note that the intensity reaches a maximum in the near-IR region of the spectrum ($\approx$1200 nm in this case).

Figure 22-5 (a) A deuterium lamp of the type used in spectrophotometers and (b) its spectrum. Note that the maximum intensity occurs $\approx$225 nm. Typically instruments switch from deuterium to tungsten at $\approx$350 nm.

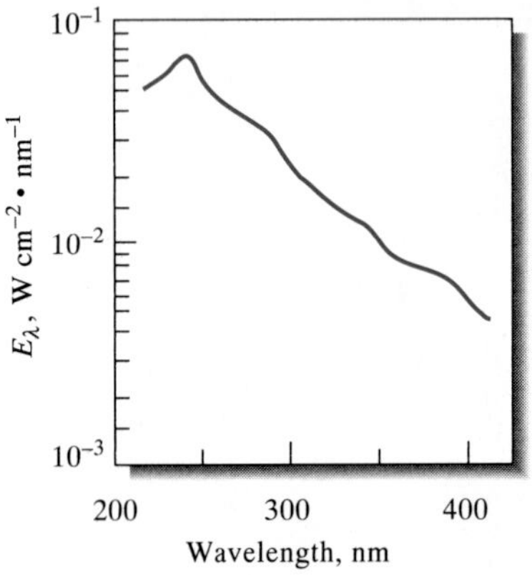

Deuterium (and also hydrogen) lamps are most often used to provide continuum radiation in the UV region. A deuterium lamp consists of a cylindrical tube, containing deuterium at low pressure, with a quartz window from which the radiation exits as shown in Figure 22-5. Excitation is carried out by applying about 40 V between a heated oxide-coated electrode and a metal electrode. Excited deuterium dissociates in the resulting plasma to give atomic species plus a UV photon. The energy of the emitted radiation can vary in a continuous manner. The result is a continuum spectrum from about 160 nm to about 350 to 400 nm.

Other UV/Visible Sources

In addition to the continuum sources discussed earlier, line sources are also important for use in the UV/visible region. Low-pressure mercury arc lamps are very common sources for use in liquid chromatography detectors. The dominant line emitted by these sources is at 253.7 nm. Hollow cathode lamps are also common line sources used specifically for atomic absorption spectroscopy as discussed in Section 23F. Lasers have also been used in molecular and atomic spectroscopy, both for single wavelength and for scanning applications. Tunable dye lasers can be scanned over wavelength ranges of several hundred nanometers when more than one dye is used.

Continuum Sources in the IR Region

The continuum sources for IR radiation are normally heated inert solids. A *Globar* source consists of a silicon carbide rod. Infrared radiation is emitted when the Globar is heated to about 1500°C by the passage of electricity; Table 22-1 gives the wavelength range of these sources.

A *Nernst glower* is a cylinder of zirconium and yttrium oxides which emits IR radiation when heated to a high temperature by an electric current. Electrically heated spirals of nichrome wire also serve as inexpensive IR sources.

22A-3   Selecting the Desired Wavelength

Spectroscopic instruments in the UV and visible regions are usually equipped with one or more devices to restrict the radiation being measured to a narrow band that is absorbed or emitted by the analyte. Such devices greatly enhance both the selectivity and the sensitivity of an instrument. In addition, for absorption measurements, as was shown in Section 21C-2, narrow bands of radiation greatly diminish the chance for Beer's law deviations due to polychromatic radiation. Many instruments employ a *monochromator* or *filter* to isolate the desired wavelength band so that only the band of interest is detected and measured. Others use a *spectrograph* to spread out, or disperse, the wavelengths so that they can be detected with a multichannel detector.

Monochromators generally employ a diffraction grating (see Feature 22-1) to disperse the radiation into its component wavelengths as shown in Figure 22-6a. Older instruments used prisms for this purpose, as seen in Figure 22-6b. By rotating the grating, different wavelengths can be made to pass through an exit slit. The output wavelength of a monochromator is thus continuously variable over a considerable spectral range. The wavelength range passed by a monochromator, called the *spectral bandpass* or *effective bandwidth,* can be less than 1 nm for moderately expensive instruments to greater than 20 nm for inexpensive systems. Because of the ease with which the wavelength can be changed with a monochromator-based instrument, these systems are widely used for spectral scanning applications as well as for applications requiring a fixed wavelength. With an instrument containing a spectrograph, the sample and wavelength selector are reversed from the configuration shown in Figure 22-1. Like the monochromator, the spectrograph contains a diffraction grating to disperse the spectrum. The spectrograph has no exit slit, however, which thus allows the dispersed spectrum to strike a multiwavelength detector. Still other instruments used for emission spectroscopy employ a device called a *polychromator,* which contains multiple exit slits and multiple detectors that allow many discrete wavelengths to be measured simultaneously.

Filters used for absorption measurements are typically *interference filters.* These filters transmit radiation over a bandwidth of 5 to 20 nm. Radiation outside the transmitted bandpass is removed by destructive interference. Filters have the advantages of simplicity, ruggedness, and low cost. One filter, however, can only isolate one band of wavelengths; a new filter must be used for a different band. Therefore, interference filter instruments are used only for measurements that are made at fixed or infrequently changed wavelength.

A *spectrograph* is a device that uses a grating to disperse a spectrum. It contains an entrance slit to define the area of the source to be viewed. A large opening at its exit allows a wide range of wavelengths to strike a multiwavelength detector. A *monochromator* is a device that contains an entrance slit and an exit slit. The latter is used to isolate a small band of wavelengths. One band at a time is isolated, and different bands can be transmitted sequentially by rotating the grating. A *polychromator* contains multiple exit slits so that several wavelength bands can be isolated simultaneously.

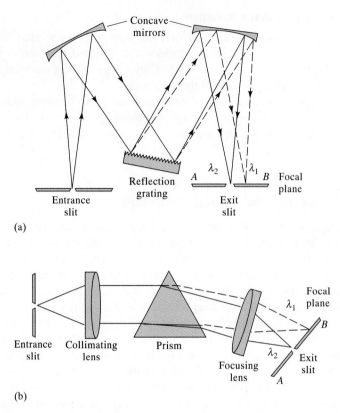

Figure 22-6 Types of monochromators: (a) grating monochromator and (b) prism monochromator. The monochromator design in (a) is a Czerny–Turner design, whereas the prism monochromator in (b) is a Bunsen design. In both cases, $\lambda_1 > \lambda_2$.

In the IR region of the spectrum, most modern instruments do not disperse the spectrum at all, although dispersion was common with older instruments. Instead, an *interferometer* is used, and the constructive and destructive interference of electromagnetic waves are used to obtain spectral information through a technique called Fourier transformation. These IR instruments are further discussed in Section 22C-2.

22A-4   Detecting and Measuring Radiant Energy

To obtain spectroscopic information, the radiant power transmitted, fluoresced, or emitted must be detected in some manner and converted into a measurable quantity. A *detector* is a device that indicates the existence of some physical phenomenon. Familiar examples of detectors include photographic film for indicating the presence of electromagnetic or radioactive radiation, the pointer of a balance for indicating mass differences, and the mercury level in a thermometer for indicating temperature. The human eye is also a detector; it converts visible radiation into an electrical signal that is passed to the brain via a chain of neurons in the optic nerve and produces vision.

Invariably in modern instruments, the information of interest is encoded and processed as an electrical signal. The term *transducer* is used to indicate the type

A *transducer* is a type of detector that converts various types of chemical and physical quantities into electrical signals such as electrical charge, current, or voltage.

Dispersion of UV/visible radiation can be brought about by directing a polychromatic beam through a *transmission grating* or onto the surface of a *reflection grating.* Reflection gratings are far more widely used than transmission gratings. *Replica gratings,* which are used in many monochromators, are manufactured from a *master grating.* The master grating consists of a large number of parallel and closely spaced grooves ruled on a hard, polished surface with a suitably shaped diamond tool. For the UV/visible region, a grating will contain from 50 to 6000 grooves per mm, with 1200 to 2400 grooves the most common. Master gratings are ruled by a diamond ruling tool operated by a ruling engine. The construction of a good master grating is tedious, time consuming, and expensive; the grooves must be identical in size, exactly parallel, and equally spaced over the typical 3- to 10-cm length of the grating. Because of the difficulty in construction, few master gratings are produced.

The modern era of gratings dates back to the 1880s when Henry A. Rowland constructed an engine capable of ruling gratings up to 6 inches wide with over 100,000 grooves. A simplified drawing of the Rowland engine is shown in Figure 22F-1. With this machine, a high-precision screw moves the grating carriage while a diamond stylus cuts the tiny parallel grooves. Imagine ruling a grating with 100,000 grooves in 6 in. manually! The engine required around 5 hr just to warm up to a nearly uniform temperature. After this, nearly 15 hr more were needed to obtain a uniform layer of lubricant on the surface. Only after this time was the diamond lowered to begin the ruling process. Large gratings required almost a week to produce. Two important improvements were made by John Strong in the 1930s. The most important of these was to vacuum deposit aluminum onto glass blanks as a medium. The thin layer of aluminum gave a much smoother surface and reduced wear on the diamond tool. Strong's second improvement was to reciprocate the grating blank instead of the diamond tool.

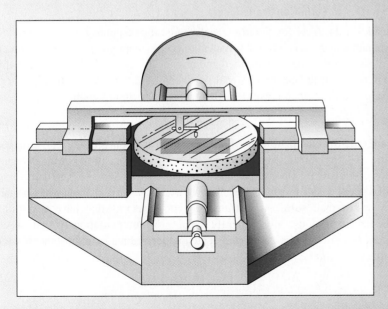

Figure 22F-1   Simplified diagram of the Rowland ruling engine. A single high-precision screw moves the grating carriage. A diamond point then travels over the grating, which is ruled on a concave mirror surface. Machines of this type were the models for many of the ruling engines made.

Today, ruling engines use interferometric control over the ruling process. Fewer than 50 ruling engines are in use around the world, however. Even if all these engines were operated 24 hr a day, they could not begin to meet the demand for gratings. Fortunately, modern coating and resin technology have made it possible to produce replica gratings of very high quality. Replica gratings are formed from the master grating by vacuum deposition of aluminum onto a ruled master grating. The aluminum layer is subsequently coated with an epoxy-type material. The material is then polymerized, and the replica is separated from the master. The replica gratings of today are superior to the master gratings produced in the past.

Another way in which gratings are made is a result of laser technology. These *holographic gratings* are made by coating a flat glass plate with a photosensitive material. Beams from a pair of identical lasers then strike the coated glass surface. The resulting interference fringes from the two beams sensitize the photoresist, producing areas that can be dissolved away to leave a grooved structure. Aluminum is then vacuum deposited to produce a reflection grating. The spacing of the grooves can be changed by changing the angle of the two laser beams with respect to one another. Nearly perfect gratings with as many as 6000 lines per mm can be manufactured in this way at a relatively low cost. Holographic gratings are not quite as efficient in terms of their light output as ruled gratings, but they can eliminate false lines, called *grating ghosts,* and reduce scattered light that results from ruling errors.

of detector that converts quantities, such as light intensity, pH, mass, and temperature, into such *electrical signals* that can be subsequently amplified, manipulated, and finally converted into numbers proportional to the magnitude of the original quantity. All the detectors discussed here are radiation transducers.

Properties of Radiation Transducers

The ideal transducer for electromagnetic radiation responds rapidly to low levels of radiant energy over a broad wavelength range. In addition, it produces an electrical signal that is easily amplified and has a low electrical noise level. Finally, it is essential that the electrical signal produced by the transducer be directly proportional to the radiant power P of the beam as shown in Equation 22-1:

$$G = KP + K' \qquad (22\text{-}1)$$

where G is the electrical response of the detector in units of current, voltage, or charge. The proportionality constant K measures the sensitivity of the detector in terms of electrical response per unit of radiant power input. Many detectors exhibit a small constant response K', known as a *dark current,* even when no radiation strikes their surfaces. Instruments with detectors that have a significant dark-current response are ordinarily capable of compensation so that the dark current is automatically subtracted. Thus, under ordinary circumstances, we can simplify Equation 22-1 to

$$G = KP \qquad (22\text{-}2)$$

Types of Transducers

As shown in Table 22-2, we encounter two general types of transducers: one type responds to photons, the other to heat. All photon detectors are based on the interaction of radiation with a reactive surface to produce electrons (*photoemis-*

Table 22-2

Common Detectors for Absorption Spectroscopy

Type	Wavelength Range, nm	Type of Spectroscopy
Photon Detectors		
Phototubes	150–1000	UV/visible and near-IR absorption
Photomultiplier tubes	150–1000	UV/visible and near IR absorption, molecular fluorescence
Silicon photodiodes	350–1100	Visible and near IR absorption
Photoconductive cells	1000–50,000	IR absorption
Thermal Detectors		
Thermocouples	600–20,000	IR absorption
Bolometers	600–20,000	IR absorption
Pneumatic cells	600–40,000	IR absorption
Pyroelectric cells	1000–20,000	IR absorption

sion) or to promote electrons to energy states in which they can conduct electricity (*photoconduction*). Only UV, visible, and near-IR radiation possess enough energy to cause photoemission to occur; thus, photoemissive detectors are limited to wavelengths shorter than about 2 μm (2000 nm). Photoconductors can be used in the near-, mid-, and far-IR regions of the spectrum.

Generally, we detect IR radiation by measuring the temperature rise of a blackened material located in the path of the beam or by measuring the increase in electrical conductivity of a photoconducting material when it absorbs IR radiation. Because the temperature changes resulting from the absorption of the IR energy are minute, close control of the ambient temperature is required if large errors are to be avoided. Usually, the detector system limits the sensitivity and precision of an IR instrument.

Photon Detectors

Widely used types of photon detectors include phototubes, photomultiplier tubes, silicon photodiodes, and photodiode arrays.

Phototubes and Photomultiplier Tubes. The response of a phototube or a photomultiplier tube is based on the photoelectric effect. As shown in Figure 22-7, a phototube consists of a semicylindrical photocathode and a wire anode sealed inside an evacuated transparent glass or quartz envelope. The concave surface of the cathode supports a layer of photoemissive material, such as an alkali metal or metal oxide, that emits electrons when irradiated with light of the appropriate energy. When a voltage is applied across the electrodes, the emitted *photoelectrons* are attracted to the positively charged wire anode. In the complete circuit shown in Figure 22-7, a *photocurrent* then results that is readily amplified and measured. The number of photoelectrons ejected from the photocathode per unit time is directly proportional to the radiant power of the beam striking the surface. With an applied voltage of about 90 V or more, all these

Photoelectrons are electrons that are ejected from a photosensitive surface by electromagnetic radiation. A photocurrent is the current in an external circuit that is limited by the rate of ejection of photoelectrons.

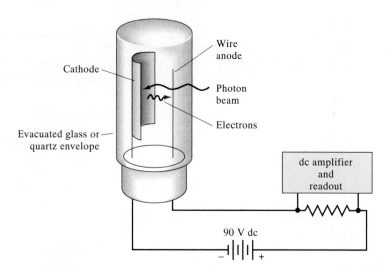

Figure 22-7   A phototube and accompanying circuit. The photocurrent induced by the radiation causes a voltage across the measuring resistor; this voltage is then amplified and measured.

photoelectrons are collected at the anode to give a photocurrent that is also proportional to the radiant power of the beam.

The *photomultiplier tube* (PMT) is similar in construction to the phototube, but it is significantly more sensitive. The photocathode is similar to that of the phototube with electrons being emitted upon exposure to radiation. In place of a single wire anode, however, the PMT has a series of electrodes called *dynodes* as shown in Figure 22-8. The electrons emitted from the cathode are accelerated toward the first dynode that is maintained 90–100 V positive with respect to the cathode. Each accelerated photoelectron that strikes the dynode surface produces several electrons, called secondary electrons, that are then accelerated to dynode 2, which is held 90–100 V more positive than dynode 1. Again electron amplification results. By the time this process has been repeated at each of the dynodes, 10^5 to 10^7 electrons have been produced for each incident photon. This cascade of electrons is finally collected at the anode to provide an average current that is further amplified electronically and measured.

Photoconductive Cells.   Photoconductive cells are transducers that consist of a thin film of a semiconductor material, such as lead sulfide, mercury cadmium telluride (MCT), or indium antimonide, deposited often on a nonconducting glass surface and sealed in an evacuated envelope. Absorption of radiation by these materials promotes nonconducting valence electrons to a higher energy state, which decreases the electrical resistance of the semiconductor. Typically, a photoconductor is placed in series with a voltage source and load resistor, and the voltage drop across the load resistor serves as a measure of the radiant power of the beam of radiation. The PbS and InSb detectors are quite popular in the near-IR region of the spectrum. The MCT detector is useful in the mid- and far-IR regions when cooled with liquid N_2 to minimize thermal noise.

Silicon Photodiodes and Photodiode Arrays.   Photodiodes are semiconductor *pn* junction devices that respond to incident light by forming electron–hole pairs (a hole is a mobile positive charge in a semiconductor). When a voltage is ap-

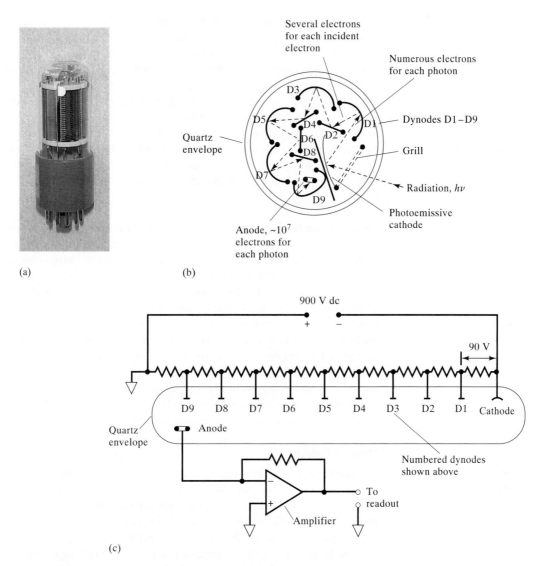

Figure 22-8 Diagram of a photomultiplier tube: (a) pictorial view, (b) cross-sectional view, (c) electrical diagram illustrating dynode polarization and photocurrent measurement. Radiation striking the photosensitive cathode (b) gives rise to photoelectrons by the photoelectric effect. Dynode D1 is held at a positive voltage with respect to the photocathode. Electrons emitted by the cathode are attracted to the first dynode and accelerated in the field. Each electron striking dynode D1 thus gives rise to 2 to 4 secondary electrons. These secondary electrons are attracted to dynode D2, which is again positive with respect to dynode D1. The resulting amplification at the anode can be 10^6 or greater. The exact amplification factor depends on the number of dynodes and the voltage difference between each dynode. This automatic internal amplification is one of the major advantages of photomultiplier tubes. With modern instrumentation, the arrival of individual photocurrent pulses can be detected and counted instead of being measured as an average current. This technique, called *photon counting,* is advantageous at very low light levels.

plied to the *pn* diode such that the *p*-type semiconductor is negative with respect to the *n*-type semiconductor, the diode is said to be *reverse biased.* If the number of light-induced charges per unit time is large compared with thermally produced charge carriers, the current in an external circuit, under reverse bias conditions, is directly related to the incident radiant power. Silicon photodiode detectors respond extremely rapidly, usually in nanoseconds. They are more sensitive than vacuum phototubes, but considerably less sensitive than photomultiplier tubes. A special type of photodiode, called an *avalanche photodiode,* provides some internal amplification, although not nearly as large as that of a PMT.

Silicon photodiodes can also be fabricated in arrays of 1000 or more detectors that are arranged linearly on a single silicon integrated circuit (the width of individual diodes is about 0.02 mm). One or two of these *diode-array detectors* can be used in conjunction with a spectrograph to monitor many wavelengths simultaneously. Diode arrays can also be obtained commercially with front-end devices called *image intensifiers* to provide gain and allow the detection of low light levels. Multiwavelength instruments based on diode arrays are discussed in Section 22B-3.

Photodiode arrays are used not only in spectroscopic instruments but also in optical scanners and bar-code readers seen at supermarket checkouts.

Charge-Transfer Devices.   Photodiode arrays cannot match the performance of photomultiplier tubes with respect to sensitivity, dynamic range, and signal-to-noise ratio. In contrast, performance characteristics of charge-transfer devices approach those of photomultipliers. As a consequence, *charge-coupled devices* (CCDs) and *charge-injection devices* (CIDs) are appearing in ever-increasing numbers in modern spectroscopic instruments. Unlike the one-dimensional photodiode array, these charge-transfer devices are two-dimensional arrays that can produce images. They are the solid state equivalent of the photographic plate. In both the CCD and the CID, photogenerated charges are collected and then measured. In spectroscopic applications, charge-transfer devices are used in conjunction with multichannel instruments as discussed in Section 22B-3. In addition to spectroscopic applications, charge-transfer devices find widespread applications in solid state television cameras and microscopy.

Thermal Detectors

Thermal detectors are often used in the infrared region of the spectrum because photon energies are insufficient to cause photoemission of electrons. Unfortunately, the performance characteristics of most thermal detectors are inferior to those of phototubes, PMTs, and silicon photodiodes.

A thermal detector consists of a tiny blackened surface that absorbs IR radiation and increases in temperature as a consequence. The temperature rise is converted to an electrical signal that is amplified and measured. Under the best of circumstances, the temperature changes involved are tiny, amounting to a few thousandths of a degree Celsius. The measurement problem is compounded because of thermal radiation from the surroundings, which is always a source of uncertainty. To minimize this background radiation, or noise, thermal detectors are housed in a vacuum and are carefully shielded from their surroundings. To further reduce the effects of this external noise, the beam from the source is modulated (chopped) by a rotating disk inserted between source and detector. Chopping produces a beam that fluctuates regularly from zero to a maximum intensity. The transducer converts this periodic radiation signal to an alternating electrical current that can be amplified and separated from the dc signal resulting from the background radiation. As a consequence, measurements with older IR instruments were usually less precise than UV/visible measurements. Modern IR instruments, however, overcome the limitations by averaging the results of multiple scans and by employing sophisticated data processing techniques. As shown in Table 22-2, four types of thermal detectors are commonly used for IR spectroscopy.

22A-5 Containing the Sample

Sample containers, usually called *cells* or *cuvettes,* must have windows that are transparent in the spectral region of interest. Thus, as shown in Figure 22-2, quartz or fused silica is required for the UV region (wavelengths less than 350 nm) and may be used in the visible region and out to about 3000 nm (3 μm) in the IR region. Silicate glass is ordinarily used for the region of 375 to 2000 nm because of its low cost compared with quartz. Plastic cells are also used in the visible region. The most common window material for IR studies is crystalline sodium chloride, which is soluble in water and in some other solvents.

The best cells have windows that are perpendicular to the direction of the beam to minimize reflection losses. The most common cell pathlength for studies in the UV and visible regions is 1 cm; matched, calibrated cells of this size are available from several commercial sources. Many other shorter and longer pathlength cells can also be purchased. Some typical UV/visible cells are shown in Figure 22-9.

For reasons of economy, cylindrical cells are sometimes encountered. Particular care must be taken to duplicate the position of such cells with respect

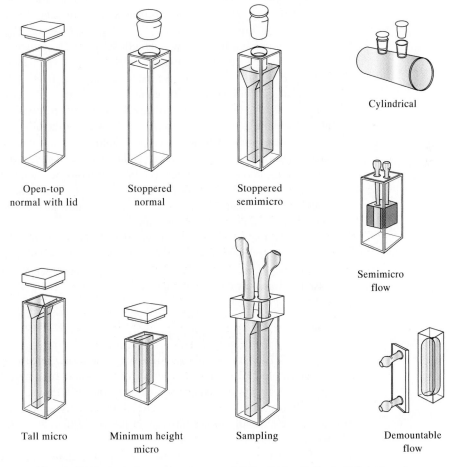

Open-top normal with lid   Stoppered normal   Stoppered semimicro   Cylindrical

Tall micro   Minimum height micro   Sampling   Semimicro flow   Demountable flow

Figure 22-9 Typical examples of commercially available cells for the UV/visible region.

to the beam; otherwise, variations in pathlength and reflection losses at the curved surfaces can cause significant error as discussed in Section 21C-2.

The quality of spectroscopic data is critically dependent on the way that cells are used and maintained. Fingerprints, grease, or other deposits on the walls markedly alter the transmission characteristics of a cell. Thus, thorough cleaning before and after use is imperative, and care must be taken to avoid touching the windows after cleaning is complete. Matched cells should never be dried by heating in an oven or over a flame because of potential physical damage or a change in pathlength. Matched cells should be calibrated against each other regularly with an absorbing solution.

22B   UV/VISIBLE PHOTOMETERS AND SPECTROPHOTOMETERS

The optical components described in Figure 22-1 have been combined in various ways to produce two types of instruments for absorption measurements. Several common terms are used to describe complete instruments. Thus, a *spectrometer* is a spectroscopic instrument that employs a monochromator or polychromator in conjunction with a transducer to convert the radiant intensities into electrical signals. *Spectrophotometers* are spectrometers that allow measurement of the ratio of the radiant powers of two beams, a requirement to measure absorbance (recall from Equation 21-6 that $A = \log P_0/P \approx \log P_{solvent}/P_{solution}$). *Photometers* employ a filter for wavelength selection in conjunction with a suitable radiation transducer. Spectrophotometers offer the considerable advantage that the wavelength used can be varied continuously, thus making it possible to record absorption spectra. Photometers have the advantages of simplicity, ruggedness, and low cost. Several dozen models of spectrophotometers are available commercially. Most spectrophotometers cover the UV/visible and occasionally the near-infrared region, whereas photometers are most often used for the visible region. Photometers find considerable use as detectors for chromatography, electrophoresis, immunoassays, or continuous flow analysis. Both photometers and spectrophotometers can be obtained in single- and double-beam varieties. In addition, multichannel instruments are now available; these systems can acquire an entire spectrum at one time.

Because the radiant power output of UV/visible sources is a strong function of wavelength (recall Figures 22-4 and 22-5), single-beam instruments typically measure $P_{solvent}$ and $P_{solution}$ at each wavelength. Alternatively, today's computer-based instruments can acquire and store an entire solvent spectrum for later calculation of absorbance.

22B-1   Single-Beam Instruments

Figure 22-10 shows the design of a simple and inexpensive spectrophotometer, the Spectronic 20, which is designed for the visible region of the spectrum. This instrument first appeared on the market in the mid-1950s, and the modified version shown is still being manufactured and widely sold. More of these instruments are currently in use throughout the world than any other single model of spectrophotometer.

The Spectronic 20 reads out linearly in transmittance or logarithmically in absorbance on a $5\frac{1}{2}$-in. scale. The instrument is equipped with an *occluder,* which is

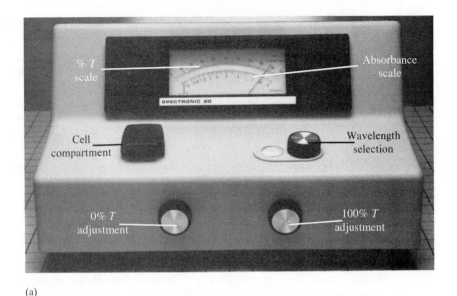

(a)

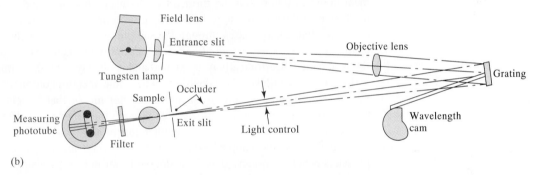

(b)

Figure 22-10   The Spectronic 20 spectrophotometer. A photograph of the instrument is shown in (a), and the optical diagram is seen in (b). (Courtesy of Spectronic Instruments, Inc., 820 Linden Avenue, Rochester, NY 14625). Radiation from the tungsten filament source passes through an entrance slit into the monochromator. A reflection grating diffracts the radiation, and the selected wavelength band passes through the exit slit into the sample chamber. A phototube converts the light intensity into a related electrical signal that is amplified and displayed on an analog meter.

a vane that automatically falls between the beam and the detector whenever the cylindrical cell is removed from its holder. The light-control device consists of a V-shaped aperture that is moved in and out of the beam to control the amount of light reaching the exit slit.

To obtain a percent transmittance reading, the pointer of the meter is first zeroed with the sample compartment empty so that the occluder blocks the beam and no radiation reaches the detector. This process is called the *0% T calibration* or *adjustment*. A cell containing the blank (often the solvent) is then inserted into the cell holder, and the pointer is brought to the 100% *T* mark by adjusting the position of the light-control aperture and thus the amount of light reaching the detector. This adjustment is called the *100% T calibration* or *adjustment*. Finally, the sample is placed in the cell compartment, and the percent transmittance or the absorbance is read directly from the scale of the meter.

Be sure to select the appropriate wavelength prior to making transmittance or absorbance measurements.

The 0% *T* and 100% *T* adjustments should be made immediately before each transmittance or absorbance measurement. To obtain reproducible transmittance measurements, it is essential that the radiant power of the source remain constant during the time that the 100% *T* adjustment is made and the % *T* is read from the meter.

The spectral range of the Spectronic 20 is 340 to 625 nm, although an accessory phototube can extend the range to 950 nm. Other specifications include a spectral bandpass of 20 nm and a wavelength accuracy of ± 2.5 nm.

Single-beam instruments of the type described are well suited for quantitative absorption measurements at a single wavelength. Here, simplicity of instrumentation, low cost, and ease of maintenance offer distinct advantages. Several instrument manufacturers offer single-beam spectrophotometers and photometers of the single-wavelength type. Prices for these instruments range from one thousand to a few thousand dollars. In addition, simple single-beam multichannel instruments based on array detectors are widely available, as discussed in Section 22B-3.

22B-2   Double-Beam Instruments

Many modern photometers and spectrophotometers are based on a double-beam design. Figure 22-11 shows two double-beam designs (b and c) compared with a single-beam system (a). Figure 22-11b illustrates a double-beam-in-space instrument in which two beams are formed by a V-shaped mirror called a *beam splitter*. One beam passes through the reference solution to a photodetector, and the other simultaneously passes through the sample to a second, matched photodetector. The two outputs are amplified, and their ratio, or the logarithm of their ratio, is obtained electronically or computed and is displayed on the output device.

Figure 22-11c illustrates a double-beam-in-time spectrophotometer. Here the beams are separated in time by a rotating sector mirror that directs the entire beam through the reference cell and then through the sample cell. The pulses of radiation are then recombined by another mirror that transmits the reference beam and reflects the sample beam to the detector. The double-beam-in-time approach is generally preferred over the double-beam-in-space because of the difficulty in matching two detectors.

Double-beam instruments offer the advantage that they compensate for all but the most short-term fluctuations in the radiant output of the source. They also compensate for wide variations of source intensity with wavelength. Furthermore, the double-beam design is well suited for continuous recording of absorption spectra.

22B-3   Multichannel Instruments

Photodiode arrays and charge-transfer devices, discussed in Section 22A-4, are the basis of multichannel instruments for UV/visible absorption. These instruments are usually of the single-beam design illustrated in Figure 22-12. With multichannel systems, the dispersive system is a grating spectrograph placed after the sample or reference cell. The photodiode array is placed in the focal plane of the spectrograph. These detectors allow the measurement of an entire spectrum in less than one second. With single-beam designs, the array dark current is acquired and stored in computer memory. Next, the spectrum of the source is obtained and stored in memory after dark current subtraction. Finally, the raw spectrum of the sample is obtained and, after dark current subtraction, the sam-

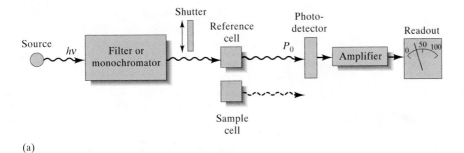

(a)

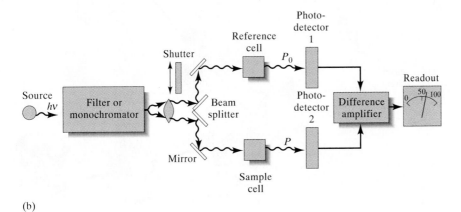

(b)

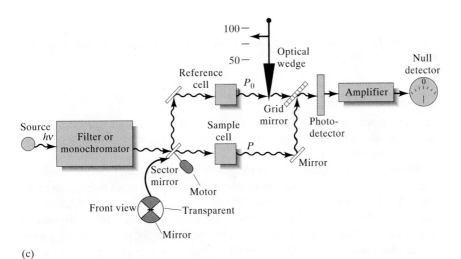

(c)

Figure 22-11   Instrumental designs for UV/visible photometers or spectrophotometers. In (a), a single-beam instrument is shown. Radiation from the filter or monochromator passes through either the reference cell or the sample cell before striking the photodetector. In (b), a double-beam-in-space instrument is shown. Here, radiation from the filter or monochromator is split into two beams that simultaneously pass through the reference and sample cells before striking two matched photodetectors. In the double-beam-in-time instrument (c), the beam is alternately sent through reference and sample cells before striking a single photodetector. Only a matter of milliseconds separate the beams as they pass through the two cells.

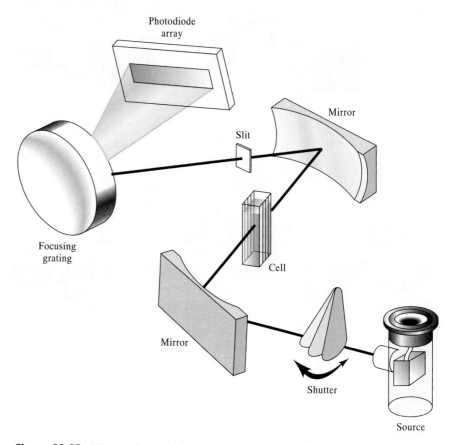

Figure 22-12   Diagram of a multichannel spectrometer based on a grating spectrograph with a photodiode array detector.

ple values are divided by the source values at each wavelength to give absorbance. Multichannel instruments can also be configured as double-beam-in-time spectrophotometers.

The spectrophotometer shown in Figure 22-12 can be controlled by most personal computers. The instrument (without the computer) can be purchased for about $10,000. Several instrument companies are combining array detector systems with fiber-optic probes that transport the light to and from the sample. These instruments allow measurements in convenient locations that are remote to the spectrometer. More CCD and CID detectors are beginning to be used in multichannel systems, particularly when advantage can be taken of the two-dimensional nature of these detectors for imaging purposes.

22C   INFRARED SPECTROPHOTOMETERS

Two types of spectrometers are used in IR spectroscopy: the dispersive type and the Fourier transform variety.

22C-1   Dispersive Infrared Instruments

Older IR instruments were invariably dispersive double-beam designs. These instruments were often of the double-beam-in-time variety shown in Figure 22-11c except that the location of the cell compartment with respect to the monochromator was reversed. In most UV/visible instruments, the cell is located between the monochromator and the detector to avoid photodecomposition of the sample, which may occur if samples are exposed to the full power of the source. Note that photodiode array instruments avoid this problem because of the short exposure time of the sample to the beam. Infrared radiation, in contrast, is not sufficiently energetic to bring about photodecomposition. Also, most samples are good emitters of IR radiation. Therefore, the cell compartment is usually located between the source and the monochromator in an IR instrument.

The components of IR instruments differ significantly from those in UV/visible instruments. Thus, IR sources are heated solids and IR detectors respond to heat rather than to photons. Furthermore, the optical components (lenses and mirrors) of IR instruments are constructed from polished salts, such as sodium chloride or potassium bromide.

22C-2   Fourier Transform Instruments

When Fourier transform infrared (FTIR) spectrometers first appeared on the market in the early 1970s, they were bulky, expensive (more than $100,000), and required frequent mechanical adjustments. For these reasons, their use was limited to special applications in which their unique characteristics (great speed, high resolution, high sensitivity, and excellent wavelength precision and accuracy) were essential. Today, however, FTIR spectrometers have been reduced to benchtop size and have become very reliable and easy to maintain. Furthermore, the simple models are now similarly priced to simple dispersive spectrometers. Hence, FTIR spectrometers have displaced dispersive instruments in most laboratories.

Fourier transform IR instruments contain no dispersing element, and all wavelengths are detected and measured simultaneously. Instead of a monochromator, an interferometer is used to produce interference patterns that contain the infrared spectral information. The same types of sources used in dispersive instruments are used in FTIR spectrometers. Transducers are typically triglycine sulfate, a pyroelectric transducer, or mercury cadmium telluride, a photoconductive transducer. To obtain radiant power as a function of wavelength, the interferometer modulates the source signal in such a way that it can be decoded by the mathematical technique of Fourier transformation. This operation requires a computer to perform the necessary calculations. The theory of Fourier transform measurements is beyond the scope of this book.[1]

Most benchtop FTIR spectrometers are of the single-beam type. To obtain the spectrum of a sample, the background spectrum is first obtained by Fourier transformation of the interferogram from the background (solvent, ambient

---

[1]For further information see J. D. Ingle, Jr. and S. R. Crouch, *Spectrochemical Analysis* (Englewood Cliffs, N.J.: Prentice-Hall, 1988); D. A. Skoog, F. J. Holler, and T. A. Nieman, *Principles of Instrumental Analysis,* 5th ed. (Philadelphia: Saunders College Publishing, 1998).

Fourier transform spectrometers detect all the wavelengths all the time. They have greater light-gathering power than dispersive instruments and consequently better precision. Although computation of the Fourier transform is somewhat complex, it is readily accomplished with inexpensive, high-speed personal computers.

water, and carbon dioxide). Next, the sample spectrum is obtained. Finally, the ratio of the single-beam sample spectrum to that of the background spectrum is calculated and absorbance or transmittance versus wavelength or wavenumber is plotted. Often, benchtop instruments purge the spectrometer with an inert gas or dry, CO_2-free air to reduce the background absorption from water vapor and CO_2.

The major advantages of FTIR instruments over dispersive spectrometers include better speed and sensitivity, better light-gathering power, more accurate wavelength calibration, simpler mechanical design, and the virtual elimination of the contribution from stray light and IR emission. Because of these advantages, nearly all the new IR instruments are FTIR systems.

WEB WORKS

Set your browser to *http://www.spectroscopymag.com* and click on the buyers guide. When on the buyers guide page, search for companies that manufacture monochromators. Open a few of the Web pages for these companies and see if you can find a UV/visible monochromator of the Czerny-Turner design that has better than 0.1 nm resolution.

22D   QUESTIONS AND PROBLEMS

22-1. Describe the differences between the following and list any particular advantages possessed by one over the other:

(a) filters and monochromators as wavelength selectors.

*(b) photoconductive cells and phototubes as detectors for electromagnetic radiation.

(c) phototubes and photomultiplier tubes.

*(d) conventional and diode-array spectrophotometers.

22-2. Define the term *effective bandwidth of a filter*.

*22-3. Why are photomultiplier tubes unsuited for the detection of infrared radiation?

22-4. Why do quantitative and qualitative analyses often require different monochromator slit widths?

*22-5. Why is iodine sometimes introduced into a tungsten lamp?

22-6. Describe the differences between the following and list any particular advantages possessed by one over the other:

(a) spectrophotometers and photometers.

(b) spectrographs and polychromators.

(c) monochromators and polychromators.

(d) single-beam and double-beam instruments for absorbance measurements.

*22-7. What minimum requirement is needed to obtain reproducible results with a single-beam spectrophotometer?

22-8. What is the purpose of

(a) the 0% T adjustment of a spectrophotometer?

(b) the 100% T adjustment of a spectrophotometer?

*22-9. What experimental variables must be controlled to ensure reproducible absorbance data?

22-10. What are the major advantages of Fourier transform IR instruments over dispersive IR instruments?

22-11. A photometer with a linear response to radiation gave a reading of 685 mV with a blank in the light path and 179 mV when the blank was replaced by an absorbing solution. Calculate

*(a) the percent transmittance and absorbance of the absorbing solution.

(b) the expected transmittance if the concentration of absorber is one-half that of the original solution.

*(c) the transmittance to be expected if the light path through the original solution is doubled.

22-12. A portable photometer with a linear response to radiation registered 73.6 μA with a blank solution in the light path. Replacement of the blank with an absorbing solution yielded a response of 24.9 μA. Calculate

  (a) the percent transmittance of the sample solution.

  *(b)** the absorbance of the sample solution.

  (c) the transmittance to be expected for a solution in which the concentration of the absorber is one-third that of the original sample solution.

  *(d)** the transmittance to be expected for a solution that has twice the concentration of the sample solution.

22-13. The following data were taken from a diode array spectrophotometer in an experiment to measure the spectrum of the Co(II)-EDTA complex. The column labeled $P_{solution}$ is the relative signal obtained with sample solution in the cell after subtraction of the dark signal. The column labeled $P_{solvent}$ is the reference signal obtained with only solvent in the cell after subtraction of the dark signal. Find the transmittance at each wavelength, and the absorbance at each wavelength. Plot the spectrum of the compound.

Wavelength, nm	$P_{solvent}$	$P_{solution}$
350	0.002689	0.002560
375	0.006326	0.005995
400	0.016975	0.015143
425	0.035517	0.031648
450	0.062425	0.024978
475	0.095374	0.019073
500	0.140567	0.023275
525	0.188984	0.037448
550	0.263103	0.088537
575	0.318361	0.200872
600	0.394600	0.278072
625	0.477018	0.363525
650	0.564295	0.468281
675	0.655066	0.611062
700	0.738180	0.704126
725	0,813694	0.777466
750	0.885979	0.863224
775	0.945083	0.921446
800	1.000000	0.977237

Chapter 23

Applying Molecular and Atomic Spectroscopic Methods: Shedding More Light on the Subject

Molecular and atomic spectroscopic methods are among the most widely used of all instrumental analytical methods. Molecular spectroscopy based on ultraviolet, visible, and infrared radiation is used for the identification and determination of a huge number of inorganic, organic, and biochemical species.[1] Molecular ultraviolet/visible absorption spectroscopy is employed primarily for quantitative analysis and is probably more widely used in chemical and clinical laboratories than any other single method. Infrared absorption spectroscopy is one of the most powerful tools for determining the structure of both inorganic and organic compounds. In addition, this technique is now assuming a role in quantitative analysis, particularly for determining environmental pollutants. Molecular fluorescence spectroscopy is an analytically important emission process in which molecules are excited by absorption of UV or visible radiation. One of the most attractive features of fluorescence methods is their inherent sensitivity, which can be many orders of magnitude better than for absorption methods. Molecular fluorescence methods are also widely used for qualitative and quantitative analysis. Atomic spectroscopic methods are widely used for elemental analysis. At present, some 70 elements are conveniently determined by atomic spectroscopy. Atomic spectroscopic methods can detect trace amounts of elemental constituents in the parts-per-million to parts-per-billion range.

---

[1]For a more detailed treatment, see J. D. Ingle, Jr. and S. R. Crouch, *Spectrochemical Analysis* (Englewood Cliffs, N.J.: Prentice-Hall, 1988); E. J. Meehan in *Treatise on Analytical Chemistry,* P. J. Elving, E. J. Meehan, and I. M. Kolthoff, Eds. (New York: Wiley, 1981), Part I, Vol. 7, Chapter 2; *Techniques in Visible and Ultraviolet Spectrometry,* C. Burgess and A. Knowles, Eds. (London: Chapman and Hall, 1981), Vol. 1.

In this chapter, we first describe the applications of UV/visible absorption spectroscopy for determining inorganic, organic, and biochemical compounds. We then turn to infrared absorption methods for qualitative and quantitative analysis as well as for structural elucidation. The principles and applications of molecular luminescence methods, including molecular fluorescence, molecular phosphorescence, and chemiluminescence, are then described. The chapter concludes with a brief overview of atomic spectroscopic methods with an emphasis on atomic absorption spectroscopy.

23A   APPLYING ULTRAVIOLET/VISIBLE MOLECULAR ABSORPTION METHODS

The applications of UV/visible molecular absorption spectroscopy are numerous and varied. The reader can obtain a notion of the scope of spectrophotometric methods by consulting the fundamental review articles published biennially in *Analytical Chemistry*[2] as well as monographs on the subject.[3]

23A-1   Overview

In the UV/visible region, many types of organic and inorganic compounds absorb radiation directly. Others can be converted to absorbing species by means of a chemical reaction.

Molecules Absorbing UV/Visible Radiation

Absorption measurements in the UV/visible region of the spectrum provide qualitative and quantitative information about organic, inorganic, and biochemical molecules.

Absorption by Organic Compounds.   Two types of electrons are responsible for the absorption of ultraviolet and visible radiation by organic molecules: (1) shared electrons that participate directly in bond formation and are thus associated with more than one atom and (2) unshared outer electrons that are largely localized about such atoms as oxygen, the halogens, sulfur, and nitrogen.

The wavelength at which an organic molecule absorbs depends on how tightly its several electrons are bound. The shared electrons in such single bonds as carbon/carbon or carbon/hydrogen are so firmly held that absorption occurs only with photons more energetic than normal UV photons. Electrons involved

---

[2]The 1998 article is J. A. Howell and R. E. Sutton, "Ultraviolet and Absorption Light Spectrometry," *Anal. Chem.* **1998**, *70*, 107R–118R.

[3]See for example, E. B. Sandell and H. Onishi, *Photometric Determination of Traces of Metals,* 4th ed. (New York: Wiley, 1978); D. F. Boltz and J. A. Howell, *Colorimetric Determination of Nonmetals,* 2nd ed. (New York: Wiley-Interscience, 1978); F. D. Snell, *Photometric and Fluorometric Methods of Analysis: Metals* (New York: Wiley-Interscience, 1978); F. D. Snell, *Photometric and Fluorometric Methods of Analysis: Non-metals* (New York: Wiley-Interscience, 1981).

A *chromophore* is, strictly speaking, the functional group in a molecule that gives rise to color. In common usage, however, it has come to mean a functional group that absorbs in the UV/visible region. The word comes from the Greek *chromo* or *chromus,* meaning color, and *phorus,* meaning bearing or a carrier of.

in double and triple bonds of organic molecules are more loosely held and are therefore more easily excited than electrons in single bonds. Thus, species with unsaturated bonds generally absorb in the UV. Unsaturated organic functional groups that absorb in the UV/visible region are known as *chromophores*. Table 23-1 list common chromophores and the approximate wavelengths at which they absorb. The data for position and absorptivity are both influenced by solvent effects as well as other structural details of the molecule. Moreover, conjugation between two or more chromophores tends to cause shifts in peak absorbances to longer wavelengths. Typical spectra for some organic compounds are shown in Figure 23-1.

The unshared electrons in such elements as sulfur, bromine, and iodine are less strongly held than the shared electrons of a saturated bond. Organic molecules incorporating these elements frequently exhibit useful absorption bands in the ultraviolet region as well.

For many nonabsorbing analytes, reagents are available that react with organic functional groups to produce species that absorb in the UV or visible regions. For example, carbonyl containing compounds react with dinitrophenylhydrazine to form colored species that absorb at 480 nm. Likewise, the red color of the 1:1 complexes that form between low-molecular-weight aliphatic alcohols and cerium(IV) can be used for quantitative estimation of such alcohols. In some cases, organic compounds can be determined by very selective enzyme re-

Table 23-1

Absorption Characteristics of Some Common Organic Chromophores

Chromophore	Example	Solvent	λ_{max}, nm	ε_{max}
Alkene	$C_6H_{13}CH{=}CH_2$	*n*-Heptane	177	13,000
Conjugate alkene	$CH_2{=}CHCH{=}CH_2$	*n*-Heptane	217	21,000
Alkyne	$C_5H_{11}C{\equiv}C{-}CH_3$	*n*-Heptane	178	10,000
			196	2,000
			225	160
Carbonyl	$CH_3\overset{\overset{O}{\|}}{C}CH_3$	*n*-Hexane	186	1,000
			280	16
	$CH_3\overset{\overset{O}{\|}}{C}H$	*n*-Hexane	180	Large
			293	12
	$CH_3\overset{\overset{O}{\|}}{C}NH_2$	Water	214	60
Carboxyl	CH_3COOH	Ethanol	204	41
Azo	$CH_3N{=}NCH_3$	Ethanol	339	5
Nitro	CH_3NO_2	Isooctane	280	22
Nitrate	$C_2H_5ONO_2$	Dioxane	270	12
Nitroso	C_4H_9NO	Ethyl ether	300	100
			665	20
Aromatic	Benzene	*n*-Hexane	204	7,900
			256	200

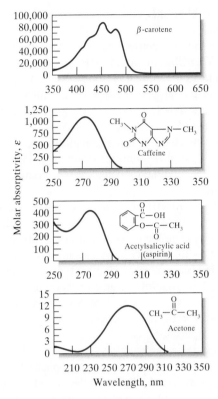

Figure 23-1   Ultraviolet spectra for several typical organic compounds.

actions for which they act as substrates. For example, the determination of lactate can be based on the reaction of L-lactate with the coenzyme nicotinamide adenine dinucleotide (NAD^+), catalyzed by L-lactate dehydrogenase. The reaction produces the reduced form of NAD^+, NADH, which absorbs at 340 nm.

Absorption by Inorganic Species.   In general, the ions and complexes of elements in the first two transition series absorb broad bands of visible radiation in at least one of their oxidation states and are, as a consequence, colored. Figure 23-2 shows some typical spectra of colored aquo-complexes of transition metals. With these complexes, absorption involves transitions between filled and unfilled *d*-orbitals of the metal ion with energies that depend on the bonded ligands. The energy differences between these *d*-orbitals and thus the position of the corresponding absorption maximum depend on the position of the element in the periodic table, its oxidation state, and the nature of the ligand bonded to it.

Many other inorganic species absorb in the UV/visible region. Ions such as nitrate, nitrite, and chromate show characteristic UV/visible absorption. Many nonabsorbing analytes can be determined photometrically by causing them to react with chromophoric reagents to produce products that absorb.

Charge-Transfer Absorption.   For quantitative purposes, charge-transfer absorption is particularly important because molar absorptivities are unusually large, a circumstance that leads to high sensitivity. Many inorganic and

Note the vast differences in molar absorptivities in the spectra shown in Figures 23-1 and 23-2.

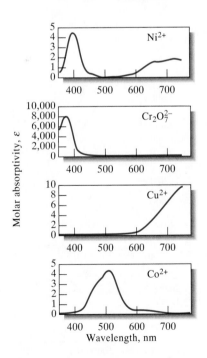

Figure 23-2   Absorption spectra of aqueous solutions of several transition metal ions. These ions exist as aquo-complexes in aqueous solutions.

Charge-transfer spectra result when the absorption of a photon promotes an electron from ligand to metal or from metal to ligand in transition metal complexes.

The 1,10-phenanthroline compound is an excellent complexing agent for Fe(II). Three 1,10-phenanthroline molecules combine with one Fe(II) molecule. The complex is used in the colorimetric determination of iron and also as a redox indicator.

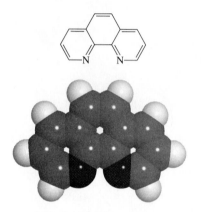

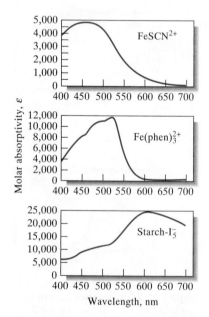

1,10-phenanthroline

Figure 23-3 Absorption spectra of aqueous charge-transfer complexes.

organic complexes exhibit this type of absorption and are therefore called *charge-transfer complexes.*

A charge-transfer complex consists of an electron-donor group bonded to an electron acceptor. When this product absorbs radiation, an electron from the donor is transferred to an orbital that is largely associated with the acceptor. The excited state is thus the product of a kind of internal oxidation/reduction process. This behavior differs from that of an organic chromophore in which the excited electron is in a *molecular* orbital that is shared by two or more atoms.

Familiar examples of charge-transfer complexes include the phenolic complex of iron(III), the 1,10-phenanthroline complex of iron(II), the iodide complex of molecular iodine, and the ferro/ferricyanide complex responsible for the color of Prussian blue.

The red color of the iron(III)/thiocyanate complex is a further example of charge-transfer absorption. Absorption of a photon results in the transfer of an electron from the thiocyanate ion to an orbital that is largely associated with the iron(III) ion. The product is an excited species involving predominantly iron(II) and the thiocyanate radical SCN. As with other types of electronic excitation, the electron in this complex ordinarily returns to its original state after a brief period. Occasionally, however, an excited complex may dissociate and produce photochemical oxidation/reduction products. Figure 23-3 shows spectra for typical charge-transfer complexes.

In most charge-transfer complexes involving a metal ion, the metal serves as the electron acceptor. Exceptions are the 1,10-phenanthroline complexes of iron(II) and copper(I), where the ligand is the acceptor and the metal ion the donor. A few other examples of this type of complex are known.

Biochemical Species.   Many species of biochemical importance also exhibit strong absorption in the UV/visible region. For example, NADH, the reduced form of the coenzyme nicotinamide adenine dinucleotide (NAD^+), absorbs at 340 nm. Many enzyme reactions involve the NAD^+/NADH couple and can be monitored by following the absorbance of NADH. Proteins are also strongly absorbing in the UV region near 280 nm. Molecules containing heme groups often show strong absorption in the 400-nm region.

Qualitative Analysis

Qualitative applications of UV/visible spectroscopy are limited because the spectra of most compounds in solution consist of one or, at most, a few broad bands with no fine structure that would be desirable for unambiguous identification. The spectral position of an absorption band is, however, an indication of the presence or absence of certain structural features or functional groups in a molecule. Examples of such groups include phenyl, conjugated double bonds, and the nitro group. Compilations of UV/visible absorption spectra are available to compare with the absorption spectrum of a pure unknown compound. Usually, UV/visible absorption spectroscopy is only used for confirmation in conjunction with a more useful qualitative technique, such as NMR, IR (see Section 23B-2), and mass spectrometry.

Fundamental Studies

Spectrophotometry in the UV/visible region is one of the major tools for studying chemical equilibria and kinetics. Wavelengths are chosen to allow monitoring of one or more reactants, products, or intermediate species. The concentrations are then obtained by using Beer's law with known or previously determined molar absorptivities. A wide variety of reaction types have been studied in this way. Whenever feasible with such studies, adherence to Beer's law should be tested by making absorption measurements on standard solutions and plotting A versus c; it is never wise to assume linearity.

For equilibrium studies, known concentrations of reactants are mixed, and the absorption spectrum of the reaction mixture is obtained after equilibrium has been established. From Beer's law, the final concentrations of reactants and products are obtained and equilibrium constants determined from known stoichiometric relationships. Spectrophotometry in the UV/visible region is also widely used to determine the stoichiometry of reactions; in particular, it has been widely employed for studies of metal complexes.

In kinetic studies, spectrophotometry is used to monitor the appearance of a product or intermediate, or the disappearance of a reactant. Beer's law is then used to determine the reaction rate or the concentration versus time profile. Such studies can be used to obtain rate laws and mechanisms for reactions as well as reaction rate constants.[4]

Spectrophotometric studies are also used to determine fundamental molecular parameters such as molar absorptivities and the molecular weights of compounds. The shape, width, and wavelength of the absorption bands of pure compounds can give information about the energy-level structure of a molecule.

Quantitative Analysis

Ultraviolet/visible spectrophotometry is one of the most powerful and widely used tools for quantitative analysis.[5] Important characteristics of UV/visible spectrophotometry include wide applicability to organic, inorganic, and biochemical systems; good sensitivity; detection limits of 10^{-4} to 10^{-7} M; moderate to high selectivity; reasonable accuracy and precision (relative errors in the 1 to 3% range and with special techniques, as low as a few tenths of a percent); and speed and convenience. In addition, spectrophotometric methods are readily automated.

23A-2   Making UV/Visible Spectrophotometry Quantitative

Spectrophotometry can be applied to quantitative analysis in several different ways. We will focus here on conventional, or direct, spectrophotometric methods as well as on methods in which spectrophotometry is used in an indirect manner to monitor the progress of a titration or a reaction used in a kinetic study.

[4]D. A. Skoog, D. M. West, and F. J. Holler, *Fundamentals of Analytical Chemistry* (Philadelphia: Saunders College Publishing, 1996), Chapter 27.
[5]See, for example, E. J. Meehan, in *Treatise on Analytical Chemistry,* P. J. Elving, E. J. Meehan, and I. M. Kolthoff, Eds. (New York: Wiley, 1981), Part I, Vol. 7, Chapter 2; *Techniques in Visible and Ultraviolet Spectrometry,* C. Burgess and A. Knowles, Eds. (London: Chapman and Hall, 1981), Vol. 1; J. R. Edisbury, *Practical Hints on Absorption Spectrometry* (New York: Plenum Press, 1968).

Conventional Spectrophotometry

Conventional spectrophotometric methods involve either a direct measurement of the analyte absorbance or an indirect measurement after reacting the analyte with a reagent to form an absorbing product. The former approach is more restricted, particularly with complex samples, because only rarely can a wavelength be found at which only the analyte absorbs. In some cases, separations will eliminate potential absorbing species before spectrophotometric detection. In other cases, the interference can be corrected by a second absorbance measurement. As an example of this latter approach, nitrate in natural waters can be determined by measuring the absorbance of the ion at 220 nm. Dissolved organic matter can also absorb at the same wavelength; however, a second absorbance measurement at 270 nm, where nitrate does not absorb, can be used to correct for the interference from organic species. Usually, a clean separation, although it can be time consuming, is the preferred method to eliminate interferences.

Quite commonly, a chemical reaction with a "selective" reagent can be used to produce a species that absorbs in a region free from most interferences. Alternatively, the product can be separated from interferences by extraction or some other separation method. For example, phosphate can be determined by the molybdenum blue method, which involves the reaction of PO_4^{3-} with Mo(VI) in acidic solution to form yellow 12-molybdophosphoric acid ($[H_2PMo_{12}O_{40}]^-$, or 12-MPA):

$$12\,MoO_4^{2-} + H_2PO_4^- + 24\,H^+ \rightleftharpoons [H_2PMo_{12}O_{40}]^- + 12\,H_2O$$

The 12-MPA can then be reduced by ascorbic acid or stannous chloride to produce molybdenum blue, which is strongly absorbing in the 650 to 890 nm region. As long as all the phosphate is converted to 12-MPA and reduced to molybdenum blue, the absorbance of the blue product is directly proportional to the phosphate concentration. An alternative method measures the yellow 12-MPA species itself after extraction from water into a polar nonaqueous solvent.

External Standards and the Calibration Curve. In most spectrophotometric methods, calibration is achieved by the method of *external standards*. Here, a series of standard solutions of the analyte is used to construct a calibration curve of absorbance versus concentration or to produce a linear regression equation as shown in Figure 23-4. The slope of the calibration curve or regression equation is the product of absorptivity and pathlength. Thus, using external standards is a way of determining the proportionality factor between absorbance and concentration under the same conditions and with the same instrument as is used for the samples. In quantitative analyses, it is seldom if ever safe to assume adherence to Beer's law and only use a single standard to determine the molar absorptivity; it is even less prudent to base the results of an analysis on a literature value of molar absorptivity.

The ideal standards for the determination would be identical in composition to the samples except that the analyte concentrations would be known. Real standards, however, are usually made from pure chemicals and do not contain most of the matrix components present in the sample. To compensate for some of the interferences caused by the matrix, the overall composition of the standards should be made to resemble as closely as feasible the composition of the sample. This process is called *matrix matching*.

The measured absorbance of a given solution will usually vary somewhat from instrument to instrument. Thus, quantitative determinations should *never* be based on molar absorptivities found in the literature.

The term *matrix* refers to the collection of all the constituents in a sample. The matrix includes the analyte as well as the other constituents, often called *concomitants*. In determining trace amounts, we tend to think of the analyte and the matrix separately because the analyte is present in such small amounts.

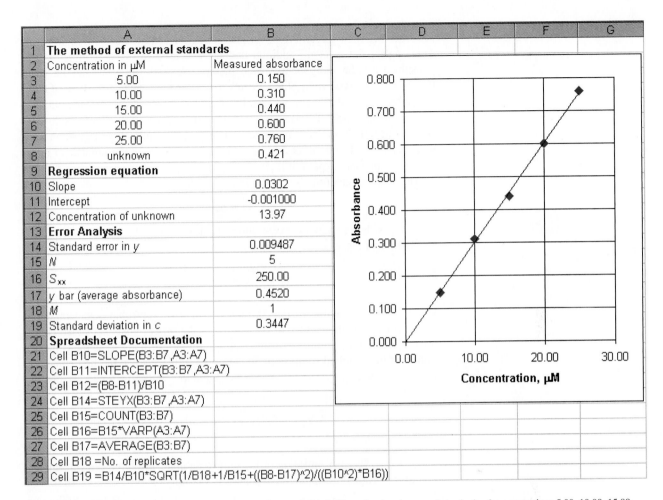

The following spreadsheet content is shown in the figure:

	A	B
1	**The method of external standards**	
2	Concentration in μM	Measured absorbance
3	5.00	0.150
4	10.00	0.310
5	15.00	0.440
6	20.00	0.600
7	25.00	0.760
8	unknown	0.421
9	**Regression equation**	
10	Slope	0.0302
11	Intercept	-0.001000
12	Concentration of unknown	13.97
13	**Error Analysis**	
14	Standard error in y	0.009487
15	N	5
16	S_{xx}	250.00
17	y bar (average absorbance)	0.4520
18	M	1
19	Standard deviation in c	0.3447
20	**Spreadsheet Documentation**	
21	Cell B10=SLOPE(B3:B7,A3:A7)	
22	Cell B11=INTERCEPT(B3:B7,A3:A7)	
23	Cell B12=(B8-B11)/B10	
24	Cell B14=STEYX(B3:B7,A3:A7)	
25	Cell B15=COUNT(B3:B7)	
26	Cell B16=B15*VARP(A3:A7)	
27	Cell B17=AVERAGE(B3:B7)	
28	Cell B18 =No. of replicates	
29	Cell B19 =B14/B10*SQRT(1/B18+1/B15+((B8-B17)^2)/((B10^2)*B16))	

Figure 23-4 The method of external standards (see also Example 21-2). Here, the absorbances of standards of concentrations 5.00, 10.00, 15.00, and 25.00 μM are measured and used to construct the calibration curve and regression equation. If a blank exists, the resulting equation is of the form $A = mc + b$, where m is the slope of the regression equation and b is the intercept (absorbance of the blank). If a proper blank has been used, the value of b will be close to zero. Next, the absorbance of the analyte solution of unknown concentration is obtained from the calibration curve or from the regression equation as shown. Note above the formula for calculating the concentration of the unknown and the error analysis to find the standard deviation in the unknown concentration.

The Standard Addition Method. The difficulties associated with production of standards with an overall composition closely resembling that of the sample can be formidable, if not insurmountable. Under such circumstances, the *method of standard additions* may prove useful. The standard addition method requires adherence to Beer's law (and this must be confirmed experimentally) and the absence of any background or the use of an accurate blank measurement. In the single-point standard addition method, a known amount of analyte is introduced into a second aliquot of the sample (see Example 23-1), and the difference in absorbance is used to calculate the analyte concentration of the sample. Alternatively, multiple additions can be made to several aliquots of the sample and a multiple standard addition calibration curve obtained. The multiple additions method can give verification that Beer's law is obeyed (see Example 23-2).

Example 23-1

The single-point standard addition method was used in the determination of phosphate by the molybdenum blue method. A 2.00-mL urine sample was treated with molybdenum blue reagents to produce a species absorbing at 820 nm, after which the sample was diluted to 100.00 mL. A 25.00-mL aliquot gave an absorbance of 0.428 (solution 1). Addition of 1.00 mL of a solution containing 0.0500 mg of phosphate to a second 25.0-mL aliquot gave an absorbance of 0.517 (solution 2). Use these data to calculate the number of milligrams of phosphate per milliliter of the sample.

The absorbance of the first solution is given by

$$A_1 = \varepsilon b c_x$$

where c_x is the unknown concentration of phosphate in the first solution. The absorbance of the second solution is given by

$$A_2 = \frac{\varepsilon b V_x c_x}{V_t} + \frac{\varepsilon b V_s c_s}{V_t}$$

where V_x is the volume of the solution of unknown phosphate concentration (25.00 mL), V_s is the volume of the standard solution of phosphate added (1.00 mL), V_t is the total volume after the addition (26.00 mL), and c_s is the concentration of the standard solution (0.500 mg mL^{-1}). If we solve the first equation for εb, substitute the result into the second equation, and solve for c_x, we obtain

$$c_x = \frac{A_1 c_s V_s}{A_2 V_t - A_1 V_x} = \frac{0.428 \times 0.0500 \text{ mg mL}^{-1} \times 1.00 \text{ mL}}{0.517 \times 26.00 \text{ mL} - 0.428 \times 25.00 \text{ mL}}$$

$$= 0.00780 \text{ mg mL}^{-1}$$

This result is the concentration of the diluted sample. To obtain the concentration of the original urine sample, we need to multiply by 100.00/2.00. Thus,

$$\text{concentration of phosphate} = 0.00780 \text{ mg mL}^{-1} \times \frac{100.00 \text{ mL}}{2.00 \text{ mL}}$$

$$= 0.390 \text{ mg mL}^{-1}$$

Example 23-2

A multiple standard addition method was used for determining Fe^{3+} in a natural water sample. Ten-milliliter aliquots of the sample were pipetted into 50.00-mL volumetric flasks. Exactly 0.00, 5.00, 10.00, 15.00, and 20.00 mL of a standard solution containing 11.1 ppm of Fe^{3+} were added to each followed by an excess of thiocyanate ion to give the red complex $Fe(SCN)^{2+}$. After dilution to volume, absorbances of the five solutions were measured in a 0.982-cm cell at 480 nm and found to be 0.240, 0.437, 0.621, 0.809, and 1.009, respectively. What is the concentration of Fe^{3+} in the water sample?

For several identical aliquots of a standard solution having a known concentration c_s, the absorbance after each addition, A_s, is given by

$$A_s = \frac{\varepsilon b V_s c_s}{V_t} + \frac{\varepsilon b V_x c_x}{V_t} = k V_s c_s + k V_x c_x$$

where V_t is the total volume of each flask, V_s is the variable volume of standard added, c_x is the unknown concentration, and k is a constant equal to $\varepsilon b/V_t$. A plot of A_s as a function of V_s should yield a straight line of the form

$$A_s = mV_s + b$$

where the slope m and intercept b are given by

$$m = kc_s \quad \text{and} \quad b = kV_x c_x$$

By combining these equations for the slope and intercept, the concentration of the unknown can be found as

$$c_x = \frac{bc_s}{mV_x}$$

The spreadsheet and graph shown in Figure 23-5 show how these data can be plotted and used to find the concentration of Fe^{3+} in the sample.

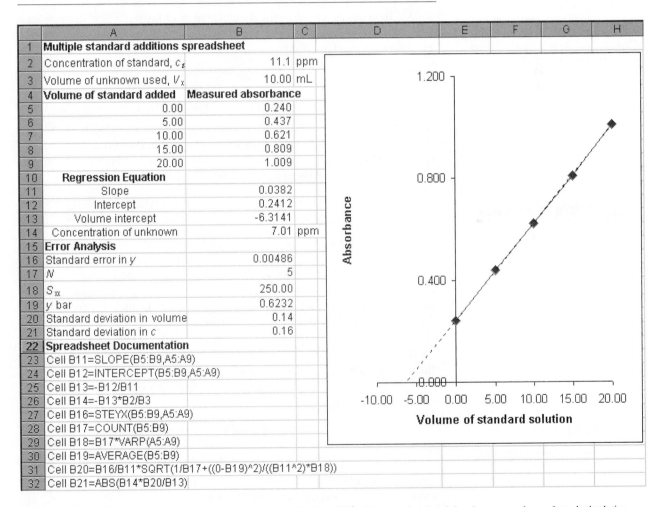

Figure 23-5   Multiple standard additions method for the determination of Fe^{3+}. Note that the plot of absorbance vs. volume of standard solution added gives a straight line and verifies that the system obeys Beer's law. The unknown concentration can be obtained from the volume intercept $(-b/m)$ by the equation given in Example 23-2 and shown in the documentation section, cell A25. The standard deviation in c is calculated by an equation similar to that given in Figure 23-4, except that the $1/M$ term is absent and the value for the absorbance of the unknown is entered as 0.00 since this is the absorbance at the x axis intercept.

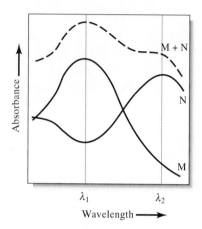

Figure 23-6 Absorption spectrum of a two-component mixture (M + N), with spectra of the individual components. Wavelengths λ_1 and λ_2 are chosen for the analysis because the individual component spectra are significantly different at these two wavelengths.

Analysis of Mixtures. The total absorbance of a solution at any given wavelength is equal to the sum of the absorbances of the individual components in the solution, as shown by Equation 21-9. This relationship makes it possible in principle to determine the concentrations of the individual components in a mixture even if there is strong overlap in their spectra. For example, Figure 23-6 shows the spectrum of a solution containing a mixture of species M and species N as well as absorption spectra for the individual components. We can see that there is no wavelength at which the absorbance is due to just one of these components. To analyze the mixture, molar absorptivities for M and N are first determined at wavelengths λ_1 and λ_2. Enough standards must be used to ensure that Beer's law is obeyed over an absorbance range that encompasses the absorbance of the sample. Note that the wavelengths selected are ones at which the two spectra differ significantly. In some cases, the wavelengths chosen may not correspond to the absorption maxima as they do in Figure 23-6. Thus, at λ_1, the molar absorptivity of component M is much larger than that for component N. The reverse is true for λ_2. To complete the analysis, the absorbance of the mixture is determined at the same two wavelengths. From the known molar absorptivities and pathlength, the following equations hold:

$$A_1 = \varepsilon_{M1}bc_M + \varepsilon_{N1}bc_N \tag{23-1}$$

$$A_2 = \varepsilon_{M2}bc_M + \varepsilon_{N2}bc_N \tag{23-2}$$

where the subscript 1 indicates measurement at λ_1, and the subscript 2 indicates measurement at λ_2. With the known values of ε and b, Equations 23-1 and 23-2 represent two equations in two unknowns (c_M and c_N), which can be readily solved by several methods as demonstrated in Example 23-3.

Spreadsheet Exercise

Example 23-3

Titanium and vanadium can be measured simultaneously as their peroxide complexes. Measurements on standard solutions containing titanium indicate the following molar absorptivities at 400 and 460 nm for the peroxy complexes: $\varepsilon_{Ti(400)} = 644$ L mol^{-1} cm^{-1}: $\varepsilon_{Ti(460)} = 321$ L mol^{-1} cm^{-1}. Similar measurements on vanadium standards gave $\varepsilon_{V(400)} = 145$ L mol^{-1} cm^{-1}: $\varepsilon_{V(460)} = 232$ L mol^{-1} cm^{-1}. A steel sample containing both titanium and vanadium was dissolved and treated with H_2O_2. The absorbance at 400 nm in a 1.00-cm cell was 0.172, while that at 460 nm was 0.116. What were the concentrations of the titanium and vanadium peroxy complexes in the solution measured?

From Equations 23-1 and 23-2 we can write the two simultaneous equations that must be solved:

$$A_{400} = \varepsilon_{Ti(400)}bc_{Ti} + \varepsilon_{V(400)}bc_V \tag{23-3}$$

$$A_{460} = \varepsilon_{Ti(460)}bc_{Ti} + \varepsilon_{V(460)}bc_V \tag{23-4}$$

These two equations can be solved simultaneously for c_{Ti} and c_V in terms of known quantities. Solving Equation 23-3 for c_{Ti} gives

$$c_{Ti} = \frac{A_{400} - \varepsilon_{V(400)}bc_V}{\varepsilon_{Ti(400)}b} \tag{23-5}$$

Solving Equation 23-4 for c_V gives,

$$c_V = \frac{A_{460} - \varepsilon_{Ti(460)}bc_{Ti}}{\varepsilon_{V(460)}b} \qquad (23\text{-}6)$$

If we now substitute the value of c_{Ti} from Equation 23-5 in this latter equation, we obtain

$$c_V = \frac{A_{460} - \varepsilon_{Ti(460)}b\left(\dfrac{A_{400} - \varepsilon_{V(400)}bc_V}{\varepsilon_{Ti(400)}b}\right)}{\varepsilon_{V(460)}b} \qquad (23\text{-}7)$$

Equation 23-7 can be solved in Excel by iteration or by matrix methods as discussed in Chapter 8. Alternatively, some further algebra can be done to develop an explicit solution as shown later in this exercise.

In the spreadsheet, we will first solve Equation 23-7 by iteration. Let us begin by typing the known quantities into a blank spreadsheet as shown in the following. Note that we have multiplied the absorptivities by the 1.00-cm pathlength to obtain εb values in L mol^{-1}.

	A	B	C	D	E
1	Spreadsheet to solve two-component absorption problem by iteration and by exact solution				
2	Known parameters	$\varepsilon_{Ti(400)}b$ in L mol^{-1}	$\varepsilon_{Ti(460)}b$ in L mol^{-1}	$\varepsilon_{V(400)}b$ in L mol^{-1}	$\varepsilon_{V(460)}b$ in L mol^{-1}
3		644	321	145	232
4					
5	Data	A(400) for mixture	A(460) for mixture		
6		0.172	0.116		

Now let us type the spreadsheet equivalent of Equation 23-7 into cell B9. The formula documentation is shown in cell B17.

	A	B	C	D	E
1	Spreadsheet to solve two-component absorption problem by iteration and by exact solution				
2	Known parameters	$\varepsilon_{Ti(400)}b$ in L mol^{-1}	$\varepsilon_{Ti(460)}b$ in L mol^{-1}	$\varepsilon_{V(400)}b$ in L mol^{-1}	$\varepsilon_{V(460)}b$ in L mol^{-1}
3		644	321	145	232
4					
5	Data	A(400) for mixture	A(460) for mixture		
6		0.172	0.116		
7					
8	Solution	Iteration			
9	Conc. V	0.000189495			
10					
11					
12					
13					
14					
15					
16	Spreadsheet Documentation				
17	Cell B9=(C6-C3*(B6-D3*B9)/B3)/E3				

Since cell B9 contains a circular reference, be sure you have checked Iteration in Tools/Options. Choose 1000 iterations with a maximum step-size of 0.000001.

By substituting the value of c_V from Equation 23-6 into Equation 23-5, we can solve for c_{Ti} and obtain

$$c_{Ti} = \frac{A_{400} - \varepsilon_{V(400)}b\left(\dfrac{A_{460} - \varepsilon_{Ti(460)}bc_{Ti}}{\varepsilon_{V(460)}b}\right)}{\varepsilon_{Ti(400)}b} \tag{23-8}$$

The spreadsheet formula for this equation can be entered into cell B10 to obtain the concentration of the titanium peroxide complex as shown below. It is useful at this point to check on the validity of the results obtained by using the calculated concentrations in Equations 23-3 and 23-4 to predict the absorbance values A_{400} and A_{460}. If the predicted values are equal to the actual values, the formulas have been entered correctly and the iteration done appropriately. The predicted values are calculated in cells B12 and B13 by the equations listed in the documentation section.

	A	B	C	D	E
1	Spreadsheet to solve two-component absorption problem by iteration and by exact solution				
2	Known parameters	$\varepsilon_{Ti(400)}b$ in L mol^{-1}	$\varepsilon_{Ti(460)}b$ in L mol^{-1}	$\varepsilon_{V(400)}b$ in L mol^{-1}	$\varepsilon_{V(460)}b$ in L mol^{-1}
3		644	321	145	232
4					
5	Data	A(400) for mixture	A(460) for mixture		
6		0.172	0.116		
7					
8	Solution	Iteration			
9	Conc. V	0.000189495			
10	Conc. Ti	0.000224415			
11					
12	Predicted A(400)	0.172			
13	Predicted A(460)	0.116			
14					
15					
16	Documentation				
17	Cell B9=(C6-C3*(B6-D3*B9)/B3)/E3				
18	Cell B10=(B6-D3*B9)/B3				
19	Cell B12=C2*B9+E2*B8				
20	Cell B13=C3*B10+E3*B9				

Note that the predicted values match the measured absorbances.

Equations 23-7 and 23-8 can also be solved exactly with some further algebra. The values obtained are:

$$c_V = \frac{\varepsilon_{Ti(400)}A_{460} - \varepsilon_{Ti(460)}A_{400}}{\varepsilon_{Ti(400)}\varepsilon_{V(460)} - \varepsilon_{Ti(460)}\varepsilon_{V(400)}} \tag{23-9}$$

and

$$c_{Ti} = \frac{\varepsilon_{V(460)} A_{400} - \varepsilon_{V(400)} A_{460}}{\varepsilon_{Ti(400)} \varepsilon_{V(460)} - \varepsilon_{Ti(460)} \varepsilon_{V(400)}} \tag{23-10}$$

These exact solutions are entered into cells D8 and D9 with the formulas shown in the documentation in cells D16 and D17. Since the same values are obtained as by iteration, there is no need here to find predicted values for the absorbances at the two wavelengths.

	A	B	C	D	E
1	Spreadsheet to solve two-component absorption problem by iteration and by exact solution				
2	*Known parameters*	$\varepsilon_{Ti(400)}b$ in L mol^{-1}	$\varepsilon_{Ti(460)}b$ in L mol^{-1}	$\varepsilon_{V(400)}b$ in L mol^{-1}	$\varepsilon_{V(460)}b$ in L mol^{-1}
3		644	321	145	232
4					
5	Data	A(400) for mixture	A(460) for mixture		
6		0.172	0.116		
7					
8	Solution	Iteration		Exact Solution	
9	Conc. V	0.000189495		0.000189495	
10	Conc. Ti	0.000224415		0.000224415	
11					
12	Predicted A(400)	0.172			
13	Predicted A(460)	0.116			
14					
15					
16	Documentation				
17	Cell B9=(C6-C3*(B6-D3*B9)/B3)/E3			Cell D9=(B3*C6-C3*B6)/(B3*E3-C3*D3)	
18	Cell B10=(B6-D3*B9)/B3			Cell D10=(E3*B6-D3*C6)/(B3*E3-C3*D3)	
19	Cell B12=B3*B10+D3*B9				
20	Cell B13=C3*B10+E3*B9				

This same problem can also be readily solved by the matrix operations introduced on page 198. Substituting the known values into Equations 23-3 and 23-4, gives

$$0.172 = 644c_{Ti} + 145c_V$$

and

$$0.116 = 321c_{Ti} + 232c_V$$

We can enter the εb values as the coefficient matrix and the absorbance values as the constant matrix and solve the two equations simultaneously as shown in the following spreadsheet. Note that the values returned in cells D8 and D9 are identical to those obtained by the methods previously described.

The matrix methods can readily be extended to more than two equations in two unknowns but in many cases errors in the parameters and in the absorbances can lead to large errors in the calculated concentrations.

	A	B	C	D	E	F	G	H
1	Spreadsheet to solve simultaneous absorbance equations by matrix operations							
2								
3	Coefficient Matrix			Constant Matrix				
4	644	145		0.172				
5	321	232		0.116				
6								
7	Inverse Matrix			Solution Matrix				
8	0.002255	-0.00141		0.000224415				
9	-0.00312	0.006261		0.000189495				
10								
11								
12	Spreadsheet Documentation							
13	Cell A8:B9=MINVERSE(A4:B5)				Cell D8:D9=MMULT(A8:B9,D4:D5)			
14	entered as an array formula with				entered as an array formula with			
15	control+shift+enter				control+shift+enter			

In principle, determinations of more than two absorbing species can be made by incorporating at least one additional absorbance reading for each added component. The uncertainties in the resulting data become greater, however, as the number of components increases. Some of the newer data processing techniques, called *multivariate calibration methods,* can lessen these uncertainties by overdetermining the system; that is, these techniques use essentially the entire spectrum with many more data points than unknowns.

Following Titrations by Measuring UV/Visible Absorption

Ultraviolet/visible spectrophotometric and photometric measurements are useful for locating the end points of titrations.[6] The method requires that one or more of the reactants or products absorb radiation or that an absorbing indicator be present. In spectrophotometric titrations, the spectrophotometer serves as the detector that monitors the transmittance or absorbance of the solution at a suitable wavelength during the addition of increments of the titrant. For example, an acid/base titration can be monitored spectrophotometrically by adding a small amount of an indicator that is colored in either the acidic or basic form such as methyl red or phenolphthalein. The sample cell serves as the titration vessel.

Titration Curves.   The plot of absorbance versus titrant volume is called a *spectrophotometric titration curve;* the shapes depend on the species that absorbs radiation, as shown in Figure 23-7. Normally, the absorbances are corrected for dilution by the titrant by multiplying the measured values by $(V_T + V_A)/V_A$, where V_A and V_T are the volumes of the analyte solution and titrant, respectively. Ideally, the end point is located by a sharp change in absorbance; often, conditions are arranged so that two straight-line regions of differing slopes intersect at the end point. If the reaction is not quantitative near the

---

[6]See, for example, J. B. Headridge, *Photometric Titrations* (New York: Pergamon Press, 1961).

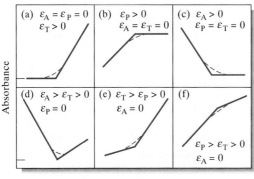

Figure 23-7 Shapes of spectrophotometric titration curves. In (a), only the titrant absorbs or the excess titrant reacts with an indicator. The absorbance increases sharply after the titrant becomes in excess. In (b), only the product absorbs and the absorbance increases until the analyte has completely reacted. In (c), only the analyte absorbs and its absorbance decreases steadily until the end point. The curves in (d) through (f) are typical of those in which more than one species absorbs. The molar absorptivities of the analyte, product, and titrant are labeled ε_A, ε_P, and ε_T.

equivalence point, the linear segments before and after the end point can be extrapolated to locate the end point; adherence to Beer's law is a necessity.

Spectrophotometric titrations have several advantages over titrations based on visual end-point detection. More precision can be obtained with spectrophotometric detection because a smaller change in absorbance can be measured than with the eye. Also, the wavelength selectivity of spectrophotometric detection allows the use of turbid or colored solutions, which cause difficulty with visual methods. Because of better sensitivity than with the eye, titration reactions that have unfavorable equilibrium constants can also be more readily employed with spectrophotometric detection.

Spectrophotometric titrations are also advantageous over conventional spectrophotometric measurements because they are relative measurements. In the curves shown in Figure 23-7, the absolute value of the absorbance is not critical in locating the end point. Hence, calibration with absorbance standards is not needed, and the presence of other species that absorb or scatter the radiation does not cause error as long as these species do not react with the titrant. For all these reasons, instrumentation for spectrophotometric titrations can often be much simpler than scanning spectrophotometric equipment. Commercial titration systems usually include the spectrophotometer or photometer, a titration vessel, an automated buret or syringe pump, and a computer-based data handling and plotting system.

Applications of Spectrophotometric Titrations. Spectrophotometric or photometric titrations have been applied to many types of reactions.[7] For example, most standard oxidizing agents have characteristic absorption spectra and thus produce photometrically detectable end points. Although standard acids or bases do not absorb, the introduction of acid/base indicators permits the use of spec-

> Spectrophotometric titration curves are often made up of linear segments that intersect at the end point.

> Spectrophotometric titrations are more selective and precise than visual methods. The absolute absorbance is not important in locating the end point; only the relative changes in absorbance are critical.

[7]A. L. Underwood, in *Advances in Analytical Chemistry and Instrumentation,* C. N. Reilley, Ed., (New York: Wiley Interscience, 1964), vol. 3, pp. 31–104.

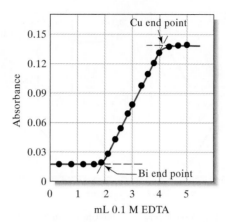

Figure 23-8   Photometric titration curve at 745 nm for 100 mL of a solution that was 2.0×10^{-3} M in Bi^{3+} and Cu^{2+}. The titrant was 0.1 M EDTA (A. L. Underwood, *Anal. Chem.,* **1954,** *26,* 1322. With permission of the American Chemical Society.)

trophotometric end points in neutralization titrations. The photometric end point has also been used to great advantage in titrations with EDTA and other complexing agents. Figure 23-8 illustrates the application of this technique to the successive titration of Bi(III) and Cu(II). At 745 nm, the cations, the reagent, and the bismuth complex formed in the first part of the titration do not absorb, but the copper complex does. Thus, the solution exhibits no absorbance until essentially all the Bi(III) has been titrated. With the first formation of the Cu(II) complex, an increase in absorbance occurs. The increase continues until the copper end point is reached. Additional reagent causes no further absorbance change. Clearly, two well-defined end points result.

Using Spectrophotometry for Kinetic Methods

In *kinetic determinations,* the rate of the analytical reaction or the concentration of reactant or product during the reaction is measured and related to the analyte concentration.[8] Spectrophotometry is the most common method for obtaining reaction rates. Often, the initial rate in the first few percent of the reaction is sufficient. For example, when phosphate reacts with Mo(VI) in acidic solution in the presence of ascorbic acid, the initial rate of formation of the resulting molybdenum blue species is directly proportional to the phosphate concentration. With spectrophotometry, the absorbance change during this initial portion of the reaction is measured. A calibration curve is made of initial rate, or a related quantity, versus the analyte concentration in standard solutions. The analyte concentration in the unknown sample is then obtained from this calibration curve.

Kinetic methods have several advantages over conventional spectrophotometric methods in which the analytical reaction is allowed to proceed to equilibrium before measurements are made. First, the analysis time can be considerably shorter, particularly for reactions that take several minutes or more to reach equilibrium. Kinetic methods can also exhibit greater selectivity over conventional

---

[8]D. A. Skoog, D. M. West, and F. J. Holler, *Fundamentals of Analytical Chemistry,* 7th ed. (Philadelphia: Saunders College Publishing, 1996), pp. 637–658.

spectrophotometric methods because they involve relative changes in absorbance rather than absolute values of absorbance and because many interferences can be discriminated against if they react at a significantly different rate than the analyte. With kinetic methods, it is possible to use reactions that do not exhibit exact stoichiometry or that have unstable reactants or products. Also, catalysts, such as enzymes, can be determined uniquely by kinetic methods since they influence the reaction rate, but not the equilibrium concentrations.

Instrumentation for kinetic methods resembles that for conventional spectrophotometry. A temperature-controlled sample compartment is an added requirement because the rates of most reactions are highly temperature dependent. Many conventional spectrophotometers have accessories for kinetic measurements. Some instruments allow several sample cells to be positioned sequentially in the light beam for monitoring several reactions nearly simultaneously.

Spectrophotometric kinetic methods are usually the preferred method for determining enzymes in clinical samples. For example, the enzymes L-lactate dehydrogenase and hexokinase are determined by measuring the rate of increase in absorbance of the product NADH in enzyme-catalyzed reactions. Many enzyme substrates are also conveniently determined by kinetic methods. Inorganic and organic species can also be determined by rate methods. Phosphate in blood serum can be determined by measuring the rate of formation of molybdenum blue by the reaction described earlier in this section. Inorganic or organic catalysts can be determined at very low levels by kinetic methods. As one example, trace quantities (ppb or lower) of iodide can be determined by catalysis of the Ce(IV)/As(III) reaction. At one time, this kinetic procedure was widely used for the determination of protein-bound iodine; the method has now largely been replaced by radioimmunoassay methods.

23B   ABSORBING INFRARED RADIATION: GOOD VIBRATIONS

Infrared absorption spectroscopy is also widely employed in analytical chemistry. Its scope is nearly as broad as that of UV/visible methods. In the IR region, absorption of radiation can give information about the identity of compounds, the presence or absence of functional groups, and the structure of molecules. Although not nearly as widely used in quantitative analysis as UV/visible absorption, IR absorption is one of the premier techniques for qualitative analysis and functional group identification.

23B-1   Molecules That Absorb Infrared Radiation

With the exception of homonuclear diatomic molecules, such as O_2, Cl_2, and N_2, all molecules, organic and inorganic, absorb infrared radiation. Thus, IR spectroscopy is one of the most generally applicable of all analytical methods. As noted in Chapter 21, absorption of IR radiation involves transitions among the vibrational energy levels of the lowest excited electronic energy levels of molecules. The number of ways a molecule can vibrate is related to the number of bonds it contains and thus the number of atoms making up the molecule. The number of vibrations is large even for a simple molecule. For example, *n*-butanal

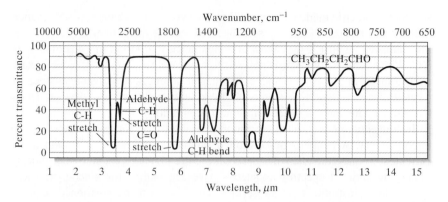

Figure 23-9   Infrared spectrum for *n*-butanal (*n*-butyraldehyde). Note that transmittance rather than absorbance is plotted in this figure.

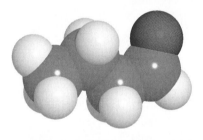

n-Butanal ($CH_3CH_2CH_2CHO$). This molecule, also called *n*-butyraldehyde, shows a complex IR spectrum.

Molecular model of *n*-butanal. The acetal formed from this compound and polyvinyl alcohol is widely used in preparing safety glass.

($CH_3CH_2CH_2CHO$) has 33 vibrational modes, most differing from each other in energy. Not all these vibrational modes produce IR absorptions; nevertheless, as shown in Figure 23-9, the spectrum of *n*-butanal is complex.

Infrared absorption occurs not only with organic molecules but also with simple inorganic compounds such as CO_2, CO, H_2S, NO_2, and SO_2 as well as covalently bonded metal complexes of various types. Several of the possible IR vibrational modes for water are illustrated in Figure 23-10.

23B-2   Qualitative and Structural Applications

Infrared absorption spectroscopy is one of the most powerful and important tools available to the chemist for identifying and determining the structure of organic, inorganic, and biochemical species. As can be seen in Figure 23-9, even simple molecules give complex spectra that provide numerous absorption bands; many of these are useful for identification purposes. Indeed, the IR spectrum of a compound in the range of 2.5 to 15 μm provides a unique fingerprint that is readily distinguished from absorption spectra of other compounds; only optical isomers have identical spectra. Because techniques for the identification of organic compounds from their IR spectra are treated in detail in most organic laboratory courses, we do not discuss the subject further here.[9]

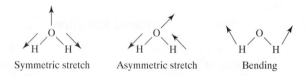

Figure 23-10   Illustrations of possible vibrational modes for water.

---

[9]See, for instance, R. M. Silverstein, G. C. Bassler, and T. C. Morrill, *Spectrometric Identification of Organic Compounds,* 5th ed. (New York: Wiley, 1991).

23B-3 Making Infrared Absorption Spectroscopy Quantitative

Quantitative applications of IR spectroscopy are much more limited than are such applications in the UV/visible region. This limitation results from low molar absorptivities in the IR region, the presence of significant solvent or background absorption, and the narrowness of IR absorption bands, which can lead to Beer's law deviations due to polychromatic radiation.

Measuring Absorbance

The use of matched cuvettes for solvent and analyte, as is common in UV/visible methods, is seldom practical for IR measurements because of the difficulty in obtaining cells with identical transmission characteristics. Part of this difficulty results from degradation of the transparency of IR cell windows (typically polished NaCl) with use due to attack by traces of moisture in the atmosphere and in samples. Furthermore, pathlengths are hard to reproduce because the thickness of IR cells is often less than 1 mm to permit the transmission of measurable intensities of IR radiation. Because of the thin cells used for solutions and their deterioration with age, frequent pathlength calibrations must be made. Measurements of dilute analyte solutions, as done in the UV/visible region, are frequently precluded by the lack of good solvents that transmit over appreciable regions of the IR spectrum.

With Fourier transform systems, it is common to store the spectrum of the solvent in memory and to obtain the ratio of the sample spectrum to that of the solvent. With older dispersive systems, it was common to eliminate the reference absorber entirely and to compare the intensity passing through the sample with that of the unobstructed beam or the beam containing only a salt plate reference. Because of solvent absorption, the resulting transmittance is often less than 100% even in regions of the IR spectrum where the sample is totally transparent (see Figure 23-9). The presence of significant solvent and other background absorption can make the measurement of the absorbance of a single component somewhat difficult in the IR region. Often, it is necessary to estimate the background absorption under the analyte absorption peak by extrapolation of the baseline on either side of the peak. Such background subtraction methods can lead to significant imprecision in IR absorbance measurements.

Applying IR Spectroscopy

Infrared spectrophotometry has the potential for determining an unusually large number of substances because nearly all species absorb in this region. Moreover, the uniqueness of the IR spectrum provides a high degree of specificity that is matched or exceeded by few analytical methods.

The recent increase in government regulations on atmospheric contaminants has demanded the development of sensitive, rapid, and highly specific methods for a variety of chemical compounds. Infrared absorption procedures appear to meet this need better than any other single analytical method.

Table 23-2 illustrates the variety of atmospheric pollutants that can be determined with the simple IR filter photometer shown in Feature 23-1; a separate interference filter is used for each analyte species. Of the more than 400 chemicals for which maximum tolerable limits have been set by the Occupational Safety and Health Administration (OSHA), more than half have absorption characteristics that make them amenable to determination by IR absorption.

Infrared absorption is used widely to determine atmospheric pollutants.

Feature 23-1

Infrared Photometers for
Routine Determination of
Atmospheric Pollutants

Figure 23F-1 is an optical diagram of a portable infrared photometer that has been designed for the routine quantitative determination of organic pollutants in the atmosphere. The source is a ceramic rod wound with a nichrome wire; the transducer is a pyroelectric detector. A variety of interference filters that transmit in the region of 3000 to 750 cm^{-1} (3.3 to 13 μm) are available; each is designed for the determination of a specific compound. The filters are readily interchangeable. Gaseous samples are introduced into the cell by means of a battery-operated pump. The pathlength of the cell as shown is 0.5 m; a series of reflecting mirrors (not shown) permits increases in the cell length to 20.5 m in increments of 0.5 m. This feature greatly enhances the concentration range of the instrument. The photometer is reported to be sensitive to a few tenths of a part per million of such substances as acrylonitrile, chlorinated hydrocarbons, carbon monoxide, phosgene, and hydrogen cyanide.

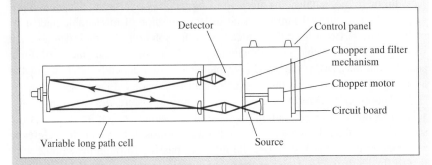

Figure 23F-1   A portable IR gas analyzer for air pollutants. (Courtesy of Wilks Scientific Corporation.)

Table 23-2

Examples of Infrared Vapor Analysis for OSHA Compliance

Compound	Allowable Exposure, ppm*	Wavelength, μm	Minimum Detectable Concentration, ppm†
Carbon disulfide	4	4.54	0.5
Chloroprene	10	11.4	4
Diborane	0.1	3.9	0.05
Ethylenediamine	10	13.0	0.4
Hydrogen cyanide	4.7‡	3.04	0.4
Methyl mercaptan	0.5	3.38	0.4
Nitrobenzene	1	11.8	0.2
Pyridine	5	14.2	0.2
Sulfur dioxide	2	8.6	0.5
Vinyl chloride	1	10.9	0.3

Source: Courtesy of the Foxboro Company, Foxboro, MA 02035.
*1992 OSHA exposure limits for 8-hr time-weighted average.
†For 20.25-m cell.
‡Short-term exposure limit; 15-min time-weighted average that shall not be exceeded at any time during the work day.

Infrared absorption remains a good quantitative technique for mixtures of organic compounds. With simple mixtures, simultaneous equations (recall Equations 23-1 and 23-2) can be solved to obtain the concentrations of each component. For more complex mixtures, IR spectrophotometry is increasingly used as a detector for compounds that have been separated by chromatographic methods.

In summary, the primary use of IR methods is for identification, determination, or confirmation of molecular structure. In the last application, IR methods are frequently used in conjunction with such techniques as mass spectrometry and NMR spectroscopy.

23C   ASSESSING ERRORS IN SPECTROPHOTOMETRY[10]

In most of the analytical methods we have considered so far, the relative error in the result can be directly related to the relative instrumental error associated with the measurement. Thus, in a volumetric analysis, an error of 0.03 mL in a 10-mL titration will lead to a relative uncertainty in the concentration of the analyte of 0.03 mL/10 mL, or 3 ppt. This straightforward relationship between instrumental error and concentration error is not found in spectrophotometric methods as shown by the experimental data plotted in Figure 23-11. To obtain these plots, a series of standard solutions was analyzed and the relative standard deviation of the results computed for various concentrations of the analyte. In the figures, these relative standard deviations are plotted as a function of absorbance for two spectrophotometers. One was a Spectronic 20 like the one shown in Figure 22-10; the other was a Cary 118 instrument, which is a research-quality spectrophotometer.

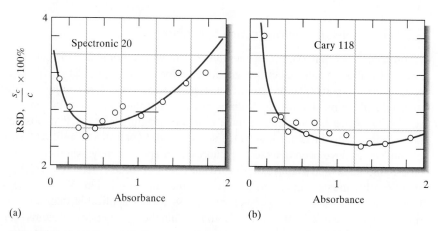

Figure 23-11   Experimental curves relating relative concentration uncertainties to absorbance for two spectrophotometers. The relative standard deviation (%RSD) in concentration (RSD = s_c/c) is plotted versus absorbance for (a) a Spectronic 20, a low-cost instrument, and (b) a Cary 118, a research-quality instrument. [From W. E. Harris and B. Kratochvil, *An Introduction to Chemical Analysis* (Philadelphia: Saunders College Publishing, 1981), p. 384. With permission.]

[10]For more information, see L. D. Rothman, S. R. Crouch, and J. D. Ingle, Jr., *Anal. Chem.* **1972,** *44,* 1375.

It is evident from both plots that concentration measurements at absorbances lower than about 0.1 are not very reliable and should be avoided. The reason for the poor precision in this region can be understood by writing Beer's law in the form

$$A = abc = \log \frac{P_0}{P} = \log P_0 - \log P$$

Note that concentration is directly proportional to the *difference* between the two measured quantities, $\log P_0$ and $\log P$. At low concentrations, $\log P$ is nearly as large as $\log P_0$ and A is a small difference between large numbers. Therefore, the relative uncertainties in A and c are large.

Note in Figure 23-11a that large errors are also encountered with the Spectronic 20 when the measured absorbance is above about 1.2. Here, the power of the beam is so very low after it has passed through the analyte solution that it cannot be measured accurately. As shown in Figure 23-11b, this type of error is much less severe with a highly sensitive research instrument. The user must ensure that concentrations of unknowns and standards fall in the absorbance range giving low relative error for the instrument used. As a rule of thumb, absorbances in the range of 0.1 to 1.2 are appropriate for many general-purpose instruments.

23D   MOLECULAR LUMINESCENCE SPECTROSCOPY

Luminescence phenomena include *fluorescence, phosphorescence,* and *chemiluminescence.*

As discussed in Section 21B-2, several spectroscopic methods are based on luminescence phenomena. Fluorescence spectroscopy is a very sensitive technique for determining some molecules. Likewise, phosphorescence and chemiluminescence methods find applications in the modern analytical laboratory.

23D-1   Molecular Fluorescence Spectroscopy

Fluorescence is a photoluminescence process in which atoms or molecules are excited by absorption of electromagnetic radiation (recall Figure 21-5). The excited species then relax to the ground state, giving up their excess energy as photons. One of the most attractive features of molecular fluorescence is its inherent sensitivity, which is often one to three orders of magnitude better than absorption spectroscopy. In fact, for selected species under controlled conditions, single molecules have been detected by fluorescence spectroscopy. Another advantage is the large linear concentration range of fluorescence methods, which is significantly greater than those encountered in absorption spectroscopy. Fluorescence methods are, however, much less widely applicable than absorption methods because of the relatively limited number of chemical systems that show appreciable fluorescence. Fluorescence is also subject to many more environmental interference effects than absorption methods. We will consider here a few of the most important aspects of molecular fluorescence methods.

Principles of Molecular Fluorescence

Molecular fluorescence is measured by exciting the sample at the absorption wavelength, also called the *excitation wavelength,* and measuring the emission at a longer wavelength called the emission or *fluorescence wavelength.* For example, the reduced form of the coenzyme nicontinamide adenine dinucleotide (NADH) can absorb radiation at 340 nm. The molecule exhibits fluorescence with an emission maximum at 465 nm. Usually, fluorescence emission is measured at right angles to the incident beam so as to avoid measuring the incident radiation (recall Figure 22-1b). The short-lived emission that occurs is called *fluorescence,* whereas luminescence that is much longer lasting is called *phosphorescence.*

Relaxation Processes.   Figure 23-12 shows a partial energy level diagram for a hypothetical molecular species. Three electronic energy states are shown, E_0, E_1, and E_2; E_0 is the ground state, and E_1 and E_2 are excited states. Each of the electronic states is shown as having four excited vibrational levels. Irradiation of this species with a band of radiation made up of wavelengths λ_1 to λ_5 (Figure 23-12a) results in the momentary population of the five vibrational levels of the first excited electronic state, E_1. Similarly, when the molecules are irradiated with a more energetic band of radiation made up of shorter wavelengths λ_1' to λ_5', the five vibrational levels of the higher energy electronic state E_2 become briefly populated.

Once the molecule is excited to E_1 or E_2, several processes can occur that cause the molecule to lose its excess energy. Two of the most important of these mechanisms, nonradiative relaxation and fluorescence emission, are illustrated in Figure 23-12b and c.

Fluorescence emission occurs in 10^{-5} s or less. In contrast, phosphorescence may persist for several minutes or even hours. Fluorescence is much more widely used for determinations than phosphorescence.

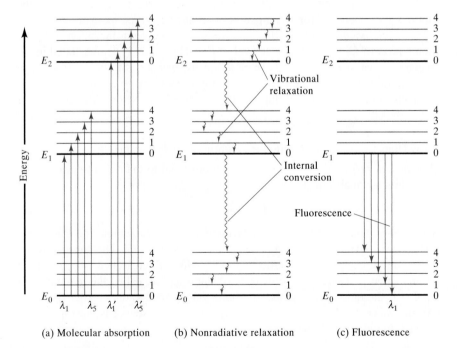

(a) Molecular absorption   (b) Nonradiative relaxation   (c) Fluorescence

Figure 23-12   Energy-level diagram shows some of the processes that occur during (a) absorption of incident radiation, (b) nonradiative relaxation, and (c) fluorescence emission by a molecular species. Absorption typically occurs in 10^{-15} s, whereas vibrational relaxation occurs in the 10^{-10} to 10^{-11} s time scale. Internal conversion between different electronic states is also very rapid (10^{-12} s), whereas the lifetime of fluorescence is typically 10^{-5} to 10^{-10} s.

The two most important nonradiative relaxation methods that compete with fluorescence are illustrated in Figure 23-12b. *Vibrational relaxation,* depicted by the short wavy arrows between vibrational energy levels, takes place during collisions between excited molecules and molecules of the solvent. Nonradiative relaxation between the lower vibrational levels of an excited electronic state and the higher vibrational levels of another electronic state can also occur. This type of relaxation, sometimes called *internal conversion,* is depicted by the two longer wavy arrows in Figure 23-12b. The exact mechanism by which these two relaxational processes occur is under study today, but the net result is a tiny increase in the temperature of the medium.

Figure 23-12c illustrates the relaxation process that is desired, fluorescence. Almost always, fluorescence involves transitions from the lowest-lying excited electronic state E_1 to the ground state, E_0. Also, the fluorescence usually occurs only from the lowest vibrational level of E_1 to various vibrational levels of E_0, because the internal conversion and vibrational relaxation processes are very rapid compared with fluorescence. Hence, a fluorescence spectrum usually consists of only one band with many closely spaced lines that represent transitions from the lowest vibrational level of E_1 to the many different vibrational levels of E_0. Note also that fluorescence emission is usually of lower energy (longer wavelength) than the energy of excitation. This shift to longer wavelength is sometimes called the *Stokes shift.*

> Stokes-shifted fluorescence is longer in wavelength than the radiation that caused the excitation.

Relationship between Excitation Spectra and Fluorescence Spectra. Because the energy differences between vibrational states is about the same for both ground and excited states, the absorption, or *excitation spectrum,* and the fluorescence spectrum for a compound often appear as approximate mirror images of one another with overlap occurring near the origin transition (0 vibrational level of E_1 to 0 vibrational level of E_0). This effect is demonstrated in the spectra shown in Figure 23-13. There are many exceptions to this mirror-image rule, particularly when the excited and ground states have different molecular geometries or when different fluorescence bands originate from different parts of the molecule.

How Does Concentration Influence Fluorescence Intensity?

The radiant power of fluorescence F is proportional to the radiant power of the excitation beam absorbed:

$$F = K(P_0 - P) \tag{23-11}$$

where P_0 is the radiant power of the beam incident on the sample and P is the radiant power after it traverses a pathlength b of the medium. The constant K depends on geometry and the efficiency with which fluorescence competes with nonradiative processes. The efficiency of fluorescence, called the *quantum efficiency,* is the ratio of the number of photons fluoresced to the number absorbed. At low concentrations (see Feature 23-2), where fluorescence is most often employed, Equation 23-11 becomes

$$F = K'P_0 c \tag{23-12}$$

where c is the concentration of the fluorescent species and K' is a new proportionality constant. At constant incident radiant power, Equation 23-12 illustrates that F is directly proportional to analyte concentration. Thus, a plot of the fluo-

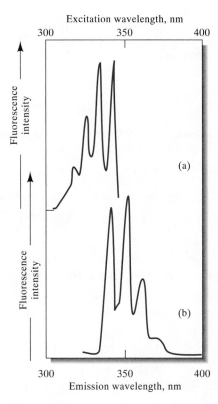

Figure 23-13   Fluorescence spectra for 1 ppm anthracene in alcohol: (a) excitation spectrum and (b) emission spectrum. Note that the two spectra are very nearly mirror images.

rescent radiant power versus the concentration of the emitting species should be, and ordinarily is, linear at low concentrations. When c becomes great enough that the absorbance is larger than about 0.05, linearity is lost and F begins to reach a plateau with concentration. This effect is known as a *primary absorption inner filter effect*. In fact, at high concentrations, fluorescence radiant power can even begin to decrease with increasing concentration.

Fluorescence Instrumentation

The several different types of fluorescence instruments all follow the general block diagram of Figure 22-1b. Optical diagrams of typical instruments are shown in Figure 23-14. If the two wavelength selectors are both filters, the instrument is called a *fluorometer*. If both wavelength selectors are monochromators, the instrument is a *spectrofluorometer*. Some instruments are hybrids and use an excitation filter along with an emission monochromator. Fluorescence instruments can incorporate a double-beam design to compensate for changes in the source radiant power with time and wavelength. Instruments that correct for the source spectral distribution are called *corrected spectrofluorometers*.

Sources for fluorescence are usually more powerful than typical absorption sources. In fluorescence, the radiant power emitted is directly proportional to the source intensity (Equation 23-12), whereas absorbance, being related to the ratio of radiant powers, is essentially independent of source intensity. Hence, mercury

Feature 23-2

How Is Fluorescence Related
to Concentration?

To relate F to the concentration c of the fluorescing species, we write Beer's law in the form

$$\frac{P}{P_0} = 10^{-\varepsilon bc}$$

where ε is the molar absorptivity of the fluorescing species and εbc is the absorbance A. By substituting this equation into Equation 23-11, we obtain

$$F = KP_0(1 - 10^{-\varepsilon bc})$$

A series expansion of the exponential term in this equation leads to

$$F = KP_0\left[2.3\varepsilon bc - \frac{(-2.3\varepsilon bc)^2}{2!} - \frac{(-2.3\varepsilon bc)^3}{3!} - \cdots \right]$$

where the factor of 2.3 arises in converting from base 10 to base e. Provided $\varepsilon bc = A < 0.05$, all the subsequent terms in the brackets are small with respect to the first, and we can write

$$F = 2.3K\varepsilon bcP_0 \tag{23-13}$$

or

$$F = K'P_0c$$

which is Equation 23-12.

arc lamps, xenon arc lamps, xenon-mercury arc lamps, and lasers are typical fluorescence sources. Monochromators and transducers are similar to those used in absorption spectrophotometers except that photomultipliers are invariably used in high-sensitivity spectrofluorometers. Array detectors, particularly charge coupled device (CCD) arrays are also becoming popular for fluorescence measurements. The sophistication, performance characteristics, and cost of fluorometers and spectrofluorometers vary as widely as the characteristics of absorption spectrophotometers. Generally, fluorescence instruments are more expensive than absorption instruments of corresponding quality.

Applying Molecular Fluorescence Spectroscopy

Fluorescence spectroscopy is not considered a major structural or qualitative analysis tool because molecules with subtle structural differences often have similar fluorescence spectra. Also, fluorescence bands in solution are relatively broad at room temperature. Fluorescence, however, has proved to be a valuable tool in oil spill identification. The source of such a spill can often be identified by comparing the fluorescence emission spectrum of the spill sample with that of a suspected source. The vibrational structure of polycyclic hydrocarbons present in the oil makes this type of identification possible.

Fluorescence methods are used to study chemical equilibria and kinetics in much the same way as absorption spectrophotometry. Often, it is possible to study chemical reactions at lower concentrations because of the higher sensitivity of fluorescence methods. In many cases where fluorescence monitoring is or-

Some typical polycyclic aromatic hydrocarbons found in oil spills are chrysene, perylene, pyrene, fluorene, and 1,2-benzofluorene. Most of these compounds are carcinogenic.

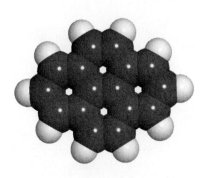

pyrene

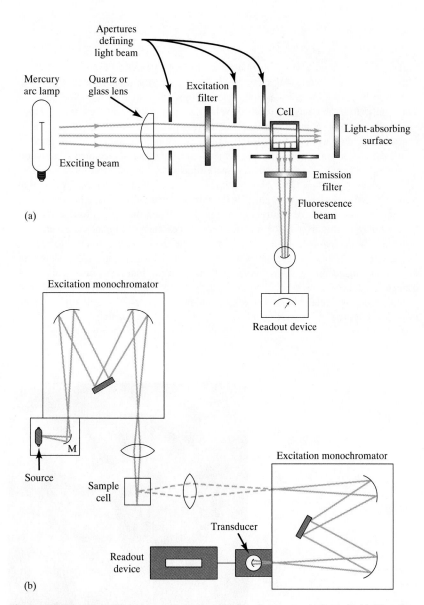

Figure 23-14   Typical fluorescence instruments. A filter fluorometer is shown in (a). Note that the emission is measured at right angles to the mercury arc lamp source. Fluorescence radiation is emitted in all directions, and the 90 deg geometry avoids the detector viewing the source. The spectrofluorometer (b) uses two grating monochromators and also views the emission at right angles. The two monochromators allow the scanning of excitation spectra (excitation wavelength scanned at a fixed emission wavelength), emission spectra (emission wavelength scanned at a fixed excitation wavelength), or synchronous spectra (both wavelengths scanned with a fixed wavelength or wavenumber offset between the two monochromators).

dinarily not feasible, *fluorescent probes* or tags can be bound covalently to specific sites in molecules such as proteins, thus making them detectable via fluorescence. These tags can be used to provide information about energy transfer processes, the polarity of the protein, and the distances between reactive sites (see, for example, Feature 23-3).

Feature 23-3

Use of Fluorescence Probes
in Neurobiology: Probing the
Enlightened Mind

Fluorescent indicators have been widely used to probe biological events in individual cells. A particularly interesting example is the so-called ion probe, which changes its excitation or emission spectrum upon binding to specific ions such as Ca^{2+} or Na^+. These indicators can be used to record events that take place in different parts of individual neurons or to monitor simultaneously the activity of a collection of neurons. In neurobiology, for example, the dye Fura-2 has been used to monitor the free intracellular calcium concentration following some pharmacological or electrical stimulation. By following the fluorescence changes with time at specific sites in the neuron, researchers can determine when and where a calcium-dependent electrical event took place. One cell that has been studied is the Purkinje neuron in the cerebellum, which is one of the largest in the central nervous system. When this cell is loaded with the Fura-2 fluorescent indicator, sharp changes in fluorescence can be measured that correspond to individual calcium action potentials. The changes are correlated to specific sites in the cell by means of fluorescence imaging techniques. Figure 23F-3 shows the fluorescent image on the right along with fluorescent transients, recorded as the change in fluorescence relative to the steady fluorescence $\Delta F/F$, correlated with sodium action potential spikes. The interpretation of these kinds of patterns can have important implications in understanding the details of synaptic activity.

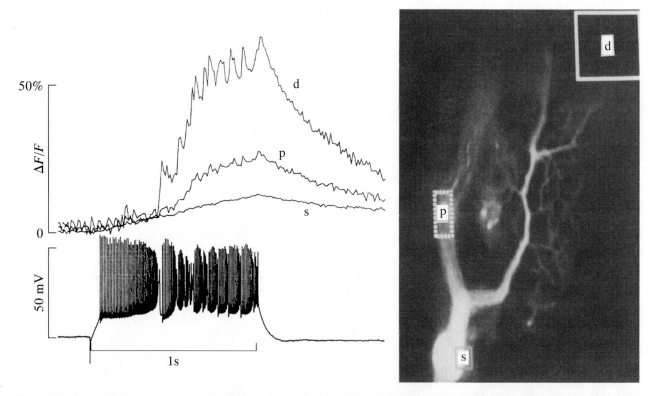

Figure 23F-3   Calcium transients in a cerebellar Purkinje cell. The image on the right is of the cell filled with a fluorescent dye that responds to the calcium concentration. Fluorescent transients are shown on the top left recorded at areas d, p, and s in the cell. The transients in region d correspond to the dendrite region of the cell. Specific calcium signals can be correlated to the action potentials shown on the bottom left. [From V. Lev-Ram, H. Miyakawa, N. Lasser-Ross and W.N. Ross, *J. Neurophysiol.* **1992,** *68,* 1170. With Permission of The American Physiological Society.]

Quantitative fluorescence methods have been developed for inorganic, organic, and biochemical species. Inorganic fluorescence methods can be divided into two classes: direct methods that are based on the reaction of the analyte with a complexing agent to form a fluorescent complex and indirect methods that depend on the decrease in fluorescence, also called *quenching,* as a result of interaction between the analyte and a fluorescent reagent. Quenching methods are primarily used for the determination of anions and dissolved oxygen.

Nonradiative relaxation of transition-metal chelates is so efficient that fluorescence of these species is seldom encountered. It is worth noting that most transition metals absorb in the UV or visible region, whereas nontransition metal ions do not. For this reason, fluorescence is often considered complementary to absorption for the determination of cations (see Figure 23-15).

The number of applications of fluorescence methods to organic and biochemical problems is impressive. Among the compound types that can be determined by fluorescence are amino acids, proteins, coenzymes, vitamins, nucleic acids, alkaloids, porphyrins, steroids, flavonoids, and many metabolites.[11] Because of its sensitivity, fluorescence is widely used as a detection technique for liquid chromatographic methods (see Section 25B-1), for flow analysis methods, and for electrophoresis. In addition to methods that are based on measurements of fluorescence intensity, many methods are based on measurements of fluorescence lifetimes. Several instruments have been developed that provide microscopic images of specific species based on fluorescence lifetimes.[12]

The most important applications of fluorescence spectroscopy are for the analysis of food products, pharmaceuticals, clinical samples, and natural products. The sensitivity and selectivity of fluorescence make it a particularly valuable tool in these fields.

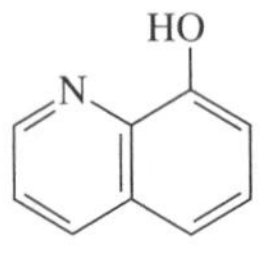

23D-2   Molecular Phosphorescence Spectroscopy

Phosphorescence is a photoluminescence phenomenon that is quite similar to fluorescence. Understanding the difference between these two phenomena requires an understanding of electron spins and the difference between a singlet state and a triplet state. Ordinary molecules that are not free radicals exist in the ground state with their electron spins paired. A molecular electronic state in which all electron spins are paired is said to be a *singlet state.* The ground state of a free radical, on the other hand, is a *doublet state,* because the odd electron can assume two orientations in a magnetic field.

When one of a pair of electrons in a molecule is excited to a higher-energy level, a singlet or a *triplet* state can be produced. In the excited singlet state, the spin of the promoted electron is still opposite that of the remaining electron. In the *triplet state,* however, the spins of the two electrons become unpaired and are thus parallel. These states can be represented as illustrated in Figure 23-16.

8-hydroxyquinoline
(reagent for Al, Be, and
other metal ions)

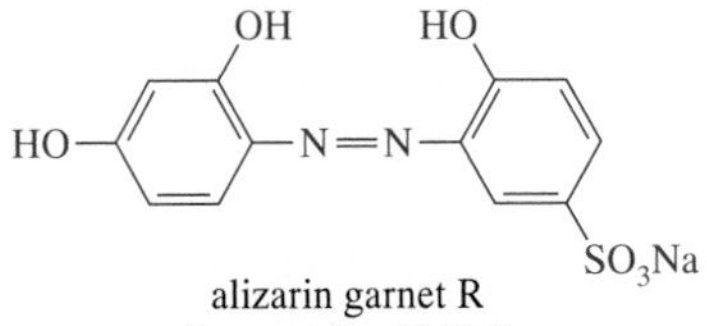

alizarin garnet R
(reagent for Al, F⁻)

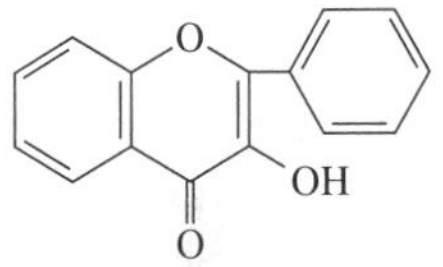

flavanol
(reagent for Zr and Sn)

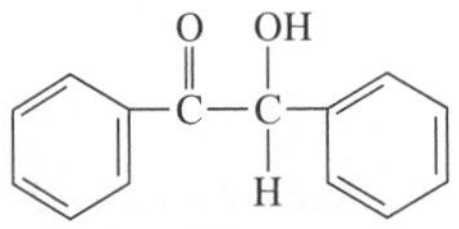

benzoin
(reagent for B, Zn, Ge, and Si)

Figure 23-15   Some fluorometric chelating agents for metal cations. Alizarin garnet R can detect Al^{3+} at levels as low as 0.007 $\mu g\ mL^{-1}$. Detection of F^- with alizarin garnet R is based on quenching of the Al^{3+} complex. Flavanol can detect Sn^{4+} at the 0.1 $\mu g\ mL^{-1}$ level.

[11]See, for example, O. S. Wolfbeis, in *Molecular Luminescence Spectroscopy: Methods and Applications, Part I,* S. G. Schulman, Ed. (New York: Wiley-Interscience, 1985), Chapter 3.

[12]See J. R. Lakowicz, H. Szmacinski, K. Nowacyzk, K. Berndt, and M. L. Johnson, in *Fluorescence Spectroscopy: New Methods and Applications,* O. S. Wolfbeis, Ed. (Berlin: Springer-Verlag, 1993), Chapter 10.

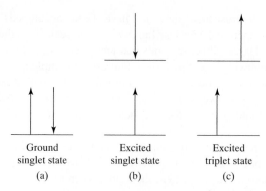

Ground singlet state	Excited singlet state	Excited triplet state
(a)	(b)	(c)

Figure 23-16   Electronic spin states of molecules. In (a), the ground electronic state is shown. In the lowest energy or ground state, the spins are always paired, and the state is said to be a singlet state. In (b) and (c), excited electronic states are shown. If the spins remain paired in the excited state, the molecule is in an excited singlet state (b). If the spins become unpaired, the molecule is in an excited triplet state (c).

The excited triplet state is less energetic than the corresponding excited singlet state.

Fluorescence of molecules involves a transition from an excited singlet state to the ground singlet state. This transition is highly probable; thus, the lifetime of an excited singlet state is very short (10^{-5} s or less). Molecular phosphorescence, on the other hand, involves a transition from an excited triplet state to the ground singlet state. Because this transition produces a change in electron spin, it is much less probable. Hence, the triplet state has a much longer lifetime (typically 10^{-4} to 10^{4} s). As an example, solid state phosphors coated on the screen of a cathode-ray tube make it possible to observe the action of electron beams in many oscilloscopes, television sets, and computer monitors.

The long lifetime of phosphorescence is also one of its drawbacks. As a result of this long lifetime, nonradiational processes can compete with phosphorescence to deactivate the excited state. Thus, the efficiency of the phosphorescence process, and the corresponding phosphorescence intensity, is relatively low. To increase this efficiency, phosphorescence is commonly observed at low temperatures in rigid media such as glasses. In recent years, room-temperature phosphorescence has become popular. In this technique, the molecule is either adsorbed on a solid surface or enclosed in a molecular cavity (micelle or cyclodextrin cavity), which protects the fragile triplet state.

Because of its weak intensity, phosphorescence is much less widely applicable than fluorescence. Phosphorimetry, however, has been used for the determination of a variety of organic and biochemical species including nucleic acids, amino acids, pyrine and pyrimidine, enzymes, polycyclic hydrocarbons, and pesticides. Many pharmaceutical compounds exhibit measurable phosphorescence signals. The instrumentation for phosphorescence is also somewhat more complex than that for fluorescence. Usually, the phosphorescence instrument allows discrimination of phosphorescence from fluorescence by delaying the phosphorescence measurement until the fluorescence has decayed to nearly zero. Many fluorescence instruments have attachments, called *phosphoroscopes,* that allow the same instrument to be used for phosphorescence measurements.

In room-temperature phosphorescence, the triplet state of the analyte can be protected by being incorporated into a surfactant aggregate called a micelle. In aqueous solutions, the aggregate has a nonpolar core due to repulsion of the polar head groups. The opposite occurs in nonpolar solvents.

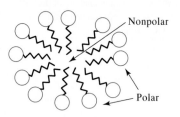

Micelle in aqueous solvent

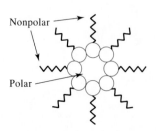

Micelle in nonaqueous solvent

23D-3 Chemiluminescence Methods: Reactions That Produce Light

Chemiluminescence is produced when a chemical reaction yields an electronically excited molecule, which emits light as it returns to the ground state. Chemiluminescence reactions are encountered in a number of biological systems, where the process is often called *bioluminescence.* Examples of species exhibiting bioluminescence include the firefly, the sea pansy, certain jellyfish, bacteria, protozoa, and crustacea.

One attractive feature of chemiluminescence for analytical uses, is the very simple instrumentation. Since no external source of radiation is needed for excitation, the instrument may consist of only a reaction vessel and a photomultiplier tube. Generally, no wavelength selection device is needed because the only source of radiation is the chemical reaction.

Chemiluminescence methods are known for their high sensitivities. Typical detection limits range from parts per million to parts per billion or lower. Applications include the determination of gases, such as oxides of nitrogen, ozone, and sulfur compounds; determination of inorganic species, such as hydrogen peroxide and some metal ions; immunoassay techniques; DNA probe assays; and polymerase chain reaction methods.[13]

The firefly produces light by the phenomenon of *bioluminescence.* Different species of fireflies flash with different on–off cycle times. Fireflies mate only with their own species. The familiar bioluminescence reaction occurs when the firefly is looking for a mate.

Several commercial analyzers for the determination of gases are based on chemiluminescence. Nitrous oxide (NO) can be determined by reaction with ozone (O_3). The reaction converts the NO to excited NO_2 with the subsequent emission of light. In addition, ozone can be determined by reaction with NO.

23E MOLECULAR SCATTERING METHODS

Thus far, we have considered the absorption and emission (luminescence) of radiation by molecules. Several additional processes, however, can occur when radiation interacts with matter. One of the most important of these is scattering of electromagnetic radiation. Scattering can be divided into two classes: *elastic scattering,* in which the scattered radiation is of the same energy as the incident radiation, and *inelastic scattering,* in which the scattered radiation has higher or lower energy than the incident radiation. Both of these types of phenomena have useful analytical applications.

The elastic scattering methods of *turbidimetry* and *nephelometry* have long been used to measure the concentration of particulate matter in suspensions. Turbidimetric methods measure the decrease that occurs in the transmitted radiation as a result of scattering from particles, whereas nephelometric methods directly measure the intensity of the scattered radiation. Laser light-scattering methods are widely used to determine molecular weights of compounds and to determine particle sizes. Low-angle laser light-scattering methods are used for the former, whereas *quasi-elastic* or *dynamic* light-scattering methods are used for the latter.

Inelastic light-scattering methods have become very important in recent years, particularly those methods based on the *Raman effect.* Raman spectroscopy involves inelastic scattering of radiation caused by vibrational and rotational transitions. The Raman method is complementary to IR spectroscopy and can be used to obtain qualitative, structural, and quantitative information about molecular species. Some molecules show no infrared and Raman bands in com-

---

[13]See, for example, T. A. Nieman, in *Handbook of Instrumental Techniques for Analytical Chemistry,* F. A. Settle, Ed. (Upper Saddle River, NJ: Prentice-Hall, 1997), Chapter 27.

mon, whereas others show some similarity in spectra. The differences are related to the symmetry properties of the molecules, and a combination of the two techniques is very important in revealing molecular symmetry. The introduction of inexpensive Raman spectrometers with CCD detectors is responsible for tremendous growth in the routine use of Raman spectroscopy.

23F   ATOMIC SPECTROSCOPY

Atomic spectroscopy is used for the qualitative and quantitative determination of 70 to 80 elements. Detection limits for many of these lie in the sub-parts-per-million range. The methods can be based on absorption, emission, or fluorescence. Atomic absorption (AA) spectroscopy currently is the most widely used of these techniques. Here, we briefly consider the most important of these techniques. The reader is referred elsewhere for more detailed discussions.[14]

23F-1   Atomic Absorption Spectroscopy

In AA spectroscopy, as in all atomic spectroscopic methods, the sample must be converted into an atomic vapor by a process known as *atomization*. In this process, the sample is volatilized and decomposed to produce atoms and perhaps some ions in the gas phase. Several methods are used to atomize samples. The two most important of these for AA spectroscopy are flame and furnace atomization.

Flame Atomic Absorption Spectroscopy

A block diagram of a flame AA spectrometer is shown in Figure 23-17. Here, the radiation of a line source of the element of interest, typically a hollow cathode lamp, is directed through the flame containing the atomic vapor. Solution samples are usually brought into the flame by means of a sprayer or *nebulizer,* which produces small sample droplets. The solvent from the droplets quickly vaporizes, and the resulting salt particles vaporize and decompose into atoms, ions, and electrons. Atoms in the sample will absorb radiation emitted by the same atom in the hollow cathode lamp and thus attenuate the power of the source. Usually a monochromator is used to separate a spectral line of the element of interest from any background radiation from the source or the flame. A photomultiplier tube is typically used to convert the radiant power from the source into a related electrical current.

Furnace Atomic Absorption Spectroscopy

Furnace, or electrothermal, AA uses the same instrumental setup as shown in Figure 23-17 except that a furnace atomizer is substituted for the burner. With furnace AA, discrete amounts of sample, usually microliter volumes, are de-

---

[14]J. D. Ingle, Jr., and S. R. Crouch, *Spectrochemical Analysis* (Englewood Cliffs, N.J.: Prentice-Hall, 1988); D. A. Skoog, F. J. Holler, and T. A. Nieman, *Principles of Instrumental Analysis,* 5th ed. (Philadelphia: Saunders College Publishing, 1998).

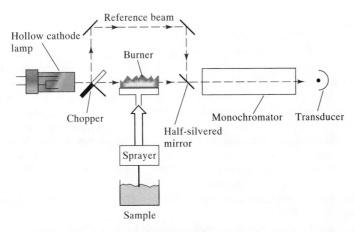

Figure 23-17   Block diagram of a double-beam atomic absorption spectrophotometer. Radiation from the hollow cathode lamp is split into two beams; one goes through the flame and one bypasses the flame as a reference beam. The sample solution is sprayed into the flame as droplets. In the flame, atoms of the element of interest are formed. These atoms can absorb the source radiation and attenuate the beam. The monochromator isolates the desired radiation from background. A photomultiplier tube acts as the radiation transducer.

posited in the furnace. A multistep heating program is usually applied to desolvate the sample, ash, or char the organic material present and then produce the atomic vapor. Furnace AA gives a transient signal that reaches a peak in a few seconds. Furnace AA is usually one to two orders of magnitude more sensitive than flame AA.

Quantitative Applications

Flame and furnace AA are quantitative techniques only, since a specific hollow cathode lamp is required for each element of interest. Quantitative analysis is based on Beer's law, which applies to the atomic absorption process just as it does to molecular absorption. The method is highly selective because of the very narrow line widths emitted by the hollow cathode lamp (0.002 to 0.005 nm). With furnace AA, measurements can be made on samples as small as a few microliters. Calibration methods use the method of external standards or the method of standard additions. Under normal conditions, flame AA can achieve standard deviations of a few percent, whereas furnace AA is usually not as precise because of the smaller volumes used. Typically, relative errors of less than 10% are achievable with furnace AA. Another technique applicable to volatile elements and compounds is the cold-vapor technique discussed in Feature 23-4.

23F-2   Some Other Types of Atomic Spectroscopy

In addition to atomic absorption spectroscopy, atomic emission and atomic fluorescence are also encountered. Of the last two, atomic emission (AE) is much more widely employed at present. Some AE methods use flames to produce

Feature 23-4

Mercury and its
Determination by
Cold-Vapor Atomic
Absorption Spectroscopy

People's fascination with mercury began when prehistoric cave dwellers discovered the mineral cinnabar (HgS) and used it as a red pigment. Our first written record of the element came from Aristotle, who described it as "liquid silver" in the fourth century B.C. Today, there are thousands of uses of mercury and its compounds in medicine, metallurgy, electronics, agriculture, and many other fields. Because it is a liquid at room temperature, mercury is used to make flexible and efficient electrical contacts in scientific, industrial, and household applications. Thermostats, silent light switches, and fluorescent light bulbs are but a few examples of its electrical applications.

A useful property of metallic mercury is that it forms amalgams with other metals that have a host of uses. For example, metallic sodium is produced as an amalgam by electrolysis of molten sodium chloride. In the past, dentists used a 50% amalgam with an alloy of silver for fillings.

The toxicological effects of mercury have been known for many years. The bizarre behavior of the Mad Hatter in Lewis Carroll's *Alice in Wonderland* was a result of the effects of mercury and mercury compounds on the Hatter's brain. Mercury that has been absorbed through the skin and lungs destroys brain cells, which are not regenerated. Hatters of the nineteenth century used mercury compounds in processing fur to make felt hats. These workers and those in other industries have suffered the debilitating symptoms of mercurialism such as loosening of teeth, tremors, muscle spasms, personality changes, depression, irritability, and nervousness.

The toxicity of mercury is complicated by its tendency to form both inorganic and organic compounds. Inorganic mercury is relatively insoluble in body tissues and fluids, so it is expelled from the body about ten times faster than organic mercury. Organic mercury, usually in the form of alkyl compounds such as methyl mercury, is somewhat soluble in fatty tissues such as the liver. Methyl mercury accumulates to toxic levels and is expelled from the body quite slowly. Even experienced scientists must take extreme precautions in handling organomercury compounds. In 1997, Dr. Karen Wetterhahn of Dartmouth College died as a result of mercury poisoning, despite being one of the world's leading experts in handling methyl mercury.

Mercury tends to concentrate in the environment, as illustrated in Figure 23F-4a. Inorganic mercury is converted to organic mercury by anaerobic bacteria in sludge deposited at the bottom of lakes, streams, and other bodies of water. Small aquatic animals consume the organic mercury and are in turn eaten by larger life forms. As the element moves up the food chain from microbes, to shrimp, to fish, and ultimately to larger animals such as swordfish, the mercury becomes ever more concentrated. Some sea creatures such as oysters may concentrate mercury by a factor of 100,000. At the top of the food chain, the concentration of mercury reaches levels as high as 20 ppm. The U.S. Food and Drug Administration has set a legal limit of 1 ppm in fish for human consumption. As a result, mercury levels in some areas threaten local fishing industries. The Environmental Protection Agency has set a limit of 1 ppb of mercury in drinking water, and the Occupational Safety and Health Administration has set a limit of 0.1 mg/m^3 in air.

Analytical methods for the determination of mercury play an important role in monitoring the safety of food and water supplies. One of the most useful methods is based on the atomic absorption by mercury of 253.7 nm of radiation. Figure 23F-4b shows an apparatus that is used to determine mercury by atomic absorption at room temperature.[15]

---

[15]W. R. Hatch and W. L. Ott, *Anal. Chem.* **1968,** *40,* 2085.

(a)

Figure 23F-4A Biological concentration of mercury in the environment.

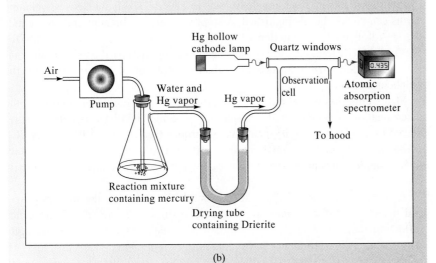

(b)

Figure 23F-4B Apparatus for cold-vapor atomic absorption determination of mercury.

A sample suspected of containing mercury is decomposed in a hot mixture of nitric acid and sulfuric acid, which converts the mercury to compounds of Hg^{2+}. The Hg(II) is reduced to the metal with a mixture of hydroxylamine sulfate and tin(II) sulfate. Air is then pumped through the solution to carry the resulting mercury-containing vapor through the drying tube and into the observation cell. Water vapor is trapped by Drierite® in the drying tube so that only mercury vapor and air pass through the cell. The monochromator of the atomic absorption spectrophotometer is tuned to a band around 254 nm. Radiation from the 253.7-nm line of the mercury hollow cathode lamp passes through the quartz windows of the observation cell, which is placed in the light path of the instrument. The absorbance is directly proportional to the concentration of mercury in the cell, which in turn is proportional to the concentration of mercury in the sample. Solutions of known mercury concentration are treated in a similar way to calibrate the apparatus. The method depends on the low solubility of mercury in the reaction mixture and its appreciable vapor pressure, which is 2×10^{-3} torr at 25°C. The sensitivity of the method is about 1 ppb, and it is used to determine mercury in foods, metals, ores, and environmental samples. The method has the advantages of sensitivity, simplicity, and room-temperature operation.

excited atoms, which then emit characteristic radiation as they return to the ground state. Other AE methods use more energetic atomizers, such as inductively coupled plasmas (ICPs), arcs, and sparks. In contrast to atomic absorption, atomic emission methods can be used for qualitative analysis. Complete spectra can be recorded that are then used to identify the elements present based on the wavelengths of the lines emitted. For example, Figure 23-18 shows the emission spectrum of a brine sample obtained using an oxyhydrogen flame. Emission lines due to several alkali metals and alkaline earths can clearly be distinguished. Also present are the broader emission bands that result from the excitation of molecular species such as MgOH, CaOH, MgO, and OH. Here, the vibrational transitions superimposed on electronic transitions produce closely spaced lines that are not completely resolved by the monochromator.

Flame emission was at one time widely used in the clinical laboratory for the determination of sodium and potassium. These methods have largely been replaced by methods using ion-selective electrodes (see Section 19D). Today, ICP emission spectrometry is much more important than flame emission. The inductively coupled plasma is also an important ion source for mass spectrometric analysis.

Atomic fluorescence (AF) spectroscopy is, in principle, the most sensitive and selective of the atomic spectrometric methods. In AF, the fluorescence following radiational excitation is measured, usually at the same wavelength. At present, AF instruments have not enjoyed commercial success, and AF is used only in specialized situations. Atomic absorption methods remain the most widely used of the atomic spectrometric techniques for quantitative analysis.

Gustav Robert Kirchhoff (1824–1877) was a German physicist who along with his colleague, chemist Robert Wilhelm Bunsen (1811–1899), discovered spectroscopic analysis. Bunsen's burner was used to atomize samples of the elements, and Kirchhoff's prism spectroscope was used to analyze the light given off by the incandescent samples. Using this technique, they were able to identify several new elements and demonstrate the presence of a number of elements in the sun by analyzing sunlight. This discovery is depicted on this Vatican stamp. The sun's corona, one of Kirchhoff's spectroscopes, and the absorption spectrum of hydrogen are also illustrated.

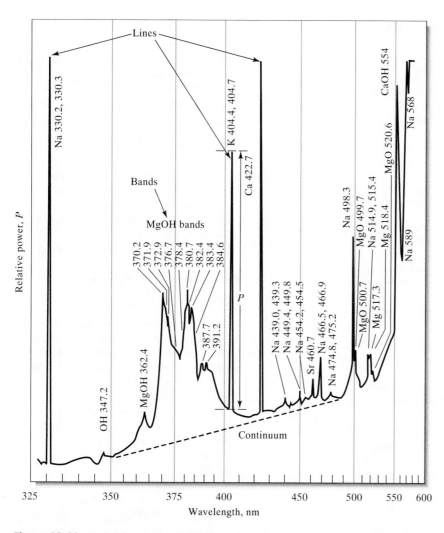

Figure 23-18   Emission spectrum of a brine sample obtained with an oxyhydrogen flame.
[R. Hermann and C. T. J. Alkemade, *Chemical Analysis by Flame Photometry,* 2nd ed. (New York: Interscience, 1979), p. 484. With permission.]

WEB WORKS

One of the developing areas in which spectroscopy is providing important information is in the field of spectroscopic imaging, particularly in biomedicine. Fluorescence and infrared spectroscopy are being employed to provide images that can be used in diagnostic and screening methods. These methods usually involve either a microscope or a fiber optic probe to give spatial and wavelength-resolved images. For one biomedical research institute interested in infrared imaging and *in-vivo* spectroscopy, point your browser to *http://www.ibd.nrc.ca/spectroscopy/ir_imaging.html.* Investigate the research programs in IR pathology, IR clinical chemistry, and IR imaging at the Institute for Biodiagnostics.

23G   QUESTIONS AND PROBLEMS

23-1. Briefly describe or define
  *(a) chromophore.
  (b) an electronic transition of a molecule.
  *(c) monochromatic radiation.
  (d) the ground state of a molecule.
  *(e) absorption spectrum.

23-2. Briefly describe or define
  *(a) fluorescence.
  (b) vibrational relaxation.
  *(c) internal conversion.
  (d) nonradiative relaxation.
  *(e) Stokes shift.
  (f) quantum yield.
  *(g) inner filter effect.

23-3. How do vibrational relaxation and electronic relaxation resemble one another? How do they differ?

***23-4.** Why do some absorbing compounds fluoresce and others not?

23-5. What structural features appear to favor fluorescence?

***23-6.** Briefly explain why
  (a) fluorescence emission ordinarily occurs at wavelengths that are longer than that of the excitation radiation.
  (b) fluorescence measurements have the capability of greater sensitivity than absorbance measurements.

23-7. What advantage can be claimed for the standard addition method? What minimum condition is needed for the successful application of this method?

***23-8.** What is the mechanism of charge-transfer absorption? Why is this type of absorption of interest in analytical chemistry?

23-9. Why is UV/visible absorption spectrometry not often used by itself for qualitative analysis?

***23-10.** Describe the basic differences between atomic emission and atomic absorption spectroscopy.

***23-11.** Why is atomic emission more sensitive to flame instability than atomic absorption or fluorescence?

23-12. Why has atomic fluorescence spectrometry not enjoyed the popularity of atomic absorption spectrometry?

23-13. The equilibrium constant for the conjugate acid/base pair

$$HIn + H_2O \rightleftharpoons H_3O^+ + In^-$$

is 8.00×10^{-5}. From the additional information,

Species	Absorption Maximum, nm	Molar Absorptivity	
		430 nm	600 nm
HIn	430	8.04×10^3	1.23×10^3
In$^-$	600	0.775×10^3	6.96×10^3

***(a)** calculate the absorbance at 430 nm and 600 nm for the following indicator concentrations: 3.00×10^{-4} M, 2.00×10^{-4} M, 1.00×10^{-4} M, 0.500×10^{-4} M, and 0.250×10^{-4} M.

(b) plot absorbance as a function of indicator concentration.

23-14. The equilibrium constant for the reaction

$$2CrO_4^{2-} + 2H^+ \rightleftharpoons Cr_2O_7^{2-} + H_2O$$

is 4.2×10^{14}. The molar absorptivities for the two principal species in a solution of $K_2Cr_2O_7$ are as shown in the table that follows. Four solutions were prepared by dissolving 4.00×10^{-4}, 3.00×10^{-4}, 2.00×10^{-4}, and 1.00×10^{-4} moles of $K_2Cr_2O_7$ in water and diluting to 1.00 L with a pH 5.60 buffer. Derive theoretical absorbance values (1.00-cm cells) for each solution and plot the data for
  (a) 345 nm.
  (b) 370 nm.
  (c) 400 nm.

λ, nm	$\varepsilon_1(CrO_4^{2-})$	$\varepsilon_2(Cr_2O_7^{2-})$
345	1.84×10^3	10.7×10^2
370	4.81×10^3	7.28×10^2
400	1.88×10^3	1.89×10^2

***23-15.** The molar absorptivity for the complex formed between bismuth(III) and thiourea is 9.32×10^3 L mol^{-1} cm^{-1} at 470 nm. Calculate the range of permissible concentrations for the complex if the absorbance is to be no less than 0.15 nor greater than 0.80 when the measurements are made in 1.00-cm cells.

23-16. The molar absorptivity for aqueous solutions of phenol at 211 nm is 6.17×10^3 L mol^{-1} cm^{-1}. Calculate the permissible range of phenol concentrations that can be used if the transmittance is to be less than 80% and greater than 5% when the measurements are made in 1.00-cm cells.

***23-17.** The logarithm of the molar absorptivity for acetone in ethanol is 2.75 at 366 nm. Calculate the range of acetone concentrations that can be used if the transmittance is to be greater than 10% and less than 90% in a 1.50-cm cell.

23-18. The logarithm of the molar absorptivity of phenol in aqueous solution is 3.812 at 211 nm. Calculate the range of phenol concentrations that can be used if the absorbance is to be greater than 0.100 and less than 2.000 with a 1.25-cm cell.

23-19. Sketch a photometric titration curve for the titration of Sn^{2+} with MnO_4^-. What color radiation should be used for this titration? Explain.

23-20. Iron(III) reacts with SCN^- to form the red complex, $FeSCN^{2+}$. Sketch a photometric titration curve for Fe(III) with SCN^- ion when a photometer with a green filter is used to collect data. Why is a green filter used?

*23-21. Ethylenediaminetetraacetic acid abstracts bismuth(III) from its thiourea complex:

$$Bi(tu)_6^{3+} + H_2Y^{2-} \rightarrow BiY^- + 6tu + 2H^+$$

where tu is the thiourea molecule, $(NH_2)_2CS$. Predict the shape of a photometric titration curve based on this process, given that the Bi(III)/thiourea complex is the only species in the system that absorbs at 465 nm, the wavelength selected to monitor the titration.

23-22. The accompanying data (1.00-cm cells) were obtained for the spectrophotometric titration of 10.00 mL of Pd(II) with 2.44 10^{-4} M Nitroso R (O. W. Rollins and M. M. Oldham, *Anal. Chem.,* **1971,** *43,* 262). Calculate the concentration of the Pd(II) solution, given that the ligand-to-cation ratio in the colored product is 2:1.

Volume of Nitroso R, mL	A_{500}
0	0
1.00	0.147
2.00	0.271
3.00	0.375
4.00	0.371
5.00	0.347
6.00	0.325
7.00	0.306
8.00	0.289

*23-23. A 4.97-g petroleum specimen was decomposed by wet-ashing and subsequently diluted to 500 mL in a volumetric flask. Cobalt was determined by treating 25.00-mL aliquots of this diluted solution as follows:

Reagent Volume			
Co(II), 3.00 ppm	Ligand	H_2O	Absorbance
0.00	20.00	5.00	0.398
5.00	20.00	0.00	0.510

Assume that the Co(II)/ligand chelate obeys Beer's law, and calculate the percentage of cobalt in the original sample.

23-24. Predict the shape of photometric titration curves (after correction for volume change) if, at the wavelength selected, the molar absorptivities for the analyte A, the titrant T, and the product P are as follows:

	ε_A	ε_T	ε_P
*(a)	0	>0	0
(b)	>0	0	0
*(c)	0	0	>0
(d)	>0	>A	0
*(e)	0	>0	<T
(f)	>0	0	<A
(g)	>0	<A	0

23-25. Iron(III) forms a complex with thiocyanate ion that has the formula $Fe(SCN)^{2+}$. The complex has an absorption maximum at 580 nm. A specimen of well water was assayed according to the scheme shown in the table at the bottom of the page.

Calculate the concentration of iron in parts per million (1.00-cm cells).

23-26. A. J. Mukhedkar and N. V. Deshpande (*Anal. Chem.,* **1963,** *35,* 47) report on a simultaneous determination of cobalt and nickel based on absorption by their 8-quinolinol complexes. Molar absorptivities are $\varepsilon_{Co} = 3529$ and $\varepsilon_{Ni} = 3228$ at 365 nm and $\varepsilon_{Co} = 428.9$ and $\varepsilon_{Ni} = 0$ at 700 nm. Calculate the concentration of nickel and cobalt

		Volumes, mL				
Sample	Sample Volume	Oxidizing Reagent	Fe(II) 2.75 ppm	KSCN 0.050 M	H_2O	Absorbance, 580 nm
1	50.00	5.00	5.00	20.00	20.00	0.549
2	50.00	5.00	0.00	20.00	25.00	0.231

in each of the following solutions (1.00-cm cells):

Solution	A_{700}	A_{365}
*1	0.0235	0.617
2	0.0714	0.755
*3	0.0945	0.920
4	0.0147	0.592
5	0.0540	0.685

23-27. Molar absorptivity data for the cobalt and nickel complexes with 2,3-quinoxalinedithiol are $\varepsilon_{Co} = 36,400$ and $\varepsilon_{Ni} = 5520$ at 510 nm and $\varepsilon_{Co} = 1240$ and $\varepsilon_{Ni} = 17,500$ at 656 nm. A 0.425-g sample was dissolved and diluted to 50.0 mL. A 25.0-mL aliquot was treated to eliminate interferences; after addition of 2,3-quinoxalinedithiol, the volume was adjusted to 50.0 mL. This solution had an absorbance of 0.446 at 510 nm and 0.326 at 656 nm in a 1.00-cm cell. Calculate the parts per million of cobalt and nickel in the sample.

23-28. The indicator HIn has an acid dissociation constant of 4.80×10^{-6} at ordinary temperatures. The accompanying absorbance data are for 8.00×10^{-5} M solutions of the indicator measured in 1.00-cm cells in strongly acidic and strongly alkaline media.

	Absorbance	
λ, nm	pH 1.00	pH 13.00
420	0.535	0.050
445	0.657	0.068
450	0.658	0.076
455	0.656	0.085
470	0.614	0.116
510	0.353	0.223
550	0.119	0.324
570	0.068	0.352
585	0.044	0.360
595	0.032	0.361
610	0.019	0.355
650	0.014	0.284

Estimate the wavelength at which absorption by the indicator becomes independent of pH (called the isosbestic point).

23-29. Calculate the absorbance (1.00-cm cells) at 450 nm of a solution in which the total molar concentration of the indicator described in Problem 23-28 is 8.00×10^{-5} and the pH is
*(a) 4.92   (b) 5.46
*(c) 5.93   (d) 6.16

23-30. What is the absorbance at 595 nm (1.00-cm cells) of a solution that is 1.25×10^{-4} M in the indicator of Problem 23-28 and has a pH of
*(a) 5.30?
 (b) 5.70?
*(c) 6.10?

23-31. A standard solution was put through appropriate dilutions to give the concentrations of iron shown in the table that follows. The iron (II)-1,10-phenanthroline complex was then formed in 25.0-mL aliquots of these solutions, following which each was diluted to 50.0 mL. The following absorbances (1.00-cm cells) were recorded at 510 nm:

Fe(II) Concentration in Original Solutions, ppm	A_{510}
4.00	0.160
10.0	0.390
16.0	0.630
24.0	0.950
32.0	1.260
40.0	1.580

 (a) Construct a spreadsheet and input these data. Use the spreadsheet to plot a calibration curve from these data.
*(b) Use the method of least squares to derive an equation relating absorbance and the concentration of iron(II).
*(c) Calculate the standard deviation of the slope and intercept.

23-32. The method developed in Problem 23-31 was used for the routine determination of iron in 25.0-mL aliquots of ground water. Express the concentration (as ppm Fe) in samples that yielded the accompanying absorbance data (1.00-cm cell). Calculate the relative standard deviations of the resulting concentrations. Repeat the calculation assuming the absorbance data are means of three measurements.
*(a) 0.143   (b) 0.675   *(c) 0.068
 (d) 1.009   *(e) 1.512   (f) 0.546

23-33. Quinine in a 1.664-g antimalarial tablet was dissolved in sufficient 0.10 M HCl to give 500 mL of solution. A 15.00-mL aliquot was then diluted to 100.0 mL with the acid. The fluorescent intensity for the diluted sample at 347.5 nm provided a reading of 288 on an arbitrary scale. A standard 100-ppm quinine solution registered 180 when measured under conditions identical to those for the diluted sample. Calculate the milligrams of quinine in the tablet.

23-34. The determination in Problem 23-33 was modified to make use of a standard addition. As before, a 2.196-g tablet was dissolved in sufficient 0.10 M HCl to give 1.000 L. Dilution of a 20.00-mL aliquot to 100 mL gave a solution that gave a reading of 540 at 347.5 nm. A second 20.00-mL aliquot was mixed with 10.0 mL of 50-ppm quinine solution before dilution to 100 mL. The fluorescent intensity of this solution was 600. Calculate the parts per million of quinine in the tablet.

23-35. The reduced form of nicotinamide adenine dinucleotide (NADH) is an important and highly fluorescent coenzyme. It has an absorption maximum of 340 nm and an emission maximum at 465 nm. Standard solutions of NADH gave the following fluorescence intensities:

Concn NADH, μmol/L	Relative Intensity
0.100	2.24
0.200	4.52
0.300	6.63
0.400	9.01
0.500	10.94
0.600	13.71
0.700	15.49
0.800	17.91

(a) Construct a spreadsheet and use it to draw a calibration curve for NADH.

***(b)** Find the least-squares slope and intercept for the plot in part (a).

(c) Calculate the standard deviation of the slope and about regression for the curve.

***(d)** An unknown exhibits a relative fluorescence of 12.16. Use the spreadsheet to calculate the concentration of NADH.

***(e)** Calculate the relative standard deviation for the result in part (d).

(f) Calculate the relative standard deviation for the result in part (d) if the reading of 12.16 was the mean of three measurements.

Section V

Separation Methods

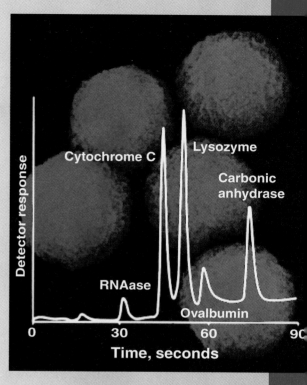

For many analytical methods, separating the analyte from potential interferents is a vital step in the procedure. The three chapters in this section deal with the most common methods for achieving analytical separations. Chapter 24 describes the classic methods for handling chemical interference problems. These methods are masking, precipitation, extraction, distillation, and ion exchange. In this chapter, we also include a discussion of the general principles of chromatography that apply to all types of this powerful separation technique.

Chapter 25 deals with the two most widely used chromatographic methods, namely gas chromatography and high performance liquid chromatography. Chapter 26 completes this section with a brief treatment of three relatively new analytical separation methods: supercritical-fluid chromatography, capillary electrochromatography, and capillary electrophoresis.

The graph above illustrates rapid separation of important biomolecules by reversed-phase high performance liquid chromatography using Poroshell silica particles as the stationary phase. The particles, shown in the figure above, consist of solid 5 μm-diameter silica spheres, each enclosed in a 1 μm thick shell of porous silica. The favorable mass-transfer characteristics of these particles permit efficient separation of mixtures of macromolecules. (Data from J. J. Kirkland, *Anal. Chem.*, 1992, *64*(11), 1239. Reproduced by permission of the American Chemical Society.)

Chapter 24

An Introduction to Analytical Separations

F ew, if any, measurement techniques used for chemical analysis are specific for a single chemical species; as a consequence, an important part of most analyses is dealing with foreign species that either attenuate the signal from the analyte or produce a signal that is indistinguishable from that of the analyte. A substance that effects an analytical signal is called an *interference* or an *interferent*.

As shown in Table 24-1, several general methods are used for dealing with interferences in an analysis, including (1) masking, (2) chemical or electrolytic precipitation, (3) distillation, (4) solvent extraction, (5) ion exchange, (6) chromatography, and (7) electrophoresis. Sections 24-A through 24-E of this chapter deal with the first five of these methods. In addition, an introduction to chromatographic separations is found in Section 24-F. Chapters 25 and 26 then deal with the applications of chromatography and electrophoresis.

> An *interferent* is a chemical species that causes a systematic error in an analysis by enhancing or attenuating the analytical signal.

24A MASKING

In masking, a reagent is added to the solution of the sample to immobilize, or chemically bind, the interferent as a complex that no longer contributes to or attenuates the signal from the analyte.[1] Clearly, a masking agent must not affect the behavior of the analyte significantly. An example of masking is the use of fluoride ion to prevent iron(III) from interfering in the iodometric determination of copper(II). In this analysis, the analyte solution is treated with an excess of iodide ion, which reacts with copper(II) ions to produce a stoichiometric amount of iodine. The liberated iodine is then titrated with a standard solution of sodium thiosulfate. Iron(III) interferes in this determination because it also oxidizes iodide ion to some extent. This interference is avoided by introducing fluoride ion

> A *masking agent* is a reagent that chemically binds an interferent as a complex and prevents it from causing errors in an analysis.

---

[1]For a monograph on masking agents, see D. D. Perrin, *Masking and Demasking Reactions* (New York: Wiley-Interscience, 1970).

Table 24-1

Methods for Eliminating Interferences
in a Chemical Analysis

Method	Basis of Method
1. Masking	Immobilization of interferent as a nonreactive complex
2. Mechanical phase separation	
a. Precipitation and filtration	Difference in solubility of compounds formed
b. Distillation	Difference in volatility of compounds
c. Extraction	Difference in solubility in two immiscible liquids
d. Ion exchange	Difference in stability of reactants with an ion-exchange resin
3. Chromatography	Difference in rate of movement of a solute through a stationary phase
4. Electrophoresis	Difference in migration rate in an electrical field gradient

in excess. Here, masking results from the strong tendency of fluoride ions to complex iron(III) but not copper(II). The consequence is a decrease in the electrode potential of the iron(III) system to the point where only copper(II) ions from the sample oxidize iodide to iodine.

A second example of masking is shown in Feature 15-4.

24B   PRECIPITATION AND FILTRATION

Precipitation, in which the analyte or an interferent is removed from a solution selectively as an insoluble species, is one of the oldest methods for dealing with interferences in an analytical procedure. In earlier chapters, we gave several examples of the use of a precipitating agent for coping with the effects of an interferent. Two examples include the use of dimethylglyoxime to separate nickel(II) from most other heavy metal cations and silver ion to separate chloride from several other anions. In Table 20-2 several cations are shown that can be separated from interferents by electrolytic deposition at a controlled cathode potential.

24C   SEPARATING SPECIES BY DISTILLATION

Distillation is widely used to separate volatile analytes from nonvolatile interferents. A common example is the separation of nitrogen analytes from many other species by converting the nitrogen to ammonia, which is then distilled from basic solution. Other examples include separating carbon as carbon dioxide and sulfur as sulfur dioxide from acidic solution.

24D   SEPARATING SOLUTES BY EXTRACTION

The extent to which solutes, both inorganic and organic, distribute themselves between two immiscible liquids differs enormously, and these differences have been used for decades to accomplish separations of chemical species. This section considers applications of the distribution phenomenon to analytical separations.

24D-1   Principles

The partition of a solute between two immiscible phases is an equilibrium phenomenon that is governed by the *distribution law*. If the solute species A is allowed to distribute itself between water and an organic phase, the resulting equilibrium may be written as

$$A_{aq} \rightleftharpoons A_{org}$$

where the subscripts refer to the aqueous and the organic phases, respectively. Ideally, the ratio of activities for A in the two phases will be constant and independent of the total quantity of A; that is, at any given temperature,

$$K = \frac{(a_A)_{org}}{(a_A)_{aq}} \approx \frac{[A]_{org}}{[A]_{aq}} \tag{24-1}$$

where $(a_A)_{org}$ and $(a_A)_{aq}$ are the activities of A in each of the phases and the bracketed terms are molar concentrations of A. The equilibrium constant K is known as the *distribution constant*. As with many other equilibria, under many conditions, molar concentrations can be substituted for activities without serious error. Generally, the numerical value for K approximates the ratio of the solubility of A in each solvent.

Distribution constants are useful because they permit us to calculate the concentration of an analyte remaining in a solution after a certain number of extractions. They also provide guidance as to the most efficient way to perform an extractive separation. Thus, we can show (see Feature 24-1) that for the simple system described by Equation 24-1, the concentration of A remaining in an aqueous solution after i extractions with an organic solvent ($[A]_i$) is given by the equation

$$[A]_i = \left(\frac{V_{aq}}{V_{org}K + V_{aq}} \right)^i [A]_0 \tag{24-2}$$

where $[A]_i$ is the concentration of A remaining in the aqueous solution after extracting V_{aq} mL of the solution having an original concentration of $[A]_0$ with i portions of the organic solvent, each with a volume of V_{org}. Example 24-1 illustrates how this equation can be used to decide on the most efficient way to perform an extraction.

Example 24-1

It is always better to use several small portions of solvent to extract a sample than to extract with one large portion.

The distribution constant for iodine between an organic solvent and H_2O is 85. Find the concentration of I_2 remaining in the aqueous layer after extraction of 50.0 mL of 1.00×10^{-3} M I_2 with the following quantities of the organic solvent: (a) 50.0 mL, (b) two 25.0-mL portions, (c) five 10.0-mL portions.

Substitution into Equation 24-2 gives the following results.

(a)

$$[I_2]_1 = \left(\frac{50.0}{50.0 \times 85 + 50.0}\right)^1 \times 1.00 \times 10^{-3} = 1.16 \times 10^{-5}$$

(b)

$$[I_2]_2 = \left(\frac{50.0}{25.0 \times 85 + 50.0}\right)^2 \times 1.00 \times 10^{-3} = 5.28 \times 10^{-7}$$

(c)

$$[I_2]_5 = \left(\frac{50.0}{10.0 \times 85 + 50.0}\right)^5 \times 1.00 \times 10^{-3} = 5.29 \times 10^{-10}$$

Note the increased extraction efficiencies that result from dividing the original 50 mL of solvent into two 25-mL or five 10-mL portions.

Figure 24-1 shows that the improved efficiency of multiple extractions falls off rapidly as a total fixed volume is subdivided into smaller and smaller portions. Clearly, little is to be gained by dividing the extracting solvent into more than five or six portions.

Feature 24-1

Derivation of Equation 24-2

Consider the simple system described by Equation 24-1. Suppose n_0 mmol of the solute A in V_{aq} mL of aqueous solution is extracted with V_{org} mL of an immiscible organic solvent. At equilibrium, n_1 mmol of A will remain in the aqueous layer, and $(n_0 - n_1)$ mmol will have been transferred to the organic layer. The concentrations of A in the two layers will then be

$$[A]_1 = \frac{n_1}{V_{aq}}$$

and

$$[A]_{org} = \frac{(n_0 - n_1)}{V_{org}}$$

Substitution of these quantities into Equation 24-1 and rearrangement gives

$$n_1 = \left(\frac{V_{aq}}{V_{org}K + V_{aq}}\right) n_0$$

Similarly, the number of millimoles, n_2, remaining after a second extraction with the same volume of solvent will be

$$n_2 = \left(\frac{V_{aq}}{V_{org}K + V_{aq}}\right) n_1$$

Substitution of the previous equation into this expression gives

$$n_2 = \left(\frac{V_{aq}}{V_{org}K + V_{aq}}\right)^2 n_0$$

By the same argument, the number of millimoles, n_i, that remain after i extractions will be given by the expression

$$n_i = \left(\frac{V_{aq}}{V_{org}K + V_{aq}} \right)^i n_0$$

Finally, this equation can be written in terms of the initial and final concentrations of A in the aqueous layer by substituting the relationships

$$n_i = [A]_i V_{aq} \quad \text{and} \quad n_0 = [A]_0 V_{aq}$$

Thus,

$$[A]_i = \left(\frac{V_{aq}}{V_{org}K + V_{aq}} \right)^i [A]_0$$

which is Equation 24-2.

24D-2   Applying Extraction to Inorganic Separations

An extraction is frequently more attractive than a precipitation method for separating inorganic species. The processes of equilibration and separation of phases in a separatory funnel are less tedious and time consuming than conventional precipitation, filtration, and washing.

Separating Metal Ions as Chelates

Many organic chelating agents are weak acids that react with metal ions to give uncharged complexes that are very soluble in organic solvents, such as ethers, hydrocarbons, ketones, and chlorinated species (including chloroform and carbon tetrachloride.)[2] Most metal chelates, on the other hand, are nearly insoluble in water. Similarly, the chelating agents themselves are often quite soluble in organic solvents but have limited solubility in water.

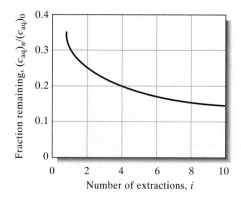

Figure 24-1   Plot of Equation 24-2 assuming that $K = 2$ and $V_{aq} = 100$ mL. The total volume of the organic solvent was assumed to be 100 mL also, so that $V_{org} = 100/n_i$.

---

[2]The use of chlorinated solvents is decreasing because of concerns about their health effects and their possible role in ozone layer depletion.

Figure 24-2 shows the equilibria that develop when an aqueous solution of a divalent cation, such as zinc(II), is extracted with an organic solution containing a large excess of 8-hydroxyquinoline (see Section 8D-3 for the structure and reactions of this chelating agent). Four equilibria are shown. The first involves distribution of the 8-hydroxyquinoline, HQ, between the organic and aqueous layers. The second is the acid dissociation of the HQ to give H^+ and Q^- ions in the aqueous layer. The third equilibrium is the complex-formation reaction giving MQ_2. Fourth is the distribution of the chelate between the two solvents (but for the fourth equilibrium, MQ_2 would precipitate out of the aqueous solution). The overall equilibrium is the sum of these four reactions, or

$$2HQ(org) + M^{2+}(aq) \rightleftharpoons MQ_2(org) + 2H^+(aq)$$

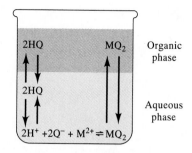

Figure 24-2 Equilibria in the extraction of an aqueous cation M^{2+} into an immiscible organic solvent containing 8-hydroxy-quinoline.

The equilibrium constant for this reaction is

$$K' = \frac{[MQ_2]_{org}[H^+]_{aq}^2}{[HQ]_{org}^2[M^{2+}]_{aq}}$$

Ordinarily, HQ is present in the organic layer in large excess with respect to M^{2+} in the aqueous phase so that $[HQ]_{org}$ remains essentially constant during the extraction. Therefore, the equilibrium-constant expression can be simplified to

$$K'[HQ]_{org}^2 = K = \frac{[MQ_2]_{org}[H^+]_{aq}^2}{[M^{2+}]_{aq}}$$

or

$$\frac{[MQ_2]_{org}}{[M^{2+}]_{aq}} = \frac{K}{[H^+]_{aq}^2}$$

Thus, we see that the ratio of concentration of the metal species in the two layers is inversely proportional to the square of the hydrogen ion concentration of the aqueous layer. Equilibrium constants K vary widely from metal ion to metal ion, and these differences often make it possible to selectively extract one cation from another by buffering the aqueous solution at a level where one is extracted nearly completely and the second remains largely in the aqueous phase.

Several useful extractive separations with 8-hydroxyquinoline have been developed. Furthermore, numerous chelating agents that behave in a similar way are described in the literature.[3] As a consequence, pH-controlled extractions provide a powerful method for separating metallic ions.

Extracting Metal Chlorides and Nitrates

A number of inorganic species can be separated by extraction with suitable solvents. For example, a single ether extraction of a 6 M hydrochloric acid solution will cause better than 50% of several ions to be transferred to the organic phase; included among these are iron(III), antimony(V), titanium(III), gold(III), molybdenum(VI), and tin(IV). Other ions, such as aluminum(III) and the divalent cations of cobalt, lead, manganese, and nickel, are not extracted.

Uranium(VI) can be separated from such elements as lead and thorium by ether extraction of a solution that is 1.5 M in nitric acid and saturated with ammonium nitrate. Bismuth and iron(III) are also extracted to some extent from this medium.

[3]For example, see J. A. Dean, in *Analytical Chemistry Handbook* (New York: McGraw-Hill, 1995), p. 2.24.

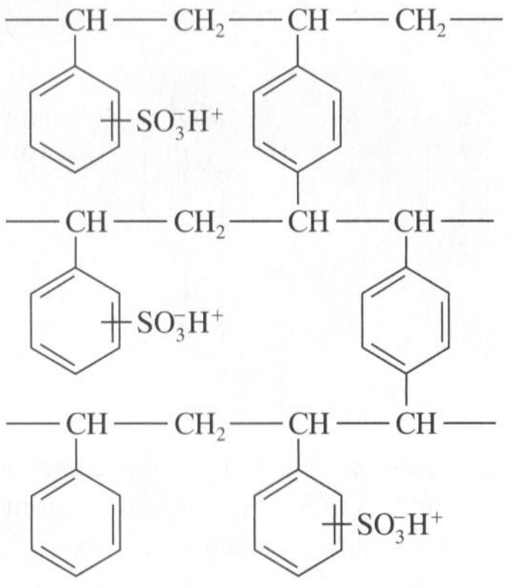

Figure 24-3 Structure of a cross-linked polystyrene ion-exchange resin. Similar resins are used in which the $-SO_3^-H^+$ group is replaced by $-COO^-H^+$, $-NH_3^+OH^-$, and $-N(CH_3)_3^+OH^-$ groups.

24E   SEPARATING IONS BY ION EXCHANGE

Ion exchange is a process by which ions held on a porous, essentially insoluble solid are exchanged for ions in a solution that is brought in contact with the solid. The ion-exchange properties of clays and zeolites have been recognized and studied since the late nineteenth century. Synthetic ion-exchange resins were first produced in 1935 and have since found widespread application in water softening, water deionization, solution purification, and ion separation.

24E-1   Ion-Exchange Resins

Synthetic ion-exchange resins are high-molecular-weight polymers that contain large numbers of an ionic functional group per molecule. Cation-exchange resins contain acidic groups, whereas anion-exchange resins have basic groups. Exchangers of the strong-acid type have sulfonic acid groups ($-SO_3^-H^+$) attached to the polymeric matrix (see Figure 24-3) and have wider application than exchangers of the weak-acid type, which owe their action to carboxylic acid ($-COOH$) groups. Similarly, strong-base anion exchangers contain quaternary amine [$-N(CH_3)_3^+OH^-$] groups, whereas weak-base types contain secondary or tertiary amines.

Cation exchange is illustrated by the equilibrium

$$x\text{RSO}_3^-\text{H}^+ + \text{M}^{x+} \rightleftharpoons (\text{RSO}_3^-)_x\text{M}^{x+} + x\text{H}^+$$
$$\quad\text{solid}\qquad\quad\text{soln}\qquad\quad\text{solid}\qquad\quad\text{soln}$$

where M^{x+} represents a cation and R represents *that part of a resin molecule that contains one sulfonic acid group*. The analogous equilibrium involving a strong-base anion exchanger and an anion A^{x-} is

$$x\text{RN}(CH_3)_3^+\text{OH}^- + \text{A}^{x-} \rightleftharpoons [\text{RN}(CH_3)_3^+]_x\text{A}^{x-} + x\text{OH}^-$$
$$\quad\text{solid}\qquad\qquad\text{soln}\qquad\qquad\text{solid}\qquad\qquad\text{soln}$$

24E-2   Ion-Exchange Equilibria

Ion-exchange equilibria can be treated by the law of mass action. For example, when a dilute solution containing calcium ions is passed through a column packed with a sulfonic acid resin, the following equilibrium is established:

$$\text{Ca}^{2+}(aq) + 2\text{H}^+(res) \rightleftharpoons \text{Ca}^{2+}(res) + 2\text{H}^+(aq)$$

for which an equilibrium constant K' is given by

$$K' = \frac{[\text{Ca}^{2+}]_{res}[\text{H}^+]_{aq}^2}{[\text{Ca}^{2+}]_{aq}[\text{H}^+]_{res}^2} \tag{24-3}$$

As usual, the bracketed terms are molar concentrations (strictly, activities) of the species in the two phases. Note that $[\text{Ca}^{2+}]_{res}$ and $[\text{H}^+]_{res}$ are molar concentrations of the two ions *in the solid phase*. In contrast to most solids, these concentrations can vary from zero to some maximum value when all the negative sites on the resin are occupied by one species only.

Ion-exchange separations are ordinarily performed under conditions in which one ion predominates in *both* phases. Thus, in the removal of calcium ions from a dilute and somewhat acidic solution, the calcium ion concentration will be much smaller than that of hydrogen ion in both the aqueous and resin phases; that is,

$$[Ca^{2+}]_{res} \ll [H^+]_{res}$$

and

$$[Ca^{2+}]_{aq} \ll [H^+]_{aq}$$

As a consequence, the hydrogen ion concentration is essentially constant in both phases, and Equation 24-3 can be rearranged to

$$\frac{[Ca^{2+}]_{res}}{[Ca^{2+}]_{aq}} = K'\frac{[H^+]_{res}^2}{[H^+]_{aq}^2} = K \qquad (24\text{-}4)$$

where K is a distribution constant analogous to the constant that governs an extraction equilibrium (Equation 24-1). Note that K in Equation 24-4 represents the affinity of the resin for calcium ion relative to another ion (here, H^+). In general, where K for an ion is large, a strong tendency for the stationary phase to retain that ion exists; where K is small, the opposite is true. Selection of a common reference ion (such as H^+) permits a comparison of distribution ratios for various ions on a given type of resin. Such experiments reveal that polyvalent ions are much more strongly retained than singly charged species. Within a given charge group, differences that exist among values for K appear to be related to the size of the hydrated ion as well as other properties. Thus, for a typical sulfonated cation-exchange resin, values of K for univalent ions decrease in the order $Ag^+ > Cs^+ > Rb^+ > K^+ > NH_4^+ > Na^+ > H^+ > Li^+$. For divalent cations, the order is $Ba^{2+} > Pb^{2+} > Sr^{2+} > Ca^{2+} > Ni^{2+} > Cd^{2+} > Cu^{2+} > Co^{2+} > Zn^{2+} > Mg^{2+} > UO_2^{2+}$.

24E-3   Applications of Ion-Exchange Methods

Ion-exchange resins are used to eliminate ions that would otherwise interfere with an analysis. For example, iron(III) and aluminum(III), as well as many other cations, tend to coprecipitate with barium sulfate during the determination of sulfate ion. Passage of a solution containing sulfate through a cation-exchange resin results in the retention of these cations and the release of an equivalent number of hydrogen ions. Sulfate ions pass freely through the column and can be precipitated as barium sulfate from the effluent.

Another valuable application of ion-exchange resins involves concentrating ions from a very dilute solution. Thus, traces of metallic elements in large volumes of natural waters can be collected on a cation-exchange column and subsequently liberated from the resin by treatment with a small volume of an acidic solution; the result is a considerably more concentrated solution for analysis.

The total salt content of a sample can be determined by titrating the hydrogen ion released as an aliquot of the sample passes through a cation exchanger in its acidic form. Similarly, a standard hydrochloric acid solution can be prepared by diluting to known volume the effluent resulting from treatment of a cation-exchange resin with a known weight of sodium chloride. Substitution of an anion-exchange resin in its hydroxide form will permit the preparation of a standard base solution.

As shown in Section 25B-4, ion-exchange resins are particularly useful for the chromatographic separation of both inorganic and organic ionic species.

Chromatography was invented by the Russian botanist Mikhail Tswett at the beginning of the twentieth century. He employed the technique to separate various plant pigments, such as chlorophylls and xanthophylls, by passing solutions of these species through glass columns packed with finely divided calcium carbonate. The separated species appeared as colored bands on the column, which accounts for the name he chose for the method (Greek *chroma* meaning color and *graphein* meaning to write).

The *stationary phase* in chromatography is a phase that is fixed in place either in a column or on a planar surface.

The *mobile phase* in chromatography is a phase that moves over or through the stationary phase, carrying with it the analyte mixture. The mobile phase may be a gas, a liquid, or a supercritical fluid.

Planar and column chromatography are based on the same types of equilibria.

Gas and supercritical mobile phases require the use of a column. Only liquid mobile phases can be used on planar surfaces.

24F   CHROMATOGRAPHIC SEPARATIONS

Chromatography is a widely used method for the separation, identification, and determination of the chemical components in complex mixtures. No other separation method is as powerful and generally applicable as chromatography.[4] The remainder of this chapter is devoted to the general principles of all types of chromatography. Chapters 25 and 26 deal with some applications of chromatography for analytical separations. Chapter 26 also treats separations by capillary electrophoresis.

24F-1   General Description of Chromatography

The term "chromatography" is difficult to define rigorously because the word has been applied to several systems and techniques. Common to all these methods, however, is the use of a *stationary phase* and a *mobile phase*. Components of a mixture are carried through the stationary phase by the flow of the mobile phase, and separations are based on differences in migration rates among the mobile-phase components.

24F-2   Classifying Chromatographic Methods

Chromatographic methods are of two basic types. In *column chromatography,* the stationary phase is held in a narrow tube, and the mobile phase is forced through the tube under pressure or by gravity. In *planar chromatography,* the stationary phase is supported on a flat plate or in the pores of a paper. Here the mobile phase moves through the stationary phase by capillary action or under the influence of gravity. We will deal with column chromatography only.

As shown in the first column of Table 24-2, column chromatographic methods can be further subdivided according to the nature of the mobile phase, specifically liquid, gas, and supercritical fluid. The second column of the table reveals that there are five types of liquid chromatography and two types of gas chromatography that differ in the nature of the stationary phase and the types of equilibria between phases.

24F-3   Elution in Column Chromatography

Figure 24-4 shows how two components A and B of a sample are resolved on a packed column by *elution.* The column consists of a narrow-bore tube packed with a finely divided inert solid that holds the stationary phase on its surface. The mobile phase occupies the open spaces between the particles of the packing. Initially, a solution of the sample containing a mixture of A and B in the mobile phase is introduced at the head of the column as a narrow plug, whereupon the

---

[4]General references on chromatography include E. Heftmann, Ed., *Chromatography: Fundamentals and Applications of Chromatography and Electrophoretic Methods, Part A Fundamentals, Part B Applications* (New York: Elsevier, 1983); P. Sewell and B. Clarke, *Chromatographic Separations* (New York: Wiley, 1988); J. A. Jonsson, Ed., *Chromatographic Theory and Basic Principles* (New York: Marcel Dekker, 1987); R. P. W. Scott, *Liquid Chromatography for the Analyst* (New York: Marcel Dekker, 1995).

Table 24-2

Classification of Column Chromatographic Methods

General Classification	Specific Method	Stationary Phase	Type of Equilibrium
1. Gas chromatography (GC)	a. Gas–liquid (GLC)	Liquid adsorbed or bonded to a solid surface	Partition between gas and liquid
	b. Gas–solid	Solid	Adsorption
2. Liquid chromatography (LC)	a. Liquid–liquid, or partition	Liquid adsorbed or bonded to a solid surface	Partition between immiscible liquids
	b. Liquid–solid, or adsorption	Solid	Adsorption
	c. Ion exchange	Ion-exchange resin	Ion exchange
	d. Size exclusion	Liquid in interstices of a polymeric solid	Partition/sieving
	e. Affinity	Group specific liquid bonded to a solid surface	Partition between surface liquid and mobile liquid
3. Supercritical-fluid chromatography (SFC) (mobile phase: supercritical fluid)		Organic species bonded to a solid surface	Partition between supercritical fluid and bonded surface

two components distribute themselves between the two phases as shown in Figure 24-4 at time t_0. Elution then involves forcing the sample components through the column by repeated additions of fresh mobile phase.

With the first introduction of fresh mobile phase (called the *eluent*), the portion of the sample contained in the mobile phase moves down the column, where further partitioning between the mobile phase and the stationary phase occurs (time t_1). Partitioning between the fresh mobile phase and the stationary phase takes place simultaneously at the original site of the sample.

Further additions of solvent carry solute molecules down the column in a continuous series of transfers between the two phases. Because solute movement can occur only in the mobile phase, the average *rate* at which a solute migrates *depends on the fraction of time it spends in that phase*. This fraction is small for solutes that are strongly retained by the stationary phase (component B in Figure 24-4, for example) and large where retention in the mobile phase is more likely (component A). Ideally, the resulting differences in rates cause the components in a mixture to separate into *bands,* or *zones,* along the length of the column (see Figure 24-5). Isolation of the separated species is then accomplished by passing a sufficient quantity of mobile phase through the column to cause the individual bands to pass out the end (to be *eluted* from the column), where they can be collected or detected (times t_3 and t_4 in Figure 24-4).

Elution is a process in which solutes are washed through a stationary phase by the movement of a mobile phase. The mobile phase that exits the column is called the *eluate.*

An *eluent* is a solvent used to carry the components of a mixture through a stationary phase.

24F-4 Chromatograms

If a detector that responds to solute concentration is placed at the end of the column during elution and its signal is plotted as a function of time (or of volume of added mobile phase), a series of peaks is obtained, as shown in the lower part

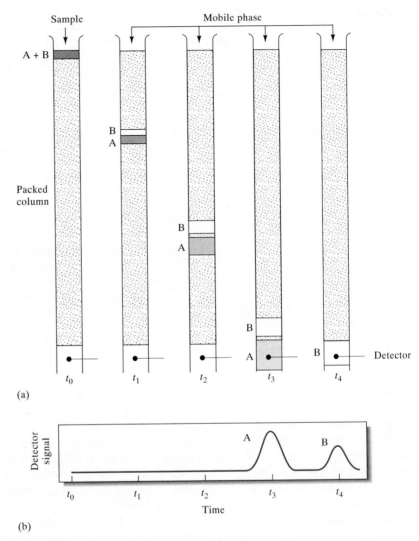

(a)

(b)

Figure 24-4 (a) Diagram showing the separation of a mixture of components A and B by column elution chromatography. (b) The output of the signal detector at the various stages of elution shown in (a).

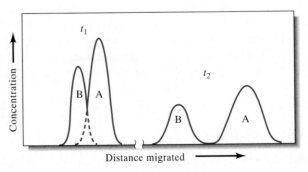

Figure 24-5 Concentration profiles of solute bands A and B at two different times in their migration down the column in Figure 24-4. The times t_1 and t_2 are indicated in Figure 24-4.

of Figure 24-4. Such a plot, called a *chromatogram,* is useful for both qualitative and quantitative analysis. The positions of the peaks on the time axis can be used to identify the components of the sample; the areas under the peaks provide a quantitative measure of the amount of each species.

> A *chromatogram* is a plot of some function of solute concentration versus elution time or elution volume.

24F-5   Improving Column Performance

Figure 24-5 shows concentration profiles for the bands containing solutes A and B on the column in Figure 24-4 at time t_1 and at a later time t_2.[5] Because B is more strongly retained by the stationary phase than A, B lags during the migration. Clearly, the distance between the two increases as they move down the column. At the same time, however, broadening of both bands takes place, which lowers the efficiency of the column as a separating device. Although band broadening is inevitable, conditions can often be found where it occurs more slowly than band separation. Thus, as shown in Figure 24-5, a clean separation of species is possible provided the column is sufficiently long.

Several chemical and physical variables influence the rates of band separation and band broadening. As a consequence, improved separations can often be realized by the control of variables that either (1) increase the rate of band separation or (2) decrease the rate of band spreading. These alternatives are illustrated in Figure 24-6.

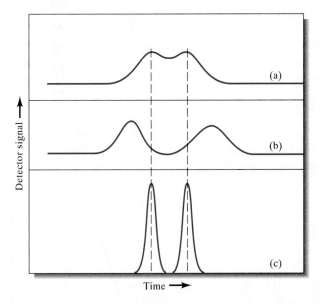

Figure 24-6   Two-component chromatograms illustrating two methods of improving separation: (a) original chromatogram with overlapping peaks, improvement brought about by (b) an increase in band separation, and (c) a decrease in bandwidth.

---

[5]Note that the relative positions of the bands for A and B in the concentration profile in Figure 24-5 appear to be reversed from their positions in the lower part of Figure 24-4. The difference is that the abscissa is distance along the column in Figure 24-5, but the abscissa is time in Figure 24-4. Thus, in Figure 24-4, the *front* of a peak lies to the left and the *tail* to the right; in Figure 24-5, the reverse is true.

The variables that influence the relative rates at which solutes migrate through a stationary phase are described in the next section. Following this discussion, we turn to those factors that play a part in zone broadening.

24F-6　Relative Migration Rates of Solutes

The effectiveness of a chromatographic column in separating two solutes depends in part on the relative rates at which the two species are eluted. These rates are in turn determined by the ratios of the solute concentrations in each of the two phases.

Distribution Constants

All chromatographic separations are based on differences in the extent to which solutes are distributed between the mobile and the stationary phase. For the solute species A, the equilibrium involved is described by the equation

$$A(\text{mobile}) \rightleftharpoons A(\text{stationary}) \tag{24-5}$$

The equilibrium constant K_c for this reaction is called a *distribution constant*, which is defined as

> The *distribution constant* for a solute in chromatography is equal to the ratio of its molar analytical concentration in the stationary phase to its molar analytical concentration in the mobile phase.

$$K_c = \frac{[A]_S}{[A]_M} = \frac{c_S}{c_M} \tag{24-6}$$

where the bracketed terms are activities of a solute A in the two phases. We shall often substitute c_S, the molar analytical concentrations of the solute in the stationary phase, and c_M, its molar analytical concentration in the mobile phase. Ideally, the distribution constant is constant over a wide range of solute concentrations; that is, c_S is directly proportional to c_M.

Retention Times

Figure 24-7 is a simple chromatogram made up of just two peaks. The small peak on the left is for a species that is *not* retained by the stationary phase. The time t_M after sample injection for this peak to appear is sometimes called the

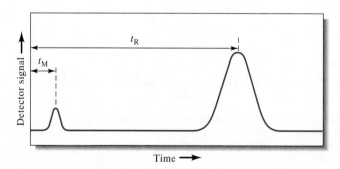

Figure 24-7 A typical chromatogram for a two-component mixture. The small peak on the left represents a solute that is not retained on the column and so reaches the detector almost immediately after elution is started. Thus, its retention time t_M is approximately equal to the time required for a molecule of the mobile phase to pass through the column.

dead time. The dead time provides a measure of the average rate of migration of the mobile phase and is an important parameter in identifying analyte peaks. Often, the sample or the mobile phase will contain an unretained species. When it does not, such a species may be added to aid in peak identification. The larger peak on the right in Figure 24-7 is that of an analyte species. The time required for this peak to reach the detector after sample injection is called the *retention time* and is given the symbol t_R.

The average linear rate of solute migration, $\bar{v}$, in centimeters per second is

$$\bar{v} = \frac{L}{t_R} \tag{24-7}$$

where L is the length of the column packing. Similarly, the average linear velocity, u, of the molecules of the mobile phase is

$$u = \frac{L}{t_M} \tag{24-8}$$

Relating Migration Rates to Distribution Constants

To relate the rate of migration of a solute to its distribution constant, we express the rate as a fraction of the velocity of the mobile phase:

$$\bar{v} = u \times \text{fraction of time solute spends in mobile phase}$$

This fraction, however, equals the average number of moles of solute in the mobile phase at any instant divided by the total number of moles of solute in the column:

$$\bar{v} = u \times \frac{\text{moles of solute in mobile phase}}{\text{total moles of solute}}$$

The total number of moles of solute in the mobile phase is equal to the molar concentration, c_M, of the solute in that phase multiplied by its volume, V_M. Similarly, the number of moles of solute in the stationary phase is given by the product of the concentration, c_S, of the solute in the stationary phase and its volume, V_S. Therefore,

$$\bar{v} = u \times \frac{c_M V_M}{c_M V_M + c_S V_S} = u \times \frac{1}{1 + c_S V_S / c_M V_M}$$

Substitution of Equation 24-6 into this equation gives an expression for the rate of solute migration as a function of its distribution constant as well as a function of the volumes of the stationary and mobile phases:

$$\bar{v} = u \times \frac{1}{1 + K_c V_S / V_M} \tag{24-9}$$

The two volumes can be estimated from the method by which the column is prepared.

The Retention Factor, *k*

The retention factor is an important experimental parameter that is widely used to compare the migration rates of solutes on columns.[6] For solute A, the retention factor k_A is defined as

$$k_A = \frac{K_A V_S}{V_M} \tag{24-10}$$

The *retention factor* k_A for solute A is related to the rate at which A migrates through a column.

where K_A is the distribution constant for solute A. Substitution of Equation 24-10 into Equation 24-9 yields

$$\bar{v} = u \times \frac{1}{1 + k_A}$$

To show how k_A can be derived from a chromatogram, we substitute Equations 24-7 and 24-8 into Equation 24-9:

$$\frac{L}{t_R} = \frac{L}{t_M} \times \frac{1}{1 + k_A} \tag{24-11}$$

This equation rearranges to

$$k_A = \frac{t_R - t_M}{t_M} \tag{24-12}$$

As shown in Figure 24-7, t_R and t_M are readily obtained from a chromatogram. When the retention factor for a solute is much less than unity, elution occurs so rapidly that accurate determination of the retention times is difficult. When the retention factor is larger than perhaps 20 to 30, elution times become inordinately long. Ideally, separations are performed under conditions in which the retention factors for the solutes in a mixture lie in the range between 1 and 5.

Ideally, the *retention factors* for analytes in a sample are between 1 and 5.

Retention factors in gas chromatography can be varied by changing the temperature and the column packing. In liquid chromatography, retention factors can often be manipulated to give better separations by varying the composition of the mobile phase and the stationary phase.

The Selectivity Factor

The *selectivity factor* α for solutes A and B is defined as the ratio of the distribution constant of the more strongly retained solute B to the distribution constant for the less strongly held solute A.

The *selectivity factor* α of a column for the two solutes A and B is defined as

$$\alpha = \frac{K_B}{K_A} \tag{24-13}$$

where K_B is the distribution constant for the more strongly retained species B and K_A is the constant for the less strongly held or more rapidly eluted species A. According to this definition, α *is always greater than unity.*

The selectivity factor for two analytes in a column provides a measure of how well the column will separate the two.

Substitution of Equation 24-11 and the analogous equation for solute B into Equation 24-13 provides a relationship between the selectivity factor for two solutes and their capacity factors:

---

[6]In the older literature, this constant was called the capacity factor and was symbolized by k'. In 1993, however, the IUPAC Committee on Analytical Nomenclature recommended that this constant be termed the *retention factor* and be symbolized by k.

$$\alpha = \frac{k_B}{k_A} \qquad (24\text{-}14)$$

where k_B and k_A are the capacity factors for B and A, respectively. Substitution of Equation 24-11 for the two solutes into Equation 24-13 gives an expression that permits the determination of α from an experimental chromatogram:

$$\alpha = \frac{(t_R)_B - t_M}{(t_R)_A - t_M} \qquad (24\text{-}15)$$

In Section 24F-9, we show how we use the retention factor to compute the resolving power of a column.

24F-7 Band Broadening and Column Efficiency

The efficiency of a chromatographic column is affected by the amount of band broadening that occurs when a compound passes through the column. Before defining column efficiency in more quantitative terms, let us examine the reasons that bands become broader as they move down a column.

The Rate Theory of Elution Chromatography

The *rate theory* of chromatography describes the shapes and breadths of elution peaks in quantitative terms based on a random-walk mechanism for the migration of molecules through a column. A detailed discussion of the rate theory is beyond the scope of this text. We can, however, give a qualitative picture of why bands broaden and what variables improve column efficiency.

If you examine the chromatograms shown in this and the next chapter, you will see that the elution peaks look very much like the Gaussian or normal error curves that you encountered in Chapters 5 and 6. As shown in Section 5A-2, normal error curves are rationalized by assuming that the uncertainty associated with any single measurement is the summation of a much larger number of small, individually undetectable and random uncertainties, each of which has an equal probability of being positive or negative. In a similar way, the typical Gaussian shape of a chromatographic band can be attributed to the additive combination of the random motions of the various molecules making up a band as it moves down the column.

It is instructive to consider a single solute molecule as it undergoes many thousands of transfers between the stationary and the mobile phases during elution. Residence time in either phase is highly irregular. Transfer from one phase to the other requires energy, and the molecule must acquire this energy from its surroundings. Thus, the residence time in a given phase may be transitory after some transfers and relatively long after others. Recall that movement down the column can occur *only while the molecule is in the mobile phase*. As a consequence, certain particles travel rapidly by virtue of their accidental inclusion in the mobile phase for a majority of the time, whereas others lag because they happen to be incorporated in the stationary phase for a greater-than-average length of time. The result of these random individual processes is a symmetric spread of velocities around the mean value, which represents the behavior of the average analyte molecule.

As shown in Figure 24-8, some chromatographic peaks are nonideal and exhibit *tailing* or *fronting*. In tailing, the tail of the peak, appearing to the right on the chromatogram, is drawn out, and the front is steepened. With fronting, the reverse is the case. A common cause of tailing and fronting is a distribution constant that varies with concentration. Fronting also occurs when the amount of sample introduced onto a column is too large. Distortions of this kind are undesirable because they lead to poorer separations and less reproducible elution times. In our discussion here, tailing and fronting are assumed to be minimal.

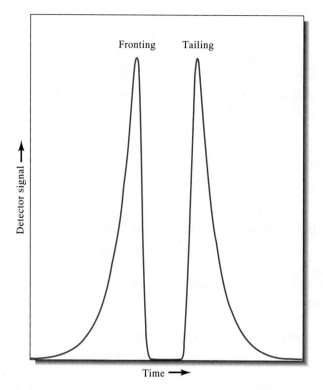

Figure 24-8   Illustration of fronting and tailing in chromatographic peaks.

Quantitative Measures of Column Efficiency

Two related terms are widely used as quantitative measures of chromatographic column efficiency: (1) *plate height H* and (2) *plate count* or *number of theoretical plates N*. The two are related by the equation

$$N = \frac{L}{H} \tag{24-16}$$

where L is the length (usually in centimeters) of the column packing. The efficiency of chromatographic columns increases as the plate count N becomes greater and as the plate height H becomes smaller. Enormous differences in efficiencies are encountered in columns as a result of differences in column type and in mobile and stationary phases. Efficiencies in terms of plate numbers can vary from a few hundred to several hundred thousand; plate heights ranging from a few tenths to one thousandth of a centimeter or smaller are not uncommon.

In Section 6B-2, we pointed out that the breadth of a Gaussian curve is described by the standard deviation σ and the variance σ^2. Because chromatographic bands are usually Gaussian and because the efficiency of a column is reflected in the breadth of chromatographic peaks, the variance per unit length of column is used by chromatographers as a measure of column efficiency. That is, the column efficiency H is defined as

$$H = \frac{\sigma^2}{L} \qquad (24\text{-}17)$$

This definition of column efficiency is illustrated in Figure 24-9a, which shows a column having a packing L cm in length. Above this schematic is a plot (Figure 24-9b) showing the distribution of molecules along the length of the column at the moment the analyte peak reaches the end of the packing (that is, at the retention time t_R). The curve is Gaussian, and the locations of $L + 1\sigma$ and $L - 1\sigma$ are indicated as broken vertical lines. Note that L carries units of centimeters and σ^2 units of centimeters squared; thus, H represents a linear distance in centimeters as well (Equation 24-17). In fact, the plate height can be thought of as the length of column that contains the fraction of the analyte that lies between L and $L - \sigma$. Because the area under a normal error curve bounded by $\pm \sigma$ is about 68% of the total area (page 123), the plate height, as defined, contains 34% of the analyte.

Determining the Number of Plates in a Column

The number of theoretical plates, N, and the plate height, H, are widely used in the literature and by instrument manufacturers as measures of column performance. Figure 24-10 shows how N can be determined from a chromatogram. Here, the retention time of a peak t_R and the width of the peak at its base W (in units of time) are measured. The number of plates can then be computed by the simple relationship[7]

$$N = 16\left(\frac{t_R}{W}\right)^2 \qquad (24\text{-}18)$$

To obtain H, the length of the column L is measured and Equation 24-16 is applied.

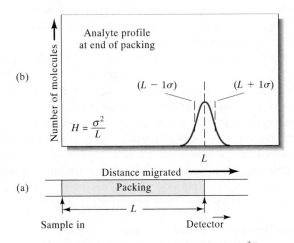

Figure 24-9 Definition of plate height $H = \sigma^2/L$.

[7]For a derivation of this relationship, see D. A. Skoog, F. J. Holler, and T. A. Nieman, *Principles of Instrumental Analysis,* 5th ed. (Philadelphia: Saunders College Publishing, 1998), pp. 682–683.

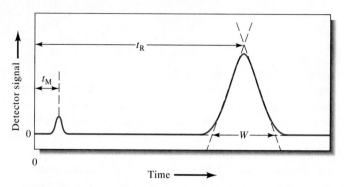

Figure 24-10 Determining the number of plates $N = 16 \left(\dfrac{t_R}{W} \right)^2$.

24F-8   Variables Affecting Column Efficiency

Band broadening, and thus loss of column efficiency, is the consequence of the finite rate at which several mass-transfer processes occur during migration of a solute down a column. Some of the variables that affect these rates are controllable and can be exploited to improve separations.

The Effect of Mobile-Phase Flow Rate

The extent of band broadening depends on the length of time the mobile phase is in contact with the stationary phase, which in turn depends on the flow rate of the mobile phase. For this reason, efficiency studies have generally been carried out by determining H as a function of mobile-phase velocity. The data obtained from such studies are typified by the two plots shown in Figure 24-11, the one for liquid chromatography and the other for gas chromatography. Although both show a minimum in H (or a maximum in efficiency) at low linear flow rates, the minimum for liquid chromatography usually occurs at flow rates that are well below those for gas chromatography and are often so low that they are not observed under normal operating conditions.

Linear flow rate and *volumetric flow rate* are two different but related quantities. The linear flow rate is the volumetric flow rate divided by the cross-sectional area of the column.

Generally, liquid chromatograms are obtained at lower flow rates than gas chromatograms. Furthermore, as shown in Figure 24-11, plate heights for liquid chromatographic columns are an order of magnitude or more smaller than those encountered with gas chromatographic columns. Offsetting this advantage is that it is impractical to employ liquid chromatographic columns longer than about 25 to 50 cm (because of high-pressure drops), whereas gas chromatographic columns may be 50 m or more in length. Consequently, the total number of plates and thus overall column efficiency are usually superior with gas chromatographic columns.

Other Variables That Influence Plate Heights

It has been found that plate heights can be decreased, and thus column efficiency increased, by decreasing the particle size of column packings, by employing thinner layers of film (where the stationary phase is a liquid adsorbed on a solid), and by lowering the viscosity of the mobile phase. Increases in temperature also reduce band broadening in most cases.

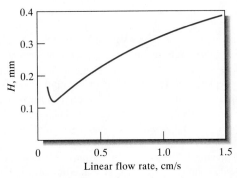

(a) Liquid chromatography

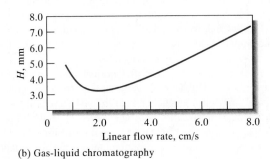

(b) Gas-liquid chromatography

Figure 24-11 Effect of mobile phase flow rate on plate height for (a) liquid chromatography and (b) gas chromatography.

24F-9   Column Resolution

The *resolution* R_s of a column provides a quantitative measure of its ability to separate two analytes. The significance of this term is illustrated in Figure 24-12, which consists of chromatograms for solute species A and B on three columns with different resolving powers. The resolution of each column is defined as

$$R_s = \frac{2\,\Delta Z}{W_A + W_B} = \frac{2[(t_R)_B - (t_R)_A]}{W_A + W_B} \qquad (24\text{-}19)$$

where all the terms on the right side are as defined in the figure.

It is evident from Figure 24-12 that a resolution of 1.5 gives an essentially complete separation of A and B, whereas a resolution of 0.75 does not. At a resolution of 1.0, zone A contains about 4% B and zone B contains about 4% A. At a resolution of 1.5, the overlap is about 0.3%. The resolution for a given stationary phase can be improved by lengthening the column and thus increasing the number of plates. An adverse consequence of the added plates, however, is an increase in the time required for separating the components.

> The *resolution* of a chromatographic column is a quantitative measure of its ability to separate analytes A and B.

Effect of Retention Factor and Selectivity Factor on Resolution

A useful equation is readily derived that relates the resolution of a column to the number of plates it contains as well as to the retention and selectivity factors of a pair of solutes on the column. Thus, it can be shown[8] that for the two solutes A and B in Figure 24-11, the resolution is given by the equation

Feature 24-2

What Is the Source of
the Terms Plate and
Plate Height?

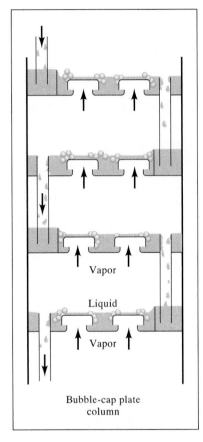

Figure 24F-1 Plates in a fractionating
column.

The 1952 Nobel Prize in chemistry was awarded to two Englishmen, A. J. P. Martin and R. L. M. Synge, for their work in the development of modern chromatography. In their theoretical studies, they adapted a model that was first developed in the early 1920s to describe separations on fractional distillation columns. Fractionating columns, which were first used in the petroleum industry for separating closely related hydrocarbons, consisted of numerous interconnected bubble-cap plates (see Figure 24F-1) at which vapor-liquid equilibria were established when the column was operated under reflux conditions.

Martin and Synge treated a chromatographic column as if it were made up of series of contiguous bubble-cap-like plates within which equilibrium conditions always prevail. This plate model successfully accounts for the Gaussian shape of chromatographic peaks as well as for factors that influence differences in solute-migration rates. The plate model is totally incapable of accounting for zone broadening, however, because of its basic assumption that equilibrium conditions prevail throughout a column during elution. This assumption can never be valid in the dynamic state that exists in a chromatographic column, where phases are moving past one another at such a pace that sufficient time is not available for equilibration.

Because the plate model is an inadequate representation of a chromatographic column, you are strongly urged (1) to avoid attaching any real or imaginary significance to the terms "plate" and "plate height" and (2) to view these terms as designators of column efficiency that are retained for historic reasons only and not because they have physical significance. Unfortunately, the terms are so well entrenched in the chromatographic literature that their replacement by more appropriate designations seems unlikely, at least in the near future.

$$R_s = \frac{\sqrt{N}}{4}\left(\frac{\alpha - 1}{\alpha}\right)\left(\frac{k_B}{1 + k_B}\right) \tag{24-20}$$

where k_B is the retention factor of the slower-moving species and α is the selectivity factor. This equation can be rearranged to give the number of plates needed to realize a given resolution:

$$N = 16R_s^2\left(\frac{\alpha}{\alpha - 1}\right)^2\left(\frac{1 + k_B}{k_B}\right)^2 \tag{24-21}$$

Effect of Resolution on Retention Time

As mentioned earlier, the goal in chromatography is the highest possible resolution in the shortest possible elapsed time. Unfortunately, these goals tend to be incompatible, and a compromise between the two is usually necessary. The time $(t_R)_B$ required to elute the two species in Figure 24-12 with a resolution of R_s is given by

[8]See D. A. Skoog, F. J. Holler, and T. A. Nieman, *Principles of Instrumental Analysis,* 5th ed. (Philadelphia: Saunders College Publishing, 1998), p. 689.

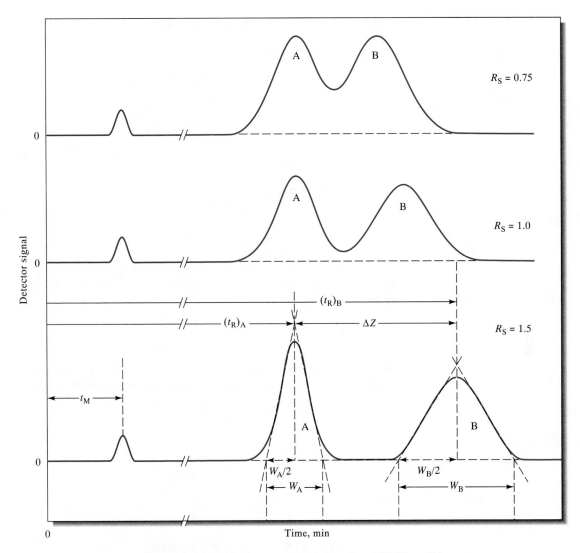

Figure 24-12 Separation at three resolutions: $R_s = 2 \Delta Z/(W_A + W_B)$.

$$(t_R)_B = \frac{16R_s^2 H}{u} \left(\frac{\alpha}{\alpha - 1} \right)^2 \frac{(1 + k_B)^3}{(k_B)^2} \qquad (24\text{-}22)$$

where u is the linear velocity of the mobile phase.

Example 24-2

Substances A and B have retention times of 16.40 and 17.63 min, respectively, on a 30.0-cm column. An unretained species passes through the column in 1.30 min. The peak widths (at base) for A and B are 1.11 and 1.21 min, respectively. Calculate (a) the column resolution, (b) the average number of plates in the column, (c) the plate height, (d) the length of column required to achieve a resolution of 1.5, and (e) the time required to elute substance B on the column that gives $R_s = 1.5$.

(a) Employing Equation 24-19, we find

$$R_s = \frac{2(17.63 - 16.40)}{1.11 + 1.21} = 1.06$$

(b) Equation 24-18 permits computation of N:

$$N = 16\left(\frac{16.40}{1.11}\right)^2 = 3493 \quad \text{and} \quad N = 16\left(\frac{17.63}{1.21}\right)^2 = 3397$$

$$N_{av} = \frac{3493 + 3397}{2} = 3445 = 3.4 \times 10^3$$

(c) $H = \dfrac{L}{N} = \dfrac{30.0}{3445} = 8.7 \times 10^{-3}$ cm

(d) k and α do not change greatly with increasing N and L. Thus, substituting N_1 and N_2 into Equation 24-20 and dividing one of the resulting equations by the other yield

$$\frac{(R_s)_1}{(R_s)_2} = \frac{\sqrt{N_1}}{\sqrt{N_2}}$$

where the subscripts 1 and 2 refer to the original and longer columns, respectively. Substituting the appropriate values for N_1, $(R_s)_1$, and $(R_s)_2$ gives

$$\frac{1.06}{1.5} = \frac{\sqrt{3445}}{\sqrt{N_2}}$$

$$N_2 = 3445\left(\frac{1.5}{1.06}\right)^2 = 6.9 \times 10^3$$

But

$$L = NH = 6.9 \times 10^3 \times 8.7 \times 10^{-3} = 60 \text{ cm}$$

(e) Substituting $(R_s)_1$ and $(R_s)_2$ into Equation 24-22 and dividing yield

$$\frac{(t_R)_1}{(t_R)_2} = \frac{(R_s)_1^2}{(R_s)_2^2} = \frac{17.63}{(t_R)_2} = \frac{(1.06)^2}{(1.5)^2}$$

$$(t_R)_2 = 35 \text{ min}$$

Thus, to obtain the improved resolution, the column length and consequently the separation time must be doubled.

24F-10   Applications of Chromatography

Chromatography is a powerful and versatile tool for separating closely related chemical species. In addition, it can be employed for the qualitative identification and quantitative determination of separated species. Examples of the applications of the various types of chromatography are given in Chapters 25 and 26.

Simulating a Chromatogram[9]

Spreadsheet Exercise

Microsoft
Excel

In this spreadsheet exercise, we use the properties of the Gaussian curve and the equations describing chromatographic peaks to simulate a chromatogram with Excel. In Feature 6-2, we explored the properties of the Gaussian curve and showed how the areas under the curve could be calculated. The equation for a Gaussian curve was presented as

$$y = \frac{1}{\sigma\sqrt{2\pi}}\, e^{-(x-\mu)^2/2\sigma^2}$$

where y is the value corresponding to a given value of x on the Gaussian curve, μ is the mean of the distribution, and σ is its standard deviation. For a chromatographic peak with a Gaussian shape, we may write a similar equation.

$$c_t = \frac{A}{\sigma}\, e^{-(t-t_R)^2/2\sigma^2}$$

where σ is the standard deviation of the Gaussian-shaped chromatographic peak, t_R is the retention time, A is an amplitude function containing the initial concentration of the solute, and c_t is the concentration of the solute at time t. It can be shown that the width of a peak $W = 4\sigma$. If we substitute this value into Equation 24-18 and solve for σ, we find that

$$\sigma = \frac{t_R}{\sqrt{N}}$$

where t_R is the retention time, and N is the number of theoretical plates for the chromatographic separation. Armed with this equation and the equation for the Gaussian chromatographic peak, we can calculate the concentration of a solute at any time t for given values of t_R, N, and A.

Another important figure of merit for chromatographic separations is the resolution R_s, which is given by Equation 24-19 for solutes A and B. Since for component A, $W_A = 4\sigma_A$ and for component B, $W_B = 4\sigma_B$, we can substitute these values into Equation 24-19 to arrive at the following equation for the resolution between the peaks for solutes A and B.

$$R_s = \frac{(t_R)_B - (t_R)_A}{2(\sigma_A + \sigma_B)}$$

We now have sufficient equations to simulate a chromatogram containing three components: A, B, and C, and compute the resolution for adjacent pairs of peaks. The spreadsheet below is designed to accomplish this task and allow you to change the various chromatographic variables to observe the effects on the resulting chromatogram.

---

[9]Based on an unpublished Mathcad® exercise by A. C. Censullo, California Polytechnic State University; P. Kucera and J. Lepore, *Chromatography Magazine,* **1987**(*4*), 23–27; F. J. Holler, *Mathcad® Applications for Analytical Chemistry,* pp. 129–133. Philadelphia: Saunders College Publishing, 1994.

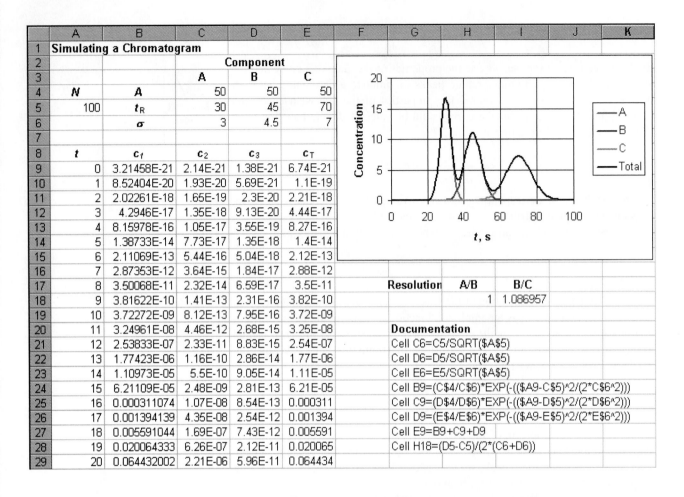

The spreadsheet table:

	A	B	C	D	E	F	G	H	I	J	K
1	**Simulating a Chromatogram**										
2				**Component**							
3			**A**	**B**	**C**						
4	N	A	50	50	50						
5	100	t_R	30	45	70						
6		σ	3	4.5	7						
7											
8	t	c_1	c_2	c_3	c_T						
9	0	3.21458E-21	2.14E-21	1.38E-21	6.74E-21						
10	1	8.52404E-20	1.93E-20	5.69E-21	1.1E-19						
11	2	2.02261E-18	1.65E-19	2.3E-20	2.21E-18						
12	3	4.2946E-17	1.35E-18	9.13E-20	4.44E-17						
13	4	8.15978E-16	1.05E-17	3.55E-19	8.27E-16						
14	5	1.38733E-14	7.73E-17	1.35E-18	1.4E-14						
15	6	2.11069E-13	5.44E-16	5.04E-18	2.12E-13						
16	7	2.87353E-12	3.64E-15	1.84E-17	2.88E-12						
17	8	3.50068E-11	2.32E-14	6.59E-17	3.5E-11		**Resolution**	**A/B**	**B/C**		
18	9	3.81622E-10	1.41E-13	2.31E-16	3.82E-10			1	1.086957		
19	10	3.72272E-09	8.12E-13	7.95E-16	3.72E-09						
20	11	3.24961E-08	4.46E-12	2.68E-15	3.25E-08		**Documentation**				
21	12	2.53833E-07	2.33E-11	8.83E-15	2.54E-07		Cell C6=C5/SQRT($A$5)				
22	13	1.77423E-06	1.16E-10	2.86E-14	1.77E-06		Cell D6=D5/SQRT($A$5)				
23	14	1.10973E-05	5.5E-10	9.05E-14	1.11E-05		Cell E6=E5/SQRT($A$5)				
24	15	6.21109E-05	2.48E-09	2.81E-13	6.21E-05		Cell B9=(C$4/C$6)*EXP(-(($A9-C$5)^2/(2*C$6^2)))				
25	16	0.000311074	1.07E-08	8.54E-13	0.000311		Cell C9=(D$4/D$6)*EXP(-(($A9-D$5)^2/(2*D$6^2)))				
26	17	0.001394139	4.35E-08	2.54E-12	0.001394		Cell D9=(E$4/E$6)*EXP(-(($A9-E$5)^2/(2*E$6^2)))				
27	18	0.005591044	1.69E-07	7.43E-12	0.005591		Cell E9=B9+C9+D9				
28	19	0.020064333	6.26E-07	2.12E-11	0.020065		Cell H18=(D5-C5)/(2*(C6+D6))				
29	20	0.064432002	2.21E-06	5.96E-11	0.064434						

Let us begin our discussion of the spreadsheet with the chromatographic variables that you can change in cells A5 and C4 through E5. Cell A5 contains the number of theoretical plates N for the column. Cells C4, D4, and E4 contain the factors A_A, A_B, and A_C for the three peaks in our chromatogram, and cells C5, D5, and E5 contain the retention times $(t_R)_A$, $(t_R)_B$, and $(t_R)_C$. The spreadsheet calculates σ_A, σ_B, and σ_C in cells C6, D6, and E6 from the retention times and the number of theoretical plates. Note that once you have entered the correct formula in cell C6, you can simply copy the formula into cells D6 and E6.

The formulas for peaks A, B, and C begin in cells B9, C9, and D9. After you have entered the formula for peak A in cell B9, you can just copy it to cells C9 and D9. Cell E9 contains the sum of cells B9, C9, and D9, and the sequence of numbers in column E beneath cell E9 represent the entire chromatogram. In other words, columns B through D represent individual chromatograms for components A, B, and C, and column E is the sum of all three, which represents a chromatogram of a mixture of the three components. Although we have shown only the first 29 rows of the spreadsheet, the formulas in cells A9 through E9 must be copied to row 109 in order to calculate a complete chromatogram.

Before you copy the cells, be sure to enter 0 in cell A9 and 1 in cell A10 so that cells A9 through A109 will contain one hundred consecutive times at 1-second intervals beginning at zero and ending in 100. The resolution between peaks A and B is computed by the formula in cell H18, which can be copied into cell I18 to determine the resolution between peaks B and C. To create the graph, highlight cells A9:E109, and use the Graph Wizard to set the plot type to XY Scatter. Once the plot has been created, you may double-click on an individual point on each peak to view the Format Data Series menu. Click on the Patterns tab, then set the options to plot smooth lines rather than points by clicking on the Custom button and the Smoothed Line checkbox. Select a different color for each peak and for the overall chromatogram in column E using the Color pull-down menu. Under Marker, click on the None button. It is also a good idea to double-click on the axes and manually set the scale on the x-axis to a minimum of 0 and a maximum of 100. Set the y-axis to a minimum of 0 and allow the maximum to scale automatically.

Begin your study of chromatographic variables with the numbers shown in the spreadsheet. Note the resolutions calculated between pairs of peaks, and then change the number of theoretical plates N to 200, 300, 400, 800, etc., noting the resolutions under each set of conditions. What resolution constitutes a good separation? Reset N to 100, vary the retention times, and note the results both visually and by inspection of the calculated resolutions. It is instructive to choose one peak, say peak A, and systematically change its retention time $(t_R)_A$ in cell C5 through the entire range of the time scale, noting the results; for example, 10, 20, 30, etc. Note especially how the peak shape changes as the retention time increases. Does the area under the Gaussian peak appear to change? Now return the retention times to their original settings, systematically vary the amplitudes in cells C4 and E4, and note the effect on resolution. As you can see, the number of theoretical plates has a powerful influence on the resolving ability of a particular column. By working with this spreadsheet, you will develop a feel for and a visual sense of the significance of chromatographic resolution.

WEB WORKS

Set your Web browser to http://www.chemicalanalysis.com/chromatography.htm for links to noncommercial chromatography resources on the Web. Look up information on the care and maintenance of capillary gas chromatography columns. Alternatively, look at the hypertextbook on liquid chromatography and see what information is available on the basics of liquid chromatography.

24G QUESTIONS AND PROBLEMS

*24-1. What is a masking agent and how does it function?

24-2. How do strong and weak acid synthetic ion-exchange resins differ in structure?

*24-3. The distribution constant for X between n-hexane and water is 9.6. Calculate the concentration of X remaining in the aqueous phase after 50.0 mL of 0.150

M X is treated by extraction with the following quantities of n-hexane:

(a) one 40.0-mL portion.

(b) two 20.0-mL portions.

(c) four 10.0-mL portions.

(d) eight 5.00-mL portions.

24-4. The distribution constant for Z between n-hexane and water is 6.25. Calculate the percent of Z remaining in 25.0 mL of water that was originally 0.0600 M in Z after extraction with the following volumes of n-hexane:

(a) one 25.0-mL portion.

(b) two 12.5-mL portions.

(c) five 5.00-mL portions.

(d) ten 2.50-mL portions.

*24-5. What volume of n-hexane is required to decrease the concentration of X in Problem 24-3 to 1.00×10^{-4} M if 25.0 mL of 0.0500 M X are extracted with

(a) 25.0-mL portions?

(b) 10.0-mL portions?

(c) 2.0-mL portions?

24-6. What volume of n-hexane is required to decrease the concentration of Z in Problem 24-4 to 1.00×10^{-5} M if 40.0 mL of 0.0200 M Z are extracted with

(a) 50.0-mL portions of n-hexane?

(b) 25.0-mL portions?

(c) 10.0-mL portions?

*24-7. What is the minimum distribution constant that permits removal of 99% of a solute from 50.0 mL of water with

(a) two 25.0-mL extractions with toluene?

(b) five 10.0-mL extractions with toluene?

24-8. If 30.0 mL of water that is 0.0500 M in Q are to be extracted with four 10.0-mL portions of an immiscible organic solvent, what is the minimum distribution constant that allows transfer of all but the following percentages of the solute to the organic layer:

*(a) 1.00×10^{-4}

(b) 1.00×10^{-2}

(c) 1.00×10^{-3}

*24-9. A 0.150 M aqueous solution of the weak organic acid HA was prepared from the pure compound, and three 50.0-mL aliquots were transferred to 100.0-mL volumetric flasks. Solution 1 was diluted to 100.0 mL with 1.0 M $HClO_4$, solution 2 was diluted to the mark with 1.0 M NaOH, and solution 3 was diluted to the mark with water. A 25.0-mL aliquot of each was extracted with 25.0-mL of n-hexane. The extract from solution 2 contained no detectable trace of A-containing species, indicating that A^- is not soluble in the organic solvent. The extract from solution 1 contained no ClO_4^- or $HClO_4$ but was found to be 0.0454 M in HA (by extraction with standard NaOH

and back-titration with standard HCl). The extract from solution 3 was found to be 0.0225 M in HA. Assume that HA does not associate or dissociate in the organic solvent, and calculate

(a) the distribution constant for HA between the two solvents.

(b) the concentration of the *species* HA and A^- in aqueous solution 3 after extraction.

(c) the dissociation constant of HA in water.

24-10. To determine the equilibrium constant for the reaction

$$I_2 + 2SCN^- \rightleftharpoons I(SCN)_2^- + I^-$$

25.0 mL of a 0.0100 M aqueous solution of I_2 were extracted with 10.0 mL of $CHCl_3$. After extraction, spectrophotometric measurements revealed that the I_2 concentration *of the aqueous layer* was 1.12×10^{-4} M. An aqueous solution that was 0.0100 M in I_2 and 0.100 M KSCN was then prepared. After extraction of 25.0 mL of this solution with 10.0 mL of $CHCl_3$, the concentration of I_2 in the $CHCl_3$ layer was found from spectrophotometric measurement to be 1.02×10^{-3} M.

(a) What is the distribution constant for I_2 between $CHCl_3$ and H_2O?

(b) What is the formation constant for $I(SCN)_2^-$?

*24-11. The total cation content of natural water is often determined by exchanging the cations for hydrogen ions on a strong-acid ion-exchange resin. A 25.0-mL sample of a natural water was diluted to 100 mL with distilled water, and 2.0 g of a cation-exchange resin were added. After stirring, the mixture was filtered and the solid remaining on the filter paper was washed with three 15.0-mL portions of water. The filtrate and washings required 15.3 mL of 0.0202 M NaOH to give a bromocresol green end point.

(a) Calculate the number of milliequivalents of cation present in exactly 1 L of sample. (Here, the equivalent weight of a cation is its formula weight divided by its charge.)

(b) Report the results in terms of milligrams of $CaCO_3$ per liter.

24-12. An organic acid was isolated and purified by recrystallization of its barium salt. To determine the equivalent weight of the acid, a 0.393-g sample of the salt was dissolved in about 100 mL of water. The solution was passed through a strong-acid ion-exchange resin, and the column was then washed with water; the eluate and washings were titrated with 18.1 mL of 0.1006 M NaOH to a phenolphthalein end point.

(a) Calculate the equivalent weight of the organic acid.

(b) A potentiometric titration curve of the solution resulting when a second sample was treated in the same way revealed two end points, one at pH 5 and the other at pH 9. What is the molecular weight of the acid?

*24-13. Describe the preparation of 2.00 L of 0.1500 M HCl from primary-standard grade NaCl using a cation-exchange resin.

24-14. An aqueous solution containing $MgCl_2$ and HCl was analyzed by first titrating a 25.00-mL aliquot to a bromocresol green end point with 18.96 mL of 0.02762 M NaOH. A 10.00-mL aliquot was then diluted to 50.00 mL with distilled water and passed through a strong-acid ion-exchange resin. The eluate and washings required 36.54 mL of the NaOH solution to reach the same end point. Report the molar concentrations of HCl and $MgCl_2$ in the sample.

24-15. Define
  *(a) elution.   **(b)** mobile phase.
  *(c) stationary phase.   **(d)** partition ratio.
  *(e) retention time.   **(f)** retention factor.
  *(g) selectivity factor.   **(h)** plate height.

24-16. List the variables that lead to zone broadening.

*24-17. What is the difference between gas-liquid and liquid-liquid chromatography?

24-18. What is the difference between liquid-liquid and liquid-solid chromatography?

*24-19. Describe a method for determining the number of plates in a column.

24-20. Name two general methods for improving the resolution of two substances on a chromatographic column.

*24-21. The following data are for a liquid chromatographic column:

Length of Packing	24.7 cm
Flow rate	0.313 mL/min
V_M	1.37 mL
V_S	0.164 mL

A chromatogram of a mixture of species A, B, C, and D provided the following data:

	Retention Time, min	Width of Peak Base (W), min
Nonretained	3.1	—
A	5.4	0.41
B	13.3	1.07
C	14.1	1.16
D	21.6	1.72

Calculate
(a) the number of plates from each peak.
(b) the mean and the standard deviation for N.
(c) the plate height for the column.

*24-22. From the data in Problem 24-21, calculate for A, B, C, and D
(a) the retention factor.
(b) the distribution constant.

*24-23. From the data in Problem 24-21, calculate for species B and C
(a) the resolution.
(b) the selectivity factor.
(c) the length of column necessary to separate the two species with a resolution of 1.5.
(d) the time required to separate the two species on the column in part (c).

*24-24. From the data in Problem 24-21, calculate for species C and D
(a) the resolution.
(b) the length of column necessary to separate the two species with a resolution of 1.5.

24-25. The following data were obtained by gas-liquid chromatography on a 40-cm packed column:

Compound	t_R, min	W, min
Air	1.9	—
Methylcyclohexane	10.0	0.76
Methylcyclohexene	10.9	0.82
Toluene	13.4	1.06

Calculate
(a) an average number of plates from the data.
(b) the standard deviation for the average in (a).
(c) an average plate height for the column.

24-26. Referring to Problem 24-25, calculate the resolution for
(a) methylcyclohexene and methylcyclohexane.
(b) methylcyclohexene and toluene.
(c) methylcyclohexane and toluene.

24-27. If a resolution of 1.5 is desired in separating methylcyclohexane and methylcyclohexene in Problem 24-25
(a) how many plates are required?
(b) how long must the column be if the same packing is employed?
(c) what is the retention time for methylcyclohexene on the column in Problem 24-27b?

*24-28. If V_S and V_M for the column in Problem 24-25 are 19.6 and 62.6 mL, respectively, and a nonretained peak appears after 1.9 min, calculate the
(a) retention factor for each compound.
(b) distribution constant for each compound.
(c) selectivity factor for methylcyclohexane and methylcyclohexene.

*24-29. From distribution studies, species M and N̄ are known to have water/hexane distribution constants of 6.01 and 6.20 ($K = [M]_{H_2O}/[M]_{hex}$). The two species are to be separated by elution with hexane in a column packed with silica gel containing adsorbed water. The ratio V_S/V_M for the packing is 0.422.

(a) Calculate the retention factor for each solute.

(b) Calculate the selectivity factor.

(c) How many plates are needed to provide a resolution of 1.5?

(d) How long a column is needed if the plate height of the packing is 2.2×10^{-3} cm?

(e) If a flow rate of 7.10 cm/min is employed, how long will it take to elute the two species?

24-30. Repeat the calculations in Problem 24-29 assuming $K_M = 5.81$ and $K_N = 6.20$.

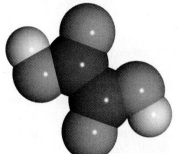

Chapter 25

Gas-Liquid and High-Performance Liquid Chromatography

I n this chapter, we describe how analytical separations are carried out with two of the most common and widely used types of chromatography: gas-liquid chromatography (GLC) and high-performance liquid chromatography (HPLC).

25A   GAS-LIQUID CHROMATOGRAPHY

In gas chromatography, the components of a vaporized sample are fractionated as a consequence of being partitioned between a mobile gaseous phase and a liquid or a solid stationary phase held in a column.[1] In performing a gas chromatographic separation, the sample is vaporized and injected onto the head of a chromatographic column. Elution is brought about by the flow of an inert gaseous mobile phase. In contrast to most other types of chromatography, the mobile phase does not interact with molecules of the sample; its only function is to transport the sample species through the column. Two types of gas chromatography are encountered: *gas-liquid chromatography* (GLC) and *gas-solid chromatography* (GSC). Gas-liquid chromatography finds widespread use in all fields of science, where its name is usually shortened to *gas chromatography* (GC). Gas-solid chromatography has limited application owing to semipermanent retention of active or polar molecules and severe tailing of elution peaks. We will not discuss GSC further.

Gas-liquid chromatography is based on partitioning of the analyte between a gaseous mobile phase and a liquid phase immobilized on the surface of an inert solid packing or on the walls of a capillary tubing. The concept of gas-liquid chromatography was first enunciated in 1941 by A. J. P. Martin and R. L. M. Synge, who were also responsible for the development of liquid-liquid partition

---

[1]For detailed treatment of gas chromatography, see J. Willet, *Gas Chromatography* (New York: Wiley, 1987); R. L. Grob, Ed. *Modern Practice of Gas Chromatography,* 3rd ed. (New York: Wiley, 1995); R. P. W. Scott, *Introduction to Analytical Gas Chromatography,* 2nd ed. (New York: Marcel Dekker, 1997).

In *gas-liquid chromatography,* the mobile phase is a gas, and the stationary phase is a liquid that is retained on the surface of an inert solid by adsorption or chemical bonding.

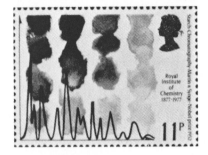

Stamp honoring biochemists Archer J. P. Martin (1910–   ) and Richard L. M. Synge (1914–   ), who were awarded the 1952 Nobel Prize in chemistry for their contributions to the development of modern chromatography.

chromatography. More than a decade was to elapse, however, before the value of gas-liquid chromatography was demonstrated experimentally and this technique began to be used as a routine laboratory tool. In 1955, the first commercial apparatus for gas-liquid chromatography appeared on the market. Since that time, the growth in applications of this technique has been phenomenal. In 1985, it was estimated that as many as 200,000 gas chromatographs were in use in the world.[2] Undoubtedly, by now the number is significantly larger.

25A-1   Instruments for Gas Chromatography

Many changes and improvements in gas chromatographic instruments have appeared in the marketplace since their commercial introduction. In the 1970s, electronic integrators and computer-based data processing equipment became common. The 1980s saw computers being used for automatic control of most instrument parameters, such as column temperature, flow rates, and sample injection; development of very high performance instruments at moderate costs; and perhaps most important, the development of open tubular columns that are capable of separating components of complex mixtures in relatively short times. In the 1990s, capillary GC was further developed along with techniques such as high-speed GC. Today, well over 30 instrument manufacturers offer some 130 different models of gas-chromatographic equipment at costs that range from $1500 to $40,000. The basic components of a typical instrument for performing gas chromatography are shown in Figure 25-1 and are described briefly in this section.

25A-2   Carrier Gas System

Helium is the most common mobile phase in gas chromatography.

The gaseous mobile phase in gas chromatography must be chemically inert. Helium is the most common mobile phase, although argon, nitrogen, and hydrogen are also used. These gases are available in pressurized tanks. Pressure regulators, gauges, and flow meters are required to control the flow rate of the gas.

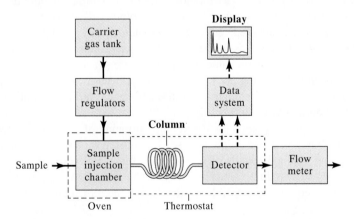

Figure 25-1   Block diagram of a gas-chromatographic apparatus.

[2]R. Schill and R. R. Freeman, in *Modern Practice of Gas Chromatography,* 2nd ed., R. L. Grob, Ed. (New York: Wiley, 1985), p. 294.

Pressures at the column inlet usually range from 10 to 50 psi (lb/in.² above room pressure) and provide flow rates of 25 to 50 mL/min. Flow rates are sometimes measured by a simple soap-bubble meter, such as that shown in Figure 25-2. A soap film is formed in the path of the gas when the rubber bulb containing a solution of soap is squeezed; the time required for this film to move between two graduations on the buret is measured and converted to a flow rate. Many modern commercial chromatographs are equipped with electronic flow meters that are computer controlled to maintain the flow rate at any desired level.

Figure 25-2   A soap-bubble flow meter. (Courtesy of Varian, Inc., Walnut Creek, CA)

25A-3   Sample Injection System

Column efficiency requires that the sample be of a suitable size and be introduced as a "plug" of vapor; slow injection or oversized samples cause band spreading and poor resolution. Calibrated microsyringes, such as those shown in Figure 25-3, are used to inject liquid samples through a rubber or silicone diaphragm or septum into a heated sample port located at the head of the column. The sample port (Figure 25-4) is ordinarily about 50°C above the boiling point of the least volatile component of the sample. For ordinary packed analytical columns, sample sizes range from a few tenths of a microliter to 20 μL. Capillary columns require samples that are smaller by a factor of 100 or more. Here, a sample splitter is often needed to deliver only a small known fraction (1:100 to 1:500) of the injected sample, with the remainder going to waste. Commercial gas chromatographs intended for use with capillary columns incorporate such splitters and also allow for sample injection without splitting when packed columns are used (split/splitless injection systems).

Figure 25-3   A set of microsyringes for sample injection. (Courtesy of Varian, Inc., Walnut Creek, CA)

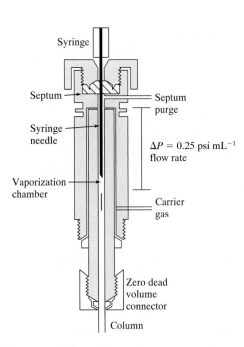

Syringe

Septum

Syringe needle

Vaporization chamber

Septum purge

$\Delta P = 0.25$ psi mL^{-1} flow rate

Carrier gas

Zero dead volume connector

Column

Figure 25-4   Cross-sectional view of a microflash vaporizer direct injector.

25A-4   Detectors

Detection devices for gas chromatography must respond rapidly to minute concentrations of solutes as they exit the column. The solute concentration in the carrier gas at any instant is no more than a few parts per thousand and often is even smaller by one or two orders of magnitude. Moreover, the time during which a peak passes the detector is typically one second (or less), which requires that the device be capable of exhibiting its full response during this brief period.

Other desirable properties for a detector include linear response, stability, and uniform response for a wide variety of chemical species or, alternatively, a predictable and selective response toward one or more classes of solutes. No single detector fulfills all these requirements. Some of the more common detectors are listed in Table 25-1. Four of the most widely used detectors are discussed in the sections that follow.

Flame-Ionization Detectors

The flame-ionization detector (FID) is the most widely used and generally applicable of all detectors for gas chromatography. Most organic compounds, when pyrolyzed in a hot flame, produce ionic intermediates that conduct electricity through the flame. Hydrogen is added to the carrier gas with this detector, and the eluent is mixed with oxygen and combusted in a burner equipped with a pair of electrodes (see Figure 25-5). Detection involves monitoring the current produced by collecting electrons and ions produced by the combustion process at biased electrodes. The FID exhibits a high sensitivity, a large linear response, and low noise. It is also rugged and easy to use. A limitation of the FID is that it destroys the sample during the combustion step.

Thermal Conductivity Detectors

The *thermal conductivity detector,* which was one of the earliest detectors for gas chromatography, still finds wide application. This device consists of an electrically heated source whose temperature at constant electric power depends on

Table 25-1

Gas Chromatography Detectors

Type	Applicable Samples	Typical Detection Limit
Thermal conductivity	Universal detector	500 pg/mL
Flame ionization	Hydrocarbons	2 pg/s
Electron capture	Halogenated compounds	5 fg/s
Thermionic (Nitrogen-phosphorus)	Nitrogen and phosphorus compounds	0.1 pg/s (P) 1 pg/s (N)
Electrolytic conductivity (Hall)	Compounds containing halogens, sulfur, or nitrogen	0.5 pg Cl/s 2 pg S/s 4 pg N/s
Photoionization	Compounds ionized by UV radiation	2 pg C/s
Fourier transform IR (FTIR)	Organic compounds	0.2 to 40 ng
Mass spectrometer (MS)	Tunable for any species	0.25 to 100 pg

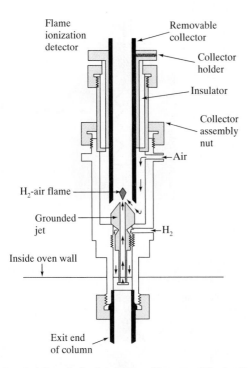

Flame
ionization
detector

Removable
collector

Collector
holder

Insulator

Collector
assembly
nut

Air

H_2-air flame

Grounded
jet

H_2

Inside oven wall

Exit end
of column

Figure 25-5 A typical flame ionization detector. (Courtesy of Hewlett-Packard Company.)

the thermal conductivity of the surrounding gas. The heated element may be a fine platinum, gold, or tungsten wire (see Figure 25-6a) or, alternatively, a small thermistor. The electrical resistance of this element depends on the thermal conductivity of the gas. Twin detectors are ordinarily used, one being located ahead of the sample injection chamber and the other immediately beyond the column; alternatively, the gas stream can be split. The detectors are incorporated in two arms of a simple bridge circuit (see Figure 25-6b) such that the thermal conductivity of the carrier is canceled. In addition, the effects of variations in temperature, pressure, and electric power are minimized. The thermal conductivities of helium and hydrogen are roughly six to ten times greater than those of most organic compounds. Thus, even small amounts of organic species cause relatively large decreases in the thermal conductivity of the column effluent, which results in a marked rise in the temperature of the detector. Detection by thermal conductivity is less satisfactory with carrier gases whose conductivities more closely resemble those of most sample components.

The advantages of the thermal conductivity detector are its simplicity, its large linear dynamic range (about five orders of magnitude), its general response to both organic and inorganic species, and its nondestructive character, which permits collection of solutes after detection. The chief limitation of the thermal conductivity detector is its relatively low sensitivity. Other detectors exceed this sensitivity by factors of 10^4 to 10^7.

Electron-Capture Detectors

The electron-capture detector (ECD) has become one of the most widely used detectors for environmental samples because this detector selectively responds to halogen-containing organic compounds, such as pesticides and polychlorinated

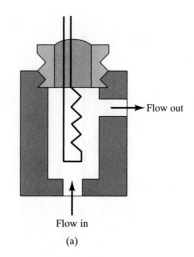

Flow out

Flow in

(a)

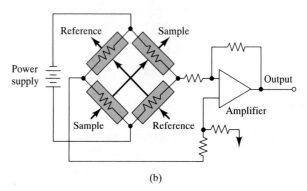

(b)

Figure 25-6   A typical thermal conductivity detector. (Courtesy of Varian Instrument Division, Palo Alto, CA.)

biphenyls. In this detector, the sample eluate from a column is passed over a radioactive β emitter, usually nickel-63. An electron from the emitter causes ionization of the carrier gas (often nitrogen) and the production of a burst of electrons. In the absence of organic species, a constant standing current between a pair of electrodes results from this ionization process. The current decreases markedly, however, in the presence of organic molecules containing electronegative functional groups that tend to capture electrons. Compounds, such as halogens, peroxides, quinones, and nitro groups, are detected with high sensitivity. The detector is insensitive to functional groups such as amines, alcohols, and hydrocarbons.

Electron-capture detectors are highly sensitive and have the advantage of not altering the sample significantly (in contrast to the flame ionization detector, which consumes the sample). The linear response of the detector, however, is limited to about two orders of magnitude.

Hyphenated Methods of Detection

Other important GC detectors include the thermionic detector, the electrolytic conductivity or Hall detector, and the photoionization detector. The thermionic detector is similar in construction to the FID. With the thermionic detector, nitrogen- and phosphorous-containing compounds produce increased currents in a flame in which an alkali metal salt is vaporized. The thermionic detector is widely used for organophosphorus pesticides and pharmaceutical compounds. With the electrolytic conductivity detector, compounds containing halogens, sulfur, or nitrogen are mixed with a reaction gas in a small reactor tube. The products are then dissolved in a liquid, which produces a conductive solution. The change in conductivity as a result of the presence of the active compound is then measured. In the photoionization detector, molecules are photoionized by UV radiation. The ions and electrons produced are then collected with a pair of biased electrodes, and the resulting current is measured. The detector is often used for aromatic and other molecules that are easily photoionized.

The effluent from chromatographic columns is often monitored continuously by the selective techniques of spectroscopy or electrochemistry. Mass spectrometry (MS) and Fourier Transform IR spectroscopy (FTIR) are widely used. The resulting so-called hyphenated methods (for instance, GC/MS and GC/FTIR) provide the chemist with powerful tools for identifying the components of complex mixtures.[3] Modern computer-based instruments incorporate large databases for comparing spectra and identifying compounds eluting from the GC column.

Hyphenated methods couple the separation capabilities of chromatography with the qualitative and quantitative detection capabilities of spectral methods.

25A-5   Gas Chromatographic Columns and Stationary Phases

Two types of columns are encountered in gas-liquid chromatography, *capillary,* or *open tubular,* and *packed.* The latter can accommodate larger samples and are generally more convenient to use than the former. Capillary columns have become of considerable importance because of their unparalleled resolution. To date, the vast majority of gas chromatography has been carried out on packed columns. This situation is changing rapidly, however, and it seems probable in the near future that packed columns will be largely replaced by the faster open tubular columns for most applications.

Capillary, or Open Tubular, Columns[4]

Currently, the most widely used capillary columns are *fused-silica open tubular columns* (FSOT columns). Fused-silica capillaries are drawn from specially purified silica that contains minimal amounts of metal oxides. Their inside diameters are typically from 0.1 to 0.5 mm. The tubes are given added strength by an outside protective polyimide coating, which is applied as the capillary tubing is being drawn. The resulting columns are quite flexible and can be bent into coils having diameters of a few centimeters. Silica open tubular columns are available commercially and offer several important advantages such as physical strength, much lower reactivity toward sample components, and flexibility. Figure 25-7 is a picture of a fused-silica open tubular column.

In 1987, a column manufacturer reported drawing a fused-silica column 1300 m in length with over 2 million theoretical plates.

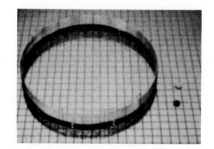

Figure 25-7   A 25-m fused-silica capillary column. (Courtesy of Varian, Inc., Walnut Creek, CA)

---

[3]For a review on hyphenated methods, see C. L. Wilkins, *Science,* **1983,** *222,* 251; *Anal. Chem.,* **1989,** *59,* 571A.

[4]For a detailed description of open tubular columns, see M. L. Lee, F. J. Yang, and K. D. Bartle, *Open Tubular Column Gas Chromatography* (New York: Wiley, 1984).

Packed columns offer the advantages of large sample size and ease and convenience of use. Capillary columns have the advantages of high resolution, short analysis time, and high sensitivity.

Packed Columns

Present-day packed columns are fabricated from glass or metal tubing; they are typically 2 to 3 m long and have inside diameters of 2 to 4 mm. The tubes are ordinarily formed as coils with diameters of roughly 15 cm to permit convenient thermostating in an oven.

The packing, or support, for a column holds the liquid stationary phase in place so that the surface area exposed to the mobile phase is as large as possible. The ideal solid packing consists of small, uniform, spherical particles with good mechanical strength and with a specific surface of at least 1 m^2/g. In addition, the material should be inert at elevated temperatures and be uniformly wetted by the liquid phase. No available substance meets all these criteria perfectly.

The earliest, and still the most widely used, packings for gas chromatography were prepared from naturally occurring diatomaceous earth, which consists of the skeletons of thousands of species of single-celled plants that inhabited ancient lakes and seas (Figure 25-8 is an enlarged photo of a diatom obtained with a scanning electron microscope). These support materials are often treated chemically with dimethylchlorosilane, which gives a surface layer of methyl groups. This treatment reduces the tendency of the packing to adsorb polar molecules.

The particle size of packings for gas chromatography typically fall in the range of 60 to 80 mesh (250 to 170 μm) or 80 to 100 mesh (170 to 149 μm). The use of smaller particles is not practical because the pressure drop within the column becomes prohibitively high.

Figure 25-8 A photomicrograph of a diatom. Magnification 5000×.

Column Thermostating

Reproducible retention times require control of the column temperature to within a few tenths of a degree. For this reason, the coiled column is ordinarily housed in a thermostated oven. The optimum temperature depends on the boiling points of the sample components. A temperature that is roughly equal to or slightly above the average boiling point of a sample results in a reasonable elution period (2 to 30 min). For samples with a broad boiling range, it may be necessary to employ temperature programming, whereby the column temperature is increased either continuously or in steps as the separation proceeds.

In general, optimum resolution is associated with minimal temperature. Low temperatures, however, result in longer elution times and hence slower analyses. Figure 25-9 illustrates this principle.

Temperature programming is a technique in which the temperature of a gas-chromatographic column is increased continuously or in steps during elution.

25A-6   Liquid Phases for Gas-Liquid Chromatography

Desirable properties for the immobilized liquid phase in a gas-liquid chromatographic column include (1) *low volatility* (ideally, the boiling point of the liquid should be 100°C higher than the maximum operating temperature for the column); (2) *thermal stability;* (3) *chemical inertness;* and (4) *solvent characteristics* such that k and α (Section 24F-6) values for the solutes to be resolved fall within a suitable range.

Many liquids have been proposed as stationary phases in the development of gas-liquid chromatography. By now, only a handful — perhaps a dozen or fewer — suffice for most applications. The proper choice of stationary phase is often crucial to the success of a separation. Qualitative guidelines exist for mak-

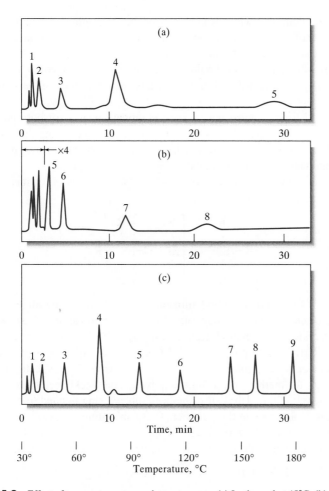

Figure 25-9 Effect of temperature on gas chromatograms. (a) Isothermal at 45°C, (b) isothermal at 145°C, (c) programmed at 30°C to 180°C. [From W. E. Harris and H. W. Habgood, *Programmed Temperature Gas Chromatography* (New York: Wiley, 1966), p. 10. Reprinted with permission.]

ing this choice, but in the end, the best stationary phase can only be determined in the laboratory.

The retention time for an analyte on a column depends on its distribution constant, which in turn is related to the chemical nature of the liquid stationary phase. Clearly, to be useful in gas-liquid chromatography, the immobilized liquid must generate different partition ratios for different sample components. In addition, however, these ratios must not be extremely large or extremely small because the former leads to prohibitively long retention times and the latter results in such short retention times that separations are incomplete.

To have a reasonable residence time in the column, an analyte must show some degree of compatibility (solubility) with the stationary phase. Here, the principle of "like dissolves like" applies, where "like" refers to the polarities of the analyte and the immobilized liquid. Polarity is the electrical field effect in the immediate vicinity of a molecule and is measured by the dipole moment of the species. Polar stationary phases contain functional groups such as —CN, —CO, and —OH. Hydrocarbon-type stationary phases and dialkyl

The polarities of common organic functional groups in increasing order are as follows: aliphatic hydrocarbons < olefins < aromatic hydrocarbons < halides < sulfides < ethers < nitro compounds < esters ≈ aldehydes ≈ ketones < alcohols ≈ amines < sulfones < sulfoxides < amides < carboxylic acids < water.

siloxanes are nonpolar, whereas polyester phases are highly polar. Polar analytes include alcohols, acids, and amines; solutes of medium polarity include ethers, ketones, and aldehydes. Saturated hydrocarbons are nonpolar. Generally, the polarity of the stationary phase should match that of the sample components. When the match is good, the order of elution is determined by the boiling point of the eluents.

Table 25-2 lists the most widely used stationary phases for both packed and open-tubular column gas chromatography in order of increasing polarity. These six liquids can probably provide satisfactory separations for 90% or more of the samples encountered by the scientist.

Five of the liquids listed in Table 25-2 are polydimethyl siloxanes that have the general structure

$$R-\underset{\underset{R}{|}}{\overset{\overset{R}{|}}{Si}}-O\left[\underset{\underset{R}{|}}{\overset{\overset{R}{|}}{Si}}-O\right]_n\underset{\underset{R}{|}}{\overset{\overset{R}{|}}{Si}}-R$$

In the first of these, polydimethyl siloxane, the —R groups are all —CH_3, giving a liquid that is relatively nonpolar. In the other polysiloxanes shown in the table, a fraction of the methyl groups is replaced by functional groups such as phenyl (—C_6H_5), cyanopropyl (—C_3H_6CN), and trifluoropropyl (—$C_3H_6CF_3$). The percentage description in each case gives the amount of substitution of the named group for methyl groups on the polysiloxane backbone. Thus, for example, 5% phenyl polydimethyl siloxane has a phenyl ring bonded to 5% by number of the silicon atoms in the polymer. These substitutions increase the polarity of the liquids to various degrees.

The fifth entry in Table 25-2 is a polyethylene glycol that has the structure

$$HO-CH_2-CH_2-(O-CH_2-CH_2)_n-OH$$

It finds widespread use for separating polar species.

Table 25-2

Some Common Liquid Stationary Phases for Gas-Liquid Chromatography

Stationary Phase	Common Trade Name	Maximum Temperature, °C	Common Applications
Polydimethyl siloxane	OV-1, SE-30	350	General-purpose nonpolar phase, hydrocarbons, polynuclear aromatics, drugs, steroids, PCBs
5% Phenyl-polydimethyl siloxane	OV-3, SE-52	350	Fatty acid methyl esters, alkaloids, drugs, halogenated compounds
50% Phenyl-polydimethyl siloxane	OV-17	250	Drugs, steroids, pesticides, glycols
50% Trifluoropropyl-polydimethyl siloxane	OV-210	200	Chlorinated aromatics, nitroaromatic, alkyl substituted benzenes
Polyethylene glycol	Carbowax 20M	250	Free acids, alcohols, ethers, essential oils, glycols
50% Cyanopropyl-polydimethyl siloxane	OV-275	240	Polyunsaturated fatty acids, rosin acids, free acids, alcohols

25A-7   Applications of Gas-Liquid Chromatography

Gas-liquid chromatography is applicable to species that are appreciably volatile and thermally stable at temperatures up to a few hundred degrees Celsius. An enormous number of compounds of interest to humans possess these qualities. Consequently, gas chromatography has been widely applied to the separation and determination of the components in a variety of sample types. Figure 25-10 shows chromatograms for a few such applications.

Qualitative Analysis

Gas chromatography is widely used for recognizing the presence or absence of components in mixtures that contain a limited number of species whose identities are known. For example, 30 or more amino acids in a protein hydrolysate can be detected with a reasonable degree of certainty by means of a chromatogram. On the other hand, because a chromatogram provides but a single piece of information about each species in a mixture (the retention time), the application of the technique to the qualitative analysis of complex samples of unknown composition is limited. This limitation has been largely overcome by linking chromatographic columns directly with ultraviolet, infrared, and mass

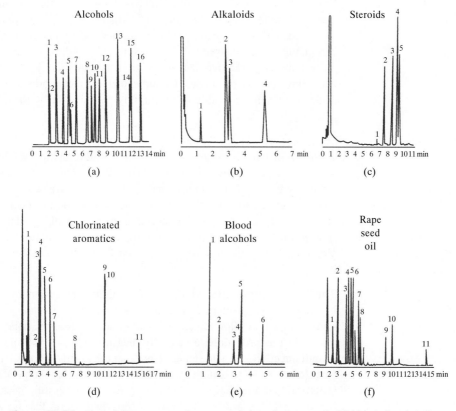

Figure 25-10   Typical chromatograms from open tubular columns coated with (a) polydimethyl siloxane, (b) 5%(phenylmethyldimethyl) siloxane, (c) 50%(phenylmethyldimethyl) siloxane, (d) 50% poly(trifluoropropyl-dimethyl) siloxane, (e) polyethylene glycol, (f) 50% poly(cyanopropyl-dimethyl) siloxane. (Courtesy of J & W Scientific.)

spectrometers. The resulting *hyphenated* instruments are powerful tools for identifying the components of complex mixture. An example of the use of mass spectroscopy combined with gas chromatography for the identification of constituents in blood is given in Feature 25-1.

Although a chromatogram may not lead to positive identification of the species in a sample, it often provides sure evidence of the *absence* of species. Thus, failure of a sample to produce a peak at the same retention time as a standard obtained under identical conditions is strong evidence that the compound in question is absent (or present at a concentration below the detection limit of the procedure).

Quantitative Analysis

Gas chromatography owes its enormous growth in part to its speed, simplicity, relatively low cost, and wide applicability to separations. It is doubtful, however, that GC would have become so widely used were it not able to provide quantitative information about separated species as well.

Quantitative GC is based on comparison of either the height or the area of an analyte peak with that of one or more standards. If conditions are properly controlled, both of these parameters vary linearly with concentration. Peak area is independent of the broadening effects discussed earlier. From this standpoint, therefore, area is a more satisfactory analytical parameter than peak height. On the other hand, peak heights are more easily measured and, for narrow peaks, more accurately determined. Most modern chromatographic instruments are equipped with computers that provide measurements of relative peak areas. If such equipment is not available, a manual estimate must be made. A simple method that works well for symmetric peaks of reasonable widths is to multiply peak height by the width at one-half peak height.

Calibration with Standards.   The most straightforward method for quantitative gas-chromatographic analyses involves the preparation of a series of standard solutions that approximate the composition of the unknown. Chromatograms for the standards are then obtained, and peak heights or areas are plotted as a function of concentration. A plot of the data should yield a straight line passing through the origin; quantitative analyses are based on this plot. Frequent standardization is necessary for highest accuracy.

The Internal-Standard Method.   The highest precision for quantitative GC is obtained using internal standards because the uncertainties introduced by sample injection, flow rate, and variations in column conditions are minimized. In this procedure, a carefully measured quantity of an internal standard is introduced into each standard and sample, and the ratio of analyte peak area (or height) to internal-standard peak area (or height) is used as the analytical parameter. For this method to be successful, it is necessary that the internal-standard peak be well separated from the peaks of all other components in the sample. It must, however, appear close to the analyte peak. With a suitable internal standard, precisions of 0.5 to 1% relative are attainable.

Feature 25-1

Use of Gas
Chromatography/Mass
Spectrometry to Identify a
Drug Metabolite in Blood[5]

A comatose patient was suspected of taking an overdose of a prescription drug glutethimide (Doriden™) because an empty prescription bottle had been found near where the patient was found. A gas chromatogram was obtained of a blood plasma extract, and two peaks were found as shown in Figure 25F-1a. The retention time for

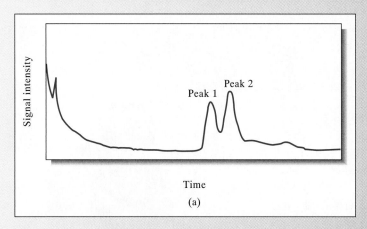

(a)

Figure 25F-1a   Gas chromatogram of a blood plasma extract from a drug overdose victim. Peak 1 was at the appropriate retention time to be glutethimide, but the compound responsible for peak 2 was unknown until GC/MS was done.

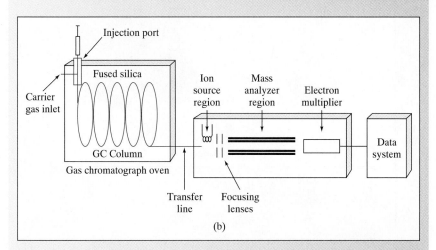

(b)

Figure 25F-1b   Schematic of a typical capillary gas chromatograph/mass spectrometer. The effluent from the gas chromatograph is passed into the inlet of a mass spectrometer, where the molecules in the gas are fragmented, ionized, and analyzed using one of a variety of different types of mass analyzers. For further details of mass spectrometry, see D. A. Skoog, F. J. Holler, and T. A. Nieman, *Principles of Instrumental Analysis,* 5th ed. (Philadelphia, Saunders College Publishing, 1998), pp. 718–720.

[5]From J. T. Watson, *Introduction to Mass Spectrometry,* 3rd ed. (New York: Lippincott-Raven, 1997), pp. 22–25.

peak 1 corresponded to the retention time of glutethimide, but the compound responsible for peak 2 was not known. The possibility that the patient had taken another drug was considered. The retention time for peak 2 under the conditions used, however, did not correspond to any other drug available to the patient or to a known drug of abuse. Hence, gas chromatography/mass spectrometry (GC/MS) was called on to establish the identity of peak 2 and confirm the identity of peak 1 before treating the patient. A schematic diagram of a typical gas chromatograph/mass spectrometer is shown in Figure 25F-1b.

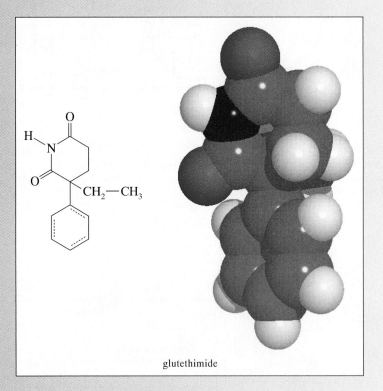

glutethimide

The plasma extract was subjected to GC/MS analysis, and the mass spectrum depicted at the top of Figure 25F-1c confirmed that peak 1 was due to glutethimide. A peak in the mass spectrum at a mass-to-charge ratio of 217 is the correct ratio for the glutethimide molecular ion, and the mass spectrum was identical to that from a known sample of glutethimide. The mass spectrum of peak 2, however, showed a molecular-ion peak at a mass-to-charge ratio of 233, as shown at the bottom of Figure 25F-1c. This peak differs from the molecular ion of glutethimide by 16 mass units. Several other peaks in the mass spectrum from GC peak 2 differed from those of glutethimide by 16 mass units indicating incorporation of oxygen into the glutethimide molecule. These differences led the scientists to believe that peak 2 was due to a 4-hydroxy metabolite of the parent drug.

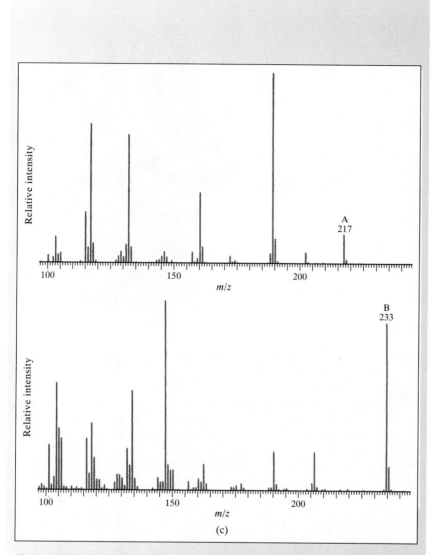

(c)

Figure 25F-1c   Top: Mass spectrum obtained during peak 1 of the gas chromatogram in Figure 25F-1a. This mass spectrum is identical with that of glutethimide. Bottom: Mass spectrum obtained during peak 2 of the gas chromatogram in Figure 25F-1a. The fragmentation of the two compounds produces ions that are separated in the mass spectrometer. Each peak in the mass spectra appears at a position (m/z) corresponding to the mass of the fragment. Peak A at $m/z = 217$ in the upper spectrum corresponds to the molar mass of glutethimide, and the mass spectrum is identical to a pure sample of the compound. The mass spectrum thus conclusively identifies the suspect compound as glutethimide. Peak B in the lower spectrum appears at $m/z = 233$, exactly 16 units more massive than glutethimide. This evidence suggests the presence of an extra oxygen atom in the molecule, which corresponds to the 4-hydroxy metabolite shown in the text. [From J. T. Watson, *Introduction to Mass Spectrometry*, 3rd ed. (Philadelphia: Lippincott-Raven, 1997), p. 24. With permission.]

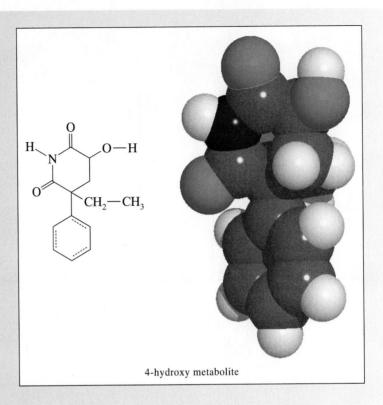

4-hydroxy metabolite

An acetic anhydride derivative of the peak 2 material was then prepared and found to be identical to the acetate derivative of 4-hydroxy-2-ethyl-2-phenylglutarimide, the metabolite previously shown. This metabolite was known to exhibit toxicity in animals. The patient was then subjected to hemodialysis, which removed the polar metabolite more rapidly than the less polar parent drug. Soon thereafter, the patient regained consciousness.

FOR THE CHEMIST ON THE GO:
LAPTOP GAS CHROMATOGRAPH/MASS SPECTROMETER

| Example 25-1 | Spreadsheet Exercise |

Gas chromatographic peaks can be influenced by a variety of instrumental factors. We can often compensate for variations in these factors by using the internal standard method. Here we add the same amount of internal standard to mixtures containing known amounts of the analyte and to the samples of unknown analyte concentration. We then ratio the peak height or area for the analyte to that of the internal standard. The internal standard should be absent in the sample to be analyzed.

The following data were obtained during the determination of a C_7 hydrocarbon with a closely related compound added to each standard and to the unknown as an internal standard.

Percent analyte	Peak height analyte	Peak height, internal standard
0.05	18.8	50.0
0.10	48.1	64.1
0.15	63.4	55.1
0.20	63.2	42.7
0.25	93.6	53.8
Unknown	58.9	49.4

Construct a spreadsheet to determine the peak height ratio of the analyte to internal standard and plot this versus the analyte concentration. Determine the concentration of the unknown and its standard deviation.

The spreadsheet is shown in Figure 25-11. The data are entered into columns A through C as shown. In cells D4 through D9, the peak height ratio is calculated by the formula shown in documentation cell A22. A plot of the calibration curve is also shown in Figure 25-11. The linear regression statistics are calculated in cells B11 through B20 using the same approach as described in Section 23A-2. The statistics and error analysis are calculated by the formulas in documentation cells A23 through A31. The percentage of the analyte in the unknown is found to be 0.16 ± 0.01.

25B   HIGH-PERFORMANCE LIQUID CHROMATOGRAPHY

High-performance liquid chromatography (HPLC) is the most versatile and widely used type of elution chromatography. The technique is used by chemists for separating and determining species in a variety of organic, inorganic, and biological materials. In liquid chromatography, the mobile phase is a liquid solvent containing the sample as a mixture of solutes. There are several types of high-performance liquid chromatography, depending on the nature of the stationary phase. They include (1) *partition*, or *liquid-liquid, chromatography;* (2) *adsorption*, or *liquid-solid, chromatography;* (3) *ion-exchange*, or *ion, chromatography;* (4) *size-exclusion chromatography;* and (5) *affinity chromatography.*

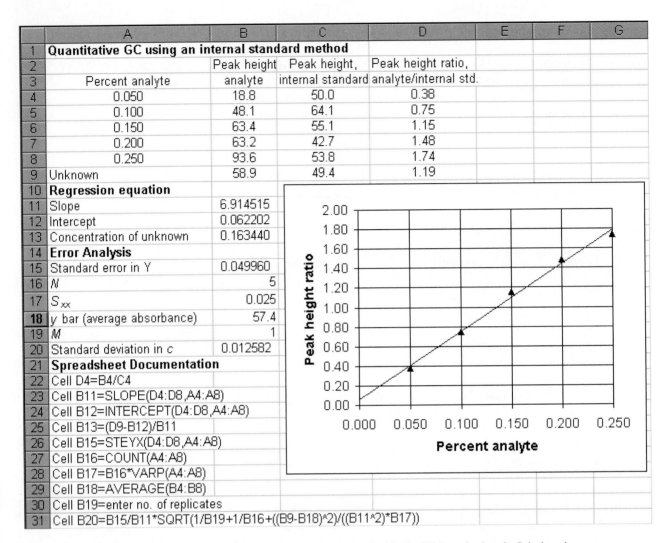

Figure 25-11 Spreadsheet to illustrate the internal standard method for the GC determination of a C_7 hydrocarbon.

Early liquid chromatography was performed in glass columns with inside diameters of 10 to 50 mm packed with particles having diameters of 150 to 200 μm. It was soon realized, however, that large decreases in plate heights would be realized if the particle size of packings were reduced. This effect is shown in the data plotted in Figure 25-12. It is of interest to note that in none of these plots is the minimum shown in Figure 24-11a (page 657) reached. The reason for this difference is that diffusion in liquids is much slower than in gases; consequently, its effect on plate heights is observed only at extremely low flow rates.

Not until the late 1960s, was the technology for producing and using packings with particle diameters as small as 3 to 10 μm developed. This technology required sophisticated instruments that contrasted markedly with the simple devices that preceded them. The name *high-performance liquid chromatography* is

High-performance liquid chromatography (HPLC) is a type of chromatography that employs a liquid mobile phase and a very finely divided stationary phase. To obtain satisfactory flow rates, the liquid must be pressurized to several hundred or more pounds per square inch.

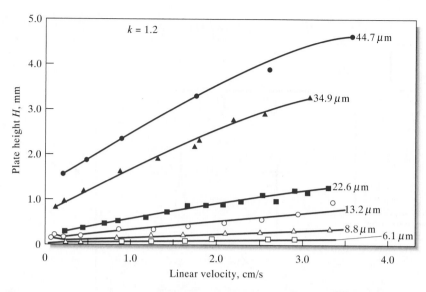

Figure 25-12   Effect of particle size of packing and flow rate on plate height in liquid chromatography. (From R. E. Majors, *J. Chromatogr. Sci.*, **1973**, *11*, 92, with permission.)

often employed to distinguish these newer procedures from their predecessors, which still find considerable use for preparative purposes.[6]

25B-1   Instruments for High-Performance Liquid Chromatography

Pumping pressures of several hundred atmospheres are required to achieve reasonable flow rates with packings in the 3- to 10-μm size range, which are common in modern liquid chromatography. As a consequence of these high pressures, the equipment for high-performance liquid chromatography tends to be considerably more elaborate and expensive than that encountered in other types of chromatography. Figure 25-13 shows the important components of a typical high-performance chromatographic instrument.

Mobile-Phase Reservoirs

A modern HPLC apparatus is equipped with one or more glass or stainless steel reservoirs, each of which contains 500 mL or more of a solvent. Provisions are often included to remove dissolved gases and dust from the liquids.

An elution with a single solvent of constant composition is called *isocratic*. In *gradient elution,* two (and sometimes more) solvent systems that differ significantly in polarity are employed. The ratio of the two solvents is varied in a preprogrammed way, sometimes continuously and sometimes in a series of steps.

An *isocratic elution* in HPLC is one in which the composition of the solvent remains constant.

A *gradient elution* in HPLC is one in which the composition of the solvent is changed continuously or in a series of steps.

[6]For a detailed discussion of HPLC systems, see N. A. Parris, *Instrumental Liquid Chromatography,* 2nd ed. (New York: Elsevier, 1984); H. Engelhardt, Ed., *Practice of High Performance Liquid Chromatography: Applications, Equipment, and Quantitative Analysis* (New York: Springer-Verlag, 1986), Chapter 1; A. Katritzky and R. J. Offerman, *Crit. Rev. Anal. Chem.*, **1989**, *21* (4), 83.

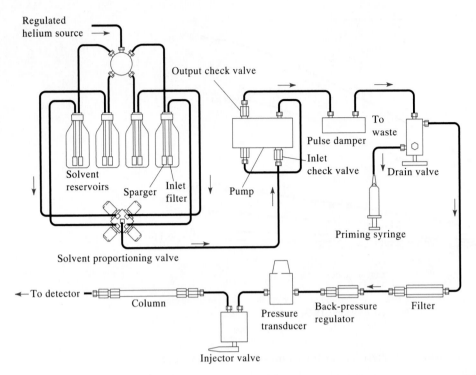

Figure 25-13   Block diagram showing components of a typical apparatus for HPLC. (Courtesy of Perkin-Elmer Corp., Norwalk, CT.)

As shown in Figure 25-14, gradient elution frequently improves separation efficiency, just as temperature programming helps in gas chromatography. Modern HPLC instruments are often equipped with proportioning valves that introduce liquids from two or more reservoirs at rates that vary continuously.

Pumping Systems

The requirements for liquid-chromatographic pumps include (1) the generation of pressures of up to 6000 psi (lb/in.^{2}), (2) pulse-free output, (3) flow rates ranging from 0.1 to 10 mL/min, (4) flow reproducibilities of 0.5% relative or better, and (5) resistance to corrosion by a variety of solvents.

Three types of pumps are employed: a screw-driven syringe type, a reciprocating pump, and a pneumatic or constant-pressure pump. Reciprocating pumps, which are the most widely used, consist of a small cylindrical chamber that is filled and then emptied by the back-and-forth motion of a piston. The pumping motion produces a pulsed flow that must be subsequently damped. Advantages of reciprocating pumps include small internal volume, high output pressure (up to 10,000 psi), ready adaptability to gradient elution, and constant flow rates, which are largely independent of column back-pressure and solvent viscosity. Most modern commercial chromatographs employ a reciprocating pump.

The high pressures generated by liquid-chromatographic pumps do not constitute an explosion hazard because liquids are not very compressible. Thus, rupture of a component results only in solvent leakage. To be sure, such leakage may constitute a fire or environmental hazard with some solvents.

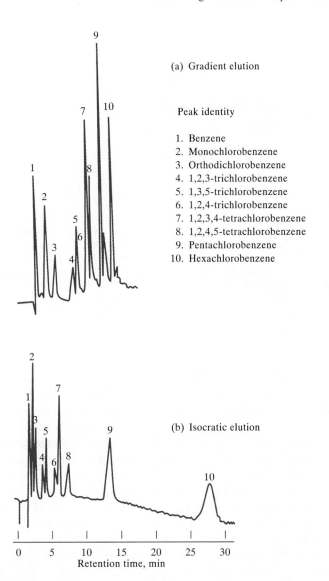

(a) Gradient elution

Peak identity

1. Benzene
2. Monochlorobenzene
3. Orthodichlorobenzene
4. 1,2,3-trichlorobenzene
5. 1,3,5-trichlorobenzene
6. 1,2,4-trichlorobenzene
7. 1,2,3,4-tetrachlorobenzene
8. 1,2,4,5-tetrachlorobenzene
9. Pentachlorobenzene
10. Hexachlorobenzene

(b) Isocratic elution

Retention time, min

Figure 25-14 Improvement in separation efficiency by gradient elution. [From J. J. Kirkland, Ed., *Modern Practice of Liquid Chromatography* (New York: Interscience, 1971), p. 88. Reprinted by permission of John Wiley & Sons, Inc.]

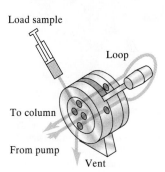

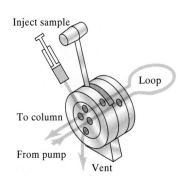

Sample Injection System

Although syringe injection through an elastomeric septum is often used in liquid chromatography, this procedure is not very reproducible and is limited to pressures less than about 1500 psi. The most widely used method of sample introduction in liquid chromatography is based on sampling loops such as that shown in Figure 25-15. These devices are often an integral part of modern liquid-chromatography equipment and have interchangeable loops that provide a choice of sample sizes ranging from 5 to 500 μL. The reproducibility of injections with a typical sampling loop is a few tenths of a percent relative.

Figure 25-15 A sampling loop for liquid chromatography. (Courtesy of Beckman Instruments, Fullerton, CA.)

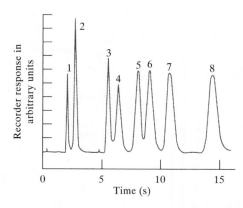

Figure 25-16   High-speed isocratic separation. Column dimensions: 4-cm length, 0.4-cm inside diameter; packing: 3-μm sperisorb; Mobile phase: 4.1% ethyl acetate in n-hexane. Compounds: (1) p-xylene, (2) anisole, (3) benzyl acetate, (4) dioctyl phthalate, (5) dipentyl phthalate, (6) dibutyl phthalate, (7) dipropyl phthalate, (8) diethyl phthalate. [From R. P. W. Scott, *Small Bore Liquid Chromatography Columns: Their Properties and Uses* (New York: Wiley, 1984), p. 156. Reprinted with permission of John Wiley & Sons, Inc.]

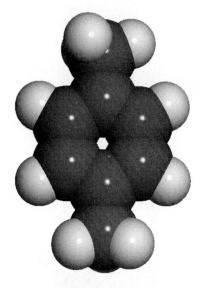

Molecular model of p-xylene. There are three xylene isomers, ortho, meta and para. Para-xylene is used for the production of artificial fibers. Xylol is a mixture of the three isomers and is used as a solvent.

Columns for High-Performance Liquid Chromatography

Liquid-chromatographic columns are usually constructed from stainless steel tubing, although glass or Tygon tubing is sometimes employed for lower pressure applications ($<$ 600 psi). Most columns range in length from 10 to 30 cm and have inside diameters of 4 to 10 mm. Column packings typically have particle sizes of 5 or 10 μm. Columns of this type often contain 40,000 to 60,000 plates/m. Recently, high-performance microcolumns with inside diameters of 1 to 4.6 mm and lengths of 3 to 7.5 cm have become available. These columns, which are packed with 3- or 5-μm particles, contain as many as 100,000 plates/m and have the advantages of speed and minimal solvent consumption. The latter property is of considerable importance because the high-purity solvents required for liquid chromatography are expensive to purchase and to dispose of after use. Figure 25-16 illustrates the speed with which a separation can be performed on this type of column. Here, eight components of diverse type are separated in about 15 s. The column is 4 cm in length and has an inside diameter of 4 mm; it is packed with 3-μm particles.

Column Thermostats

For many applications, close control of column temperature is not necessary and columns are operated at room temperature. Often, however, better chromatograms are obtained by maintaining column temperatures constant to a few tenths of a degree Celsius. Most modern commercial instruments are now equipped with heaters that control column temperatures to a few tenths of a degree from near ambient to 150°C. Columns may also be fitted with water jackets fed from a constant-temperature bath to give precise temperature control.

Table 25-3

Performances of HPLC Detectors

HPLC Detector	Commercially Available	Mass LOD (commercial detectors)*	Mass LOD (state of the art)†
Absorbance	Yes‡	100 pg–1 ng	1 pg
Fluorescence	Yes‡	1–10 pg	10 fg
Electrochemical	Yes‡	10 pg–1 ng	100 fg
Refractive index	Yes	100 ng–1 μg	10 ng
Conductivity	Yes	500 pg–1 ng	500 ng
Mass spectrometry	Yes	100 pg–1 ng	1 pg
FTIR	Yes	1 μg	100 ng
Light scattering	Yes	10 μg	500 ng

Source: From E. S. Yeung and R. E. Synovec, *Anal. Chem.,* **1986,** *58,* 1238. With permission. Copyright American Chemical Society.

*Mass LOD (limit of detection) is calculated for injected mass that yields a signal equal to five times the standard deviation of the noise, using a molar mass of 200 g/mol, 10 μL injected for conventional or 1 μL injected for microbore HPLC.

†Same definition as above, but the injected volume is generally smaller.

‡Commercially available for microbore HPLC also.

Detectors

No highly sensitive, universal detector system, such as those for gas chromatography, is available for HPLC. Thus, the detector used will depend on the nature of the sample. Table 25-3 lists some of the common detectors and their properties.

The most widely used detectors for liquid chromatography are based on absorption of ultraviolet or visible radiation (see Figure 25-17). Both photometers and spectrophotometers, specifically designed for use with chromatographic columns, are available from commercial sources. The former often make use of the 254- and 280-nm lines from a mercury source because many organic functional groups absorb in this region. Inexpensive instruments can use deuterium or tungsten filament sources in combination with interference filters to provide a simple means of detecting absorbing species. Spectrophotometric detectors are considerably more versatile than photometers and are widely used in high-performance instruments. Modern instruments use diode-array instruments that can display an entire spectrum as an analyte exits the column. The combination of HPLC with a mass spectrometry detector is currently receiving a great deal of attention. Such HPLC/MS systems can identify the analytes exiting from the HPLC column.

Another detector, that was used in early HPLC instruments, is based on the changes in the refractive index of the solvent that is caused by analyte molecules. In contrast to most of the other detectors listed in Table 25-3, the refractive index detector is general rather than selective and responds to the presence of all solutes. It is not widely used in current instruments because of its low sensitivity. Several electrochemical detectors have also been introduced that are based on potentiometric, conductometric, and voltammetric measurements. An example of the last detector is shown in Figure 25-18.

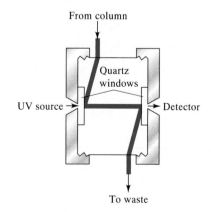

Figure 25-17   A UV detector for HPLC.

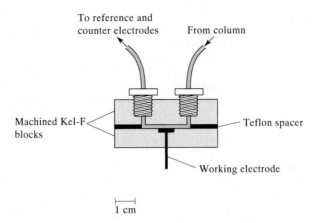

Figure 25-18   Amperometric thin-layer detector cell for HPLC.

25B-2   High-Performance Partition Chromatography

The most widely used type of HPLC is partition chromatography, in which the stationary phase is a second liquid that is immiscible with the liquid mobile phase. This liquid is held in place on the surface of a finely divided solid by either adsorption or by chemical bonding.

Column Packing

The most common packing for partition chromatography is prepared from silica particles, which are synthesized by agglomerating submicron silica particles under conditions that lead to larger particles with highly uniform diameters. The resulting particles are coated with thin organic films, which are chemically or physically bonded to the surface.

Two types of partition chromatography are distinguishable based on the relative polarities of the mobile and stationary phases. Early work in liquid chromatography was based on highly polar stationary phases such as triethylene glycol or water; a relatively nonpolar solvent such as hexane or *i*-propyl ether then served as the mobile phase. For historic reasons, this type of chromatography is now called *normal-phase chromatography*. In *reversed-phase chromatography*, the stationary phase is nonpolar, often a hydrocarbon, and the mobile phase is a relatively polar solvent (such as water, methanol, or acetonitrile).[7] In normal-phase chromatography, the *least* polar component is eluted first; *increasing* the polarity of the mobile phase then *decreases* the elution time. In contrast, in the reversed-phase method, the *most* polar component elutes first, and *increasing* the mobile phase polarity *increases* the elution time.

It has been estimated that more than three-quarters of all HPLC separations are currently performed with reversed-phase, bonded, octyl-, or octyldecyl siloxane packings. With such preparations, the long-chain hydrocarbon groups are aligned parallel to one another and perpendicular to the surface of the particle, giving a brushlike, nonpolar, hydrocarbon surface. The mobile phase used with

In *normal-phase partition chromatography,* the stationary phase is polar and the mobile phase nonpolar. In *reversed-phase partition chromatography,* the polarity of these phases is reversed.

In normal-phase chromatography, the least polar analyte is eluted first. In reversed-phase chromatography, the least polar analyte is eluted last.

---

[7]For a detailed discussion of reversed-phase HPLC, see A. M. Krstulovic and P. R. Brown, *Reversed-Phase High-Performance Liquid Chromatography* (New York: Wiley, 1982).

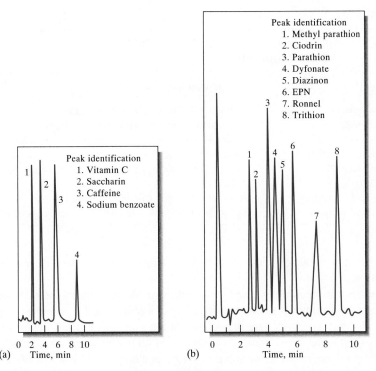

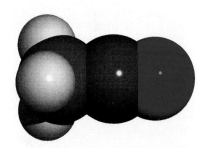

Figure 25-19 Typical applications of bonded-phase chromatography. (a) Soft-drink additives. Column: 4.6 × 250 mm paced with polar (nitrile) bonded-phase packings. Isocratic solvent: 6% HOAC/94% H_2O. Flow rate: 1.0 cm³/min. (Courtesy of BTR Separations, a DuPont/ConAgra affiliate.) (b) Organophosphate insecticides. Column: 4.5 × 250 mm packed with 5-μm, C_8, bonded-phase particles. Gradient: 67% CH_3OH/33% H_2O to 80% CH_3/20% H_2O. Flow rate: 2 mL/min. (Courtesy of IBM Instruments, Inc., Danbury, CT.) Both used 254-nm UV detectors.

these packings is often an aqueous solution containing various concentrations of such solvents as methanol, acetonitrile, or tetrahydrofuran.

Applications of High-Performance Partition Chromatography

Figure 25-19 illustrates typical applications of partition chromatography. Table 25-4 further illustrates the variety of samples to which the technique is applicable.

25B-3   High-Performance Adsorption Chromatography

All the pioneering work in chromatography was based on adsorption of analyte species on a solid surface. Here, the stationary phase is the surface of a finely divided polar solid. With such a packing, the analyte competes with the mobile phase for sites on the surface of the packing, and retention is the result of adsorption forces.

Finely divided silica and alumina are the only stationary phases that find extensive use for adsorption chromatography. Silica is preferred for most (but not all) applications because of its higher sample capacity and its wider range of useful forms. The adsorption characteristics of the two substances parallel one

Molecular model of acetonitrile. Acetonitrile ($CH_3C\equiv N$) is a widely used organic solvent. Its use as an LC mobile phase stems from its being more polar than ethanol and methanol, but less polar than water.

In *adsorption chromatography,* analyte species are adsorbed onto the surface of a polar packing.

Table 25-4

Typical Applications of High-Performance
Partition Chromatography

Field	Typical Mixtures
Pharmaceuticals	Antibiotics, sedatives, steroids, analgesics
Biochemicals	Amino acids, proteins, carbohydrates, lipids
Food products	Artificial sweeteners, antioxidants, aflatoxins, additives
Industrial chemicals	Condensed aromatics, surfactants, propellants, dyes
Pollutants	Pesticides, herbicides, phenols, polychlorinated biphenyls (PCBs)
Forensic chemistry	Drugs, poisons, blood alcohol, narcotics
Clinical medicine	Bile acids, drug metabolites, urine extracts, estrogens

In *adsorption chromatography*, the mobile phase is usually an organic solvent or a mixture of organic solvents; the stationary phase is finely divided particles of silica or alumina.

another. For both, retention times become longer as the polarity of the analyte increases.

In adsorption chromatography, the only variable that affects the partition coefficient of analytes is the composition of the mobile phase (in contrast to partition chromatography, where the polarity of the stationary phase can also be varied). Fortunately, enormous variations in retention and thus resolution accompany variations in the solvent system, and only rarely is a suitable mobile phase not available.

Currently, liquid-solid HPLC is used extensively for the separations of relatively nonpolar, water-insoluble organic compounds with molecular weights that are less than about 5000. A particular strength of adsorption chromatography not shared by other methods is its ability to resolve isomeric mixtures such as meta- and para-substituted benzene derivatives.

25B-4   High-Performance Ion-Exchange Chromatography

In Section 24E, we described some of the applications of ion-exchange resins to analytical separations. In addition, these materials are useful as stationary phases for liquid chromatography, where they are used to separate charged species.[8] In most cases, conductivity measurements are used for detecting eluents.

Two types of ion chromatography are currently in use: *suppressor based* and *single column*. They differ in the method used to prevent the conductivity of the eluting electrolyte from interfering with the measurement of analyte conductivities.

Ion Chromatography Based on Suppressors

Conductivity detectors have many of the properties of the ideal detector. They can be highly sensitive, they are universal for charged species, and as a general rule, they respond in a predictable way to concentration changes. Furthermore,

---

[8]For a brief review of ion chromatography, see J. S. Fritz, *Anal. Chem.*, **1987,** *59,* 335A. For a detailed description of the method, see H. Small, *Ion Chromatography* (New York: Plenum Press, 1989); D. T. Gjerde and J. S. Fritz, *Ion Chromatography,* 2nd ed. (New York: A. Heuthig, 1987).

such detectors are simple to operate, inexpensive to construct and maintain, easy to miniaturize, and ordinarily give prolonged, trouble-free service. The only limitation to the use of conductivity detectors, which delayed their general application to ion chromatography until the mid-1970s, was due to the high electrolyte concentrations required to elute most analyte ions in a reasonable time. The conductivity from the mobile-phase components tends to swamp that from the analyte ions, thus greatly reducing the detector sensitivity.

In 1975, the problem created by the high conductance of eluents was solved by the introduction of an *eluent suppressor column* immediately following the ion-exchange column.[9] The suppressor column is packed with a second ion-exchange resin that effectively converts the ions of the eluting solvent to a molecular species of limited ionization without affecting the conductivity due to analyte ions. For example, when cations are being separated and determined, hydrochloric acid is chosen as the eluting reagent, and the suppressor column is an anion-exchange resin in the hydroxide form. The product of the reaction in the suppressor is water. That is,

$$H^+(aq) + Cl^-(aq) + resin^+OH^-(s) \rightarrow resin^+Cl^-(s) + H_2O$$

The analyte cations are not retained by this second column.

For anion separations, the suppressor packing is the acid form of a cation-exchange resin, and sodium bicarbonate or carbonate is the eluting agent. The reaction in the suppressor is

$$Na^+(aq) + HCO_3^-(aq) + resin^-H^+(s) \rightarrow resin^-Na^+(s) + H_2CO_3(aq)$$

The largely undissociated carbonic acid does not contribute significantly to the conductivity.

Single-Column Ion Chromatography

Recently, equipment has become available commercially for ion chromatography in which no suppressor column is used. This approach depends on the small differences in conductivity between sample ions and the prevailing eluent ions. To amplify these differences, low-capacity exchangers are used to permit elution with solutions with low electrolyte concentrations. Furthermore, eluents with low conductivity are chosen.[10]

Single-column ion chromatography offers the advantage of not requiring special equipment for suppression. It is, however, a somewhat less sensitive method for determining anions than suppressor column methods.

Applications

Figures 25-20 and 25-21 illustrate two applications of ion chromatography based on a suppressor column and conductometric detection. In each instance, the ions were present in the parts-per-million range. The method is particularly important for anion analysis.

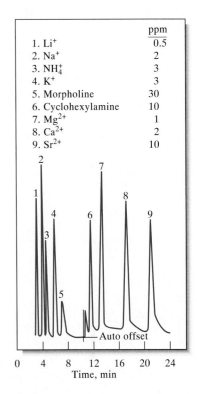

		ppm
1. Li^+		0.5
2. Na^+		2
3. NH_4^+		3
4. K^+		3
5. Morpholine		30
6. Cyclohexylamine		10
7. Mg^{2+}		1
8. Ca^{2+}		2
9. Sr^{2+}		10

Figure 25-20 Ion chromatogram of a mixture of cations. (Courtesy of Dionex, Sunnyvale, CA.)

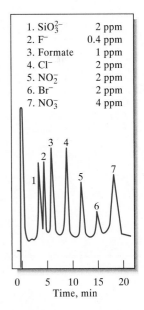

1. SiO_3^{2-}	2 ppm
2. F^-	0.4 ppm
3. Formate	1 ppm
4. Cl^-	2 ppm
5. NO_2^-	2 ppm
6. Br^-	2 ppm
7. NO_3^-	4 ppm

Figure 25-21 Ion chromatogram of a mixture of anions. (Courtesy of Dionex, Sunnyvale, CA.)

[9]H. Small, T. S. Stevens, and W. C. Bauman, *Anal. Chem.*, **1975**, *47*, 1801.

[10]See R. M. Becker, *Anal Chem.*, **1980**, *52*, 1510; J. R. Benson, *Amer. Lab.*, **1985**, *17*(6), 30; T. Jupille, *Amer. Lab.*, **1986**, *18*(5), 114.

25B-5   High-Performance Size-Exclusion Chromatography

Size-exclusion, or gel, chromatography is the newest of the liquid-chromatographic procedures. It is a powerful technique that is particularly applicable to high-molecular-weight species.[11]

Packings

Packings for size-exclusion chromatography consist of small ($\sim$10 μm) silica or polymer particles containing a network of uniform pores into which solute and solvent molecules can diffuse. While in the pores, molecules are effectively trapped and removed from the flow of the mobile phase. The average residence time of analyte molecules depends on their effective size. Molecules that are significantly larger than the average pore size of the packing are excluded and thus suffer no retention; that is, they travel through the column at the rate of the mobile phase. Molecules that are appreciably smaller than the pores can penetrate throughout the pore maze and are thus entrapped for the greatest time; they are last to be eluted. Between these two extremes are intermediate-size molecules whose average penetration into the pores of the packing depends on their diameters. The fractionation that occurs within this group is directly related to molecular size and, to some extent, molecular shape. Note that size-exclusion separations differ from the other chromatographic procedures in the respect that no chemical or physical interactions between analytes and the stationary phase are involved. Indeed, every effort is made to avoid such interactions because they lead to impaired column efficiencies.

Numerous size-exclusion packings are on the market. Some are hydrophilic for use with aqueous mobile phases; others are hydrophobic and are used with nonpolar organic solvents. Chromatography based on the hydrophilic packings is sometimes called *gel filtration,* whereas techniques based on hydrophobic packings are called *gel permeation.* With both types of packings, many pore diameters are available. Ordinarily, a given packing will accommodate a 2- to 2.5-decade range of molecular weight. The average molecular weight suitable for a given packing may be as small as a few hundred or as large as several million.

Applications

Figures 25-22 and 25-23 illustrate typical applications of size-exclusion chromatography. In Figure 25-22, a hydrophilic packing was used to exclude molecular weights greater than 1000. The chromatogram in this figure was obtained with a hydrophobic packing in which the eluent was tetrahydrofuran. The sample was a commercial epoxy resin in which each monomer unit had a molecular weight of 280 (n = number of monomer units).

Another important application of size-exclusion chromatography is the rapid determination of the molecular weight or the molecular weight distribution of

---

[11]For monographs on this subject, see B. J. Hunt and S. R. Holding, Eds., *Size Exclusion Chromatography* (New York: Chapman and Hall, 1988); P. L. Dubin, Ed., *Aqueous Size-Exclusion Chromatography* (Amsterdam: Elsevier, 1988); C. S. Wu, Ed., *Handbook of Size Exclusion Chromatography* (New York: Marcel Dekker, 1995).

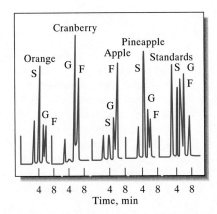

Figure 25-22   Gel-filtration chromatogram for glucose (G), fructose (F), and sucrose (S) in canned juices. (Courtesy of BTR Separations, a DuPont/ConAgra affiliate.)

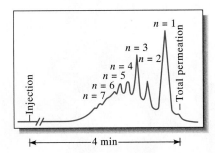

Figure 25-23   Gel-permeation separation of components in an epoxy resin. (Courtesy of BTR Separations, a DuPont/ConAgra affiliate.)

large polymers or natural products. Here, the elution volumes of the sample are compared with elution volumes for a series of standard compounds that have the same chemical characteristics.

25C   HIGH-PERFORMANCE LIQUID-CHROMATOGRAPHY VERSUS GAS-LIQUID CHROMATOGRAPHY

Table 25-5 provides a comparison between high-performance liquid chromatography and gas-liquid chromatography. When either is applicable, gas-liquid chromatography offers the advantage of speed and simplicity of equipment. On the other hand, high-performance liquid chromatography is applicable to non-volatile substances (including inorganic ions) and thermally unstable materials, whereas gas-liquid chromatography is not. Often the two methods are complementary.

Feature 25-2

Buckyballs: The
Chromatographic Separation
of Fullerenes

Our ideas about the nature of matter are often profoundly influenced by chance discoveries. No event in recent memory has captured the imagination of both the scientific community and the public as did the serendipitous discovery in 1985 of the soccerball-shaped molecule C_{60}. This molecule, illustrated in Figure 25F-2a, its cousin C_{70}, and other similar molecules discovered since 1985 are called *fullerenes*, or more commonly, *buckyballs*.[12] The compounds are named in honor of the famous architect R. Buckminster Fuller (1895–1983), who designed many geodesic dome buildings having the same hexagonal/pentagonal structure as buckyballs. Since their discovery, thousands of research groups throughout the world have studied various chemical and physical properties of these highly stable molecules. They represent a third allotropic form of carbon besides graphite and diamond.

The preparation of buckyballs is almost trivial. When an ac arc is established between two carbon electrodes in a flowing helium atmosphere, the soot that is collected is rich in C_{60} and C_{70}. Although the preparation is easy, the separation and purification of more than a few milligrams of C_{60} proved tedious and expensive. Recently, it was

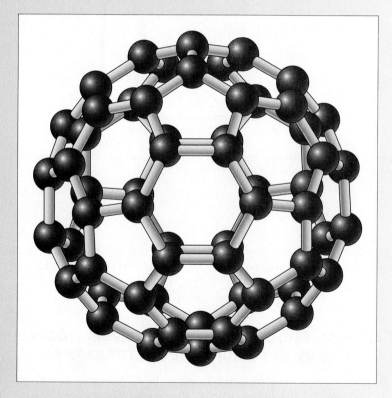

Figure 25F-2a   Buckminster fullerene, C_{60}.

[12]R. F. Curl and R. E. Smalley, *Scientific American,* **1991,** *265* (4), 54.

found that relatively large quantities of buckyballs can be separated using size-exclusion chromatography.[13] Fullerenes are extracted from soot prepared as mentioned earlier and injected on a 19 mm × 500 Å Ultratyragel column (Waters Chromatography Division of Millipore Corporation), using toluene as the mobile phase and UV/visible detection following separation. A typical chromatogram is shown in Figure 25F-2b. The peaks in the chromatogram are labeled with their identities and retention times.

Note that C_{60} elutes before C_{70} and the higher fullerenes. This situation is contrary to what we expect; the smallest molecule, C_{60}, should be retained more strongly than C_{70} and the higher fullerenes. It has been suggested that the interaction between the solute molecules and the gel is on the surface of the gel rather than in its pores. Since C_{70} and the higher fullerenes have larger surface areas than C_{60}, the higher fullerenes are retained more strongly on the surface of the gel and are thus eluted after C_{60}. Using automated apparatus, this method of separation may be used to prepare several grams of 99.8% pure C_{60} from 5 to 19 g of C_{60} to C_{70} mixture in a 24-hr period. These quantities of C_{60} can then be used to prepare and study the chemistry and physics of derivatives of this interesting and unusual form of carbon.

As an example, C_{60} forms stoichiometric compounds with the alkali metals that have the general formula M_3C_{60}, where M is potassium, rubidium, or cesium. Researchers have found that these compounds are superconductors at temperatures below 30 K. In the future, these or similar compounds may help lead the way to practical high-temperature superconductors, which would revolutionize the electrical, electronics, and telecommunications industries and would conserve vast amounts of energy.

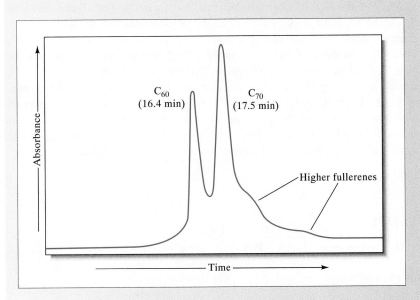

Figure 25F-2b   Separation of fullerenes.

---

[13]M. S. Meier and J. P. Selegue, *J. Org. Chem.,* **1992,** *57,* 1924; A. Gugel and K. Mullen, *J. Chromatogr.,* **1993,** *628,* 23.

Table 25-5

Comparison of High-Performance Liquid Chromatography and Gas-Liquid Chromatography

Characteristics of both methods
- Efficient, highly selective, widely applicable
- Only small sample required
- May be nondestructive of sample
- Readily adapted to quantitative analysis

Advantages of HPLC
- Can accommodate nonvolatile and thermally unstable samples
- Generally applicable to inorganic ions

Advantages of GC
- Simple and inexpensive equipment
- Rapid
- Unparalleled resolution (with capillary columns)
- Easily interfaced with mass spectroscopy

WEB WORKS

Use your Web browser to connect to the LC-GC Online Web site at http://www.lcgcmag.com/. *LC-GC* is a free magazine that contains interesting and timely articles for chromatographers and other users of chromatography equipment. From the LC-GC Online home page, browse to the Tips, Tricks, and Other Resources page and click on the link to the GC Troubleshooting Guide. Read the general guidelines for diagnosing and solving problems with gas chromatography. Then, click on the links to specific problems and read the solutions that you find. Now, find the page that provides links to the tables of contents of LC-GC, and locate the table of contents of the December 1998 issue. Find and read the summary of the article by M. W. Dong on HPLC determination of organic acids in juice and wine. Do you notice anything unusual about the title of the article? Finally, browse around the site to get an idea of the wealth of useful information that you can find there. You may wish to create a bookmark for the LC-GC Online site.

25D   QUESTIONS AND PROBLEMS

*25-1.   What are the effects of slow sample injection on a gas chromatogram?

 25-2.   What is meant by temperature programming in gas chromatography?

*25-3.   What is a chromatogram?

 25-4.   What variables must be controlled if satisfactory quantitative data are to be obtained from chromatograms?

*25-5.   Describe the physical differences between open tubular and packed columns. What are the advantages and disadvantages of each?

25-6.   Define the following terms used in HPLC:
  (a)   gradient elution
  (b)   isocratic elution
  (c)   reversed-phase packing
  (d)   sampling loops
  (e)   bonded-phase packing
  (f)   gel filtration

25-7.   Indicate the order in which the following compounds would be eluted from an HPLC column containing a reversed-phase packing:
  *(a)   benzene, diethyl ether, *n*-hexane
  (b)   acetone, dichloroethane, acetamide

25-8. Indicate the order of elution of the following compounds from a normal-phase packed HPLC column:

  *(a) ethyl acetate, acetic acid, dimethylamine

  (b) propylene, hexane, benzene, dichlorobenzene

*25-9. Describe the fundamental difference between adsorption and partition chromatography.

25-10. Describe the fundamental difference between ion-exchange and size-exclusion chromatography.

*25-11. Describe the difference between gel-filtration and gel-permeation chromatography.

25-12. What types of species can be separated by HPLC but not by GLC?

*25-13. One method for quantitative determination of the concentration of constituents in a sample analyzed by gas chromatography is the area normalization method. Here, complete elution of all the sample constituents is necessary. The area of each peak is then measured and corrected for differences in detector response to the different eluates. This correction involves dividing the area by an empirically determined correction factor. The concentration of the analyte is found from the ratio of its corrected area to the total corrected area of all peaks. For a chromatogram containing three peaks, the relative areas were found to be 16.4, 45.2, and 30.2 in the order of increasing retention time. Calculate the percentage of each compound if the relative detector responses were 0.60, 0.78, and 0.88, respectively.

25-14. Peak areas and relative detector responses are to be used to determine the concentrations of the five species in a sample. The area-normalization method described in Problem 25-13 is to be used. The relative areas for the five gas chromatographic peaks are given below. Also shown are the relative responses of the detector. Calculate the percentage of each component in the mixture.

Compound	Relative Peak Area	Relative Detector Response
A	32.5	0.70
B	20.7	0.72
C	60.1	0.75
D	30.2	0.73
E	18.3	0.78

25-15. For the data given in Spreadsheet Example 25-1, compare the method of external standards to the internal standard method. Plot the analyte peak height versus percent analyte and determine the unknown without using the internal standard results. Are your results any more precise using the internal standard method?

*25-16. Describe the various kinds of pumps used in high-performance liquid chromatography. What are the advantages and disadvantages of each?

25-17. Describe the differences between single-column and suppressor-column ion chromatography.

Chapter 26

Supercritical-Fluid Chromatography, Capillary Electrophoresis, and Capillary Electrochromatography

I n this chapter, we discuss three relatively new methods for performing analytical separations, namely: supercritical-fluid chromatography, capillary electrophoresis, and capillary electrochromatography.

26A SUPERCRITICAL-FLUID CHROMATOGRAPHY

Supercritical-fluid chromatography (SFC), in which the mobile phase is a supercritical fluid, is a hybrid of gas and liquid chromatography that combines some of the best features of each. For certain applications, it appears to be clearly superior to both gas-liquid and high-performance liquid chromatography.[1]

26A-1 Important Properties of Supercritical Fluids

A *supercritical fluid* is the physical state of a substance when it is held above its critical temperature.

The *critical temperature* is the temperature above which a distinct liquid phase for a substance cannot exist.

The density of a supercritical fluid is 200 to 400 times that of its gaseous state, and it is nearly as dense as its liquid state.

A *supercritical fluid* is formed whenever a substance is heated above its *critical temperature*. At the critical temperature, a substance can no longer be condensed into its liquid state through the application of pressure. For example, carbon dioxide becomes a supercritical fluid at temperatures above 31°C. In this state, the molecules of carbon dioxide act independently of one another just as they do in a gas.

As shown by the data in Table 26-1, the physical properties of a substance in the supercritical-fluid state can be remarkably different from the same properties in either the liquid or the gaseous state. For example, the density of a supercritical fluid is typically 200 to 400 times greater than that of the corresponding gas and approaches that of the substance in its liquid state. The properties compared

---

[1]T. L. Chester, J. D. Pinkston, and D. B. Raynie, *Anal. Chem.,* **1998,** *70,* 301R; M. D. Palmieri, *J. Chem. Educ.,* **1988,** *65,* A254; M. D. Palmieri, *J. Chem. Educ.,* **1989,** *66,* A141; P. R. Griffiths, *Anal. Chem.,* **1988,** *60,* 593A; R. D. Smith, B. W. Wright, and C. R. Yonker, *Anal. Chem.,* **1988,** *60,* 1323A.

Table 26-1

Comparison of Properties of Supercritical Fluids with Liquids and Gases

Property	Gas (STP)	Supercritical Fluid	Liquid
Density (g/cm³)	$(0.6-2) \times 10^{-3}$	$0.2-0.5$	$0.6-2$
Diffusion coefficient (cm²/s)	$(1-4) \times 10^{-1}$	$10^{-3}-10^{-4}$	$(0.2-2) \times 10^{-5}$
Viscosity (g cm^{-1} s^{-1})	$(1-3) \times 10^{-4}$	$(1-3) \times 10^{-4}$	$(0.2-3) \times 10^{-2}$

Note: All the data are order-of-magnitude only.

in Table 26-1 are those that are important in gas, liquid, and supercritical-fluid chromatography.

An important property of supercritical fluids related to their high densities (0.2 to 0.5 g/cm³) is their ability to dissolve large nonvolatile molecules. For example, supercritical carbon dioxide readily dissolves n-alkanes containing from 5 to 22 carbon atoms, di-n-alkylphthalates in which the alkyl groups contain 4 to 16 carbon atoms, and various polycyclic aromatic hydrocarbons consisting of several rings.[2]

Critical temperatures for fluids used in chromatography vary widely, from about 30°C to above 200°C. Lower critical temperatures are advantageous in chromatography from several standpoints. For this reason, much of the work to date has focused on such supercritical fluids as shown in Table 26-2. Note that these temperatures, and the pressures at these temperatures, are well within the operating conditions of ordinary high-performance liquid chromatography.

Supercritical fluids tend to dissolve large, nonvolatile molecules.

26A-2   Instrumentation and Operating Variables

Instruments for supercritical-fluid chromatography are similar in design to high-performance liquid chromatographs except that provision is made in the former for controlling and measuring the column pressure. Several manu-

Table 26-2

Properties of Some Supercritical Fluids

Fluid	Critical Temperature, °C	Critical Pressure, atm	Critical Point Density, g/mL	Density at 400 atm, g/mL
CO_2	31.3	72.9	0.47	0.96
N_2O	36.5	71.7	0.45	0.94
NH_3	132.5	112.5	0.24	0.40
n-Butane	152.0	37.5	0.23	0.50

Source: From M. L. Lee and K. E. Markides, *Science,* **1987,** *235,* 1345. With permission. Data taken form *Matheson Gas Data Book* and *CRC Handbook of Chemistry and Physics.*

[2]Certain important industrial processes are based on the high solubility of organic species in supercritical carbon dioxide. For example, this medium has been employed in extracting caffeine from coffee beans to give decaffeinated coffee and in extracting nicotine from cigarette tobacco.

facturers began to offer apparatus for supercritical-fluid chromatography in the mid-1980s.[3]

The Effect of Pressure

The density of a supercritical fluid increases rapidly and nonlinearly with pressure increases. Density increases also alter retention factors (k) and thus elution times. For example, the elution time for hexadecane is reported to decrease from 25 to 5 min as the pressure of carbon dioxide is raised from 70 to 90 atm. Gradient elution can thus be achieved through linear increases in column pressure or through regulation of pressure to obtain linear density increases. A clear analogy exists between gradient elution through adjustment of pressure or density of a supercritical fluid and by temperature-gradient elution in gas-liquid chromatography and solvent-gradient elution in liquid chromatography. Figure 26-1 illustrates the improvement in chromatograms realized by pressure programming.

> Gradient elution can be achieved in SFC by systematically changing the column pressure or the density of the supercritical fluid.

Columns

Both packed columns and open tubular columns are used in supercritical fluid chromatography. Packed columns have the advantages of greater efficiency per unit time and the capability of handling larger sample volumes. Packed columns

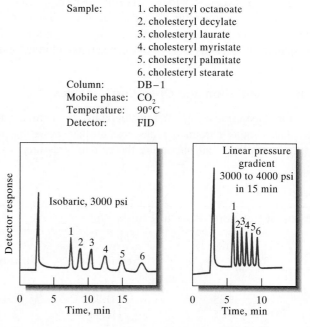

Sample:	1. cholesteryl octanoate
	2. cholesteryl decylate
	3. cholesteryl laurate
	4. cholesteryl myristate
	5. cholesteryl palmitate
	6. cholesteryl stearate
Column:	DB–1
Mobile phase:	CO_2
Temperature:	90°C
Detector:	FID

Figure 26-1 Effect of pressure programming in supercritical-fluid chromatography. Note the shorter time for the pressure-gradient chromatogram on the right compared with the isobaric chromatogram on the left. (Courtesy of Brownlee Labs, Santa Clara, CA.)

---

[3]For descriptions of several commercial instruments for SFC, see F. Wach, *Anal. Chem.,* **1994,** *66,* 372A.

for SFC can be much longer than those for HPLC, providing well over 100,000 plates. Open tubular columns are similar to the fused-silica open tubular (FSOT) columns described on page 673. Because of the low viscosity of supercritical media, columns can be much longer than those used in liquid chromatography, and column lengths of 10 to 20 m and inside diameters of 50 or 100 μm are common. For difficult separations, columns 60 m in length and longer have been used.

Many of the column coatings used in liquid chromatography have been applied to supercritical-fluid chromatography as well. Typically, these are polysiloxanes (see Section 25A-3) that are chemically bonded to the surface of silica particles or to the inner silica wall of capillary tubing. Film thicknesses are 0.05 to 0.4 μm.

Very long columns can be used in SFC because the viscosity of supercritical fluids is so low.

Mobile Phases

The most widely used mobile phase for supercritical-fluid chromatography is carbon dioxide. It is an excellent solvent for a variety of nonpolar organic molecules. In addition, it transmits in the ultraviolet and is odorless, nontoxic, readily available, and remarkably inexpensive relative to other chromatographic solvents. Its critical temperature of 31°C and its pressure of 73 atm at the critical temperature permit a wide selection of temperatures and pressures without exceeding the operating limits of modern high-performance liquid-chromatographic equipment. In some applications, polar organic modifiers, such as methanol, are introduced in small concentrations ($\approx 1\%$) to modify alpha values for analytes.

A number of other substances have served as mobile phases in supercritical-fluid chromatography, including ethane, pentane, dichlorodifluoromethane, diethyl ether, and tetrahydrofuran.

Detectors

A major advantage of supercritical-fluid chromatography is that the sensitive and universal detectors of gas-liquid chromatography are applicable to this technique as well. For example, the convenient flame ionization detector of gas-liquid chromatography can be applied by simply allowing the supercritical carrier to expand through a restrictor and into a hydrogen flame, where ions formed from the analytes are collected at biased electrodes, giving rise to an electrical current.

26A-3   Supercritical-Fluid Chromatography versus Other Column Methods

The information in Table 26-1 and other data as well reveal that several physical properties of supercritical fluids are intermediate between the properties of gases and liquids. As a consequence, this new type of chromatography combines some of the characteristics of both gas and liquid chromatography. Thus, like gas chromatography, supercritical-fluid chromatography is inherently faster than liquid chromatography because of the lower viscosity and higher diffusion rates in the mobile phase. High diffusivity, however, leads to longitudinal band spreading, which is a significant factor with gas but not with liquid chromatography. Thus, the intermediate diffusivities and viscosities of supercritical fluids result in faster

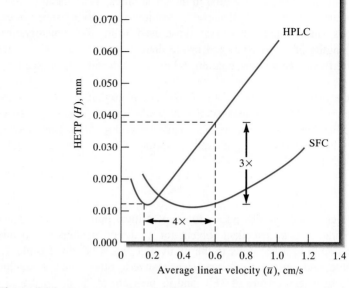

Figure 26-2   Performance characteristics of a 5-μm ODS column when elution is carried out with a conventional mobile phase (HPLC) and supercritical carbon dioxide SFC. [From D. R. Gere, *Application Note 800-3.* (Hewlett-Packard Company, Palo Alto, CA, 1983). With permission.]

biphenyl

Molecular model of biphenyl, a hazardous aromatic hydrocarbon. It is used an intermediate in the production of emulsifiers, brighteners, plastics and many other compounds. Biphenyl has been used as a heat transfer medium in heating fluids, as a dye carrier for textiles and copying paper, and as a solvent in pharmaceutical preparations. Paper impregnated with biphenyl is used in citrus fruit packaging to reduce fruit damage by fungus. Biphenyl is extremely hazardous to humans. Short-term exposure causes eye and skin irritation and toxic effects on the liver, kidneys and nervous system. Long-term exposure causes kidney damage to laboratory animals and may effect the central nervous system in humans.

SFC with flame ionization detection works very well for nonvolatile or thermally unstable compounds that have no chromophores for photometric detection.

separations than are achieved with liquid chromatography accompanied by less zone spreading than is encountered in gas chromatography.

Figure 26-2 shows plots of plate heights H as a function of average linear velocity $\bar{u}$ in cm/s for high-performance liquid chromatography and supercritical-fluid chromatography. In both cases, the solute was pyrene and the stationary phase a reversed-phase octyldecyl silane (ODS) maintained at 40°C. The mobile phase for HPLC was acetonitrile and water, and for SFC it was carbon dioxide. These conditions yielded about the same retention factor (k) for both mobile phases. Note that the minimum in plate height occurred at a flow rate of 0.13 cm/s with the HPLC and 0.40 cm/s for the SFC. The consequence of this difference is shown in Figure 26-3, where these same conditions are used for the separation of pyrene from biphenyl. Note that the HPLC separation required more than twice the time of the SFC separation.

26A-4   Applications

Supercritical-fluid chromatography appears to have a potential niche in the spectrum of column-chromatographic methods because it is applicable to a class of compounds that is not readily amenable to either gas-liquid or liquid chromatography. These compounds include species that are nonvolatile or thermally unstable and, in addition, contain no chromophoric groups that can be used for photometric detection. Separation of these compounds is possible with supercritical-fluid chromatography at temperatures below 100°C; furthermore, detection is readily carried out by means of the highly sensitive flame ionization detector. It is also noteworthy that supercritical columns have the added advantage of being much easier to interface with mass spectrometers than liquid-chromatographic columns.

26B   CAPILLARY ELECTROPHORESIS[4]

Electrophoresis is a separation method based on the differential rates of migration of charged species in an applied dc electric field. This separation technique for macrosize samples was first developed by Swedish chemist Arne Tiselius in the 1930s for the study of serum proteins; he was awarded the 1948 Nobel Prize in chemistry for his work.

Electrophoresis on a macro scale has been applied to a variety of difficult analytical separation problems: inorganic anions and cations, amino acids, catecholamines, drugs, vitamins, carbohydrates, peptides, proteins, nucleic acids, nucleotides, polynucleotides, and numerous other species. A particular strength of electrophoresis is its unique ability to separate charged macromolecules of interest to biochemists, biologists, and clinical chemists. For many years, electrophoresis has been the powerhouse method for separating proteins (enzymes, hormones, antibodies) and nucleic acids (DNA, RNA), for which it offers unparalleled resolution.

Until the appearance of capillary electrophoresis, electrophoretic separations were not carried out in columns but were performed in a flat stabilized medium such as paper or a porous semisolid gel. Remarkable separations were realized in such media, but the technique was slow, tedious, and required a good deal of operator skill. In the early 1980s, scientists began to explore the feasibility of performing these same separations on micro amounts of sample in fused-silica capillary tubes. Their results proved promising in terms of resolution, speed, and potential for automation. As a consequence, capillary electrophoresis (CE) has become an important tool for a wide variety of analytical separation problems and is the only type of electrophoresis we will consider.

26B-1   Instrumentation for Capillary Electrophoresis

As shown in Figure 26-4, the instrumentation for capillary electrophoresis is simple.[5] A buffer-filled fused-silica capillary, typically 10 to 100 μm in internal diameter and 40 to 100 cm long, extends between two buffer reservoirs that also hold platinum electrodes. Sample introduction takes place at one end and detection at the other. A potential difference of 5- to 30-kV dc is applied between the two electrodes. The polarity of this high voltage can be as indicated in Figure 26-4 or can be reversed to allow rapid separation of anions.

Sample introduction is often accomplished by pressure injection in which one end of the capillary is inserted into a vessel containing the sample. The vessel is then raised briefly above the level of the capillary to force sample into the tube.

---

[4]For additional discussion on the principles, instrumentation, and applications of capillary electrophoresis, see M. G. Khaledi, Ed., *High-Performance Capillary Electrophoresis, Theory, Techniques and Applications,* (New York: Wiley, 1998); P. Camilleri, Ed., *Capillary Electrophoresis: Theory and Practice,* (Boca Raton, FL: CRC Press, 1997).

[5]For a review of current commercially available capillary electrophoresis instruments, see S. C. Beale, *Anal. Chem.,* **1998,** *70,* 279R; C. Henry, *Anal. Chem.,* **1996,** *68,* 747A.

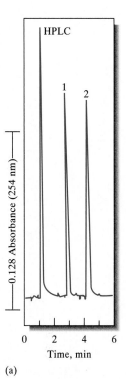

(a)

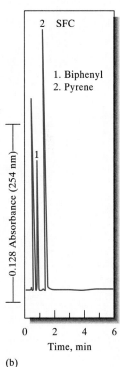

(b)

Figure 26-3   Separation of pyrene and biphenyl by (a) HPLC and (b) SFC. (From D. R. Gere, *Science,* **1983,** *222,* 255. With permission.)

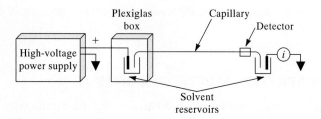

Figure 26-4   Schematic of a capillary zone electrophoresis system.

Alternatively, a vacuum is applied at the detector end of the tubing. Introduction may also be carried out by electroosmosis, described in the next section.

Because the separated analytes move past a common point in most types of capillary electrophoresis, detectors are similar in design and function to those described for HPLC. Table 26-3 lists several of the detection methods that have been reported through 1988 for capillary electrophoresis. The second column of the table shows representative detection limits for these detectors.

26B-2   Electroosmotic Flow

A unique feature of capillary electrophoresis is *electroosmotic flow.* When a high voltage is applied across a fused-silica capillary tube containing a buffer solution, electroosmotic flow usually occurs in which the solvent migrates toward the cathode. The rate of migration can be appreciable. For example, a 50-mM pH 8 buffer has been found to flow through a 50-cm capillary toward the cathode at approximately 5 cm/min with an applied voltage of 25 kV.[6]

Table 26-3

Detection Modes Developed for Capillary Electrophoresis

Detection Principle	Representative Detection Limit* (moles detected)
Spectrometry	
Absorption†	10^{-15}–10^{-13}
Fluorescence	10^{-17}–10^{-20}
Thermal lens†	4×10^{-17}
Raman†	2×10^{-15}
Mass spectrometry	1×10^{-17}
Electrochemical	
Conductivity†	1×10^{-16}
Potentiometry	Not reported
Amperometry	7×10^{-19}

Source: From A. G. Ewing, R. A. Wallingford, and T. M. Olefirowicz, *Anal. Chem.,* **1989,** *61,* 298A. With permission.

*Detection limits quoted have been determined with a wide variety of injection volumes that range from 18 pL to 10 nL.

†Mass detection limit converted from concentration detection limit using a 1-nL injection volume.

[6] J. D. Olechno, J. M. Y. Tso, J. Thayer, and A. Wainright, *Amer. Lab.,* **1990** (Nov.), *22* (17), 51.

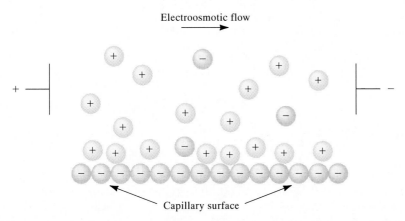

Figure 26-5   Charge distribution at a silica/capillary interface and resulting electroosmotic flow. (From A. G. Ewing, R. A. Wallingford, and T. M. Olefirowicz, *Anal. Chem.,* **1989,** *61,* 294A. With permission.)

As shown in Figure 26-5, the cause of electroosmotic flow is the electric double layer that develops at the silica/solution interface. At pH values higher than 3, the inside wall of a silica capillary is negatively charged due to ionization of the surface silanol groups (Si—OH). Buffer cations congregate in an electrical double layer adjacent to the negative surface of the silica capillary. The cations in the diffuse outer layer of the double layer are attracted toward the cathode, or negative electrode. Since these cations are solvated, they drag the bulk solvent along with them. As shown in Figure 26-6, electroosmosis leads to bulk solution flow that has a flat profile across the tube because flow originates at the walls. This profile is in contrast to the laminar (parabolic) flow profile observed with the pressure-driven flow encountered in HPLC. Because the profile is essentially flat, electroosmotic flow does not contribute significantly to band broadening the way pressure-driven flow does in liquid chromatography.

The rate of electroosmotic flow is generally greater than the electrophoretic migration velocities of the individual ions and effectively becomes the mobile-phase pump of capillary zone electrophoresis. Even though analytes migrate according to their charges within the capillary, the electroosmotic flow rate is usually sufficient to sweep all positive, neutral, and even negative species toward the same end of the capillary, so all can be detected as they pass by a common point (see Figure 26-7). The resulting *electropherogram* looks like a chromatogram but with narrower peaks.

Electroosmosis is often desirable in certain types of capillary electrophoresis, but in other types it is not. Electroosmotic flow can be eliminated by coating the inside capillary wall with a reagent like trimethylchlorosilane to eliminate the surface silanol groups.

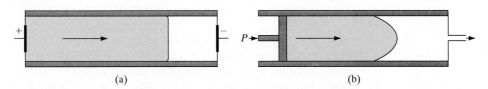

Figure 26-6   Flow profiles for liquids under (a) electroosmotic flow and (b) pressure-induced flow.

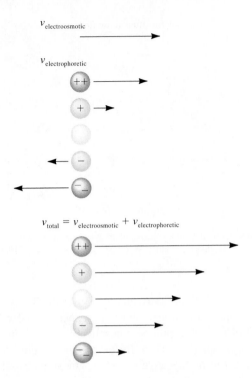

Figure 26-7 Velocities in the presence of electroosmotic flow. The length of the arrow next to an ion indicates the magnitude of its velocity; the direction of the arrow indicates the direction of motion. The negative electrode would be to the right and the positive electrode to the left of this section of solution.

26B-3 The Basis for Electrophoretic Separations

The migration rate v of an ion in an electric field is given by

$$v = \mu_e E = \mu_e \cdot \frac{V}{L} \tag{26-1}$$

where E is the electric field strength in volts per centimeter, V is the applied voltage, L is the length of the tube between electrodes, and μ_e is the *electrophoretic mobility,* which is proportional to the charge on the ion and inversely proportional to the frictional retarding force on the ion. The frictional retarding force on an ion is determined by the size and shape of the ion and the viscosity of the medium.

It has been shown that the plate count N of a capillary electrophoresis column is given by

$$N = \frac{\mu_e V}{2D} \tag{26-2}$$

where D is the diffusion coefficient of the solute ($cm^2 s^{-1}$). Because resolution increases with plate count, it is desirable to use high applied voltages to achieve high-resolution separations. Note that for electrophoresis, contrary to the situation in chromatography, the plate count does not increase with the column length.

Typically, capillary electrophoresis plate counts are 100,000 to 200,000 at the usual applied voltages.

26B-4   Applications of Capillary Electrophoresis[7]

Capillary electrophoretic separations are performed in several ways called modes. These include *isoelectric focusing, isotachophoresis,* and *capillary zone electrophoresis* (CZE). We shall consider only capillary zone electrophoresis in which the buffer composition is constant throughout the region of the separation. The applied field causes each of the different ionic components of the mixture to migrate according to its own mobility and to separate into zones that may be completely resolved or may be partially overlapped. Completely resolved zones have regions of buffer between them. The situation is analogous to elution column chromatography, where regions of mobile phase are located between zones containing separated analytes.

Separation of Small Ions

For most electrophoretic separations of small ions, it has been found best from the standpoint of decreased time to have the analyte ions move in the same direction as the electroosmotic flow. Thus, for cation separations, the walls of the capillary are untreated, and the electroosmotic flow and the cation movement is toward the cathode. For the separation of anions, on the other hand, the electroosmotic flow is usually reversed by treating the walls of the capillary with an alkyl ammonium salt, such as cetyl trimethylammonium bromide. The positively charged ammonium ions become attached to the negatively charged silica surface and in turn create a negatively charged double layer of solution, which is attracted toward the anode, reversing the electroosmotic flow.

In the past, the most common method for analysis of small anions has been ion-exchange chromatography. For cations, the preferred techniques have been atomic absorption spectroscopy and inductively coupled plasma emission spectroscopy. Recently, however, capillary electrophoretic methods have begun to compete with these traditional methods for small ion analysis. Several major reasons for adoption of electrophoretic methods have been recognized: lower equipment costs, smaller sample size requirements, much greater speed, and better resolution.

The initial cost of equipment and the expense of maintenance for electrophoresis is generally significantly lower than those for ion chromatographic and atomic spectroscopic instruments. Thus, commercial electrophoretic instruments are marketed in the range of $15,000 to $40,000.[8]

Sample sizes for electrophoresis are in the nanoliter range, whereas microliter or larger samples are usually needed for other types of small ion analysis. Thus, electrophoretic methods are more sensitive than the other methods on a mass basis (but usually not on a concentration basis).

---

[7]For excellent reviews of applications of capillary electrophoresis and electrochromatography, see B. L. Karger, A. S. Cohen, and A. Gutman, *J. Chromatogr.,* **1989,** *492,* 585; J. P. Landers, Ed., *Handbook of Capillary Electrophoresis,* 2nd ed. (Boca Raton, FL: CRC Press, 1997).
[8]See M. Warner, *Anal. Chem.,* **1994,** *66,* 1137A.

Figure 26-8 illustrates the unsurpassed quickness and resolution of electrophoretic separations of small anions. Here, 30 anions were separated cleanly in just over 3 min. Typically, an ion-exchange separation of only three or four anions could be accomplished in this brief time. Figure 26-9 further illustrates the speed at which separations can be carried out. Here, 19 cations were separated in less than 2 min.

Separation of Molecular Species

A variety of small synthetic herbicides, pesticides, and pharmaceuticals that are ions or can be derivatized to yield ions have been separated and analyzed by CZE. Figure 26-10 is illustrative of this type of application in which three anti-inflammatory drugs, which are carboxylic derivatives, are separated in less than 15 min.

Proteins, amino acids, and carbohydrates have all been separated very rapidly by CZE. In the case of neutral carbohydrates, the separations are preceded by formation of negatively charged borate complexes. The separation of protein mixtures is illustrated in Figure 26-11.

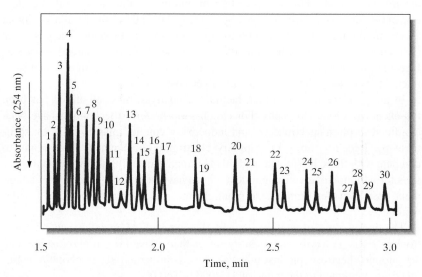

Figure 26-8   Electropherogram showing the separation of 30 anions. Capillary internal diameter: 50 μm (fused silica). Detection: indirect UV, 254 nm. Peaks: 1 = thiosulfate (4 ppm), 2 = bromide (4 ppm), 3 = chloride (2 ppm), 4 = sulfate (4 ppm), 5 = nitrite (4 ppm), 6 = nitrate (4 ppm), 7 = molybdate (10 ppm), 8 = azide (4 ppm), 9 = tungstate (10 ppm), 10 = monofluorophosphate (4 ppm), 11 = chlorate (4 ppm), 12 = citrate (2 ppm), 13 = fluoride (1 ppm), 14 = formate (2 ppm), 15 = phosphate (4 ppm), 16 = phosphite (4 ppm), 17 = chlorite (4 ppm), 18 = galactarate (5 ppm), 19 = carbonate (4 ppm), 20 = acetate (4 ppm), 21 = ethanesulfonate (4 ppm), 22 = propionate (5 ppm), 23 = propanesulfonate (4 ppm), 24 = butyrate (5 ppm), 25 = butanesulfonate (4 ppm), 26 = valerate (5 ppm), 27 = benzoate (4 ppm), 28 = *l*-glutamate (5 ppm), 29 = pentanesulfonate (4 ppm), 30 = *d*-gluconate (5 ppm). (From W. A. Jones and P. Jandik, *J. Chromatogr.,* **1991,** *546,* 445. With permission.)

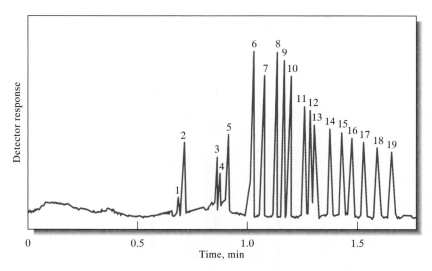

Figure 26-9   Separation of alkali, alkaline earths, and lanthanides. Capillary: 36.5 cm × 75 μm fused silica, +30 kV. Injection: hydrostatic, 20 s at 10 cm. Detection: indirect UV, 214 nm. Peaks: 1 = rubidium (2 ppm), 2 = potassium (5 ppm), 3 = calcium (2 ppm), 4 = sodium (1 ppm), 5 = magnesium (1 ppm), 6 = lithium (1 ppm), 7 = lanthanum (5 ppm), 8 = cerium (5 ppm), 9 = praseodymium (5 ppm), 10 = neodymium (5 ppm), 11 = samarium (5 ppm), 12 = europium (5 ppm), 13 = gadolinium (5 ppm), 14 = terbium (5 ppm), 15 = dysprosium (5 ppm), 16 = holmium (5 ppm), 17 = erbium (5 ppm), 18 = thulium (5 ppm), 19 = ytterbium (5 ppm). (From P. Jandik, W. R. Jones, O. Weston, and P. R. Brown, *LC-GC,* **1991,** *9,* 634. With permission.)

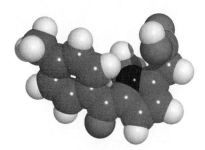

Naproxin

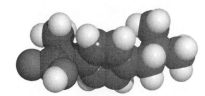

Ibuprofen

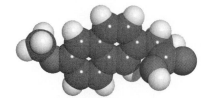

Tolmetin

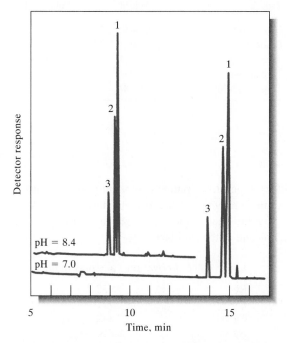

Figure 26-10   Separation of anti-inflammatory drugs by CZE. Detection: absorption at 215 nm. Analytes: (1) naproxen, (2) ibuprofen, (3) tolmetin. (Reprinted with permission from A. Wainright, *J. Microcolumn. Sep.,* **1990,** *2,* 166.)

Molecular models of anti-inflammatory drugs, naproxen, ibuprofen and tolmetin. These nonsteroidal anti-inflammatory agents are thought to relieve pain by inhibiting the synthesis of prostaglandins which are involved in the perception of pain and the production of fever and inflammation. Ibuprofen is also known as Motrin, Advil or Nuprin. Naproxen sodium is Aleve, and tolmetin is tolectin. Each of these are used to treat symptoms of arthritis and to relieve the pain caused by gout, bursitis, tendonitis, sprains, strains and other injuries and menstrual cramps. Ibuprofen and naproxen are available over-the-counter in the United States.

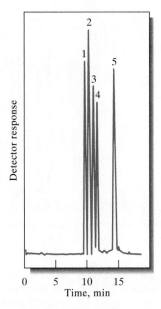

Figure 26-11   CZE separation of a model protein mixture. Conditions: pH 2.7 buffer; absorbance detection at 214 nm; 22 kV, 10 μA. Peaks are identified in the following table.

	Model Proteins Separated at pH 2.7		
Peak No.	**Proteins**	**Molecular Weight**	**Isoelectric Point, pH**
1	Cytochrome c	12,400	10.7
2	Lysozyme	14,100	11.1
3	Trypsin	24,000	10.1
4	Trypsinogen	23,700	8.7
5	Trypsin inhibitor	20,100	4.5

Spreadsheet Exercise   Example 26-1

For the capillary zone electrophoretic separation of the ions shown in Figure 26-9, accurate measurements of the arrival times of the first seven ions at the UV detector gave the results shown below. The detector was located 30.0 cm from the injection end of the capillary.

Ion	**Arrival time at detector, min**
Rb^+	0.658
K^+	0.711
Ca^{2+}	0.842
Na^+	0.868
Mg^{2+}	0.895
Li^+	1.026
La^{3+}	1.079
Ce^{3+}	1.132

The mobility of Na^+ is known from conductance data to be 5.19×10^{-4} cm^2 s^{-1} V^{-1}. Use this value to calculate the electroosmotic flow velocity (cm/s) for the data in Figure 26-9, and then use the velocity to calculate the electrophoretic mobilities of the other ions. Compare the values obtained to the limiting ionic mobilities obtained from conductance data given below.

Ion	Limiting ionic mobility, cm^2 s^{-1} V^{-1} $\times 10^{-4}$ at 25·C
Rb^+	8.06×10^{-4}
K^+	7.62×10^{-4}
Ca^{2+}	6.17×10^{-4}
Na^+	5.19×10^{-4}
Mg^{2+}	5.50×10^{-4}
Li^+	4.01×10^{-4}
La^{3+}	7.21×10^{-4}
Ce^{3+}	7.26×10^{-4}

Source: J. A. Dean, *Analytical Chemistry Handbook*, (New York: McGraw-Hill, 1995), p. 14.41

Electroosmotic Flow Velocity

The arrival of an ion at the detector t_D is given by the distance to the detector l_D divided by the velocity with which the ion migrates. The migration rate is the sum of the ions electrophoretic velocity v_e and the velocity of the electroosmotic flow v_{eo}. Hence we can write

$$t_D = \frac{l_D}{v_e + v_{eo}} \qquad (26\text{-}3)$$

From Equation 26-1, we can relate the electrophoretic velocity to the electrophoretic mobility, the applied voltage and the length of the capillary. Hence, we can write the arrival time as

$$t_D = \frac{l_D}{\dfrac{\mu_e V}{L} + v_{eo}} \qquad (26\text{-}4)$$

We can solve Equation 26-4 for v_{eo} and obtain

$$v_{eo} = \frac{l_D}{t_D} - \frac{\mu_e V}{L} \qquad (26\text{-}5)$$

For Na^+ we can write

$$v_{eo} = \frac{30.0 \text{ cm}}{0.868 \text{ min} \times 60.0 \text{ s/min}} - \frac{5.19 \times 10^{-4} \text{ cm}^2 \text{ s}^{-1} \text{ V}^{-1} \times 30{,}000 \text{ V}}{36.5 \text{ cm}}$$

$$= 0.14946 \text{ cm s}^{-1}$$

Mobilities of the Other Ions

Before entering values into the spreadsheet for calculating mobilities, let us rearrange Equation 26-4 and solve for μ_e. The result is

$$\mu_e = \frac{l_D L}{t_D V} - \frac{v_{eo} L}{V} \qquad (26\text{-}6)$$

Now we can make entries into the spreadsheet as shown in Figure 26-12. We first enter the known values in Equation 26-6 into cells B2 through B5. The table of arrival times on page 712 is entered into B8 through B15 after putting the appropriate labels in A8 through A15 and titles in A7 and B7. Equation 26-6 is then entered into cell C8 as shown in the documentation cell. The fill handle is then used to apply this equation to the remaining values in the table.

If we then add the values from the table of ionic mobilities, we can compare the capillary electrophoresis values directly to the conductance values as shown in Figure 26-12. The limiting ionic mobilities obtained from conductance values are called *limiting* because they are values obtained by extrapolating conductances at finite concentration to infinite dilution. Because the capillary electrophoresis results of Figure 26-9 were obtained in a buffer solution, we would not expect perfect correlations. Note that for the univalent ions, values compare fairly well with the trends in mobility being the same ($Rb^+ > K^+ > Na^+ > Li^+$). In all cases except for Li^+, the electrophoretic mobilities are smaller probably because the ionic atmosphere surrounding the migrating ion exhibits a drag on the ion. The Na^+ ion has an asterisk next to it, because its mobility was used to calculate the electroosmotic velocity. The electrophoretic mobilities of the divalent and trivalent ions do not correlate well with the limiting ionic mobilities obtained from conductance. The electrophoresis values are lower due not only to the ionic atmosphere, but also because of the tendency of the rare earth ions to form complexes with anions from the buffer. The disagreement seen in Figure 26-12 is not uncommon when comparing values taken under conditions of relative high ionic strength to infinite dilution values. The comparison is useful, however, because trends can often be identified and in capillary electrophoresis arrival times can be rationalized in terms of ionic mobilities.

	A	B	C	D
1	**Calculation of electrophoretic mobilities of ions**			
2	Capillary length, cm	36.5		
3	Distance to detector, cm	30.0		
4	Applied voltage, V	30000.0		
5	Electroosmotic flow, cm/s	0.14946		
6				
7	**Ion**	**Arrival time, s**	**Electrophoretic mobility, cm^2 s^{-1} V^{-1}**	**Ionic moblity, cm^2 s^{-1} V^{-1}**
8	Rb^+	0.658	7.43E-04	8.06E-04
9	K^+	0.711	6.74E-04	7.62E-04
10	Ca^{2+}	0.842	5.41E-04	6.17E-04
11	Na^+	0.868	5.19E-04	5.19E-04*
12	Mg^{2+}	0.895	4.98E-04	5.50E-04
13	Li^+	1.026	4.11E-04	4.01E-04
14	La^{3+}	1.079	3.82E-04	7.21E-04
15	Ce^{3+}	1.132	3.56E-04	7.26E-04
16				
17	**Spreadsheet documentation**			
18	Cell C8=\$B\$3*\$B\$2/(B8*60*\$B\$4)-\$B\$5*\$B\$2/\$B\$4			

Figure 26-12 Spreadsheet to calculate electrophoretic mobilities from the capillary electropherogram of Figure 26-9.

26C   CAPILLARY ELECTROCHROMATOGRAPHY

Capillary electrochromatography (CEC) is a hybrid of HPLC and capillary electrophoresis (CE) that offers some of the best features of the two methods.[9] Like HPLC, it is applicable to the separation of neutral species. Like CE, however, it provides highly efficient separations on microvolumes of sample solution without the need for the high-pressure pumping system required for HPLC. In CEC, a mobile phase is transported across a stationary phase by electroosmotic flow. As shown in Figure 26-6, electroosmotic pumping leads to a flat plug profile rather than the parabolic profile that results from mechanical pumping (pressure-induced flow). The flat profile of electroosmotic pumping leads to narrow bands and thus high separation efficiencies.

26C-1   Packed Column Electrochromatography

Electrochromatography based on packed columns is the least mature of the various electroseparation techniques. In this method, a polar solvent is usually driven by electroosmotic flow through a capillary that is packed with a reversed-phase HPLC packing. Separations depend on the distribution of the analyte species between the mobile phase and the liquid stationary phase held on the packing. Figure 26-13 shows a typical electrochromatogram for the separation of

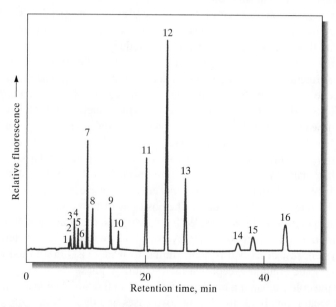

Figure 26-13   Electrochromatogram showing the electrochromatographic separation of 16 PAHs. The peaks are identified as follows: ($\sim 10^{-6}$ to 10^{-8} M of each compound): 1 = naphthalene, 2 = acenaphthylene, 3 = acenaphthene, 4 = fluorene, 5 = phenanthrene, 6 = anthracene, 7 = fluoranthene, 8 = pyrene, 9 = benz[*a*]anthracene, 10 = chrysene, 11 = benzo[*b*]fluoranthene, 12 = benzo[*k*]fluoranthene, 13 = benzo[*a*]pyrene, 14 = dibenz[*a,h*]anthracene, 15 = benzo[*ghi*]perylene, 16 = indeo[1,2,3-*cd*]pyrene. (From C. Yan, R. Dadoo, H. Zhao, D. J. Rakestraw, and R. N. Zare, *Anal. Chem.,* **1995,** *67,* 2026. With permission.)

---

[9]For a discussion of this method, see S. Terabe, *Trends Anal. Chem.,* **1989,** *8,* 129.

16 polyaromatic hydrocarbons (PAHs) in a 33-cm-long capillary having an inside diameter of 75 μm. The mobile phase consisted of acetonitrile in a 4-mM sodium borate solution. The stationary phase consisted of 3-μm octadecylsilica particles.

26-C2   Micellar Electrokinetic Capillary Chromatography

The capillary electrophoretic methods we have described thus far are not applicable to the separation of uncharged solutes. In 1984, however, Shigeru Terabe and collaborators[10] described a modification of the method that permits the separation of low-molecular-weight aromatic phenols and nitro compounds with equipment such as shown in Figure 26-4. This technique involves introduction of surfactant, at a concentration level at which *micelles* form. Micelles form in aqueous solutions when the concentration of an ionic species with a long-chain hydrocarbon tail is increased above a certain level called the *critical micelle concentration* (CMC). At this point, the surfactant begins to form spherical aggregates made up to 40 to 100 ions with their hydrocarbon tails in the interior of the aggregate and their charged ends exposed to water on the outside. Micelles constitute a stable second phase that can incorporate nonpolar compounds in the hydrocarbon interior of the particles, thus *solubilizing* the nonpolar species. Solubilization is commonly encountered when a greasy material or surface is washed with a detergent solution.

Capillary electrophoresis carried out in the presence of micelles is called *micellar electrokinetic capillary chromatography* and given the acronym MECC or MEKC. In this technique, surfactants are added to the operating buffer in amounts that exceed the critical micelle concentration. For most applications to date, the surfactant has been sodium dodecyl sulfate (SDS). The surface of an ionic micelle of this type has a large negative charge, which gives it a large electrophoretic mobility. Most buffers, however, exhibit such a high electroosmotic flow rate toward the negative electrode that the anionic micelles are carried toward that electrode also, but at a much reduced rate. Thus, during an experiment, the buffer mixture consists of a faster-moving aqueous phase and a slower-moving micellar phase. When a sample is introduced into this system, the components distribute themselves between the aqueous phase and the hydrocarbon phase in the interior of the micelles. The positions of the resulting equilibria depend on the polarity of the solutes. With polar solutes, the aqueous solution is favored; with nonpolar compounds, the hydrocarbon environment is preferred.

The system just described is quite similar to what exists in a liquid partition chromatographic column except that the "stationary phase" is moving along the length of the column but at a much slower rate than the mobile phase. The mechanism of separations is identical in the two cases and depends on differences in distribution constants for analytes between the mobile aqueous phase and the hydrocarbon *pseudostationary phase*. The process is thus true chromatography; hence, the name micellar electrokinetic capillary *chromatography*. Figure 26-14 illustrates two typical separations by MECC.

---

[10]S. Terabe, K. Otsuka, K. Ichikawa, A. Tsuchiya, and T. Ando, *Anal. Chem.,* **1984,** *56,* 111; S. Terabe, K. Otsuka, and T. Ando, *Anal. Chem.,* **1985,** *57,* 841. For a discussion of optimization of separations by this technique, see J. P. Foley, *Anal. Chem.,* **1990,** *62,* 1302.

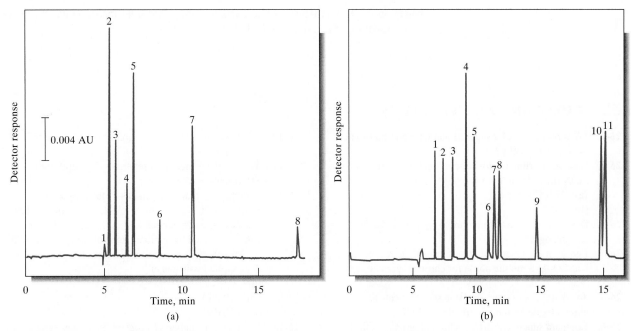

Figure 26-14   Typical separation by MECC. (a) Some test compounds: 1 = methanol, 2 = resorcinol, 3 = phenol, 4 = *p*-nitroaniline, 5 = nitrobenzene, 6 = toluene, 7 = 2-naphthol, 8 = Sudan III; capillary, 50 μm inside diameter, 500 mm to the detector; applied voltage, ca. 15 kV; detection, absorbance at 210 nm. (b) Analysis of a cold medicine: 1 = acetaminophen, 2 = caffeine, 3 = sulpyrine, 4 = naproxen, 5 = guaifenesin, 10 = noscapine, 11 = chlorpheniramine and tipepidine; applied voltage, 20 kV; capillary, as in (a); detection, absorption at 220 nm. (From S. Terabe, *Trends Anal. Chem.,* **1989,** *8,* 129. With permission.)

Capillary chromatography in the presence of micelles appears to have a promising future. One advantage that this hybrid technique has over HPLC is much higher column efficiencies (100,000 plates or more). In addition, changing the second phase in MECC is simple, involving only a change of the micellar composition of the buffer. In contrast, in HPLC, the second phase can only be altered by changing the type of column packing.

WEB WORKS

Use your Web browser to view the lecture slides by J. K. Hardy at http://ull.chemistry.uakron.edu/chemsep/super/ on supercritical-fluid chromatography. Use these slides to supplement your reading of this chapter. Also, browse to other sets of lecture slides related to other topics of interest in analytical chemistry.

Browse to the Hewlett-Packard on-line publication Web site at http://chem.external.hp.com/scripts/cag_litsearch.asp. Perform a search for documents containing the word "electrophoresis." On-line publications at this site are provided as Adobe Acrobat Portable Document Format (PDF) files, so you must be sure that the Adobe Acrobat Reader is installed as a plugin for your Web browser. Find the application brief entitled *Separation of nucleotides and related coenzymes using capillary zone electrophoresis* and determine the experimental conditions for this separation. Before you leave the site, find a few other application notes and note the kinds of information that you can obtain. Browse to other

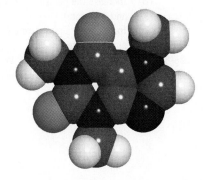

Molecular model of caffeine. Caffeine stimulates the cerebral cortex by inhibiting an enzyme that inactivates a certain form of adenosine triphosphate, the molecule that supplies energy. Caffeine occurs in coffee, tea and cola drinks.

pages on the Hewlett-Packard site and see if you can find instruments and techniques that are of use in analytical chemistry.

26D   QUESTIONS AND PROBLEMS

*26-1. What properties of a supercritical fluid are important in chromatography?

26-2. How do instruments for supercritical-fluid chromatography differ from those for
   (a) HPLC?
   (b) GC?

*26-3. Describe the effect of pressure on supercritical-fluid chromatograms.

26-4. List some of the advantageous properties of supercritical CO_2 as a mobile phase for chromatographic separations.

*26-5. To what types of compounds is supercritical-fluid chromatography particularly applicable?

26-6. Compare supercritical-fluid chromatography with other column chromatographic methods.

*26-7. For supercritical carbon dioxide, predict the effect that the following changes will have on the elution time in an SFC experiment.
   (a) Increase the flow rate (at constant temperature and pressure).
   (b) Increase the pressure (at constant temperature and flow rate).
   (c) Increase the temperature (at constant pressure and flow rate).

26-8. What is electroosmotic flow? Why does it occur?

*26-9. Suggest a way in which electroosmotic flow might be repressed.

26-10. Why does pH affect separation of amino acids by electrophoresis?

*26-11. What is the principle of separation by capillary zone electrophoresis?

26-12. A certain inorganic cation has a electrophoretic mobility of 4.31×10^{-4} cm^2 s^{-1} V^{-1}. This same ion has a diffusion coefficient of 9.8×10^{-6} cm^2 s^{-1}. If this ion is separated by capillary zone electrophoresis with a 50.0-cm capillary, what is the expected plate count N at applied voltages of
   (a) 5.0 kV?
   (b) 10.0 kV?
   (c) 30.0 kV?

*26-13. The cationic analyte of Problem 26-12 was separated by capillary zone electrophoresis with a 50-cm capillary at 10.0 kV. Under the separation conditions, the electroosmotic flow rate was 0.85 mm s^{-1} toward the cathode. If the detector were placed 40.0 cm from the injection end of the capillary, how long does it take in minutes for the analyte cation to reach the detector after the field is applied?

26-14. What is the principle of micellar electrokinetic capillary chromatography? How does it differ from capillary zone electrophoresis?

*26-15. Describe a major advantage of micellar electrokinetic capillary chromatography over conventional liquid chromatography.

Section VI

Selected Methods
of Analysis

This section contains but one chapter with detailed instructions for performing many chemical analyses. After an introductory experiment, which includes a sampling exercise, two gravimetric experiments are presented. The instructions for several neutralization titrations are next given including an experiment to determine amine nitrogen by the Kjeldahl method. In addition to neutralization titrations, instructions are presented for precipitation titrations, complexometric titrations with EDTA, and redox titrations involving permanganate, iodine, thiosulfate, and potassium bromate. Potentiometric methods described include the determination of chloride and iodide in a mixture, a determination involving an acid/base titration, and the determination of fluoride with an ion-selective electrode. Electrogravimetry is illustrated by an experiment to determine copper and lead in brass. Absorption spectrophotometry experiments include the determination of iron in a natural water sample, the determination of manganese in steel and the spectrophotometric determination of pH. An atomic absorption determination and an atomic emission determination complete analyses involving spectrochemical methods. Several separation experiments are presented including an ion-exchange experiment and the gas chromatographic determination of ethanol in beverages.

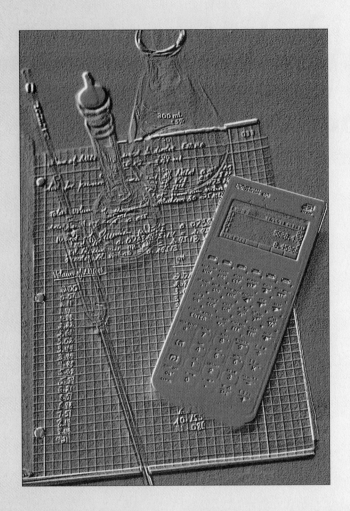

Some of the common laboratory tools of the analytical chemist are shown in this image including the laboratory notebook, a pipet, a volumetric flask, an Erlenmeyer flask, and a calculator. Other useful tools include pH meters, balances, spectrophotometers, chromatographs and the computers that control these instruments or manipulate the data produced.

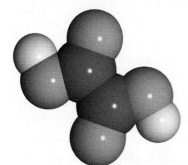

Chapter 27

Selected Methods of Analysis

This chapter contains detailed directions for performing a variety of chemical analyses. The methods have been chosen to introduce you to analytical techniques that are widely used by chemists. For most of these analyses, the composition of the samples is known to the instructor. Thus, you will be able to judge how well you are mastering these techniques.

Your chances of success in the laboratory will greatly improve if you take time before beginning any analysis to read carefully and *understand* each step in the method and to develop a plan for how and when you will perform each step. For greatest efficiency, such study and planning should take place *before you enter the laboratory.*

The discussion in this introduction is aimed at helping you develop efficient work habits in the laboratory; it also provides you with some general information about an analytical chemistry laboratory.

Background Information

Before you start an analysis, you should understand the significance of each step in the procedure to avoid the pitfalls and potential sources of error that exist in all analytical methods. Information about these steps can usually be found in preliminary discussion sections, in earlier chapters that are referred to in the discussion section, and in the "Notes" that follow many of the procedures. If, after reading these materials, you still do not understand the reason for one or more of the steps in the method, consult your instructor *before you begin laboratory work.*

The Accuracy of Measurements

In looking over an analytical procedure, you should decide which measurements must be made with maximum precision, and thus with maximum care, as opposed to those that can be carried out rapidly with little concern for precision.

Generally, measurements that appear in the equation used to compute the results must be performed with maximum precision. The remaining measurements can *and should* be made less carefully to conserve time. The words "about" and "approximately" are frequently used to indicate that a measurement does not have to be done carefully. You should not waste time and effort to measure, let us say, a volume to ± 0.02 mL when an uncertainty of ± 0.5 mL or even ± 5 mL will have no discernible effect on the results.

In some procedures, a statement such as "weigh three 0.5-g samples to the nearest 0.1 mg" is encountered. Here, samples of perhaps 0.4 to 0.6 g are acceptable, but their weights must be known to the nearest 0.1 mg. The number of significant figures in the specification of a volume or a weight is also a guide as to the care that should be taken in making a measurement. For example, the statement "add 10.00 mL of a solution to the beaker" indicates that you should measure the volume carefully with a buret or a pipet with the aim of limiting the uncertainty to perhaps ± 0.02 mL. In contrast, if the directions read "add 10 mL," the measurement can be made with a graduated cylinder.

Planning Your Time

You should study carefully the time requirements of the several unit operations involved in an analysis *before work is started*. Such study will reveal operations that require considerable elapsed, or clock, time but little or no operator time. Examples of such operations include drying a sample in an oven, cooling a sample in a desiccator, or evaporating liquid on a hot plate. The experienced chemist plans to use such periods of waiting to perform other operations or perhaps to begin a new analysis. Some people find it worthwhile to prepare a written time schedule for each laboratory period to avoid periods when no work can be done.

Time planning is also needed to identify places where an analysis can be interrupted for overnight or longer as well as those operations that must be completed without a break.

Reagents

Directions for the preparation of reagents accompany many of the procedures. Before preparing such reagents, be sure and check to see if they are already prepared and available on a side shelf for general use.

If a reagent is known to pose a hazard, you should plan *in advance of the laboratory period* the steps that you should take to minimize injury or damage. Furthermore, you must acquaint yourself with the rules that apply in your laboratory for the disposal of waste liquids and solids. These rules vary from one location to another and even among laboratories in the same locale.

Water

Some laboratories use deionizers to purify water; others employ stills for this purpose. The terms "distilled water" and "deionized water" are used interchangeably in the directions that follow. Either type is satisfactory for the procedures in this chapter.

You should use tap water only for preliminary cleaning of glassware. The cleaned glassware is then rinsed with at least three small portions of distilled or deionized water.

27A AN INTRODUCTORY EXPERIMENT

The purpose of this experiment is to introduce several of the tools, techniques, and skills necessary for work in the analytical chemistry laboratory. The techniques are considered one at a time, as unit operations. It is important to learn proper techniques and acquire individual skills before attempting additional laboratory experiments.

27A-1 Using the Analytical Balance

Discussion

In this experiment, you will obtain the mass of five new pennies, first by determining the mass of each penny individually. Then you will determine the mass of all five pennies at once, remove one penny at a time, and calculate the individual masses of the pennies by difference. The pair of masses determined for a particular penny by the two different methods should agree to within a few tenths of a milligram. From the data, you will determine the mean and median values, the standard deviation, and the relative standard deviation of the masses of the pennies.

You will then weigh an unknown aluminum cylinder and report the mass of this unknown.

Procedure

1. After you have been instructed in the use of the balance and have become familiar with its use, obtain a set of pennies, an unknown aluminum cylinder, and a pair of tweezers from the instructor.
2. Do not handle the pennies or the cylinder with your fingers; always use tweezers. If you are using a mechanical balance, be sure to have the balance in the "off" or "complete arrest" position whenever you remove anything from or add anything to the balance pan.

3. Before you begin to determine masses, zero your analytical balance carefully. Select five pennies at random from the vial containing the pennies, and weigh each penny on your balance. Enter the data in your laboratory notebook. Keep track of the identity of each penny by placing them on a labeled piece of paper.
4. Check the zero setting on your balance. Place these same five pennies on the balance pan, determine their total mass, and record it.
5. Then, remove one of the pennies from the balance, obtain the mass of the remaining four, and record the mass.
6. Repeat this process, removing one penny at a time. Obtain the individual masses by subtraction. This process is known as *weighing by difference,* which is the way nearly all mass determinations are done in the analytical laboratory.
7. Finally, check the zero on your balance, and find the mass of the unknown aluminum cylinder.

27A-2   Making Quantitative Transfers

Discussion

The following experiment is designed to provide experience in the correct use of the volumetric flask.

Procedure

1. Weigh a 50-mL beaker on a triple beam balance.
2. Adjust the balance for an additional 0.4 g, and add solid $KMnO_4$ to the beaker until the beam is again balanced. Note that chemicals should never be returned to stock bottles, because they may contaminate the bottle.
3. Dissolve the potassium permanganate in about 20 mL of distilled water, stirring gently to avoid loss. This solution is nearly saturated, and some care is required to dissolve the crystals completely.
4. Quantitatively transfer the solution to a 100-mL volumetric flask fitted with a small funnel. To prevent solution from running down the outside of the beaker, pour it down the stirring rod, and then touch the rod to the spout of the beaker to remove the last drop. Add more water to the beaker, stir, and repeat the procedure.
5. Repeat the procedure until no trace of the color of the permanganate remains in the beaker. Note the number of washings required to quantitatively transfer the permanganate from the beaker to the flask.
6. Rinse the last portions of solution from the stirring rod into the volumetric flask with a stream of water from the wash bottle. Rinse the funnel and remove it. Dilute the solution in the flask until the bottom of the meniscus is even with the graduation mark. Stopper, invert, and shake the flask. Return it to the upright position, and allow the air bubble to return all the way to the top of the neck.
7. Repeat until the solution is completely homogeneous; about 10 inversions and shakings are required. Save the solution for Section 27A-3.

27A-3   Delivering an Aliquot

Discussion

Whenever a buret or pipet is used to deliver a measured volume of solution, the liquid it contains before measurement should have the same composition as the solution to be dispensed. The following operations are designed to illustrate how to rinse and fill a pipet and how to deliver an aliquot of solution.

Procedure

1. Fill a pipet with the solution of potassium permanganate, and let it drain.
2. Draw a few milliliters of distilled water from a 50-mL beaker into the pipet, rinse all internal surfaces of the pipet, and discard the rinse solution. Do not fill the pipet completely; this step is wasteful, time consuming, and inefficient. Just draw in a small amount, tilt the pipet horizontally, and turn it to rinse the sides.
3. Determine the minimum number of such rinsings required to completely remove the permanganate color from the pipet. If your technique is efficient, three rinsings will suffice.
4. Again fill the pipet with permanganate solution, and proceed as before. This time, determine the minimum volume of rinse water required to remove the color by collecting the rinsings in a graduated cylinder. Less than 5 mL is enough with efficient technique. In the rinsing operations, was the water in the 50-mL beaker contaminated with permanganate? If a pink color shows that it was, repeat the exercise with more care.
5. As a test of your technique, ask the laboratory instructor to observe and comment on the following operation: Rinse a 10-mL pipet several times with the solution of potassium permanganate you prepared.
6. Pipet 10 mL of the permanganate solution into a 250-mL volumetric flask.
7. Carefully dilute the solution to volume, trying to mix the contents as little as possible.
8. Mix the solution by repeatedly inverting and shaking the flask. Note the effort required to disperse the permanganate color uniformly throughout the solution.
9. Rinse the pipet with the solution in the volumetric flask. Pipet a 10-mL aliquot of the solution into a conical flask.

27A-4   Calibrating a Pipet

Discussion

The proper technique for calibrating an analytical transfer pipet is readily learned with practice, care, and attention to detail. Note that this manual technique can only be learned through practice. With the possible exception of a mass determination, this experiment has the potential of being the most accurate and precise set of measurements that you will ever make.

Buret section constructed from a discarded buret. Broken burets are carefully cleaned and cut into pieces about 10 cm in length. The upper end of each section is carefully sealed by glassblowing, and the opposite end is drawn out to a tip. The tipped end is then cut so that there is approximately a 1-mm opening in the tipped end of the buret section. A hypodermic syringe fitted with a large-bore needle is then used to add distilled water to each section until it is about half full. The tipped end of each section is then sealed by glassblowing, and the sections are stored upside down in a test-tube rack or a wooden block with holes drilled to accommodate the sections. Each buret section should be permanently marked with a unique number.

Procedure

1. Clean a 10-mL pipet. When a pipet, buret, or other piece of volumetric glassware is cleaned properly, no droplets of reagent remain on the internal surfaces when they are drained, which is very important for accurate and reproducible results. If reagent is retained inside a pipet, you cannot deliver the nominal volume. If you clean a pipet or any other glassware with alcoholic KOH, use the bottle of cleaning solution only inside the sink and rinse it off thoroughly before putting it back on the plastic matting. Do not put the bottle of cleaning solution directly on a benchtop; it may ruin the surface. The solution is very corrosive. If your fingers feel slippery after use, or if some part of your body develops an itch, wash the area thoroughly with water.
2. Obtain a pipetting bulb, a 50-mL Erlenmeyer flask with dry stopper, a 400-mL beaker of distilled water equilibrated to room temperature, and a thermometer.
3. Determine the mass of the flask and stopper, and record it to the nearest 0.1 mg. Do not touch the flask with your fingers after this weighing. Use tongs or a folded strip of waxed paper.
4. Determine and record the temperature of the water.
5. Pipet 10.00 mL of the distilled water into the flask using the technique described on page 45. Stopper the flask, determine the mass of the flask and the water that it contains, and record the mass.
6. In the same way, add a second pipet of water to the flask; remove the stopper just before the addition. Replace the stopper, and once again determine and record the mass of the flask and the water. Following each trial, determine the mass of water added to the flask by the pipet.
7. Repeat this process until you have determined four consecutive masses of water that agree within a range of 0.02 g. If the determinations of the mass of water delivered by the pipet do not agree within this range, your pipetting technique may be suspect. Consult your instructor for assistance in finding the source of your error, and then repeat the experiment until you are able to deliver four consecutive volumes of water with the precision cited above.
8. Correct the mass for buoyancy as described on page 29, and calculate the volume of the pipet in milliliters.
9. Report the mean, the standard deviation, and the relative standard deviation of the volume of your pipet. Calculate and report the 95% confidence interval for the volume of your pipet.

27A-5   Reading Buret Sections

Discussion

The following exercise will give you practice in reading a buret and will confirm the accuracy of your readings.

Procedure

1. Obtain a set of five buret sections from your instructor.
2. Invert each section, and tap the section lightly to remove any solvent that might remain in the sealed tip.

3. Record the number and reading of each buret section on the form provided. Use a buret reading card to observe the readings to the nearest 0.01 mL.
4. Check your readings of the buret sections against the known values provided by your instructor.

27A-6 Reading a Buret

Discussion

The following exercise demonstrates the proper way to use a buret.

Procedure

1. Mount a buret in a buret stand, and fill the buret with distilled water.
2. Wait at least 30 s before taking the initial reading. Use a buret reading card to take readings. A buret reading card can be easily constructed by applying a piece of black electrical tape to a three-by-five-inch card. Never adjust the volume of solution in a buret to exactly 0.00 mL. Attempting to do so will introduce bias into the measurement process and will waste time.
3. Now let about 5 mL run into a 250-mL Erlenmeyer flask. Wait at least 30 s, and take the "final reading." The amount of solution in the Erlenmeyer flask is equal to the difference between the final reading and the initial reading. Record the final reading in your laboratory notebook, and then ask your instructor to take the final reading. Compare the two readings. They should agree within 0.01 mL. Notice that the final digit in the buret reading is the estimation of the distance between two consecutive 0.1-mL marks on the buret.
4. Refill the buret, and take a new zero reading. Now add 30 drops to the Erlenmeyer flask, and take the final reading. Calculate the mean volume of one drop, then repeat this step except with 40 drops, and again calculate the mean volume of a drop. Record these results and compare them.
5. Finally, practice adding half-drops to the flask. Calculate the mean volume of several half-drops, and compare your results with those that you obtained with full drops. When you perform titrations, you should attempt to determine end points to within half a drop to achieve good precision.

27A-7 Sampling[1]

Discussion

In most analytical methods, only a small fraction of the entire population is analyzed. The results from the determination of an analyte in a laboratory sample are assumed to be similar to the concentration of the analyte in the whole population. Consequently, a laboratory sample taken from the entire batch must be representative of the population.

In this experiment, you will investigate how the sample size influences the uncertainty associated with the sampling step. Generally, the required sample

---

[1]J. E. Vitt and R. C. Engstrom, *J. Chem. Educ.,* **1999,** 76 (1), 99.

size must increase as the sample heterogeneity increases, as the fraction of the analyte decreases, or as the desired uncertainty decreases. The model system used in this experiment consists of a collection of plastic beads that are identical in size, shape, and density but that are different in color. If p represents the fraction of the particles of the analyte (beads of the first color), then $1 - p$ is the fraction of the second type of particles (beads of the second color). If a sample of n particles is drawn from the population, the number of particles of the analyte in the sample should be np. It can be shown that the standard deviation of the number of particles of analyte np obtained from a sample of the two-component mixture is $\sqrt{np(1 - p)}$. The relative standard deviation (RSD) is then

$$\text{RSD} = \frac{\sqrt{np(1 - p)}}{np} = \sqrt{\frac{1 - p}{np}}$$

This equation suggests that as the number of particles sampled increases, the relative uncertainty decreases. Using a mixture of beads of two colors, you will determine the uncertainty of sampling as a function of sample size.

Procedure

1. Stir the container of beads thoroughly, and withdraw a sample of beads using a small beaker. Make sure that the beaker is full to the top but not overflowing.
2. Empty the beads into a counting tray, and count the number of beads of each color.
3. Repeat step 1 using a medium-size beaker and then the larger beaker. Record the total number of beads in your sample and the percentage of beads of a particular color specified by your instructor. Each student in your class will collect and count three similar samples and enter the data on a class chart that will be provided by your instructor. After all data are entered, the chart will be copied and distributed to all students in your class.

Calculations

1. Using the compiled class data, calculate the mean percentage of beads of the specified color and the relative standard deviation of that percentage for each sample size.
2. Using the equation given above based on sampling theory, calculate the theoretical relative standard deviation using the values of p and the mean number of particles for each of the three sample sizes.
3. Compare your class data with the theoretical result. Does the relative standard deviation decrease as the sample size increases as predicted by sampling theory?
4. Use the equation for the relative standard deviation to find the number of beads that would have to be sampled to achieve a relative standard deviation of 0.002.
5. Suggest two reasons why the theory above might not be adequate to describe the sampling of many materials for chemical analysis.

27B   GRAVIMETRIC METHODS OF ANALYSIS

General aspects, calculations, and typical applications of gravimetric analysis are discussed in Chapter 8.

27B-1   The Gravimetric Determination of Chloride in a Soluble Sample

Discussion

The chloride content of a soluble salt can be determined by precipitation as silver chloride:

$$Ag^+ + Cl^- \rightarrow AgCl(s)$$

The precipitate is collected in a weighed filtering crucible and washed. After the precipitate has been dried to constant mass at 110°C, its mass is then determined.

The solution containing the sample is kept slightly acidic during the precipitation to eliminate possible interference from anions of weak acids (such as CO_3^{2-}) that form sparingly soluble silver salts in a neutral environment. A moderate excess of silver ion is needed to diminish the solubility of silver chloride, but a large excess is avoided to minimize coprecipitation of silver nitrate.

Silver chloride forms first as a colloid and is subsequently coagulated with heat. Nitric acid and the small excess of silver nitrate promote coagulation by providing a moderately high electrolyte concentration. Nitric acid in the wash solution maintains the electrolyte concentration and eliminates the possibility of peptization during the washing step; the acid subsequently decomposes to give volatile products when the precipitate is dried. See Section 8A-2 for additional information concerning the properties and treatment of colloidal precipitates.

In common with other silver halides, finely divided silver chloride undergoes photodecomposition:

$$2AgCl(s) \xrightarrow{h\nu} 2Ag(s) + Cl_2(g)$$

The elemental silver produced in this reaction is responsible for the violet color that develops in the precipitate. In principle, this reaction leads to low results for chloride ion. In practice, however, its effect is negligible provided you avoid direct and prolonged exposure of the precipitate to sunlight.

If photodecomposition of silver chloride occurs before filtration, the additional reaction

$$3Cl_2(aq) + 3H_2O + 5Ag^+ \rightarrow 5AgCl(s) + ClO_3^- + 6H^+$$

tends to cause high results.

Some photodecomposition of silver chloride is inevitable as the analysis is ordinarily performed. It is worthwhile to minimize exposure of the solid to intense sources of light as far as possible.

Because silver nitrate is expensive, any unused reagent should be collected in a storage container; similarly, precipitated silver chloride should be retained after the analysis is complete.[2]

---

[2]Silver can be recovered from silver chloride and from surplus reagent by reduction with ascorbic acid; see J. W. Hill and L. Bellows, *J. Chem. Educ.,* **1986,** *63* (4), 357; see also J. P. Rawat and S. Iqbal M. Kamoonpuri, *J. Chem. Educ.,* **1986,** *63* (6), 537 for recovery (as $AgNO_3$) based on ion exchange. For a potential hazard in the recovery of silver nitrate, see D. D Perrin, W. L. F. Armarego, and D. R. Perrin, *Chemistry International,* **1987,** *9* (1), 3.

Procedure

Clean three medium-porosity sintered-glass or porcelain filtering crucibles by allowing about 5 mL of concentrated HNO_3 to stand in each crucible for about 5 min. Use a vacuum (Figure 2-16) to draw the acid through the crucible. Rinse each crucible with three portions of tap water, and then discontinue the vacuum. Next, add about 5 mL of 6 M NH_3 and wait about 5 min before drawing it through the filter. Finally, rinse each crucible with six to eight portions of distilled or deionized water. Provide each crucible with an identifying mark. Dry the crucibles to constant mass by heating at 110°C while the other steps in the analysis are being carried out. The first drying should be for at least 1 hr; subsequent heating periods can be somewhat shorter (30 to 40 min). This process of heating and drying should be repeated until the mass becomes constant to within 0.2 to 0.3 mg.

Transfer the unknown to a weighing bottle and dry it at 110°C (Figure 2-9) for 1 to 2 hr; allow the bottle and contents to cool to room temperature in a desiccator. Weigh (to the nearest 0.1 mg) individual samples by difference into 400-mL beakers (note 1). Dissolve each sample in about 100 mL of distilled water to which 2 to 3 mL of 6 M HNO_3 has been added.

Slowly, and with good stirring, add 0.2 M $AgNO_3$ to each of the cold sample solutions until AgCl is observed to coagulate (notes 2 and 3), and then introduce an additional 3 to 5 mL. Heat almost to boiling, and digest the solids for about 10 min. Add a few drops of $AgNO_3$ to confirm that precipitation is complete. If more precipitate forms, add about 3 mL of $AgNO_3$, digest, and again test for completeness of precipitation. Pour any unused $AgNO_3$ into a waste container (*not* into the original reagent bottle). Cover each beaker, and store in a dark place for at least 2 hr and preferably until the next laboratory period.

Read the instructions for filtration in Section 2F-2. Decant the supernatant liquids through weighed filtering crucibles. Wash the precipitates several times (while they are still in the beaker) with a solution consisting of 2 to 5 mL of 6 M HNO_3 per liter of distilled water; decant these washings through the filters. Quantitatively transfer the AgCl from the beakers to the individual crucibles with fine streams of wash solution; use rubber policemen to dislodge any particles that adhere to the walls of the beakers. Continue washing until the filtrates are essentially free of Ag^+ ion (note 4).

Dry the precipitate at 110°C for at least 1 hr. Store the crucibles in a desiccator while they cool. Determine the mass of the crucibles and their contents. Repeat the cycle of heating, cooling, and weighing until consecutive weighings agree to within 0.2 mg. Calculate the percentage of Cl^- in the sample.

Upon completion of the analysis, remove the precipitates by gently tapping the crucibles over a piece of glazed paper. Transfer the collected AgCl to a container for silver wastes. Remove the last traces of AgCl by filling the crucibles with 6 M NH_3 and allowing them to stand.

Notes

1. Consult with the instructor concerning an appropriate sample size.
2. Determine the approximate amount of $AgNO_3$ needed by calculating the volume that would be required if the unknown were pure NaCl.
3. Use a separate stirring rod for each sample, and leave it in its beaker throughout the determination.
4. To test the washings for Ag^+, collect a small volume in a test tube and add a few drops of HCl. Washing is judged complete when little or no turbidity develops.

Be sure to label your beakers and crucibles.

Digest means to heat an unstirred precipitate in the *mother liquid*, that is, the solution from which it is formed.

27B-2 The Gravimetric Determination of Nickel in Steel

Discussion

The nickel in a steel sample can be precipitated from a slightly alkaline medium with an alcoholic solution of dimethylglyoxime (Section 8D-3). Interference from iron(III) is eliminated by masking with tartaric acid. The product is freed of moisture by drying at 110° C.

The bulky character of nickel dimethylglyoxime limits the mass of nickel that can be accommodated conveniently and thus the sample mass. Care must also be taken to control the excess of alcoholic dimethylglyoxime used. If too much is added, the alcohol concentration becomes sufficient to dissolve appreciable amounts of the nickel dimethylglyoxime, which leads to low results. If the alcohol concentration becomes too low, however, some of the reagent may precipitate and cause a positive error.

Iron(III) forms a highly stable complex with tartrate ion, which prevents it from precipitating as $Fe_2O_3 \cdot xH_2O$ in slightly alkaline solutions.

Preparation of Solutions

1. *Dimethylglyoxime, 1% (w/v)*. Dissolve 10 g of dimethylglyoxime in 1 L of ethanol. (This solution is sufficient for about 50 precipitations.)
2. *Tartaric acid, 15% (w/v)*. Dissolve 225 g of tartaric acid in sufficient water to give 1500 mL of solution. Filter before use if the solution is not clear. (This solution is sufficient for about 50 precipitations.)

Procedure

Clean and mark three medium-porosity sintered-glass crucibles (note 1); bring them to constant mass by drying at 110°C for at least 1 hr.

Weigh (to the nearest 0.1 mg) samples containing between 30 and 35 mg of nickel into individual 400-mL beakers (note 2). In the hood, dissolve each sample in about 50 mL of 6 M HCl with gentle warming. Carefully add approximately 15 mL of 6 M HNO_3, and boil gently to expel any oxides of nitrogen that may have been produced. Dilute to about 200 mL, and heat to boiling. Introduce about 30 mL of 15% tartaric acid and sufficient concentrated NH_3 to produce a faint odor of NH_3 in the vapors over the solutions (note 3); then add another 1 to 2 mL of NH_3. If the solutions are not clear at this stage, proceed as directed in note 4. Make the solutions acidic with HCl (no odor of NH_3), heat to 60 to 80°C, and add about 20 mL of the 1% dimethylglyoxime solution. With good stirring, add 6 M NH_3 until a slight excess exists (faint odor of NH_3) plus an additional 1 to 2 mL. Digest the precipitates for 30 to 60 min, cool for at least 1 hr, and filter.

Wash the solids with water until the washings are free of Cl^- (note 5). Bring the crucibles and their contents to constant mass at 110°C. Report the percentage of nickel in the sample. The dried precipitate has the composition $Ni(C_4H_7O_2N_2)_2$ (288.92 g/mol).

Notes

1. Medium-porosity porcelain filtering crucibles or Gooch crucibles with glass pads can be substituted for sintered-glass crucibles in this determination.
2. Use a separate stirring rod for each sample, and leave it in the beaker throughout.

3. The presence or absence of excess NH_3 is readily established by odor; use a waving motion with your hand to waft the vapors toward your nose.
4. If $Fe_2O_3 \cdot xH_2O$ forms upon addition of NH_3, acidify the solution with HCl, introduce additional tartaric acid, and neutralize again. Alternatively, remove the solid by filtration. Thorough washing with a hot NH_3/NH_4Cl solution is required; the washings are combined with the solution containing the bulk of the sample.
5. Test the washings for Cl^- by collecting a small portion in a test tube, acidifying with HNO_3, and adding a drop or two of 0.1 M $AgNO_3$. Washing is judged complete when little or no turbidity develops.

27C NEUTRALIZATION TITRATIONS

Neutralization titrations are performed with standard solutions of strong acids or bases. Although a single solution (of either acid or base) is sufficient for the titration of a given type of analyte, it is convenient to have standard solutions of both acid and base available in the event that a back-titration is needed to locate end points more exactly. The concentration of one solution is established by titration against a primary standard; the concentration of the other is then determined from the acid/base ratio (that is, the volume of acid needed to neutralize 1.000 mL of the base).

27C-1 The Effect of Atmospheric Carbon Dioxide on Neutralization Titrations

Water in equilibrium with the atmosphere is about 1×10^{-5} M in carbonic acid as a consequence of the equilibrium

$$CO_2(g) + H_2O \rightleftharpoons H_2CO_3(aq)$$

At this concentration level, the amount of 0.1-M base consumed by the carbonic acid in a typical titration is negligible. With more dilute reagents (<0.05 M), however, the water used as a solvent for the analyte and in the preparation of reagents must be freed of carbonic acid by boiling for a brief period.

Water that has been purified by distillation rather than by deionization is often supersaturated with carbon dioxide and may thus contain sufficient acid to affect the results of an analysis.[3] The instructions that follow are based on the assumption that the amount of carbon dioxide in the water supply can be neglected without causing serious error. For further discussion on the effects of carbon dioxide in neutralization titrations, see Section 14A-3.

---

[3]Water that is to be used for neutralization titrations can be tested by adding 5 drops of phenolphthalein to a 500-mL portion. Less than 0.2 to 0.3 mL of 0.1 M OH^- should suffice to produce the first faint pink color of the indicator. If a larger volume is needed, the water should be boiled and cooled before it is used to prepare standard solutions or to dissolve samples.

27C-2 Preparation of Indicator Solutions for Neutralization Titrations

Discussion

The theory of acid/base indicators is discussed in Section 12A-2. An indicator exists for virtually any pH range between 1 and 13.[4] Directions follow for the preparation of indicator solutions suitable for most neutralization titrations.

Procedure

Stock solutions ordinarily contain between 0.5 and 1.0 g of indicator per liter. (One liter of indicator is sufficient for hundreds of titrations.)

1. *Bromocresol green.* Dissolve the sodium salt directly in distilled water.
2. *Phenolphthalein, thymolphthalein.* Dissolve the solid indicator in a solution consisting of 800 mL of ethanol and 200 mL of distilled or deionized water.

27C-3 Preparation of Dilute Hydrochloric Acid Solutions

Discussion

The preparation and standardization of acids are considered in Sections 14A-1 and 14A-2.

Procedure

For a 0.1-M solution, add about 8 mL of concentrated HCl to about 1 L of distilled water (note). Mix thoroughly, and store in a glass-stoppered bottle.

Note

It is advisable to eliminate CO_2 from the water by a preliminary boiling if very dilute solutions (<0.05 M) are being prepared.

27C-4 Preparation of Carbonate-Free Sodium Hydroxide

Discussion

See Sections 14A-3 and 14A-4 for information concerning the preparation and standardization of bases.

---

[4]See, for example, J. Beukenkemp and W. Rieman III, in *Treatise on Analytical Chemistry,* I. M. Kolthoff and P. J. Elving, Eds., Part I, Vol. 11 (New York: Wiley, 1974), pp. 6987–7001.

Procedure

Transfer 1 L of distilled water to a polyethylene storage bottle (see the note in Section 27C-3). Decant 4 to 5 mL of 50% NaOH into a small container (note 2), add it to the water, and *mix thoroughly. Use extreme care in handling* 50% NaOH, which is highly corrosive. If the reagent comes into contact with skin, *immediately* flush the area with *copious* amounts of water.

Protect the solution from unnecessary contact with the atmosphere.

Notes

1. A solution of base that will be used up within two weeks can be stored in a tightly capped polyethylene bottle. After each removal of base, squeeze the bottle while tightening the cap to minimize the air space above the reagent. The bottle will become embrittled after extensive use as a container for bases.
2. Be certain that any solid Na_2CO_3 in the 50% NaOH has settled to the bottom of the container and that the decanted liquid is absolutely clear. If necessary, filter the base through a glass mat in a Gooch crucible; collect the clear filtrate in a test tube inserted in the filter flask.

27C-5   The Determination of the Acid/Base Ratio

Discussion

If both acid and base solutions have been prepared, it is useful to determine their volumetric combining ratio. Knowledge of this ratio and the concentration of one solution permits calculation of the molarity of the other.

Procedure

Instructions for placing a buret into service are given in Sections 2G-4 and 2G-6; consult these instructions if necessary. Place a test tube or small beaker over the top of the buret that holds the NaOH solution to minimize contact between the solution and the atmosphere.

Record the initial volumes of acid and base in the burets to the nearest 0.01 mL. Deliver 35 to 40 mL of the acid into a 250-mL conical flask. Touch the tip of the buret to the inside wall of the flask, and rinse down with a little distilled water. Add two drops of phenolphthalein (note 1) and then sufficient base to render the solution a definite pink. Introduce acid dropwise to discharge the color, and again rinse down the walls of the flask. Carefully add base until the solution again acquires a faint pink hue that persists for at least 30 s (notes 2, 3). Record the final buret volumes (again, to nearest 0.01 mL). Repeat the titration. Calculate the acid/base volume ratio. The ratios for duplicate titrations should agree to within 1 to 2 ppt. Perform additional titrations, if necessary, to achieve this order of precision.

Notes

1. The volume ratio can also be determined with an indicator that has an acid transition range, such as bromocresol green. If the NaOH is contaminated with carbonate, the ratio obtained with this indicator will differ significantly from the value obtained with phenolphthalein. In general, the acid/base ratio should be evaluated with the indicator that is to be used in subsequent titrations.

2. Fractional drops can be formed on the buret tip, touched to the wall of the flask, and then rinsed down with a small amount of water.
3. The phenolphthalein end point fades as CO_2 is absorbed from the atmosphere.

27C-6   Standardization of Hydrochloric Acid against Sodium Carbonate

Discussion

See Section 14A-2.

Procedure

Dry a quantity of primary-standard Na_2CO_3 for about 2 hr at 110°C (Figure 2-9), and cool in a desiccator. Weigh individual 0.20- to 0.25-g samples (to the nearest 0.1 mg) into 250-mL conical flasks, and dissolve each in about 50 mL of distilled water. Introduce 3 drops of bromocresol green, and titrate with HCl until the solution just begins to change from blue to green. Boil the solution for 2 to 3 min, cool to room temperature (note 1), and complete the titration (note 2).

Determine an indicator correction by titrating approximately 100 mL of 0.05 M NaCl and 3 drops of indicator. Boil briefly, cool, and complete the titration. Subtract any volume needed for the blank from the titration volumes. Calculate the concentration of the HCl solution.

Notes

1. The indicator should change from green to blue as CO_2 is removed during heating. If no color change occurs, an excess of acid was added originally. This excess can be back-titrated with base provided that the acid/base combining ratio is known; otherwise, the sample must be discarded.
2. It is permissible to back-titrate with base to establish the end point with greater certainty.

27C-7   Standardization of Sodium Hydroxide against Potassium Hydrogen Phthalate

Discussion

See Section 14A-4.

Procedure

Dry a quantity of primary-standard potassium hydrogen phthalate (KHP) for about 2 hr at 110°C (Figure 2-9), and cool in a desiccator. Weigh individual 0.7- to 0.8-g samples (to the nearest 0.1 mg) into 250-mL conical flasks, and dissolve each in 50 to 75 mL of distilled water. Add 2 drops of phenolphthalein; titrate with base until the pink color of the indicator persists for 30 s (note). Calculate the concentration of the NaOH solution.

Note

It is permissible to back-titrate with acid to establish the end point more precisely. Record the volume used in the back-titration. Use the acid/base ratio to calculate the net volume of base used in the standardization.

27C-8   Determining Potassium Hydrogen Phthalate in an Impure Sample

Discussion

The unknown is a mixture of KHP and a neutral salt. This analysis is conveniently performed concurrently with the standardization of the base.

Procedure

Consult with the instructor concerning an appropriate sample size. Then follow the directions in Section 27C-7.

27C-9   Determining the Acid Content of Vinegars and Wines

Discussion

The total acid content of a vinegar or a wine is readily determined by titration with a standard base. It is customary to report the acid content of vinegar in terms of acetic acid, the principal acidic constituent, even though other acids are present. Similarly, the acid content of a wine is expressed as percent tartaric acid, even though there are other acids in the sample. Most vinegars contain about 5% acid (w/v) expressed as acetic acid; wines ordinarily contain somewhat under 1% acid (w/v) expressed as tartaric acid.

Procedure

1. *If the unknown is a vinegar* (note 1), pipet 25.00 mL into a 250-mL volumetric flask, and dilute to the mark with distilled water. Mix thoroughly, and pipet 50.00-mL aliquots into 250-mL conical flasks. Add about 50 mL of water and 2 drops of phenolphthalein (note 2) to each, and titrate with standard 0.1 M NaOH to the first permanent ($\approx$30 s) pink color.

   Report the acidity of the vinegar as percent (w/v) CH_3COOH (60.053 g/mol).

2. *If the unknown is a wine,* pipet 50.00-mL aliquots into 250-mL conical flasks, add about 50 mL of distilled water and 2 drops of phenolphthalein to each (note 2), and titrate to the first permanent ($\approx$30 s) pink color.

   Express the acidity of the sample as percent (weight/volume) tartaric acid $C_2H_4O_2$ $(COOH)_2$ (150.09 g/mol) (note 3).

Notes

1. The acidity of bottled vinegar tends to decrease on exposure to air. It is recommended that unknowns be stored in individual vials with snug covers.
2. The amount of indicator used should be increased as necessary to make the color change visible in colored samples.
3. Tartaric acid has two acidic hydrogens, both of which are titrated at a phenolphthalein end point.

27C-10   The Determination of Sodium Carbonate in an Impure Sample

Discussion

The titration of sodium carbonate is discussed in Section 14A-2 in connection with its use as a primary standard; the same considerations apply for the determination of carbonate in an unknown that has no interfering contaminants.

Procedure

Dry the unknown at 110°C for 2 hr, and then cool in a desiccator. Consult with the instructor on an appropriate sample size. Then follow the instructions in Section 27C-6.
Report the percentage of Na_2CO_3 in the sample.

27C-11   The Determination of Amine Nitrogen by the Kjeldahl Method

Discussion

These directions are suitable for the Kjeldahl determination of protein in materials such as blood meal, wheat flour, pasta products, dry cereals, and pet foods. A simple modification permits the analysis of unknowns that contain more highly oxidized forms of nitrogen.[5]

In the Kjeldahl method (see Section 14B-1), the organic sample is digested in hot concentrated sulfuric acid, which converts amine nitrogen in the sample to ammonium sulfate. After cooling, the sulfuric acid is neutralized by the addition of an excess of concentrated sodium hydroxide. The ammonia liberated by this treatment is then distilled into a measured excess of a standard solution of acid; the excess is determined by back-titration with standard base.

Figure 27-1 illustrates typical equipment for a Kjeldahl distillation. The long-necked container, which is used for both digestion and distillation, is called a *Kjeldahl flask*. In the apparatus in Figure 27-1a, the base is added slowly by

---

[5]See *Official Methods of Analysis,* 14th ed. (Washington, D.C.: Association of Official Analytical Chemists, 1984), p. 16.

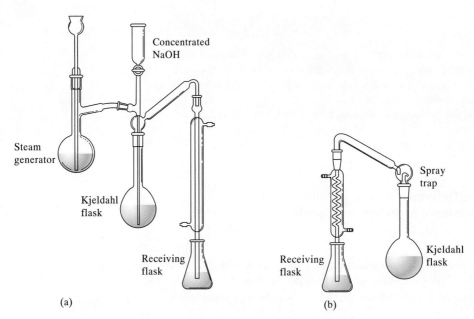

(a)

(b)

Figure 27-1 Kjeldahl distillation apparatus.

partially opening the stopcock from the NaOH storage vessel, the liberated ammonia is then carried to the receiving flask by steam distillation.

In an alternative method (Figure 27-1b), a dense, concentrated sodium hydroxide solution is carefully poured down the side of the Kjeldahl flask to form a second, lower layer. The flask is then quickly connected to a spray trap and an ordinary condenser before loss of ammonia can occur. Only then are the two layers mixed by gentle swirling of the flask.

Quantitative collection of ammonia requires the tip of the condenser to extend into the liquid in the receiving flask throughout the distillation step. The tip must be removed before heating is discontinued, however. Otherwise, the liquid will be drawn back into the apparatus.

Two methods are commonly used for collecting and determining the ammonia liberated from the sample. In one, the ammonia is distilled into a measured volume of standard acid. After the distillation is complete, the excess acid is back-titrated with standard base. An indicator with an acidic transition range is required because of the acidity of the ammonium ions present at equivalence. A convenient alternative, which requires only one standard solution, involves the collection of the ammonia in an unmeasured excess of boric acid, which retains the ammonia by the reaction

$$H_3BO_3 + NH_3 \rightarrow NH_4^+ + H_2BO_3^-$$

The dihydrogen borate ion produced is a reasonably strong base that can be titrated with a standard solution of hydrochloric acid:

$$H_2BO_3^- + H_3O^+ \rightarrow H_3BO_3 + H_2O$$

At the equivalence point, the solution contains boric acid and ammonium ions; an indicator with an acidic transition interval (such as bromocresol green) is again required.

Procedure

Preparing Samples

Consult with the instructor on sample size. *If the unknown is powdered* (such as blood meal), weigh samples onto individual 9-cm filter papers (note 1). Fold the paper around the sample, and drop each into a Kjeldahl flask (the paper keeps the samples from clinging to the neck of the flask). *If the unknown is not powdered* (such as breakfast cereals or pasta), the samples can be weighed by difference directly into the Kjeldahl flasks.

Add 25 mL of concentrated H_2SO_4, 10 g of powdered K_2SO_4, and the catalyst (note 2) to each flask.

Digestion

Clamp the flasks in a slanted position in a hood or vented digestion rack. Heat carefully to boiling. Discontinue heating briefly if foaming becomes excessive; never allow the foam to reach the neck of the flask. Once foaming ceases and the acid is boiling vigorously, the samples can be left unattended; prepare the distillation apparatus during this time. Continue digestion until the solution becomes colorless or faint yellow; 2 to 3 hr may be needed for some materials. If necessary, *cautiously* replace the acid lost by evaporation.

When digestion is complete, discontinue heating and allow the flasks to cool to room temperature; swirl the flasks if the contents show signs of solidifying. Cautiously add 250 mL of water to each flask and again allow the solution to cool to room temperature.

Distillation of Ammonia

Arrange a distillation apparatus similar to that shown in Figure 27-1. Pipet 50.00 mL of standard 0.1 M HCl into the receiver flask (note 3). Clamp the flask so that the tip of the adapter extends below the surface of the standard acid. Circulate water through the condenser jacket.

Hold the Kjeldahl flask at an angle, and gently introduce about 60 mL of 50% (w/v) NaOH solution, taking care to minimize mixing with the solution in the flask. *The concentrated caustic solution is highly corrosive and should be handled with great care* (note 4). Add several pieces of granulated zinc (note 5) and a small piece of litmus paper. *Immediately* connect the Kjeldahl flask to the spray trap. Cautiously mix the contents by gentle swirling. The litmus paper should be blue after mixing is complete, indicating that the solution is basic.

Bring the solution to a boil, and distill at a steady rate until one-half to one-third of the original volume remains. Control the rate of heating to prevent the liquid in the receiver flask from being drawn back into the Kjeldahl flask. After distillation is judged complete, lower the receiver flask to bring the adapter well clear of the liquid. Discontinue heating, disconnect the apparatus, and rinse the inside of the condenser with small portions of distilled water, collecting the washings in the receiver flask. Add 2 drops of bromocresol green to the receiver flask, and titrate the residual HCl with standard 0.1 M NaOH to the color change of the indicator.

Report the percentage of nitrogen and the percentage of protein (note 6) in the unknown.

Notes

1. If filter paper is used to hold the sample, carry a similar piece through the analysis as a blank. Acid-washed filter paper is frequently contaminated with measurable amounts of ammonium ion and should be avoided if possible.

2. Any of the following catalyze the digestion: a crystal of $CuSO_4$, 0.1 g of selenium, or 0.2 g of $CuSeO_3$. The catalyst can be omitted, if desired.

3. A modification of this procedure uses about 50 mL of 4% boric acid solution in lieu of the standard HCl in the receiver flask. After distillation is complete, the ammonium borate produced is titrated with standard 0.1 M HCl, with 2 to 3 drops of bromocresol green as indicator.

4. If any sodium hydroxide solution comes into contact with your skin, wash the affected area *immediately* with copious amounts of water.

5. Granulated zinc (10 to 20 mesh) is added to minimize bumping during the distillation; it reacts slowly with the base to give small bubbles of hydrogen that prevent superheating of the liquid.

6. The percentage of protein in the unknown is calculated by multiplying the % N by an appropriate factor: 5.70 for cereals, 6.25 for meats, and 6.38 for dairy products.

27D   PRECIPITATION TITRATIONS

As noted in Section 15B-2, most precipitation titrations make use of a standard silver nitrate solution as titrant. Directions follow for the volumetric titration of chloride ion using an adsorption indicator.

27D-1   Preparing a Standard Silver Nitrate Solution

Procedure

Use a top-loading balance to transfer the approximate mass of $AgNO_3$ to a weighing bottle (note 1). Dry at 110°C for about 1 hr but not much longer (note 2), and then cool to room temperature in a desiccator. Weigh the bottle and contents (to the nearest 0.1 mg). Transfer the bulk of the $AgNO_3$ to a volumetric flask using a powder funnel. Cap the weighing bottle and reweigh it and any solid that remains. Rinse the powder funnel thoroughly. Dissolve the $AgNO_3$, dilute to the mark with water, and mix well (note 3). Calculate the molar concentration of this solution.

Notes

1. Consult with the instructor concerning the volume and concentration of $AgNO_3$ to be prepared. The mass of $AgNO_3$ to be taken is as follows:

Silver Ion Concentration, M	Approximate Mass (g) of $AgNO_3$ Needed to Prepare		
	1000 mL	500 mL	250 mL
0.10	16.9	8.5	4.2
0.05	8.5	4.2	2.1
0.02	3.4	1.8	1.0

2. Prolonged heating causes partial decomposition of $AgNO_3$. Some discoloration may occur, even after only 1 hr at 110°C; the effect of this decomposition on the purity of the reagent is ordinarily imperceptible.

3. Silver nitrate solutions should be stored in a dark place when not in use.

27D-2   The Determination of Chloride by Titration with an Adsorption Indicator

Discussion

In this titration, the anionic adsorption indicator dichlorofluorescein is used to locate the end point. With the first excess of titrant, the indicator becomes incorporated in the counter-ion layer surrounding the silver chloride and imparts color to the solid (page 357). To obtain a satisfactory color change, it is desirable to maintain the particles of silver chloride in the colloidal state. Dextrin is added to the solution to stabilize the colloid and prevent its coagulation.

Preparation of Solutions

Dichlorofluorescein indicator (sufficient for several hundred titrations). Dissolve 0.2 g of dichlorofluorescein in a solution prepared by mixing 75 mL of ethanol and 25 mL of water.

Procedure

Dry the unknown at 110°C for about 1 hr; allow it to return to room temperature in a desiccator. Weigh individual samples (to the nearest 0.1 mg) into individual conical flasks, and dissolve them in appropriate volumes of distilled water (note 1). To each, add about 0.1 g of dextrin and 5 drops of indicator. Titrate (note 2) with $AgNO_3$ to the first permanent pink color of silver dichlorofluoresceinate. Report the percentage of Cl^- in the unknown.

Notes

1. Use 0.25-g samples for 0.1 M $AgNO_3$ and about half that amount for 0.05 M reagent. Dissolve the former in about 200 mL of distilled water and the latter in about 100 mL. If 0.02 M $AgNO_3$ is to be used, weigh a 0.4-g sample into a 500-mL volumetric flask, and take 50-mL aliquots for titration.
2. Colloidal AgCl is sensitive to photodecomposition, particularly in the presence of the indicator; attempts to perform the titration in direct sunlight will fail. If photodecomposition appears to be a problem, establish the approximate end point with a rough preliminary titration, and use this information to estimate the volumes of $AgNO_3$ needed for the other samples. For each subsequent sample, add the indicator and dextrin only after most of the $AgNO_3$ has been added, and then complete the titration without delay.

27E   COMPLEX-FORMATION TITRATIONS WITH EDTA

See Section 15D for a discussion of the analytical uses of EDTA as a chelating reagent. Directions follow for a direct titration of magnesium and a determination of the hardness of a natural water.

27E-1   Preparation of Solutions

Procedure

A pH-10 buffer and an indicator solution are needed for these titrations.

1. *Buffer solution, pH 10* (sufficient for 80 to 100 titrations). Dilute 57 mL of concentrated NH_3 and 7 g of NH_4Cl in sufficient distilled water to give 100 mL of solution.
2. *Eriochrome Black T (Erio T) indicator* (sufficient for about 100 titrations). Dissolve 100 mg of the solid in a solution containing 15 mL of ethanolamine and 5 mL of absolute ethanol. This solution should be freshly prepared every two weeks; refrigeration slows its deterioration.

27E-2   Preparing a Standard 0.01 M EDTA Solution

Discussion

See Section 15D-1 for a description of the properties of reagent-grade $Na_2H_2Y \cdot 2H_2O$ and its use in the direct preparation of standard EDTA solutions.

Procedure

Dry about 4 g of the purified dihydrate $Na_2H_2Y \cdot 2H_2O$ (note 1) at 80°C to remove superficial moisture. Cool to room temperature in a desiccator. Weigh (to the nearest milligram) about 3.8 g into a 1-L volumetric flask (note 2). Use a powder funnel to ensure quantitative transfer; rinse the funnel well with water before removing it from the flask. Add 600 to 800 mL of water (note 3) and swirl periodically. Dissolution may take 15 min or longer. When all the solid has dissolved, dilute to the mark with water and mix well (note 4). In calculating the molarity of the solution, correct the weight of the salt for the 0.3% moisture it ordinarily retains after drying at 80°C.

Notes

1. Directions for the purification of the disodium salt are described by W. J. Blaedel and H. T. Knight, *Anal. Chem.*, **1954,** *26* (4), 741.
2. The solution can be prepared from the anhydrous disodium salt, if desired. The weight taken should be about 3.6 g.
3. Water used in the preparation of standard EDTA solutions must be totally free of polyvalent cations. If any doubt exists concerning its quality, pass the water through a cation-exchange resin before use.
4. As an alternative, an EDTA solution that is approximately 0.01 M can be prepared and standardized by direct titration against a Mg^{2+} solution of known concentration (using the directions in Section 27E-3).

27E-3   The Determination of Magnesium by Direct Titration

Discussion

See Section 15D-7.

Procedure

Submit a clean 500-mL volumetric flask to receive the unknown, dilute to the mark with water, and mix thoroughly. Transfer 50.00-mL aliquots to 250-mL conical flasks, add 1 to 2 mL of pH 10 buffer and 3 to 4 drops of Erio T indicator to each. Titrate with 0.01 M EDTA until the color changes from red to pure blue (notes 1, 2).

Express the results as parts per million of Mg^{2+} in the sample.

Notes

1. The color change tends to be slow in the vicinity of the end point. Care must be taken to avoid overtitration.
2. Other alkaline earths, if present, are titrated along with the Mg^{2+}; removal of Ca^{2+} and Ba^{2+} can be accomplished with $(NH_4)_2CO_3$. Most polyvalent cations are also titrated. Precipitation as hydroxides or the use of a masking reagent may be needed to eliminate this source of interference.

27E-4   The Determination of Hardness in Water

Discussion

See Section 15D-9.

Procedure

Acidify 100.0-mL aliquots of the sample with a few drops of HCl, and boil gently for a few minutes to eliminate CO_2. Cool, add 3 to 4 drops of methyl red, and neutralize with 0.1 M NaOH. Introduce 2 mL of pH 10 buffer, 3 to 4 drops of Erio T, and titrate with standard 0.01 M Na_2H_2Y to a color change from red to pure blue (note).

Report the results in terms of milligrams of $CaCO_3$ per liter of water.

Note

The color change is sluggish if Mg^{2+} is absent. In this event, add 1 to 2 mL of 0.1 M MgY^{2-} before starting the titration. This reagent is prepared by adding 2.645 g of $MgSO_4 \cdot 7H_2O$ to 3.722 g of $Na_2H_2Y \cdot 2H_2O$ in 50 mL of distilled water. This solution is rendered faintly alkaline to phenolphthalein and diluted to 100 mL. A small portion, mixed with pH 10 buffer and a few drops of Erio T indicator, should have a dull violet color. A single drop of 0.01 M EDTA solution should cause a color change to blue, whereas an equal volume of 0.01 Mg^{2+} should cause a change to red. If necessary, adjust the composition with EDTA or with Mg^{2+} until these criteria are met.

27F   TITRATIONS WITH POTASSIUM PERMANGANATE

The properties and uses of potassium permanganate are described in Section 18C-1. Directions follow for the determination of iron in an ore and calcium in a limestone.

27F-1   Preparation of 0.02 M Potassium Permanganate

Discussion

See page 452 for a discussion of the precautions needed in the preparation and storage of permanganate solutions.

Procedure

Dissolve about 3.2 g of $KMnO_4$ in 1 L of distilled water. Keep the solution at a gentle boil for about 1 hr. Cover and let stand overnight. Remove MnO_2 by filtration (note 1) through a fine-porosity filtering crucible (note 2) or through a Gooch crucible fitted with glass mats. Transfer the solution to a clean glass-stoppered bottle; store in the dark when not in use.

Notes

1. The heating and filtering can be omitted if the permanganate solution is standardized and used on the same day.
2. Remove the MnO_2 that collects on the fritted plate with 1 M H_2SO_4 containing a few milliliters of 3% H_2O_2, followed by a rinse with copious quantities of water.

27F-2   Standardization of Potassium Permanganate Solutions

Discussion

See Section 18C-1 for a discussion of primary standards for permanganate solutions. Directions follow for standardization with sodium oxalate.

Procedure

Dry about 1.5 g of primary-standard $Na_2C_2O_4$ at 110°C for at least 1 hr. Cool in a desiccator; weigh (to the nearest 0.1 mg) individual 0.2- to 0.3-g samples into 400-mL beakers. Dissolve each in about 250 mL of 1 M H_2SO_4. Heat each solution to 80 to 90°C, and titrate with $KMnO_4$ while stirring with a thermometer. The pink color imparted by one addition should be permitted to disappear before any further titrant is introduced (notes 1, 2). Reheat if the temperature drops below 60°C. Take the first persistent ($\approx$30 s) pink color as the end point (notes 3, 4). Determine a blank by titrating an equal volume of the 1 M H_2SO_4.

Correct the titration data for the blank, and calculate the concentration of the permanganate solution (note 5).

Notes

1. Promptly wash any $KMnO_4$ that spatters on the walls of the beaker into the bulk of the liquid with a stream of water.
2. Finely divided MnO_2 will form along with Mn^{2+} if the $KMnO_4$ is added too rapidly and will cause the solution to acquire a faint brown discoloration. Precipitate formation is not a serious problem so long as sufficient oxalate remains to reduce the MnO_2 to Mn^{2+}; the titration is simply discontinued until the brown color disappears. The solution must be free of MnO_2 at the end point.

3. The surface of the permanganate solution rather than the bottom of the meniscus can be used to measure titrant volumes. Alternatively, backlighting with a flashlight or a match will permit reading of the meniscus in the conventional manner.

4. A permanganate solution should not be allowed to stand in a buret any longer than necessary because partial decomposition to MnO_2 may occur. Freshly formed MnO_2 can be removed from a glass surface with 1 M H_2SO_4 containing a small amount of 3% H_2O_2.

5. As noted on page 454, this procedure yields molarities that are a few tenths of a percent low. For more accurate results, introduce from a buret sufficient permanganate to react with 90 to 95% of the oxalate (about 40 mL of 0.02 M $KMnO_4$ for a 0.3-g sample). Let the solution stand until the permanganate color disappears. Then warm to about 60°C and complete the titration, taking the first permanent pink ($\approx$30 s) as the end point. Determine a blank by titrating an equal volume of the 1 M H_2SO_4.

27F-3   Determining Calcium in a Limestone

Discussion

In common with a number of other cations, calcium is conveniently determined by precipitation with oxalate ion. The solid calcium oxalate is filtered, washed free of excess precipitating reagent, and dissolved in dilute acid. The oxalic acid liberated in this step is then titrated with standard permanganate or some other oxidizing reagent. This method is applicable to samples that contain magnesium and the alkali metals. Most other cations must be absent since they either precipitate or coprecipitate as oxalates and cause positive errors in the analysis.

Factors Affecting the Composition of Calcium Oxalate Precipitates.   It is essential that the mole ratio between calcium and oxalate be exactly unity in the precipitate and thus in solution at the time of titration. A number of precautions are needed to ensure this condition. For example, the calcium oxalate formed in a neutral or ammoniacal solution is likely to be contaminated with calcium hydroxide or with a basic calcium oxalate, either of which will cause low results. The formation of these compounds is prevented by adding the oxalate to an acidic solution of the sample and slowly forming the desired precipitate by the dropwise addition of ammonia. The coarsely crystalline calcium oxalate that is produced under these conditions is readily filtered. Losses resulting from the solubility of calcium oxalate are negligible above pH 4, provided washing is limited to freeing the precipitate of excess oxalate.

Coprecipitation of sodium oxalate becomes a source of positive error in the determination of calcium whenever the concentration of sodium in the sample exceeds that of calcium. The error from this source can be eliminated by reprecipitation.

Magnesium, if present in high concentration, may also be a source of contamination. An excess of oxalate ion helps prevent this interference through the formation of soluble oxalate complexes of magnesium. Prompt filtration of the calcium oxalate can also help prevent interference because of the pronounced tendency of magnesium oxalate to form supersaturated solutions from which precipitate formation occurs only after an hour or more. These measures do not

Reprecipitation is a method for minimizing coprecipitation errors by dissolving the initial precipitate and then reforming the solid.

suffice for samples that contain more magnesium than calcium. Here, reprecipitation of the calcium oxalate becomes necessary.

The Composition of Limestones.   Limestones are composed principally of calcium carbonate; dolomitic limestones contain large amounts of magnesium carbonate as well. Calcium and magnesium silicates, along with the carbonates and silicates of iron, aluminum, manganese, titanium, sodium, and other metals, are also present in smaller amounts.

Hydrochloric acid is an effective solvent for most limestones. Only silica, which does not interfere with the analysis, remains undissolved. Some limestones are more readily decomposed after they have been ignited; a few yield only to a carbonate fusion.

The method that follows is remarkably effective for determining calcium in most limestones. Iron and aluminum, in amounts equivalent to that of calcium, do not interfere. Small amounts of manganese and titanium can also be tolerated.

Procedure

Sample Preparation

Dry the unknown for 1 to 2 hr at 110°C, and cool in a desiccator. If the material is readily decomposed in acid, weigh 0.25- to 0.30-g samples (to the nearest 0.1 mg) into 250-mL beakers. Add 10 mL of water to each sample, and cover with a watch glass. Add 10 mL of concentrated HCl dropwise, taking care to avoid losses due to spattering as the acid is introduced.

Precipitation of Calcium Oxalate

Add 5 drops of saturated bromine water to oxidize any iron in the samples, and boil gently (HOOD) for 5 min to remove the excess Br_2. Dilute each sample solution to about 50 mL, heat to boiling, and add 100 mL of hot 6% (w/v) $(NH_4)_2C_2O_4$ solution. Add 3 to 4 drops of methyl red, and precipitate CaC_2O_4 by slowly adding 6 M NH_3. As the indicator starts to change color, add the NH_3 at a rate of one drop every 3 to 4 s. Continue until the solutions turn to the intermediate yellow-orange color of the indicator (pH 4.5 to 5.5). Allow the solutions to stand for no more than 30 min (note) and filter; medium-porosity filtering crucibles or Gooch crucibles with glass mats are satisfactory. Wash the precipitates with several 10-mL portions of cold water. Rinse the outside of the crucibles to remove residual $(NH_4)_2C_2O_4$, and return them to the beakers in which the CaC_2O_4 was formed.

Titration

Add 100 mL of water and 50 mL of 3 M H_2SO_4 to each of the beakers containing the precipitated calcium oxalate and the crucible. Heat to 80 to 90°C, and titrate with 0.02 M permanganate. The temperature should be above 60°C throughout the titration; reheat if necessary.

Report the percentage of CaO in the unknown.

Note

The period of standing can be longer if the unknown contains no Mg^{2+}.

27F-4   The Determination of Iron in an Ore

Discussion

The common ores of iron are hematite (Fe_2O_3), magnetite (Fe_3O_4), and limonite ($2Fe_2O_3 \cdot 3H_2O$). Steps in the analysis of these ores are (1) dissolution of the sample, (2) reduction of iron to the divalent state, and (3) titration of iron(II) with a standard oxidant.

The Decomposition of Iron Ores.   Iron ores often decompose completely in hot concentrated hydrochloric acid. The rate of attack by this reagent is increased by the presence of a small amount of tin(II) chloride. The tendency of iron(II) and iron(III) to form chloro complexes accounts for the effectiveness of hydrochloric acid over nitric or sulfuric acid as a solvent for iron ores.

Many iron ores contain silicates that may not be entirely decomposed by treatment with hydrochloric acid. Incomplete decomposition is indicated by a dark residue that remains after prolonged treatment with the acid. A white residue of hydrated silica, which does not interfere in any way, is indicative of complete decomposition.

The Prereduction of Iron.   Because part of or all the iron is in the trivalent state after decomposition of the sample, prereduction to iron(II) must precede titration with the oxidant. Any of the methods described in Section 18A-1 can be used. Perhaps the most satisfactory prereductant for iron is tin(II) chloride:

$$2Fe^{3+} + Sn^{2+} \rightarrow 2Fe^{2+} + Sn^{4+}$$

The only other common species reduced by this reagent are the high oxidation states of arsenic, copper, mercury, molybdenum, tungsten, and vanadium.

The excess reducing agent is eliminated by the addition of mercury(II) chloride:

$$Sn^{2+} + 2HgCl_2 \rightarrow Hg_2Cl_2(s) + Sn^{4+} + 2Cl^-$$

The slightly soluble mercury(I) chloride (Hg_2Cl_2) does not reduce permanganate, nor does the excess mercury(II) chloride ($HgCl_2$) reoxidize iron(II). Care must be taken, however, to prevent the occurrence of the alternative reaction

$$Sn^{2+} + HgCl_2 \rightarrow Hg(l) + Sn^{4+} + 2Cl^-$$

Elemental mercury reacts with permanganate and causes the results of the analysis to be high. The formation of mercury, which is favored by an appreciable excess of tin(II), is prevented by careful control of this excess and by the rapid addition of excess mercury(II) chloride. A proper reduction is indicated by the appearance of a small amount of a silky white precipitate after the addition of mercury(II). Formation of a gray precipitate at this juncture indicates the presence of metallic mercury; the total absence of a precipitate indicates that an insufficient amount of tin(II) chloride was used. In either event, the sample must be discarded.

The Titration of Iron(II).   The reaction of iron(II) with permanganate is smooth and rapid. The presence of iron(II) in the reaction mixture, however, *induces* the oxidation of chloride ion by permanganate, a reaction that does not ordinarily proceed rapidly enough to cause serious error. High results are obtained if this parasitic reaction is not controlled. Its effects can be eliminated through removal of the hydrochloric acid by evaporation with sulfuric acid or by introduction of *Zimmermann–Reinhardt reagent,* which contains manganese(II) in a fairly concentrated mixture of sulfuric and phosphoric acids.

The oxidation of chloride ion during a titration is believed to involve a direct reaction between this species and the manganese(III) ions that form as an intermediate in the reduction of permanganate ion by iron(II). The presence of manganese(II) in the Zimmermann–Reinhardt reagent is believed to inhibit the formation of chlorine by decreasing the potential of the manganese(III)/manganese(II) couple. Phosphate ion is believed to exert a similar effect by forming stable manganese(III) complexes. Moreover, phosphate ions react with iron(III) to form nearly colorless complexes so that the yellow color of the iron(II)/chloro complexes does not interfere with the end point.[6]

Preparation of Reagents

The following solutions suffice for about 100 titrations.

1. *Tin(II)chloride, 0.25 M.* Dissolve 60 g of iron-free $SnCl_2 \cdot 2H_2O$ in 100 mL of concentrated HCl; warm if necessary. After the solid has dissolved, dilute to 1 L with distilled water and store in a well-stoppered bottle. Add a few pieces of mossy tin to help preserve the solution.
2. *Mercury(II) chloride, 5% (w/v).* Dissolve 50 g of $HgCl_2$ in 1 L of distilled water.
3. *Zimmermann–Reinhardt reagent.* Dissolve 300 g of $MnSO_4 \cdot 4H_2O$ in 1 L of water. Cautiously add 400 mL of concentrated H_2SO_4, add 400 mL of 85% H_3PO_4, and dilute to 3 L.

Procedure

Sample Preparation

Dry the ore at 110°C for at least 3 hr, and then allow it to cool to room temperature in a desiccator. Consult with the instructor for a sample size that will require from 25 to 40 mL of standard 0.02 M $KMnO_4$. Weigh samples into 500-mL conical flasks. To each, add 10 mL of concentrated HCl and about 3 mL of 0.25 M $SnCl_2$ (note 1). Cover each flask with a small watch glass or Tuttle flask cover. Heat the flasks in a hood at just below boiling until the samples are decomposed and the undissolved solid, if any, is pure white (note 2). Use another 1 to 2 mL of $SnCl_2$ to eliminate any yellow color that may develop as the solutions are heated. Heat a blank consisting of 10 mL of HCl and 3 mL of $SnCl_2$ for the same amount of time.

After the ore has been decomposed, remove the excess Sn(II) by the dropwise addition of 0.02 M $KMnO_4$ until the solutions become faintly yellow. Dilute to about 15 mL. Add sufficient $KMnO_4$ solution to impart a faint pink color to the blank, and then decolorize with one drop of the $SnCl_2$ solution.

---

[6]The mechanism by which Zimmermann–Reinhardt reagent acts has been the subject of much study. For a discussion of this work, see H. A. Laitinen, *Chemical Analysis* (New York: McGraw-Hill, 1960), pp. 369–372.

Take samples and blank individually through subsequent steps to minimize air oxidation of iron(II).

Reduction of Iron

Heat the sample solution nearly to boiling, and make dropwise additions of 0.25 M $SnCl_2$ until the yellow color just disappears; then add two more drops (note 3). Cool to room temperature, and *rapidly* add 10 mL of 5% $HgCl_2$ solution. A small amount of silky white Hg_2Cl_2 should precipitate (note 4). The blank should be treated with the $HgCl_2$ solution.

Titration

Following addition of the $HgCl_2$, wait 2 to 3 min. Then add 25 mL of Zimmermann–Reinhardt reagent and 300 mL of water. Titrate *immediately* with standard 0.02 M $KMnO_4$ to the first faint pink that persists for 15 to 20 s. Do not add the $KMnO_4$ rapidly at any time. Correct the titrant volume for the blank.

Report the percentage of Fe_2O_3 in the sample.

Notes

1. The $SnCl_2$ hastens decomposition of the ore by reducing iron(III) oxides to iron(II). Insufficient $SnCl_2$ is indicated by the appearance of yellow iron(III)/chloride complexes.
2. If dark particles persist after the sample has been heated with acid for several hours, filter the solution through ashless paper, wash the residue with 5 to 10 mL of 6 M HCl, and retain the filtrate and washings. Ignite the paper and its contents in a small platinum crucible. Mix 0.5 to 0.7 g of Na_2CO_3 with the residue, and heat until a clear melt is obtained. Cool, add 5 mL of water, and then cautiously add a few milliliters of 6 M HCl. Warm the crucible until the melt has dissolved, and combine the contents with the original filtrate. Evaporate the solution to 15 mL and continue the analysis.
3. The solution may not become entirely colorless but instead may acquire a faint yellow-green hue. Further additions of $SnCl_2$ will not alter this color. If too much $SnCl_2$ is added, it can be removed by adding 0.2 M $KMnO_4$ and repeating the reduction.
4. The absence of precipitate indicates that insufficient $SnCl_2$ was used and that the reduction of iron(III) was incomplete. A gray residue indicates the presence of elemental mercury, which reacts with $KMnO_4$. The sample must be discarded in either event.
5. These directions can be used to standardize a permanganate solution against primary standard iron. Weigh (to the nearest 0.1 mg) 0.2-g lengths of electrolytic iron wire into 250-mL conical flasks, and dissolve in about 10 mL of concentrated HCl. Dilute each sample to about 75 mL. Then take each individually through the reduction and titration steps.

27G   TITRATIONS WITH IODINE

The oxidizing properties of iodine, the composition and stability of triiodide solutions, and the applications of this reagent in volumetric analysis are discussed in Section 18C-3. Starch is ordinarily employed as an indicator for iodimetric titrations.

27G-1   Preparation of Reagents

Procedure

(a) *Iodine approximately 0.05 M.* Weigh about 40 g of KI into a 100-mL beaker. Add 12.7 g of I_2 and 10 mL of water. Stir for several minutes (note 1). Introduce an additional 20 mL of water, and stir again for several minutes. Carefully decant the bulk of the liquid into a storage bottle containing 1 L of distilled water. It is essential that any undissolved iodine remain in the beaker (note 2).

(b) *Starch indicator* (sufficient for about 100 titrations). Rub 1 g of soluble starch and 15 mL of water into a paste. Dilute to about 500 mL with boiling water, and heat until the mixture is clear. Cool; store in a tightly stoppered bottle. For most titrations, 3 to 5 mL of the indicator is used.

The indicator is readily attacked by airborne organisms and should be freshly prepared every few days.

Notes

1. Iodine dissolves slowly in the KI solution. Thorough stirring is needed to hasten the process.
2. Any solid I_2 inadvertently transferred to the storage bottle will cause the concentration of the solution to increase gradually. Filtration through a sintered-glass crucible eliminates this potential source of difficulty.

27G-2   Standardization of Iodine Solutions

Discussion

Arsenic(III) oxide, long a favored primary standard for iodine solutions, is now seldom used because of the elaborate federal regulations governing the use of even small amounts of arsenic-containing compounds. Barium thiosulfate monohydrate and anhydrous sodium thiosulfate have been proposed as alternative standards.[7] Perhaps the most convenient method for determining the concentration of an iodine solution is the titration of aliquots with a sodium thiosulfate solution that has been standardized against pure potassium iodate. Instructions for this method follow:

Preparation of Reagents

1. *Sodium thiosulfate, 0.1 M.* Follow the directions in Sections 27H-1 and 27H-2 for the preparation and standardization of this solution.
2. *Starch indicator.* See Section 27G-1(b).

---

[7]W. M. McNevin and O. H. Kriege, *Anal. Chem.,* **1953,** *25* (5), 767; A. A. Woolf, *Anal. Chem.,* **1982,** *54* (12), 2134.

Procedure

Transfer 25.00-mL aliquots of the iodine solution to 250-mL conical flasks, and dilute to about 50 mL. *Take each aliquot individually through subsequent steps.* Introduce approximately 1 mL of 3 M H_2SO_4, and titrate immediately with standard sodium thiosulfate until the solution becomes a faint straw yellow. Add about 5 mL of starch indicator, and complete the titration, taking as the end point the change in color from blue to colorless (note).

Note

The blue color of the starch/iodine complex may reappear after the titration has been completed, owing to the air oxidation of iodide ion.

27G-3   The Determination of Antimony in Stibnite

Discussion

The analysis of stibnite, a common antimony ore, is a typical application of iodimetry and is based on the oxidation of Sb(III) to Sb(V):

$$SbO_3^{3-} + I_2 + H_2O \rightleftharpoons SbO_4^{3-} + 2I^- + 2H^+$$

The position of this equilibrium is strongly dependent on the hydrogen ion concentration. To force the reaction to the right, it is common practice to carry out the titration in the presence of an excess of sodium hydrogen carbonate, which consumes the hydrogen ions as they are produced.

Stibnite is an antimony sulfide ore containing silica and other contaminants. Provided that the material is free of iron and arsenic, the analysis of stibnite for its antimony content is straightforward. Samples are decomposed in hot concentrated hydrochloric acid to eliminate sulfide as gaseous hydrogen sulfide. Care is needed to prevent loss of volatile antimony(III) chloride during this step. The addition of potassium chloride helps by favoring formation of nonvolatile chloro complexes such as $SbCl_4^-$ and $SbCl_6^{3-}$.

Sparingly soluble basic antimony salts, such as SbOCl, often form when the excess hydrochloric acid is neutralized; these react incompletely with iodine and cause low results. The difficulty is overcome by adding tartaric acid, which forms a soluble complex ($SbOC_4H_4O_6^-$) from which antimony is rapidly oxidized by the reagent.

Procedure

Dry the unknown at 110°C for 1 hr, and allow it to cool in a desiccator. Weigh individual samples (note 1) into 500-mL conical flasks. Introduce about 0.3 g of KCl and 10 mL of concentrated HCl to each flask. Heat the mixtures (HOOD) just below boiling until only white or slightly gray residues of SiO_2 remain.

Add 3 g of tartaric acid to each sample, and heat for an additional 10 to 15 min. Then, with good swirling, add water (note 2) from a pipet or buret until the volume is about 100 mL. If reddish Sb_2S_3 forms, discontinue dilution and heat further to eliminate H_2S; add more HCl if necessary.

Add 3 drops of phenolphthalein, and neutralize with 6 M NaOH to the first faint pink of the indicator. Discharge the color by the dropwise addition of 6 M HCl, and then add 1 mL in excess. Introduce 4 to 5 g of $NaHCO_3$, taking care to avoid losses of solution by spattering during the addition. Add 5 mL of starch indicator, rinse down the inside of the flask, and titrate with standard 0.05 M I_2 to the first blue color that persists for 30 s.

Report the percentage of Sb_2S_3 in the unknown.

Notes

1. Samples should contain between 1.5 and 2 mmol of antimony; consult with the instructor for an appropriate sample size. Weighings to the nearest milligram are adequate for samples larger than 1 g.
2. The slow addition of water, with efficient stirring, is essential to prevent the formation of SbOCl.

27H   TITRATIONS WITH SODIUM THIOSULFATE

Numerous methods are based upon the reducing properties of iodide ion:

$$2I^- \rightarrow I_2 + 2e^-$$

Iodine, the reaction product, is ordinarily titrated with a standard sodium thiosulfate solution, with starch serving as the indicator:

$$I_2 + 2S_2O_3^{2-} \rightarrow 2I^- + S_4O_6^{2-}$$

A discussion of thiosulfate methods is found in Section 18B-2.

27H-1   Preparation of 0.1 M Sodium Thiosulfate

Procedure

Boil about 1 L of distilled water for 10 to 15 min. Allow the water to cool to room temperature; then add about 25 g of $Na_2S_2O_3 \cdot 5H_2O$ and 0.1 g of Na_2CO_3. Stir until the solid has dissolved. Transfer the solution to a clean glass or plastic bottle, and store in a dark place.

27H-2   Standardizing Sodium Thiosulfate against Potassium Iodate

Discussion

Solutions of sodium thiosulfate are conveniently standardized by titration of the iodine produced when an unmeasured excess of potassium iodide is added to a known volume of an acidified standard potassium iodate solution. The reaction is

$$IO_3^- + 5I^- + 6H^+ \rightarrow 3I_2 + 3H_2O$$

Note that each mole of iodate results in the production of three moles of iodine. The procedure that follows is based on this reaction.

Preparation of Solutions

1. *Potassium iodate, 0.0100 M.* Dry about 1.2 g of primary-standard KIO_3 at 110°C for at least 1 hr, and cool in a desiccator. Weigh (to the nearest 0.1 mg) about 1.1 g into a 500-mL volumetric flask; use a powder funnel to ensure quantitative transfer of the solid. Rinse the funnel well, dissolve the KIO_3 in about 200 mL of distilled water, dilute to the mark, and mix thoroughly.
2. *Starch indicator.* See Section 27G-1(b).

Procedure

Pipet 50.00-mL aliquots of standard iodate solution into 250-mL conical flasks. *Treat each sample individually from this point to minimize error resulting from the air oxidation of iodide ion.* Introduce 2 g of iodate-free KI, and swirl the flask to hasten solution. Add 2 mL of 6 M HCl, and immediately titrate with thiosulfate until the solution becomes pale yellow. Introduce 5 mL of starch indicator, and titrate with constant stirring to the disappearance of the blue color. Calculate the molarity of the thiosulfate solution.

27H-3   Standardization of Sodium Thiosulfate against Copper

Discussion

Thiosulfate solutions can also be standardized against pure copper wire or foil. This procedure is advantageous when the solution is to be used for the determination of copper because any systematic error in the method tends to be canceled.

Copper(II) is reduced quantitatively to copper(I) by iodide ion:

$$2Cu^{2+} + 4I^- \rightarrow 2CuI(s) + I_2$$

The importance of CuI formation in forcing this reaction to completion can be seen from the following standard electrode potentials:

$$Cu^{2+} + e^- \rightleftharpoons Cu^+ \qquad E^0 = 0.15 \text{ V}$$

$$I_2 + 2e^- \rightleftharpoons 2I^- \qquad E^0 = 0.54 \text{ V}$$

$$Cu^{2+} + I^- + e^- \rightleftharpoons CuI(s) \qquad E^0 = 0.86 \text{ V}$$

The first two potentials suggest that iodide should have no tendency to reduce copper(II); the formation of CuI, however, favors the reduction. The solution must contain at least 4% excess iodide to force the reaction to completion. Moreover, the pH must be less than 4 to prevent the formation of basic copper species that react slowly and incompletely with iodide ion. The acidity of the solution cannot be greater than about 0.3 M, however, because of the tendency of iodide ion to undergo air oxidation, a process catalyzed by copper salts. Nitrogen oxides also catalyze the air oxidation of iodide ion. A common source of these oxides is the nitric acid ordinarily used to dissolve metallic copper and the other copper-containing solids. Urea is used to scavenge nitrogen oxides from solutions:

$$(NH_2)_2CO + 2HNO_2 \rightarrow 2N_2(g) + CO_2(g) + 3H_2O$$

The titration of iodine by thiosulfate tends to yield slightly low results owing to the adsorption of small but measurable quantities of iodine upon solid CuI. The adsorbed iodine is released only slowly, even when thiosulfate is in excess; transient and premature end points result. This difficulty is largely overcome by the addition of thiocyanate ion. The sparingly soluble copper(I) thiocyanate replaces part of the copper iodide at the surface of the solid:

$$CuI(s) + SCN^- \rightarrow CuSCN(s) + I^-$$

Accompanying this reaction is the release of the adsorbed iodine, which thus becomes available for titration. The addition of thiocyanate must be delayed until most of the iodine has been titrated to prevent interference from a slow reaction between the two species, possibly

$$2SCN^- + I_2 \rightarrow 2I^- + (SCN)_2$$

Preparation of Solutions

1. *Urea, 5% (w/v).* Dissolve about 5 g of urea in sufficient water to give 100 mL of solution. Approximately 10 mL will be needed for each titration.
2. *Starch indicator.* See Section 27G-1(b).

Procedure

Use scissors to cut copper wire or foil into 0.20- to 0.25-g portions. Wipe the metal free of dust and grease with a filter paper; do not dry it. The pieces of copper should be handled with paper strips, cotton gloves, or tweezers to prevent contamination by contact with the skin.

Use a weighed watch glass or weighing bottle to obtain the mass of individual copper samples (to the nearest 0.1 mg) by difference. Transfer each sample to a 250-mL conical flask. Add 5 mL of 6 M HNO_3, cover with a small watch glass, and warm gently (HOOD) until the metal has dissolved. Dilute with about 25 mL of distilled water, add 10 mL of 5% (w/v) urea, and boil briefly to eliminate nitrogen oxides. Rinse the watch glass, collecting the rinsings in the flask. Cool.

Add concentrated NH_3 dropwise and with thorough mixing to produce the intensely blue $Cu(NH_3)_4^{2+}$; the solution should smell faintly of ammonia (note). Make dropwise additions of 3 M H_2SO_4 until the color of the complex just disappears, and then add 2.0 mL of 85% H_3PO_4. Cool to room temperature.

Treat each sample individually from this point on to minimize the air oxidation of iodide ion. Add 4.0 g of KI to the sample, and titrate immediately with $Na_2S_2O_3$ until the solution becomes pale yellow. Add 5 mL of starch indicator, and continue the titration until the blue color becomes faint. Add 2 g of KSCN; swirl vigorously for 30 s. Complete the titration, using the disappearance of the blue starch/I_2 color as the end point.

Calculate the molarity of the $Na_2S_2O_3$ solution.

Note

Do not sniff vapors directly from the flask; instead, waft them toward your nose with a waving motion of your hand.

27H-4 The Determination of Copper in Brass

Discussion

The standardization procedure described in Section 27H-3 is readily adapted to the determination of copper in brass, an alloy that also contains appreciable amounts of tin, lead, and zinc (and perhaps minor amounts of nickel and iron). The method is relatively simple and applicable to brasses with less than 2% iron. A weighed sample is treated with nitric acid, which causes the tin to precipitate as a hydrated oxide of uncertain composition. Evaporation with sulfuric acid to the appearance of sulfur trioxide eliminates the excess nitrate, redissolves the tin compound, and possibly causes the formation of lead sulfate. The pH is adjusted through the addition of ammonia, followed by acidification with a measured amount of phosphoric acid. An excess of potassium iodide is added, and the liberated iodine is titrated with standard thiosulfate. See Section 27H-3 for additional discussion.

Procedure

If so directed, free the metal of oils by treatment with an organic solvent; briefly heat in an oven to drive off the solvent. Weigh (to the nearest 0.1 mg) 0.3-g samples into 250-mL conical flasks, and introduce 5 mL of 6 M HNO_3 into each; warm (HOOD) until solution is complete. Add 10 mL of concentrated H_2SO_4, and evaporate (HOOD) until copious white fumes of SO_3 are given off. Allow the mixture to cool. Cautiously add 30 mL of distilled water, boil for 1 to 2 min, and again cool.

Follow the instructions in the third and fourth paragraphs of the procedure in Section 27H-3.

Report the percentage of Cu in the sample.

27I TITRATIONS WITH POTASSIUM BROMATE

Applications of standard bromate solutions to the determination of organic functional groups are described in Section 18C-4. Directions follow for the determination of ascorbic acid in vitamin C tablets.

27I-1 Preparation of Solutions

Procedure

1. *Potassium bromate, 0.015 M.* Transfer about 1.5 g of reagent-grade potassium bromate to a weighing bottle, and dry at 110°C for at least 1 hr. Cool in a desiccator. Weigh approximately 1.3 g (to the nearest 0.1 mg) into a 500-mL volumetric flask; use a powder funnel to ensure quantitative transfer of the solid. Rinse the

funnel well, and dissolve the KBrO$_3$ in about 200 mL of distilled water. Dilute to the mark, and mix thoroughly.

Solid potassium bromate can cause a fire if it comes into contact with damp organic material (such as paper towels in a waste container). Consult with the instructor concerning the disposal of any excess.

2. *Sodium thiosulfate, 0.05 M.* Follow the directions in Section 27H-1; use about 12.5 g of Na$_2$S$_2$O$_3 \cdot$5H$_2$O per liter of solution.

3. *Starch indicator.* See Section 27G-1(b).

27I-2   Standardizing Sodium Thiosulfate against Potassium Bromate

Discussion

Iodine is generated by the reaction between a known volume of standard potassium bromate and an unmeasured excess of potassium iodide:

$$BrO_3^- + 6I^- + 6H^+ \rightarrow Br^- + 3I_2 + 3H_2O$$

The iodine produced is titrated with the sodium thiosulfate solution.

Procedure

Pipet 25.00-mL aliquots of the KBrO$_3$ solution into 250-mL conical flasks and rinse the interior wall with distilled water. *Treat each sample individually beyond this point.* Introduce 2 to 3 g of KI and about 5 mL of 3 M H$_2$SO$_4$. Immediately titrate with Na$_2$S$_2$O$_3$ until the solution is pale yellow. Add 5 mL of starch indicator, and titrate to the disappearance of the blue color.

Calculate the concentration of the thiosulfate solution.

27I-3   The Determination of Ascorbic Acid in Vitamin C Tablets by Titration with Potassium Bromate

Discussion

Ascorbic acid, C$_6$H$_8$O$_6$, is cleanly oxidized to dehydroascorbic acid by bromine:

An unmeasured excess of potassium bromide is added to an acidified solution of the sample. The solution is titrated with standard potassium bromate to the first permanent appearance of excess bromine; this excess is then determined iodometrically with standard sodium thiosulfate. The entire titration must be performed without delay to prevent air oxidation of the ascorbic acid.

Procedure

Weigh (to the nearest milligram) 3 to 5 vitamin C tablets (note 1). Pulverize them thoroughly in a mortar, and transfer the powder to a dry weighing bottle. Weigh individual 0.40- to 0.50-g samples (to the nearest 0.1 mg) into dry 250-mL conical flasks. *Treat each sample individually beyond this point.* Dissolve the sample (note 2) in 50 mL of 1.5 M H_2SO_4; then add about 5 g of KBr. Titrate immediately with standard $KBrO_3$ to the first faint yellow due to excess Br_2. Record the volume of $KBrO_3$ used. Add 3 g of KI and 5 mL of starch indicator; back-titrate (note 3) with standard 0.05 M $Na_2S_2O_3$.

Calculate the average mass (in milligrams) of ascorbic acid (176.12 g/mol) in each tablet.

Notes

1. This method is not applicable to chewable vitamin C tablets.
2. The binder in many vitamin C tablets remains in suspension throughout the analysis. If the binder is starch, the characteristic color of the complex with iodine appears upon the addition of KI.
3. The volume of thiosulfate needed for the back-titration seldom exceeds a few milliliters.

27J POTENTIOMETRIC METHODS

Potentiometric measurements provide a highly selective method for the quantitative determination of numerous cations and anions. A discussion of the principles and applications of potentiometric measurements is found in Chapter 19. Detailed instructions are given in this section on the use of potentiometric measurements to locate end points in volumetric titrations. In addition, a procedure for the direct potentiometric determination of fluoride ion in drinking water and in toothpaste is described.

27J-1 General Directions for Performing a Potentiometric Titration

The procedure that follows is applicable to the titrimetric methods described in this section. With the proper choice of indicator electrode, it can also be applied to most of the volumetric methods given in Sections 27C through 27I.

Procedure

1. Dissolve the sample in 50 to 250 mL of water. Rinse a suitable pair of electrodes with deionized water, and immerse them in the sample solution. Provide magnetic

(or mechanical) stirring. Position the buret so that reagent can be delivered without splashing.

2. Connect the electrodes to the meter, commence stirring, and record the initial buret volume and the initial potential (or pH).

3. Record the meter reading and buret volume after each addition of titrant. Introduce fairly large volumes (about 5 mL) at the outset. Withhold a succeeding addition until the meter reading remains constant within 1 to 2 mV (or 0.05 pH unit) for at least 30 s (note). Judge the volume of reagent to be added by estimating a value for $\Delta E/\Delta V$ after each addition. In the immediate vicinity of the equivalence point, introduce the reagent in 0.1-mL increments. Continue the titration 2 to 3 mL beyond the equivalence point, increasing the volume increments as $\Delta E/\Delta V$ again becomes smaller.

Note

Stirring motors occasionally cause erratic meter readings; it may be advisable to turn off the motor while meter readings are being made.

27J-2   The Potentiometric Titration of Chloride and Iodide in a Mixture

Discussion

Figure 27-2 is a theoretical argentometric titration curve for a mixture of iodide and chloride ions. Initial additions of silver nitrate result in formation of silver iodide exclusively because the solubility of that salt is only about 5×10^{-7} that of silver chloride. It can be shown that this solubility difference is great enough so that formation of silver chloride is delayed until all but $7 \times 10^{-5}\%$ of the iodide has precipitated. Thus, short of the equivalence point, the curve is essentially indistinguishable from that for iodide alone (Figure 27-2). Just beyond the iodide equivalence point, the silver ion concentration is determined by the concentration of chloride ion in the solution, and the titration curve becomes essentially identical to that for chloride ion by itself.

Curves resembling Figure 27-2 can be obtained experimentally by measuring the potential of a silver electrode immersed in the analyte solution. Hence, a chloride/iodide mixture can be analyzed for each of its components. This technique is not as applicable to analyzing iodide/bromide or bromide/chloride mixtures, however, because the solubility differences between the silver salts are not great enough. Thus, the more soluble salt begins to form in significant amounts before precipitation of the less soluble salt is complete. The silver indicator electrode can be a commercial billet type or simply a polished wire. A calomel electrode can be used as reference, although diffusion of chloride ion from the salt bridge may cause the results of the titration to be measurably high. This source of error can be eliminated by placing the calomel electrode in a potassium nitrate solution that is in contact with the analyte solution by means of a KNO_3 salt bridge. Alternatively, the analyte solution can be made slightly acidic with several drops of nitric acid; a glass electrode can then serve as the reference electrode because the pH of the solution and thus its potential will remain essentially constant throughout the titration.

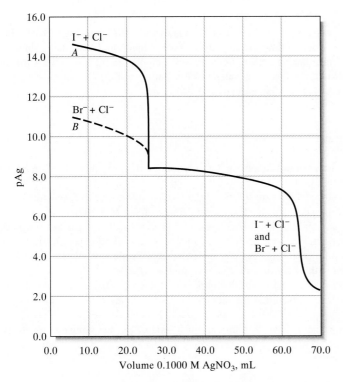

Figure 27-2   Titration curves for 50.00 mL of a solution 0.0800 M in Cl^- and 0.0500 M in I^- or Br^-.

The titration of I^-/Cl^- mixtures demonstrates how a potentiometric titration can have multiple end points. The potential of the silver electrode is proportional to pAg. Thus, a plot of E_{Ag} against titrant volume yields an experimental curve with the same shape as theoretical curve shown in Figure 27-2 (the ordinate units will be different, of course).

Experimental curves for the titration of I^-/Cl^- mixtures do not show the sharp discontinuity that occurs at the first equivalence point of the theoretical curve (Figure 27-2). More important, the volume of silver nitrate needed to reach the I^- end point is generally somewhat greater than theoretical. This effect is the result of coprecipitation of the more soluble AgCl during formation of the less soluble AgI. An overconsumption of reagent thus occurs in the first part of the titration. The total volume closely approaches the correct amount.

Despite this coprecipitation error, the potentiometric method is useful for the analysis of halide mixtures. With approximately equal quantities of iodide and chloride, relative errors can be kept within 2% relative.

Preparation of Reagents

1. *Silver nitrate, 0.05 M.* Follow the instructions in Section 27D-1.
2. *Potassium nitrate salt bridge.* Bend an 8-mm glass tube into a U-shape with arms that are long enough to extend nearly to the bottom of two 100-mL beakers. Heat 50 mL of water to boiling, and stir in 1.8 g of powdered agar; continue to heat

and stir until a uniform suspension is formed. Dissolve 12 g of KNO_3 in the hot suspension. Allow the mixture to cool somewhat. Clamp the U-tube with the openings facing up, and use a medicine dropper to fill it with the warm agar suspension. Cool the tube under a cold-water tap to form the gel. When the bridge is not in use, immerse the ends in 2.5 M KNO_3.

Procedure

Obtain the unknown in a clean 250-mL volumetric flask; dilute to the mark with water, and mix well.

Transfer 50.00 mL of the sample to a clean 100-mL beaker, and add 1 or 2 drops of concentrated HNO_3. Place about 25 mL of 2.5 M KNO_3 in a second 100-mL beaker, and make contact between the two solutions with the agar salt bridge. Immerse a silver electrode in the analyte solution and a calomel reference electrode in the second beaker. Titrate with $AgNO_3$ as described in Section 27J-1. Use small increments of titrant in the vicinity of the two end points.

Plot the data, and establish end points for the two analyte ions. Plot a theoretical titration curve, assuming the measured concentrations of the two constituents to be correct.

Report the number of milligrams of I^- and Cl^- in the sample or as otherwise instructed.

27J-3   The Potentiometric Determination of Solute Species in a Carbonate Mixture

Discussion

A glass/calomel electrode system can be used to locate end points in neutralization titrations and to estimate dissociation constants. As a preliminary step to the titrations, the electrode system is standardized against a buffer of known pH.

The unknown is issued as an aqueous solution prepared from one or perhaps two adjacent members of the following series: $NaHCO_3$, Na_2CO_3, and NaOH (see Section 14B-2). The object is to determine which of these components were used to prepare the unknown as well as the weight percent of each solute.

Most unknowns require a titration with either standard acid or standard base. A few may require separate titrations, one with acid and one with base. The initial pH of the unknown provides guidance concerning the appropriate titrant(s); a study of Figure 14-3 and Table 14-2 may be helpful in interpreting the data.

Preparation of Solutions

Standardized 0.1 M HCl and/or 0.1 M NaOH. Follow the directions in Sections 27C-3 through 27C-7.

Procedure

Obtain the unknown in a clean 250-mL volumetric flask. Dilute to the mark, and mix well. Transfer a small amount of the diluted unknown to a beaker, and determine its pH. Titrate a 50.00-mL aliquot with standard acid or standard base (or perhaps both). Use the resulting titration curves to select indicator(s) suitable for end-point detection, and perform duplicate titrations with these indicators.

Identify the solute species in the unknown, and report the mass/volume percent of each. Calculate the approximate dissociation constant that can be obtained for any carbonate-containing species from the titration data. Estimate the ionic strength of the solution and correct the calculated constant to give an approximate thermodynamic dissociation constant.

27J-4   The Direct Potentiometric Determination of Fluoride Ion

Discussion

The solid-state fluoride electrode (Section 19D-6) has found extensive use in the determination of fluoride in a variety of materials. Directions follow for the determination of this ion in drinking water and in toothpaste. A total ionic strength adjustment buffer (TISAB) is used to adjust all unknowns and standards to essentially the same ionic strength; when this reagent is used, the *concentration* of fluoride, rather than its activity, is measured. The pH of the buffer is about 5, a level at which F^- is the predominant fluorine-containing species. The buffer also contains cyclohexylaminedinitrilotetraacetic acid, which forms stable chelates with iron(III) and aluminum(III), thus freeing fluoride ion from its complexes with these cations.

Review Sections 19D and 19F before undertaking these experiments.

Preparation of Solutions

....

1. *Total ionic strength adjustment buffer (TISAB).* This solution is marketed commercially under the trade name TISAB.[8] Sufficient buffer for 15 to 20 determinations can be prepared by mixing (with stirring) 57 mL of glacial acetic acid, 58 g of NaCl, 4 g of cyclohexylaminedinitrilotetraacetic acid, and 500 mL of distilled water in a 1-L beaker. Cool the contents in a water or ice bath, and carefully add 6 M NaOH to a pH of 5.0 to 5.5. Dilute to 1 L with water, and store in a plastic bottle.
2. *Standard fluoride solution, 100 ppm.* Dry a quantity of NaF at 110°C for 2 hr. Cool in a desiccator; then weigh (to the nearest milligram) 0.22 g into a 1-L volumetric flask. (*Caution! NaF is highly toxic. Immediately* wash any skin touched by this compound with copious quantities of water.) Dissolve in water, dilute to the mark, mix well, and store in a plastic bottle. Calculate the exact concentration of fluoride in parts per million.

    A standard F^- solution can be purchased from commercial sources.

Procedure

....

The apparatus for this experiment consists of a solid-state fluoride electrode, a saturated calomel electrode, and a pH meter. A sleeve-type calomel electrode is needed for

---

[8]Orion Research, Boston, MA.

the toothpaste determination because the measurement is made on a suspension that tends to clog the liquid junction. The sleeve must be loosened momentarily to renew the interface after each series of measurements.

Determining Fluoride in Drinking Water

Transfer 50.00-mL portions of the water to 100-mL volumetric flasks, and dilute to the mark with TISAB solution.

Prepare a 5-ppm F⁻ solution by diluting 25.0 mL of the 100-ppm standard to 500 mL in a volumetric flask. Transfer 5.00-, 10.0-, 25.0-, and 50.0-mL aliquots of the 5-ppm solution to 100-mL volumetric flasks, add 50 mL of TISAB solution, and dilute to the mark. (These solutions correspond to 0.5, 1.0, 2.5, and 5.0 ppm F⁻ in the sample.)

After thorough rinsing and drying with paper tissue, immerse the electrodes in the 0.5-ppm standard. Stir mechanically for 3 min; then record the potential. Repeat with the remaining standards and samples.

Plot the measured potential against the log of the concentration of the standards. Use this plot to determine the concentration in parts per million of fluoride in the unknown.

Determining Fluoride in Toothpaste[9]

Weigh (to the nearest milligram) 0.2 g of toothpaste into a 250-mL beaker. Add 50 mL of TISAB solution, and boil for 2 min with good mixing. Cool and then transfer the suspension quantitatively to a 100-mL volumetric flask, dilute to the mark with distilled water, and mix well. Follow the directions for the analysis of drinking water, beginning with the second paragraph.

Report the parts per million of F⁻ in the sample.

27K   ELECTROGRAVIMETRIC METHODS

A convenient example of an electrogravimetric method of analysis is the simultaneous determination of copper and lead in a sample of brass. Additional information concerning electrogravimetric methods is found in Section 20C.

27K-1   The Electrogravimetric Determination of Copper and Lead in Brass

Discussion

This procedure is based on the deposition of metallic copper on a cathode and of lead as PbO_2 on an anode. As a first step, the hydrous oxide of tin ($SnO_2 \cdot xH_2O$) that forms when the sample is treated with nitric acid must be removed by filtration. Lead dioxide is deposited quantitatively at the anode from a solution with a high nitrate ion concentration; copper is only partially deposited on the cathode under these conditions. Therefore, it is necessary to eliminate the excess nitrate after deposition of the PbO_2 is complete. Removal is accomplished through the addition of urea:

---

[9]From T. S. Light and C. C. Cappuccino, *J. Chem. Educ.*, **1975**, *52*, 247.

$$6NO_3^- + 6H^+ + 5(NH_2)_2CO \rightarrow 8N_2(g) + 5CO_2(g) + 13H_2O$$

Copper then deposits quantitatively from the solution after the nitrate ion concentration has been decreased.

Procedure

Preparation of Electrodes

Immerse the platinum electrodes in hot 6 M HNO_3 for about 5 min (note 1). Wash them thoroughly with distilled water, rinse with several small portions of ethanol, and dry in an oven at 110°C for 2 to 3 min. Cool and weigh both anodes and cathodes to the nearest 0.1 mg.

Preparation of Samples

It is not necessary to dry the unknown. Weigh (to the nearest 0.1 mg) 1-g samples into 250-mL beakers. Cover the beakers with watch glasses. Cautiously add about 35 mL of 6 M HNO_3 (HOOD). Digest for at least 30 min; add more acid if necessary. Evaporate to about 5 mL but never to dryness (note 2).

To each sample, add 5 mL of 3 M HNO_3, 25 mL of water, and one-quarter of a tablet of filter paper pulp; digest without boiling for about 45 min. Filter off the $SnO_2 \cdot xH_2O$, using a fine-porosity filter paper (note 3); collect the filtrates in tall-form electrolysis beakers. Use many small washes with hot 0.3 M HNO_3 to remove the last traces of copper; test for completeness with a few drops of NH_3. The final volume of filtrate and washings should be between 100 and 125 mL; either add water or evaporate to attain this volume.

Electrolysis

With the current switch off, attach the cathode to the negative terminal and the anode to the positive terminal of the electrolysis apparatus. Briefly turn on the stirring motor to be sure the electrodes do not touch. Cover the beakers with split watch glasses, and commence the electrolysis. Maintain a current of 1.3 A for 35 min.

Rinse the cover glasses, and add 10 mL of 3 M H_2SO_4 followed by 5 g of urea to each beaker. Maintain a current of 2 A until the solutions are colorless. To test for completeness of the electrolysis, remove one drop of the solution with a medicine dropper, and mix it with a few drops of NH_3 in a small test tube. If the mixture turns blue, rinse the contents of the tube back into the electrolysis vessel and continue the electrolysis for an additional 10 min. Repeat the test until no blue $Cu(NH_3)_4^{2+}$ is produced.

When electrolysis is complete, discontinue stirring but leave the current on. Rinse the electrodes thoroughly with water as they are removed from the solution. After rinsing is complete, turn off the electrolysis apparatus (note 4), disconnect the electrodes, and dip them in acetone. Dry the cathodes for about 3 min and the anodes for about 15 min at 110°C. Allow the electrodes to cool in air, and then weigh them.

Report the percentages of lead (note 5) and copper in the brass.

Notes

1. Alternatively, grease and organic materials can be removed by heating platinum electrodes to redness in a flame. Electrode surfaces should not be touched with your fingers after cleaning because grease and oil cause nonadherent deposits that can flake off during washing and weighing.
2. Chloride ion must be totally excluded from this determination because it attacks the platinum anode during electrolysis. This reaction is not only destructive but

also causes positive errors in the analysis by codepositing platinum with copper on the cathode.

3. If desired, the tin content can be determined gravimetrically by ignition of the $SnO_2 \cdot xH_2O$ to SnO_2.

4. It is important to maintain a potential between the electrodes until they have been removed from the solution and washed. Some copper may redissolve if this precaution is not observed.

5. Experience has shown that a small amount of moisture is retained by the PbO_2 and that better results are obtained if 0.8643 is used instead of 0.8662, the stoichiometric factor.

27L   METHODS BASED ON THE ABSORPTION OF RADIATION

Molecular absorption methods are discussed in Chapter 23. Directions follow for (1) the use of a calibration curve for the determination of iron in water, (2) the use of a standard-addition procedure for the determination of manganese in steel, and (3) a spectrophotometric determination of the pH of a buffer solution.

27L-1   The Cleaning and Handling of Cells

The accuracy of spectrophotometric measurements is critically dependent on the availability of good-quality matched cells. These cells should be calibrated against one another at regular intervals to detect differences resulting from scratches, etching, and wear. Equally important is the proper cleaning of the exterior sides (the *windows*) just before the cells are inserted into a photometer or spectrophotometer. The preferred method is to wipe the windows with a lens paper soaked in methanol; the methanol is then allowed to evaporate, leaving the windows free of contaminants. It has been shown that this method is far superior to the usual procedure of wiping the windows with a dry lens paper, which tends to leave a residue of lint and a film on the window.[10]

27L-2   The Determination of Iron in a Natural Water

Discussion

The red-orange complex that forms between iron(II) and 1,10-phenanthroline (orthophenanthroline) is useful for determining iron in water supplies. The reagent is a weak base that reacts to form phenanthrolinium ion, $PhenH^+$, in acidic media. Complex formation with iron is thus best described by the equation

$$Fe^{2+} + 3PhenH^+ \rightleftharpoons Fe(Phen)_3^{2+} + 3H^+$$

---

[10]For further information, see J. O. Erickson and T. Surles, *Amer. Lab.,* **1976,** *8* (6), 50.

The structure of the complex is shown in Section 17D-1. The formation constant for this equilibrium is 2.5×10^6 at 25°C. Iron(II) is quantitatively complexed in the pH range between 3 and 9. A pH of about 3.5 is ordinarily recommended to prevent precipitation of iron salts, such as phosphates.

An excess of a reducing reagent, such as hydroxylamine or hydroquinone, is needed to maintain iron in the +2 state. The complex, once formed, is very stable.

This determination can be performed with a spectrophotometer set at 508 nm or with a photometer equipped with a green filter.

Preparation of Solutions

1. *Standard iron solution, 0.01 mg/mL.* Weigh (to the nearest 0.2 mg) 0.0702 g of reagent-grade $Fe(NH_4)_2(SO_4)_2 \cdot 6H_2O$ into a 1-L volumetric flask. Dissolve in 50 mL of water that contains 1 to 2 mL of concentrated sulfuric acid; dilute to the mark, and mix well.
2. *Hydroxylamine hydrochloride* (sufficient for 80 to 90 measurements). Dissolve 10 g of $H_2NOH \cdot HCl$ in about 100 mL of distilled water.
3. *Orthophenanthroline solution* (sufficient for 80 to 90 measurements). Dissolve 1.0 g of orthophenanthroline monohydrate in about 1 L of water. Warm slightly if necessary. Each milliliter is sufficient for no more than about 0.09 mg of Fe. Prepare no more reagent than needed; it darkens on standing and must then be discarded.
4. *Sodium acetate, 1.2 M* (sufficient for 80 to 90 measurements). Dissolve 166 g of $NaOAc \cdot 3H_2O$ in 1 L of distilled water.

Procedure

Preparation of a Calibration Curve

Transfer 25.00 mL of the standard iron solution to a 100-mL volumetric flask and 25 mL of distilled water to a second 100-mL volumetric flask. Add 1 mL of hydroxylamine, 10 mL of sodium acetate, and 10 mL of orthophenanthroline to each flask. Allow the mixtures to stand for 5 min; dilute to the mark, and mix well.

Clean a pair of matched cells for the instrument. Rinse each cell with at least three portions of the solution it is to contain. Determine the absorbance of the standard with respect to the blank.

Repeat this procedure with at least three other volumes of the standard iron solution; attempt to encompass an absorbance range between 0.1 and 1.0. Plot a calibration curve.

Determination of Iron

Transfer 10.00 mL of the unknown to a 100-mL volumetric flask; treat in exactly the same way as the standards, measuring the absorbance with respect to the blank. Alter the volume of unknown taken to obtain absorbance measurements for replicate samples that are within the range of the calibration curve.

Report the parts per million of iron in the unknown.

27L-3   The Determination of Manganese in Steel

Discussion

Small quantities of manganese are readily determined photometrically by the oxidation of Mn(II) to the intensely colored permanganate ion. Potassium periodate is an effective oxidizing reagent for this purpose. The reaction is

$$5IO_4^- + 2Mn^{2+} + 3H_2O \rightarrow 5IO_3^- + 2MnO_4^- + 6H^+$$

Permanganate solutions that contain an excess of periodate are quite stable.

Interferences to the method are few. The presence of most colored ions can be compensated for with a blank. Cerium(III) and chromium(III) are exceptions; these yield oxidation products with periodate that absorb to some extent at the wavelength used for the measurement of permanganate.

The method given here is applicable to steels that do not contain large amounts of chromium. The sample is dissolved in nitric acid. Any carbon in the steel is oxidized with peroxodisulfate. Iron(III) is eliminated as a source of interference by complexation with phosphoric acid. The standard-addition method (page 601) is used to establish the relationship between absorbance and amount of manganese in the sample.

A spectrophotometer set at 525 nm or a photometer with a green filter can be used for the absorbance measurements.

Preparation of Solutions

Standard manganese(II) solution (sufficient for several hundred analyses). Weigh 0.1 g (to the nearest 0.1 mg) of manganese into a 50-mL beaker, and dissolve in about 10 mL of 6 M HNO_3 (HOOD). Boil gently to eliminate oxides of nitrogen. Cool; then transfer the solution quantitatively to a 1-L volumetric flask. Dilute to the mark with water, and mix thoroughly. The manganese in 1 mL of the standard solution, after being converted to permanganate, causes a volume of 50 mL to increase in absorbance by about 0.09.

Procedure

The unknown does not require drying. If there is evidence of oil, rinse with acetone and dry briefly. Weigh (to the nearest 0.1 mg) duplicate samples (note 1) into 150-mL beakers. Add about 50 mL of 6 M HNO_3, and boil gently (HOOD); heating for 5 to 10 min should suffice. Cautiously add about 1 g of ammonium peroxodisulfate, and boil gently for an additional 10 to 15 min. If the solution is pink or has a deposit of MnO_2, add 1 mL of NH_4HSO_3 (or 0.1 g of $NaHSO_3$) and heat for 5 min. Cool; transfer quantitatively (note 2) to 250.0-mL volumetric flasks. Dilute to the mark with water, and mix well. Use a 20.0-mL pipet to transfer three aliquots of each sample to individual beakers. Treat as follows:

Aliquot	Volume of 85% H_3PO_4, mL	Volume of Standard Mn, mL	Weight of KIO_4, g
1	5	0.00	0.4
2	5	5.00 (note 3)	0.4
3	5	0.00	0.0

Boil each solution gently for 5 min, cool, and transfer it quantitatively to a 50-mL volumetric flask. Mix well. Measure the absorbance of aliquots 1 and 2 using aliquot 3 as the blank (note 4).

Report the percentage of manganese in the unknown.

Notes

1. The sample size depends on the manganese content of the unknown; consult with the instructor.
2. If there is evidence of turbidity, filter the solutions as they are transferred to the volumetric flasks.
3. The volume of the standard addition may be dictated by the absorbance of the sample. It is useful to obtain a rough estimate by generating permanganate in about 20 mL of sample, diluting to about 50 mL, and measuring the absorbance.
4. A single blank can be used for all measurements, provided the samples weigh within 50 mg of one another.

27L-4   The Spectrophotometric Determination of pH

Discussion

The pH of an unknown buffer is determined by addition of an acid/base indicator and spectrophotometric measurement of the absorbance of the resulting solution. Because overlap exists between the spectra for the acid and base forms of the indicator, it is necessary to evaluate individual molar absorptivities for each form at two wavelengths. See page 604 for further discussion.

The relationship between the two forms of bromocresol green in an aqueous solution is described by the equilibrium

$$HIn + H_2O \rightleftharpoons H_3O^+ + In^-$$

for which

$$K_a = \frac{[H_3O^+][In^-]}{[HIn]} = 1.6 \times 10^{-5}$$

The spectrophotometric evaluation of $[In^-]$ and $[HIn]$ permits the calculation of $[H_3O^+]$ and thus pH.

Preparation of Solutions

1. *Bromocresol green, $1.0 \times 10^{-4} M$* (sufficient for about five determinations). Dissolve 40.0 mg (to the nearest 0.1 mg) of the sodium salt of bromocresol green (720 g/mol) in water, and dilute to 500 mL in a volumetric flask.
2. *HCl, 0.5 M.* Dilute about 4 mL of concentrated HCl to approximately 100 mL with water.
3. *NaOH, 0.4 M.* Dilute about 7 mL of 6 M NaOH to about 100 mL with water.

Procedure

Determination of Individual Absorption Spectra

Transfer 25.00-mL aliquots of the bromocresol green indicator solution to two 100-mL volumetric flasks. To one, add 25 mL of 0.5 M HCl; to the other, add 25 mL of 0.4 M NaOH. Dilute to the mark and mix well.

Obtain the absorption spectra for the acid and conjugate-base forms of the indicator between 400 and 600 nm, using water as a blank. Record absorbance values at 10-nm intervals routinely and at closer intervals as needed to define maxima and minima. Evaluate the molar absorptivity for HIn and In$^-$ at wavelengths corresponding to their absorption maxima.

Determination of the pH of an Unknown Buffer

Transfer 25.00 mL of the stock bromocresol green indicator to a 100-mL volumetric flask. Add 50.0 mL of the unknown buffer, dilute to the mark, and mix well. Measure the absorbance of the diluted solution at the wavelengths for which absorptivity data were calculated.

Report the pH of the buffer.

27M   ATOMIC SPECTROSCOPY

Several methods of analysis based on atomic spectroscopy are discussed in Section 23F. One such application is atomic absorption, which is demonstrated in the experiment that follows.

27M-1   The Determination of Lead in Brass by Atomic Absorption Spectroscopy

Discussion

Brasses and other copper-based alloys contain from 0 to about 10% lead as well as tin and zinc. Atomic absorption spectroscopy permits the quantitative estimation of these elements. The accuracy of this procedure is not as great as that obtainable with gravimetric or volumetric measurements, but the time needed to acquire the analytical information is considerably less.

A weighed sample is dissolved in a mixture of nitric and hydrochloric acids, the latter being needed to prevent the precipitation of tin as metastannic acid, $SnO_2 \cdot xH_2O$. After suitable dilution, the sample is aspirated into a flame, and the absorption of radiation from a hollow cathode lamp is measured.

Preparation of Solutions

Standard lead solution, 100 mg/L. Dry a quantity of reagent-grade $Pb(NO_3)_2$ for about 1 hr at 110°C. Cool; weigh (to the nearest 0.1 mg) 0.17 g into a 1-L volumetric flask. Dissolve in a solution of 5 mL water and 1 to 3 mL of concentrated HNO_3. Dilute to the mark with distilled water, and mix well.

Procedure

Weigh duplicate samples of the unknown (note 1) into 150-mL beakers. Cover with watch glasses; then dissolve (HOOD) in a mixture consisting of about 4 mL of concentrated HNO_3 and 4 mL of concentrated HCl (note 2). Boil gently to remove oxides of nitrogen. Cool; transfer the solutions quantitatively to a 250-mL volumetric flask, dilute to the mark with water, and mix well.

Use a buret to deliver 0-, 5-, 10-, 15-, and 20-mL portions of the standard lead solution to individual 50-mL volumetric flasks. Add 4 mL of concentrated HNO_3 and 4 mL of concentrated HCl to each, and dilute to the mark with water.

Transfer 10.00-mL aliquots of each sample to 50-mL volumetric flasks, add 4 mL of concentrated HCl and 4 mL of concentrated HNO_3, and dilute to the mark with water.

Set the monochromator at 283.3 nm, and measure the absorbance for each standard and the sample at that wavelength. Take at least three, and preferably more, readings for each measurement.

Plot the calibration data. Report the percentage of lead in the brass.

Notes

1. The weight of sample depends on the lead content of the brass and on the sensitivity of the instrument used for the absorption measurements. A sample containing 6 to 10 mg of lead is reasonable. Consult with the instructor.
2. Brasses that contain a large percentage of tin require additional HCl to prevent the formation of metastannic acid. The diluted samples may develop some turbidity on prolonged standing; a slight turbidity has no effect on the determination of lead.

27M-2 The Determination of Sodium, Potassium, and Calcium in Mineral Waters by Atomic Emission Spectroscopy

Discussion

A convenient method for the determination of alkali and alkaline-earth metals in water and in blood serum is based on the characteristic spectra these elements emit upon being aspirated into a natural gas/air flame. The accompanying directions are suitable for the analysis of the three elements in water samples. Radiation buffers are used to minimize the effect of each element on the emission intensity of the others.

A radiation buffer is a substance that is added in large excess to both samples and standards to swamp out the effect of matrix species and thus minimize interferences.

Preparation of Solutions

1. *Standard calcium solution, approximately 500 ppm.* Dry a quantity of $CaCO_3$ for about 1 hr at 110°C. Cool in a desiccator; weigh (to the nearest milligram) 1.25 g into a 600-mL beaker. Add about 200 mL of distilled water and about 10 mL of concentrated HCl; cover the beaker with a watch glass during the addition of the acid to avoid loss due to spattering. After reaction is complete, transfer the solution quantitatively to a 1-L volumetric flask, dilute to the mark, and mix well.
2. *Standard potassium solution, approximately 500 ppm.* Dry a quantity of KCl for about 1 hr at 110°C. Cool; weigh (to the nearest milligram) about 0.95 g into a 1-L volumetric flask. Dissolve in distilled water, and dilute to the mark.

3. *Standard sodium solution, approximately 500 ppm.* Proceed as in part 2, using 1.25 g (to the nearest milligram) of dried NaCl.
4. *Radiation buffer for the determination of calcium.* Prepare about 100 mL of a solution that has been saturated with NaCl, KCl, and $MgCl_2$, in that order.
5. *Radiation buffer for the determination of potassium.* Prepare about 100 mL of a solution that has been saturated with NaCl, $CaCl_2$, and $MgCl_2$, in that order.
6. *Radiation buffer for the determination of sodium.* Prepare about 100 mL of a solution that has been saturated with $CaCl_2$, KCl, and $MgCl_2$, in that order.

Procedure

Preparation of Working Curves

Add 5.00 mL of the appropriate radiation buffer to each of a series of 100-mL volumetric flasks. Add volumes of standard that will produce solutions that range from 0 to 10 ppm in the cation to be determined. Dilute to the mark with water, and mix well.

Measure the emission intensity for each solution, taking at least three readings for each. Aspirate distilled water between each set of measurements. Correct the average values for background luminosity, and prepare a working curve from the data.

Repeat for the other two cations.

Analysis of a Water Sample

Prepare duplicate aliquots of the unknown as directed for measurement with the working curves. If necessary, use a standard to adjust the response of the instrument to the working curve; then measure the emission intensity of the unknown. Correct the data for background. Determine the cation concentration in the unknown by comparison with the working curve.

27N   SEPARATING CATIONS BY ION EXCHANGE

The application of ion-exchange resins to the separation of ionic species of opposite charge is discussed in Section 24E. Directions follow for the ion-exchange separation of nickel(II) from zinc(II) based on converting the zinc ions to negatively charged chloro complexes. After separation, each of the cations is determined by EDTA titration.

Discussion

The separation of the two cations is based on differences in their tendency to form anionic complexes. Stable chlorozincate(II) complexes (such as $ZnCl_3^-$ and $ZnCl_4^{2-}$) are formed in 2-M hydrochloric acid and retained on an anion-exchange resin. In contrast, nickel(II) is not complexed appreciably in this medium and passes rapidly through such a column. After separation is complete, elution with water effectively decomposes the chloro complexes and permits removal of the zinc.

Both nickel and zinc are determined by titration with standard EDTA at pH 10. Eriochrome Black T is the indicator for the zinc titration. Bromopyrogallol Red or murexide is used for the nickel titration.

Preparation of Solutions

1. *Standard EDTA.* See Section 27E-2.
2. *pH 10 buffer.* See Section 27E-1.
3. *Eriochrome Black T indicator.* See Section 27E-1.
4. *Bromopyrogallol Red indicator* (sufficient for 100 titrations). Dissolve 0.5 g of the solid indicator in 100 mL of 50% (v/v) ethanol.
5. *Murexide indicator.* The solid is approximately 0.2% indicator by weight in NaCl. Approximately 0.2 g is needed for each titration. The solid preparation is used because solutions of the indicator are quite unstable.

Preparation of Ion-Exchange Columns

A typical ion-exchange column is a cylinder 25 to 40 cm in length and 1 to 1.5 cm in diameter. A stopcock at the lower end permits adjustment of liquid flow through the column. A buret makes a convenient column. It is recommended that two columns be prepared to permit the simultaneous treatment of duplicate samples.

Insert a plug of glass wool to retain the resin particles. Then introduce sufficient strong-base anion-exchange resin (note) to give a 10- to 15-cm column. Wash the column with about 50 mL of 6 M NH_3, followed by 100 mL of water and 100 mL of 2 M HCl. At the end of this cycle, the flow should be stopped so that the liquid level remains about 1 cm above the resin column. *At no time should the liquid level be allowed to drop below the top of the resin.*

Note

Amberlite® CG 400 or its equivalent can be used.

Procedure

Obtain the unknown, which should contain between 2 and 4 mmol of Ni^{2+} and Zn^{2+}, in a clean 100-mL volumetric flask. Add 16 mL of 12 M HCl, dilute to the mark with distilled water, and mix well. The resulting solution is approximately 2 M in acid. Transfer 10.00 mL of the diluted unknown onto the column. Place a 250-mL conical flask beneath the column, and slowly drain until the liquid level is barely above the resin. Rinse the interior of the column with several 2- to 3-mL portions of the 2 M HCl; lower the liquid level to just above the resin surface after each washing. Elute the nickel with about 50 mL of 2 M HCl at a flow rate of 2 to 3 mL/min.

Elute the Zn(II) by passing about 100 mL of water through the column, using the same flow rate; collect the liquid in a 500-mL conical flask.

Titration of Nickel

Evaporate the solution containing the nickel to dryness to eliminate excess HCl. Avoid overheating; the residual $NiCl_2$ must not be permitted to decompose to NiO. Dissolve the residue in 100 mL of distilled water, and add 10 to 20 mL of pH 10 buffer. Add 15 drops of Bromopyrogallol Red indicator or 0.2 g of murexide. Titrate to the color change (blue to purple for Bromopyrogallol Red, yellow to purple for murexide).

Calculate the milligrams of nickel in the unknown.

Titration of Zinc

Add 10 to 20 mL of pH 10 buffer and 1 to 2 drops of Eriochrome Black T to the eluate. Titrate with standard EDTA solution to a color change from red to blue.
  Calculate the milligrams of zinc in the unknown.

27O   GAS-LIQUID CHROMATOGRAPHY

As noted in Chapter 25, gas-liquid chromatography enables the analyst to resolve the components of complex mixtures. The accompanying directions are for the determination of ethanol in beverages.

27O-1   The Gas-Chromatographic Determination of Ethanol in Beverages[11]

Discussion

Ethanol is conveniently determined in aqueous solutions by means of gas chromatography. The method is readily extended to measurement of the *proof* of alcoholic beverages. By definition, the proof of a beverage is two times its volume percent of ethanol at 60°F.

  The operating instructions pertain to a 1/4-in. (o.d.) × 0.5-m Poropack column containing 80- to 100-mesh packing. A thermal conductivity detector is needed. (Flame ionization is not satisfactory because of its insensitivity to water.)

  The determination is based on a calibration curve in which the ratio of the area under the ethanol peak to the area under the ethanol-plus-water peak is plotted as a function of the volume percent of ethanol:

$$\text{vol \% EtOH} = \frac{\text{vol EtOH}}{\text{vol soln}} \times 100\%$$

This relationship is not strictly linear. At least two reasons can be cited to account for the curvature. First, the thermal conductivity detector responds linearly to mass ratios rather than volume ratios. Second, at the high concentrations involved, the volumes of ethanol and water are not strictly additive, as would be required for linearity. That is,

$$\text{vol EtOH} + \text{vol H}_2\text{O} \neq \text{vol soln}$$

Preparation of Standards

Use a buret to deliver 10.00, 20.00, 30.00, and 40.00 mL of absolute ethanol into separate 50-mL volumetric flasks (note). Dilute to volume with distilled water, and mix well.

---

[11]Adapted from J. J. Leary, *J. Chem. Educ.*, **1983,** *60,* 675.

Note

The coefficient of thermal expansion for ethanol is approximately five times that for water. It is thus necessary to keep the temperature of the solutions used in this experiment constant to $\pm 1°C$ during volume measurements.

Procedure

The following operating conditions have yielded satisfactory chromatograms for this determination:

Column temperature	100°C
Detector temperature	130°C
Injection-port temperature	120°C
Bridge current	100 mA
Flow rate	60 mL/min

Inject a 1 μL sample of the 20% (v/v) standard, and record the chromatogram. Obtain additional chromatograms, adjusting the recorder speed until the water peak has a width of about 2 mm at half-height. Then vary the volume of sample injected and the attenuation until peaks with a height of at least 40 mm are produced. Obtain chromatograms for the remainder of the standards (including pure water and pure ethanol) in the same way. Measure the area under each peak, and plot $area_{EtOH}/(area_{EtOH} + area_{H_2O})$ as a function of the volume percentage of ethanol.

Obtain chromatograms for the unknown. Report the volume percentage of ethanol.

WEB WORKS

Material Safety Data Sheets (MSDSs) are an important resource for anyone using chemicals. These documents provided essential information on the properties and toxicity of chemicals that are used in the laboratory. Set your browser to http://www.ilpi.com/msds/index.chtml and find a comprehensive listing of most of the internet sites that present MSDSs. Browse to one of the sites and look up the MSDS for oxalic acid. Browse the entire MSDS, read the reactivity data, note its chemical properties, and find the first aid treatment for ingestion of oxalic acid. Chemical manufacturers are required by law to furnish an MSDS for every chemical that they sell, and many of them can be found on the internet. It is a good idea to examine the MSDS for any substance that you use in the laboratory.

Appendix 1

Solubility Product Constants at 25°C

Compound	Formula	K_{sp}	Notes
Aluminum hydroxide	$Al(OH)_3$	3×10^{-34}	
Barium carbonate	$BaCO_3$	5.0×10^{-9}	
Barium chromate	$BaCrO_4$	2.1×10^{-10}	
Barium hydroxide	$Ba(OH)_2 \cdot 8H_2O$	3×10^{-4}	
Barium iodate	$Ba(IO_3)_2$	1.57×10^{-9}	
Barium oxalate	BaC_2O_4	1×10^{-6}	
Barium sulfate	$BaSO_4$	1.1×10^{-10}	
Cadmium carbonate	$CdCO_3$	1.8×10^{-14}	
Cadmium hydroxide	$Cd(OH)_2$	4.5×10^{-15}	
Cadmium oxalate	CdC_2O_4	9×10^{-8}	
Cadmium sulfide	CdS	1×10^{-27}	
Calcium carbonate	$CaCO_3$	4.5×10^{-9}	Calcite
	$CaCO_3$	6.0×10^{-9}	Aragonite
Calcium fluoride	CaF_2	3.9×10^{-11}	
Calcium hydroxide	$Ca(OH)_2$	6.5×10^{-6}	
Calcium oxalate	$CaC_2O_4 \cdot H_2O$	1.7×10^{-9}	
Calcium sulfate	$CaSO_4$	2.4×10^{-5}	
Cobalt(II) carbonate	$CoCO_3$	1.0×10^{-10}	
Cobalt(II) hydroxide	$Co(OH)_2$	1.3×10^{-15}	
Cobalt(II) sulfide	CoS	5×10^{-22}	α
	CoS	3×10^{-26}	β
Copper(I) bromide	$CuBr$	5×10^{-9}	
Copper(I) chloride	$CuCl$	1.9×10^{-7}	
Copper(I) hydroxide*	Cu_2O	2×10^{-15}	
Copper(I) iodide	CuI	1×10^{-12}	
Copper(I) thiocyanate	$CuSCN$	4.0×10^{-14}	
Copper(II) hydroxide	$Cu(OH)_2$	4.8×10^{-20}	
Copper(II) sulfide	CuS	8×10^{-37}	
Iron(II) carbonate	$FeCO_3$	2.1×10^{-11}	
Iron(II) hydroxide	$Fe(OH)_2$	4.1×10^{-15}	
Iron(II) sulfide	FeS	8×10^{-19}	
Iron(III) hydroxide	$Fe(OH)_3$	2×10^{-39}	
Lanthanum iodate	$La(IO_3)_3$	1.0×10^{-11}	
Lead carbonate	$PbCO_3$	7.4×10^{-14}	
Lead chloride	$PbCl_2$	1.7×10^{-5}	
Lead chromate	$PbCrO_4$	3×10^{-13}	

Compound	Formula	K_{sp}	Notes
Lead hydroxide	PbO†	8×10^{-16}	Yellow
	PbO†	5×10^{-16}	Red
Lead iodide	PbI_2	7.9×10^{-9}	
Lead oxalate	PbC_2O_4	8.5×10^{-9}	$\mu = 0.05$
Lead sulfate	$PbSO_4$	1.6×10^{-8}	
Lead sulfide	PbS	3×10^{-28}	
Magnesium ammonium phosphate	$MgNH_4PO_4$	3×10^{-13}	
Magnesium carbonate	$MgCO_3$	3.5×10^{-8}	
Magnesium hydroxide	$Mg(OH)_2$	7.1×10^{-12}	
Manganese carbonate	$MnCO_3$	5.0×10^{-10}	
Manganese hydroxide	$Mn(OH)_2$	2×10^{-13}	
Manganese sulfide	MnS	3×10^{-11}	Pink
	MnS	3×10^{-14}	Green
Mercury(I) bromide	Hg_2Br_2	5.6×10^{-23}	
Mercury(I) carbonate	Hg_2CO_3	8.9×10^{-17}	
Mercury(I) chloride	Hg_2Cl_2	1.2×10^{-18}	
Mercury(I) iodide	Hg_2I_2	4.7×10^{-29}	
Mercury(I) thiocyanate	$Hg_2(SCN)_2$	3.0×10^{-20}	
Mercury(II) hydroxide	HgO‡	3.6×10^{-26}	
Mercury(II) sulfide	HgS	2×10^{-53}	Black
	HgS	5×10^{-54}	Red
Nickel carbonate	$NiCO_3$	1.3×10^{-7}	
Nickel hydroxide	$Ni(OH)_2$	6×10^{-16}	
Nickel sulfide	NiS	4×10^{-20}	α
	NiS	1.3×10^{-25}	β
Silver arsenate	Ag_3AsO_4	6×10^{-23}	
Silver bromide	AgBr	5.0×10^{-13}	
Silver carbonate	Ag_2CO_3	8.1×10^{-12}	
Silver chloride	AgCl	1.82×10^{-10}	
Silver chromate	Ag_2CrO_4	1.2×10^{-12}	
Silver cyanide	AgCN	2.2×10^{-16}	
Silver iodate	$AgIO_3$	3.1×10^{-8}	
Silver iodide	AgI	8.3×10^{-17}	
Silver oxalate	$Ag_2C_2O_4$	3.5×10^{-11}	
Silver sulfide	Ag_2S	8×10^{-51}	
Silver thiocyanate	AgSCN	1.1×10^{-12}	
Strontium carbonate	$SrCO_3$	9.3×10^{-10}	
Strontium oxalate	SrC_2O_4	5×10^{-8}	
Strontium sulfate	$SrSO_4$	3.2×10^{-7}	
Thallium(I) chloride	TlCl	1.8×10^{-4}	
Thallium(I) sulfide	Tl_2S	6×10^{-22}	
Zinc carbonate	$ZnCO_3$	1.0×10^{-10}	
Zinc hydroxide	$Zn(OH)_2$	3.0×10^{-16}	Amorphous
Zinc oxalate	ZnC_2O_4	8×10^{-9}	
Zinc sulfide	ZnS	2×10^{-25}	α
	ZnS	3×10^{-23}	β

*$Cu_2O(s) + H_2O \rightleftharpoons 2Cu^+ + 2OH^-$

†$PbO(s) + H_2O \rightleftharpoons Pb^{2+} + 2OH^-$

‡$HgO(s) + H_2O \rightleftharpoons Hg^{2+} + 2OH^-$

Source: Most of these data were taken from A. E. Martell and R. M. Smith, *Critical Stability Constants*, Vol. 3–6 (New York: Plenum, 1976–1989). In most cases, the ionic strength was 0.0 and the temperature was 25°C.

Appendix 2

Acid Dissociation Constants at 25°C

Acid	Formula	K_1	K_2	K_3
Acetic acid	CH_3COOH	1.75×10^{-5}		
Ammonium ion	NH_4^+	5.70×10^{-10}		
Anilinium ion	$C_6H_5NH_3^+$	2.51×10^{-5}		
Arsenic acid	H_3AsO_4	5.8×10^{-3}	1.1×10^{-7}	3.2×10^{-12}
Arsenous acid	H_3AsO_3	5.1×10^{-10}		
Benzoic acid	C_6H_5COOH	6.28×10^{-5}		
Boric acid	H_3BO_3	5.81×10^{-10}		
1-Butanoic acid	$CH_3CH_2CH_2COOH$	1.52×10^{-5}		
Carbonic acid	H_2CO_3	1.5×10^{-4}	4.69×10^{-11}	
Chloroacetic acid	$ClCH_2COOH$	1.36×10^{-3}		
Citric acid	$HOOC(OH)C(CH_2COOH)_2$	7.45×10^{-4}	1.73×10^{-5}	4.02×10^{-7}
Dimethyl ammonium ion	$(CH_3)_2NH_2^+$	1.68×10^{-11}		
Ethanol ammonium ion	$HOC_2H_4NH_3^+$	3.18×10^{-10}		
Ethyl ammonium ion	$C_2H_5NH_3^+$	2.31×10^{-11}		
Ethylene diammonium ion	$^+H_3NCH_2CH_2NH_3^+$	1.42×10^{-7}	1.18×10^{-10}	
Formic acid	$HCOOH$	1.80×10^{-4}		
Fumaric acid	*trans*-$HOOCCH:CHCOOH$	8.85×10^{-4}	3.21×10^{-5}	
Glycolic acid	$HOCH_2COOH$	1.47×10^{-4}		
Hydrazinium ion	$H_2NNH_3^+$	1.05×10^{-8}		
Hydrazoic acid	HN_3	2.2×10^{-5}		
Hydrogen cyanide	HCN	6.2×10^{-10}		
Hydrogen fluoride	HF	6.8×10^{-4}		
Hydrogen peroxide	H_2O_2	2.2×10^{-12}		
Hydrogen sulfide	H_2S	9.6×10^{-8}	1.3×10^{-14}	
Hydroxyl ammonium ion	$HONH_3^+$	1.10×10^{-6}		
Hypochlorous acid	$HOCl$	3.0×10^{-8}		
Iodic acid	HIO_3	1.7×10^{-1}		
Lactic acid	$CH_3CHOHCOOH$	1.38×10^{-4}		
Maleic acid	*cis*-$HOOCCH:CHCOOH$	1.3×10^{-2}	5.9×10^{-7}	
Malic acid	$HOOCCHOHCH_2COOH$	3.48×10^{-4}	8.00×10^{-6}	
Malonic acid	$HOOCCH_2COOH$	1.42×10^{-3}	2.01×10^{-6}	
Mandelic acid	$C_6H_5CHOHCOOH$	4.0×10^{-4}		
Methyl ammonium ion	$CH_3NH_3^+$	2.3×10^{-11}		
Nitrous acid	HNO_2	7.1×10^{-4}		
Oxalic acid	$HOOCCOOH$	5.60×10^{-2}	5.42×10^{-5}	
Periodic acid	H_5IO_6	2×10^{-2}	5×10^{-9}	

Acid	Formula	K_1	K_2	K_3
Phenol	C_6H_5OH	1.00×10^{-10}		
Phosphoric acid	H_3PO_4	7.11×10^{-3}	6.32×10^{-8}	4.5×10^{-13}
Phosphorous acid	H_3PO_3	3×10^{-2}	1.62×10^{-7}	
o-Phthalic acid	$C_6H_4(COOH)_2$	1.12×10^{-3}	3.91×10^{-6}	
Picric acid	$(NO_2)_3C_6H_2OH$	4.3×10^{-1}		
Piperidinium ion	$C_5H_{11}NH^+$	7.50×10^{-12}		
Propanoic acid	CH_3CH_2COOH	1.34×10^{-5}		
Pyridinium ion	$C_5H_5NH^+$	5.90×10^{-6}		
Pyruvic acid	$CH_3COCOOH$	3.2×10^{-3}		
Salicylic acid	$C_6H_4(OH)COOH$	1.06×10^{-3}		
Sulfamic acid	H_2NSO_3H	1.03×10^{-1}		
Succinic acid	$HOOCCH_2CH_2COOH$	6.21×10^{-5}	2.31×10^{-6}	
Sulfuric acid	H_2SO_4	Strong	1.02×10^{-2}	
Sulfurous acid	H_2SO_3	1.23×10^{-2}	6.6×10^{-8}	
Tartaric acid	$HOOC(CHOH)_2COOH$	9.20×10^{-4}	4.31×10^{-5}	
Thiocyanic acid	HSCN	0.13		
Thiosulfuric acid	$H_2S_2O_3$	0.3	2.5×10^{-2}	
Trichloroacetic acid	Cl_3CCOOH	3		
Trimethyl ammonium ion	$(CH_3)_3NH^+$	1.58×10^{-10}		

Note: Most data are for zero ionic strength.

Source: From A. E. Martell and R. M. Smith, *Critical Stability Constants,* Vol. 1–6 (New York: Plenum Press, 1974–1989).

Appendix 3

Formation Constants of Complex Compounds at 25°C

Ligand	Cation	$\log K_1$	$\log K_2$	$\log K_3$	$\log K_4$	Ionic Strength
Acetate (CH_3COO^-)	Ag^+	0.73	-0.9			0.0
	Ca^{2+}	1.18				0.0
	Cd^{2+}	1.93	1.22			0.0
	Cu^{2+}	2.21	1.42			0.0
	Fe^{3+}	3.38*	3.1*	1.8*		0.1
	Hg^{2+}	$\log K_1K_2 = 8.45$				0.0
	Mg^{2+}	1.27				0.0
	Pb^{2+}	2.68	1.40			0.0
Ammonia (NH_3)	Ag^+	3.31	3.91			0.0
	Cd^{2+}	2.55	2.01	1.34	0.84	0.0
	Co^{2+}	1.99*	1.51	0.93	0.64	0.0
		$\log K_5 = 0.06$		$\log K_6 = -0.74$		0.0
	Cu^{2+}	4.04	3.43	2.80	1.48	0.0
	Hg^{2+}	8.8	8.6	1.0	0.7	0.5
	Ni^{2+}	2.72	2.17	1.66	1.12	0.0
		$\log K_5 = 0.67$		$\log K_6 = -0.03$		0.0
	Zn^{2+}	2.21	2.29	2.36	2.03	0.0
Bromide (Br^-)	Ag^+	$Ag^+ + 2Br^- \rightleftharpoons AgBr_2^-$		$\log K_1K_2 = 7.5$		0.0
	Hg^{2+}	9.00	8.1	2.3	1.6	0.5
	Pb^{2+}	1.77				0.0
Chloride (Cl^-)	Ag^+	$Ag^+ + 2Cl^- \rightleftharpoons AgCl_2^-$		$\log K_1K_2 = 5.25$		0.0
		$AgCl_2^- + Cl^- \rightleftharpoons AgCl_3^{2-}$		$\log K_3 = 0.37$		0.0
	Cu^+	$Cu^+ + 2Cl^- \rightleftharpoons CuCl_2^-$		$\log K_1K_2 = 5.5*$		0.0
	Fe^{3+}	1.48	0.65			0.0
	Hg^{2+}	7.30	6.70	1.0	0.6	0.0
	Pb^{2+}	$Pb^{2+} + 3Cl^- \rightleftharpoons PbCl_3^-$		$\log K_1K_2K_3 = 1.8$		0.0
	Sn^{2+}	1.51	0.74	-0.3	-0.5	0.0
Cyanide (CN^-)	Ag^+	$Ag^+ + 2CN^- \rightleftharpoons Ag(CN)_2^-$		$\log K_1K_2 = 20.48$		0.0
	Cd^{2+}	6.01	5.11	4.53	2.27	0.0
	Hg^{2+}	17.00	15.75	3.56	2.66	0.0
	Ni^{2+}	$Ni^{2+} + 4CN^- \rightleftharpoons Ni(CN)_4^-$		$\log K_1K_2K_3K_4 = 30.22$		0.0
	Zn^{2+}	$\log K_1K_2 = 11.07$		4.98	3.57	0.0
EDTA	See Table 15-5, page 365.					
Fluoride (F^-)	Al^{3+}	7.0	5.6	4.1	2.4	0.0
	Fe^{3+}	5.18	3.89	3.03		0.0

Ligand	Cation	$\log K_1$	$\log K_2$	$\log K_3$	$\log K_4$	Ionic Strength
Hydroxide (OH⁻)	Al^{3+}	$Al^{3+} + 4OH^- \rightleftharpoons Al(OH)_4^-$		$\log K_1K_2K_3K_4 = 33.4$		0.0
	Cd^{2+}	3.9	3.8			0.0
	Cu^{2+}	6.5				0.0
	Fe^{2+}	4.6				0.0
	Fe^{3+}	11.81	11.5			0.0
	Hg^{2+}	10.60	11.2			0.0
	Ni^{2+}	4.1	4.9	3		0.0
	Pb^{2+}	6.4	$Pb^{2+} + 3OH^- \rightleftharpoons Pb(OH)_3^-$		$\log K_1K_2K_3 = 13.9$	0.0
	Zn^{2+}	5.0	$Zn^{2+} + 4OH^- \rightleftharpoons Zn(OH)_4^{2-}$		$\log K_1K_2K_3K_4 = 15.5$	0.0
Iodide (I⁻)	Cd^{2+}	2.28	1.64	1.0	1.0	0.0
	Cu^+	$Cu^+ + 2I^- \rightleftharpoons CuI_2^-$		$\log K_1K_2 = 8.9$		0.0
	Hg^{2+}	12.87	10.95	3.8	2.2	0.5
	Pb^{2+}	$Pb^{2+} + 3I^- \rightleftharpoons PbI_3^-$		$\log K_1K_2K_3 = 3.9$		0.0
		$Pb^{2+} + 4I^- \rightleftharpoons PbI_4^{2-}$		$\log K_1K_2K_3K_4 = 4.5$		0.0
Oxalate ($C_2O_4^{2-}$)	Al^{3+}	5.97	4.96	5.04		0.1
	Ca^{2+}	3.19				0.0
	Cd^{2+}	2.73	1.4	1.0		1.0
	Fe^{3+}	7.58	6.23	4.8		1.0
	Mg^{2+}	3.42(18°C)				
	Pb^{2+}	4.20	2.11			1.0
Sulfate (SO_4^{2-})	Al^{3+}	3.89				0.0
	Ca^{2+}	2.13				0.0
	Cu^{2+}	2.34				0.0
	Fe^{3+}	4.04	1.34			0.0
	Mg^{2+}	2.23				0.0
Thiocyanate (SCN⁻)	Cd^{2+}	1.89	0.89	0.1		0.0
	Cu^+	$Cu^+ + 3SCN^- \rightleftharpoons Cu(SCN)_3^{2-}$		$\log K_1K_2K_3 = 11.60$		0.0
	Fe^{3+}	3.02	0.62*			0.0
	Hg^{2+}	$\log K_1K_2 = 17.26$		2.7	1.8	0.0
	Ni^{2+}	1.76				0.0
Thiosulfate ($S_2O_3^{2-}$)	Ag^+	8.82*	4.7	0.7		0.0
	Cu^{2+}	$\log K_1K_2 = 6.3$				0.0
	Hg^{2+}	$\log K_1K_2 = 29.23$		1.4		0.0

*20°C.

Source: Data from A. E. Martell and R. M. Smith, *Critical Stability Constants,* Vol. 3–6 (New York: Plenum Press, 1974–1989).

Appendix 4

Standard and Formal Electrode Potentials

Half-Reaction	E^0, V*	Formal Potential, V†
Aluminum		
$Al^{3+} + 3e^- \rightleftharpoons Al(s)$	-1.662	
Antimony		
$Sb_2O_5(s) + 6H^+ + 4e^- \rightleftharpoons 2SbO^+ + 3H_2O$	$+0.581$	
Arsenic		
$H_3AsO_4 + 2H^+ + 2e^- \rightleftharpoons H_3AsO_3 + H_2O$	$+0.559$	0.577 in 1 M HCl, $HClO_4$
Barium		
$Ba^{2+} + 2e^- \rightleftharpoons Ba(s)$	-2.906	
Bismuth		
$BiO^+ + 2H^+ + 3e^- \rightleftharpoons Bi(s) + H_2O$	$+0.320$	
$BiCl_4^- + 3e^- \rightleftharpoons Bi(s) + 4Cl^-$	$+0.16$	
Bromine		
$Br_2(l) + 2e^- \rightleftharpoons 2Br^-$	$+1.065$	1.05 in 4 M HCl
$Br_2(aq) + 2e^- \rightleftharpoons 2Br^-$	$+1.087‡$	
$BrO_3^- + 6H^+ + 5e^- \rightleftharpoons \frac{1}{2}Br_2(l) + 3H_2O$	$+1.52$	
$BrO_3^- + 6H^+ + 6e^- \rightleftharpoons Br^- + 3H_2O$	$+1.44$	
Cadmium		
$Cd^{2+} + 2e^- \rightleftharpoons Cd(s)$	-0.403	
Calcium		
$Ca^{2+} + 2e^- \rightleftharpoons Ca(s)$	-2.866	
Carbon		
$C_6H_4O_2 \text{ (quinone)} + 2H^+ + 2e^- \rightleftharpoons C_6H_4(OH)_2$	$+0.699$	0.696 in 1 M HCl, $HClO_4$, H_2SO_4
$2CO_2(g) + 2H^+ + 2e^- \rightleftharpoons H_2C_2O_4$	-0.49	
Cerium		
$Ce^{4+} + e^- \rightleftharpoons Ce^{3+}$		$+1.70$ in 1 M $HClO_4$; $+1.61$ in 1 M HNO_3; $+1.44$ in 1 M H_2SO_4
Chlorine		
$Cl_2(g) + 2e^- \rightleftharpoons 2Cl^-$	$+1.359$	
$HClO + H^+ + e^- \rightleftharpoons \frac{1}{2}Cl_2(g) + H_2O$	$+1.63$	
$ClO_3^- + 6H^+ + 5e^- \rightleftharpoons \frac{1}{2}Cl_2(g) + 3H_2O$	$+1.47$	
Chromium		
$Cr^{3+} + e^- \rightleftharpoons Cr^{2+}$	-0.408	
$Cr^{3+} + 3e^- \rightleftharpoons Cr(s)$	-0.744	
$Cr_2O_7^{2-} + 14H^+ + 6e^- \rightleftharpoons 2Cr^{3+} + 7H_2O$	$+1.33$	
Cobalt		
$Co^{2+} + 2e^- \rightleftharpoons Co(s)$	-0.277	
$Co^{3+} + e^- \rightleftharpoons Co^{2+}$	$+1.808$	

Half-Reaction	E^0, V*	Formal Potential, V†
Copper		
$Cu^{2+} + 2e^- \rightleftharpoons Cu(s)$	+ 0.337	
$Cu^{2+} + e^- \rightleftharpoons Cu^+$	+ 0.153	
$Cu^+ + e^- \rightleftharpoons Cu(s)$	+ 0.521	
$Cu^{2+} + I^- + e^- \rightleftharpoons CuI(s)$	+ 0.86	
$CuI(s) + e^- \rightleftharpoons Cu(s) + I^-$	− 0.185	
Fluorine		
$F_2(g) + 2H^+ + 2e^- \rightleftharpoons 2HF(aq)$	+ 3.06	
Hydrogen		
$2H^+ + 2e^- \rightleftharpoons H_2(g)$	0.000	− 0.005 in 1 M HCl, $HClO_4$
Iodine		
$I_2(s) + 2e^- \rightleftharpoons 2I^-$	+ 0.5355	
$I_2(aq) + 2e^- \rightleftharpoons 2I^-$	+ 0.615‡	
$I_3^- + 2e^- \rightleftharpoons 3I^-$	+ 0.536	
$ICl_2^- + e^- \rightleftharpoons \frac{1}{2}I_2(s) + 2Cl^-$	+ 1.056	
$IO_3^- + 6H^+ + 5e^- \rightleftharpoons \frac{1}{2}I_2(s) + 3H_2O$	+ 1.196	
$IO_3^- + 6H^+ + 5e^- \rightleftharpoons \frac{1}{2}I_2(aq) + 3H_2O$	+ 1.178‡	
$IO_3^- + 2Cl^- + 6H^+ + 4e^- \rightleftharpoons ICl_2^- + 3H_2O$	+ 1.24	
$H_5IO_6 + H^+ + 2e^- \rightleftharpoons IO_3^- + 3H_2O$	+ 1.601	
Iron		
$Fe^{2+} + 2e^- \rightleftharpoons Fe(s)$	− 0.440	
$Fe^{3+} + e^- \rightleftharpoons Fe^{2+}$	+ 0.771	0.700 in 1 M HCl; 0.732 in 1 M $HClO_4$; 0.68 in 1 M H_2SO_4
$Fe(CN)_6^{3-} + e^- \rightleftharpoons Fe(CN)_6^{4-}$	+ 0.36	0.71 in 1 M HCl; 0.72 in 1 M $HClO_4$, H_2SO_4
Lead		
$Pb^{2+} + 2e^- \rightleftharpoons Pb(s)$	− 0.126	− 0.14 in 1 M $HClO_4$; − 0.29 in 1 M H_2SO_4
$PbO_2(s) + 4H^+ + 2e^- \rightleftharpoons Pb^{2+} + 2H_2O$	+ 1.455	
$PbSO_4(s) + 2e^- \rightleftharpoons Pb(s) + SO_4^{2-}$	− 0.350	
Lithium		
$Li^+ + e^- \rightleftharpoons Li(s)$	− 3.045	
Magnesium		
$Mg^{2+} + 2e^- \rightleftharpoons Mg(s)$	− 2.363	
Manganese		
$Mn^{2+} + 2e^- \rightleftharpoons Mn(s)$	− 1.180	
$Mn^{3+} + e^- \rightleftharpoons Mn^{2+}$		1.51 in 7.5 M H_2SO_4
$MnO_2(s) + 4H^+ + 2e^- \rightleftharpoons Mn^{2+} + 2H_2O$	+ 1.23	
$MnO_4^- + 8H^+ + 5e^- \rightleftharpoons Mn^{2+} + 4H_2O$	+ 1.51	
$MnO_4^- + 4H^+ + 3e^- \rightleftharpoons MnO_2(s) + 2H_2O$	+ 1.695	
$MnO_4^- + e^- \rightleftharpoons MnO_4^{2-}$	+ 0.564	
Mercury		
$Hg_2^{2+} + 2e^- \rightleftharpoons 2Hg(l)$	+ 0.788	0.274 in 1 M HCl; 0.776 in 1 M $HClO_4$; 0.674 in 1 M H_2SO_4
$2Hg^{2+} + 2e^- \rightleftharpoons Hg_2^{2+}$	+ 0.920	0.907 in 1 M $HClO_4$
$Hg^{2+} + 2e^- \rightleftharpoons Hg(l)$	+ 0.854	
$Hg_2Cl_2(s) + 2e^- \rightleftharpoons 2Hg(l) + 2Cl^-$	+ 0.268	0.244 in sat'd KCl; 0.282 in 1 M KCl; 0.334 in 0.1 M KCl
$Hg_2SO_4(s) + 2e^- \rightleftharpoons 2Hg(l) + SO_4^{2-}$	+ 0.615	
Nickel		
$Ni^{2+} + 2e^- \rightleftharpoons Ni(s)$	− 0.250	
Nitrogen		
$N_2(g) + 5H^+ + 4e^- \rightleftharpoons N_2H_5^+$	− 0.23	
$HNO_2 + H^+ + e^- \rightleftharpoons NO(g) + H_2O$	+ 1.00	
$NO_3^- + 3H^+ + 2e^- \rightleftharpoons HNO_2 + H_2O$	+ 0.94	0.92 in 1 M HNO_3
Oxygen		
$H_2O_2 + 2H^+ + 2e^- \rightleftharpoons 2H_2O$	+ 1.776	
$HO_2^- + H_2O + 2e^- \rightleftharpoons 3OH^-$	+ 0.88	

Half-Reaction	E^0, V*	Formal Potential, V†
$O_2(g) + 4H^+ + 4e^- \rightleftharpoons 2H_2O$	$+ 1.229$	
$O_2(g) + 2H^+ + 2e^- \rightleftharpoons H_2O_2$	$+ 0.682$	
$O_3(g) + 2H^+ + 2e^- \rightleftharpoons O_2(g) + H_2O$	$+ 2.07$	
Palladium		
$Pd^{2+} + 2e^- \rightleftharpoons Pd(s)$	$+ 0.987$	
Platinum		
$PtCl_4^{2-} + 2e^- \rightleftharpoons Pt(s) + 4Cl^-$	$+ 0.73$	
$PtCl_6^{2-} + 2e^- \rightleftharpoons PtCl_4^{2-} + 2Cl^-$	$+ 0.68$	
Potassium		
$K^+ + e^- \rightleftharpoons K(s)$	$- 2.925$	
Selenium		
$H_2SeO_3 + 4H^+ + 4e^- \rightleftharpoons Se(s) + 3H_2O$	$+ 0.740$	
$SeO_4^{2-} + 4H^+ + 2e^- \rightleftharpoons H_2SeO_3 + H_2O$	$+ 1.15$	
Silver		
$Ag^+ + e^- \rightleftharpoons Ag(s)$	$+ 0.799$	0.228 in 1 M HCl; 0.792 in 1 M HClO$_4$; 0.77 in 1 M H$_2$SO$_4$
$AgBr(s) + e^- \rightleftharpoons Ag(s) + Br^-$	$+ 0.073$	
$AgCl(s) + e^- \rightleftharpoons Ag(s) + Cl^-$	$+ 0.222$	0.228 in 1 M KCl
$Ag(CN)_2^- + e^- \rightleftharpoons Ag(s) + 2CN^-$	$- 0.31$	
$Ag_2CrO_4(s) + 2e^- \rightleftharpoons 2Ag(s) + CrO_4^{2-}$	$+ 0.446$	
$AgI(s) + e^- \rightleftharpoons Ag(s) + I^-$	$- 0.151$	
$Ag(S_2O_3)_2^{3-} + e^- \rightleftharpoons Ag(s) + 2S_2O_3^{2-}$	$+ 0.017$	
Sodium		
$Na^+ + e^- \rightleftharpoons Na(s)$	$- 2.714$	
Sulfur		
$S(s) + 2H^+ + 2e^- \rightleftharpoons H_2S(g)$	$+ 0.141$	
$H_2SO_3 + 4H^+ + 4e^- \rightleftharpoons S(s) + 3H_2O$	$+ 0.450$	
$SO_4^{2-} + 4H^+ + 2e^- \rightleftharpoons H_2SO_3 + H_2O$	$+ 0.172$	
$S_4O_6^{2-} + 2e^- \rightleftharpoons 2S_2O_3^{2-}$	$+ 0.08$	
$S_2O_8^{2-} + 2e^- \rightleftharpoons 2SO_4^{2-}$	$+ 2.01$	
Thallium		
$Tl^+ + e^- \rightleftharpoons Tl(s)$	$- 0.336$	$- 0.551$ in 1 M HCl; $- 0.33$ in 1 M HClO$_4$, H$_2$SO$_4$
$Tl^{3+} + 2e^- \rightleftharpoons Tl^+$	$+ 1.25$	0.77 in 1 M HCl
Tin		
$Sn^{2+} + 2e^- \rightleftharpoons Sn(s)$	$- 0.136$	$- 0.16$ in 1 M HClO$_4$
$Sn^{4+} + 2e^- \rightleftharpoons Sn^{2+}$	$+ 0.154$	0.14 in 1 M HCl
Titanium		
$Ti^{3+} + e^- \rightleftharpoons Ti^{2+}$	$- 0.369$	
$TiO^{2+} + 2H^+ + e^- \rightleftharpoons Ti^{3+} + H_2O$	$+ 0.099$	0.04 in 1 M H$_2$SO$_4$
Uranium		
$UO_2^{2+} + 4H^+ + 2e^- \rightleftharpoons U^{4+} + 2H_2O$	$+ 0.334$	
Vanadium		
$V^{3+} + e^- \rightleftharpoons V^{2+}$	$- 0.256$	0.21 in 1 M HClO$_4$
$VO^{2+} + 2H^+ + e^- \rightleftharpoons V^{3+} + H_2O$	$+ 0.359$	
$V(OH)_4^+ + 2H^+ + e^- \rightleftharpoons VO^{2+} + 3H_2O$	$+ 1.00$	1.02 in 1 M HCl, HClO$_4$
Zinc		
$Zn^{2+} + 2e^- \rightleftharpoons Zn(s)$	$- 0.763$	

*G. Milazzo, S. Caroli, and V. K. Sharma, *Tables of Standard Electrode Potentials* (London: Wiley, 1978).

†E. H. Swift and E. A. Butler, *Quantitative Measurements and Chemical Equilibria* (New York: Freeman, 1972).

‡These potentials are hypothetical because they correspond to solutions that are 1.00 M in Br$_2$ or I$_2$. The solubilities of these two compounds at 25°C are 0.18 M and 0.0020 M, respectively. In saturated solutions containing an excess of Br$_2$(l) or I$_2$(s), the standard potentials for the half-reaction Br$_2$(l) + 2e$^-$ $\rightleftharpoons$ 2Br$^-$ or I$_2$(s) + 2e$^-$ $\rightleftharpoons$ 2I$^-$ should be used. In contrast, at Br$_2$ and I$_2$ concentrations less than saturation, these hypothetical electrode potentials should be employed.

Appendix 5

Use of Exponential Numbers and Logarithms

Scientists frequently find it necessary (or convenient) to use exponential notation to express numerical data. A brief review of this notation follows.

EXPONENTIAL NOTATION

An exponent is used to describe the process of repeated multiplication or division. For example, 3^5 means

$$3 \times 3 \times 3 \times 3 \times 3 = 3^5 = 243$$

The power 5 is the exponent of the number (or base) 3; thus, 3 raised to the fifth power is equal to 243.

A negative exponent represents repeated division. For example, 3^{-5} means

$$\frac{1}{3} \times \frac{1}{3} \times \frac{1}{3} \times \frac{1}{3} \times \frac{1}{3} = \frac{1}{3^5} = 3^{-5} = 0.00412$$

Note that changing the sign of the exponent yields the *reciprocal* of the number; that is,

$$3^{-5} = \frac{1}{3^5} = \frac{1}{243} = 0.00412$$

It is important to note that a number raised to the first power is the number itself, and any number raised to the zero power has a value of 1. For example,

$$4^1 = 4$$

$$4^0 = 1$$

$$67^0 = 1$$

Fractional Exponents

A fractional exponent symbolizes the process of extracting the root of a number. The fifth root of 243 is 3; this process is expressed exponentially as

$$(243)^{1/5} = 3$$

Other examples are

$$25^{1/2} = 5$$

$$25^{-1/2} = \frac{1}{25^{1/2}} = \frac{1}{5}$$

The Combination of Exponential Numbers in Multiplication and Division

Multiplication and division of exponential numbers having the same base are accomplished by adding and subtracting the exponents. For example,

$$3^3 \times 3^2 = (3 \times 3 \times 3)(3 \times 3) = 3^{(3+2)} = 3^5 = 243$$

$$3^4 \times 3^{-2} \times 3^0 = (3 \times 3 \times 3 \times 3)\left(\frac{1}{3} \times \frac{1}{3}\right) \times 1 = 3^{(4-2+0)} = 3^2 = 9$$

$$\frac{5^4}{5^2} = \frac{5 \times 5 \times 5 \times 5}{5 \times 5} = 5^{(4-2)} = 5^2 = 25$$

$$\frac{2^3}{2^{-1}} = \frac{(2 \times 2 \times 2)}{1/2} = 2^4 = 16$$

Note that in the last equation the exponent is given by the relationship

$$3 - (-1) = 3 + 1 = 4$$

Extraction of the Root of an Exponential Number

To obtain the root of an exponential number, the exponent is divided by the desired root. Thus,

$$(5^4)^{1/2} = (5 \times 5 \times 5 \times 5)^{1/2} = 5^{(4/2)} = 5^2 = 25$$

$$(10^{-8})^{1/4} = 10^{(-8/4)} = 10^{-2}$$

$$(10^9)^{1/2} = 10^{(9/2)} = 10^{4.5}$$

THE USE OF EXPONENTS IN SCIENTIFIC NOTATION

Scientists and engineers are frequently called upon to use very large or very small numbers for which ordinary decimal notation is either awkward or impossible. For example, to express Avogadro's number in decimal notation would require 21 zeros following the number 602. In scientific notation, the number is written as a multiple of two numbers, the one number in decimal notation and the other expressed as a power of 10. Thus, Avogadro's number is written as 6.02×10^{23}. Other examples are

$$4.32 \times 10^3 = 4.32 \times 10 \times 10 \times 10 = 4320$$

$$4.32 \times 10^{-3} = 4.32 \times \frac{1}{10} \times \frac{1}{10} \times \frac{1}{10} = 0.00432$$

$$0.002002 = 2.002 \times \frac{1}{10} \times \frac{1}{10} \times \frac{1}{10} = 2.002 \times 10^{-3}$$

$$375 = 3.75 \times 10 \times 10 = 3.75 \times 10^{2}$$

It should be noted that the scientific notation for a number can be expressed in any of several equivalent forms. Thus,

$$4.32 \times 10^{3} = 43.2 \times 10^{2} = 432 \times 10^{1} = 0.432 \times 10^{4} = 0.0432 \times 10^{5}$$

The number in the exponent is equal to the number of places the decimal must be shifted to convert a number from scientific to purely decimal notation. The shift is to the right if the exponent is positive and to the left if it is negative. The process is reversed when decimal numbers are converted to scientific notation.

ARITHMETIC OPERATIONS WITH SCIENTIFIC NOTATION

The use of scientific notation is helpful in preventing decimal errors in arithmetic calculations. Some examples follow.

Multiplication

Here, the decimal parts of the numbers are multiplied and the exponents are added; thus,

$$420,000 \times 0.0300 = (4.20 \times 10^{5})(3.00 \times 10^{-2})$$
$$= 12.60 \times 10^{3} = 1.26 \times 10^{4}$$

$$0.0060 \times 0.000020 = 6.0 \times 10^{-3} \times 2.0 \times 10^{-5}$$
$$= 12 \times 10^{-8} = 1.2 \times 10^{-7}$$

Division

Here, the decimal parts of the numbers are divided; the exponent in the denominator is subtracted from that in the numerator. For example,

$$\frac{0.015}{5000} = \frac{15 \times 10^{-3}}{5.0 \times 10^{3}} = 3.0 \times 10^{-6}$$

Addition and Subtraction

Addition or subtraction in scientific notation requires that all numbers be expressed to a common power of 10. The decimal parts are then added or subtracted, as appropriate. Thus,

$$2.00 \times 10^{-11} + 4.00 \times 10^{-12} - 3.00 \times 10^{-10}$$
$$= 2.00 \times 10^{-11} + 0.400 \times 10^{-11} - 30.0 \times 10^{-11}$$
$$= -2.76 \times 10^{-10} = -27.6 \times 10^{-11}$$

Raising to a Power a Number Written in Exponential Notation

Here, each part of the number is raised to the power separately. For example,

$$(2 \times 10^{-3})^4 = (2.0)^4 \times (10^{-3})^4 = 16 \times 10^{-(3 \times 4)}$$
$$= 16 \times 10^{-12} = 1.6 \times 10^{-11}$$

Extraction of the Root of a Number Written in Exponential Notation

Here, the number is written in such a way that the exponent of 10 is evenly divisible by the root. Thus,

$$(4.0 \times 10^{-5})^{1/3} = \sqrt[3]{40 \times 10^{-6}} = \sqrt[3]{40} \times \sqrt[3]{10^{-6}}$$
$$= 3.4 \times 10^{-2}$$

LOGARITHMS

In this discussion, we will assume that the reader has available an electronic calculator for obtaining logarithms and antilogarithms of numbers. (The key for the antilogarithm function on most calculators is designated as 10^x.) It is desirable, however, to understand what a logarithm is as well as some of its properties. The discussion that follows provides this information.

A logarithm (or log) of a number is the power to which some base number (usually 10) must be raised to give the desired number. Thus, a logarithm is an exponent of the base 10. From the discussion in the previous paragraphs about exponential numbers, we can draw the following conclusions with respect to logs:

1. The logarithm of a product is the sum of the logarithms of the individual numbers in the product.

$$\log (100 \times 1000) = \log 10^2 + 10^3 = 2 + 3 = 5$$

2. The logarithm of a quotient is the difference between the logarithms of the individual numbers.

$$\log (100/1000) = \log 10^2 - \log 10^3 = 2 - 3 = -1$$

3. The logarithm of a number raised to some power is the logarithm of the number multiplied by that power.

$$\log (1000)^2 = 2 \times \log 10^3 = 2 \times 3 = 6$$

$$\log (0.01)^6 = 6 \times \log 10^{-2} = 6 \times (-2) = -12$$

4. The logarithm of a root of a number is the logarithm of that number divided by the root.

$$\log (1000)^{1/3} = \frac{1}{3} \times \log 10^3 = \frac{1}{3} \times 3 = 1$$

The following examples illustrate these statements:

$$\log 40 \times 10^{20} = \log 4.0 \times 10^{21} = \log 4.0 + \log 10^{21}$$
$$= 0.60 + 21 = 21.60$$

$$\log 2.0 \times 10^{-6} = \log 2.0 + \log 10^{-6} = 0.30 + (-6) = -5.70$$

For some purposes, it is helpful to dispense with the subtraction step shown in the last example and report the log as a *negative* integer and a *positive* decimal number; that is,

$$\log 2.0 \times 10^{-6} = \log 2.0 + \log 10^{-6} = \bar{6}.30$$

The last two examples demonstrate that the logarithm of a number is the sum of two parts, a *characteristic* located to the left of the decimal point and a *mantissa* that lies to the right. The characteristic is the logarithm of 10 raised to a power and serves to indicate the location of the decimal point in the original number when that number is expressed in decimal notation. The mantissa is the logarithm of a number in the range between 0.00 and 9.99. . . . Note that the mantissa is *always positive*. As a consequence, the characteristic in the last example is -6 and the mantissa is $+0.30$.

Appendix 6

Volumetric Calculations Using Normality and Equivalent Weight

The *normality* of a solution expresses the number of equivalents of solute contained in 1 L of solution or the number of milliequivalents in 1 mL. The equivalent and milliequivalent, like the mole and millimole, are units for describing the amount of a chemical species. The former two, however, are defined so that we may state that, at the equivalence point in *any* titration,

$$\text{no. meq analyte present} = \text{no. meq standard reagent added} \qquad \text{(A6-1)}$$

or

$$\text{no. eq analyte present} = \text{no. eq standard reagent added} \qquad \text{(A6-2)}$$

As a consequence, stoichiometric ratios such as those described in Section 3C-2 (page 73) need not be derived every time a volumetric calculation is performed. Instead, the stoichiometry is taken into account by how the equivalent or milliequivalent weight is defined.

A6-1 THE DEFINITIONS OF EQUIVALENT AND MILLIEQUIVALENT

In contrast to the mole, the amount of a substance contained in one equivalent can vary from reaction to reaction. Consequently, the weight of one equivalent of a compound can never be computed *without reference to a chemical reaction* in which that compound is, directly or indirectly, a participant. Similarly, the normality of a solution can never be specified *without knowledge about how the solution will be used.*

Equivalent Weights in Neutralization Reactions

One equivalent weight of a substance participating in a neutralization reaction is that amount of substance (molecule, ion, or paired ion such as NaOH) that either reacts with or supplies 1 mol of hydrogen ions *in that reaction.*[1] A milliequivalent is simply 1/1000 of an equivalent.

---

[1] An alternative definition, proposed by the International Union of Pure and Applied Chemistry, is as follows: An equivalent is "that amount of substance, which, in a specified reaction, releases or replaces that amount of hydrogen that is combined with 3 g of carbon-12 in methane $^{12}CH_4$" (see *Information Bulletin* No. 36, International Union of Pure and Applied Chemistry, August 1974). This definition applies to acids. For other types of reactions and reagents, the amount of hydrogen referred to may be replaced by the equivalent amount of hydroxide ions, electrons, or cations. The reaction to which the definition is applied must be specified.

Once again, we find ourselves using the term *weight* when we really mean *mass*. The term "equivalent weight" is so firmly engrained in the literature and vocabulary of chemistry that we retain it in this discussion.

The relationship between equivalent weight (eqw) and the molar mass ($\mathcal{M}$) is straightforward for strong acids or bases and for other acids or bases that contain a single reactive hydrogen or hydroxide ion. For example, the equivalent weights of potassium hydroxide, hydrochloric acid, and acetic acid are equal to their molar masses because each has but a single reactive hydrogen ion or hydroxide ion. Barium hydroxide, which contains two identical hydroxide ions, reacts with two hydrogen ions in any acid/base reaction, and so its equivalent weight is one-half its molar mass:

$$\text{eqw } Ba(OH)_2 = \frac{\mathcal{M}_{Ba(OH)_2}}{2}$$

The situation becomes more complex for acids or bases that contain two or more reactive hydrogen or hydroxide ions with different tendencies to dissociate. With certain indicators, for example, only the first of the three protons in phosphoric acid is titrated:

$$H_3PO_4 + OH^- \rightarrow H_2PO_4^- + H_2O$$

With certain other indicators, a color change occurs only after two hydrogen ions have reacted:

$$H_3PO_4 + 2OH^- \rightarrow HPO_4^{2-} + 2H_2O$$

For a titration involving the first reaction, the equivalent weight of phosphoric acid is equal to the molar mass; for the second, the equivalent weight is one-half the molar mass. (Because it is not practical to titrate the third proton, an equivalent weight that is one-third the molar mass is not generally encountered for H_3PO_4.) If it is not known which of these reactions is involved, an unambiguous definition of the equivalent weight for phosphoric acid *cannot be made.*

Equivalent Weights in Oxidation/Reduction Reactions

The equivalent weight of a participant in an oxidation/reduction reaction is that amount that directly or indirectly produces or consumes 1 mol of electrons. The numerical value for the equivalent weight is conveniently established by dividing the molar mass of the substance of interest by the change in oxidation number associated with its reaction. As an example, consider the oxidation of oxalate ion by permanganate ion:

$$5C_2O_4^{2-} + 2MnO_4^- + 16H^+ \rightarrow 10CO_2 + 2Mn^+ + 8H_2O \quad \text{(A6-3)}$$

In this reaction, the change in oxidation number of manganese is 5 because the element passes from the $+7$ to the $+2$ state; the equivalent weights for MnO_4^- and Mn^{2+} are therefore one-fifth their molar masses. Each carbon atom in the oxalate ion is oxidized from the $+3$ to the $+4$ state, leading to the production of two electrons by that species. Therefore, the equivalent weight of sodium oxalate is one-half its molar mass. It is also possible to assign an equivalent weight to the carbon dioxide produced by the reaction. Since this molecule contains but a single carbon atom and since that carbon undergoes a change in oxidation number of 1, the molar mass and equivalent weight of the two are identical.

It is important to note that in evaluating the equivalent weight of a substance, *only* its change in oxidation number during the titration is considered. For example, suppose the manganese content of a sample containing Mn_2O_3 is to be determined by a titration based on the reaction given in Equation A6-3. That each manganese in the Mn_2O_3 has an oxidation number of $+3$ plays no part in determining equivalent weight. Thus, we must assume that, by suitable treatment, all the manganese is oxidized to the $+7$ state before the titration is begun. Each manganese from the Mn_2O_3 is then reduced from the $+7$ to the $+2$ state in the titration step. The equivalent weight is thus the molar mass of Mn_2O_3 divided by $2 \times 5 = 10$.

As in neutralization reactions, the equivalent weight for a given oxidizing or reducing agent is not invariant. Potassium permanganate, for example, reacts under some conditions to give MnO_2:

$$MnO_4^- + 3e^- + 2H_2O \rightarrow MnO_2(s) + 4OH^-$$

The change in the oxidation state of manganese in this reaction is from $+7$ to $+4$, and the equivalent weight of potassium permanganate is now equal to its molar mass divided by 3 (instead of 5 as in the earlier example).

Equivalent Weights in Precipitation and Complex-Formation Reactions

The equivalent weight of a participant in a precipitation or a complex-formation reaction is the weight that reacts with or provides one mole of the *reacting* cation if it is univalent, one-half mole if it is divalent, one-third mole if it is trivalent, and so on. It is important to note that the cation referred to in this definition is always *the cation directly involved in the analytical reaction* and not necessarily the cation contained in the compound whose equivalent weight is being defined.

Example A6-1

Define equivalent weights for $AlCl_3$ and $BiOCl$ if the two compounds are determined by a precipitation titration with $AgNO_3$:

$$Ag^+ + Cl^- \rightarrow AgCl(s)$$

In this instance, the equivalent weight is based on the number of moles of *silver ions* involved in the titration of each compound. Since 1 mol of Ag^+ reacts with 1 mol of Cl^- provided by one-third mole of $AlCl_3$, we can write

$$\text{eqw } AlCl_3 = \frac{\mathcal{M}_{AlCl_3}}{3}$$

Because each mole of $BiOCl$ reacts with only 1 Ag^+ ion,

$$\text{eqw } BiOCl = \frac{\mathcal{M}_{BiOCl}}{1}$$

Note that Bi^{3+} (or Al^{3+}) being trivalent has no bearing because the definition is based *on the cation involved in the titration:* Ag^+.

A6-2 THE DEFINITION OF NORMALITY

The normality c_N of a solution expresses the number of milliequivalents of solute contained in 1 mL of solution or the number of equivalents contained in 1 L. Thus, a 0.20 N hydrochloric acid solution contains 0.20 meq of HCl in each milliliter of solution or 0.20 eq in each liter.

The normal concentration of a solution is defined by equations analogous to Equations 3-1 and 11-3. Thus, for a solution of the species A, the normality $c_{N(A)}$ is given by the equations

$$c_{N(A)} = \frac{\text{no. meq A}}{\text{no. mL solution}} \qquad (A6\text{-}4)$$

$$c_{N(A)} = \frac{\text{no. eq A}}{\text{no. L solution}} \qquad (A6\text{-}5)$$

A6-3 SOME USEFUL ALGEBRAIC RELATIONSHIPS

Two pairs of algebraic equations, analogous to Equations 11-1 and 11-2 as well as 11-3 and 11-4 in Chapter 11, apply when normal concentrations are used:

$$\text{amount A} = \text{no. meq A} = \frac{\text{mass A (g)}}{\text{meqw A (g/meq)}} \qquad (A6\text{-}6)$$

$$\text{amount A} = \text{no. eq A} = \frac{\text{mass A (g)}}{\text{eqw A (g/eq)}} \qquad (A6\text{-}7)$$

$$\text{amount A} = \text{no. meq A} = V\,(\text{mL}) \times c_{N(A)}(\text{meq/mL}) \qquad (A6\text{-}8)$$

$$\text{amount A} = \text{no. eq A} = V\,(\text{L}) \times c_{N(A)}(\text{eq/L}) \qquad (A6\text{-}9)$$

A6-4 CALCULATION OF THE NORMALITY OF STANDARD SOLUTIONS

Example A6-2 shows how the normality of a standard solution is computed from preparatory data. Note the similarity between this example and Example 11-1 (Chapter 11).

Example A6-2

Describe the preparation of 5.000 L of 0.1000 N Na_2CO_3 (105.99 g/mol) from the primary-standard solid, assuming that the solution is to be used for titrations in which the reaction is

$$CO_3^{2-} + 2H^+ \rightarrow H_2O + CO_2$$

Applying Equation A6-9 gives

$$\text{amount } Na_2CO_3 = V \text{ soln (L)} \times c_{N(Na_2CO_3)}(\text{eq/L})$$
$$= 5.000 \text{ L} \times 0.1000 \text{ eq/L} = 0.5000 \text{ eq } Na_2CO_3$$

Rearranging Equation A6-7 gives

$$\text{mass } Na_2CO_3 = \text{no. eq } Na_2CO_3 \times \text{eqw } Na_2CO_3$$

But 2 eq of Na_2CO_3 are contained in each mole of the compound; therefore,

$$\text{mass } Na_2CO_3 = 0.5000 \text{ eq } Na_2CO_3 \times \frac{105.99 \text{ g } Na_2CO_3}{2 \text{ eq } Na_2CO_3} = 26.50 \text{ g}$$

Therefore, dissolve 26.50 g in water and dilute to 5.000 L.

It is worth noting that when the carbonate ion reacts with two protons, the weight of sodium carbonate required to prepare a 0.10 N solution is just one-half that required to prepare a 0.10 M solution.

A6-5   THE USE OF NORMALITIES TO TREAT TITRATION DATA

Calculating Normalities from Titration Data

Examples A6-3 and A6-4 illustrate how normality is obtained from standardization data. Note that these examples are similar to Examples 11-4 and 11-5 in Chapter 11.

Example A6-3

A 50.00 mL volume of an HCl solution required 29.71 mL of 0.03926 N $Ba(OH)_2$ to give an end point with bromocresol green indicator. Calculate the normality of the HCl.

Note that the molarity of $Ba(OH)_2$ is one-half its normality. That is,

$$c_{Ba(OH)_2} = 0.03926 \frac{\text{meq}}{\text{mL}} \times \frac{1 \text{ mmol}}{2 \text{ meq}} = 0.01963 \text{ M}$$

Because we are basing our calculations on the milliequivalent, we write

$$\text{no. meq HCl} = \text{no. meq } Ba(OH)_2$$

The number of milliequivalents of standard is obtained by substituting into Equation A6-8:

$$\text{amount } Ba(OH)_2 = 29.71 \text{ mL } \cancel{Ba(OH)_2} \times 0.03926 \frac{\text{meq } Ba(OH)_2}{\text{mL } \cancel{Ba(OH)_2}}$$

To obtain the number of milliequivalents of HCl, we write

$$\text{amount HCl} = (29.71 \times 0.03926) \text{ meq } \cancel{Ba(OH)_2} \times \frac{1 \text{ meq HCl}}{1 \text{ meq } \cancel{Ba(OH)_2}}$$

Equating this result to Equation A6-8 yields

$$\text{amount HCl} = 50.00 \text{ mL} \times c_{N(HCl)}$$
$$= (29.71 \times 0.03926 \times 1) \text{ meq HCl}$$

$$c_{N(HCl)} = \frac{(29.71 \times 0.03926 \times 1) \text{ meq HCl}}{50.00 \text{ mL HCl}} = 0.02333 \text{ N}$$

Example A6-4

A 0.2121-g sample of pure $Na_2C_2O_4$ (134.00 g/mol) was titrated with 43.31 mL of $KMnO_4$. What is the normality of the $KMnO_4$ solution? The chemical reaction is

$$2MnO_4^- + 5C_2O_4^{2-} + 16H^+ \rightarrow 2Mn^{2+} + 10CO_2 + 8H_2O$$

By definition, at the equivalence point in the titration,

$$\text{no. meq } Na_2C_2O_4 = \text{no. meq } KMnO_4$$

Substituting Equations A6-6 and A6-8 into this relationship gives

$$V_{KMnO_4} \times c_{N(KMnO_4)} = \frac{\text{mass } Na_2C_2O_4(g)}{\text{meqw } Na_2C_2O_4 \text{ (g/meq)}}$$

$$43.31 \text{ mL } KMnO_4 \times c_{N(KMnO_4)} = \frac{0.2121 \text{ g } Na_2C_2O_4}{0.13400 \text{ g } Na_2C_2O_4/2 \text{ meq}}$$

$$c_{N(KMnO_4)} = \frac{0.2121 \text{ g } Na_2C_2O_4}{43.31 \text{ mL } KMnO_4 \times 0.1340 \text{ g } Na_2C_2O_4/2 \text{ meq}}$$
$$= 0.073093 \text{ meq/mL } KMnO_4 = 0.07309 \text{ N}$$

Note that the normality found here is five times the molarity computed in Example 11-5.

CALCULATING THE QUANTITY OF ANALYTE FROM TITRATION DATA

The examples that follow illustrate how analyte concentrations are calculated when normalities are involved. Note that these examples are similar to Examples 11-6 and 11-8.

Example A6-5

A 0.8040-g sample of an iron ore was dissolved in acid. The iron was then reduced to Fe^{2+} and titrated with 47.22 mL of 0.1121 N (0.02242 M) $KMnO_4$ solution. Calculate the results of this analysis in terms of (a) percent Fe (55.847 g/mol) and (b) percent Fe_3O_4 (231.54 g/mol). The reaction of the analyte with the reagent is described by the equation

$$MnO_4^- + 5Fe^{2+} + 8H^+ \rightarrow Mn^{2+} + 5Fe^{3+} + 4H_2O$$

(a) At the equivalence point, we know that

$$\text{no. meq } KMnO_4 = \text{no. meq } Fe^{2+} = \text{no. meq } Fe_3O_4$$

Substituting Equations A6-8 and A6-6 leads to

$$V_{KMnO_4}(\text{mL}) \times c_{N(KMnO_4)}(\text{meq/mL}) = \frac{\text{mass } Fe^{2+}(g)}{\text{meqw } Fe^{2+}(g/\text{meq})}$$

Substituting numerical data into the equation gives, after rearranging,

$$\text{mass Fe}^{2+} = 47.22 \text{ mL KMnO}_4 \times 0.1121 \frac{\text{meq}}{\text{mL KMnO}_4} \times \frac{0.055847 \text{ g}}{1 \text{ meq}}$$

Note that the milliequivalent weight of the Fe^{2+} is equal to its millimolar mass. The percentage of iron is

$$\text{percent Fe}^{2+} = \frac{(47.22 \times 0.1121 \times 0.055847) \text{ g Fe}^{2+}}{0.8040 \text{ g sample}} \times 100\%$$

$$= 36.77\%$$

(b) Here,

$$\text{no. meq KMnO}_4 = \text{no. meq Fe}_3\text{O}_4$$

and

$$V_{\text{KMnO}_4}(\text{mL}) \times c_{\text{N(KMnO}_4)}(\text{meq/mL}) = \frac{\text{mass Fe}_3\text{O}_4(g)}{\text{meqw Fe}_3\text{O}_4(g/\text{meq})}$$

Substituting numerical data and rearranging give

$$\text{mass Fe}_3\text{O}_4 = 47.22 \text{ mL} \times 0.1121 \frac{\text{meq}}{\text{mL}} \times 0.23154 \frac{\text{g Fe}_3\text{O}_4}{3 \text{ meq}}$$

Note that the milliequivalent weight of Fe_3O_4 is one-third its millimolar mass because each Fe^{2+} undergoes a one-electron change and the compound is converted to $3Fe^{2+}$ before titration. The percentage of Fe_3O_4 is then

$$\text{percent Fe}_3\text{O}_4 = \frac{(47.22 \times 0.1121 \times 0.23154/3) \text{ g Fe}_3\text{O}_4}{0.8040 \text{ g sample}} \times 100\%$$

$$= 50.81\%$$

Note that the answers to this example are identical to those in Example 11-6.

Example A6-6

A 0.4755-g sample containing $(NH_4)_2C_2O_4$ and inert compounds was dissolved in water and made alkaline with KOH. The liberated NH_3 was distilled into 50.00 mL of 0.1007 N (0.05035 M) H_2SO_4. The excess H_2SO_4 was back-titrated with 11.13 mL of 0.1214 N NaOH. Calculate the percentage of N (14.007 g/mol) and of $(NH_4)_2C_2O_4$ (124.10 g/mol) in the sample.

At the equivalence point, the number of milliequivalents of acid and base are equal. In this titration, however, two bases are involved: NaOH and NH_3. Thus,

$$\text{no. meq H}_2\text{SO}_4 = \text{no. meq NH}_3 + \text{no. meq NaOH}$$

After rearranging,

$$\text{no. meq NH}_3 = \text{no. meq N} = \text{no. meq H}_2\text{SO}_4 - \text{no. meq NaOH}$$

Substituting Equations A6-6 and A6-8 for the number of milliequivalents of N and H_2SO_4, respectively, yields

$$\frac{\text{mass N (g)}}{\text{meqw N (g/meq)}} = 50.00 \text{ mL } H_2SO_4 \times 0.1007 \frac{\text{meq}}{\text{mL } H_2SO_4}$$

$$- 11.13 \text{ mL NaOH} \times 0.1214 \frac{\text{meq}}{\text{mL NaOH}}$$

$$\text{mass N} = (50.00 \times 0.1007 - 11.13 \times 0.1214) \text{ meq} \times 0.014007 \text{ g N/meq}$$

$$\text{percent N} = \frac{(50.00 \times 0.1007 - 11.13 \times 0.1214) \times 0.014007 \text{ g N}}{0.4755 \text{ g sample}} \times 100\%$$

$$= 10.85\%$$

The number of milliequivalents of $(NH_4)_2C_2O_4$ is equal to the number of milliequivalents of NH_3 and N, but the milliequivalent weight of the $(NH_4)_2C_2O_4$ is equal to one-half its molar mass. Thus,

$$\text{mass } (NH_4)_2C_2O_4 = (50.00 \times 0.1007 - 11.13 \times 0.1214) \text{ meq}$$
$$\times 0.12410 \text{ g/2 meq}$$

percent $(NH_4)_2C_2O_4$

$$= \frac{(50.00 \times 0.1007 - 11.13 \times 0.1214) \times 0.06205 \text{ g } (NH_4)_2C_2O_4}{0.4755 \text{ g sample}} \times 100\%$$

$$= 48.07\%$$

Note that the results obtained here are identical to those obtained in Example 11-8.

Appendix 7

Acronyms and Abbreviations of Significance in Analytical Chemistry

Acronym or Abbreviation	Name
AAS	Atomic Absorption Spectroscopy
ACS	American Chemical Society
ADC	Analog-to-Digital Converter
AES	Atomic Emission Spectroscopy
AFS	Atomic Fluorescence Spectroscopy
amu	atomic mass unit
AOAC	Association of Official Analytical Chemists
ASTM	American Society for Testing Materials
BUN	Blood Urea Nitrogen
CCD	Charge-Coupled Device
CE	Capillary Electrophoresis
CGE	Capillary Gel Electrophoresis
CHC	Chlorinated HydroCarbons
CL	Confidence Level
CL	Confidence Limit
CZE	Capillary Zone Electrophoresis
DAC	Digital-to-Analog Converter
DCP	Direct Current Plasma
DMCS	DiMethylChloroSilane
DME	Dropping Mercury Electrode
DNA	DeoxyriboNucleic Acid
DVM	Digital VoltMeter
EAAS	Electrothermal Atomic Absorption Spectroscopy
EC	ElectroChromatography
EDTA	EthyleneDiamineTetraacetic Acid
EMR	ElectroMagnetic Radiation
EPA	Environmental Protection Agency
FAAS	Flame Atomic Absorption Spectroscopy
FDA	Food and Drug Administration
FES	Flame Emission Spectroscopy
FIA	Flow Injection Analysis
FID	Flame Ionization Detector
FSOT	Fused-Silica Open Tubular column
FTIR	Fourier Transform Infrared spectroscopy
FTNMR	Fourier Transform Nuclear Magnetic Resonance
GC	Gas Chromatography

Acronym or Abbreviation	Name
GC/IR	Gas Chromatography/InfraRed spectroscopy
GC/MS	Gas Chromatography/Mass Spectroscopy
GLC	Gas-Liquid Chromatography
HETP	Height Equivalent to a Theoretical Plate
HPLC	High-Performance Liquid Chromatography
IC	Ion Chromatography
ICP	Inductively Coupled Plasma
IEC	Ion Exchange Chromatography
IR	InfraRed
ISE	Ion Selective Electrode
IUPAC	International Union of Pure and Applied Chemistry
KHP	Potassium (K) Hydrogen Phthalate
LC	Liquid Chromatography
LED	Light-Emitting Diode
LIMS	Laboratory Information Management System
LOD	Limit Of Detection
MS	Mass Spectrometry
NBS	National Bureau of Standards (the predecessor to NIST)
NIH	National Institutes of Health
NIR	Near InfraRed spectroscopy
NIST	National Institute of Standards and Technology
NMR	Nuclear Magnetic Resonance
NSF	National Science Foundation
OSHA	Occupational Safety and Health Administration
PC	Paper Chromatography
PMT	PhotoMultiplier Tube
ppm	parts per million
ROM	Read Only Memory
RSD	Relative Standard Deviation
SCE	Saturated Calomel Electrode
SCF	SuperCritical Fluid
SCOT	Support Coated Open Tubular column
SEC	Size Exclusion Chromatography
SFC	Supercritical Fluid Chromatography
SHE	Standard Hydrogen Electrode
SN	Signal-to-Noise ratio
STP	Standard Temperature and Pressure
TA	Thermal Analysis
TCD	Thermal Conductivity Detector
TG	ThermoGravimetry
TGA	ThermoGravimetric Analysis
TID	ThermIonic Detector
TISAB	Total Ionic Strength Adjusting Buffer
TLC	Thin Layer Chromatography
USDA	United States Department of Agriculture
USGS	United States Geological Survey
UV	UltraViolet
WCOT	Wall-Coated Open Tubular column
XFS	X-ray Fluorescence Spectroscopy

Answers to Selected Questions and Problems

Chapter 3

3-1. **(a)** The *millimole* is the amount of an elementary species, such as an atom, an ion, a molecule, or an electron. A millimole contains

$$6.02 \times 10^{23} \; \frac{\text{particles}}{\cancel{\text{mole}}} \times 10^{-3} \; \frac{\cancel{\text{mole}}}{\text{millimole}}$$

$$= 6.02 \times 10^{20} \; \frac{\text{particles}}{\text{millimole}}$$

(c) The millimolar mass of a species is the mass in grams of one millimole of the species.

3-3. $1 \text{ L} = 10^{-3} \text{ m}^3$

$$1 \text{ M} = \frac{1 \text{ mol}}{\text{L}} = \frac{1 \text{ mol}}{10^{-3} \text{ m}^3}$$

$1 \text{ Å} = 10^{-10} \text{ m}$

3-4. **(a)** 25 kHz **(c)** 843 mmol **(e)** 74.4 μm

3-5. $5.08 \times 10^{22} \text{ Na}^+$ ions

3-7. **(a)** 0.0983 mol **(b)** 7.76×10^{-4} mol
(c) 3.82×10^{-2} mol **(d)** 5.26×10^{-4} mol

3-9. **(a)** 5.52 mmol **(b)** 31.2 mmol
(c) 6.58×10^{-3} mmol **(d)** 966 mmol

3-11. **(a)** 4.20×10^4 mg **(b)** 1.21×10^4 mg
(c) 1.52×10^6 mg **(d)** 2.92×10^6 mg

3-13. **(a)** 1.33×10^3 mg **(b)** 519 mg

3-14. **(a)** 2.51 g **(b)** 2.89×10^{-3} g

3-15. **(a)** pNa = 1.132 pCl = 1.629 pOH = 1.298
(c) pH = 0.097 pCl = −0.001 pZn = 0.996
(e) pK = 5.715 pOH = 6.385
pFe(CN)$_6$ = 6.421

3-16. **(a)** 6.2×10^{-10} M **(c)** 0.35 M
(e) 4.8×10^{-8} M **(g)** 1.6 M

3-17. **(a)** pNa = pBr = 2.000 pH = pOH = 7.00
(c) pBa = 2.46 pOH = 2.15 pH = 11.85
(e) pCa = 2.28 pBa = 2.44 pCl = 1.75
pH = pOH = 7.00

3-18. **(a)** 2.14×10^{-9} M **(c)** 0.925 M **(e)** 2.09 M
(g) 0.994 M

3-19. **(a)** $[\text{Na}^+] = 4.79 \times 10^{-2}$ M,
$[\text{SO}_4^{2-}] = 2.87 \times 10^{-2}$ M
(b) pNa = 1.320 pSO$_4$ = 2.543

3-21. **(a)** 1.141×10^{-2} M **(b)** 1.141×10^{-2} M
(c) 3.423×10^{-2} M **(d)** 0.317% (w/v)
(e) 0.856 mmol Cl$^-$ **(f)** 446 ppm K$^+$
(g) 1.943 **(h)** 1.466

3-23. **(a)** 0.281 M **(b)** 0.843 M **(c)** 68.0 g/L

3-25. **(a)** Dissolve 23.8 mL of ethanol in water and make up to 500 mL with water.
(b) Mix 23.8 g of ethanol with 476.2 g of water.
(c) Dissolve 23.8 g of ethanol in water and dilute to 500 mL.

3-27. Dilute 307 mL of the H$_3$PO$_4$ to 750 mL.

3-29. **(a)** Dissolve 5.52 g AgNO$_3$ in water and dilute to 500 mL.
(b) Dilute 47.5 mL of 6.00 M HCl to 1.00 L.
(c) Dissolve 3.04 g K$_4$Fe(CN)$_6$ in water and dilute to 400 mL.
(d) Dilute 216 mL of 0.400 M BaCl$_2$ to 600 mL.
(e) Dilute 25.1 mL of the reagent to 2.00 L.
(f) Dissolve 1.67 g Na$_2$SO$_4$ in water and dilute to 9.00 L

3-31. 8.30 g La(IO$_3$)$_3$

3-33. **(a)** 0.04652 g CO$_2$ **(b)** 0.0286 M HClO$_4$

3-35. **(a)** 1.602 g SO$_2$ **(b)** 0.0386 M HClO$_4$

3-37. 3184 mL AgNO$_3$

Chapter 4

4-1. **(a)** A weak electrolyte is a substance that ionizes only partially in a solvent.
(c) The conjugate acid of a Brønsted-Lowry base is the species formed when the base has combined with a proton.
(e) An amphiprotic solute is one that can act either as an acid or as a base when dissolved in a solvent.
(g) Autoprotolysis is self-ionization of a solvent to give a conjugate acid and a conjugate base.
(i) The Le Châtelier principle states that the position of equilibrium in a system always shifts in a direction that tends to relieve an applied stress to the system.

4-2. **(a)** An amphiprotic solvent is a solvent that acts as a base with acidic solutes and as an acid with basic solutes.
(c) A leveling solvent is one in which a series of acids (or bases) all dissociate completely.

4-3. For an aqueous equilibrium in which water is a reactant or a product, the concentration of water is normally so much larger than the concentrations of the reactants and products that its concentration can be assumed to be constant and independent of the position of the equilibrium. Thus, its concentration is assumed to be constant and is included in the equilibrium constant. For a solid reactant or product, it is the concentration of that reactant in the solid phase that would influence the position of the equilibrium. However, the concentration of a species in the solid phase is constant. Thus as long as some solid exists as a second phase, its effect on the equilibrium is constant, and its concentration is included in the equilibrium constant.

4-4.

Acid	Conjugate Base
(a) HOCl	OCl$^-$
(c) NH$_4^+$	NH$_3$
(e) H$_2$PO$_4^-$	HPO$_4^{2-}$

4-5.

Base	Conjugate Acid
(a) H$_2$O	H$_3$O$^+$
(c) H$_2$O	H$_3$O$^+$
(e) PO$_4^{3-}$	HPO$_4^{2-}$

4-6. (a) $2H_2O \rightleftharpoons H_3O^+ + OH^-$
(c) $2CH_3NH_2 \rightleftharpoons CH_3NH_3^+ + CH_3NH^-$

4-7. (a) $C_2H_5NH_2 + H_2O \rightleftharpoons C_2H_5NH_3^+ + OH^-$

$$K_b = \frac{[C_2H_5NH_3^+][OH^-]}{[C_2H_5NH_2]} = 4.33 \times 10^{-4}$$

(c) $C_5H_5NH^+ + H_2O \rightleftharpoons C_5H_5N + H_3O^+$

$$K_a = \frac{[H_3O^+][C_5H_5N]}{[C_5H_5NH^+]} = 5.90 \times 10^{-6}$$

(e) $H_3AsO_4 + 3H_2O \rightleftharpoons 3H_3O^+ + AsO_4^{3-}$

$$\frac{[H_3O^+]^3[AsO_4^{3-}]}{[H_3AsO_4]} = 2.0 \times 10^{-21} = \beta_3$$

4-8. (a) $K_{sp} = [Ag^+][IO_3^-]$
(b) $K_{sp} = [Ag^+]^2[SO_3^{2-}]$
(c) $K_{sp} = [Ag^+]^3[AsO_4^{3-}]$
(d) $K_{sp} = [Pb^{2+}][Cl^-][F^-]$

4-9. (a) $K_{sp} = S^2$　　(b) $K_{sp} = 4S^3$
(c) $K_{sp} = 27S^4$　　(d) $K_{sp} = S^3$

4-10. (a) $K_{sp} = 5.0 \times 10^{-7}$　(c) $K_{sp} = 3.2 \times 10^{-13}$
(e) $K_{sp} = 1.0 \times 10^{-15}$

4-11. (a) 1.0×10^{-5} mol/L　(c) 1.3×10^{-6} mol/L
(e) 1.5×10^{-3} mol/L

4-12. (a) 1.0×10^{-5} mol/L　(c) 1.3×10^{-10} mol/L
(e) 1.6×10^{-10} mol/L

4-13. (a) 2.7×10^{-7} mol/L　(b) 1.2 mol/L

4-15. (a) 0.0250 M　　(b) 1.7×10^{-2} M
(c) 1.9×10^{-3} M　(d) 7.6×10^{-7} M

4-17. (a) PbI$_2$(1.2×10^{-3} M) > TlI(2.5×10^{-4} M) > BiI$_3$(1.3×10^{-5} M) > AgI(9.1×10^{-9} M)
(b) PbI$_2$(7.1×10^{-7} M) > TlI(6.5×10^{-7} M) > AgI(8.3×10^{-16} M) > BiI$_3$(8.1×10^{-16} M)
(c) PbI$_2$(4.2×10^{-4} M) > TlI(6.5×10^{-6} M) > BiI$_3$(1.4×10^{-6} M) > AgI(8.3×10^{-15} M)

4-20.

	[H$_3$O$^+$]	[OH$^-$]
(a)	2.4×10^{-5} M	4.1×10^{-10} M
(c)	1.1×10^{-12} M	9×10^{-3} M
(e)	5.0×10^{-11} M	2.0×10^{-4} M
(g)	3.32×10^{-4} M	3.02×10^{-11} M

4-22. (a) 1.10×10^{-2} M　(c) 5.33×10^{-12} M
(e) 1.46×10^{-4} M

4-23.

Percent	(a) [H$_3$O$^+$]	(c) [H$_3$O$^+$]	(e) [H$_3$O$^+$]
10.00	3.07×10^{-3}	2.10×10^{-11}	3.91×10^{-5}
20.00	4.58×10^{-3}	1.35×10^{-11}	5.94×10^{-5}
30.00	5.74×10^{-3}	1.06×10^{-11}	7.51×10^{-5}
40.00	6.73×10^{-3}	8.94×10^{-12}	8.84×10^{-5}
50.00	7.59×10^{-3}	7.86×10^{-12}	1.00×10^{-4}
60.00	8.38×10^{-3}	7.08×10^{-12}	1.11×10^{-4}
70.00	9.10×10^{-3}	6.49×10^{-12}	1.21×10^{-4}
80.00	9.77×10^{-3}	6.02×10^{-12}	1.30×10^{-4}
90.00	1.04×10^{-2}	5.64×10^{-12}	1.38×10^{-4}
100	1.10×10^{-2}	5.32×10^{-12}	1.46×10^{-4}
200	1.58×10^{-2}	3.65×10^{-12}	2.12×10^{-4}
300	1.95×10^{-2}	2.94×10^{-12}	2.62×10^{-4}
400	2.27×10^{-2}	2.53×10^{-12}	3.05×10^{-4}

Chapter 5

5-1. (a) *Constant errors* are the same magnitude regardless of the sample size. *Proportional errors* are proportional in size to the sample size.
(c) The *mean* is the sum of the measurements in a set divided by the number of measurements. The *median* is the central value for a set of data; half of the measurements are larger and half are smaller than the median.

5-2. (1) Random temperature fluctuations causing random changes in the length of the metal rule; (2) uncertainties arising from having to move and position the rule twice; (3) personal judgment in reading the rule; (4) vibrations in the table and/or rule; (5) uncertainty in locating the rule perpendicular to the edge of the table.

5-3. The three types of systematic error are *instrumental error, method error,* and *personal error.*

5-5. Constant errors.

5-6. (a) -0.04%　(c) -0.3%

5-7. (a) 17 g　(b) 4 g

5-8. (a) 0.08%　(c) 0.16%

5-9. (a) -1.0%　(c) -0.10%

5-10.

	Mean	Median	Deviation from Mean	Mean Deviation
(a)	0.0106	0.0105	0.0004, 0.0002, 0.0001	0.0002
(c)	190	189	2, 0, 4, 3	2
(e)	39.59	39.64	0.24, 0.02, 0.34, 0.090	0.17

Chapter 6

6-1. (a) The *spread* or *range* for a set of replicate data is the numerical difference between the highest and lowest value.

(c) *Significant figures* include all of the digits in a number that are known with certainty plus the first uncertain digit.

6-2. (a) The *sample variance*, s^2, is given by the expression

$$s^2 = \frac{\sum_{i=1}^{N} (x_i - \bar{x})^2}{N - 1}$$

where $\bar{x}$ is the sample mean.

The *sample standard deviation* is given by

$$s = \sqrt{\frac{\sum_{i=1}^{N} (x_i - \bar{x})^2}{N - 1}}$$

(c) *Accuracy* represents the agreement between an experimentally measured value and the true value. *Precision* describes the agreement among measurements that have been performed in exactly the same way.

6-3. (a) In statistics, a sample is a small set of replicate measurements. In chemistry, a sample is a portion of a material that is used for analysis.

6-5.

	(a) Mean	(b) Median	(c) Spread	(d) Std Dev	(e) CV
A	3.1	3.1	1.0	0.4	12%
C	0.825	0.803	0.108	0.051	6.2%
E	70.53	70.64	0.44	0.22	0.31%

6-6.

	Absolute Error	Relative Error, ppt
A	0.10	33
C	−0.005	−7
E	0.48	6.9

6-7.

	s_y	CV	y
(a)	0.03	7%	0.44(± 0.03)
(c)	0.14×10^{-16}	1.8%	7.4(± 0.1) $\times 10^{-16}$
(e)	0.5×10^{-2}	6.9%	7.6(± 0.5) $\times 10^{-2}$

6-8.

	s_y	CV	y
(a)	3×10^{-10}	2%	$-1.37(\pm 0.03) \times 10^{-8}$
(c)	3	25%	12 ± 3
(e)	25	50%	50 ± 25

6-9.

Sample	s, % K
1	0.10
2	0.12
3	0.12
4	0.10
5	0.11

$s_{\text{pooled}} = 0.11\%$ K

6-11. $s_{\text{pooled}} = 0.29\%$ heroin

Chapter 7

7-1.

	$\bar{x}$	s	95% CL
A	2.08	0.35	2.1 ± 0.4
C	0.0918	0.0055	0.092 ± 0.009
E	69.53	0.22	69.5 ± 0.3

7-2.

A	2.08 ± 0.18	or	2.1 ± 0.2
C	0.0918 ± 0.0087	or	0.092 ± 0.009
E	69.53 ± 0.35	or	69.5 ± 0.4

7-3. A Retain.

C Reject.

E Reject.

7-4. (a) 80% CL, 18 ± 3 μg/mL; 95% CL, 18 ± 5 μg/mL

(b) 80% CL, 18 ± 2 μg/mL; 95% CL, 18 ± 3 μg/mL

(c) 80% CL, 18 ± 2 μg/mL; 95% CL, 18 ± 2 μg/mL

7-6. 95%, 10 measurements; 99%, 17 measurements

7-8. (a) 3.22 ± 0.15 mmol/L   (b) 3.22 ± 0.06 mmol/L

7-10. (a) 12 replications

7-11. (a) Systematic error indicated at 95% confidence.

(b) No systematic error demonstrated.

7-13. At the 99% confidence level, the difference between the cobalt results should be no greater than ± 0.053 ppm. The actual difference was -0.07 ppm. Similarly, the difference in the thorium results should be no greater than ± 0.09 ppm. The actual difference was -0.12 ppm. Therefore, reasonable doubt exists as to the driver's guilt.

7-15. Difference is significant even at the 99.9% confidence level. Probability = 0.00013 or 0.013%.

7-16. (a) Outlier retained.

(b) Outlier rejected.

7-19. (b) $E = -29.74$ pCa $+ 92.86$

(c) pCa = 2.44; $s_c = 0.080$; $(s_c)_r = $ CV = 3.3%

(d) For 2 replicates, $s_c = 0.061$; CV = 2.5% For 8 replicates, $s_c = 0.043$; CV = 1.8%

Chapter 8

8-1. (a) The individual particles of a *colloid* are smaller than about 10^{-5} mm in diameter, while those of a *crystalline precipitate* are larger. As a consequence, crystalline precipitates settle out of solution relatively rapidly, whereas colloidal particles do not unless they can be caused to agglomerate.

(c) *Precipitation* is the process by which a solid phase forms and is carried out of solution when the solubility product of a species is exceeded. *Coprecipitation* is the process in which a normally soluble species is carried out of solution during the formation of a precipitate.

(e) *Occlusion* is a type of coprecipitation in which an impurity is entrapped in a pocket formed by a rapidly growing crystal. *Mixed-crystal formation* is a type of coprecipitation in which a foreign ion is incorporated into a growing crystal in a lattice position that is ordinarily occupied by one of the ions of the precipitate.

8-2. **(a)** *Digestion* is a process for improving the purity and filterability of a precipitate by heating the solid in contact with the solution from which it is formed (the *mother liquor*).

(c) In *reprecipitation,* a precipitate is filtered, washed, redissolved, and then reformed from the new solution. Because the concentration of contaminant is lower in this new solution than in the original, the second precipitate contains less coprecipitated impurity.

(e) A *counter-ion layer* is a charged layer of solution that surrounds a colloidal particle.

(g) *Supersaturation* is a phenomenon in which a solvent contains temporarily a concentration of solute that is greater than the equilibrium concentration of a saturated solution of the solute.

8-3. A *chelating agent* is an organic compound that contains two or more electron-donor groups located in such a configuration that five- or six-membered rings are formed when the donor groups form a complex with a cation.

8-5. **(a)** Positive charge   **(b)** adsorbed Ag^+ **(c)** NO_3^-

8-7. *Peptization* is the process in which a coagulated colloid returns to its original dispersed state as a consequence of a decrease in the electrolyte concentration of the solution in contact with the precipitate. Peptization of a coagulated colloid can be avoided by washing with an electrolyte solution rather than with pure water.

8-9. **(a)** mass SO_3 = mass of $BaSO_4 \times \dfrac{\text{molar mass } SO_3}{\text{molar mass } BaSO_4}$

(c) mass In = mass of $In_2O_3 \times \dfrac{2 \times \text{atomic mass In}}{\text{molar mass } In_2O_3}$

(e) mass CuO

$= \text{mass } Cu_2(SCN)_2 \times \dfrac{2 \times \text{molar mass CuO}}{\text{molar mass } Cu_2(SCN)_2}$

(i) mass $Na_2B_4O_7 \cdot 10H_2O$

$= \text{mass } B_2O_3 \times \dfrac{\text{molar mass } Na_2B_4O_7 \cdot 10H_2O}{2 \times \text{molar mass } B_2O_3}$

8-10. 95.35% KCl
8-12. 0.662 g
8-14. 0.124 g
8-16. 14.03%
8-18. 18.99%
8-20. 38.82%
8-22. 46.40%
8-24. 0.617 g sample and 0.824 g $BaSO_4$
8-26. **(a)** 0.239 g sample   **(b)** 0.494 g AgCl
(c) 0.4065 g sample
8-28. 4.72% Cl^- and 27.05% I^-
8-30. 0.395 g CO_2
8-32. **(a)** 0.369 g $Ba(IO_3)_2$ **(b)** 0.0149 g $BaCl_2 \cdot 2\,H_2O$

Chapter 9

9-1. **(a)** *Activity, a,* is the effective concentration of a species A in solution. The *activity coefficient,* γ_A, is the factor needed to convert a molar concentration to activity: $\alpha_A = \gamma_A[A]$.

(b) The *thermodynamic equilibrium constant* refers to an ideal system within which each species is unaffected by any others. A *concentration equilibrium constant* takes account of the influence exerted by solute species upon one another. A thermodynamic constant is based upon activities of reactants and products; a concentration constant is based upon molar concentrations of reactants and products.

9-3. **(a)** Ionic strength should decrease.
(b) Ionic strength should be unchanged.
(c) Ionic strength should increase.

9-5. For a given ionic strength, activity coefficients for ions with multiple charge show greater departures from ideality.

9-7. **(a)** 0.16   **(c)** 1.2
9-8. **(a)** 0.20   **(c)** 0.073
9-9. **(a)** 0.21   **(c)** 0.078
9-10. **(a)** 1.7×10^{-12} **(c)** 7.6×10^{-11}
9-11. **(a)** 5.2×10^{-6} M   **(b)** 6.2×10^{-6} M
(c) 9.5×10^{-12} M   **(d)** 1.5×10^{-7} M
9-12. **(a)** (1) 1.4×10^{-6} M   (2) 1.0×10^{-6} M
(b) (1) 2.1×10^{-3} M   (2) 1.3×10^{-3} M
(c) (1) 2.9×10^{-5} M   (2) 1.0×10^{-5} M
(d) (1) 1.4×10^{-5} M   (2) 2.0×10^{-6} M
9-13. **(a)** (1) 2.2×10^{-4} M   (2) 1.8×10^{-4} M
(b) (1) 1.7×10^{-4} M   (2) 1.2×10^{-4} M
(c) (1) 3.3×10^{-8} M   (2) 6.6×10^{-9} M
(d) (1) 1.3×10^{-3} M   (2) 8.0×10^{-4} M

Chapter 10

10-3. A charge-balance equation is derived by relating the concentration of cations and anions in such a way that no. mol/L positive charge = no. mol/L negative charge. For a doubly charged ion such as Ba^{2+}, the concentration of electrons for each mole is twice the molar *concentration* of the Ba^{2+}. That is, mol/L positive charge = $2[Ba^{2+}]$. Thus, the molar concentration of all multiply charged species is always multiplied by the charge in a charge-balance equation.

10-4. **(a)** $0.10 = [H_3PO_4] + [H_2PO_4^-] + [HPO_4^{2-}] +$
(c) $0.100 + 0.0500 = [HNO_2] + [NO_2^-]$
(e) $0.100 = [Na^+] = [OH^-] + 2[Zn(OH)_4^{2-}]$
(g) $[Ca^{2+}] = \frac{1}{2}([F^-] + [HF])$

10-5. **(a)** $[H_3O^+] = [OH^-] + [HPO_4^-] + 2[HPO_4^{2-}] + 3[PO_4^{3-}]$
(c) $[H_3O^+] + [Na^+] = [OH^-] + [NO_2^-]$
(e) $[H_3O^+] + [Na^+] + 2[Zn^{2+}] = [OH^-] + 2[Zn(OH)_4^{2-}]$
(g) $[H_3O^+] + 2[Ca^{2+}] = [OH^-] + [F^-]$

10-6. (a) 3.5×10^{-3} M (c) 3.6×10^{-4} M

10-7. (a) 1.47×10^{-4} M (c) 7.42×10^{-5} M

10-8. (a) 2.5×10^{-9} M (b) 2.5×10^{-12} M

10-10. (a) 5×10^{-2} M

10-11. 1.26×10^{-5} M

10-13. (a) $Cu(OH)_2$ forms first

(b) 9.8×10^{-10} M (c) 9.6×10^{-9} M

10-15. (a) 8.3×10^{-11} M (b) 1.6×10^{-11} M

(c) 5.2×10^6 M (d) 1.3×10^4 M

10-17. 1.877 g

10-19. (a) 0.0095 M; 48% (b) 6.6×10^{-3} M; 70%

Chapter 11

11-2. (a) The *millimole* is the amount of an elementary species, such as an atom, an ion, a molecule, or an electron. A millimole contains

$$6.02 \times 10^{23} \frac{\text{particles}}{\text{mole}} \times 10^{-3} \frac{\text{mole}}{\text{millimole}}$$

$$= 6.02 \times 10^{20} \frac{\text{particles}}{\text{millimole}}$$

(c) The *stoichiometric ratio* is the molar ratio of two species that appear in a balanced chemical equation.

11-3. (a) The *equivalence point* in a titration is that point at which sufficient titrant has been added so that stoichiometrically equivalent amounts of analyte and titrant are present. The *end point* in a titration is the point at which an observable physical change signals the equivalence point.

(c) A *primary standard* is a highly purified substance that serves as the basis for a titrimetric method. It is used either (1) to prepare a standard solution directly by mass or (2) to standardize a solution to be used in a titration.

A *secondary standard* is a material or solution whose concentration is determined from the stoichiometry of its reaction with a primary-standard material. Secondary standards are employed when a reagent is not available in primary-standard quality. For example, solid sodium hydroxide is hygroscopic and cannot be used to prepare a standard solution directly. A secondary-standard solution of the reagent is readily prepared, however, by standardizing a solution of sodium hydroxide against a primary-standard reagent such as potassium hydrogen phthalate.

11-4. $\dfrac{\text{mg}}{\text{L}} = \dfrac{10^{-3} \text{ g solute}}{10^3 \text{ g solution}} = \dfrac{1 \text{ g solute}}{10^6 \text{ g solution}} = 1 \text{ ppm}$

11-5. (a) $\dfrac{1 \text{ mol } H_2NNH_2}{2 \text{ mol } I_2}$

(c) $\dfrac{1 \text{ mol } Na_2B_4O_7 \cdot 10H_2O}{2 \text{ mol } H^+}$

11-6. (a) 5.52 mmol

(b) 31.2 mmol

(c) 6.58×10^{-3} mmol

(d) 966 mmol

11-8. (a) 1.33×10^3 mg (b) 520 mg

11-9. (a) 2.51 g (b) 2.88×10^{-3} g

11-10. 19.0 M

11-12. (a) 6.161 M (c) 4.669 M

11-13. (a) Dilute 80 g ethanol to 500 mL with water.

(b) Dilute 80.0 mL ethanol to 500 mL with water.

(c) Dilute 80.0 g ethanol with 420 g water.

11-15. Dilute 26 mL of the concentrated reagent to 2.0 L.

11-17. (a) Dissolve 6.37 g of $AgNO_3$ in water and dilute to 500 mL.

(b) Dilute 52.5 mL of 6.00 M HCl to 1.00 L.

(c) Dissolve 4.56 g $K_4Fe(CN)_6$ in water and dilute to 600 mL.

(d) Dilute 144 mL of 0.400 M $BaCl_2$ to a volume of 400 mL.

(e) Dilute 25 mL of the commercial reagent to a volume of 2.0 L.

(f) Dissolve 1.67 g Na_2SO_4 in water and dilute to 9.00 L.

11-19. 9.151×10^{-2} M

11-21. 0.06114 M $BaCl_2$

11-22. 0.2970 M $HClO_4$; 0.3259 M NaOH

11-24. 0.08411 M

11-25. 345.8 ppm S

11-27. 5.471% As_2O_3

11-28. 7.317% $(NH_2)_2CS$

11-30. (a) 9.36×10^{-3} M $Ba(OH)_2$

(b) 1.9×10^{-5} M

(c) -3 ppt

11-32. (a) $\dfrac{0.02966 \text{ mmol } KMnO_4}{\text{mL soln}}$

(b) 47.59% Fe_2O_3

11-34. (a) 1.821×10^{-2} M

(b) 1.821×10^{-2} M

(c) 5.463×10^{-2} M

(d) 0.506% (w/v)

(e) 1.366 mmol Cl^-

(f) 712 ppm K^+

Chapter 12

In the answers to this chapters, (Q) indicates that the answer was obtained by solving a quadratic equation.

12-1. (a) The initial pH of the NH_3 solution will be less than that for the solution containing NaOH. With the first addition of titrant, the pH of the NH_3 solution will decrease rapidly and then level off and become nearly constant throughout the middle part of the titration. In contrast, additions of standard acid to the NaOH solution will cause the pH of the NaOH solution to decrease gradually and nearly linearly until the equivalence point is approached. The equivalence-point pH for the NH_3 solution will be well below 7, whereas for the NaOH solution it will be exactly 7.

(b) Beyond the equivalence point, the pH is determined by the excess titrant. Thus, the curves become identical in this region.

12-3. The limited sensitivity of the eye to small color differences requires that there be a roughly tenfold excess of one or the other form of the indicator to be present in order for the color change to be seen. This change corresponds to a pH range of ± 1 pH unit about the pK of the indicator.

12-5. The standard reagents in neutralization titrations are always strong acids or strong bases because the reactions with this type of reagent are more complete than with those of their weaker counterparts. Sharper end points are the consequence of this difference.

12-7. The *buffer capacity* of a solution is the number of moles of hydronium ion or hydroxide ion needed to cause 1.00 L of the buffer to undergo a unit change in pH.

12-9. The three solutions will have the same pH, since the ratios of the amounts of weak acid to conjugate base are identical. They will differ in buffer capacity, however, with (a) having the greatest and (c) the least.

12-10. (a) Malic acid/sodium hydrogen malate
(c) NH_4Cl/NH_3

12-11. (a) NaOCl   (c) CH_3NH_2

12-12. (a) HIO_3 (c) $CH_3COCOOH$

12-14. 3.24

12-16. (a) 14.94

12-17. (a) 12.94

12-18. -0.607

12-20. 7.04(Q)

12-23. (a) 1.05   (b) 1.05   (c) 1.81
(d) 1.81   (e) 12.60

12-24. (a) 1.30   (b) 1.37

12-26. (a) 4.26   (b) 4.76   (c) 5.76

12-28. (a) 11.12   (b) 10.62   (c) 9.53(Q)

12-30. (a) 12.04(Q)   (b) 11.48(Q)   (c) 9.97(Q)

12-32. (a) 1.94   (b) 2.45   (c) 3.52

12-34. (a) 2.41(Q)   (b) 8.35   (c) 12.35   (d) 3.84

12-37. (a) 3.85   (b) 4.06   (c) 2.63(Q)   (d) 2.10(Q)

12-39. (a) 0.000   (c) -1.000
(e) -0.500 (g) 0.000

12-40. (a) -5.00 (c) -0.097
(e) -3.369 (g) -0.017

12-41. (a) 5.00   (c) 0.079   (e) 3.272   (g) 0.017

12-42. (b) -0.141

12-43. 15.5 g sodium formate

12-45. 194 mL HCl

12-47.

V_{HCl}	pH	V_{HCl}	pH
0.00	13.00	49.00	11.00
10.00	12.82	50.00	7.00
25.00	12.52	51.00	3.00
40.00	12.05	55.00	2.32
45.00	11.72	60.00	2.04

12-48. The theoretical pH at 24.95 mL is 6.44; at 25.05 mL, it is 9.82. Thus, the indicator should change color in the range of pH 6.5 to 9.8. Cresol purple (range: 7.6 to 9.2) (Table 12-1) would be quite suitable.

12-50.

Vol, mL	(a) pH	(c) pH
0.00	2.09(Q)	3.12
5.00	2.38(Q)	4.28
15.00	2.82(Q)	4.86
25.00	3.17(Q)	5.23
40.00	3.76(Q)	5.83
45.00	4.11(Q)	6.18
49.00	4.85(Q)	6.92
50.00	7.92	8.96
51.00	11.00	11.00
55.00	11.68	11.68
60.00	11.96	11.96

12-51.

Vol HCl, mL	(a) pH
0.00	11.12
5.00	10.20
15.00	9.61
25.00	9.24
40.00	8.64
45.00	8.29
49.00	7.55
50.00	5.27
51.00	3.00
55.00	2.32
60.00	2.04

12-52.

Vol, mL	(a) pH	(c) pH
0.00	2.80	4.26
5.00	3.65	6.57
15.00	4.23	7.15
25.00	4.60	7.52
40.00	5.20	8.12
49.00	6.29	9.21
50.00	8.65	10.11
51.00	11.00	11.00
55.00	11.68	11.68
60.00	11.96	11.96

12-53. (a) $\alpha_0 = 0.215$; $\alpha_1 = 0.785$
(c) $\alpha_0 = 0.769$; $\alpha_1 = 0.231$
(e) $\alpha_0 = 0.917$; $\alpha_1 = 0.083$

12-54. 6.61×10^{-2} M

Chapter 13

13-1. Not only is NaHA a proton donor, it is also the conjugate base of the parent acid H_2A.

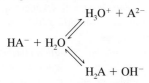

Solutions of acid salts are acidic or alkaline, depending upon which of these equilibria predominates. In order to compute the pH of this type, it is necessary to take both equilibria into account.

13-3. The HPO_4^{2-} is such a weak acid ($K_a = 4.5 \times 10^{-13}$) that the change in pH in the vicinity of the third equivalence point is too small to be observable.

13-4. (a) Since the Ks are essentially identical, the solution should be approximately neutral.

(c) neutral   (e) basic   (g) acidic

13-6. phenolphthalein

13-8. (a) cresol purple   (c) cresol purple
(e) bromocresol green   (g) phenolphthalein

13-9. (a) 1.86(Q)   (c) 1.64(Q)   (e) 4.21

13-10. (a) 4.71   (c) 4.28   (e) 9.80

13-11. (a) 12.32(Q)   (c) 9.70   (e) 12.58(Q)

13-12. (a) 2.07(Q)   (b) 7.18   (c) 10.63
(d) 2.55(Q)   (e) 2.06

13-14. (a) 1.54(Q)   (b) 1.99(Q)
(c) 12.07(Q)   (d) 12.01(Q)

13-16. (a) $[SO_3^{2-}]/[HSO_3^-] = 15.2$
(b) $[HCit^{2-}]/[Cit^{3-}] = 2.5$
(c) $[HM^-]/[M^{2-}] = 0.498$
(d) $[HT^-]/[T^{2-}] = 0.0232$

13-18. 50.2 g

13-20. (a) 2.11(Q)   (b) 7.38

13-22. Mix 442 mL of 0.300 M Na_2CO_3 with 558 mL of 0.200 M HCl.

13-24. Mix 704 mL of the HCl with 296 mL of the Na_3AsO_4.

13-28.

Vol reagent, mL	(a) pH	(c) pH
0.00	11.66	0.96
12.50	10.33	1.28
20.00	9.73	1.50
24.00	8.95	1.63
25.00	7.08	1.67
26.00	5.21	1.70
37.50	3.83	2.19
45.00	3.25	2.70
49.00	2.74	3.46
50.00	2.57	7.35
51.00	2.42	11.30
60.00	1.74	12.26

13-29.   0.00 mL, pH = 13.00;   26.00 mL, pH = 9.26
10.00 mL, pH = 12.70;   35.00 mL, pH = 7.98
20.00 mL, pH = 12.15;   44.00 mL, pH = 6.70
24.00 mL, pH = 11.43;   45.00 mL, pH = 4.68
25.00 mL, pH = 10.35;   46.00 mL, pH = 2.68
50.00 mL, pH = 2.00

13-31. (a) $\dfrac{[H_3AsO_4][HAsO_4^{2-}]}{[H_2AsO_4^-]^2} = 1.9 \times 10^{-5}$

13-32. $\dfrac{[NH_3][HOAc]}{[NH_4^+][OAc^-]} = 3.26 \times 10^{-5}$

13.33.

	pH	D	α_0	α_1	α_2	α_3
(a)	2.00	1.112×10^{-4}	0.899	0.101	3.94×10^{-5}	
	6.00	5.500×10^{-9}	1.82×10^{-4}	0.204	0.796	
	10.00	4.379×10^{-9}	2.28×10^{-12}	2.56×10^{-5}	1.000	
(c)	2.00	1.075×10^{-6}	0.931	6.93×10^{-2}	1.20×10^{-4}	4.82×10^{-9}
	6.00	1.882×10^{-14}	5.31×10^{-5}	3.96×10^{-2}	0.685	0.275
	10.00	5.182×10^{-15}	1.93×10^{-16}	1.44×10^{-9}	2.49×10^{-4}	1.000
(e)	2.00	4.000×10^{-4}	0.250	0.750	1.22×10^{-5}	
	6.00	3.486×10^{-9}	2.87×10^{-5}	0.861	0.139	
	10.00	4.863×10^{-9}	2.06×10^{-12}	6.17×10^{-4}	0.999	

Chapter 14

14-1. Carbon dioxide is not strongly bonded by water molecules, and thus is readily volatilized from aqueous media. Gaseous HCl molecules, on the other hand, are fully dissociated into H_3O^+ and Cl^- when dissolved in water; neither of these species is volatile.

14-3. Primary-standard Na_2CO_3 can be obtained by heating primary-standard-grade $NaHCO_3$ for about an hour at 270 to 300°C. The reaction is

$$2NaHCO_3(s) \rightarrow Na_2CO_3(s) + H_2O(g) + CO_2(g)$$

14-5. For, let us say, a 40-mL titration

$$\text{mass } KH(IO_3)_2 \text{ required} = 0.16 \text{ g}$$

$$\text{mass } HBz \text{ required} = 0.049 \text{ g}$$

The $KH(IO_3)_2$ is preferable because the relative weighing error would be less with a 0.16-g sample than with a

0.049-g sample. A second advantage of $KH(IO_3)_2$ is that it acts as a strong acid, which makes the choice of indicator simpler.

14-8. **(a)** Dissolve 17 g KOH in water and dilute to 2.0 L.

     (b) Dissolve 9.5 g of $Ba(OH)_2 \cdot 2 H_2O$ in water and dilute to 2.0 L.

     (c) Dilute about 120 mL of the reagent HCl to 2.0 L.

14-10. **(a)** 0.1026 M

     (b) $s = 0.00039$ and CV = 0.38%

     (c) 0.0009, not an outlier

14-12. **(a)** 0.1388 M   **(b)** 0.1500 M

14-14. **(a)** 0.08387 M   **(b)** 0.1007 M   **(c)** 0.1311 M

14-16. **(a)** 0.28 to 0.36 g Na_2CO_3 **(c)** 0.85 to 1.1 g HBz

     (e) 0.17 to 0.22 g TRIS

14-17.

mL HCl	SD TRIS	SD Na_2CO_3	SD $Na_2B_4O_7 \cdot H_2O$
20.00	0.00004	0.00009	0.00003
30.00	0.00003	0.00006	0.00002
40.00	0.00002	0.00005	0.00001
50.00	0.00002	0.00004	0.00001

14-19. 0.1217 g H_2T/100 mL

14-21. **(a)** 46.25% $Na_2B_4O_7$

     (b) 87.67% $Na_2B_4O_7 \cdot 10H_2O$

     (c) 32.01% B_2O_3

     (d) 9.94% B

14-23. 24.3% HCHO

14-25. 7.079% active ingredient.

14-27. $MgCO_3$ with a molar mass of 84.31 seems a likely candidate.

14-29. 3.35×10^3 ppm CO_2

14-31. 6.333% P

14-32. 13.86% analyte

14-33. 22.08% RN_4

14-35. 3.885% N

14-37. **(a)** 10.09% N   **(c)** 47.61% $(NH_4)_2SO_4$

14-39. 19.04% $(NH_4)_2SO_4$ and 30.49% NH_4NO_3

14-40. 69.84% KOH; 21.04% K_2CO_3; 9.12% H_2O

14-42.

(a)	(b)	(c)	(d)
9.07 mL HCl	18.15 mL HCl	19.14 mL HCl	9.21 mL HCl
13.61 mL HCl	27.22 mL HCl	23.93 mL HCl	12.27 mL HCl
22.68 mL HCl	36.30 mL HCl	28.71 mL HCl	21.48 mL HCl
36.30 mL HCl	45.37 mL HCl	38.28 mL HCl	24.55 mL HCl

14-44. **(a)** 4.314 mg NaOH/mL

     (b) 7.985 mg Na_2CO_3/mL and 4.358 mg $NaHCO_3$/mL

     (c) 3.455 mg Na_2CO_3/mL and 4.396 mg NaOH/mL

     (d) 8.215 mg Na_2CO_3/mL

     (e) 13.46 mg $NaHCO_3$/mL

Chapter 15

15-1. **(a)** A *chelate* is a cyclic complex consisting of a metal ion and a reagent that contains two or more electron donor groups located in such a position that they can bond with the metal ion to form a heterocyclic ring structure.

     (c) A *ligand* is a species that contains one or more electron pair donor groups that tend to form bonds with metal ions.

     (e) A *conditional formation constant* is an equilibrium constant for the reaction between a metal ion and a complexing agent that applies only when the pH and/or the concentration of other complexing ions are carefully specified.

     (g) *Water hardness* is the concentration of calcium carbonate that is equivalent to the total concentration of all of the multivalent metal carbonates in the water.

15-2. Three general methods for performing EDTA titrations are (1) direct titration, (2) back-titration, and (3) displacement titration. Method (1) is simple, rapid, and requires but one standard reagent. Method (2) is advantageous for those metals that react so slowly with EDTA as to make direct titration inconvenient. In addition, the procedure is useful for cations for which satisfactory indicators are not available. Finally, it is useful for analyzing samples that contain anions that form sparingly soluble precipitates with the analyte under the analytical conditions. Method (3) is particularly useful in situations where no satisfactory indicators are available for direct titration.

15-4. **(a)** $Ag^+ + S_2O_3^{2-} \rightleftharpoons AgS_2O_3^-$ $K_1 = \dfrac{[AgS_2O_3^-]}{[Ag^+][S_2O_3^{2-}]}$

$$AgS_2O_3^- + S_2O_3^{2-} \rightleftharpoons Ag(S_2O_3)_2^{3-}$$

$$K_2 = \frac{[Ag(S_2O_3)_2^{3-}]}{[AgS_2O_3^-][S_2O_3^{2-}]}$$

15-5. The overall formation constant β_n is equal to the product of the individual stepwise constants. Thus, the overall constant for formation of $Ni(SCN)_3^-$ in Problem 15-4(b) is

$$\beta_3 = K_1K_2K_3 = \frac{[Ni(SCN)_3^-]}{[Ni^{2+}][SCN^-]^3}$$

which is the equilibrium constant for the reaction

$$Ni^{2+} + 3SCN^- \rightleftharpoons Ni(SCN)_3^-$$

and

$$\beta_2 = K_1K_2 = \frac{Ni(SCN)_2}{[Ni^{2+}][SCN^-]^2}$$

where the overall constant β_2 is for the reaction

$$Ni^{2+} + 2SCN^- \rightleftharpoons Ni(SCN)_2$$

15-6. The Fajan determination of chloride involves a direct titration, while the Volhard approach requires two standard solutions and a filtration step to eliminate AgCl.

15-9. Potassium is determined by precipitation with an excess of a standard solution of sodium tetraphenylboron. An excess of standard $AgNO_3$ is then added, which precipitates the excess tetraphenylboron ion. The excess $AgNO_3$ is then titrated with a standard solution of SCN^-. The reactions are

$$K^+ + B(C_6H_5)_4^- \rightleftharpoons KB(C_6H_5)_4(s)$$

[measured excess $B(C_6H_5)_4^-$]

$$Ag^+ + B(C_6H_5)_4^- \rightleftharpoons AgB(C_6H_5)_4(s)$$

[measured excess $AgNO_3$]

The excess $AgNO_3$ is then determined by a Volhard titration with KSCN.

15-12. 0.01032 M EDTA

15-14. (a) 39.1 mL EDTA   (c) 41.6 mL EDTA
(e) 31.2 mL EDTA

15-16. 3.028% Zn

15-17. 0.998 mg Cr/cm^2

15-18. (a) 51.78 mL $AgNO_3$ (c) 10.64 mL $AgNO_3$
(e) 46.24 mL $AgNO_3$

15-20. (a) 44.70 mL $AgNO_3$ (c) 14.87 mL $AgNO_3$

15-22. 1.228% Tl_2SO_4

15-24. 184.0 ppm Fe^{2+}; 213.0 ppm Fe^{2+}

15-26. 55.16% Pb; 44.86% Cd

15-28. 99.7% ZnO; 0.256% Fe_2O_3

15-30. 15.75% Cr; 60.88% Ni; 23.18% Fe

15-32. (a) 4.6×10^9 (b) 1.1×10^{12} (c) 7.4×10^{13}

15-34.

Vol, mL	pSr	Vol, mL	pSr
0.00	2.00	25.00	5.37
10.00	2.30	25.10	6.16
24.00	3.57	26.00	7.16
24.90	4.57	30.00	7.86

15-36.

(a)

V_{NH_4SCN} (mL)	$[Ag^+]$	$[SCN^-]$	pAg
30.00	9.09×10^{-3}	1.2×10^{-10}	2.04
40.00	3.85×10^{-3}	2.9×10^{-10}	2.42
49.00	3.38×10^{-4}	3.3×10^{-9}	3.47
50.00	1.05×10^{-6}	1.05×10^{-6}	5.98
51.00	3.3×10^{-9}	3.3×10^{-4}	8.48
60.00	3.7×10^{-10}	2.94×10^{-3}	9.43
70.00	2.1×10^{-10}	5.26×10^{-3}	9.68

(c)

V_{NaCl} (mL)	$[Ag^+]$	$[Cl^-]$	pAg
10.00	3.75×10^{-2}	4.85×10^{-9}	1.43
20.00	1.50×10^{-2}	1.21×10^{-8}	1.82
29.00	1.27×10^{-3}	1.43×10^{-7}	2.90
30.00	1.35×10^{-5}	1.35×10^{-5}	4.87
31.00	1.48×10^{-7}	1.23×10^{-3}	6.83
40.00	1.70×10^{-8}	1.07×10^{-2}	7.77
50.00	9.71×10^{-9}	1.88×10^{-2}	8.01

(e)

$V_{Na_2SO_4}$ (mL)	$[Ba^{2+}]$	$[SO_4^{2-}]$	pBa
0.00	2.50×10^{-2}	0.0	1.60
10.00	1.00×10^{-2}	1.1×10^{-8}	2.00
19.00	8.48×10^{-4}	1.3×10^{-7}	3.07
20.00	1.05×10^{-5}	1.05×10^{-5}	4.98
21.00	1.3×10^{-7}	8.20×10^{-4}	6.87
30.00	1.5×10^{-8}	7.14×10^{-3}	7.81
40.00	8.8×10^{-9}	1.25×10^{-2}	8.06

15-38. (b) 339 ppm $CaCO_3$

Chapter 16

16-1. (a) *Oxidation* is a process in which a species loses one or more electrons.

(c) A *salt bridge* is a device that provides electrical contact but prevents mixing of dissimilar solutions in an electrochemical cell.

(e) The *Nernst equation* relates the potential to the concentrations (strictly, activities) of the participants in an electrochemical reaction.

16-2. (a) The *electrode potential* is the potential of an electrochemical cell in which a standard hydrogen electrode acts as the reference electrode (left-hand electrode) and the half-cell of interest is the right-hand electrode.

(c) The *standard electrode potential* for a half-reaction is the potential of a *cell* consisting of the half-reaction of interest on the right and a standard hydrogen electrode behaving as the reference. The activities of all of the participants in the half-reaction are specified as having a value of unity. The additional specification that the standard hydrogen electrode is the left-hand electrode implies that the standard potential for a half-reaction is always a *reduction potential.*

(e) An *oxidation potential* is the potential of an electrochemical cell in which the right electrode is a standard hydrogen electrode and the half-cell of interest is on the left.

16-3. (a) *Reduction* is the process whereby a substance acquires electrons; a *reducing agent* is a supplier of electrons.

(c) The *anode* of an electrochemical cell is the electrode at which oxidation occurs. The *cathode* is the electrode at which reduction occurs.

(e) The *standard electrode potential* is the potential of an electrochemical cell in which the standard hydrogen electrode acts as the reference electrode (left) and all participants in the right-hand electrode process have unit activity. The formal potential differs in that the molar *concentrations* of the reactants and products are unity and the concentration of other species in the solution are carefully specified.

16-4. The first standard potential is for a solution that is saturated with I_2, which has an $I_2(aq)$ activity significantly less than one. The second potential is for a *hypothetical* half-cell in which the $I_2(aq)$ activity is unity. Such a half-cell, if it existed, would have a greater potential, since the driving force for the reduction would be greater at the higher I_2 concentration. The second half-cell potential, although hypothetical, is nevertheless useful for calculating electrode potentials for solutions that are undersaturated in I_2.

16-5. It is necessary to bubble hydrogen through the electrolyte in a hydrogen electrode in order to keep the solution saturated with the gas. Only under these circumstances is the hydrogen activity constant so that the electrode potential is constant and reproducible.

16-7. **(a)** $2Fe^{3+} + Sn^{2+} \rightarrow 2Fe^{2+} + Sn^{4+}$

(c) $2NO_3^- + Cu(s) + 4H^+ \rightarrow$
$$2NO_2(g) + 2H_2O + Cu^{2+}$$

(e) $Ti^{3+} + Fe(CN)_6^{3-} + H_2O \rightarrow$
$$TiO^{2+} + Fe(CN)_6^{4-} + 2H^+$$

(g) $2Ag(s) + 2I^- + Sn^{4+} \rightarrow 2AgI(s) + Sn^{2+}$

(i) $5HNO_2 + 2MnO_4^- + H^+ \rightarrow$
$$5NO_3^- + 2Mn^{2+} + 3H_2O$$

16-8. **(a)** Oxidizing agent Fe^{3+}; $Fe^{3+} + e^- \rightleftharpoons Fe^{2+}$
Reducing agent Sn^{2+}; $Sn^{2+} \rightleftharpoons Sn^{4+} + 2e^-$

(c) Oxidizing agent NO_3^-; $NO_3^- + 2H^+ + e^- \rightleftharpoons$
$$NO_2(g) + H_2O$$
Reducing agent Cu; $Cu(s) \rightleftharpoons Cu^{2+} + 2e^-$

(e) Oxidizing agent $Fe(CN)_6^{3-}$; $Fe(CN)_6^{3-} + e^- \rightleftharpoons$
$$Fe(CN)_6^{4-}$$
Reducing agent Ti^{3+}; $Ti^{3+} + H_2O \rightleftharpoons$
$$TiO^{2+} + 2H^+ + e^-$$

(g) Oxidizing agent Sn^{4+}; $Sn^{4+} + 2e^- \rightleftharpoons Sn^{2+}$
Reducing agent Ag; $Ag(s) + I^- \rightleftharpoons AgI(s) + e^-$

(i) Oxidizing agent MnO_4^-; $MnO_4^- + 8H^+ + 5e^- \rightleftharpoons$
$$Mn^{2+} + 4H_2O$$
Reducing agent HNO_2; $HNO_2 + H_2O \rightleftharpoons$
$$NO_3^- + 3H^+ + 2e^-$$

16-9. **(a)** $MnO_4^- + 5VO^{2+} + 11H_2O \rightarrow$
$$Mn^{2+} + 5V(OH)_4^+ + 2H^+$$

(c) $Cr_2O_7^{2-} + 3U^{4+} + 2H^+ \rightarrow$
$$2Cr^{3+} + 3UO_2^{2+} + H_2O$$

(e) $IO_3^- + 5I^- + 6H^+ \rightarrow 3I_2 + H_2O$

(g) $HPO_3^{2-} + 2MnO_4^- + 3OH^- \rightarrow$
$$PO_4^{3-} + 2MnO_4^{2-} + 2H_2O$$

(i) $V^{2+} + 2V(OH)_4^+ + 2H^+ \rightarrow 3VO^{2+} + 5H_2O$

16-10. **(a)** Oxidizing agent MnO_4^-; $MnO_4^- + 5e^- + 8H^+ \rightleftharpoons$
$$Mn^{2+} + 4H_2O$$
Reducing agent VO^{2+}; $VO^{2+} + 3H_2O \rightleftharpoons$
$$V(OH)_4^+ + 2H^+ + e^-$$

(c) Oxidizing agent $Cr_2O_7^{2-}$;
$$Cr_2O_7^{2-} + 6e^- + 14H^+ \rightleftharpoons 2Cr^{3+} + 7H_2O$$
Reducing agent U^{4+}; $U^{4+} + 2H_2O \rightleftharpoons$
$$UO_2^{2+} + 2e^- + 4H^+$$

(e) Oxidizing agent IO_3^-; $IO_3^- + 5e^- + 6H^+ \rightleftharpoons$
$$\tfrac{1}{2}I_2 + 3H_2O$$
Reducing agent I^-; $I^- \rightleftharpoons \tfrac{1}{2}I_2 + e^-$

(g) Oxidizing agent MnO_4^-; $MnO_4^- + e^- \rightleftharpoons MnO_4^{2-}$
Reducing agent HPO_3^{2-}; $HPO_3^{2-} + H_2O \rightleftharpoons$
$$PO_4^{3-} + 2e^- + 3H^+$$

(i) Oxidizing agent
$V(OH)_4^+$; $V(OH)_4^+ + e^- + 2H^+ \rightleftharpoons VO^{2+} + 3H_2O$
Reducing agent V^{2+}; $V^{2+} + H_2O \rightleftharpoons$
$$VO^{2+} + 2e^- + 2H^+$$

16-11. **(a)** $AgBr(s) + e^- \rightleftharpoons Ag(s) + Br^-$; $V^{2+} \rightleftharpoons V^{3+} + e^-$
$Tl^{3+} + 2e^- \rightleftharpoons Tl^+$; $Fe(CN)_6^{4-} \rightleftharpoons Fe(CN)_6^{3-} + e^-$
$V^{3+} + e^- \rightleftharpoons V^{2+}$; $Zn \rightleftharpoons Zn^{2+} + 2e^-$
$Fe(CN)_6^{3-} + e^- \rightleftharpoons Fe(CN)_6^{4-}$; $Ag(s) + Br^- \rightleftharpoons AgBr(s) + e^-$
$S_2O_8^{2-} + 2e^- \rightleftharpoons 2SO_4^{2-}$; $Tl^+ \rightleftharpoons Tl^{3+} + 2e^-$

(b), (c)	E^0
$S_2O_8^{2-} + 2e^- \rightleftharpoons 2SO_4^{2-}$	2.01
$Tl^{3+} + 2e^- \rightleftharpoons Tl^+$	1.25
$Fe(CN)_6^{3-} + e^- \rightleftharpoons Fe(CN)_6^{4-}$	0.36
$AgBr(s) + e^- \rightleftharpoons Ag(s) + Br^-$	0.073
$V^{3+} + e^- \rightleftharpoons V^{2+}$	-0.256
$Zn^{2+} + 2e^- \rightleftharpoons Zn(s)$	-0.763

16-13. **(a)** 0.297 V   **(b)** 0.190 V   **(c)** -0.152 V
(d) 0.047 V   **(e)** 0.007 V

16-15. -0.121 V

16-16. **(a)** 0.78 V   **(b)** 0.198 V   **(c)** -0.355 V
(d) 0.210 V   **(e)** 0.177 V   **(f)** 0.86 V

16-18. **(a)** -0.280 anode   **(b)** -0.009 anode
(c) 1.003 cathode   **(d)** 0.171 V cathode
(e) -0.009 V anode

16-20. 0.390 V

16-22. -0.964 V

16-24. -1.25 V

16-25. 0.13 V

Chapter 17

17-1. The electrode potential of a system is the electrode potential of all half-cell processes at equilibrium in the system.

17-2. **(a)** *Equilibrium* is the state that a system assumes after each addition of reagent. *Equivalence* refers to a particular equilibrium state when a stoichiometric amount of titrant has been added.

17-4. For points before equivalence, potential data are computed from the analyte standard potential and the analytical concentrations of the analyte and its product. Post-equivalence point data are based upon the standard potential for the titrant and its analytical concentrations. The equivalence point potential is computed from the two standard potentials and the stoichiometric relation between the analyte and titrant.

17-6. An asymmetric titration curve will be encountered whenever the titrant and the analyte react in a ratio that is not $1:1$.

17-8. (a) 0.452 V $\quad Co^{2+} + Zn \rightarrow Zn^{2+} + Co$
(b) 0.031 V $\quad Hg^{2+} + 2Fe^+ \rightarrow Hg + 2Fe^{3+}$
(c) 0.414 V $\quad O_2 + 4H^+ + 4Ag \rightarrow 2H_2O + 4Ag^+$
(d) -0.401 V $\quad 2Ag + Cu^{2+} + 2I^- \rightarrow 2AgI + Cu$
(e) -0.208 V
$\quad 2H^+(1.00\ M) + H_2 \rightarrow H_2 + 2H^+(3.07 \times 10^{-4}\ M)$
(f) 0.724 V
$\quad 2Fe^{3+} + U^{4+} + 2H_2O \rightarrow 2Fe^{2+} + UO_2^{2+} + 4H^+$

17-9. (a) 0.631 V (c) 0.331 V

17-11. (a) 2.2×10^{17} (c) 3.2×10^{22}
(e) 9×10^{37} (g) 2.4×10^{10}

17-12. (a) 0.258 V (c) 0.444 V (e) 0.951 V
(g) -0.008 V

17-14. (a) Phenosafranine (c) Indigo tetrasulfonate or Methylene blue
(e) Erioglaucin A (g) None

17-15.

Vol, mL	E, V		
	(a)	(c)	(e)
10.00	-0.292	0.32	0.316
25.00	-0.256	0.36	0.334
49.00	-0.156	0.46	0.384
49.90	-0.097	0.52	0.414
50.00	0.017	0.95	1.17
50.10	0.074	1.17	1.48
51.00	0.104	1.20	1.49
60.00	0.133	1.23	1.50

Chapter 18

18-1. (a) $2Mn^{2+} + 5S_2O_8^{2-} + 8H_2O \rightarrow$
$\qquad\qquad 10SO_4^{2-} + 2MnO_4^- + 16H^+$
(b) $NaBiO_3(s) + 2Ce^{3+} + 4H^+ \rightarrow$
$\qquad\qquad BiO^+ + 2Ce^{4+} + 2H_2O + Na^+$
(c) $H_2O_2 + U^{4+} \rightarrow UO_2^{2+} + 2H^+$
(d) $V(OH)_4^+ + Ag(s) + Cl^- + 2H^+ \rightarrow$
$\qquad\qquad VO^{2+} + AgCl(s) + 3H_2O$
(e) $2MnO_4^- + 5H_2O_2 + 6H^+ \rightarrow$
$\qquad\qquad 5O_2 + 2Mn^{2+} + 8H_2O$
(f) $ClO_3^- + 6I^- + 6H^+ \rightarrow 3I_2 + Cl^- + 3H_2O$

18-3. Only in the presence of Cl^- ion is Ag a sufficiently good reducing agent to be very useful for prereductions. In the presence of Cl^- ion, the half-reaction occurring in the Walden reductor is

$$Ag(s) + Cl^- \rightarrow AgCl(s) + e^-$$

The excess HCl increases the tendency of this reaction to occur by the common ion effect.

18-5. $UO_2^{2+} + 2Ag(s) + 4H^+ + 2Cl^- \rightleftharpoons$
$\qquad\qquad U^{4+} + 2AgCl(s) + H_2O$

18-7. Standard solutions of reductants find somewhat limited use because of their susceptibility to air oxidation.

18-8. Standard $KMnO_4$ solutions are seldom used to titrate solutions containing HCl because of the tendency of MnO_4^- to oxidize Cl^- to Cl_2, thus causing an overconsumption of MnO_4^-.

18-10. $2MnO_4^- + 3Mn^{2+} + 2H_2O \rightarrow 5MnO_2(s) + 4H^+$

18-13. $4MnO_4^- + 2H_2O \rightarrow 4MnO_2(s) + 3O_2 + 4OH^-$
$\qquad\qquad$ brown

18-15. Iodine is not sufficiently soluble in water to produce a useful standard reagent. It is quite soluble in solutions that contain an excess of iodide, however, as a consequence of the formation of the triiodide complex. The rate at which iodine dissolves in iodide solutions increases as the concentration of iodide ion becomes greater. For this reason, iodine is always dissolved in a very concentrated solution of potassium iodide and diluted only after solution is complete.

18-17. $S_2O_3^{2-} + H^+ \rightarrow HSO_3^- + S(s)$

18-19. $BrO_3^- + 6I^- + 6H^+ \rightarrow Br^- + 3I_2 + 3H_2O$
$\qquad\qquad$ excess

$$I_2 + 2S_2O_3^{2-} \rightarrow 2I^- + S_4O_6^{2-}$$

18-21. $2I_2 + N_2H_4 \rightarrow N_2 + 4H^+ + 4I^-$

18-23. (a) 0.1122 M Ce^{4+} (c) 0.02245 M MnO_4^-
(e) 0.02806 M IO_3^-

18-24. Dissolve 2.574 g $K_2Cr_2O_7$ in sufficient water to give 250.0 mL of solution.

18-26. Dissolve about 24 g of $KMnO_4$ in 1.5 L of water.

18-28. 0.01518 M $KMnO_4$

18-30. 0.06711 M $Na_2S_2O_3$

18-32. (a) 14.72% Sb (b) 20.54% Sb_2S_3

18-34. 9.38% $CS(NH_2)_2$

18-35. (a) 32.08% Fe (b) 45.86% Fe_2O_3

18-37. 0.03074 M

18-39. 56.53% $KClO_3$

18-41. 0.701% As_2O_3

18-43. 3.64% C_2H_5SH

18-45. 2.635% KI

18-46. 69.07% Fe and 21.07% Cr

18-48. 0.5554 g Tl

18-49. 10.4 ppm SO_2

18-51. 19.5 ppm H_2S

18-53. 0.0397 mg O_2/mL

Chapter 19

19-1. (a) An indicator electrode is an electrode used in potentiometry that responds to variations in the activity of an analyte ion or molecule.
(c) An electrode of the first kind is a metal electrode that is used to determine the concentration of the cation of that metal in a solution.

19-2. **(a)** A liquid-junction potential is the potential that develops across the interface between two solutions having different electrolyte compositions.

19-3. **(a)** An electrode of the first kind for Hg(II) would take the form

$$\|\,Hg^{2+}(x\mathrm{M})\,|\,Hg$$

$$E_{Hg} = E^0_{Hg} - \frac{0.0592}{2}\log\frac{1}{[Hg^{2+}]} = E^0_{Hg} - \frac{0.0592}{2}\,pHg$$

(b) An electrode of the second kind for EDTA would take the form

$$\|\,HgY^{2-}(y\,\mathrm{M}),\,Y^{4-}\,(x\,\mathrm{M})\,|\,Hg$$

where a small and fixed amount of HgY^{2-} is introduced into the analyte solution so that its concentration is xM. Here the potential of the mercury electrode is given by

$$E_{Hg} = K - \frac{0.0592}{2}\log[Y^{4-}] = K + \frac{0.0592}{2}\,pY$$

where

$$K = E^0_{HgY^{2-}} - \frac{0.0592}{2}\log\frac{1}{a_{HgY^{2-}}}$$

$$\approx 0.21 - \frac{0.0592}{2}\log\frac{1}{[HgY^{2-}]}$$

19-5. The pH-dependent potential that develops across a glass membrane arises from the difference in positions of dissociation equilibria that arise on each of the two surfaces. These equilibria are described by the equation

$$H^+Gl^- \rightleftharpoons H^+ + Gl^-$$
membrane　soln　membrane

The surface exposed to the solution having the higher hydrogen ion activity then becomes positive with respect to the other surface. This charge difference, or potential, serves as the analytical parameter when the pH of the solution on one side of the membrane is held constant.

19-7. Uncertainties that may be encountered in pH measurements include (1) the acid error in highly acidic solutions, (2) the alkaline error in strongly basic solutions, (3) the error that arises when the ionic strength of the calibration standards differ from that of the analyte solution, (4) uncertainties in the pH of the standard buffers, (5) nonreproducible junction potentials when samples of low ionic strength are measured, and (6) dehydration of the working surface.

19-9. The alkaline error arises when a glass electrode is employed to measure the pH of solutions having pH values in the 10 to 12 range or greater. In the presence of alkali ions, the glass surface becomes responsive to both hy-

drogen and alkali ions. Low pH values arise as a consequence.

19-11. **(b)** The *boundary potential* for a membrane electrode is a potential that develops when the membrane separates two solutions that have different concentrations of a cation or an anion that the membrane binds selectively. For an aqueous solution, the following equilibria develop when the membrane is positioned between two solutions of A^-:

$$A^+M^- \rightleftharpoons A^+ + M^-$$
membrane₁　soln₁　membrane₁

$$A^+M^- \rightleftharpoons A^+ + M^-$$
membrane₂　soln₂　membrane₂

where the subscripts refer to the two sides of the membrane. A potential develops across this membrane if one of these equilibria proceeds further to the right than the other, and this potential is the boundary potential. For example if the concentration of A^+ is greater in solution 1 than in solution 2, the negative charge on side 1 of the membrane will be less than that on side 2 because the equilibrium on side 1 still lies further to the left. Thus, a greater fraction of the negative charge on side 1 will be neutralized by A^+.

(d) The membrane in a solid state electrode for F^- is crystalline LaF_3, which when immersed in aqueous solution LaF_3, dissociates according to the equation

$$LaF_3(s) \rightleftharpoons La^{3+} + 3F^-$$

Thus, a boundary potential develops across this membrane when it separates two solutions of different F^- ion concentration. The source of this potential is described in part (b) of this answer.

19-12. The direct potentiometric measurement of pH provides a measure of the equilibrium activity of hydronium ions present in a solution of the sample. A potentiometric titration provides information on the amount of reactive protons, both ionized and nonionized, that are present in a sample.

19-15. **(a)** 0.354 V

(b) $SCE\,\|\,IO_3^-(x\mathrm{M}),\,AgIO_3(\mathrm{sat'd})\,|\,Ag$

(c) $pIO_3 = \dfrac{E_{cell} - 0.110}{0.0592}$

(d) 3.11

19-17. **(a)** $SCE\,\|\,SCN^-(x\mathrm{M}),\,AgSCN(\mathrm{sat'd})\,|\,Ag$

(c) $SCE\,\|\,SO_3^{2-}(x\mathrm{M}),\,Ag_2SO_3(\mathrm{sat'd})\,|\,Ag$

19-18. **(a)** $pSCN = \dfrac{E_{cell} + 0.153}{0.0592}$

(c) $pSO_3 = \dfrac{2(E_{cell} - 0.146)}{0.0592}$

19-19. (a) 4.65 (c) 5.20

19-20. 6.76

19-21. (a) pH = 12.629; $a_{H^+} = 2.35 \times 10^{-13}$
(b) pH = 5.579; $a_{H^+} = 2.64 \times 10^{-6}$
(c) pH = 12.596; $a_{H^+} = 2.54 \times 10^{-13}$
pH = 12.663; $a_{H^+} = 2.17 \times 10^{-13}$
and
pH = 5.545 and 5.612
$a_{H^+} = 2.85 \times 10^{-6}$ and 2.44×10^{-6}

19-22. 136 g HA/mol

19-24. (a)

mL Ce(IV)	E vs. SCE, V	mL Ce(IV)	E vs. SCE, V
5.00	0.58	50.00	0.80
10.00	0.59	50.01	0.98
15.00	0.60	50.05	1.02
25.00	0.61	50.10	1.04
40.00	0.63	50.20	1.05
49.00	0.66	50.30	1.06
49.50	0.67	50.40	1.07
49.60	0.67	50.50	1.08
49.70	0.67	51.00	1.10
49.80	0.68	60.00	1.15
49.90	0.69	75.00	1.18
49.95	0.70	90.00	1.19
49.99	0.72		

19-26. 3.2×10^{-4} M

19-28. (b) Unknown 1 = 0.020 M; Unknown 2 = 0.008 M

Chapter 20

20-1. (a) *Concentration polarization* is a condition in which the current in an electrochemical cell is limited by the rate at which reactants are brought to or removed from the surface of one or both electrodes. *Kinetic polarization* is a condition in which the current in an electrochemical cell is limited by the rate at which electrons are transferred between the electrode surfaces and reactants in solution. For either type of polarization, the current is no longer proportional to the cell potential.

(c) Both the *coulomb* and the *faraday* are units of quantity of charge, or electricity. The former is the quantity transported by one ampere of current in one second; the latter is equal to 96,485 coulombs or one mole of electrons.

(e) The *electrolysis circuit* consists of a working electrode and a counter electrode. The *control circuit* regulates the applied potential such that the potential between the working electrode and a reference electrode in the control circuit is constant and at a desired level.

20-2. (a) *Current density* is the current at an electrode divided by the surface area of that electrode. Ordinarily, it has units of amperes per square centimeter.

(c) A *coulometric titration* is an electroanalytical method in which a constant current of known magnitude generates a reagent that reacts with the analyte. The time required to generate enough reagent to complete the reaction is measured.

(e) *Current efficiency* is a measure of agreement between the number of faradays of current and the number of moles of reactant oxidized or reduced at a working electrode.

20-3. Mass transport in an electrochemical cell results from one or more of the following: (1) *diffusion,* which arises from the concentration difference between the solution immediately adjacent to the electrode surface and the bulk of the solution; (2) *migration,* which results from electrostatic attraction or repulsion between a species and an electrode; and (3) *convection,* which results from stirring, vibration, or temperature difference.

20-5. Both kinetic and concentration polarization cause the potential of a cell to be more negative than the thermodynamic potential. Concentration polarization arises from the slow rate at which reactants or products are transported to or away from the electrode surfaces. Kinetic polarization arises from the slow rate of the electrochemical reactions at the electrode surfaces.

20-7. Kinetic polarization is often encountered when the product of a reaction is a gas, particularly when the electrode is a soft metal such as mercury, zinc, or copper. It is likely to occur at low temperatures and high current densities.

20-9. Temperature, current density, complexation of the analyte, and codeposition of a gas influence the physical properties of an electrogravimetric deposit.

20-11. (a) An *amperostat* is an instrument that provides a constant current to an electrolysis cell.

(b) A *potentiostat* controls the applied potential to maintain a constant potential between the working electrode and a reference electrode.

20-12. In *amperostatic coulometry,* the cell is operated so that the current in the cell is held constant. In *potentiostatic coulometry,* the potential of the working electrode is maintained constant.

20-13. The species produced at the counter electrode is a potential interference by reacting with the product at the working electrode. Isolation of the one from the other is ordinarily necessary.

20-15. (a) *Voltammetry* is an analytical technique that is based on measuring the current that develops in a cell containing a microelectrode as the applied potential is varied. *Polarography* is a particular type of voltammetry in which the microelectrode is a dropping mercury electrode.

20-16. (a) A *voltammogram* is a plot of current as a function of voltage applied to a microelectrode.

(e) The *half-wave potential* is the potential on a voltammetric wave at which the current is one half the limiting current.

20-18. **(b)** 6.2×10^{16} cations

20-19. **(a)** -0.738 V   **(c)** -0.337 V

20-20. -0.913 V

20-21. -0.493 V

20-24. **(a)** -0.94 V   **(b)** -0.35 V
      (c) -2.09 V   **(d)** -2.37 V

20-25. **(a)** 4.2×10^{-6} M   **(b)** -0.425 V

20-28. **(a)** Separation is impossible.
      (b) Separation is feasible.
      (c) Separation would be feasible if cell potential held between 0.099 and -0.14 V.

20-29. **(a)** 0.237 V   **(c)** 0.0395 V   **(e)** 0.118 V
      (g) 0.276 V   **(i)** 0.0789 V

20-30. **(a)** 28.4 min   **(b)** 9.47 min

20-32. 112.6 g/eq

20-34. 79.5 ppm $CaCO_3$

20-36. 4.06% $C_6H_5NO_2$

20-38. 2.471% CCl_4 and 1.854% $CHCl_3$

20-40. **(a)** 0.067%   **(b)** 0.13%   **(c)** 0.40%

20-42. Stripping methods are generally more sensitive than other voltammetric procedures because the analyte can be removed from a relatively large volume of solution and concentrated on a tiny electrode for a long period (often many minutes). After concentration, the potential is reversed and all of the analyte that has been stored on the electrode is rapidly oxidized or reduced, producing a large current.

Chapter 21

21-1. Because it absorbs yellow light in the 580 to 595-nm region of the spectrum and transmits blue light.

21-2. **(a)** Absorbance $= -\log$ transmittance

21-3. Interactions at high concentrations, shift of equilibria, polychromatic radiation, stray light, mismatched cells.

21-6. **(a)** 1.13×10^{18} Hz   **(c)** 4.318×10^{14} Hz
      (e) 1.53×10^{13} Hz

21-7. **(a)** 252.8 cm   **(c)** 286 cm

21-9. 1.00×10^{14} Hz to 1.62×10^{15} Hz
      3.33×10^3 cm^{-1} to 5.41×10^4 cm^{-1}

21-11. 136 cm and 1.46×10^{-25} J

21-12. **(a)** 436 nm

21-13. **(a)** cm^{-1} ppm^{-1} **(c)** cm^{-1} %$^{-1}$

21-14. **(a)** 88.9% **(c)** 41.8%   **(e)** 32.7%

21-15. **(a)** 0.593   **(c)** 0.484   **(e)** 1.07

21-18. **(a)** $\%T = 67.3$, $c = 4.07 \times 10^{-5}$ M, $c_{ppm} = 8.13$ ppm,
               $a = 2.11 \times 10^{-2}$ cm^{-1} ppm^{-1}
      (c) $\%T = 30.2$, $c = 6.54 \times 10^{-5}$ M, $c_{ppm} = 13.1$ ppm,
               $a = 0.0397$ cm^{-1} ppm^{-1}
      (e) $A = 0.638$, $\%T = 23.0$, $c_{ppm} = 342$ ppm,
               $a = 1.87 \times 10^{-2}$ cm^{-1} ppm^{-1}

(g) $\%T = 15.9$, $c = 1.68 \times 10^{-4}$ M,
      $\varepsilon = 3.17 \times 10^3$ L mol^{-1} cm^{-1},
      $a = 0.0158$ cm^{-1} mol^{-1}

(i) $c = 2.62 \times 10^{-5}$ M, $A = 1.281$, $b = 5.00$ cm,
      $a = 0.0489$ cm^{-1} ppm^{-1}

21-19. 1.80×10^4 L cm^{-1} mol^{-1}

21-21. **(a)** 0.175   **(b)** 0.350
      (c) 66.8% and 44.7%   **(d)** 0.476

21-22. 0.530

21-24. **(a)** 0.595   **(b)** 25.4%
      (c) 1.70×10^{-5} M   **(d)** 2.50 cm

21-26. Slope $= 0.12$; intercept $= 0.18$; $c_X = 2.8$ ppm; standard error in $y = 0.017$; standard deviation for $c_X = 0.2$ μM.

21-28. **(a)** slope $= 0.033$; intercept $= 0.002$; standard error in $y = 0.0049$;
      (c) concentration of unknown $= 11.98$ μM; standard deviation $= 0.16$ μM.

Chapter 22

22-1. **(b)** A *phototube* is a vacuum tube equipped with a photoemissive cathode. When photons strike the photocathode, photoelectrons are emitted by the photoelectric effect. When a voltage of 90 V or more is applied between cathode and anode, the photoelectrons are attracted to the anode and collected to give a small photocurrent. A *photoconductive cell* consists of a thin film of a semiconductor material. Absorption of radiation decreases the electrical resistance of the semiconductor. When placed in series with a voltage source and load resistor, the voltage drop across the load resistor is measured.

      (d) A diode array spectrometer detects the entire spectral range simultaneously and can produce a spectrum in less than one second. Conventional spectrometers must scan the spectrum by mechanically rotating a grating. Consequently they take minutes to obtain the spectrum. Diode array instruments can be used to obtain spectra when changes are occurring rapidly such as at the output of a liquid chromatograph or a fast kinetics apparatus. Conventional instruments are normally too slow for such tasks.

22-3. Photons from the infrared region of the spectrum do not have sufficient energy to cause photoemission from the cathode of a photomultipler tube.

22-5. *Tungsten/halogen lamps* contain a small amount of iodine in the evacuated quartz envelope that contains the tungsten filament. The iodine prolongs the life of the lamp and permits it to operate at a higher temperature. The iodine combines with gaseous tungsten that sublimes from the filament and causes the metal to be redeposited, thus adding to the life of the lamp.

22-7. As a minimum, the radiation emitted by the source of a single-beam instrument must be stable long enough to make the 0% T adjustment, the 100% T adjustment, and the measurement of T for the sample.

22-9. pH, electrolyte concentration, temperature.

22-11. (a) $T = 26.1\%, A = 0.583$
        (c) $T = 6.82\%, A = 1.166$

22-12. (b) 0.471   (d) 11.4%

Chapter 23

23-1. (a) A *chromophore* is an organic functional group that absorbs radiation in the ultraviolet/visible regions.
        (c) Radiation that consists of a single wavelength is said to be *monochromatic*.
        (e) An *absorption spectrum* is a plot of a spectral property (absorbance, log absorbance, absorptivity, transmittance) as the ordinate and wavelength, wavenumber, or frequency as the abscissa.

23-2. (a) *Fluorescence* is a process by which an excited singlet species relaxes by emitting electromagnetic radiation.
        (c) *Internal conversion* is the nonradiative relaxation of a molecule from the lowest vibrational level of an excited electronic state to the highest vibrational level of a lower electronic state.
        (e) The *Stokes shift* is the difference in wavelength between the radiation used to excite fluorescence and the wavelength of the emitted radiation.
        (g) An *inner-filter effect* occurs when the fluorescent radiation from an excited analyte molecule is absorbed by an unexcited analyte molecule. This process results in a decrease in fluorescence intensity. Also, excessive absorption of the incident beam causes a primary inner-filter effect.

23-4. Compounds that fluoresce have structures that slow the rate of nonradiative relaxation to the point where there is time for fluorescence to occur. Compounds that do not fluoresce have structures that permit rapid relaxation by nonradiative processes.

23-6. (a) Excitation of fluorescence usually involves transfer of an electron to a high vibrational state of an upper electronic state. Relaxation to a lower vibrational state of this electronic state goes on much more rapidly than fluorescence relaxation. When fluorescence relaxation occurs it is to a high vibrational state of the ground state or to a high vibrational state of an electronic state that is above the ground state. Such transitions involve less energy than the excitation energy. Therefore, the emitted radiation is longer in wavelength than the excitation wavelength.
        (b) For spectrofluorometry, the analytical signal F is given by $F = 2.3\,K'\varepsilon bcP_0$. The magnitude of F, and thus sensitivity, can be enhanced by increasing

the source intensity P_0 or the transducer sensitivity.

For spectrophotometry, the analytical signal A is given by $A = \log P_0/P$. Increasing P_0 or the detector's response to P_0 is accompanied by a corresponding increase in P. Thus, the ratio does not change nor does the analytical signal. Consequently, no improvement in sensitivity accompanies such changes.

23-8. Charge-transfer absorption occurs in species that contain both an electron donor and an electron acceptor group. The absorbed energy results in an electron being transferred from an orbital of the donor group to an orbital that is largely associated with the acceptor group. This type of absorption is analytically important because the molar absorptivities associated with this type of transition are usually very high, which leads to high sensitivities and low detection limits.

23-10. In *atomic emission spectroscopy* the radiation source is the sample itself. The energy for excitation of analyte atoms comes from a flame, a furnace, or a plasma. The signal is the measured intensity of the flame at the wavelength of interest. In *atomic absorption spectroscopy* the radiation source is usually a line source such as a hollow cathode lamp, and the output signal is the absorbance calculated from the incident power of the source and the resulting power after the light has passed through the atomized sample in the heated source.

23-11. In atomic emission spectroscopy, the analytical signal is produced by *excited* atoms or ions, whereas in atomic absorption the signal results from absorption by *unexcited* species. Typically, the number of unexcited species exceeds the excited by several orders of magnitude. The ratio of unexcited to excited atoms in a hot medium varies exponentially with temperature. Thus a small change in temperature brings about a large change in the number of excited atoms. The number of unexcited atoms changes very little, however, because they are present in an enormous excess. Therefore, emission spectroscopy can be more sensitive to temperature changes that is absorption spectroscopy.

23-13. (a)

c_{ind}, M	A_{430}	A_{530}
3.00×10^{-4}	1.54	1.06
2.00×10^{-4}	0.935	0.777
1.00×10^{-4}	0.383	0.455
0.500×10^{-4}	0.149	0.261
0.250×10^{-4}	0.056	0.145

23-15. 1.6×10^{-5} M to 8.6×10^{-5} M

23-17. 5.4×10^{-5} M to 1.2×10^{-3} M

23-21. The absorbance will decrease linearly before the equivalence point and reach a constant value of approximately zero after the equivalence point.

23-23. 0.0214% Co

23-24. (a) No absorbance until the equivalence point after which the absorbance increases linearly.

(c) Absorbance increases linearly until the equivalence point after which it stays relatively constant.

(e) Absorbance increases linearly until the equivalence point after which it also increases linearly, but with a large slope.

23-26. (1) 5.48×10^{-5} M Co and 1.31×10^{-4} M Ni

(3) 2.20×10^{-4} M Co and 4.41×10^{-5} M Ni

23-29. (a) 0.492 (c) 0.190

23-30. (a) 0.301 (c) 0.491

23-31. (b) $A_{510} = 0.03949 c_{Fe} - 0.001008$

(c) $s_m = 1.1 \times 10^{-4}$ and $s_b = 2.7 \times 10^{-3}$

23-32.

	c_{Fe}, ppm	s_c, rel % 1 Result	3 Results
(a)	3.65	2.8	2.1
(c)	1.75	6.1	4.6
(e)	38.3	0.27	0.20

23-35. (b) Slope = 22.34; intercept = 4×10^{-4}

(d) Unknown = 0.54 μM

(e) Standard deviation = 0.008; RSD = 0.015

Chapter 24

24-1. A *masking agent* is a complexing reagent that reacts selectively with one or more components of a solution to prevent them from interfering in an analysis.

24-3. (a) 1.73×10^{-2} M   (b) 6.40×10^{-3} M

(c) 2.06×10^{-3} M   (d) 6.89×10^{-4} M

24-5. (a) 75 mL   (b) 40 mL   (c) 22 mL

24-7. (a) 18.0   (b) 7.56

24-8. (a) 91.9

24-9. (a) $K_d = 1.53$

(b) $[HA]_{aq} = 0.0147$ M; $[A^-]_{aq} = 0.0378$ M

(c) $K_a = 9.7 \times 10^{-2}$

24-11 (a) 12.4 meq Ca^{2+}/L   (b) 6.19×10^2 mg $CaCO_3$/L

24-13. Dissolve 17.53 g of NaCl in about 100 mL of water and pass the solution through a column packed with a cation exchange resin in its acid form. Wash the column with several hundred milliliters of water, collecting the liquid from the original solution and the washings in a 2-L volumetric flask. Dilute to the mark and mix well.

24-15. (a) *Elution* is a process in which species are washed through a chromatographic column by additions of fresh mobile phase.

(c) The *stationary phase* in a chromatographic column is a solid or liquid that is fixed in place. A mobile phase then passes over or through the stationary phase.

(e) The *retention time* for an analyte is the time interval between its injection onto a column and the appear-

ance of its peak at the other end of the column.

(g) The *selectivity factor* α of a column toward two species is given by the equation $\alpha = K_B/K_A$ where K_B is the distribution constant of the more strongly held species B and K_A is the corresponding constant for the less strongly held solute A.

24-17. In gas-liquid chromatography, the mobile phase is a gas, whereas in liquid-liquid chromatography it is a liquid.

24-19. The number of plates in a column can be determined by measuring the retention time t_R and width of a peak at its base W. The number of plates N is then given by the equation $N = 16(t_R/W)^2$.

24-21. (a)

Peak	N
A	2775
B	2472
C	2364
D	2523

(b) $\bar{N} = 2.5 \times 10^3$; $s = 0.2 \times 10^3$

(c) $H = 0.0097$ cm

24-22.

	(a) k	(b) K
A	0.74	6.2
B	3.3	2.7
C	3.5	30
D	6.0	50

24-23. (a) $R_S = 0.72$

(b) $\alpha_{C,B} = 1.1$

(c) L = 108 cm

(d) $(t_R)_S = 62$ min

24-24. (a) $R_S = 5.2$ (b) 2.0 cm

24-28.

	(a) k	(b) K
Methylcyclohexane	4.3	14
Methylcyclohexene	4.7	15
Toluene	6.1	19

24-29. (a) $k_M = 2.54$ $k_N = 2.62$

(b) $\alpha = 1.03$

(c) $N = 8.1 \times 10^4$ plates

(d) $L = 1.8 \times 10^2$ cm

(e) 91 min

Chapter 25

25-1. *Slow sample injection* in gas chromatography leads to band broadening and lowered resolution.

25-3. A chromatogram is a plot of the response of the detector of a chromatographic column versus the time of elution or the volume of eluent.

25-5. In *open tubular columns,* the stationary phase is held on the inner surface of a capillary, whereas in packed columns, the stationary phase is supported on particles that are contained in a glass or metal tube. Open tubular columns, which are only applicable in gas chromatography, contain an enormous number of plates that permit rapid separation of closely related species. They suffer from small sample capacities.

25.7. (a) Diethyl ether, benzene, *n*-hexane

25.8. (a) Ethyl acetate, dimethylamine, acetic acid

25.9. In *adsorption chromatography,* separations are based upon *adsorption equilibria* between the components of the sample and a solid surface. In partition chromatography, separations are based upon *distribution equilibria* between two immiscible liquids.

25-11. *Gel filtration* is a type of size-exclusion chromatography in which the packings are hydrophilic, and eluents are aqueous. It is used for separating high-molecular-weight polar compounds. *Gel-permeation chromatography* is a type of size-exclusion chromatography in which the packings are hydrophobic, and eluents are nonaqueous. It is used for separation high-molecular-weight nonpolar species.

25-13. 22.9% methyl acetate   48.5% methyl propionate
28.7% methyl *n*-butyrate

25-16. The simplest type of pump for liquid chromatography is a *pneumatic pump,* which consists of a collapsible solvent container housed in a vessel that can be pressurized by a compressed gas. This type of pump is simple, inexpensive, and pulse-free. It has limited capacity and pressure output, it is not adaptable to gradient elution, and its pumping rate depends upon the viscosity of the solvent.

A *screw-driven syringe pump* consists of a large syringe in which the piston is moved in or out by means of a motor-driven screw. It also is pulse-free and the rate of delivery is easily varied. It suffers from lack of capacity and is inconvenient to use when solvents must be changed.

The most versatile and widely used pump is the *reciprocating pump* that usually consists of a small cylindrical chamber that is filled and then emptied by the back-and-forth motion of a piston. Advantages of the reciprocating pump include small internal volume, high output pressures, adaptability to gradient elution, and flow rates that are constant and independent of viscosity and back pressure. The main disadvantage is pulsed output, which must be damped.

Chapter 26

26-1. The properties of a supercritical fluid that are of particular importance to its application to chromatography are its density, its viscosity, and the rates at which solutes diffuse in it. The magnitude of each of these properties lies intermediate between a typical gas and a typical liquid.

26-3. Pressure increases cause the density of a supercritical fluid to increase which causes the k for analytes to change. Generally increases in pressure reduce the elution times of solutes.

26-6. Supercritical fluid chromatography is ordinarily faster than HPLC and exhibits less band spreading than is encountered in GLC. In contrast to GLC, SFC separations can be carried out at relatively low temperatures thus making the technique applicable to thermally unstable and nonvolatile species. Unlike HPLC, but like GLC, flame ionization can be used for detection, which makes the method applicable to more analytes.

26-8. Electroosmotic flow occurs when a mobile phase in a chromatographic column is subjected to a high potential difference between one end of the tube and the other. For a silica tube, the flow is generally away from the positive electrode towards the negative. The flow occurs because of the attraction of positive charged species toward the negative silica surface. This ring of positive charge is mobile and is attracted toward the negative electrode carrying with it the mobile phase molecules.

26-10. In solution, amino acids exist as zwitterions that bear both a positive and a negative charge. At low pH values, the net effective charge is positive owing to the interaction of hydrogen ions with the amine groups of the amino acid, whereas at high pH values the net charge is negative due to the dissociation of the carboxylic acid groups. Thus, at low pH values the protein molecules will be strongly attracted toward the negative electrode, while in basic solutions the reverse will be the case.

26-13. 3.9 min.

26-15. Micellar electrokinetic capillary chromatography serves to separate neutral species under the influence of a high potential difference. In this technique, neutral analyte molecules are solubilized in micelles that usually bear a negative charge owing to the presence of sulfonic acid anions on their surface. This charge repels the micelles from the negative electrode. Normally, however, this repulsion is more than offset by the electroosmotic flow. Thus, the micelles move toward the negative electrode at a rate that is lower than that of the mobile phase. As a consequence, the micelles act like a slowly moving stationary phase. Separations then occur in the same way as on an HPLC column.

Glossary

Absolute error   An accuracy measurement based on the numerical difference between an experimental measurement and its true (or accepted) value.

Absolute standard deviation   A precision estimate based on the deviations between individual members in a set and the mean of that set.

Absorbance, A   The logarithm of the ratio between the initial power of a beam of radiation P_0 and its power after it has traversed an absorbing medium P, $A = \log(P_0/P)$.

Absorption   A process in which a substance is incorporated or assimilated within another; also, a process in which a beam of electromagnetic radiation is attenuated during passage through a medium.

Absorption filter   A colored medium (usually glass) that transmits a relatively narrow band of the visible spectrum.

Absorption spectrum   A plot of absorbance as a function of wavelength or frequency.

Absorptivity, a   The proportionality constant in the Beer's law equation, $A = abc$, where A is the absorbance, b is the pathlength of radiation (usually in centimeters), and c is the concentration of the absorbing species.

Accuracy   The numerical difference between an analytical result and the true or accepted value for the measured quantity; this agreement is measured in terms of error.

Acid dissociation constant, K_a   The equilibrium constant for the dissociation reaction of a weak acid.

Acid error   The tendency of a glass electrode to register anomalously high pH response in highly acidic media.

Acid rain   Rainwater that has been rendered acidic from absorption of airborne nitrogen- and sulfur oxides produced mainly by humans.

Acids   Species that are capable of donating protons to other species.

Activity, a   The effective concentration of a participant in a chemical equilibrium; the activity of a species is given by the product of the equilibrium concentration of the species and its activity coefficient.

Activity coefficient, γ_X   A unitless quantity whose numerical value depends on the ionic strength of a solution. The activity for a species is equal to the product of its equilibrium concentration and its activity coefficient.

Adsorption   A process in which a substance becomes physically bound to the surface of a solid.

Agar   A polysaccharide that forms a conducting gel with electrolyte solutions. Used in salt bridges to provide electric contact between dissimilar solutions while preventing mixing.

Air damper   A device that hastens achievement of equilibrium by the beam of a mechanical analytical balance; also called a dashpot.

Aliquot   A volume of liquid that is a known fraction of a larger volume.

Alkaline error   The tendency of many glass electrodes to provide an anomalously low pH response in basic solutions containing alkali metal ions.

α-values   The ratio between the molar concentration of a particular species and the analytical concentration of the solute from which it is derived.

Alumina   The common name for aluminum oxide.

Amines   Derivatives of ammonia with one or more organic groups replacing hydrogen.

Amino acids   Weak organic carboxylic acids that also contain basic amine groups.

Amperometric titration   A titrimetric method based on applying a constant potential to a working electrode and recording the resulting current; a linear segment curve is obtained.

Amperostat   An instrument that maintains a constant current; used for coulometric titrations.

Amphiprotic substances   Species that can either donate protons or accept protons, depending on the environment.

Amylose   A component of starch, the β form of which is a specific indicator for iodine.

Analyte   The species in the sample about which analytical information is sought.

Analytical balance   An instrument for the accurate determination of mass.

Analytical molarity, c_X   The moles of solute X in 1.000 liter of solution. Also numerically equal to the number of millimoles of solute per milliliter of solution. Compare with species molarity.

Angstrom unit (Å)   A unit of length equal to 1×10^{-10} meter.

Anhydrone® Trade name for magnesium perchlorate, a drying agent.

Anion exchange resins High-molecular-weight polymers to which numerous amine groups are bonded.

Anode The electrode of an electrochemical cell at which oxidation occurs.

Aqua regia A mixture containing three volumes of concentrated hydrochloric acid and one volume nitric acid; a potent oxidizing solution.

Argentometric titration A titration in which the reagent is a standard solution of $AgNO_3$.

Arithmetic mean Synonymous with mean or average of a set of numbers.

Ashing The process whereby an organic material is combusted in air.

Ashless filter paper Paper produced from cellulose fibers that have been treated to eliminate inorganic species, thus leaving no residue when ashed.

Aspirator A device for sucking fluid through a medium.

Assay The process of determining how much of a given sample is the material indicated by its name.

Asymmetry potential A small potential that results from slight differences between the two surfaces of a glass membrane electrode.

Atomic absorption spectroscopy An analytical method that is based on absorption of light by a gaseous mixture of atoms formed in a flame, furnace, or a cold vapor cell.

Atomic emission spectroscopy An analytical method based on emission of light by atoms excited in a flame, a furnace, an inductively coupled plasma, or an electric arc or spark.

Atomic fluorescence spectroscopy An analytical method based on the fluorescent radiation produced when an atomic vapor is excited by a beam of electromagnetic radiation.

Atomic mass unit A unit of mass based on $\frac{1}{12}$ of the mass of the most abundant isotope of carbon, ^{12}C.

Atomization The process of producing an atomic gas by applying energy to a sample.

Attenuation In absorption spectroscopy, a decrease in the power of a beam of radiant energy.

Autocatalysis A condition where the product of a reaction catalyses the reaction itself.

Autoprotolysis A self-dissociation process in which molecules of a substance react to give a pair of ions.

Auxiliary balance A generic term for a balance that is less sensitive but more rugged than an analytical balance; synonymous with laboratory balance.

Average A number obtained by summing the data in a set and dividing the sum by the number of data in the set.

Azo indicators A group of acid/base indicators that have in common the structure R—N=N—R'.

Back-titration The titration of an excess of a standard solution that has reacted completely with an analyte.

Band Ideally, a Gaussian shaped distribution of (1) adjacent wavelengths encountered in spectroscopy and (2) of the amount of a compound as it exits from a chromatograph or an electrophoretic column.

Band spectrum A molecular spectrum made up of one or more wavelength regions in which spectral components are numerous and close together due to rotational and vibrational transitions.

Bandwidth Usually, the range of wavelengths or frequencies of a spectral absorption or emission peak at half the height of the peak; the range passed by a wavelength isolation device.

Base dissociation constant, K_b The equilibrium constant for the reaction of a weak base with water.

Bases Species that are capable of accepting protons from donor (acid) species.

Beam arrest A mechanism that lifts the beam from its bearing surface when an analytical balance is not in use or when the load is being changed.

Beam splitter A device for dividing radiation from a monochromator such that one portion passes through the sample while the other passes through the blank.

Beer's law The fundamental relationship for the absorption of radiation by matter; that is, $A = abc$, where A is the absorbance, a is the absorptivity, b is the pathlength of the beam of radiation, and c is the concentration of the absorbing species.

β-amylose That component of starch that serves as a specific indicator for iodine.

Bias The tendency to skew estimates in the direction that favors the anticipated result; also used to describe the effect of a systematic error on a set of measurements; also a dc voltage that is used to polarize a circuit element.

Blackbody radiation Continuum radiation produced by a heated solid.

Blank determination The process of performing all steps of an analysis in the absence of sample. Used to detect and compensate for systematic errors in an analysis.

Brønsted–Lowry acids and bases An acid of this type is defined as a proton donor and a base as a proton acceptor. The loss of a proton by an acid results in the formation of a species that is a potential proton acceptor, or a conjugate base of the parent acid.

Buffer capacity The number of moles of strong acid (or strong base) needed to alter the pH of 1.00 L of a buffer solution by 1.00 unit.

Buffer solutions Solutions that tend to resist changes in pH as the result of dilution or the addition of small amounts of acids or bases.

Bumping The sudden and often violent boiling of a liquid that results from local overheating.

Buoyancy The displacement of the medium (ordinarily air) by an object, thus producing an apparent loss of mass; a significant source of error when the densities of the object and the comparison standards (weights) differ.

Buret A graduated tube from which accurately known volumes can be dispensed.

Calomel The compound Hg_2Cl_2.

Calomel electrode A versatile reference electrode based on the half-reaction $Hg_2Cl_2(s) + 2e^- \rightleftharpoons 2Hg(l) + 2Cl^-$.

Capillary column A small-diameter chromatographic column for GLC, fabricated of metal, glass, or fused silica. The stationary phase is a thin coating of liquid on the interior wall of the tube.

Capillary electrophoresis High-speed, high-resolution electrophoresis performed in capillary tubes.

Carbonate error A systematic error caused by absorption of carbon dioxide by standard solutions of base that will be used in the titration of weak acids.

Carrier gas The mobile phase for gas-liquid chromatography.

Catalytic reaction A reaction whose progress toward the equilibrium state is hastened by a substance that is not consumed in the overall process.

Cathode The electrode in an electrochemical cell at which reduction takes place.

Cathode depolarizer A substance that is more easily reduced than hydrogen ion; used to prevent codeposition of hydrogen during an electrolysis.

Cation exchange resins High-molecular-weight polymers to which acidic groups are bonded; these resins permit the substitution of cations in solution for hydrogen ions from the exchanger.

Cell A term with several meanings. In statistics, the combination of adjacent data for display in a histogram. In electrochemistry, an array consisting of a pair of electrodes immersed in solutions that are in electrical contact; the electrodes are connected externally by a metallic conductor. In spectroscopy, the container that holds the sam-

ple in the light path of an optical instrument. In an electronic balance, a system of constraints that assure alignment of the pan.

Cells without liquid junction Electrochemical cells in which both electrodes are immersed in a common electrolyte solution.

Charge-balance equation An expression relating the concentrations of anions and cations based on charge neutrality in a given solution.

Charge-coupled device (CCD) A solid-state two-dimensional detector array used for spectroscopy and imaging.

Charge injection device (CID) A solid-state photodetector array used in spectroscopy.

Charge transfer complexes Complexes that are made up of an electron donor group and an electron acceptor group. Absorption of radiation by these complexes involves a transfer of electrons from the donor to the acceptor.

Chelating agents Substances with multiple sites available for coordination bonding with metal ions. Such bonding typically results in the formation of five- or six-membered rings.

Chelation The reaction between a metal ion and a chelating reagent.

Chemical deviations from Beer's law Deviations from Beer's law that result from association or dissociation of the absorbing species or reaction with the solvent that produce a product that absorbs differently from the analyte.

Chemical equilibrium A dynamic state in which the rates of forward and reverse processes are identical. A system in equilibrium will not spontaneously depart from this condition.

Chemiluminescence The emission of energy as electromagnetic radiation during a chemical reaction.

Chopper A mechanical device that alternately transmits and blocks radiation from a source.

Chromatogram A plot of an analyte concentration signal as a function of elution time or elution volume.

Chromatographic bands The distribution (ideally Gaussian) of the concentration of eluted species about a central value; the result of variations in the time an analyte species resides in the mobile phase.

Chromatographic zones Synonymous with chromatographic bands.

Chromatography A term for methods of separation based on the partition of analyte species between a stationary phase and a mobile phase.

Clark oxygen sensor A voltammetric sensor for dissolved oxygen.

Coagulation The process whereby particles with colloidal dimensions are caused to form larger aggregates or to coagulate.

Coefficient of variation, CV   The relative standard deviation, expressed as a percentage.

Colloidal suspension   A mixture of a solid in a liquid in which the particles are so finely divided that they have little or no tendency to settle.

Colorimeter   An optical instrument for the measurement of electromagnetic radiation in the visible region of the spectrum.

Column chromatography   A chromatographic method in which the stationary phase is held within or on the surface of a narrow tube and the mobile phase is forced through the tube where compound separation occurs.

Column efficiency   A measure of the degree of broadening of a chromatographic band; often expressed in terms of plate height H or the number of theoretical plates N. Insofar as the distribution of analyte is Gaussian within the band, the plate height is given by the variance σ^2 divided by the length L of the column packing.

Column resolution, R   Measures the capability of a column to separate two analyte bands.

Common-ion effect   The shift in the position of equilibrium caused by the addition of a participating ion.

Complex formation   The process whereby a species with one or more unshared electron pairs forms coordinate bonds with metal ions.

Concentration-based equilibrium constant, K'   The equilibrium constant based on molar equilibrium concentrations. The numerical value of K' depends on the ionic strength of the medium.

Concentration polarization   A condition in an electrochemical cell where the transport of species to and from electrode surfaces is insufficient to maintain the current at a desired level.

Conduction of electricity   The movement of charge by ions in solution, by electrochemical reaction at the surface of electrodes, and by movement of electrons in metals.

Conductometric detector   A detector for charged species; finds use in ion chromatography.

Confidence interval   Defines bounds about an experimental value within which, with a given probability, the true value should be located.

Confidence limits   The values that define the confidence interval.

Conjugate acid/base pairs   Species that differ from one another by one proton.

Constant-boiling HCl   A solution of hydrochloric acid having a constant concentration that depends on the atmospheric pressure.

Constant error   A systematic error that is independent of the size of the sample taken for analysis. Its effect on the results of an analysis increases as the sample size decreases.

Constructive interference   Increase in amplitude of a wave in regions where two or more wave fronts are in phase with one another.

Continuous source   A source that emits radiation continuously in time.

Continuum source, absorption spectroscopy   Sources that emit a continuum of wavelengths; examples include tungsten filament lamps and deuterium lamps.

Continuum spectrum   Radiation consisting of a broad band of wavelengths as opposed to discrete lines. Incandescent solids provide a continuum output (blackbody radiation) in the visible and IR regions; deuterium- and hydrogen-lamps yield continuum spectra in the ultraviolet region.

Controlled potential methods   Coulometric and electrogravimetric methods that involve the continuous variation of the potential applied to the cell to maintain a constant potential between the working electrode and a reference electrode.

Convection   The transport of a species in a liquid or gaseous medium by stirring, by mechanical agitation, or by temperature gradients.

Coordination compounds   Species formed between metal ions and electron-pair donating groups. The product may be anionic, neutral, or cationic.

Coprecipitation   The carrying down of otherwise soluble species either within a solid or on the surface of a solid as it precipitates.

Coulomb, C   The quantity of charge provided by a constant current of one ampere in one second.

Coulometer   A device that permits measurement of the quantity of charge. Electronic coulometers evaluate the integral of the current/time curve; chemical coulometers are based on the extent of reaction in an auxiliary cell.

Coulometric titration   A type of coulometric analysis that involves measurement of the time needed for a constant current to produce enough reagent to react completely with an analyte.

Counter electrode   The electrode that, with the working electrode, forms the electrolysis circuit in a three-electrode cell.

Counter-ion layer   A region of solution surrounding a colloidal particle within which there exists a quantity of ions sufficient to balance the charge on the surface of the particle.

Creeping The tendency of some precipitates to spread over a wetted surface.

Critical temperature The temperature above which a fluid can no longer exist in two phases, regardless of pressure.

Cross-linked stationary phase A polymer stationary phase in a chromatographic column in which covalent bonds link different strands of the polymer, thus creating a more stable phase.

Crystalline membrane electrode Electrodes in which the sensing element is a crystalline solid that responds selectively to the activity of an ionic analyte.

Crystalline precipitates Solids that tend to form as large-particulate, easily filtered solids.

Current, i The amount of electric charge in amperes A that passes through an electrical circuit per unit time.

Current density The electrical current per unit area of an electrode (A/m^2).

Current efficiency Measures of the effectiveness of a quantity of electricity to bring about an equivalent amount of chemical change in an analyte.

Current maxima Anomalous peaks in the current of a polarographic cell; eliminated by the introduction of surface active agents.

Current-to-voltage converter A device for converting an electric current into a voltage that is proportional to the current.

Cuvette The container that holds the analyte in the light path, in absorption spectroscopy.

Dalton Synonymous with atomic mass unit.

Dark currents Small currents that occur even when no radiation is reaching a photometric transducer.

Dashpot Synonymous with air damper in an analytical balance.

dc Plasma (DCP) spectroscopy A method that makes use of an electrically induced argon plasma to excite the emission spectra of analyte species.

Dead-stop end point The end point of a biamperometric titration characterized by a linear increase or decrease in current.

Dead time In column chromatography, the time, t_M, required for an unretained species to traverse the column.

Debye–Hückel equation An expression that permits calculation of activity coefficients in media with ionic strengths less that 0.1.

Debye–Hückel limiting law A simplified form of the Debye–Hückel equation, applicable to solutions in which the ionic strength is less than 0.01.

Decantation The transfer of supernatant liquid and washings from a container to a filter without disturbing the precipitated solid in the container.

Decrepitation The shattering of a crystalline solid as it is heated; caused by vaporization of occluded water.

Degrees of freedom The number of members in a statistical sample that provide an independent measure of the precision of the set.

Dehydration The loss of water by a solid.

Dehydrite® Trade name for magnesium perchlorate, a drying agent.

Density The ratio between the mass of an object and its volume.

Derivative titration curve A plot of the change in the quantity measured per unit volume against the volume of titrant added. A derivative curve displays a maximum where there is a point of inflection in a conventional titration curve.

Desiccants Drying agents.

Desiccator A container that provides a dry atmosphere for the storage of samples, crucibles, and precipitates.

Destructive interference Decrease in amplitude of waves resulting from the superposition of two or more wave fronts that are not in phase with one another.

Detection limit The minimum amount of analyte that a system or method is capable of measuring.

Detector A device that responds to some characteristic of the system under observation and converts that response into a measurable signal.

Determinate error A class of errors that — in principle, at least — has a known cause. Synonymous with systematic error.

Deuterium lamp A source that provides a spectral continuum in the ultraviolet region of the spectrum. Radiation results from application of about 40 V to a pair of electrodes housed in a deuterium atmosphere.

Devarda's alloy An alloy of copper, aluminum, and zinc. Used to reduce nitrates and nitrites to ammonia in a basic medium.

Deviation The difference between an individual measurement and the mean (or median) value for a set of data.

Diatomaceous earth The siliceous skeletons of unicellular algae that are used as a solid support in GLC.

Differentiating solvents Solvents in which differences in the strength of solute acids or bases are enhanced. Compare with leveling solvents.

Diffraction order, n Integer multiples of a wavelength at which constructive interference occurs.

Diffusion The migration of species from a region of high concentration in a solution to a more dilute region.

Diffusion coefficient (polarographic, D; chromatographic, D_M) A measure of the mobility of a species in units cm^2/s.

Diffusion current, i_d The limiting current in voltammetry when diffusion is the major form of mass transfer.

Digestion The practice of maintaining an unstirred mixture of freshly formed precipitate and solution from which it was formed at temperatures just below boiling. Results in improved purity and particle size.

Dimethylglyoxime A precipitating reagent that is specific for nickel(II). Its formula is $CH_3(C=NOH)_2CH_3$.

Diode array detector A silicon chip that accommodates numerous photodiodes; provides the capability to collect data from entire spectral regions simultaneously.

Diphenylthiocarbazide A chelating reagent, also known as dithizone; adducts with cations are sparingly soluble in water but are readily extracted with organic solvents.

Dissociation The splitting of ions or molecules of a substance, commonly into two simpler entities.

Distribution constant The ratio of the equilibrium molar concentrations of an analyte in two immiscible solvents.

Dithizone Synonymous with diphenylthiocarbazide.

Doping The intentional introduction of traces of group III or group V elements to increase the semiconductor properties of a silicon or germanium crystal.

Double-beam instrument An optical instrument design that eliminates the need to alternate blank and analyte solutions manually in the light path. A beam splitter partitions the radiation in a double beam in space spectrometer; a chopper directs the beam alternately between blank and analyte in a double beam in time instrument.

Double precipitation Synonymous with reprecipitation.

Drierite® Trade name for calcium sulfate, a drying agent.

Dropping mercury electrode An electrode in which mercury is forced through a capillary giving regular drops of mercury.

Dry ashing The elimination of organic matter from a sample by direct heating in air.

Dumas method A method of analysis based on the combustion of nitrogen-containing organic samples by CuO to convert the nitrogen to N_2, which is then measured volumetrically.

Dynode An intermediate electrode in a photomultiplier tube.

EDTA An abbreviation of ethylenediaminetetraacetic acid, which is a chelating agent that is widely used for complex formation titrations. Its formula is $(HOOCCH_2)_2NCH_2CH_2N(CH_2COOH)_2$.

Effective bandwidth See bandwidth.

Electric double layer Refers to the charge on the surface of a colloidal particle and counter-ion layer in the surrounding solution that balances this charge.

Electroanalytical methods A large group of methods that have in common the measurement of an electrical property of the system that is related to the amount of analyte in the sample.

Electrochemical cell An array consisting of two or three electrodes, each of which is in contact with an electrolyte solution. Typically, the electrolytes are in electrical contact through a salt bridge. An external metal conductor connects the electrodes.

Electrode A conductor at the surface of which electron transfer to or from the surrounding solution takes place.

Electrodeless discharge tube A source of atomic line spectra that is powered by radio frequency or microwave radiation.

Electrode of the first kind A metallic electrode whose potential is proportional to the logarithm of the concentration (strictly activity) of a cation (or the ratio of cations) derived from the electrode metal.

Electrode of the second kind A metallic electrode whose response is proportional to the logarithm of the concentration (strictly activity) of an anion that forms a sparingly soluble species or stable complexes with a cation (or the ratio of cations) derived from the electrode metal.

Electrode potential The potential of an electrochemical cell in which the potential of the electrode of interest is the right-hand electrode and the standard hydrogen electrode is the left-hand electrode.

Electrogravimetric analysis A branch of gravimetric analysis that involves measuring the mass of species deposited on an electrode of an electrochemical cell.

Electrolyte effect The dependence of numerical values for equilibrium constants on the ionic strength of the solution.

Electrolytes Solute species whose aqueous solutions conduct electricity.

Electrolytic cell An electrochemical cell that requires an external source of energy to drive the cell reaction. Compare with galvanic cell.

Electromagnetic radiation A form of energy with properties that can be described in terms of waves, or alternatively as particulate photons, depending on the method of observation.

Electromagnetic spectrum The power or intensity of

electromagnetic radiation plotted as a function of wavelength or frequency.

Electronic balance A balance in which an electromagnetic field supports the pan and its contents; the current needed to restore the loaded pan to its original position is proportional to the mass on the pan.

Electronic transitions The promotion of an electron from one electronic state to a second electronic state, and conversely.

Electroosmotic flow The net flow of bulk liquid in an applied electric field.

Electrophoresis A separation method based on differential rates of migration of charged species in an electric field.

Electrothermal analyzer Any of several devices that form an atomized gas containing an analyte in the light path of an instrument by electrical heating. Used for atomic absorption and atomic fluorescence measurements.

Eluent A mobile phase in chromatography that is used to carry solutes through a stationary phase.

Eluent suppressor column In ion chromatography, a column downstream from the analytical column where ionic eluents are converted to nonconducting species while analyte ions remain unaffected.

Elution chromatography Describes processes in which analytes are separated from one another on a column owing to differences in the time they are retained in the column.

Emission spectrum The collection of spectral lines or bands that are observed when species in excited states relax by giving off their excess energy as electromagnetic radiation.

Empirical formula The simplest whole-number combination of atoms in a molecule.

End point An observable change during titration that signals that the amount titrant added is chemically equivalent to the amount of analyte in the sample.

Eppendorf pipet A type of micropipet that delivers adjustable volumes of liquid.

Equal-arm balance An analytical balance equipped with a beam that supports two pans equidistant from the fulcrum, one for the load and the other to accommodate an equal mass of known weights.

Equilibrium-constant expression An algebraic statement that describes the equilibrium relationship among the activities of participants of a chemical reaction.

Equilibrium molarity The concentration of a solute *species* (in mol/L or mmol/mL). Synonymous with species molarity.

Equivalence point That point in a titration where the amount of standard titrant added is chemically equivalent to the amount of analyte in the sample.

Equivalence-point potential The electrode potential of the system in an oxidation/reduction titration when the amount of titrant that has been added is chemically equivalent to the amount of analyte in the sample.

Equivalent For an oxidation/reduction reaction, that weight of a species that can donate or accept 1 mole of electrons. For an acid/base reaction, that weight of a species that can donate or accept 1 mole of protons.

Equivalent of chemical change The mass of a species that is directly or indirectly equivalent to one faraday (6.02×10^{23} electrons).

Equivalent weight A specialized basis for expressing mass in chemical terms. Similar to, but different from, molar mass. As a consequence of definition, one equivalent of an analyte reacts with one equivalent of a reagent, even if the stoichiometry of the reaction is not one to one.

Error The difference between an experimental measurement and its accepted value.

Essential water Water in a solid that exists in a fixed amount, either within the molecular structure (water of constitution) or within the crystalline structure (water of crystallization).

Ethylenediaminetetraacetic acid Probably the most versatile reagent for complex-formation titrations; forms chelates with most cations. See EDTA.

Excitation The promotion of an atom, an ion, or a molecule to a state that is more energetic than a lower energy state.

Excitation spectrum In fluorescence spectroscopy, a plot of fluorescence intensity as a function of excitation wavelength.

Faradaic current An electrical current produced by oxidation/reduction processes in an electrochemical cell.

Faraday, F The quantity of electricity associated with 6.022×10^{23} electrons.

Ferroin A common name for the 1,10-phenanthroline-iron(II) complex, which is a versatile redox indicator. Its formula is $(C_{12}H_8N_2)_3Fe^{2+}$.

Flame emission spectroscopy Methods that use a flame to cause an atomized analyte to emit its characteristic emission spectrum. Also known as flame photometry.

Flame ionization detector (FID) A detector for gas chromatography based on the collection of ions produced during the pyrolysis of organic analytes in a flame.

Fluorescence   Radiation produced by an atom or a molecule that has been excited by photons to a singlet excited state.

Fluorescence bands   Groups of fluorescence lines that originate from the same excited electronic state.

Fluorescence spectrum   A plot of fluorescence intensity versus wavelength where the excitation wavelength is held constant.

Fluorometer   A filter instrument for quantitative fluorescence measurements.

Formality, F   The number of formula masses of solute contained in each liter of solution. Synonymous with analytical molarity.

Formal potential, $E^{0'}$ The electrode potential for a couple when the analytical concentrations of all participants are unity and the concentrations of other species in the solution are defined.

Formula weight   The summation of atomic masses in the chemical formula of a substance. Synonymous with gram formula weight and molar mass.

Fourier transform spectrometer   A spectrometer in which an interferometer and Fourier transformation are used to obtain a spectrum.

Frequency, of electromagnetic radiation   The number of oscillations per second; has units of hertz (Hz), which is one oscillation per second.

Fritted-glass crucible   A filtering crucible equipped with a porous glass bottom; also called a sintered-glass crucible.

Fused-silica open tubular column   A wall-coated gas chromatography column that has been fabricated from purified silica.

Galvanic cell   An electrochemical cell that provides energy during its operation. Synonymous with voltaic cell.

Galvanostat   Synonymous with amperostat.

Gas chromatography   Methods that make use of a gaseous mobile phase and a liquid or a solid stationary phase.

GC/MS   A combined technique in which a mass spectrometer is used as a detector for gas chromatography.

Gas electrode   An electrode that involves the formation or consumption of a gas during its operation.

Gas-sensing probe   An indicator/reference electrode system that is isolated from the analyte solution by a hydrophobic membrane. The membrane is permeable to a gas; the composition of the internal solution, and thus the potential, is proportional to the gas content of the analyte solution.

Gaussian distribution   A theoretical bell-shaped distribution of results obtained for replicate measurements that are affected by random errors.

Gel filtration chromatography   A type of size exclusion chromatography that makes use of a hydrophilic packing. Used for separating polar species.

Gel permeation chromatography   A type of size exclusion chromatography that makes use of a hydrophobic packing. Used for separating nonpolar species.

Glass electrode   An electrode in which a potential develops across a thin glass membrane that provides a measure of the pH of a solution in which the electrode is immersed.

Gooch crucible   A porcelain filtering crucible; filtration is accomplished by a glass fiber mat or a layer of asbestos fiber.

Gradient elution   The systematic alteration of mobile-phase composition in liquid chromatography to optimize the chromatographic resolution of the components in a mixture.

Grating   A device consisting of closely spaced grooves that is used to disperse polychromatic radiation by diffraction into its component wavelengths.

Gravimetric analysis   A group of analytical methods in which the amount of analyte is established through the measurement of the mass of a pure substance containing the analyte.

Gravimetric factor, GF   The stoichiometric ratio between the analyte and the solid weighed in a gravimetric analysis.

Gravimetric titrimetry   Titrations in which the mass of standard titrant is measured rather than volume. Concentration of titrant is expressed in mmol/g of solution (rather than the more familiar mmol/mL).

Gross error   An occasional error, neither random nor systematic, that results in the occurrence of a questionable outlier result.

Ground state   The lowest energy state of an atom or molecule.

Guard column   A precolumn located ahead of an HPLC column; composition of the packing in the guard column is selected to extend the useful lifetime of the analytical column by removing particulate matter and contaminants and by saturating the eluent with the stationary phase.

Half-cell potential   The potential of an electrochemical half-cell measured with respect to the standard hydrogen anode.

Half reaction   A method of portraying the oxidation or the reduction of a species. A balanced equation that shows the oxidized and reduced form of a species, any H_2O and H^+ needed to balance the hydrogen and oxygen atoms in the system, and the number of electrons required to balance the charge.

Half-wave potential, $E_{1/2}$ The potential (ordinarily versus SCE) at which the current of a voltammetric wave is one half the limiting current.

Height equivalent of a theoretical plate (HETP) A measure of chromatographic column efficiency; equal to the length of a column divided by the number of theoretical plates in the column.

Henderson–Hasselbalch equation An expression used by biochemists to calculate the pH of a buffer solution; $pH = pK_a + \log (c_{NaA}/c_{HA})$, where pK_a is the negative log of the dissociation constant for the acid and c_{NaA} and c_{HA} are the molar concentrations of the compounds making up the buffer.

High-performance liquid chromatography, HPLC Column chromatography in which the stationary phase is made up of small particles and the mobile phase is forced through the particles by high pressure.

Histogram A bar graph in which replicate results are grouped according to ranges of magnitude along the horizontal axis and by frequency of occurrence on the vertical axis.

Hollow cathode lamp A source used in atomic absorption spectroscopy that emits sharp lines for a single element or sometimes for several elements.

Homogeneous precipitation A technique in which a precipitating agent is generated slowly throughout a solution of an analyte to yield a dense and easily filtered precipitate for gravimetric analysis.

Hundred percent T adjustment Adjustment of an optical absorption instrument to register 100% T with a suitable blank in the light path.

Hydrodynamic voltammetry Voltammetry performed with the analyte solution in constant motion relative to the electrode surface; produced by pumping the solution past a stationary electrode or by moving the electrode through the solution.

Hydrogen lamp A continuum source of radiation in the ultraviolet that is similar in structure to a deuterium lamp.

Hydronium ion The hydrated proton whose symbol is H_3O^+.

8-Hydroxyquinoline A versatile chelating reagent; used in gravimetric analysis, in volumetric analysis, as a protective reagent in atomic spectroscopy, and as an extracting reagent. Also known as oxine. Its formula is HOC_9H_6N.

Hyphenated methods Methods involving the combination of two or more types of instrumentation. The product is an instrument with greater capabilities than any one method alone.

Indeterminate error Synonymous with random error.

Indicator electrode An electrode whose potential is related to the activity of one or more species in contact with the electrode.

Inductively coupled plasma (ICP) spectroscopy A method that makes use of an argon plasma formed by the absorption of radio-frequency radiation to atomize a sample for atomic emission spectroscopy.

Infrared radiation Electromagnetic radiation in the 0.78 to 300 μm range.

Inner-filter effect Phenomenon causing nonlinear fluorescence calibration curves due to excessive absorption of the incident beam or the emitted beam.

Instrumental deviations from Beer's law Departures from linearity between absorbance and concentration that are attributable to the measuring device.

Interference filter An optical filter that provides narrow bandwidths owing to constructive interference.

Interference order, n An integer that, along with the thickness and the refractive index of the dielectric material, determines the wavelength transmitted by an interference filter.

Interferences Species that affect the signal upon which an analysis is based.

Interferometer A nondispersive device that obtains spectral information through constructive and destructive interference. Used in Fourier transform IR instruments.

Internal standard A known quantity of a species with properties similar to an analyte that is introduced into solutions of the standard and the unknown. The ratio of the signal from the internal standard to the signal from the analyte then serves as the basis for the analysis.

Ion chromatography An HPLC technique based on the partition of ionic species between a liquid mobile phase and a solid polymeric ionic exchanger.

Ion exchange resin A high-molecular-weight polymer to which a large number of acidic or basic functional groups have been bonded. Cathodic resins permit the exchange of hydrogen ion for cations in solution; anodic resins substitute hydroxide ion for anions.

Ionic strength, μ A property of a solution that depends on the total concentration of ions in the solution as well as by the charge carried by each of these ions. That is, $\mu = \frac{1}{2}\Sigma c_i Z_i^2$, where c_i is the molar concentration of each ion and Z_i is its charge.

IR drop The potential drop across a cell due to

resistance to the movement of charge. Also known as the ohmic potential.

Irreversible cell   An electrochemical cell in which the chemical reaction as a galvanic cell is different from that which occurs when the current is reversed.

Isocratic elution   Elution with a single solvent.

Isoelectric point   The pH at which an amino acid has no tendency to migrate under the influence of an electric field.

IUPAC convention   A set of definitions relating to electrochemical cells and their potentials. Also known as the Stockholm convention.

Jones reductor   A column packed with amalgamated zinc; used for the prereduction of analytes.

Junction potential   The potential that develops at the interface between solutions with dissimilar composition. Synonymous with liquid junction potential.

Karl Fischer reagent   A reagent for the titrimetric analysis of water.

Kilogram   The SI base unit of mass.

Kinetic determination   Analytical method based on relating the kinetics of a reaction to the analyte concentration.

Kinetic polarization   Nonlinear behavior of an electrochemical cell caused by the slowness of reaction at the surface of one or both electrodes.

Kjeldahl flask   A long-necked flask used for the digestion of samples with hot, concentrated sulfuric acid.

Kjeldahl method   A titration method for the determination of nitrogen in organic compounds in which the nitrogen is converted to ammonia, which is then distilled and determined by a neutralization titration.

Knife edge   The nearly friction-free contact between the moving components of a mechanical analytical balance.

Laboratory balance   Synonymous with auxiliary balance.

Laminar flow   Flow in a region near the surface of a solid within which parallel layers of liquid slide by one another.

Least-squares method   A statistical method for fitting a mathematical function (such as the equation for a straight line) to a set of experimental data by minimizing the sum of the square of the differences between the experimental points and the points predicted by the mathematical functions.

Le Châtelier principle   A statement that the application of a stress to a chemical system at equilibrium will result in a shift in the position of the equilibrium that tends to relieve the stress.

Leveling solvents   Solvents in which the strength of solute acids or bases tend to be the same.

Levitation   As applied to electronic balances, the suspension of the pan of the balance in air by a magnetic field.

Ligand   A molecule or ion with at least one pair of unshared electrons available for coordinate bonding with cations.

Limiting current   Current plateau reached in voltammetry when rate of mass transfer is at its maximum value.

Linear segment curve   A titration curve in which the end point is obtained from the interaction of linear extrapolations of data from regions well before and after the equivalence point. Useful for reactions that do not strongly favor the formation of products.

Line source   In atomic absorption spectroscopy, a radiation source that emits sharp atomic lines characteristic of the analyte atoms.

Liquid junction   The interface between two liquids with different composition.

Liter   One cubic decimeter.

Luminescence   Radiation resulting from photoexcitation (photoluminescence), chemical excitation (chemiluminescence), or thermal excitation (thermoluminescence).

Macrobalance   An analytical balance with a capacity of 160 to 200 g and a precision of 0.1 mg.

Majority carrier   The species principally responsible for the transport of electricity in a semiconductor.

Masking agent   A reagent that combines with and inactivates matrix species that would otherwise interfere with the determination of an analyte.

Mass   An invariant measure of the amount of matter in an object.

Mass action effect   The shift in the position of equilibrium through the addition or removal of a participant in the equilibrium.

Mass balance equation   An expression that relates the equilibrium concentration of various species in a solution to one another and to the analytical concentration of the various solutes.

Mass spectrometry   Methods based on forming ions in the gas phase and separating them on the basis of mass-to-charge ratio.

Mass transport   The movement of species through a solution caused by diffusion, convection, and electrostatic forces.

Matrix   The medium that contains an analyte.

Mean Synonymous with arithmetic mean and average. Used to report what is considered the most representative value for a set of measurements.

Mean activity coefficient An experimentally measured activity coefficient for an ionic compound. It is not possible to resolve the mean activity coefficient into values for the individual participants.

Measuring pipet A pipet calibrated to deliver any desired volume up to its maximum capacity.

Mechanical entrapment The incorporation of impurities within a growing crystal.

Median The central value in a set of replicate measurements. For an odd number of data, there are an equal number of data above and below the median. For an even number of data, the median is the average of the central pair.

Membrane electrode An indicator electrode whose response is due to ion-exchange processes on each side of a thin membrane.

Meniscus The curved surface displayed by a liquid held in a vessel.

Microanalytical balance An analytical balance with a capacity of 1 to 3 g with precision of 0.001 mg.

Microporous membrane A hydrophobic membrane with pore size that permits the passage of gases and is impermeable to other species; the sensing element of a gas-sensing probe.

Microgram, μg 1×10^{-6} g.

Microliter, μL 1×10^{-6} L.

Millimole, mmol 1×10^{-3} mol.

Mixed crystal formation A type of coprecipitation encountered in crystalline precipitates in which some of the ions in the analyte crystals are replaced by nonanalyte ions.

Mobile phase In chromatography, a liquid or a gas that carries analytes through a liquid or solid stationary phase.

Mohr's salt A common name for iron(II) ammonium sulfate hexahydrate.

Molar absorptivity, ε The proportionality constant in the Beer's law expression when b is in centimeters and c is in moles per liter; characteristic of the absorbing species.

Molarity, M The number of moles of a species contained in one liter of solution or the number of millimoles contained in one milliliter.

Molar mass, $\mathcal{M}$ The mass, in grams, of one mole of a chemical substance.

Mole The amount of substance contained in 6.022×10^{23} particles of that substance.

Molecular absorption The absorption of ultraviolet, visible, and infrared radiation brought about by quantized transitions in molecules.

Molecular fluorescence The process whereby singlet excited-state electrons in molecules return a lower quantum state, with the resulting energy being given off as electromagnetic radiation.

Molecular formula A formula that includes structural information in addition to the number and identity of atoms in a molecule.

Molecular weight Synonymous with molecular mass.

Monochromator A device for resolving polychromatic radiation into its component wavelengths.

Monochromatic radiation Ideally, electromagnetic radiation that consists of a single wavelength; in practice, a very narrow band of wavelengths.

Mother liquor The solution that remains following the precipitation of a solid.

Muffle furnace A heavy-duty oven, capable of maintaining temperatures in excess of 1100°C.

Nanometer, nm 1×10^{-9} m.

National Institute of Standards and Technology, NIST An agency of the U.S. Department of Commerce; formerly the National Bureau of Standards (NBS).

Nebulization The transformation of a liquid into myriad tiny droplets.

Nernst equation A mathematical expression that relates the potential of an electrode to a logarithmic function of those species in solution that are responsible for the potential.

Nernst glower A source of infrared radiation that consists of a cylinder of zirconium and yttrium oxides heated to a high temperature by passage of an electrical current.

Nichrome A nickel/chromium alloy; when incandescent, a source of infrared radiation.

Noise Random fluctuations of an analytical signal that result from a large number of uncontrolled variables that affect the signal.

Nonessential water Water that is retained in or on a solid by physical, rather than chemical, forces.

Normal error curve A plot of a Gaussian distribution of the frequency of results from random errors in a measurement.

Normal hydrogen electrode, NHE Synonymous with standard hydrogen electrode.

Normality, c_N The number of equivalent weights of a species in one liter of solution.

Normal-phase chromatography A type of partition HPLC that involves a polar stationary phase and a nonpolar mobile phase.

Nucleation A process involving formation of very small aggregates of a solid during formation of a precipitate.

Null hypothesis A claim that a characteristic of a single population is equal to some specified value or that two or more population characteristics are identical.

Number of theoretical plates A characteristic of a chromatographic column used to describe its efficiency.

Occlusion The physical entrainment of soluble impurities in a growing crystalline precipitate.

Oesper's salt Common name for iron(II) ethylenediamine sulfate tetrahydrate.

Ohmic potential Synonymous with IR drop.

Open tubular column A capillary column of glass or fused silica used in gas chromatography. The walls of the tube are coated with a thin layer of the stationary phase.

Optical instruments A broad term for instruments that measure absorption, emission, or fluorescence by analyte species based on ultraviolet, visible, or infrared radiation.

Optical wedge A device used in optical spectroscopy whose transmission decreases linearly along its length.

Outlier A result that appears at odds with the other members in a data set.

Overpotential, overvoltage Excess voltage necessary to produce current in a polarized electrochemical cell.

Oxidant Synonymous with oxidizing agent.

Oxidation The loss of electrons by a species in an oxidation/reduction reaction.

Oxidation potential The potential of an electrode process that is written as an oxidation.

Oxidizing agent A substance that acquires electrons in an oxidation/reduction reaction.

Oxine A common name for 8-hydroxyquinoline.

Pan arrest A device to support the pans of a balance when a load is being placed on them.

Parallax Apparent change in position of an object as a result of the movement of the observer. Results in systematic errors in reading laboratory scales such as those on burets, pipets, and meters with pointers.

Partition chromatography A type of chromatography based on the distribution of solutes between a liquid mobile phase and a liquid stationary phase retained on the surface of a solid.

Partition coefficient An equilibrium constant for the distribution of a solute between two immiscible liquid phases. See distribution constant.

Partition ratio Synonymous with distribution constant.

Parts per million, ppm A convenient method for expressing the concentration of a solute species that exists in trace amounts. For dilute aqueous solutions, ppm is synonymous with mg solute/L solution.

Peptization A process in which a coagulated colloid returns to its dispersed state.

Period of electromagnetic radiation The time required for successive peaks of an electromagnetic wave to pass a fixed point in space.

pH The negative logarithm of the hydrogen-ion activity of a solution.

Phosphorescence Emission of light from an excited triplet state. Phosphorescence is slower than fluorescence and may occur over several minutes.

Phosphorus pentoxide, P_2O_5 A drying agent.

Photoconductive cell A detector for electromagnetic radiation whose electrical conductivity increases with the intensity of radiation impinging upon it.

Photodecomposition The formation of new species from molecules excited by radiation; one of several ways by which excitation energy is dissipated.

Photodiode (1) A vacuum tube consisting of a wire anode and a photosensitive surface that produces an electron for each photon absorbed on the surface. (2) A reverse-biased silicon semiconductor that produces electrons and holes when irradiated by electromagnetic radiation. The resulting current provides a measure of the number of photons per second striking the device.

Photoelectric colorimeter A photometer that responds to visible radiation.

Photoelectron An electron released by the absorption of a photon striking a photoemissive surface.

Photoionization detector A chromatographic detector that uses intense ultraviolet radiation to ionize analyte species. The resulting currents, which are amplified and recorded, are proportional to analyte concentration.

Photometer An instrument for the measurement of absorbance that incorporates a filter for wavelength selection and a photon detector.

Photomultiplier tube A sensitive detector for electromagnetic radiation; amplification is accomplished by a series of dynodes that produce a cascade of electrons for each photon received by the tube.

Photon detector A generic term for transducers that convert an optical signal to an electrical signal.

Photons Energy packets of electromagnetic radiation; also known as quanta.

Phototube A transducer consisting of a photoemissive cathode, a wire anode, and a power supply to maintain a suitable potential between the electrodes.

pIon meter An instrument that directly measures the concentration (strictly, activity) of an analyte. Consists of a specific ion indicator electrode, a reference electrode, and a potential-measuring device.

Plasma A gaseous medium that owes its conductivity to appreciable amounts of ions and electrons.

Plate height A quantity describing the efficiency of a chromatographic column.

Pneumatic detector A heat detector that is based on changes in the pressure that a gas exerts on a flexible diaphragm.

pn Junction diode A junction between electron-rich and electron-deficient regions of a semiconductor; permits currents in one direction only.

Polarization A phenomenon in an electrochemical cell in which the magnitude of the current is limited by the low rate of the electrode reactions or the slowness of transport of reactants to the electrode surface.

Polarography Voltammetry at the dropping mercury electrode.

Polychromatic radiation Electromagnetic radiation consisting of more than one wavelength. Compare with monochromatic radiation.

Polyfunctional acids and bases Species that contain more than one acidic or basic functional group.

Population mean, μ The mean value for an infinite set of measurements. The true value for a quantity that is free of systematic error.

Population of data A theoretical aggregate of the infinite number of values that a measurement can take. Also referred to as a universe of data.

Population standard deviation, σ A precision parameter based in principle on a population containing an infinite number of measurements.

Potentiometric titration A titrimetric method involving measurement of the potential between a reference electrode and an indicator electrode as a function of titrant volume.

Potentiometry That branch of electrochemistry concerned with the relation between the potential of an electrochemical cell and the concentration of the contents of the cell.

Potentiostat An electronic device that alters the applied potential so that the potential between a working electrode and a reference electrode is maintained at a fixed value.

Potentiostatic methods Coulometric and gravimetric methods that use a constant potential between the working electrode and a reference electrode.

Power, P, of electromagnetic radiation The energy that reaches a given area per second; often used synonymously with intensity, although the two are not precisely the same.

Precision A measure of internal agreement among a set of replicate observations.

Primary adsorption layer Charged layer of ions on the surface of a solid, resulting from the attraction of lattice ions for ions of opposite charge in the solution.

Primary standard A highly pure chemical compound that is used to prepare or determine the concentrations of standard solutions for titrimetry.

Prism A transparent, prism-shaped solid that disperses polychromatic radiation into its component wavelengths.

Proportional error An error whose magnitude increases as the sample size increases.

p-Values An expression of the concentration of a solute species as its negative logarithm. The use of p-values permits expression of enormous ranges of concentration in terms of relatively small numbers.

Quanta Synonymous with photons.

Quenching (1) Process by which molecules in an excited state lose energy to other species without fluorescing. The diminution in the intensity of a fluorescing reagent as a result of reaction with the analyte may be used for determining the analyte. (2) An action that brings about the cessation of a reaction.

Random error Uncertainties resulting from the operation of small uncontrolled variables that are inevitable as measurement systems are extended to and beyond their limits.

Range, w, of data The difference between extreme values in a set of data; synonymous with spread.

Reagent-grade chemicals Highly pure chemicals that meet the standards of the Reagent Chemical Committee of the American Chemical Society.

Redox Synonymous with oxidation/reduction.

Redox electrode An inert electrode that responds to the electrode potential of the system.

Reducing agent The species that supplies electrons in an oxidation/reduction reaction.

Reduction The process whereby a species acquires electrons.

Reduction potential The potential of an electrode process expressed as a reduction. Synonymous with electrode potential.

Reductor A column packed with a granular metal through which a sample is passed to prereduce an analyte.

Reference electrode An electrode whose potential relative to the standard hydrogen electrode is known and against which potentials of unknown electrodes may be measured. A reference electrode potential is completely independent of the analyte concentration.

Reference standards Complex materials that have been extensively analyzed. A prime source for these standards is the National Institute of Standards and Technology (NIST).

Reflection grating An optical body that disperses polychromatic radiation into its component wavelengths. Consists of lines ruled on a reflecting surface; dispersion is the result of constructive and destructive interference.

Refractive index The ratio of the velocity of electromagnetic radiation in a vacuum to the velocity in some other medium.

Relative electrode potential The potential of an electrode with respect to another (ordinarily the standard hydrogen electrode).

Relative error The error in a measurement divided by the true (or accepted) value for the measurement. Often expressed as a percentage.

Relative standard deviation (RSD) The standard deviation divided by the mean value for a set of data. When expressed as a percentage, the relative standard deviation is referred to as the coefficient of variation.

Relative supersaturation The difference between the instantaneous (Q) and the equilibrium (S) concentrations of a solute in a solution divided by S. Provides general guidance as to the particle size of a precipitate formed by addition of reagent to an analyte solution.

Relaxation The return of excited species to a lower energy level; the process is accompanied by the release of excitation energy as heat, fluorescence, or phosphorescence.

Replica grating An impression of a master grating; used as the dispersing element in most grating instruments, owing to the high cost of a master grating.

Replicate samples Portions of a material of approximately the same size that are carried through an analysis at the same time and in the same way.

Reprecipitation A method for improving the purity of precipitates, involving formation and filtration of the solid, followed by redissolving and again forming the precipitate.

Resolution, R_S Measures the ability of a chromatographic column to separate two analytes. Defined as the difference between the retention times for the two peaks divided by their average widths.

Resonance fluorescence Fluorescence emission at a wavelength that is identical with the excitation wavelength.

Resonance line The wavelength of radiation responsible for both the excitation and the fluorescent emission of an atom.

Retention factor, k A term used to describe the migration of a species through a chromatographic column. Its numerical value is given by $k = (t_R - t_M)/t_M$, where t_R is the retention time for a peak and t_M is the dead time; also called the *capacity factor*.

Retention time, t_R In chromatography, the time between sample injection on a chromatographic column and the arrival of an analyte peak at the detector.

Reversed-phase partition chromatography Liquid-liquid chromatography that makes use of a nonpolar stationary phase and a polar liquid phase.

Reversible cell An electrochemical cell in which the oxidation/reduction process is reversed when the direction of current is reversed.

Rotational transition A change in quantized rotational energy states in a molecule.

Rubber policeman A small length of rubber tubing that has been crimped on one end; used to dislodge adherent particles of precipitate from beaker walls.

Salt An ionic solid formed by the reaction of an acid and a base.

Salt bridge A device in an electrochemical cell that conducts current between the two electrolyte solutions surrounding the electrodes while minimizing mixing of the two.

Salt effect Synonymous with electrolyte effect.

Sample matrix Refers to the medium that contains an analyte.

Sample mean, $\bar{x}$ The mean of a finite set of measurements.

Sample of data A finite group of replicate measurements.

Sample splitter A device that permits the introduction of small and reproducible portions of sample to a chromatographic column.

Sample standard deviation, s A precision estimate based on deviations of individual data from the mean, $\bar{x}$, of a finite sample. Also referred to as the standard deviation.

Sampling The process of collecting a small part of a material whose composition is representative of the bulk of material from which it was taken.

Sampling loop A type of sample splitter.

Saturated calomel electrode (SCE) A reference electrode that can be formulated as $Hg|Hg_2Cl_2(sat'd)$, $KCl(sat'd)\|$. Its half-reaction is $Hg_2Cl_2(s) + 2e^- \rightleftharpoons 2Hg(l) + 2Cl^-$.

Secondary standard A substance used in volumetric analysis whose purity has been established and verified by chemical analysis.

Selectivity The tendency for a reagent or an instrumental method to respond similarly to only a few species.

Selectivity coefficient, $k_{A,B}$ The selectivity coefficient for a specific ion electrode is a measure of the relative response of the electrode to ions A and B.

Selectivity factor, α In chromatography, $\alpha = K_B/K_A$, where K_B is the distribution constant for a less strongly retained species and K_A is the constant for a more strongly retained species.

Self-absorption A process in which analyte molecules absorb radiation emitted by other analyte molecules.

Semiconductor A material with electrical conductivity that is intermediate between a metal and an insulator.

Semimicroanalytical balance A balance with a capacity of about 30 g and a precision of 0.01 mg.

Servo system A device in which a small electrical signal provided by a mechanical displacement causes the system to return to a null position.

Sigmoid curve An S-shaped curve that is typical of plots of p-function of an analyte versus volume of reagent in titrimetry.

Signal-to-noise ratio (S/N) The ratio of the mean analyte output signal to the standard deviation of the inherently random fluctuations of an electronic device used to measure the signal.

Significant figure convention A system of imparting to the reader information concerning the reliability of numerical data. In general, all digits known with certainty, plus the first uncertain digit, are significant.

Silicon photodiode A photon detector based on a reverse-biased silicon diode; exposure to radiation creates new holes and electrons, thereby increasing conductivity.

Silver-silver chloride electrode A widely encountered reference electrode, which can be formulated as $Ag|AgCl(sat'd), KCl(xM)||$. The half-reaction for the electrode is $AgCl(s) + e^- \rightleftharpoons Ag(s) + Cl^-(xM)$.

Single-beam instruments Photometric instruments that require the operator to position the sample and the blank alternately in a single light path.

Single-pan balance An unequal arm balance with the pan and weights on one side of the fulcrum and an air damper on the other; the weighing operation involves removal of standard weights in an amount equal to the mass of the object on the pan.

Sintered-glass crucible Synonymous with fritted-glass crucible.

SI units An international system of measurement that makes use of seven base units. All other units are derived from these seven units.

Size-exclusion chromatography A type of chromatography in which the packing is a finely divided solid having a uniform pore size. Separation is based on the size of analyte molecules.

Soap-bubble meter A device for measuring gas flow rates in gas chromatography.

Solubility-product constant, K_{sp} A numerical constant that describes equilibrium in a saturated solution of a sparingly soluble ionic salt.

Soluble starch, β-amylose An aqueous suspension is a specific indicator for iodine.

Solvent programming The systematic alteration of mobile-phase composition to optimize migration rates of solutes in a chromatographic column.

Sorbed water Nonessential water that is retained in the interstices of solid materials.

Sparging The removal of an unwanted dissolved gas by aeration with an inert gas.

Species molarity The equilibrium concentration of a species expressed in moles per liter and symbolized with square brackets [ ]. Synonymous with equilibrium molarity.

Specific gravity, sp gr The ratio of the density of a substance to that of water at a specified temperature (ordinarily 4°C).

Specific indicator A species that reacts with a particular species in an oxidation/reduction titration.

Specificity Refers to methods or reagents that respond or react with one and only one analyte.

Specific surface area The ratio between the surface area of a solid and its mass.

Spectra Plots of absorbance, transmittance, or emission intensity as a function of wavelength, frequency, or wavenumber.

Spectrochemical methods Synonymous with spectrometric methods.

Spectrofluorometer An instrument that incorporates a monochromator to disperse, isolate, and determine the intensity of fluorescent radiation. Some also have a second monochromator to provide control over the wavelength of the excitation radiation.

Spectrograph An optical instrument equipped with a dispersing element such as a grating or prism that allows a range of wavelengths to strike a spatially selective detector such as a diode array or photographic plate.

Spectrometer An instrument equipped with a monochromator or polychromator, a photodetector, and an electronic readout to display a number that is proportional to the intensity of an isolated spectral band.

Spectrometric methods Methods based on the absorption, emission, or fluorescence of electromagnetic radiation that is related to the amount of analyte in the sample.

Spectrophotometer A spectrometer designed for the measurement of the absorption of ultraviolet, visible, or infrared radiation. The instrument includes a source of radiation, a monochromator, and an electrical means of measuring radiation intensity.

Spectrophotometric titration A titration monitored by UV/visible spectrometry.

Spectroscope An optical instrument similar to a spectrometer except that a movable eyepiece is used in place of an electronic detector to allow visual detection of spectral lines.

Spectroscopy A general term to describe techniques based on the measurement of absorption, emission, or fluorescence of electromagnetic radiation.

Spread, *w*, of data A precision estimate. Synonymous with range.

Standard addition method A method for determining the concentration of an analyte in a solution. Small measured increments of the analyte are added to the sample solution, and instrument readings are recorded after one or more additions. The method compensates for matrix interferences.

Standard deviation, *s* or σ A measure of how closely replicate data clusters around the mean.

Standard electrode potential, E^0 The potential (relative to the standard hydrogen electrode) of a half reaction written as a reduction when the activities of all reactants and products are unity.

Standard error of the mean, σ_m The standard deviation divided by the square root of the number of measurements in the set.

Standard hydrogen electrode, SHE A gas electrode consisting of a platinized platinum electrode immersed in a solution that has a hydrogen ion activity of 1.00 and is kept saturated with hydrogen at a pressure of 1.00 atm. Its potential is assigned a value of 0.000 V at all temperatures.

Standardization Determination of the concentration of a solution through reaction, directly or indirectly, with a primary standard.

Standard reference materials, SRM Samples of various materials in which the concentration of one or more species is known.

Standard solution A solution in which the concentration of a solute is known with high reliability.

Stationary phase A solid or immobilized liquid in chromatography upon which analyte species are partitioned during passage of a mobile phase.

Stirrup The link between the beam of a mechanical balance and its pan (or pans).

Stockholm convention A set of conventions relating to electrochemical cells and their potentials. Also known as the IUPAC convention.

Stoichiometry Refers to the combining ratios among molar quantities of species in a chemical reaction.

Stokes shift Differences in wavelengths of incident and emitted or scattered radiation.

Stop-flow injection In HPLC, introduction of the sample at the head of the column while solvent flow is temporarily discontinued.

Stray radiation Radiation of wavelength other than the wavelength selected for optical measurement.

Strong acids and strong bases Acids and bases that are completely dissociated in a particular solvent.

Strong electrolytes Solutes that are completely dissociated into ions in a particular solvent.

Successive approximations A procedure for solving higher-order equations through the use of intermediate estimates of the quantity sought.

Sulfonic acid group $-RSO_3H$.

Supercritical fluid A substance that is maintained above its critical temperature. Its properties are intermediate between a liquid and a gas.

Supercritical fluid chromatography HPLC involving a supercritical fluid as the mobile phase.

Supersaturation A solution that temporarily contains an amount of solute that exceeds its equilibrium solubility.

Surface adsorption The retention of a normally soluble species on the surface of a solid.

Systematic error Errors that have a known source, affect measurements in one and only one way, and can, in principle, be accounted for. Also called determinate error or bias.

0% *T* adjustment A calibration step to eliminate dark current from the response of a spectrophotometer.

100% *T* adjustment Adjustment of a spectrophotometer to register 100% transmittance with a blank in the light path.

Tailing A nonideal condition in a chromatographic peak where the latter portions are drawn out.

Tare A counterweight used on an analytical balance to compensate for the mass of a container.

Temperature programming The systematic adjustment

of column temperature in gas chromatography to optimize migration rates for solutes.

THAM A primary standard for bases. Its formula is $(HOCH_2)_3CNH_2$ or tris-(hydroxymethyl) aminomethane.

Thermal conductivity detector A detector used in gas chromatography that depends on measuring the thermal conductivity of the column eluent.

Thermal detector An infrared detector that produces heat as a result of absorption of radiation.

Thermionic detector, TID A detector for gas chromatography, similar to a flame ionization detector; particularly sensitive for analytes that contain nitrogen or phosphorus.

Thermodynamic equilibrium constant, K The equilibrium constant expressed in terms of the activities for all reactants and products.

TISAB (total ionic strength adjustment buffer) A solution used to swamp the effect of electrolytes on direct potentiometric analyses.

Titration The procedure whereby a measured volume of a standard solution reacts with an analyte to the point of chemical equivalence.

Titration error The difference between the titrant volume needed to reach an end point in a titration and the theoretical volume required to obtain an equivalence point.

Titrimetry The process of systematically introducing an amount of titrant that is chemically equivalent to the quantity of analyte in a sample.

Transducer A device that converts a chemical or physical phenomenon into an electrical signal.

Transfer pipet Synonymous with volumetric pipet.

Transmittance, T The ratio of the power P of a beam of radiation after it has traversed an absorbing medium to its original power P_0; often expressed as a percentage: $\%T = (P/P_0) \times 100\%$.

Transverse wave A wave motion in which the direction of displacement is perpendicular to the direction of propagation.

Triple beam balance A rugged laboratory balance that is used to weigh approximate amounts.

TRIS Synonymous with THAM.

t-test A statistical test used to decide whether an experimental value equals a known or theoretical value or whether two or more experimental values are identical with a given level of confidence; used with s and $\bar{x}$ when estimated σ and μ are not available.

Tungsten-halogen lamp A tungsten lamp that contains a small amount of I_2 within a quartz envelope that allows the lamp to operate at a higher temperature.

Tyndall effect The scattering of radiation by particles in a solution or a gas that have colloidal dimensions.

Ultraviolet/visible region The region of the electromagnetic spectrum between 180 and 780 nm; associated with electronic transitions in atoms and molecules.

Valinomycin An antibiotic that has also found application in liquid membrane electrodes for potassium.

Variance, s^2 A precision estimate consisting of the square of the standard deviation.

Vernier An aid for making estimates between graduation marks on a scale.

Vibrational deactivation A very efficient process in which excited molecules relax to the lowest vibrational level of an electronic state.

Vibrational relaxation Synonymous with vibrational deactivation.

Vibrational transition Transitions between vibrational states of an electronic state that are responsible for infrared absorption.

Visible radiation That portion of the electromagnetic spectrum (380 to 780 nm) to which the human eye is perceptive.

Volatilization methods of analysis A variant of the gravimetric method based on mass loss caused by heating or ignition.

Voltage divider A device that provides voltages ranging from zero to the maximum of the power supply.

Voltaic cell Synonymous with galvanic cell.

Voltammetry A group of electroanalytical methods that measure current as a function of applied voltage.

Volume percent (v/v) The ratio between the volume of a liquid and the volume of its solution, multiplied by 100%.

Volumetric flask A device containing precise volumes of solution.

Volumetric methods Methods of analysis in which the final measurement is a volume of a standard titrant needed to react with the analyte in a known quantity of sample.

Volumetric pipet A device that will deliver a precise volume from one container to another; also called a measuring pipet.

Walden reductor A column packed with finely divided silver granules; used to prereduce HCl solutions of analytes.

Wavelength, λ, of electromagnetic radiation The distance between successive maxima or minima of an electromagnetic wave.

Wavelength selector A device that limits the range of wavelengths used for an optical measurement.

Wavenumber, $\bar{\nu}$ The reciprocal of frequency of radiation, which usually has units of cm^{-1}.

Weak acids and weak bases Acids and bases that are only partially dissociated in a particular solvent.

Weak electrolytes Solutes that are incompletely dissociated into ions in a particular solvent.

Weighing by difference The process of weighing a container plus the sample, followed by weighing the container after the sample has been removed.

Weighing form In gravimetric analysis, the species collected whose mass is proportional to the amount of analyte in the sample.

Weight The attraction between an object and its surroundings, terrestrially, the earth.

Weight molarity, M_w The concentration of titrant expressed as mmol/g.

Weight percent (w/w) The ratio of the mass of a solute to the mass of its solution, multiplied by 100%.

Weight titrimetry Synonymous with gravimetric titrimetry.

Weight/volume percent (w/v) The ratio of the mass of a solute to the volume of solution in which it is dissolved, multiplied by 100%.

Wet ashing The use of strong liquid oxidizing reagents to decompose the organic matter in a sample.

Zero percent T adjustment A calibration step to compensate for dark current in the response of a spectrophotometer.

Zones, chromatographic Synonymous with chromatographic bands.

Zimmermann–Reinhardt reagent A solution of manganese(II) in concentrated H_2SO_4 and H_3PO_4 that prevents the induced oxidation of chloride ion by permanganate during the titration of iron(II).

Zwitterion The species that results from the transfer in solution of a proton from an acidic group to an acceptor site on an amino acid molecule.

Index

Entries in **bold face** refer to specific laboratory directions; *t* refers to a table

INTERNATIONAL ATOMIC MASSES

Element	Symbol	Atomic Number	Atomic Mass	Element	Symbol	Atomic Number	Atomic Mass
Actinium	Ac	89	227	Mercury	Hg	80	200.59
Aluminum	Al	13	26.981539	Molybdenum	Mo	42	95.94
Americium	Am	95	243	Neodymium	Nd	60	144.24
Antimony	Sb	51	121.757	Neon	Ne	10	20.1797
Argon	Ar	18	39.948	Neptunium	Np	93	237
Arsenic	As	33	74.92159	Nickel	Ni	28	58.6934
Astatine	At	85	210	Niobium	Nb	41	92.90638
Barium	Ba	56	137.327	Nitrogen	N	7	14.00674
Berkelium	Bk	97	247	Nobelium	No	102	259
Beryllium	Be	4	9.012182	Osmium	Os	76	190.2
Bismuth	Bi	83	208.98037	Oxygen	O	8	15.9994
Boron	B	5	10.811	Palladium	Pd	46	106.42
Bromine	Br	35	79.904	Phosphorus	P	15	30.973762
Cadmium	Cd	48	112.411	Platinum	Pt	78	195.08
Calcium	Ca	20	40.078	Plutonium	Pu	94	244
Californium	Cf	98	251	Polonium	Po	84	210
Carbon	C	6	12.011	Potassium	K	19	39.0983
Cerium	Ce	58	140.115	Praseodymium	Pr	59	140.90765
Cesium	Cs	55	132.90543	Promethium	Pm	61	145
Chlorine	Cl	17	35.4527	Protactinium	Pa	91	231.03588
Chromium	Cr	24	51.9961	Radium	Ra	88	226
Cobalt	Co	27	58.93320	Radon	Rn	86	221
Copper	Cu	29	63.546	Rhenium	Re	75	186.207
Curium	Cm	96	247	Rhodium	Rh	45	102.90550
Dysprosium	Dy	66	162.50	Rubidium	Rb	37	85.4678
Einsteinium	Es	99	252	Ruthenium	Ru	44	101.07
Erbium	Er	68	167.26	Samarium	Sm	62	150.36
Europium	Eu	63	151.965	Scandium	Sc	21	44.955910
Fermium	Fm	100	257	Selenium	Se	34	78.96
Fluorine	F	9	18.9984032	Silicon	Si	14	28.0855
Francium	Fr	87	223	Silver	Ag	47	107.8682
Gadolinium	Gd	64	157.25	Sodium	Na	11	22.989768
Gallium	Ga	31	69.723	Strontium	Sr	38	87.62
Germanium	Ge	32	72.61	Sulfur	S	16	32.066
Gold	Au	79	196.96654	Tantalum	Ta	73	180.9479
Hafnium	Hf	72	178.49	Technetium	Tc	43	98
Helium	He	2	4.002602	Tellurium	Te	52	127.60
Holmium	Ho	67	164.93032	Terbium	Tb	65	158.92534
Hydrogen	H	1	1.00794	Thallium	Tl	81	204.3833
Indium	In	49	114.82	Thorium	Th	90	232.0381
Iodine	I	53	126.90447	Thulium	Tm	69	168.93421
Iridium	Ir	77	192.22	Tin	Sn	50	118.710
Iron	Fe	26	55.847	Titanium	Ti	22	47.88
Krypton	Kr	36	83.80	Tungsten	W	74	183.85
Lanthanum	La	57	138.9055	Uranium	U	92	238.0289
Lawrencium	Lr	103	262	Vanadium	V	23	50.9415
Lead	Pb	82	207.2	Xenon	Xe	54	131.29
Lithium	Li	3	6.941	Ytterbium	Yb	70	173.04
Lutetium	Lu	71	174.967	Yttrium	Y	39	88.90585
Magnesium	Mg	12	24.305	Zinc	Zn	30	65.39
Manganese	Mn	25	54.93805	Zirconium	Zr	40	91.224
Mendelevium	Md	101	258				